Solutions Manual

Jan William Simek
California Polytechnic State University

ORGANIC CHEMISTRY

NINTH EDITION

LEROY G. WADE

JAN WILLIAM SIMEK

PEARSON

Editor-in-Chief: Jeanne Zalesky
Executive Editor: Terry Haugen
Product Marketing Manager: Elizabeth Ellsworth
Executive Field Marketing Manager: Chris Barker
Director of Development: Jennifer Hart
Project Manager: Elisa Mandelbaum
Program Manager: Lisa Pierce
Editorial Assistant: Fran Falk
Team Lead, Project Management: David Zielonka
Team Lead, Program Management: Kristen Flathman
Compositor: GEX Publishing Services
Project Manager: GEX Publishing Services
Illustrator: Lachina
Operations Specialist: Maura Zaldivar-Garcia
Cover Photo Credit: Klaus Rein / image/BROKER / Alamy Stock Photo
Spectra: ©Sigma-Aldrich Co.
Supplement Cover Designer: Seventeenth Street Studios

www.pearsonhighered.com

ISBN 10: 0-13-416037-1; ISBN 13: 978-0-13-416037-5

TABLE OF CONTENTS

NOTES TO THE STUDENT, AND CONCEPT MAPS

PREFACE

Hints for Passing Organic Chemistry

Do you want to pass your course in organic chemistry? Here is my best advice, based on over thirty-five years of observing students learning organic chemistry:

Hint #1: *Do the problems.* It seems straightforward, but humans, including students, try to take the easy way out until they discover there is no shortcut. Unless you have a measured IQ above 200 and comfortably cruise in the top 1% of your class, *do the problems.* Usually your teacher (professor or teaching assistant) will recommend certain ones; try to do all those recommended. If you do half of them, you will be half-prepared at test time. (Do you want your surgeon coming to your appendectomy having practiced only *half* the procedure?) And when you do the problems, keep this Solutions Manual CLOSED. Avoid looking at *my* answer before you write *your* answer—your trying and struggling with the problem is the most valuable part of the problem. Discovery is a major part of learning. Remember that the primary goal of doing these problems is *not* just getting the right answer, but understanding the material well enough to get right answers to the questions you haven't seen yet.

Hint #2: *Keep up.* Getting behind in your work in a course that moves as quickly as this one is the Kiss of Death. For most students, organic chemistry is the most rigorous intellectual challenge they have faced so far in their studies. Some are taken by surprise at the diligence it requires. Don't think that you can study all of the material in the couple of days before the exam—well, you can, but you won't get a passing grade. Study organic chemistry like a foreign language: try to do some every day so that the freshly trained neurons stay sharp.

Hint #3: *Get help when you need it.* Use your teacher's office hours when you have difficulty. Many schools have tutoring centers (in which organic chemistry is a popular offering). Here's a secret: absolutely the best way to cement this material in your brain is to get together with a few of your fellow students and make up problems for each other, then correct and discuss them. When *you* write the problems, you will gain great insight into what this is all about.

Hint #3.5: When you write answers to problems, *write* them. Use the old-fashioned method of a pen or pencil on paper. Keep a notebook of your work; show your instructor; he/she will be impressed.

Purpose of This Solutions Manual

What is the point of this Solutions Manual? First, I can't do your studying for you. Second, since I am not leaning over your shoulder as you write your answers, I can't give you direct feedback on what you write and think—the print medium is limited in its usefulness. What I *can* do for you is: 1) provide correct answers: the publishers, Professor Wade, Professor Palandoken (my reviewer), and I have gone to great lengths to assure that what I have written is correct, for we all understand how it can shake a student's confidence to discover that the answer book flubbed up; 2) provide a considerable degree of rigor: beyond the fundamental requirement of correctness, I have tried to flesh out these answers, being complete but succinct; 3) provide insight into how to solve a problem and into where the sticky intellectual points are. Insight is the toughest to accomplish, but over the years, I have come to understand where students have trouble, so I have tried to anticipate your questions and to add enough detail so that the concept, as well as the answer, is clear.

It is difficult for students to understand or acknowledge that their teachers are human (some are more human than others). Since I am human (despite what my students might report), I can and do make mistakes. If there are mistakes in this book, they are my sole responsibility, and I am sorry. If you find one, PLEASE let me know so that it can be corrected in future printings. Nip it in the bud.

What's New in This Edition?

Better answers! Part of my goal in this edition has been to add more explanatory material to clarify how to arrive at the answer. In many problems, the possibility of more than one answer to a problem has been noted. Concept maps have been added at appropriate places to demonstrate the logic of particular concepts.

Better graphics! A second color has been added to correlate with the graphics in the text, and for emphasis. The official name of this color is magenta, but in the text and here, it is called red.

Better jokes? Like, seriously?

Some Web Stuff

Here I am: http://www.chemistry.calpoly.edu/content/faculty/simek_jan . Or read my bio on the Amazon page for the solutions manual.

Two essential web sites providing spectra are listed on the bottom of p. 320.

Acknowledgments

No project of this scope is ever done alone. These are team efforts, and several people who have assisted and facilitated in one fashion or another deserve my thanks.

Professor L. G. Wade, Jr., first author on the textbook, is a remarkable person. He has gone to extraordinary lengths to make the textbook as clear, organized, informative, and insightful as possible. He has solicited and followed many suggestions on his text, and his comments on my solutions have been perceptive and valuable. We agreed early on that our primary goal is to help the students learn a fascinating and challenging subject, and all of our efforts have been directed toward that goal. I continue to appreciate our collaboration.

My friend and colleague, Dr. Hasan Palandoken, has reviewed the entire manuscript for accuracy and style. His extraordinary diligence, attention to detail, and chemical wisdom have made this a better manual. Hasan stands on the shoulders of previous reviewers who scoured earlier editions for errors: Dr. Kristen Meisenheimer, Dr. Jessica Gilman Ernakovich, Dr. Eric Kantorowski, and Dr. Dan Mattern. Mr. Richard King of Pasadena, Texas, Editorial Adviser, has offered numerous suggestions on how to clarify murky explanations. I am grateful to them all.

The people at Pearson have made this project possible. Good books would not exist without their dedication, professionalism, and experience. Among the many people who contributed are: Lee Englander, who originally connected me with this project; Jeanne Zalesky, Executive Editor in Chemistry; Lisa Pierce, Program Manager in Chemistry; and Elisa Mandelbaum, Project Manager in Chemistry. Many thanks.

The entire manuscript was produced using *ChemDraw*®, the remarkable software for drawing chemical structures developed by CambridgeSoft Corp., now part of PerkinElmer Informatics.

Finally, I appreciate my friends and family who supported me throughout this project, most notably my wife and friend of over fifty years, Judy Lang. The students are too numerous to list, but it is for them that all this happens.

Jan William Simek, Professor Emeritus
Department of Chemistry and Biochemistry
Cal Poly State University
San Luis Obispo, CA 93407
Email: jsimek@calpoly.edu

DEDICATION

To my inspirational chemistry teachers:
Joe Plaskas, who made the batter;
Kurt Kaufman, who baked the cake;
Carl Djerassi, who put on the icing;

and to my parents:
Ervin J. and Imilda B. Simek,
who had the original concept.

SYMBOLS AND ABBREVIATIONS

Below is a list of symbols and abbreviations used in this Solutions Manual, consistent with those used in the textbook by Wade; see the inside front cover of the text. (Do not expect all of these to make sense to you now. You will learn them throughout your study of organic chemistry.)

BONDS

—— a single bond

═ a double bond

≡ a triple bond

 a bond in three dimensions, coming out of the paper toward the reader

 a bond in three dimensions, going behind the paper away from the reader

-------- a stretched bond, in the process of forming or breaking

ARROWS

→ in a reaction, shows direction from reactants to products

⇌ signifies equilibrium (not to be confused with resonance)

↔ signifies resonance (not to be confused with equilibrium)

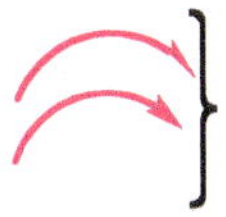 shows direction of electron movement:
the arrowhead with one barb shows movement of one electron;
the arrowhead with two barbs shows movement of a pair of electrons

 shows polarity of a bond or molecule, the arrowhead signifying the more negative end of the dipole

SUBSTITUENT GROUPS

Me a methyl group, CH_3

Et an ethyl group, CH_2CH_3

Pr a propyl group, a three-carbon group (two possible arrangements)

Bu a butyl group, a four-carbon group (four possible arrangements)

R the general abbreviation for an alkyl group (or any substituent group bonded at carbon)

Ph a phenyl group, the name of a benzene ring as a substituent, represented:

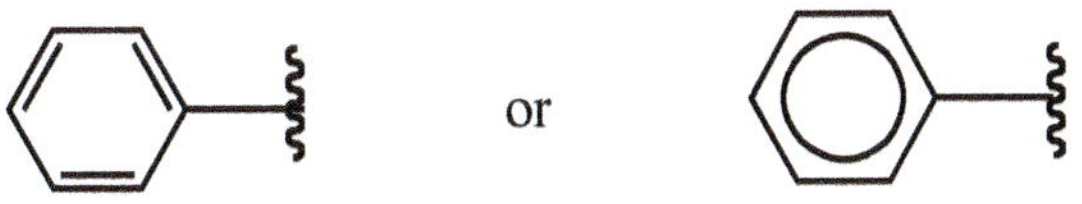

Ar the general abbreviation for an aromatic group, for example, benzene

continued on next page

Symbols and Abbreviations, continued

SUBSTITUENT GROUPS, continued

Ac an **ac**etyl group $CH_3-C(=O)-$

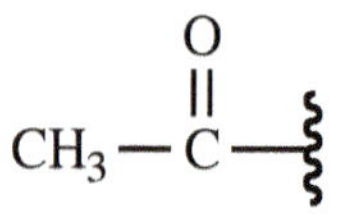

c-Hx a **c**yclo**hex**yl group

TIPS **tri**iso**p**ropyl**s**ilyl

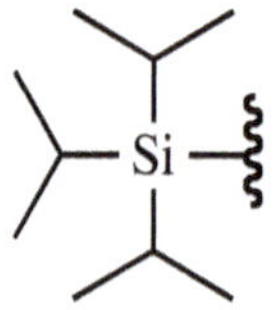

Ts **t**o**s**yl, or *p*-**t**oluene**s**ulfonyl group $CH_3-C_6H_4-SO_2-$

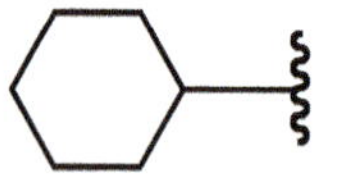

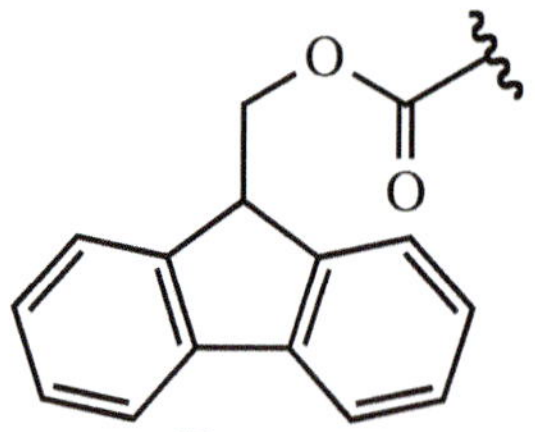

Fmoc

9-(**f**luorenyl**m**eth**o**xy**c**arbonyl)

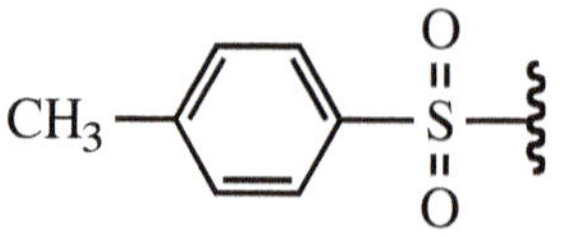

REAGENTS AND SOLVENTS

DCC **d**i**c**yclohexyl**c**arbodiimide N=C=N

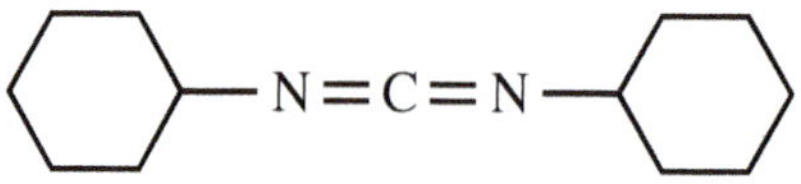

DMP **D**ess-**M**artin **p**eriodinane

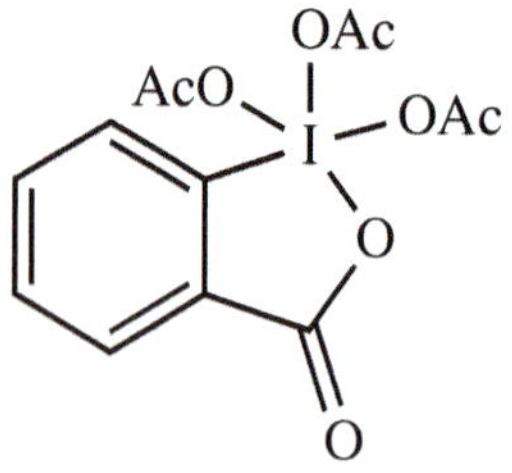

OAc AcO I OAc O O

DMSO 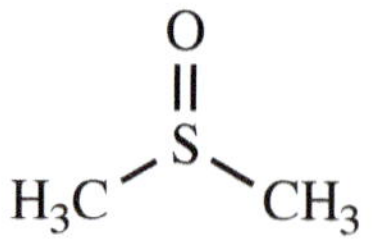 **d**i**m**ethyl**s**ulf**o**xide $H_3C-S(=O)-CH_3$

ether diethyl ether, $CH_3CH_2OCH_2CH_3$

HA or H—A is a generic acid; the conjugate base may appear as: A^- $A^{\ominus}$ $:A^{\ominus}$

LG **l**eaving **g**roup

mCPBA *meta*-**c**hloro**p**eroxy**b**enzoic **a**cid Cl C—O—OH

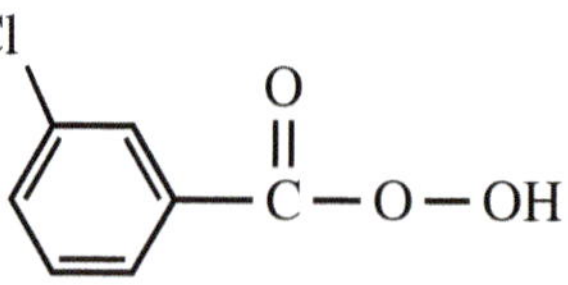

MVK **m**ethyl **v**inyl **k**etone $H_3C-C(=O)-CH=CH_2$

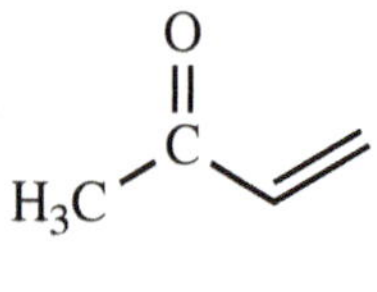

NBS *N*-**b**romo**s**uccinimide O N O Br

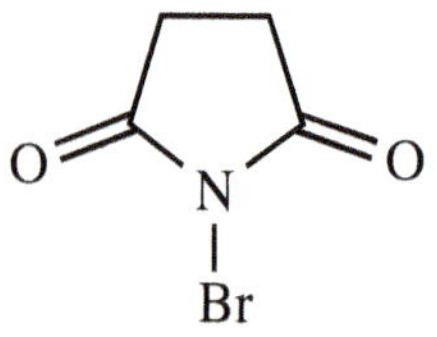

continued on next page

Symbols and Abbreviations, continued

REAGENTS AND SOLVENTS, continued

Nuc or **:**Nuc or Nuc^- is a generic nucleophile, a Lewis base; E or E^+ is a generic electrophile, a Lewis acid

PCC	**p**yridinium **c**hloro**c**hromate,	$CrO_3 \cdot HCl \cdot$ N(pyridine)
Sia_2BH	di**sia**mylborane	$(CH_3)_2CH{-}CH(CH_3){-}BH{-}CH(CH_3){-}CH(CH_3)_2$
THF	**t**etra**h**ydro**f**uran	
TEMPO	2,2,6,6-**te**tra**m**ethyl**p**iperidinyl-1-**o**xy	N−O•

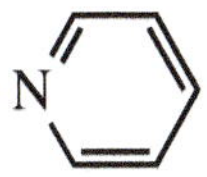

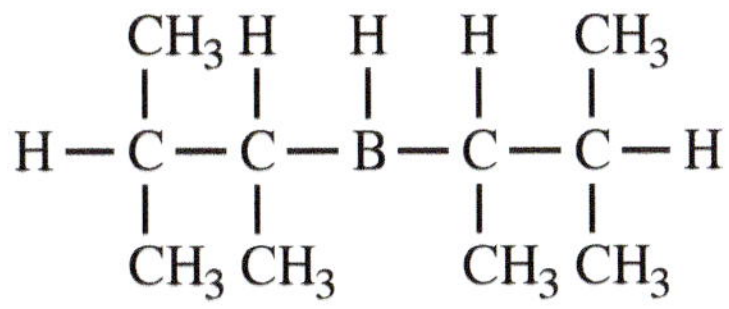

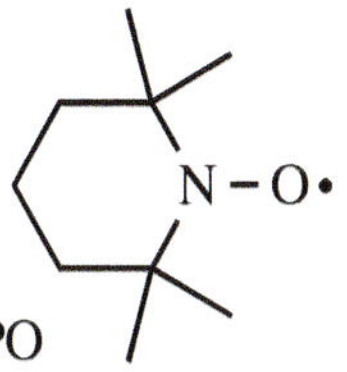

SPECTROSCOPY

IR	**i**nfra**r**ed spectroscopy
NMR	**n**uclear **m**agnetic **r**esonance spectroscopy
MS	**m**ass **s**pectrometry
UV	**u**ltra**v**iolet spectroscopy
ppm	**p**arts **p**er **m**illion, a unit used in NMR
Hz	hertz, cycles per second, a unit of frequency
MHz	megahertz, millions of cycles per second
TMS	**t**etra**m**ethyl**s**ilane, $(CH_3)_4Si$, the reference compound in NMR
s, d, t, q, m	**s**inglet, **d**oublet, **t**riplet, **q**uartet, **m**ultiplet: the number of peaks an NMR absorption gives
nm	nanometers, 10^{-9} meters (usually used as a unit of wavelength)
m/z	mass-to-charge ratio, in mass spectrometry
δ	in NMR, chemical shift value, measured in ppm (Greek lower case delta)
λ	wavelength (Greek lambda)
ν	frequency (Greek nu)
J	coupling constant in NMR

OTHER

•• or **:**	unshared electron pair
a, ax	axial (in chair forms of cyclohexane)
e, eq	equatorial (in chair forms of cyclohexane)
HOMO	**h**ighest **o**ccupied **m**olecular **o**rbital
LUMO	**l**owest **u**noccupied **m**olecular **o**rbital
NR	no reaction
o, m, p	*ortho, meta, para* (positions on an aromatic ring)
Δ	when written over an arrow: "heat"; when written before a letter: "change in"
δ^+, δ^-	partial positive charge, partial negative charge
$h\nu$	energy from electromagnetic radiation (light)
$[\alpha]_D$	specific rotation at the D line of sodium (589 nm)

Students: Add your own notes on symbols and abbreviations.

CHAPTER 1—STRUCTURE AND BONDING

1-1

(a) Nitrogen has atomic number 7, so all nitrogen atoms have 7 protons. The mass number is the total number of neutrons and protons; therefore, ^{13}N has 6 neutrons, ^{14}N has 7 neutrons, ^{15}N has 8 neutrons, ^{16}N has 9 neutrons, and ^{17}N has 10 neutrons.

(b)

Na	$1s^2 2s^2 2p^6 3s^1$	P	$1s^2 2s^2 2p^6 3s^2 3p_x^1 3p_y^1 3p_z^1$
Mg	$1s^2 2s^2 2p^6 3s^2$	S	$1s^2 2s^2 2p^6 3s^2 3p_x^2 3p_y^1 3p_z^1$
Al	$1s^2 2s^2 2p^6 3s^2 3p_x^1$	Cl	$1s^2 2s^2 2p^6 3s^2 3p_x^2 3p_y^2 3p_z^1$
Si	$1s^2 2s^2 2p^6 3s^2 3p_x^1 3p_y^1$	Ar	$1s^2 2s^2 2p^6 3s^2 3p_x^2 3p_y^2 3p_z^2$

1-2 Lines between atom symbols represent covalent bonds between those atoms. Nonbonding electrons are indicated with dots.

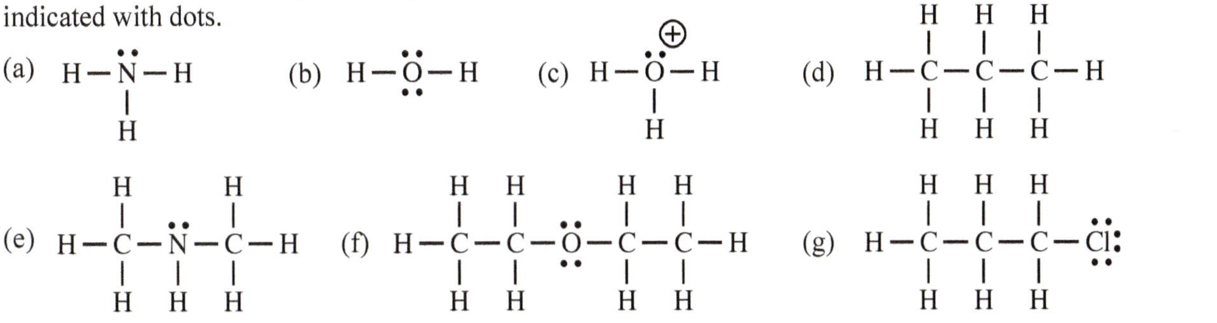

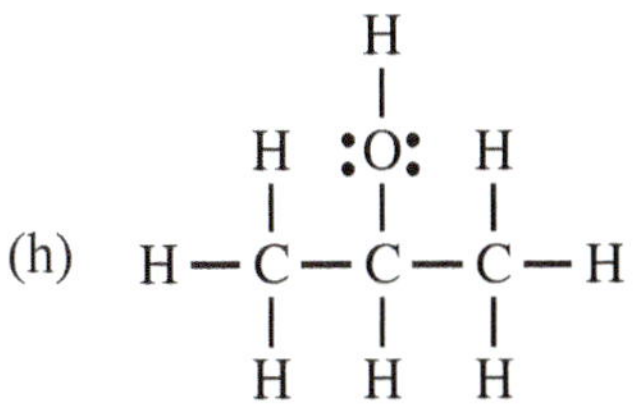

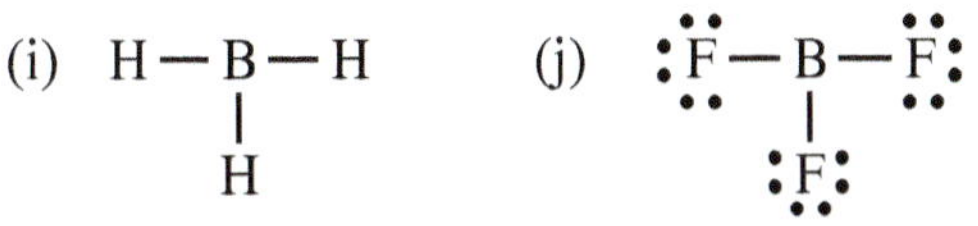

(a) H—N̈—H with H below N
(b) H—Ö—H (two lone pairs on O)
(c) H—Ö—H with H below O, ⊕ on O
(d) H—C—C—C—H

(e) H—C—N̈—C—H; first C with H above and below, N with H below, last C with H above and below
(f) H—C—C—Ö—C—C—H, each C bearing H above and below; O with two lone pairs
(g) H—C—C—C—C̈l: , each C bearing H above and below; Cl with three lone pairs

(h) H—C—C—C—H, each C bearing H below; first and last C bearing H above; middle C bearing :O: above, with H on O

(i) H—B—H with H below B
(j) :F̤̈—B—F̤̈: with :F̤̈: below B

The compounds in (i) and (j) are unusual in that boron does not have an octet of electrons—normal for boron because it has only three valence electrons.

1-3

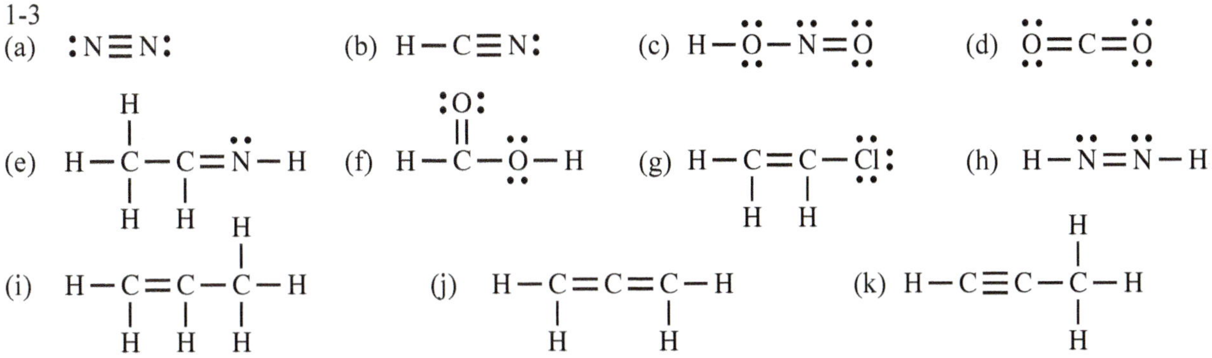

(a) :N≡N:
(b) H—C≡N:
(c) H—Ö—N̈=Ö (lone pairs on O, N, O)
(d) Ö=C=Ö

(e) H—C—C=N̈—H; first C with H above and below, second C with H below
(f) H—C—Ö—H with :O: double-bonded above C
(g) H—C=C—C̈l: ; each C with H below
(h) H—N̈=N̈—H

(i) H—C=C—C—H; each C with H below, last C also with H above
(j) H—C=C=C—H; first and last C with H below
(k) H—C≡C—C—H; last C with H above and below

1-4 There are no unshared electron pairs in parts (i), (j), and (k).

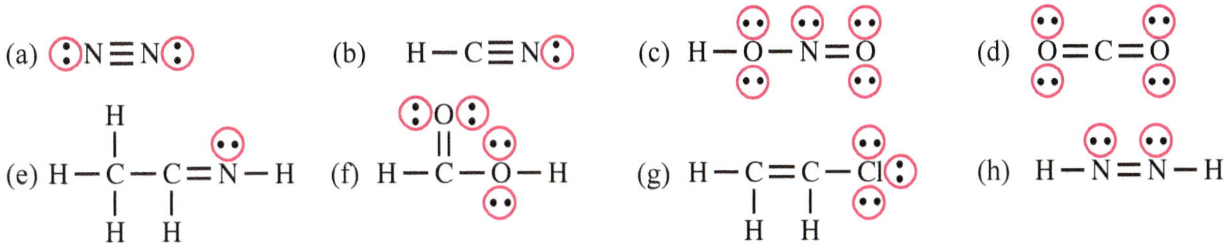

(a) :N≡N: (unshared pairs circled)
(b) H—C≡N: (unshared pair circled)
(c) H—Ö—N̈=Ö (unshared pairs circled)
(d) Ö=C=Ö (unshared pairs circled)

(e) H—C—C=N̈—H; first C with H above and below, second C with H below (unshared pair circled)
(f) H—C—Ö—H with :O: double-bonded above C (unshared pairs circled)
(g) H—C=C—C̈l: ; each C with H below (unshared pairs circled)
(h) H—N̈=N̈—H (unshared pairs circled)

1-5 The symbols "δ^{+}" and "δ^{-}" indicate bond polarity by showing partial charge. (In the arrow symbolism, the arrow should point to the partial negative charge.)

(a) $\overset{\delta+}{C}-\overset{\delta-}{Cl}$ (b) $\overset{\delta+}{C}-\overset{\delta-}{O}$ (c) $\overset{\delta+}{C}-\overset{\delta-}{N}$ (d) $\overset{\delta+}{C}-\overset{\delta-}{S}$ (e) $\overset{\delta-}{C}-\overset{\delta+}{B}$

(f) $\overset{\delta+}{N}-\overset{\delta-}{Cl}$ (g) $\overset{\delta+}{N}-\overset{\delta-}{O}$ (h) $\overset{\delta-}{N}-\overset{\delta+}{S}$ (i) $\overset{\delta-}{N}-\overset{\delta+}{B}$ (j) $\overset{\delta+}{B}-\overset{\delta-}{Cl}$

1-6 Non-zero formal charges are shown beside the atoms, circled for clarity.

(a) $H_3C-\overset{\oplus}{O}H_2$ (b) $H_4N^{\oplus}\ :\!Cl\!:^{\ominus}$ (c) $(CH_3)_4N^{\oplus}\ :\!Cl\!:^{\ominus}$

In (b) and (c), the chlorine is present as chloride ion. There is no covalent bond between chloride and other atoms in the formula.

(d) $Na^{\oplus}\ {}^{\ominus}:\!O-CH_3$ (e) $H-\overset{\oplus}{C}H-H$ (f) $H-\overset{\ominus}{\ddot{C}}H-H$ (g) $Na^{\oplus}\ H-\overset{\ominus}{B}H_2-H$

(h) $Na^{\oplus}\ H-\overset{\ominus}{B}H_2-C\equiv N:$ (i) $H_3C-\overset{\oplus}{O}(-\overset{\ominus}{B}F_3)-CH_3$ (j) $H-\ddot{O}-\overset{\oplus}{N}H_2-H$

(k) $K^{\oplus}\ {}^{\ominus}:\!O-C(CH_3)_3$ (l) $H-C(H)=\overset{\oplus}{\ddot{O}}-H$

As shown in (d), (g), (h), and (k), alkali metals like sodium and potassium form only ionic bonds, never covalent bonds.

1-7 Resonance forms in which all atoms have full octets are the most significant contributors. In resonance forms, ALL ATOMS KEEP THEIR POSITIONS—ONLY ELECTRONS ARE SHOWN IN DIFFERENT POSITIONS. (In this Solutions Manual, braces {} are commonly used to denote resonance forms.)

One more helpful hint when drawing resonance forms of structures with charges: *move electrons toward (+) charges and away from (–) charges.* Arrows show how to alter one structure to make it into the next one. In Chapter 1, arrows in resonance forms are the same as the green arrows in the text.

(a) $\left\{ {}^{\ominus}O-C(=O)-O^{\ominus} \longleftrightarrow {}^{\ominus}O-C(-O^{\ominus})=O \longleftrightarrow O=C(-O^{\ominus})-O^{\ominus} \right\}$ Move electrons away from (–) charge.

(b) $\left\{ H-CH=CH-\overset{\oplus}{C}H-H \longleftrightarrow H-\overset{\oplus}{C}H-CH=CH-H \right\}$ Move electrons toward (+) charge.

1-7 continued

(c) 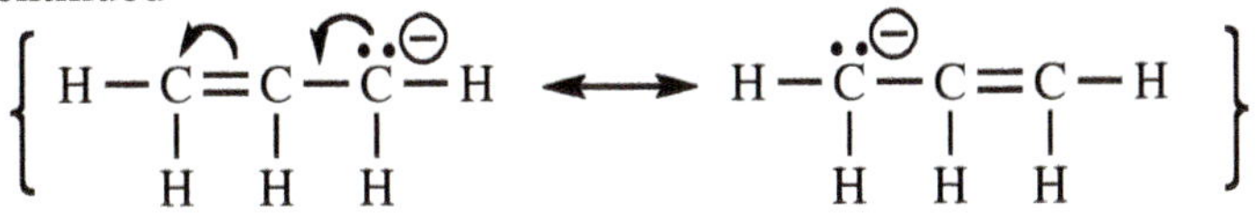 Move electrons away from (–) charge.

(d) 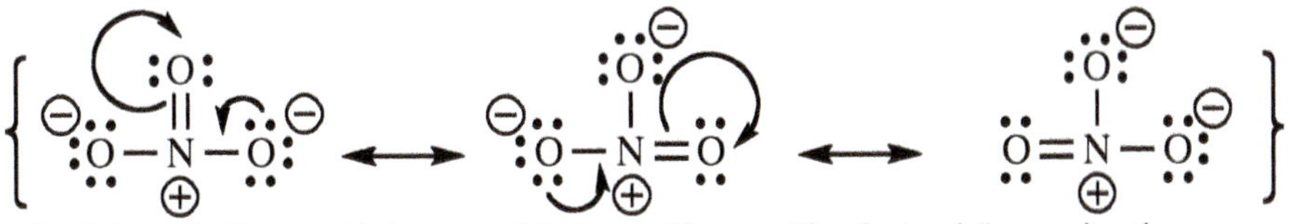 Move electrons away from (–) charge.

Look how similar part (d) is to part (a) above. They are "isoelectronic", meaning the same number of electrons.

(e)  Move electrons away from (–) charge.

(f)

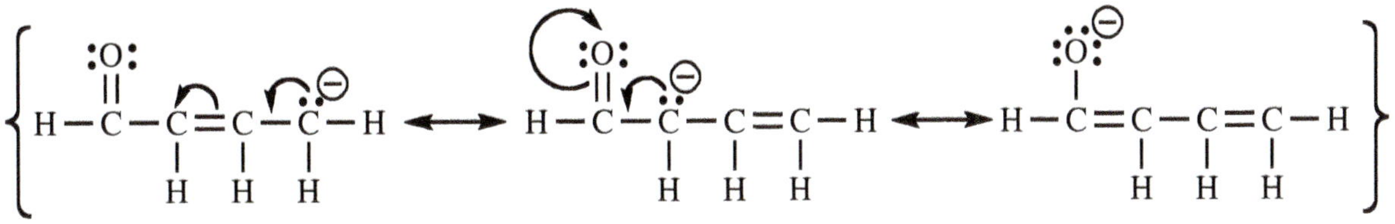

Move electrons away from (–) charge.

(g) There are three resonance forms shown here; the structure given in the problem ("*original*") is shown twice because the electrons can be delocalized in two different directions; each of the other structures is derived from the original.

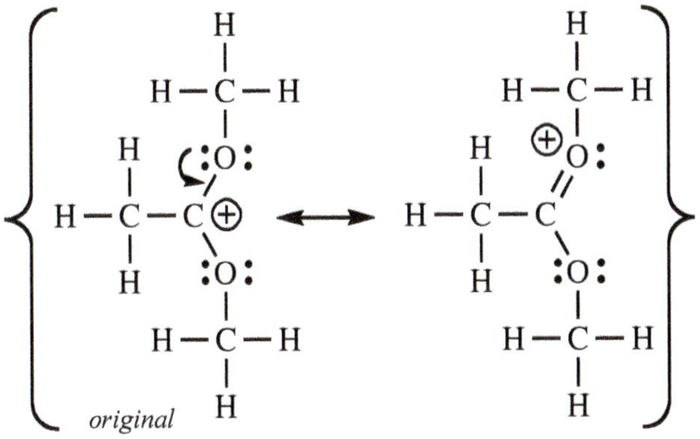

original | *original*

Move electrons toward (+) charge. 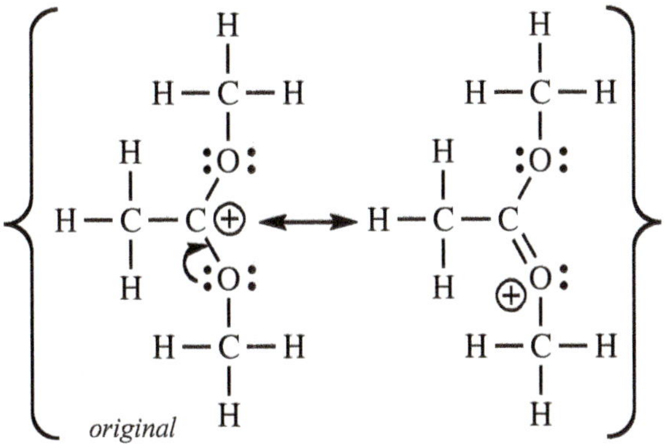 | Move electrons toward (+) charge.

(h) In similar fashion to part (g), the three other resonance forms are derived from the first. In this solution, however, for simplicity, only the first structure has the arrow showing how the first structure can be made into the second. It should be apparent to you to see how to make the first structure into the third and fourth. Alternatively, you can transform the second into the third directly by using two arrows.

It is very important to keep in mind that these arrows are simply helpful devices to make one picture into another. *They do not mean that electrons are moving!* There is one real chemical species that is an average of all the individual resonance forms which are just pictures.

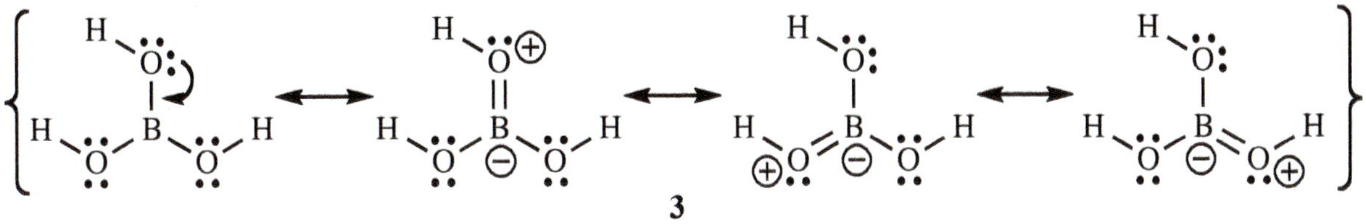

1-8 Major resonance contributors would have the lowest energy. The most important factors are: maximize full octets; maximize number of bonds; put negative charge on electronegative atoms; minimize charge separation— see the Problem-Solving Hint above Solved Problem 1-2. Part (a) has been solved in the text.

(b) H−C−C≡N: ⟷ H−C=C=N:

minor — major

The second structure has negative charge on the more electronegative atom.

(c) H−C=O−C−H ⟷ H−C−O−C−H

major — minor

The first structure has full octets and one more bond.

(d) H−C−N=O ⟷ H−C−N−O: ⟷ H−C=N−O:

minor — minor — major

All atoms have octets; same number of bonds; third structure has both (−) charges on the more electronegative oxygen atoms instead of carbon.

(e)

minor — major — major

The latter two structures have equivalent energy and are major because they have full octets and more bonds.

(f) H−C=C−N−H ⟷ H−C−C=N−H

major — minor

The first structure has the negative charge on the more electronegative nitrogen atom.

(g) There are three resonance forms shown here; the structure given in the problem ("*original*") is shown twice in order to demonstrate how each of the other structures is derived from the original.

H−C−C−C−H ⟷ H−C=C−C−H

original

minor — major

H−C−C−C−H ⟷ H−C−C=C−H

original

minor — major

The latter two structures have equivalent energy and are major because the negative charge is on the more electronegative oxygen atom rather than on the less electronegative carbon atom.

1-8 continued

(h) There are three resonance forms shown here; the structure given in the problem ("*original*") is shown twice in order to demonstrate how each of the other structures is derived from the original.

All three resonance forms have full octets and the same number of bonds. The second structure is the most significant contributor because the negative charge is on the oxygen (most electronegative); the third structure with negative charge on nitrogen is next; the least significant contributor is the first structure ("*original*") with negative charge on the least electronegative carbon atom.

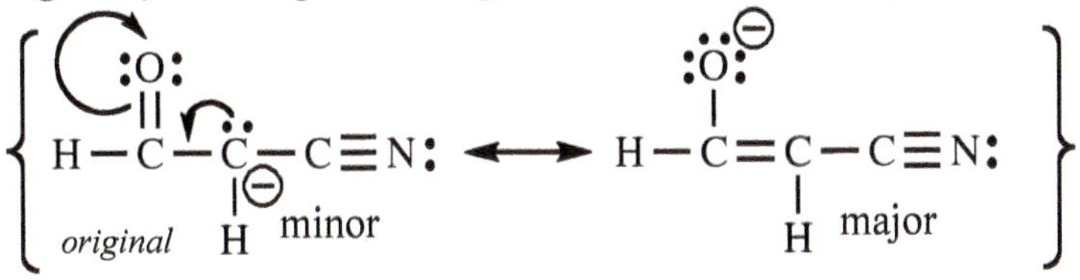

H—C(=O)—C^{-}(H)—C≡N: ⟷ H—C(O^{-})=C(H)—C≡N:

original minor ⟷ major

Move electrons away from (–) charge.

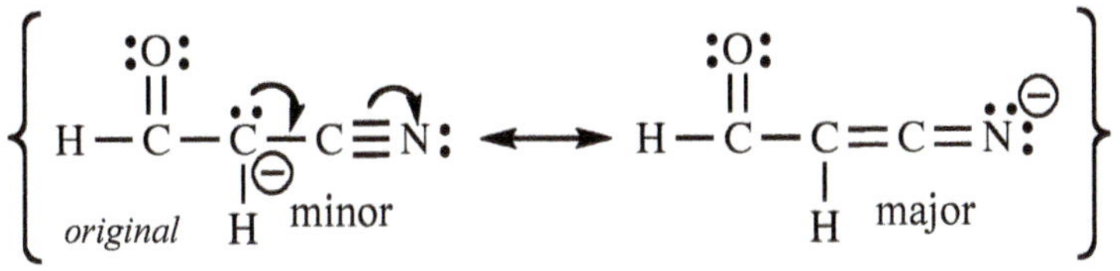

H—C(=O)—C^{-}(H)—C≡N: ⟷ H—C(=O)—C(H)=C=N:$^{-}$

original minor ⟷ major

1-9

(a) There are four resonance forms shown here; the structure given in the problem ("*original*") is shown twice in order to demonstrate how each of the other structures is derived from the original.

H_3C—O—C^{+}(H)—C(H)=C(H)—N(OH)—H ⟷ H_3C—O^{+}=C(H)—C(H)=C(H)—N(OH)—H

original major—full octets

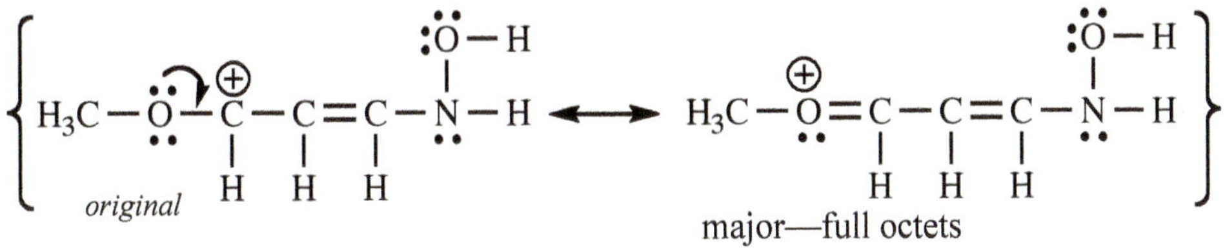

Move electrons toward (+) charge.

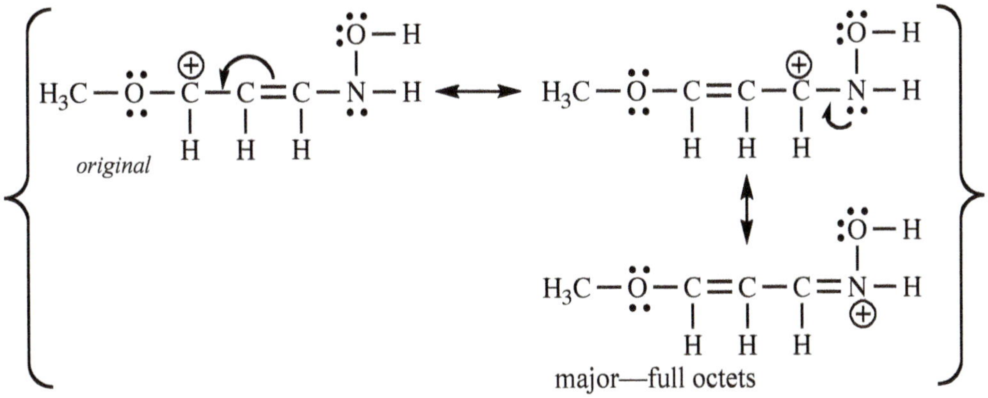

H_3C—O—C^{+}(H)—C(H)=C(H)—N(OH)—H ⟷ H_3C—O—C(H)=C(H)—C^{+}(H)—N(OH)—H

⟷ H_3C—O—C(H)=C(H)—C(H)=N^{+}(OH)—H

original major—full octets

Although the O and N have (+) charges in these major resonance forms, their full octets are more important than charge in determining the significance of their contribution to the resonance hybrid.

1-9 continued

(b) There are five resonance forms shown here; one of the structures is duplicated to show how the other resonance forms are derived from it. Resonance forms with all atoms having their full octets are the most significant contributors. In this case, the ones with full octets also have more bonds.

major—full octets

this structure repeated below

Move electrons toward (+) charge.

major—full octets

this structure repeated from above

major—full octets

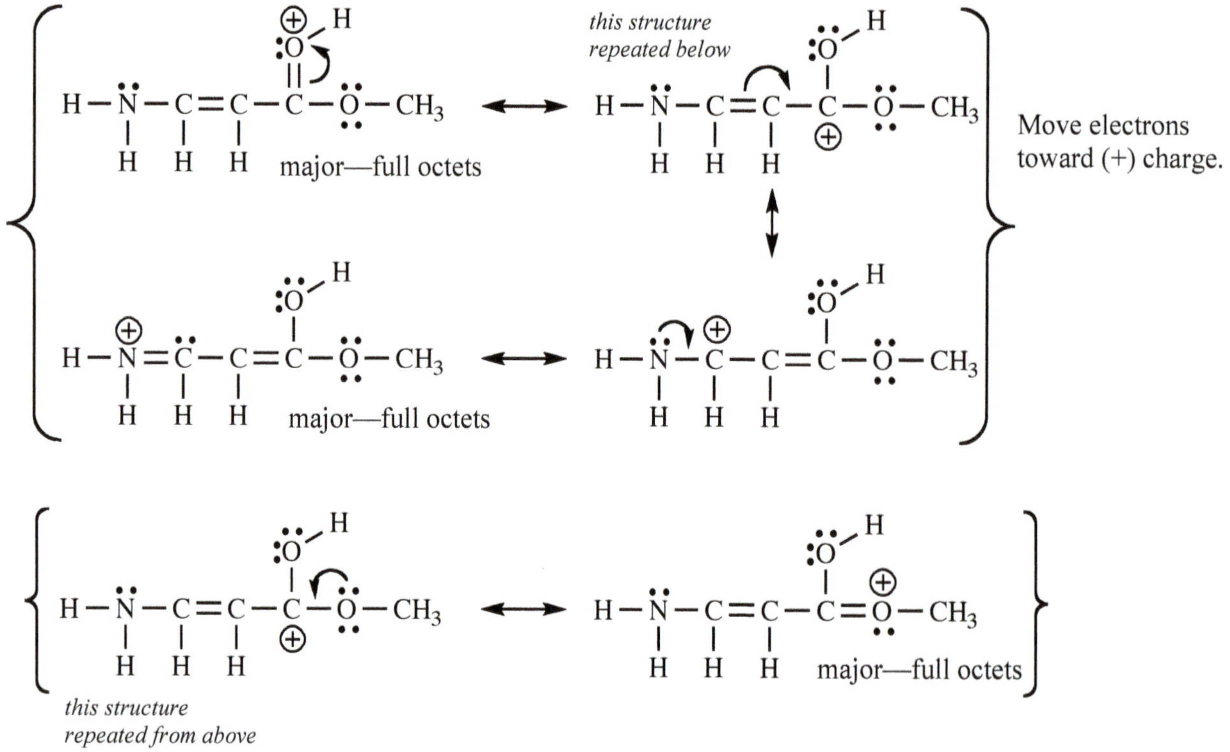

(c) There are four resonance forms shown here; one of the structures is duplicated to show how the other resonance forms are derived from it. Resonance forms with all atoms having their full octets are the most significant contributors. In this case, the ones with full octets also have more bonds.

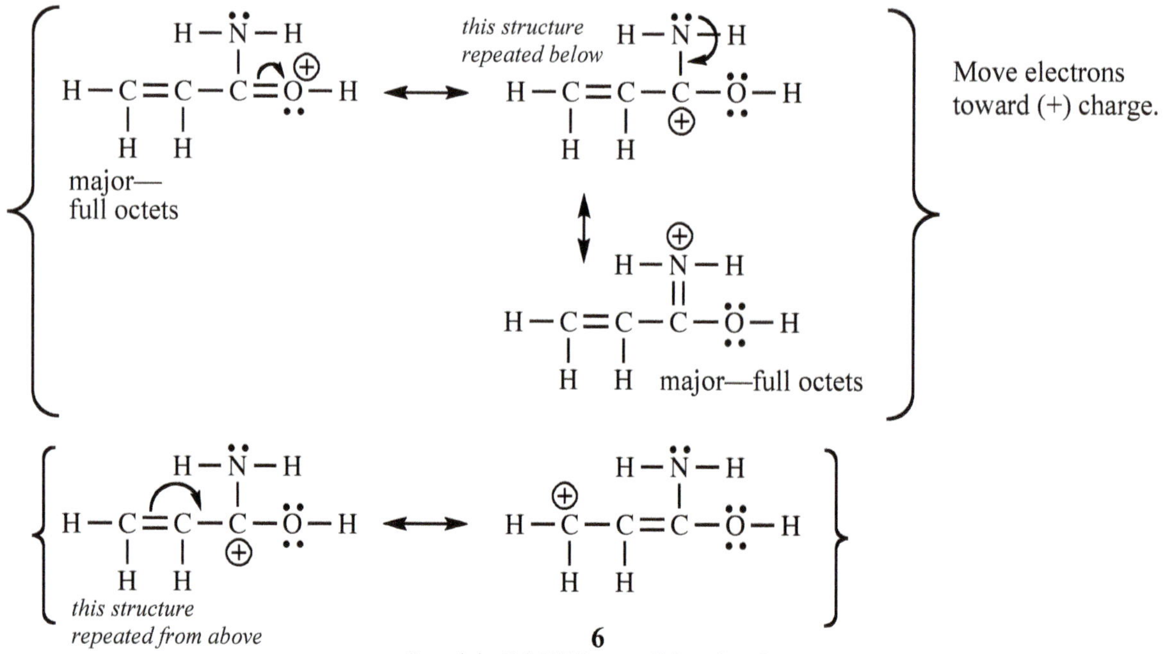

major—full octets

this structure repeated below

Move electrons toward (+) charge.

major—full octets

this structure repeated from above

1-9 continued

(d) The lone pair of electrons on C cannot be delocalized to the C=C or the triple bond.

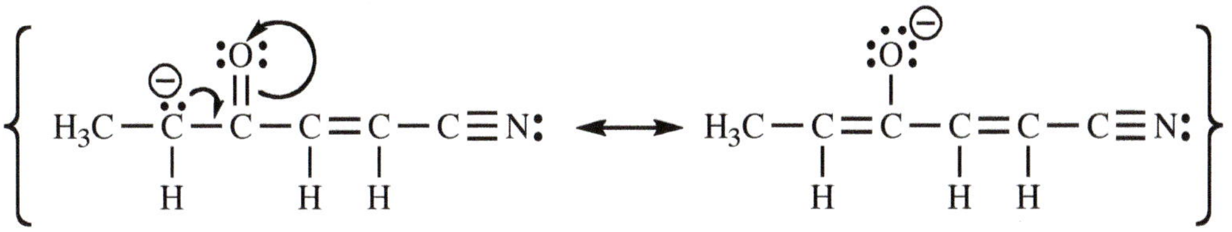

$$\left\{ H_3C-\overset{\ominus}{\ddot{C}}H-C(=O)-CH=CH-C\equiv N: \longleftrightarrow H_3C-CH=C(-\ddot{O}:^{\ominus})-CH=CH-C\equiv N: \right\}$$

major—negative charge on the more electronegative atom

Move electrons away from (–) charge.

(e) There are four resonance forms shown here; the original structure is duplicated to show how the other resonance forms are derived from it.

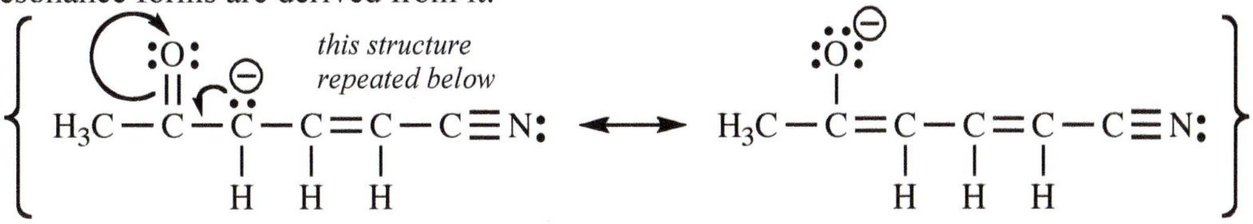

$$\left\{ H_3C-C(=O)-\overset{\ominus}{\ddot{C}}H-CH=CH-C\equiv N: \longleftrightarrow H_3C-C(-\ddot{O}:^{\ominus})=CH-CH=CH-C\equiv N: \right\}$$

this structure repeated below

major—negative charge on the more electronegative atom

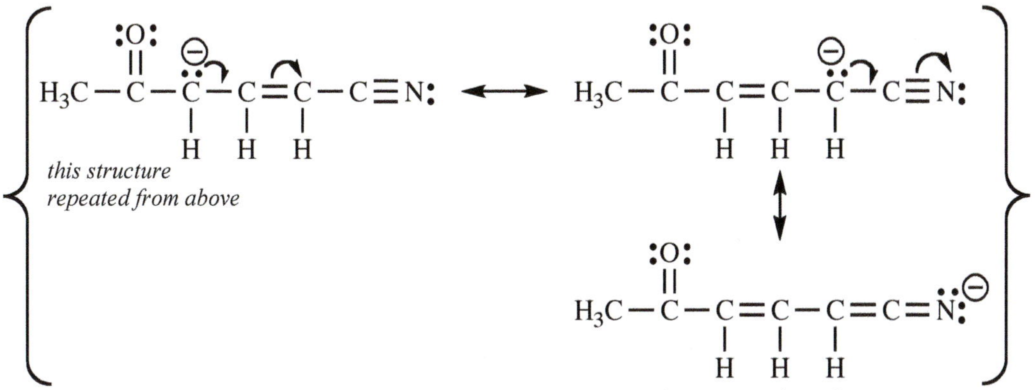

Move electrons away from (–) charge.

$$\left\{ H_3C-C(=O)-\overset{\ominus}{\ddot{C}}H-CH=CH-C\equiv N: \longleftrightarrow H_3C-C(=O)-CH=CH-\overset{\ominus}{\ddot{C}}H-C\equiv N: \longleftrightarrow H_3C-C(=O)-CH=CH-CH=C=\ddot{N}:^{\ominus} \right\}$$

this structure repeated from above

major—negative charge on the more electronegative atom

(f) The lone pair of electrons on C cannot be delocalized to the C=C.

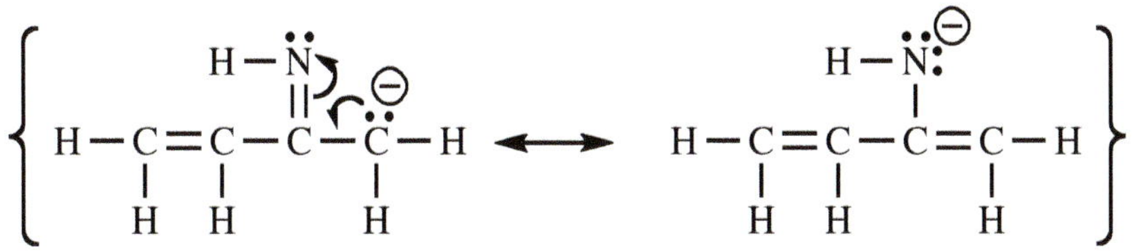

$$\left\{ H-CH=CH-C(=\ddot{N}-H)-\overset{\ominus}{\ddot{C}}H-H \longleftrightarrow H-CH=CH-C(-\ddot{N}:^{\ominus}-H)=CH-H \right\}$$

major—negative charge on the more electronegative atom

Move electrons away from (–) charge.

1-10 Resonance structures depict the distribution of electrons around the molecule. The significant resonance contributors suggest the areas of highest and lowest electron density.

(a)

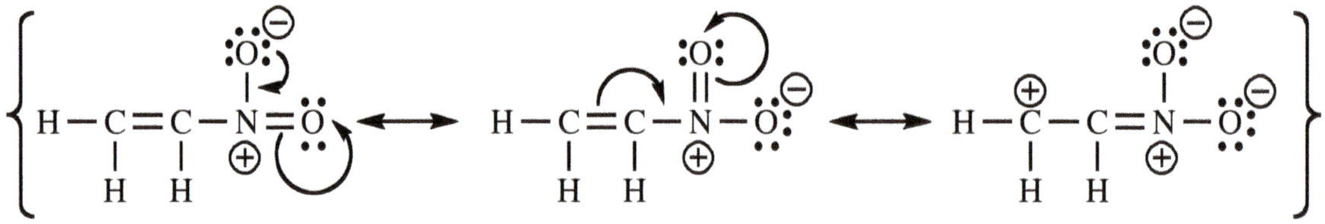

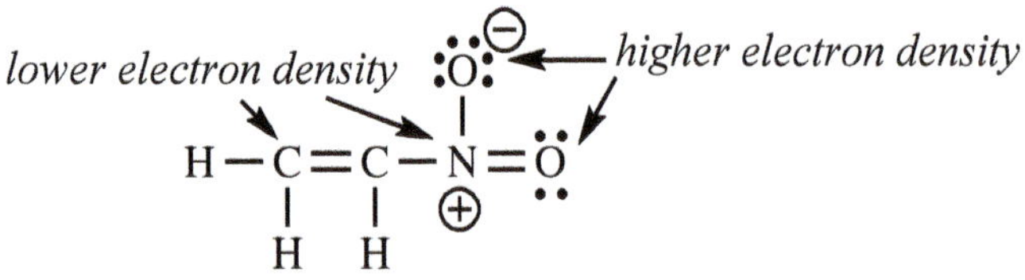

lower electron density

higher electron density

Although the N has a (+) charge in all resonance forms, it still has an octet and is not as electron-deficient as the CH_2 in the form where the CH_2 bears a (+) charge.

(b)

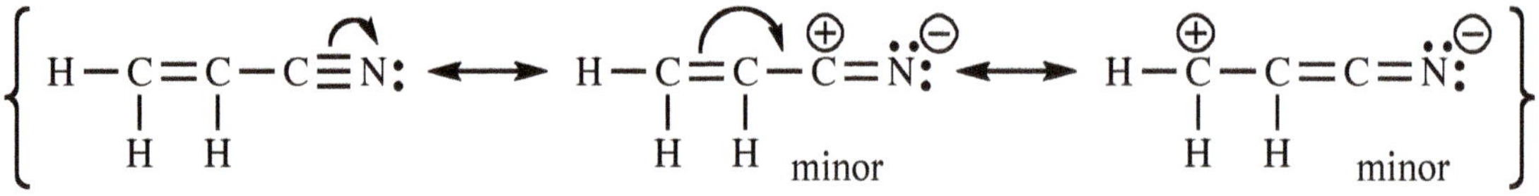

minor

minor

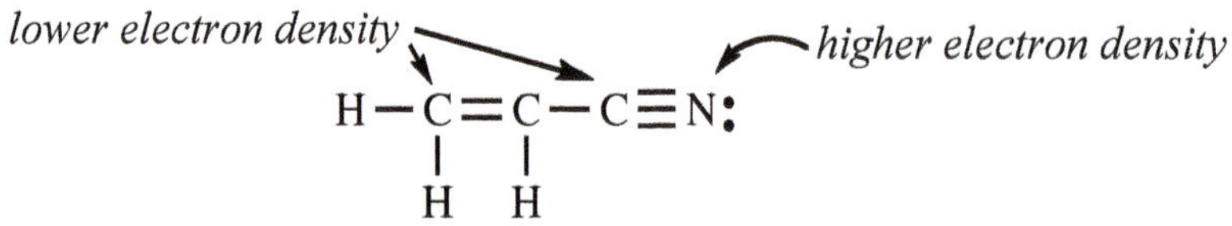

lower electron density

higher electron density

(c)

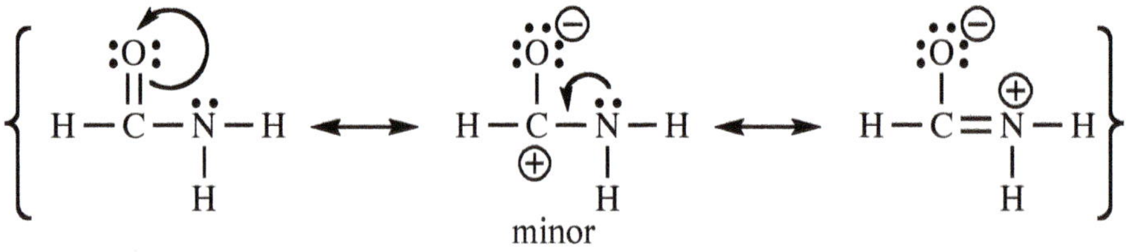

minor

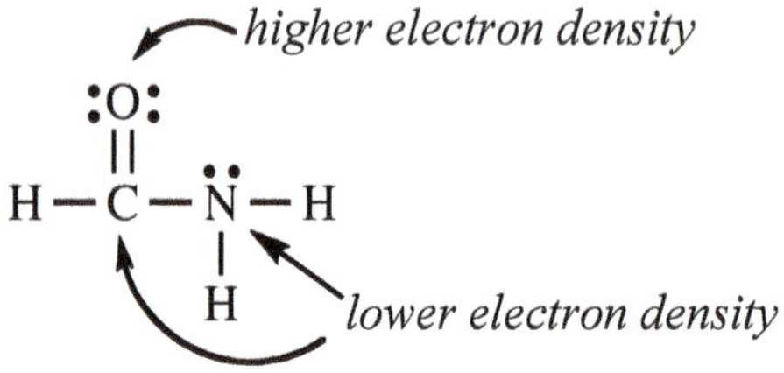

higher electron density

lower electron density

(d)

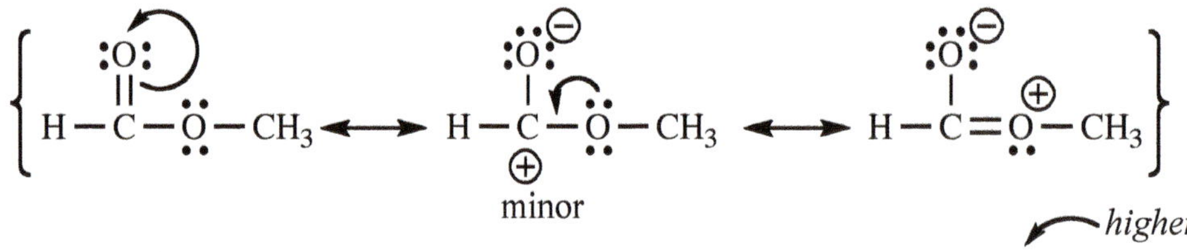

minor

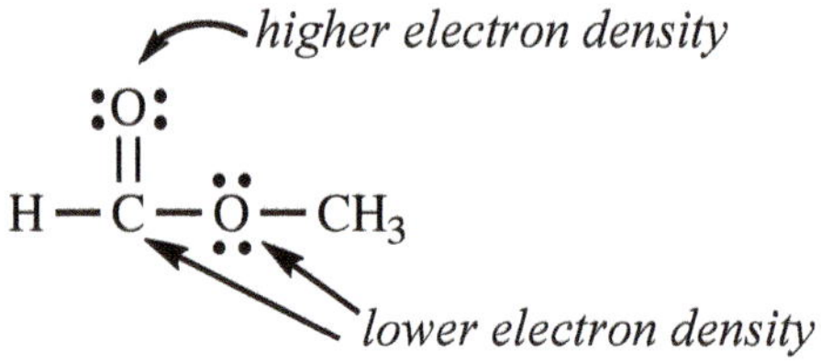

higher electron density

lower electron density

1-10 continued

(e)

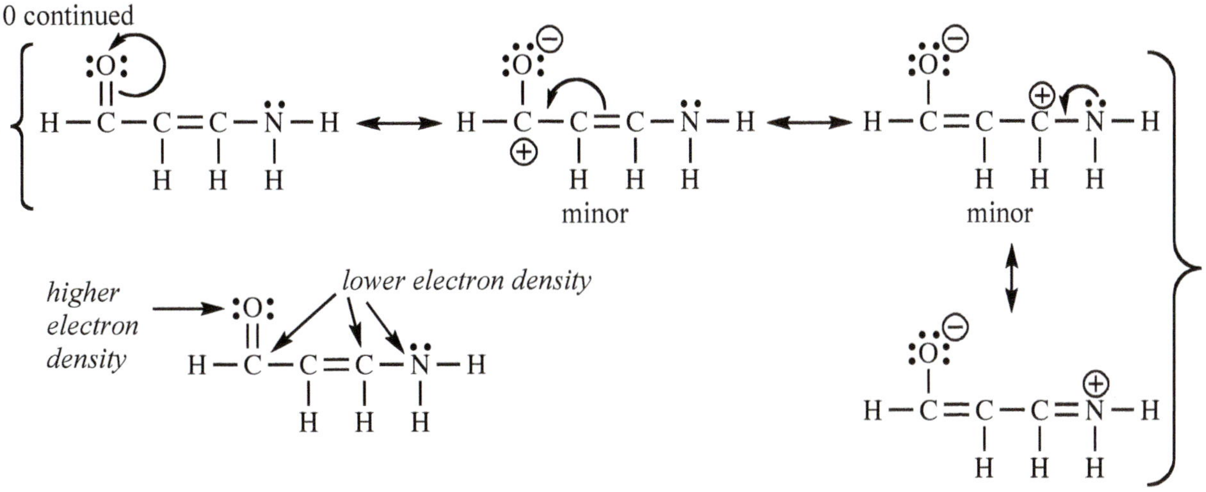

minor

minor

higher electron density

lower electron density

(f)

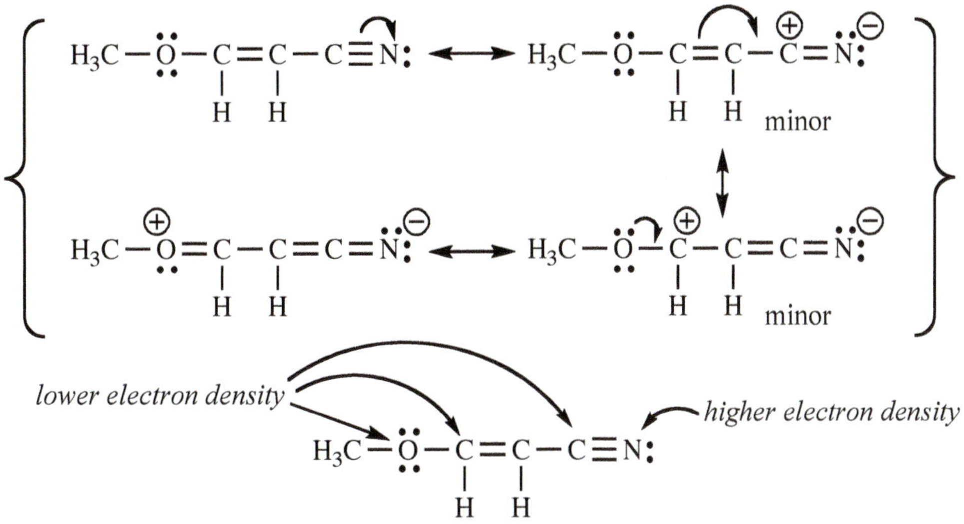

minor

minor

lower electron density

higher electron density

1-11 Your Lewis structures may *appear* different from how these are drawn. As long as the atoms are connected in the same order and by the same type of bond, they are equivalent structures. For now, the exact placement of the atoms on the page is not significant. A Lewis structure is "complete" with unshared electron pairs shown.

(a) 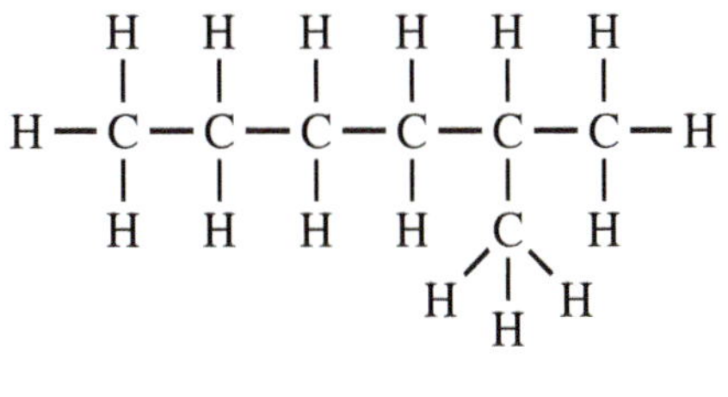 (b) 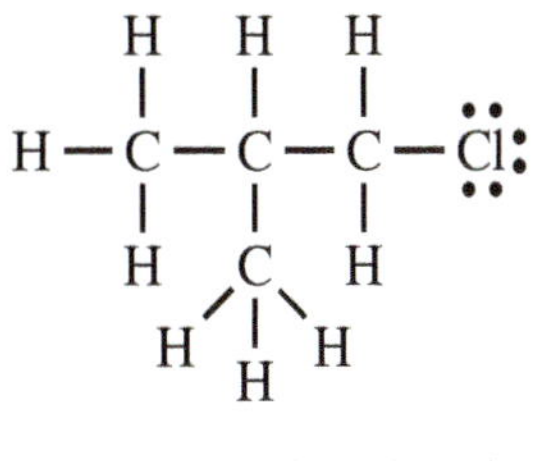 (c)

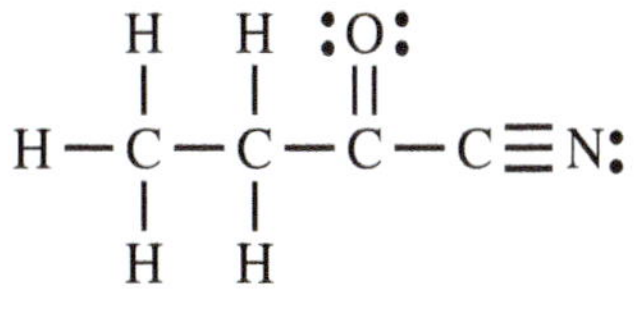

Always be alert for the implied double or triple bond. Remember that the normal valence of C is four bonds, nitrogen has three bonds, oxygen has two bonds, and hydrogen has one bond. The only exceptions to these valence rules are structures with formal charges. (We will see other unusual exceptions in later chapters.)

1-11 continued

(d) (e) (f)

(g) (h)

1-12 *Complete* Lewis structures display all atoms, bonds, and unshared electron pairs.

(a) $C_6H_{13}N$

(b) $C_8H_{16}O$

(c) C_4H_5N

(d) $C_5H_{10}O$

(e) $C_7H_{10}O$

(f) C_6H_8O

(g) C_8H_8O

(h) $C_5H_{10}O$

1-13 Line-angle structures, sometimes called "stick" figures, usually omit unshared electron pairs.

(a) C_7H_{16}

(b) C_4H_9Cl

(c) C_4H_5NO

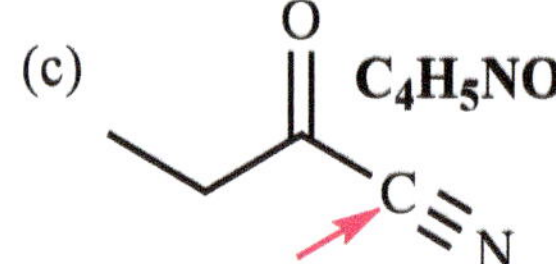

C is usually not shown but this clarifies what ends this triple bond.

(d) C_3H_4O

H on C is usually not shown but this is an exception; it clarifies what ends this chain.

1-13 continued

(e) $C_7H_{12}O$

better placement

not as good

(f) $C_3H_4O_3$

(g) $C_5H_{10}O$

(h) $C_4H_{10}O$

not as good

1-14 If the percent values do not sum to 100%, the remainder must be oxygen. Assume 100 g of sample; percents then translate directly to grams of each element.

There are usually MANY possible structures for a molecular formula. Yours may be different from the examples shown here and they could still be correct.

(a) $\frac{40.0 \text{ g C}}{12.0 \text{ g/mole}} = 3.33 \text{ moles C} \div 3.33 \text{ moles} = 1 \text{ C}$

$\frac{6.67 \text{ g H}}{1.01 \text{ g/mole}} = 6.60 \text{ moles H} \div 3.33 \text{ moles} = 1.98 \approx 2 \text{ H}$

$\frac{53.33 \text{ g O}}{16.0 \text{ g/mole}} = 3.33 \text{ moles O} \div 3.33 \text{ moles} = 1 \text{ O}$

empirical formula = CH_2O ⇨ empirical weight = 30.02

molecular weight = 90, three times the empirical weight ⇨

three times the empirical formula = molecular formula = $C_3H_6O_3$

some possible structures:

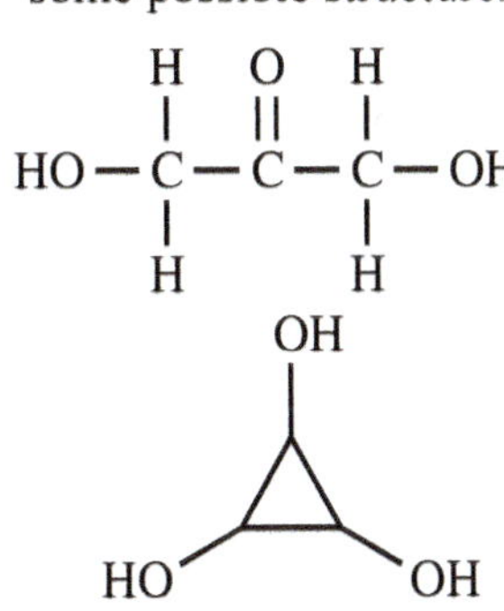

Other structures are possible.

(b) $\frac{32.0 \text{ g C}}{12.0 \text{ g/mole}} = 2.67 \text{ moles C} \div 1.34 \text{ moles} = 1.99 \approx 2 \text{ C}$

$\frac{6.67 \text{ g H}}{1.01 \text{ g/mole}} = 6.60 \text{ moles H} \div 1.34 \text{ moles} = 4.93 \approx 5 \text{ H}$

$\frac{18.7 \text{ g N}}{14.0 \text{ g/mole}} = 1.34 \text{ moles N} \div 1.34 \text{ moles} = 1 \text{ N}$

$\frac{42.6 \text{ g O}}{16.0 \text{ g/mole}} = 2.66 \text{ moles O} \div 1.34 \text{ moles} = 1.99 \approx 2 \text{ O}$

empirical formula = $C_2H_5NO_2$ ⇨ empirical weight = 75.05

molecular weight = 75, same as the empirical weight ⇨

empirical formula = molecular formula = $C_2H_5NO_2$

some possible structures:

$CH_3CH_2NO_2$

$H_2N-C(=O)-O-CH_3$

MANY other structures are possible.

1-14 continued

(c) $\frac{25.6 \text{ g C}}{12.0 \text{ g/mole}} = 2.13 \text{ moles C} \div 1.07 \text{ moles} = 1.99 \approx 2 \text{ C}$

$\frac{4.32 \text{ g H}}{1.01 \text{ g/mole}} = 4.28 \text{ moles H} \div 1.07 \text{ moles} = 4 \text{ H}$

$\frac{37.9 \text{ g Cl}}{35.45 \text{ g/mole}} = 1.07 \text{ moles Cl} \div 1.07 \text{ moles} = 1 \text{ Cl}$

$\frac{15.0 \text{ g N}}{14.0 \text{ g/mole}} = 1.07 \text{ moles N} \div 1.07 \text{ moles} = 1 \text{ N}$

$\frac{17.2 \text{ g O}}{16.0 \text{ g/mole}} = 1.07 \text{ moles O} \div 1.07 \text{ moles} = 1 \text{ O}$

empirical formula = **C_2H_4ClNO** ⇨ empirical weight = 93.49

molecular weight = 93, same as the empirical weight ⇨

empirical formula = molecular formula = **C_2H_4ClNO**

some possible structures:

MANY other structures are possible.

(d) $\frac{38.4 \text{ g C}}{12.0 \text{ g/mole}} = 3.20 \text{ moles C} \div 1.60 \text{ moles} = 2 \text{ C}$

$\frac{4.80 \text{ g H}}{1.01 \text{ g/mole}} = 4.75 \text{ moles H} \div 1.60 \text{ moles} = 2.97 \approx 3 \text{ H}$

$\frac{56.8 \text{ g Cl}}{35.45 \text{ g/mole}} = 1.60 \text{ moles Cl} \div 1.60 \text{ moles} = 1 \text{ Cl}$

empirical formula = **C_2H_3Cl** ⇨ empirical weight = 62.45

molecular weight = 125, twice the empirical weight ⇨

twice the empirical formula = molecular formula = **$C_4H_6Cl_2$**

some possible structures:

MANY other structures are possible.

1-15 *The fundamental principle of organic chemistry is that a molecule's chemical and physical properties depend on the molecule's structure:* the structure-function or structure-reactivity correlation. It is essential that you understand the three-dimensional nature of organic molecules, and there is no better device to assist you than a molecular model set. You are strongly encouraged to use models regularly when reading the text and working the problems.

(a) Requires use of models.

(b)

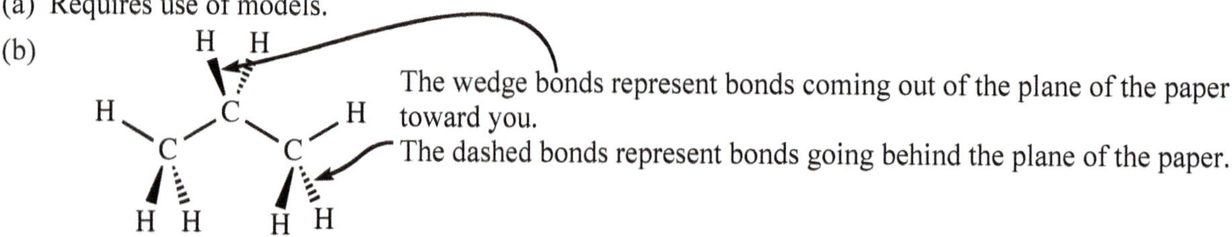

The wedge bonds represent bonds coming out of the plane of the paper toward you.
The dashed bonds represent bonds going behind the plane of the paper.

1-16 (a) The hybridization of oxygen is sp^3 since it has two sigma bonds and two pairs of nonbonding electrons. The reason that the bond angle of 104.5° is less than the perfect tetrahedral angle of 109.5° is that the lone pairs in the two sp^3 orbitals are repelling each other more strongly than the electron pairs in the sigma bonds, thereby compressing the bond angle.

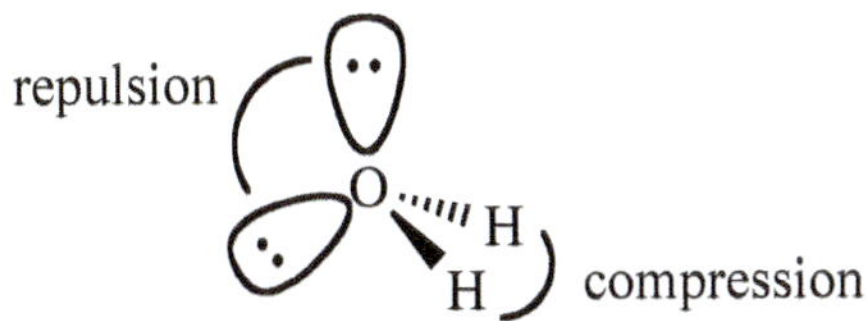

(b) The electrostatic potential map for water shows that the hydrogens have low electron potential (blue), and the area of the unshared electron pairs in sp^3 orbitals has high electron potential (red).

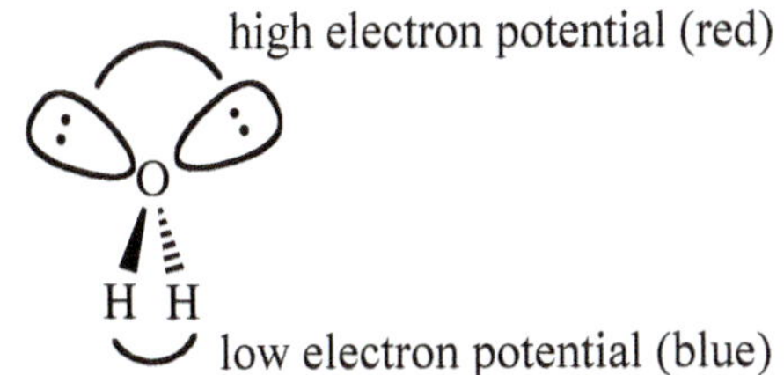

1-17 Each double-bonded atom is sp^2 hybridized with bond angles about 120°; geometry around sp^2 atoms is trigonal planar. In (a), all four carbons and the two hydrogens on the sp^2 carbons are all in one plane. Each carbon on the end is sp^3 hybridized with tetrahedral geometry and bond angles about 109°. In (b), the two carbons, the nitrogen, and the two hydrogens on the sp^2 carbon and nitrogen are all in one plane. The CH_3 carbon is sp^3 hybridized with tetrahedral geometry and bond angles about 109°.

(a) sp^2; 120°; 120°; 109.5°; sp^3

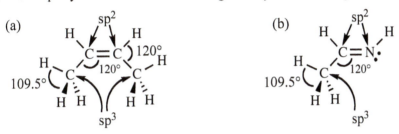

(b) sp^2; 120°; 109.5°; sp^3

1-18 The hybridization of the nitrogen and the triple-bonded carbon are sp, giving linear geometry (C—C—N are linear) and a bond angle around the triple-bonded carbon of 180°. The CH_3 carbon is sp^3 hybridized, tetrahedral, with bond angles about 109°.

sp; 109.5°; 180°; sp^3

1-19

(a) linear, bond angle 180°

O=C=O: sp^2 sp sp^2

(b) All atoms are sp^3; tetrahedral geometry and bond angles of 109° around each atom.

not a bond—shows lone pair coming out of paper

not a bond—shows lone pair going behind paper

1-19 continued

(c) All atoms are sp^3; tetrahedral geometry and bond angles of 109° around each atom.

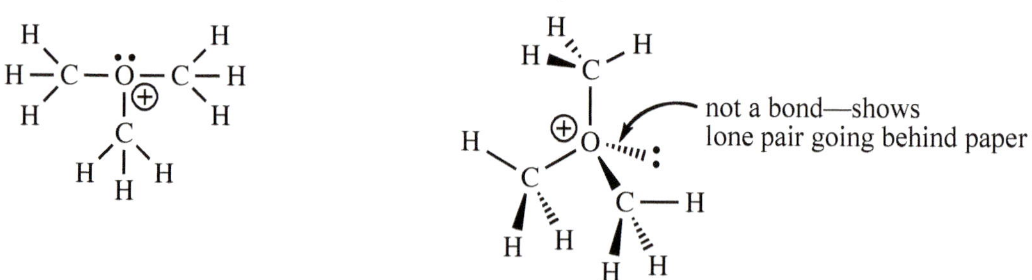

(d) trigonal planar around the carbonyl carbon (C=O), bond angles 120°; tetrahedral around the single-bonded oxygen and the CH_3, bond angles 109°

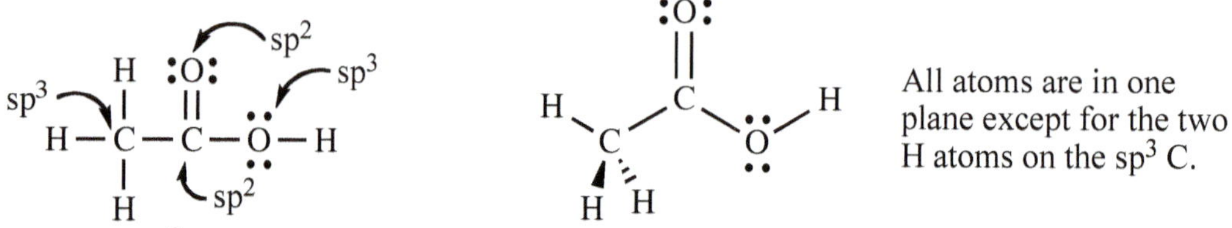

All atoms are in one plane except for the two H atoms on the sp^3 C.

(e) tetrahedral around the sp^3 carbon; the other two carbons both sp, linear, bond angle 180°

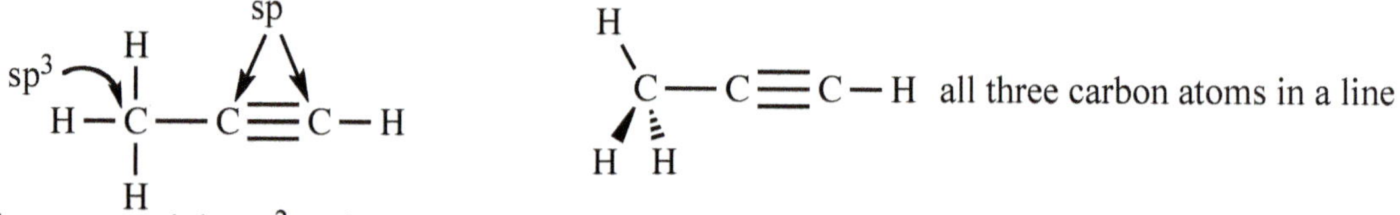

(f) trigonal planar around the sp^2 carbon and nitrogen, bond angles 120°; tetrahedral geometry and 109° bond angles around the sp^3 carbons

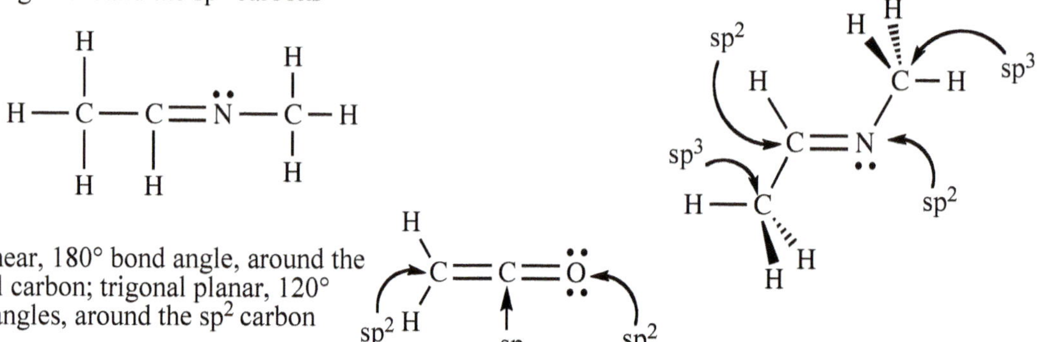

(g) linear, 180° bond angle, around the central carbon; trigonal planar, 120° bond angles, around the sp^2 carbon

1-20 Carbon-2 is sp hybridized. If the p orbitals making the pi bond between C-1 and C-2 are in the plane of the paper (putting the hydrogens in front of and behind the paper), then the other p orbital on C-2 must be perpendicular to the plane of the paper, making the pi bond between C-2 and C-3 perpendicular to the paper. This necessarily places the hydrogens on C-3 in the plane of the paper. (Models will surely help.)

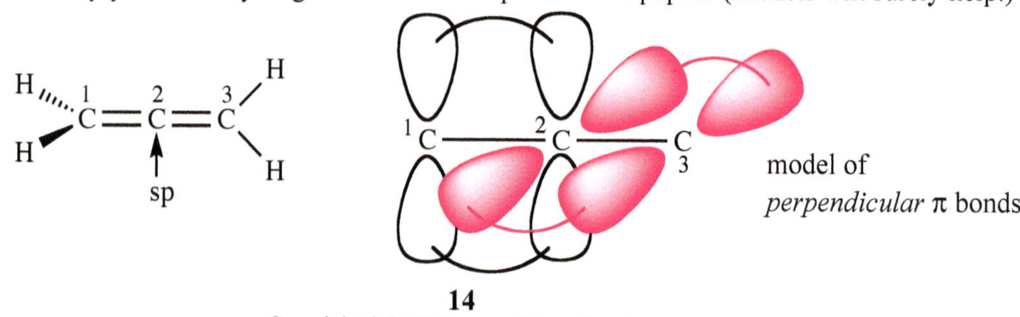

1-21 For clarity, electrons in sigma bonds are not shown. Part (a) has been solved in the text.

(b) Carbon and oxygen are both sp^2 hybridized.

(top view)

(side view)

One pair of electrons on oxygen is always in an sp^2 orbital. The other pair of electrons is shown in a p orbital in the first resonance form, and in a pi bond in the second resonance form.

(c) Oxygen and both carbons are sp^2 hybridized.

(top view)

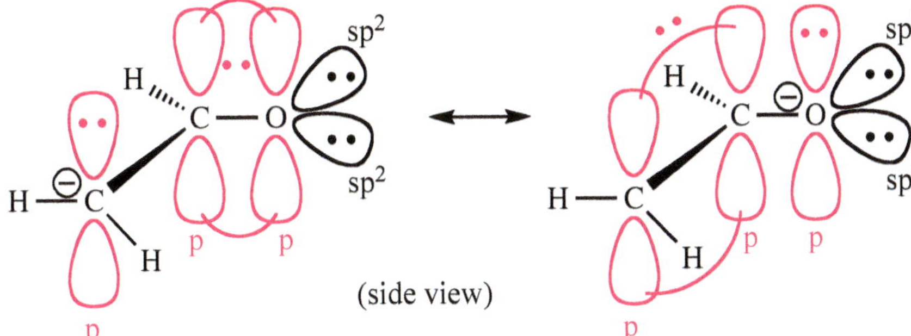

1-21 continued

(d) All atoms are sp^2 hybridized except the C labeled as sp^3 (and H which is NEVER hybridized); all bond angles around the sp^3 carbon are 109°.

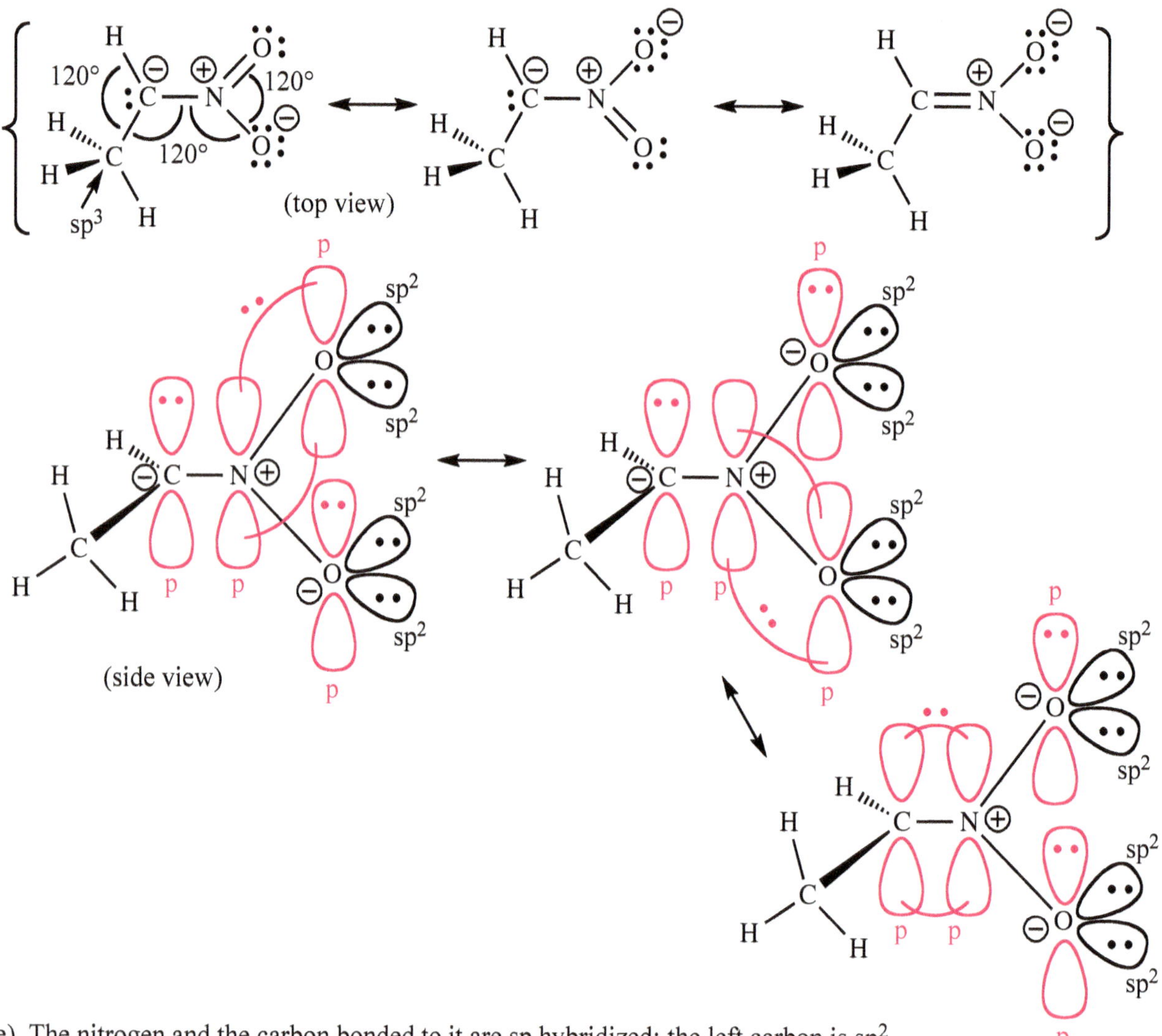

(e) The nitrogen and the carbon bonded to it are sp hybridized; the left carbon is sp^2.

1-21 continued

(f) The boron and the oxygens bonded to it are sp^2 hybridized.

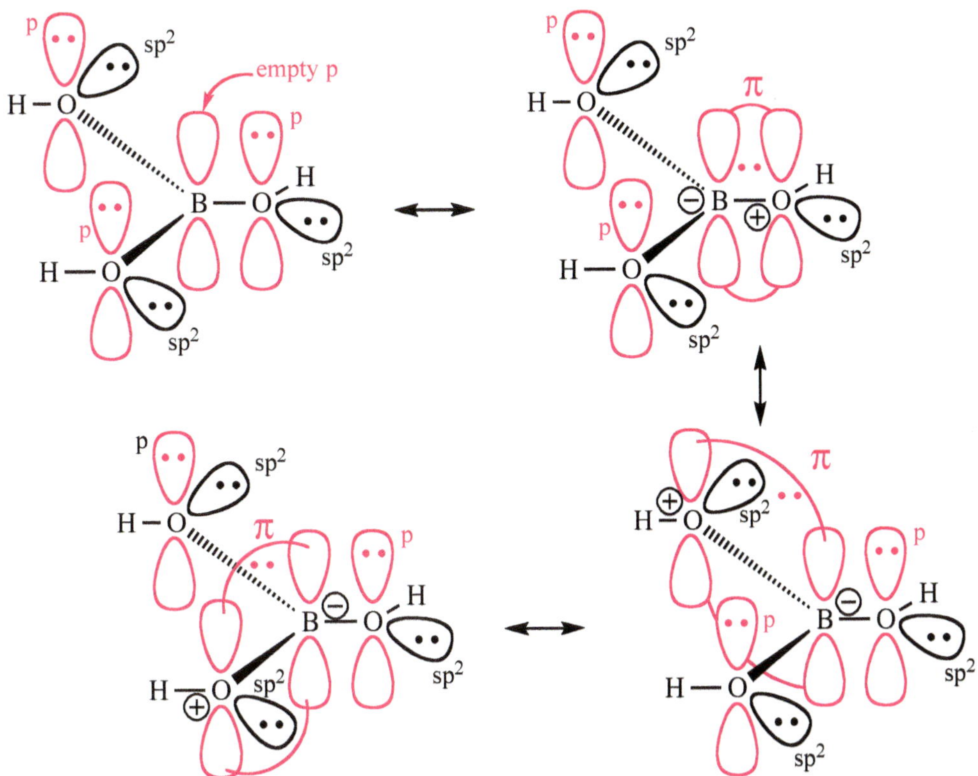

(g)

All oxygens are sp^2 with bond angle 120°.

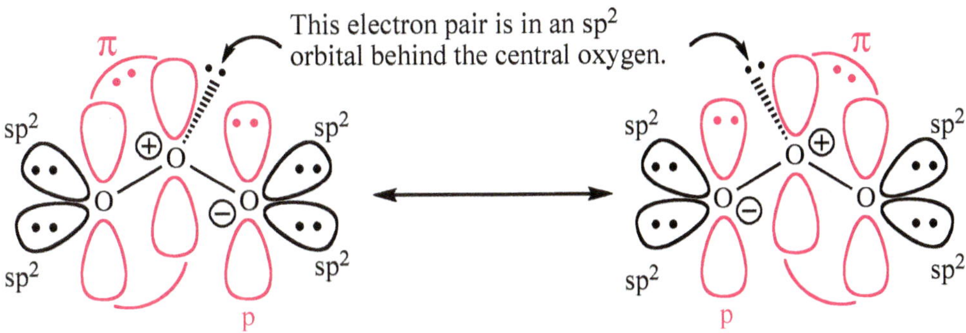

1-22 Very commonly in organic chemistry, we have to determine whether two structures are the same or different, and if they are different, what structural features are different. In order for two structures to be the same, all bonding connections have to be identical, and in the case of double bonds, the groups must be on the same side of the double bond in both structures. (A good exercise to do with your study group is to draw two structures and ask if they are the same; or draw one structure and ask how to draw a different compound.)

(a) Different compounds; H and CH_3 on one carbon of the double bond, and CH_3 and CH_2CH_3 on the other carbon—same in both structures. Drawing a plane through the p orbitals shows the H and CH_3 are on the same side of the double bond in the first structure, and the H and the CH_2CH_3 are on the same side in the second structure, so they are DIFFERENT compounds.

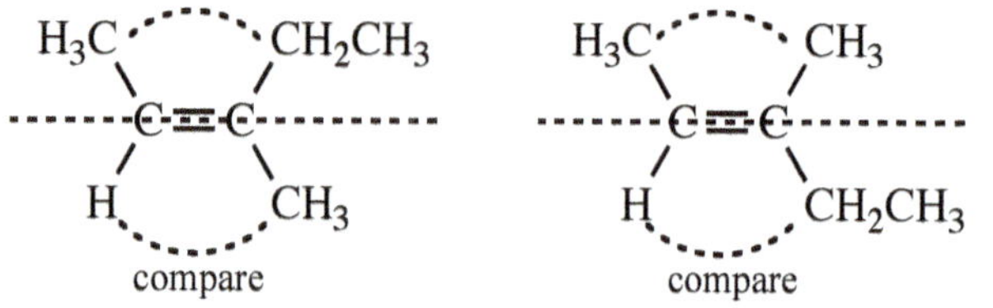

H_3C CH_2CH_3 / H CH_3 compare — H_3C CH_3 / H CH_2CH_3 compare — These are DIFFERENT.

(b) Same compound; in the structure on the right, the right carbon has been rotated, but the bonding is identical between the two structures.

(c) Different compounds; H and Br on one carbon, F and Cl on the other carbon in both structures; H and Cl on the same side of the plane through the C=C in the first structure, and H and F on the same side of the plane through the C=C in the second structure, so they are DIFFERENT compounds.

(d) Same compound: in the structure on the right, the right carbon has been rotated 120°.

1-23

(a) H—C—C=N—C—H: sp^3, sp^2, sp^3

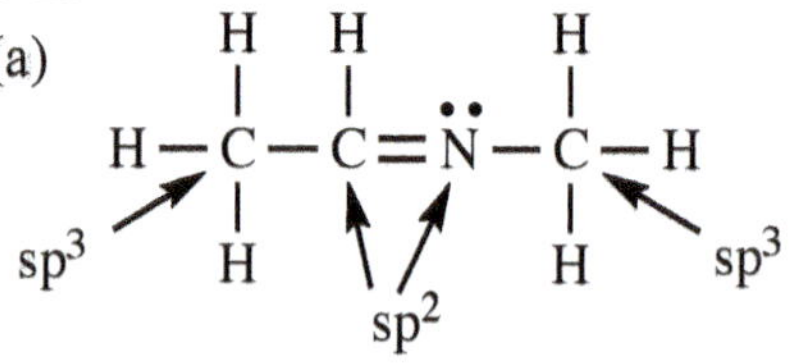

(b) NOT INTER-CONVERTIBLE

two CH_3 on opposite sides of the C=N

two CH_3 on the same side of the C=N

(c) The CH_3 on the N is on the same side as another CH_3 no matter how it is drawn—only one possible structure.

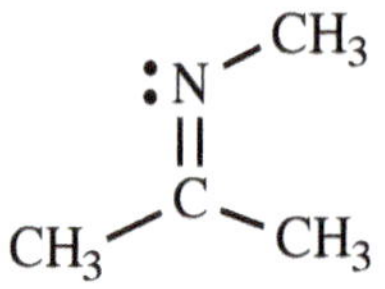

1-24

(a) F, H / F, H: *cis* and H, F / F, H: *trans*

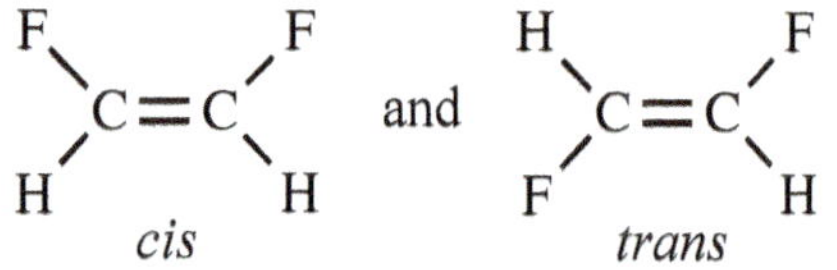

(b) no *cis-trans* isomerism
(c) no *cis-trans* isomerism
(d) no *cis-trans* isomerism

} two identical groups on one carbon of the double bond

(e) *cis* and *trans*

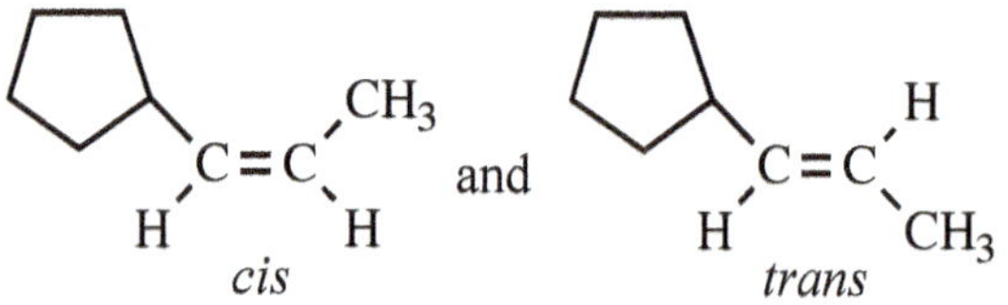

(f) These structures show cis-trans isomerism (also known as geometric isomerism), although "*cis*" and "*trans*" are not defined for this particular case. A different, unambiguous system for naming geometric isomers will be described in Chapter 7.

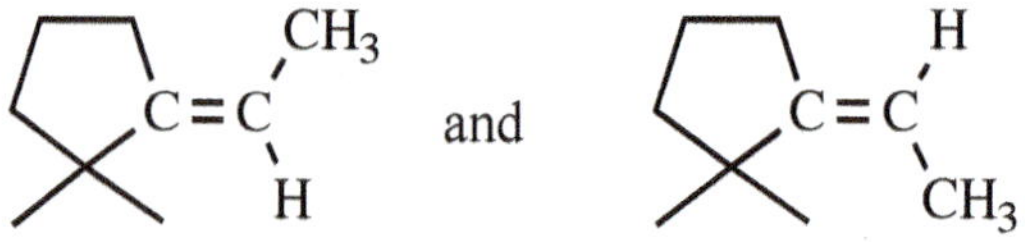

1-25 Models will be helpful here.

(a) Constitutional isomers—the carbon skeleton is different.

(b) *Cis-trans* isomers—the first is *trans,* the second is *cis.*

(c) Constitutional isomers—the bromines are on different carbons in the first structure, on the same carbon in the second structure.

1-25 continued

(d) same compound—just flipped over
(e) same compound—just rotated
(f) same compound—just rotated
(g) not isomers—different molecular formulas
(h) Constitutional isomers—the double bond has changed position.
(i) same compound—just reversed
(j) Constitutional isomers—the CH_3 groups are in different relative positions.
(k) Constitutional isomers—the double bond is in a different position relative to the CH_3 and an H has moved.

1-26

(a) { $O{=}\overset{\oplus}{S}{-}O^{\ominus}$ ⟷ $^{\ominus}O{-}\overset{\oplus}{S}{=}O$ ⟷ $O{=}S{=}O$ }

more than 8 electrons; see part (c)

(b) { $O{=}\overset{\oplus}{O}{-}O^{\ominus}$ ⟷ $^{\ominus}O{-}\overset{\oplus}{O}{=}O$ }

(c) The last resonance form of SO_2 has no equivalent form in O_3. Sulfur, a third-row element, can have more than eight electrons around it because of d orbitals, whereas oxygen, a second-row element, must adhere strictly to the octet rule.

1-27 (a) CARBON! (the best element) (b) oxygen (c) phosphorus (d) chlorine

1-28

valence e^- →	1	2	3	4	5	6	7	8
	H							He ($2e^-$)
	Li	Be	B	C	N	O	F	Ne
					P	S	Cl	
							Br	
							I	

1-29

(a) ionic only (b) covalent ($H{-}O^-$) and ionic (Na^+ ^-OH)

(c) covalent ($H{-}C$ and $C{-}Li$), but the $C{-}Li$ bond is strongly polarized

(d) covalent only (e) covalent ($H{-}C$ and $C{-}O^-$) and ionic (Na^+ $^-OCH_3$)

(f) covalent ($H{-}C$ and $C{=}O$ and $C{-}O^-$) and ionic (HCO_2^- Na^+) (g) covalent only

1-30

(a)

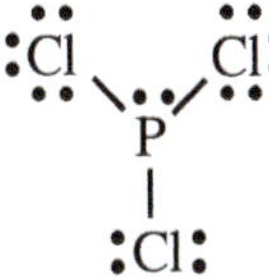

(b)

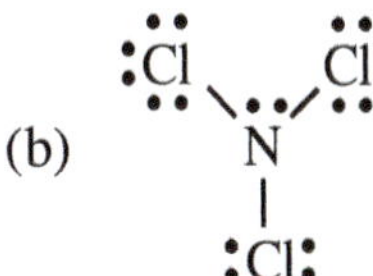

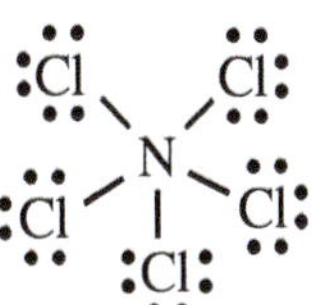

CANNOT EXIST

NCl_5 violates the octet rule; nitrogen can have no more than eight electrons (or four atoms) around it. Phosphorus, a third-row element, can have more than eight electrons because phosphorus can use d orbitals in bonding, so PCl_5 is a stable, isolable compound.

1-31 Your Lewis structures may look different from these. As long as the atoms are connected in the same order and by the same type of bond, they are equivalent structures. For now, the exact placement of the atoms on the page is not significant.

(a) 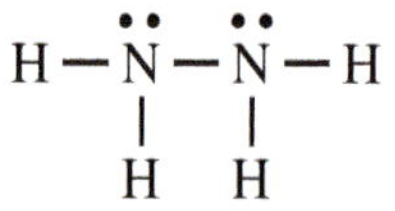 H—N—N—H

(b)  H—N=N—H

(c) 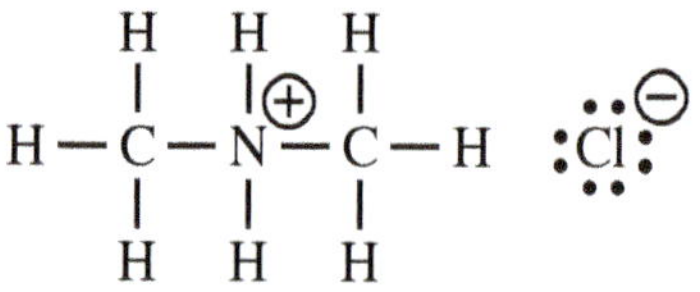 $(CH_3)_2NH_2^+\ Cl^-$

(d) 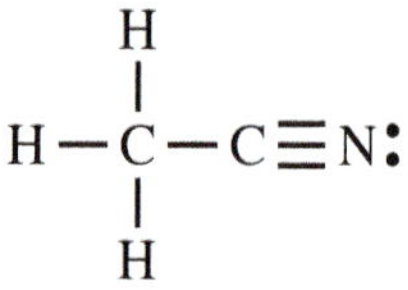 H—C(H)(H)—C≡N:

(e) 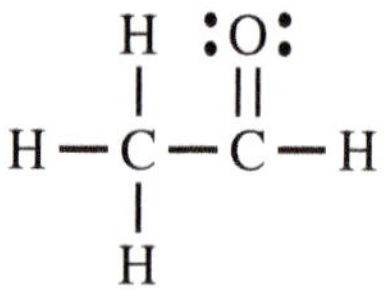 CH_3CHO

(f) 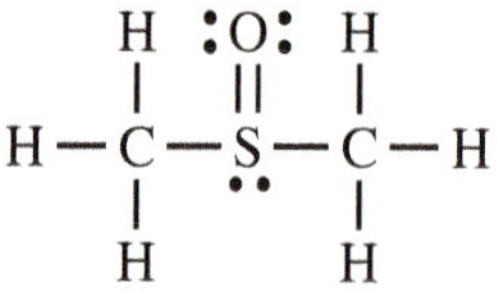 $CH_3S(=O)CH_3$

(g) H_2SO_4

(h) $CH_3N=C=O$

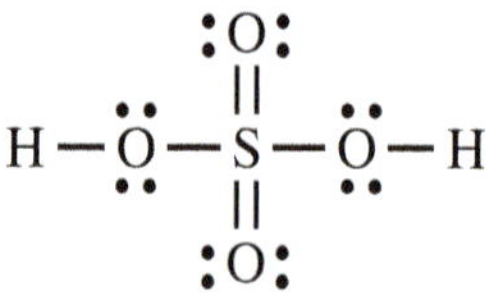

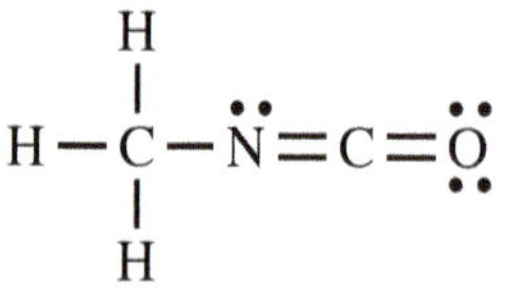

(i) 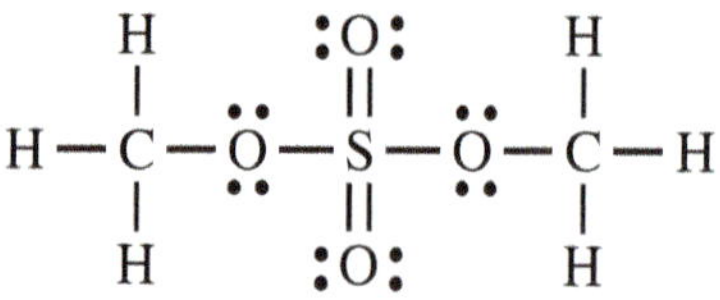 $CH_3OS(=O)_2OCH_3$

(j) 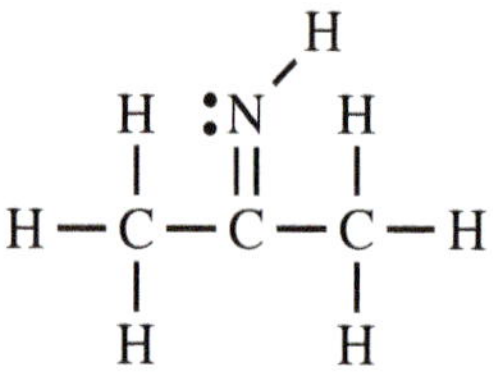 $CH_3C(=NH)CH_3$

(k) 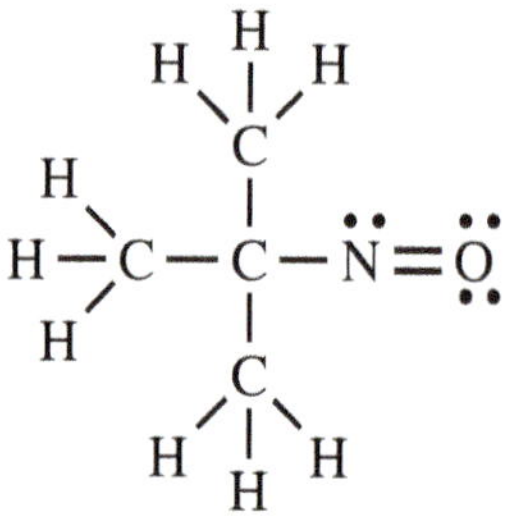 $(CH_3)_3C–N=O$

1-32

(a) 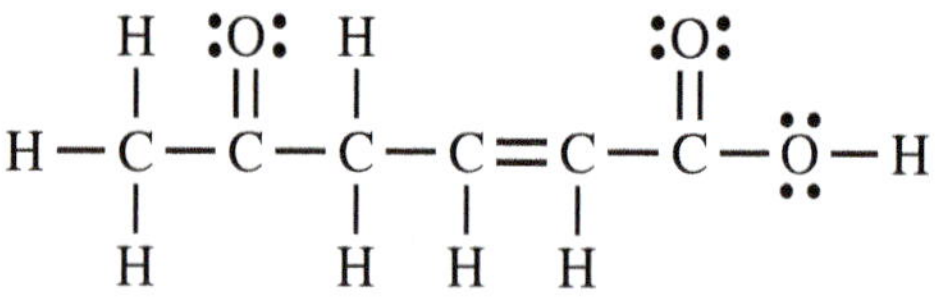 $CH_3C(=O)CH_2CH=CHCOOH$

(b) 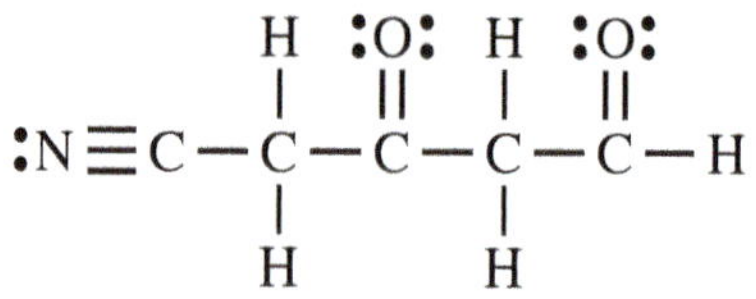 $N≡C–CH_2–C(=O)–CH_2–CHO$

(c) $CH_2=CH–CH(OH)–CH_2–COOH$

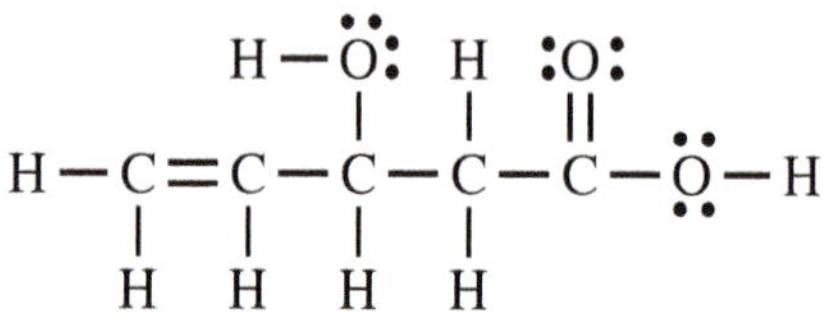

(d) 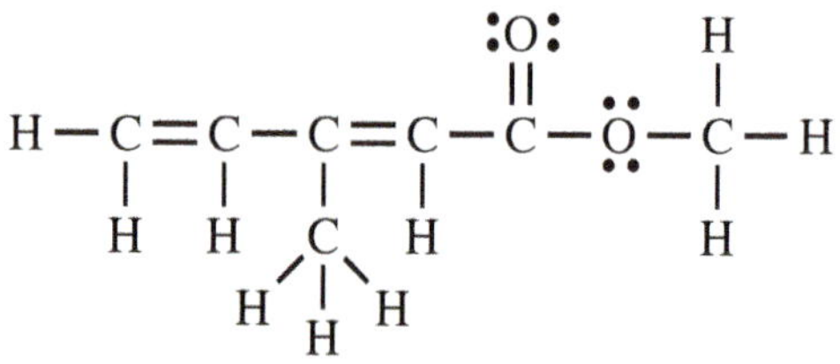 $CH_2=CH–C(CH_3)=CH–COOCH_3$

1-33 (Structures from problem 1-32 are redrawn in line-angle format.) In each set below, the second structure is a more correct line formula. Since chemists are human (surprise!), they will take shortcuts where possible; the first structure in each pair uses a common abbreviation, either COOH or CHO. Make sure you understand that COOH does not stand for C—O—O—H. Likewise for CHO.

(a) 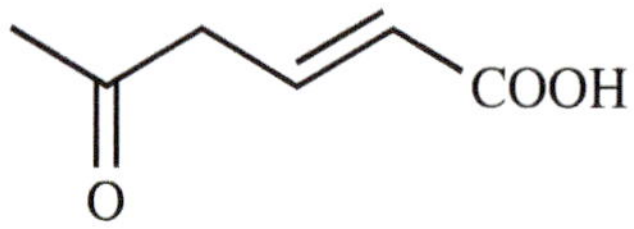 COOH

OR

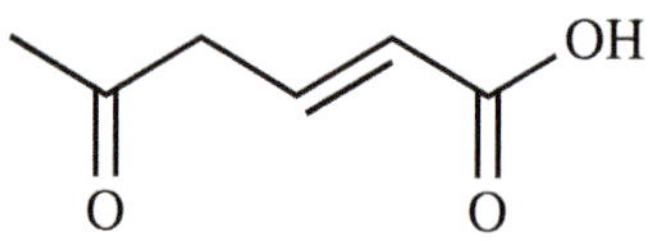

OH

(b) 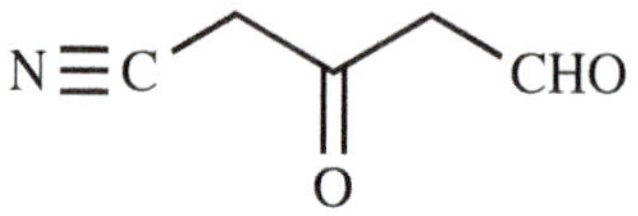 N≡C CHO

OR

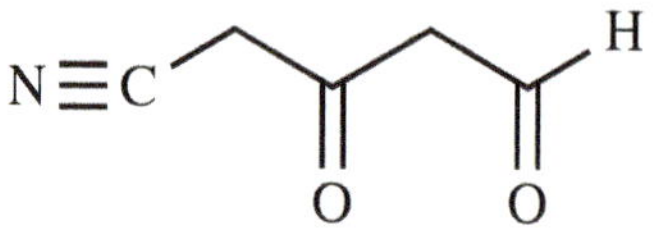

N≡C H

1-33 continued

(c) 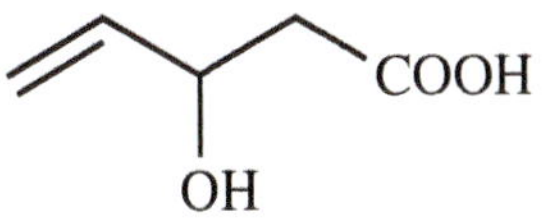 COOH, OH

OR

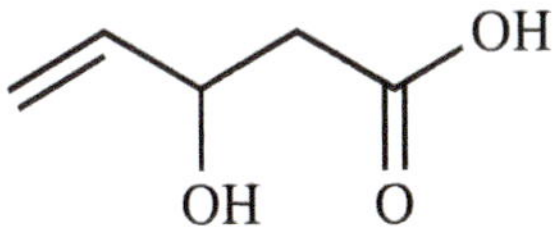

OH, OH, O

(d) 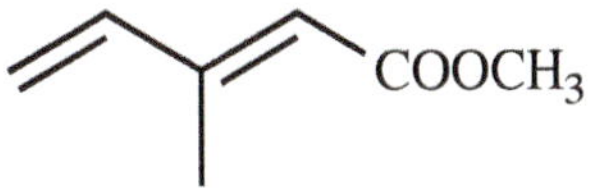 $COOCH_3$

OR

O, O

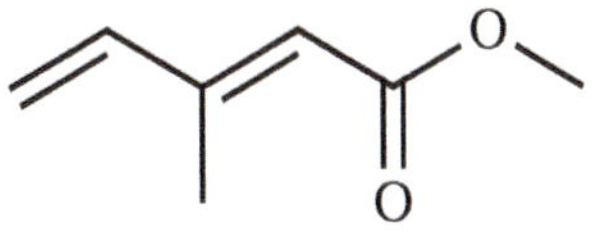

1-34

In some cases, you are asked for the maximum number of structures possible for a molecular formula containing a small number of atoms. In other cases, however, there may be many more structures possible, and your structures can be correct even if they are not the ones shown here.

(a)

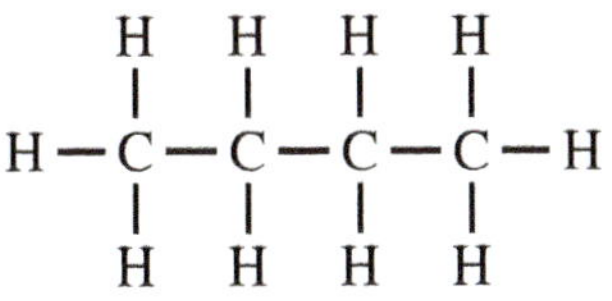

```
    H   H   H   H             H   H   H
    |   |   |   |             |   |   |
H — C — C — C — C — H   and   H — C — C — C — H
    |   |   |   |             |   |   |
    H   H   H   H             H   C   H
                                 /|\
                                H H H
```

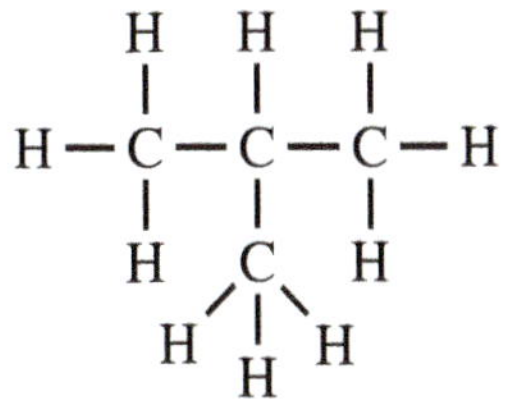

These are the only two possibilities, but your structures may appear different—making models will help you visualize these structures.

(b)

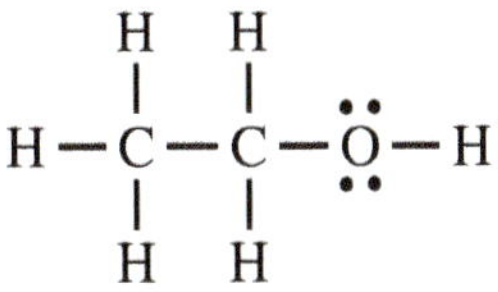

```
    H   H                      H         H
    |   |   ••                 |    ••   |
H — C — C — O — H      and H — C — O — C — H
    |   |   ••                 |    ••   |
    H   H                      H         H
```

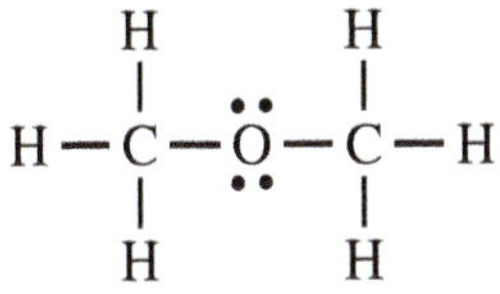

These are the only two possibilities.

(c)

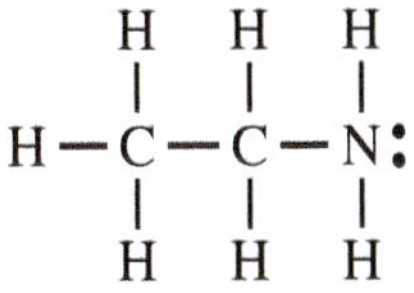

```
    H   H   H                 H   H   H
    |   |   |                 |   |   |
H — C — C — N:       and  H — C — N — C — H
    |   |   |                 |   ••  |
    H   H   H                 H       H
```

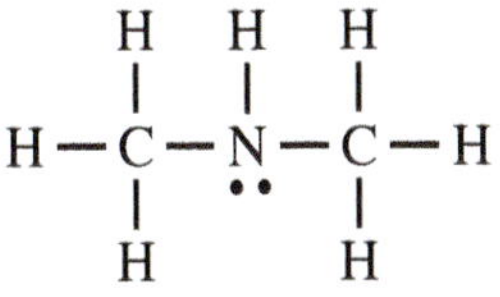

These are the only two possibilities, but your structures may appear different—making models will help you visualize these structures.

(d)

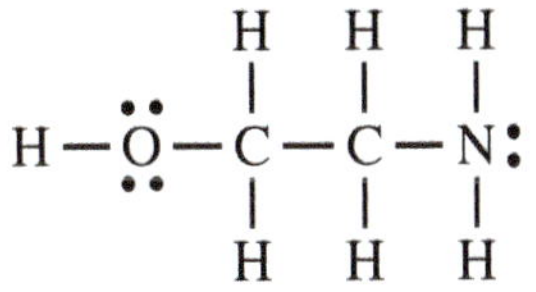

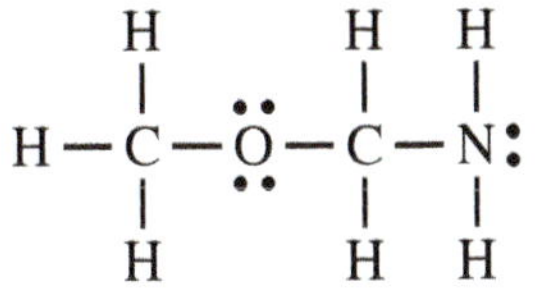

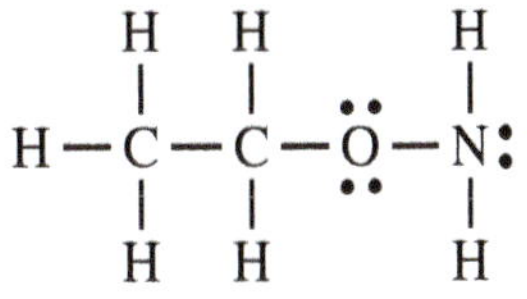

```
             H   H   H              H         H   H              H   H         H
    ••       |   |   |              |    ••   |   |              |   |    ••   |
H — O — C — C — N:           H — C — O — C — N:           H — C — C — O — N:
    ••       |   |   |              |    ••   |   |              |   |    ••   |
             H   H   H              H         H   H              H   H         H
```

Think of all the ways an oxygen could be added to the structures in (c). There are many more!

(e) There are several other possibilities as well. Your answer may be correct even if your structures do not appear here. Check with others in your study group.

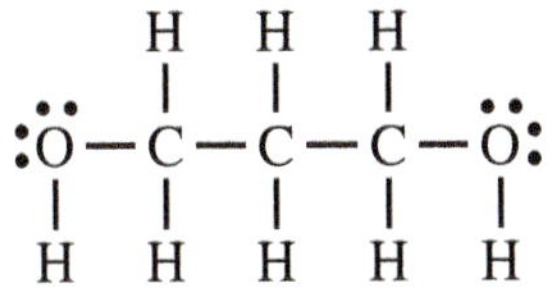

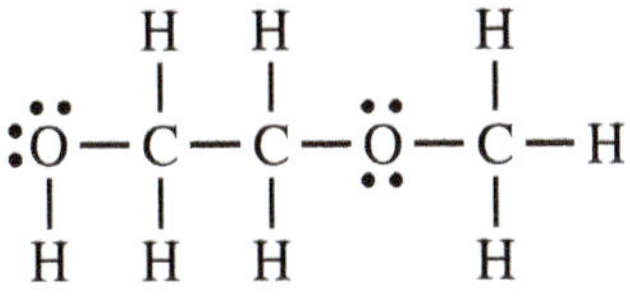

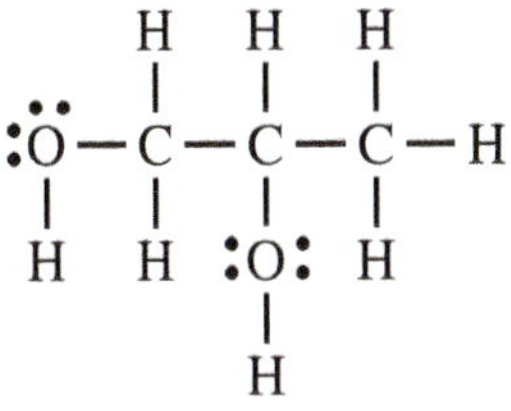

```
      H   H   H                  H   H         H               H   H   H
 ••   |   |   |   ••        ••   |   |   ••    |          ••   |   |   |
:O — C — C — C — O:        :O — C — C — O — C — H        :O — C — C — C — H
 |    |   |   |   |         |    |   |   ••    |          |    |   |   |
 H    H   H   H   H         H    H   H         H          H    H  :O:  H
                                                                    |
                                                                    H
```

(f)

```
    H  :O:                                  :O:
    |   ||                         ••       / \
H — C — C — H        H — C = C — O — H     H — C — C — H
    |                    |   |     ••          |   |
    H                    H   H                 H   H
```

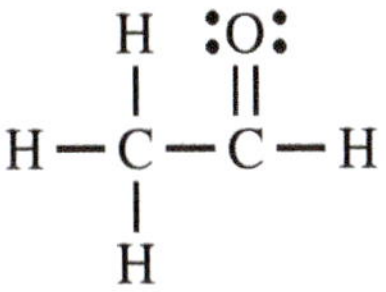

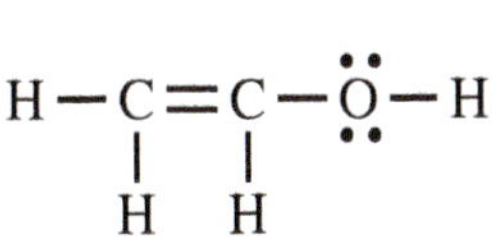

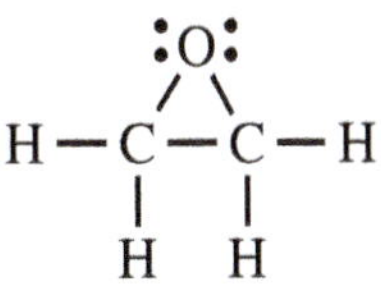

These are the only three structures with this molecular formula.

1-35

(a) only three possible structures

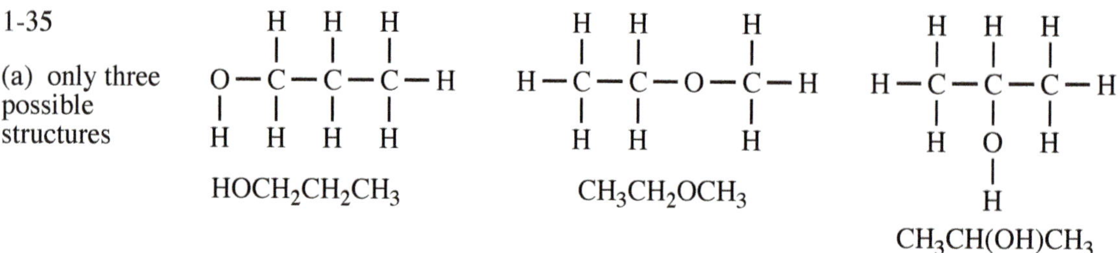

(b) This is most (maybe all) of the possible structures.

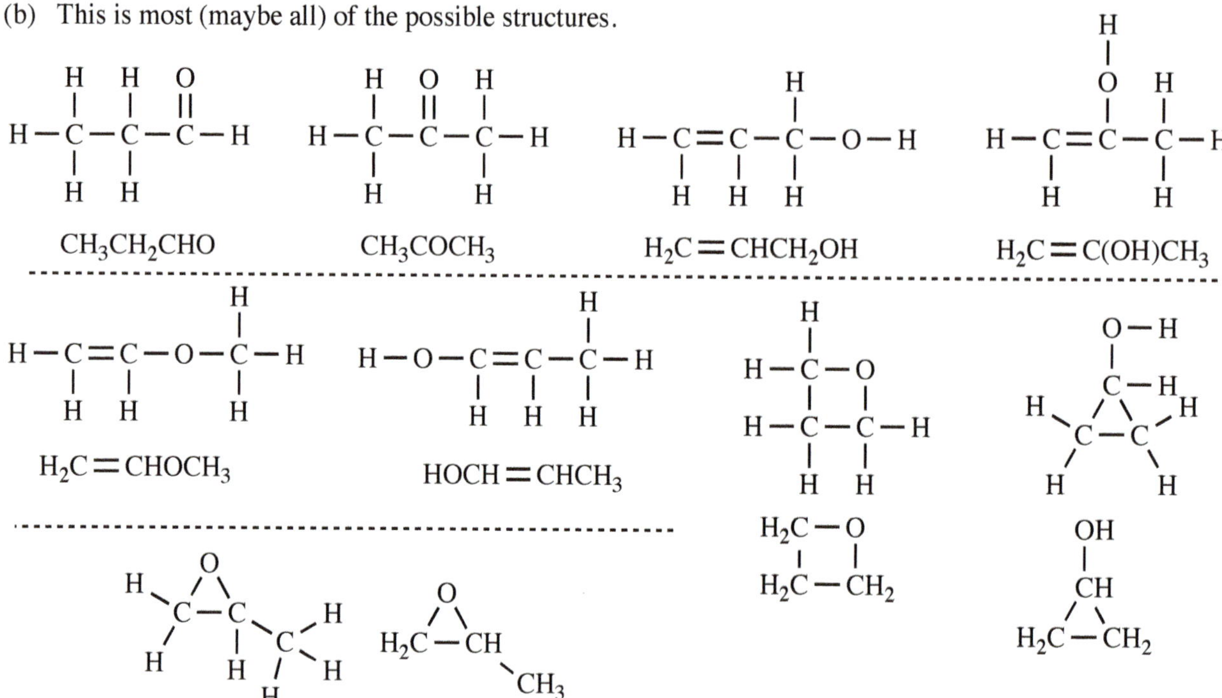

1-36 General rule: *molecular formulas of stable hydrocarbons must have an even number of hydrogens.* The formula CH_2 does not have enough atoms to bond with the four orbitals of carbon.

one carbon: **CH_4**

two carbons: **C_2H_2** **C_2H_4** **C_2H_6**

three carbons: **C_3H_4** **C_3H_6** **C_3H_8**

1-37

(a)

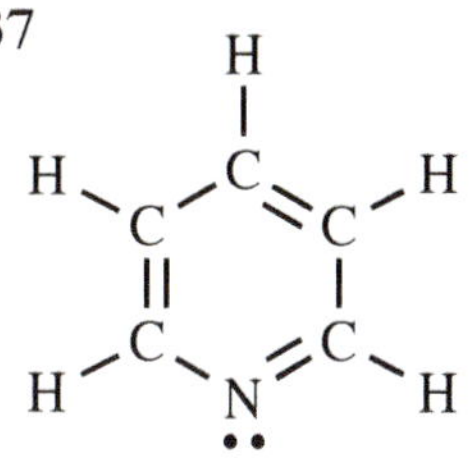

(b)

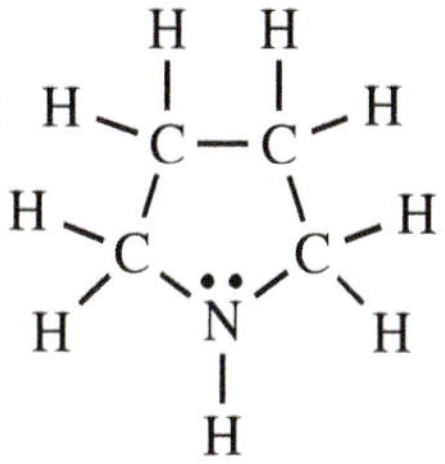

(c)

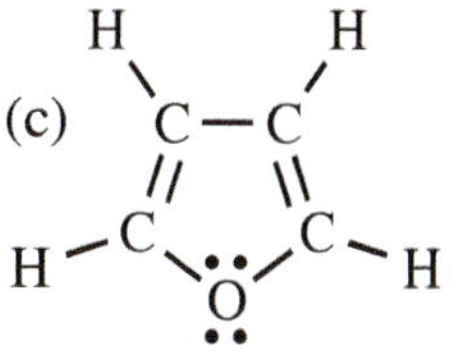

(d)

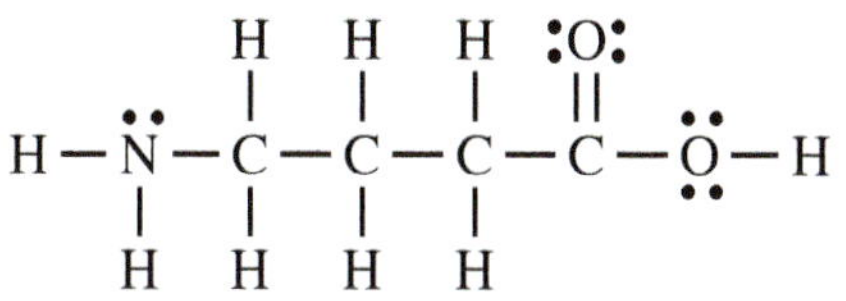

(e)

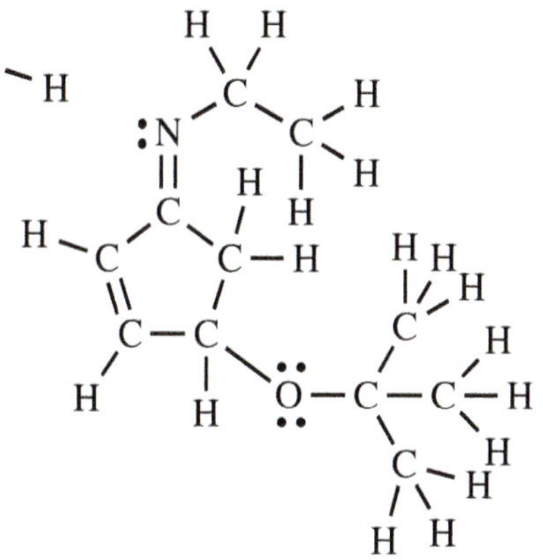

(f)

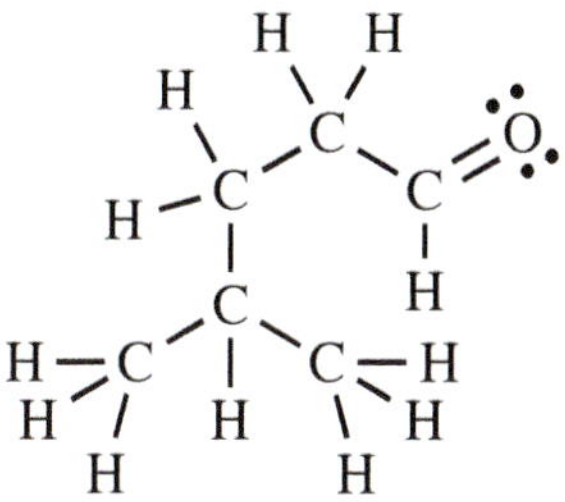

(g)

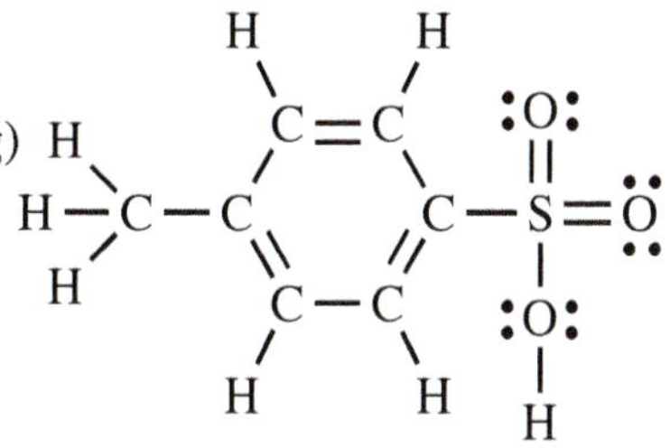

(h)

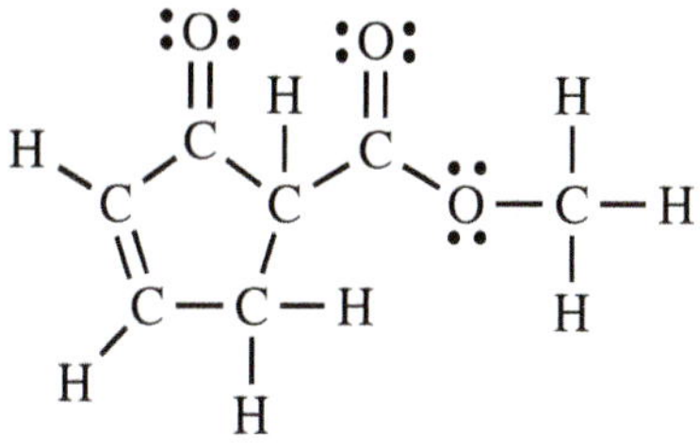

1-38 Molecular formulas of the structures in problem 1-37.

(a) C_5H_5N (b) C_4H_9N (c) C_4H_4O (d) $C_4H_9NO_2$ (e) $C_{11}H_{19}NO$

(f) $C_6H_{12}O$ (g) $C_7H_8O_3S$ (h) $C_7H_8O_3$

1-39 The symbols "δ^+" and "δ^-" indicate bond polarity by showing partial charge. Electronegativity differences greater than or equal to 0.5 are considered large.

	Bond	Polarity	Difference
(a)	C—Cl	δ^+ δ^-	large
(b)	C—H	δ^- δ^+	small
(c)	C—Li	δ^- δ^+	large
(d)	C—N	δ^+ δ^-	small
(e)	C—O	δ^+ δ^-	large
(f)	C—B	δ^- δ^+	large
(g)	C—Mg	δ^- δ^+	large
(h)	N—H	δ^- δ^+	large
(i)	O—H	δ^- δ^+	large
(j)	C—Br	δ^+ δ^-	small

1-40 Non-zero formal charges are shown by the atoms.

(a) 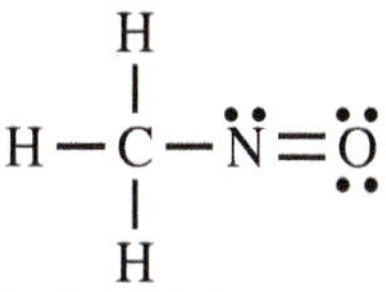 No formal charges.

(b) 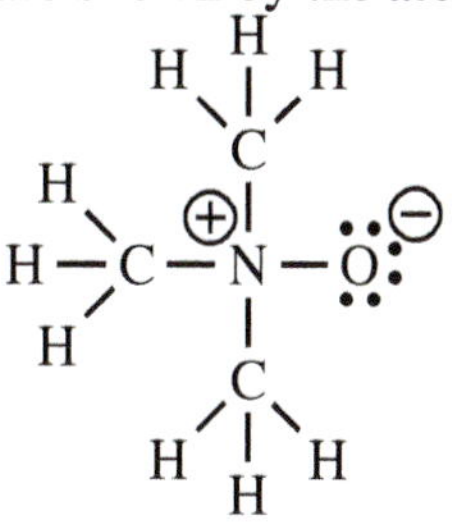 Formal charges sum to 0.

(c) 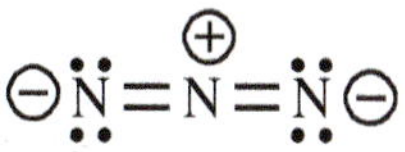 Formal charges sum to –1.

(d)

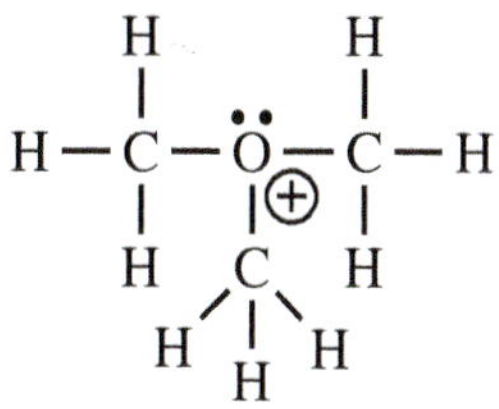

(e) 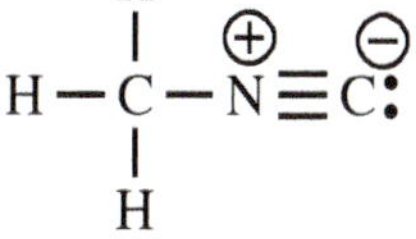 Formal charges sum to 0.

(f) 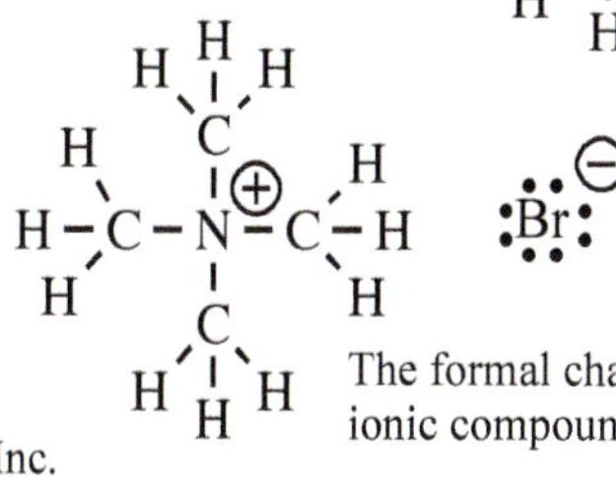 The formal charges in this ionic compound sum to 0.

1-41 Resonance forms must have atoms in identical positions. If any atom moves position, it is a different structure. It is particularly helpful in this problem to draw in all of the H atoms and bonds.

(a) Different compounds—a hydrogen atom has changed position.
(b) Resonance forms—only the position of electrons is different.
(c) Different compounds—a hydrogen atom has changed position.
(d) Different compounds—a hydrogen atom has changed position.
(e) Resonance forms—only the position of electrons is different.
(f) Resonance forms—only the position of electrons is different.
(g) Resonance forms—only the position of electrons is different.
(h) Different compounds—a hydrogen atom has changed position.
(i) Different compounds—a hydrogen atom has changed position.
(j) Resonance forms—only the position of electrons is different.
(k) Resonance forms—only the position of electrons is different.
(l) Resonance forms—only the position of electrons is different.

1-42 When drawing resonance forms with charges on ring atoms, it helps to keep track of the charge by writing the C or N or O with the charge. (In Chapter 1, arrows will be shown when drawing resonance forms, equivalent to the green arrows in the text. This important skill should become automatic by Chapter 2 when the arrows will not be shown.)

(a)

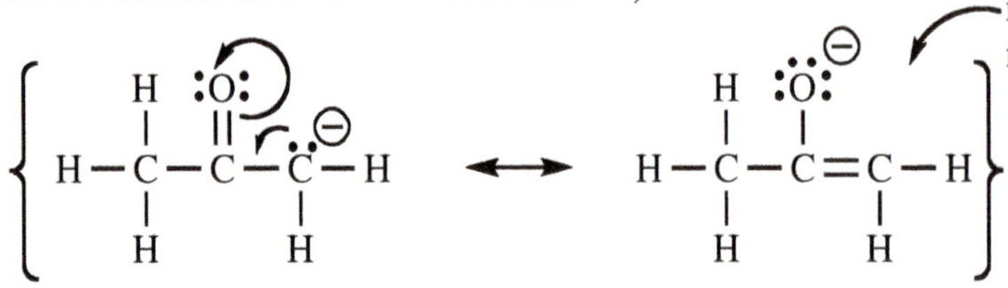

major—negative charge on the more electronegative atom

(b)

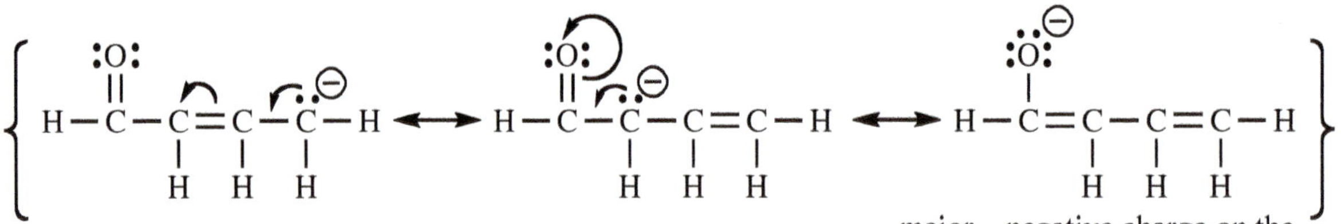

major—negative charge on the more electronegative atom

(c)

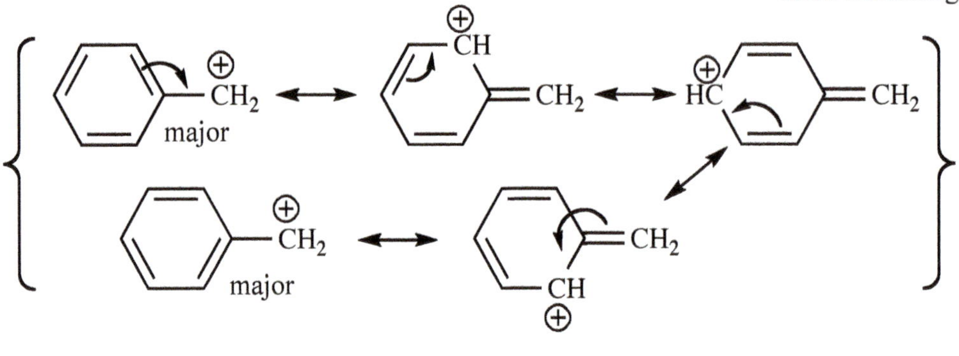

major

major

For reasons to be discussed in a later chapter, the structure with all double bonds in the ring is the major contributor.

(d)

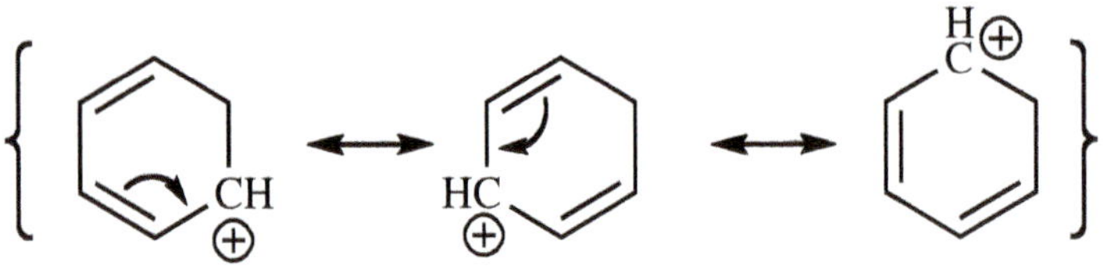

All resonance forms are of equal energy.

1-42 continued

(e) major* major*

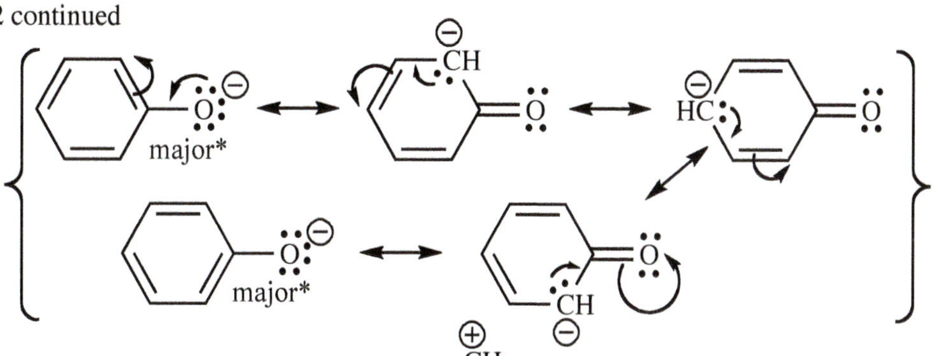

*major—negative charge on the more electronegative atom

(f) major—all atoms have octets

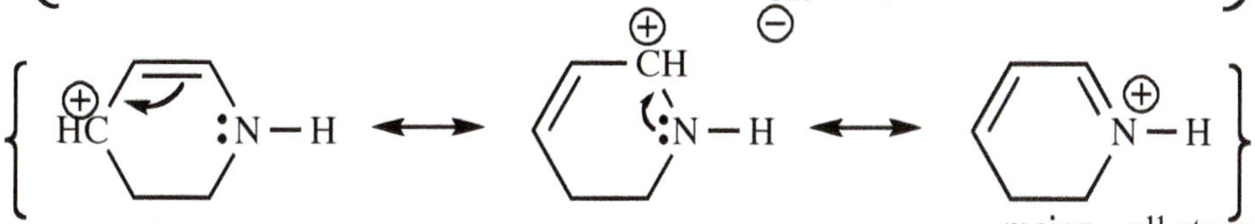

(g) major—all atoms have octets

(h) major—negative charge on the more electronegative atom

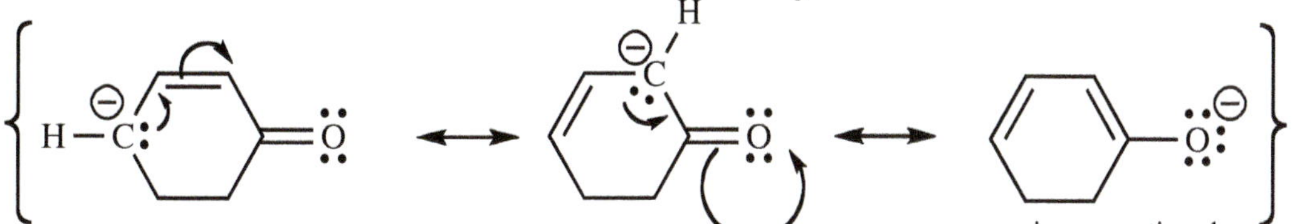

(i) All resonance forms are of equal energy.

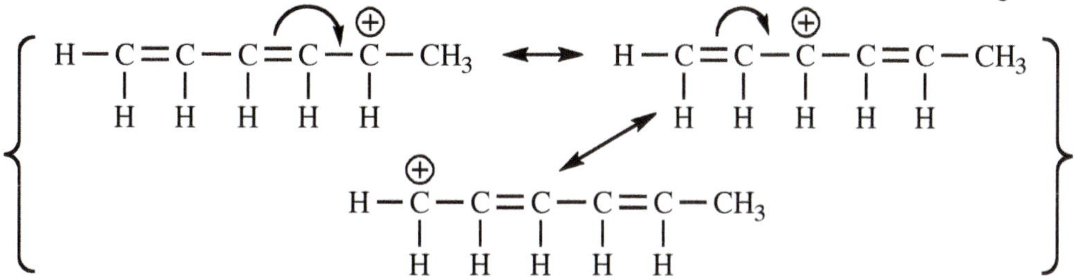

(j) No resonance forms—the charge must be on an atom next to a double or triple bond, or next to a non-bonded pair of electrons, in order for resonance to delocalize the charge.

(k) major—all atoms have octets

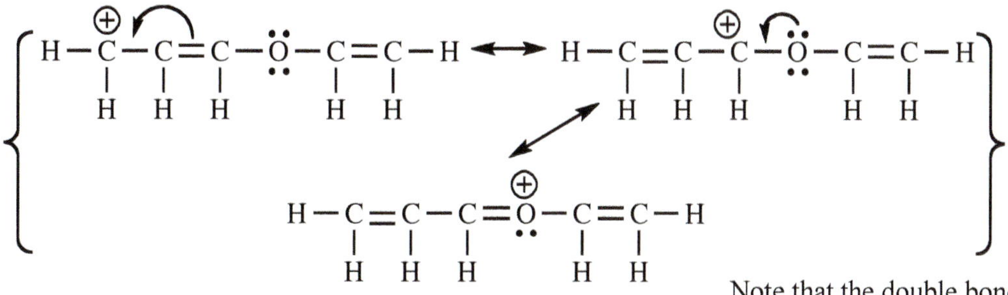

Note that the double bond on the right does not contribute its electrons to the resonance forms.

1-43

(a) $\left\{ CH_3-\overset{\ominus}{\underset{H}{\ddot{C}}}-C\equiv N: \longleftrightarrow CH_3-\underset{H}{C}=C=\overset{\ominus}{\ddot{N}}: \right\}$

minor — major (negative charge on electronegative atom)

(b) $\left\{ CH_3-\overset{:\ddot{O}:^{\ominus}}{C}=\underset{H}{C}-\overset{\oplus}{\underset{H}{C}}-CH_3 \longleftrightarrow CH_3-\overset{:\ddot{O}:^{\ominus}}{\underset{\oplus}{C}}-\underset{H}{C}=\underset{H}{C}-CH_3 \longleftrightarrow CH_3-\overset{:O:}{\overset{\|}{C}}-\underset{H}{C}=\underset{H}{C}-CH_3 \right\}$

minor — minor — major—full octets, no charge separation

(c) $\left\{ CH_3-\overset{:O:}{\overset{\|}{C}}-\overset{\ominus}{\underset{H}{\ddot{C}}}-\overset{:O:}{\overset{\|}{C}}-CH_3 \longleftrightarrow CH_3-\overset{:\ddot{O}:^{\ominus}}{C}=\underset{H}{C}-\overset{:O:}{\overset{\|}{C}}-CH_3 \longleftrightarrow CH_3-\overset{:O:}{\overset{\|}{C}}-\underset{H}{C}=\overset{:\ddot{O}:^{\ominus}}{C}-CH_3 \right\}$

minor — major — major

negative charge on electronegative atoms—equal energy

(d) $\left\{ CH_3-\overset{\ominus}{\underset{H}{\ddot{C}}}-\underset{H}{C}=\underset{H}{C}-\overset{\oplus}{\underset{:\ddot{O}:^{\ominus}}{N}}=\ddot{O} \longleftrightarrow CH_3-\underset{H}{C}=\underset{H}{C}-\overset{\ominus}{\underset{H}{\ddot{C}}}-\overset{\oplus}{\underset{:\ddot{O}:^{\ominus}}{N}}=\ddot{O} \longleftrightarrow CH_3-\underset{H}{C}=\underset{H}{C}-\underset{H}{C}=\overset{\oplus}{\underset{:\ddot{O}:^{\ominus}}{N}}-\ddot{O}:^{\ominus} \right\}$

minor — minor — major—negative charge on electronegative atoms

NOTE: The two structures below are resonance forms, varying from the first two structures in part (d) by the different positions of the double bonds in the NO_2. Usually, chemists omit drawing the second form of the NO_2 group although we all understand that its presence is implied. It is a good idea to draw all of the resonance forms until they become second nature. The importance of understanding resonance forms cannot be overemphasized.

$\left\{ CH_3-\overset{\ominus}{\underset{H}{\ddot{C}}}-\underset{H}{C}=\underset{H}{C}-\overset{\oplus}{\underset{:O:}{\underset{\|}{N}}}-\ddot{O}:^{\ominus} \longleftrightarrow CH_3-\underset{H}{C}=\underset{H}{C}-\overset{\ominus}{\underset{H}{\ddot{C}}}-\overset{\oplus}{\underset{:O:}{\underset{\|}{N}}}-\ddot{O}:^{\ominus} \right\}$

(e) $\left\{ CH_3CH_2-\overset{\ddot{N}H_2}{\underset{\oplus}{C}}-\ddot{N}H_2 \longleftrightarrow CH_3CH_2-\overset{\overset{\oplus}{N}H_2}{\overset{\|}{C}}-\ddot{N}H_2 \longleftrightarrow CH_3CH_2-\overset{\ddot{N}H_2}{C}=\overset{\oplus}{N}H_2 \right\}$

minor — major—full octets — major—full octets

equal energy

1-44 In Chapter 1, arrows will be shown when drawing resonance forms. This important skill should become automatic by Chapter 2 when the arrows will not be shown.

(a)

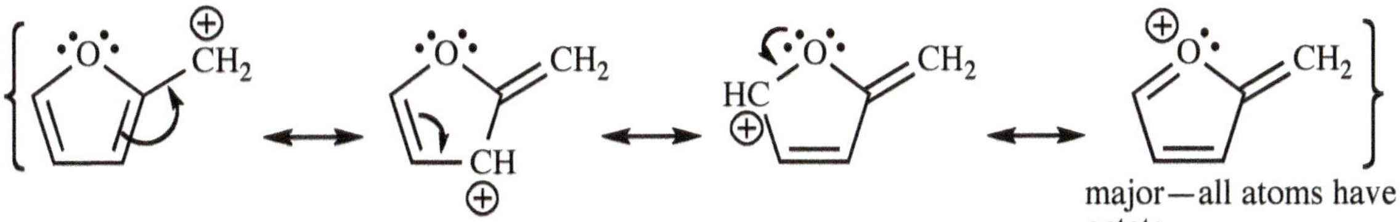

major—all atoms have octets

(b)

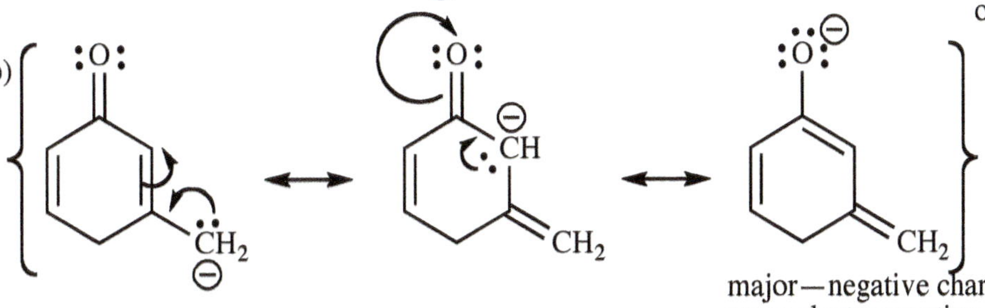

major—negative charge on more electronegative atom

(c)

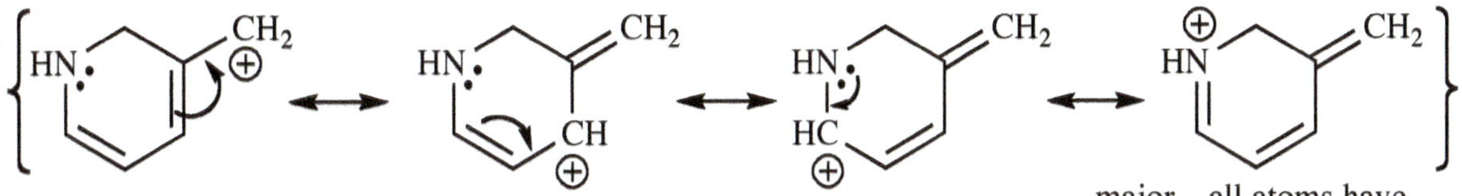

major—all atoms have octets

(d)

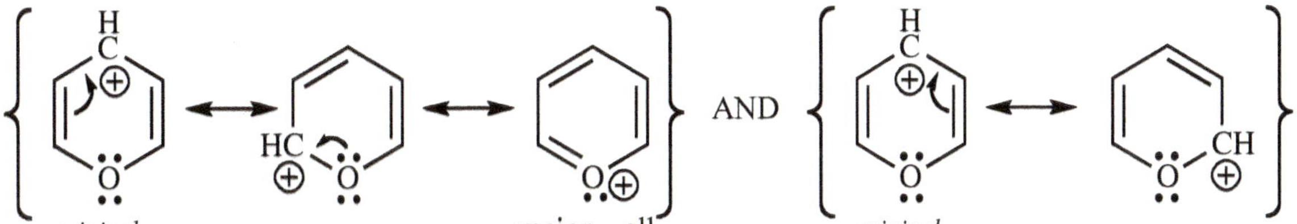

original

major—all atoms have octets

AND

original

(e) This remarkable anion has five equivalent resonance forms, distributing the negative charge equally on all five carbons. (Braces omitted to save space.)

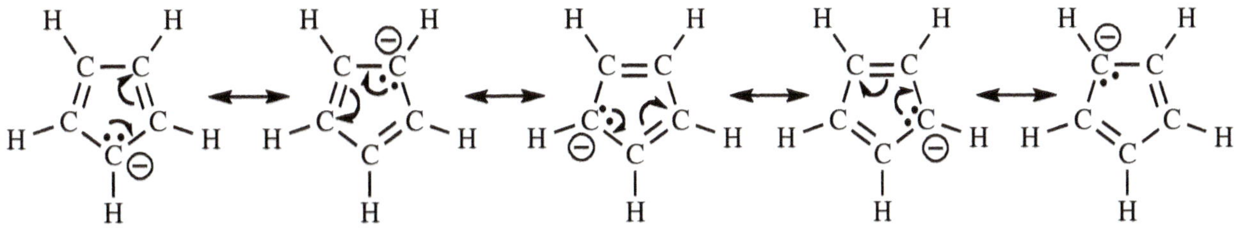

(f)

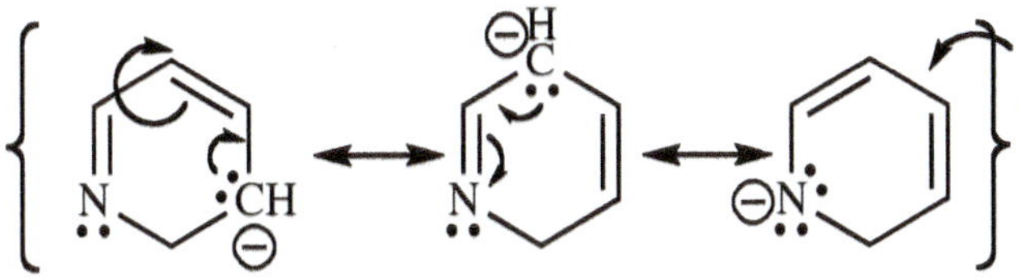

major—negative charge on more electronegative atom

1-45

(a) $CH_3-\overset{\oplus}{C}H-CH_3$

no resonance stabilization

$\left\{ CH_3-\overset{\oplus}{C}H-\ddot{\underset{..}{O}}-CH_3 \longleftrightarrow CH_3-CH=\overset{\oplus}{\underset{..}{O}}-CH_3 \right\}$

more stable—resonance stabilized

(b) $\left\{ (CH_3)_2\ddot{N}-\overset{\oplus}{C}(CH_3)_2 \longleftrightarrow (CH_3)_2\overset{\oplus}{N}=C(CH_3)_2 \right\}$

more stable—resonance stabilized

$(CH_3)_2CH-\overset{\oplus}{C}(CH_3)_2$

no resonance stabilization

(c) $\left\{ CH_2=CH-\overset{\oplus}{C}H-CH_3 \longleftrightarrow \overset{\oplus}{C}H_2-CH=CH-CH_3 \right\}$

more stable—resonance stabilized

$CH_2=CH-CH_2-\overset{\oplus}{C}H_2$

no resonance stabilization

(d) $H-\overset{\ominus}{\ddot{C}}H-CH_3$

no resonance stabilization

$\left\{ H-\overset{\ominus}{\ddot{C}}H-C\equiv N: \longleftrightarrow H-CH=C=\overset{\ominus}{\ddot{N}}: \right\}$

more stable—resonance stabilized

(e) {cyclohexenyl–$\overset{\oplus}{C}H_2$ ⟷ methylenecyclohexane with $\oplus$ on ring C–H}

more stable—resonance stabilized

{non-conjugated cyclohexenyl–$\overset{\oplus}{C}H_2$}

no resonance stabilization

(f) {cyclohexenone carbanion ⟷ carbanion on the other ring carbon}

some resonance stabilization, but negative charge is never on an electronegative atom

{conjugated cyclohexenone carbanion ⟷ carbanion α to C=O ⟷ enolate, $\ddot{O}:^{\ominus}$}

more stable—more resonance stabilization, and negative charge is on an electronegative atom

1-46 Resonance structures depict the distribution of electrons around the molecule. The significant resonance contributors suggest the areas of highest and lowest electron density.

(a)

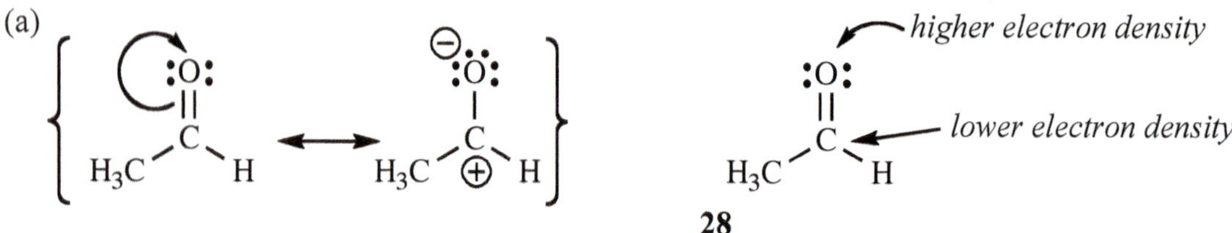

1-46 continued

(b) $\left\{ H_3C{-}\overset{\overset{\large :O:}{\|}}{C}{-}\ddot{N}{-}H \longleftrightarrow H_3C{-}\overset{\overset{:\ddot{O}:^{\ominus}}{|}}{\underset{\oplus}{C}}{-}\ddot{N}{-}H \longleftrightarrow H_3C{-}\overset{\overset{:\ddot{O}:^{\ominus}}{|}}{C}{=}\overset{\oplus}{N}{-}H \right\}$

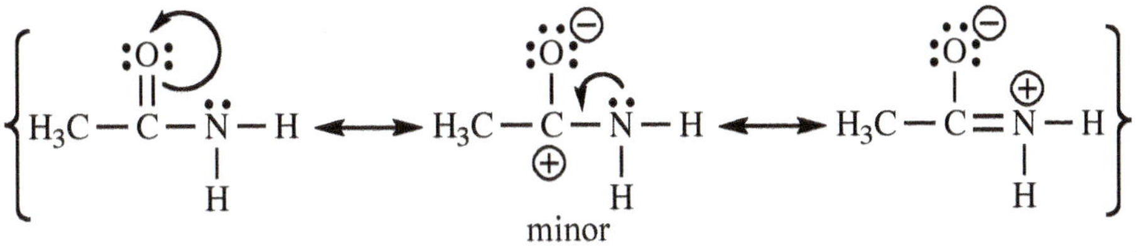

minor

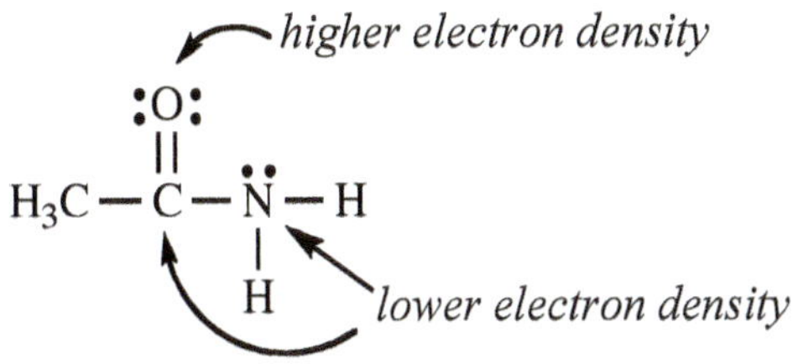

higher electron density

lower electron density

(c)

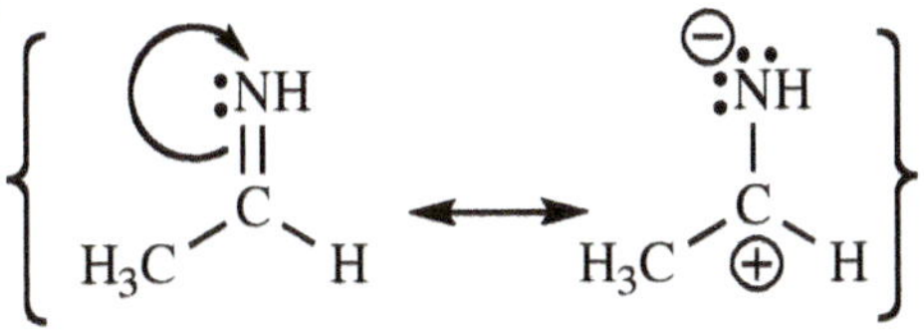

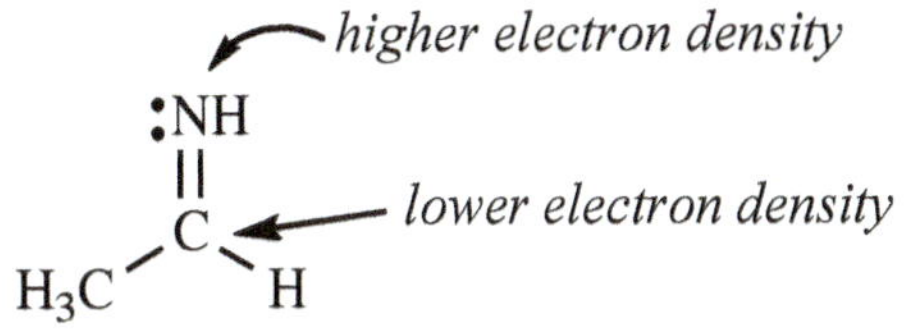

higher electron density

lower electron density

(d) $\left\{ H_3C{-}\overset{\overset{:O:}{\|}}{C}{-}\ddot{\underset{..}{O}}{-}CH_3 \longleftrightarrow H_3C{-}\overset{\overset{:\ddot{O}:^{\ominus}}{|}}{\underset{\oplus}{C}}{-}\ddot{\underset{..}{O}}{-}CH_3 \longleftrightarrow H_3C{-}\overset{\overset{:\ddot{O}:^{\ominus}}{|}}{C}{=}\overset{\oplus}{\underset{..}{O}}{-}CH_3 \right\}$

minor

higher electron density

lower electron density

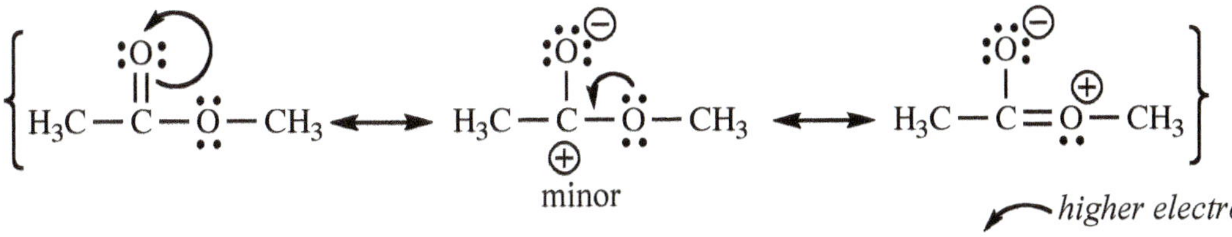

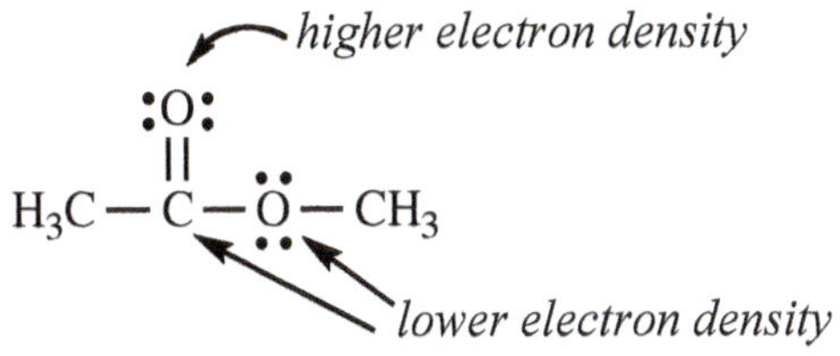

(e) $\left\{ H_3C{-}\ddot{\underset{..}{O}}{-}\overset{\overset{:O:}{\|}}{C}{-}\ddot{N}{-}H \longleftrightarrow H_3C{-}\ddot{\underset{..}{O}}{-}\overset{\overset{:\ddot{O}:^{\ominus}}{|}}{\underset{\oplus}{C}}{-}\ddot{N}{-}H \longleftrightarrow H_3C{-}\ddot{\underset{..}{O}}{-}\overset{\overset{:\ddot{O}:^{\ominus}}{|}}{C}{=}\overset{\oplus}{N}{-}H \right\}$

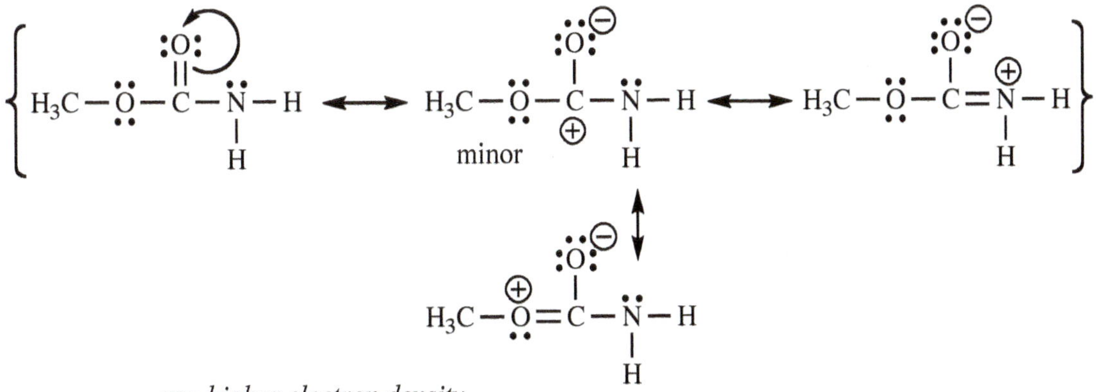

minor

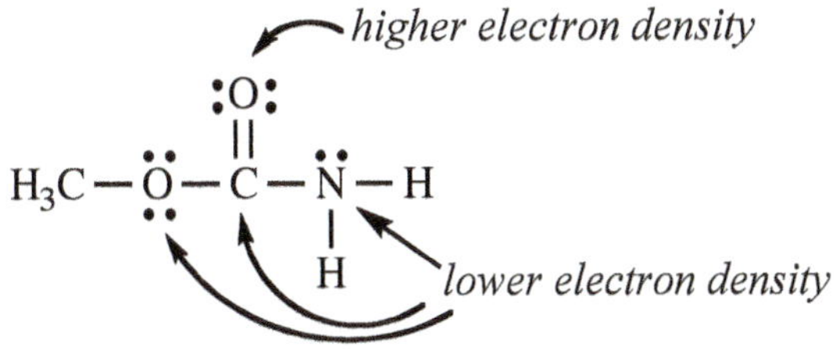

$H_3C{-}\overset{\oplus}{\underset{..}{O}}{=}\overset{\overset{:\ddot{O}:^{\ominus}}{|}}{C}{-}\ddot{N}{-}H$

higher electron density

lower electron density

1-46 continued

(f)

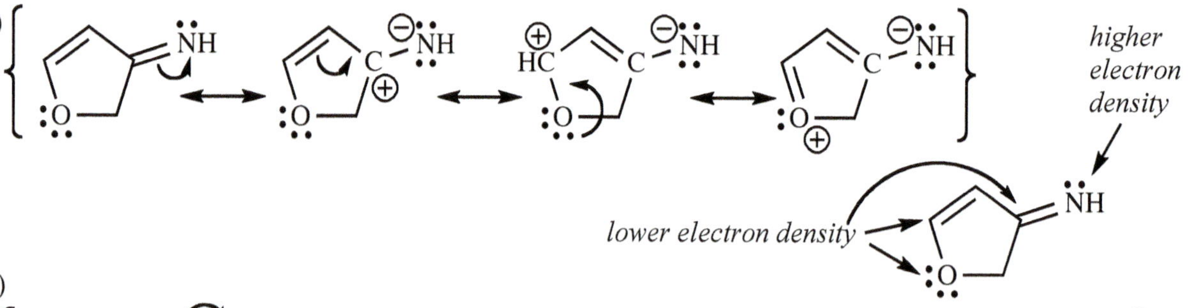

higher electron density

lower electron density

(g)

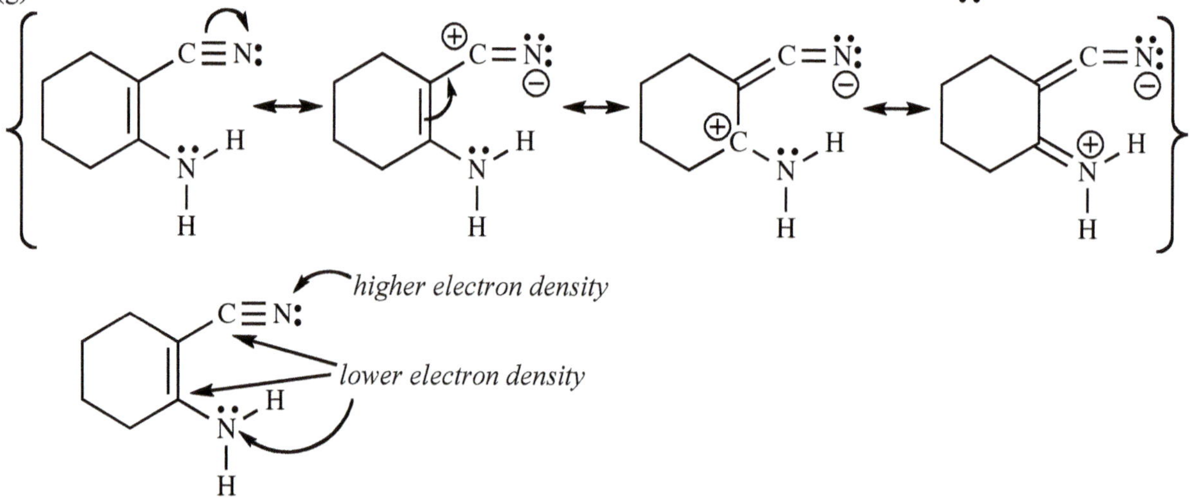

higher electron density

lower electron density

(h)

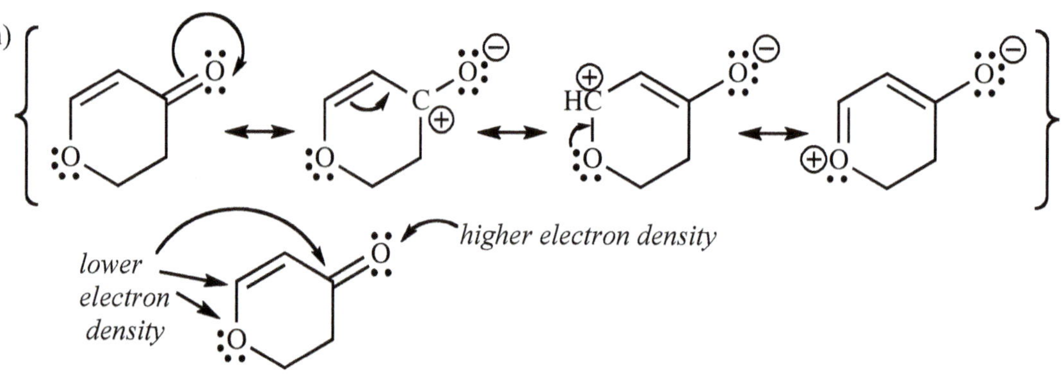

higher electron density

lower electron density

(i)

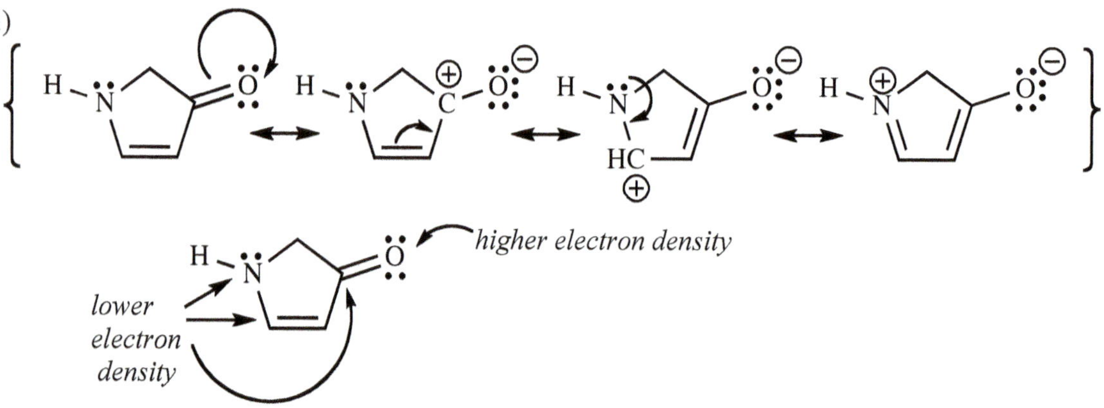

higher electron density

lower electron density

1-46 continued

(j)

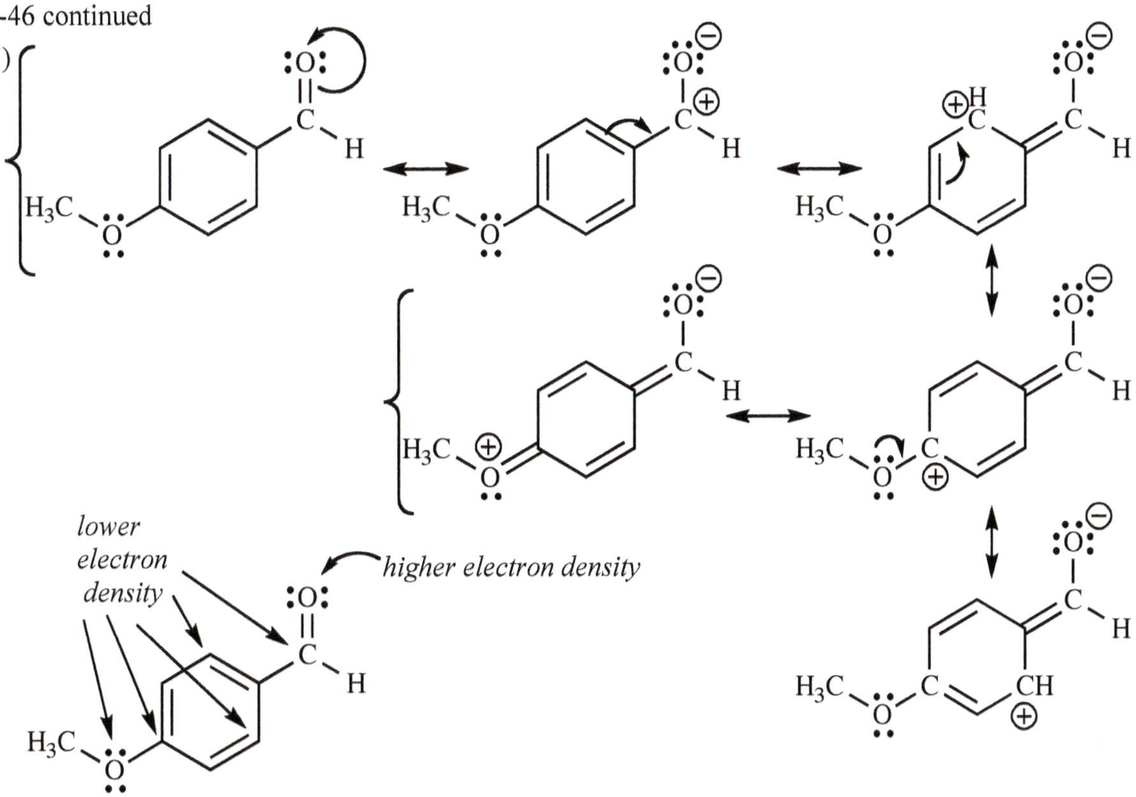

lower electron density

higher electron density

1-47

(a) 100% − 62.0% C − 10.4% H = 27.6% oxygen

$$\frac{62.0 \text{ g C}}{12.0 \text{ g/mole}} = 5.17 \text{ moles C} \div 1.73 \text{ moles} = 2.99 \approx 3 \text{ C}$$

$$\frac{10.4 \text{ g H}}{1.01 \text{ g/mole}} = 10.3 \text{ moles H} \div 1.73 \text{ moles} = 5.95 \approx 6 \text{ H}$$

$$\frac{27.6 \text{ g O}}{16.0 \text{ g/mole}} = 1.73 \text{ moles O} \div 1.73 \text{ moles} = 1 \text{ O}$$

(b) empirical formula = C_3H_6O ⇨ empirical weight = 58

molecular weight = 117, about double the empirical weight

⇨ double the empirical formula = molecular formula = $C_6H_{12}O_2$

(c) some possible structures—MANY other structures are possible:

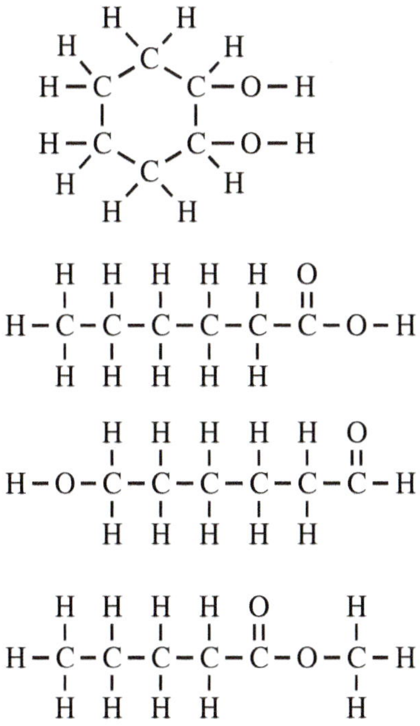

1-48 From the amounts of CO_2 and H_2O generated, the milligrams of C and H in the original sample can be determined, thus giving by difference the amount of oxygen in the 5.00-mg sample. From these values, the empirical formula and empirical weight can be calculated.

(a) how much carbon in 14.54 mg CO_2

$$14.54 \text{ mg } CO_2 \times \frac{1 \text{ mmole } CO_2}{44.01 \text{ mg } CO_2} \times \frac{1 \text{ mmole C}}{1 \text{ mmole } CO_2} \times \frac{12.01 \text{ mg C}}{1 \text{ mmole C}} = 3.968 \text{ mg C}$$

how much hydrogen in 3.97 mg H_2O

$$3.97 \text{ mg } H_2O \times \frac{1 \text{ mmole } H_2O}{18.016 \text{ mg } H_2O} \times \frac{2 \text{ mmoles H}}{1 \text{ mmole } H_2O} \times \frac{1.008 \text{ mg H}}{1 \text{ mmole H}} = 0.444 \text{ mg H}$$

how much oxygen in 5.00 mg estradiol

5.00 mg estradiol – 3.968 mg C – 0.444 mg H = 0.59 mg O

calculate empirical formula

$$\frac{3.968 \text{ mg C}}{12.01 \text{ mg/mmole}} = 0.3304 \text{ mmoles C} \div 0.037 \text{ mmoles} = 8.93 \approx 9 \text{ C}$$

$$\frac{0.444 \text{ mg H}}{1.008 \text{ mg/mmole}} = 0.440 \text{ mmoles H} \div 0.037 \text{ mmoles} = 11.9 \approx 12 \text{ H}$$

$$\frac{0.59 \text{ mg O}}{16.00 \text{ mg/mmole}} = 0.037 \text{ mmoles O} \div 0.037 \text{ mmoles} = 1 \text{ O}$$

empirical formula = $\boxed{C_9H_{12}O}$ ⟹ empirical weight = 136

(b) molecular weight = 272, exactly twice the empirical weight

twice the empirical formula = molecular formula = $\boxed{C_{18}H_{24}O_2}$

1-49 Imagine that you had four tennis balls, two yellow and two green. If you are asked to arrange them on a flat surface, how many different ways could you do it? There are two possible "square" arrangements (diamonds are the same, just oriented differently): one has the two yellow spheres touching and the two green spheres touching, while the other arrangement has the two yellow diagonal to each other, and the two greens diagonal. If this represented CH_2Cl_2, with carbon at the center, there should be two isomers of CH_2Cl_2 in square planar geometry—and yet only one CH_2Cl_2 has ever been isolated. The geometry must be something different from square planar, and the only other way of arranging the five atoms in CH_2Cl_2, while satisfying carbon's valence of 4, is tetrahedral. In tetrahedral geometry, the relationship of each sphere (tennis ball) to the other three is equivalent, no matter how the spheres are arranged.

1-50

(a) 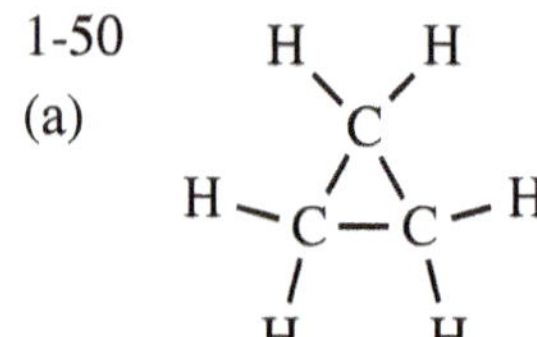

(b) Cyclopropane must have 60° bond angles compared with the usual sp^3 bond angle of 109.5° in an acyclic molecule.

(c) Like a bent spring, bonds that deviate from their normal angles or positions are highly strained. Cyclopropane is reactive because breaking the ring relieves the strain.

1-51

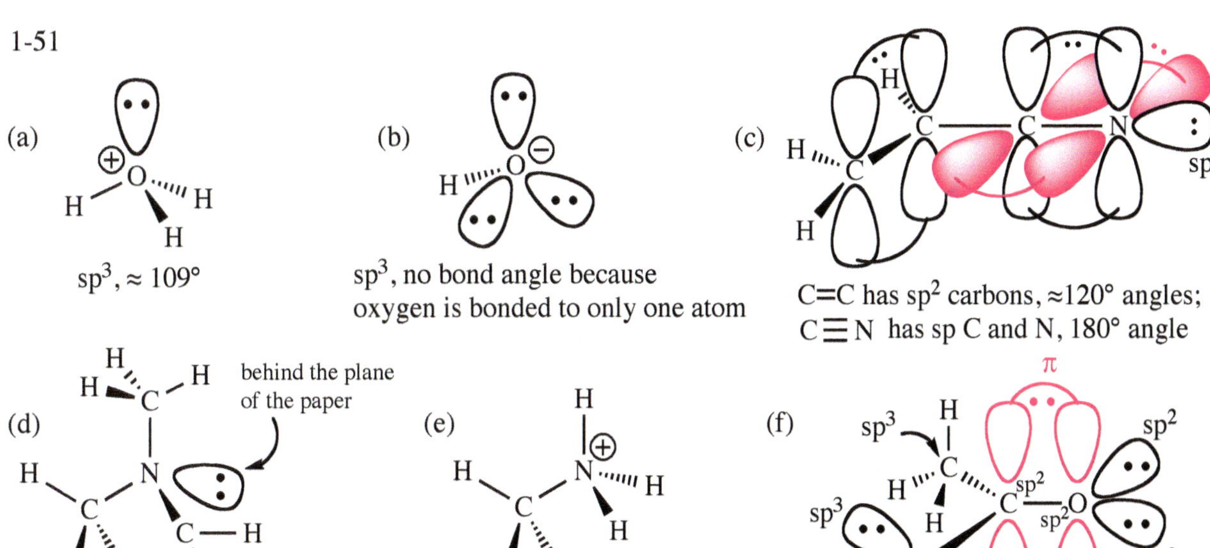

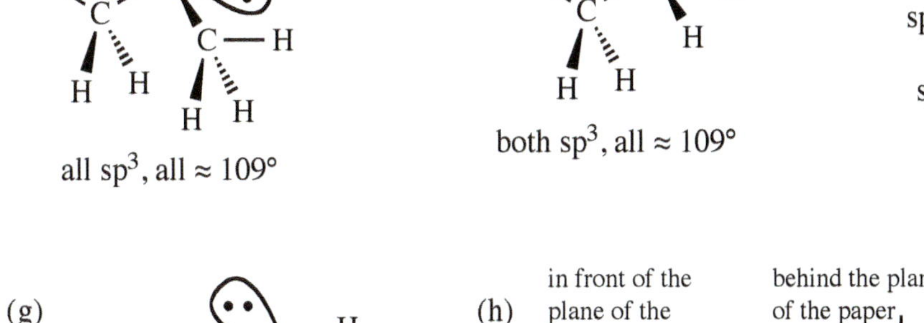

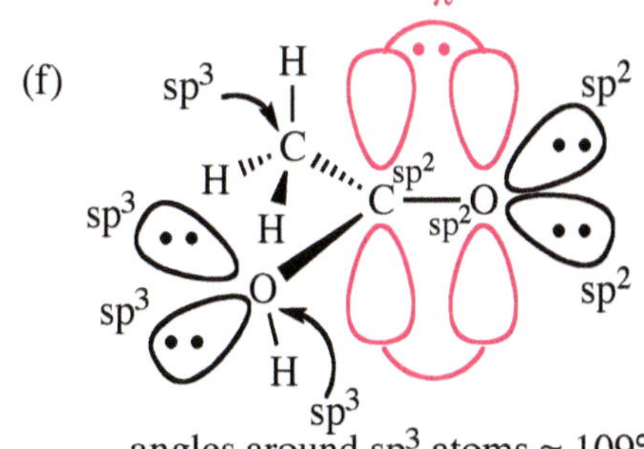

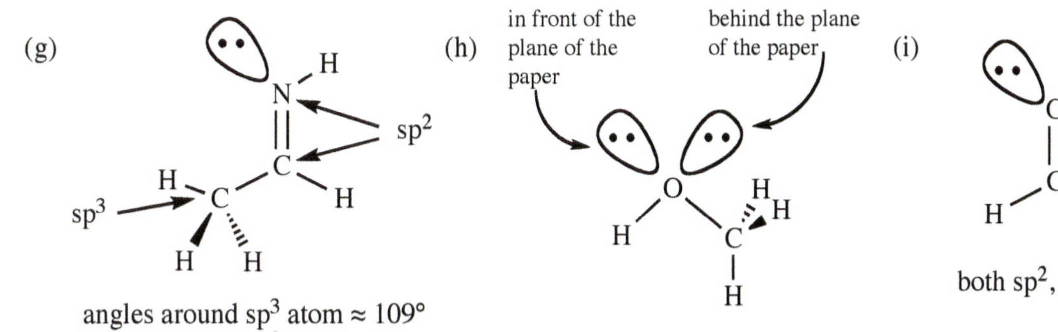

1-52 For clarity in these pictures, bonds between hydrogen and an sp^3 atom are not labeled; these bonds are s-sp^3 overlap.

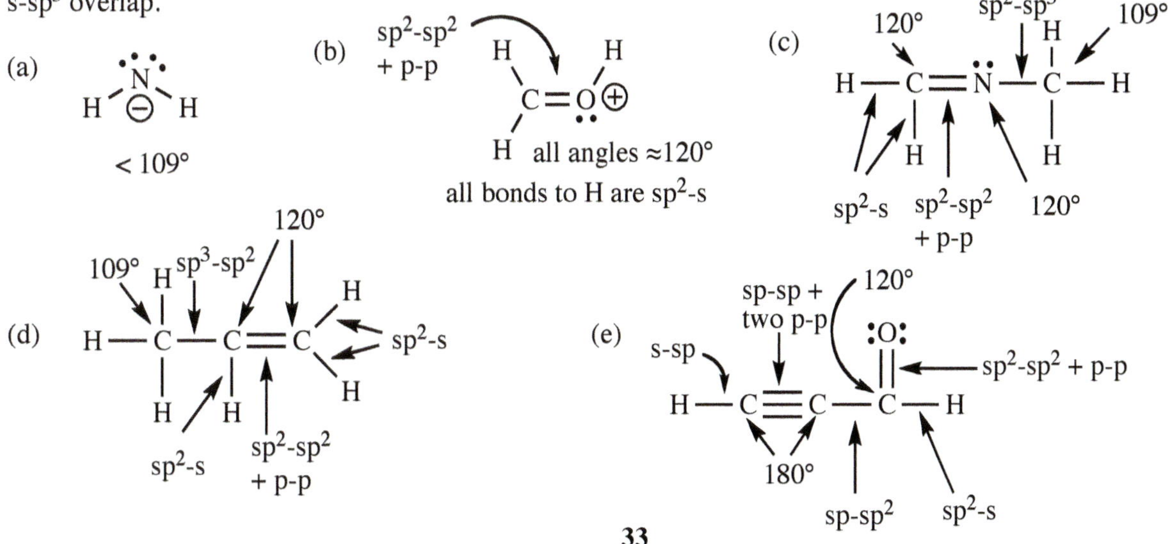

1-52 continued

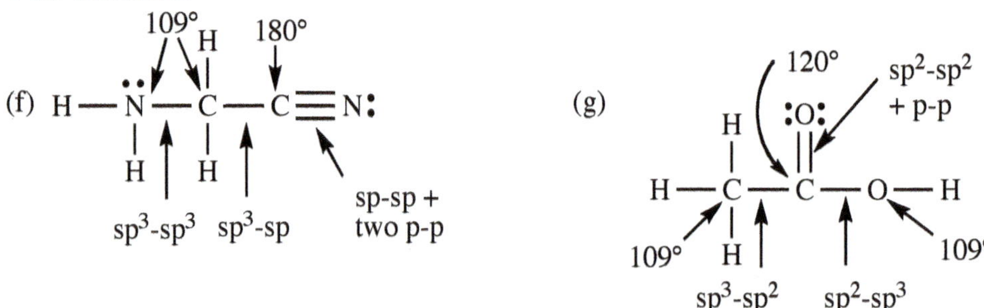

(h) sp³-sp³, sp³-sp², sp²-sp² + p-p, 109°, 120°, sp²-sp², sp²-sp² + p-p, sp²-s, sp³-sp², sp²-s, ≈120° around C=C

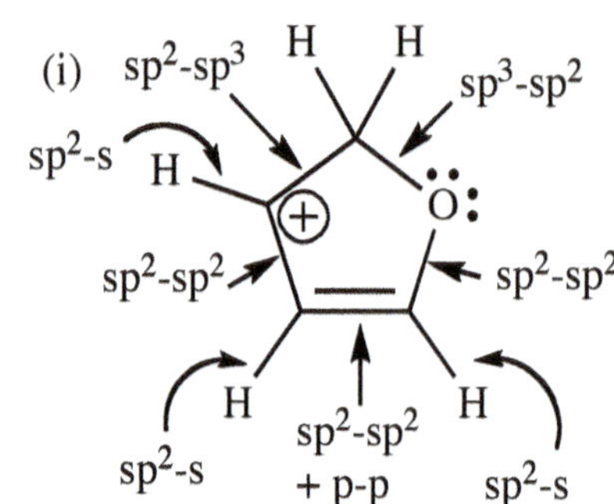

Resonance forms show that the O is sp^2. All atoms are sp^2 except the top C which is sp^3. Angles ≈120° around sp^2, ≈109° around the sp^3 C.

1-53 The second and third resonance forms of urea are minor but still significant. They show that the nitrogen-carbon bonds have some double bond character, requiring that the nitrogens be sp^2 hybridized with bond angles approaching 120°.

1-54

(a) The major resonance contributor shows a carbon-carbon double bond, suggesting that both carbons are sp^2 hybridized with trigonal planar geometry. The CH_3 carbon is sp^3 hybridized with tetrahedral geometry.

minor major

(b) The major resonance contributor shows a carbon-nitrogen double bond, suggesting that all three carbons and the nitrogen are sp^2 hybridized with trigonal planar geometry.

minor minor major

1-54 continued

(c) The nitrogen and the carbon bonded to it are sp hybridized; the other carbon is sp^2. This ion has linear geometry. See the solution to 1-21(e) in this manual for an orbital picture.

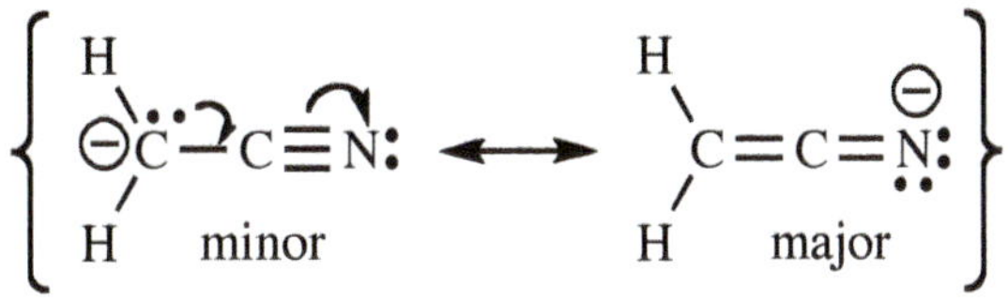

minor ⟷ major

(d) The carbon of the CH_2 is sp^3 hybridized. The other carbons and the nitrogen are all sp^2.

minor ⟷ minor ⟷ major—negative charge on more electronegative atom

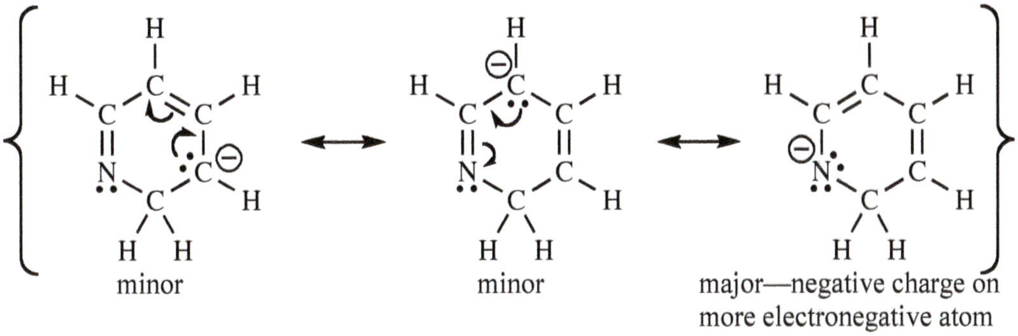

(e) The two carbons of CH_2 are sp^3 hybridized. The other carbons and the nitrogen are all sp^2.

minor ⟷ minor ⟷ major—all atoms have octets

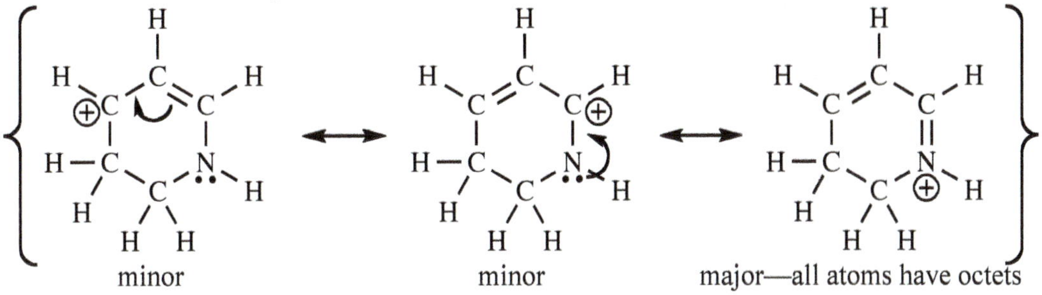

(f) All of the non-hydrogen atoms in this molecule are sp^2 hybridized.

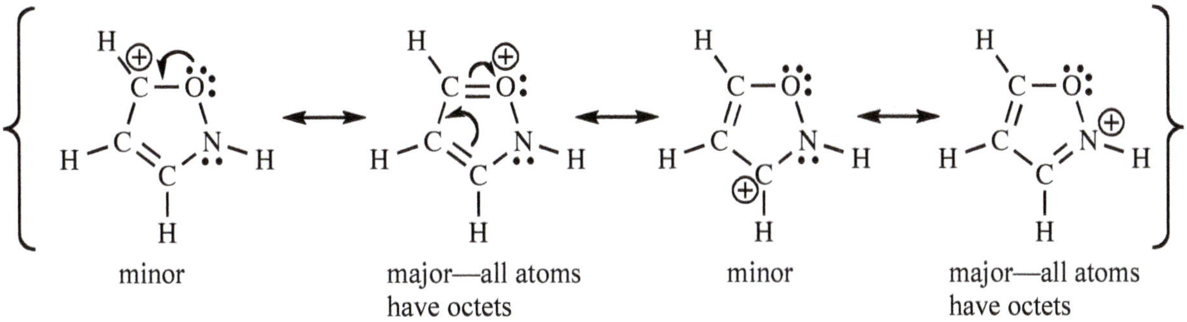

minor ⟷ major—all atoms have octets ⟷ minor ⟷ major—all atoms have octets

1-54 continued

(g) The carbon of the CH_2 is sp^3 hybridized. The other carbons and the nitrogen and oxygen are all sp^2.

major—all atoms have octets; no charges — minor — middle—all atoms have octets

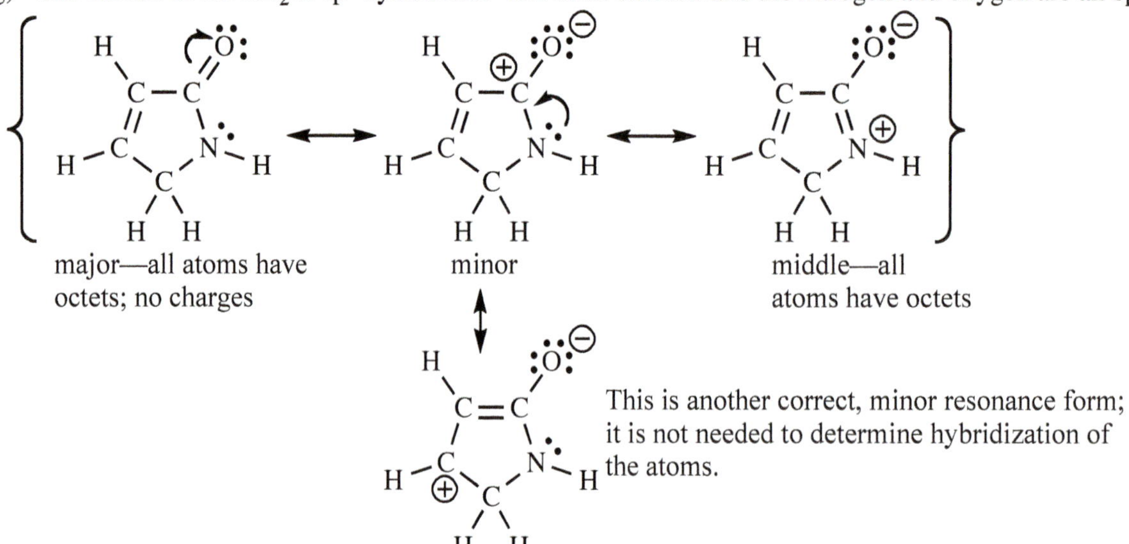

This is another correct, minor resonance form; it is not needed to determine hybridization of the atoms.

(h) The carbon of the CH_2 is sp^3 hybridized. The other carbons and the nitrogen and oxygen are all sp^2.

major—all atoms have octets; no charges — minor — minor — middle—all atoms have octets

(i) Only the C and O of the C=O are sp^2 hybridized. All other non-hydrogen atoms are sp^3 hybridized.

major—all atoms have octets; no charges — minor

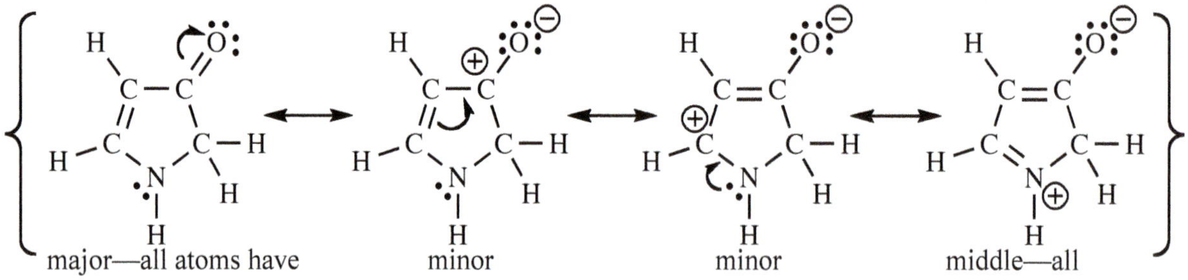

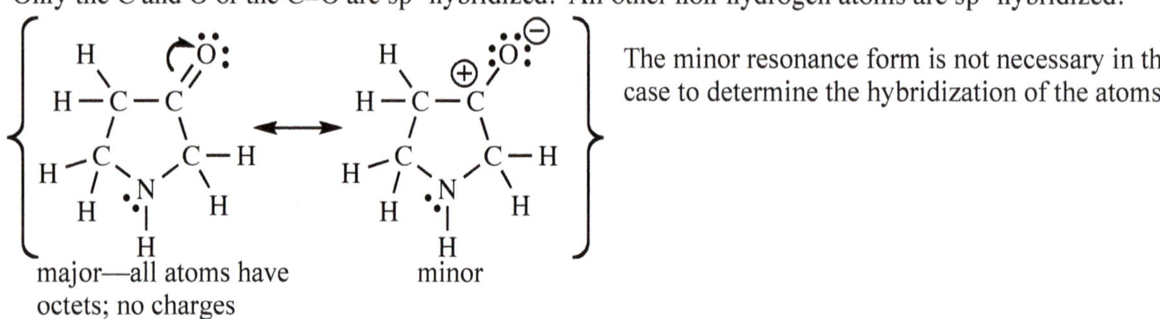

The minor resonance form is not necessary in this case to determine the hybridization of the atoms.

1-55 In (c), (d), (e) and (f), the unshadowed p orbitals are vertical and parallel; the shadowed p orbitals are perpendicular to the paper and horizontal.

(a)

(b)

(c)

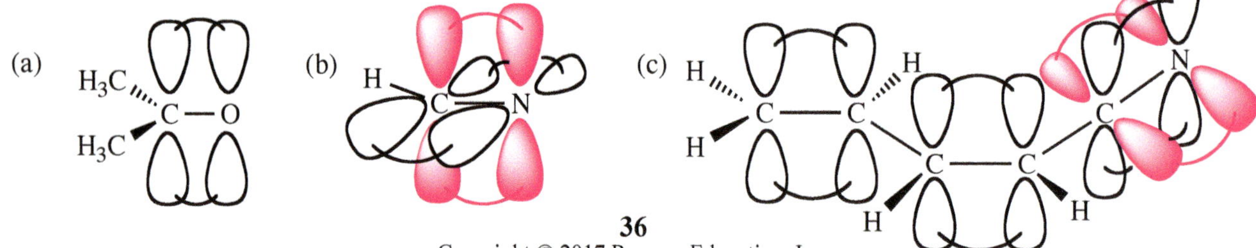

1-55 continued

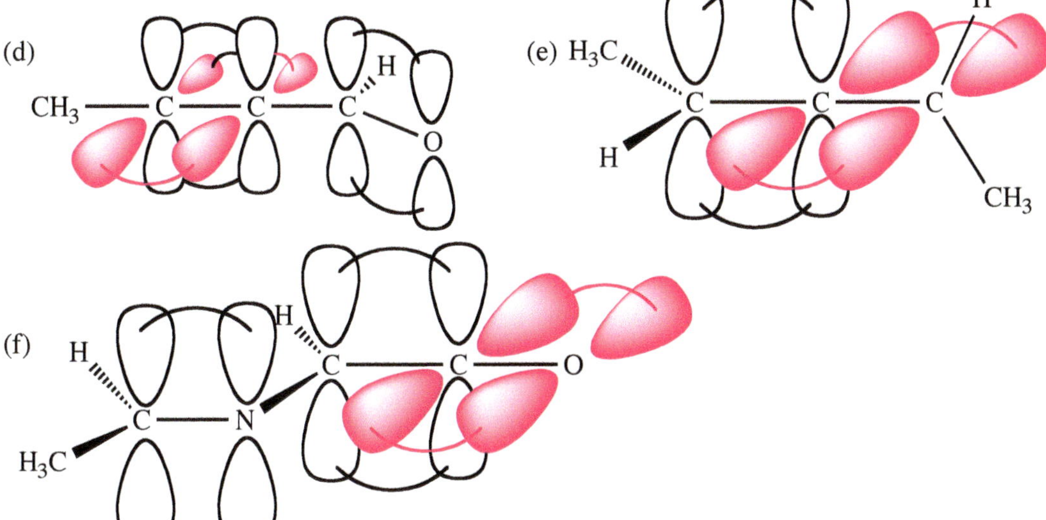

1-56

(a) *cis*

(b) The coplanar atoms in the structures to the left and below are marked with asterisks.

(c) *trans*

There are still six coplanar atoms.

(d)

1-57 Collinear atoms are marked with asterisks.

1-58

Cis and trans isomers are determined by comparing two identical groups on the two carbons of the double bond, and determining their relative position: same side of the double bond = *cis*; opposite side = *trans*. Drawing in H atoms is a big help!

(a) $H_3C(H)C{=}C(H)CH_3$ *cis* and $H(H_3C)C{=}C(H)CH_3$ *trans*

(b) no *cis-trans* isomerism around a triple bond

(c) no *cis-trans* isomerism; two groups on each carbon are the same

(d) Theoretically, cyclopentene could show *cis-trans* isomerism. In reality, the *trans* form is too unstable to exist because of the necessity of stretched bonds and deformed bond angles. *trans*-Cyclopentene has never been detected.

(e) $H_3C(H)C{=}C(CH_2CH_3)(CH_2CH_2CH_3)$ and $H(H_3C)C{=}C(CH_2CH_3)(CH_2CH_2CH_3)$ These are *cis-trans* isomers, but the designation of *cis* and *trans* to specific structures is not defined because of four different groups on the double bond.

(f) $H_3C(H)C{=}\ddot{N}CH_3$ *cis* and $H(H_3C)C{=}\ddot{N}CH_3$ *trans*

1-59

(a) constitutional isomers—The carbon skeletons are different.

(b) constitutional isomers—The position of the chlorine atom has changed.

(c) *cis-trans* isomers—The first is *cis*, the second is *trans*.

(d) constitutional isomers—The carbon skeletons are different.

(e) *cis-trans* isomers—The first is *trans*, the second is *cis*. (The $-CH_2CH_3$ groups decide.)

(f) same compound—Rotation of the first structure gives the second.

(g) *cis-trans* isomers—The first is *cis*, the second is *trans*.

(h) constitutional isomers—The position of the double bond relative to the ketone has changed (while it is true that the first double bond is *cis* and the second is *trans*, in order to have *cis-trans* isomers, the rest of the structure must be identical).

1-60

$CH_3-C(=\ddot{O}:)-CH_3$ — sp^2—planar

$\{ CH_3-\ddot{S}(=\ddot{O}:)-CH_3 \longleftrightarrow CH_3-\ddot{S}^{\oplus}(-\ddot{O}:^{\ominus})-CH_3 \}$

a misleading picture — sp^3—tetrahedral

The key to this problem is understanding that sulfur has a *lone pair of electrons*. The second resonance form shows four pairs of electrons around the sulfur atom, an electronic configuration requiring sp^3 hybridization. Sulfur in DMSO cannot be sp^2 like carbon in acetone, so we would expect sulfur's geometry to be pyramidal (the four electron pairs around sulfur require tetrahedral geometry, but the three atoms around sulfur define its shape as pyramidal). The first resonance form is a misleading picture because it suggests a p-p pi bond in DMSO that does not exist. Sulfur might use a d orbital for some pi bonding but it is definitely not a p-p pi bond.

CHAPTER 2—ACIDS AND BASES; FUNCTIONAL GROUPS

2-1

(a) $2.4 \text{ D} = 4.8 \times \delta \times 1.23 \text{ Å}$

$\delta = 0.41$, or 41% of a positive charge on carbon and 41% of a negative charge on oxygen

(b)

Resonance form **A** must be the major contributor. If **B** were the major contributor, the value of the charge separation would be between 0.5 and 1.0, meaning >50% of a positive charge on C and >50% of a negative charge on O. Even though **B** is "minor", it is quite significant, explaining in part the high polarity of the C=O.

2-2

Both NH_3 and NF_3 have a pair of nonbonding electrons on the nitrogen. In NH_3, the *direction* of polarization of the N—H bonds is *toward* the nitrogen; thus, all three bond polarities and the lone pair polarity reinforce each other. In NF_3, on the other hand, the direction of polarization of the N—F bonds is *away* from the nitrogen; the three bond polarities cancel the lone pair polarity, so the net result is a very small *molecular* dipole moment.

polarities reinforce; large dipole moment (1.50)

polarities oppose; small dipole moment (0.20)

2-3 Some magnitudes of dipole moments are difficult to predict; however, the direction of the dipole should be straightforward, in most cases. Actual values of molecular dipole moments are given in parentheses. (Each halogen atom has three nonbonded electron pairs, not shown below.) The C—H is usually considered nonpolar. (Parts (a) and (b) are solved in the text.)

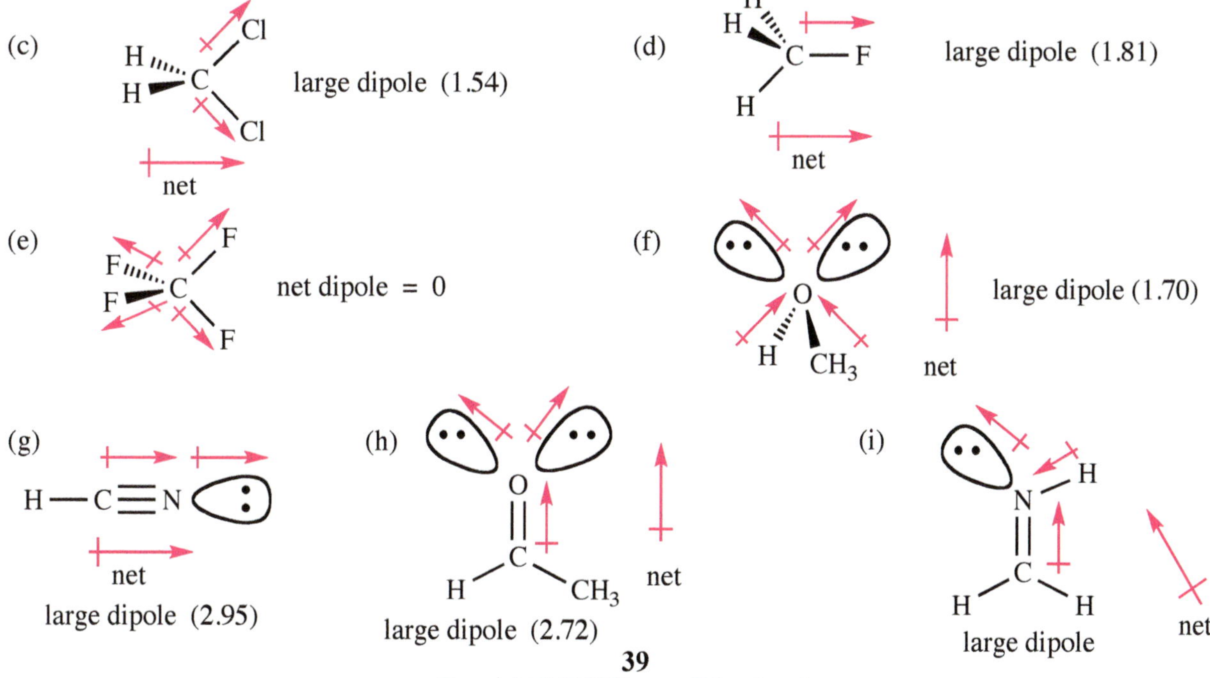

2-3 continued

(j)

(k)

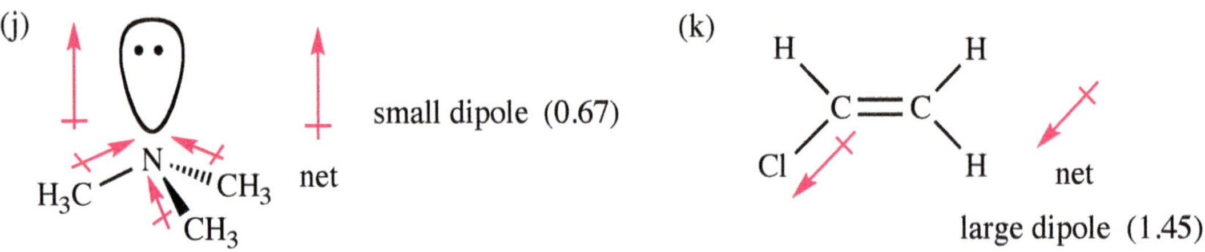

(l)

F B F F

net dipole = 0

(m)

Cl — Be — Cl

net dipole = 0

In (l) and (m), the symmetry of the molecule allows the individual bond dipoles to cancel.

2-4 With chlorines on the same side of the double bond, the bond dipole moments reinforce each other, resulting in a large net dipole. With chlorines on opposite sides of the double bond, the bond dipole moments exactly cancel each other, resulting in a zero net dipole.

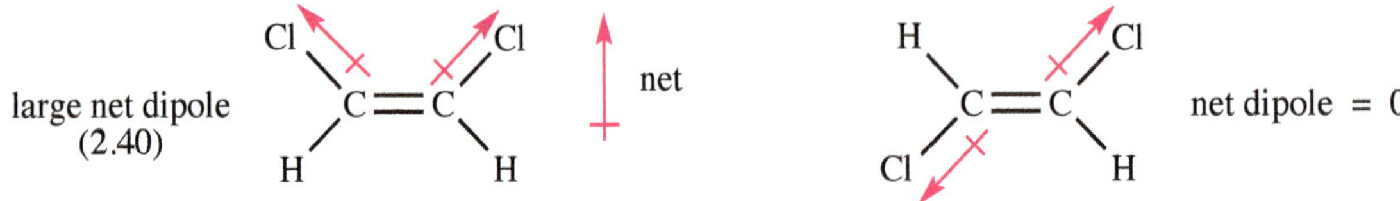

2-5 Hydrogen bonds are shown as wavy bonds.

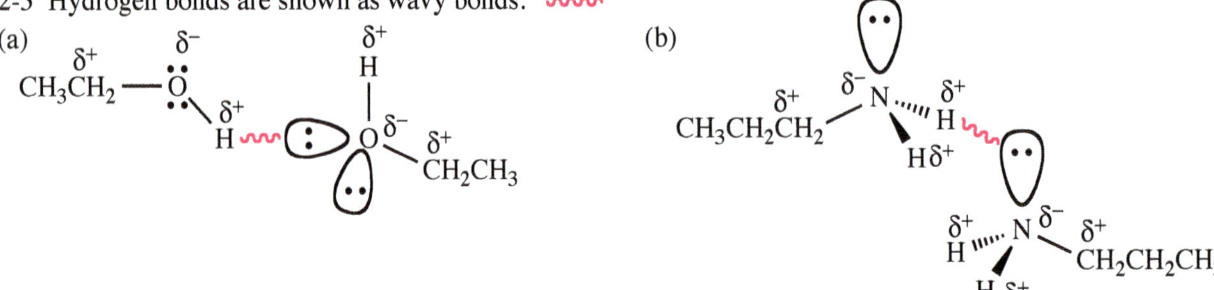

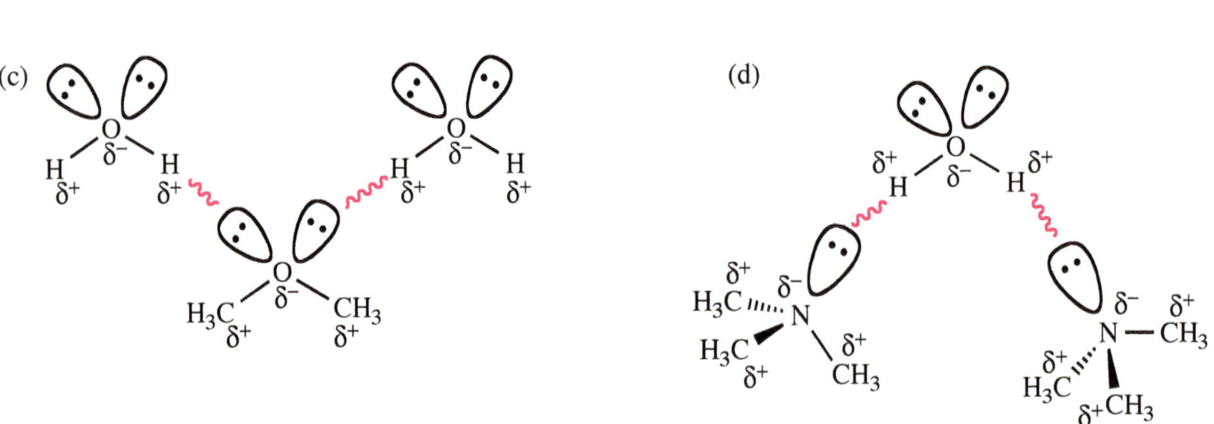

2-6

(a) $(CH_3)_2CHCH_2CH_2CH(CH_3)_2$ has less branching and boils at a higher temperature than $(CH_3)_3CC(CH_3)_3$.

(b) $CH_3(CH_2)_5CH_2OH$ can form hydrogen bonds and will boil at a much higher temperature than $CH_3(CH_2)_6CH_3$, which cannot form hydrogen bonds.

(c) $CH_3CH_2CH_2CH_2OH$ can form hydrogen bonds and boils at a higher temperature than $CH_3CH_2OCH_2CH_3$.

(d) $HOCH_2(CH_2)_4CH_2OH$ can form hydrogen bonds at both ends and has no branching; it will boil at a much higher temperature than $(CH_3)_3CCH(OH)CH_3$.

(e) $(CH_3CH_2CH_2)_2NH$ has an N—H bond and can form hydrogen bonds; it will boil at a higher temperature than $(CH_3CH_2)_3N$, which cannot form hydrogen bonds.

(f) The second compound shown (**B**) has the higher boiling point for two reasons: **B** has a higher molecular weight than **A**; and **B**, a primary amine with two N—H bonds, has more opportunity for forming hydrogen bonds than **A**, a secondary amine with only one N—H bond.

A N-H B N H H

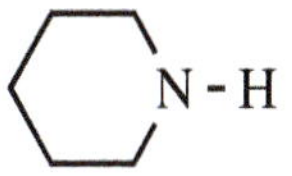

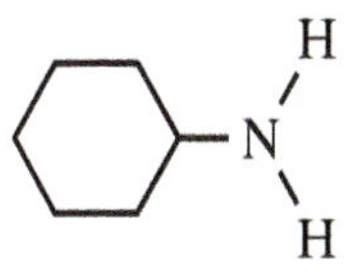

2-7

(a) $CH_3CH_2OCH_2CH_3$ can form hydrogen bonds with water and is more soluble than $CH_3CH_2CH_2CH_2CH_3$, which cannot form hydrogen bonds with water.

(b) $CH_3CH_2CH_2OH$ is more soluble in water because it has one fewer carbon than $CH_3CH_2OCH_2CH_3$.

(c) $CH_3CH_2NHCH_3$ is more water soluble because it can form hydrogen bonds with water; $CH_3CH_2CH_2CH_3$ cannot form hydrogen bonds.

(d) CH_3CH_2OH is more soluble in water. The polar O—H group forms hydrogen bonds with water, overcoming the resistance of the nonpolar CH_3CH_2 group toward entering the water. In $CH_3CH_2CH_2CH_2OH$, however, the hydrogen bonding from only one OH group cannot carry a four-carbon chain into the water; this substance is only slightly soluble in water.

(e) Both compounds form hydrogen bonds with water at the double-bonded oxygen, but only the smaller molecule (CH_3COCH_3) dissolves. The cyclic compound has too many nonpolar CH_2 groups to dissolve.

2-8

(a) $$5.00 \text{ g HBr} \times \frac{1 \text{ mole HBr}}{80.9 \text{ g HBr}} = 0.0618 \text{ moles HBr}$$

$$0.0618 \text{ moles HBr} \Longrightarrow 0.0618 \text{ moles } H_3O^+ \text{ (100\% dissociated)}$$

$$\frac{0.0618 \text{ moles } H_3O^+}{100. \text{ mL}} \times \frac{1000 \text{ mL}}{1 \text{ L}} = \frac{0.618 \text{ moles } H_3O^+}{1 \text{ L solution}}$$

$$pH = -\log_{10}[H_3O^+] = -\log_{10}(0.618) = \boxed{0.209}$$

(b) $$1.50 \text{ g NaOH} \times \frac{1 \text{ mole NaOH}}{40.0 \text{ g NaOH}} = 0.0375 \text{ moles NaOH}$$

$$0.0375 \text{ moles NaOH} \Longrightarrow 0.0375 \text{ moles } ^-OH \text{ (100\% dissociated)}$$

$$\frac{0.0375 \text{ moles } ^-OH}{50. \text{ mL}} \times \frac{1000 \text{ mL}}{1 \text{ L}} = \frac{0.75 \text{ moles } ^-OH}{1 \text{ L solution}} = 0.75 \text{ M}$$

$$[H_3O^+] = \frac{1 \times 10^{-14}}{[^-OH]} = \frac{1 \times 10^{-14}}{0.75} = 1.33 \times 10^{-14}$$

$$pH = -\log_{10}[H_3O^+] = -\log_{10}(1.33 \times 10^{-14}) = \boxed{13.88}$$

(The number of decimal places in a pH value is the number of significant figures.)

2-9

(a) By definition, an acid is any species that can donate a proton. Ammonia has a proton bonded to nitrogen, so ammonia can be an acid (although a very weak one). A base is a proton acceptor; that is, it must have a pair of electrons to share with a proton. In theory, any atom with an unshared electron pair can be a base. The nitrogen in ammonia has an unshared electron pair so ammonia is basic. In water, ammonia is too weak an acid to give up its proton; instead, it acts as a base and pulls a proton from water to a small extent.

(b) water as an acid: $H_2O + NH_3 \rightleftharpoons {}^-OH + NH_4^+$

water as a base: $H_2O + HCl \rightleftharpoons H_3O^+ + Cl^-$

(c) Hydronium acting as an acid in water solution will have this chemical equation:

$$\underset{HA}{H_3O^+} + H_2O \rightleftharpoons H_3O^+ + \underset{A^-}{H_2O} \Longrightarrow K_a = \frac{[H_3O^+]\,[A^-]}{[HA]} = \frac{[H_3O^+]\,[H_2O]}{[H_3O^+]} = [H_2O]$$

$$[H_2O] = \frac{1000 \text{ g } H_2O}{1 \text{ L } H_2O} \times \frac{1 \text{ mole } H_2O}{18.0 \text{ g } H_2O} = \underline{55.55 \text{ M}} = K_a \Longrightarrow pK_a = -\log(55.55) = \underline{-1.74}$$

(d) methanol as an acid: $CH_3OH + NH_3 \rightleftharpoons CH_3O^- + NH_4^+$

methanol as a base: $CH_3OH + H_2SO_4 \rightleftharpoons CH_3OH_2^+ + HSO_4^-$

2-10 Values are from Table 2-2, or, as noted, from Appendix 4 in the text.

(a) $HCOOH + {}^-CN \rightleftharpoons HCOO^- + HCN$ — FAVORS **PRODUCTS** (Part (a) is already solved in the text.)

- HCOOH: stronger acid, pK_a 3.76
- ${}^-CN$: stronger base
- $HCOO^-$: weaker base
- HCN: weaker acid, pK_a 9.22

(b) $CH_3COO^- + CH_3OH \rightleftharpoons CH_3COOH + CH_3O^-$ — FAVORS **REACTANTS**

- CH_3COO^-: weaker base
- CH_3OH: weaker acid, pK_a 15.5 (Appendix 4)
- CH_3COOH: stronger acid, pK_a 4.74
- CH_3O^-: stronger base

(c) $(CH_3)_2CHOH + NaNH_2 \rightleftharpoons (CH_3)_2CHO^-\ Na^+ + NH_3$ — FAVORS **PRODUCTS**

- $(CH_3)_2CHOH$: stronger acid, $pK_a \approx 15.9$ (estimated from Table 2-2)
- $NaNH_2$: stronger base
- $(CH_3)_2CHO^-\ Na^+$: weaker base
- NH_3: weaker acid, pK_a 36

(d) $Na^+\ {}^-OCH_3 + HCN \rightleftharpoons HOCH_3 + Na^+\ {}^-CN$ — FAVORS **PRODUCTS**

- $Na^+\ {}^-OCH_3$: stronger base
- HCN: stronger acid, pK_a 9.22
- $HOCH_3$: weaker acid, pK_a 15.5 (Appendix 4)
- $Na^+\ {}^-CN$: weaker base

(e) $HCl + CH_3CH_2OH \rightleftharpoons CH_3CH_2OH_2^+ + Cl^-$ — FAVORS **PRODUCTS**

- HCl: stronger acid, pK_a −7
- CH_3CH_2OH: stronger base
- $CH_3CH_2OH_2^+$: weaker acid, pK_a −2.4 (Appendix 4)
- Cl^-: weaker base

2-10 continued

(f) H_3O^+ + CH_3O^- ⇌ H_2O + CH_3OH FAVORS **PRODUCTS**

stronger acid, pK_a −1.7 | stronger base | weaker base | weaker acid, pK_a 15.5 (Appendix 4)

(g) pyrrolidine (N−H) + phenol (OH) ⇌ pyrrolidinium (N^+H_2) + phenoxide (O^-) FAVORS **PRODUCTS**

stronger base | stronger acid, pK_a 10.0 (Appendix 4) | weaker acid, pK_a 11.3 (Appendix 4) | weaker base

(h) pyrrolidine (N−H) + cyclopentane−COOH ⇌ pyrrolidinium (N^+H_2) + cyclopentane−COO^- FAVORS **PRODUCTS**

stronger base | stronger acid, $pK_a \approx 4.7$ (Appendix 4) similar to CH_3COOH | weaker acid, pK_a 11.3 (Appendix 4) | weaker base

(i) benzene−SH + $CH_3CH_2O^-$ ⇌ CH_3CH_2OH + benzene−S^- FAVORS **PRODUCTS**

stronger acid, pK_a 7.8 (Appendix 4) | stronger base | weaker acid, pK_a 15.9 (Appendix 4) | weaker base

(j) benzene−$C\equiv C^-$ + CH_3CH_3OH ⇌ benzene−$C\equiv CH$ + $CH_3CH_2O^-$ FAVORS **PRODUCTS**

stronger base | stronger acid, pK_a 15.9 (Appendix 4) | weaker acid, $pK_a \approx 25$ (Appendix 4) similar to $HC\equiv CH$ | weaker base

2-11

(a) $H_3C-C(=O)-OH$ + $:B^-$ ⇌ $H-B$ + $H_3C-C(=O)-O^-$

strongest acid pK_a 4.74

CH_3CH_2OH + $:B^-$ ⇌ $H-B$ + $CH_3CH_2O^-$

pK_a 15.9

CH_3NH_2 + $:B^-$ ⇌ $H-B$ + CH_3NH^-

weakest acid pK_a 40

2-11 continued

(b)

$$CH_3NH_2 + H{-}A \rightleftharpoons A^{\ominus} + CH_3NH_3^{\oplus}$$

strongest base — $CH_3NH_3^{\oplus}$: pK_a 10.7 *weakest conjugate acid*

$$CH_3CH_2OH + H{-}A \rightleftharpoons A^{\ominus} + CH_3CH_2OH_2^{\oplus}$$

$CH_3CH_2OH_2^{\oplus}$: pK_a –2.4

$$H_3C{-}C({=}O){-}OH + H{-}A \rightleftharpoons A^{\ominus} + H_3C{-}C({=}OH^{\oplus}){-}OH$$

weakest base — $H_3C{-}C({=}OH^{\oplus}){-}OH$: pK_a –6.1 *strongest conjugate acid*

2-12

First, review the structures and properties of the solvents in question.

$CH_3CH_2CH_2CH_2CH_3$	$CH_3CH_2OCH_2CH_3$	CH_3CH_2OH	H_2O	NH_3
pentane	diethyl ether	ethanol	water	ammonia

•Pentane is a hydrocarbon, neither acidic nor basic. Any acid-base reaction can occur in hydrocarbons without interacting with the solvent.

•Diethyl ether is not acidic, so reactions with strong bases are compatible, but it is weakly basic, so reactions with very strong acids (stronger than pK_a –3.6) will be leveled by the solvent.

•Ethanol is both acidic (pK_a 15.9) and basic (pK_a of the conjugate acid –2.4), so only acids and bases within this range are possible without interaction with the solvent.

•Water is both acidic (pK_a 15.7) and basic (pK_a of the conjugate acid –1.7), so only acids and bases within this range are possible without interaction with the solvent.

•Ammonia is both acidic (pK_a 36, very weak) and basic (pK_a of the conjugate acid 9.2), so only acids and bases within this range are possible without interaction with the solvent.

This question is asking, "Are the ultimate products compatible with the solvents?" The reactants may be leveled by the solvent, but as long as the products can exist within the pK_a limits of the solvent, the solvent is an acceptable choice.

Reaction	$CH_3CH_2CH_2CH_2CH_3$ pentane	$CH_3CH_2OCH_2CH_3$ diethyl ether	CH_3CH_2OH ethanol	H_2O water	NH_3 ammonia
(a) very strong base	OK	OK	NO	NO	OK
$HC{\equiv}CLi$ is a strong base from an acid of pK_a 25. It is not compatible with alcohols and water.					
(b) strong base	OK	OK	NO	NO	OK
$(CH_3)_3COLi$ is a strong base from an acid of pK_a 18. It is not compatible with alcohols and water.					
(c) very strong acid	OK	NO	NO	NO	NO
The protonated ring compound has pK_a –3.8. It is compatible only with pentane.					
(d) strong base	OK	OK	NO	NO	OK also a reactant
$(CH_3)_3CONa$ is a strong base from an acid of pK_a 18. It is not compatible with alcohols and water.					
(e) very strong acid	OK	NO	NO	NO	NO
The protonated oxygen compound has pK_a –7.3. It is compatible only with pentane.					
(e) strong acid	OK	OK	OK	OK	NO
The protonated oxygen compound has pK_a 0.0. It is compatible with the four solvents except ammonia.					

2-13

(a) $CH_3CH_2-\ddot{O}-H$ + $CH_3-\ddot{N}^{-}-H$ ⇌ $CH_3CH_2-\ddot{O}:^{-}$ + $CH_3-\ddot{N}(H)-H$ *favors PRODUCTS*

stronger acid; stronger base; conjugate base, weaker base (more stable anion—negative charge on the more electronegative atom); conjugate acid, weaker acid

(b) $F_3CCOONa$ + Br_3CCOOH ⇌ F_3CCOOH + $Br_3CCOONa$ *favors REACTANTS*

weaker base (more stable anion—F is more electronegative than Br); weaker acid; conjugate acid, stronger acid; conjugate base, stronger base

(c) $CH_3-\ddot{O}-H$ + H_2SO_4 ⇌ $CH_3-\ddot{O}^{+}(H)-H$ + HSO_4^- *favors PRODUCTS*

stronger base; stronger acid, pK_a −5; conjugate acid, weaker acid, pK_a −2.5; conjugate base, weaker base

(d) Na^{+} $^{-}:\ddot{O}-H$ + $H-\ddot{S}-H$ ⇌ $H-\ddot{O}-H$ + Na^{+} $^{-}:\ddot{S}-H$ *favors PRODUCTS*

stronger base; stronger acid; conjugate acid, weaker acid; conjugate base, weaker base (larger anion—size matters!)

(e) $CH_3-N^{+}H_3$ + $CH_3-\ddot{O}:^{-}$ ⇌ $CH_3-\ddot{N}H_2$ + $CH_3-\ddot{O}-H$ *favors PRODUCTS*

stronger acid, pK_a 10.7; stronger base; conjugate base, weaker base; conjugate acid, weaker acid, pK_a 15.5

(f) $BrCH_2CH_2OH$ + $F_3CCH_2O^-$ ⇌ $BrCH_2CH_2O^-$ + F_3CCH_2OH *favors REACTANTS*

weaker acid; weaker base (more stable anion—F is more electronegative than Br, especially 3 F); conjugate base, stronger base; conjugate acid, stronger acid

(g) $NaOCH_2CH_3$ + Cl_2CHCH_2OH ⇌ $HOCH_2CH_3$ + Cl_2CHCH_2ONa *favors PRODUCTS*

stronger base; stronger acid; conjugate acid, weaker acid; conjugate base, weaker base (more stable anion—2 Cl stabilize anion)

(h) H_2Se + $NaNH_2$ ⇌ $NaHSe$ + NH_3 *favors PRODUCTS*

stronger acid, pK_a 7.0; stronger base; conjugate base, weaker base (more stable anion—Se much larger than N); conjugate acid, weaker acid, pK_a 36

(i) $CH_3CHFCOOH$ + $FCH_2CH_2COO^-$ ⇌ $CH_3CHFCOO^-$ + FCH_2CH_2COOH *favors PRODUCTS*

stronger acid; stronger base; conjugate base, weaker base (more stable anion—F is closer to COO^-); conjugate acid, weaker acid

(j) $CF_3CH_2O^-$ + FCH_2CH_2OH ⇌ CF_3CH_2OH + $FCH_2CH_2O^-$ *favors REACTANTS*

weaker base (more stable anion—3 F stabilizes anion better than 1 F); weaker acid; conjugate acid, stronger acid; conjugate base, stronger base

2-14

Three principles guide the effect of electron-withdrawing substituents on acid strength. Acidity is increased by:

1. increasing strength of electron-withdrawing substituents; for single atoms, this means more electronegative;
2. increasing number of electron-withdrawing substituents;
3. closer proximity of the electron-withdrawing substituents to the acidic group.

"Electron-Withdrawing Group" is abbreviated "EWG".

2,2-dichlorobutanoic acid > 2-chlorobutanoic acid > 2-bromobutanoic acid > 3-bromobutanoic acid

strongest acid two, strong EWG next to acid group	one, strong EWG next to acid group	one, moderate EWG next to acid group (Cl more electronegative than Br)	weakest acid one, moderate EWG farther from acid group

2-15

(a) $H_3C(H)C{=}O^{+}H$ (stronger conjugate acid, pK_a −8.0) + H_2O ⇌ H_3O^+ + $H_3C(H)C{=}\ddot{O}$: sp^2 weaker base

(b) $H_3C(H)_2C{-}O^{+}H_2$ (weaker conjugate acid, pK_a −2.4) + H_2O ⇌ H_3O^+ + $H_3C(H)_2C{-}\ddot{O}H$ sp^3 stronger base

An atom with sp^2 hybridization is less basic than the same atom with sp^3 hybridization. Therefore, the conjugate acid of the sp^2 atom must be a stronger acid than the atom when sp^3 hybridized.

2-16

cyclopentyl–$CH(CH_3)$–$\ddot{N}HCH_3$ > cyclopentyl–$C(CH_3){=}\ddot{N}CH_3$ > cyclopentyl–$C{\equiv}N$:

strongest base, sp^3	sp^2	weakest base, sp

↓ H−A ↓ H−A ↓ H−A

cyclopentyl–$CH(CH_3)$–$N^{+}H_2CH_3$ < cyclopentyl–$C(CH_3){=}N^{+}HCH_3$ < cyclopentyl–$C{\equiv}N^{+}{-}H$

weakest conjugate acid pK_a 10.7 for $H{-}N^{+}H(CH_3)_2$	sp^2 pK_a 5.5 for $(H_3C)_2C{=}N^{+}HCH_3$	strongest conjugate acid pK_a −10.1 for $H_3CC{\equiv}N^{+}H$ really strong!

2-17

When comparing basic atoms, sp^3 atoms are stronger bases than sp^2, which are stronger than sp. Conjugate acids have the opposite order: H on sp is stronger than H on sp^2, which is stronger than H on sp^3.

(a)

C≡C:

weaker base, sp

stronger base, sp^2

H—A

H—A

H

C≡C—H

stronger conjugate acid

weaker conjugate acid

(b)

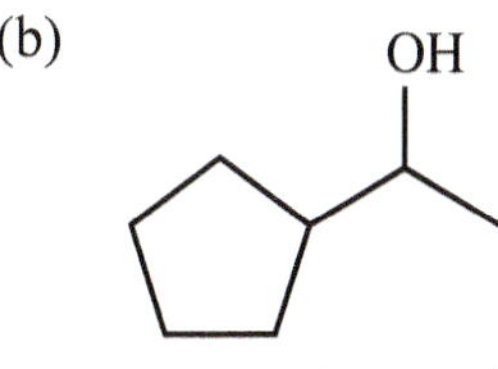

stronger base, sp^3

H—A

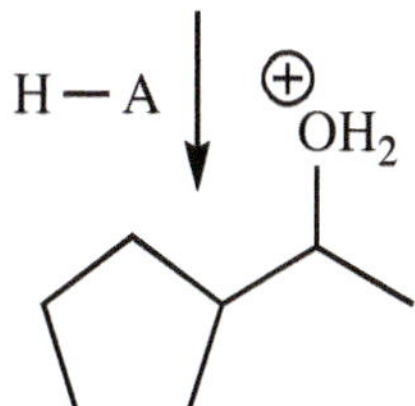

weaker conjugate acid

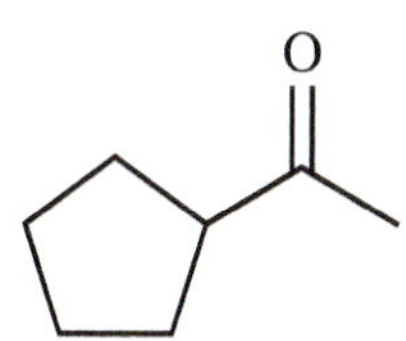

weaker base, sp^2

H—A

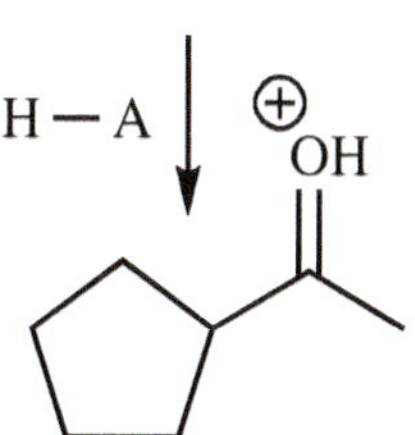

stronger conjugate acid

(c)

NH

N

stronger base, sp^3

weaker base, sp^2

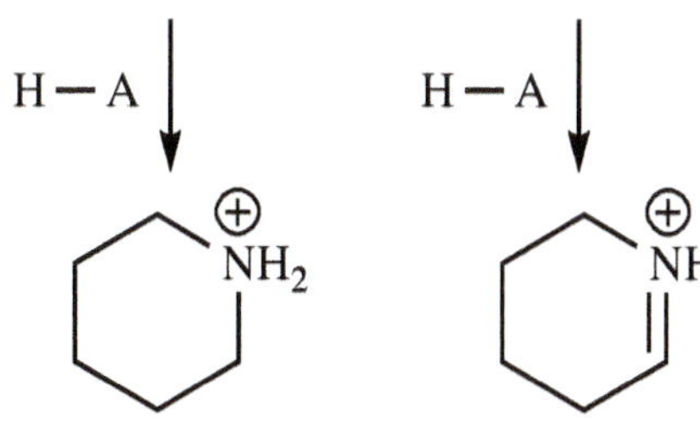

weaker conjugate acid

stronger conjugate acid

(d) These structures are not the same comparison as in the other problems. They differ by a more important feature than hybridization: an anion at an sp^3 carbon is always going to be a stronger base than a neutral nitrogen.

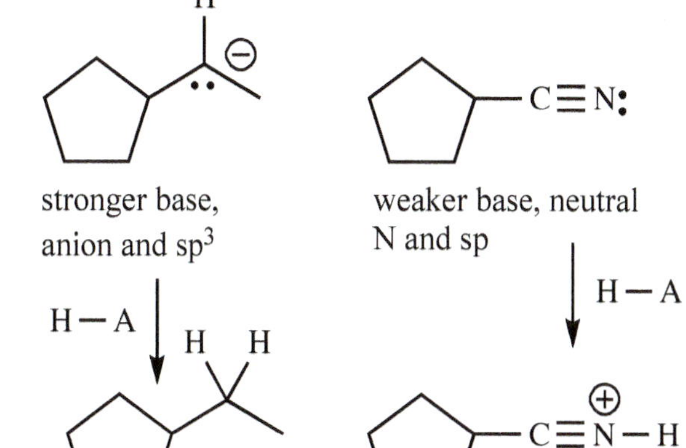

weaker conjugate acid

stronger conjugate acid

2-18

Recall that for any conjugate acid-base pair, $pK_a + pK_b = 14$. Applying this to the structures in question:

CH_3CH_2OH ⟹ $CH_3CH_2O^{\ominus}$

ethanol, pK_a 15.9 — ethoxide ion, pK_b –1.9

weaker acid — stronger conjugate base

CH_3COOH ⟹ $CH_3COO^{\ominus}$

acetic acid, pK_a 4.74 — acetate ion, pK_b 9.26

stronger acid — weaker conjugate base

The strength of an acid is determined by the stability of its conjugate base. In this case, the resonance stabilization of the acetate ion explains its weaker basicity.

2-19

The resonance forms of acetamide show that the oxygen, not the nitrogen, is the likely site of protonation because the oxygen has a higher electron density than nitrogen.

$pK_b = 14$

Protonation on oxygen also maintains resonance stabilization, unlike protonation on nitrogen which interrupts resonance stabilization.

$pK_a = 14 - pK_b = 14 - 14 = 0$

2-20

Protonation could occur at either oxygen.

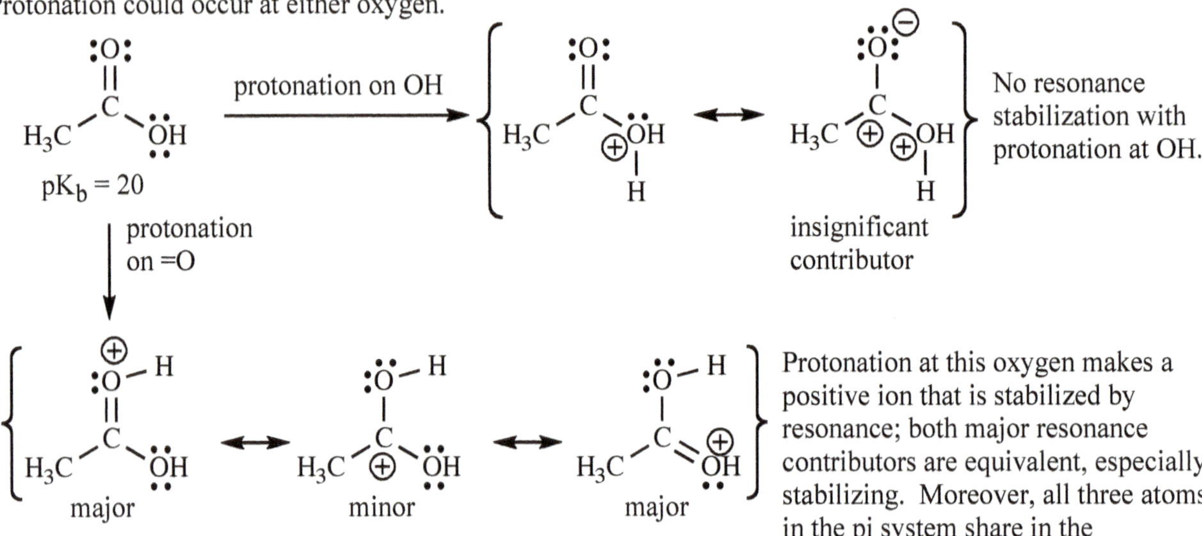

protonation on OH
$pK_b = 20$
protonation on =O
insignificant contributor
No resonance stabilization with protonation at OH.
major
minor
major
Protonation at this oxygen makes a positive ion that is stabilized by resonance; both major resonance contributors are equivalent, especially stabilizing. Moreover, all three atoms in the pi system share in the delocalization of the positive charge.
$pK_a = 14 - pK_b = 14 - 20 = -6$

2-21 *The strength of an acid is determined by the stability of its conjugate base.*

This problem analyzes resonance stabilization of conjugate bases, which in this problem are anions except for part (g). The conjugate base that has more resonance forms that are significant contributors will be more stable, and therefore its conjugate acid is the stronger acid.

Part (a) was solved in the textbook.

(b) The anion of the first structure has three significant resonance contributors, two of which are equivalent (particularly stable) and with negatve charge on the more electronegative atoms; high stabilization of the anion makes the first structure the stronger acid.

major *minor* *major*

See the next page for the resonance forms of the other conjugate base.

2-21(b) continued

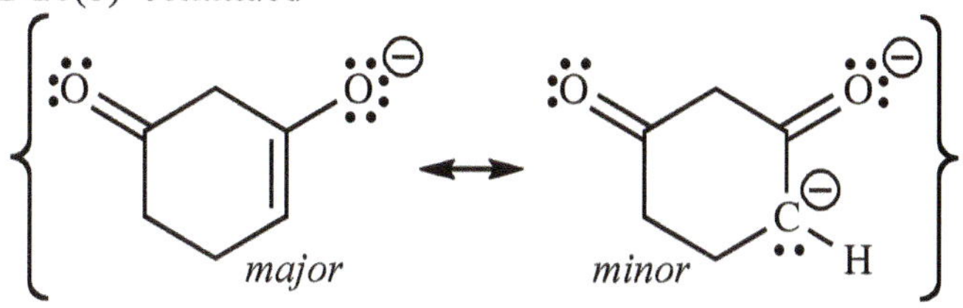

major minor

The anion obtained by removing the H^+ from the OH in the second structure gives only two resonance forms, and only one has negative charge on oxygen—not as stable as the first structure.

However, there is another choice of which H^+ to remove. Removing an H from the CH_2 between the two oxygens also gives a resonance-stabilized anion:

remove

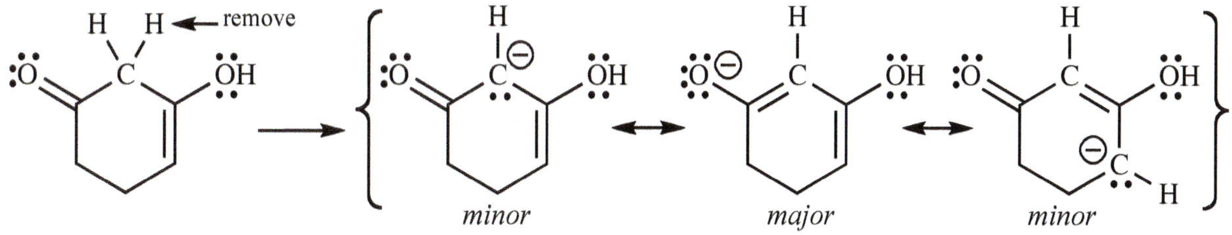

minor major minor

Arguably, the CH is a better proton to remove than the OH because of greater resonance stabilization, but either of the anions from the second structure is less stable than the anion from the first structure. **The first structure is still the stonger acid.**

(c) The most acidic H^+ from the first structure is the OH. The anion from removing this H^+ is not stabilized by resonance, and the F is too far away to have any significant inductive effect.

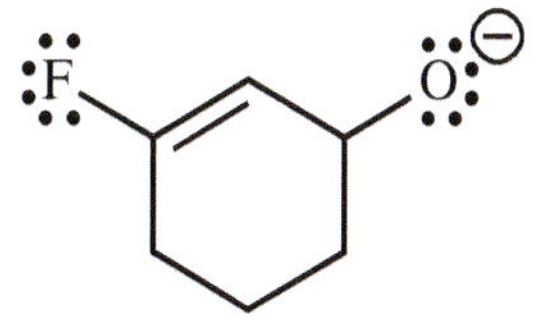

no resonance forms; F is too far away to stabilize anion

The anion from the second structure has two resonance forms, the second of which is even more stable because of the electronegativity of F. Therefore, **the second structure is the stronger acid.**

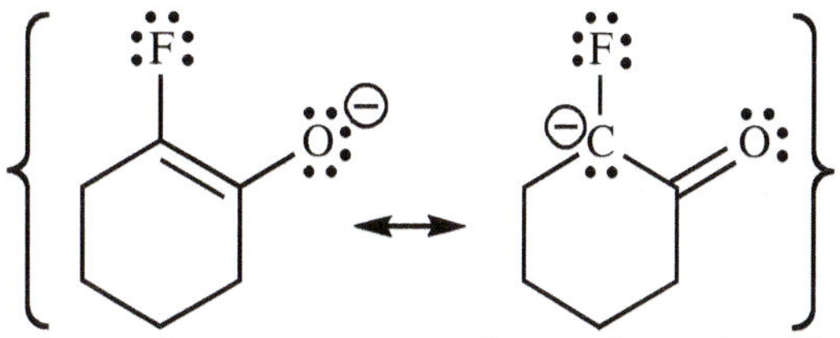

major *better than minor because of F electronegativity*

(d) Removing an H^+ from NH_2 gives a highly stabilized anion:

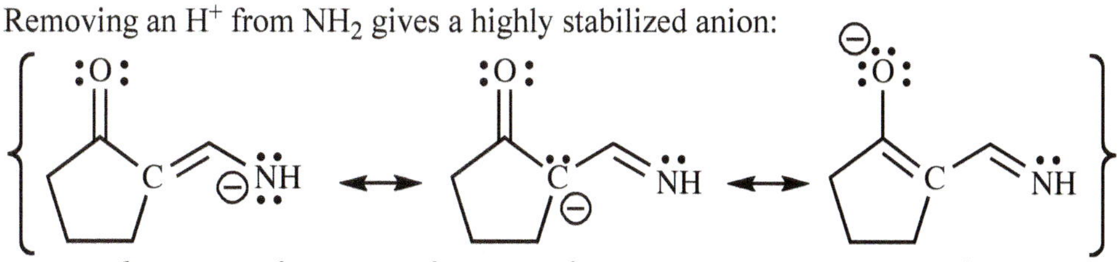

second most significant *least significant* *most significant*

There are two choices for which H^+ to remove from the second structure. If we want a direct comparison with the resonance-stabilized anion directly above, we should look at removing an H^+ from the N, giving this conjugate base with no resonance stabilization.

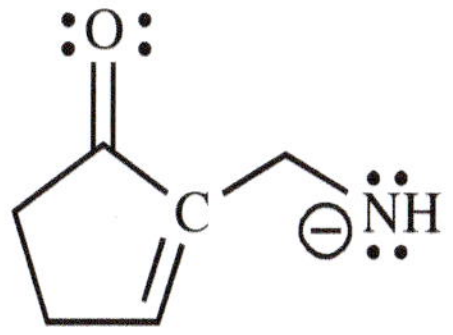

See the next page for the resonance forms of the other conjugate base.

2-21(d) continued

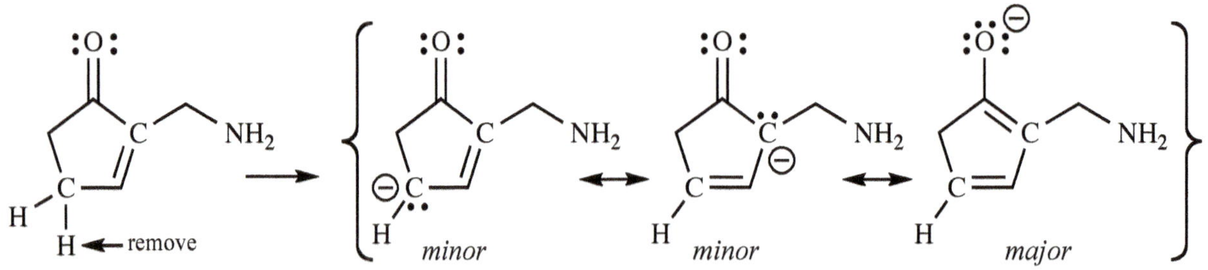

Although removing the CH_2 proton gives a more stable conjugate base than removing the NH_2 proton, neither is better than the anion from the first structure. **The first structure is the stronger acid.**

(e) Remove the proton from the OH in each case. The first structure gives a resonance-stabilized conjugate base, but only one of the resonance forms has the negative charge on an electronegative atom.

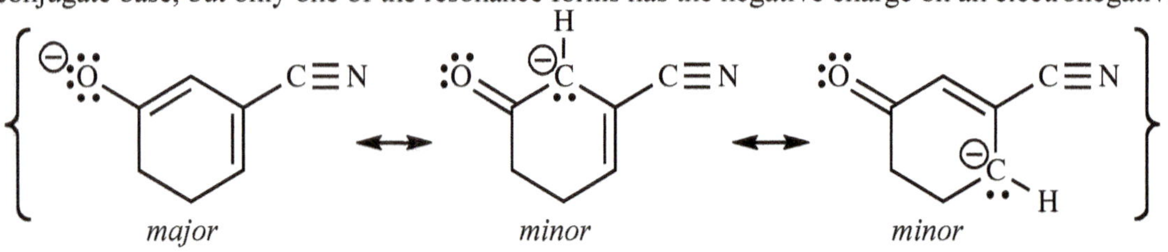

From the second structure:

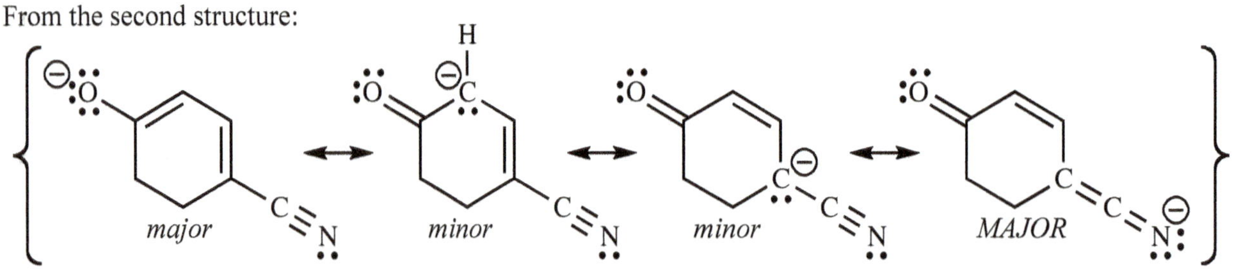

Woo hoo! With the proper placement of the CN group, the negative charge can be delocalized onto the N as well as the O, making this the more stable conjugate base. **The second structure is the stronger acid.**

(f) Remove the proton from the OH in each case. The first structure gives a resonance-stabilized conjugate base in which the negative charge is delocalized onto both oxygen atoms—hard to beat.

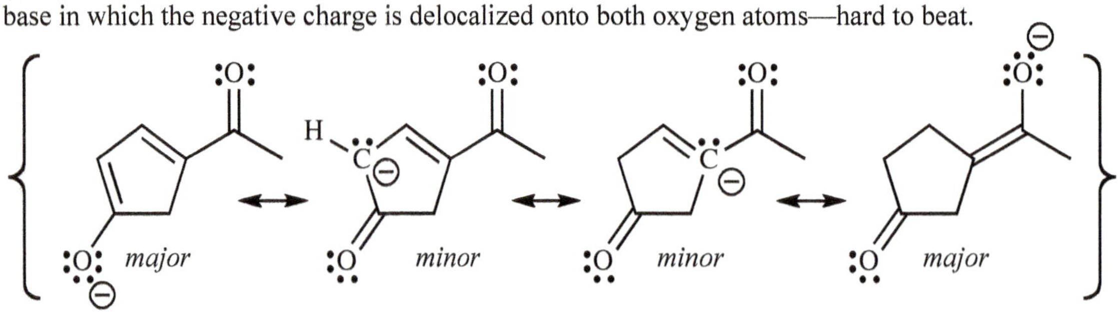

See the next page for the structure of the other conjugate base.

2-21(f) continued

The conjugate base of the second structure shows that there is no resonance stabilization of the anion. Therefore, **the first structure is the stronger acid.**

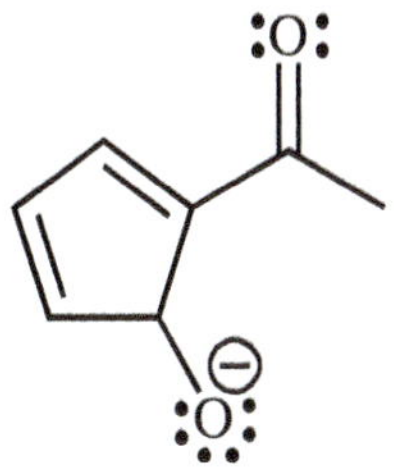

(g) The conjugate bases of these structures are not anions, yet the same principle applies: *the stability of the conjugate base determines the strength of the acid.*

After one of the NH protons is removed from the first structure, the conjugate base is a structure that has no charge and the NH_2 is not stabilized by resonance.

NH_2 NH_2 HC NH_2

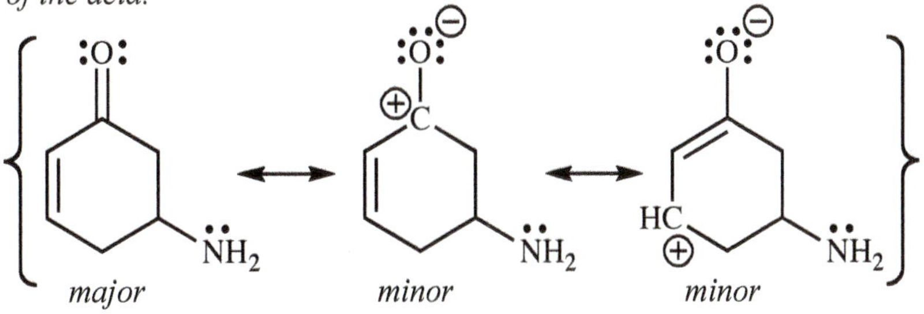

major *minor* *minor*

After one of the NH protons is removed from the second structure, however, the conjugate base is highly stabilized by resonance.

NH_2 NH_2 NH_2 NH_2

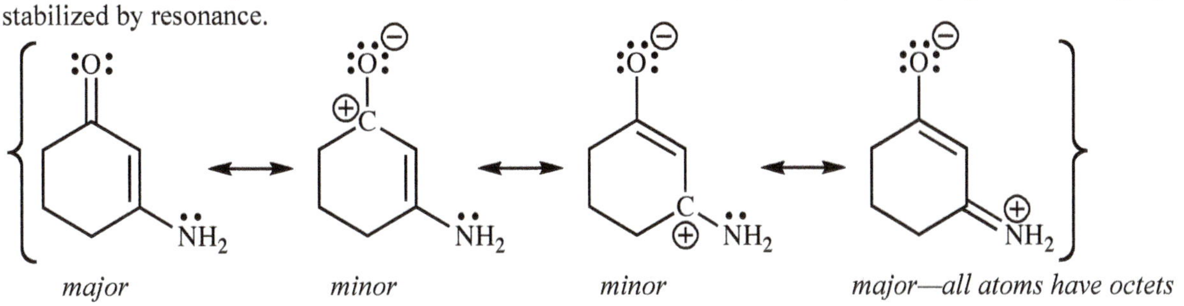

major *minor* *minor* *major—all atoms have octets*

Because it has the more stable conjugate base, **the second structure is the stronger acid**.

(h)

After the OH proton is removed, the first structure is stabilized by resonance only to a small degree.

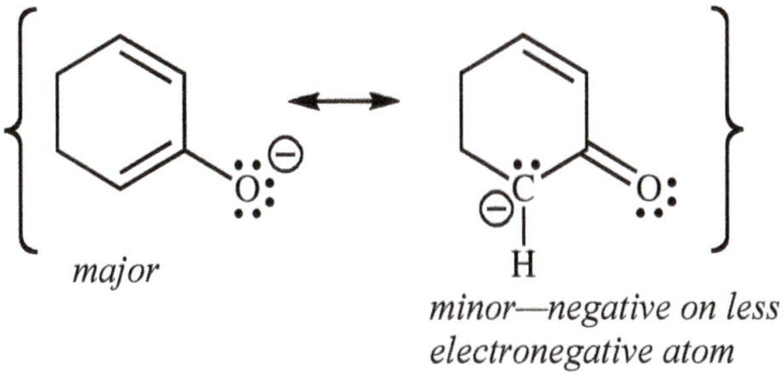

major

H

minor—negative on less electronegative atom

After one of the OH protons is removed from the second structure, however, the conjugate base is highly stabilized by resonance.

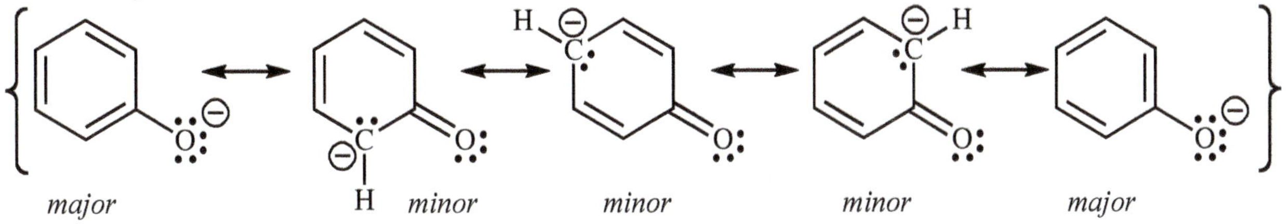

major H *minor* *minor* *minor* *major*

Because it has the more stable conjugate base, **the second structure is the stronger acid**.

2-21 continued

(i) In both of these structures, removal of the OH proton results in a highly stabilized anion.

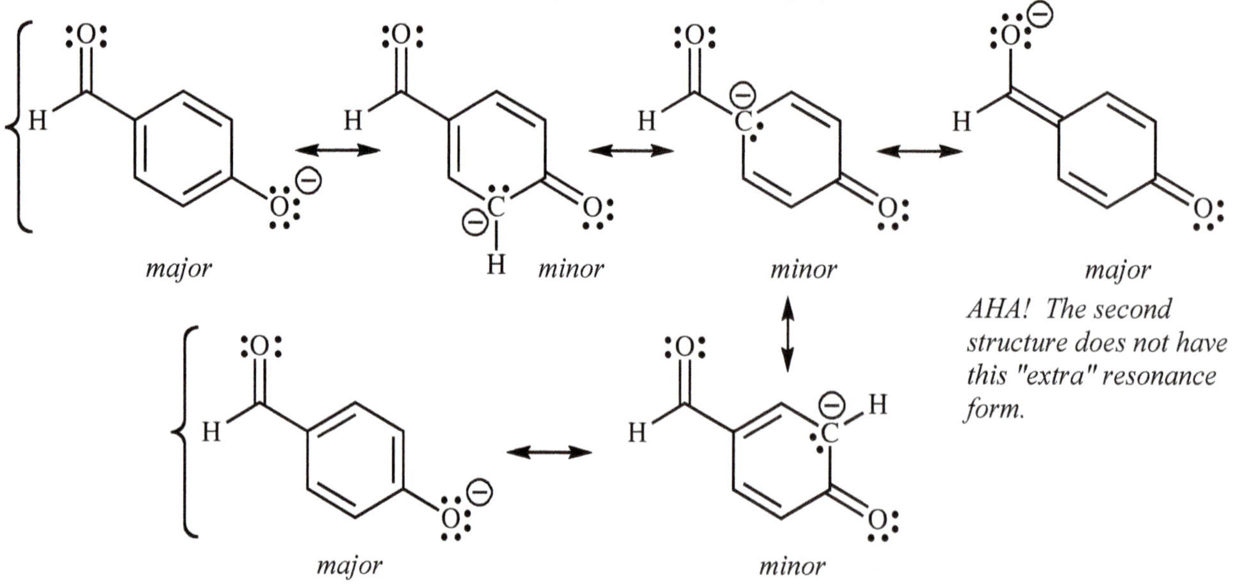

major ↔ minor ↔ minor ↔ major

AHA! The second structure does not have this "extra" resonance form.

major ↔ minor

Resonance forms of the anion of the second structure show that the negative charge is never delocalized on the C=O group.

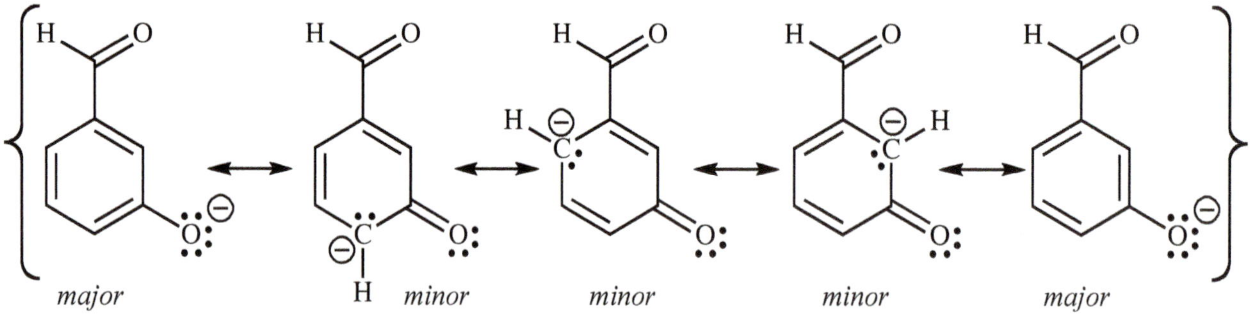

major ↔ minor ↔ minor ↔ minor ↔ major

The first structure has the extra resonance form where the negative charge is delocalized onto the C=O, resulting in greater stability. **The first structure is the stronger acid.**

> Note to the student: Resonance is your friend. Many, if not most, of the effects of structure on stability and reactivity are based on resonance, and understanding how to draw resonance forms, and when to apply the concept of resonance, is a critical tool in your success in solving problems.

2-22

The more stable the anion, the less basic it will be. This is the corollary of: a stronger acid has a weaker conjugate base, because a stronger acid has a more stable conjugate base.

(b) The first structure has its negative charge delocalized by resonance.

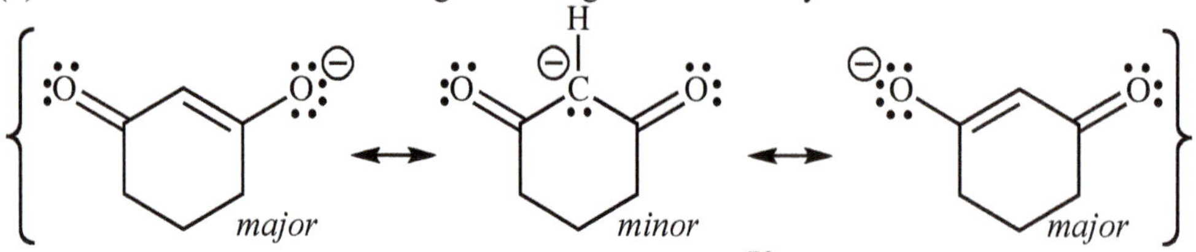

major ↔ minor ↔ major

2-22(b) continued

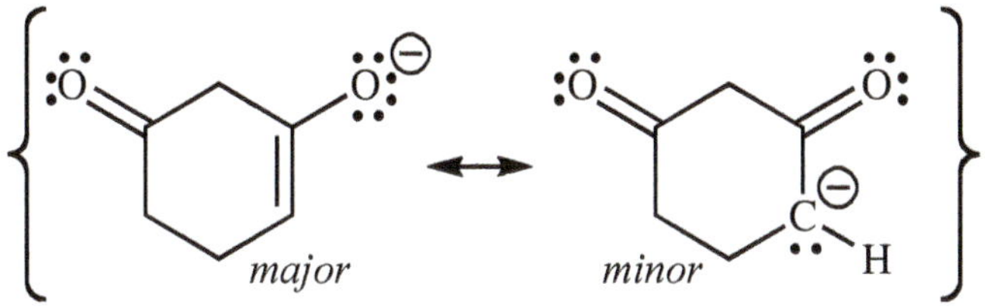

The negative charge in the second structure is only slightly stabilized by one minor resonance contributor.

major ⟷ minor

Because the first structure is more stable because of greater resonance delocalization of the negative charge, **the second structure will be the stronger base**. Note that this is the same as problem 2-21(b), just from the other side of the equation.

(c) The first structure has the negative charge delocalized onto two oxygens and a nitrogen:

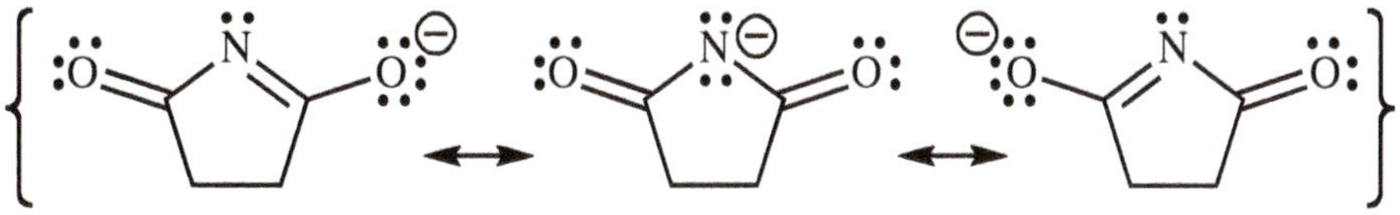

The second structure has the negative charge delocalized onto one oxygen and a carbon:

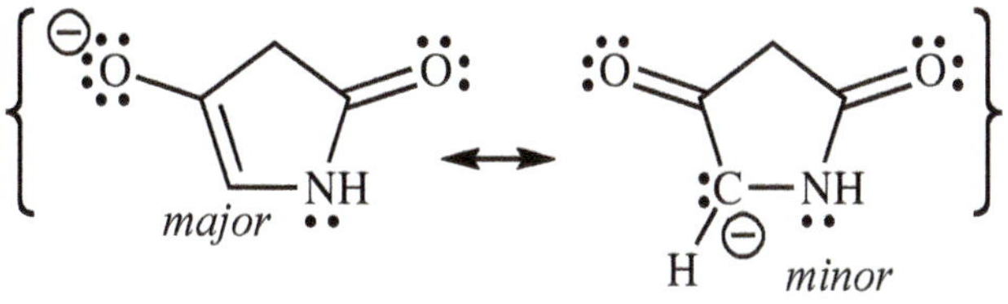

major ⟷ minor

Because the first structure is more stable because of greater resonance delocalization of the negative charge, **the second structure will be the stronger base**.

(d) Even though there is resonance between the C=C and the NO_2 group, the first structure has the negative charge on oxygen that is not delocalized by resonance.

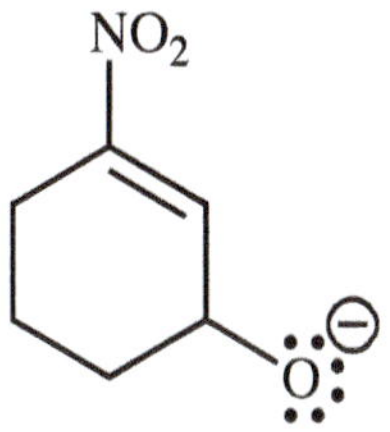

NO_2

The second structure has the negative charge delocalized onto two oxygens:

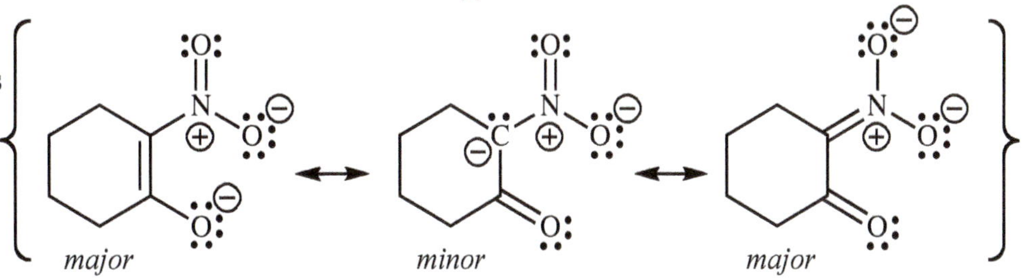

major ⟷ minor ⟷ major

Because the second structure is more stable because of greater resonance delocalization of the negative charge, **the first structure will be the stronger base**.

(e) As shown on the next page, the negative charge in both structures is highly delocalized by resonance, but only the second structure has the negative charge on the oxygen and the nitrogen. Because the second structure is more stable, **the first structure will be the stronger base.**

2-22(e) continued

First structure:

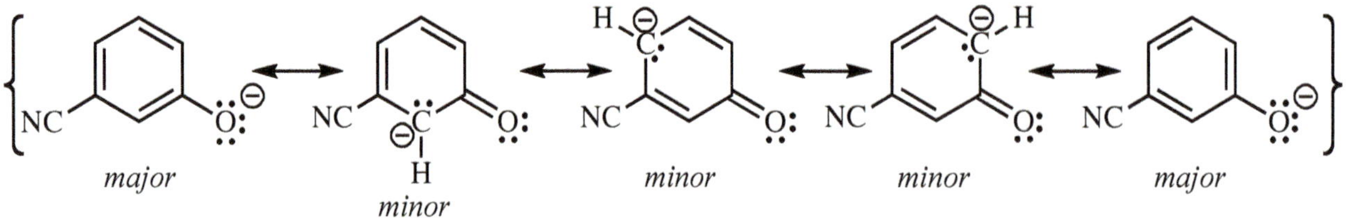

major *minor* *minor* *minor* *major*

Second structure:

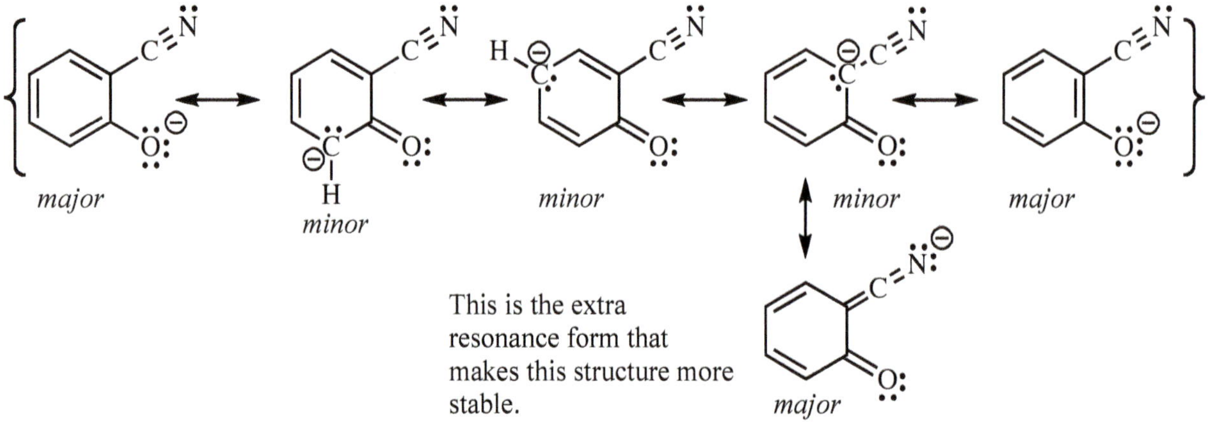

major *minor* *minor* *minor* *major*

This is the extra resonance form that makes this structure more stable. *major*

The first structure will be the stronger base.

2-23 Solutions for (a) and (b) are presented in the Solved Problem in the text. Here, the newly formed bonds are shown in bold.

Curved arrows that show electron movement are shown here and in the text in red. Arrows within resonance forms that show *imaginary* movement are in green in the text, and in black here.

(c) $H-BH_2$ + $CH_3-\ddot{O}-CH_3$ ⇌ $H_3B^{-}-O^{+}(CH_3)_2$

Lewis acid—electrophile; Lewis base—nucleophile

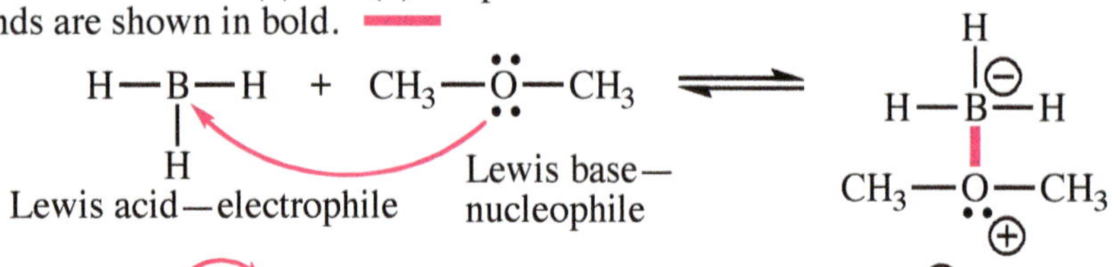

(d) $CH_3-C(=O)-H$ + $^{-}:\ddot{O}-H$ ⇌ $CH_3-CH(O^{-})-OH$

Lewis acid—electrophile; Lewis base—nucleophile

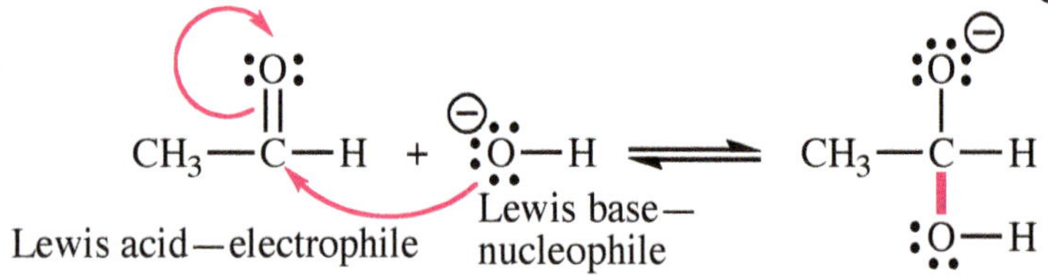

The problem asks for the *imaginary* movement of electrons in the resonance forms.

(e) Brønsted-Lowry—proton transfer

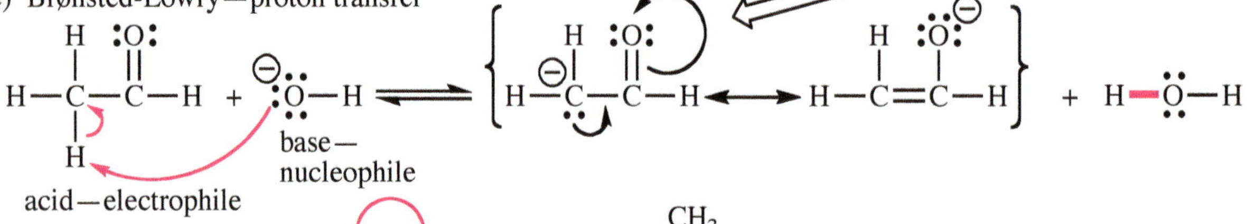

$H-CH_2-C(=O)-H$ + $^{-}:\ddot{O}-H$ ⇌ $\{ H-\bar{C}H-C(=O)-H \longleftrightarrow H-CH=C(O^{-})-H \}$ + $H-\ddot{O}-H$

acid—electrophile; base—nucleophile

(f) $CH_3-\ddot{N}H-H$ + $CH_3-\ddot{Cl}:$ ⇌ $CH_3-N^{+}H_2-CH_3$ + $:\ddot{Cl}:^{-}$

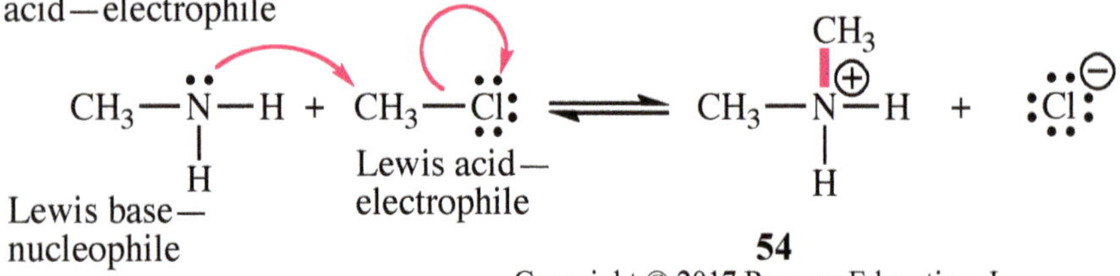

Lewis base—nucleophile; Lewis acid—electrophile

2-23 continued

(g) $CH_3-CH=CH_2$ + BF_3 ⇌ $CH_3-\overset{+}{C}H-CH_2-\overset{-}{B}F_3$

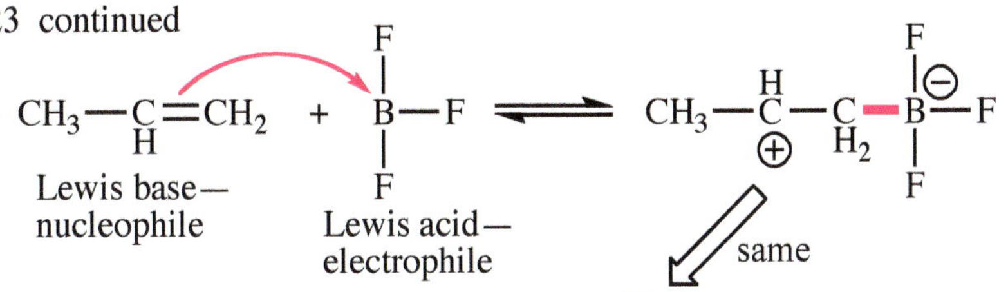

Lewis base—nucleophile

Lewis acid—electrophile

same

(h) $CH_3-CH=CH_2$ + $CH_3-\overset{+}{C}H-CH_2-\overset{-}{B}F_3$ ⇌ $H_3C-\overset{+}{C}H-CH_2-CH(CH_3)-CH_2-\overset{-}{B}F_3$

Lewis base—nucleophile

Lewis acid—electrophile

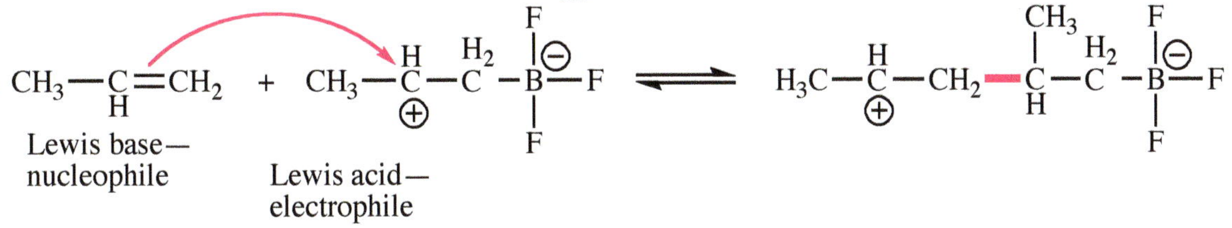

(i) Cl OH ⇌ O⁻ Cl ⊕O H

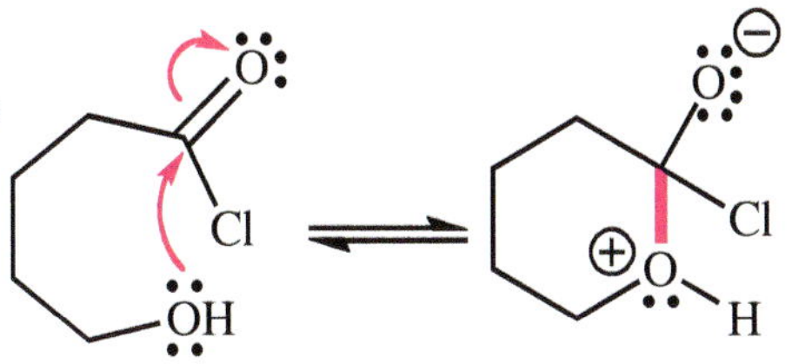

The problem asks for the *imaginary* movement of electrons in the resonance forms.

(j) OCH_3 $O=N=O$ ⊕ ⇌ { ⊕C OCH_3 ⊕N O⁻ O ↔ C=OCH_3⊕ ⊕N O⁻ O }

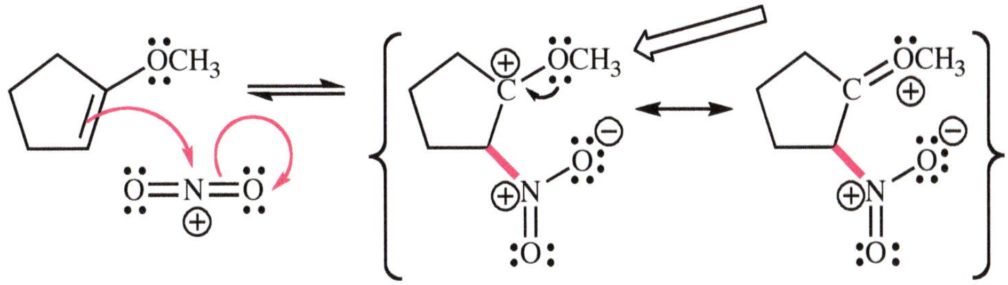

2-24

(a) H–C–C–C–C–H chain structure: alkane

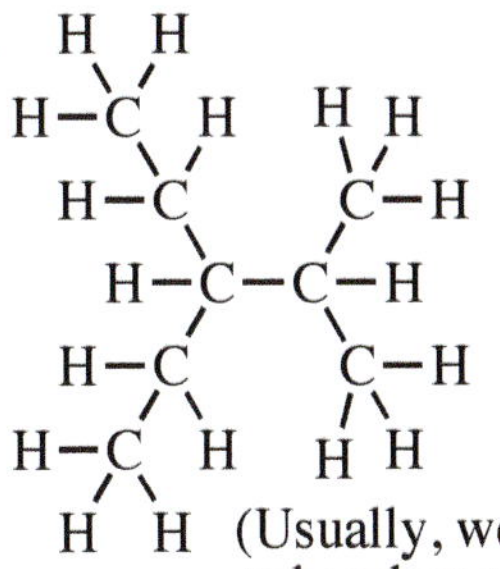

(Usually, we use the term "alkane" only when no other groups are present.)

(b) H–C–C=C–C–C–H — alkene

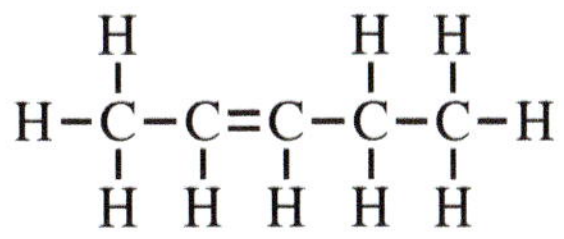

(c) H–C–C≡C–C–C–C–H — alkyne

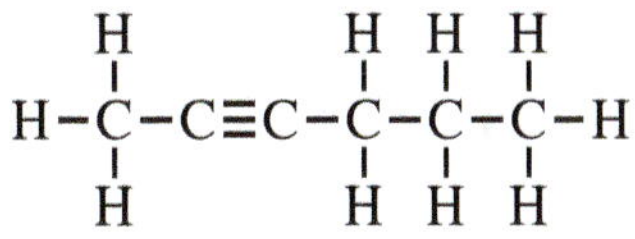

(d) cycloalkyne and cycloalkene

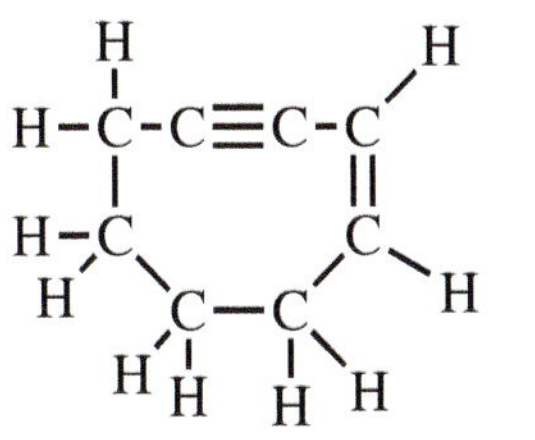

(e) cycloalkane and alkene (a "cycloalkene" would have the C=C in the ring)

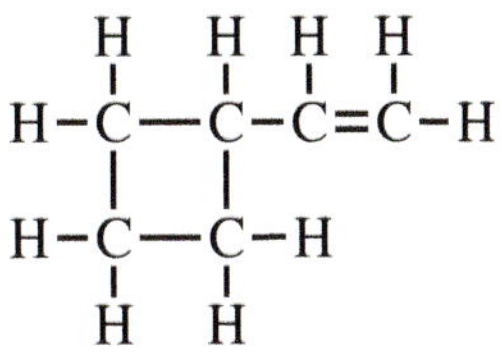

(f) aromatic hydrocarbon and alkyne

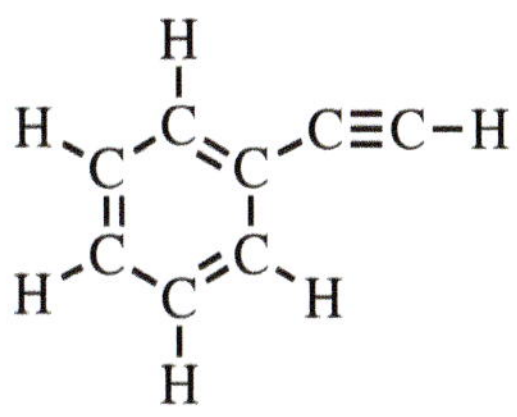

continued on page 57

Note to the student: One of the most fundamental skills in organic chemistry is to identify what functional groups are present in a molecular structure. This page is called a "concept map" that prompts you with a series of questions about a structure, leading to the determination of the functional group. Do not memorize this chart! Use it as an aid as you are becoming familiar with the functional groups.

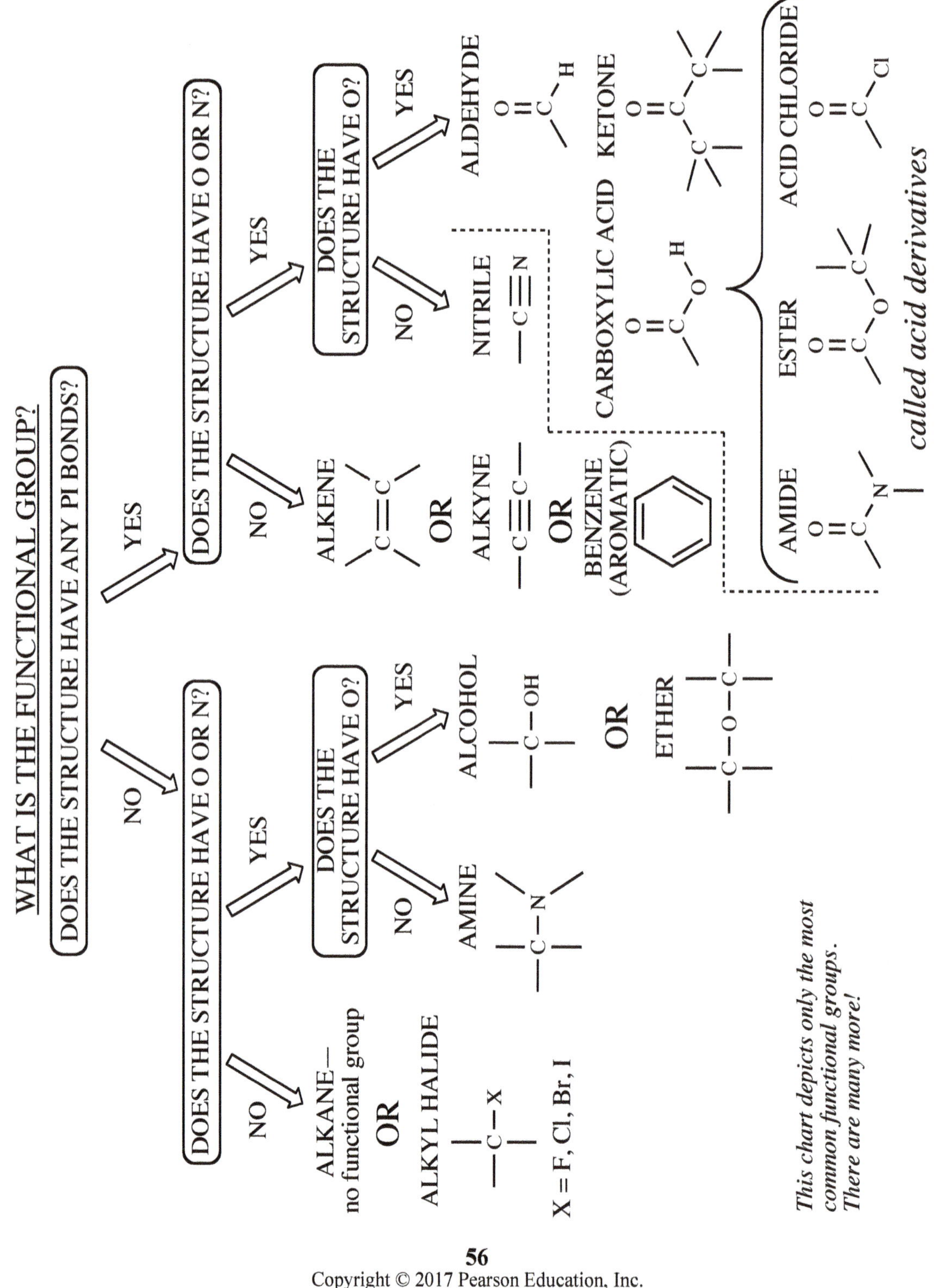

2-24 continued

(g) cycloalkene and alkene

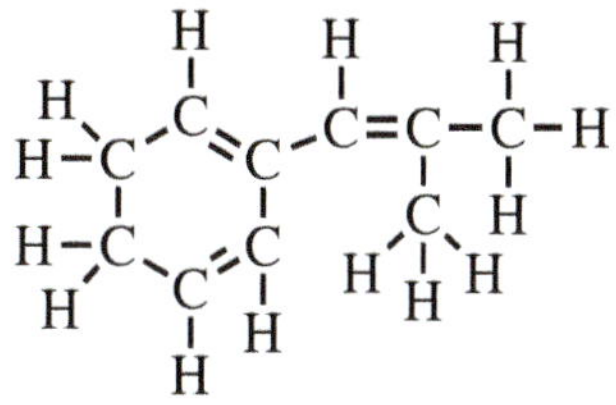

(h) cycloalkane and alkane

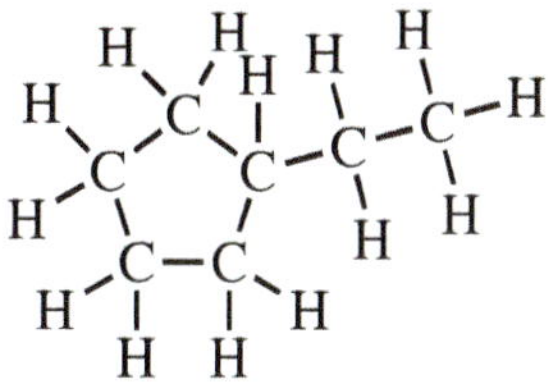

(i) aromatic hydrocarbon and cycloalkene

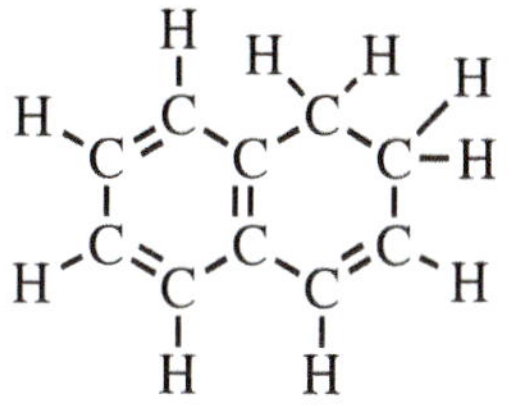

2-25

(a) alkene and aldehyde

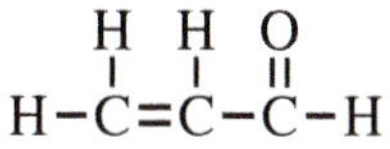

(b) alcohol

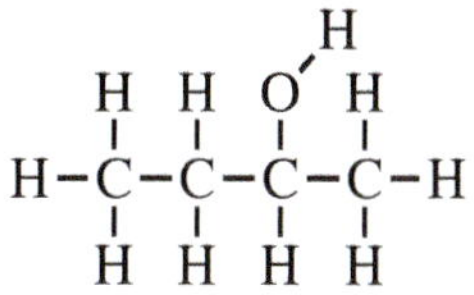

(c) ketone

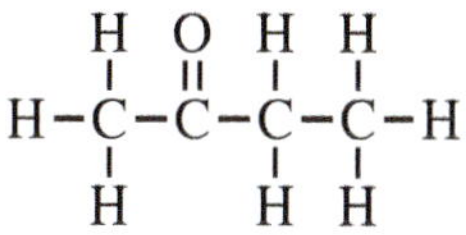

(d) ether and alkene

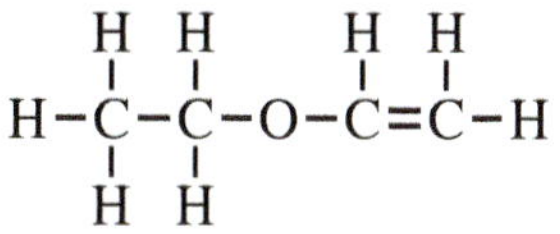

(e) carboxylic acid

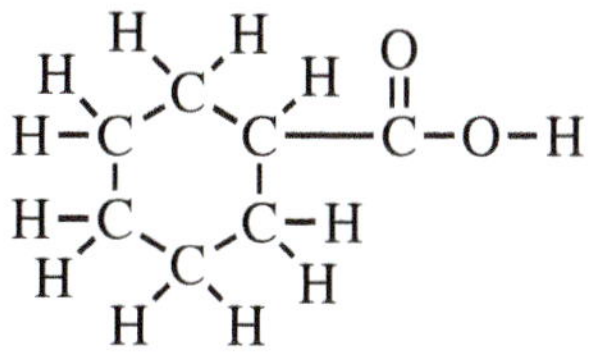

(f) ether and alkene

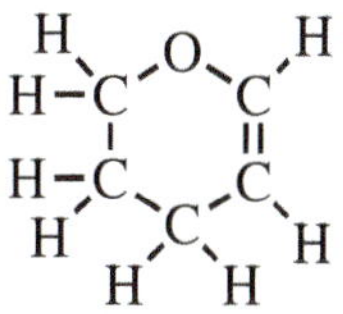

(g) alkene and ketone

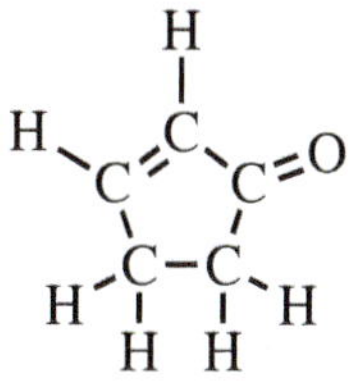

(h) aldehyde

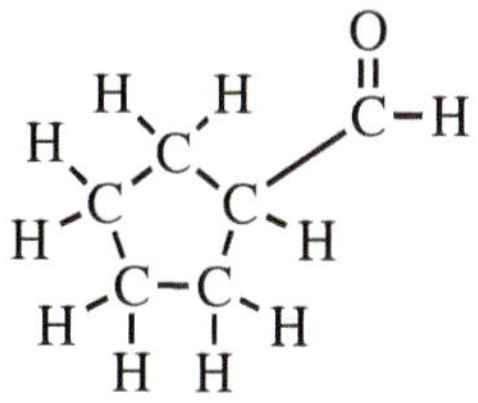

(i) alcohol

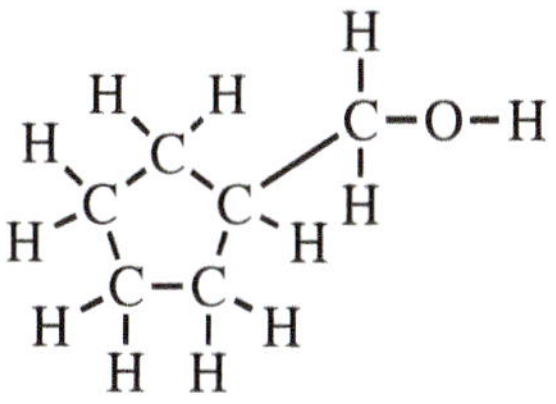

2-26

(a) amide

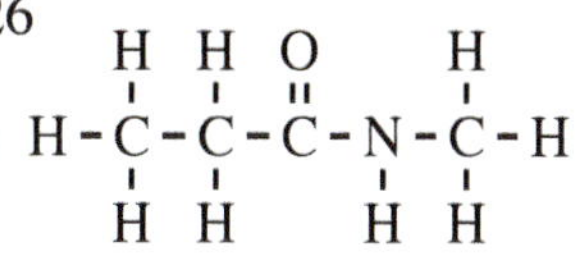

(b) amine

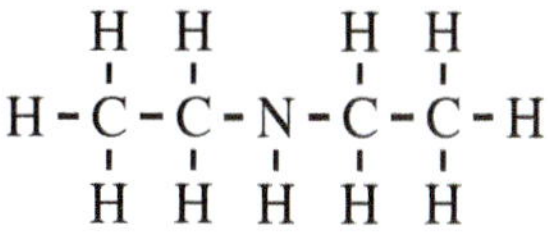

(c) ester

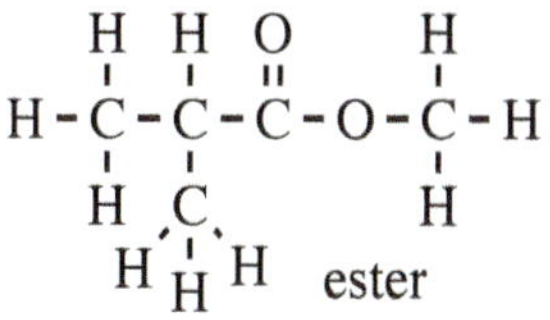

(d) alkene and acid chloride

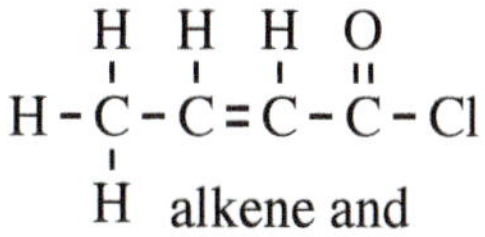

(e) ether

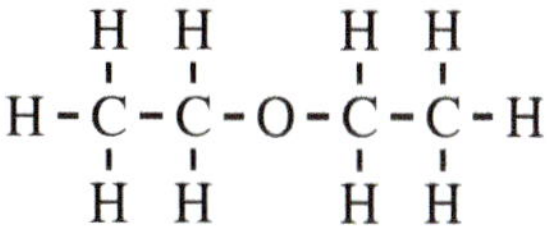

(f) nitrile

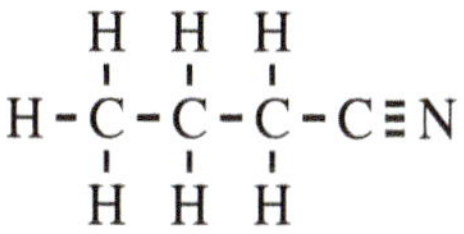

(g) carboxylic acid

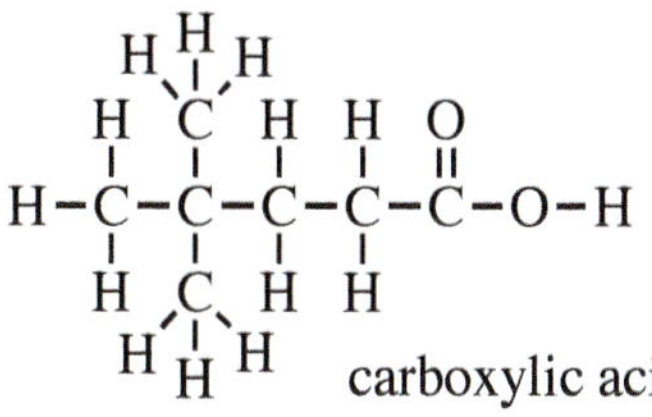

(h) cycloalkene and cyclic ester

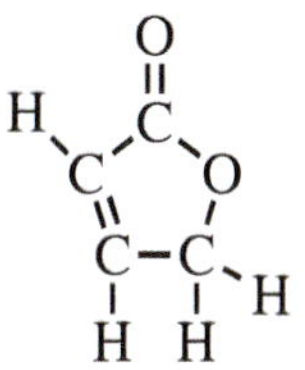

(i) cyclic ketone and ether

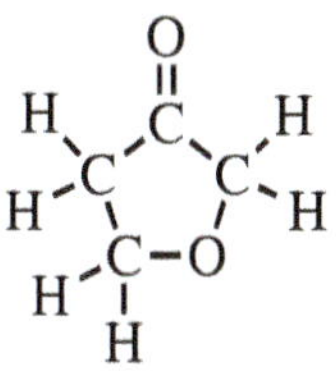

(j) cyclic amine

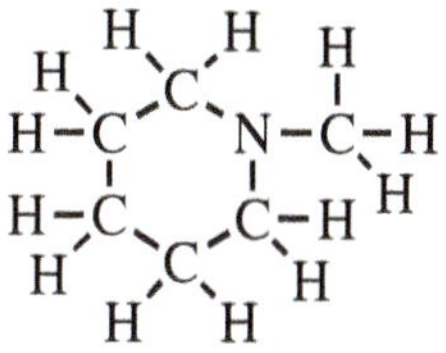

2-26 continued

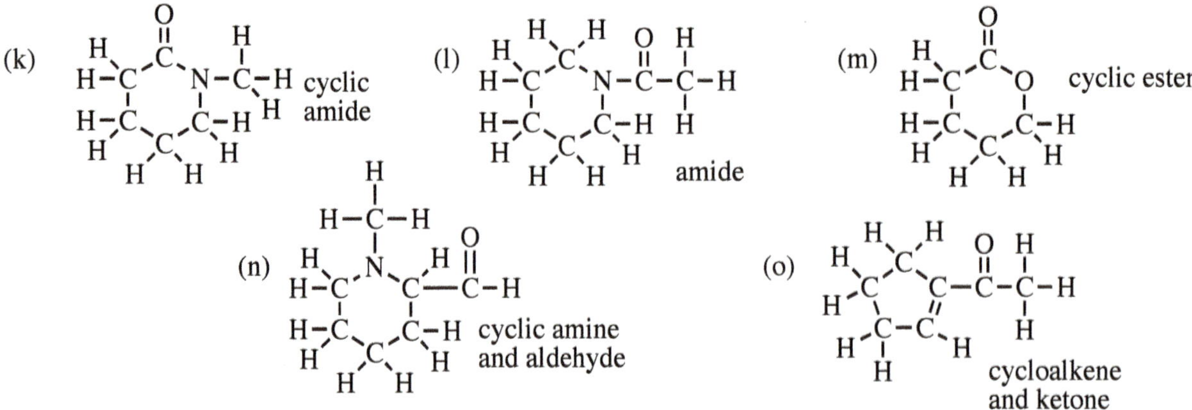

2-27 When the identity of a functional group depends on several atoms, all of those atoms should be circled. For example, an ether is an oxygen between two carbons, so the oxygen and both carbons should be circled. A ketone is a carbonyl group between two other carbons, so all those atoms should be circled.

(a) 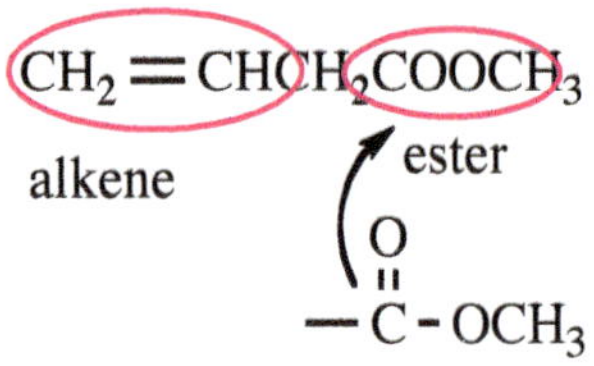 $CH_2{=}CHCH_2COOCH_3$ alkene; ester

(b) 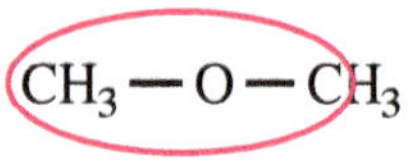 $CH_3{-}O{-}CH_3$ ether

(c) 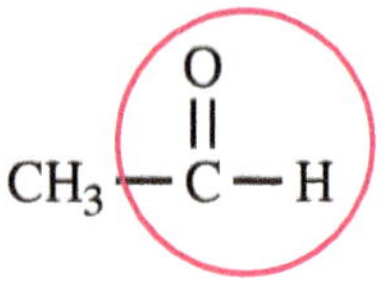 $CH_3{-}C(=O){-}H$ aldehyde

(d) 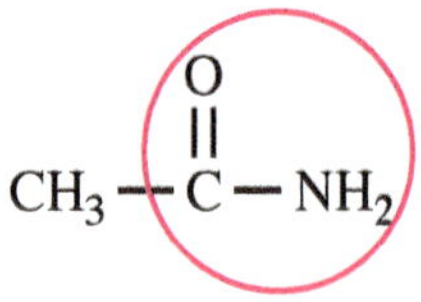 $CH_3{-}C(=O){-}NH_2$ amide

(e) 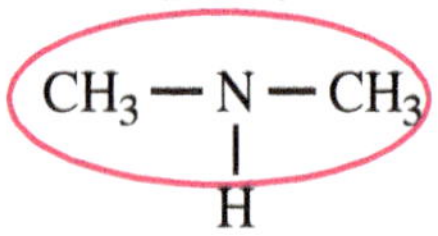 $CH_3{-}N(H){-}CH_3$ amine

(f) 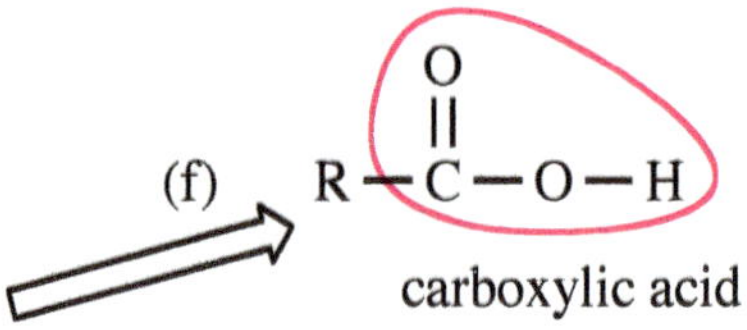 $R{-}C(=O){-}O{-}H$ carboxylic acid

R is the symbol that organic chemists use to represent alkyl groups. Sometimes, when the identity of the group does not matter, aryl groups or others can also be included in the R abbreviation.

(g) 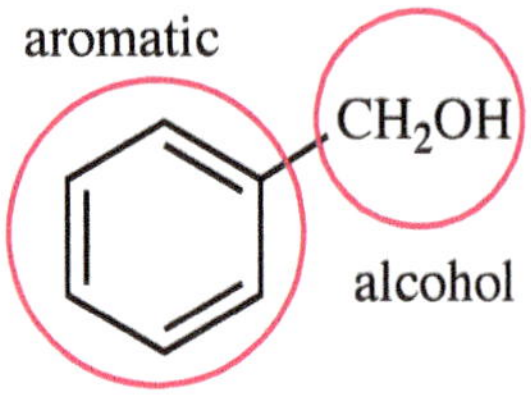 aromatic; CH_2OH alcohol

(h) 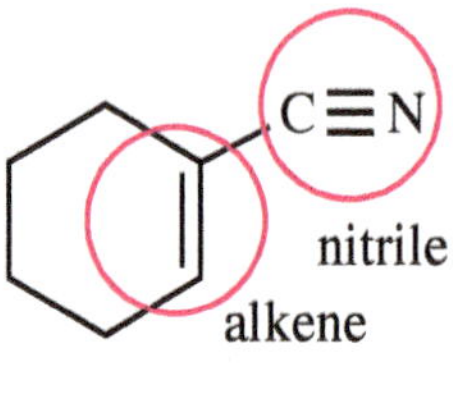 $C{\equiv}N$ nitrile; alkene

(i) 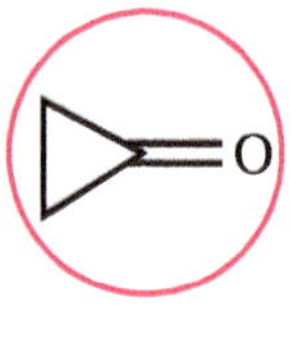 ketone

(j) 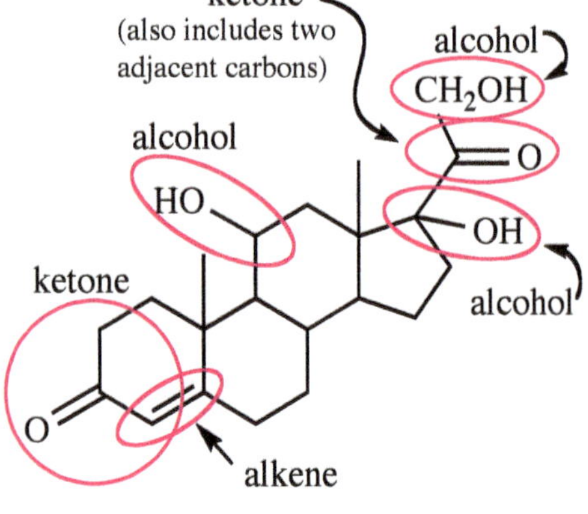 ketone (also includes two adjacent carbons); alcohol CH_2OH; alcohol HO; OH alcohol; ketone; alkene

(k) (In later chapters, you will learn that the OH group on a benzene ring is a special functional group called a "phenol". For now, it fits the broad definition of an alcohol.)

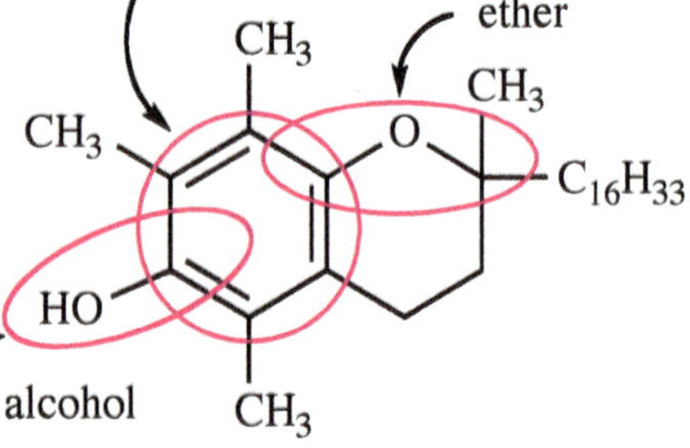

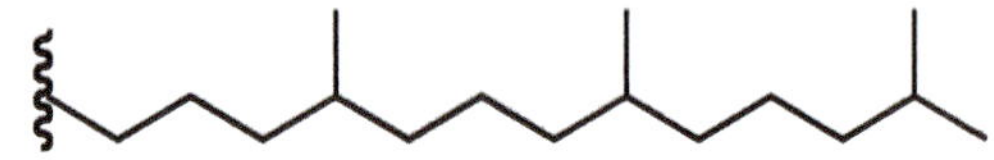

aromatic; ether; CH_3; CH_3; CH_3; $C_{16}H_{33}$; HO alcohol; CH_3

This group is represented by $C_{16}H_{33}$ in the problem.

2-28
The triple bond between C and N is highly polarized: $\left\{ H_3C-C\equiv N\text{:} \longleftrightarrow H_3C-\overset{\oplus}{C}=\ddot{N}\text{:}^{\ominus} \right\}$

A B

$3.6\ D = 4.8 \times \delta \times 1.16\ \text{Å}$

$\delta = 0.65$, or 65% of a positive charge on carbon and 65% of a negative charge on nitrogen

Resonance form **B** must be the major contributor. If **A** were the major contributor, the value of the charge separation would be between 0.0 and 0.5, meaning <50% of a positive charge on C and <50% of a negative charge on N.

2-29 Some magnitudes of dipole moments are difficult to predict; however, the direction of the dipole should be straightforward in most cases. Actual values of molecular dipole moments are given in parentheses. (The C—H bond is usually considered nonpolar.)

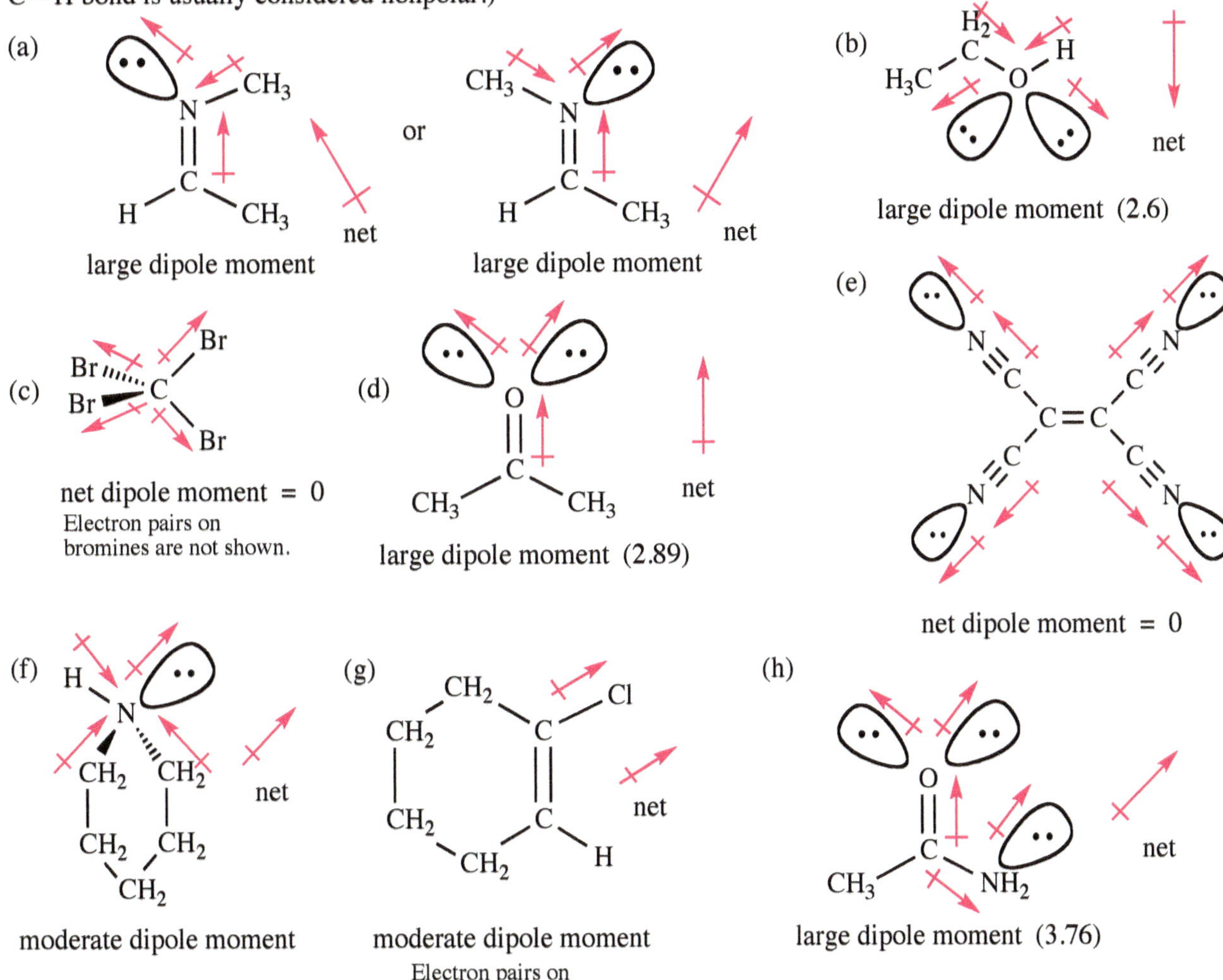

(a) large dipole moment or large dipole moment

(b) large dipole moment (2.6)

(c) net dipole moment = 0
Electron pairs on bromines are not shown.

(d) large dipole moment (2.89)

(e) net dipole moment = 0

(f) moderate dipole moment

(g) moderate dipole moment
Electron pairs on chlorine are not shown.

(h) large dipole moment (3.76)

2-30 CO_2 is linear; its bond dipoles cancel, so it has no net dipole. SO_2 is bent, so its bond dipoles do not cancel.

$\ddot{O}=C=\ddot{O}$

net dipole moment = 0

$\left\{ :\ddot{O}=\overset{\oplus}{\ddot{S}}-\ddot{O}:^{\ominus} \longleftrightarrow {}^{\ominus}:\ddot{O}-\overset{\oplus}{\ddot{S}}=\ddot{O}: \longleftrightarrow :\ddot{O}=\ddot{S}=\ddot{O}: \right\}$

net dipole moment

2-31

(a) Can hydrogen bond with itself and with water.
(b) Can hydrogen bond only with water.
(c) Can hydrogen bond with itself and with water.
(d) Can hydrogen bond only with water.
(e) Cannot hydrogen bond: no N or O.
(f) Cannot hydrogen bond: no N or O.
(g) Can hydrogen bond only with water.
(h) Can hydrogen bond with itself and with water.
(i) Can hydrogen bond only with water.
(j) Can hydrogen bond only with water.
(k) Can hydrogen bond only with water.
(l) Can hydrogen bond with itself and with water.

Water solubility depends on the number of carbons and the number of hydrogen-bonding groups in a molecule, and to a lesser extent, on the shape of the molecule and its ability to form hydrogen bonds. A structure with one hydrogen-bonding group can carry in three carbons for sure, called "miscible", meaning completely soluble; four carbons will be partially soluble—these are borderline cases and are the most difficult to predict; five or more carbons will be only slightly soluble, which is sometimes called "insoluble".

•completely soluble, miscible: (c) 3C; (g) 3C; (h) 3C; (i) 3C; (l) 2C;
•moderately soluble, possibly miscible: (a), (j), (k)—each has one H-bonding group and 4C;
•slightly soluble, "insoluble": (b) 6C; (d) 6C; (e) and (f) have no H-bonding group.

2-32

As summarized in the solution to problem 2-31, water solubility depends on the number of carbons and the number of hydrogen-bonding groups in a molecule. A structure with one hydrogen-bonding group can carry in three carbons for sure, called "miscible", meaning completely soluble; four carbons will be partially soluble; five or more carbons will be only slightly soluble, which is sometimes called "insoluble".

(a) The first structure with 5 carbons and no hydrogen-bonding group is completely insoluble in water. The second structure has 4 carbons and 1 oxygen; it is miscible with water because oxygen's electron pairs can form hydrogen bonds with water.
(b) The first structure with 5 carbons and 1 oxygen is partially soluble in water (tetrahydropyran, 8 g/100 mL water). The second structure has 4 carbons and 2 oxygens; it is miscible with water.
(c) The first structure with 4 carbons and 1 oxygen is miscible with water. The second structure with 5 carbons and 1 oxygen is only partially soluble in water (2-methyltetrahydrofuran, 14 g/100 mL).
(d) The first structure with 5 carbons and no hydrogen-bonding group is completely insoluble in water. The second structure has 6 carbons and 1 oxygen; it is partially soluble in water.
(e) The first structure with 5 carbons and 1 oxygen is slightly soluble in water. The second structure has 5 carbons and two bromines—which do not form hydrogen bonds—and is completely insoluble in water.
(f) The first structure with 5 carbons and 1 oxygen is partially soluble in water (tetrahydropyran, 8 g/100 mL water). The second structure has 4 carbons and 1 oxygen; it is miscible with water because the electron pairs on oxygen are exposed and can readily form hydrogen bonds with water.

2-33 Diethyl ether and butan-1-ol each have one oxygen, so each can form hydrogen bonds with water (water supplies the H for hydrogen bonding with diethyl ether); their water solubilities should be similar. The boiling point of butan-1-ol is much higher because these molecules can hydrogen bond with each other, thus requiring more energy to separate one molecule from another. Diethyl ether molecules cannot hydrogen bond with each other, so it is relatively easy to separate them.

$CH_3CH_2—O—CH_2CH_3$	$CH_3CH_2CH_2CH_2—OH$
diethyl ether	butan-1-ol
Can hydrogen bond with water; cannot hydrogen bond with itself.	Can hydrogen bond with water; can hydrogen bond with itself.

2-34 N-methylpyrrolidine b.p. 81 °C; piperidine b.p. 106 °C; tetrahydropyran b.p. 88 °C; cyclopentanol b.p. 141 °C

(a) Piperidine has an N—H bond, so it can hydrogen bond with other molecules of itself. *N*-Methylpyrrolidine has no N—H, so it cannot hydrogen bond and will require less energy (lower boiling point) to separate one molecule from another.

(b) Two effects need to be explained: 1) Why does cyclopentanol have a higher boiling point than tetrahydropyran? and 2) Why do the oxygen compounds have a greater difference in boiling points than the analogous nitrogen compounds?

The answer to the first question is the same as in (a): cyclopentanol can hydrogen bond with its neighbors while tetrahydropyran cannot.

The answer to the second question lies in the text, Table 2-1, that shows the bond dipole moments for C—O and H—O are much greater than C—N and H—N; bonds to oxygen are more polarized, with greater charge separation than bonds to nitrogen.

How is this reflected in the data? The boiling points of tetrahydropyran (88 °C) and *N*-methylpyrrolidine (81 °C) are close; tetrahydropyran molecules would have a slightly stronger dipole-dipole attraction, and tetrahydropyran is a little less "branched" than *N*-methylpyrrolidine, so it is reasonable that tetrahydropyran boils at a slightly higher temperature. The large difference comes when comparing the boiling points of cyclopentanol (141 °C) and piperidine (106 °C). The greater polarity of O—H versus N—H is reflected in a more negative oxygen (more electronegative than nitrogen) and a more positive hydrogen, resulting in a much stronger intermolecular attraction. The conclusion is that hydrogen bonding due to O—H is much stronger than that due to N—H.

(c) Resonance plays a large role in the properties of amides.

N,N-dimethylformamide, b.p. 150 °C: $H-C(=O)-N(CH_3)_2 \longleftrightarrow H-C(-O^-)=N^+(CH_3)_2$

N-methylacetamide, b.p. 206 °C: $H_3C-C(=O)-N(H)CH_3 \longleftrightarrow H_3C-C(-O^-)=N^+(H)CH_3$

The first structure has no hydrogen bonding because it has no O—H or N—H bond, but it is a highly polar structure and has a large dipole moment, so its dipole-dipole intermolecular force is very high. The second structure has hydrogen bonding because of the N—H bond in addition to the strong dipole-dipole interaction, reflected in its higher boiling point. Both amides boil higher than the four at the beginning of the problem because of the dipole-dipole interactions that are accentuated by the resonance forms.

2-35 Higher-boiling compounds are the ones described.

(a) $CH_3CH(OH)CH_3$ can form hydrogen bonds with other molecules of itself.

(b) $CH_3CH_2CH_2CH_2CH_3$ has a higher molecular weight than $CH_3CH_2CH_2CH_3$.

(c) $CH_3CH_2CH_2CH_2CH_3$ has less branching than $(CH_3)_2CHCH_2CH_3$.

(d) $CH_3CH_2CH_2CH_2CH_2Cl$ has a higher molecular weight AND a dipole-dipole interaction compared with $CH_3CH_2CH_2CH_2CH_3$.

(e) [piperidine, NH] can form hydrogen bonds with other molecule of itself (actual bp 106°C; the other structure has bp 81°C).

(f) [2-pyrrolidinone, O, NH] has two sites for hydrogen bonding, not just one, and a very strong dipole because of resonance (actual bp 245°C; the other structure has bp ≈102°C).

2-36 *The strength of the acid depends on the stability of the conjugate base*. The two primary factors governing the strength of organic acids are resonance and inductive effects; of these two, resonance is usually the stronger and more important effect.

Any organic structure with an $-SO_3H$ in it is a very strong acid because the anion has three significant resonance contributors. An organic structure with $-COOH$ is moderately strong since the conjugate base has two significant resonance contributors. A structure with a simple $-OH$ does not have any resonance stabilization of the conjugate base, so it is the weakest acid. Within each group, inductive effects from an electronegative atom like Cl will have a small effect.

CH_3CH_2OH	>	CH_3CH_2COOH	>	$ClCH_2CH_2COOH$	>	$CH_3CHClCOOH$	>	$CH_3CH_2SO_3H$
(b)		(c)		(e)		(d)		(a)
weakest acid				Cl further from COOH		Cl closer to COOH		*strongest acid*

2-37 These pK_a values from the text, Table 2-2 and Appendix 4, provide the answers. The lower the pK_a, the stronger the acid. Water and CH_3OH are very close.

least acidic												most acidic
NH_3	<	H_2O	~	CH_3OH	<	CH_3COOH	<	HF	<	H_3O^+	<	H_2SO_4
36		15.7		15.5		4.74		3.2		−1.7		−5

2-38 Conjugate bases of the weakest acids will be the strongest bases. The pK_a values of the conjugate acids are listed here. (The relative order of some bases was determined from the pK_a values in Appendix 4 of the textbook.)

least basic												most basic
HSO_4^-	<	H_2O	<	CH_3COO^-	<	NH_3	<	CH_3O^-	~	NaOH	<	$^-NH_2$
from −5		from −1.7		from 4.74		from 9.4		from 15.5		from 15.7		from 36

2-39

(a) $pK_a = -\log_{10} K_a = -\log_{10}(5.2 \times 10^{-5}) = \mathbf{4.3}$ for phenylacetic acid

for propionic acid, pK_a 4.87: $K_a = 10^{-4.87} = \mathbf{1.35 \times 10^{-5}}$

(b) Phenylacetic acid is 3.9 times stronger than propionic acid.

$$\frac{5.2 \times 10^{-5}}{1.35 \times 10^{-5}} = 3.9$$

(c) $C_6H_5-CH_2COO^- + CH_3CH_2COOH \rightleftharpoons C_6H_5-CH_2COOH + CH_3CH_2COO^-$

weaker acid (CH₃CH₂COOH); stronger acid (C₆H₅CH₂COOH)

Equilibrium favors the weaker acid and base. In this reaction, **reactants** are favored.

2-40

(a)

$$CH_3-\ddot{N}H-C(=\ddot{N}H)-\ddot{N}H_2$$

#1 #2 #3

H⁺ to #1: $CH_3-\overset{\oplus}{N}H_2-C(=\ddot{N}H)-\ddot{N}H_2$

no other significant resonance forms

H⁺ to #3: $CH_3-\ddot{N}H-C(=\overset{\oplus}{N}H_2)-\ddot{N}H_2$

H⁺ to #2: $CH_3-\ddot{N}H-C(=\ddot{N}H)-\overset{\oplus}{N}H_3$

no other significant resonance forms

$$\left\{ CH_3-\overset{\oplus}{N}H=C(\ddot{N}H_2)-\ddot{N}H_2 \longleftrightarrow CH_3-\ddot{N}H-\overset{\oplus}{C}(\ddot{N}H_2)-\ddot{N}H_2 \longleftrightarrow CH_3-\ddot{N}H-C(\ddot{N}H_2)=\overset{\oplus}{N}H_2 \right\}$$

(b) Protonation at nitrogen #3 gives four resonance forms that delocalize the positive charge over all three nitrogens and a carbon—a very stable condition. Nitrogen #3 will be protonated preferentially, which we interpret as being more basic.

2-41 The critical principle: *the strength of an acid is determined by the stability of its conjugate base.*

(a) conjugate bases

W $C_6H_5-C(=\ddot{O}:)-\ddot{C}H_2^{\ominus} \longleftrightarrow C_6H_5-C(-\ddot{O}:^{\ominus})=CH_2$

X $C_6H_5-C(=\ddot{O}:)-\ddot{N}H^{\ominus} \longleftrightarrow C_6H_5-C(-\ddot{O}:^{\ominus})=\ddot{N}H$

Y $C_6H_5-C(=\ddot{O}:)-\ddot{N}^{\ominus}-OH \longleftrightarrow C_6H_5-C(-\ddot{O}:^{\ominus})=\ddot{N}-OH$

Z $C_6H_5-C(=\ddot{O}:)-\ddot{O}:^{\ominus} \longleftrightarrow C_6H_5-C(-\ddot{O}:^{\ominus})=\ddot{O}:$

(b) **X** is a stronger acid than **W** because the more electronegative N in **X** can support the negative charge better than carbon, so the anion of **X** is more stable than the anion of **W**.

(c) **Y** is a stronger acid than **X** because the negative charge in **Y** is stabilized by the *inductive effect* from the electronegative oxygen substituent, the OH.

(d) **Z** is a stronger acid than **Y** because of two effects: O is more electronegative than N and can support the negative charge of the anion better, plus the anion of **Z** has two EQUIVALENT resonance forms, which is particularly stable.

2-42

(a) $H_2SO_4 + CH_3COO^- \rightleftharpoons HSO_4^- + CH_3COOH$

(b) $CH_3COOH + (CH_3)_3N\!: \rightleftharpoons CH_3COO^- + (CH_3)_3\overset{\oplus}{N}-H$

(c) $C_6H_5-\overset{O}{\overset{\|}{C}}-O-H + {}^-OH \rightleftharpoons C_6H_5-\overset{O}{\overset{\|}{C}}-O^- + H_2O$

(d) $HO-\overset{O}{\overset{\|}{C}}-OH + \mathbf{2}\,{}^-OH \rightleftharpoons {}^-O-\overset{O}{\overset{\|}{C}}-O^- + \mathbf{2}\,H_2O$

(e) $H_2O + NH_3 \rightleftharpoons HO^- + {}^+NH_4$

(f) $(CH_3)_3\overset{\oplus}{N}-H + {}^-OH \rightleftharpoons (CH_3)_3N\!: + H_2O$

(g) $HCOOH + CH_3O^- \rightleftharpoons HCOO^- + CH_3OH$

(h) $\overset{\oplus}{N}H_3CH_2COOH + \mathbf{2}\,{}^-OH \rightleftharpoons NH_2CH_2COO^- + \mathbf{2}\,H_2O$

2-43 Values are from Appendix 4 in the text. The side of the equation with the weaker acid and base is favored.

(a) pyridine + 4-nitrophenol ($O_2N-C_6H_4-OH$) $\rightleftharpoons$ pyridinium ($C_5H_5\overset{\oplus}{N}-H$) + 4-nitrophenoxide ($O_2N-C_6H_4-O^{\ominus}$) FAVORS **REACTANTS**

weaker base; weaker acid pK_a 7.1; stronger acid pK_a 5.2; stronger base

(b) pyridine + 4-nitrobenzoic acid ($O_2N-C_6H_4-COOH$) $\rightleftharpoons$ pyridinium ($C_5H_5\overset{\oplus}{N}-H$) + 4-nitrobenzoate ($O_2N-C_6H_4-COO^{\ominus}$) FAVORS **PRODUCTS**

stronger base; stronger acid pK_a 3.4; weaker acid pK_a 5.2; weaker base

2-43 continued

(c) pyrrole (N–H) + $NaNH_2$ ⇌ pyrrolide anion ($N^{\ominus}$ Na^+) + NH_3 FAVORS **PRODUCTS**

stronger acid pK_a 17; stronger base; weaker base; weaker acid pK_a 36

(d) $H_3C-C(=O)-NH_2$ + $NaOCH_2CH_3$ ⇌ $H_3C-C(=O)-NH^{\ominus}$ Na^+ + $HOCH_2CH_3$ FAVORS **REACTANTS** but not by much

weaker acid pK_a 16; weaker base; stronger base; stronger acid pK_a 15.9

The reaction in part (d) merits comment. Only 0.1 pK_a units separates the two acids in this equation. When pK_a values are measured in solvents other than water (any pK_a value higher than 15.7), the experimental error increases. Without specifying that these two pK_a values were measured in the same solvent, it is difficult to give the difference of 0.1 pK_a units much credence. Conclusion: the two sides of the equation are fairly equally matched, and the products and reactants would probably be close to 50:50. In one solvent, reactants may be favored, and in a different solvent, products favored.

(e) $H_3C-C(=O)-NH_2$ + HCl (aq) ⇌ $H_3C-C(=\overset{\oplus}{O}H)-NH_2$ + $Cl^{\ominus}$ FAVORS **PRODUCTS**

stronger base; stronger acid: in water, leveled to H_3O^+ pK_a −1.7; weaker acid pK_a 0.0; weaker base

Recall from problem 2-19 that the site of protonation of the reactant is the O, not the N.

(f) $H_3C-C(=O)-CH_2-C(=O)-OEt$ + NaOH (aq) ⇌ $H_3C-C(=O)-\overset{\ominus}{C}H-C(=O)-OEt$ Na^+ + H_2O FAVORS **PRODUCTS**

stronger acid pK_a 10.7; Et = ethyl, CH_2CH_3; stronger base; weaker base; weaker acid pK_a 15.7

The most acidic proton in the reactant is the CH between the two C=O because of resonance stabilization of the resultant anion by both C=O. See the solutions to problems 1-8(g) or 1-43(c) for similar examples.

2-44 Note that this question is asking for the strength of the aqueous solution (pH), not just the strength of the acid.

H_2SO_4 > HCl = HBr > F_3CCOOH > $BrCH_2COOH$ > CH_3COOH > NH_4Cl

H_2SO_4	HCl = HBr	F_3CCOOH	$BrCH_2COOH$	CH_3COOH	NH_4Cl
lowest pH, strongest acid	equivalent—both leveled to H_3O^+	pK_a 0.2	pK_a 2.9	pK_a 4.74	pK_a 9.4 highest pH, weakest acid

Explanation: For acids weaker than water, the pH of the aqueous solution will follow the sequence of the pK_a values: the higher the pK_a, the less they ionize in water. For acids stronger than water, they will be leveled to hydronium ion, H_3O^+, so 1 molar solutions of HCl and HBr would be completely ionized, giving $[H^+] = 1M$. The exception is H_2SO_4 because it is a diprotic acid: the first proton is 100% ionized, but the second proton (HSO_4^-) has pK_a 2.0, higher than water but still a strong acid, so some portion of it will also be ionized in excess of the first proton, making this aqueous solution the strongest acid = lowest pH, with [H+] > 1M.

2-45 (a) conjugate bases

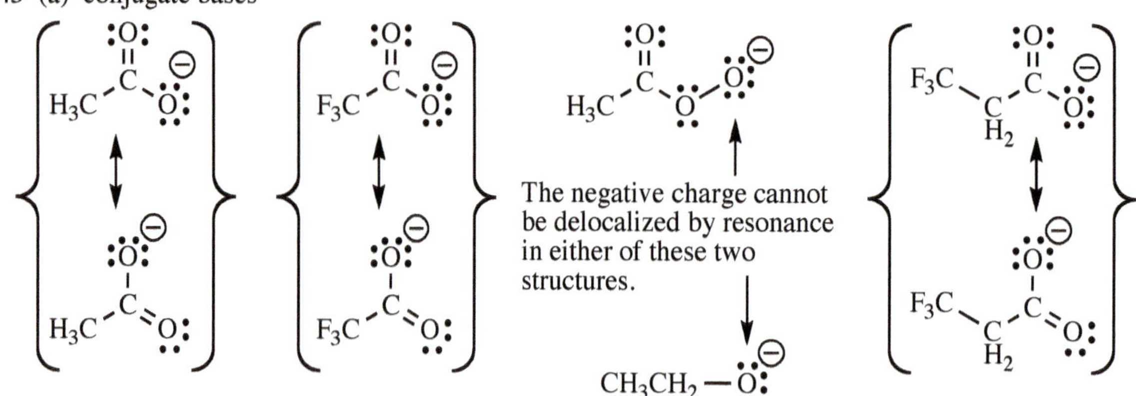
The negative charge cannot be delocalized by resonance in either of these two structures.
$CH_3CH_2—O^-$

(b)

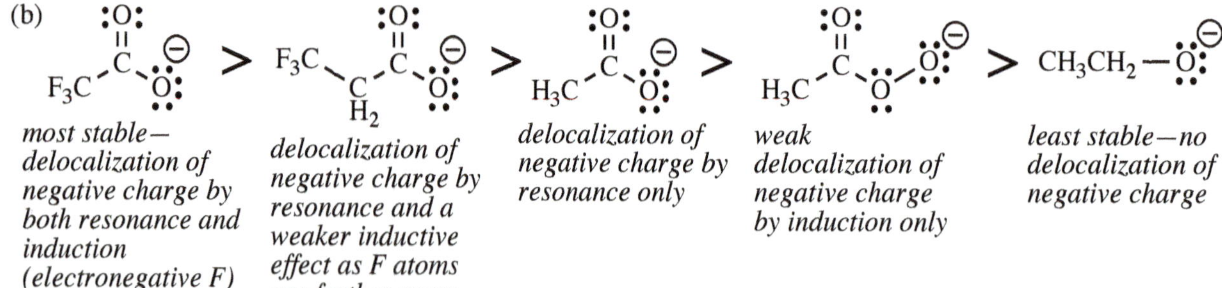
most stable—delocalization of negative charge by both resonance and induction (electronegative F)
delocalization of negative charge by resonance and a weaker inductive effect as F atoms are farther away
delocalization of negative charge by resonance only
weak delocalization of negative charge by induction only
$CH_3CH_2—O^-$
least stable—no delocalization of negative charge

(c) The strongest acid will have the most stable conjugate base. The actual pK_a values (some from text Appendix 4) are listed beneath each acid.

$F_3C\text{–}COOH$ > $F_3C\text{–}CH_2\text{–}COOH$ > $H_3C\text{–}COOH$ > $H_3C\text{–}C(=O)\text{–}O\text{–}OH$ > CH_3CH_2OH

F_3CCOOH	F_3CCH_2COOH	H_3CCOOH	$H_3CC(=O)OOH$	CH_3CH_2OH
strongest acid pK_a 0.2	pK_a 3.07	pK_a 4.74	pK_a 8.2	weakest acid pK_a 15.9

2-46

(a) conjugate acids

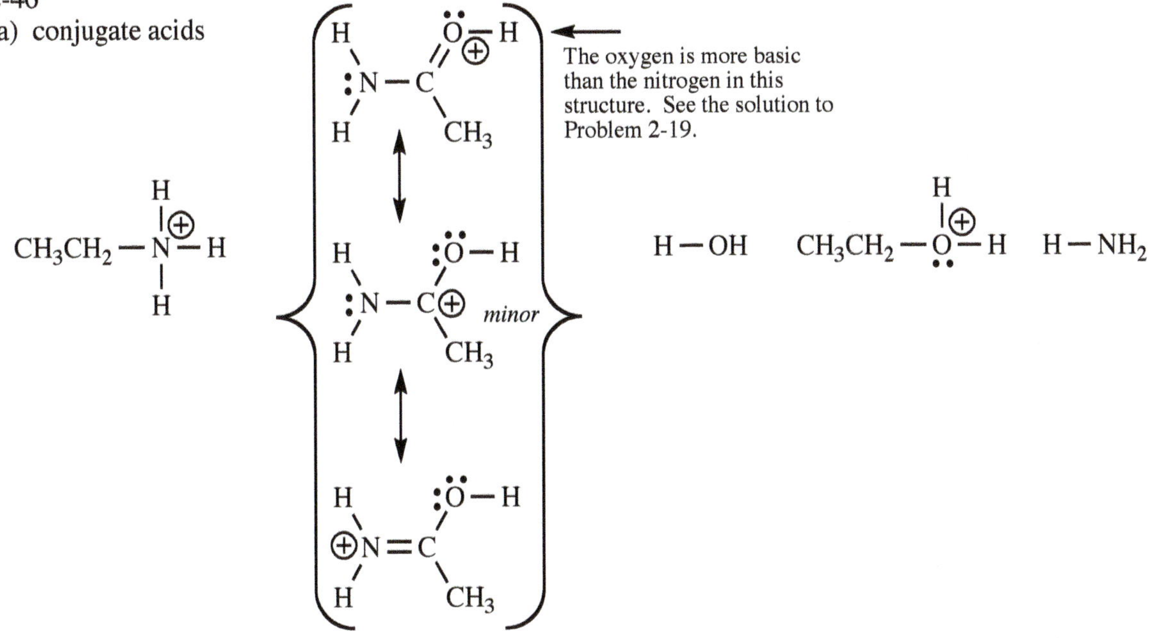
The oxygen is more basic than the nitrogen in this structure. See the solution to Problem 2-19.
minor
H—OH
H—NH$_2$

(b) order of decreasing stability (pK$_a$ values from Appendix 4 in the text)

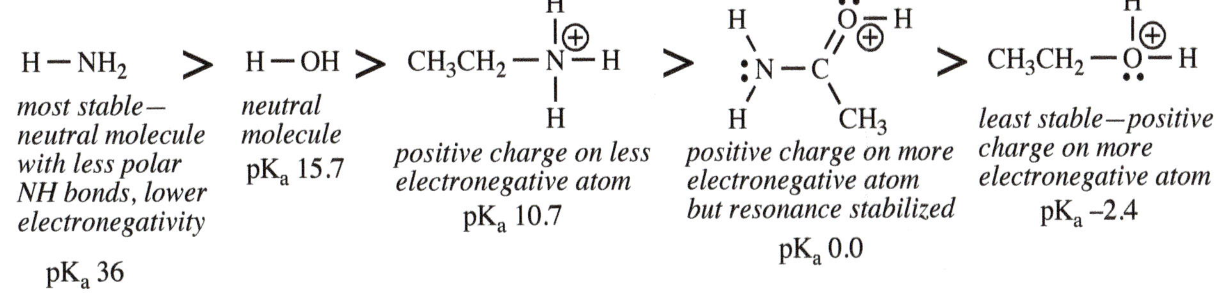
H—NH$_2$ > H—OH > CH$_3$CH$_2$—NH$_3^+$ > protonated amide > CH$_3$CH$_2$—OH$_2^+$
most stable—neutral molecule with less polar NH bonds, lower electronegativity
pK$_a$ 36
neutral molecule
pK$_a$ 15.7
positive charge on less electronegative atom
pK$_a$ 10.7
positive charge on more electronegative atom but resonance stabilized
pK$_a$ 0.0
least stable—positive charge on more electronegative atom
pK$_a$ −2.4

(c) The weakest conjugate acid will form the strongest conjugate base.

$^{-}$:NH$_2$ > $^{-}$:OH > CH$_3$CH$_2$—NH—H > H$_2$N—C(=O)—CH$_3$ > CH$_3$CH$_2$—O—H

strongest base ... *weakest base*

2-47

(a) conjugate bases

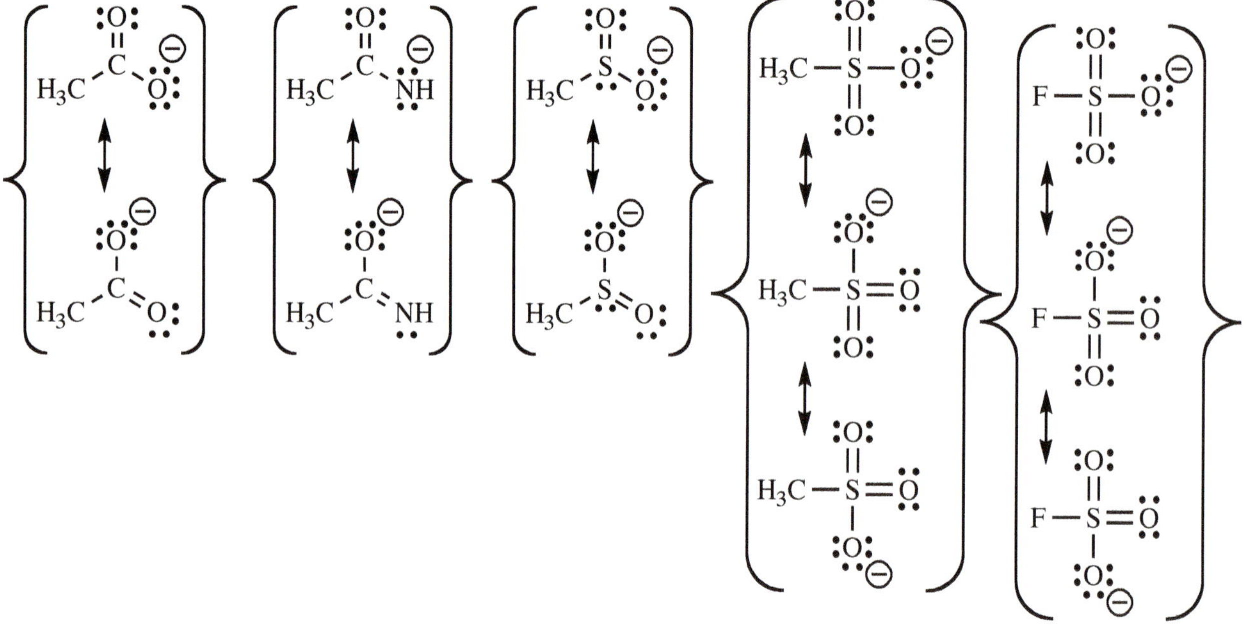

(b)

most stable—delocalization of negative charge by three resonance forms and induction (F and S more electro-negative than C) > *delocalization of negative charge by three resonance forms and induction (S more electronegative than C)* > *delocalization of negative charge by two resonance forms and induction (S more electronegative than C)* > *delocalization of negative charge by two equivalent resonance forms* > *least stable—delocalization of negative charge by two resonance forms, although non-equivalent*

(c) The strongest acid will have the most stable conjugate base. The pK_a values are listed beneath each acid.

strongest acid $pK_a < -5$ > pK_a −1.2 > pK_a 2.3 > pK_a 4.74 > *weakest acid* pK_a 16

2-48

Of all the substituent groups in organic molecules, the nitro group, NO_2, is the best electron-withdrawing group. Not only does it have three electronegative atoms that can stabilize negative charge by induction, but it also has very strong resonance stabilization. Adding just one nitro group to methane increase the acidity by 40 pK_a units!

$^{\ominus}:CH_3$
no stabilization; one of the strongest bases known

one nitro group—strong stabilization

two nitro groups—very strong stabilization

three nitro groups—maximum stabilization—essentially zero negative charge on central C

Each subsequent substitution of H by NO_2 increases the amount of the delocalization of the electron pair on C, stabilizing the anion.

2-49

(a) $CH_3CH_2—O—H \quad + \quad CH_3—Li \longrightarrow CH_3CH_2—O^- \; Li^+ \quad + \quad CH_4$

(b) The conjugate acid of CH_3Li is CH_4. Text Appendix 4 gives the pK_a of CH_4 as 50, one of the weakest acids known. The conjugate base of one of the weakest acids known must be one of the strongest bases known.

2-50 The newly formed bond is shown in bold.

(a) $CH_3—\ddot{O}:^{\ominus} \quad + \quad CH_3—\ddot{Cl}: \longrightarrow CH_3—\ddot{O}—CH_3 \quad + \quad :\ddot{Cl}:^{\ominus}$

nucleophile
Lewis base

electrophile
Lewis acid

2-50 continued

(b) $CH_3-\overset{+}{O}(CH_3)-CH_3$ + $H-\ddot{O}-H$ ⟶ $CH_3-\ddot{O}-CH_3$ + $H-\overset{+}{O}(CH_3)-H$

H_3C electrophile Lewis acid; H_2O nucleophile Lewis base

(c) $H_2C=O$ + $:NH_3$ ⟶ $H_2C(O^-)-NH_3^+$

$H_2C=O$: electrophile Lewis acid; NH_3: nucleophile Lewis base

(d) $CH_3-\ddot{N}H_2$ + CH_3CH_2-Cl ⟶ $CH_3-\overset{+}{N}H_2-CH_2CH_3$ + $:Cl:^-$

CH_3NH_2: nucleophile Lewis base; CH_3CH_2Cl: electrophile Lewis acid

(e) $CH_3-C(=O)-CH_3$ + $H-O-SO_2-OH$ ⟶ $CH_3-C(=\overset{+}{O}H)-CH_3$ + $^-O-SO_2-OH$

CH_3COCH_3: nucleophile Lewis base; H_2SO_4: electrophile Lewis acid

plus resonance forms

(f) $(CH_3)_3C-Cl$ + $AlCl_3$ ⟶ $H_3C-C^+(CH_3)_2$ + $Cl-\overset{-}{Al}Cl_3$

$(CH_3)_3CCl$: nucleophile Lewis base; $AlCl_3$: electrophile Lewis acid

This may also be written in two steps: association of the Cl with Al, and a second step where the C—Cl bond breaks.

(g) $CH_3-C(=O)-CH_2-H$ + $^-:\ddot{O}-H$ ⟶ $CH_3-C(O^-)=CH_2$ + $H-\ddot{O}-H$

CH_3COCH_3: electrophile Lewis acid; ^-OH: nucleophile Lewis base

(h) $F-B(F)-F$ + $CH_2=CH_2$ ⟶ $F-\overset{-}{B}F_2-CH_2-\overset{+}{C}H_2$

BF_3: electrophile Lewis acid; $CH_2=CH_2$: nucleophile Lewis base

(i) $\overset{-}{B}F_3-CH_2-\overset{+}{C}H_2$ + $CH_2=CH_2$ ⟶ $\overset{-}{B}F_3-CH_2-CH_2-CH_2-\overset{+}{C}H_2$

$^-BF_3CH_2CH_2^+$: electrophile Lewis acid; $CH_2=CH_2$: nucleophile Lewis base

2-51 In each product, the new bond is shown in bold.

(a) $CH_3-\ddot{N}(H)-CH_3$ + $H-Cl$ ⟶ $CH_3-N^{\oplus}(H)(H)-CH_3$ + $Cl^{\ominus}$

Lewis base—nucleophile; Lewis acid—electrophile

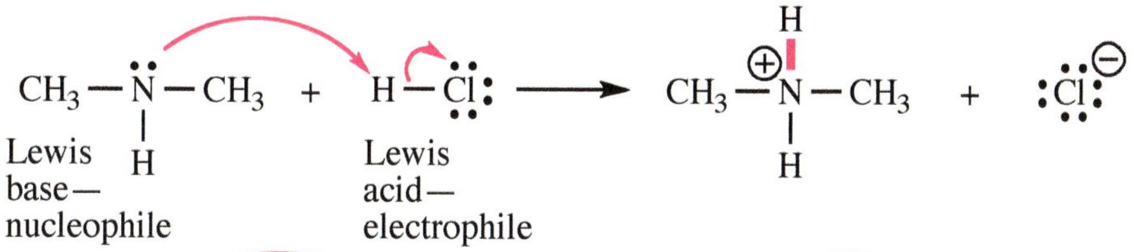

(b) $CH_3-\ddot{N}(H)-CH_3$ + H_3C-Cl ⟶ $CH_3-N^{\oplus}(CH_3)(H)-CH_3$ + $Cl^{\ominus}$

Lewis base—nucleophile; Lewis acid—electrophile

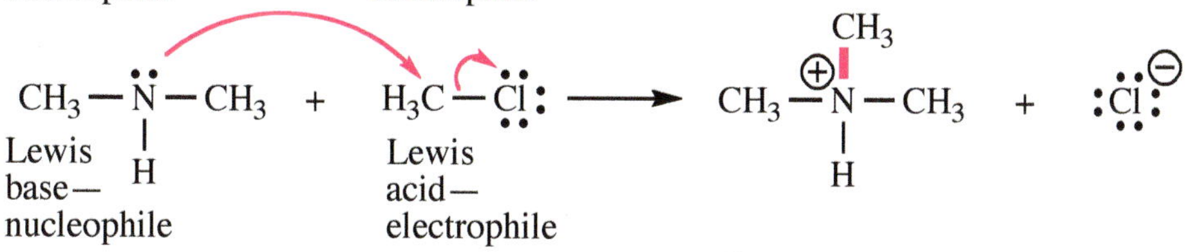

(c) $CH_3-C(=O)-H$ + $H-Cl$ ⟶ $CH_3-C(=O^{\oplus}H)-H$ + $Cl^{\ominus}$

Lewis base—nucleophile; Lewis acid—electrophile

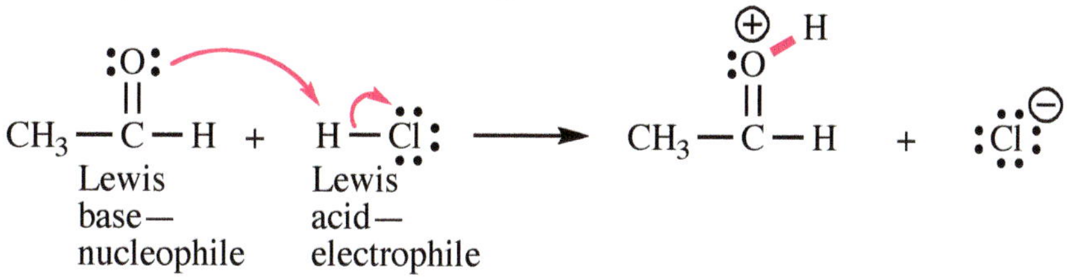

(d) $CH_3-C(=O)-H$ + $^{\ominus}O-CH_3$ ⟶ $CH_3-C(O^{\ominus})(OCH_3)-H$

Lewis acid—electrophile; Lewis base—nucleophile

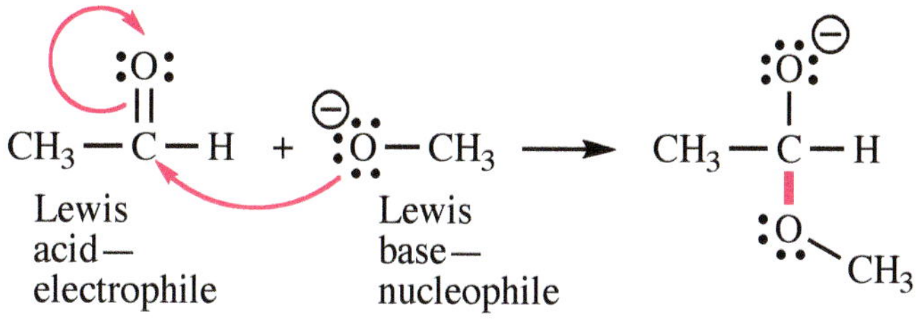

Notice in parts (d) and (e) that the identical reactants can generate two separate products.

(e) $H-C(H)(H)-C(=O)-H$ + $^{\ominus}O-CH_3$ ⟶ $H_2C=C(O^{\ominus})-H$ + $H-O-CH_3$

Lewis acid—electrophile; Lewis base—nucleophile

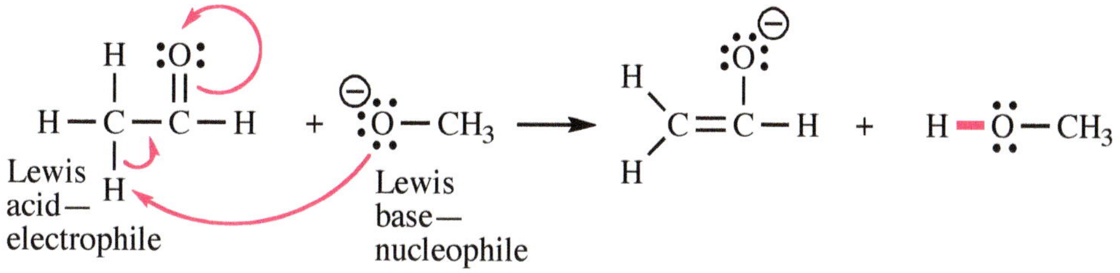

2-52 In each product, the new bond is shown in bold.

In each case, look for the atom that is the most positive because of induction or resonance, or is the most electron deficient (like boron with an empty orbital).

(a) {cyclopentanone ⟷ $C^{\oplus}-O^{\ominus}$ resonance form}

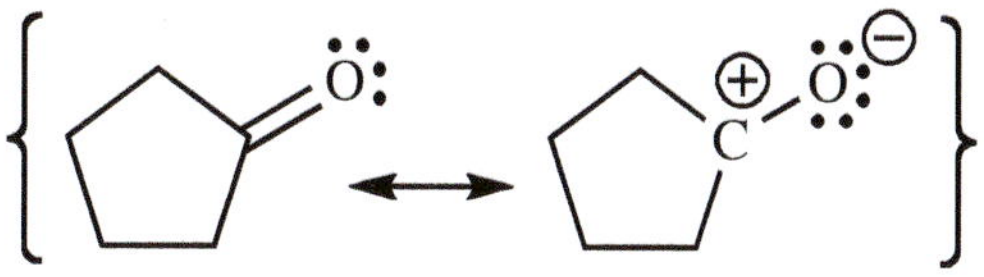

Resonance forms show that the C in a C=O group always has some (+) charge; this carbon is an electrophile. Nucleophiles will be attracted to it as shown here and in parts (e) and (f).

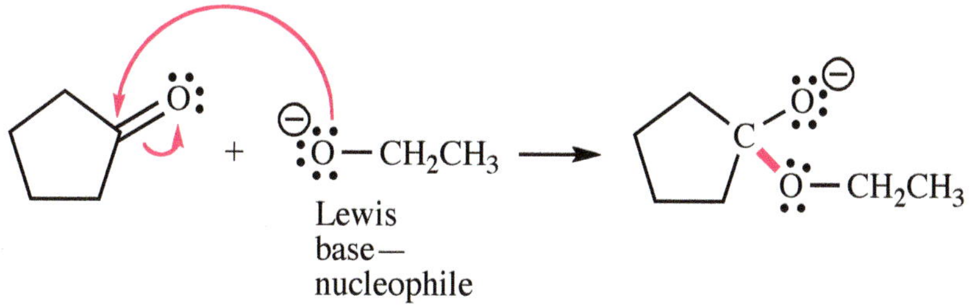

cyclopentanone + $^{\ominus}O-CH_2CH_3$ ⟶ $C(O^{\ominus})(O-CH_2CH_3)$ (cyclopentane)

Lewis base—nucleophile

The Na^+ is a spectator ion and plays no active role in these reactions. It is not shown here to avoid clutter.

2-52 continued

(b) $H_4N^+ + {}^-OCH_2CH_3 \longrightarrow NH_3 + H-O-CH_2CH_3$

(c) $H_3C-CHBr-H + {}^-OCH_2CH_3 \longrightarrow H_3C-CH_2-O-CH_2CH_3 + Br^-$

(d) $BH_3 + {}^-OCH_2CH_3 \longrightarrow H_3B^--O-CH_2CH_3$

(e) As shown in 2-51(d) and (e), there can be more than one reaction pathway. $NaOCH_2CH_3$ is a strong base and can quickly remove a proton from acetic acid, CH_3COOH. In addition, the reaction can occur at the C=O (see 2-52(a)) of acetic acid, although more slowly than proton removal which is one of the fastest reactions known.

1) $CH_3COOH + {}^-OCH_2CH_3 \longrightarrow H-O-CH_2CH_3 + CH_3COO^-$

(draw the other resonance form)

2) $CH_3COOH + {}^-OCH_2CH_3 \longrightarrow H_3C-C(O^-)(OH)-O-CH_2CH_3$

(f) As shown in 2-51(d) and (e), and 2-52(e) above, there can be more than one reaction pathway. Draw important resonance forms to reveal where the sites of (+) and (–) charge are.

$\{H_2C=CH-CH=O \longleftrightarrow H_2C=CH-CH^+-O^- \longleftrightarrow H_2C^+-CH=CH-O^-\}$

site 1

site 2

1) $H_2C=CH-CH=O + {}^-OCH_2CH_3 \longrightarrow H_2C=CH-CH(O^-)-O-CH_2CH_3$

2) $H_2C=CH-CH=O + {}^-OCH_2CH_3 \longrightarrow CH_3CH_2O-CH_2-CH=CH-O^-$

This is the most significant resonance contributor. Draw the other one.

2-53 In each product, the new bond is shown in bold.

In each case, look for an unshared electron pair, or an electron pair in a double or triple bond, to make a molecule nucleophilic. An arrow showing electron movement must always begin at an unshared electron pair or a bond.

(a)

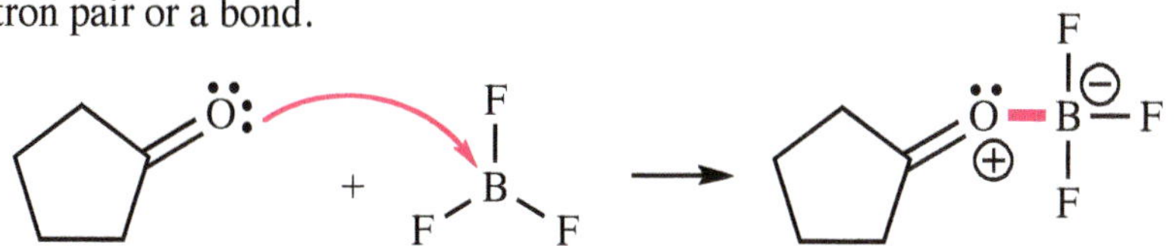

(b)

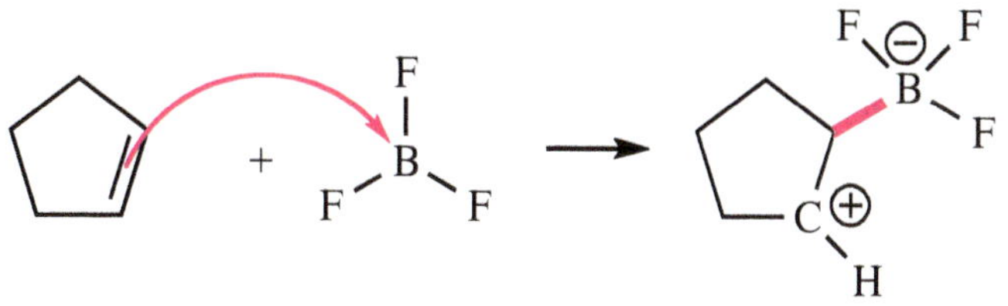

(c)

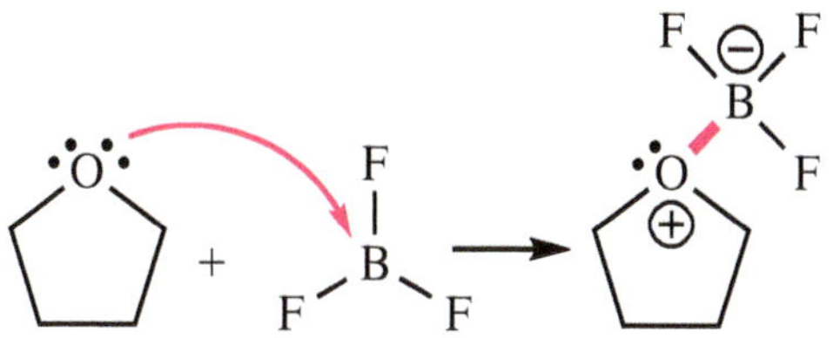

(d)

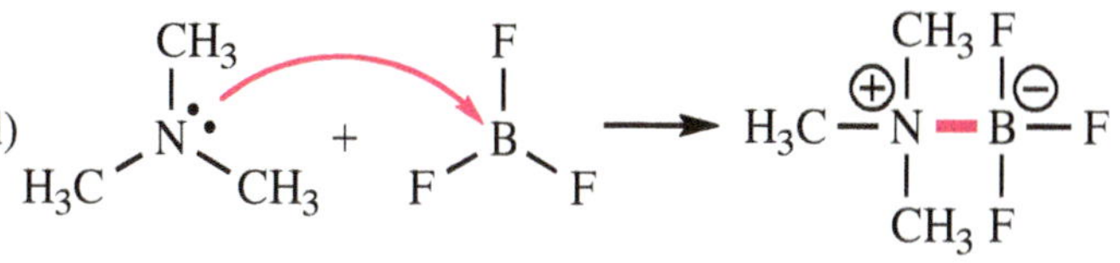

(e)

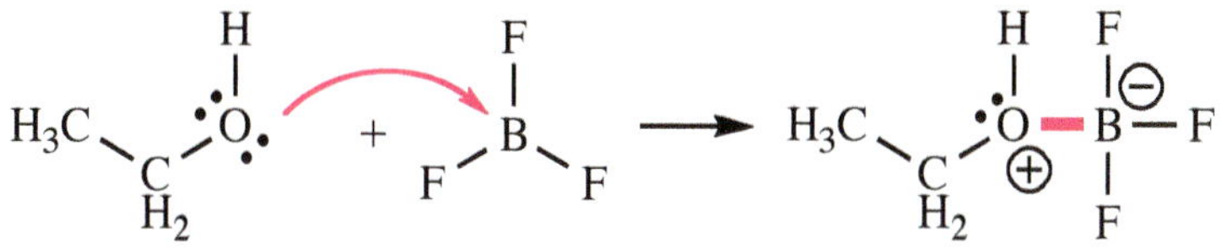

(f)

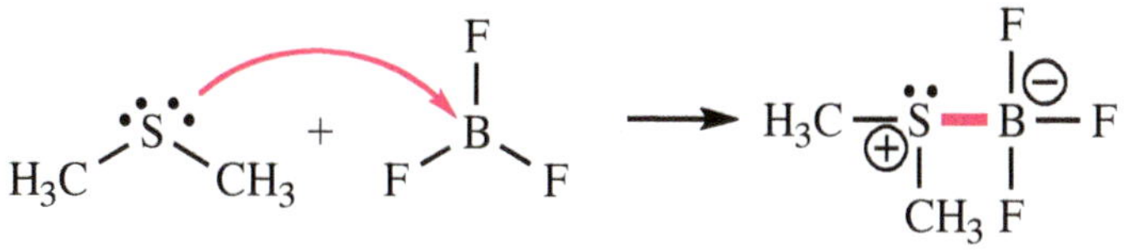

2-54 (a) Ascorbic acid has four OH groups that could act as acids. The ionization of each shows that one gives a more stable conjugate base.

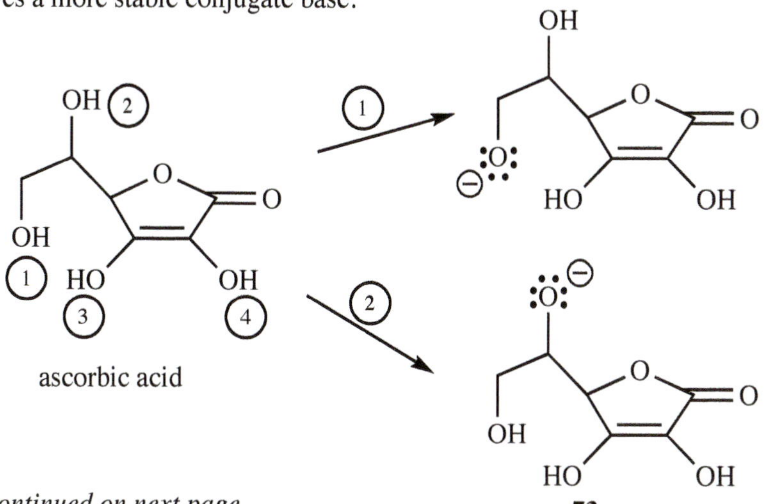

ascorbic acid

localized negative charge—
no resonance forms

localized negative charge—
no resonance forms

continued on next page

2-54(a) continued

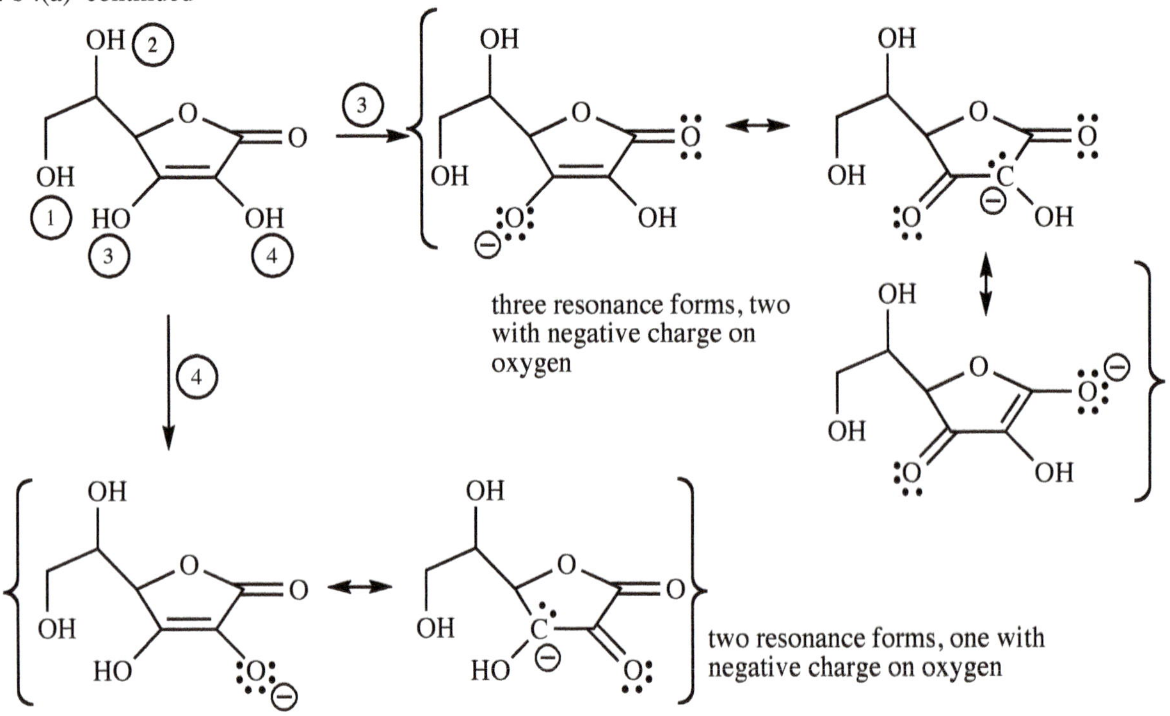

three resonance forms, two with negative charge on oxygen

two resonance forms, one with negative charge on oxygen

(b) The ionization of the OH labeled "3" produces a conjugate base with three resonance forms, two of which have negative charge on oxygen. This OH is the most acidic group in ascorbic acid.

(c) The conjugate base of acetic acid, the acetate ion, CH_3COO^- , has two resonance forms (see the solution to problem 2-47), each of which has a C=O and a negatively charged oxygen, similar to two of the resonance forms of the ascorbate ion. The acidity of these two very different molecules is similar because the stabilization of the conjugate base is so similar. *The strength of an acid is determined by the stability of its conjugate base.*

Note to the student: Organic chemistry professors will ask you to "explain" questions, that is, to explain a certain trend in organic structures or behavior of an organic reaction. The professor is trying to determine two things: 1) Does the student understand the principle underlying the behavior? 2) Does the student understand how the principle applies in this particular example?

To answer an "explain" question, somewhere in your answer should be a statement of the principle, like: "*The strength of an acid is determined by the stability of its conjugate base.*" From there, show through a series of logical steps how the principle applies, like drawing resonance forms to show which acid has the most stable conjugate base through resonance or induction. Answering these questions is like crossing a creek on stepping stones. Each phrase or sentence is a step to the next stone. When strung together, the steps bridge the gap between the principle and the observation.

2-55 The examples here are representative. Your examples may be different and still be correct. What is important in this problem is to have the same functional group.

(a) alkane: hydrocarbon with all single bonds; can be acyclic (no ring) or cyclic

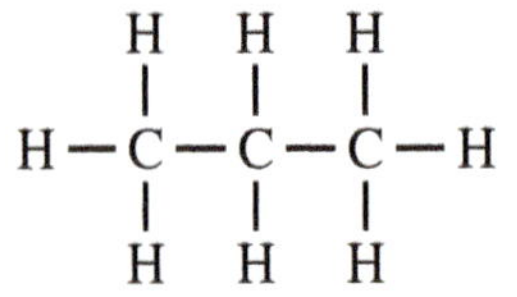

(b) alkene: contains a carbon-carbon double bond

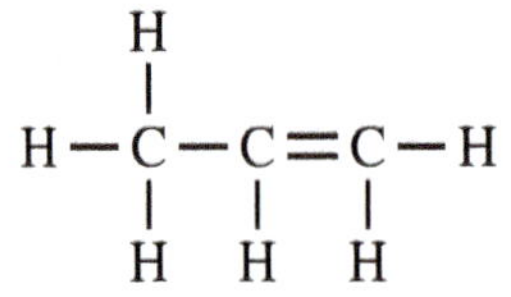

(c) alkyne: contains a carbon-carbon triple bond

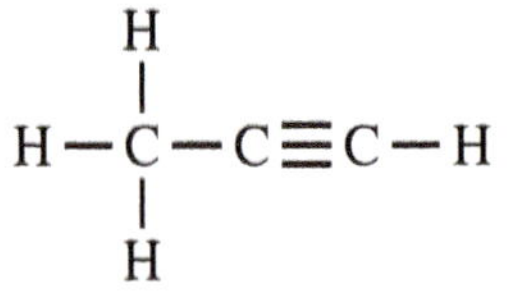

2-55 continued

(d) alcohol: contains an OH group on a carbon

```
    H  H
    |  |
H - C - C - O - H
    |  |
    H  H
```

(e) ether: contains an oxygen between two carbons

```
    H       H
    |       |
H - C - O - C - H
    |       |
    H       H
```

(f) ketone: contains a carbonyl group between two carbons

```
    H   O   H
    |   ||  |
H - C - C - C - H
    |       |
    H       H
```

(g) aldehyde: contains a carbonyl group with a hydrogen on one side

```
    H   H   O
    |   |   ||
H - C - C - C - H
    |   |
    H   H
```

(h) aromatic hydrocarbon: a cyclic hydrocarbon with alternating double and single bonds

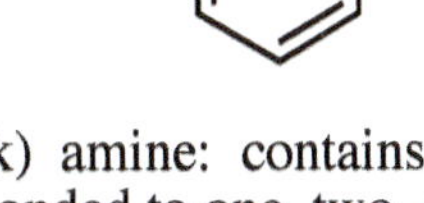

(i) carboxylic acid: contains a carbonyl group with an OH group on one side

```
    H   O
    |   ||
H - C - C - O - H
    |
    H
```

(j) ester: contains a carbonyl group with an O—C on one side

```
    H   O       H
    |   ||      |
H - C - C - O - C - H
    |           |
    H           H
```

(k) amine: contains a nitrogen bonded to one, two, or three carbons

```
    H   H or R group
    |   |
H - C - N - H or R group
    |
    H
```

(l) amide: contains a carbonyl group with a nitrogen on one side

```
    H   O   H or R group
    |   ||  |
H - C - C - N - H or R group
    |
    H
```

(m) nitrile: contains the carbon-nitrogen triple bond: $H_3C—C{\equiv}N$

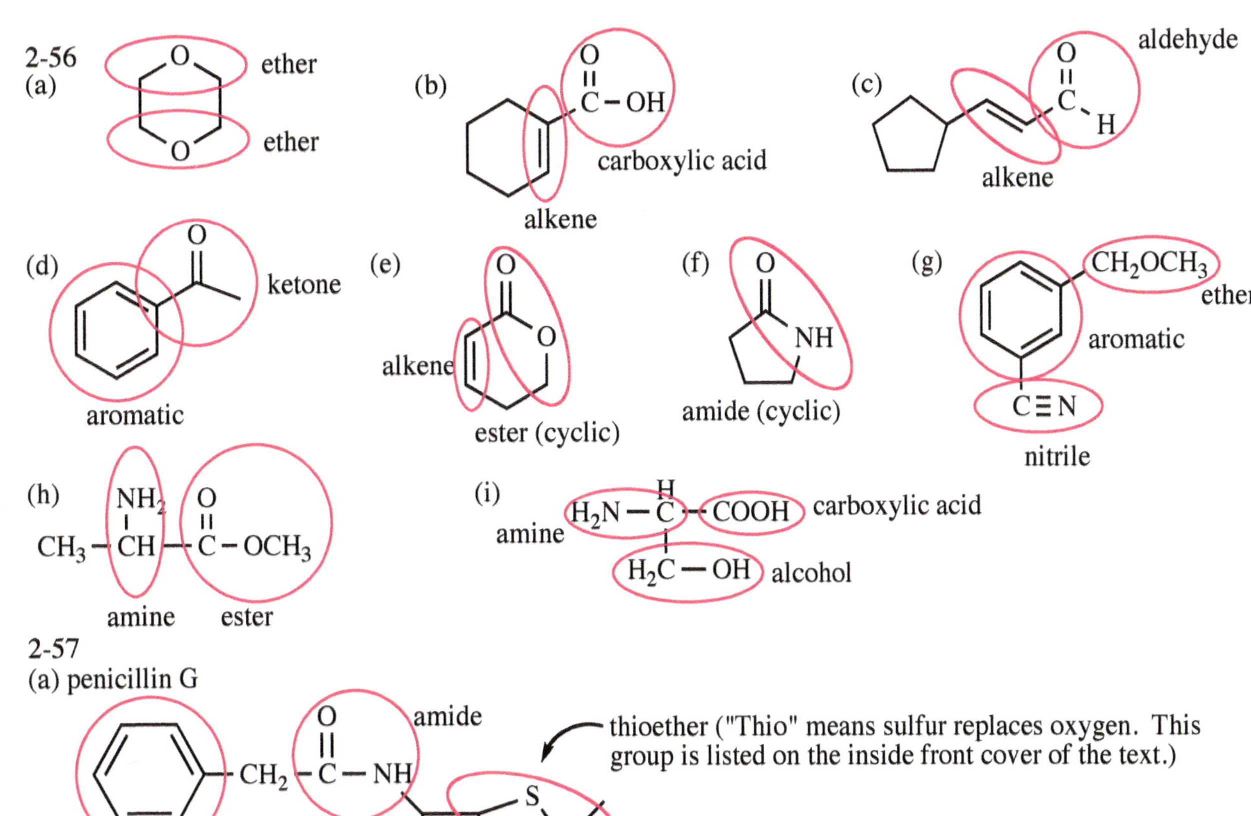

2-57 continued

(b) dopamine

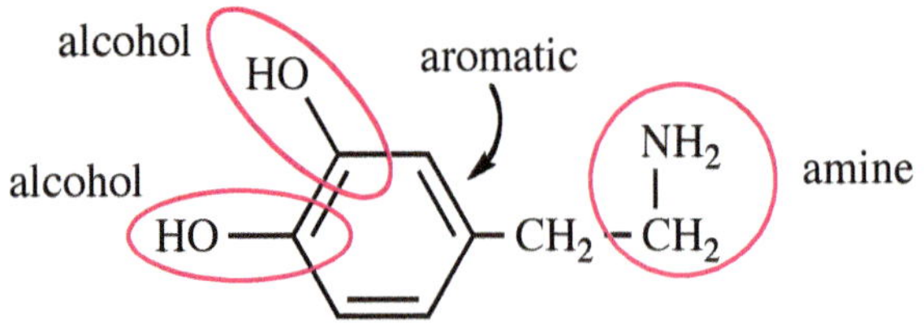

alcohol HO; alcohol HO; aromatic; NH_2; CH_2-CH_2; amine

(In later chapters, you will learn that the OH group on a benzene ring is a special functional group called a "phenol". For now, it fits the broad definition of an alcohol.)

(c) capsaicin

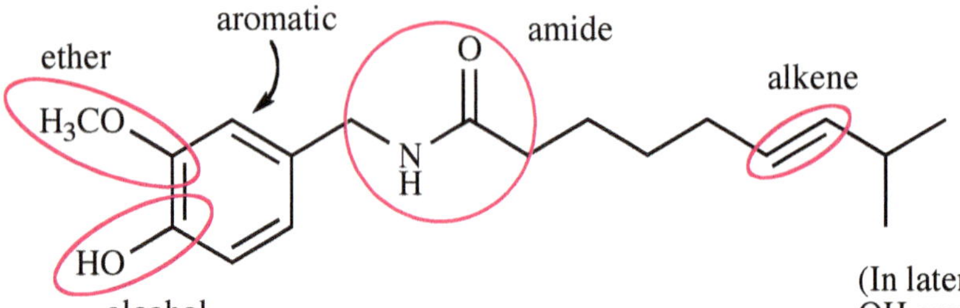

ether H_3CO; aromatic; O; N H; amide; alkene; HO; alcohol

(In later chapters, you will learn that the OH group on a benzene ring is a special functional group called a "phenol". For now, it fits the broad definition of an alcohol.)

(d) thyroxine

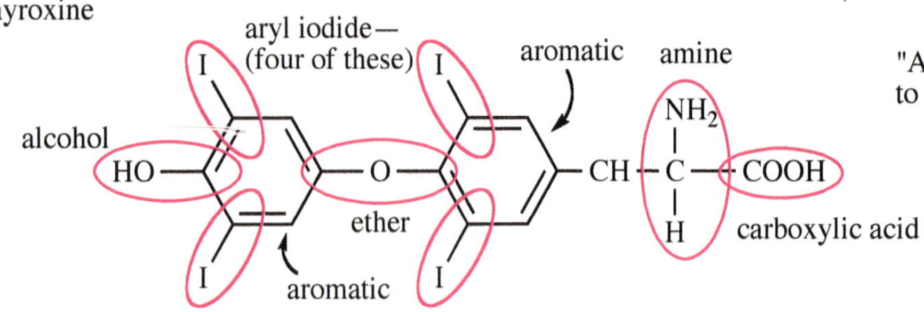

aryl iodide—(four of these); I; I; aromatic; amine; NH_2; alcohol HO; O; CH—C—COOH; H; ether; aromatic; I; I; carboxylic acid

"Aryl" iodides are attached to aromatic rings.

(e) testosterone

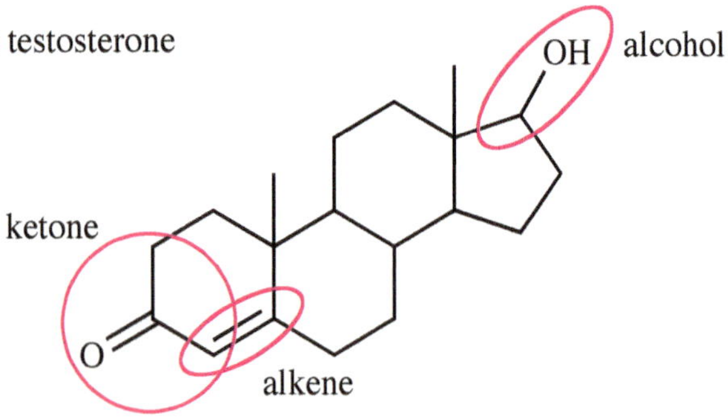

OH alcohol; ketone; O; alkene

CHAPTER 3—STRUCTURE AND STEREOCHEMISTRY OF ALKANES

3-1 (a) C_nH_{2n+2} where n = 28 gives $C_{28}H_{58}$ (b) C_nH_{2n+2} where n = 44 gives $C_{44}H_{90}$

> Note to the student: The IUPAC system of nomenclature has a well-defined set of rules determining how structures are named. You will find a summary of these rules as Appendix 5 in the textbook.

3-2 Use hyphens to separate letters from numbers. Structures might not display their longest chains left to right—you have to search for them. In the name, write substituents in alphabetical order, even though that might be different from the numerical order of position numbers.

(a) 3-methylpentane (Always find the longest chain; it may not be written in a straight line.)
(b) 2-bromo-3-methylpentane (Always find the longest chain.)
(c) 5-ethyl-2-methyl-4-propylheptane ("When there are two longest chains of equal length, use the chain with the greater number of substituents.")
(d) 4-isopropyl-2-methyldecane ("Number from the end closest to the first substituent." —also for (c))

3-3 This Solutions Manual will present line–angle formulas, or simply "line formulas", where a question asks for an answer including a structure. If you use condensed structural formulas instead, be sure that you are able to "translate" one structure type into the other.

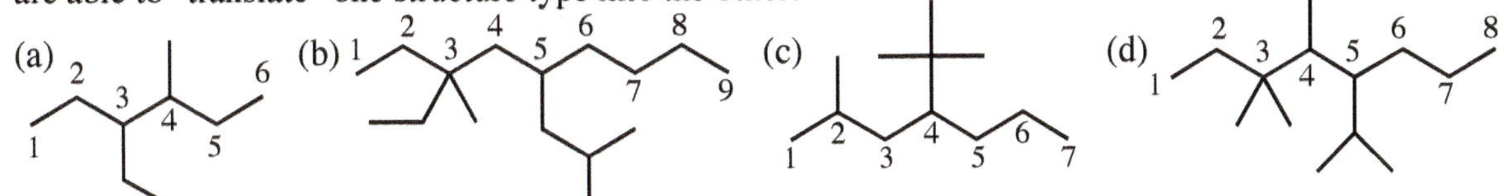

3-4 Separate numbers from numbers with commas. Separate numbers from letters with hyphens.

(a) 2-methylbutane
(b) 2,2-dimethylpropane
(c) 3-ethyl-2-methylhexane
(d) 2,4-dimethylhexane
(e) 3-ethyl-2,2,4,5-tetramethylhexane
(f) 4-*tert*-butyl-3-methylheptane

3-5 In some cases of an ambiguous or incorrect name, more than one possible structure might be implied. That is often the problem with a wrong name: it does not describe a unique structure.

(a) incorrect: *3-ethyl-4-methylpentane*

The chain is numbered in the wrong direction. The methyl should be on the second carbon, not the fourth.

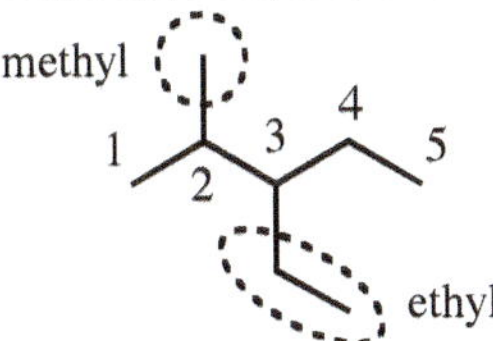

correct: *3-ethyl-2-methylpentane*

(b) incorrect: *2-ethyl-3-methylpentane*

The longest chain was not identified.

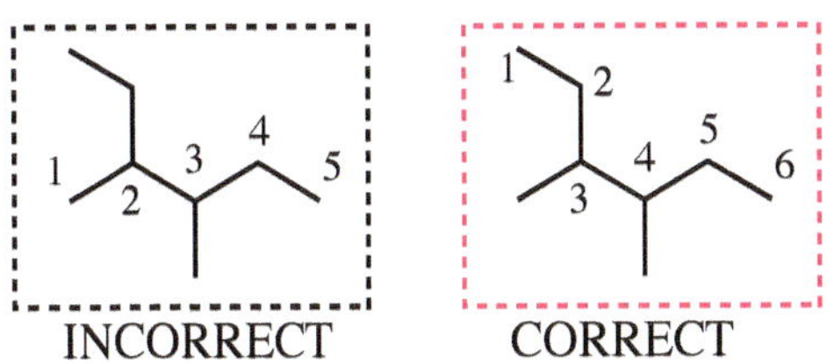

correct: *3,4-dimethylhexane*

(c) incorrect: *3-dimethylhexane*

Two substituents require two position numbers, even if they are both 3; that is the example shown below. It is also possible that a different position number was omitted, like "3,4-".

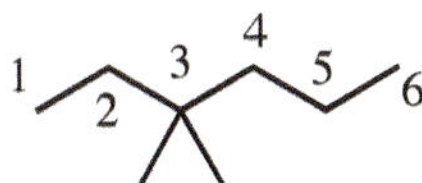

correct: *3,3-dimethylhexane*

(d) incorrect: *4-isobutylheptane*

This is a more subtle error. If two chains of equal length are possible, select the one that maximizes the number of substituents.

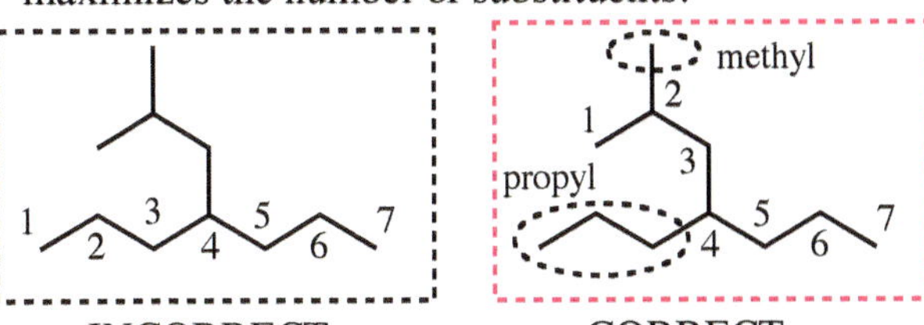

correct: *2-methyl-4-propylheptane*

3-5 continued

(e) incorrect: *2-bromo-3-ethylbutane*

The longest chain was not identified.

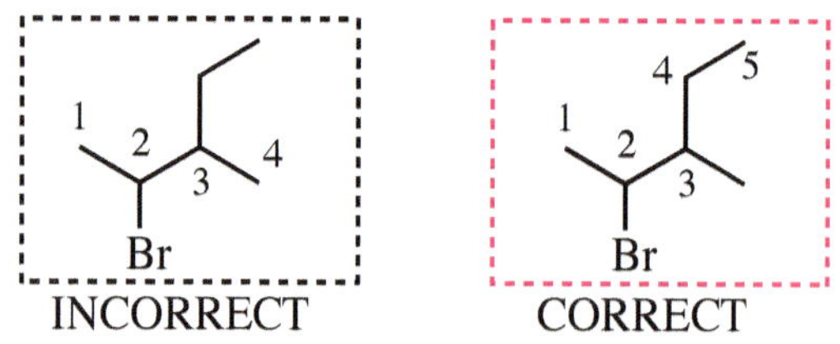

INCORRECT CORRECT

correct: *2-bromo-3-methylpentane*

(f) incorrect: *2,3-diethyl-5-isopropylheptane*

The longest chain is 8 carbons: octane. Numbering from the lower right corner as shown below puts the first substituent on carbon-2, and also maximizes the number of substituents.

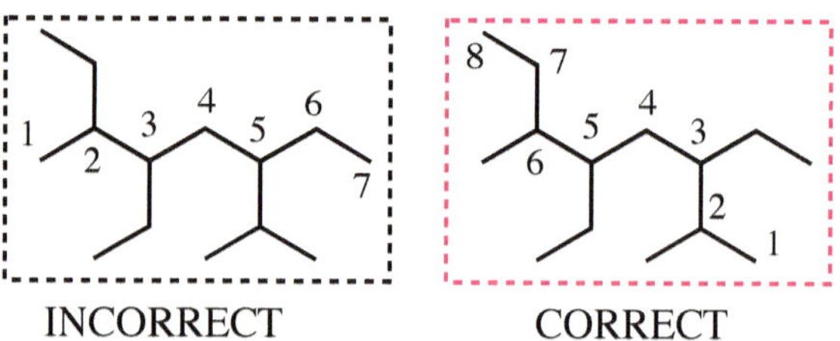

INCORRECT CORRECT

correct: *3,5-diethyl-2,6-dimethyloctane*

3-6 (Hints: *systematize* your approach to these problems. For the isomers of a six-carbon formula, for example, start with the isomer containing all six carbons in a straight chain, then the isomers containing a five-carbon chain, then a four-carbon chain, etc. Carefully check your answers to AVOID DUPLICATE STRUCTURES.)

(a)

hexane (or *n*-hexane) 2-methylpentane 3-methylpentane 2,2-dimethylbutane 2,3-dimethylbutane

(b)

heptane (or *n*-heptane) 2-methylhexane 3-methylhexane 2,2-dimethylpentane

3,3-dimethylpentane 2,3-dimethylpentane 2,4-dimethylpentane 3-ethylpentane 2,2,3-trimethylbutane

3-7 For this problem, the carbon numbers in the substituents are indicated in italics. The bold bond shows the point of attachment.

(a) $—CH(CH_3)CH_3$ (carbons *1 2*)

1-methylethyl
common name = isopropyl

(b) $—CH_2CH(CH_3)CH_3$ (carbons *1 2 3*)

2-methylpropyl
common name = isobutyl

(c) $—CH(CH_3)CH_2CH_3$ (carbons *1 2 3*; 2° at carbon 1)

1-methylpropyl
common name = *sec*-butyl, abbreviation for *secondary*-butyl

(d) $—C(CH_3)_2CH_3$ (carbons *1 2*; 3° at carbon 1)

1,1-dimethylethyl
common name = *tert*-butyl or *t*-butyl, abbreviation for *tertiary*-butyl

(e) $—CH_2CH_2CH(CH_3)CH_3$ (carbons *1 2 3 4*)

3-methylbutyl
common name = isopentyl or isoamyl

3-8

(a)

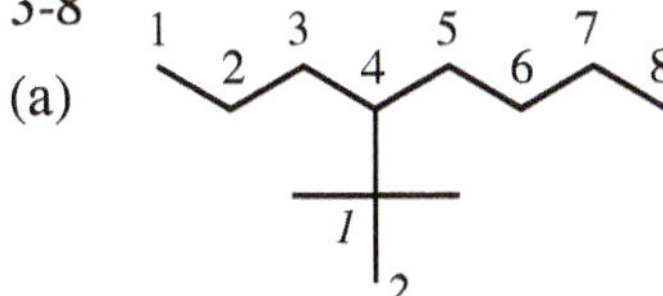

(b)

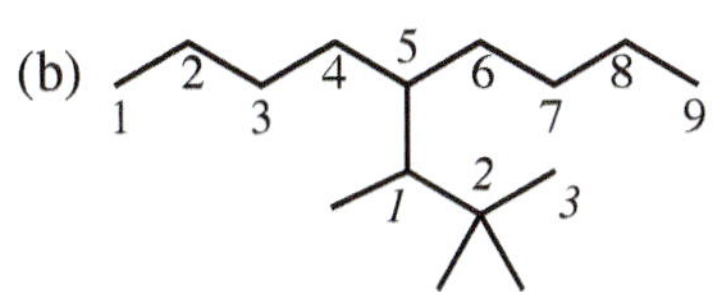

(c)

1 2 3 4 5 6 7 8
1 2 3

3-9 Once the number of carbons is determined, C_nH_{2n+2} gives the formula.

(a) octane = 8C; plus 1,1-dimethylethyl = 4C ⟹ 12 C, $C_{12}H_{26}$

(b) nonane = 9C; plus trimethyl = 3C; plus propyl = 3C ⟹ 15 C, $C_{15}H_{32}$

3-10

(a) (lowest b.p.) hexane < octane < decane (highest b.p.) —molecular weight

(b) $(CH_3)_3C{-}C(CH_3)_3$ < $CH_3CH_2C(CH_3)_2CH_2CH_2CH_3$ < octane —branching differences
(lowest b.p.) (highest b.p.)

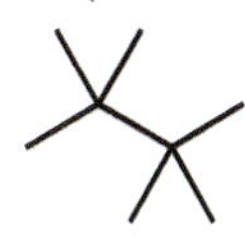

most branching
actual b.p. 107 °C

some branching
actual b.p. 112 °C

no branching
actual b.p. 126 °C

3-11 Part (a) was answered in the text.

(b) front carbon — back carbon — Newman projection

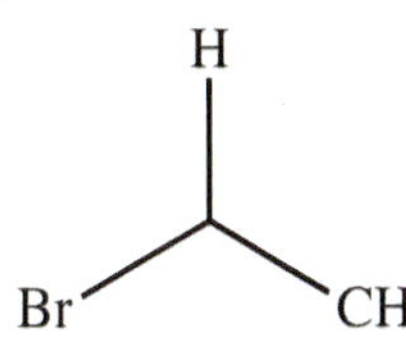

Cl H
CH_2CH_3

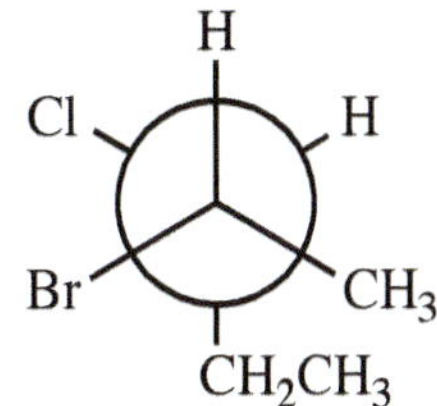

For nomenclature practice, can you name these?

2-bromo-3-chloropentane

(c) front carbon — back carbon — Newman projection

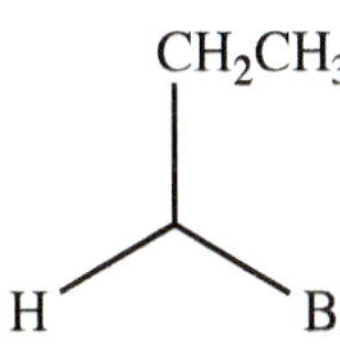

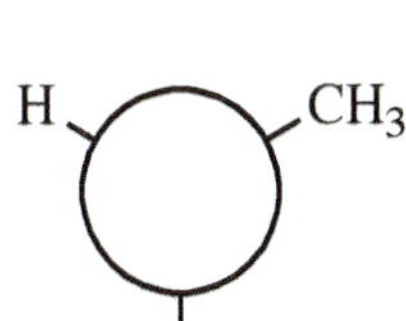

CH_2CH_3
H CH_3
H Br
CH_2CH_3

3-bromo-4-methylhexane

An excellent exercise for your study group is for one person to create a Newman projection, and for the other people in the group to translate that Newman into a line formula, and name it.

3-12

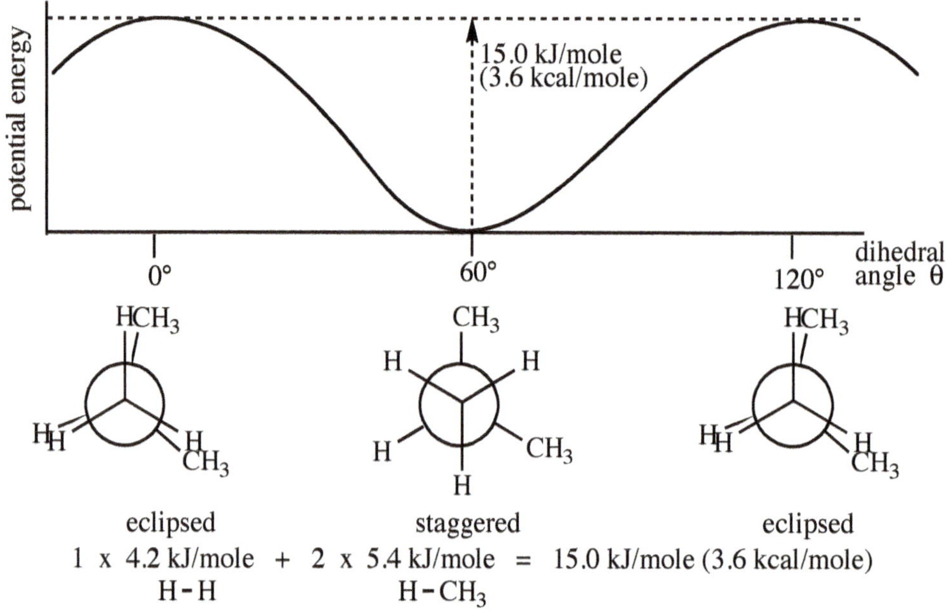

3-13

Relative energies on the graph on the next page were calculated using these values from the text: 3.8 kJ/mole (0.9 kcal/mole) for a CH_3–CH_3 gauche (staggered) interaction; 4.2 kJ/mole (1.0 kcal/mole) for a H–H eclipsed interaction; 5.4 kJ/mole (1.3 kcal/mole) for a H–CH_3 eclipsed interaction; 12.5 kJ/mole (3.0 kcal/mole) for a CH_3–CH_3 eclipsed interaction. These values in kJ/mole are noted on each structure and are summed to give the energy value on the graph. Slight variations between values here and in the text are due to rounding.

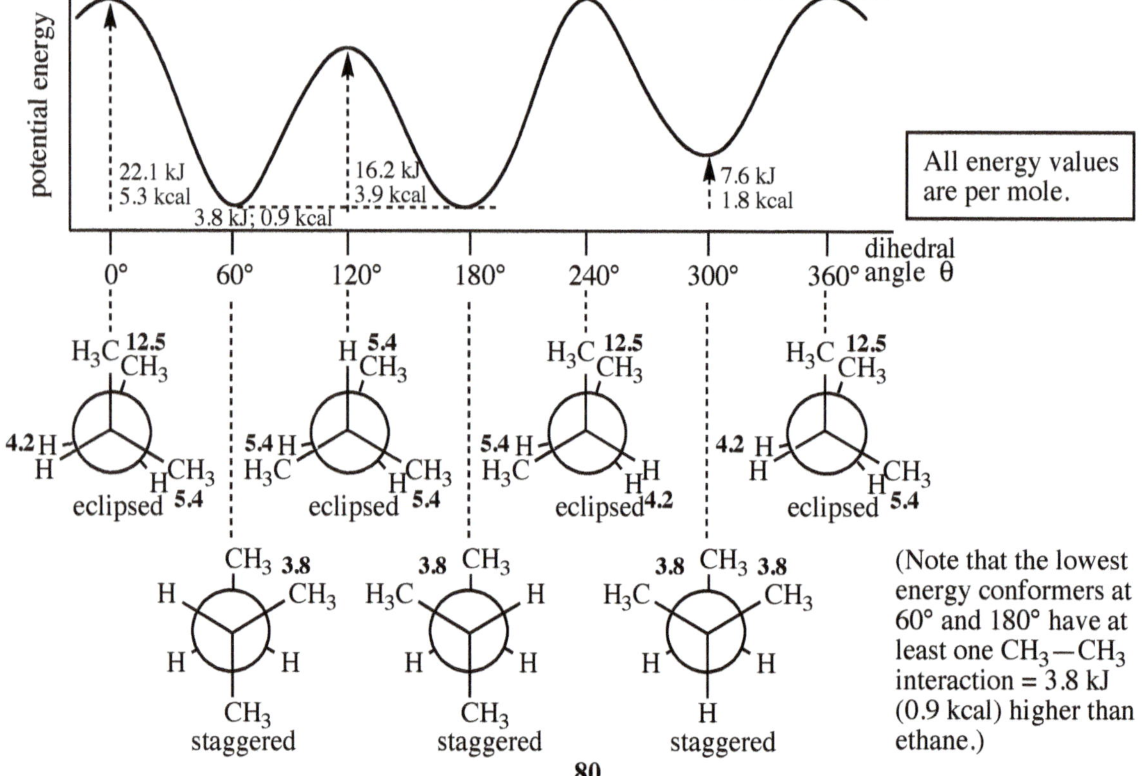

3-14

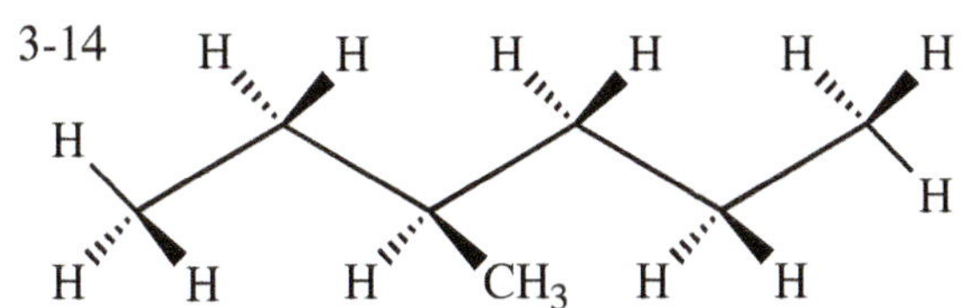

All bonds are staggered.

▬ Wedge bonds are coming toward the reader above the plane of the paper.

⋯ Dashed bonds are going away from the reader behind the plane of the paper.

3-15 Note that in rings, parts (a) and (b), two substituents on one carbon indicate where to begin numbering.
(a) 3-*sec*-butyl-1,1-dimethylcyclopentane
(b) 3-cyclopropyl-1,1-dimethylcyclohexane
(c) 4-cyclobutylnonane (The chain is longer than the ring.)

3-16

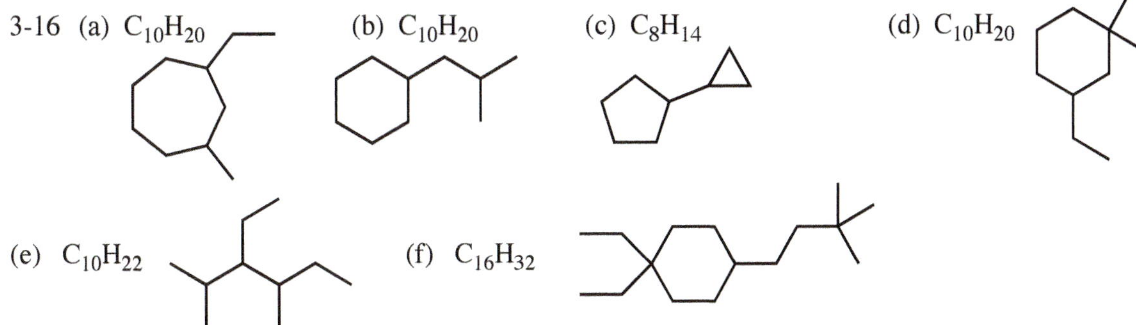

3-17 It helps to visualize *cis-trans* isomerism by putting in the H atoms on the carbons with substituents.
(a) no *cis-trans* isomerism possible

(b)

H, 3, H, 2, CH_3, 1, CH_2CH_3 — *cis*

and

CH_3, 3, H, 2, H, 1, CH_2CH_3 — *trans*

The terms *cis* and *trans* designate *relative* position. For *cis*, both substituents have to be on the same side; it does not matter if both are up or both down. For *trans*, one must be up and the other down; it does not matter which is which.

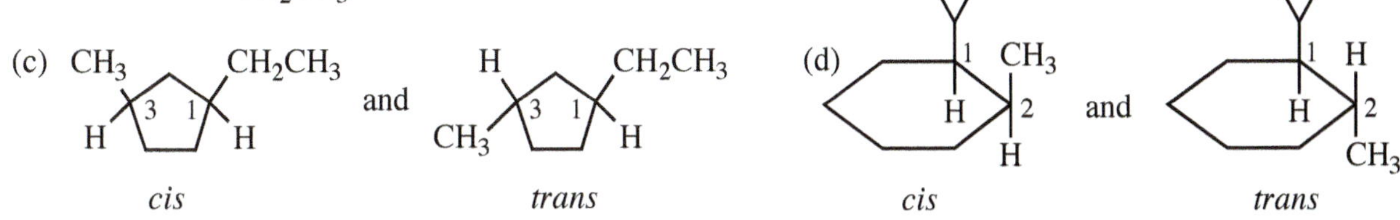

cis *trans* *cis* *trans*

3-18 In (a) and (b), numbering of the ring is determined by the first group *alphabetically* being assigned to ring carbon 1.

(a) *cis*-1-methyl-3-propylcyclobutane ("M" comes before "p"—practice that alphabet!)
(b) *trans*-1-*tert*-butyl-3-ethylcyclohexane (*Tert*, *sec*, and *n* are ignored in assigning alphabetical priority.)
(c) *trans*-1,2-dimethylcyclopropane (Either carbon with a CH_3 could be carbon-1; the same name results.)

3-19 Combustion of the *cis* isomer gives off more energy, so *cis*-1,2-dimethylcyclopropane must start at a higher energy than the *trans* isomer. The Newman projection of the *cis* isomer shows the two methyls are eclipsed with each other; in the *trans* isomer, the methyls are still eclipsed, but with hydrogens, not each other—a lower energy interaction.

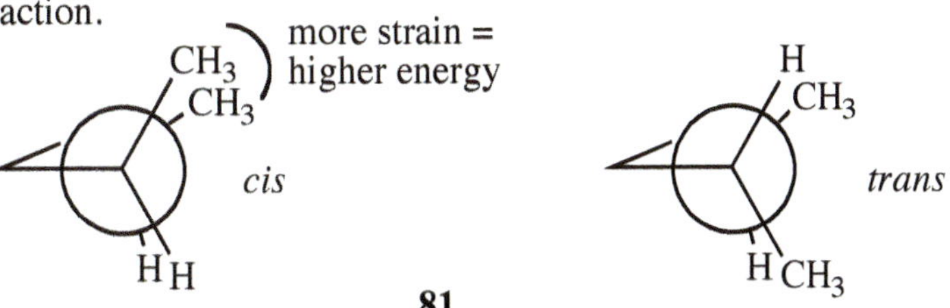

3-20 *trans*-1,2-Dimethylcyclobutane is more stable than *cis* because the two methyls can be farther apart when *trans*, as shown in the Newman projections.

more strain = higher energy

CH_3 CH_3 H H *cis*

H CH_3 H CH_3 *trans*

In the 1,3-dimethylcyclobutanes, however, the *cis* allows the methyls to be farther from other atoms and therefore more stable than the *trans*.

CH_3 CH_3 H H *cis*

H CH_3 CH_3 ↔ H *trans*

3-21

equatorial only

showing both axial and equatorial

axial only

3-22 The abbreviation for a methyl group, CH_3, is "Me". Ethyl is "Et", propyl is "Pr", and butyl is "Bu".

(a)

all methyls axial (all H equatorial)

Note that axial groups alternate up and down around the ring.

(b)

all methyls equatorial (all H axial)

3-23 Carbons 4 and 6 are the back carbons.

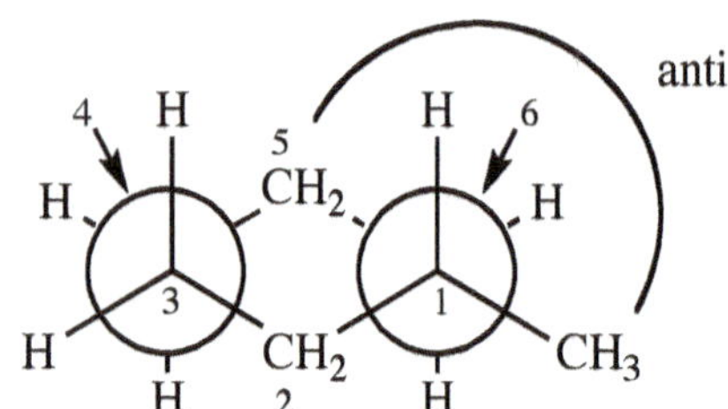

3-24 The isopropyl group can rotate so that its hydrogen is near the axial hydrogens on carbons 3 and 5, similar to a methyl group's hydrogen, and therefore similar to a methyl group in energy. The *tert*-butyl group, however, must point a methyl group toward the hydrogens on carbons 3 and 5, giving severe diaxial interactions, causing the energy of this conformer to jump dramatically.

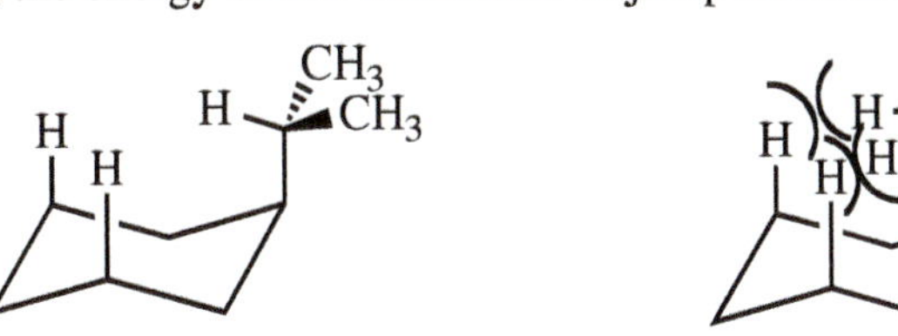

isopropylcyclohexane

tert-butylcyclohexane

3-25 The most stable conformers have larger substituents equatorial.

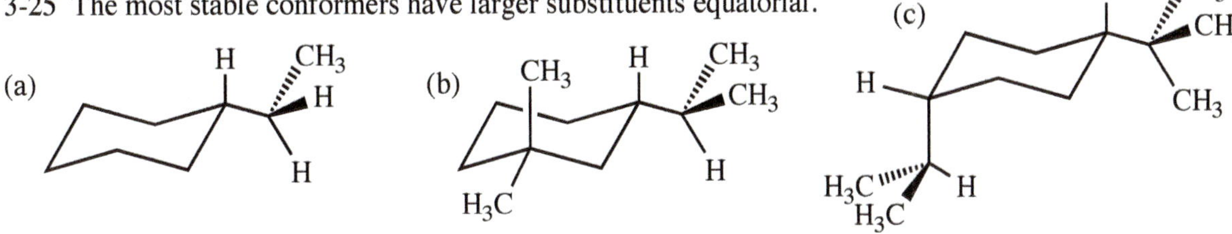

The *cis*-1,4 isomer must have one group axial and the other equatorial. The *tert*-butyl group is larger and will take the equatorial position.

3-26 The key to determining *cis* and *trans* around a cyclohexane ring is to see whether a substituent group is "up" or "down" *relative to the H at the same carbon*. Two "up" groups or two "down" groups will be *cis*; one "up" and one "down" will be *trans*. This works independent of the conformation the molecule is in!

(a) *cis*-1,3-dimethylcyclohexane
(b) *cis*-1,4-dimethylcyclohexane
(c) *trans*-1,2-dimethylcyclohexane
(d) *cis*-1,3-dimethylcyclohexane
(e) *cis*-1,3-dimethylcyclohexane
(f) *trans*-1,4-dimethylcyclohexane

3-27

(a) *cis*

(b) *trans*

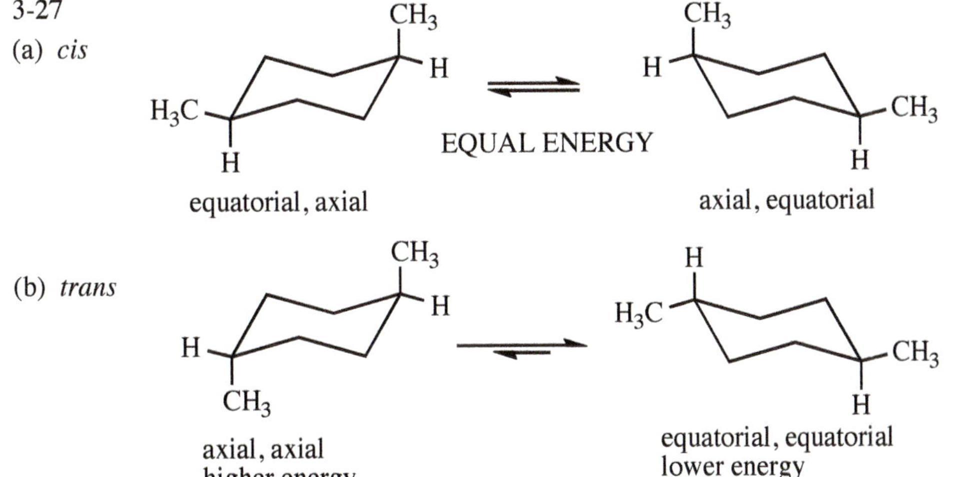

(c) The *trans* isomer is more stable because BOTH substituents can be in the preferred equatorial positions.

3-28

Positions	*cis*	*trans*
1,2	(e,a) or (a,e)	(e,e) or (a,a)
1,3	(e,e) or (a,a)	(e,a) or (a,e)
1,4	(e,a) or (a,e)	(e,e) or (a,a)

3-29 The more stable conformer places the larger group equatorial.

(a)

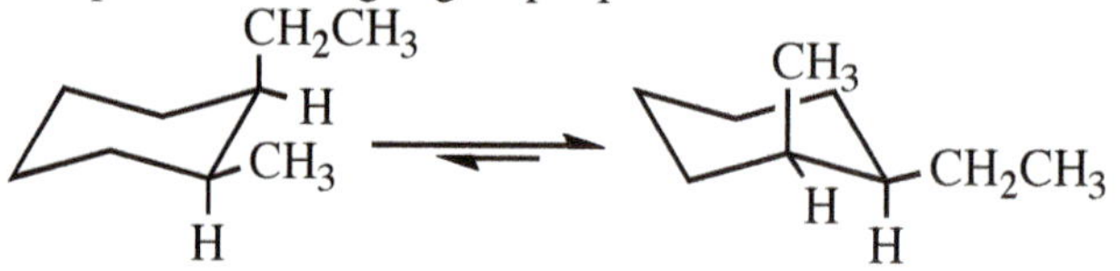

CH_2CH_3 H CH_3 H ⇌ CH_3 CH_2CH_3 H H

more stable—
larger group equatorial

(b)

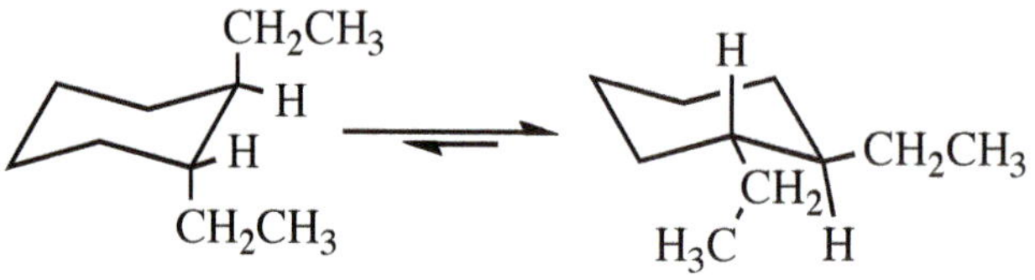

CH_2CH_3 H H CH_2CH_3 ⇌ H CH_2CH_3 CH_2 H_3C H

more stable—
both groups equatorial

(c) more stable—
larger group equatorial

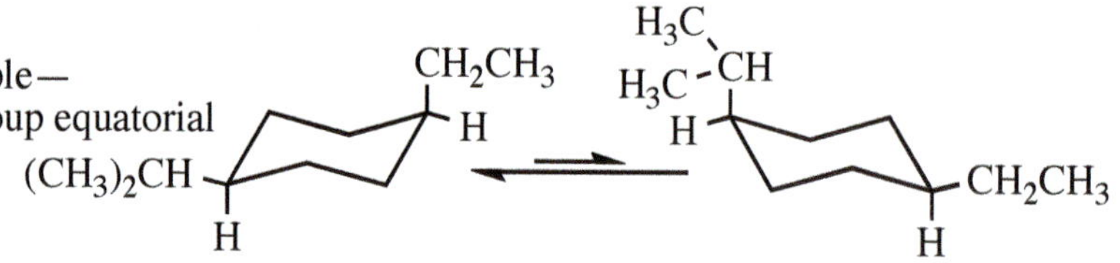

$(CH_3)_2CH$ H CH_2CH_3 H ⇌ H_3C H_3C-CH H CH_2CH_3 H

(d)

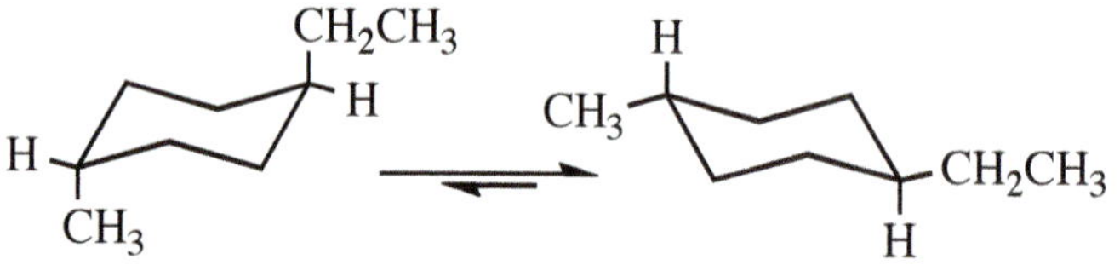

CH_2CH_3 H H CH_3 ⇌ H CH_3 CH_2CH_3 H

more stable—
both groups equatorial

3-30

(a) H H $C(CH_3)_3$ CH_2CH_3

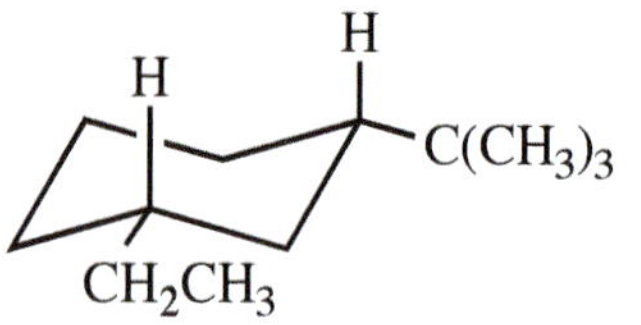

(b) H $C(CH_3)_3$ CH_3 H

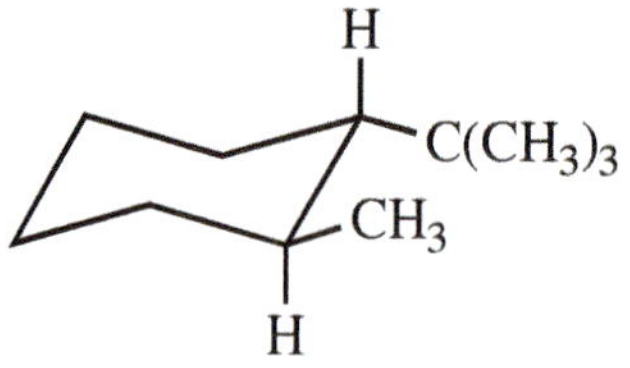

(c) Bulky substituents like *tert*-butyl adopt equatorial rather than axial positions, even if that means altering the conformation of the ring. The twist boat conformation allows both bulky substituents to be "equatorial".

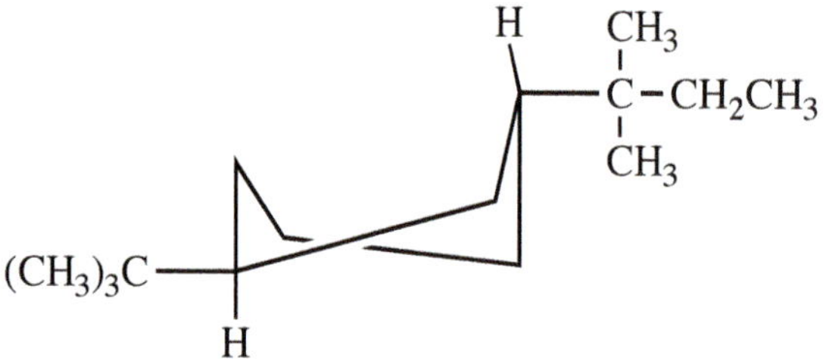

H CH_3 C–CH_2CH_3 CH_3 $(CH_3)_3C$ H

3-31 The nomenclature of bicyclic alkanes is summarized in Appendix 5 in the textbook.

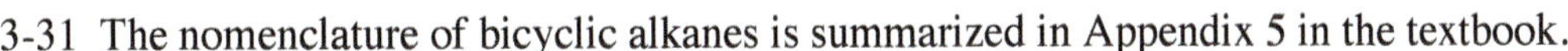

(a) bicyclo[3.1.0]hexane (b) bicyclo[3.3.1]nonane (c) bicyclo[2.2.2]octane (d) bicyclo[3.1.1]heptane

3-32 Using models is essential for this problem.

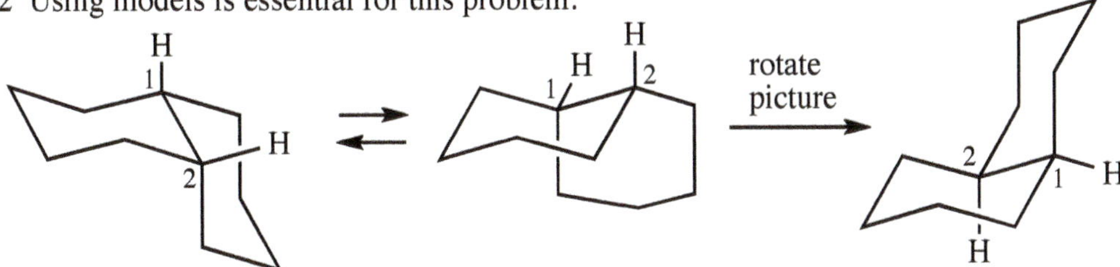

from text Figure 3-27

3-33

(a) Here are eighteen isomers of C_8H_{18}. An easy way to compare is to name yours and see if the names match. Note how these isomers were generated systematically: 8C chain (only one), all the possible 7C chains with one methyl, all the 6C chains with two methyls or 1 ethyl, etc.

8-carbon chain

octane

7-carbon chain

2-methylheptane 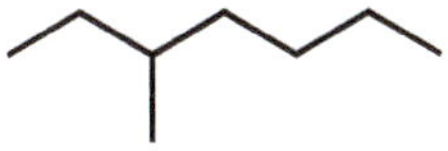 3-methylheptane  4-methylheptane

6-carbon chain

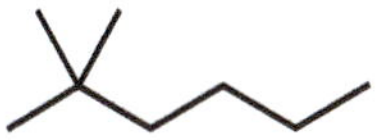

2,2-dimethylhexane 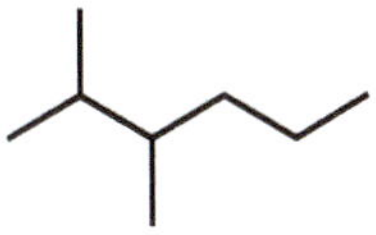 2,3-dimethylhexane 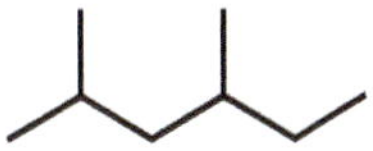 2,4-dimethylhexane 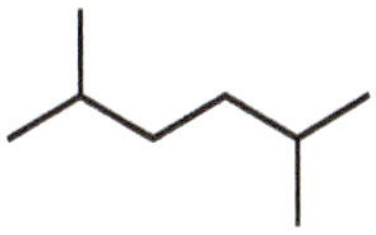 2,5-dimethylhexane

3,3-dimethylhexane 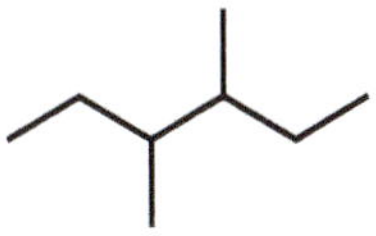 3,4-dimethylhexane 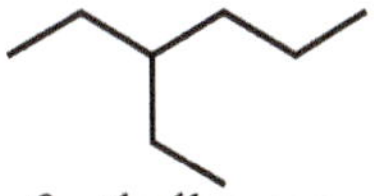 3-ethylhexane

5-carbon chain

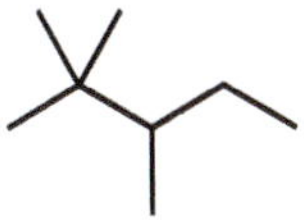

2,2,3-trimethylpentane 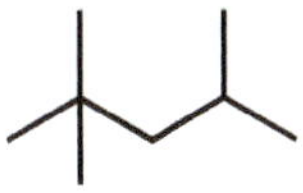 2,2,4-trimethylpentane 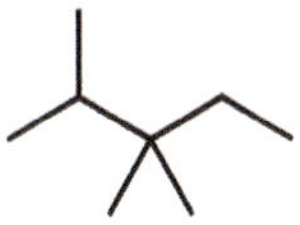 2,3,3-trimethylpentane 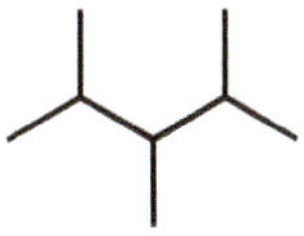 2,3,4-trimethylpentane

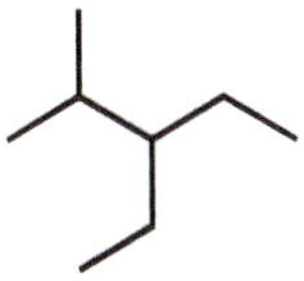

3-ethyl-2-methylpentane 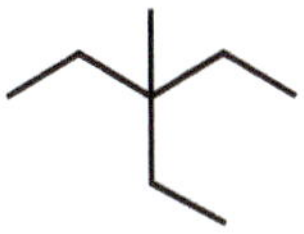 3-ethyl-3-methylpentane

4-carbon chain

2,2,3,3-tetramethylbutane

3-33 continued

(b)

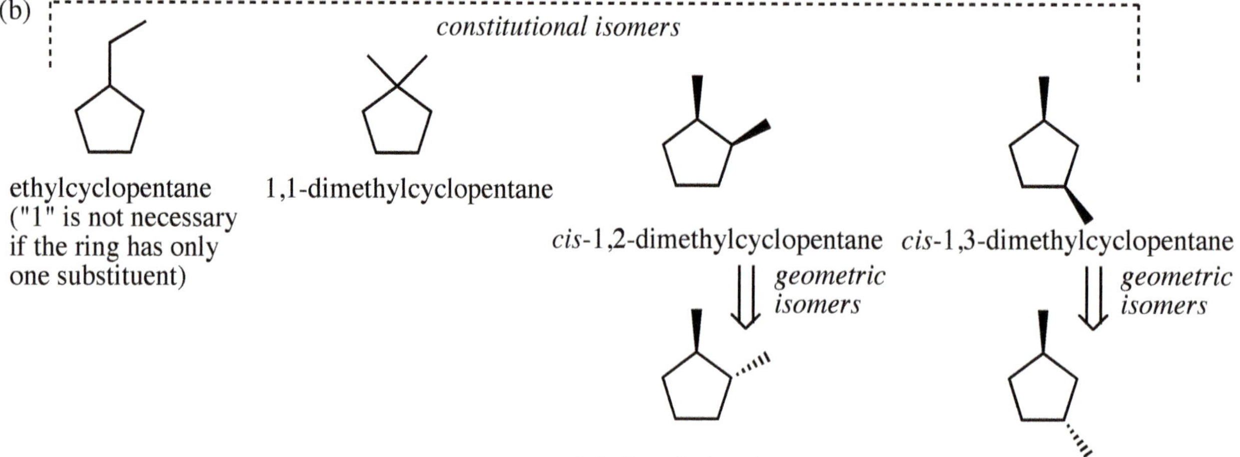

3-34

(a) The third structure is 2-methylpropane (isobutane). The other four structures are all butane (*n*-butane). Remember that a compound's identity is determined by how the atoms are connected, not by the position of the atoms when a structure is drawn on a page.

(b) The first and fourth structures are both *cis*-but-2-ene. The second and fifth structures are both but-1-ene. The third structure is *trans*-but-2-ene. The last structure is 2-methylpropene.

(c) The first two structures are both *cis*-1,2-dimethylcyclopentane. The next two structures are both *trans*-1,2-dimethylcyclopentane. The last structure, *cis*-1,3-dimethylcyclopentane, is different from all the others.

(d)

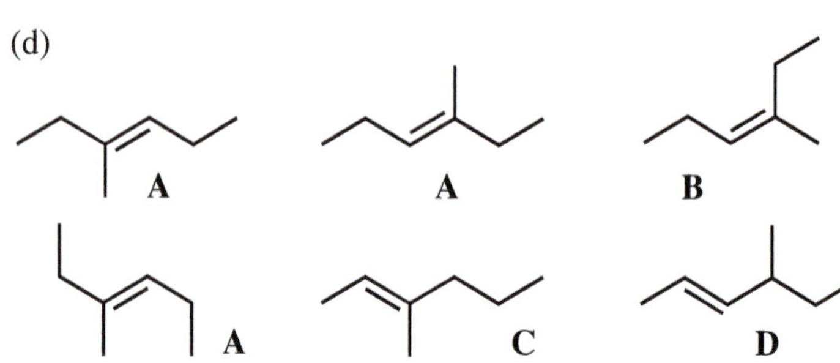

Analysis of the structures shows that some double bonds begin at carbon-2 and some at carbon-3 of the longest chain.

The three structures labeled **A** are the same, with the double bond *trans*; **B** is a geometric isomer (*cis*) of **A**. **C** and **D** are constitutional isomers of the others.

(e) Naming the structures shows that three of the structures are *trans*-1,4-dimethylcyclohexane, two are the *cis* isomer, and one is *cis*-1,3-dimethylcyclohexane. Although a structure may be shown in two different conformations, it still represents only one compound.

cis-1,4-dimethyl   *trans*-1,4-dimethyl   *cis*-1,4-dimethyl

cis-1,3-dimethyl   *trans*-1,4-dimethyl   *trans*-1,4-dimethyl

3-34 continued

(f) An important and useful skill is the ability to translate Newman projections into other structural formulas and *vice versa*. The way to do this is to start with the two carbons that represent the front and back carbons in the middle of the Newman projections, then attach the three groups emanating from each one.

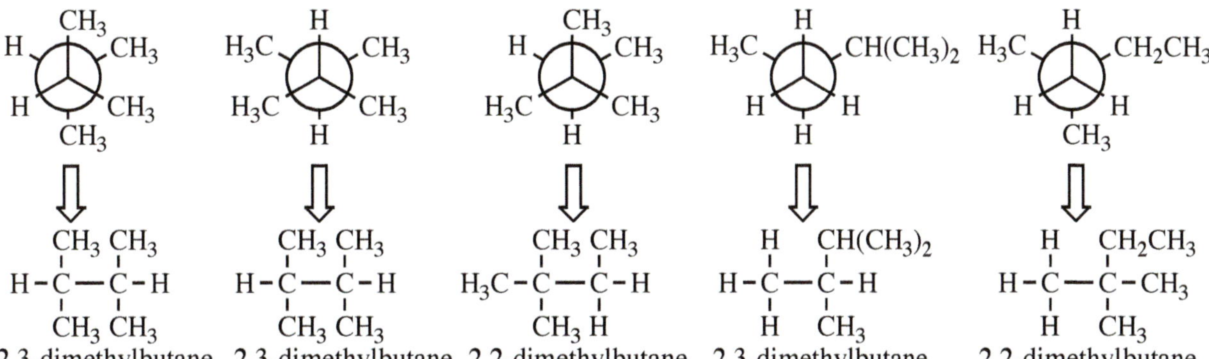

What may not be clear in the Newman projections is revealed in the structural formulas and especially in the names. There are only two compounds represented here: the first, second, and fourth are the same; the third and fifth are the same.

(g) Names have been incorporated into the other solutions.

3-35

(a) Start with a 5-carbon ring and work to smaller rings:

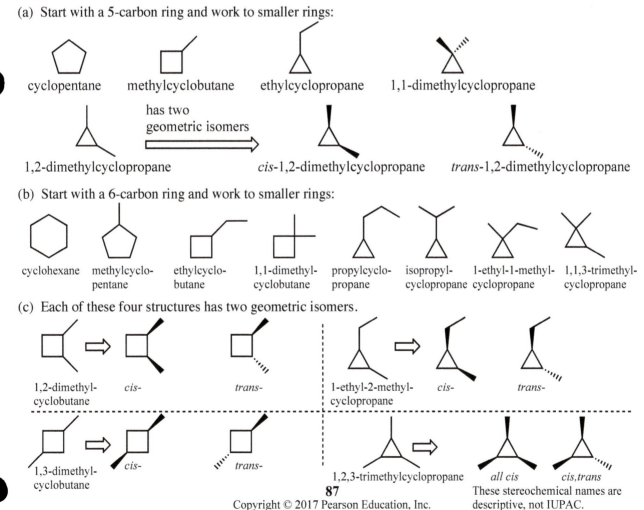

cyclopentane methylcyclobutane ethylcyclopropane 1,1-dimethylcyclopropane

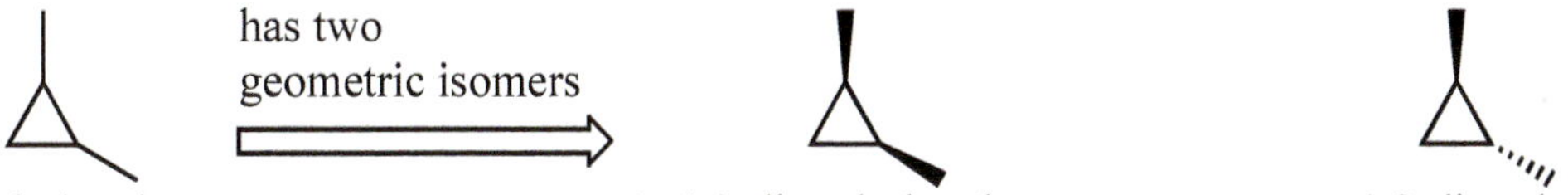

1,2-dimethylcyclopropane has two geometric isomers *cis*-1,2-dimethylcyclopropane *trans*-1,2-dimethylcyclopropane

(b) Start with a 6-carbon ring and work to smaller rings:

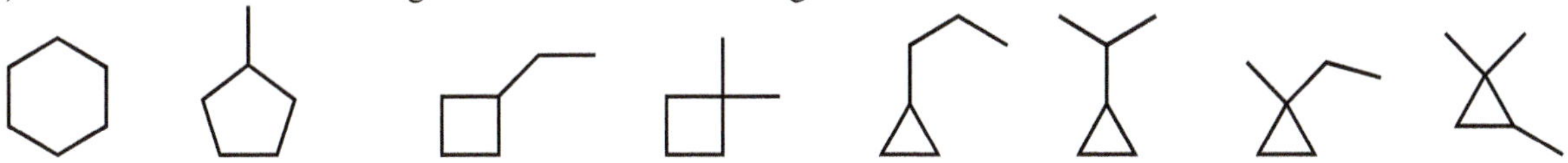

cyclohexane methylcyclopentane ethylcyclobutane 1,1-dimethylcyclobutane propylcyclopropane isopropylcyclopropane 1-ethyl-1-methylcyclopropane 1,1,3-trimethylcyclopropane

(c) Each of these four structures has two geometric isomers.

1,2-dimethylcyclobutane *cis-* *trans-* | 1-ethyl-2-methylcyclopropane *cis-* *trans-*

1,3-dimethylcyclobutane *cis-* *trans-* | 1,2,3-trimethylcyclopropane *all cis* *cis,trans*

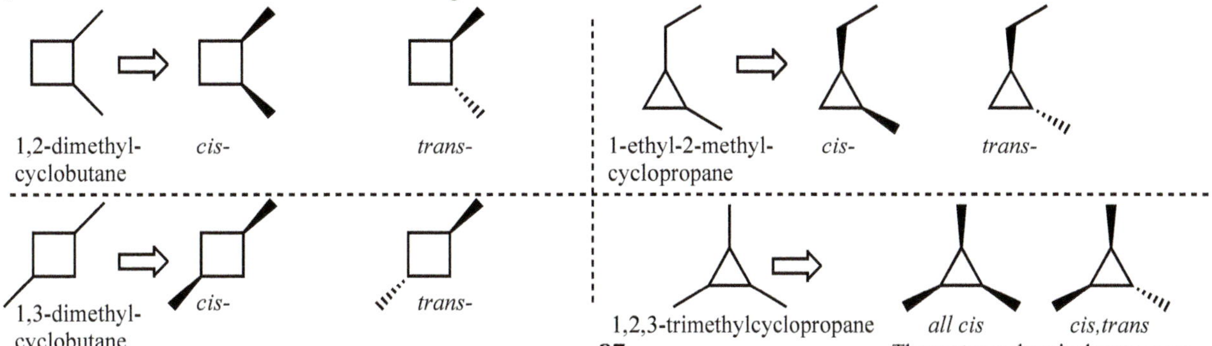

These stereochemical names are descriptive, not IUPAC.

3-36 Line–angle formulas are shown.

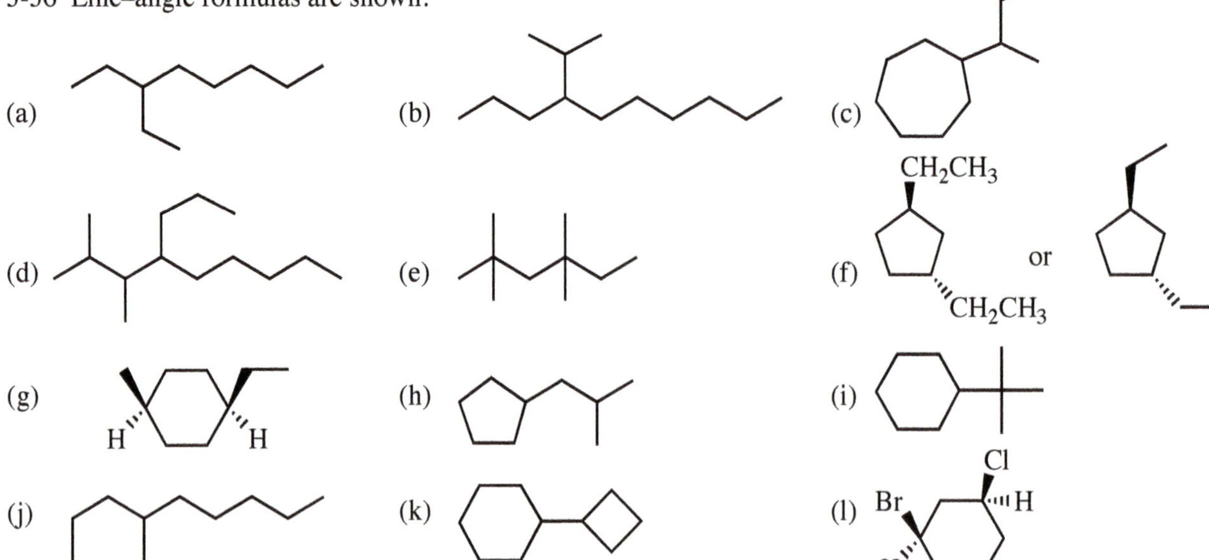

3-37 There are often many possible answers to this type of problem, although some can have unique solutions. The ones shown here are examples of correct answers. Your answers may be different AND correct. Check your answers in your study group.

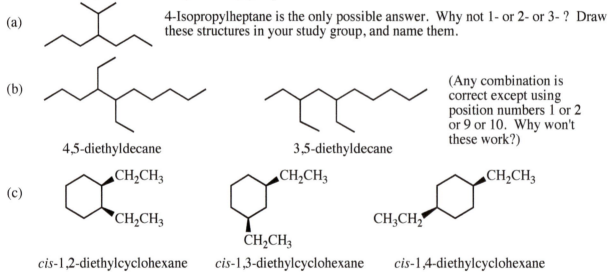

(a) 4-Isopropylheptane is the only possible answer. Why not 1- or 2- or 3- ? Draw these structures in your study group, and name them.

(b) (Any combination is correct except using position numbers 1 or 2 or 9 or 10. Why won't these work?)

4,5-diethyldecane   3,5-diethyldecane

(c) CH_2CH_3 CH_2CH_3 CH_2CH_3 CH_2CH_3 CH_2CH_3 CH_3CH_2

cis-1,2-diethylcyclohexane   *cis*-1,3-diethylcyclohexane   *cis*-1,4-diethylcyclohexane

NOT 1,5—same as 1,3; NOT 1,6—same as 1,2

(d) "Halo" is the abbreviation for any of the four halogens (F, Cl, Br, I), and "X" is the symbol used in organic structures. If unspecified, any of the four elements will satisfy the "halo" group. However, on a 5-membered ring, the substitution can be only two possibilities: 1,2- or 1,3- .

"X" stands for any of the four halogen elements.

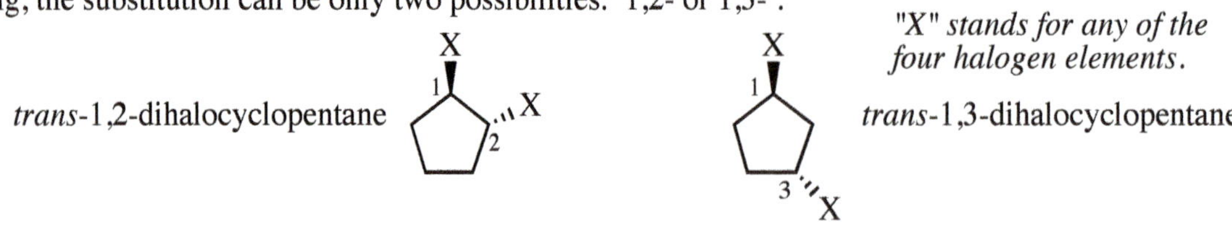

trans-1,2-dihalocyclopentane   *trans*-1,3-dihalocyclopentane

3-37 continued

(e)

(2,3-dimethylpentyl)cycloheptane (2,3-dimethylpentyl)cyclooctane

Other ring sizes are possible, although they must have 6 or more carbons to be longer than the 5 carbons of the substituent chain.

(f)

bicyclo[4.3.0]nonane

bicyclo[3.2.2]nonane bicyclo[3.3.1]nonane

Any combination where the number of carbons in the bridges sums to 7 will work. (Two carbons are the bridgehead carbons.)

3-38 $HO-CH_3$ $HO-CH_2CH_2CH_3$ $HO-CH_2CH_2CH_2CH_2CH_3$

$HO-CH_2CH_3$ $HO-CH_2CH_2CH_2CH_3$ $HO-CH_2CH_2CH_2CH_2CH_2CH_3$

3-39

(a) 3-ethyl-2,2,6-trimethylheptane
(b) 3-ethyl-2,6,7-trimethyloctane
(c) 3,7-diethyl-2,2,8-trimethyldecane
(d) 1,1-diethyl-2-methylcyclobutane
(e) bicyclo[4.1.0]heptane
(f) *cis*-1-ethyl-3-propylcyclopentane
(g) (1,1-diethylpropyl)cyclohexane
(h) *cis*-1-ethyl-4-isopropylcyclodecane

3-40 The ethyl group, $-CH_2CH_3$, is abbreviated Et. The back carbon is kept constant while the front carbon is rotated clockwise. The problem asks for relative energies, not calculations; the ethyl group is slightly larger than a methyl group, causing a higher energy than methyl in eclipsed conformations.

Interestingly, all of the minima and all of the maxima are at different energies; this arises because the three groups on the back carbon (H, CH_3, CH_2CH_3) are all different.

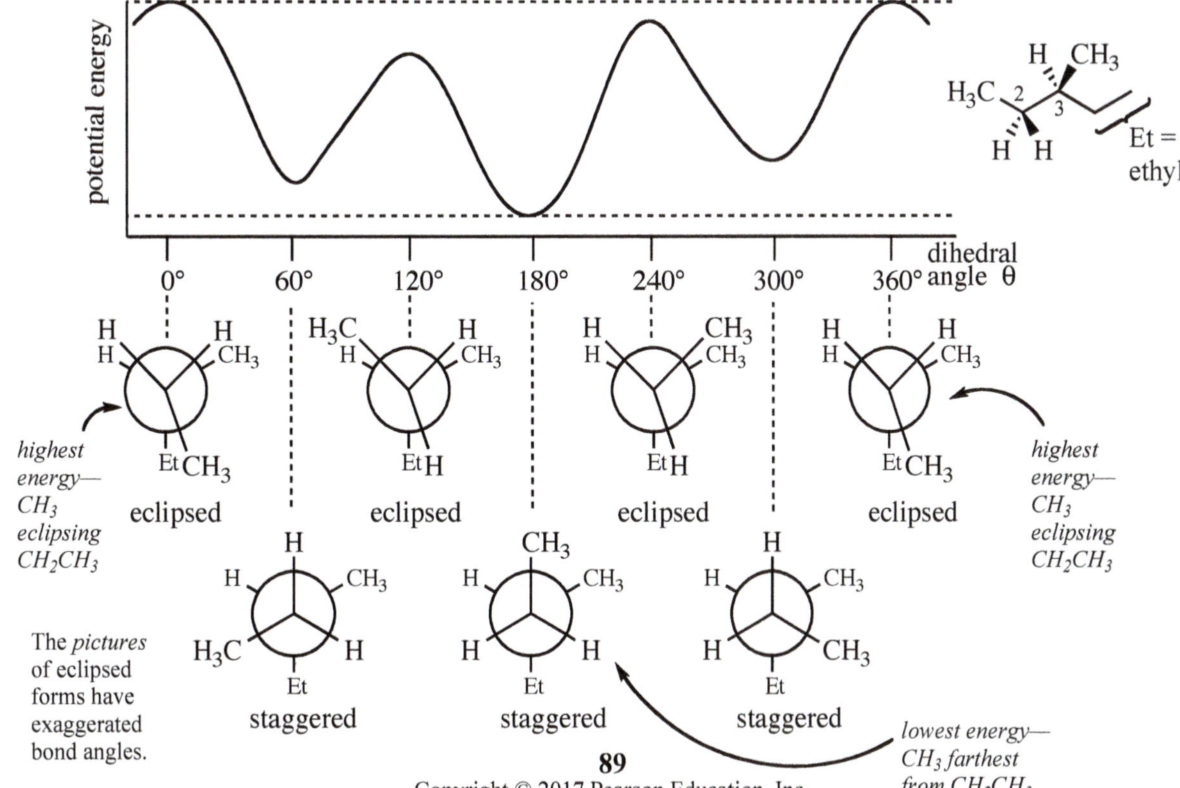

3-41

(a) correct name: 3-methylhexane
(longer chain)

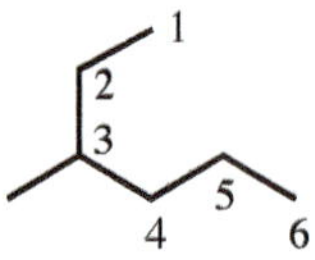

(b) correct name: 3-ethyl-2-methylhexane
(more branching with this numbering)

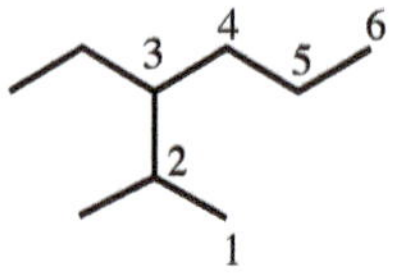

(c) correct name: 2-chloro-3-methylhexane
(Begin numbering at end closest to substituent.)

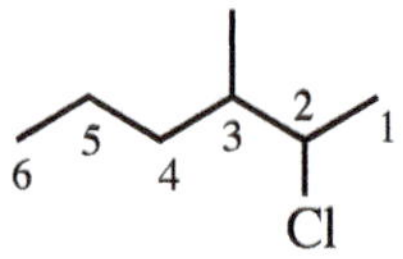

(d) correct name: 2,2-dimethylbutane (Include a position number for each substituent, regardless of redundancies.)

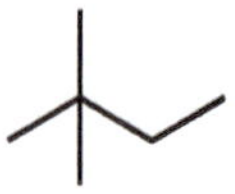

(e) correct name: *sec*-butylcyclohexane or (1-methylpropyl)cyclohexane
(The longer chain or ring is the base name.)

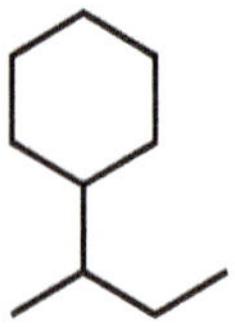

(f) CH_2CH_3 / CH_2CH_3

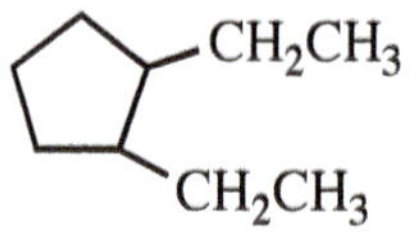

correct name: 1,2-diethylcyclopentane
(Position numbers are the lowest possible.)
cis- or *trans*- is not specified.

3-42

(a) Octane has a higher boiling point than 2,2,3-trimethylpentane because linear molecules boil higher than branched molecules of the same molecular weight (increased van der Waals interaction).
(b) Nonane has a higher boiling point than 2-methylheptane because nonane has a higher molecular weight and it is a linear structure as opposed to the lighter, branched 2-methylheptane.
(c) Nonane boils higher than 2,2,5-trimethylhexane for the same reason as in (a).

3-43 The point of attachment is shown by the bold bond at the left of each structure.

$—CH_2CH_2CH_2CH_2CH_3$ 1° pentyl (old: amyl)

$—CHCH_2CH_2CH_3$ (with CH_3 on C1) 2° 1-methylbutyl

$—CH_2CHCH_2CH_3$ (with CH_3 on C2) 1° 2-methylbutyl

$—CH_2CH_2CHCH_3$ (with CH_3 on C3) 1° 3-methylbutyl (isopentyl; old: isoamyl)

$—CHCH_2CH_3$ (with CH_2CH_3 on C1) 2° 1-ethylpropyl

$—C(CH_3)_2–CH_2CH_3$ 3° 1,1-dimethylpropyl (*tert*-pentyl; old: *tert*-amyl)

$—CHCHCH_3$ (with CH_3 on C1 and CH_3 on C2) 2° 1,2-dimethylpropyl

$—CH_2–C(CH_3)_2–CH_3$ 1° 2,2-dimethylpropyl (*neo*-pentyl)

3-44 In each case, put the largest groups on adjacent carbons in anti positions to make the most stable conformations.

(a) 3-methylpentane

C-2 is the front carbon with H, H, and CH_3
C-3 is the back carbon with H, CH_3, and CH_2CH_3

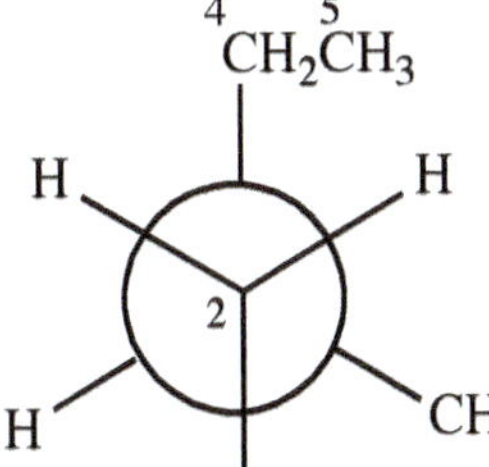

Newman projections were introduced in text section 3-7. They are used extensively to show the three-dimensional arrangement of atoms in a molecule.

Carbon-3 is behind carbon-2.

(b) 3,3-dimethylhexane

C-3 is the front carbon with CH_3, CH_3, and CH_2CH_3
C-4 is the back carbon with H, H, and CH_2CH_3

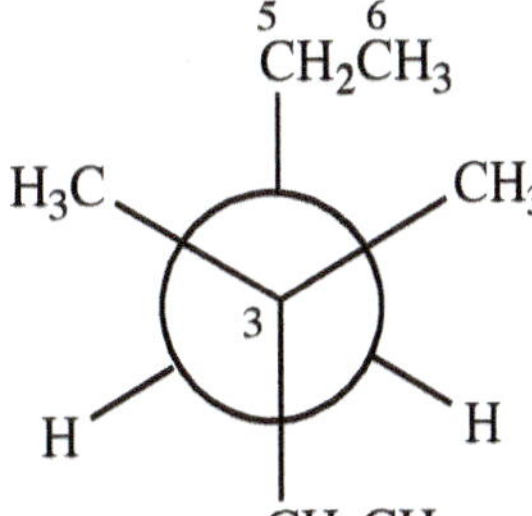

Carbon-4 is behind carbon-3.

3-45
(a)

axial H, axial H; CH_3 equatorial, CH_3 equatorial — equatorial H, H equatorial; CH_3 axial, CH_3 axial

(b) more stable (lower energy) — less stable (higher energy)

(c) From Section 3-14 of the text, each gauche interaction raises the energy 3.8 kJ/mole (0.9 kcal/mole), and each axial methyl has two gauche interactions, so the energy is:
2 methyls x 2 interactions per methyl x 3.8 kJ/mole per interaction = 15.2 kJ/mole (3.6 kcal/mole)
(d) The steric strain from the 1,3-diaxial interaction of the methyls must be the difference between the total energy and the energy due to gauche interactions: 23 kJ/mole − 15.2 kJ/mole = 7.8 kJ/mole
(5.4 kcal/mole − 3.6 kcal/mole = 1.8 kcal/mole)

3-46 The more stable conformer places the larger group equatorial.

(a) ax $CH(CH_3)_2$, H, CH_2CH_3, H eq ⇌ ax CH_2CH_3, H, $CH(CH_3)_2$, H eq
more stable—
larger group equatorial

(b) ax $CH(CH_3)_2$, H, H, CH_2CH_3 ax ⇌ H, eq $CH(CH_3)_2$, Et, eq H
Et = ethyl
more stable—
both groups equatorial

(c) ax CH_3, ax CH_2CH_3, H, H ⇌ eq CH_3, eq CH_2CH_3, H, H
more stable—
both groups equatorial

(d) H, ax CH_2CH_3, H, CH_3 eq ⇌ H, eq CH_2CH_3, ax CH_3, H
more stable—
larger group equatorial

3-46 continued

(e) ax CH_2CH_3, H; CH_3 eq, H ⇌ ax CH_3, H; eq CH_2CH_3, H

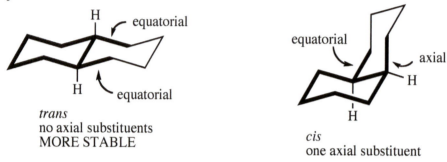

more stable—
larger group equatorial

(f) ax CH_2CH_3, H; H, CH_3 ax ⇌ H, CH_3 eq; eq CH_2CH_3, H

more stable—
both groups equatorial

3-47 (Using models is essential to this problem.)

In both *cis*- and *trans*-decalin, the cyclohexane rings can be in chair conformations. The relative energies will depend on the number of axial substituents.

H equatorial, H equatorial

trans
no axial substituents
MORE STABLE

equatorial, axial, H, H

cis
one axial substituent

3-48 Solutions are given in the boxes. The front and back carbons from the Newman projection are denoted by a dark dot.

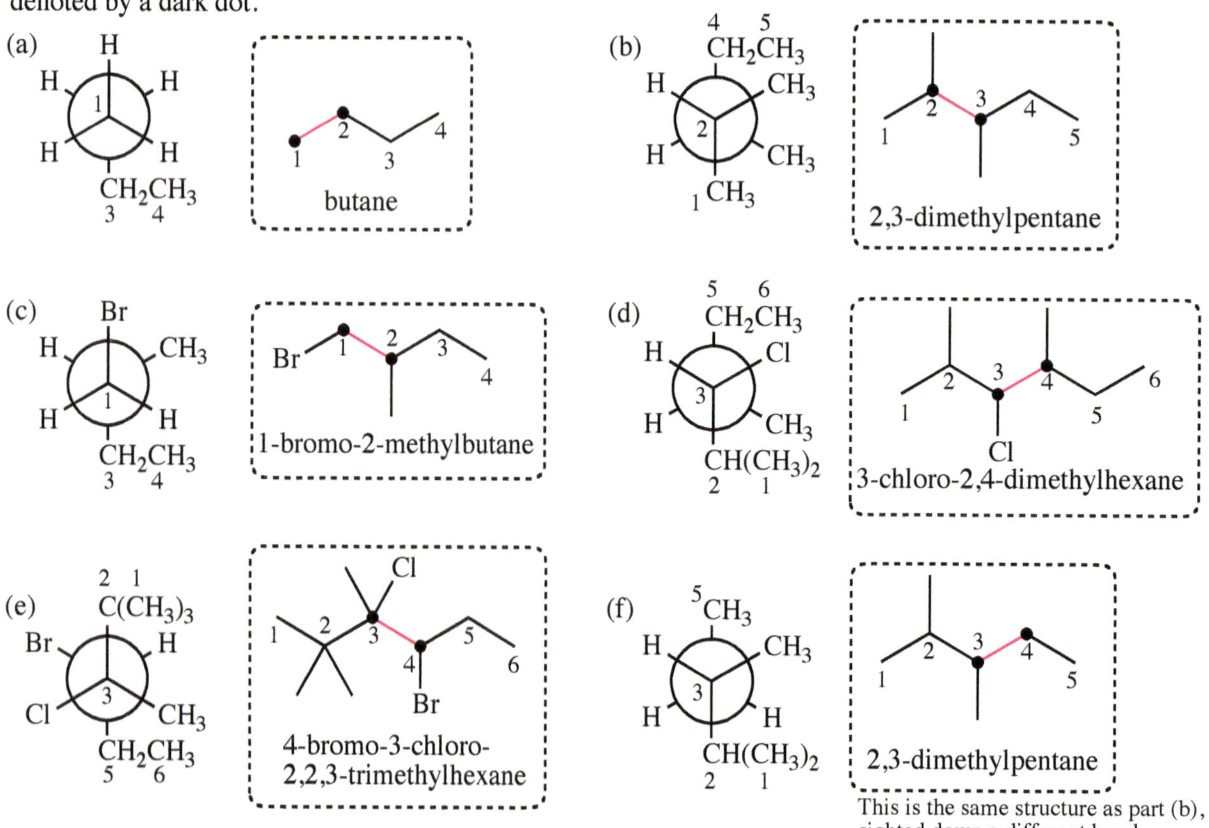

This is the same structure as part (b), sighted down a different bond.

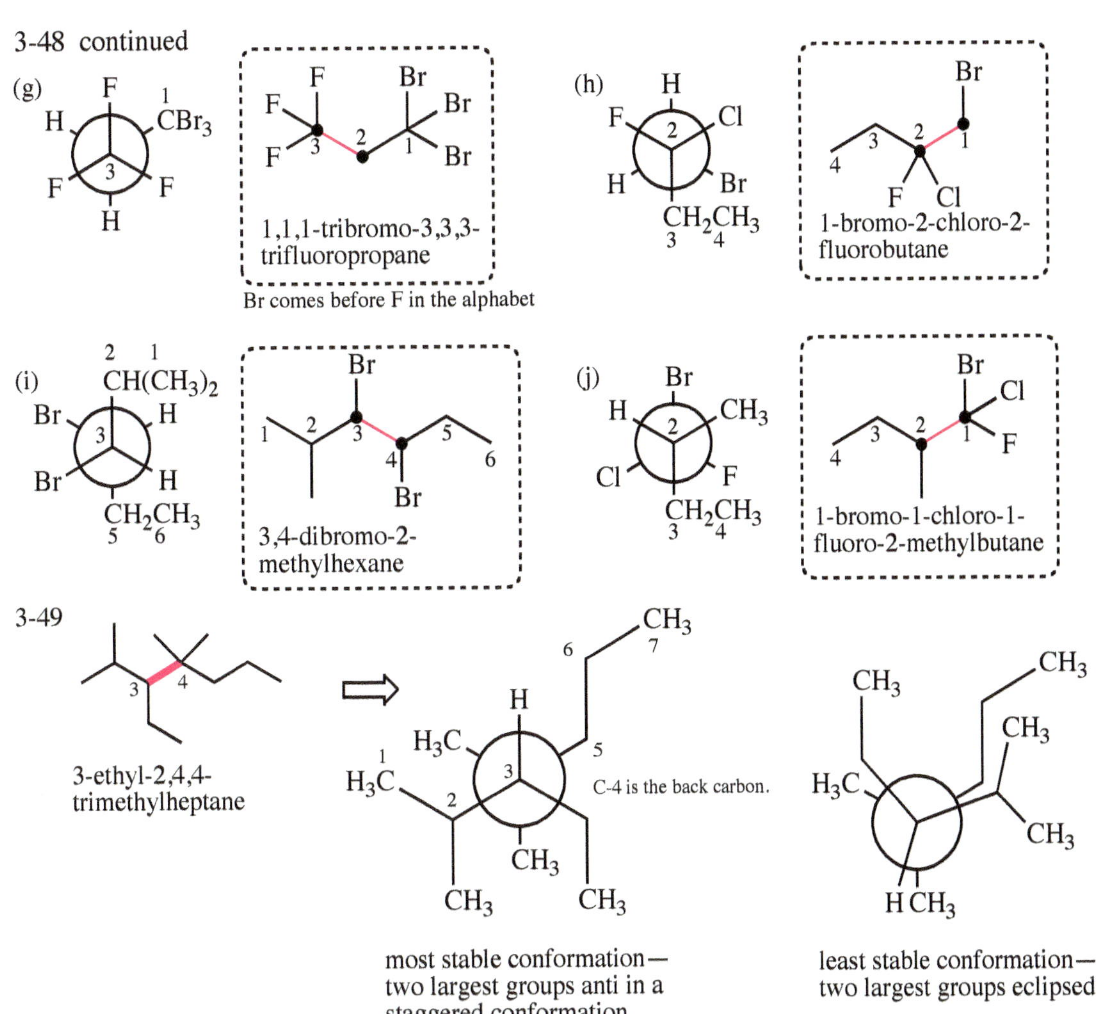

3-50 Usually, the anti conformation is more stable than the gauche because the two larger groups are farther away in the anti conformer. There must be something special about the gauche conformer, some attractive force that overcomes the normal repulsion between groups in the gauche conformation. What could it be?

anti *gauche*

Hydrogen bonding! Moving the two OH groups closer together permits the H of one to come close to the O of the other, as can be seen in the picture below. Such intramolecular (within one molecule) hydrogen bonding cannot take place in the anti conformation where the two OH groups are too far apart.

intramolecular hydrogen bonding, a strong attractive force

gauche

3-51 chair form of glucose with all substituents equatorial

(without ring H atoms shown)

3-52

Begin this problem by labeling the important parts of the structure given in the problem.

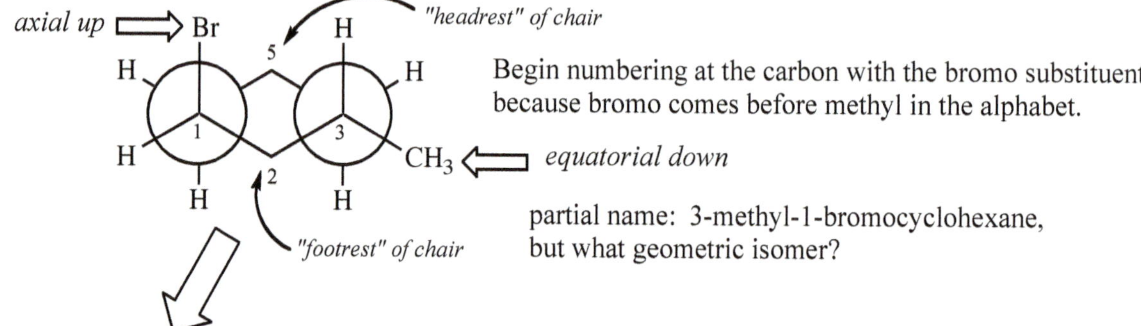

(a)

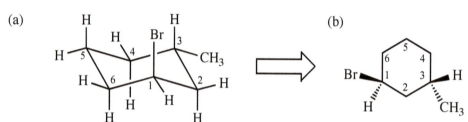

(b)

(c) *trans*-1-bromo-3-methylcyclohexane

Note to the student: Creating practice problems for your study group is very helpful in learning new material, and in some areas, it is not difficult. For example, to create nomenclature problems, draw a cyclopentane, add three substituents, and name it. Then erase one of the ring bonds and name that acyclic structure. Go around the ring, erasing one bond at a time, and you have generated five new nomenclature problems for your group to practice naming!

CHAPTER 4—THE STUDY OF CHEMICAL REACTIONS

4-1

(a) 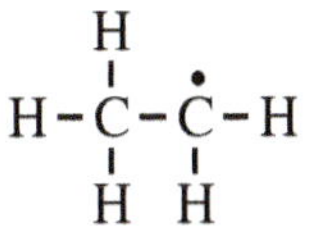
```
   H
   |  •
H–C–C–H
   |  |
   H  H
```

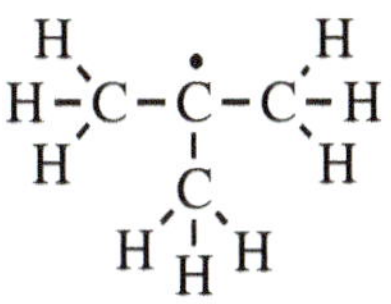

(b)
```
H       •    H
 H–C–C–C–H
H     |      H
       C
    H  H  H
```

(c)
```
   H  •  H
   |  |  |
H–C–C–C–H
   |  |  |
   H  H  H
```

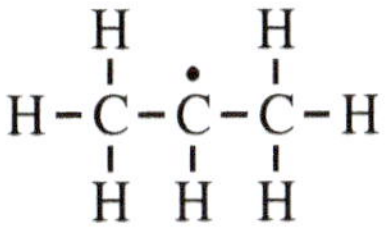

(d)  :Ï• (iodine atom with seven valence electrons)

4-2

(a)
```
     H                          H
     |                          |
Cl–C–H  +  Cl•   ——→    Cl–C•   +  H–Cl
     |                          |
     H                          H

     H                          H
     |                          |
Cl–C•   +  Cl–Cl  ——→   Cl–C–Cl   +  Cl•
     |                          |
     H                          H
```

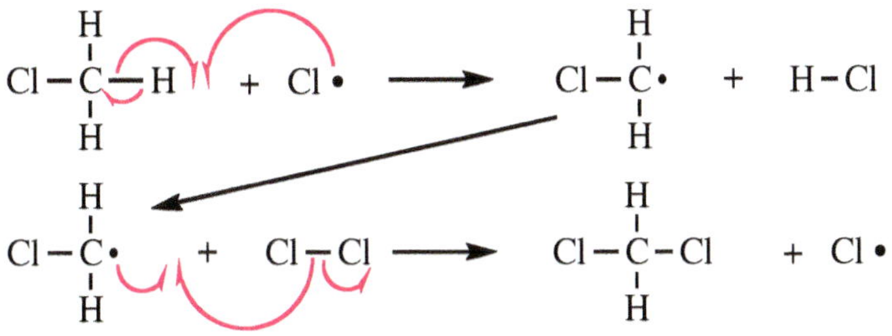

(b) Free-radical halogenation substitutes a halogen atom for a hydrogen. Even if a molecule has only one type of hydrogen, substitution of the first of these hydrogens forms a new compound. Any remaining hydrogens in this product can compete with the initial reactant for the available halogen. Thus, chlorination of methane, CH_4, produces all possible substitution products: CH_3Cl, CH_2Cl_2, $CHCl_3$, and CCl_4.

If a molecule has different types of hydrogens, the reaction can generate a mixture of the possible substitution products.

(c) Production of CCl_4 or CH_3Cl can be controlled by altering the ratio of CH_4 to Cl_2. To produce CCl_4, use an excess of Cl_2 and let the reaction proceed until all C—H bonds have been replaced with C—Cl bonds. Producing CH_3Cl is more challenging because the reaction tends to proceed past the first substitution. By using a very large excess of CH_4 to Cl_2, perhaps 100 to 1 or even more, a chlorine atom is more likely to find a CH_4 molecule than it is to find a CH_3Cl, so only a small amount of CH_4 is transformed to CH_3Cl by the time the Cl_2 runs out, with almost no CH_2Cl_2 being produced.

4-3

(a) This mechanism requires that one photon of light be added for each CH_3Cl generated, a quantum yield of 1. The actual quantum yield is several hundred or thousand. The high quantum yield suggests a chain reaction, but this mechanism is not a chain; it has no propagation steps.

(b) This mechanism conflicts with at least two experimental observations. First, the energy of light required to break a H–CH_3 bond is 439 kJ/mole (105 kcal/mole, from Table 4-2); the energy of light determined by experiment to initiate the reaction is only 242 kJ/mole of photons (58 kcal/mole of photons, from text section 4-3A), much less than the energy needed to break this H–C bond. Second, as in (a), each CH_3Cl produced would require one photon of light, a quantum yield of 1, instead of the actual number of several hundred or thousand. As in (a), there is no provision for a chain process, because all of the radicals generated are also consumed in the mechanism.

4-4

(a) The twelve hydrogens of cyclohexane are all on equivalent 2° carbons. Replacement of any one of the twelve will lead to the same product, chlorocyclohexane. Hexane, however, has hydrogens in three different positions: on carbon-1 (equivalent to carbon-6), carbon-2 (equivalent to carbon-5), and carbon-3 (equivalent to carbon-4). Monochlorination of hexane will produce a mixture of all three possible isomers: 1-, 2-, and 3-chlorohexane.

(b) The best conversion of cyclohexane to chlorocyclohexane would require the ratio of cyclohexane/chlorine to be a large number. If the ratio were small, as the concentration of chlorocyclohexane increased during the reaction, chlorine would begin to substitute for a second hydrogen of chlorocyclohexane, generating unwanted products. The goal is to have chlorine attack a molecule of cyclohexane before it ever encounters a molecule of chlorocyclohexane, so the concentration of cyclohexane should be kept high.

4-5

(a) $K_{eq} = e^{-\Delta G°/RT}$ ← -2.1 kJ/mole = -2100 J/mole

$= e^{-(-2100\ J/mole)/((8.314\ J/°K\text{-}mole) \bullet (298\ °K))}$

$= e^{2100/2478} = e^{0.847} =$ **2.3**

°K = temperature in the Kelvin scale
To be consistent with the textbook, and to avoid confusion with the symbol for an equilibrium constant, the degree symbol will be used with Kelvin temperatures.

(b) $K_{eq} = 2.3 = \dfrac{[CH_3SH]\,[HBr]}{[CH_3Br]\,[H_2S]}$

	$[CH_3Br]$	$[H_2S]$	$[CH_3SH]$	$[HBr]$
initial concentrations:	1	1	0	0
final concentrations:	1 – x	1 – x	x	x

$$K_{eq} = 2.3 = \frac{x \bullet x}{(1-x)(1-x)} = \frac{x^2}{1-2x+x^2} \Longrightarrow x^2 = 2.3x^2 - 4.6x + 2.3$$

$$0 = 1.3x^2 - 4.6x + 2.3 \Longrightarrow x = 0.60\ ,\ 1 - x = 0.40$$

(using quadratic equation)

$[CH_3SH] = [HBr] = 0.60$ M; $[CH_3Br] = [H_2S] = 0.40$ M

4-6 2 acetone ⇌ diacetone alcohol

Assume that the initial concentration of acetone is 1 molar, and 5% of the acetone is converted to diacetone alcohol. NOTE THE MOLE RATIO. The coefficients in a chemical equation become the exponents in the equilibrium expression.

	[acetone]	[diacetone alcohol]
initial concentrations:	1 M	0
final concentrations:	0.95 M	0.025 M

$$K_{eq} = \frac{[\text{diacetone alcohol}]}{[\text{acetone}]^2} = \frac{0.025}{(0.95)^2} = \boxed{0.028}$$

$\Delta G° = -2.303\ RT \log_{10} K_{eq} = -2.303\ ((8.314\ J/°K\text{-}mole) \bullet (298\ °K)) \bullet \log(0.028)$

$=$ **+ 8.9 kJ/mole (+ 2.1 kcal/mole)**

4-7 $\Delta S°$ will be negative since two molecules are combined into one, a loss of freedom of motion. Since $\Delta S°$ is negative, $-T\Delta S°$ is positive; but $\Delta G°$ is a large negative number since the reaction goes to completion. Therefore, $\Delta H°$ must also be a large negative number, necessarily larger in absolute value than $\Delta G°$. We can explain this by formation of two strong C–H sigma bonds after breaking a strong H–H sigma bond and a WEAKER C=C pi bond.

4-8

(a) $\Delta S°$ is positive—one molecule became two smaller molecules with greater freedom of motion.
(b) $\Delta S°$ is negative—two smaller molecules combined into one larger molecule with less freedom of motion.
(c) $\Delta S°$ cannot be predicted since the number of molecules in reactants and products is the same.

4-9 (a) initiation (1) $Cl-Cl \xrightarrow{h\nu} 2\ Cl\bullet$

propagation
(2) $Cl\bullet + H-CH_2CH_3 \longrightarrow H-Cl + \bullet CH_2CH_3$
(3) $Cl-Cl + \bullet CH_2CH_3 \longrightarrow Cl-CH_2CH_3 + Cl\bullet$

termination
$Cl\bullet + \bullet Cl \longrightarrow Cl-Cl$
$Cl\bullet + \bullet CH_2CH_3 \longrightarrow Cl-CH_2CH_3$
$CH_3CH_2\bullet + \bullet CH_2CH_3 \longrightarrow CH_3CH_2CH_2CH_3$

(b) step (1) break Cl–Cl $\Delta H° = +240$ kJ/mole (+ 57 kcal/mole)

step (2) break $H-CH_2CH_3$ $\Delta H° = +423$ kJ/mole (+ 101 kcal/mole)
make H–Cl $\Delta H° = -432$ kJ/mole (– 103 kcal/mole)
step (2) $\mathbf{\Delta H° = -9}$ **kJ/mole (–2 kcal/mole)**

step (3) break Cl–Cl $\Delta H° = +240$ kJ/mole (+ 57 kcal/mole)
make $Cl-CH_2CH_3$ $\Delta H° = -355$ kJ/mole (– 85 kcal/mole)
step (3) $\mathbf{\Delta H° = -115}$ **kJ/mole (–28 kcal/mole)**

(c) $\Delta H°$ for the reaction is the sum of the $\Delta H°$ values of the individual propagation steps:

– 9 kJ/mole + – 115 kJ/mole = **– 124 kJ/mole**
(–2 kcal/mole + – 28 kcal/mole = **– 30 kcal/mole)**

4-10 (a) initiation (1) $Br-Br \xrightarrow{h\nu} 2\ Br\bullet$

propagation
(2) $Br\bullet + H-CH_3 \longrightarrow H-Br + \bullet CH_3$
(3) $Br-Br + \bullet CH_3 \longrightarrow Br-CH_3 + Br\bullet$

step (1) break Br–Br $\Delta H° = +190$ kJ/mole (+ 45 kcal/mole)

step (2) break $H-CH_3$ $\Delta H° = +439$ kJ/mole (+ 105 kcal/mole)
make H–Br $\Delta H° = -366$ kJ/mole (– 88 kcal/mole)
step (2) $\mathbf{\Delta H° = +73}$ **kJ/mole (+ 17 kcal/mole)**

step (3) break Br–Br $\Delta H° = +190$ kJ/mole (+ 45 kcal/mole)
make $Br-CH_3$ $\Delta H° = -302$ kJ/mole (– 72 kcal/mole)
step (3) $\mathbf{\Delta H° = -112}$ **kJ/mole (–27 kcal/mole)**

(b) $\Delta H°$ for the reaction is the sum of the $\Delta H°$ values of the individual propagation steps:

+ 73 kJ/mole + – 112 kJ/mole = **– 39 kJ/mole**
(+ 17 kcal/mole + – 27 kcal/mole = **– 10 kcal/mole)**

4-11
(a) first order: the exponent of [$(CH_3)_3CCl$] in the rate law = 1
(b) zeroth order: [CH_3OH] does not appear in the rate law (its exponent is zero)
(c) first order: the sum of the exponents in the rate law = 1 + 0 = **1**

4-12
(a) first order: the exponent of [cyclohexene] in the rate law = 1
(b) second order: the exponent of [Br_2] in the rate law = 2
(c) third order: the sum of the exponents in the rate law = 1 + 2 = **3**

4-13

(a) The reaction rate depends on neither [ethylene] nor [hydrogen], so it is zeroth order in both species. The overall reaction must be zeroth order.

(b) rate = k_r

(c) The rate law does not depend on the concentration of the reactants. It must depend, therefore, on the only other chemical present, the catalyst. Apparently, whatever is happening on the surface of the catalyst determines the rate, regardless of the concentrations of the two gases. Increasing the surface area of the catalyst, or simply adding more catalyst, would accelerate the reaction.

4-14

(a)

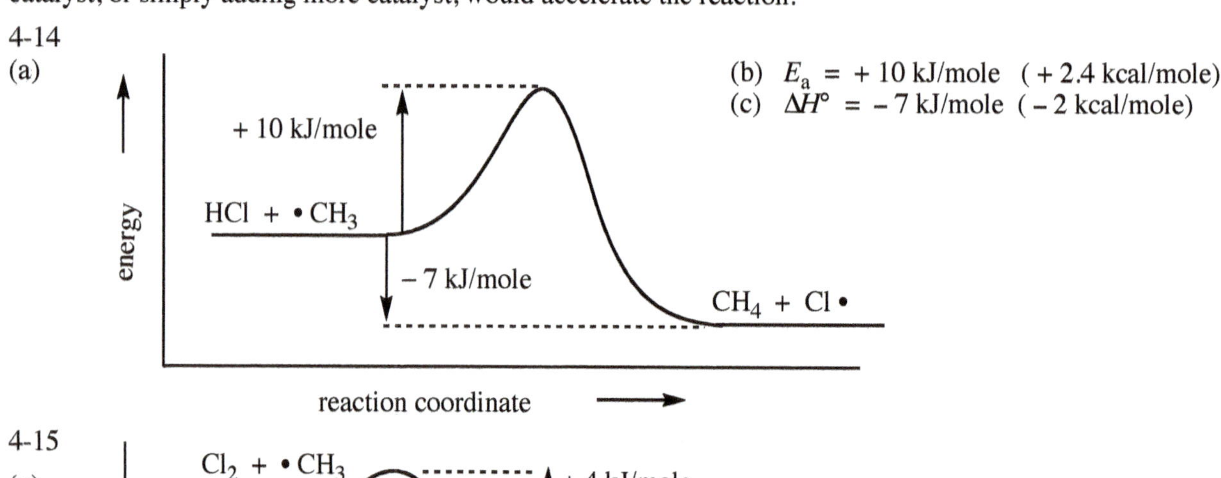

(b) E_a = + 10 kJ/mole (+ 2.4 kcal/mole)

(c) $\Delta H°$ = – 7 kJ/mole (– 2 kcal/mole)

4-15

(a)

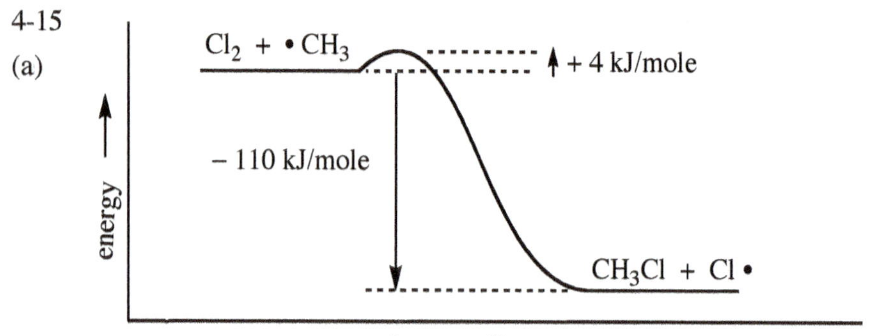

(b) reverse: CH_3Cl + Cl • ⟶ Cl_2 + • CH_3

(c) reverse: E_a = + 110 kJ/mole + + 4 kJ/mole = **+ 114 kJ/mole**
(+ 27 kcal/mole + + 1 kcal/mole = **+ 28 kcal/mole**)

4-16

(a)

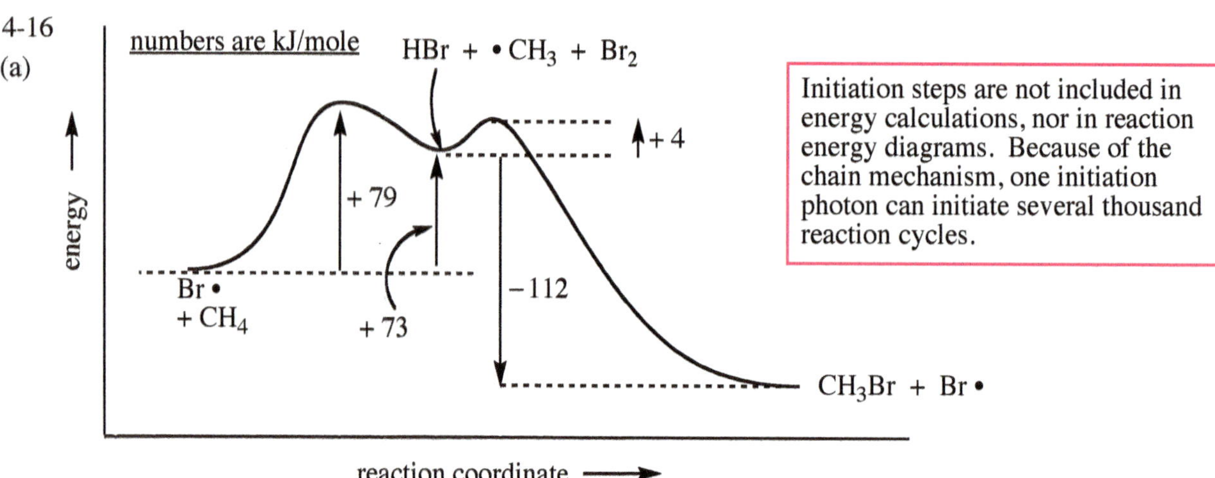

Initiation steps are not included in energy calculations, nor in reaction energy diagrams. Because of the chain mechanism, one initiation photon can initiate several thousand reaction cycles.

4-16 continued

(b) The step leading to the highest energy transition state is rate-limiting. In this mechanism, the first propagation step is rate-limiting:

$$Br\bullet + CH_4 \longrightarrow HBr + \bullet CH_3$$

(c) (1) $\left[\overset{\delta\bullet}{Br}\text{--------}\overset{\delta\bullet}{Br}\right]^{\ddagger}$ ("δ •" means partial radical character on the atom)

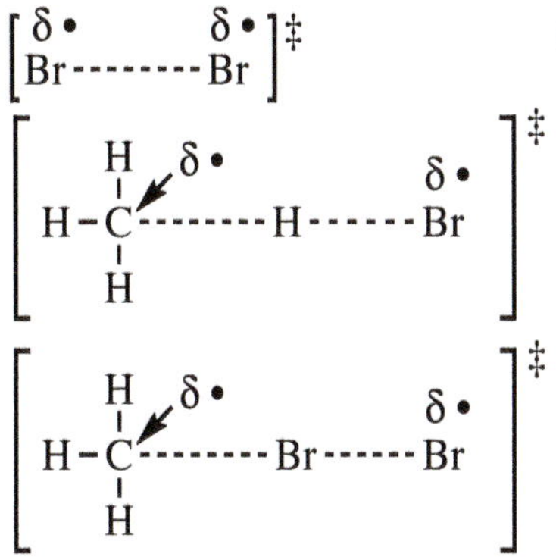

(2) $\left[H_3\overset{\delta\bullet}{C}\text{--------}H\text{------}\overset{\delta\bullet}{Br}\right]^{\ddagger}$

(3) $\left[H_3\overset{\delta\bullet}{C}\text{--------}Br\text{------}\overset{\delta\bullet}{Br}\right]^{\ddagger}$

(d) $\Delta H°$ for the reaction is the sum of the $\Delta H°$ values of the individual propagation steps (refer to the solution to 4-10 (a) and (b)):

+ 73 kJ/mole + – 112 kJ/mole = **– 39 kJ/mole**
(+ 17 kcal/mole + – 27 kcal/mole = **– 10 kcal/mole)**

4-17

(a) initiation (1) $I{-}I \xrightarrow{h\nu} 2\ I\bullet$

propagation
(2) $I\bullet + H{-}CH_3 \longrightarrow H{-}I + \bullet CH_3$
(3) $I{-}I + \bullet CH_3 \longrightarrow I{-}CH_3 + I\bullet$

step (1) break I–I $\Delta H°$ = + 149 kJ/mole (+ 36 kcal/mole)

step (2) break H–CH_3 $\Delta H°$ = + 439 kJ/mole (+ 105 kcal/mole)
make H–I $\Delta H°$ = – 298 kJ/mole (– 71 kcal/mole)
step (2) **$\Delta H°$ = + 141 kJ/mole (+ 34 kcal/mole)**

step (3) break I–I $\Delta H°$ = + 149 kJ/mole (+ 36 kcal/mole)
make I–CH_3 $\Delta H°$ = – 241 kJ/mole (– 58 kcal/mole)
step (3) **$\Delta H°$ = – 92 kJ/mole (–22 kcal/mole)**

(b) $\Delta H°$ for the reaction is the sum of the $\Delta H°$ values of the individual propagation steps:

+ 141 kJ/mole + – 92 kJ/mole = **+ 49 kJ/mole**
(+ 34 kcal/mole + – 22 kcal/mole = **+ 12 kcal/mole)**

(c) Iodination of methane is unfavorable for both kinetic and thermodynamic reasons. Kinetically, the rate of the first propagation step must be very slow because it is very endothermic; the activation energy must be at least + 138 kJ/mole. Thermodynamically, the overall reaction is endothermic, so an equilibrium would favor reactants, not products; there is no energy decrease to drive the reaction to products.

4-18 Propane has six primary hydrogens and two secondary hydrogens, a ratio of 3 : 1. If primary and secondary hydrogens were replaced by chlorine at equal rates, the chloropropane isomers would reflect the same 3 : 1 ratio, that is, 75% 1-chloropropane and 25% 2-chloropropane.

4-19 (a) $CH_3CH_2CH_2CH_3$

↑ ↑ ↑ ↑

1° 2° 2° 1°

(b) $CH_3-\underset{CH_3}{\overset{H}{C}}-CH_3$ — H ← 3°; the three CH_3 groups ← 1°, 1°, 1°

(c) $CH_3-\underset{CH_3}{\overset{H}{C}}-CH_2CH_3$ — H ← 3°; CH_3 ← 1°, CH_3 ← 1°, CH_2 ← 2°, CH_3 ← 1°

(d)

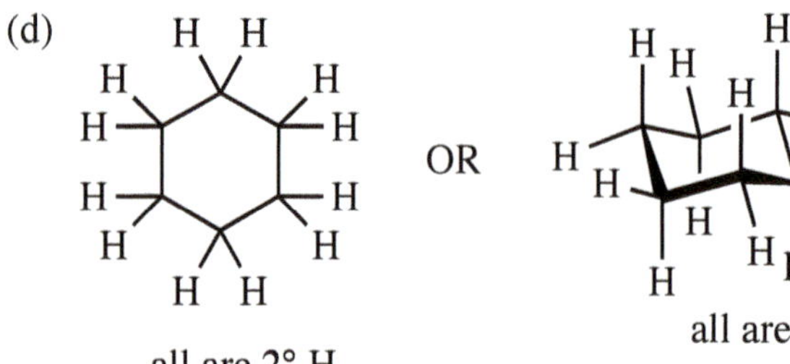

all are 2° H

all are 2° H

(e)

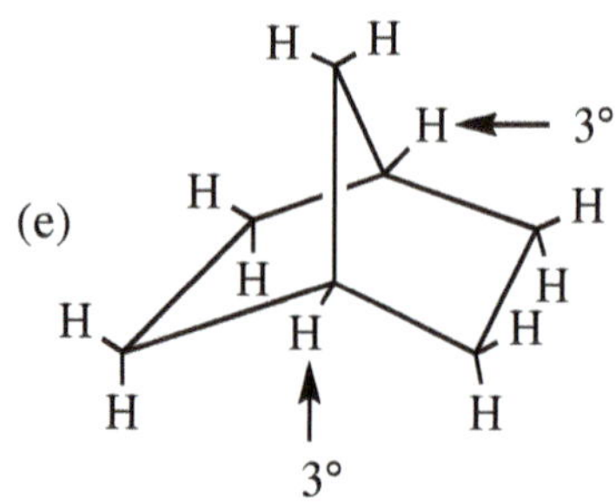

all are 2° H except at bridgehead (labeled 3°)

4-20 $\Delta H°$ for abstraction of a 1° H from both ethane and propane are + 423 kJ/mole (+ 101 kcal/mole). It is reasonable to use this same value for abstraction of the 1° H in isobutane.

3° H abstraction

$$Cl\bullet + CH_3-\underset{CH_3}{\overset{H}{C}}-CH_3 \longrightarrow H-Cl + CH_3-\underset{CH_3}{\overset{\bullet}{C}}-CH_3$$

break 3° $H-C(CH_3)_3$	$\Delta H°$ = + 403 kJ/mole	(+ 96 kcal/mole)
make H–Cl	$\Delta H°$ = – 432 kJ/mole	(– 103 kcal/mole)
overall 3° H abstraction	**$\Delta H°$ = –29 kJ/mole**	**(–7 kcal/mole)**

1° H abstraction

$$Cl\bullet + H-CH_2-\underset{CH_3}{\overset{H}{C}}-CH_3 \longrightarrow H-Cl + \bullet CH_2-\underset{CH_3}{\overset{H}{C}}-CH_3$$

break 1° $H-CH_2CH(CH_3)_2$	$\Delta H°$ = + 423 kJ/mole	(+ 101 kcal/mole)
make H–Cl	$\Delta H°$ = – 432 kJ/mole	(– 103 kcal/mole)
overall 1° H abstraction	**$\Delta H°$ = –9 kJ/mole**	**(–2 kcal/mole)**

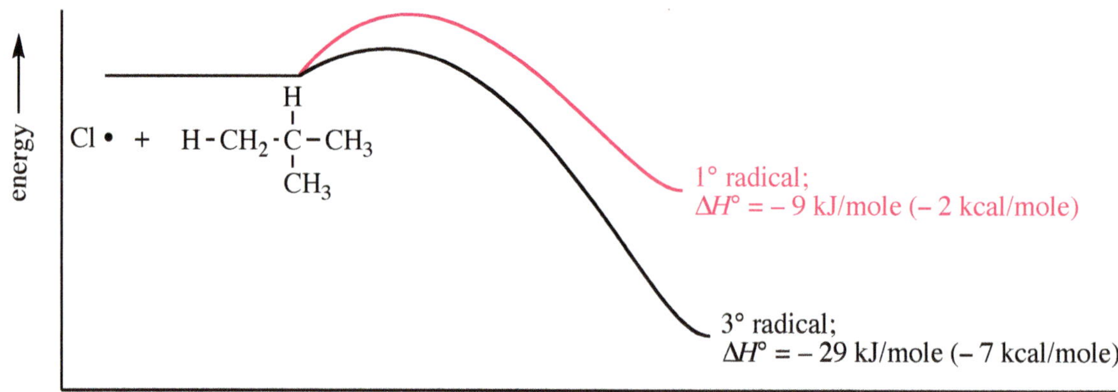

Since $\Delta H°$ for forming the 3° radical is more negative than $\Delta H°$ for forming the 1° radical, it is reasonable to infer that the activation energy leading to the 3° radical is lower than the activation energy leading to the 1° radical.

4-21

2-Methylbutane can produce four mono-chloro isomers. To calculate the relative amount of each in the product mixture, multiply the numbers of hydrogens that could lead to that product times the reactivity for that type of hydrogen. Each relative amount divided by the sum of all the amounts will provide the percent of each in the product mixture.

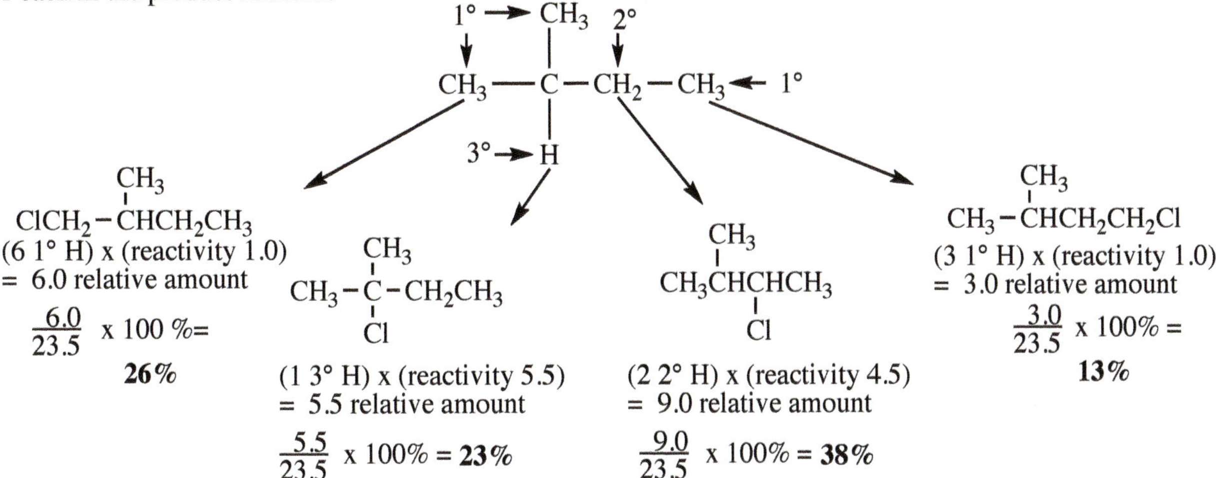

4-22

(a) When heptane is burned, only 1° and 2° radicals can be formed (from either C—H or C—C bond cleavage). These are high-energy, unstable radicals that rapidly form other products. When isooctane (2,2,4-trimethylpentane, below) is burned, 3° radicals can be formed from either C—H or C—C bond cleavage. The 3° radicals are lower in energy than 1° or 2°, relatively stable, with lowered reactivity. Slower combustion translates to less "knocking."

$CH_3-C(CH_3)_2-CH_2-CH(CH_3)-CH_3$

isooctane (2,2,4-trimethylpentane)

Any indicated bond cleavage will produce a 3° radical.

(b)

$(CH_3)_3C-O-H \; + \; \bullet R \;\longrightarrow\; (CH_3)_3C-O\bullet \; + \; R-H$

•R: an alkyl radical

When the alcohol hydrogen is abstracted from *tert*-butyl alcohol, a relatively stable *tert*-butoxy radical ($(CH_3)_3C—O\bullet$) is produced. This low-energy radical is slower to react than alkyl radicals, moderating the reaction and producing less "knocking."

(c)

BDE + 376 kJ/mol

toluene ($C_6H_5CH_3$; benzylic H) + •R (an alkyl radical) → $C_6H_5\dot{C}H_2$ (resonance stabilized) + R—H

BDE – 403 (3°) to –423 (1°) kJ/mol EXOTHERMIC

When the benzylic hydrogen is abstracted from toluene, a relatively stable, *resonance-stabilized* benzylic radical is produced. This low-energy radical is energetically "easy" to produce, making toluene a reactive molecule with a high octane rating. (If you want to peek ahead for the resonance at benzylic positions, see the solution to 4-45(b).)

4-23

(a) 1° H abstraction

$F\bullet + CH_3CH_2CH_3 \longrightarrow H{-}F + \bullet CH_2CH_2CH_3$

break 1° $H{-}CH_2CH_2CH_3$	$\Delta H° = +423$ kJ/mole (+101 kcal/mole)
make H–F	$\Delta H° = -570$ kJ/mole (−136 kcal/mole)
overall 1° H abstraction	**$\Delta H° = -147$ kJ/mole (−35 kcal/mole)**

2° H abstraction

$F\bullet + CH_3CH_2CH_3 \longrightarrow H{-}F + CH_3\dot{C}HCH_3$

break 2° $H{-}CH(CH_3)_2$	$\Delta H° = +413$ kJ/mole (+ 99 kcal/mole)
make H–F	$\Delta H° = -570$ kJ/mole (−136 kcal/mole)
overall 2° H abstraction	**$\Delta H° = -157$ kJ/mole (−37 kcal/mole)**

(b) Fluorination is extremely exothermic and is likely to be indiscriminate in which hydrogens are abstracted. (In fact, C—C bonds are also broken during fluorination.)

(c) Free-radical fluorination is extremely exothermic. In exothermic reactions, the transition states resemble the starting materials more than the products, so while the 1° and 2° radicals differ by about 10 kJ/mole (2 kcal/mole), the transition states will differ by only a tiny amount: from Figure 4-11(b), only 4 kJ/mole (1 kcal/mole). For fluorination, then, the rate of abstraction for 1° and 2° hydrogens will be virtually identical. Product ratios will depend on statistical factors only.

Fluorination is difficult to control, but if propane were monofluorinated, the product mixture would reflect the ratio of the types of hydrogens: six 1° H to two 2° H, or 3 : 1 ratio, giving 75% 1-fluoropropane and 25% 2-fluoropropane.

4-24

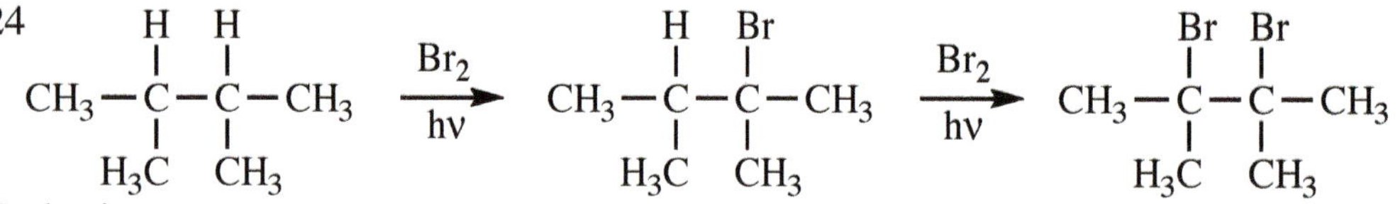

$CH_3{-}CH(CH_3){-}CH(CH_3){-}CH_3 \xrightarrow[h\nu]{Br_2} CH_3{-}CH(CH_3){-}CBr(CH_3){-}CH_3 \xrightarrow[h\nu]{Br_2} CH_3{-}CBr(CH_3){-}CBr(CH_3){-}CH_3$

Mechanism

initiation $Br{-}Br \xrightarrow{h\nu} 2\ Br\bullet$

propagation $CH_3{-}CH(CH_3){-}CH(CH_3){-}CH_3 + Br\bullet \longrightarrow HBr + CH_3{-}CH(CH_3){-}\dot{C}(CH_3){-}CH_3$

$CH_3{-}CH(CH_3){-}\dot{C}(CH_3){-}CH_3 + Br{-}Br \longrightarrow CH_3{-}CH(CH_3){-}CBr(CH_3){-}CH_3 + Br\bullet$

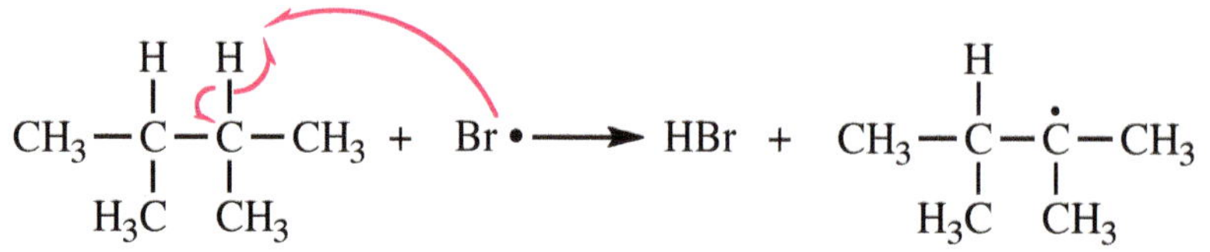

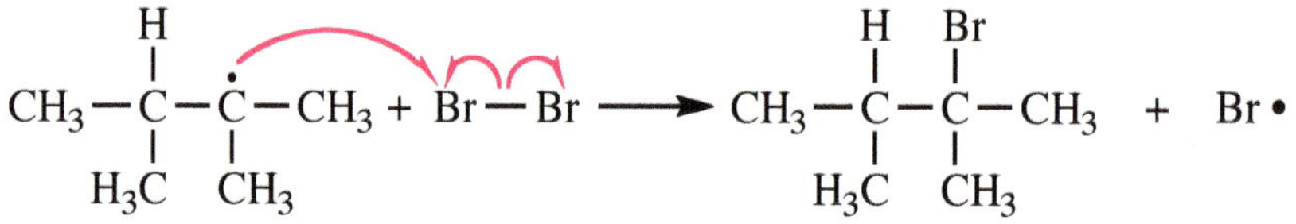

4-25 (a) Mechanism

initiation $Br{-}Br \xrightarrow{h\nu} 2\ Br\bullet$

propagation cyclohexene + $Br\bullet$ ⟶ {allylic radical ⟷ resonance form} + HBr

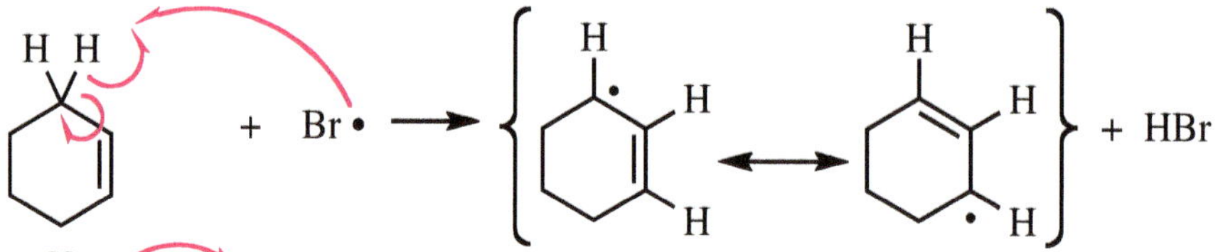

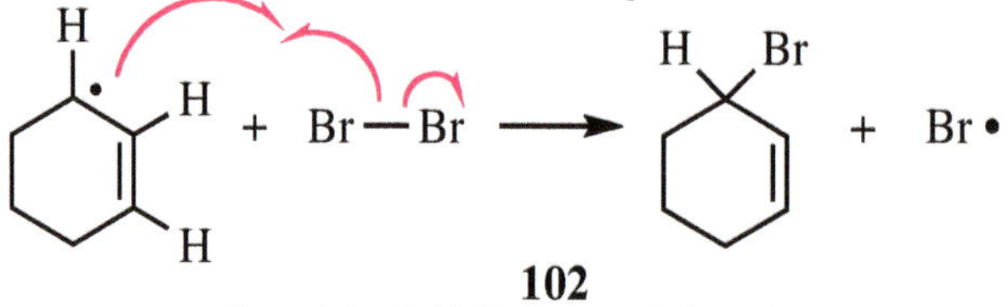

allylic radical + $Br{-}Br$ ⟶ 3-bromocyclohexene + $Br\bullet$

4-25 continued

(b) Energy calculation uses the value for the allylic C—H bond from Table 4-2.

First propagation step	break allylic H–CH[ring]	$\Delta H°$ = + 372 kJ/mole (+ 89 kcal/mole)
	make H–Br	$\Delta H°$ = – 366 kJ/mole (– 88 kcal/mole)
	overall allylic H abstraction	**$\Delta H°$ = + 6 kJ/mole (+1 kcal/mole)**
Second propagation step	break Br–Br	$\Delta H°$ = + 190 kJ/mole (+ 45 kcal/mole)
	make 2° C–Br	$\Delta H°$ = – 309 kJ/mole (– 74 kcal/mole)
	overall C–Br formation	**$\Delta H°$ = – 119 kJ/mole (–29 kcal/mole)**

The first propagation step is rate-limiting.

structure of the transition state:

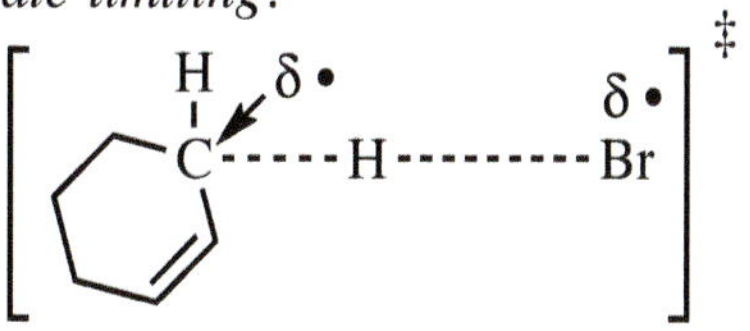

(c) Since the first propagation step is very slightly endothermic, it is reasonable to infer that the transition state is close to cyclohexene + bromine radical. This is indicated in the transition state structure by showing the H slightly closer to the C than to the Br. This one is a close call.

(d) A bromine radical will abstract the hydrogen with the lowest bond dissociation energy at the fastest rate. The allylic hydrogen of cyclohexene is more easily abstracted than a hydrogen of cyclohexane because the radical produced is stabilized by resonance. (Energy values below are per mole.)

aliphatic: 413 kJ (99 kcal) ⟶ 372 kJ (89 kcal): allylic

SLOWER *FASTER*

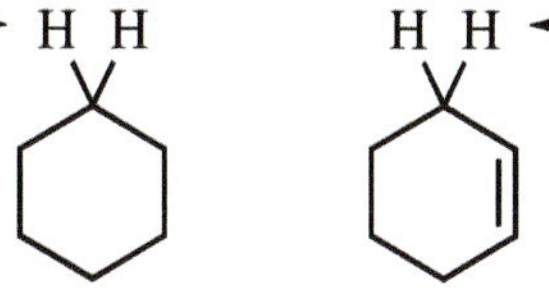

4-26

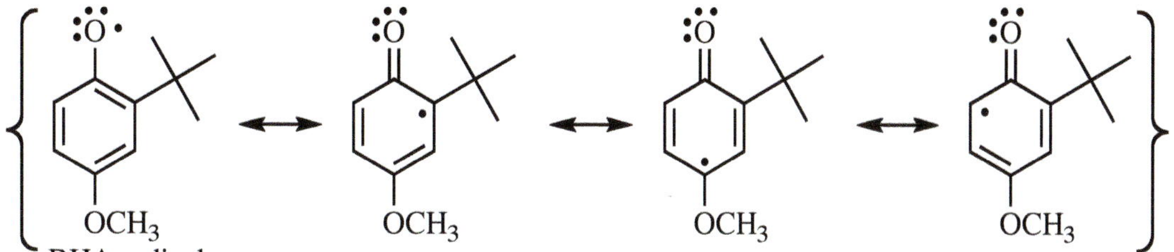

BHA radical

4-27

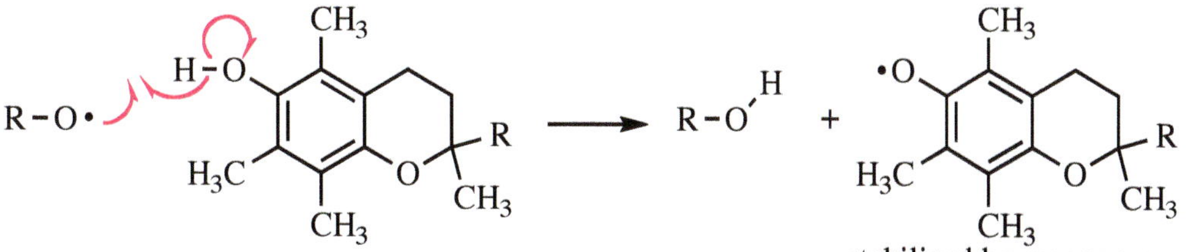

Vitamin E

stabilized by resonance—
less reactive than RO• —
similar resonance forms in 4-45(f)

4-28

The resonance forms on the next page do not include the simple benzene resonance forms as shown below; they are significant, but repetitive, so for simplicity, they are not drawn here. Braces are omitted.

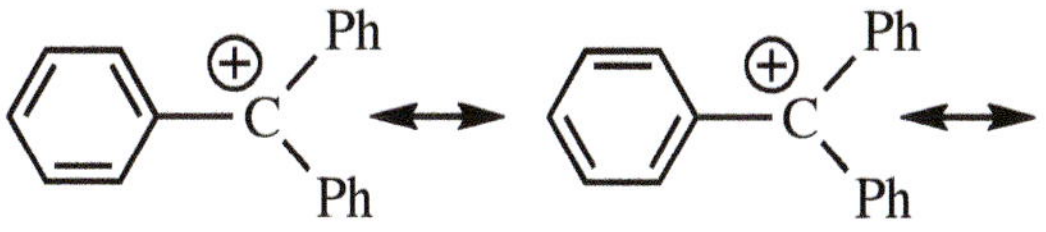

many more

Ph = phenyl =

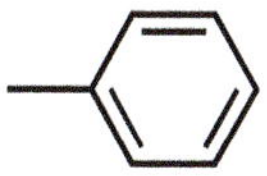

4-28 continued

The triphenylmethyl cation is so stable because of the delocalization of the charge. The more resonance forms a species has—especially equivalent resonance forms—the more stable it will be.

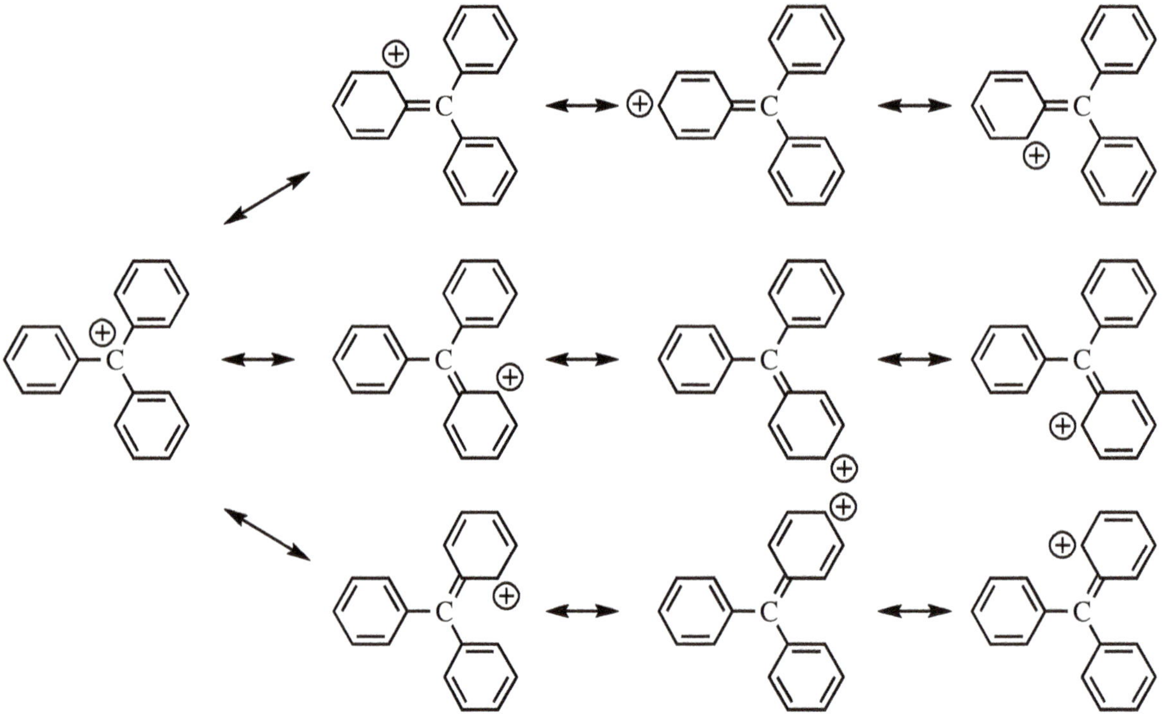

4-29 most stable (d) > (c) > (b) > (a) least stable

(d) 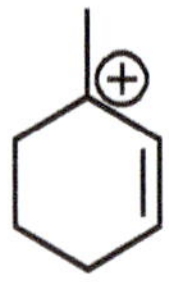 3° and resonance stabilized

(c) $CH_3-\overset{+}{C}CH_2CH_3$ with CH_3 branch — 3°

(b) $CH_3-CH\overset{+}{C}HCH_3$ with CH_3 branch — 2°

(a) $CH_3-CHCH_2\overset{+}{C}H_2$ with CH_3 branch — 1°

4-30 most stable (d) > (c) > (b) > (a) least stable

(d) 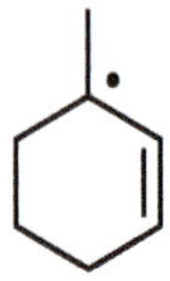 3° and resonance stabilized

(c) $CH_3-\dot{C}CH_2CH_3$ with CH_3 branch — 3°

(b) $CH_3-CH\dot{C}HCH_3$ with CH_3 branch — 2°

(a) $CH_3-CHCH_2\dot{C}H_2$ with CH_3 branch — 1°

4-31

$H_3C-C(=O)-CH_2-C(=O)-CH_3 + {}^{-}OH$ (base) $\longrightarrow H_2O +$

$\{H_3C-C(=O)-\overset{-}{C}H-C(=O)-CH_3 \longleftrightarrow H_3C-C(O^-)=CH-C(=O)-CH_3 \longleftrightarrow H_3C-C(=O)-CH=C(O^-)-CH_3\}$

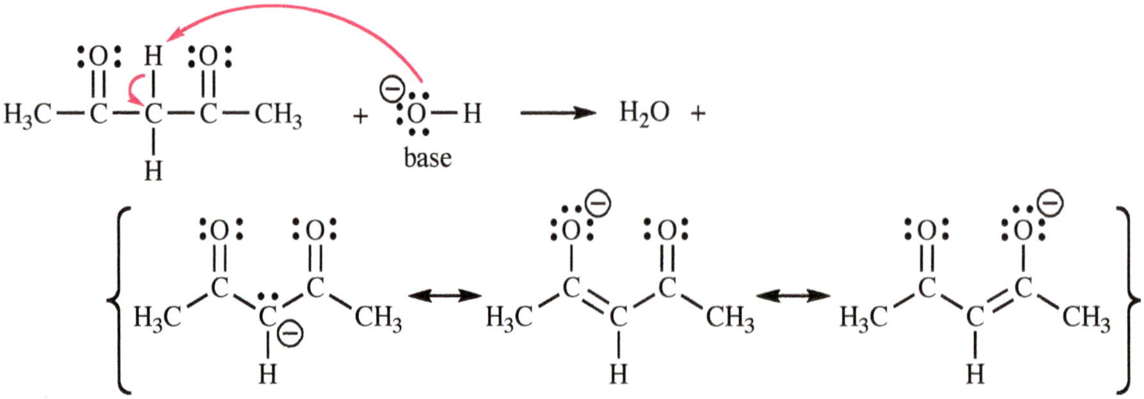

Negative charge delocalized on two oxygens lends a great deal of stability to this anion.

4-32

$$H-\underset{H}{\overset{H}{C}}-C\equiv N\!: \;+\; :Base^{\ominus} \longrightarrow H-Base \;+\; \left\{ H-\underset{H}{\ddot{C}^{\ominus}}-C\equiv N\!: \longleftrightarrow H-\underset{H}{C}=C=\ddot{N}^{\ominus}\!: \right\}$$

4-33

$$H-\ddot{C}^{-}-\overset{O}{\overset{\|}{C}}-O-CH_2CH_3$$

4-34 (a) rate = k [CH_3O^-] [C_4H_9Br] The reaction is first order in each, methoxide and 1-bromobutane. The overall order is the sum of the individual orders: 1 + 1 = 2. Second order overall.

(b) If the solvent is reduced by half with the same amount of reactants, the concentration of each doubles. Doubling the concentration of each will increase the rate by a factor of 4, i.e., four times faster.

4-35

(a) and (c)

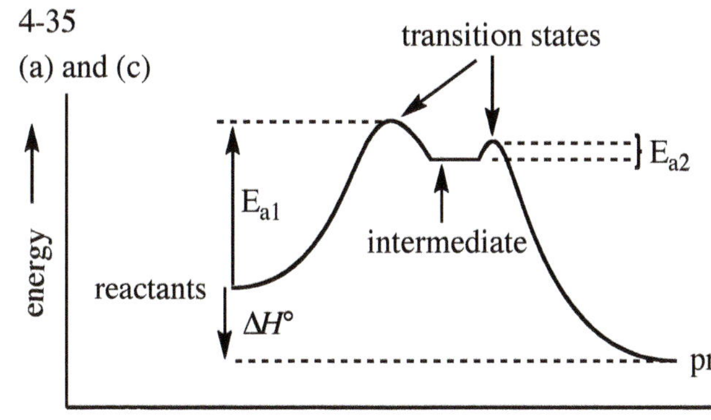

(b) $\Delta H°$ is negative (decreases), so the reaction is exothermic.

(d) The first transition state determines the rate since it is the highest energy point. The *structure* of the first transition state resembles the *structure* of the intermediate since the *energy* of the transition state is closest to the *energy* of the intermediate.

4-36

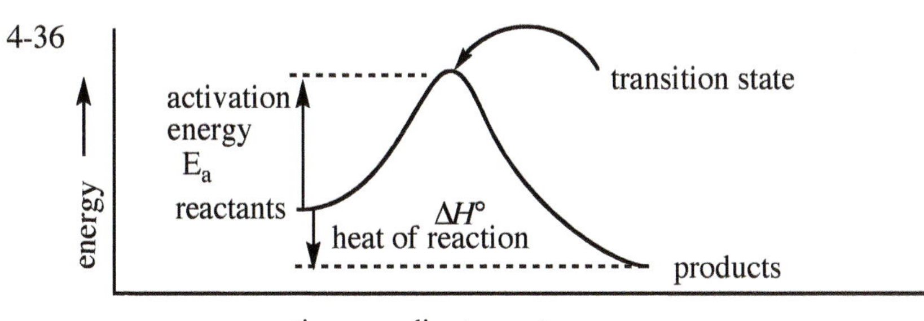

4-37

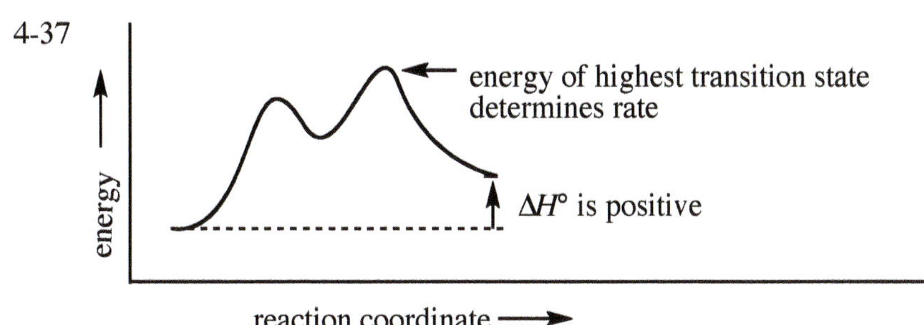

4-38

For each reaction determine whether products or reactants are favored. Because acid-base reactions are equilibria, the side that is favored will be of lower energy on the energy diagram.

(a)

phenol + NaOH ⇌ phenoxide $O^- Na^+$ + H_2O

phenol
pK_a 10.0
stronger acid

Products are favored at equilibrium.

pK_a 15.7
weaker acid

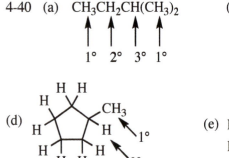

Shown in solid line on the graph below.

energy ⟶

H_3C
$H_3C-C-O^- Na^+ + H_2O$
H_3C

$O^- Na^+ + H_2O$

reaction coordinate ⟶

(b)

H_3C
$H_3C-C-OH$ + NaOH ⇌ $H_3C-C-O^- Na^+$ + H_2O
H_3C

tert-butyl alcohol
pK_a 18.0
weaker acid

Reactants are favored at equilibrium.

pK_a 15.7
stronger acid

Shown in dashed line on the graph above.

4-39

The rate law is first order with respect to the concentrations of hydrogen ion and of *tert*-butyl alcohol, zeroth order with respect to the concentration of chloride ion, second order overall.

rate = k_r $[(CH_3)_3COH][H^+]$

4-40 (a) $CH_3CH_2CH(CH_3)_2$

1° 2° 3° 1°

(b) $(CH_3)_3CCH_2CH_3$

1° 2° 1°

(c) 3° → H CH_3 ← 1°
$H_3C-C-C-CH_2CH_3$
1° → H_3C H
3° 2°

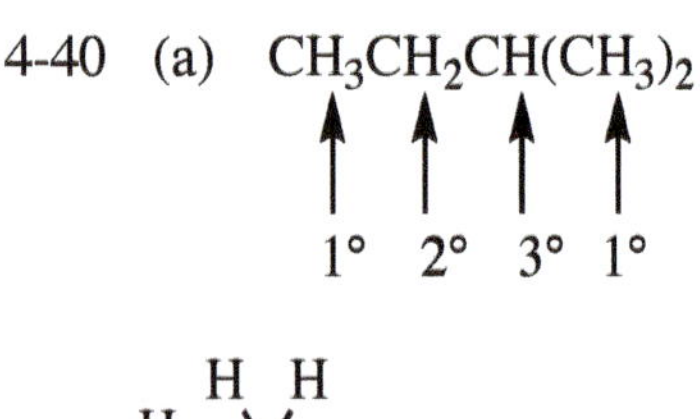

All are 2° H except for the two types labeled.

(e) H H H CH_3 ← 1°
H
H
H
H
H
H H ← 3°
H

All are 2° H except for the two types labeled.

(f)

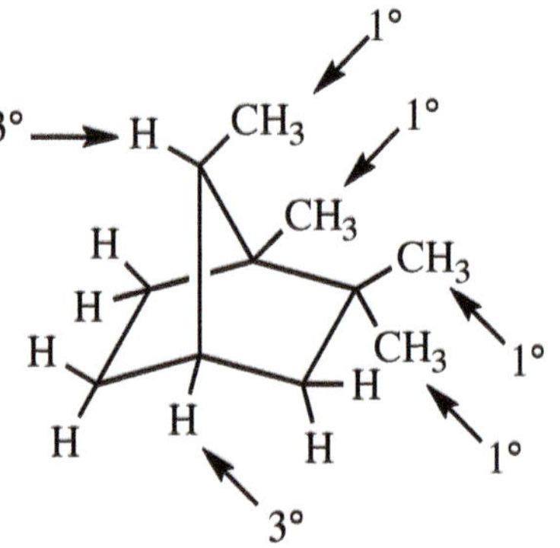

All are 2° H except as labeled.

4-41

(a) break $H–CH_2CH_3$ and I–I, make $I–CH_2CH_3$ and H–I
kJ/mole: (+ 423 + + 149) + (− 238 + − 298) = **+ 36 kJ/mole**
kcal/mole: (+ 101 + + 36) + (− 57 + − 71) = **+ 9 kcal/mole**
(b) break $CH_3CH_2–Cl$ and H–I, make $CH_3CH_2–I$ and H–Cl
kJ/mole: (+ 355 + + 298) + (− 238 + − 432) = **− 17 kJ/mole**
kcal/mole: (+ 85 + + 71) + (− 57 + − 103) = **− 4 kcal/mole**
(c) break $(CH_3)_3C–OH$ and H–Cl, make $(CH_3)_3C–Cl$ and H–OH
kJ/mole: (+ 401 + + 432) + (− 355 + − 497) = **− 19 kJ/mole**
kcal/mole: (+ 96 + + 103) + (− 85 + − 119) = **− 5 kcal/mole**
(d) break $CH_3CH_2–CH_3$ and H–H, make $CH_3CH_2–H$ and $H–CH_3$
kJ/mole: (+ 372 + + 436) + (− 423 + − 439) = **− 54 kJ/mole**
kcal/mole: (+ 89 + + 104) + (− 101 + − 105) = **− 13 kcal/mole**
(e) break $CH_3CH_2–OH$ and H–Br, make $CH_3CH_2–Br$ and H–OH
kJ/mole: (+ 393 + + 366) + (− 303 + − 497) = **− 41 kJ/mole**
kcal/mole: (+ 94 + + 88) + (− 72 + − 119) = **− 9 kcal/mole**

4-42 Numbers are bond dissociation energies in kcal/mole in the top line and kJ/mole in the bottom line.

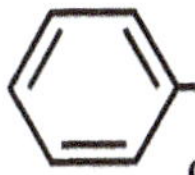

$C_6H_5\dot{C}H_2$	> $CH_2{=}CH\dot{C}H_2$	> $(CH_3)_3\dot{C}$	> $(CH_3)_2\dot{C}H$	> $CH_3\dot{C}H_2$	> $\dot{C}H_3$
90	89	96	99	101	105
376	372	403	413	423	439
benzyl	allyl	3°	2°	1°	methyl
most stable					*least stable*

Note: On an energy diagram, toluene, precursor of the benzyl radical, starts much lower than propene from which the allyl radical comes. The benzyl radical, therefore, is more stable than any of the other radicals in this list.

4-43
(a) 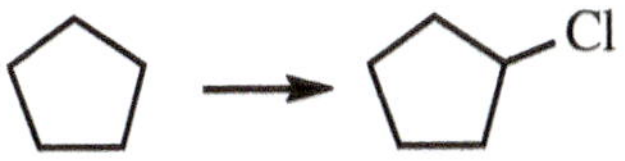 cyclopentane → chlorocyclopentane (Cl)

Only one product; chlorination would work. Bromination on a 2° carbon would not be predicted to be a high-yielding process.

(b) 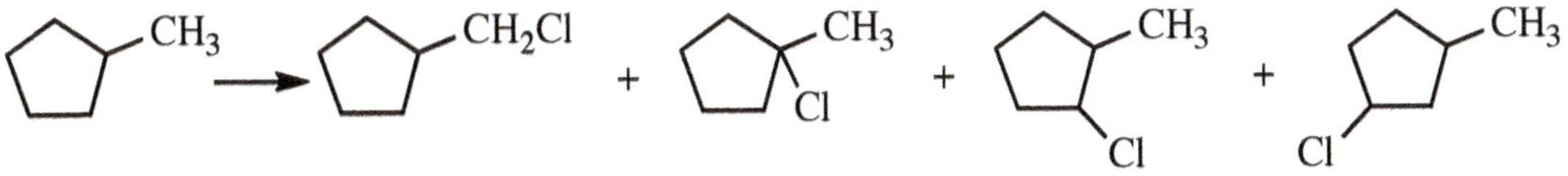 methylcyclopentane (CH_3) → (CH_2Cl) + (CH_3, Cl) + (CH_3, Cl) + (CH_3, Cl)

Chlorination would produce four constitutional isomers and would not be a good method to make only one of these. Monobromination at the 3° carbon would give a reasonable yield of the pure 3° bromide, in contrast to the results from chlorination.

(c) 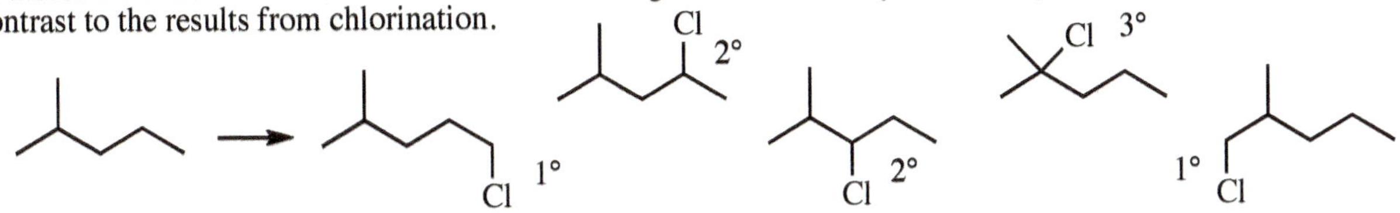 → Cl 1°; Cl 2°; Cl 2°; Cl 3°; 1° Cl

Chlorination would produce five constitutional isomers and would not be a good method to make only one of these. Monobromination would be selective for the 3° carbon and would give an excellent yield.

(d) $CH_3{-}C(CH_3)_2{-}C(CH_3)_2{-}CH_3 \longrightarrow CH_3{-}C(CH_3)_2{-}C(CH_3)_2{-}CH_2Cl$

Only one product; chlorination would give a high yield. Monobromination would be very difficult since all hydrogens are on 1° carbons.

4-44

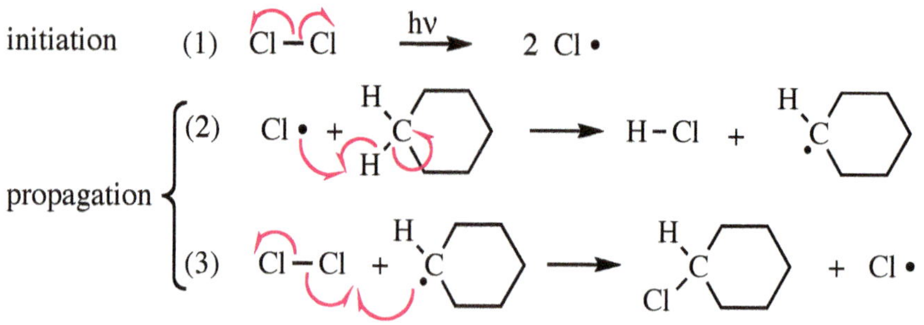

initiation (1) Cl—Cl $\xrightarrow{h\nu}$ 2 Cl•

propagation
(2) Cl• + H–C(H)(cyclohexyl) ⟶ H–Cl + H–C•(cyclohexyl)
(3) Cl—Cl + H–C•(cyclohexyl) ⟶ H–C(Cl)(cyclohexyl) + Cl•

Termination steps are any two radicals combining.

4-45

(a) { $CH_2{=}CH{-}\dot{C}H_2$ ⟷ $\dot{C}H_2{-}CH{=}CH_2$ }

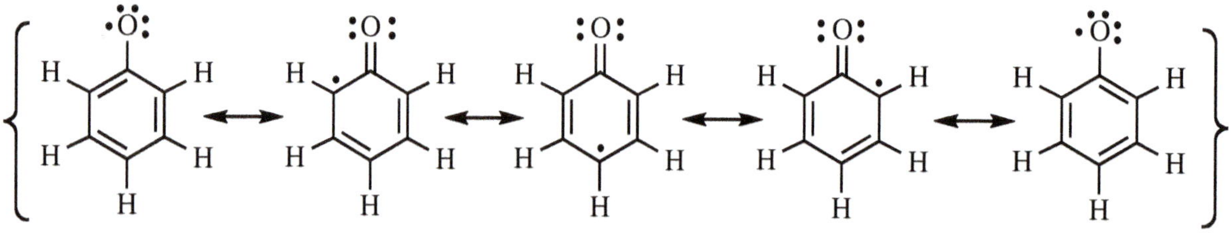

(b) { •CH_2 ⟷ CH_2 ⟷ CH_2 ⟷ CH_2 ⟷ •CH_2 }

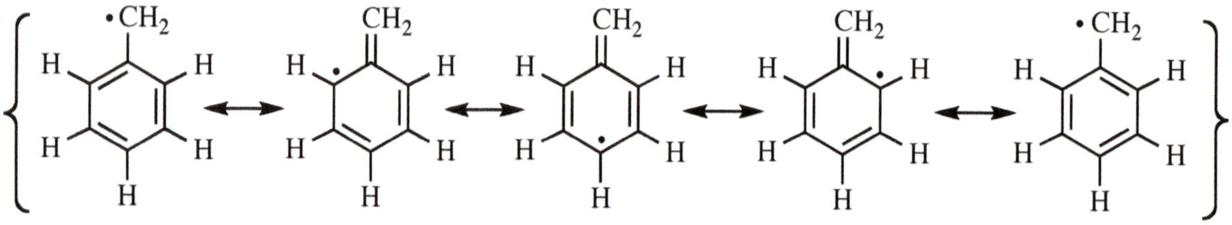

(c) { $CH_3{-}C({=}O){-}O$• ⟷ $CH_3{-}C(O•){=}O$ }

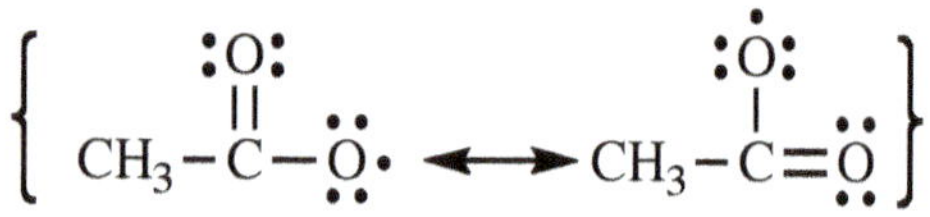

(d) { H H H ⟷ H H H }

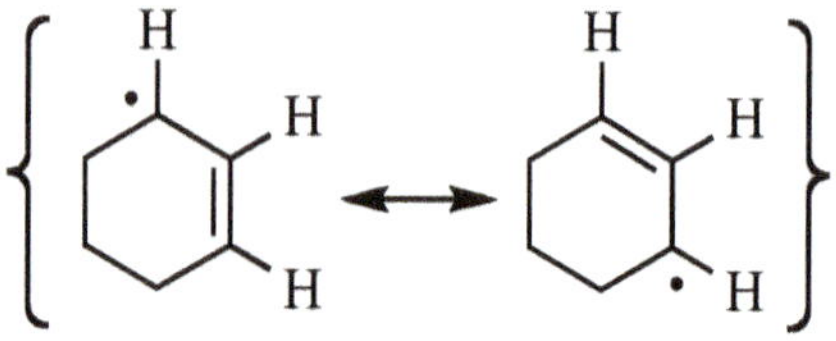

(e) { H H H ⟷ H H H }

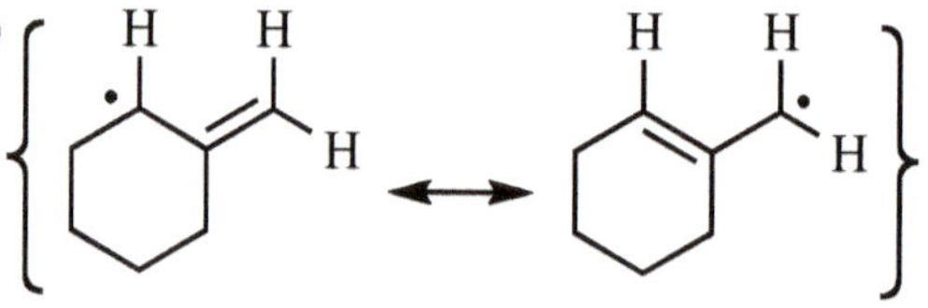

(f) { •O: ⟷ :O: ⟷ :O: ⟷ :O: ⟷ •O: }

4-46

(a) Mechanism

initiation Br—Br $\xrightarrow{h\nu}$ 2 Br•

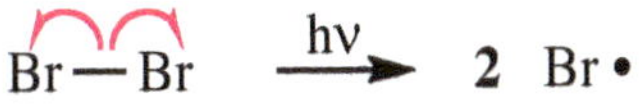

propagation

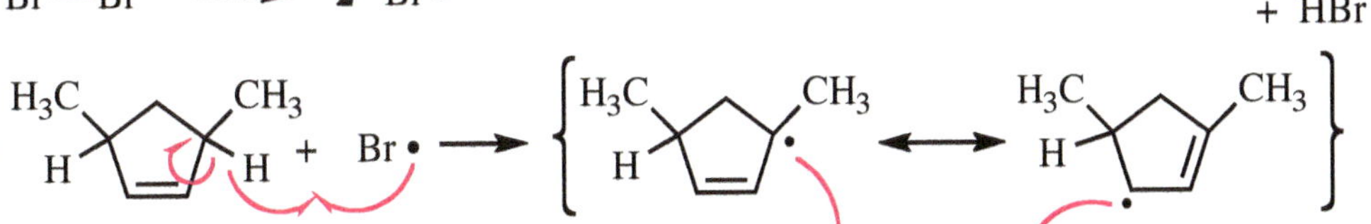

The 3° allylic H is abstracted selectively (faster than any other type in the molecule), forming an intermediate represented by two *non-equivalent* resonance forms. Partial radical character on two different carbons of the allylic radical leads to two different products.

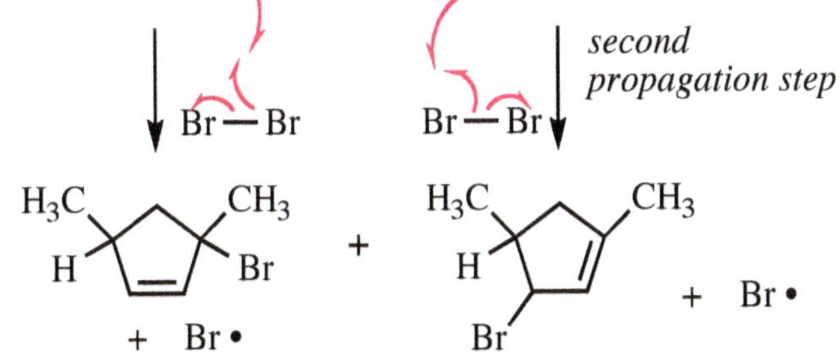

+ HBr

second propagation step

+ Br• + Br•

(b) There are two reasons why the H shown is the one that is abstracted by bromine radical: the H is 3° and it is allylic, that is, neighboring a double bond. Both of these factors stabilize the radical that is created by removing the H atom.

4-47 Where mixtures are possible, only the major product is shown.

(a) 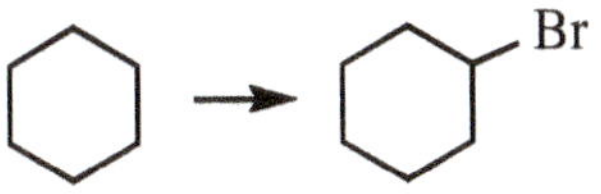 only one product possible

(b) 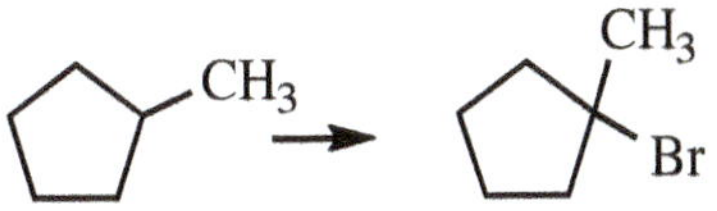 3° hydrogen abstracted selectively, faster than 2° or 1°

(c) 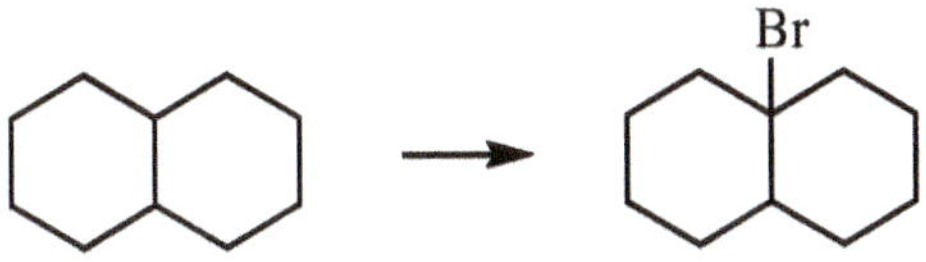 decalin — 3° hydrogen abstracted selectively, faster than 2°

(d) 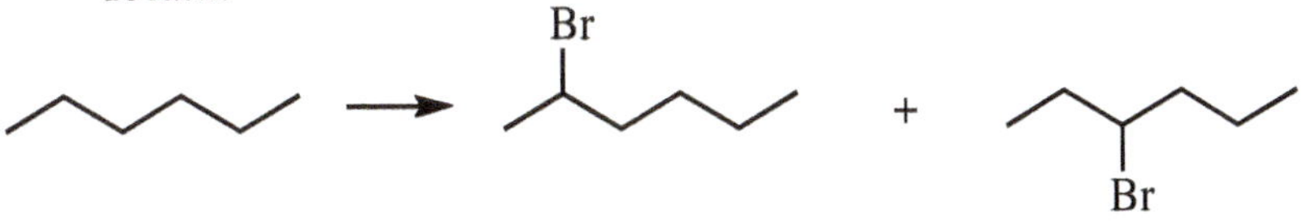 both 2°—formed in equal amounts

(e) 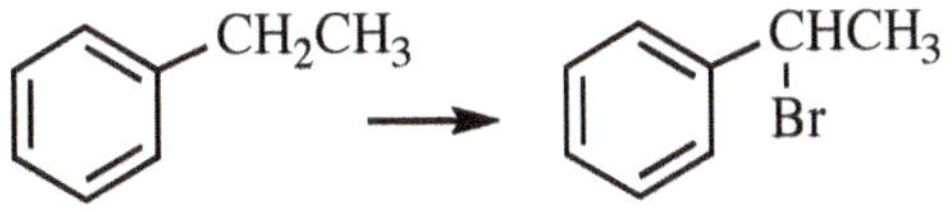 CH_2CH_3 → $CHCH_3$ (Br) from resonance-stabilized benzylic radical

(f) 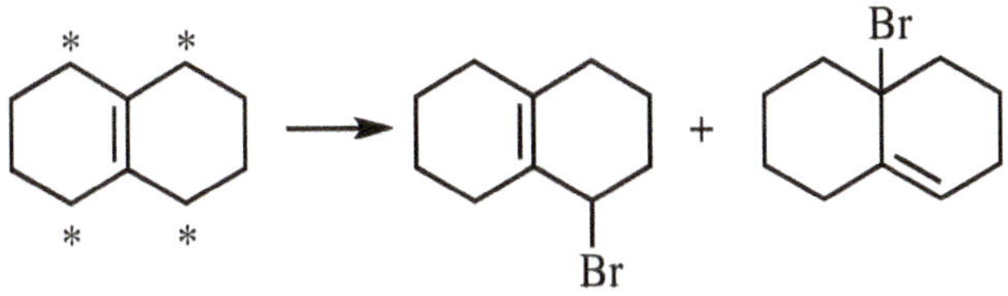 All hydrogens at the starred positions are equivalent and *allylic*. The H from the lower right carbon has been removed to make an intermediate with two non-equivalent resonance forms, giving the two products shown. *Drawing the resonance forms is the key to answering this question correctly.*

4-48

(a) As CH_3Cl is produced, it can compete with CH_4 for available Cl •, generating CH_2Cl_2. This can generate $CHCl_3$, *etc.*

propagation steps

$$CH_4 + Cl\bullet \longrightarrow HCl + \bullet CH_3$$
$$\bullet CH_3 + Cl_2 \longrightarrow ClCH_3 + Cl\bullet$$
$$ClCH_3 + Cl\bullet \longrightarrow HCl + \bullet CH_2Cl$$
$$\bullet CH_2Cl + Cl_2 \longrightarrow CH_2Cl_2 + Cl\bullet$$
$$CH_2Cl_2 + Cl\bullet \longrightarrow HCl + \bullet CHCl_2$$
$$\bullet CHCl_2 + Cl_2 \longrightarrow CHCl_3 + Cl\bullet$$
$$CHCl_3 + Cl\bullet \longrightarrow HCl + \bullet CCl_3$$
$$\bullet CCl_3 + Cl_2 \longrightarrow CCl_4 + Cl\bullet$$

(b) To maximize CH_3Cl and minimize formation of polychloromethanes, the ratio of methane to chlorine must be kept high (see solution to problem 4-2).

To guarantee that all hydrogens are replaced with chlorine to produce CCl_4, the ratio of chlorine to methane must be kept high.

4-49

(a) Pentane can produce three monochloro isomers. To calculate the relative amount of each in the product mixture, multiply the numbers of hydrogens which could lead to that product times the reactivity for that type of hydrogen. Each relative amount divided by the sum of all the amounts will provide the percent of each in the product mixture.

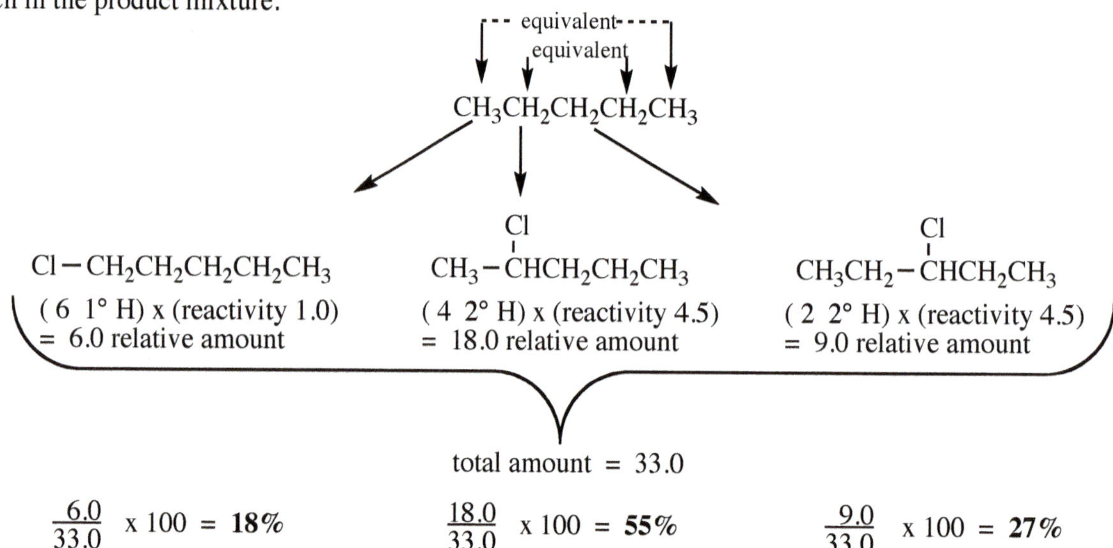

(b) $\frac{6.0}{33.0}$ x 100 = **18%** $\frac{18.0}{33.0}$ x 100 = **55%** $\frac{9.0}{33.0}$ x 100 = **27%**

4-50 (a) The second propagation step in the chlorination of methane is highly exothermic ($\Delta H°$ = −110 kJ/mole (−27 kcal/mole)). The transition state resembles the reactants; that is, the Cl–Cl bond will be slightly stretched and the Cl–CH_3 bond will just be starting to form.

$$\left[\overset{\delta\bullet}{Cl} \cdots\cdots Cl \cdots\cdots\cdots\cdots \overset{\delta\bullet}{CH_3} \right]^{\ddagger}$$

stronger weaker

(b) The second propagation step in the bromination of methane is highly exothermic ($\Delta H°$ = −112 kJ/mole (− 27 kcal/mole)). The transition state resembles the reactants; that is, the Br–Br bond will be slightly stretched and the Br–CH_3 bond will just be starting to form.

$$\left[\overset{\delta\bullet}{Br} \cdots\cdots Br \cdots\cdots\cdots\cdots \overset{\delta\bullet}{CH_3} \right]^{\ddagger}$$

stronger weaker

4-51 Two mechanisms are possible depending on whether HO• reacts with chlorine or with cyclopentane. It is reasonable to use the same bond dissociation energy for the 2° H of propane, 413 kJ/mole (from text Table 4-2), for the C—H bond in cyclopentane.

Mechanism 1 — all in kJ/mole

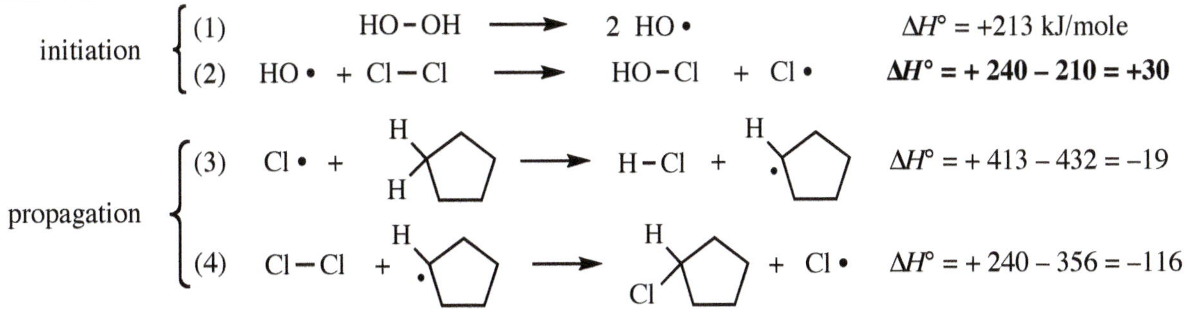

initiation
(1) HO−OH ⟶ 2 HO• $\Delta H° = +213$ kJ/mole
(2) HO• + Cl−Cl ⟶ HO−Cl + Cl• **$\Delta H° = +240 - 210 = +30$**

propagation
(3) Cl• + cyclopentane (H_2C) ⟶ H−Cl + cyclopentyl radical $\Delta H° = +413 - 432 = -19$
(4) Cl−Cl + cyclopentyl radical ⟶ chlorocyclopentane + Cl• $\Delta H° = +240 - 356 = -116$

Mechanism 2

initiation
(1) HO−OH ⟶ 2 HO• $\Delta H° = +213$ kJ/mole
(2) HO• + cyclopentane ⟶ H−OH + cyclopentyl radical **$\Delta H° = +413 - 497 = -84$**

propagation
(3) Cl−Cl + cyclopentyl radical ⟶ chlorocyclopentane + Cl• $\Delta H° = +240 - 356 = -116$
(4) Cl• + cyclopentane ⟶ H−Cl + cyclopentyl radical $\Delta H° = +413 - 432 = -19$

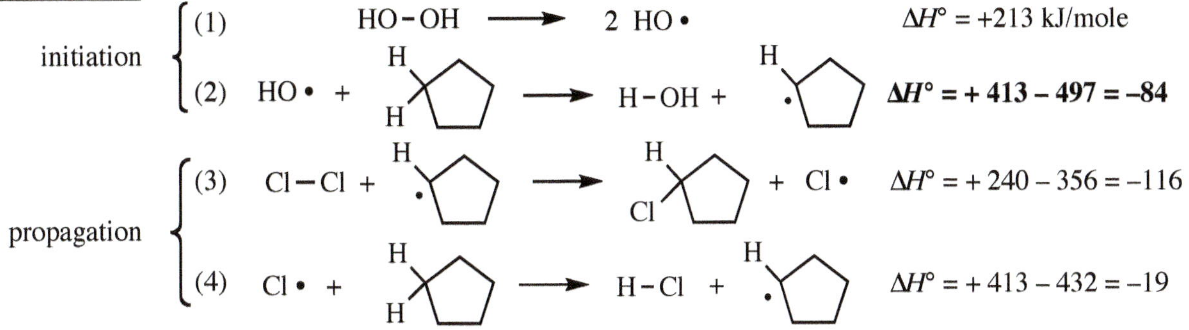

In this case, the energies of initiation steps (in **bold**) determine which mechanism is followed. The bond dissociation energy of HO—Cl is about 210 kJ/mole (about 50 kcal/mole), making initiation step (2) in mechanism 1 *endothermic* by about 30 kJ/mole (about 8 kcal/mole). In mechanism 2, initiation step (2) is *exothermic* by about 84 kJ/mole (20 kcal/mole); mechanism 2 is preferred. One strongly endothermic step can be enough to stop a mechanism.

4-52 $CHCl_3$ (Cl−C(Cl)(H)−H...) + $^{\ominus}$:Ö−H ⟶ H_2O + $Cl-\ddot{C}(Cl)(H)^{\ominus}$ ⟶ Cl−C̈−H + :C̈l:$^{\ominus}$

a carbene

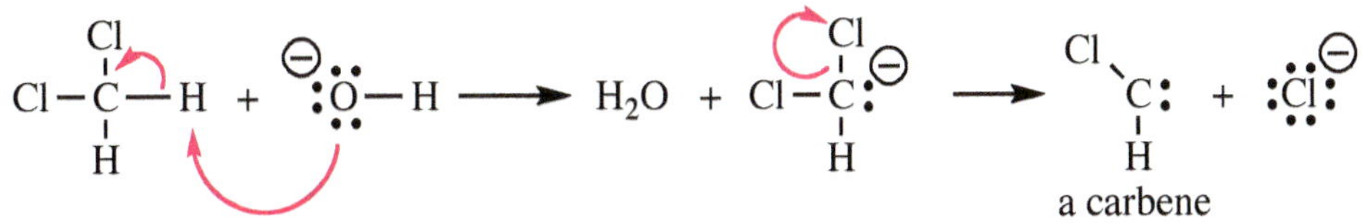

4-53 This critical equation is the key to this problem: $\Delta G = \Delta H - T\Delta S$

At 1400 °K, the equilibrium constant is 1; therefore:

$$K_{eq} = 1 \Longrightarrow \Delta G = -2.303\ RT(\log_{10}(1)) \Longrightarrow \Delta G = 0 \Longrightarrow \Delta H = T\Delta S$$

Assuming ΔH is about the same at 1400 °K as it is at calorimeter temperature:

$$\Delta S = \frac{\Delta H}{T} = \frac{-137\ \text{kJ/mole}}{1400\ °K} = \frac{-137{,}000}{1400}\ \text{J/°K-mole}$$

$$= -98\ \text{J/°K-mole}\ \ (-23\ \text{cal/°K-mole})$$

This is a large *decrease* in entropy, consistent with two molecules combining into one.

4-54
Assume that chlorine atoms (radicals) are still generated in the initiation reaction. Focus on the propagation steps. Bond dissociation energies are given below the bonds, in kJ/mole (kcal/mole).

$$\underset{}{Cl\bullet} + \underset{439\ (105)}{H-CH_3} \longrightarrow \underset{432\ (103)}{H-Cl} + \bullet CH_3 \qquad \Delta H = +7 \text{ kJ/mole } (+2 \text{ kcal/mole})$$

$$\underset{240\ (57)}{Cl-Cl} + \bullet CH_3 \longrightarrow \underset{350\ (84)}{Cl-CH_3} + Cl\bullet \qquad \Delta H = -110 \text{ kJ/mole } (-27 \text{ kcal/mole})$$

What happens when the different radical species react with iodine?

$$Cl\bullet + \underset{149\ (36)}{I-I} \longrightarrow \underset{211\ (50)}{I-Cl} + \bullet I \qquad \Delta H = -62 \text{ kJ/mole } (-14 \text{ kcal/mole})$$

$$\underset{149\ (36)}{I-I} + \bullet CH_3 \longrightarrow \underset{241\ (58)}{I-CH_3} + I\bullet \qquad \Delta H = -92 \text{ kJ/mole } (-22 \text{ kcal/mole})$$

Compare the second reaction in each pair: methyl radical reacting with chlorine is more exothermic than methyl radical reacting with iodine; this does not explain how iodine prevents the chlorination reaction. Compare the first reaction in each pair: chlorine atom reacting with iodine is very exothermic whereas chlorine atom reacting with methane is slightly endothermic. That is the key: chlorine atoms will be scavenged by iodine before they have a chance to react with methane. Without chlorine atoms, the reaction comes to a dead stop.

4-55 (a)

initiation
(1) $Br-Br \xrightarrow{h\nu} 2\ Br\bullet$
(2) $Br\bullet + H-SnBu_3 \longrightarrow H-Br + \bullet SnBu_3$

propagation
(3) (bromocyclohexane, H) $Br + \bullet SnBu_3 \longrightarrow$ (cyclohexyl radical, H) $+ Br-SnBu_3$
(4) (cyclohexyl radical, H) $+ H-SnBu_3 \longrightarrow$ (cyclohexane, H, H) $+ \bullet SnBu_3$

(b) All energies are in kJ/mole. The abbreviation "c-Hx" stands for the cyclohexane ring.

Step 2: break H—Sn, make H—Br: +310 + −366 = −56 kJ/mole

Step 3: break c-Hx—Br, make Br—Sn: +309 + −552 = −243 kJ/mole WOW!

Step 4: break H—Sn, make c-Hx—H: +310 + −413 = −103 kJ/mole

The sum of the two propagation steps is: −243 + −103 = −346 kJ/mole —a hugely exothermic reaction.

4-56

Mechanism 1:

(1a) $Cl\bullet + O_3 \longrightarrow ClO\bullet + O_2$

(1b) $\mathbf{2}\ ClO\bullet \longrightarrow Cl{-}O{-}O{-}Cl$

(1c) $Cl{-}O{-}O{-}Cl \xrightarrow{h\nu} O_2 + \mathbf{2}\ Cl\bullet$

The biggest problem in Mechanism 1 lies in step (1b). The concentration of Cl atoms is very small, so at any given time, the concentration of ClO will be very small. The probability of two ClO radicals finding each other to form ClOOCl is virtually zero. Even though this mechanism shows a catalytic cycle with Cl• (starting the mechanism and being regenerated at the end), the middle step makes it highly unlikely.

Mechanism 2:

(2a) $O_3 \xrightarrow{h\nu} O_2 + O$

Step (2a) is the "light" reaction that occurs naturally in daylight. At night, the reaction reverses and regenerates ozone.

(2b) $Cl\bullet + O_3 \longrightarrow ClO\bullet + O_2$

(2c) $ClO\bullet + O \longrightarrow O_2 + Cl\bullet$

Step (2c) is the crucial step. A low concentration of ClO *will* find a relatively high concentration of O atoms because the "light" reaction is producing O atoms in relative abundance. Cl• is regenerated and begins propagation step (2b), continuing the catalytic cycle.

Mechanism 2 is believed to be the dominant mechanism in ozone depletion. Mechanism 1 can be discounted because of the low probability of step (1b) occurring, because two species in very low, catalytic concentration are required to find each other in order for the step to occur.

4-57

(a) $\left[H_3C^{\delta\bullet} \cdots\cdots H \cdots\cdots Cl^{\delta\bullet} \right]^{\ddagger}$ $\left[D_3C^{\delta\bullet} \cdots\cdots D \cdots\cdots Cl^{\delta\bullet} \right]^{\ddagger}$

In each case, the bond from carbon to H (D) is breaking and the bond from H (D) to Cl is forming.

(b) $C_2H_5{-}D + Cl_2 \longrightarrow \underbrace{C_2H_5{-}Cl + DCl}_{7\%} + \underbrace{C_2H_4DCl + HCl}_{93\%}$

D replacement: 7% ÷ 1 D = 7 (reactivity factor)
H replacement: 93% ÷ 5 H = 18.6 (reactivity factor)
relative reactivity of H : D abstraction = 18.6 ÷ 7 = 2.7

Each hydrogen is abstracted 2.7 times faster than deuterium.

(c) In both reactions of chlorine with either methane or ethane, the first propagation step is rate-limiting. The reaction of a chlorine atom with methane is *endothermic* by 7 kJ/mole (2 kcal/mole), while for ethane this step is *exothermic* by 9 kJ/mole (2 kcal/mole). By Hammond's Postulate, differences in activation energy are most pronounced in *endothermic* reactions where the transition states most resemble the products. Therefore, a change in the methane molecule causes a greater change in its transition state energy than the same change in the ethane molecule causes in its transition state energy. Deuterium will be abstracted more slowly in both methane and ethane, but the rate effect will be more pronounced in methane than in ethane.

4-58 All energies are in kJ/mole (kcal/mole).

initiation

(1) HO—OH $\xrightarrow{\Delta}$ **2** HO• ΔH = + 213 kJ/mole (+ 51)

(2) HO• + I—CI$_3$ ⟶ HO—I + •CI$_3$ ΔH = – 46 kJ/mole (– 11)

I—CI$_3$: +188 (+45); HO—I: –234 (–56)

propagation

(3) cyclopentane–H + •CI$_3$ ⟶ cyclopentyl• + H—CI$_3$ ΔH = – 5 kJ/mole (– 1)

C—H: +413 (+99); H—CI$_3$: –418 (–100)

(4) cyclopentyl• + I—CI$_3$ ⟶ iodocyclopentane + •CI$_3$ ΔH = – 50 kJ/mole (– 12)

I—CI$_3$: +188 (+45); C—I: –238 (–57)

The sum of the two propagation steps is: – 5 + – 50 = – 55 kJ/mole —a mildly exothermic reaction.
(– 1 + – 12 = – 13 kcal/mole)

CHAPTER 5—STEREOCHEMISTRY

Note to the student: Stereochemistry is the study of molecular structure and reactions in three dimensions. Molecular models will be especially helpful in this chapter.

5-1 The best test of whether a household object is chiral is whether it would be used equally well by a left- or right-handed person. The chiral objects are the corkscrew, the writing desk, the can opener, the screw-cap bottle (only for refilling, however; in use, it would not be chiral), the rifle and the knotted rope. The corkscrew, the bottle top, and the rope each have a twist in one direction. The rifle, corkscrew, and desk are clearly made for right-handed users; the can opener is for a left-handed person. All the other objects are achiral and would feel equivalent to right- or left-handed users.

5-2

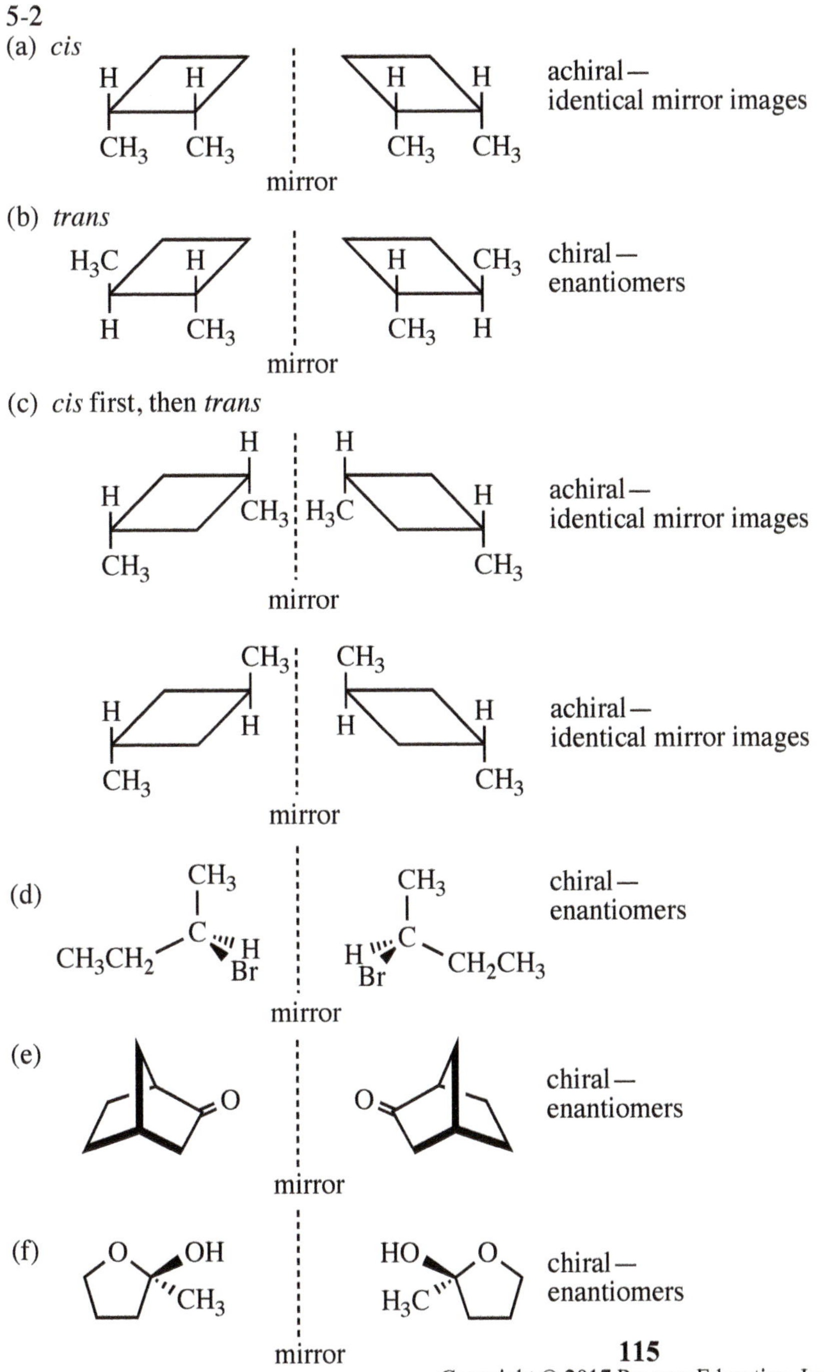

achiral—identical mirror images

chiral—enantiomers

achiral—identical mirror images

achiral—identical mirror images

chiral—enantiomers

chiral—enantiomers

chiral—enantiomers

5-3 Asymmetric carbon atoms are starred.

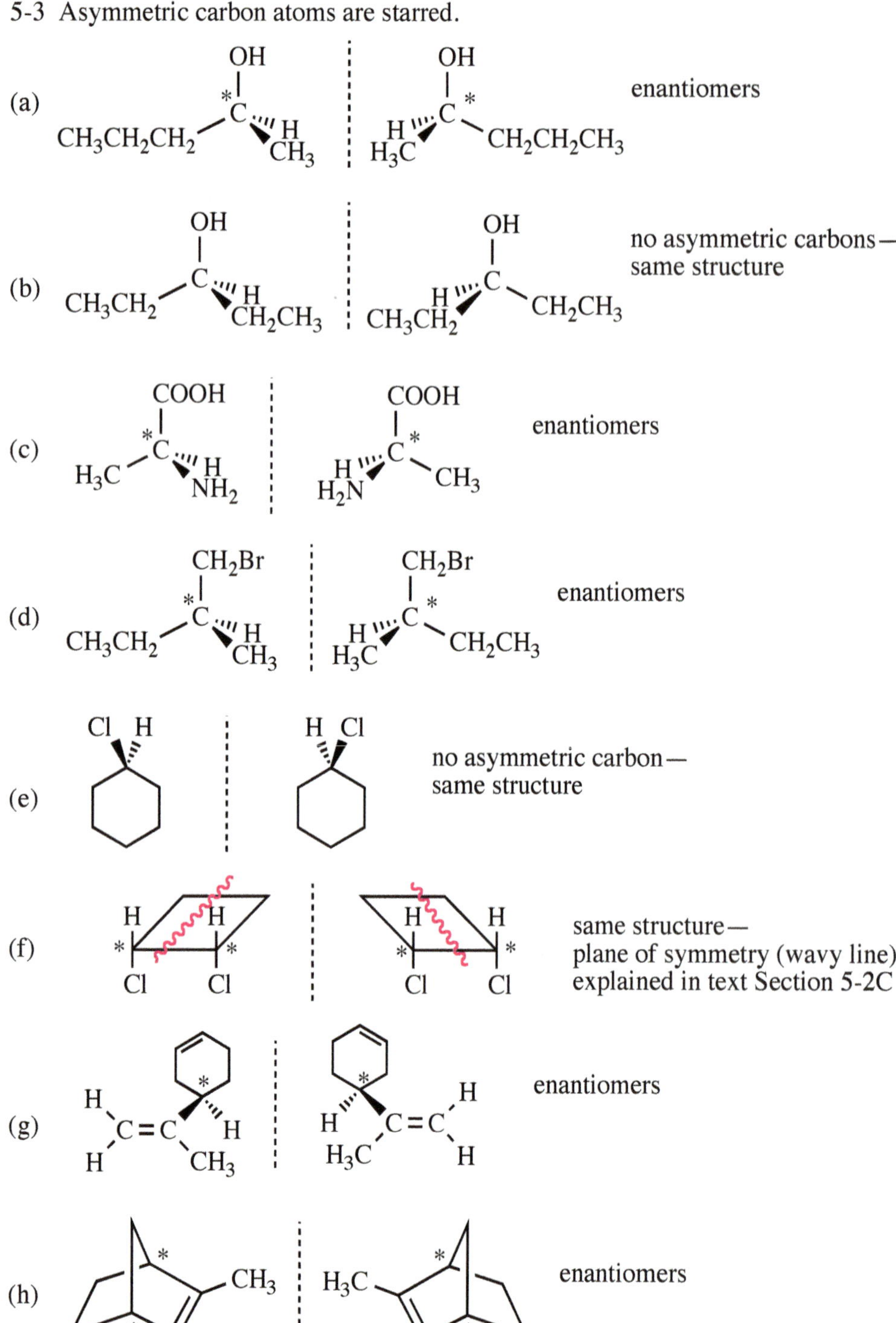

5-4 You may have chosen to interchange two groups different from the ones shown here. The type of isomer produced will still be the same as listed here.

Interchanging any two groups around a chiral center (*) will create an enantiomer of the first structure.

Interchanging the Br and the H creates an enantiomer of the structure in Figure 5-5.

Interchanging the ethyl and the isopropyl creates an enantiomer of the structure in Figure 5-5.

On a double bond, interchanging the two groups on ONE of the stereocenters will create the other geometric (*cis-trans*) isomer. However, interchanging the two groups on BOTH of the stereocenters will give the original structure.

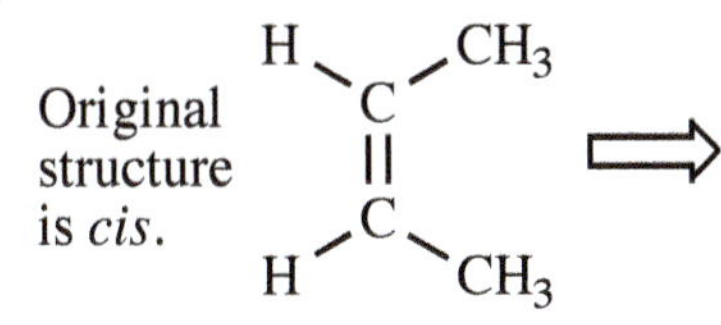

Original structure is *cis*.

Interchange H and CH_3 on top stereocenter to produce *trans* (interchanging bottom two groups will give the same structure).

5-5

(a) plane of symmetry

(b)

plane of symmetry

(c) chiral—no plane of symmetry

(d) chiral—no plane of symmetry

(e) chiral—no plane of symmetry

(f) chiral—no plane of symmetry

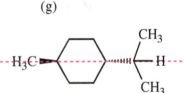

(g) plane of symmetry

(h) plane of symmetry

This view is from the right side of the structure as drawn in the text.

5-6 Place the 4th priority group away from you, where possible. Then determine if the sequence 1→2→3 is clockwise (*R*) or counter-clockwise (*S*). (There is a Problem-Solving Hint near the end of section 5-3 in the text that describes what to do when the 4th priority group is closest to you.)

(a)

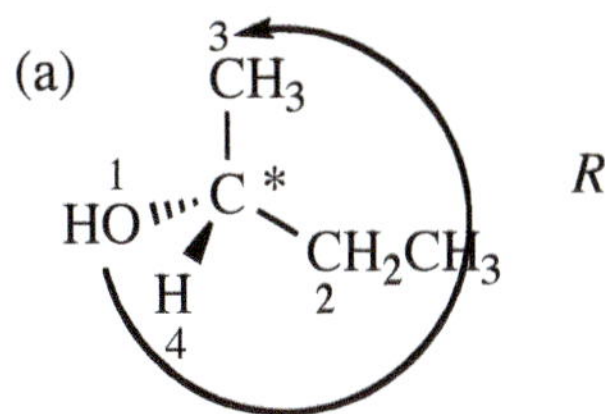

R

This appears to be *S*, but group 4 is coming toward the viewer, so the opposite chirality must be assigned; it is actually *R*.

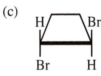

(b) *S*

Viewing from the bottom to put group 4 going away, the arrow is counterclockwise; the chirality is *S*.

(c)

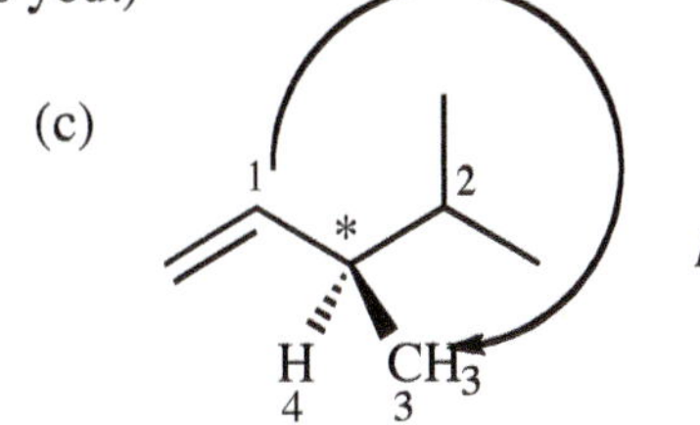

R

Viewing from the front to put group 4 going away, the arrow is clockwise; the chirality is *R*.

5-6 continued

(d)

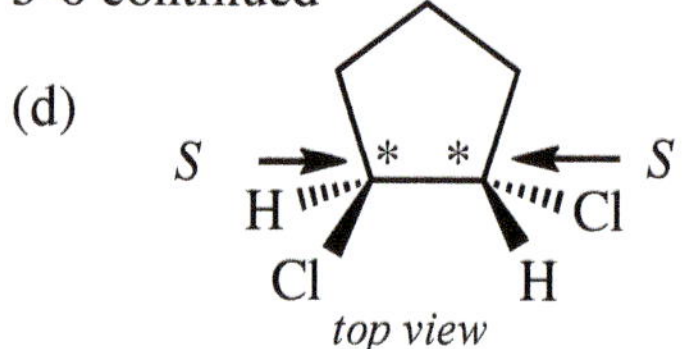

(e)

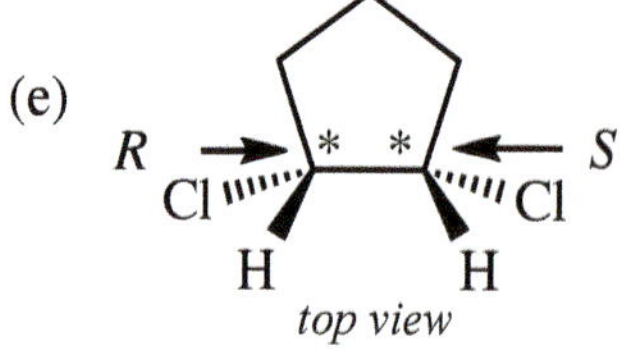

(f)

(g)

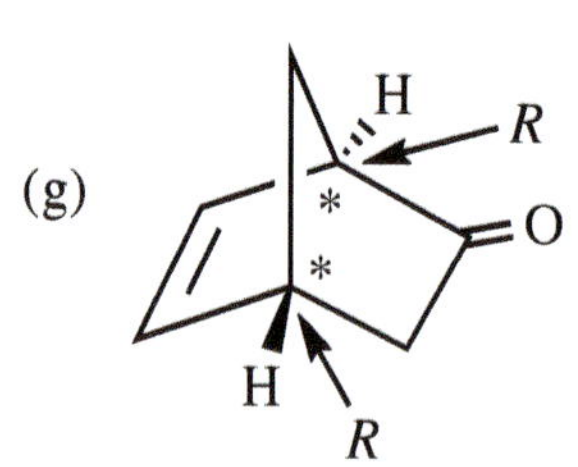

(h)

(i) S

Part (i) deserves some explanation. The difference between groups 1 and 2 hinges on what is on the "extra" oxygen.

higher priority

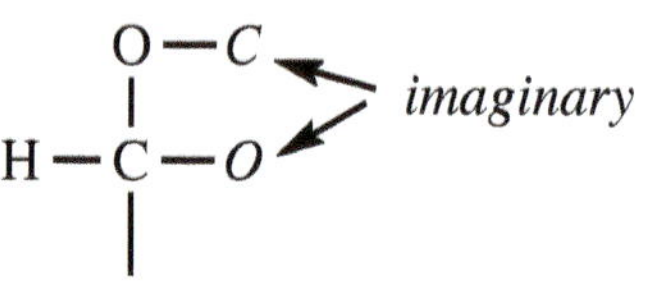

lower priority

5-7 There are no asymmetric carbons in 5-3 (b) and (e).

(a)

(c)

(d)

(f)

The identity of these two structures becomes more clear with the assignments of configuration.

(g)

(h)

(i)

What can you conclude about the designation of configuration between two enantiomers?

5-8

2.0 g / 10.0 mL = 0.20 g/mL ; 100 mm = 1 dm

$$[\alpha]_D^{25} = \frac{+1.74°}{(0.20)\,(1)} = +8.7° \text{ for (+)-glyceraldehyde}$$

5-9

0.50 g / 10.0 mL = 0.050 g/mL ; 20 cm = 2 dm

$$[\alpha]_D^{25} = \frac{-5.1°}{(0.050)\,(2)} = -51° \text{ for (–)-epinephrine}$$

5-10

Measure using a solution of about one-fourth the concentration of the first. The value will be either + 45° or – 45°, which gives the sign of the rotation.

5-11

Whether a sample is dextrorotatory (abbreviated "(+)") or levorotatory (abbreviated "(–)") is determined experimentally by a polarimeter. Except for the molecule glyceraldehyde (see text Section 5-14), there is no direct, universal correlation between direction of optical rotation ((+) and (–)) and designation of configuration (*R* and *S*). In other words, one dextrorotatory compound might have *R* configuration while a different dextrorotatory compound might have *S* configuration.

(a) Yes, both of these are determined experimentally: the (+) or (–) by the polarimeter and the smell by the nose.

(b) No, *R* or *S* cannot be determined by either the polarimeter or the nose.

(c) The drawings show that (+)-carvone from caraway has the *S* configuration and (–)-carvone from spearmint has the *R* configuration.

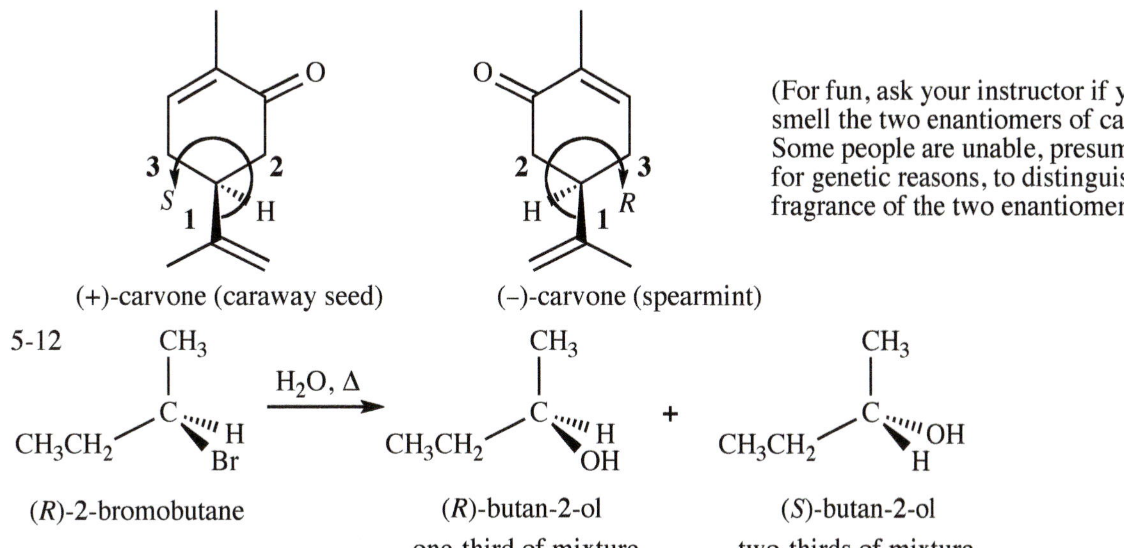

(For fun, ask your instructor if you can smell the two enantiomers of carvone. Some people are unable, presumably for genetic reasons, to distinguish the fragrance of the two enantiomers.)

one-third of mixture — two-thirds of mixture

Chapter 6 will explain how these mixtures come about. For this problem, the *S* enantiomer accounts for 66.7% of the butan-2-ol in the mixture and the rest, 33.3%, is the *R* enantiomer. Therefore, the excess of one enantiomer over the racemic mixture must be 33.3% of the *S*, the enantiomeric excess. (All of the *R* is "canceled" by an equal amount of the *S*, algebraically as well as in optical rotation.)

The optical rotation of pure (*S*)-butan-2-ol is + 13.5°. (If you had chosen the opposite sign, then the answer will be opposite. That is not the important part of this problem.) The optical rotation of this mixture is:

$$33.3\% \times (+13.5°) = +4.5°$$

see next page for an alternative solution

5-12 continued

(This algebraic approach has been suggested by Editorial Adviser Richard King.)
From the problem, the reaction produces twice as much (*S*)-(+)-butan-2-ol (the *d* isomer) as (*R*)-(–)-butan-2-ol (the *l* isomer): $d = 2\,l$

$$\text{e.e.} = \frac{d-l}{d+l} \text{ x } 100\% = \frac{2l-l}{2l+l} \text{ x } 100\% = \frac{l}{3l} \text{ x } 100\% = 33.3\%$$

The calculation of optical rotation of the mixture is the same as on the previous page.

5-13 The rotation of pure (+)-butan-2-ol is + 13.5°.

$$\frac{\text{observed rotation}}{\text{rotation of pure enantiomer}} = \frac{+0.45°}{+13.5°} \text{ x } 100\% = 3.3\% \text{ optical purity}$$
$$= 3.3\% \text{ e.e.} = \text{excess of (+) over (–)}$$

To calculate percentages of (+) and (–): (two equations in two unknowns)

$$(+) + (-) = 100\% \Longrightarrow (-) = 100\% - (+)$$

$$(+) - (-) = 3.3\% \Longrightarrow (+) - (100\% - (+)) = 3.3\%$$

$$\mathbf{2}\ (+) = 103.3\%$$

(+) = 51.6% (rounded) (–) = 48.4%

(This algebraic approach has been suggested by Editorial Adviser Richard King.)

$$\text{e.e.} = \frac{d-l}{d+l} \text{ x } 100\% = 3.33\% \Longrightarrow \frac{d-l}{d+l} = \frac{3.33\%}{100\%} \Longrightarrow$$

$$\left.\begin{aligned} d - l &= 3.33\% \\ \text{and } d + l &= 100\% \end{aligned}\right\} \text{add these two equations}$$

$$2d = 103.33\% \Longrightarrow d = 51.6\% \qquad l = 100\% - d = 100\% - 51.6\% = 48.4\%$$

5-14 Drawing Newman projections is the clearest way to determine symmetry of conformations.

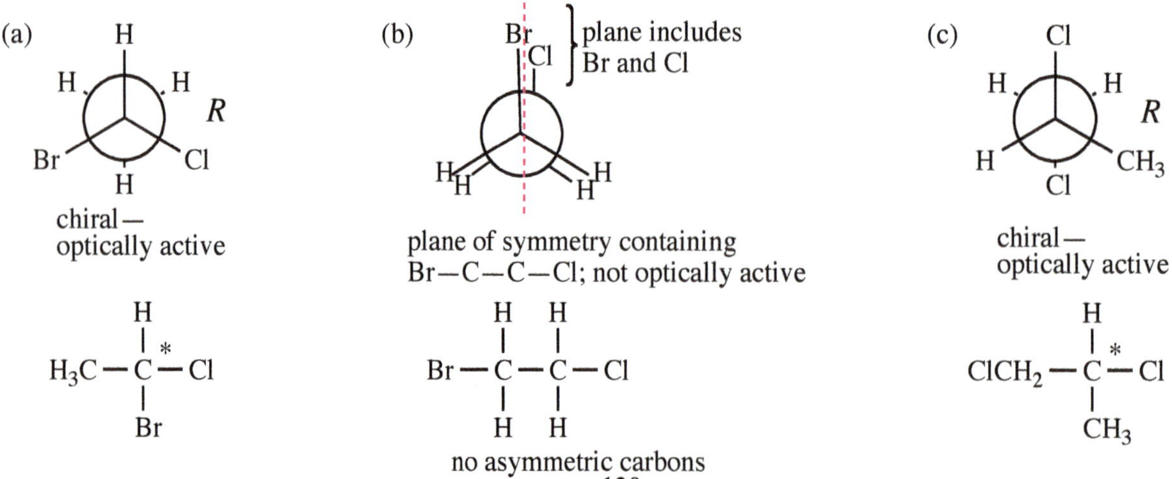

5-14 continued

(d)

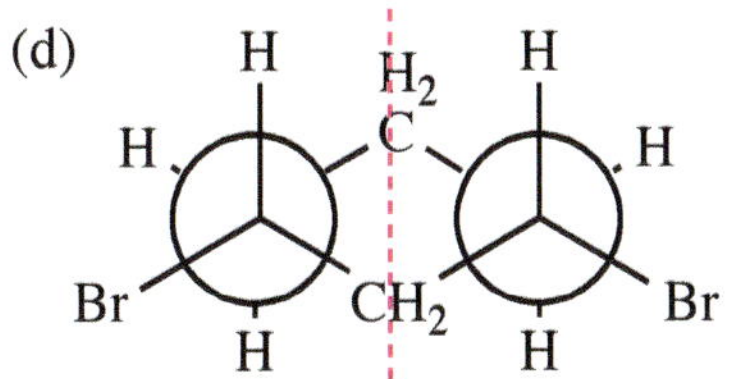

plane of symmetry—not optically active despite the presence of two asymmetric carbons

(e)

no plane of symmetry—optically active (other chair form is equivalent—no plane of symmetry)

(f)

plane of symmetry through C-1 and C-4—not optically active

no asymmetric carbons

Part (2) Predictions of optical activity based on asymmetric centers give the same answers as predictions based on the most symmetric conformation. NOTE: The assignment of *R* and *S* above might be different from your answers if you happened to draw the enantiomer.

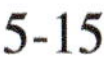

5-15

(a)

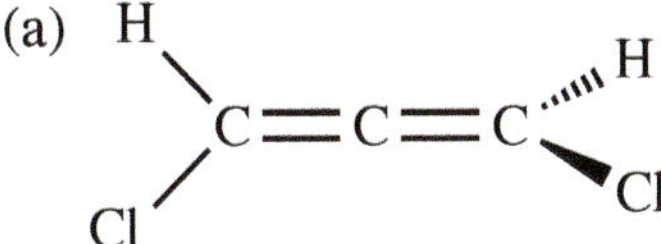

No asymmetric carbons, but the molecule is chiral (an allene); the drawing below is a three-dimensional picture of the allene in (a) showing there is no plane of symmetry because the substituents of an allene are in different planes.

(b)

No asymmetric carbons, but the molecule is chiral (an allene).

(c)

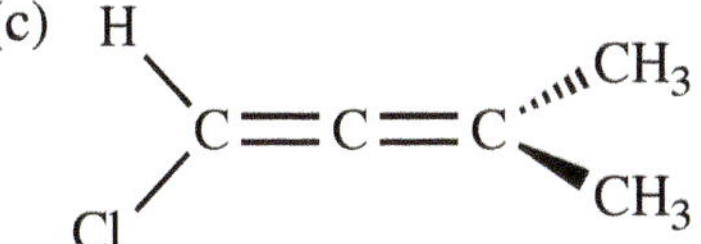

No asymmetric carbons; this allene has a plane of symmetry between the two methyls (the plane of the paper), including all the other atoms because the two pi bonds of an allene are perpendicular, the Cl is in the plane of the paper and the plane of symmetry goes through it; not a chiral molecule.

(d)

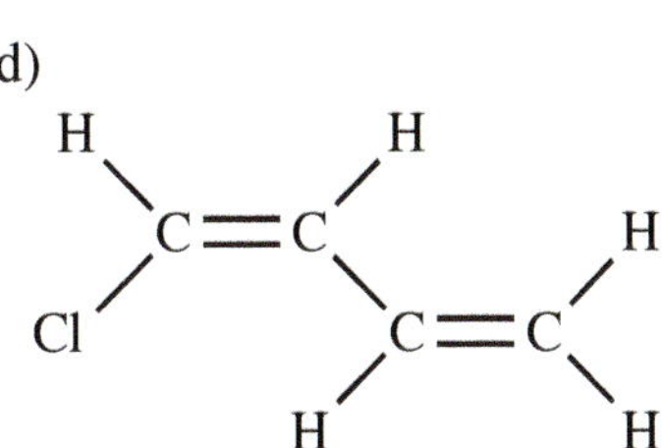

planar molecule—no asymmetric carbons; not a chiral molecule

(e)

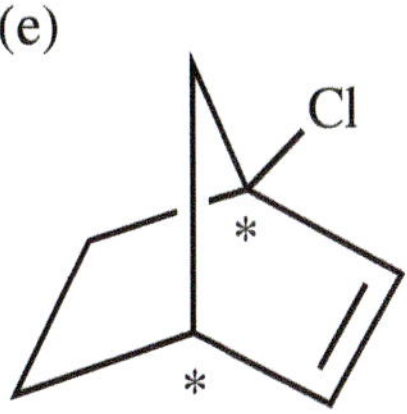

two asymmetric carbons and no plane of symmetry; a chiral compound

5-15 continued

(f)

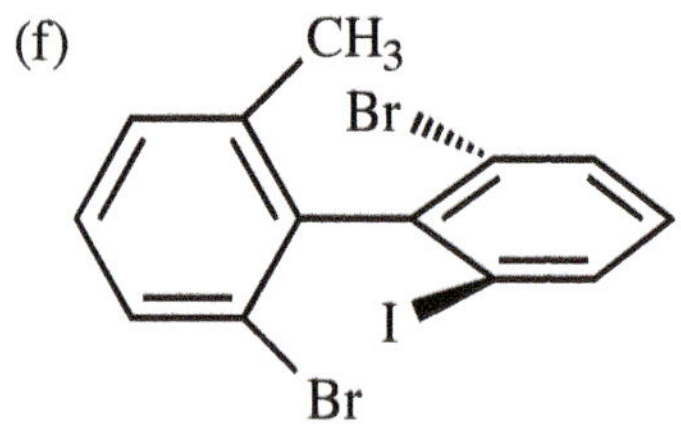

No asymmetric carbons, but the molecule is chiral due to restricted rotation that precludes a plane of symmetry; the drawing below is a three-dimensional picture showing that the rings are perpendicular (hydrogens are not shown).

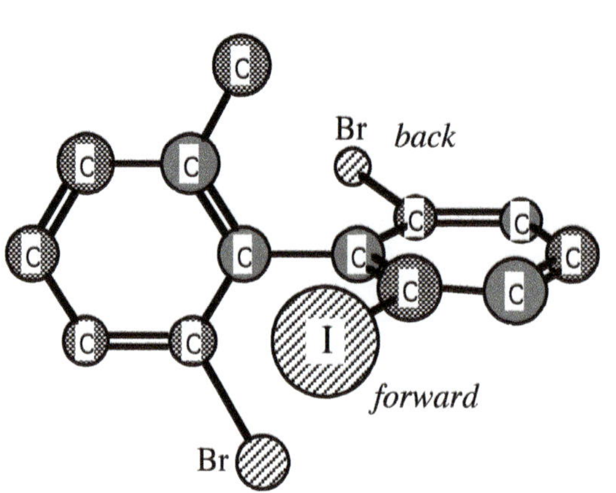

(g)

D D

H H

No asymmetric carbons, and the groups are not large enough to restrict rotation; thus, it has a plane of symmetry and is not a chiral compound.

5-16

(a)

	original			
top	COOH	COOH	H	CH_3
left — right	H — OH	HO — H	CH_3 — COOH	HO — H
bottom	CH_3	CH_3	OH	COOH
		enantiomer	enantiomer	same

(b)

	original			
top	CH_2CH_3	CH_3	CH_2CH_3	CH_3
left — right	H — Br	Br — H	Br — H	H — Br
bottom	CH_3	CH_2CH_3	CH_3	CH_2CH_3
		same	enantiomer	enantiomer

(c)

	original			
top	CH_3	CH_3	CH_3	CH_2CH_3
left — right	HO — H	H — OH	HO — H	H — OH
bottom	CH_2CH_3	CH_2CH_3	CH_2CH_3	CH_3
	(*R*)-butan-2-ol	enantiomer	same	same

Rules for Fischer projections:

1. Interchanging any two groups an odd number of times (once, three times, etc.) makes an enantiomer. Interchanging any two groups an even number of times (e.g. twice) returns to the original stereoisomer.

2. Rotating the structure by 90° makes the enantiomer. Rotating by 180° returns to the original stereoisomer. (The second rule is an application of the first. Prove this to yourself.)

5-17

	(a)	(b)	(c)	(d)	(e)
top	CH_2OH	CH_2OH	CH_2Br	CH_3	CHO
left — right	HO — H	H — Br	Br — H	HO — H	H — OH
bottom	CH_3	CH_2CH_3	CH_2CH_3	CH_2CH_3	CH_2OH

5-18

(a) Fischer projections across a mirror: left — CHO top, H–C–OH, CH_2OH bottom; right — CHO top, HO–C–H, CH_2OH bottom.

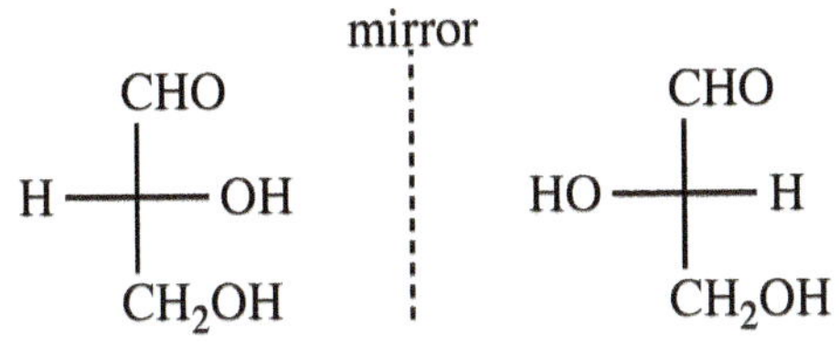

180° rotation of the right structure does not give left structure; no plane of symmetry: chiral—**enantiomers**

(b) Left — CH_2OH top, H–C–Br, CH_2OH bottom; right — CH_2OH top, Br–C–H, CH_2OH bottom.

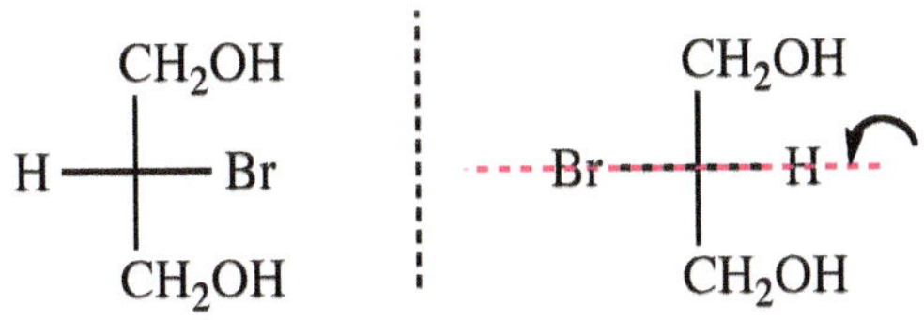

180° rotation of the right structure gives same structure as on the left; also has plane of symmetry: **same structure**

(c) Left — CH_2Br top, Br–C–Br, CH_3 bottom; right — CH_2Br top, Br–C–Br, CH_3 bottom.

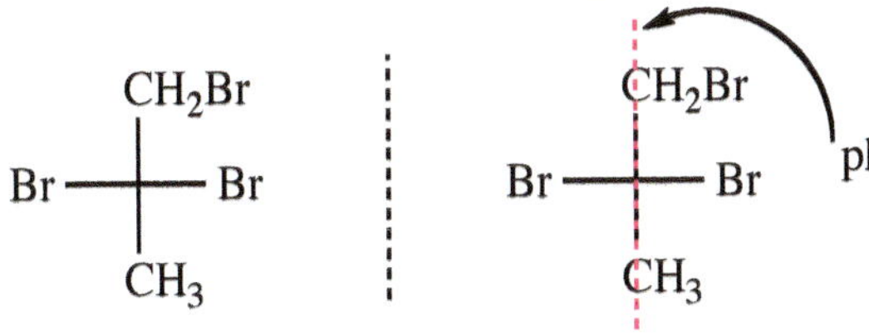

plane of symmetry: **same structure**

(d) Left — CHO top, H–C–OH, H–C–OH, CH_2OH bottom; right — CHO top, HO–C–H, HO–C–H, CH_2OH bottom.

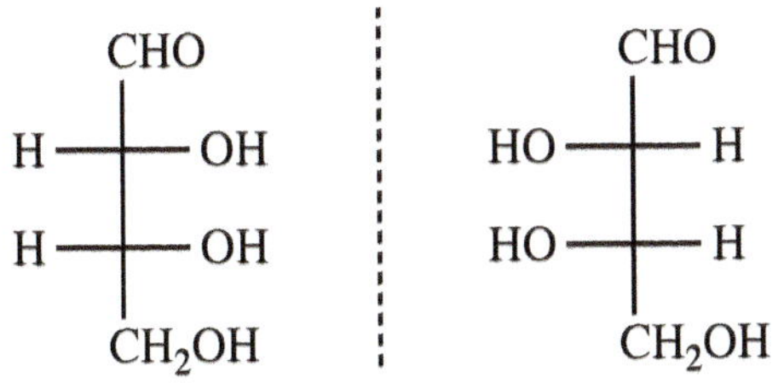

180° rotation of the right structure does not give left structure; no plane of symmetry: chiral—**enantiomers**

(e) Left — CH_2OH top, H–C–OH, H–C–OH, CH_2OH bottom; right — CH_2OH top, HO–C–H, HO–C–H, CH_2OH bottom.

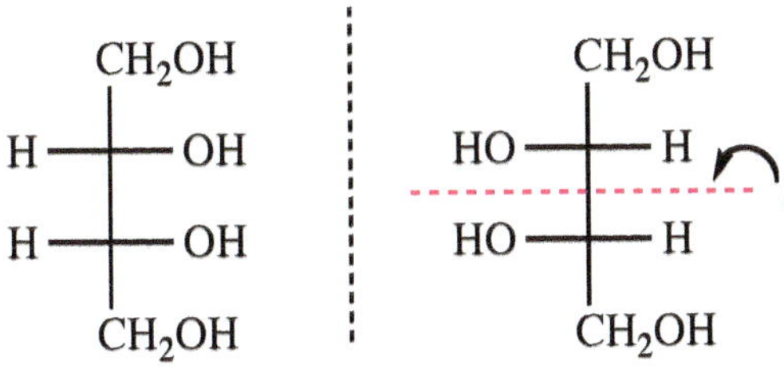

180° rotation of the right structure gives same structure as on the left; also has plane of symmetry: **same structure**

(f) Left — CH_2OH top, HO–C–H, H–C–OH, CH_2OH bottom; right — CH_2OH top, H–C–OH, HO–C–H, CH_2OH bottom.

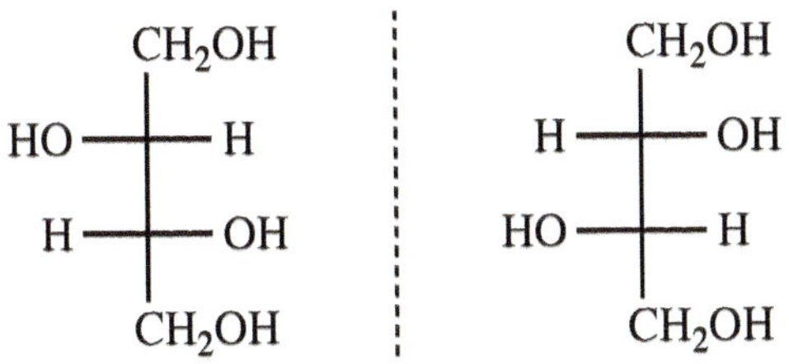

180° rotation of the right structure does not give left structure; no plane of symmetry: chiral—**enantiomers**

5-19 If the Fischer projection is drawn correctly, the most oxidized carbon (most bonds to oxygen) will be at the top; this is the carbon with the greatest number of bonds to oxygen. Then the numbering goes from the top down.

(a) *R*
(b) no chiral center
(c) no chiral center
(d) *2R,3R*
(e) *2S,3R* (numbering down)
(f) *2R, 3R*
(g) *R*
(h) *S*
(i) *S*

5-20

(a) enantiomers—configurations at both asymmetric carbons inverted
(b) diastereomers—configuration at only one asymmetric carbon inverted
(c) diastereomers—configuration at only one asymmetric carbon inverted (the left carbon)
(d) constitutional isomers—C=C shifted position
(e) enantiomers—chiral, mirror images
(f) diastereomers—configuration at only one asymmetric carbon inverted (the top one)
(g) enantiomers—configuration at all asymmetric carbons inverted
(h) enantiomers—the front C is an asymmetric carbon atom; the two Newmans are mirror images
(i) diastereomers—configuration at only one chiral center (the nitrogen) inverted

5-21 It would be excellent practice to draw Newman projections of each structure! Symmetry is quickly revealed.

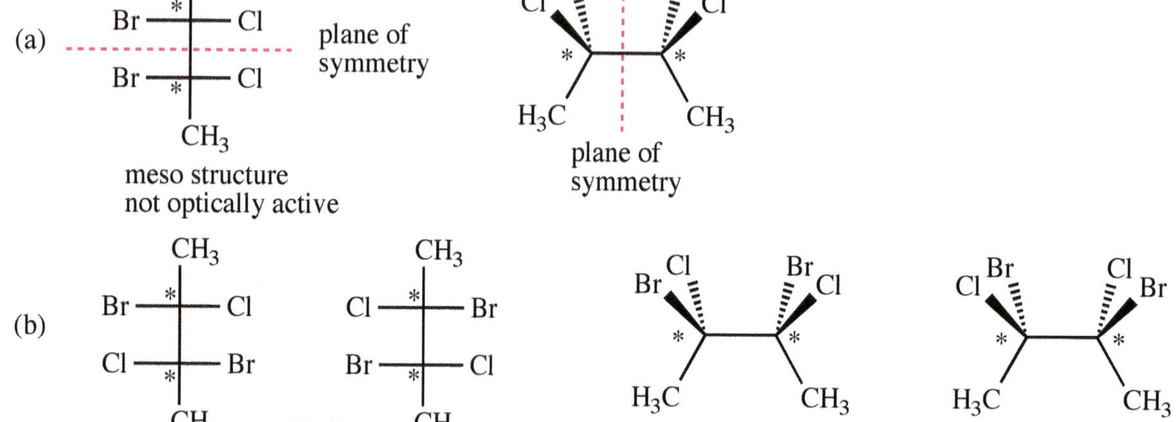

racemic mixture—not optically active, although each enantiomer by itself would be optically active

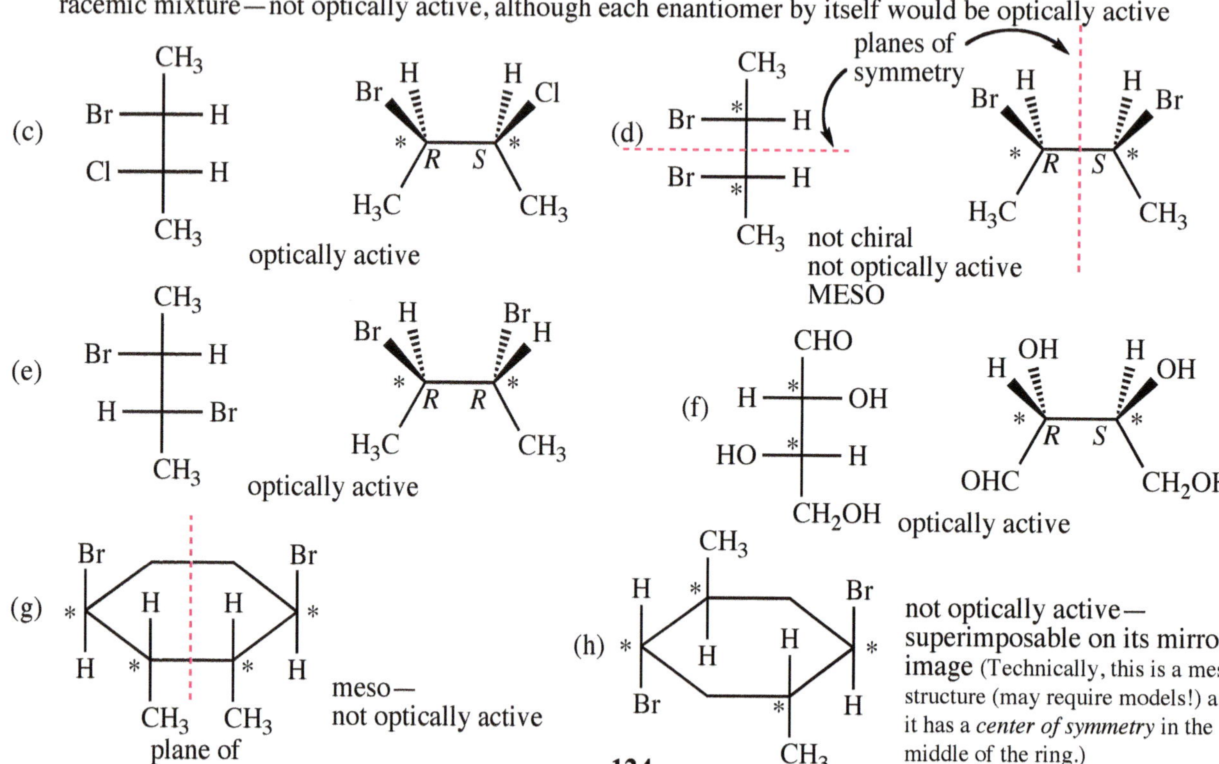

5-22

(a)

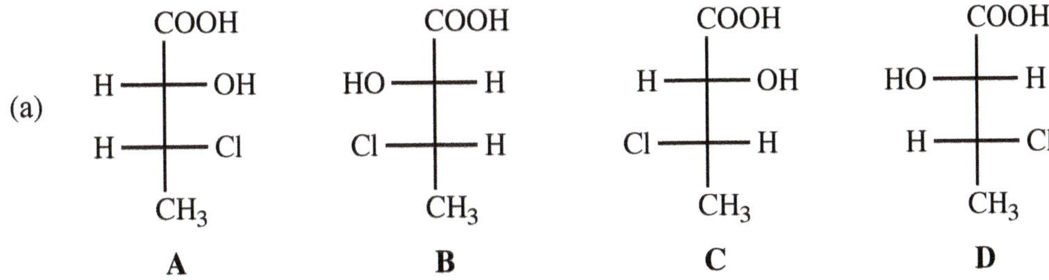

enantiomers: **A** and **B**; **C** and **D**
diastereomers: **A** and **C**; **A** and **D**; **B** and **C**; **B** and **D**

(b)

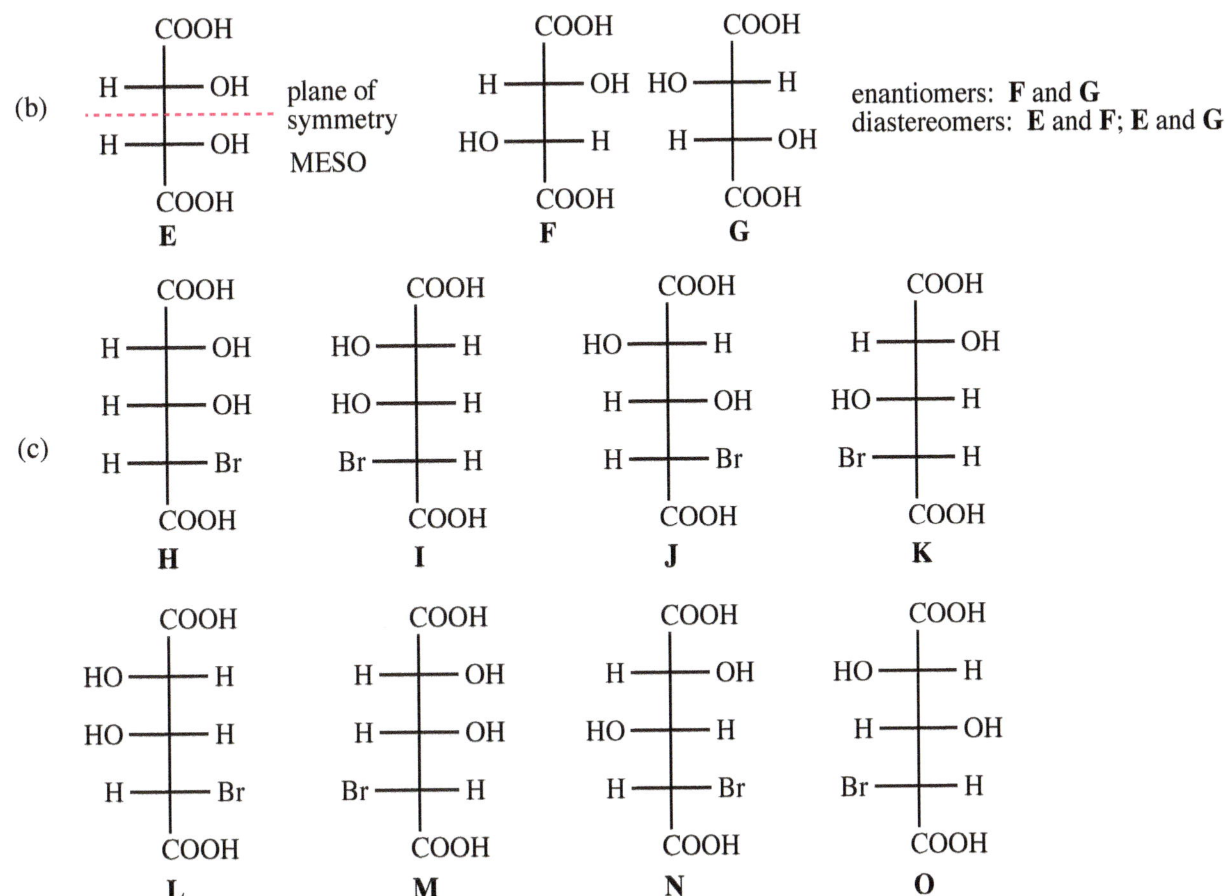

(c)

enantiomers: **H** and **I**; **J** and **K**; **L** and **M**; **N** and **O**
diastereomers: any pair that is not enantiomeric

(d)

P plane of symmetry MESO

cis ring fusion

Q **R**

trans ring fusion

enantiomers: **Q** and **R**
diastereomers: **P** and **Q**; **P** and **R**

5-22 continued

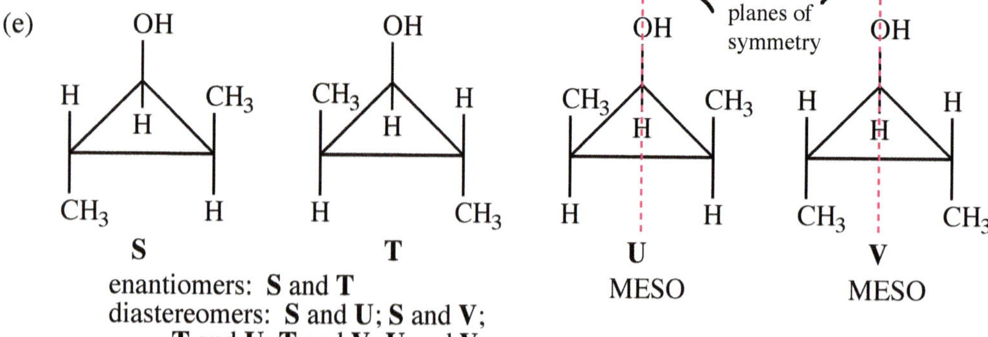

5-23 Any diastereomeric pair could be separated by a physical process like distillation or crystallization. Diastereomers are found in parts (a), (b), and (d). The structures in (c) are enantiomers; they could not be separated by normal physical means.

5-24

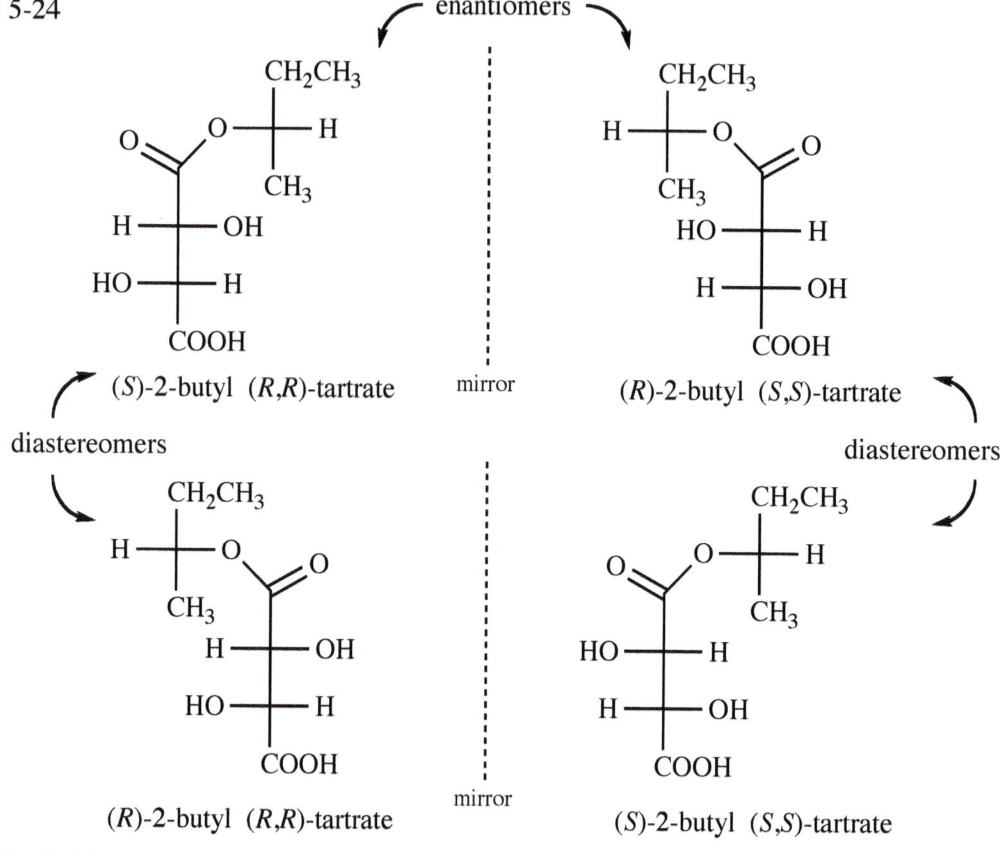

5-25 The asymmetric carbon atoms are indicated by asterisks.

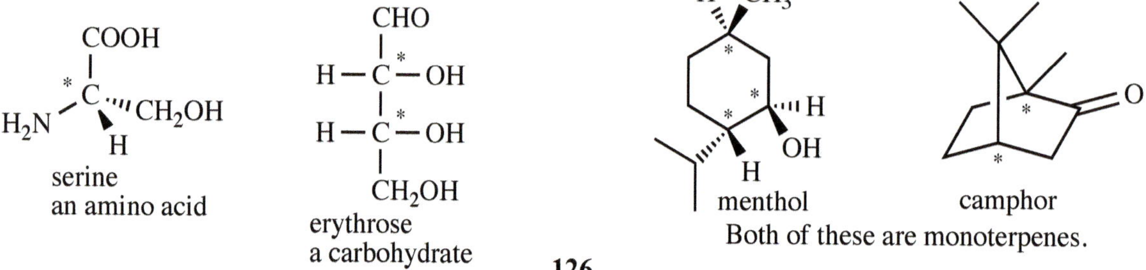

5-26

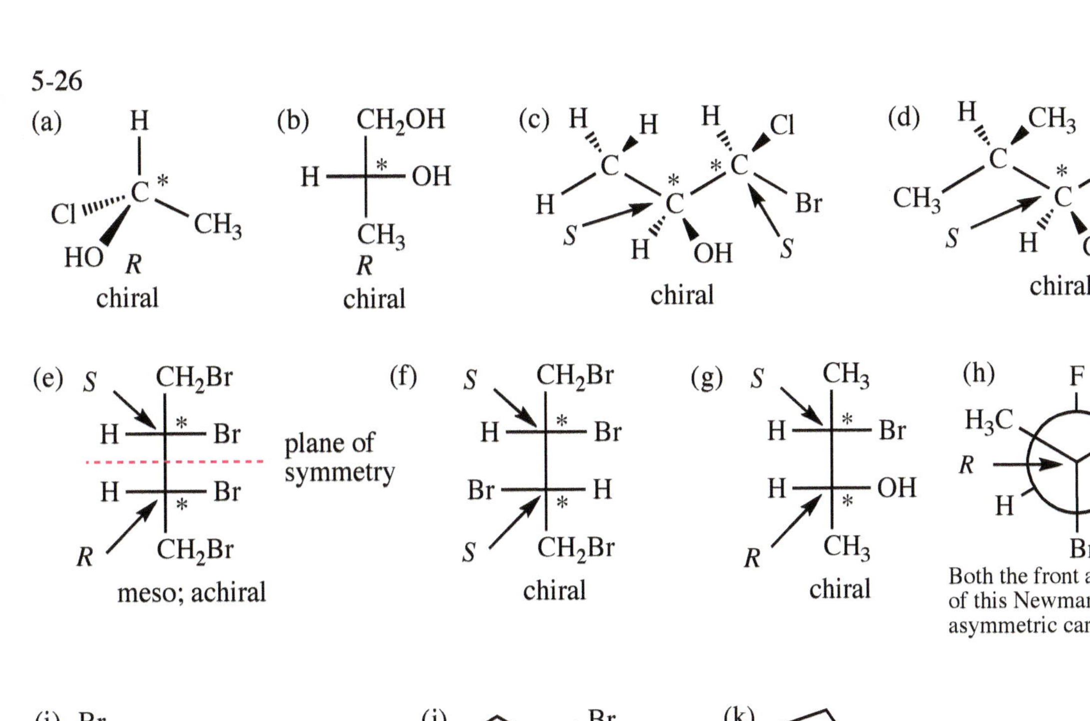

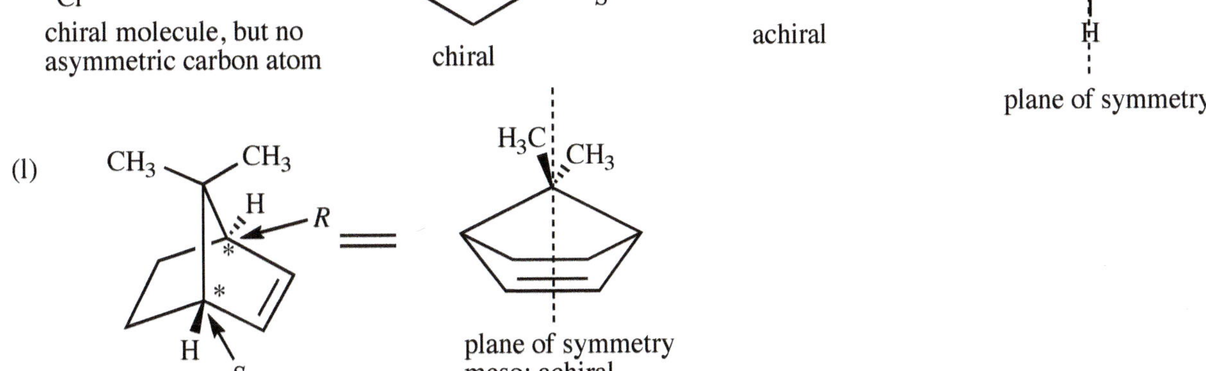

(l)

R

S

plane of symmetry
meso; achiral

5-27

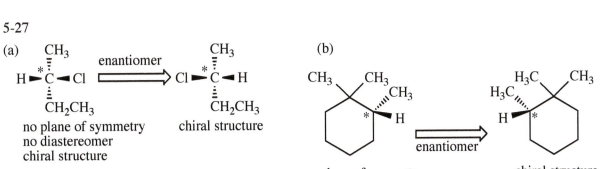

(a)

enantiomer

no plane of symmetry
no diastereomer
chiral structure

chiral structure

(b)

enantiomer

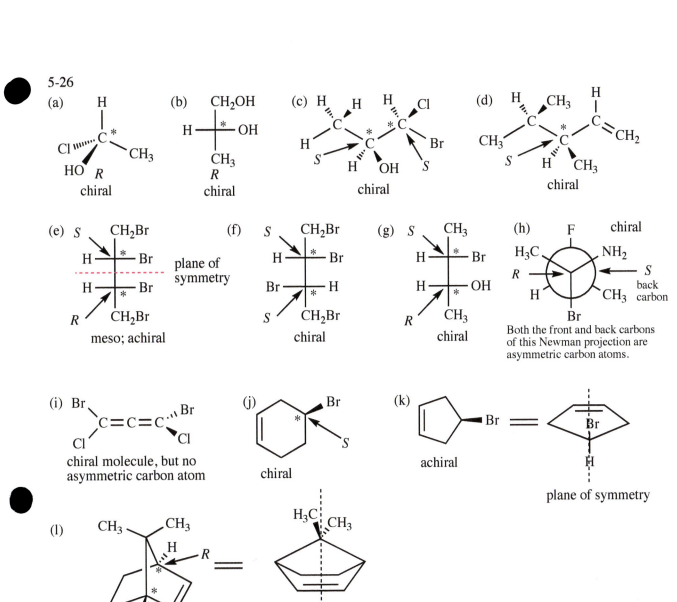

no plane of symmetry
no diastereomer
chiral structure

chiral structure

5-27 continued

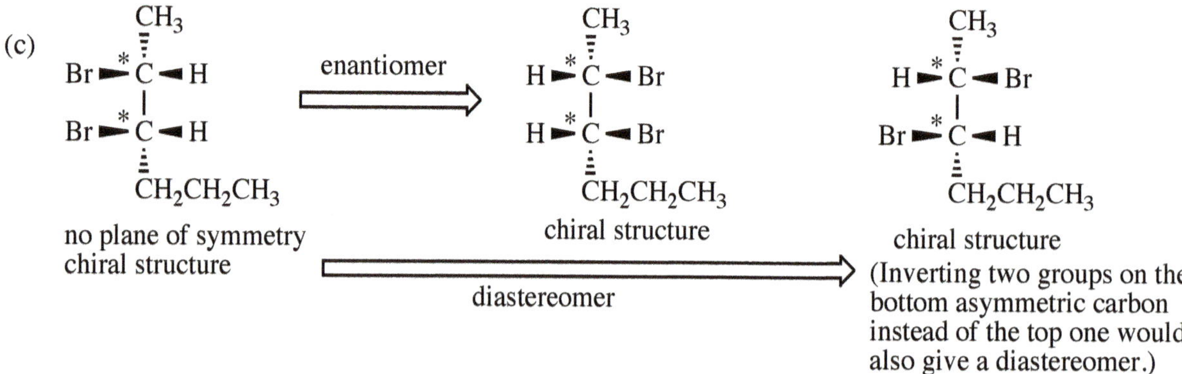

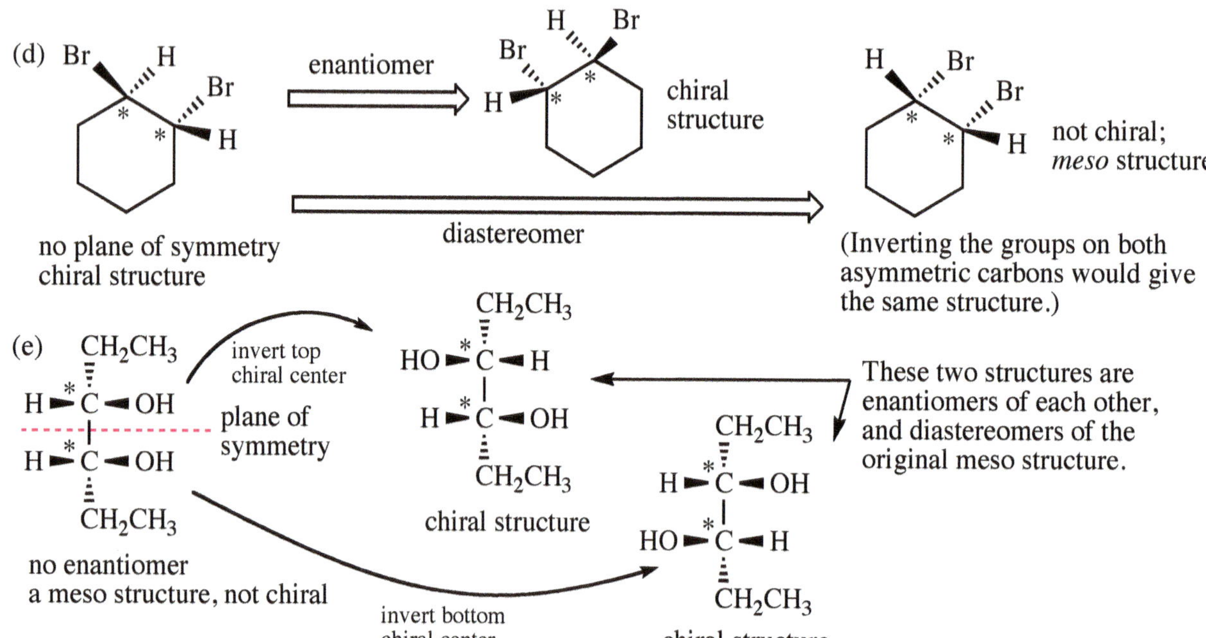

racemic mixture of enantiomers; each is chiral with no plane of symmetry

(f)

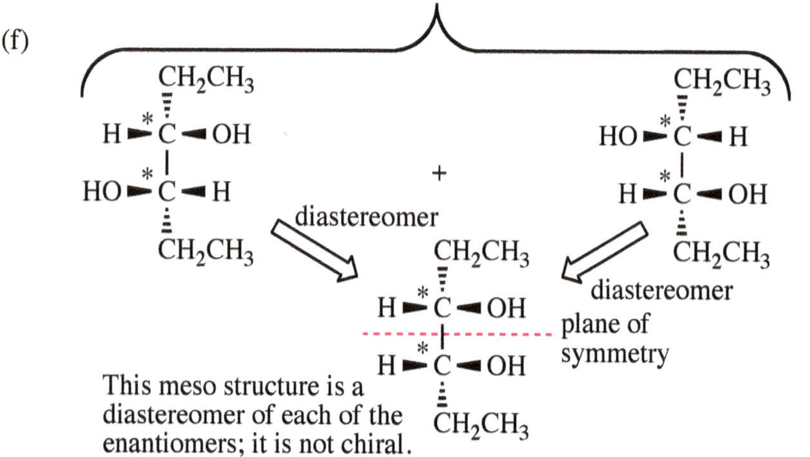

This meso structure is a diastereomer of each of the enantiomers; it is not chiral.

5-28

(a) 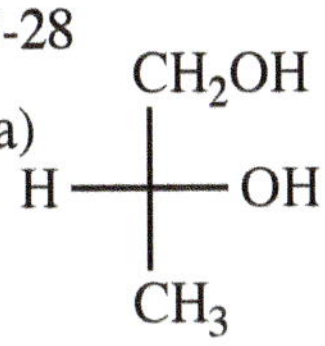 Fischer projection: top CH_2OH; H—C—OH; bottom CH_3

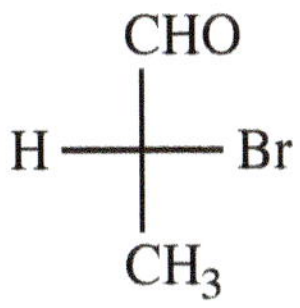

(b) Fischer projection: top CHO; H—C—Br; bottom CH_3

(c) Fischer projection: top CH_2OH; H—C—Br; HO—C—H; bottom CH_3

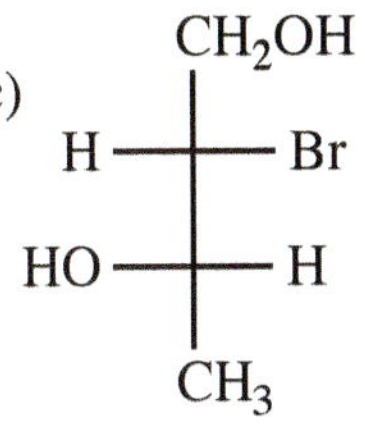

(d) 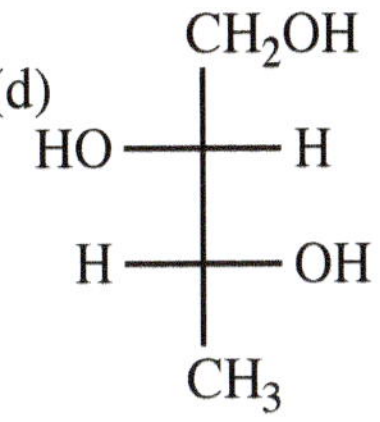 Fischer projection: top CH_2OH; HO—C—H; H—C—OH; bottom CH_3

5-29 Your drawings may look different from these and still be correct. Check configuration by assigning *R* and *S* to be sure.

(a) 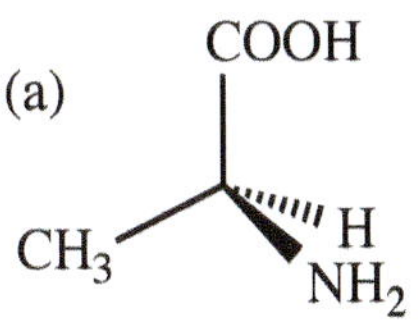 COOH, CH_3, H, NH_2

(b) 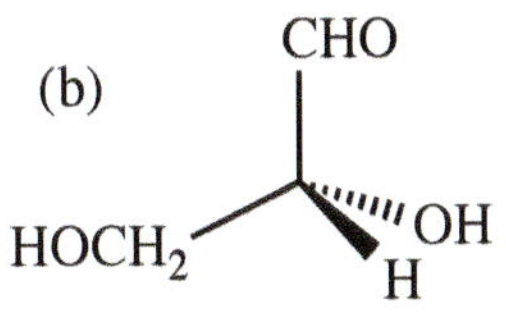 CHO, $HOCH_2$, OH, H

(c) 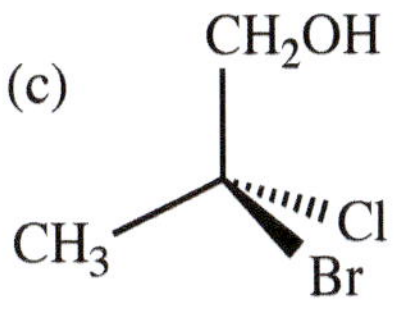 CH_2OH, CH_3, Cl, Br

(d) 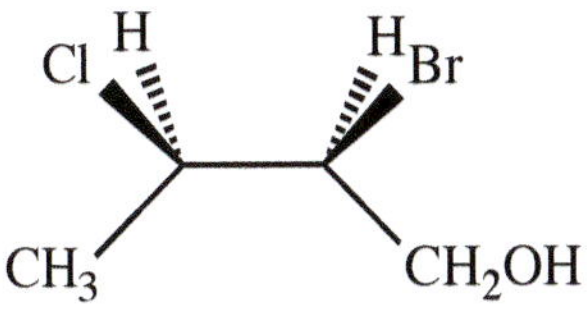 Cl, H, H, Br, CH_3, CH_2OH

5-30

(a) same (meso)—plane of symmetry, superimposable
(b) enantiomers—configuration inverted at both asymmetric carbon atoms
(c) enantiomers—configuration inverted at both asymmetric carbon atoms
(d) enantiomers—Solve this problem by switching two groups at a time to put the groups in the same positions as in the first structure; it takes three switches to make the identical compound, so they are enantiomers; an even number of switches would prove they are the same structure.
(e) diastereomers—Front carbon has same configuration (rotate Br down), back carbon is mirror image.
(f) diastereomers—configuration inverted at only one asymmetric carbon
(g) enantiomers—configuration inverted at both asymmetric carbon atoms
(h) same compound—Rotate the right structure 180° around a horizontal axis and it becomes the left structure; it might help to assign *R* and *S* to the C—Br groups: one is *R* and the other is *S*, and this makes it easier to see how they are superimposable.
(i) enantiomers—nonsuperimposable mirror images

Enantiomers have identical chemical and physical properties and cannot be separated by normal methods like distillation or crystallization. Diastereomers, however, have different properties and are therefore separable: the structures in parts (e) and (f) could be separated by normal physical methods.

5-31 Drawing the enantiomer of a chiral structure is as easy as drawing its mirror image.

(a) 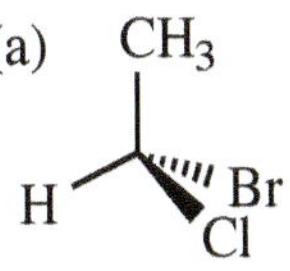 CH_3, H, Br, Cl

(b) 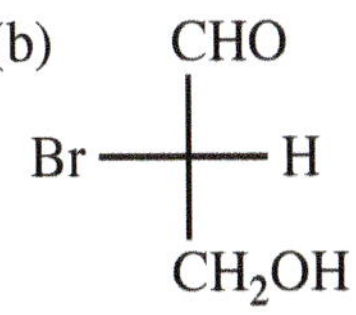 Fischer projection: top CHO; Br—C—H; bottom CH_2OH

(c) Fischer projection: top CHO; HO—C—H; HO—C—H; HO—C—H; bottom CH_2OH

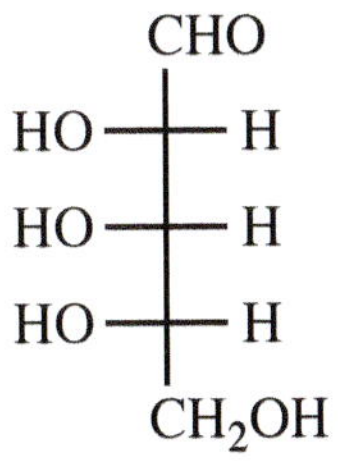

(d) 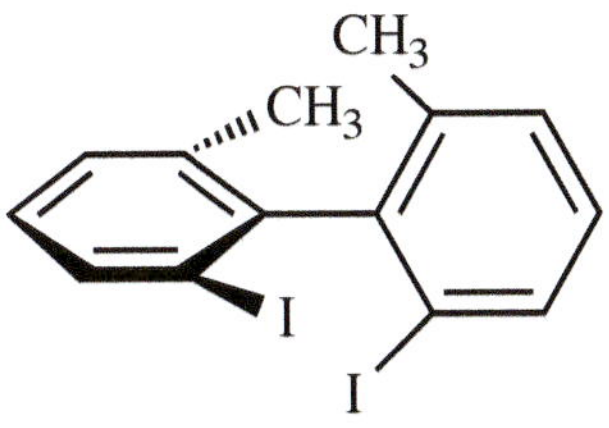 CH_3, CH_3, I, I

(e) 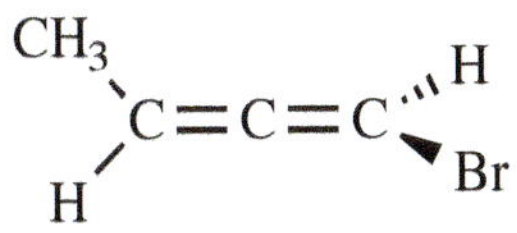 CH_3, H, C=C=C, H, Br

(f) 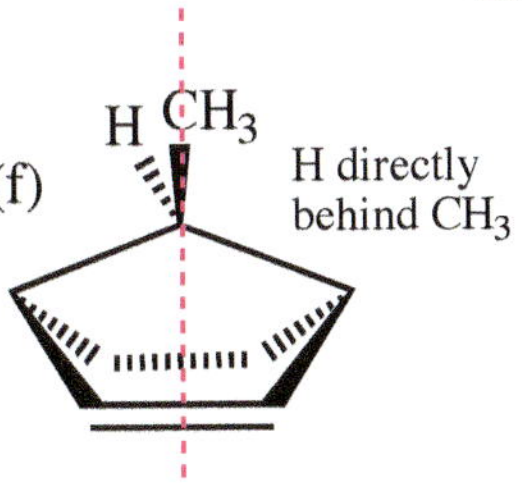 H, CH_3; H directly behind CH_3

plane of symmetry—
no enantiomer

(g) 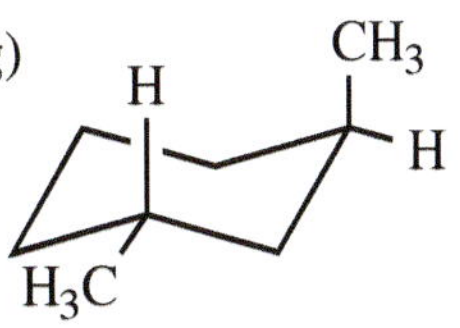 H, 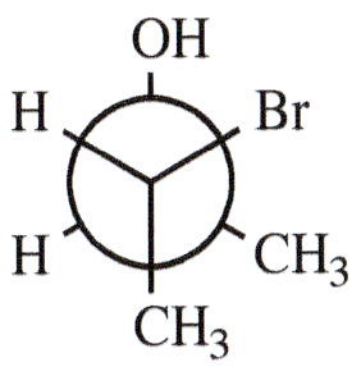 CH_3, H, H_3C

(h) OH, H, Br, H, CH_3, CH_3

5-32

(a) 1.00 g / 20.0 mL = 0.0500 g/mL ; 20.0 cm = 2.00 dm

$$[\alpha]_D^{25} = \frac{-1.25°}{(0.0500)\,(2.00)} = -12.5°$$

(b) 0.050 g / 2.0 mL = 0.025 g/mL ; 2.0 cm = 0.20 dm

$$[\alpha]_D^{25} = \frac{+0.043°}{(0.025)\,(0.20)} = +8.6°$$

5-33 The 32% of the mixture that is (–)-tartaric acid will cancel the optical rotation of the 32% of the mixture that is (+)-tartaric acid, leaving only (68 – 32) = 36% of the mixture as excess (+)-tartaric acid to give measurable optical rotation. The specific rotation will therefore be only 36% of the rotation of pure (+)-tartaric acid: (+ 12.0°) x 36% = + 4.3°

(This algebraic approach has been suggested by Editorial Adviser Richard King.)

$$\text{e.e.} = \frac{d-l}{d+l} \times 100\% = \frac{68-32}{68+32} \times 100\% = 36\%$$

The optical rotation of this mixture is: 36% x (+ 12.0°) = + 4.3°

5-34

(a) CH_3 (*S*) with H, I, CH_2CH_3

(b) Rotation of the enantiomer will be equal in magnitude, opposite in sign: – 15.90°.

(c) The rotation – 7.95° is what percent of – 15.90°?

$$\frac{-7.95°}{-15.90°} \times 100\% = 50\%\ \text{e.e.}$$

There is 50% excess of (*R*)-2-iodobutane over the racemic mixture; that is, another 25% must be *R* and 25% must be *S*. The total composition is 75% (*R*)-(–)-2-iodobutane and 25% (*S*)-(+)-2-iodobutane.

5-35 All structures in parts (a) and (b) of this problem are chiral.

(a)

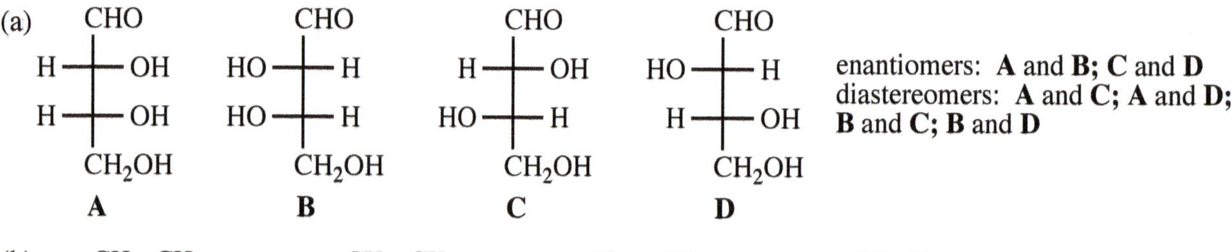

enantiomers: **A** and **B; C** and **D**
diastereomers: **A** and **C; A** and **D; B** and **C; B** and **D**

(b)

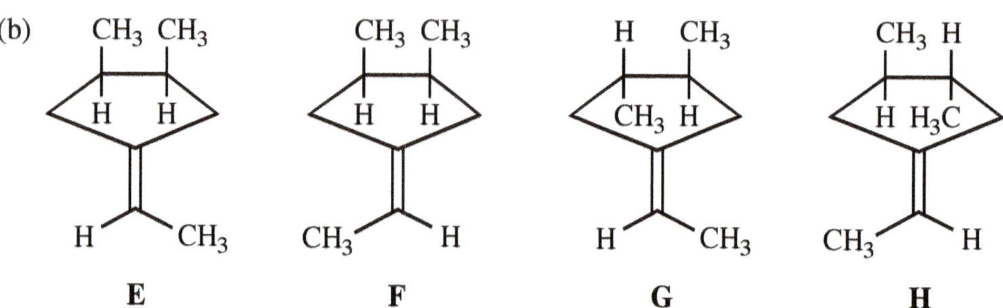

enantiomers: **E** and **F; G** and **H**
diastereomers: **E** and **G; E** and **H; F** and **G; F** and **H**

5-35 continued

(c) This structure is a challenge to visualize. A model helps. One way to approach this problem is to assign *R* and *S* configurations. Each arrow shows a change at one asymmetric carbon.

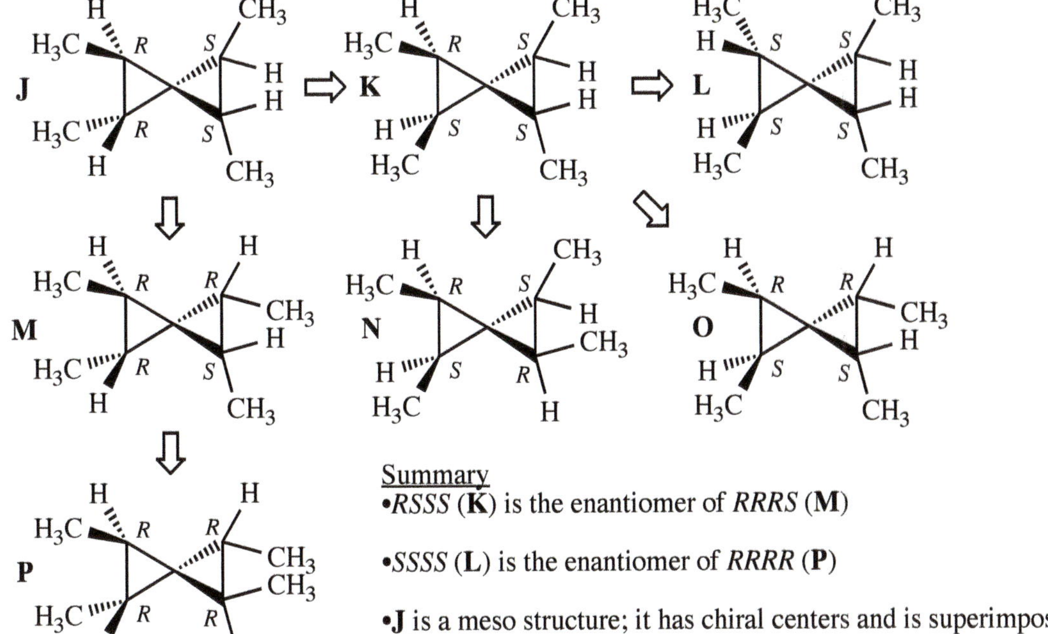

Summary

•*RSSS* (**K**) is the enantiomer of *RRRS* (**M**)

•*SSSS* (**L**) is the enantiomer of *RRRR* (**P**)

•**J** is a meso structure; it has chiral centers and is superimposable on its mirror image.

•**N** and **O** are enantiomers, and are diastereomers of all of the other structures.

Give yourself a gold star if you got this correct!

5-36

(a) Benzylic bromination follows a free-radical chain mechanism.

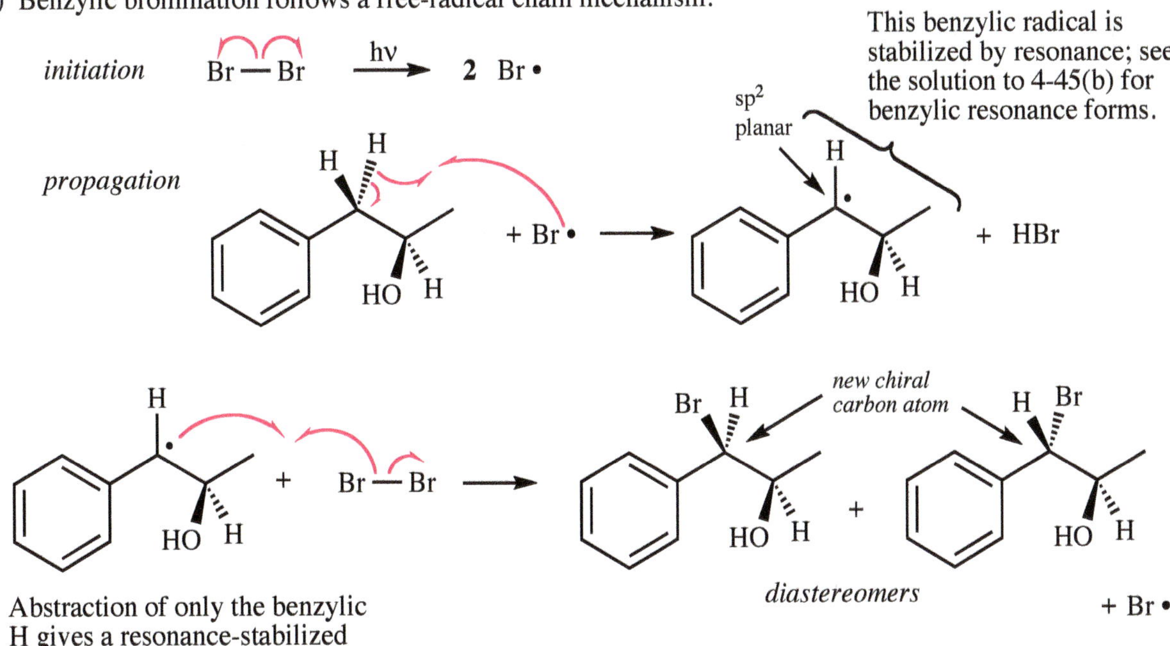

Abstraction of only the benzylic H gives a resonance-stabilized intermediate.

5-36 continued

(b) and (c)

(d) These are diastereomers. *R,R* and *S,R* have one asymmetric carbon with the same configuration and one of opposite configuration.

(e) It is possible but unlikely that they will be produced in a 50:50 mixture. Unlike racemic mixtures of enantiomers that must be 50:50, diastereomers can be and usually are unequal mixtures.

(f) Diastereomers have different physical properties like melting point and boiling point, so in theory, they could be separated by a physical method like distillation or crystallization.

5-37

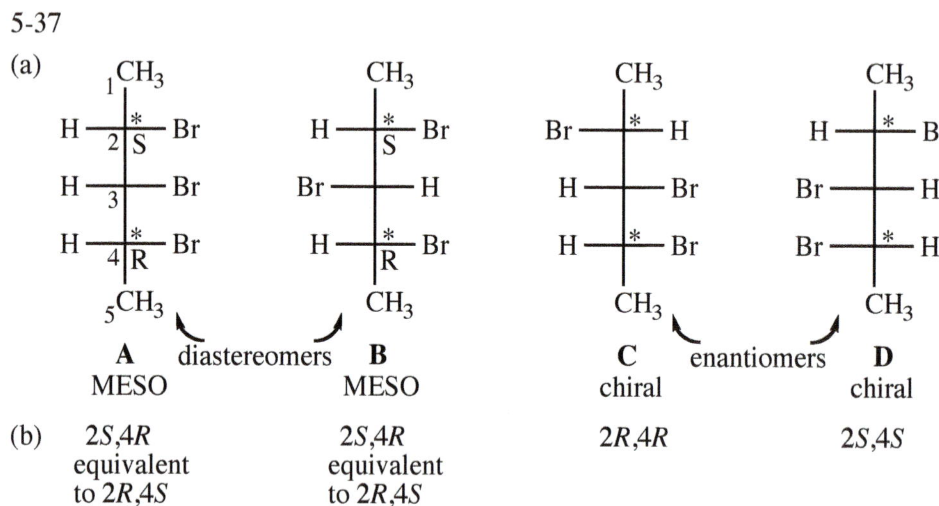

(c) According to the IUPAC designation described in text Section 5-2B, a chiral center is "any atom holding a set of ligands in a spatial arrangement which is not superimposable on its mirror image." An asymmetric carbon must have four different groups on it, but in **A** and **B**, C-3 has two groups that are identical (except for their stereochemistry). C-3 holds its groups in a spatial arrangement that is superimposable on its mirror image, so it is not a chiral center. But it is a stereocenter: in structure **A**, interchanging the H and Br at C-3 gives structure **B**, a diastereomer of **A**; therefore, C-3 is a stereocenter.

(d) In structure **C** or **D**, C-3 is not a stereocenter. Inverting the H and the Br, then rotating the structure 180°, shows that the same structure is formed. Therefore, interchanging two atoms at C-3 does *not* give a stereoisomer, so C-3 does not fit the definition of a stereocenter.

5-38 The Cahn-Ingold-Prelog priorities of the groups are the circled numbers in (a).

(a)

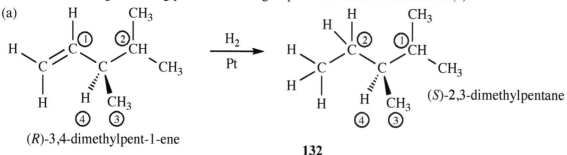

$\xrightarrow[\text{Pt}]{H_2}$

(*R*)-3,4-dimethylpent-1-ene

(*S*)-2,3-dimethylpentane

5-38 continued

(b) The reaction did not occur at the asymmetric carbon atom, so the configuration has not changed—the reaction went with retention of configuration at the asymmetric carbon.
(c) The *name* changed because the *priority* of groups in the Cahn-Ingold-Prelog system of nomenclature changed. When the alkene became an ethyl group, its priority changed from the highest priority group to priority 2. (We will revisit this anomaly in problem 6-21(c).)
(d) There is no general correlation between *R* and *S* designation and the physical property of optical rotation. Professor Wade's poetic couplet makes an important point: do not confuse an object and its properties with the *name* for that object. (Scholars of Shakespeare have come to believe that this quote from Juliet is a veiled reference to designation of *R,S* configuration versus optical rotation of a chiral molecule. Shakespeare was *way* ahead of his time.)

5-39

(a) The product has no asymmetric carbon atoms but it has three stereocenters: the carbon with the OH, plus both carbons of the double bond. Interchange of two bonds on any of these makes the enantiomer.
(b) The product is an example of a chiral compound with no asymmetric carbons. Like the allenes, it is classified as an "extended tetrahedron"; that is, it has four groups that extend from the rigid molecule in four different directions. (A model will help.) In this structure, the plane containing the COOH and carbons of the double bond is perpendicular to the plane bisecting the OH and H and carbon that they are on. Since the compound is chiral, it is capable of being optically active.

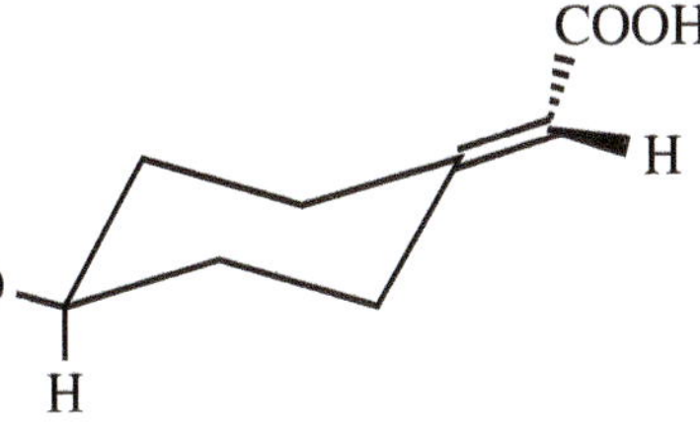

(c) As shown in text Figure 5-17, Section 5-6, catalytic hydrogenation that creates a new chiral center creates a racemic mixture (both enantiomers in a 1:1 ratio). A racemic mixture is not optically active. In contrast, by using a chiral enzyme to reduce the ketone to the alcohol (as in part (b)), an excess of one enantiomer was produced, so the product was optically active.

5-40

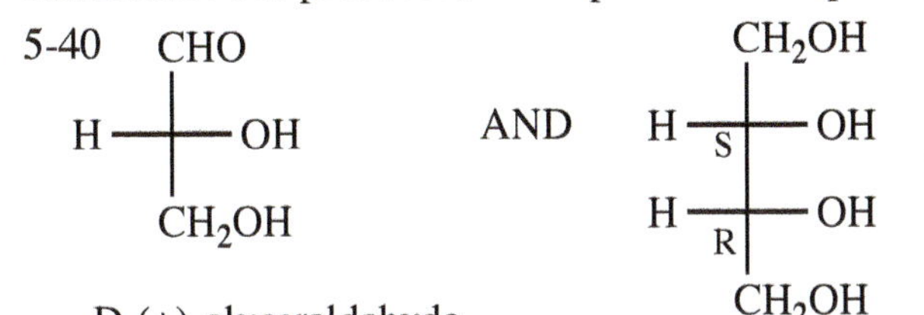

D-(+)-glyceraldehyde has *R* configuration.

Reduced form is optically inactive because it is MESO.

⟹ CHO; H–R–OH; H–R–OH; CH_2OH

The Cahn-Ingold-Prelog priorities change! The OH = #1, but the CHO = #2, so C-2 is now *R*!

D-(−)-erythrose must have the (2*R*,3*R*) configuration.

5-41 (a) The key to this problem is shown in Figure 5-19: *trans*-cyclooctene is chiral because of the twist created by the strained ring (called a chirality helix). Therefore, *cis*-cyclooctene has a chiral diastereomer (*cis* and *trans* are diastereomers) and it fits the original definition, but neither *cis* nor *trans* has chiral centers, so the *cis* isomer does not fit the working definition. (Without the information in Figure 5-19, or a model, this could not have been deduced from principles.)

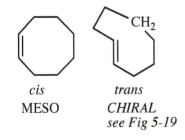

cis MESO — *trans* *CHIRAL* *see Fig 5-19*

(b) Let's try nona-2,3,6,7-tetraene:

(plane of symmetry) MESO

Consider the "twist" of an allene: in the top structure, viewing down C2 to C3, the CH_3 to CH_2 twist is clockwise; from the other end, the CH_3 to CH_2 has a counterclockwise twist.

a chiral diastereomer

However, in the bottom diastereomer, both ends have a clockwise twist; this is a chiral compound although it has no chiral centers.

Students: use this page for notes or to solve problems.

CHAPTER 6—ALKYL HALIDES; NUCLEOPHILIC SUBSTITUTION

6-1 In problems like part (a), draw out the whole structure to detect double bonds:

(a) vinyl halide (b) alkyl halide (c) alkyl halide
(d) alkyl halide (e) vinyl halide (f) aryl halide

structure in part (a)

6-2 (a) CH_2I_2 (H–C–I with I, H) (b) CBr_4 (c) Br (d) CHI_3 (e) Br

(f) 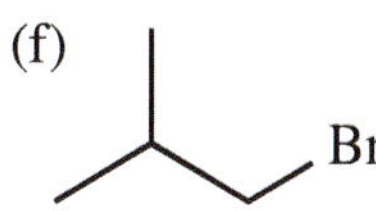

(g) H, H, CH_2F, F

You may have drawn the other enantiomer. Either is correct.

(h) Cl or $(CH_3)_3CCl$

6-3 IUPAC name; common name; degree of halogen-bearing carbon

(a) 1-chloro-2-methylpropane; isobutyl chloride; 1° halide
(b) 2-bromo-2-methylpropane; *tert*-butyl bromide; 3° halide
(c) 1-chloro-2-methylbutane; no common name; 1° halide
(d) 4-fluoro-1,1-dimethylcyclohexane; no common name; 2° halide
(e) 4-bromo-3-methylheptane; no common name; 2° halide
(f) *cis*-1-bromo-2-chlorocyclobutane; no common name; both 2° halides; also correct is (1*R*,2*S*)-1-bromo-2-chlorocyclobutane

6-4

Kepone®

Aldrin®

Chlordane®

6-5

(a) Table 6-1 shows that in all cases, the iodide has the lowest dipole moment of all four halides. Even though the C—I bond length is longer than C—Cl, the larger electronegativity of Cl makes a more significant contribution to the dipole moment. Ethyl chloride has a larger dipole moment.

(b) 1-Bromopropane has a polar C—Br bond and has a large dipole moment. Cyclopropane has no electronegative atom and has essentially zero dipole moment.

(c) The isomer with two bromine atoms *cis* to each other will have a large dipole moment. The *trans* isomer has the individual bond dipole moments pointing in opposite directions, so *trans*-2,3-dibromobut-2-ene has essentially zero dipole moment.

(d) Two chlorine atoms *cis* to each other on a ring like *cis*-1,2-dichlorocyclobutane will have a large molecular dipole moment. When two chlorine atoms point in opposite directions in a molecule as in *trans*-1,3-dichlorocyclobutane, they effectively cancel and the molecular dipole moment is essentially zero.

6-6

(a) *n*-Butyl bromide (1-bromobutane) has a higher molecular weight and less branching, and boils at a higher temperature than isopropyl bromide.

(b) *tert*-Butyl bromide has a higher molecular weight and a larger halogen, and despite its greater branching, boils at a higher temperature than isopropyl chloride.

(c) 1-Bromobutane has a higher molecular weight and a larger halogen, and boils at a higher temperature than 1-chlorobutane.

6-7 From Table 3-2, the density of hexane is 0.66; it will float on the water layer (d 1.00). From Table 6-2, the density of chloroform is 1.50; water will float on the chloroform. Water is immiscible with many organic compounds; whether water is the top layer or bottom layer depends on whether the other material is less dense or more dense than water. (This is an important consideration to remember in lab procedures.) Water and ethanol are miscible, so only one phase would appear after shaking these two together.

6-8

(a) Step (1) is initiation; steps (2) and (3) are propagation.

(1) $Br{-}Br \xrightarrow{h\nu} 2\ Br\bullet$

(2) $H_2C{=}CH{-}CH_2{-}H + Br\bullet \longrightarrow HBr + \left\{ H_2C{=}CH{-}\dot{C}H_2 \longleftrightarrow H_2\dot{C}{-}CH{=}CH_2 \right\}$

(3) $H_2C{=}CH{-}\dot{C}H_2 + Br{-}Br \longrightarrow H_2C{=}CH{-}CH_2Br + Br\bullet$

(b) Step (2): break allylic C—H, make H—Br: kJ/mole: + 372 – (+ 366) = + 6 kJ/mole

kcal/mole: + 89 – (+ 88) = + 1 kcal/mole

Step (3): break Br—Br, make allylic C—Br: kJ/mole: + 190 – (+ 280) = – 90 kJ/mole

kcal/mole: + 45 – (+ 67) = – 22 kcal/mole

ΔH^0 overall = + 6 + – 90 = – 84 kJ/mole (+ 1 + – 22 = – 21 kcal/mole)

This is a very exothermic reaction; it is reasonable to expect a small activation energy in step (2), the first propagation step, so this reaction should be very rapid.

6-9 (a) propagation steps

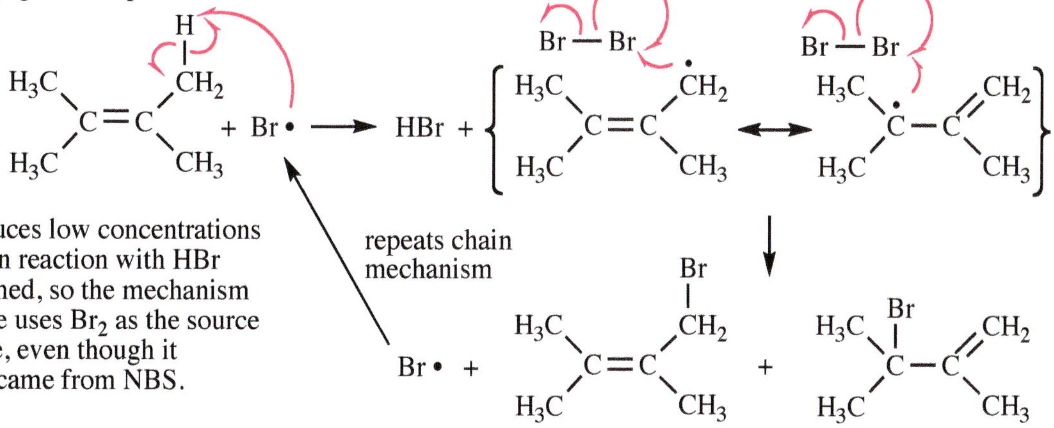

$$(CH_3)_2C{=}C(CH_3)CH_2{-}H + Br\bullet \longrightarrow HBr + \left\{ (CH_3)_2C{=}C(CH_3)\dot{C}H_2 \longleftrightarrow (CH_3)_2\dot{C}{-}C(CH_3){=}CH_2 \right\}$$

$$Br{-}Br + \text{radical} \longrightarrow Br\bullet + (CH_3)_2C{=}C(CH_3)CH_2Br + (CH_3)_2CBr{-}C(CH_3){=}CH_2$$

repeats chain mechanism

NBS produces low concentrations of Br_2 upon reaction with HBr that is formed, so the mechanism shown here uses Br_2 as the source of bromine, even though it originally came from NBS.

The resonance-stabilized allylic radical intermediate has radical character on both the 1° and 3° carbons, so bromine can bond to either of these carbons producing two isomeric products.

(b) Allylic bromination of cyclohexene gives 3-bromocyclohex-1-ene regardless of whether there is an allylic shift. Either pathway leads to the same product. If one of the ring carbons were somehow marked or labeled, then the two products could be distinguished. (We will see in following chapters how labeling is done experimentally.)

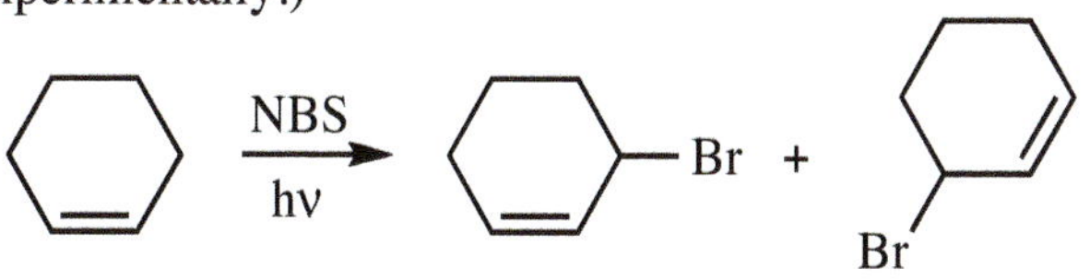

$$\text{cyclohexene} \xrightarrow[h\nu]{NBS} \text{3-bromocyclohexene (Br)} + \text{3-bromocyclohexene (Br)}$$

This second structure from the allylic shift is *identical* to the first structure—only one compound is produced here.

6-10

(a)

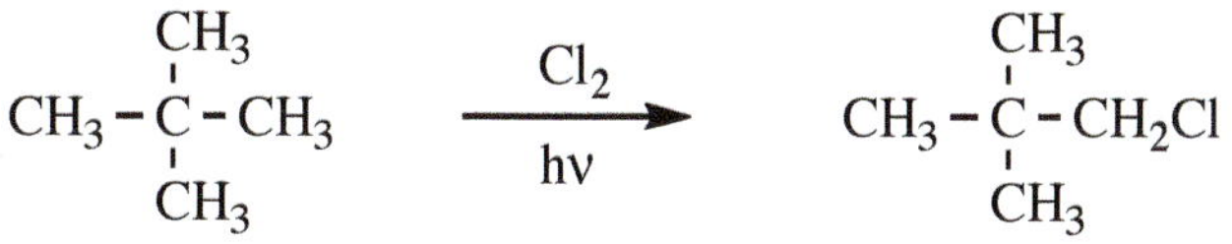

$$CH_3{-}C(CH_3)_2{-}CH_3 \xrightarrow[h\nu]{Cl_2} CH_3{-}C(CH_3)_2{-}CH_2Cl$$

This compound has only one type of hydrogen—only one monochlorine isomer can be produced.

(b)

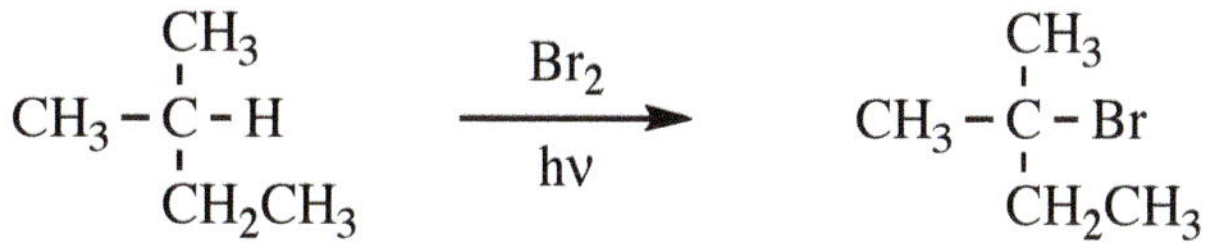

$$CH_3{-}C(CH_3)(CH_2CH_3){-}H \xrightarrow[h\nu]{Br_2} CH_3{-}C(CH_3)(CH_2CH_3){-}Br$$

Bromination has a strong preference for abstracting hydrogens that give the most stable radical intermediates, like 3° in this case.

(c)

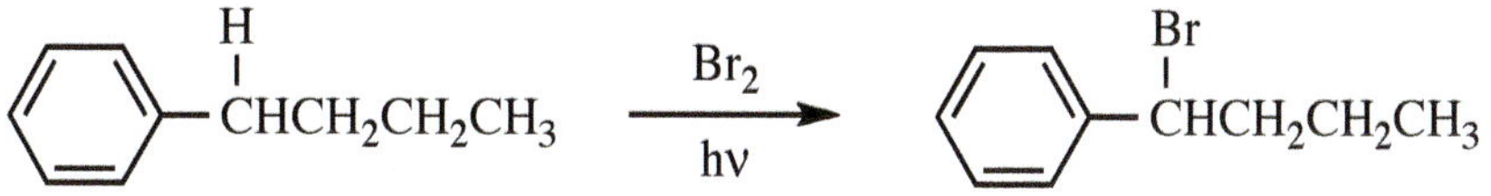

$$C_6H_5{-}CH(H)CH_2CH_2CH_3 \xrightarrow[h\nu]{Br_2} C_6H_5{-}CH(Br)CH_2CH_2CH_3$$

(NBS can also be used for benzylic bromination)

The bromine atom will abstract the hydrogen giving the most stable radical; in this case, the radical intermediate will be stabilized by resonance with the benzene ring.

(d)

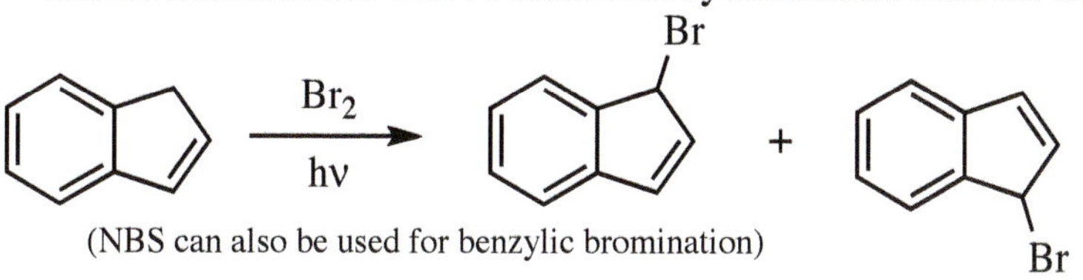

$$\text{indene} \xrightarrow[h\nu]{Br_2} \text{1-bromoindene (Br)} + \text{1-bromoindene (Br)}$$

(NBS can also be used for benzylic bromination)

This second structure from the allylic shift is *identical* to the first structure—only one compound is produced here.

The bromine atom will abstract the hydrogen giving the most stable radical; in this case, the radical intermediate will be stabilized by resonance with the benzene ring.

6-11
(a) Substitution—Br^- is the leaving group; CH_3O^- is the nucleophile.
(b) Elimination—when OH is protonated, H_2O is the leaving group.
(c) Elimination—both Br atoms are lost; iodide ion is a nucleophile that reacts at Br.

6-12 (a) $CH_3(CH_2)_4CH_2-OCH_2CH_3$ (b) $CH_3(CH_2)_4CH_2-CN$ (c) $CH_3(CH_2)_4CH_2-OH$

6-13 (a) The rate law is first order in both 1-bromobutane, C_4H_9Br, and methoxide ion. If the concentration of C_4H_9Br is lowered to one-fifth the original value, the rate must decrease to one-fifth; if the concentration of methoxide is doubled, the rate must also double. Thus, the rate must decrease to two-fifths of the original rate, 0.02 mole/L per second:

$$\text{rate} = \underbrace{(0.05 \text{ mole/L per second})}_{\text{original rate}} \times \underbrace{\frac{(0.1\text{ M})}{(0.5\text{ M})}}_{\text{change in } C_4H_9Br} \times \underbrace{\frac{(2.0\text{ M})}{(1.0\text{ M})}}_{\text{change in } NaOCH_3} = \underbrace{0.02 \text{ mole/L per second}}_{\text{new rate}}$$

A completely different way to answer this problem is to solve for the rate constant k, then put in new values for the concentrations.

$\text{rate} = k\ [C_4H_9Br]\ [NaOCH_3] \Longrightarrow 0.05 \text{ mole L}^{-1} \text{ sec}^{-1} = k\ (0.5 \text{ mol L}^{-1})\ (1.0 \text{ mol L}^{-1}) \Longrightarrow$

rate constant $k = 0.1 \text{ L mol}^{-1} \text{ sec}^{-1}$

$\text{rate} = k\ [C_4H_9Br]\ [NaOCH_3] = (0.1 \text{ L mol}^{-1} \text{ sec}^{-1})\ (0.1 \text{ mol L}^{-1})\ (2.0 \text{ mol L}^{-1}) = 0.02 \text{ mole L}^{-1} \text{ sec}^{-1}$

(b)

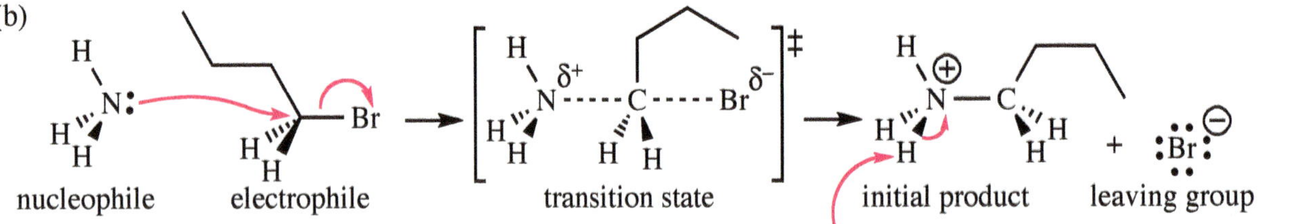

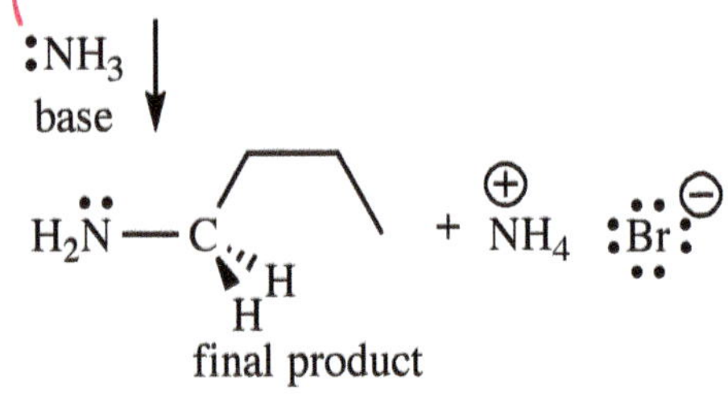

nucleophile ($:NH_3$) + electrophile (1-bromobutane) → transition state $[H_3N^{\delta+}\cdots C \cdots Br^{\delta-}]^‡$ → initial product ($R-NH_3^+$) + leaving group ($:\ddot{Br}:^-$)

$:NH_3$ base ↓

$H_2\ddot{N}-C$ (butylamine) + $\overset{\oplus}{NH_4}$ $:\ddot{Br}:^{\ominus}$

final product

(c) $NaOCH_2CH_2CH_2CH_3$ + CH_3Br ⟶
sodium butoxide, bromomethane

$CH_3CH_2CH_2CH_2O-CH_3$ + NaBr
1-methoxybutane

6-14 Organic and inorganic products are shown here for completeness.

(a) $(CH_3)_3C-O-CH_2CH_3$ + KBr

(b) $HC\equiv C-CH_2CH_2CH_2CH_3$ + NaCl

(c) $(CH_3)_2CHCH_2-\overset{\oplus}{N}H_3\ Br^- \xrightarrow{NH_3} (CH_3)_2CHCH_2-NH_2 + NH_4^+\ Br^-$

The first NH_3 replaces the Br. The second NH_3 removes H^+ from the N, leaving R-NH_2.

(d) $CH_3CH_2CH_2-CN$ + NaI

(e)  1-iodohexane (skeletal, I) + NaCl

(f) 1-fluorohexane (skeletal, F)  + KCl (18-Crown-6 is the catalyst and does not change; CH_3CN is the solvent.)

6-15 All reactions in this problem follow the same pattern; the only difference is the nucleophile ($^-$:Nuc). Only the nucleophile is listed below. (Cations like Na^+ or K^+ accompany the nucleophile but are simply spectator ions and do not take part in the reaction; they are not shown here.)

1-chlorobutane

(a) HO^- (b) F^- from KF/18-crown-6 (c) I^- (d) ^-CN (e) $HC{\equiv}C{:}^-$

(f) $^-OCH_2CH_3$ (g) excess NH_3 (or $^-NH_2$)

6-16

(a) $(CH_3CH_2)_2NH$ is a better nucleophile—less hindered.
(b) $(CH_3)_2S$ is a better nucleophile—S is larger, more polarizable than O.
(c) PH_3 is a better nucleophile—P is larger, more polarizable than N.
(d) CH_3S^- is a better nucleophile—anions are better than neutral atoms of the same element.
(e) $(CH_3)_3N$ is a better nucleophile—less electronegative than oxygen, better able to donate an electron pair.
(f) CH_3COO^- is a better nucleophile—more basic, electrons less delocalized than in CF_3COO^- because of inductive effect of F substituents.
(g) $CH_3CH_2CH_2O^-$ is a better nucleophile—less branching, less steric hindrance.
(h) I^- is a better nucleophile—larger, more polarizable than Cl.

6-17 A mechanism uses arrows to show *electron movement*. An arrow must begin at either a bond or an unshared electron pair (or a single electron in radical reactions). "Et" is the abbreviation for an ethyl group.

$$Et{-}\ddot{\underset{..}{O}}{-}Et \;+\; H{-}Br \;\rightleftharpoons\; Et{-}\overset{\oplus}{\underset{H}{O}}{-}CH_2CH_3 \;+\; :\ddot{Br}:^{\ominus} \;\longrightarrow\; Et{-}\underset{H}{\ddot{O}}: \;+\; Et{-}\ddot{Br}:$$

Protonation converts OCH_2CH_3 to a good leaving group so that bromide can effect substitution.

6-18 The type of carbon with the halide, and relative leaving group ability of the halide, determine the reactivity.

methyl iodide > methyl chloride > ethyl chloride > isopropyl bromide >> neopentyl bromide, *tert*-butyl iodide } *least reactive*

most reactive; ethyl chloride 1°; isopropyl bromide 2°; neopentyl bromide 1°, *tert*-butyl iodide 3°

Predicting the relative order of neopentyl bromide and *tert*-butyl iodide would be difficult because both would be extremely slow.

6-19 In all cases, the less hindered structure is the better S_N2 substrate.

(a) 2-methyl-1-iodopropane (1° versus 3°)

(b) cyclohexyl bromide (2° versus 3°)

(c) isopropyl bromide (no substituent on neighboring carbon)

(d) 2-chlorobutane (Even though this is a 2° halide, it is easier to attack than the 1° neopentyl type in 1-chloro-2,2-dimethylbutane—see structure and the solution to Problem 6-18.)

Cl–CH₂–C(CH₃)₂(CH₂CH₃) with $^-$:Nuc

a neopentyl halide—hindered to backside attack by neighboring methyl groups

(e) 1-iodobutane (1° versus 2°)

6-20 All S_N2 reactions occur with inversion of configuration at carbon.

(a)

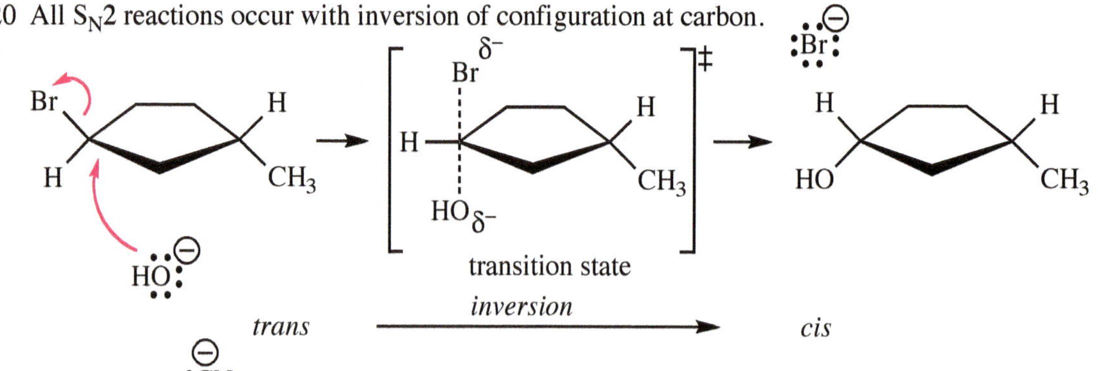

trans → inversion → cis

transition state

(b)

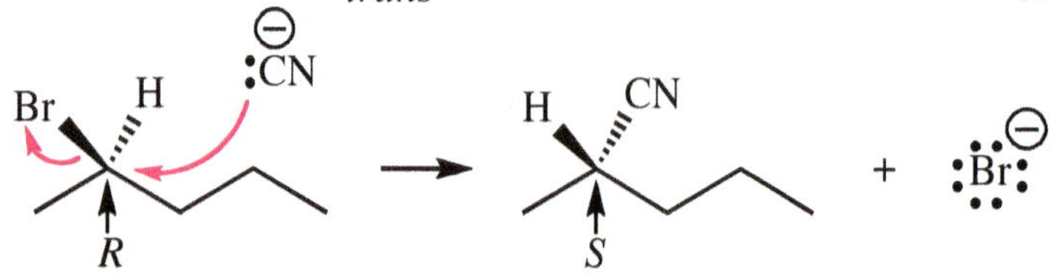

(c)

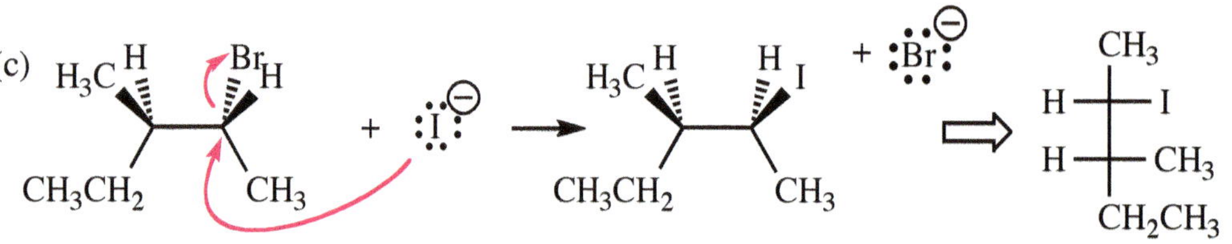

(d) 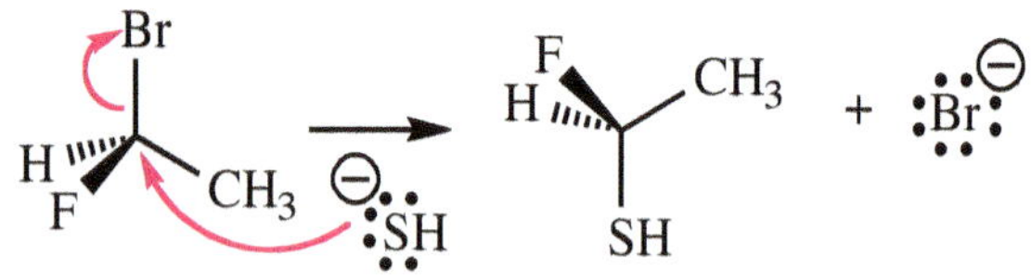 Fluoride is a bad leaving group; see the solution to 6-21(a).

(e)

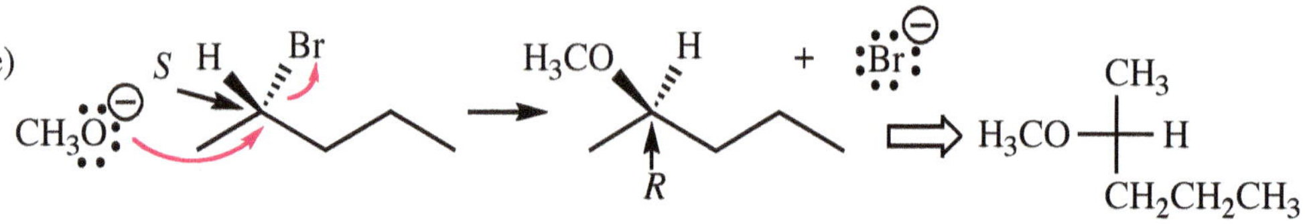

(f)

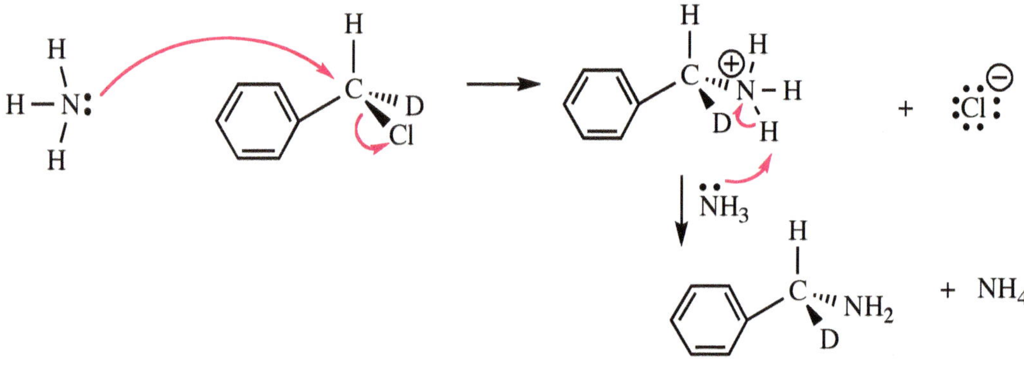

6-21

(a) The best leaving groups are the weakest bases. Bromide ion is so weak it is not considered at all basic; it is an excellent leaving group. Fluoride is moderately basic, by far the most basic of the halides. It is a terrible leaving group. Bromide is many orders of magnitude better than fluoride in leaving group ability.

6-21 continued

(b)

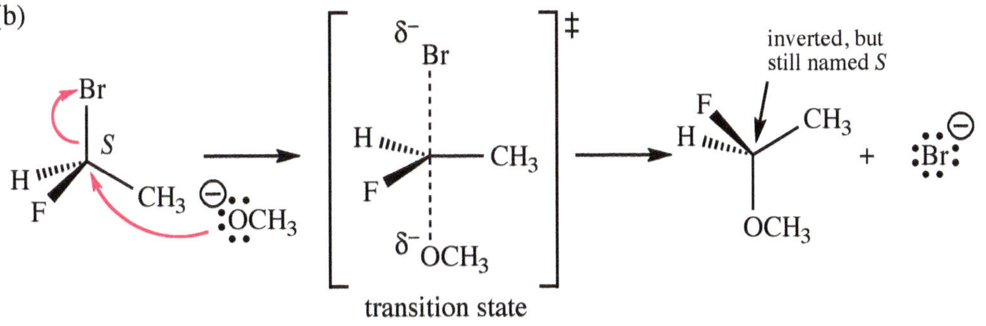

Br
S
H
F
CH3
OCH3
δ−
Br
H
CH3
F
δ−
OCH3
‡
transition state
inverted, but still named S
F
H
CH3
OCH3
+
Br

(c) As noted on the structure above, the configuration is inverted even though the designations of the configuration for both the starting material and the product are *S*; the oxygen of the product has a lower priority than the bromine it replaces. Refer to the solution to problem 5-38 for the caution about confusing absolute configuration with the *designation* of configuration.

(d) The result is perfectly consistent with the S_N2 mechanism. Even though both the reactant and the product have the *S* designation, the configuration has been inverted: the nomenclature priority of fluorine changes from second (after bromine) in the reactant to first (before oxygen) in the product. While the designation may be misleading, the structure shows with certainty that an inversion has occurred.

6-22

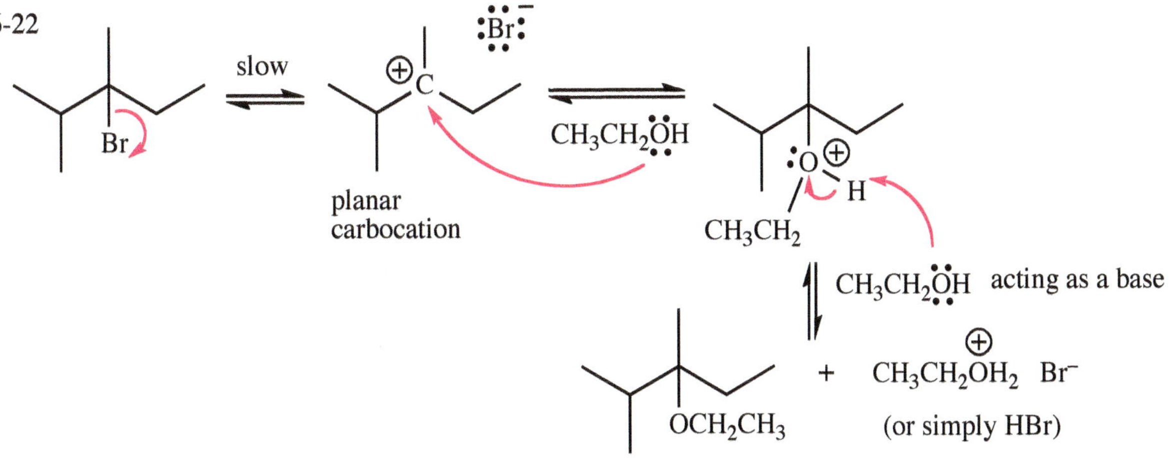

Br
slow
C
Br−
planar
carbocation
CH3CH2OH
O
H
CH3CH2
CH3CH2OH acting as a base
OCH2CH3
+
CH3CH2OH2 Br−
(or simply HBr)

6-23 The structure that can form the more stable carbocation will undergo S_N1 faster.

(a) 2-Bromopropane: will form a 2° carbocation.
(b) 2-Bromo-2-methylbutane: will form a 3° carbocation.
(c) Allyl bromide is faster than propyl bromide: allyl bromide can form a resonance-stabilized intermediate.
(d) 2-Bromopropane: will form a 2° carbocation.
(e) 2-Iodo-2-methylbutane is faster than *tert*-butyl chloride (iodide is a better leaving group than chloride).
(f) 2-Bromo-2-methylbutane (3°) is faster than ethyl iodide (1°); although iodide is a somewhat better leaving group, the difference between 3° and 1° carbocation stability dominates.

6-24 Ionization is the rate-determining step in S_N1. Anything that stabilizes the carbocation intermediate will speed the reaction. Both of these compounds form resonance-stabilized intermediates.

allylic

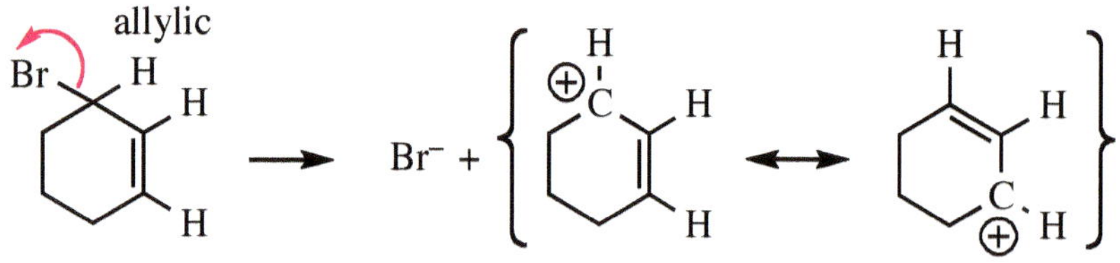

$Br^- +$ allylic carbocation

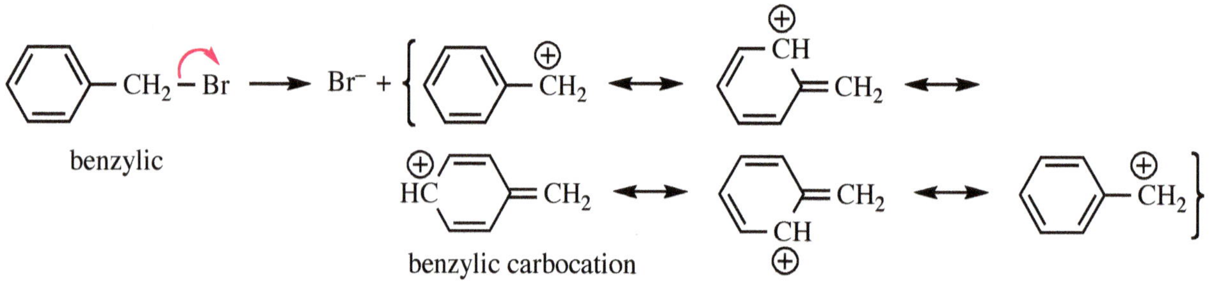

benzylic

$Br^- +$ benzylic carbocation

6-25

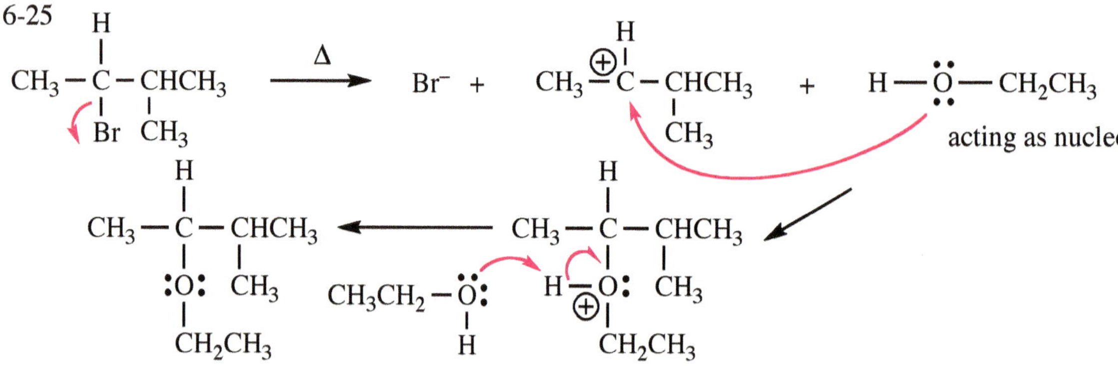

$CH_3-CH(CHCH_3)(Br)...$ $\xrightarrow{\Delta}$ $Br^- +$ $CH_3-\overset{\oplus}{C}H-CHCH_3$ $+$ $H-\ddot{O}-CH_2CH_3$ acting as nucleophile

acting as base

6-26 It is important to analyze the structure of carbocations to consider if migration of any groups from adjacent carbons will lead to a more stable carbocation. As a general rule, if rearrangement would lead to a more stable carbocation, a carbocation will rearrange. (Beginning with this problem, only those unshared electron pairs involved in a particular step will be shown.)

(a) $CH_3-C(CH_3)_2-CH(I)-CH_3$ $\xrightarrow{\Delta}$ $I^- + CH_3-C(CH_3)_2-\overset{\oplus}{C}H-CH_3$ 2° carbocation

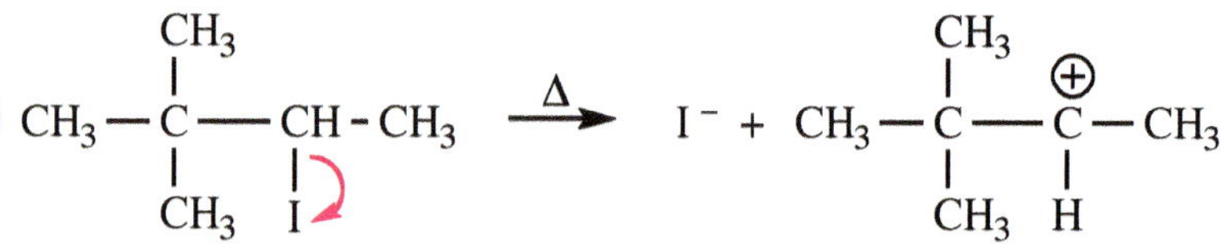

nucleophilic attack on unrearranged carbocation

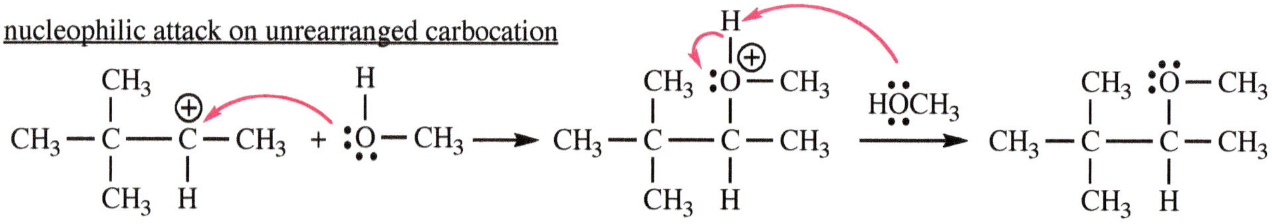

$CH_3-C(CH_3)_2-\overset{\oplus}{C}H-CH_3 + H\ddot{O}-CH_3 \rightarrow CH_3-C(CH_3)_2-CH(\overset{\oplus}{O}HCH_3)-CH_3 \xrightarrow{H\ddot{O}CH_3} CH_3-C(CH_3)_2-CH(\ddot{O}CH_3)-CH_3$

unrearranged product

continued on next page

6-26 (a) continued

nucleophilic attack after carbocation rearrangement

$CH_3-C(CH_3)_2-C^{+}H-CH_3 \longrightarrow CH_3-C^{+}(CH_3)-CH(CH_3)-CH_3$ (3°) $+ \; H\ddot{O}-CH_3 \longrightarrow CH_3-C(CH_3)(O^{+}(H)CH_3)-CH(CH_3)-CH_3$

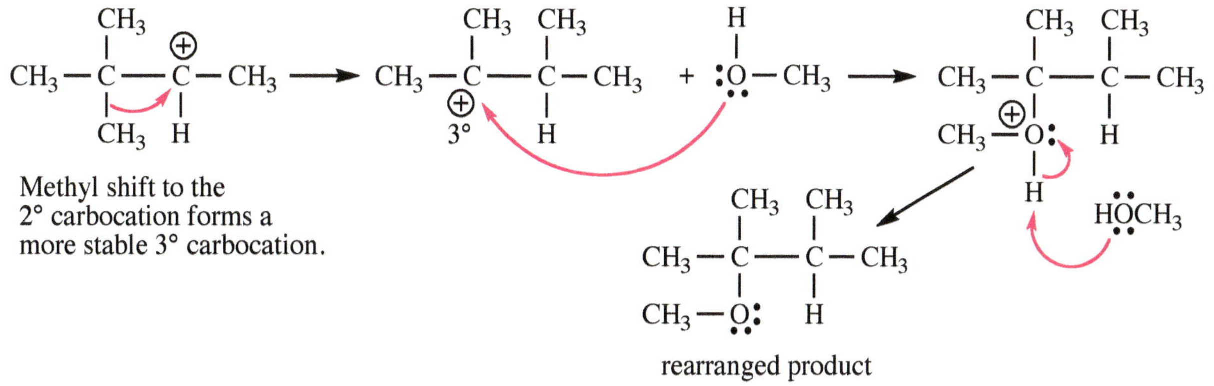

Methyl shift to the 2° carbocation forms a more stable 3° carbocation.

$\xrightarrow{HOCH_3} CH_3-C(CH_3)(\ddot{O}CH_3)-CH(CH_3)-CH_3$

rearranged product

(b)

$\xrightarrow{\Delta} Cl^{-} +$ 2° carbocation

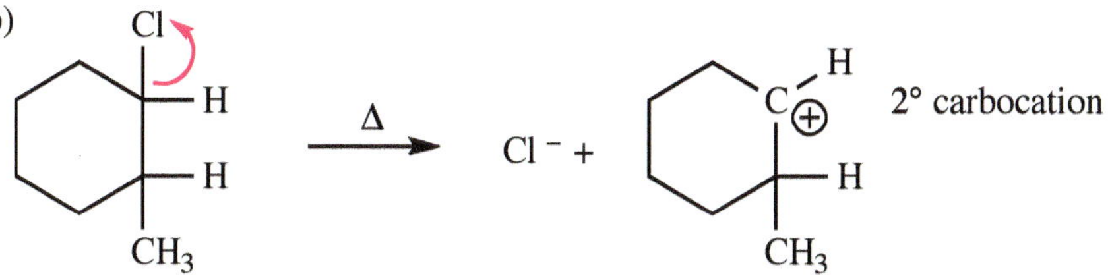

nucleophilic attack on unrearranged carbocation

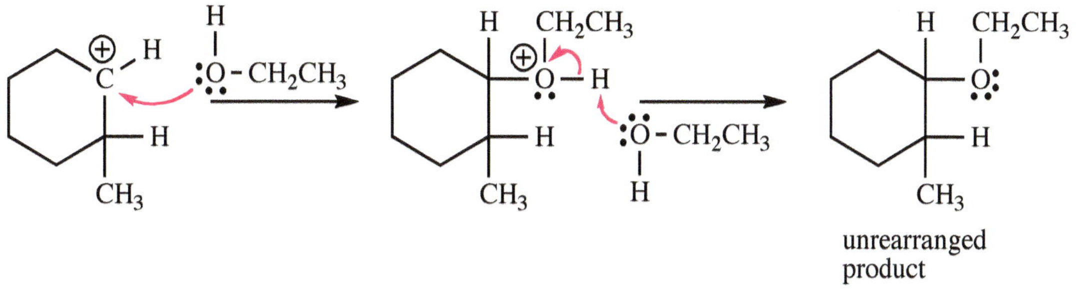

$H\ddot{O}-CH_2CH_3$

$\ddot{O}-CH_2CH_3$ (with H)

unrearranged product

nucleophilic attack after carbocation rearrangement

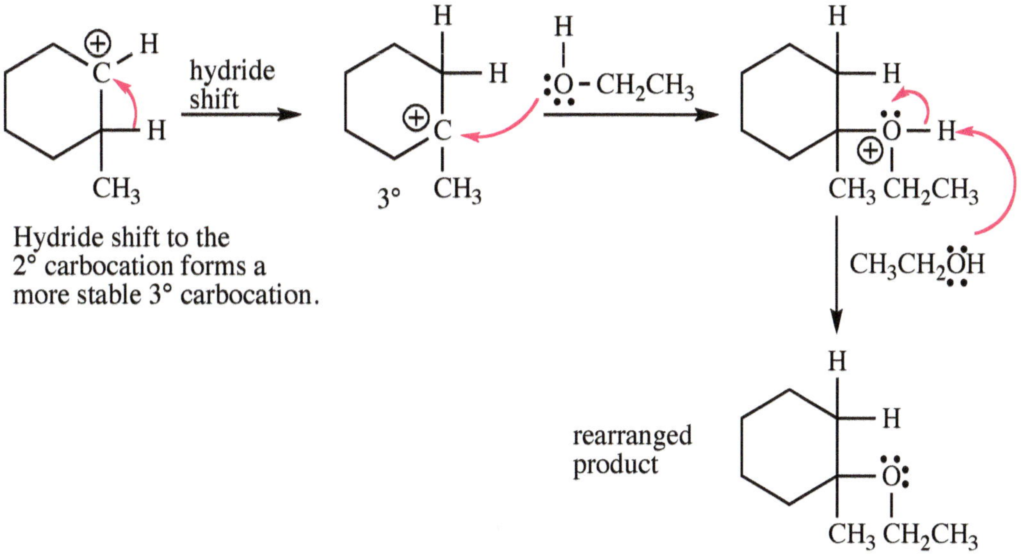

hydride shift

3°

$H\ddot{O}-CH_2CH_3$

$CH_3CH_2\ddot{O}H$

Hydride shift to the 2° carbocation forms a more stable 3° carbocation.

rearranged product

6-26 continued

(c)

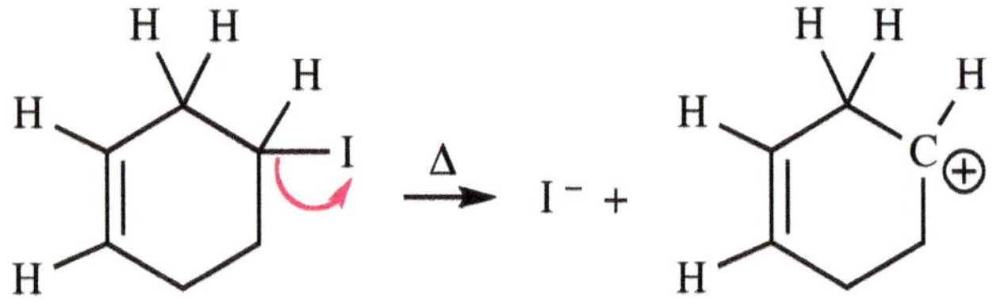

2° carbocation

Note: braces are used to indicate the ONE chemical species represented by multiple resonance forms.

nucleophilic attack on unrearranged carbocation

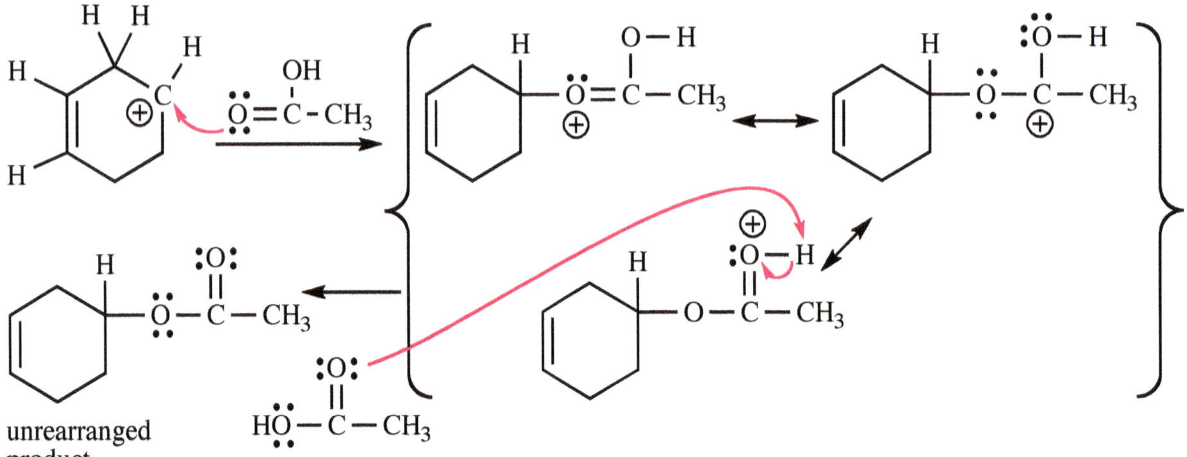

unrearranged product

The most basic species in a mixture is the most likely to remove a proton. In this reaction, acetic acid is more basic than iodide ion.

nucleophilic attack after carbocation rearrangement

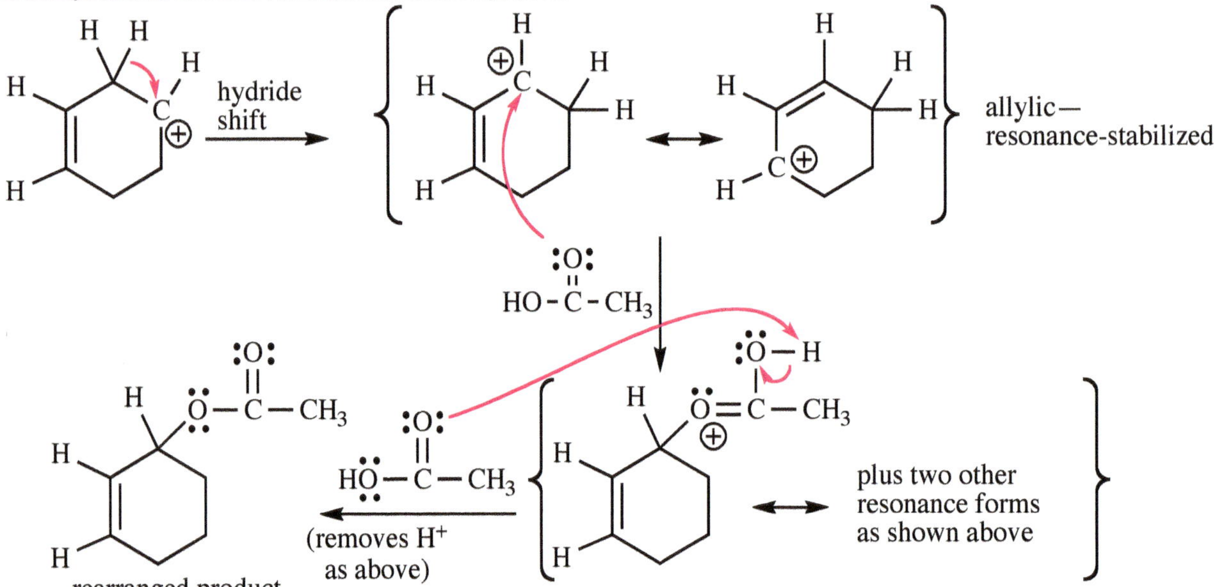

hydride shift

allylic—resonance-stabilized

plus two other resonance forms as shown above

(removes H^+ as above)

rearranged product

continued on next page

6-26(c) continued

Comments on 6-26(c)

(1) The hydride shift to a 2° carbocation generates an allylic, resonance-stabilized 2° carbocation.

(2) The double-bonded oxygen of acetic acid is more nucleophilic because of the resonance forms it can have after attack. (See the solution to Problem 2-20.)

(3) Attack on only one carbon of the allylic carbocation is shown. In reality, both positive carbons would be attacked in equal amounts, but they would give the identical product *in this case*. In other compounds, however, attack on the different carbons might give different products. ALWAYS CONSIDER ALL POSSIBILITIES.

(d) H, $CH_2—I$ $\xrightarrow{\Delta}$ I^- + [H ⊕ CH_2]

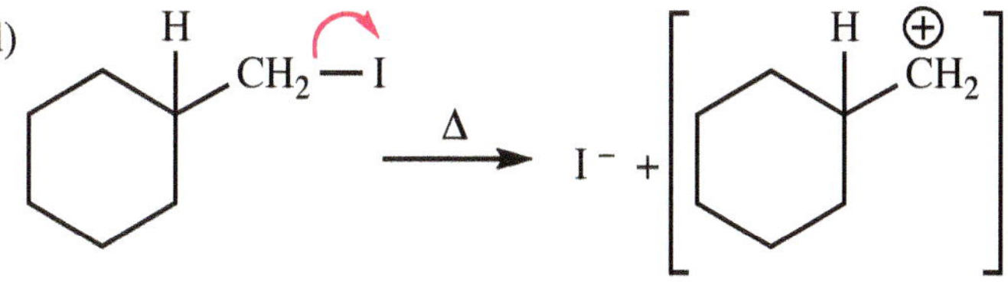

The 1° carbocation shown here would be very unstable. Rearrangement usually happens at the same time as the leaving group leaves from a 1° carbon.

hydride shift followed by nucleophilic attack

[H ⊕ CH_2] → 3° ⊕C CH_3 ; H—$\ddot{O}$—CH_2CH_3 → CH_3, $\ddot{O}$(⊕)(H)—CH_2CH_3 → CH_3, $\ddot{O}$—CH_2CH_3

H—$\ddot{O}$—CH_2CH_3 acting as base

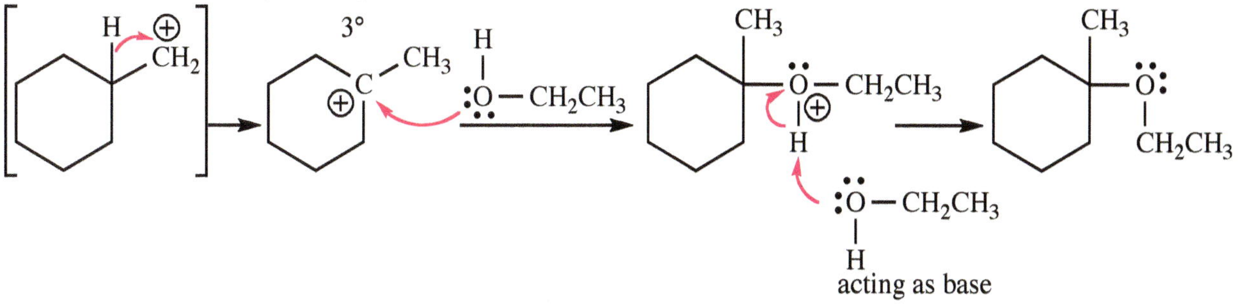

alkyl migration (ring expansion) followed by nucleophilic attack

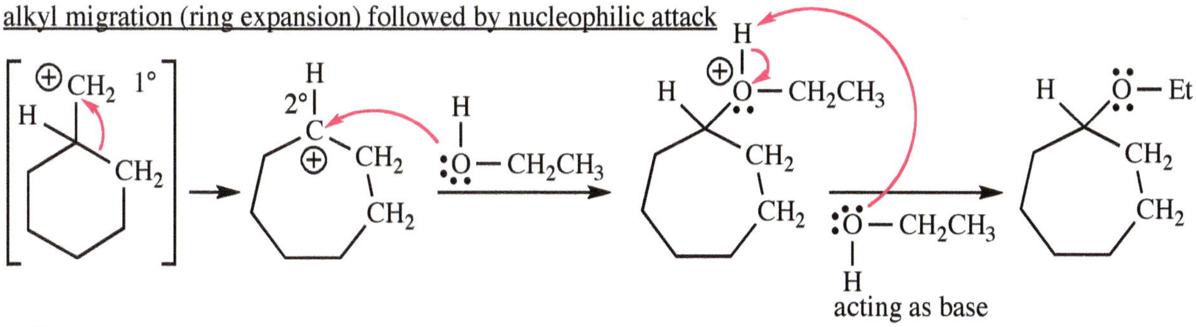

[⊕CH_2 1°, H, CH_2] → H, 2° C⊕, CH_2, CH_2 ; H—$\ddot{O}$—CH_2CH_3 → H, H—⊕O—CH_2CH_3, CH_2, CH_2 → H, $\ddot{O}$—Et, CH_2, CH_2

H—$\ddot{O}$—CH_2CH_3 acting as base

6-27

(a) $CH_3-C(CH_3)(CH_2CH_3)-O-C(=O)-CH_3$
S_N1 on 3° C, weak nucleophile

(b) $CH_3-CH(CH_3)CH_2OCH_3$
S_N2 on 1° C, strong nucleophile

(c) 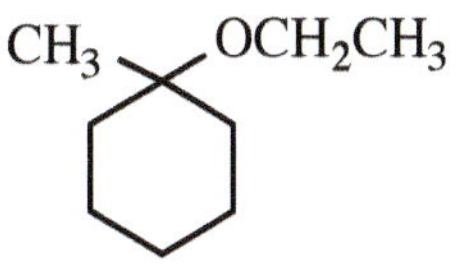 1-methyl-1-(OCH_2CH_3)cyclohexane (CH_3, OCH_2CH_3)
S_N1 on 3° C, weak nucleophile

(d) 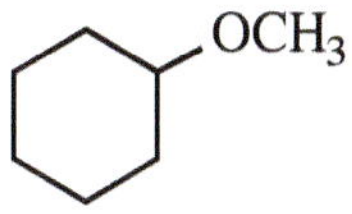 cyclohexyl OCH_3
S_N1 on 2° C, weak nucleophile

(e) 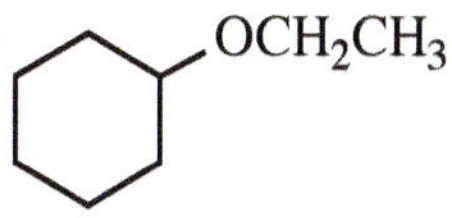 cyclohexyl OCH_2CH_3
S_N2 on 2° C, strong nucleophile

6-28

(R)-2-bromobutane $\xrightarrow[S_N2\text{—inversion}]{H_2O,\ \Delta}$ (S)-butan-2-ol

(R)-2-bromobutane $\xrightarrow[S_N1\text{—racemization}]{H_2O,\ \Delta}$ (R)-butan-2-ol + (S)-butan-2-ol

50 : 50 mixture—racemic

If S_N1, which gives racemization, occurs exactly twice as fast as S_N2, which gives inversion, then the racemic mixture (50 : 50 *R* + *S*) is 66.7% of the mixture and the rest, 33.3%, is the *S* enantiomer from S_N2. Therefore, the excess of one enantiomer over the racemic mixture must be 33.3%, the enantiomeric excess. (In the racemic mixture, the *R* and *S* "cancel" each other algebraically as well as in optical rotation.)

The optical rotation of pure (*S*)-butan-2-ol is + 13.5°. The optical rotation of this mixture is:

$$33.3\% \times +13.5° = +4.5°$$

6-29

(a) Methyl shift may occur simultaneously with ionization.

$CH_3-C(CH_3)_2-CH_2-I \xrightarrow{Ag^+} AgI + [CH_3-C(CH_3)_2-\overset{+}{C}H_2]$ $\xrightarrow{\text{methyl shift}}$ 3° $CH_3-\overset{+}{C}(CH_3)-CH_2CH_3$

This 1° carbocation would be very unstable.

$\xrightarrow{H_2\ddot{O}:}$ $CH_3-C(CH_3)(\overset{+}{O}H_2)-CH_2CH_3$ $\xrightarrow{H_2\ddot{O}:}$ $CH_3-C(CH_3)(\ddot{O}-H)-CH_2CH_3$

(b) Alkyl shift may occur simultaneously with ionization.

CH_2-I $\xrightarrow{Ag^+}$ AgI + [$\overset{+}{C}H_2$ 1°, CH_2]

This 1° carbocation would be very unstable.

$\xrightarrow{\text{alkyl shift—ring expansion}}$ 3° $\overset{+}{C}$, CH_2, CH_2 $\xrightarrow{H_2\ddot{O}:}$ $\overset{+}{O}H$ (H) $\xrightarrow{H_2\ddot{O}:}$ $\ddot{O}H$ + $\ddot{O}CH_2CH_3$

also produced from the solvent as nucleophile

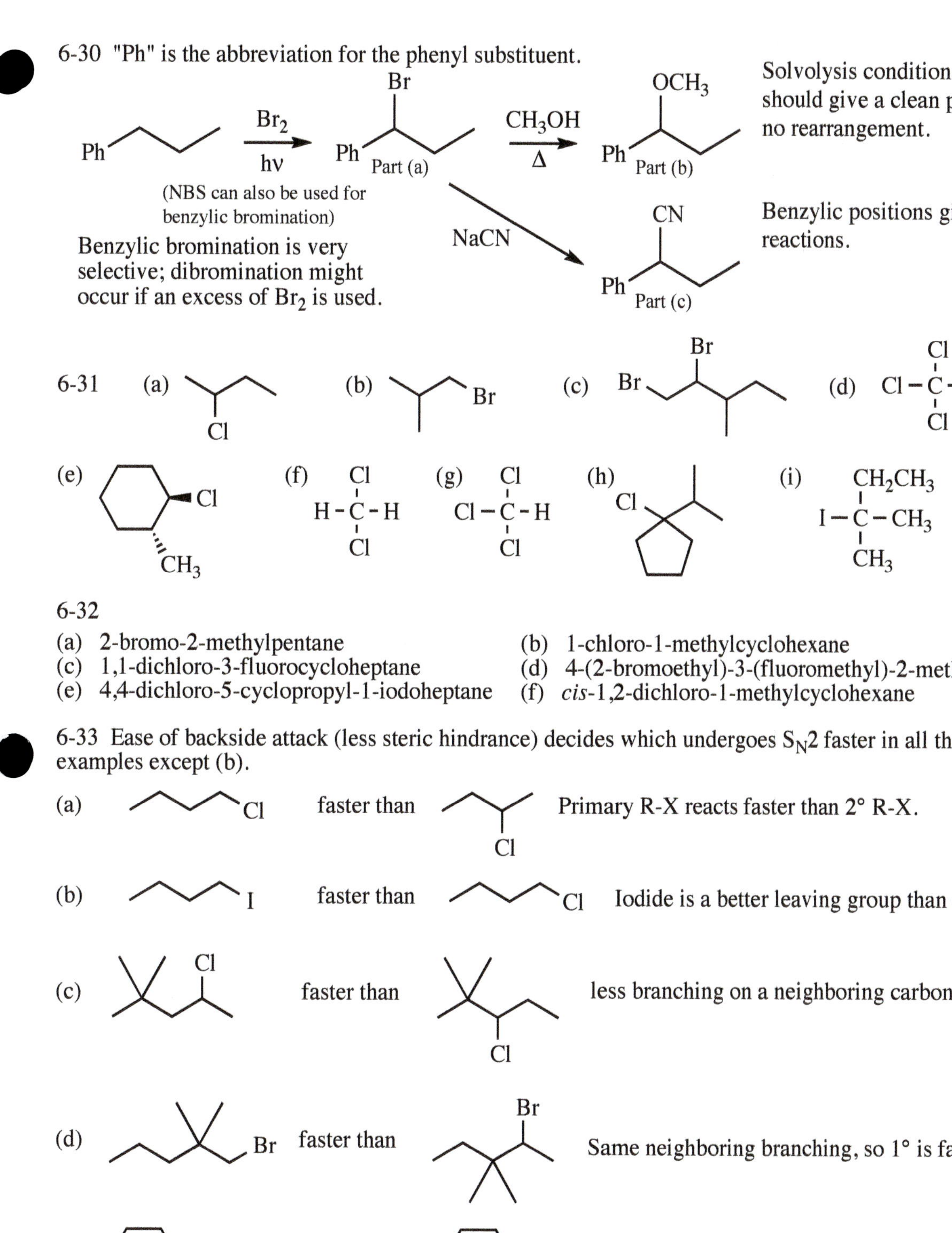

6-30 "Ph" is the abbreviation for the phenyl substituent.

(NBS can also be used for benzylic bromination)

Benzylic bromination is very selective; dibromination might occur if an excess of Br_2 is used.

Solvolysis conditions—S_N1—should give a clean product with no rearrangement.

Benzylic positions give clean S_N2 reactions.

6-31 (a) (b) (c) (d) (e) (f) (g) (h) (i)

6-32

(a) 2-bromo-2-methylpentane
(b) 1-chloro-1-methylcyclohexane
(c) 1,1-dichloro-3-fluorocycloheptane
(d) 4-(2-bromoethyl)-3-(fluoromethyl)-2-methylheptane
(e) 4,4-dichloro-5-cyclopropyl-1-iodoheptane
(f) *cis*-1,2-dichloro-1-methylcyclohexane

6-33 Ease of backside attack (less steric hindrance) decides which undergoes S_N2 faster in all these examples except (b).

(a) Cl faster than Cl Primary R-X reacts faster than 2° R-X.

(b) I faster than Cl Iodide is a better leaving group than chloride.

(c) Cl faster than Cl less branching on a neighboring carbon

(d) Br faster than Br Same neighboring branching, so 1° is faster than 2°.

(e) CH_2Cl faster than Cl Primary R-X reacts faster than 2° R-X.

(f) Br faster than Br less branching on a neighboring carbon

6-34 Formation of the more stable carbocation decides which undergoes S_N1 faster in all these examples except (d).

(a) 3° [tert-butyl chloride] faster than [3-chloropentane] 2°

(b) 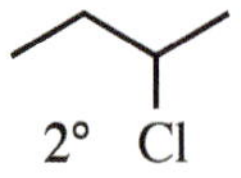 [2-chlorobutane] 2° Cl faster than [1-chlorohexane] Cl 1°

(c) 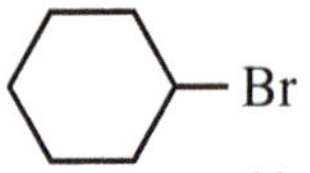 [bromocyclohexane] Br 2° faster than [cyclohexyl]-CH_2Br 1°

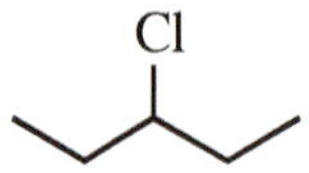

(d) 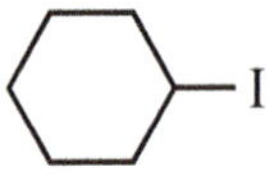 [iodocyclohexane] I faster than [chlorocyclohexane] Cl (leaving group ability)

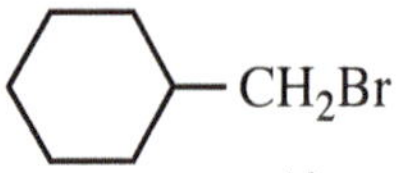

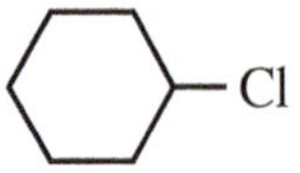

(e) 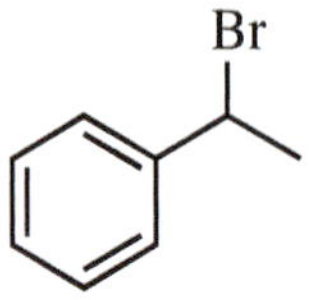 [1-bromoethylbenzene] Br faster than [branched alkyl bromide] Br 2°

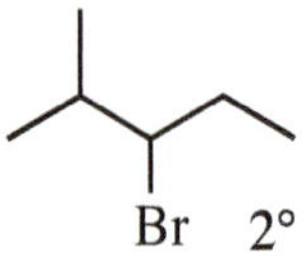

2° benzylic! (resonance!)

(f) 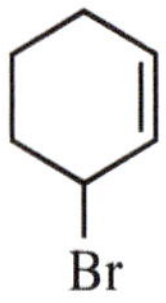 [3-bromocyclohexene] Br faster than [bromocyclohexane] Br

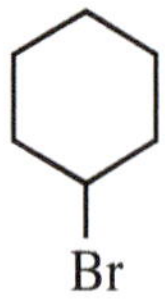

2° allylic! (resonance!) 2°

6-35 For S_N2, reactions should be designed such that the nucleophile attacks the least highly substituted alkyl halide. ("X" stands for a halide: Cl, Br, or I.)

(a)  [cyclohexyl]-CH_2X + HO^- → [cyclohexyl]-CH_2OH

(b) 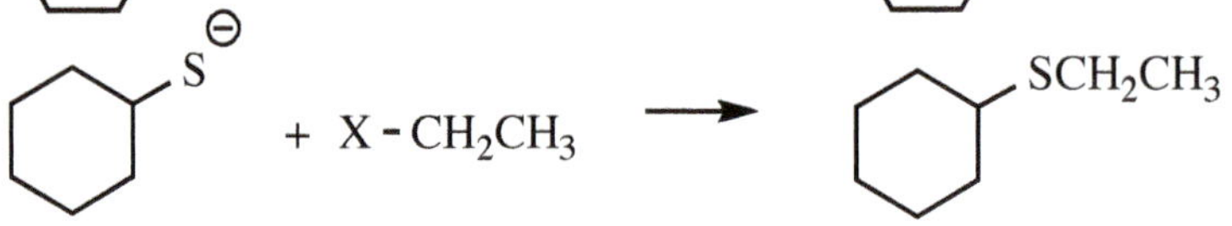 [cyclohexyl]-S^- + $X-CH_2CH_3$ → [cyclohexyl]-SCH_2CH_3

(c) 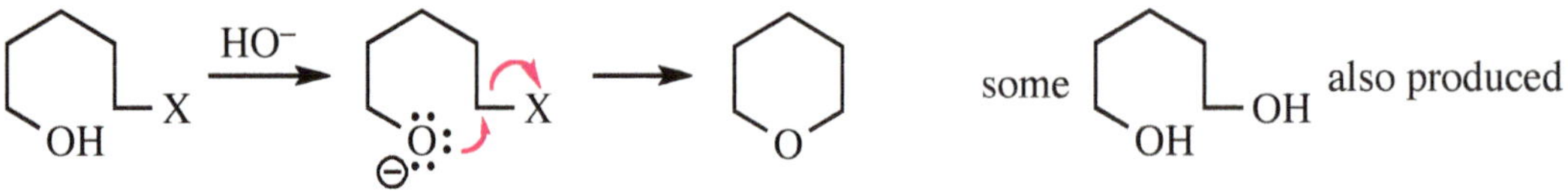 [HO-(CH₂)₅-X] OH, X —HO^-→ [alkoxide] $\ddot{O}^{:-}$, X → [tetrahydropyran] O some [HO-(CH₂)₅-OH] OH OH also produced

(d)  [cyclohexyl]-CH_2X + NH_3 excess → [cyclohexyl]-CH_2NH_2

(e) $H_2C{=}CHCH_2X$ + ^-CN → $H_2C{=}CHCH_2CN$

(f) $HC{\equiv}C^{\ominus}$ + $X-CH_2CH_2CH_3$ → $HC{\equiv}C-CH_2CH_2CH_3$

6-36

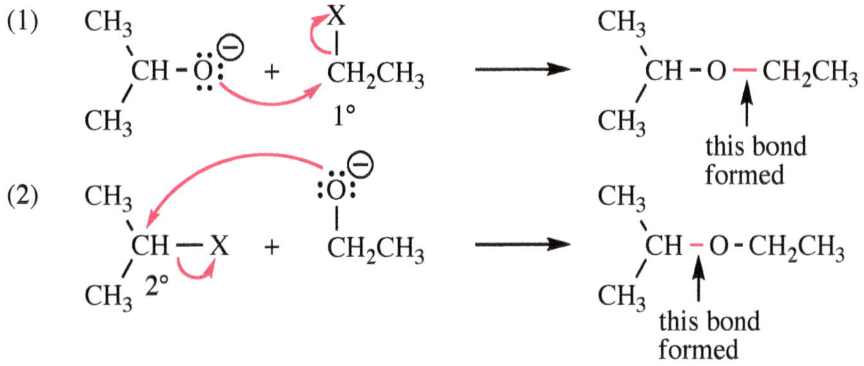

(1) CH_3 \ $CH-\ddot{O}:^{\ominus}$ / CH_3 + X | CH_2CH_3 1° ⟶ CH_3 \ $CH-O-CH_2CH_3$ / CH_3 (this bond formed: O–CH_2CH_3)

(2) CH_3 \ $CH-X$ / CH_3 2° + $:\ddot{O}:^{\ominus}$ | CH_2CH_3 ⟶ CH_3 \ $CH-O-CH_2CH_3$ / CH_3 (this bond formed: CH–O)

Synthesis (1) would give a better yield of the desired ether product. (1) uses S_N2 attack of a nucleophile on a 1° carbon, while (2) requires attack on a more hindered 2° carbon. Reaction (2) would give a lower yield of substitution, with more elimination.

6-37

(a) S_N2—second order: reaction rate doubles: rate = k [EtBr] [KO-*t*-Bu]
(b) S_N2—second order: reaction rate increases six times
(c) Virtually all reaction rates, including this one, increase with a temperature increase.

6-38 This is an S_N1 reaction; the rate law depends only on the substrate concentration, not on the nucleophile concentration: rate = k [C_4H_9Br]
(a) no change in rate
(b) the rate triples, dependent only on [*tert*-butyl bromide]
(c) Virtually all reaction rates, including this one, increase with a temperature increase.

6-39 The key to this problem is that iodide ion is both an excellent nucleophile AND leaving group. Substitution on chlorocyclohexane is faster with iodide than with cyanide (see Table 6-3 for relative nucleophilicities). Once iodocyclohexane is formed, substitution by cyanide is much faster on iodocyclohexane than on chlorocyclohexane because iodide is a better leaving group than chloride. So two fast reactions involving iodide replace a slower single reaction, resulting in an overall rate increase.

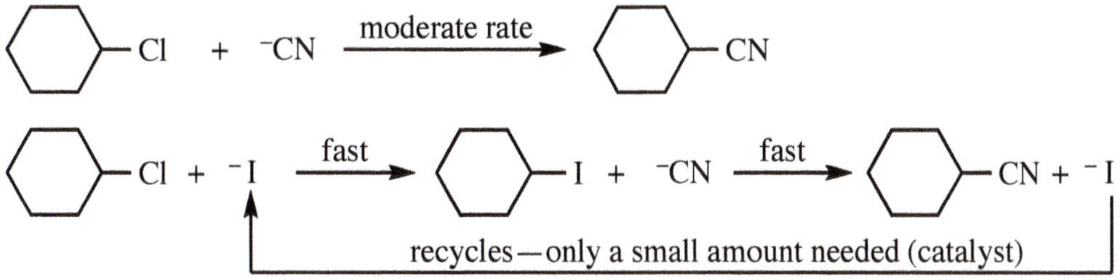

Cyclohexyl–Cl + $^-$CN —moderate rate→ Cyclohexyl–CN

Cyclohexyl–Cl + $^-$I —fast→ Cyclohexyl–I + $^-$CN —fast→ Cyclohexyl–CN + $^-$I

recycles—only a small amount needed (catalyst)

6-40 The solvolysis (S_N1) products are shown.

(a) (b) (c)

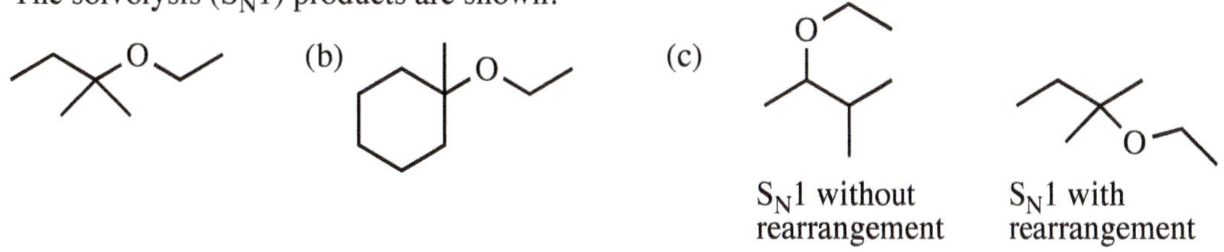

S_N1 without rearrangement

S_N1 with rearrangement

(d) rearrangement—methyl shift

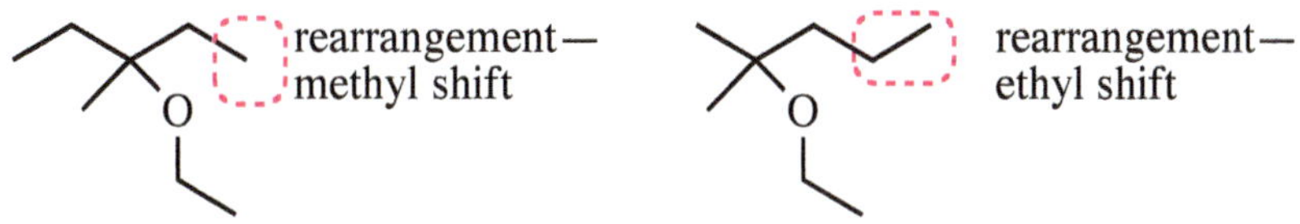

rearrangement—ethyl shift

6-41

(a) C=C–C–X → (−X⁻) {C=C–C⊕ ↔ ⊕C–C=C}

(b)

(i) equivalent resonance forms

(ii)

(iii) equivalent resonance forms

(iv) Loss of bromide gives unstable 1° carbocation that quickly rearranges to a 2° allylic carbocation.

equivalent resonance forms—same as from part (iii)

(c) Only one substitution product arises from equivalent resonance forms.

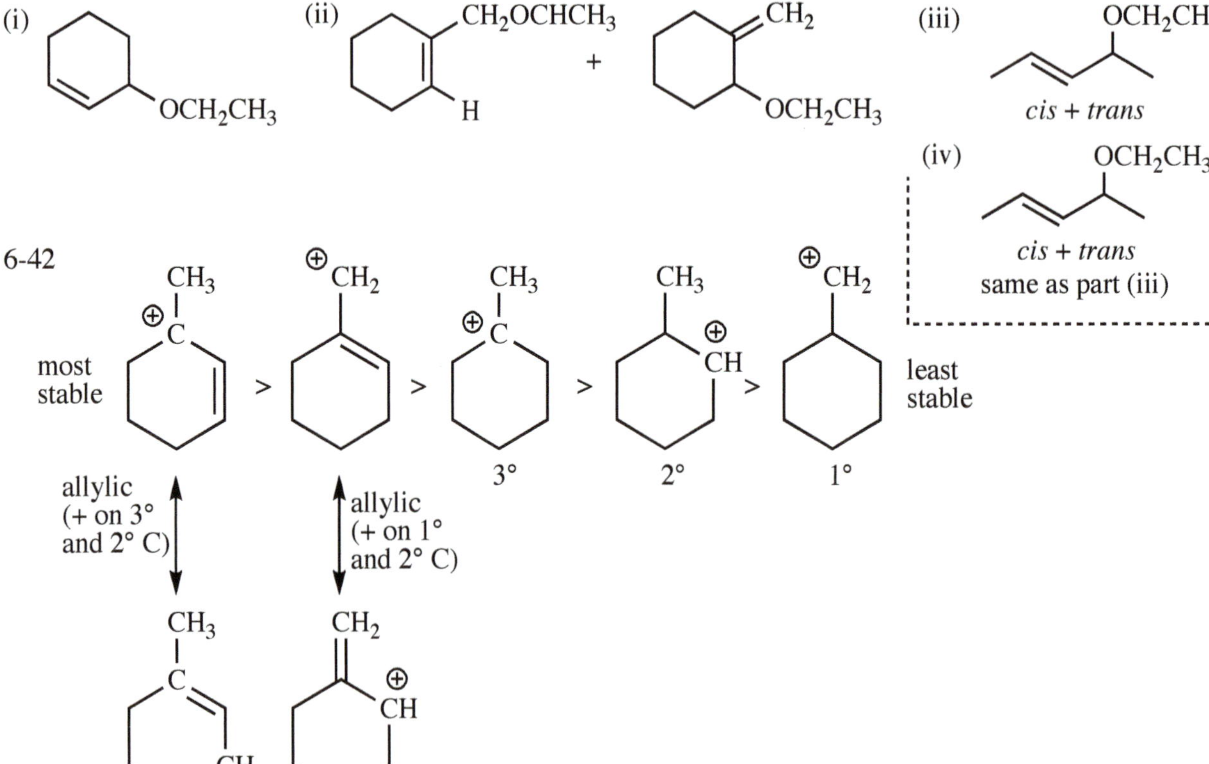

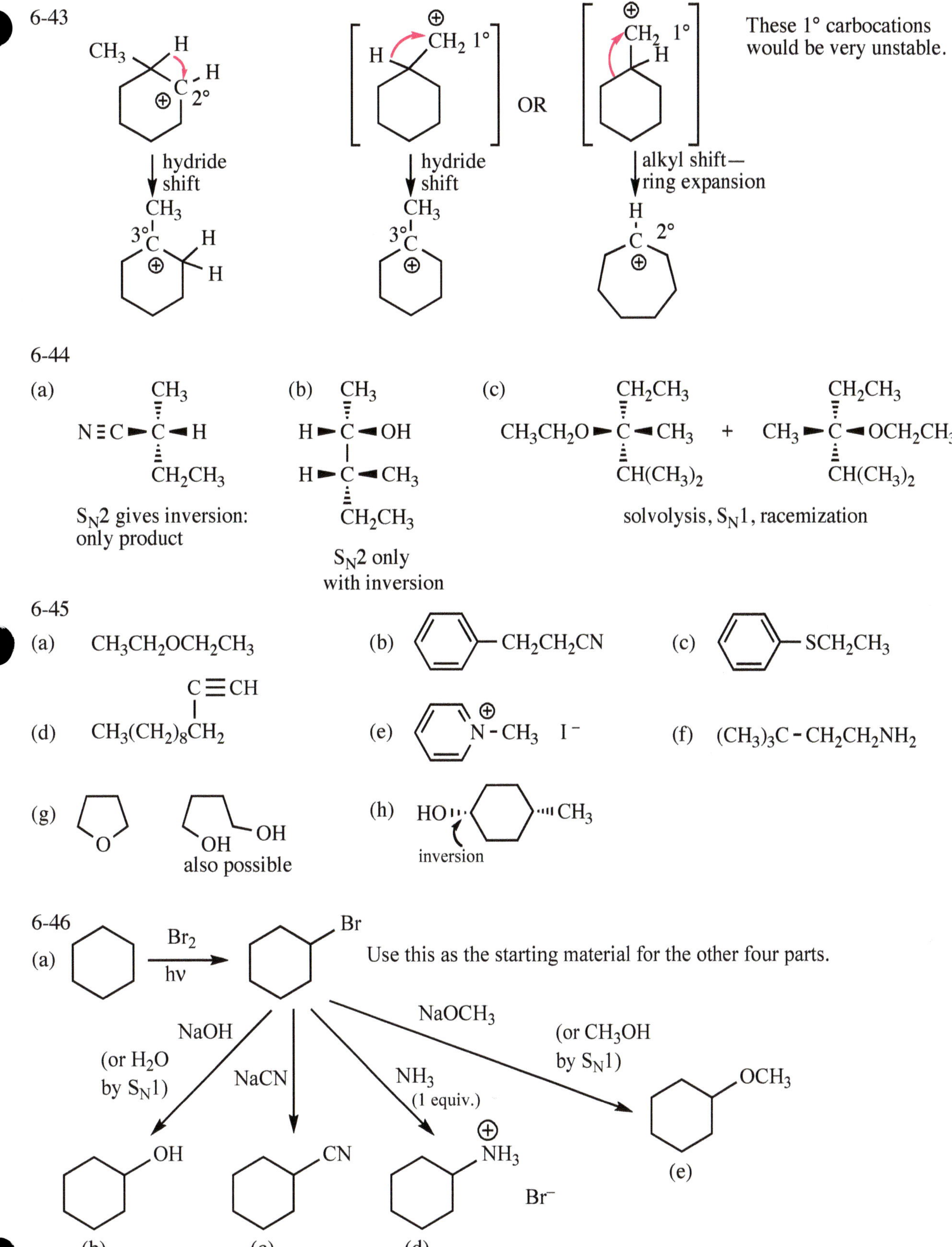

6-43
CH_3
H
H
2°
hydride shift
CH_3
3°
H
H
H
CH_2 1°
OR
CH_2 1°
H
These 1° carbocations would be very unstable.
hydride shift
CH_3
3°
alkyl shift—ring expansion
H
2°
6-44
(a)
CH_3
N≡C—C—H
CH_2CH_3
S_N2 gives inversion: only product
(b)
CH_3
H—C—OH
H—C—CH_3
CH_2CH_3
S_N2 only with inversion
(c)
CH_2CH_3
CH_3CH_2O—C—CH_3
$CH(CH_3)_2$
+
CH_2CH_3
CH_3—C—OCH_2CH_3
$CH(CH_3)_2$
solvolysis, S_N1, racemization
6-45
(a) $CH_3CH_2OCH_2CH_3$
(b) CH_2CH_2CN
(c) SCH_2CH_3
(d) C≡CH
$CH_3(CH_2)_8CH_2$
(e) N-CH_3 I^-
(f) $(CH_3)_3C-CH_2CH_2NH_2$
(g) O
OH
OH
also possible
(h) HO
CH_3
inversion
6-46
(a)
Br_2
hν
Br
Use this as the starting material for the other four parts.
NaOH
(or H_2O by S_N1)
NaCN
NH_3 (1 equiv.)
$NaOCH_3$
(or CH_3OH by S_N1)
OH
CN
NH_3
Br^-
OCH_3
(b)
(c)
(d)
(e)

6-47

The use of methyl bromide (bromomethane) as a soil sterilant has been controversial: it is effective in killing soil organisms that are pathogenic to agricultural crops, especially strawberries, but its toxicity is not specific to soil organisms. It is acutely toxic to most plants and animals, including humans, and is a potent mutagen which is typical of alkylating agents. (Alkylated DNA bases are mutations.) It is also a potent disrupter of stratospheric ozone; one estimate is that the C—Br bond is about 50 times more effective at destroying ozone than is a C—Cl bond.

Methyl iodide (iodomethane) is more toxic because iodide a better leaving group than bromide; therefore, methyl iodide is faster to alkylate any nucleophile that it contacts. On the other hand, methyl iodide is a liquid at room temperature, although it evaporates quickly; methyl bromide is a gas and diffuses more readily than methyl iodide. The other "benefit" of methyl iodide is that because it is more reactive than methyl bromide, it reacts in the soil or in the atmosphere more rapidly, so it is much less likely to reach the stratosphere than is methyl bromide—all because iodide is a better leaving group than bromide.

6-48

(a) $\dfrac{+15.58°}{+15.90°} \times 100\% = 98\%$ of original optical activity $= 98\%$ e.e.

Thus, 98% of the *S* enantiomer and 2% racemic mixture gives an overall composition of 99% *S* and 1% *R*.

(b) The 1% of radioactive iodide has produced exactly 1% of the *R* enantiomer. Each substitution must occur with inversion, a classic S_N2 mechanism.

6-49

(a) An S_N2 mechanism with inversion will convert *R* to its enantiomer, *S*. An accumulation of excess *S* does not occur because it can also react with bromide, regenerating *R*. The system approaches a racemic mixture at equilibrium.

CH3 / Br—C—H (CH2CH3), *R* + :Br:⊖ —S_N2, inversion→ CH3 / H—C—Br (CH2CH3), *S* + :Br:⊖

(b) In order to undergo substitution and therefore inversion, HO^- would have to be the leaving group, but HO^- is never a leaving group in S_N2. No reaction can occur.

(c) Once the OH is protonated, it can leave as H_2O. Racemization occurs in the S_N1 mechanism because of the planar, achiral carbocation intermediate which "erases" all stereochemistry of the starting material. Racemization occurs in the S_N2 mechanism by establishing an equilibrium of *R* and *S* enantiomers, as explained in part (a).

The symbol H—A is used for a generic acid, usually a catalyst in a mechanism. The conjugate base is A^-.

HO—C(CH3)(CH2CH3)—H —H-A→ HO^+(H)—C(CH3)(CH2CH3)—H —H_2O:, S_N2→ H—C(CH3)(CH2CH3)—O^+H_2 —H_2O:→ H—C(CH3)(CH2CH3)—OH (inverted)

S_N1 ↓

planar carbocation: CH3, CH_3CH_2, H, C^+ + H_2O: —H_2O removes H^+→ HO—C(CH3)(CH2CH3)—H + H—C(CH3)(CH2CH3)—OH (racemic mixture)

6-50 NBS generates bromine which produces bromine radical. Bromine radical abstracts an allylic hydrogen, resulting in a resonance-stabilized allylic radical. The allylic radical can bond to bromine at either of the two carbons with radical character.

$H_2C{=}C(H){-}CH(H)(CH_3){-}CH_3 + Br\bullet \longrightarrow HBr + \left\{ H_2C{=}C(H){-}\dot{C}(CH_3){-}CH_3 \longleftrightarrow H_2\dot{C}{-}C(H){=}C(CH_3){-}CH_3 \right\}$ (allylic)

HBr + NBS → $Br{-}Br$

continues propagation

$\bullet Br + H_2C{=}C(H){-}C(Br)(CH_3){-}CH_3 + H_2C(Br){-}C(H){=}C(CH_3){-}CH_3$

6-51 The bromine radical from NBS will abstract whichever hydrogen produces the most stable intermediate; in this structure, that is a benzylic hydrogen, giving the resonance-stabilized benzylic radical.

$C_6H_5{-}CH(H)CH_3 \xrightarrow{Br\bullet} HBr + \left\{ C_6H_5{-}\dot{C}HCH_3 \longleftrightarrow \cdots{=}CHCH_3 \longleftrightarrow \cdots{=}CHCH_3 \longleftrightarrow \cdot\cdots{=}CHCH_3 \right\}$

HBr + NBS → Br_2

$C_6H_5{-}CH(Br)CH_3 + Br\bullet$

(Even though three carbons of the ring have some radical character, these are minor resonance contributors. The product is most stable when the ring has all three double bonds intact, necessitating that the bromine bond to the benzylic carbon.)

6-52 Two related factors could explain this observation. First, as carbocation stability increases, the leaving group will be less tightly held by the carbocation for stabilization; the more stable carbocations are more "free" in solution, meaning more exposed. Second, more stable carbocations will have longer lifetimes, allowing the leaving group to drift off in the solvent, leading to more possibility for the incoming nucleophile to attack from the side that the leaving group just left.

The less stable carbocations hold tightly to their leaving groups, preventing nucleophiles from attacking this side. Backside attack with inversion is the preferred stereochemical route in this case.

6-53

$Nuc^{-}: \xrightarrow{S_N2} (CH_3CH_2)_3O^{\oplus}\ BF_4^{-} \longrightarrow Nuc{-}CH_2CH_3 + H_3CH_2C{-}\ddot{O}{-}CH_2CH_3$

The leaving group is a molecule of diethyl ether.

From Table 6-2, all of the ethyl halides are liquids. Triethyloxonium tetrafluoroborate is a solid which is easier to handle and often safer.

6-54 $\xrightarrow{S_N1}$ { $CH_2^{\oplus}$ ↔ ↔ $\overset{\oplus}{HC}$ ↔ $\ddot{O}^{\oplus}$ }

$\overset{\ominus}{:\ddot{O}:}$ formate nucleophile

Ionization of the 1° chloride gives a cation stabilized by resonance, leading to an S_N1 mechanism. Elimination cannot occur because no neighboring carbon has a hydrogen atom.

6-55 The energy, and therefore the structure, of the transition state determines the rate of a reaction. **Any factor that lowers the energy of the transition state will speed the reaction.**

$R_3N: \; + \; R'-X \longrightarrow \left[R_3\overset{\delta+}{N}\text{-----}R'\text{------}\overset{\delta-}{X} \right]^{\ddagger} \longrightarrow R_3\overset{\oplus}{N}-R' \quad X^-$

transition state—
developing charge

This example of S_N2 is unusual in that the nucleophile is a neutral molecule—it is not negatively charged. The transition state is beginning to show the positive and negative charges of the products (ions), so the transition state is more charged than the reactants. The polar transition state will be stabilized in a more polar solvent through dipole-dipole interactions, so the rate of reaction will be enhanced in a polar solvent.

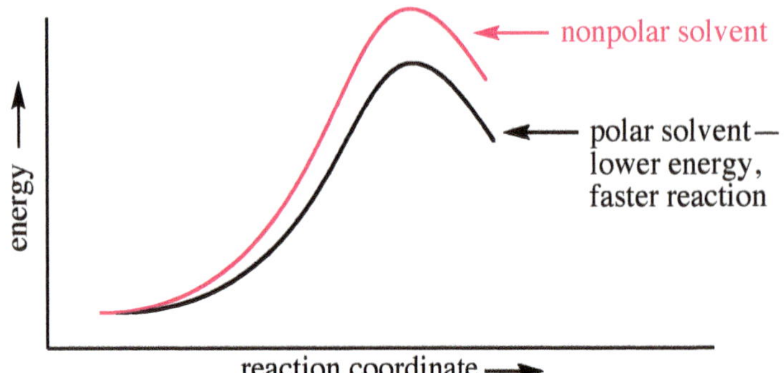

6-56 The problem is how to explain this reaction:

$$\underset{\mathbf{1}}{H_2C(NEt_2)-CH(Cl)-CH_2CH_3} \xrightarrow{HO^-} \underset{\mathbf{2}}{H_2C(OH)-CH(NEt_2)-CH_2CH_3} + Cl^-$$

<u>facts</u>

1) second order, but several thousand times faster than similar second-order reactions without the NEt_2 group
2) NEt_2 group migrates

See explanation on next page.

6-56 continued

Solution

Clearly, the NEt_2 group is involved. The nitrogen is a nucleophile and can do an internal nucleophilic substitution (S_Ni), a very fast reaction for entropy reasons because two different molecules do not have to come together.

$$\underset{\mathbf{1}}{Et_2N\text{–}H_2C\text{–}CH(Cl)\text{–}CH_2CH_3} \xrightarrow[\text{very fast}]{S_Ni} \underset{\mathbf{3}}{\text{(aziridinium: }Et_2N^{\oplus}\text{ bridging }H_2C\text{–}CH\text{)}\text{–}CH_2CH_3} + Cl^-$$

The slower step is attack of HO^- on intermediate **3**; the N is a good leaving group because it has a positive charge. Where will HO^- attack **3**? On the less substituted carbon, in typical S_N2 fashion.

$$HO^- + \mathbf{3} \xrightarrow{S_N2} \underset{\mathbf{2}}{HO\text{–}H_2C\text{–}CH(NEt_2)\text{–}CH_2CH_3}$$

This overall reaction is fast because of the *neighboring group assistance* in forming **3**. It is second order because the HO^- group and **3** collide in the slow step (not the *only* step, however). And the NEt_2 group "migrates", although in two steps.

6-57

(a) Only the propagation steps are shown. NBS provides a low concentration of Br_2 which generates bromine radical in ultraviolet light. In the starting material, all 8 allylic hydrogens are equivalent.

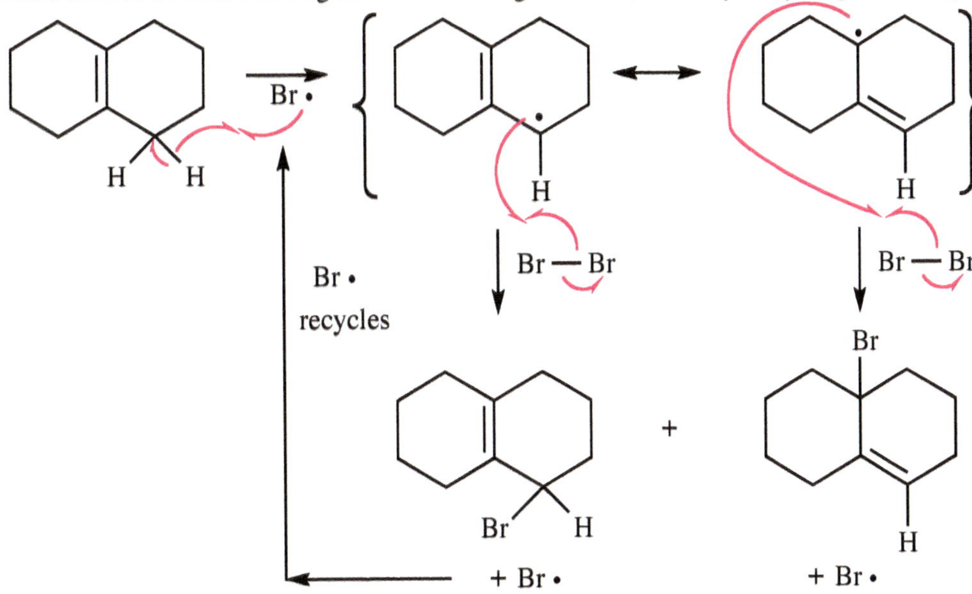

6-57 continued

(b) NBS contains traces of bromine; when combined with HBr, NBS produces more bromine. Bromine contains small amounts of bromine radical. Bromine radical abstracts an allylic hydrogen, resulting in a resonance-stabilized allylic radical. The allylic radical can bond to bromine at either of the two carbons with radical character.

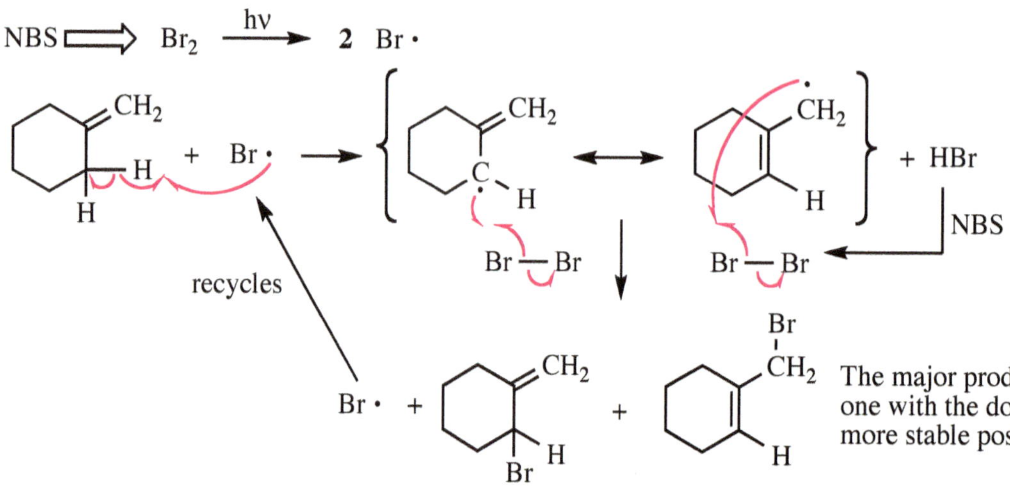

NBS ⟹ Br_2 $\xrightarrow{h\nu}$ 2 Br ·

CH_2 H H + Br · → { CH_2 C · H ⟷ · CH_2 H } + HBr

NBS

Br—Br Br—Br

recycles

Br · + CH_2 H Br + Br CH_2 H

The major product is usually the one with the double bond in the more stable position.

6-58 In each example, arrows show the allylic positions where a hydrogen atom can be removed.

(a) 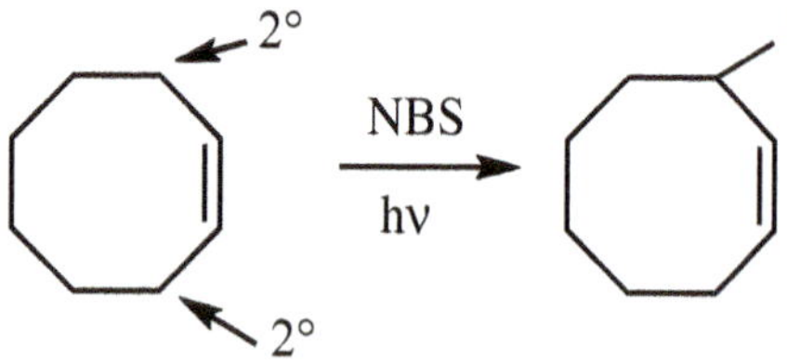 2° 2° $\xrightarrow[h\nu]{NBS}$ Br

Only one isomer is produced because of symmetry.

(b) 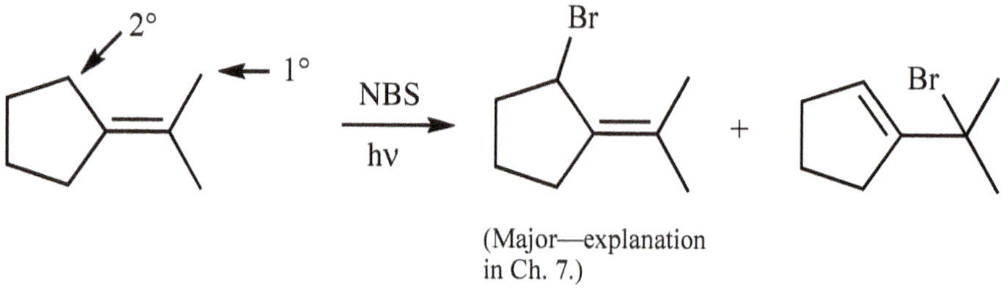 2° 1° $\xrightarrow[h\nu]{NBS}$ Br + Br

(Major—explanation in Ch. 7.)

The 2° H will be removed many times faster than the 1° H.

(c) 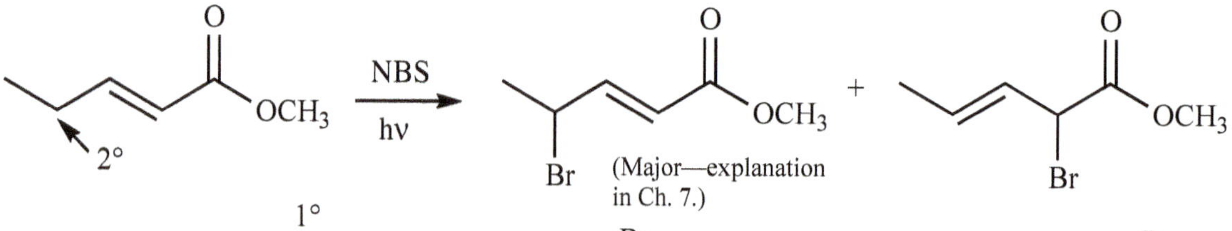 O OCH_3 2° $\xrightarrow[h\nu]{NBS}$ O OCH_3 Br + O OCH_3 Br

(Major—explanation in Ch. 7.)

(d) 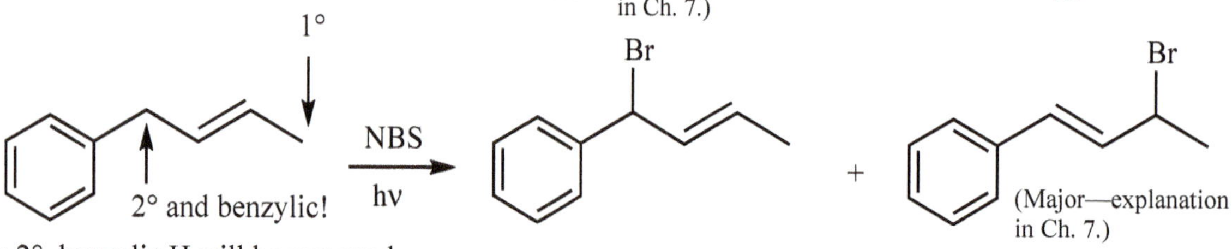 1° 2° and benzylic! $\xrightarrow[h\nu]{NBS}$ Br + Br

(Major—explanation in Ch. 7.)

The 2° benzylic H will be removed many times faster than the 1° H.

6-59

(a)

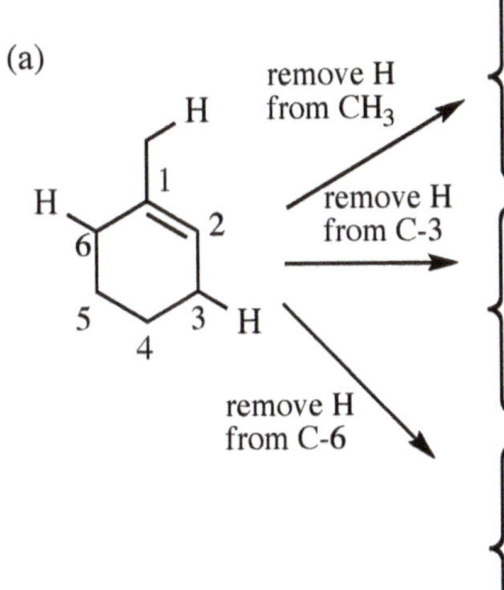

remove H from CH_3 → $\{\cdot CH_2$ (1-cyclohexenyl) $\longleftrightarrow$ CH_2 / $\cdot CH\}$

remove H from C-3 → $\{CH_3$ / $\cdot CH$ $\longleftrightarrow$ CH_3 / $\dot{C}\}$

remove H from C-6 → $\{CH_3$ / $H\dot{C}$ $\longleftrightarrow$ CH_3 / $\cdot CH\}$

(b)

Radical character on 1° and 2° carbons—least stable of these three (will be slowest to form).

Radical character on 2° and 3° carbons—most stable of these three (will be fastest to form).

Radical character on two 2° carbons, equivalent resonance forms—almost as stable as the 2°/3° radical.

(c)

Br + Br

Two products from the 1°/2° radical; so minor that only trace quantities might be detected.

Br + Br

Two products from the 2°/3° radical.

Br + Br

SAME COMPOUND

Only ONE compound is produced from the 2°/2° radical.

CHAPTER 7—STRUCTURE AND SYNTHESIS OF ALKENES; ELIMINATION

7-1 The number of elements of unsaturation in a hydrocarbon formula is given by:

$$\frac{2(\#C) + 2 - (\#H)}{2}$$

(a) Solve this equation for #H. Three double bonds plus one ring make 4 elements of unsaturation.

$$\mathbf{4} = \frac{2(9) + 2 - (\#H)}{2} \Rightarrow \boxed{\#H = \mathbf{12}}$$

C_9H_{12} has 4 elements of unsaturation.

(b) $C_6H_{12} \Rightarrow \dfrac{2(6) + 2 - (12)}{2} = \mathbf{1}$ element of unsaturation

(c) Many examples are possible. Yours may not match these structures, but all possible answers must have either one double bond or one ring, that is, one element of unsaturation.

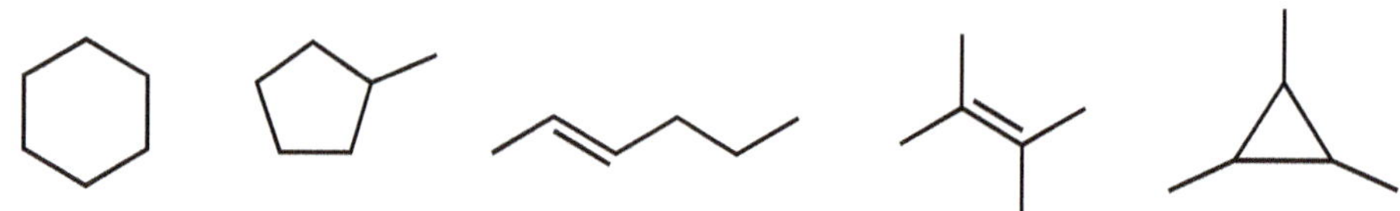

7-2 $C_4H_6 \Rightarrow \dfrac{2(4) + 2 - (6)}{2} = \mathbf{2}$ elements of unsaturation

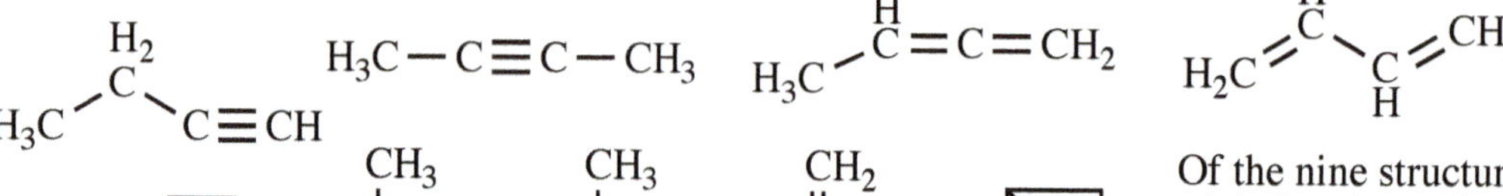

Of the nine structures, two have 1 triple bond, two have 2 double bonds, four have 1 ring plus 1 double bond, and one has 2 rings.

7-3 Hundreds of examples of C_4H_6NOCl are possible. Yours may not match these, although all must contain two elements of unsaturation.

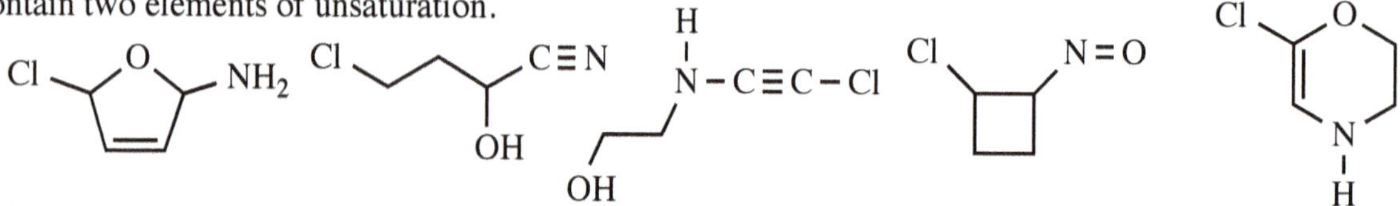

7-4 Many examples (sometimes thousands) of these formulas are possible. Yours may not match these, but correct answers must have the same number of elements of unsaturation. *Check your answers in your study group.*

(a) $C_4H_4Cl_2 \Rightarrow C_4H_6$ = *hydrocarbon equivalent*
C_4H_{10} = *saturated formula*

a 4H deficiency =
2 elements of unsaturation

(b) $C_4H_8O \Rightarrow C_4H_8$ = *hydrocarbon equivalent*
C_4H_{10} = *saturated formula*

a 2H deficiency =
1 element of unsaturation

(c) $C_6H_8O_2 \Rightarrow C_6H_8$ = *hydrocarbon equivalent*
C_6H_{14} = *saturated formula*

a 6H deficiency =
3 elements of unsaturation

$H_3C-C\equiv C-\overset{O}{\overset{||}{C}}-OCH_2CH_3$

7-4 continued

(d) $C_5H_5NO_2$ —using the formula:

$$\frac{2(5) + 2 + (1) - (5)}{2} = \mathbf{4}$$

(e) C_6H_3NClBr —using the formula:

$$\frac{2(6) + 2 + (1) - (3 + 2)}{2} = \mathbf{5}$$

Note to the student: The IUPAC system of nomenclature is undergoing many changes, most notably in the placement of position numbers. The new system places the position number close to the functional group designation, which is what this Solutions Manual will attempt to follow; however, you should be able to use and recognize names in either the old or the new style. Ask your instructor which system to use.

7-5 The *E/Z* system is unambiguous and is generally preferred for designating stereochemistry around a double bond. However, *cis* and *trans* are still used for geometric isomers from substituents on a ring.

(a) 4-methylpent-1-ene
(b) 2-ethylhex-1-ene (Number the longest chain *containing the double bond.*)
(c) penta-1,4-diene
(d) penta-1,2,4-triene
(e) 2,5-dimethylcyclopenta-1,3-diene
(f) 3-methylene-4-vinylcyclohex-1-ene ("1" is optional)
(g) 3-phenylprop-1-ene ("1" is optional)

For comparison, these are the old IUPAC names for problem 7-5. The old names will not be used in this Solutions Manual.

(a) 4-methyl-1-pentene
(b) 2-ethyl-1-hexene
(c) 1,4-pentadiene
(d) 1,2,4-pentatriene
(e) 2,5-dimethyl-1,3-cyclopentadiene
(f) 3-methylene-4-vinyl-1-cyclohexene ("1" is optional)
(g) 3-phenyl-1-propene ("1" is optional)

Note to the student: How to tell whether a double bond has geometric isomers? Look at the four groups bonded to the two carbons of the double bond: w, x, y, and z.

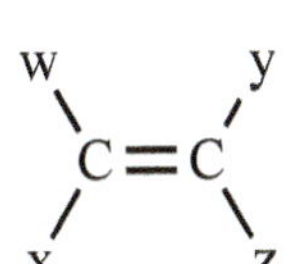

If w ≠ x AND y ≠ z, it will have geometric isomers.

If w = y (or x = z), it is the *cis* isomer.
If w = z (or x = y), it is the *trans* isomer.

7-6 Parts (b) and (e) do not show *cis,trans* isomerism.

(a)

(*Z*)-hex-3-ene
(*cis*-hex-3-ene)

(*E*)-hex-3-ene
(*trans*-hex-3-ene)

(c)

(2*Z*,4*Z*)-hexa-2,4-diene
cis,cis-hexa-2,4-diene

(2*Z*,4*E*)-hexa-2,4-diene
cis,trans-hexa-2,4-diene

(2*E*,4*E*)-hexa-2,4-diene
trans,trans-hexa-2,4-diene

7-6 continued

(d)

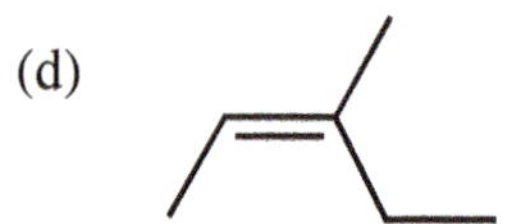

(*Z*)-3-methylpent-2-ene

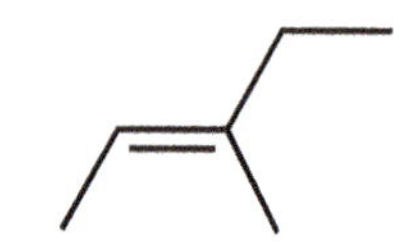

(*E*)-3-methylpent-2-ene

"*Cis*" and "*trans*" are not clear for this example; "*E*" and "*Z*" are unambiguous.

(f)

1 H H 2 4 3 Br Br

cis-3,4-dibromocyclopent-1-ene (Both Br atoms could also be up, the enantiomer.)

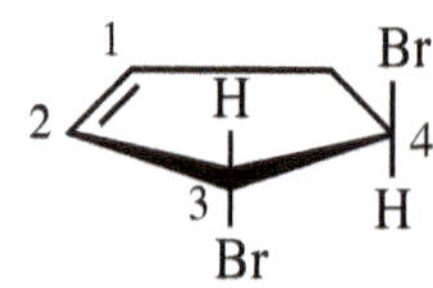

trans-3,4-dibromocyclopent-1-ene (Both Br atoms could also be inverted, the enantiomer.)

In this example, *cis* and *trans* refer to the relative positions of the Br atoms, not the double bond. All double bonds in rings of 7 or fewer atoms must be *cis*.

7-7

(a)

2 3 4 5 1

cis-3,4-dimethylpent-2-ene (The only positions to attach two methyl groups to make the double bond cis are at C-3 and C-4.)

(b)

4 3 6 5 2 1

trans-3-ethylhexa-1,4-diene (Whether *cis* or *trans* is not specified in the problem; the vinyl group is part of the main chain. This could also be named as the *E* isomer. The *cis* (*Z*) isomer is equally good.)

(c)

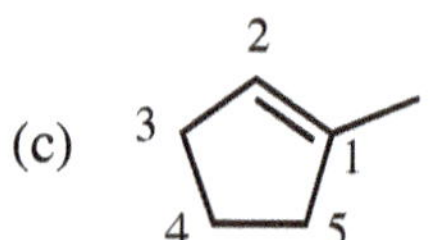

1-methylcyclopentene

(d)

4 Cl 5 3 2 1

5-chlorocyclohexa-1,3-diene (Positions of double bonds need to be specified.)

(e)

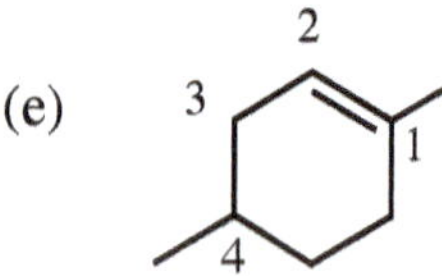

1,4-dimethylcyclohexene

(f)

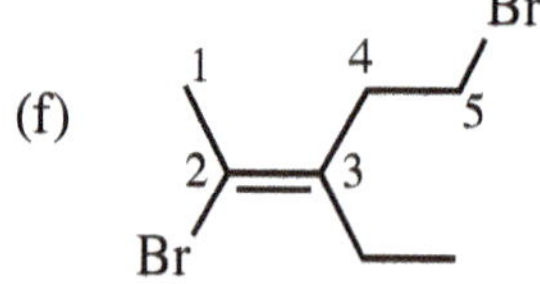

(*E*)-2,5-dibromo-3-ethylpent-2-ene (The *cis* designation does not apply—four different groups on the double bond.)

7-8

(a)

Cl Br

(*E*)-3-bromo-2-chloropent-2-ene

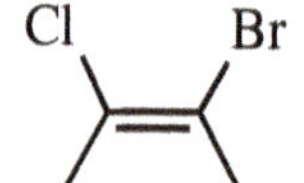

(*Z*)-3-bromo-2-chloropent-2-ene

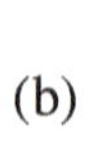

(b)

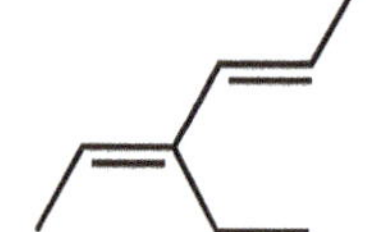

(2*E*,4*E*)-3-ethylhexa-2,4-diene

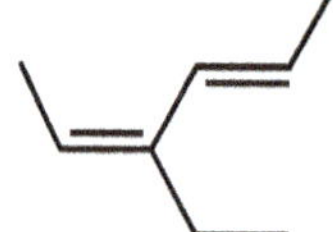

(2*Z*,4*E*)-

(2*E*,4*Z*)-

(2*Z*,4*Z*)-

(c)

H Br

(*Z*)-3-bromo-2-methylhex-3-ene

H Br

(*E*)-3-bromo-2-methylhex-3-ene

7-8 continued

(d)

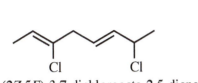

(Z)-penta-1,3-diene

(E)-penta-1,3-diene

(e)

no geometric isomer

(f)

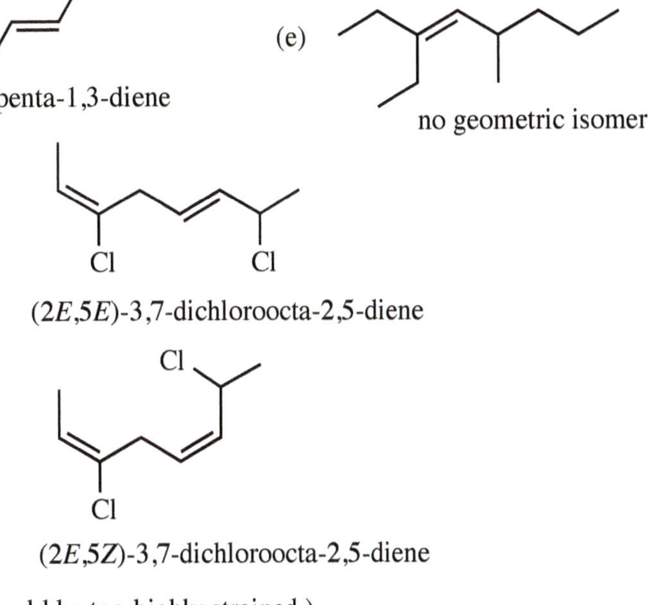

(2Z,5E)-3,7-dichloroocta-2,5-diene

(2E,5E)-3,7-dichloroocta-2,5-diene

(2Z,5Z)-3,7-dichloroocta-2,5-diene

(2E,5Z)-3,7-dichloroocta-2,5-diene

(g) no geometric isomers (An *E* double bond would be too highly strained.)

(h)

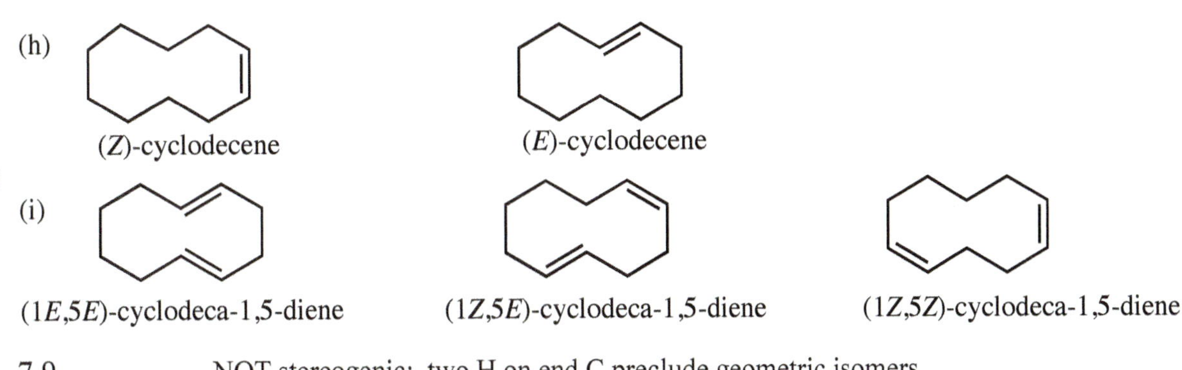

(Z)-cyclodecene

(E)-cyclodecene

(i)

(1E,5E)-cyclodeca-1,5-diene

(1Z,5E)-cyclodeca-1,5-diene

(1Z,5Z)-cyclodeca-1,5-diene

7-9

(a)

NOT stereogenic: two H on end C preclude geometric isomers.

Stereogenic: unequal groups on each side.

C=C−C=C−C=C−C−C

*Each of the four stereocenters is marked with *.

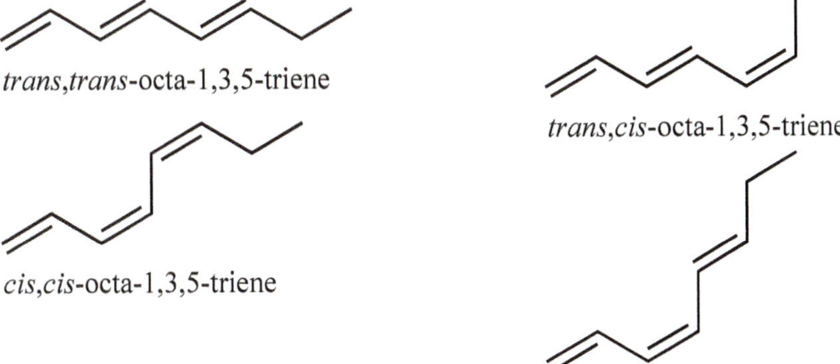

trans,trans-octa-1,3,5-triene

trans,cis-octa-1,3,5-triene

cis,cis-octa-1,3,5-triene

cis,trans-octa-1,3,5-triene

7-9 continued

(b) C—C=C—C=C—C=C—C Whereas octa-2,4,6-triene has three stereogenic double bonds, because of symmetry, the middle one has two equal and indistinguishable ends, reducing the total number of stereoisomers very much like a meso form with chiral stereocenters.

isomers with trans in the middle

trans,trans,trans-octa-2,4,6-triene

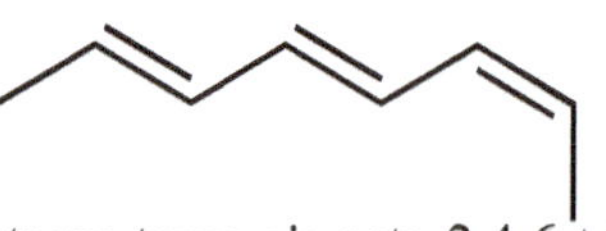

trans,trans,cis-octa-2,4,6-triene (same as *cis,trans,trans*- if numbered in the reverse direction)

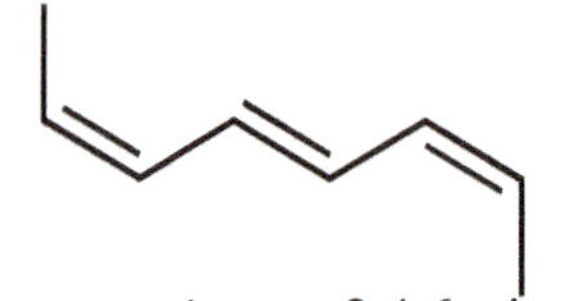

cis,trans,cis-octa-2,4,6-triene

isomers with cis in the middle

cis,cis,cis-octa-2,4,6-triene

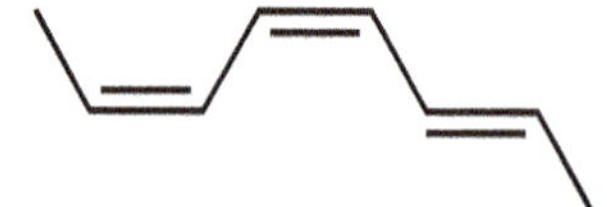

cis,cis,trans-octa-2,4,6-triene (same as *trans,cis,cis*- if numbered in the reverse direction)

trans,cis,trans-octa-2,4,6-triene

7-10

The rate-limiting step in oxidation is abstraction of an atom A by atmospheric O_2 to form A–OO• and a carbon free radical. The energies of bond breaking and bond forming in this step determine whether it will be successful, that is, whether the oxidation will occur.

It is important to note that atmospheric oxygen exists in a diradical form (•O–O•), so there is no energy required to disrupt any bond in O_2.

polyethylene

~C(H)(H)~ (400 kJ/mole) + O_2 ⟶ ~Ċ(H)~ + H—OO• (192 kJ/mole)

$\Delta H = -(192) + (400) =$ +208 kJ/mole—endothermic—requires high temperature

Teflon

~C(F)(F)~ (460 kJ/mole) + O_2 ⟶ ~Ċ(F)~ + F—OO• (63 kJ/mole)

$\Delta H = -(63) + (460) =$ +397 kJ/mole—almost twice as endothermic—requires very high temperature

Because the C-F bond is particularly strong, and the O-F bond is particularly weak, the energy required for O_2 to abstract an F atom is huge, almost twice as large as the energy required for O_2 to abstract an H from polyethylene. Both require high temperature to get over the energy barrier, but Teflon will decompose at a much higher temperature because the rate-limiting step has a higher energy barrier.

7-11

(a) The dibromo compound should boil at a higher temperature because of its much larger molecular weight.
(b) The *cis* should boil at a higher temperature than the *trans* as the *trans* has a zero dipole moment and therefore no dipole-dipole interactions.
(c) 1,2-Dichlorocyclohexene should boil at a higher temperature because of its much larger molecular weight and larger dipole moment than cyclohexene.

7-12 From Table 7-2, approximate heats of hydrogenation can be determined for similarly substituted alkenes. The energy difference is approximately 6 kJ/mole (1.4 kcal/mole), the more substituted alkene being more stable.

gem-disubstituted 116.3 kJ/mole (27.8 kcal/mole)

tetrasubstituted 110.4 kJ/mole (26.4 kcal/mole)

7-13 Use the values in Table 7-2. Alternatively, the relative values in Figure 7-8 could be used to reach the same conclusions. Minor discrepancies between kcal and kJ are due to rounding.

(a) 2 x (*cis*-disubstituted – *trans*-disubstituted) =
2 x (118.5 – 114.6) = 7.8 kJ/mole more stable for *trans, trans*
(2 x (28.3 – 27.4) = 1.8 kcal/mole)

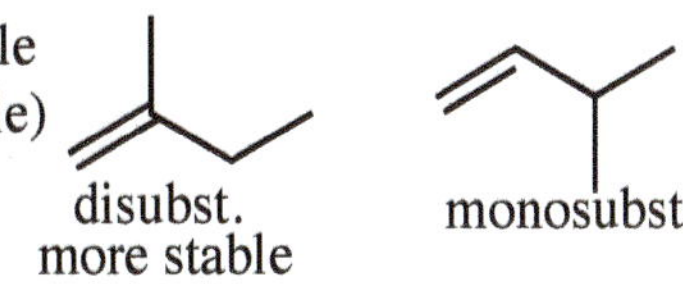

(b) monosubstituted – *gem*-disubstituted = 126.3 – 118.2 = 8.1 kJ/mole
(30.2 – 28.3 = 1.9 kcal/mole)

2-methylbut-1-ene is more stable

disubst. more stable   monosubst.

(c) *gem*-disubstituted – trisubstituted = 118.2 – 111.6 = 6.6 kJ/mole
(28.3 – 26.7 = 1.6 kcal/mole)

2-methylbut-2-ene is more stable

trisubst. more stable   disubst.

(d) *cis*-disubstituted – trisubstituted = 117.7 – 111.6 = 6.1 kJ/mole
(28.1 – 26.7 = 1.4 kcal/mole)

2-methylpent-2-ene is more stable

trisubst. more stable   *cis*-disubst.

7-14

(a) CH_3 CH_3 Slightly strained but stable with C=C in a 5-membered ring.

(b) Could not exist as the ring size must be 8 atoms or greater to include *trans* double bond—see answer to part (e).

(c) CH_3 CH_3

The *trans* refers to the two methyls since this compound can exist, rather than the *trans* referring to the alkene, a molecule that could not exist because the ring would have too much strain. A ring must have 8 or more atoms before a *trans* double bond can exist in the ring.

7-14 continued

(d)

CH3 / CH3 structure

stable—*trans* in 10-membered ring

(e) Unstable at room temperature—it cannot have *trans* alkene in 7-membered ring (possibly isolable at very low temperature—this type of experiment is one of the challenges chemists attack with gusto).
(f) Stable—the alkene is not at a bridgehead.
(g) Unstable—violates Bredt's Rule (alkene at bridgehead in 6-membered ring).
(h) Stable—alkene is at bridgehead in 8-membered ring.
(i) Unstable—violates Bredt's Rule (alkene at bridgehead in 7-membered ring).

7-15

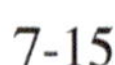

(a)

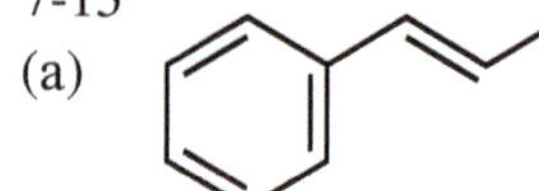

aromatic and conjugated; MOST STABLE

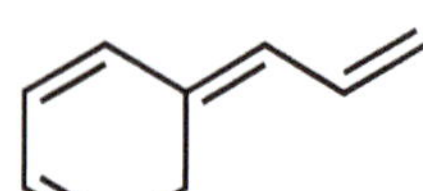

not aromatic, fully conjugated

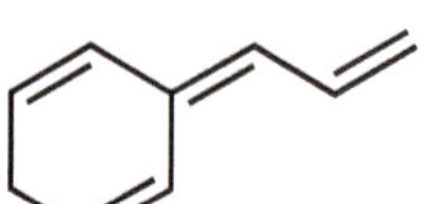

not aromatic, partially conjugated; LEAST STABLE

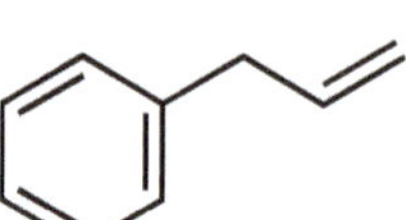

aromatic but not conjugated

(b)

only 2 pi bonds conjugated

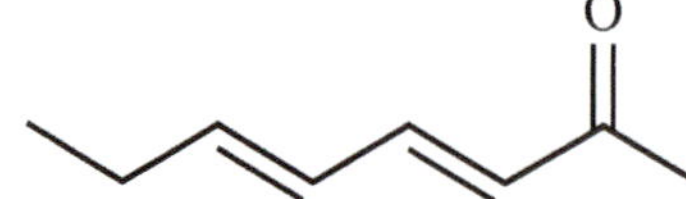

all 3 pi bonds conjugated; MOST STABLE

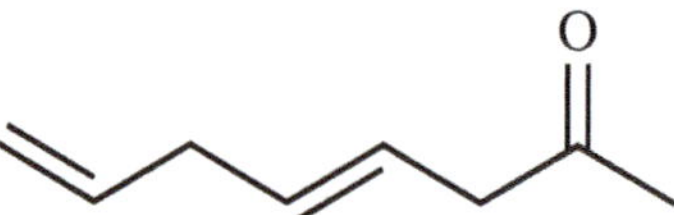

all 3 pi bonds isolated; LEAST STABLE

(c)

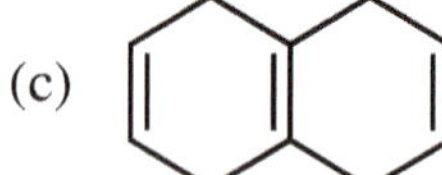

all 3 pi bonds isolated; LEAST STABLE

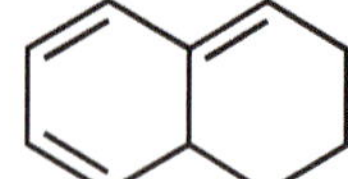

all 3 pi bonds conjugated

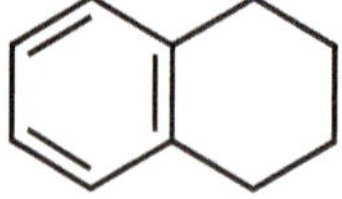

aromatic; MOST STABLE

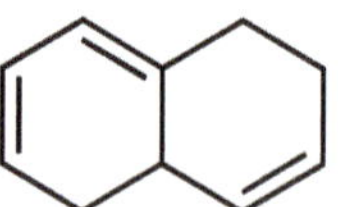

only 2 pi bonds conjugated

7-16

If the CH_3OH acts as a nucleophile, the product will be a substitution; the overall mechanism will be S_N1.

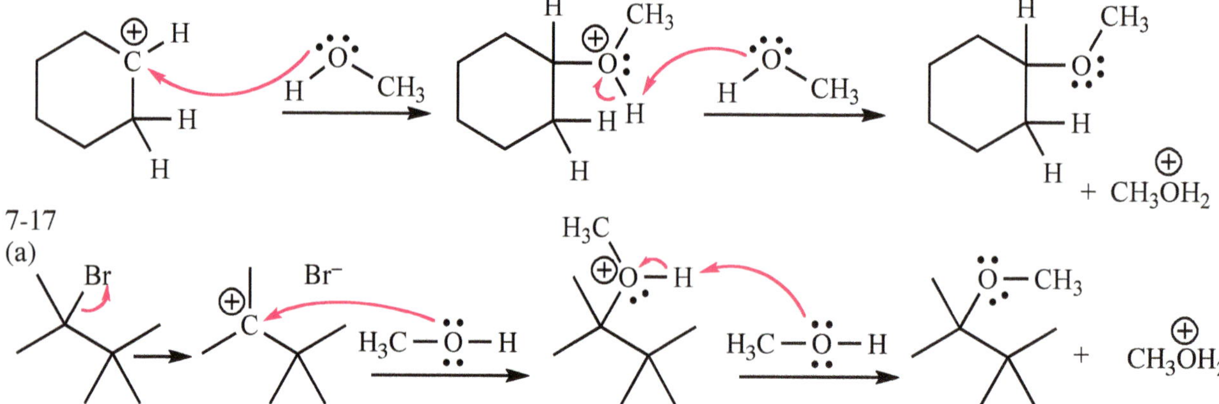

The S_N1 mechanism begins with ionization to form a carbocation, attack of a nucleophile, and in the case of ROH nucleophiles, removal of a proton by a base to form a neutral product.

7-17 continued

(b) In the E1 reaction, the solvent (methanol, in this case) serves two functions: it aids the ionization process by solvating both the leaving group (bromide) and the carbocation; and second, it serves as a base to remove the proton from a carbon adjacent to the carbocation in order to form the carbon–carbon double bond. The S_N1 mechanism adds a third function to the solvent: the first step is the same as in E1, ionization to form the carbocation; the second step has the solvent acting as a nucleophile—this step is different from E1; third, the solvent acts like a base and removes a proton, although from an oxygen (S_N1) and not a carbon (E1). Solvents are versatile!

7-18

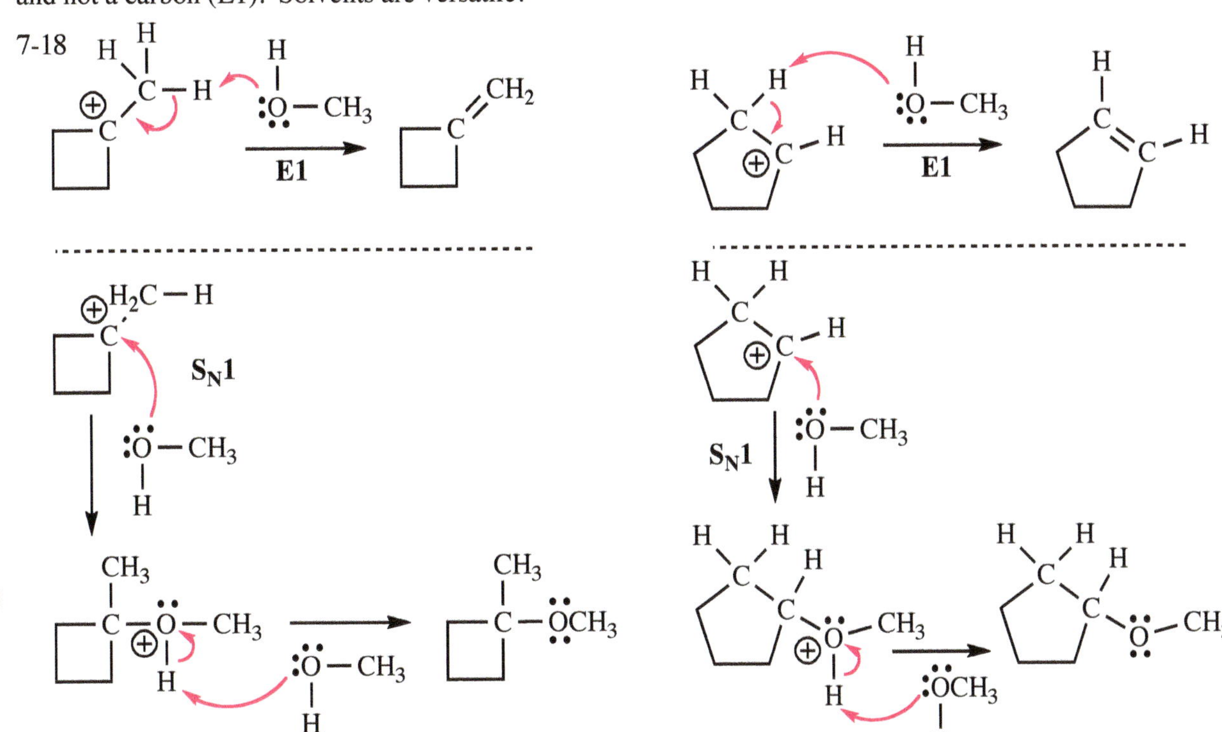

7-19 This solution does not show complete mechanisms. Rather, it shows the general sequence of all four pathways leading to five different products.

Ethanol is the solvent shown here in the substitution products; it is often used in solvolysis reactions.

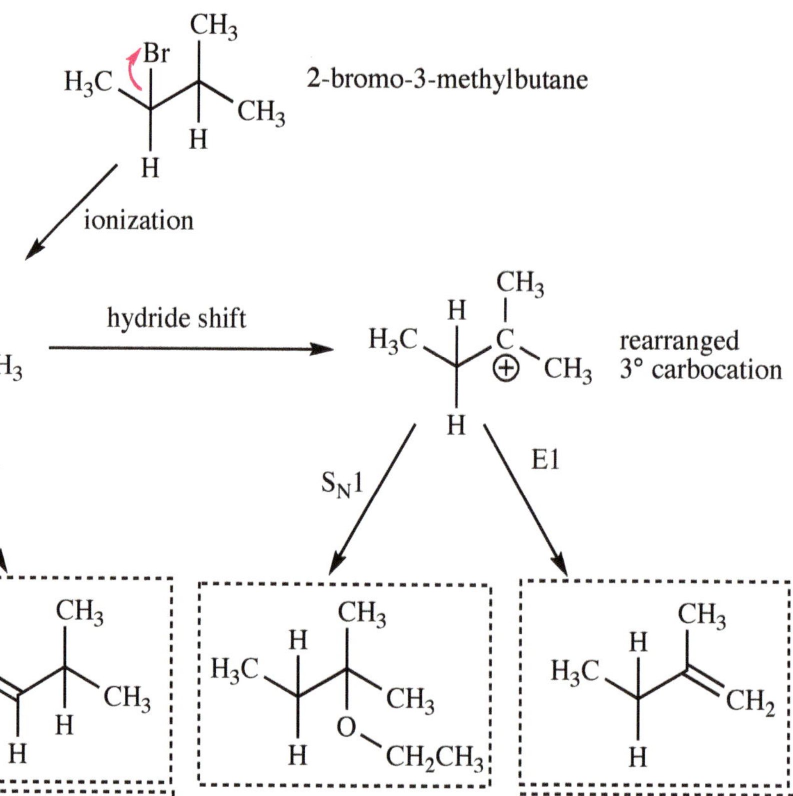

Five unique products shown in the boxes. One alkene can be produced from either the 2° or the 3° carbocation.

same compound

two possible E1 products

two possible E1 products

7-20 Substitution products are shown first, then elimination products.

(a) OCH_3 +

(b) Ph OCH_2CH_3 + Ph

Ph = phenyl

(c) OH CH_3 + OH CH_3 + CH_3 CH_3 CH_2

S_N1 product without rearrangement

S_N1 product with rearrangement

major E1 product with or without rearrangement

minor

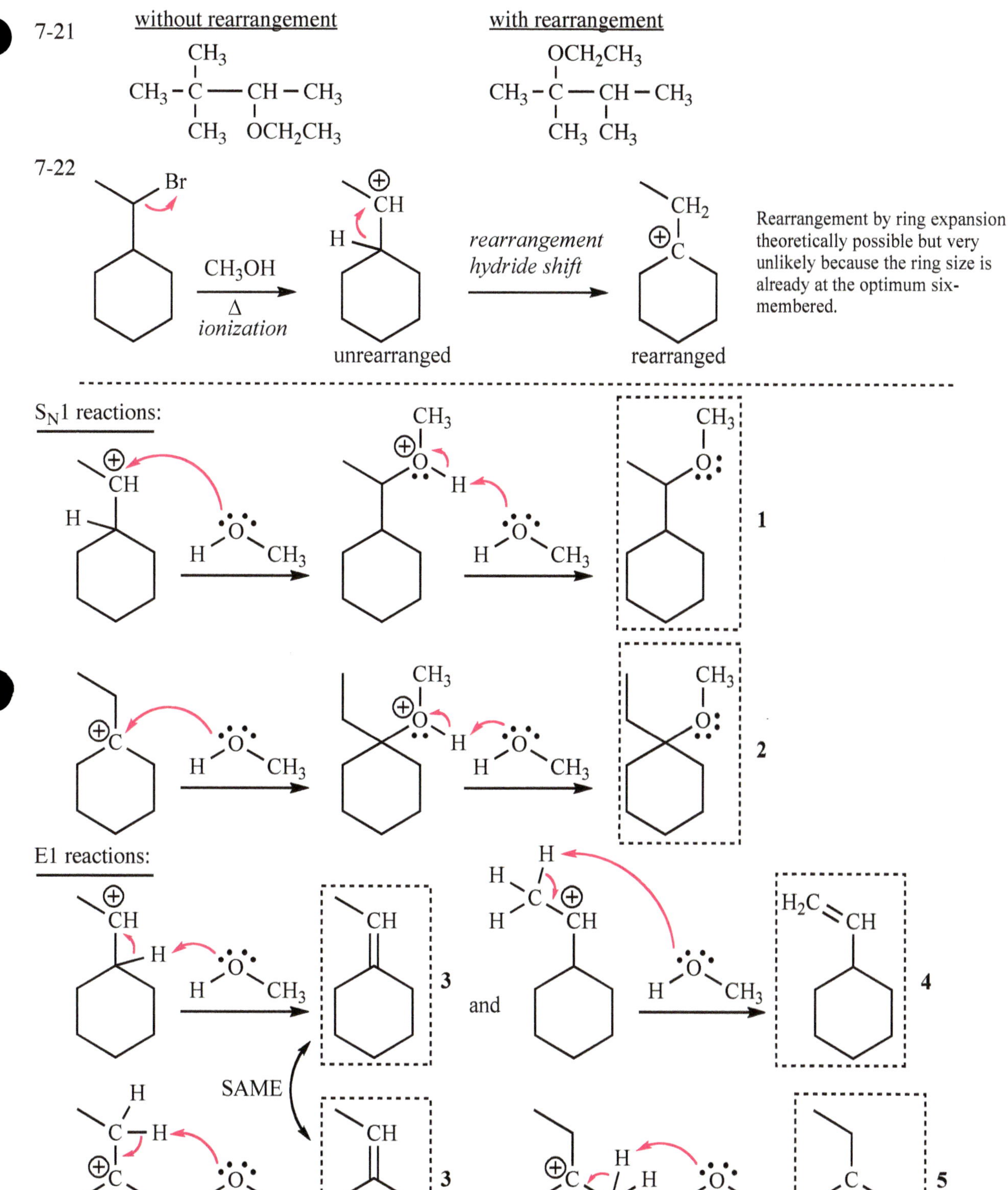

5 is the major elimination product: trisubstituted and endocyclic (in the ring), the more stable position.
3 is trisubstituted but is exocyclic, a less stable position.

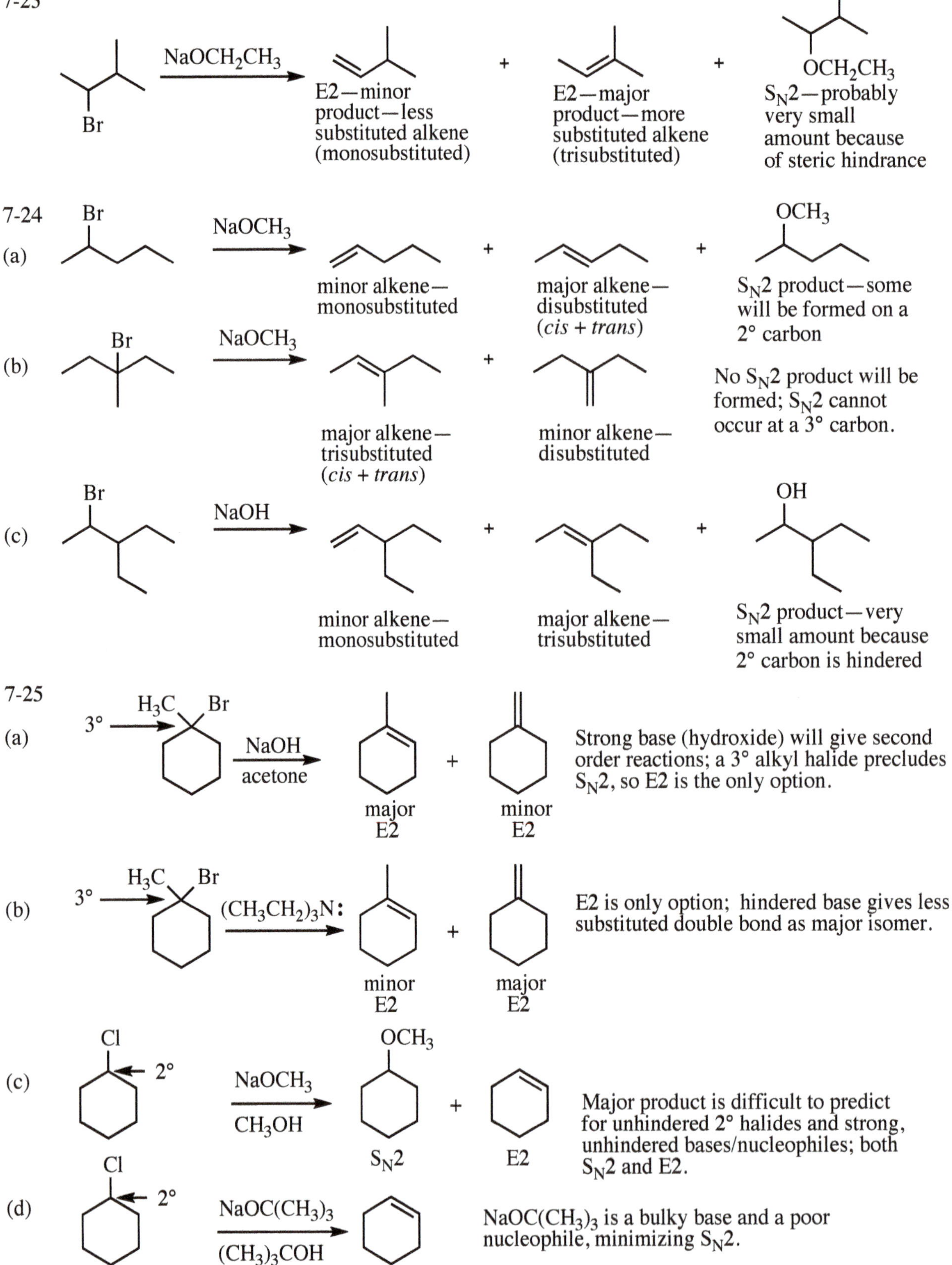
7-23
Br
NaOCH2CH3
E2—minor product—less substituted alkene (monosubstituted)
E2—major product—more substituted alkene (trisubstituted)
OCH2CH3
SN2—probably very small amount because of steric hindrance
7-24
(a)
Br
NaOCH3
minor alkene—monosubstituted
major alkene—disubstituted (cis + trans)
OCH3
SN2 product—some will be formed on a 2° carbon
(b)
Br
NaOCH3
major alkene—trisubstituted (cis + trans)
minor alkene—disubstituted
No SN2 product will be formed; SN2 cannot occur at a 3° carbon.
(c)
Br
NaOH
minor alkene—monosubstituted
major alkene—trisubstituted
OH
SN2 product—very small amount because 2° carbon is hindered
7-25
(a)
3°
H3C
Br
NaOH
acetone
major E2
minor E2
Strong base (hydroxide) will give second order reactions; a 3° alkyl halide precludes SN2, so E2 is the only option.
(b)
3°
H3C
Br
(CH3CH2)3N:
minor E2
major E2
E2 is only option; hindered base gives less substituted double bond as major isomer.
(c)
Cl
2°
NaOCH3
CH3OH
OCH3
SN2
E2
Major product is difficult to predict for unhindered 2° halides and strong, unhindered bases/nucleophiles; both SN2 and E2.
(d)
Cl
2°
NaOC(CH3)3
(CH3)3COH
E2
NaOC(CH3)3 is a bulky base and a poor nucleophile, minimizing SN2.

7-26

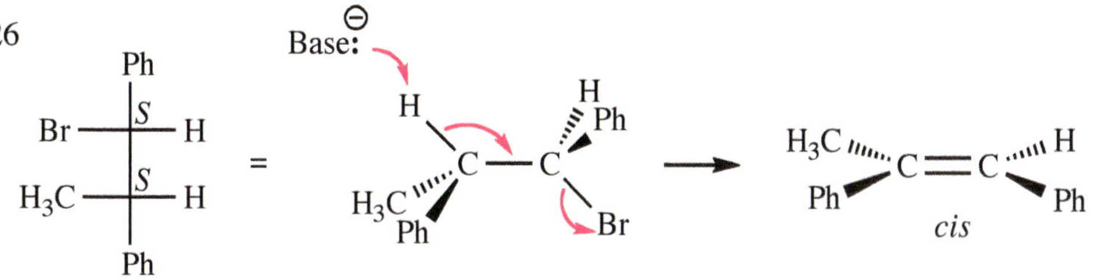

Base: ... → *cis* alkene (H_3C, Ph / H, Ph)

7-27 (a)

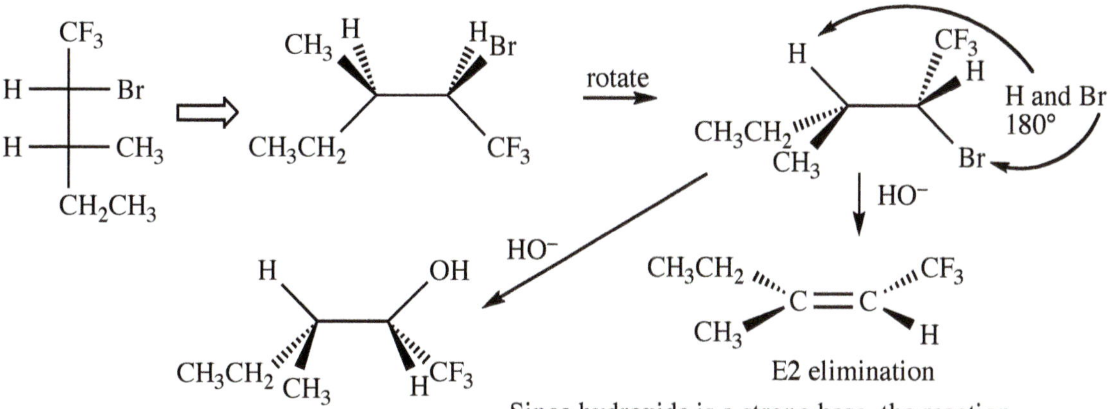

rotate → H and Br 180° → HO^- → E2 elimination; HO^- → S_N2 product

Since hydroxide is a strong base, the reaction will be second order: on a 2° alkyl bromide, both substitution and elimination will occur.

(b) meso —*hindered base* $(CH_3CH_2)_3N$: → only alkene isomer E2

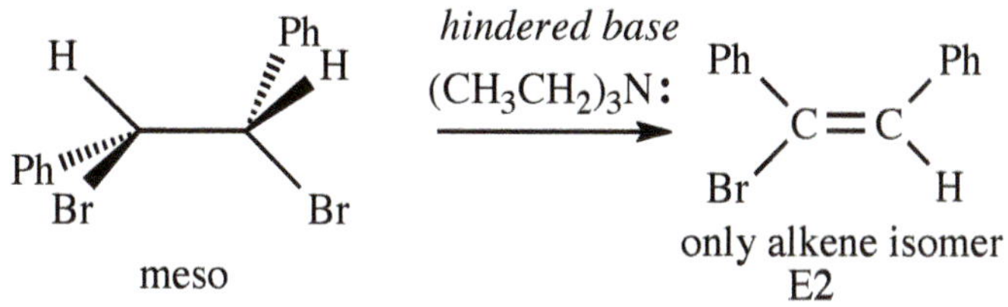

(c) one enantiomer of the *d,l* pair —*hindered base* $(CH_3CH_2)_3N$: → only alkene isomer E2

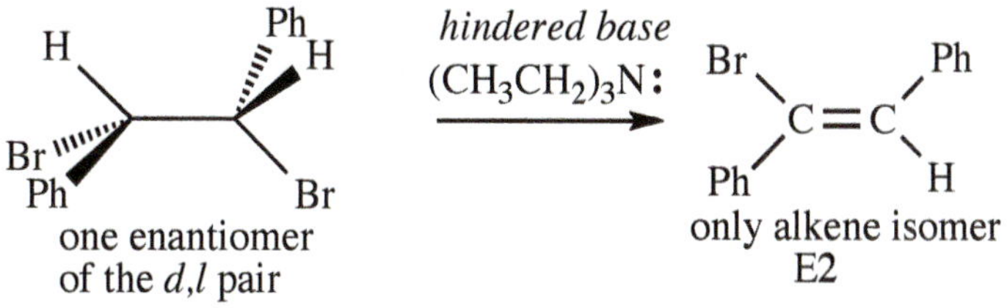

Note that these products are geometric isomers. If a reaction starts with stereoisomers and produces stereoisomers, that is the definition of a *stereospecific* reaction.

(d) NaOH, acetone →

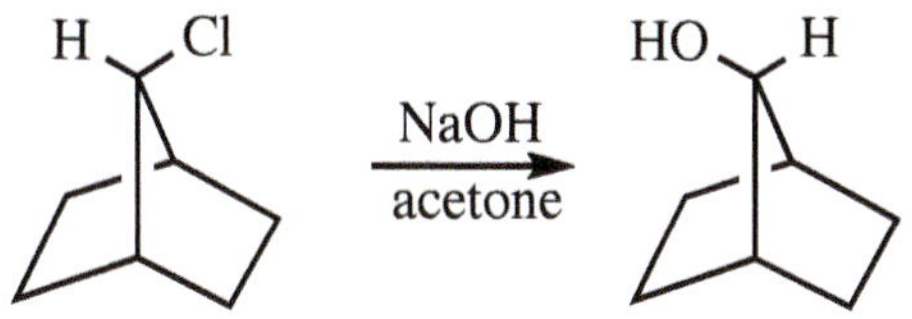

S_N2 is the only mechanism possible because no H can get coplanar with Cl; an elimination product would violate Bredt's Rule.

(e) $(CH_3)_3CO^-$ →

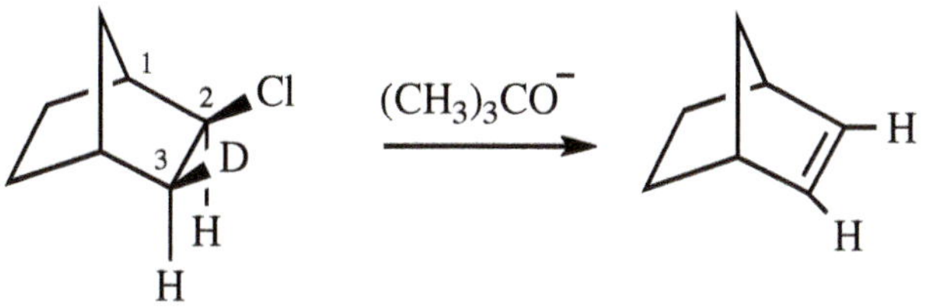

Models show that the H on C-3 cannot be anti-coplanar with the Cl on C-2. Thus, this E2 elimination must occur with a *syn*-coplanar orientation: the D must be removed as the Cl leaves.

7-27 continued

(f) [Newman projections: front carbon CH3 (top), H, H; back carbon CF3, CH2CH3, Br → rotate back carbon → H (top), H, CH3CH2; CF3, CH3, Br]

rotate back carbon

H and Br must be 180°.

hindered base: E2 only

π bond

CH_3CH_2, H, CH_3, CF_3

H, CH_2CH_3, F_3C, CH_3 only E2 product

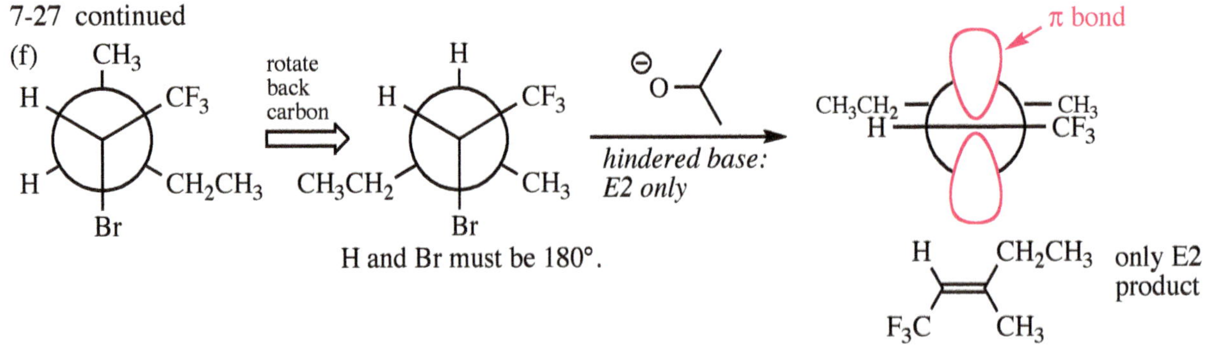

7-28 As shown in Solved Problem 7-5, the H and the Br must have *trans*-diaxial orientation for the E2 reaction to occur. In part (a), the *cis* isomer has the methyl in the equatorial position, the $NaOCH_3$ can remove a hydrogen from either C-2 or C-6, giving a mixture of alkenes where the most highly substituted isomer is the major product (Zaitsev). In part (b), the *trans* isomer has the methyl in the axial position at C-2, so no elimination can occur at C-2. The only possible elimination orientation is toward C-6.

(a) Br, H_3C, 1, 6, H, H — $NaOCH_3$ / CH_3OH → CH_3 (major) + CH_3 (minor) Zaitsev orientation

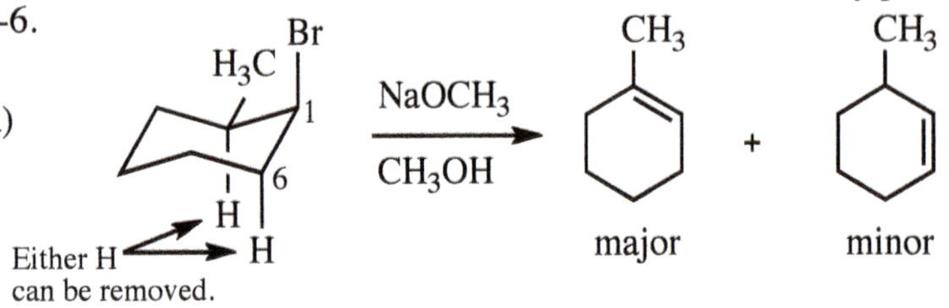

Either H can be removed.

(b) Br, H, 1, 6, H_3C, H — $NaOCH_3$ / CH_3OH → CH_3 Only alkene formed; stereochemistry of E2 precludes other isomer from forming.

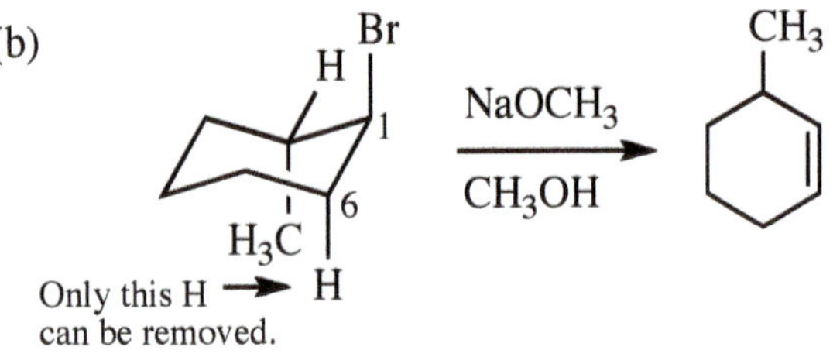

Only this H can be removed.

7-29 E2 elimination requires that the H and the leaving group be anti-coplanar; in a chair cyclohexane, this requires that the two groups be *trans* diaxial. However, when the bromine atom is in an axial position, there are no hydrogens in axial positions on adjacent carbons, so no elimination can occur.

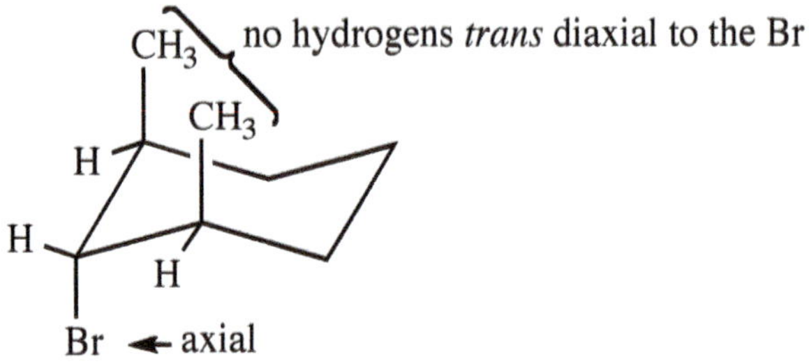

CH_3, CH_3 no hydrogens *trans* diaxial to the Br

H, H, H, Br ← axial

7-30
(a)

KOH

Showing the chair form of the decalins makes the answer clear. The top isomer locks the H and the Br that eliminate into a *trans*-diaxial conformation—optimum for E2 elimination. The bottom isomer has Br equatorial where it is exceedingly slow to eliminate.

(b)

Br must be axial to eliminate by E2, forcing the large isopropyl group axial, a very unstable conformer; this *trans* isomer will be exceedingly slow to react as the preferred conformer will be diequatorial.

KOH

In this *cis* isomer, the Br is in the optimum axial position for elimination while the large isopropyl group is in the energetically preferred equatorial position; this will be the reactive conformer.

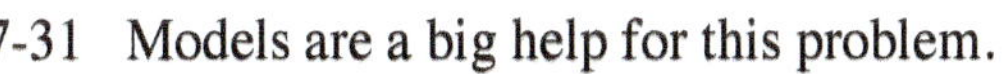

7-31 Models are a big help for this problem.

(a)

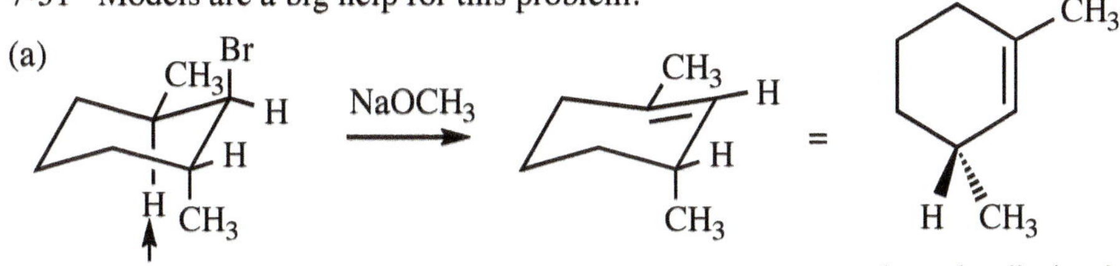

the only H *trans* diaxial

the only elimination product

(b)

NaOCH3

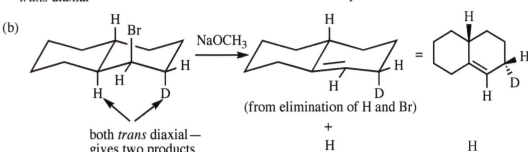

both *trans* diaxial—gives two products

(from elimination of H and Br)

+

(from elimination of D and Br)

7-32

(a) MAJOR + MINOR

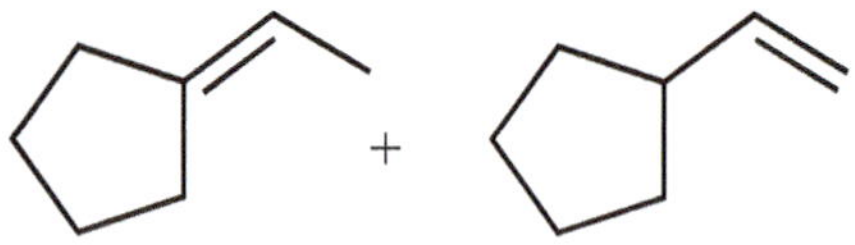

The base NaOEt is strong and not hindered, so this reaction will follow the E2 mechanism. The product ratio will follow Zaitsev's Rule: trisubstituted double bond is major, and monosubstituted double bond is minor.

(b) This E1 reaction is complicated because the carbocation intermediate in two possible ways, and each carbocation can give two possible alkene products.

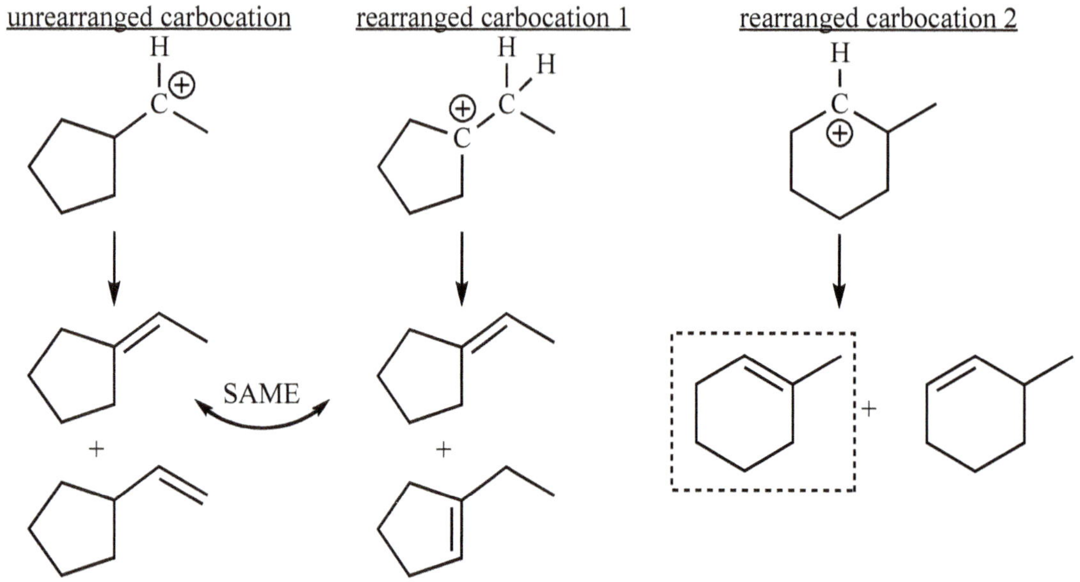

unrearranged carbocation | rearranged carbocation 1 | rearranged carbocation 2

SAME

Five different alkene products are possible in this E1 reaction. The ratio of products is difficult to predict because it depends on the rate of rearrangement and the relative amounts of the two different rearrangements. The safest prediction is that the most stable alkene is most likely the major product: of these five alkenes, the trisubstituted double bond in the six-membered ring is the most stable (the one in the box), so it is likely to be the major product.

Follow-up question to Solved Problem 7-6(a):

In the E2 transition state, the C—H bond and the C—Br bond are stretched as the O—H bond is being formed.

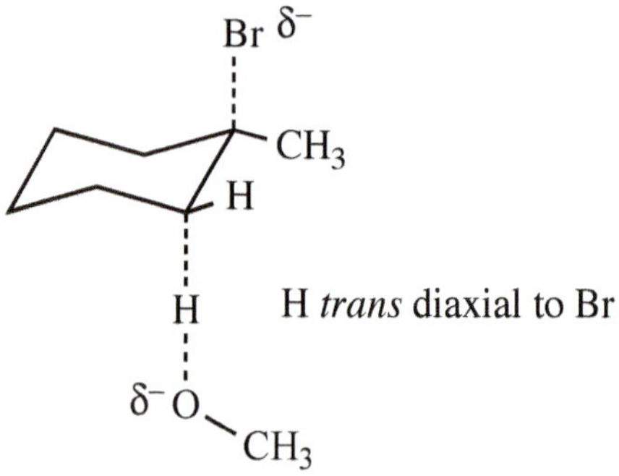

Br δ^- CH_3 H H *trans* diaxial to Br δ^-O CH_3

Follow-up question to Solved Problem 7-6(b):

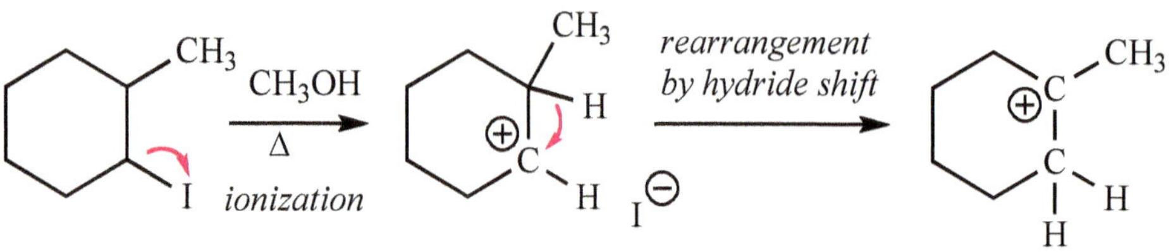

CH_3 I CH_3OH Δ *ionization* CH_3 H H $I^\ominus$ *rearrangement by hydride shift* CH_3 H H

All of the products shown below are derived from one of these two carbocations.

Follow-up question to Solved Problem 7-6(b), continued

<u>S_N1 on unrearranged carbocation</u>

<u>E1 on unrearranged carbocation</u>

major | minor

<u>S_N1 on rearranged carbocation</u>

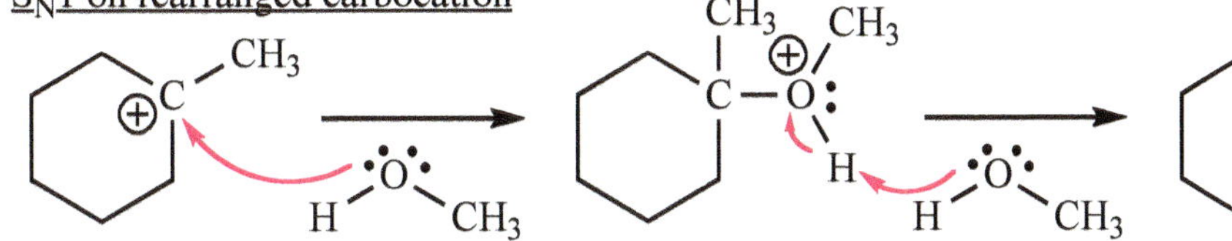

<u>E1 on rearranged carbocation</u>

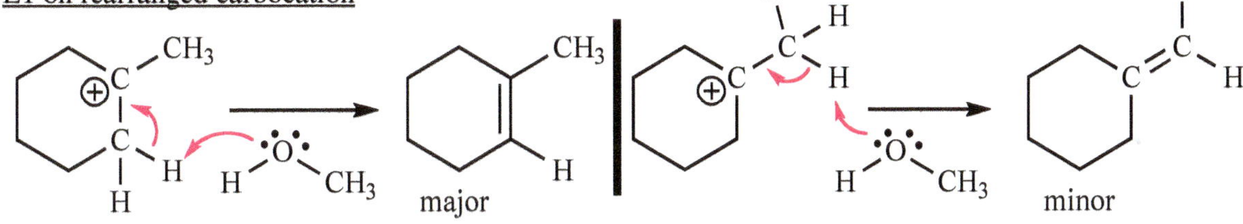

7-33

(a) Ethoxide is a strong base/nucleophile—second-order conditions. The 1° bromide favors substitution over elimination, so S_N2 will predominate over E2.

substitution—major   elimination—minor

(b) Methoxide is a strong base/nucleophile—second-order conditions. The 2° chloride will undergo S_N2 by backside attack as well as E2 to make a mixture of alkenes.

OCH_3

substitution   elimination—minor alkene   elimination—major alkene (*cis* + *trans*)

(c) Ethoxide is a strong base/nucleophile—second-order conditions. The 3° chloride is hindered and cannot undergo S_N2 by backside attack. E2 is the only route possible.

$CH_3-C(CH_3)=CHCH_3$ major $CH_2=C(CH_3)-CH_2CH_3$ minor

7-33 continued

(d) Heating in ethanol are conditions for solvolysis, an S_N1 reaction. Heat also promotes elimination.

$$CH_3-\underset{CH_3}{\overset{OCH_2CH_3}{C}}-CH_2CH_3 \quad S_N1 \qquad CH_3-\underset{CH_3}{C}=CHCH_3 \quad E1$$

major elimination product

Substitution and elimination are possible.

(e) Hydroxide is a strong base/nucleophile—second-order conditions. The 1° iodide is more likely to undergo S_N2 than E2, but both products will be observed.

$$CH_3-\underset{CH_3}{C}HCH_2OH \qquad CH_3-\underset{CH_3}{C}=CH_2$$

substitution—major elimination—minor

(f) Silver nitrate in ethanol/water are ionizing conditions for 1° alkyl halides that will lead to rearrangement followed by substitution on the 3° carbocation. (No heat, so E1 is unlikely.)

$$CH_3-\underset{CH_3}{\overset{OCH_2CH_3}{C}}-CH_3$$

from ethanol as nucleophile

$$CH_3-\underset{CH_3}{\overset{OH}{C}}-CH_3$$

from water as nucleophile

(g) Ethoxide in ethanol on a 3° halide will lead to E2 elimination; there will be no substitution.

CH_3

major alkene trisubstituted

minor alkene disubstituted

CH_2

(h) Heating a 3° halide in methanol is quintessential first-order conditions, either E1 or S_N1 (solvolysis).

CH_3 OCH_3

substitution (S_N1)

elimination (E1): CH_3 CH_2 (trace)

Solved Problem 7-7(b): removal of protons from carbocation intermediates:

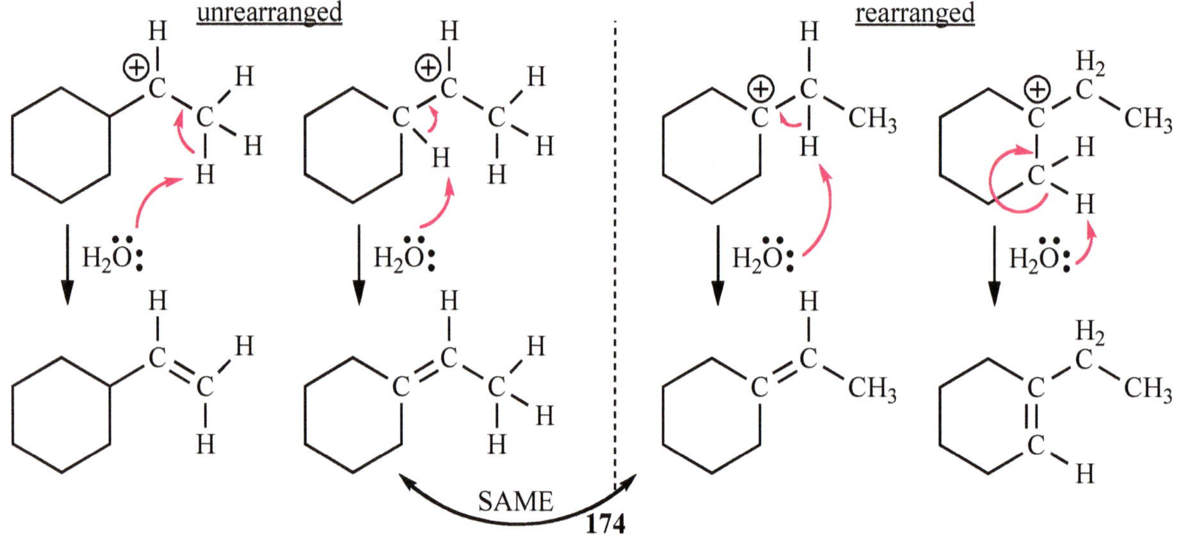

7-34 In these mechanisms, the base removing the final proton is shown as water. Other alcohols are just as likely to serve as the base.

(a)

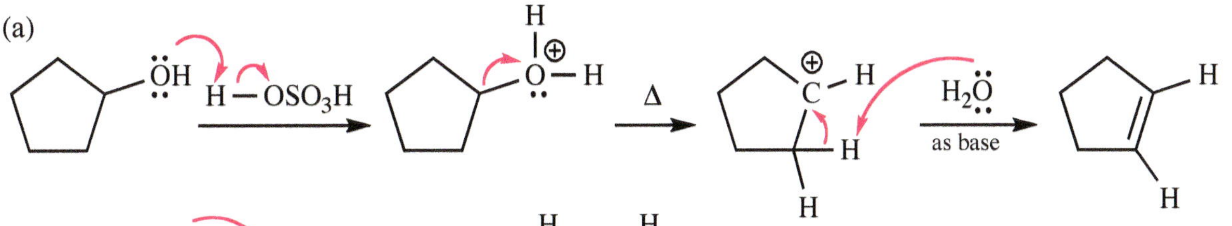

$H-OSO_3H$ Δ $H_2\ddot{O}$ as base

(b)

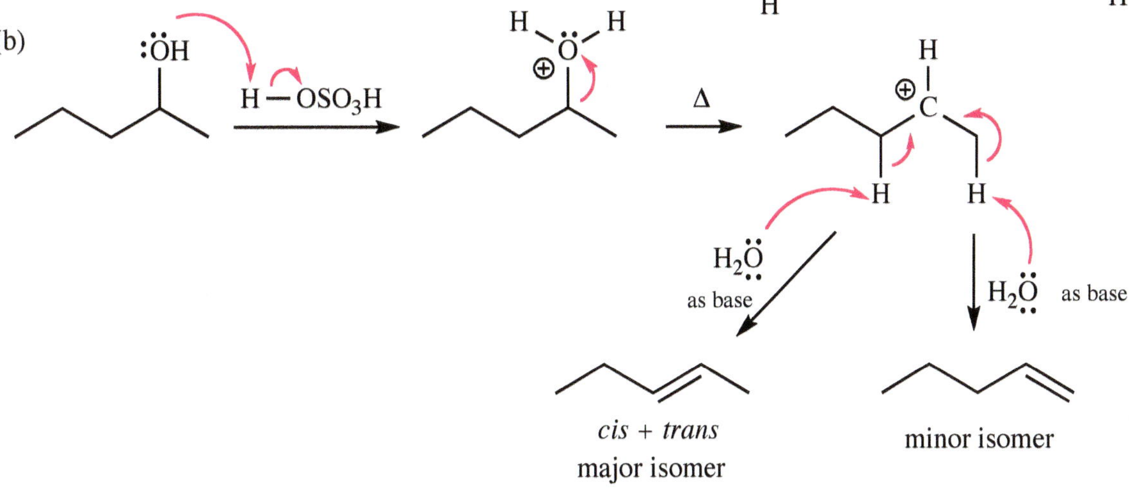

$H-OSO_3H$ Δ

$H_2\ddot{O}$ as base

$H_2\ddot{O}$ as base

cis + *trans*
major isomer

minor isomer

(c)

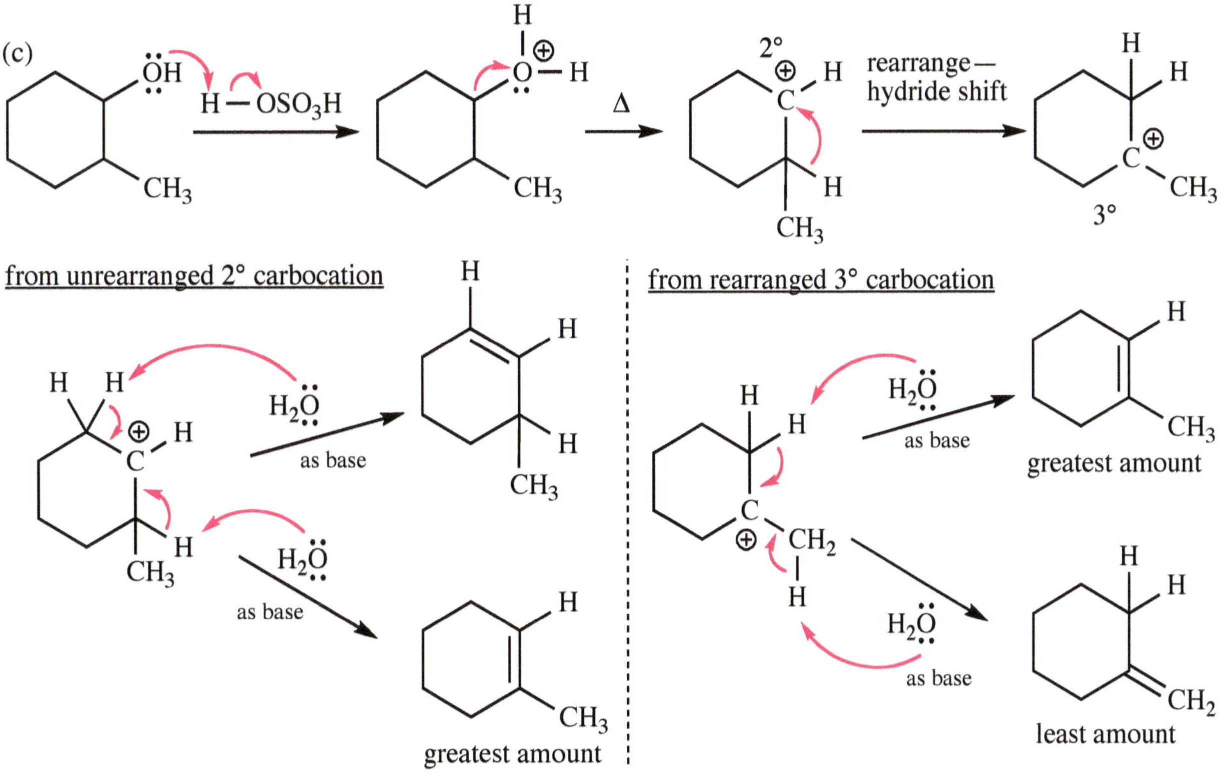

$H-OSO_3H$ Δ

2° rearrange— hydride shift 3°

from unrearranged 2° carbocation

$H_2\ddot{O}$ as base

$H_2\ddot{O}$ as base

greatest amount

from rearranged 3° carbocation

$H_2\ddot{O}$ as base

greatest amount

$H_2\ddot{O}$ as base

least amount

7-34 continued

(d)

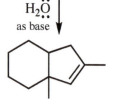

When proposing mechanisms, always look for what bonds are broken and what new bonds must be formed. In this problem, the new ring must come from the pi electrons forming a new sigma bond at a 3° carbon, which must have been a carbocation.

$H_2\ddot{O}$ as base

7-35 H—A is the generic symbol for an acid.

(a) Only one elimination product is possible.

benzylic—
resonance-stabilized

(b) Three elimination products are possible: two from an unrearranged carbocation, and two from a rearranged carbocation, one product of which is the same to both pathways.

unrearranged   rearranged

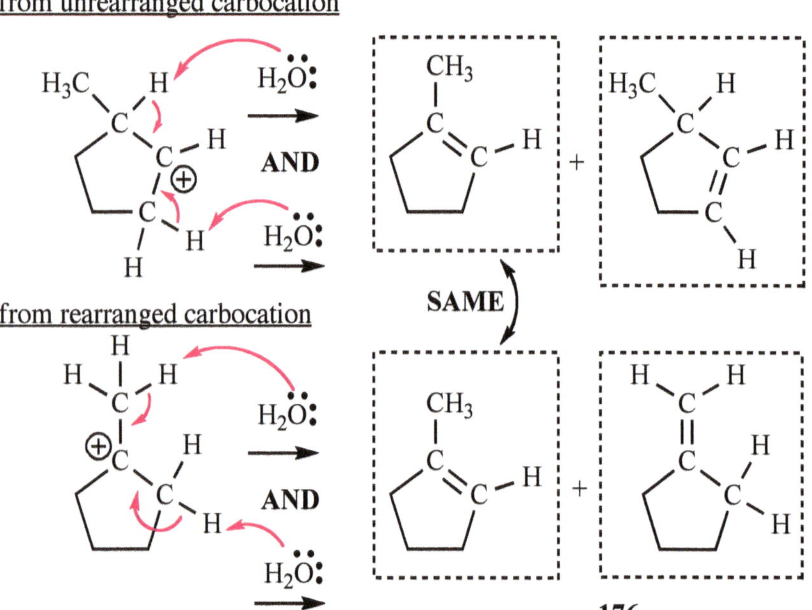

7-35 continued

(c) Three elimination products are possible: the intermediate carbocation is allylic, with positive charge shared by two different carbons. An H can be removed from any of three carbons adjacent to positively-charged carbons.

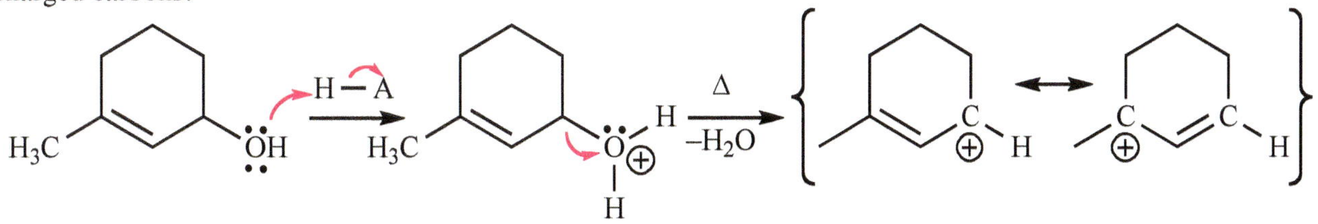

Although the two resonance forms do not represent real structures, each can be treated separately to determine the alkenes that could be produced. Resonance braces omitted for space reasons.

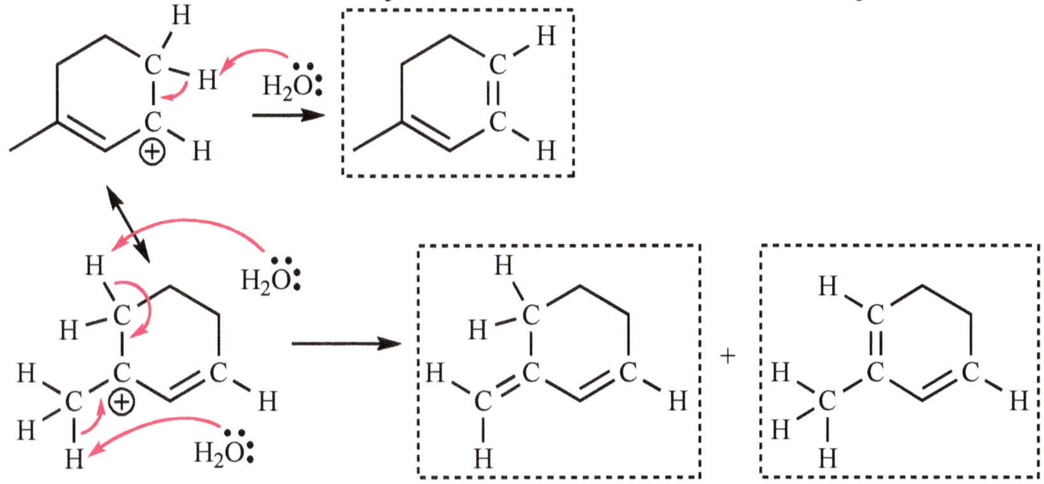

7-36

(a) $\Delta G° = \Delta H° - T\Delta S°$

$= +116{,}000 \text{ J/mol} - 298\ °\text{K}\ (117 \text{ J/°K•mol}) = +116{,}000 \text{ J/mol} - 34{,}866 \text{ J/mol}$

$= +81{,}134 \text{ J/mol} = +81.1 \text{ kJ/mol}\ (+19.3 \text{ kcal/mole})$

$\Delta G°$ is positive, the reaction is **disfavored** at 25°C.

(b) $\Delta G_{1000} = +116{,}000 \text{ J/mol} - 1273\ °\text{K}\ (117 \text{ J/°K•mol}) = +116{,}000 \text{ J/mol} - 148{,}941 \text{ J/mol}$

$= -32{,}941 \text{ J/mol} = -32.9 \text{ kJ/mol}\ (-8.0 \text{ kcal/mole})$

ΔG is negative, the reaction is **favored** at 1000°C.

7-37

(a) Basic and nucleophilic mechanism: $Ba(OH)_2$ is a strong base.
(b) Acidic and electrophilic mechanism: the catalyst is H^+.
(c) Free radical chain reaction: the catalyst is a peroxide that initiates free radical reactions.
(d) Acidic and electrophilic mechanism: the catalyst BF_3 is a strong Lewis acid.

7-38

(a)

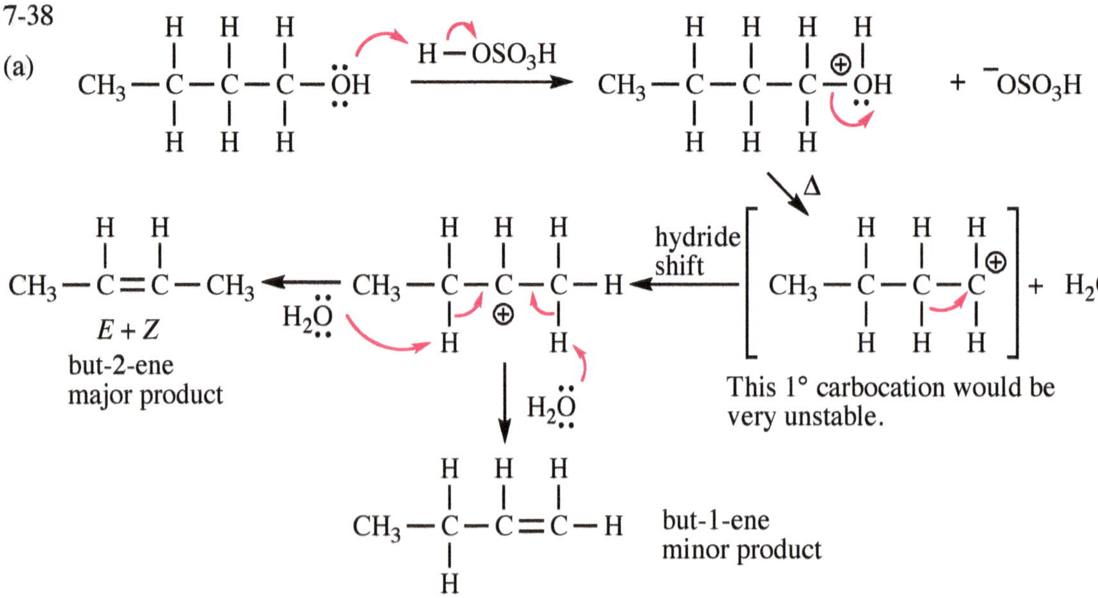

$CH_3CH_2CH_2CH_2\ddot{O}H$ + H–OSO_3H → $CH_3CH_2CH_2CH_2\overset{\oplus}{O}H_2$ + $^{-}OSO_3H$

→ Δ [$CH_3CH_2CH_2\overset{\oplus}{C}H_2$] + H_2O

This 1° carbocation would be very unstable.

hydride shift → $CH_3–CH_2–\overset{\oplus}{C}H–CH_3$

$H_2\ddot{O}$ → $CH_3–CH=CH–CH_3$

E + Z

but-2-ene

major product

$H_2\ddot{O}$ → $CH_3–CH_2–CH=CH_2$

but-1-ene

minor product

(b)

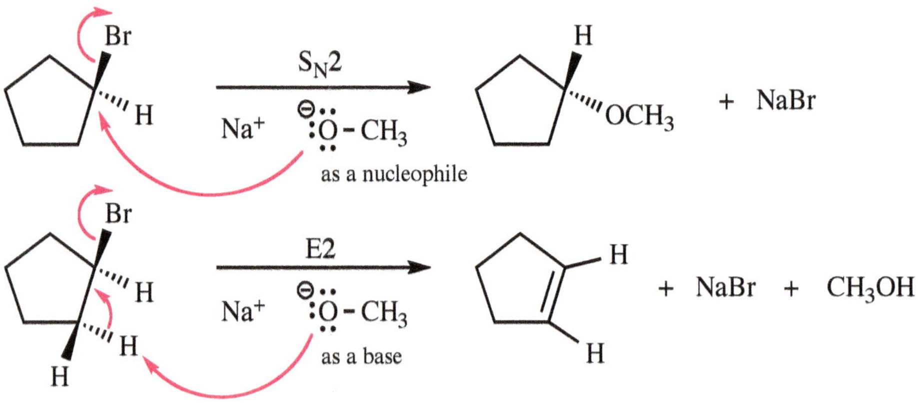

S_N2

Na^+ $^{\ominus}\ddot{O}–CH_3$ as a nucleophile

+ NaBr

E2

Na^+ $^{\ominus}\ddot{O}–CH_3$ as a base

+ NaBr + CH_3OH

(c)

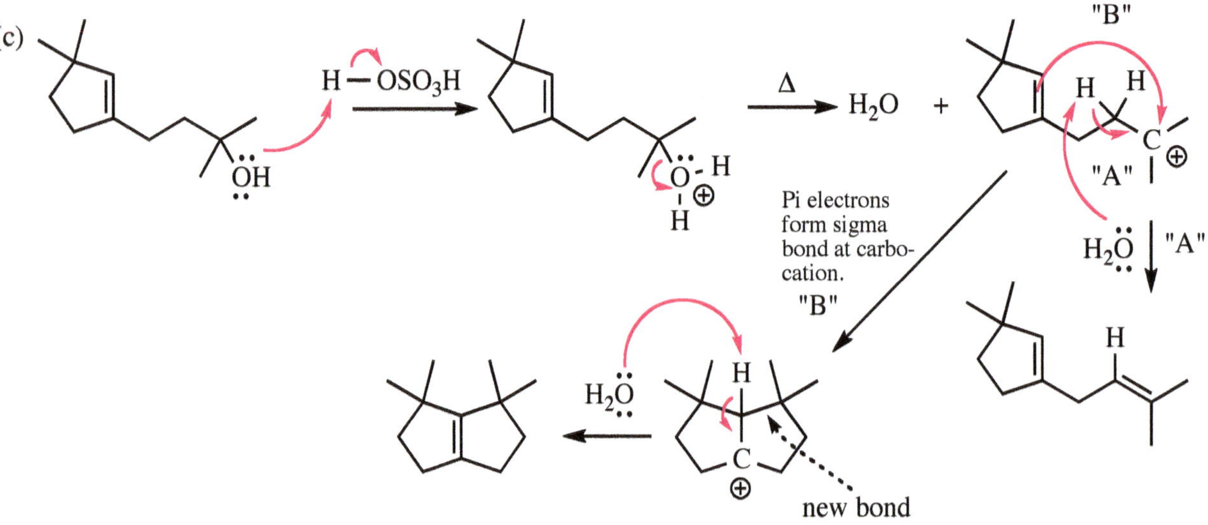

H–OSO_3H

Δ → H_2O +

"B"

"A"

Pi electrons form sigma bond at carbo-cation.

"B"

$H_2\ddot{O}$ "A"

$H_2\ddot{O}$

new bond

7-39

(a)

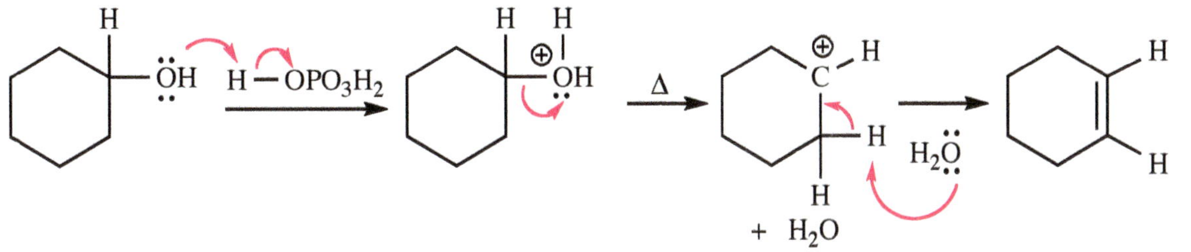

H_2O + H_2O

(b)

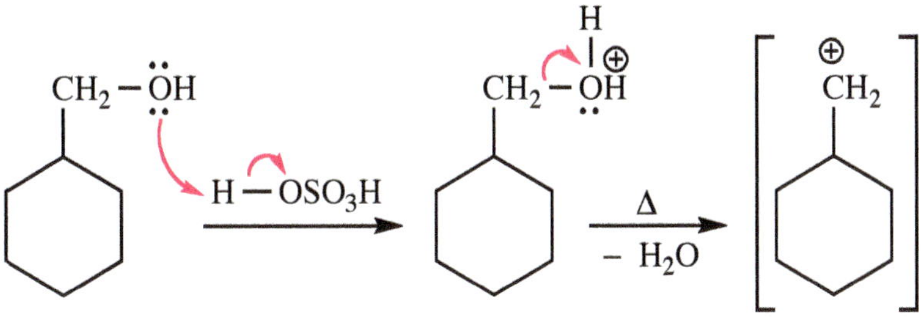

This 1° carbocation would be very unstable.

Two possible rearrangements

1. Hydride shift

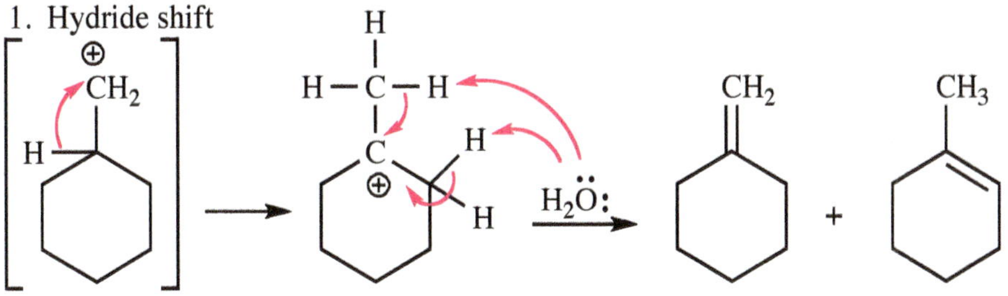

2. Alkyl shift—ring expansion

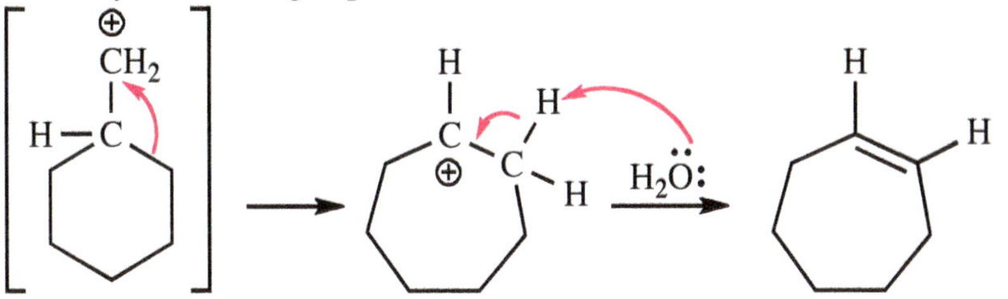

(c) carbocation formation

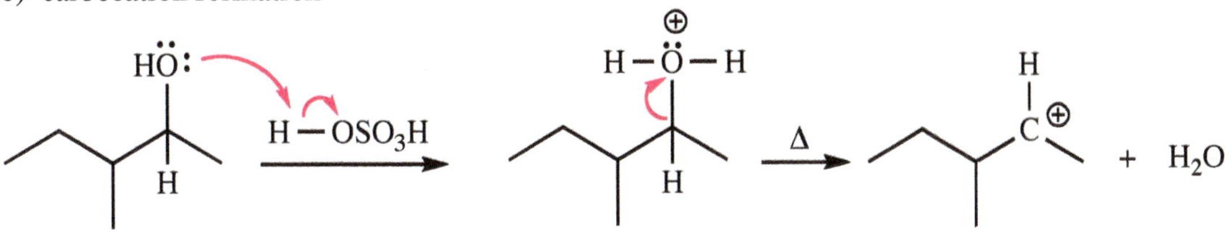

+ H_2O

continued on next page

7-39 (c) continued

without rearrangement

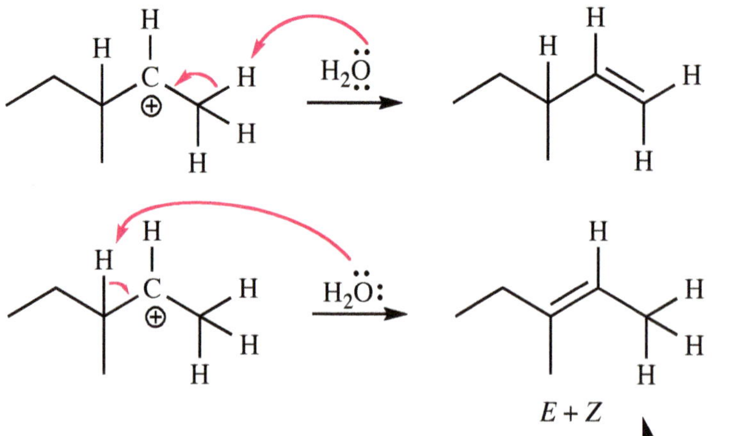

$H_2\ddot{O}$

$H_2\ddot{O}$:

$E + Z$

with hydride shift

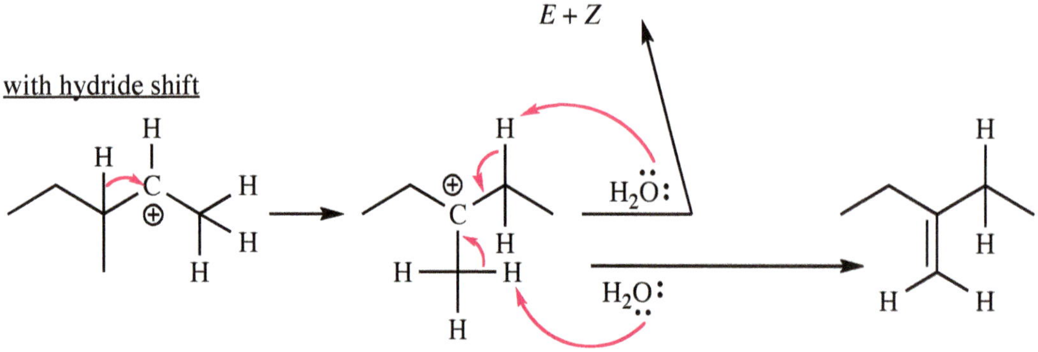

$H_2\ddot{O}$:

$H_2\ddot{O}$:

(d)

$H-OSO_3H$

Δ

$+\ H_2\ddot{O}$:

E1 product from unrearranged carbocation

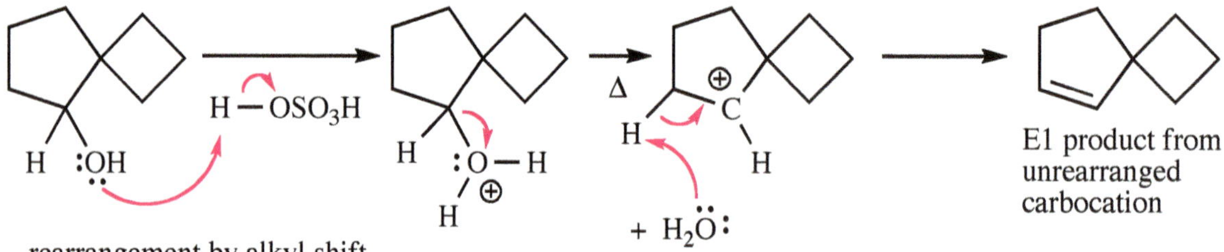

rearrangement by alkyl shift

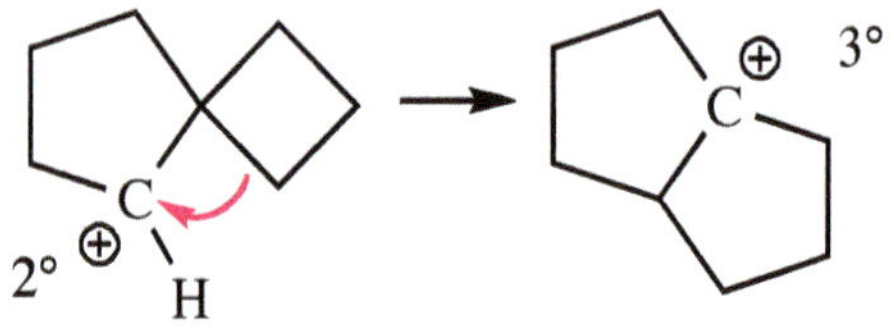

2°

3°

E1 on rearranged carbocation

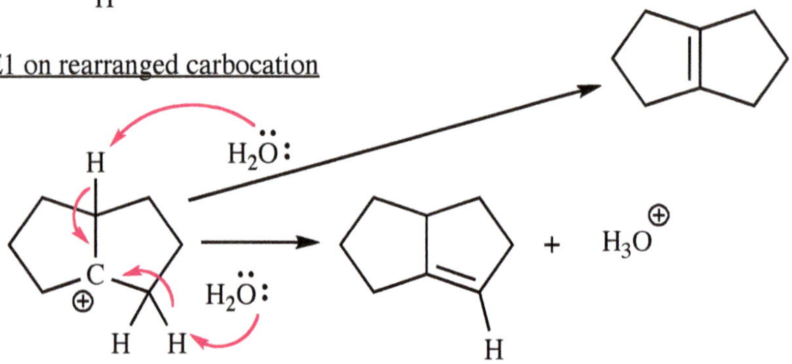

$H_2\ddot{O}$:

$H_2\ddot{O}$:

$+ \quad H_3O^{\oplus}$

7-40

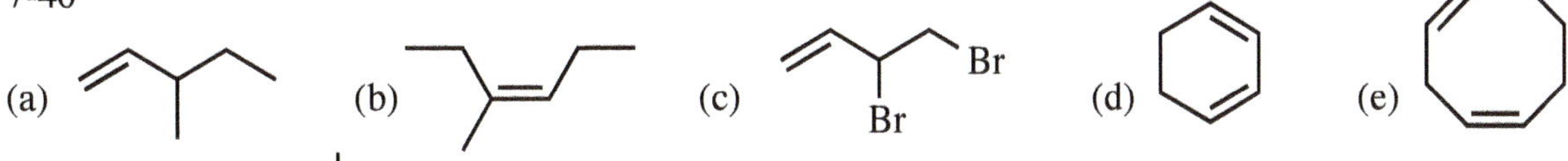

(f) Z (g) (h) Br Z (i) Z E

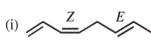

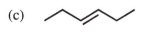

7-41 (a) *E* (b) neither—two methyl groups on one carbon (c) *Z* (d) *Z*

7-42 Parts (a) and (d) have no geometric isomers.

(b)

trans-pent-2-ene (*E*)-pent-2-ene

cis-pent-2-ene (*Z*)-pent-2-ene

(c)

trans-hex-3-ene (*E*)-hex-3-ene

cis-hex-3-ene (*Z*)-hex-3-ene

(e) Br Br

trans-1,2-dibromopropene (*E*)-1,2-dibromopropene

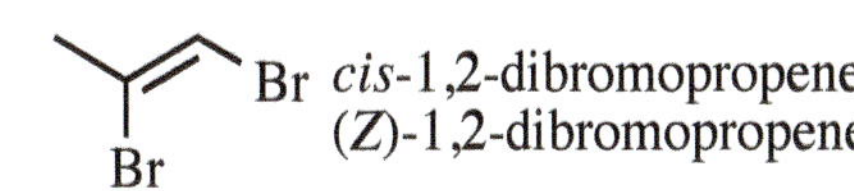

cis-1,2-dibromopropene (*Z*)-1,2-dibromopropene

(f) *Cis* and *trans* do not apply to the C=C at C-1, only the one at C-3. The *E*/*Z* system is unambiguous.

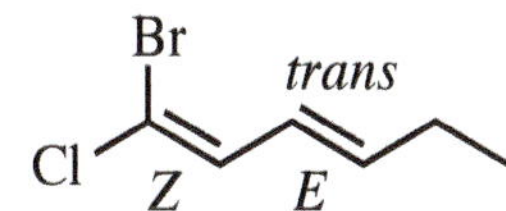

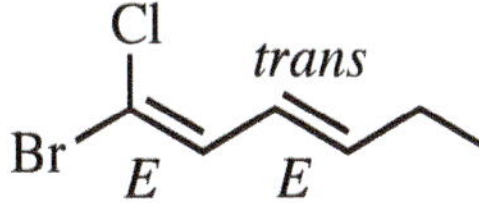

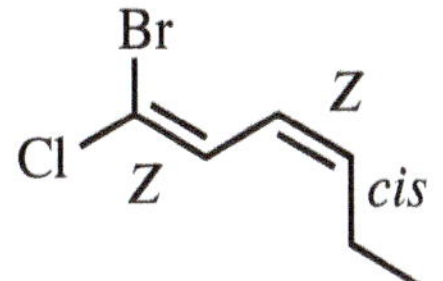

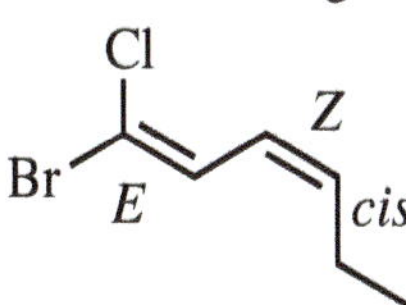

7-43 These names follow the modern IUPAC system of placement of position numbers.
(a) 2-ethylpent-1-ene (Number the longest chain *containing the double bond*.)
(b) 3-ethylpent-2-ene
(c) (3*E*,5*E*)-2,6-dimethylocta-1,3,5-triene
(d) (*E*)-4-ethylhept-3-ene
(e) 1-cyclohexylcyclohexa-1,3-diene
(f) (3*Z*,5*Z*)-6-chloro-3-(chloromethyl)octa-1,3,5-triene

7-44

(a) F

(*Z*)-1-fluoro-prop-1-ene

(*E*)-1-fluoro-prop-1-ene

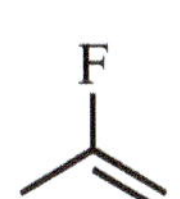

2-fluoro-prop-1-ene

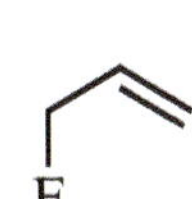

3-fluoro-prop-1-ene

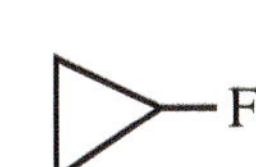

fluorocyclopropane

7-44 continued

(b) C_4H_7Br has one element of unsaturation, but no rings are permitted in the problem, so all the isomers must have one double bond. Only four isomers are possible with four carbons and one double bond, so the 12 answers must have these four skeletons with a Br substituted in all possible positions.

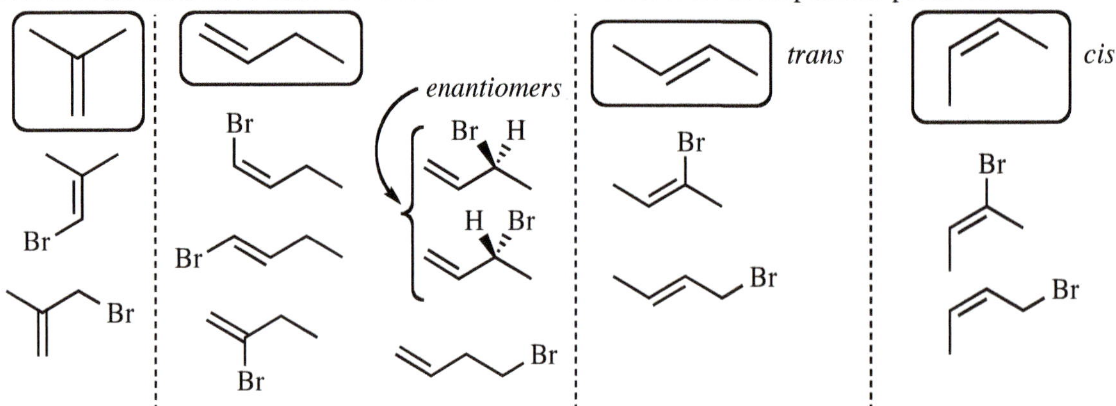

(c) Cholesterol, $C_{27}H_{46}O$, has five elements of unsaturation. If only one of those is a pi bond, the other four must be rings. (The structure of cholesterol can be found in text section 25-6.)

7-45 (a) Neither *cis* nor *trans* is defined for the double bond beginning at C-2 because none of the four groups are the same.

trans   *trans*   *cis*   *cis*

Cl   Cl   Cl   Cl

All are 3-chlorohepta-2,4-diene.

(b)   2*Z*,4*E*   2*E*,4*E*   2*Z*,4*Z*   2*E*,4*Z*

The *E*/*Z* nomenclature is unambiguous and is preferred for all four of these isomers.

7-46

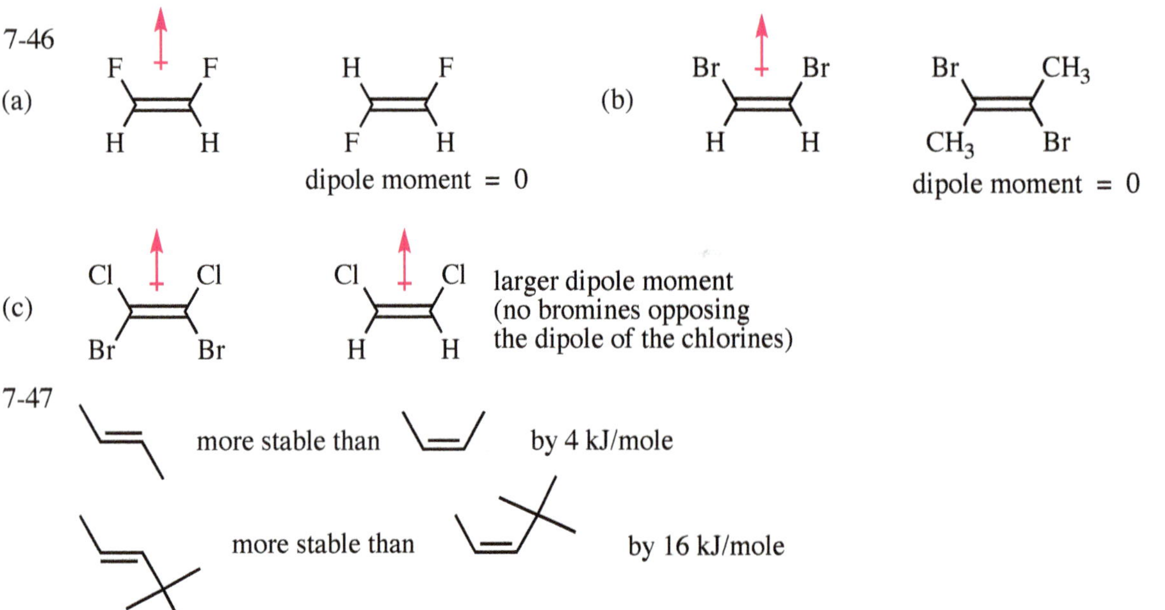

Steric crowding by the *tert*-butyl group is responsible for the energy difference. In *cis*-but-2-ene, the two methyl groups have only slight interaction. However, in the 4,4-dimethylpent-2-enes, the larger size of the *tert*-butyl group crowds the methyl group in the *cis* isomer, increasing its strain and therefore its energy.

7-48

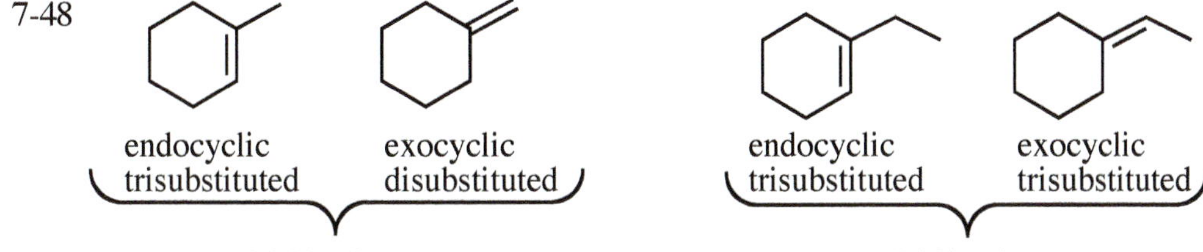

9 kJ/mole 5 kJ/mole

A standard principle of science is to compare experiments that differ by only one variable. Changing more than one variable clouds the interpretation, possibly to the point of invalidating the experiment.

The first set of structures compares endo and exocyclic double bonds, but the degree of substitution on the alkene is also different, so this comparison is not valid—we are not isolating simply the exo or endocyclic effect. The second pair is a much better measure of endo versus exocyclic stability because both alkenes are trisubstituted, so the degree of substitution plays no part in the energy values. Thus, 5 kJ/mole is a better value.

7-49

(a) + +

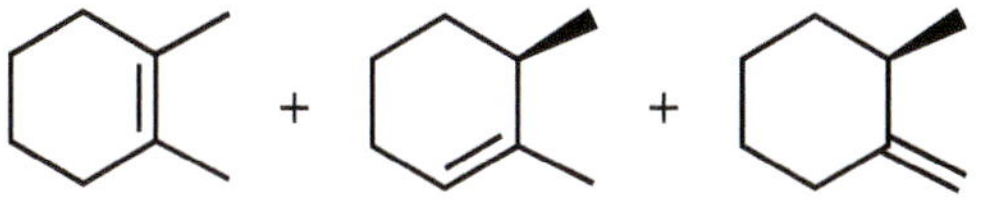

major minor trace amount

The carbocation produced in this E1 elimination is 3° and will not rearrange. The product ratio follows the Zaitsev rule.

(b) + +

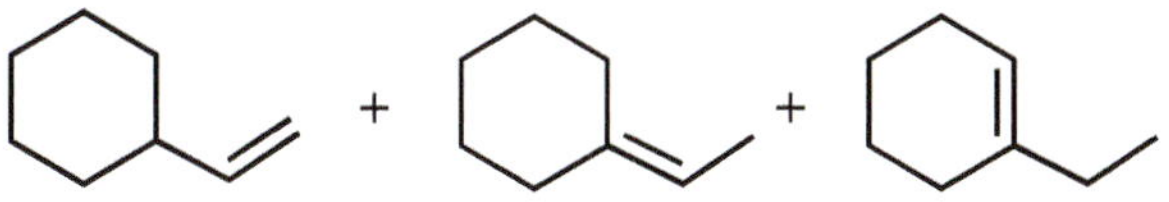

trace—from unrearranged 2° carbocation

from either unrearranged 2° carbocation or rearranged 3° carbocation

from rearranged 3° carbocation

The carbocation produced in this E1 elimination is 2° and can either eliminate to give the first two alkenes, or can rearrange by a hydride shift to a 3° carbocation which would produce the last two products. The amounts of the last two products are not predictable as they are both trisubstituted, but the first product will certainly be the least.

(c) + +

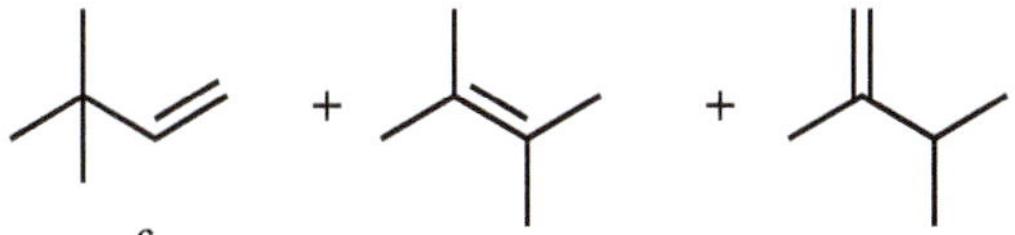

trace—from unrearranged 2° carbocation

major from rearranged 3° carbocation

minor from rearranged 3° carbocation

The carbocation produced in this E1 elimination is 2° and can either eliminate to give the first alkene, or can rearrange by a methyl shift to a 3° carbocation which would produce the last two products. The middle product is major as it is tetrasubstituted versus disubstituted for the last structure and monosubstituted for the first structure.

7-50

(a)

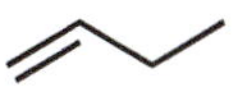

only product

(b) + +

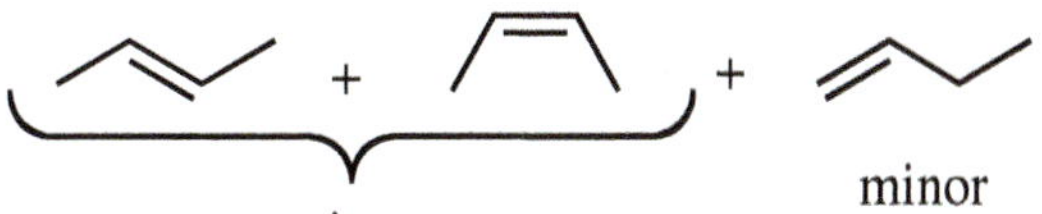

major minor

(c) +

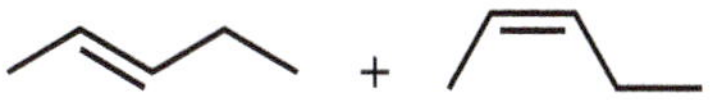

(d) +

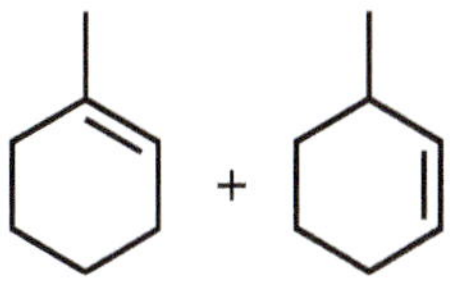

major minor

(e) CH_3 Br NaOH → CH_3

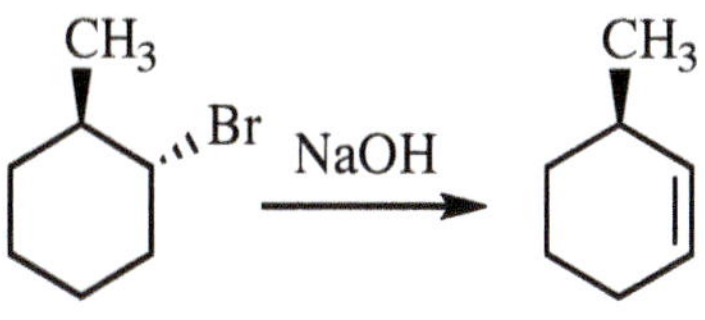

The E2 mechanism requires anti-coplanar orientation of H and Br.

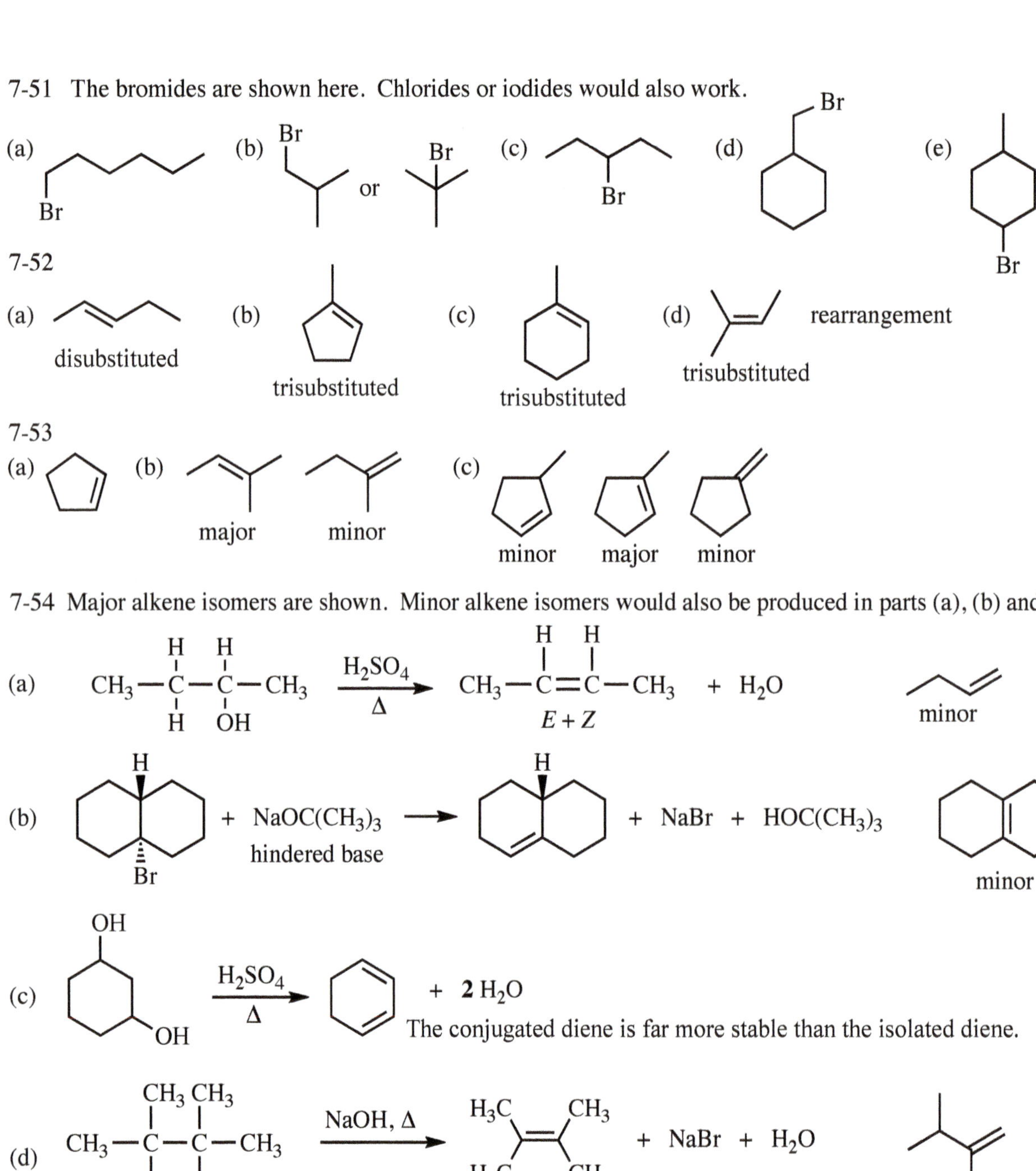

7-55

(a) major + minor

(b) major + minor

(c) major + minor

(d) CH_3 D Only product from E2—anti-coplanar is possible only from carbon without CH_3 by removing H and Cl.

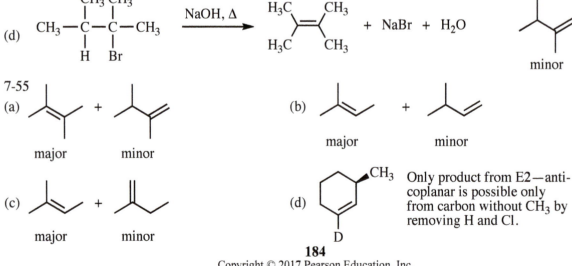

7-55 continued

(e)

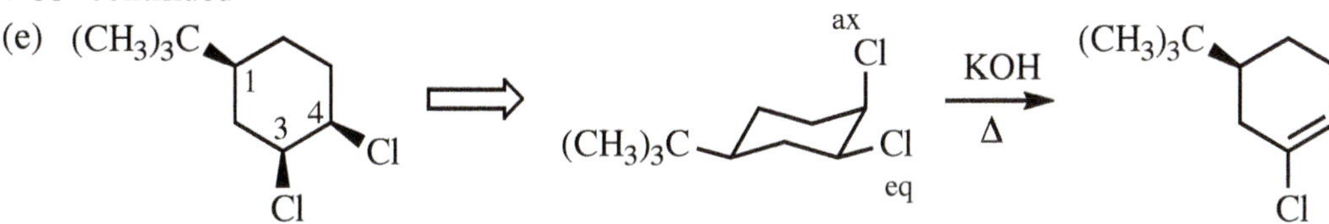

+

The conformation must be considered because the leaving group must be in an axial position. The *tert*-butyl group is so large that it must be equatorial, locking the conformation into the chair shown. As a result, only the Cl at position 4 is axial, so that is the one that must leave, not the one at position 3. Two isomers are possible but it is difficult to say which would be formed in greater amount.

(f)

Only one of the two Br atoms is in the required axial position for E2 elimination. There are axial H atoms at the two adjacent carbons, so two alkene products can be formed. Predicting major and minor would be difficult.

H Br Br H — KOH, Δ → (H, H, Br) + (H, H, Br)

7-56

(a) cyclohexane —Br_2, hν→ bromocyclohexane (Br)

all positions equivalent

(b) bromocyclohexane (from (a)) —KOH, Δ→ cyclohexene

or KO-t-Bu

(c) bromocyclohexane (from (a)) —S_N2 NaOEt, cold to avoid E2→ ethoxycyclohexane (OEt)

or S_N1: EtOH, Δ

(d) cyclohexene (from (b)) —NBS, hν→ 3-bromocyclohexene (Br)

or low conc. of Br_2 , hν

(e) 3-bromocyclohexene (from (d)) —KOH, Δ→ 1,3-cyclohexadiene

or KO-t-Bu

(f) bromocyclohexane (from (a)) —S_N2 NaOH, cold to avoid E2→ cyclohexanol (OH)

or S_N1: H_2O, Δ

7-57

(a) cyclopentanol (OH) —H_2SO_4 , Δ→ cyclopentene

(b) bromocyclopentane (Br) —$KOC(CH_3)_3$→ cyclopentene

Other bases could also be used. The hindered bases would minimize the amount of substitution that would occur.

(c) cyclopentane —Br_2 , hν→ bromocyclopentane (Br) —$KOC(CH_3)_3$→ cyclopentene

7-58 "Ph" is the abbreviation for the phenyl substituent.

(NBS can also be used for benzylic bromination)

Benzylic bromination is very selective; dibromination might occur if an excess of Br_2 is used.

solvolysis conditions—S_N1
If the reaction mixture got too warm, some E1 elimination might occur.

Using a hindered base like *tert*-butoxide will give exclusively the E2 product, with no competing S_N2 as KOH or $NaOCH_3$ might give.

7-59

E1 works well because only one carbocation and only one alkene are possible. Substitution is not a problem here. The only nucleophiles are water, which would simply form starting material by a reverse of the dehydration, and bisulfate anion. Bisulfate anion is an extremely weak base and poor nucleophile; if it did attack the carbocation, the unstable product would quickly re-ionize, with no net change, back to the carbocation.

7-60
(a)

Methoxide is a strong base and nucleophile so the reaction is second order. The Br is on a 3° carbon so S_N2 is not possible, only E2. The transition state shows bond-forming and bond-breaking with dashed bonds.

The reaction-energy diagram for a one-step mechanism like E2 has one peak.

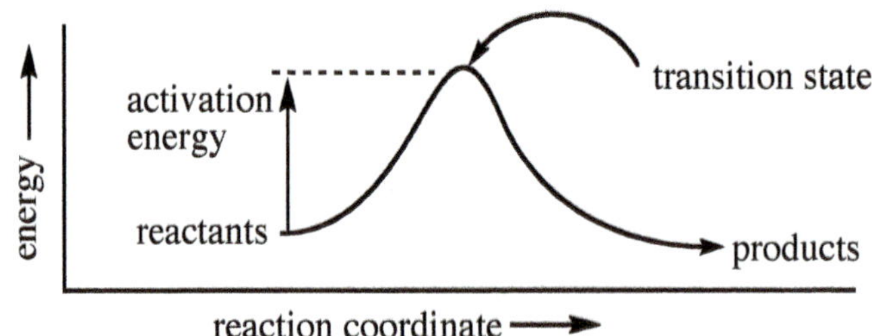

7-60 continued

(b)

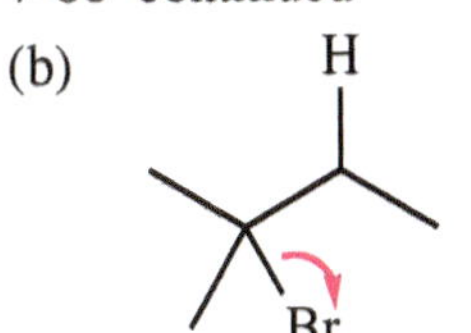

HOCH$_3$

Δ

slow

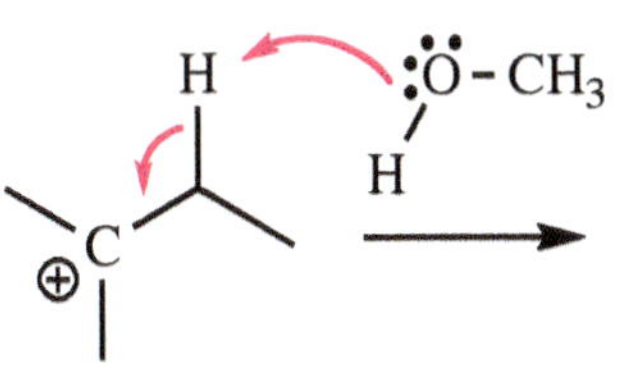

3° carbocation intermediate

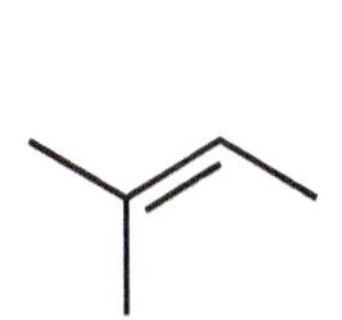

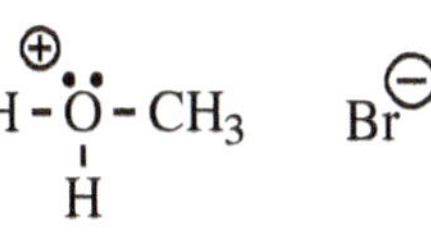

The reaction-energy diagram for a two-step mechanism like E1 has two peaks, with the first step as the slow step with higher activation energy.

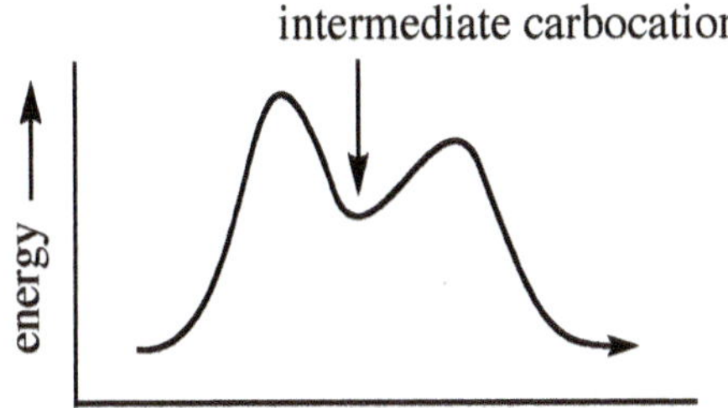

7-61

(a)

Br

HOCH$_3$

Δ

– Br$^-$

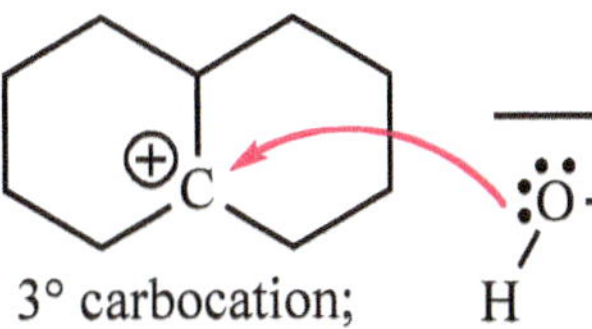

3° carbocation; no rearrangement

:Ö-CH$_3$

H

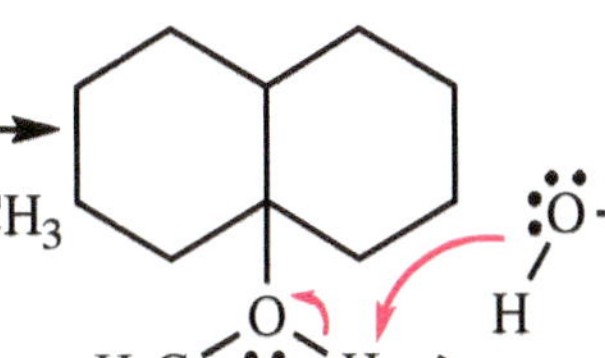

:Ö-CH$_3$

H

S$_N$1

H$_3$C

:Ö:

H

:Ö-CH$_3$

H

C

⊕C

C

C

E1

⊕C

C

H H

:Ö-CH$_3$

H

C

C

H

E1

7-61 continued

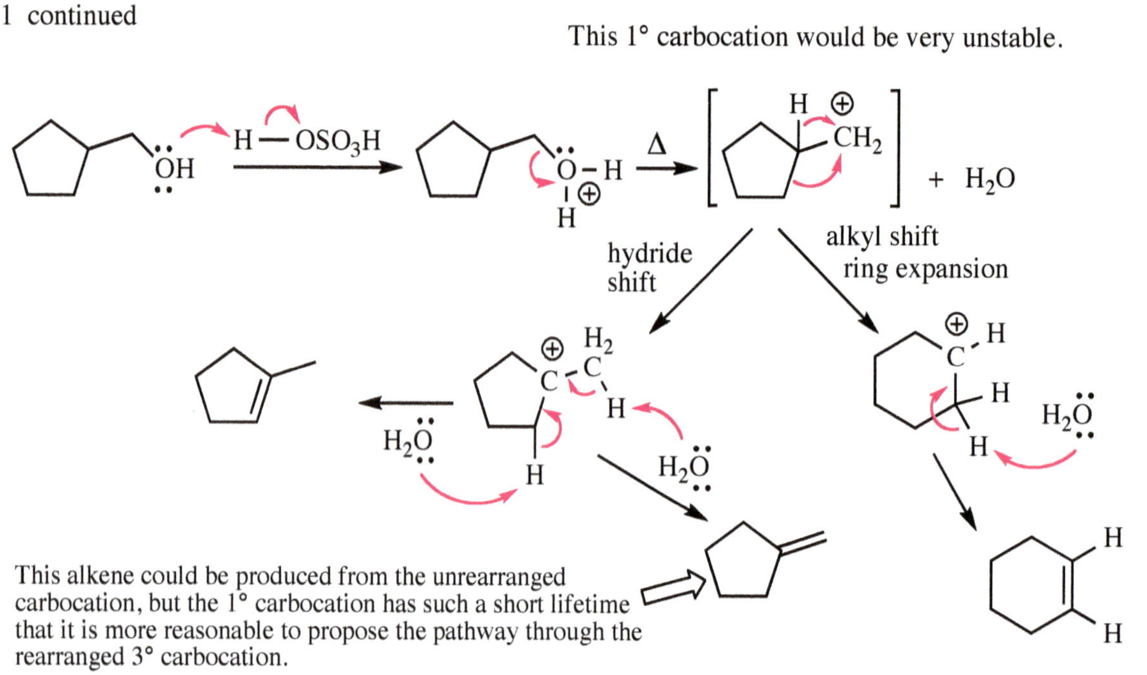

This 1° carbocation would be very unstable.

(b) OH $H—OSO_3H$ O–H H Δ   H CH_2 + H_2O

hydride shift

alkyl shift ring expansion

H_2 C   C   H H_2O H H_2O

H   C   H   H H_2O

H   H

This alkene could be produced from the unrearranged carbocation, but the 1° carbocation has such a short lifetime that it is more reasonable to propose the pathway through the rearranged 3° carbocation.

(c) EtOH Δ   H   Br   C   H   H–OEt   H   :OEt   H   H–OEt   H   OEt

H   H–OEt

C   H   H–OEt   Et   O   H   H–OEt   OEt

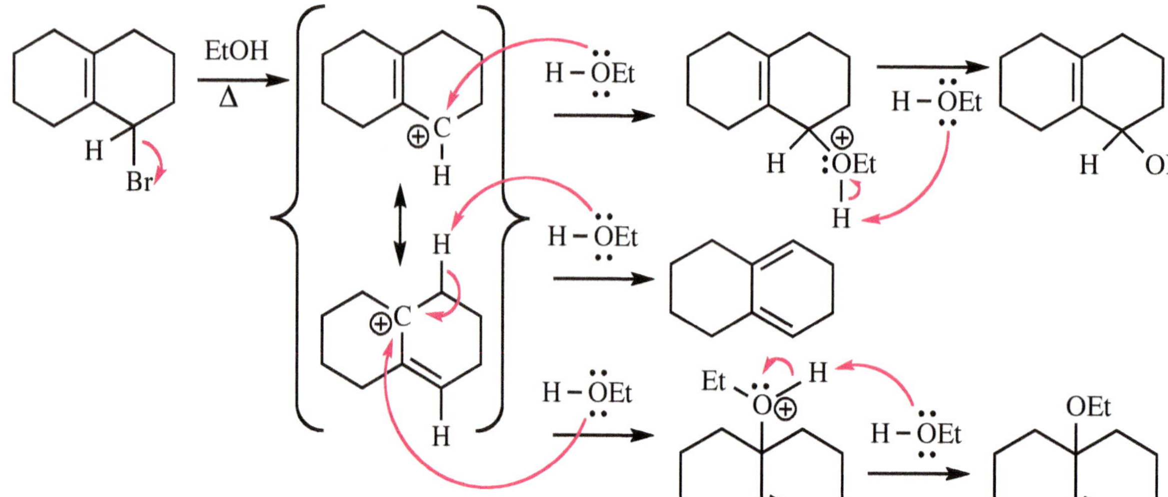

7-62

(a) The first step in this first-order solvolysis is ionization, followed by rearrangement.

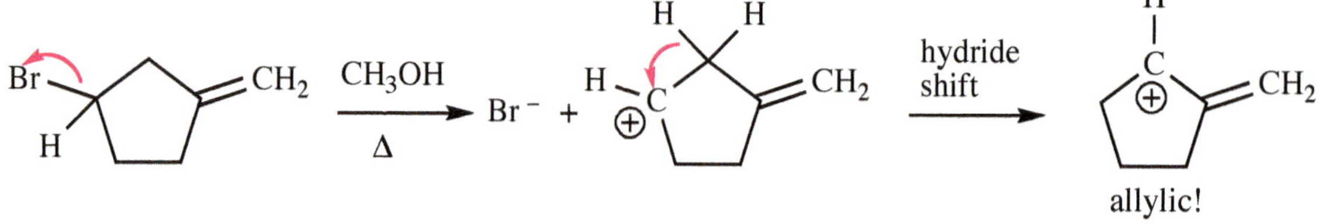

Br   H CH_2 CH_3OH Δ Br^- +   H   H   H   C CH_2 hydride shift   H   C CH_2

allylic!

continued on next page

7-62(a) continued

E1

S_N1

(b) The first step in this first-order solvolysis is ionization, followed by rearrangement.

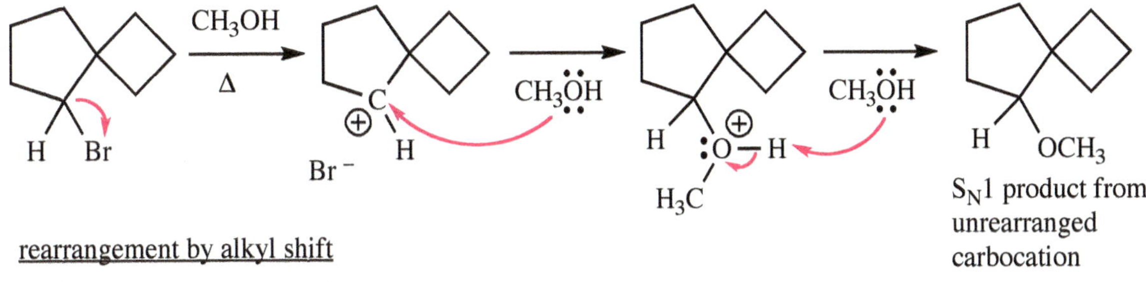

rearrangement by alkyl shift

E1 on rearranged carbocation

S_N1 on rearranged carbocation

7-63 All five products (boxed) come from rearranged carbocations. Rearrangement, which may occur simultaneously with ionization, can occur by hydride shift to the 3° methylcyclopentyl cation, or by ring expansion to the cyclohexyl cation.

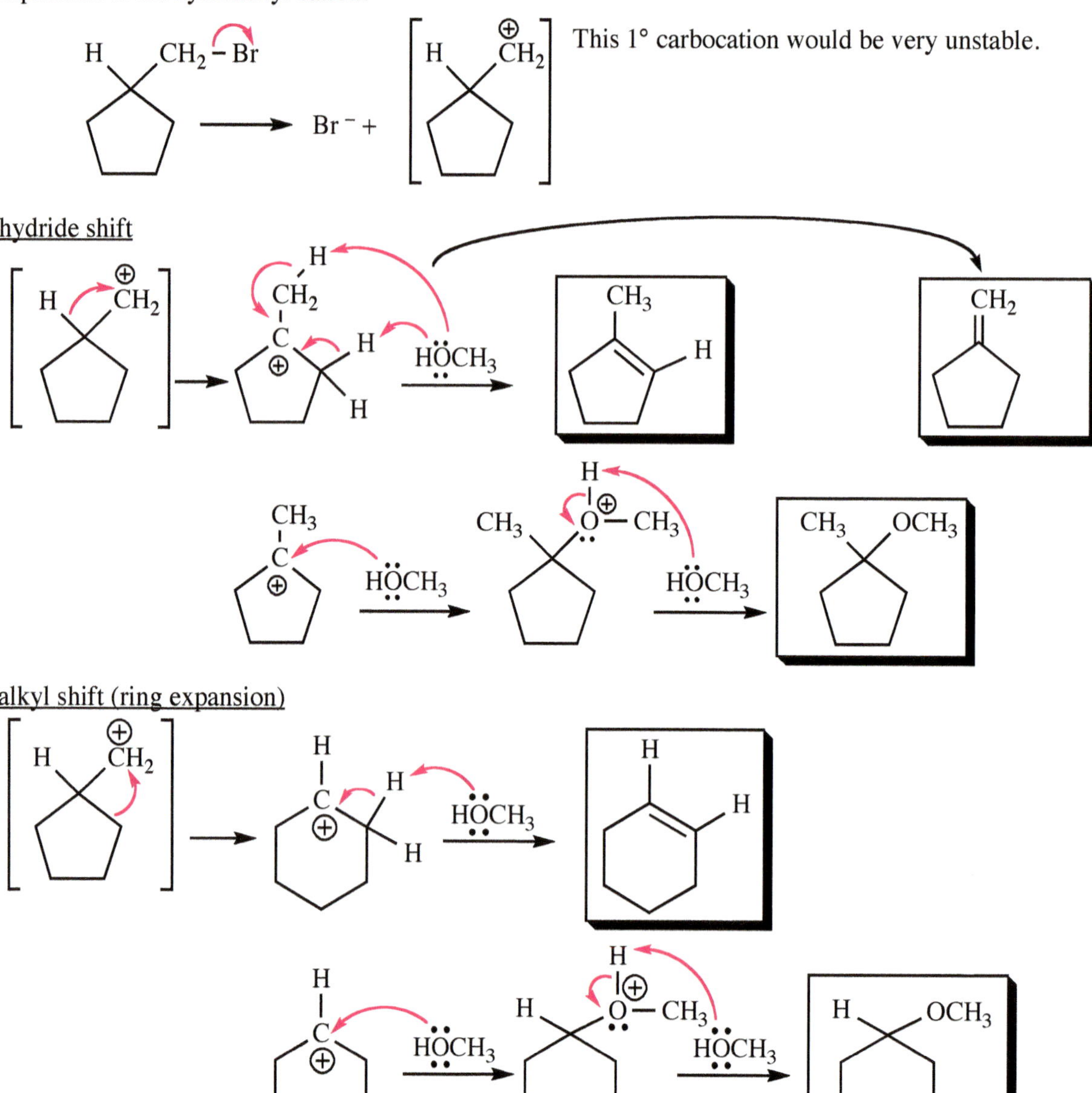

7-64

(a) S_N1

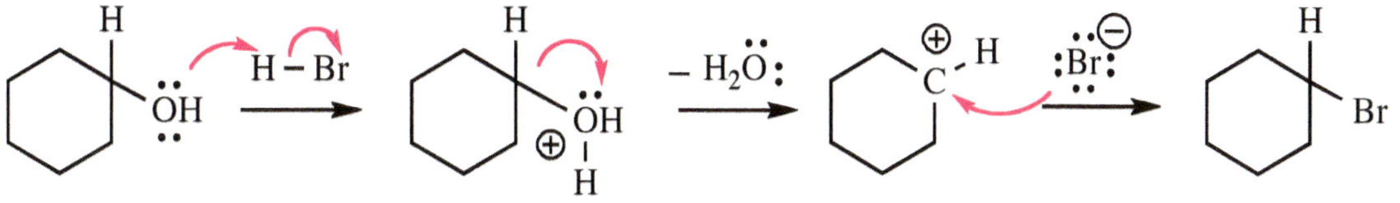

(reaction scheme: cyclohexanol + H–Br → protonated alcohol → $-H_2O$ → carbocation + Br^- → bromocyclohexane)

S_N2

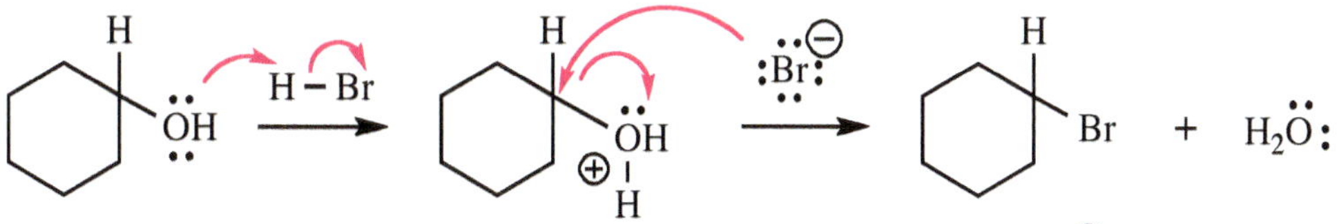

(reaction scheme: cyclohexanol + H–Br → protonated alcohol; Br^- attack → bromocyclohexane + H_2O)

(b)

(reaction scheme: H–OSO_3H protonates OH; Δ; loss of water, methyl shift, loss of H by H_2O gives 1,2,3,3-tetramethyl... cyclohexene product)

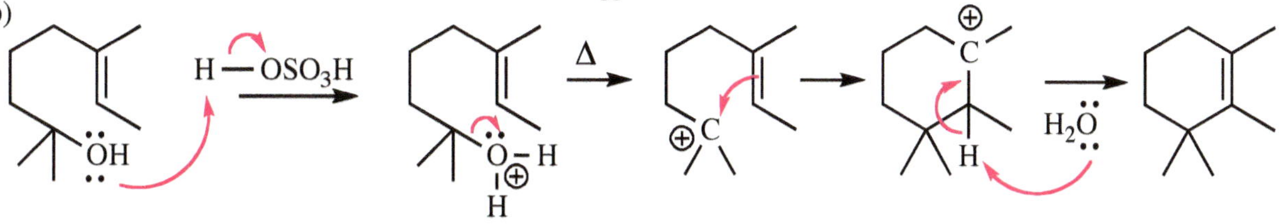

7-65 The symmetry of this molecule is crucial.

(a) $CH_3-CH(Ph)-CHBr-CH(Ph)-CH_3$

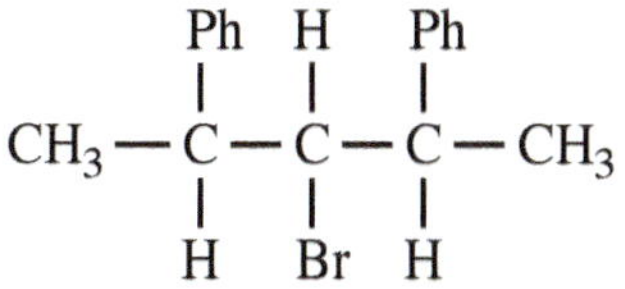

Regardless of which adjacent H is removed by *tert*-butoxide, the product will be 2,4-diphenylpent-2-ene.

(b) Here are a Newman projection, a three-dimensional representation, and a Fischer projection of the required diastereomer. On both carbons 2 and 4, the H has to be anti-coplanar with the bromine while leaving the other groups to give the same product. Not coincidentally, the correct diastereomer is a meso structure.

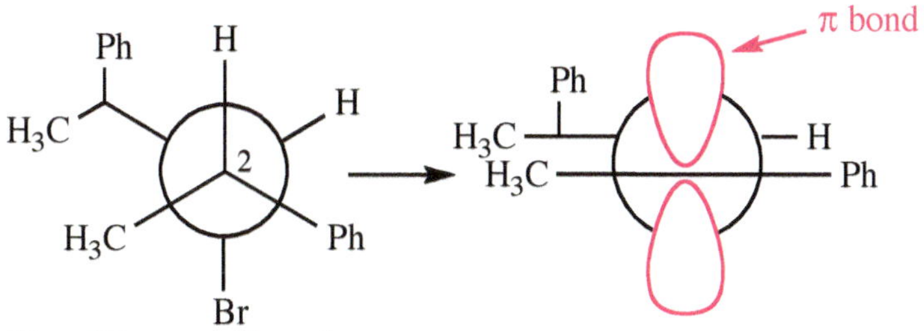

H and Br are anti-coplanar

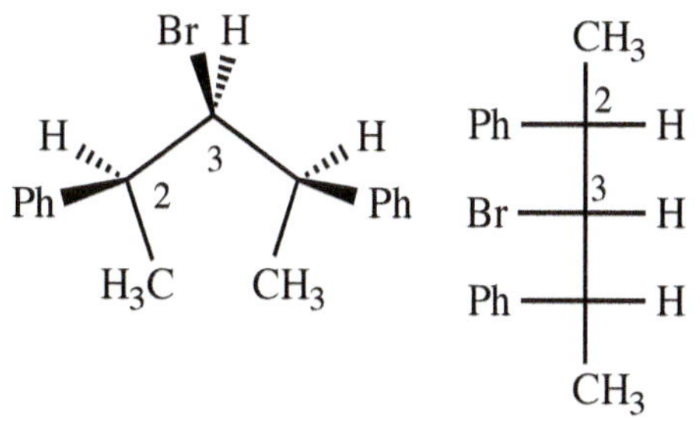

π bond

7-66

2*S*,3*R* —NaOEt→ **B** + **A** = *E*,*Z* mixture

2*S*,3*S* —NaOEt→ **C** + **A** = *E*,*Z* mixture

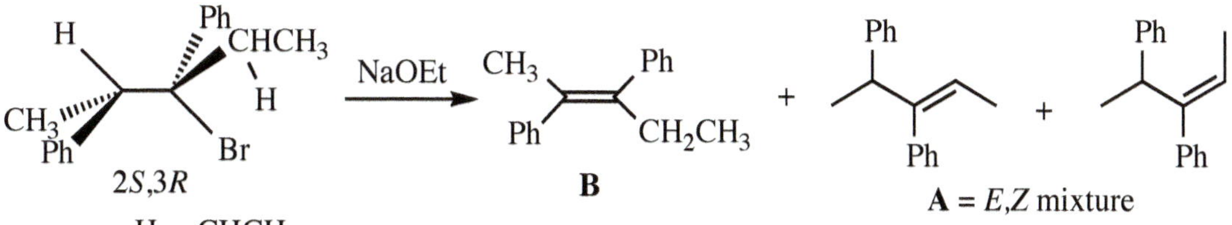

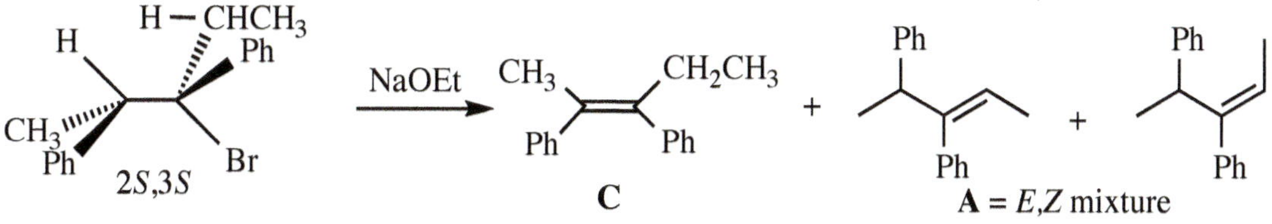

E2 dehydrohalogenation requires anti-coplanar arrangement of H and Br, so specific *cis-trans* isomers (**B** or **C**) are generated depending on the stereochemistry of the starting material. Removing a hydrogen from C-4 (achiral) will give about the same mixture of *E* and *Z* (**A**) from either diastereomer.

7-67 Begin with a structure of (*S*)-2-bromo-2-fluorobutane. Since there is no H on C-2, the lowest priority group must be the CH_3 . The Br has highest priority, then F, then CH_2CH_3, and CH_3 is fourth. Sodium methoxide is a strong base and nucleophile, so the reaction must be second order, E2 or S_N2.

(a)

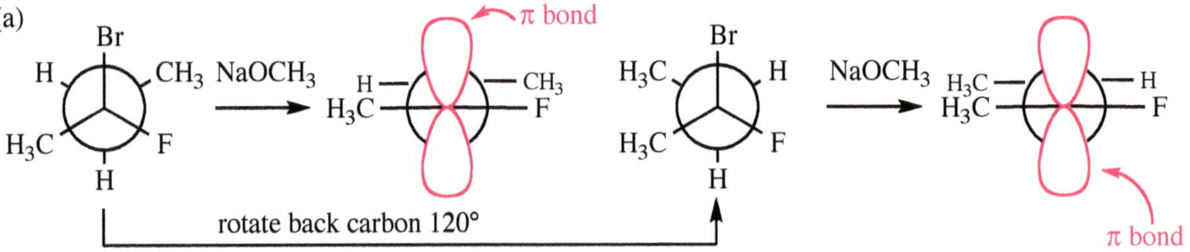

Br, H, CH_3, H_3C, F, H → $NaOCH_3$ → H, CH_3, H_3C, F (π bond)

rotate back carbon 120°

Br, H_3C, H, H_3C, F, H → $NaOCH_3$ → H_3C, H, H_3C, F (π bond)

In regular structural formulas, the reaction would give three products including the stereoisomers shown above.

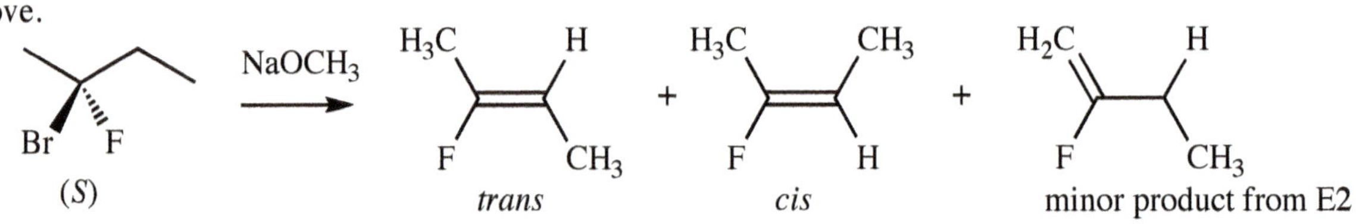

Br F (*S*) → $NaOCH_3$ → H_3C H F CH_3 *trans* + H_3C CH_3 F H *cis* + H_2C H F CH_3 minor product from E2

(b) 4 3 1 Br 2 F (*S*) → $NaOCH_3$, S_N2 → 4 3 1 F 2 OCH_3 (*S*)

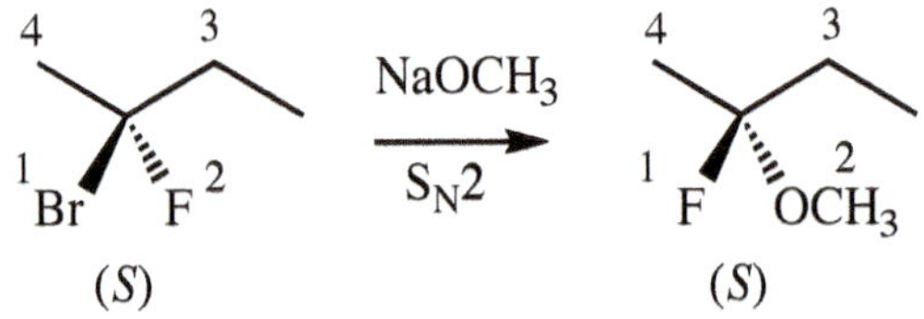

In these structures, the numbers 1 to 4 indicate the group's priority in the Cahn-Ingold-Prelog system.

A cursory analysis of the *designation* of configuration would suggest to the uncritical mind that this reaction proceeded with retention of configuration—but that would be wrong! You know by now that a careful analysis is required. In the Cahn-Ingold-Prelog system, the F in the starting material was priority group 2, but in the product, because Br has left, F is now the first priority group. So even though the *designation* of configuration suggests retention of configuration, the molecule has actually undergone inversion as would be expected with an S_N2 reaction. (See the solution to problem 6-21 for a similar example.)

7-68 substitution

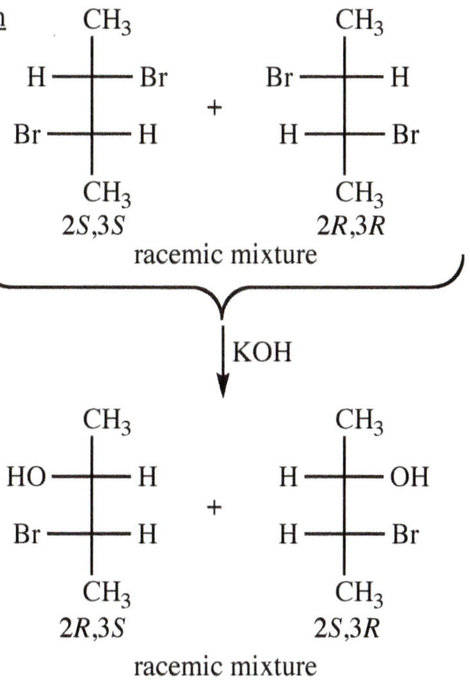

Regardless of which bromine is substituted on each molecule, the same mixture of products results.

2*S*,3*S* + 2*R*,3*R* — racemic mixture

$\xrightarrow{KOH}$

2*R*,3*S* + 2*S*,3*R* — racemic mixture

Each of the substitution products has one chiral center inverted from the starting material. The mechanism that accounts for inversion is S_N2. If an S_N1 process were occurring, the product mixture would also contain 2*R*,3*R* and 2*S*,3*S* diastereomers. Their absence argues against an S_N1 process occurring here.

elimination

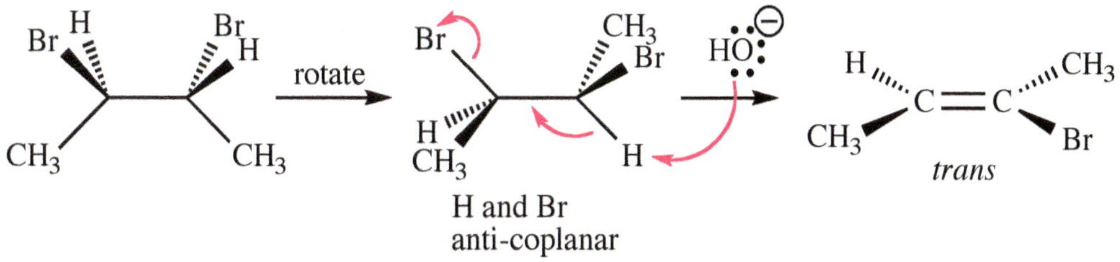

rotate → H and Br anti-coplanar → *trans*

The other enantiomer gives the same product (you should prove this to yourself).

The absence of *cis* product is evidence that only the E2 elimination is occurring in one step through an anti-coplanar transition state with no chance of rotation. If E1 had been occurring, rotation around the carbocation intermediate would have been possible, leading to both the *cis* and *trans* products.

7-69 "Ph" = phenyl

(a) $\xrightarrow[E2]{NaOCH_3}$ major *Z,E* mixture + minor

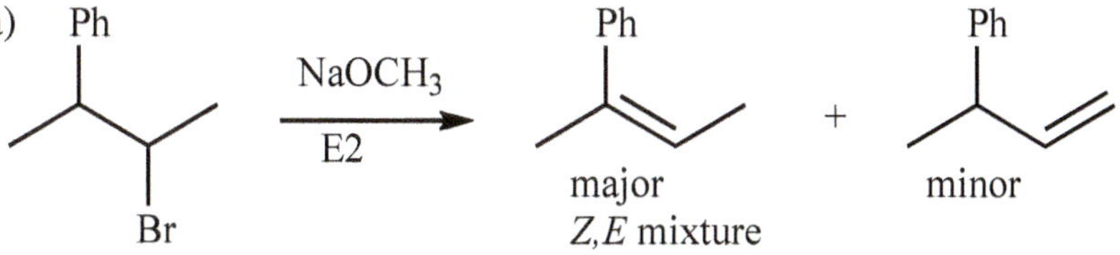

7-69 continued

(b) H and Br must be anti-coplanar in the transition state.

H, H, CH_3, Ph, CH_3, Br

2*R*,3*R*

$NaOCH_3$

Ph, H, C=C, CH_3, CH_3

methyls *cis*

H, H, CH_3, Ph, CH_3, Br

δ^- H------OCH_3 ‡

H, CH_3, Ph, CH_3, Br δ^-

transition state

H, CH_3, Ph, CH_3

(c) H, CH_3, H, Ph, CH_3, Br 2*S*,3*R*

CH_3, Ph, H, CH_3

Ph, CH_3, C=C, CH_3, H

methyls *trans*

(d) The 2*S*,3*S* is the mirror image of 2*R*,3*R*; it would give the mirror image of the alkene that 2*R*,3*R* produced (with two methyl groups *cis*). The alkene product is planar, not chiral, so its mirror image is the same: the 2*S*,3*S* and the 2*R*,3*R* give the same alkene.

7-70

A pi bond is formed from two parallel p orbitals. Rotating around a double bond breaks the pi bond by forcing the p orbitals to be perpendicular. If anything can stabilize the transition state, the rotational energy barrier will be lower.

H, H, C=C, H_3C, CH_3 — rotate right sp^2 carbon by 90° → [H, •C—C•, H, H_3C, CH_3]‡

Each radical is on a 2° carbon, not particularly stable. Two radicals on adjacent carbons are particularly unstable.

H, H, C=C, Ph, Ph — rotate right sp^2 carbon by 90° → [H, •C—C•, H]‡

Each radical is on a *benzylic* carbon; each is stabilized by resonance. Review the solution to 4-45(b) or 6-51 for benzylic resonance forms.

7-70 continued

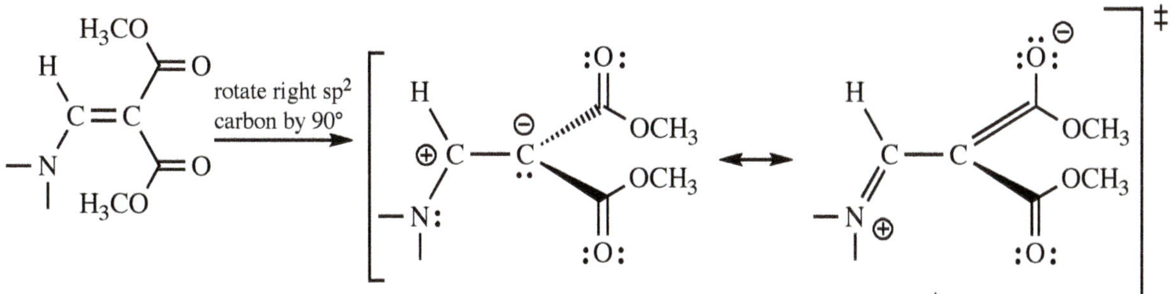

rotate right sp^2 carbon by 90°

An *unequal* distribution of electrons between the two sp^2 carbons gives resonance forms maximizing the number of pi bonds, all atoms with full octets, and negative charge on the more electronegative atoms! This large degree of stabilization explains the lowest rotational energy barrier: this transition state is the easiest to form.

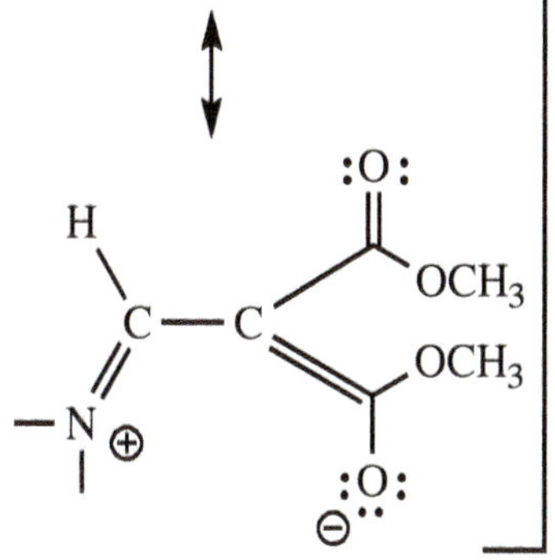

7-71 In E2, the two groups to be eliminated must be coplanar. In conformationally mobile systems like acyclic molecules, or in cyclohexanes, anti-coplanar is the preferred orientation where the H and leaving group are 180° apart. In rigid systems like norbornanes, however, SYN-coplanar (angle 0°) is the only possible orientation and E2 will occur, although at a slower rate than anti-coplanar.

The structure having the H and the Cl syn-coplanar is the *trans*, which undergoes the E2 elimination. (It is possible that the *other* H and Cl eliminate from the *trans* isomer; the results from this reaction cannot distinguish between these two possibilities.)

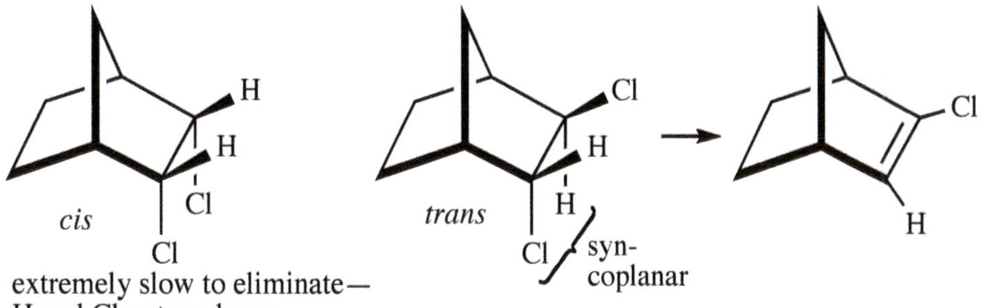

cis

extremely slow to eliminate—
H and Cl not coplanar

trans

syn-
coplanar

7-72 It is interesting to note that even though three-membered rings are more strained than four-membered rings, three-membered rings are far more common in nature than four-membered rings. Rearrangement from a four-membered ring to something else, especially a larger ring, will happen quickly.

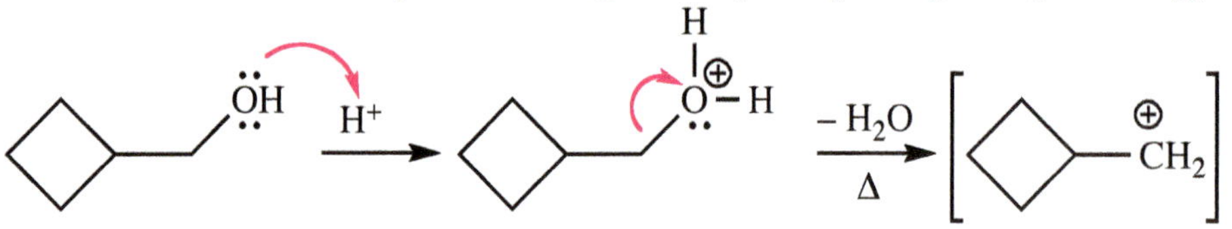

$\xrightarrow{H^+}$ $\xrightarrow[\Delta]{-H_2O}$ This 1° carbocation would be very unstable.

without rearrangement

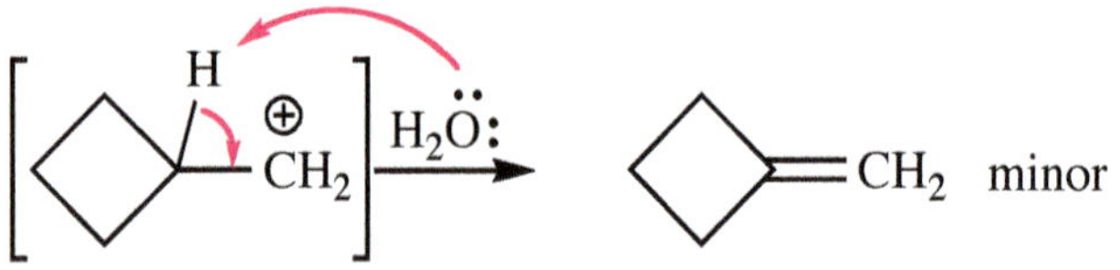

$H_2\ddot{O}$: CH_2 minor

This alkene could be produced from the unrearranged carbocation, but the 1° carbocation has such a short lifetime that it is more reasonable to propose the pathway through the rearranged 3° carbocation.

with rearrangement—hydride shift

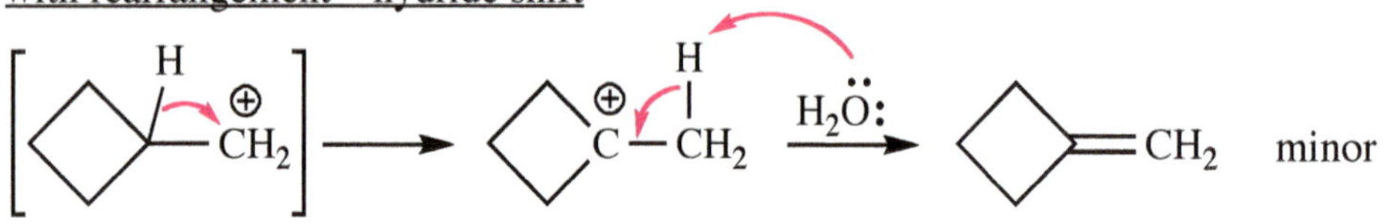

$H_2\ddot{O}$: CH_2 minor

3° carbocation, but still in a strained 4-membered ring

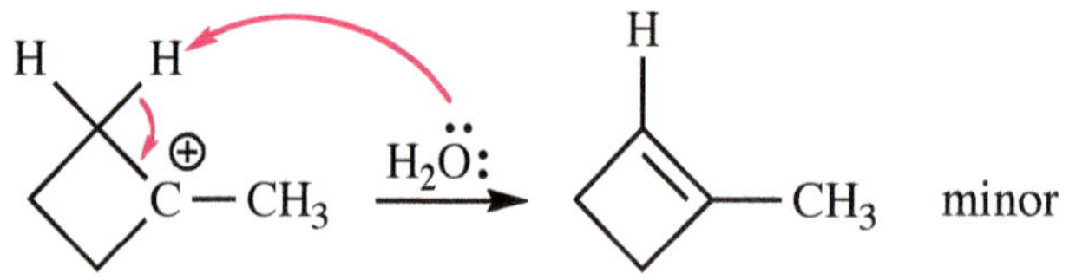

$H_2\ddot{O}$: CH_3 minor

with rearrangement—alkyl shift—ring expansion

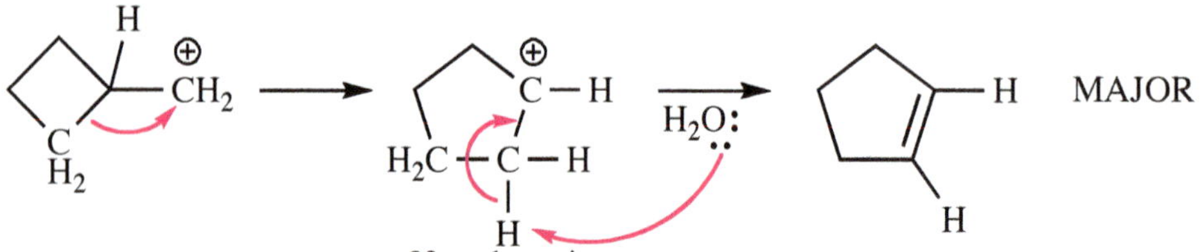

$H_2\ddot{O}$: MAJOR

2° carbocation on 5-membered ring—HOORAY!

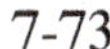

7-73

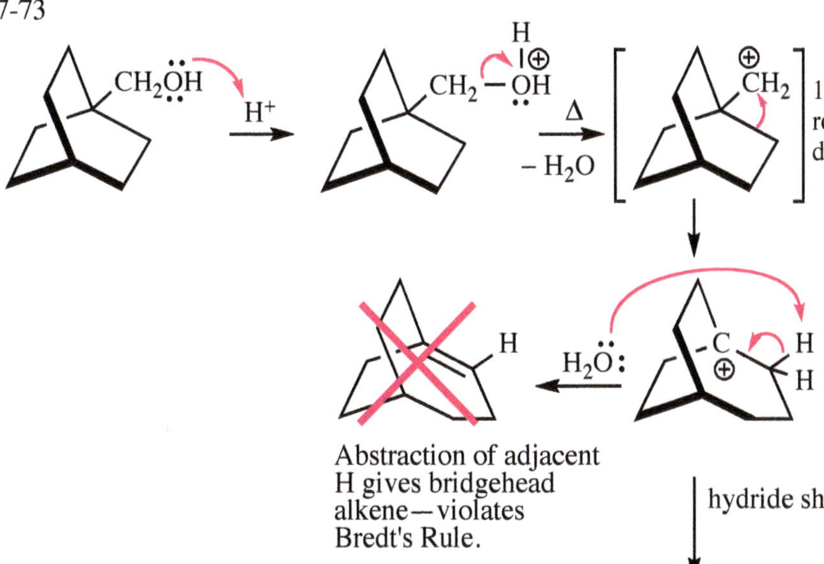

1° carbocation—terrible!—will rearrange; can't do hydride shift, must do alkyl shift = ring expansion.

This is an unstable carbocation even though it is 3°; bridgehead carbons cannot be sp^2 (planar—try to make a model), so this carbocation does a hydride shift to a 2°, *more stable* carbocation.

hydride shift

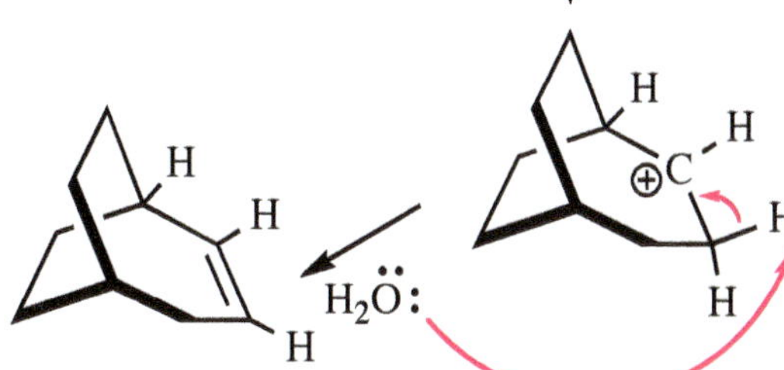

This 2° carbocation loses an adjacent H to form an alkene; can't form at bridgehead (Bredt's Rule)—only one other choice.

7-74

$CH_3-C(CH_3)(HO)-C(CH_3)(:\ddot{O}H)-CH_3 \xrightarrow{H-OSO_3H} CH_3-C(CH_3)(HO)-C(CH_3)(\overset{+}{O}H_2)-CH_3 \xrightarrow[\Delta]{-H_2O} CH_3-C(CH_3)(HO)-\overset{+}{C}(CH_3)-CH_3$

methyl shift

$CH_3-C(=:O:)-C(CH_3)_2-CH_3 \xleftarrow{H_2\ddot{O}:} \left\{ CH_3-C(=\overset{+}{O}-H)-C(CH_3)_2-CH_3 \longleftrightarrow CH_3-\overset{+}{C}(:\ddot{O}H)-C(CH_3)_2-CH_3 \right\}$

The driving force for this rearrangement is the great stability of the resonance-stabilized, protonated carbonyl group.

7-75

(a) E2— one step

$$H_2C-CH(Br)-CH_3 \;(\text{H on } H_2C,\ :\ddot{O}H^-) \longrightarrow H_2C{=}CH-CH_3 + H_2O + Br^-$$

S_N2— one step

$$CH_3-CH(Br)-CH_3 \;(:\ddot{O}H^-) \longrightarrow CH_3-CH(OH)-CH_3 + Br^-$$

(b) In the E2 reaction, a C—H bond is broken. When D is substituted for H, a C—D bond is broken, slowing the reaction. In the S_N2 reaction, no C—H (C—D) bond is broken, so the rate is unchanged.

(c) These are first-order reactions. The slow, rate-determining step is the first step in each mechanism.

E1 $H_2C(H)-CH(Br)-CH_3 \xrightarrow{\text{slow}} H_2\ddot{O}: \quad H(H_2C)-\overset{\oplus}{C}H-CH_3 \xrightarrow{\text{fast}} H_2C{=}CH-CH_3$

S_N1 $H_3C-CH(Br)-CH_3 \xrightarrow{\text{slow}} H_3C-\overset{\oplus}{C}H-CH_3 + Br^- \xrightarrow[\text{fast}]{H_2\ddot{O}:} H_3C-CH(\overset{\oplus}{O}H_2)-CH_3 \xrightarrow[\text{fast}]{H_2\ddot{O}:} H_3C-CH(OH)-CH_3$

The only mechanism of these two involving C—H bond cleavage is the E1, but the C—H cleavage does NOT occur in the slow, rate-determining step. Kinetic isotope effects are observed only when C—H (C—D) bond cleavage occurs in the rate-determining step. Thus, we would expect to observe *no change in rate* for the deuterium-substituted molecules in the E1 or S_N1 mechanisms. (In fact, this technique of measuring isotope effects is one of the most useful tools chemists have for determining what mechanism a reaction follows.)

7-76 Both products are formed through E2 reactions. The difference is whether a D or an H is removed by the base. As explained in Problem 7-75, C—D cleavage can be up to 7 times slower than C—H cleavage, so the product from C—H cleavage should be formed about 7 times as fast. This rate preference is reflected in the 7 : 1 product mixture. ("Ph" is the abbreviation for a benzene ring.)

$$\text{(D)(H)(Ph)C}-\text{C(H)(H)(Br)} \xrightarrow{-\text{D, Br}} \text{(H)(Ph)C}{=}\text{C(H)(H)}$$

requires C—D bond cleavage; slow; minor product

$$\text{(H)(Ph)(D)C}-\text{C(H)(H)(Br)} \xrightarrow{-\text{H, Br}} \text{(D)(Ph)C}{=}\text{C(H)(H)}$$

requires C—H bond cleavage; 7 times faster; major product

Note to the student: The most common question from students studying Chapter 7 material is: How do I know whether a reaction will follow an S_N1, S_N2, E1, or E2 mechanism? The answer is not always simple even for experienced chemists, and mixtures of substitution and elimination products are typical. Here are some generalizations that can provide guidelines about how to make reasonable predictions. Most instructors have their own version of this "concept map"; this one was modified from one of my colleagues, Dr. Hima Joshi, who has permitted its use here.

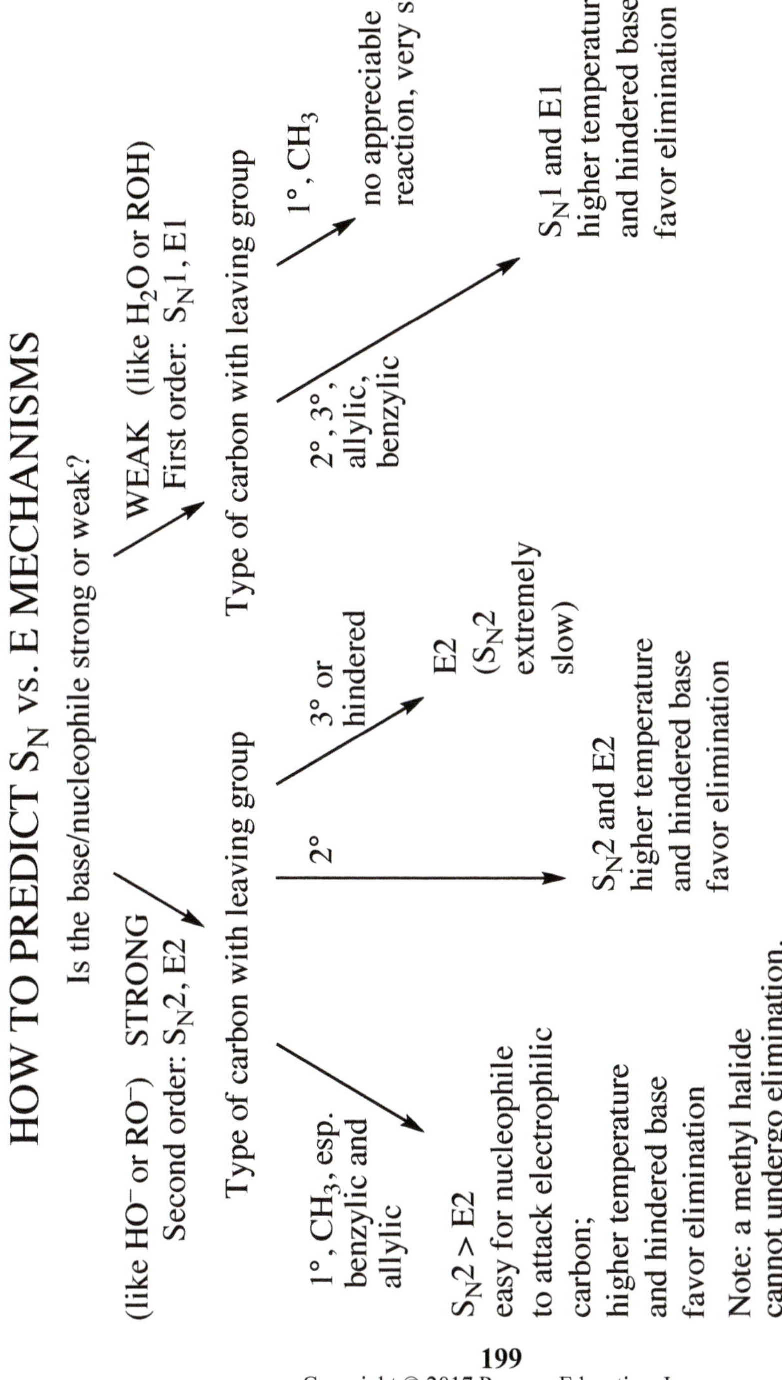

HOW TO PREDICT S_N vs. E MECHANISMS
Is the base/nucleophile strong or weak?
(like HO^- or RO^-) STRONG
Second order: S_N2, E2
WEAK (like H_2O or ROH)
First order: S_N1, E1
Type of carbon with leaving group
1°, CH_3, esp. benzylic and allylic
2°
3° or hindered
S_N2 > E2
easy for nucleophile to attack electrophilic carbon;
higher temperature and hindered base favor elimination
Note: a methyl halide cannot undergo elimination.
S_N2 and E2
higher temperature and hindered base favor elimination
E2 (S_N2 extremely slow)
Type of carbon with leaving group
2°, 3°, allylic, benzylic
1°, CH_3
S_N1 and E1
higher temperature and hindered base favor elimination
no appreciable reaction, very slow

Students: use this space for notes or to solve problems.

CHAPTER 8—REACTIONS OF ALKENES

8-1 *Major* products are produced in greatest amount; they are not necessarily the *only* products produced.

(a)

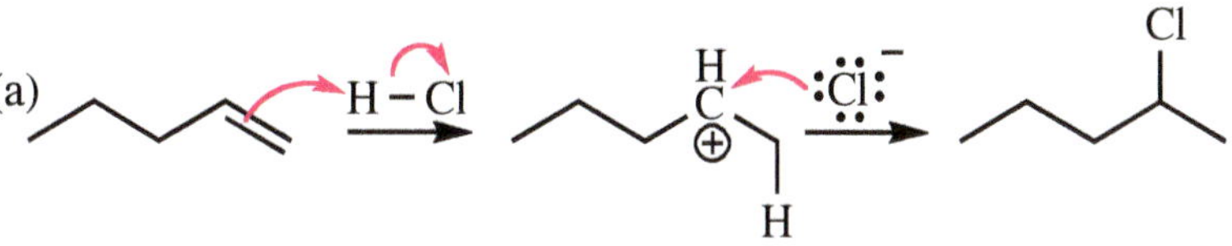

H − Cl

:Cl:⁻

Cl

(b)

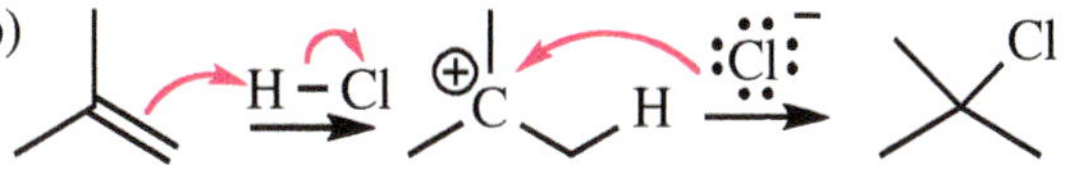

H − Cl

:Cl:⁻

Cl

(c)

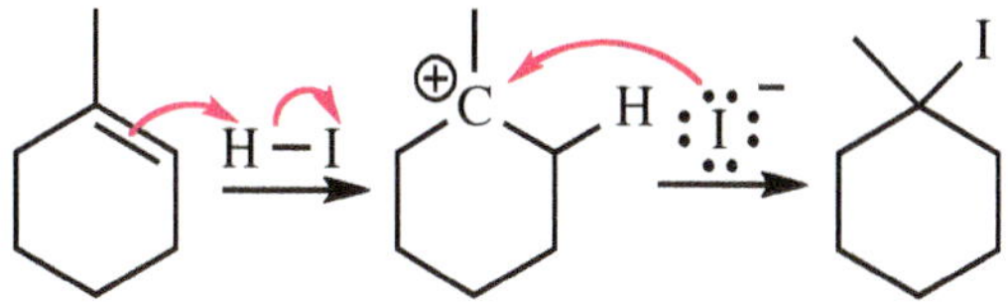

H − I

H :I:⁻

I

(d)

H − Br

H

CH

H

:Br:⁻

Br

Br

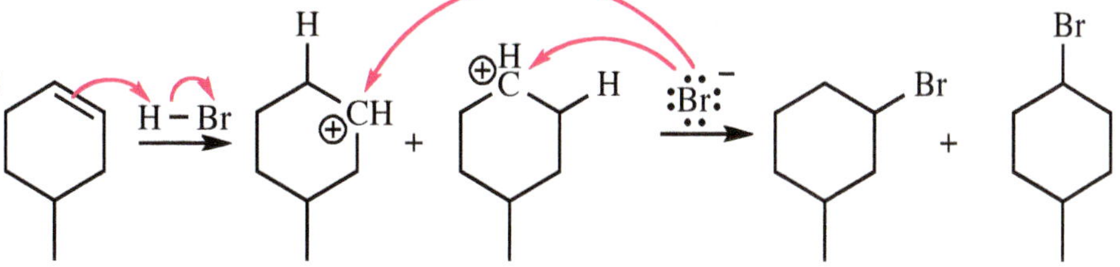

Produced in about equal amounts; each is a mixture of *cis* and *trans*.

Both 2° carbocations will form.

8-2

(a)

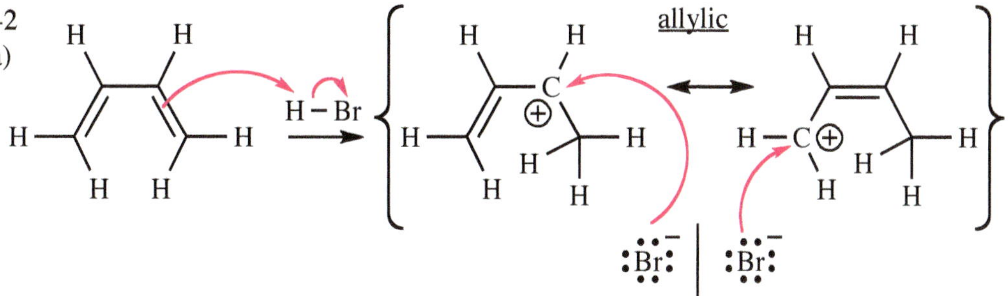

H − Br

allylic

:Br:⁻

Because the allylic carbocation has partial positive charge at two carbons, the bromide nucleophile can bond at either electrophilic carbon, giving two products.

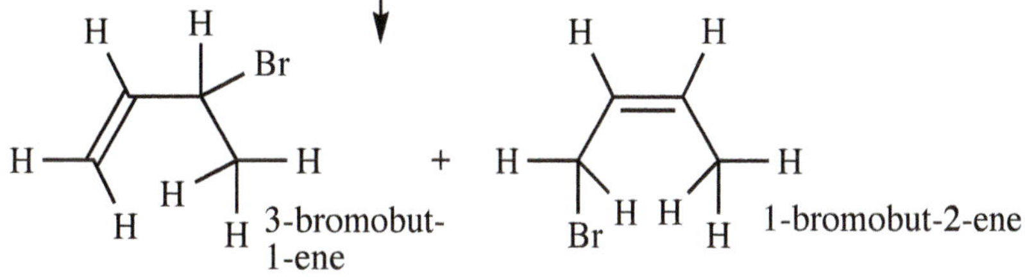

3-bromobut-1-ene + 1-bromobut-2-ene

(b)

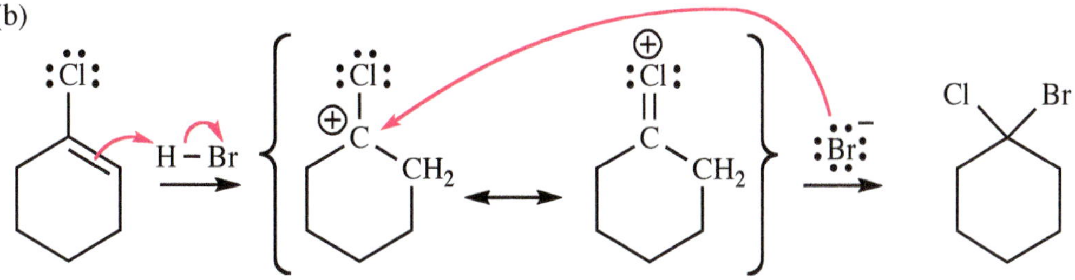

:Cl:

H − Br

CH_2

CH_2

:Br:⁻

Cl Br

This orientation produces a carbocation intermediate that is stabilized by resonance with the Cl atom.

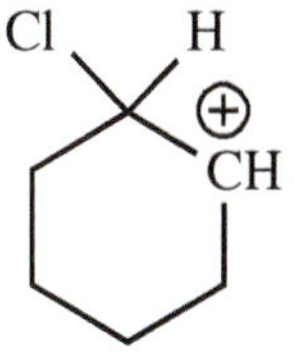

Cl H

CH

This possible intermediate is not produced; it is a 2° carbocation without any stabilization.

8-3

(a) initiation steps

$$H_3C-\overset{O}{\overset{\|}{C}}-O-O-\overset{O}{\overset{\|}{C}}-CH_3 \xrightarrow{\Delta} 2\ H_3C-\overset{O}{\overset{\|}{C}}-\ddot{\underset{\cdot\cdot}{O}}\cdot$$ radical—no charge

$$H_3C-\overset{O}{\overset{\|}{C}}-\ddot{\underset{\cdot\cdot}{O}}\cdot + H-Br \longrightarrow H_3C-\overset{O}{\overset{\|}{C}}-\ddot{\underset{\cdot\cdot}{O}}H + \cdot Br$$

propagation steps

•Br

Br

The 3° radical is more stable than the 2° radical.

Br

H—Br

H

Br

1-bromo-2-methylcyclopentane (*cis* + *trans*)

+ •Br recycles in first propagation step

(b) initiation steps

$$\xrightarrow{\Delta} 2\ (CH_3)_3C-\ddot{\underset{\cdot\cdot}{O}}\cdot$$

$$(CH_3)_3C-\ddot{\underset{\cdot\cdot}{O}}\cdot + H-Br \longrightarrow (CH_3)_3C-\ddot{\underset{\cdot\cdot}{O}}H + \cdot Br$$

propagation steps

(Recall that "Ph" is the abbreviation for "phenyl".)

Ph • Br

H C Ph Br

The benzylic radical is stabilized by resonance, and more stable than the aliphatic radical.

H C Ph Br

H—Br

H Ph H Br

+ •Br recycles in first propagation step

2-bromo-1-phenylpropane

8-4

(a) $\xrightarrow[\text{ROOR},\ \Delta]{\text{HBr}}$ Br

(b) $\xrightarrow{\text{HBr}}$ Br

(c) OH $\xrightarrow[\Delta]{H_2SO_4}$ $\xrightarrow{\text{HBr}}$ Br

(d) OH $\xrightarrow[\Delta]{H_2SO_4}$ $\xrightarrow[\text{ROOR},\ \Delta]{\text{HBr}}$ Br

Note: A good synthesis uses major products as intermediates, not minor products. Knowing orientation of addition and elimination is critical to using reactions correctly.

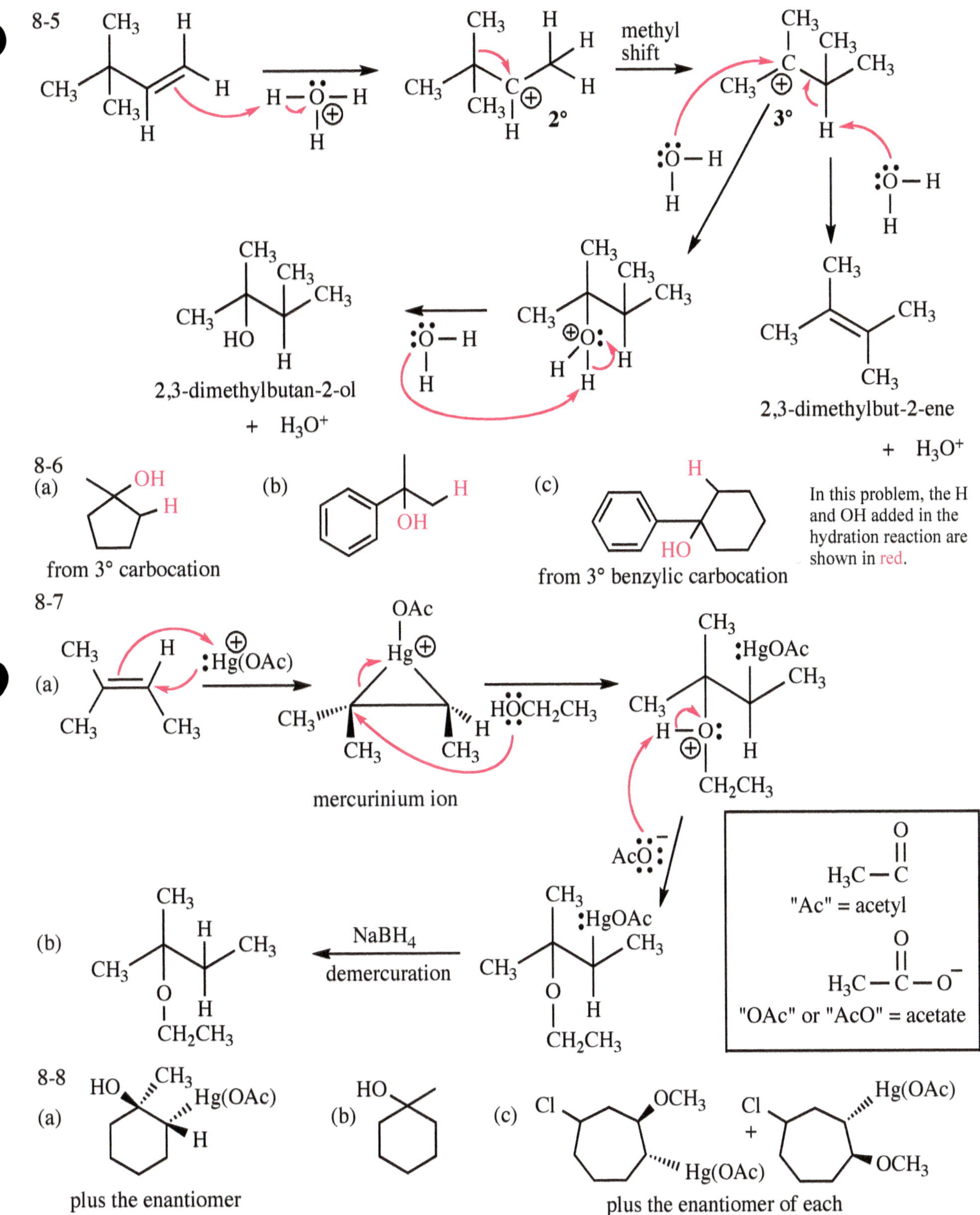

Note: When new chiral centers are generated from achiral or racemic reactants, the products are racemic mixtures. This book will indicate a racemic mixture by adding "plus the enantiomer".

8-8 continued

(d) Cl, OCH_3 + Cl, OCH_3

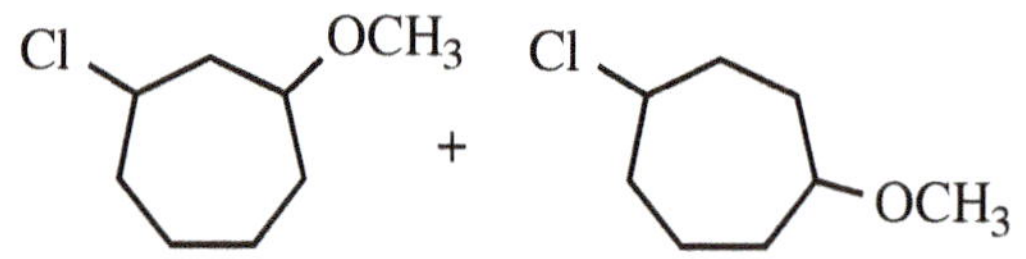

The *cis* and *trans* stereoisomers of each positional isomer would also be produced.

8-9

(a) $Hg(OAc)_2$ / CH_3OH → $NaBH_4$ → OCH_3

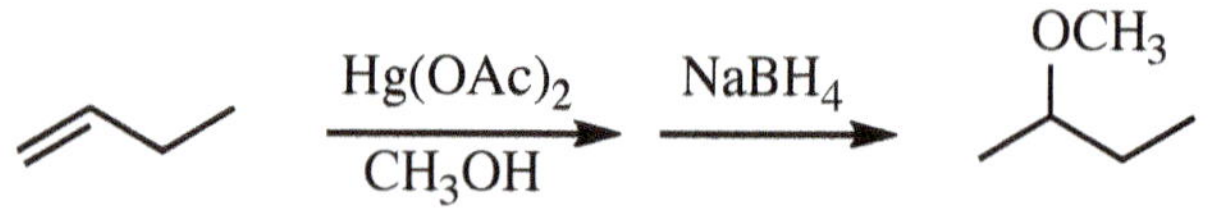

(b) I — KOH, Δ → Heat favors elimination. — $Hg(OAc)_2$ / H_2O → $NaBH_4$ → OH

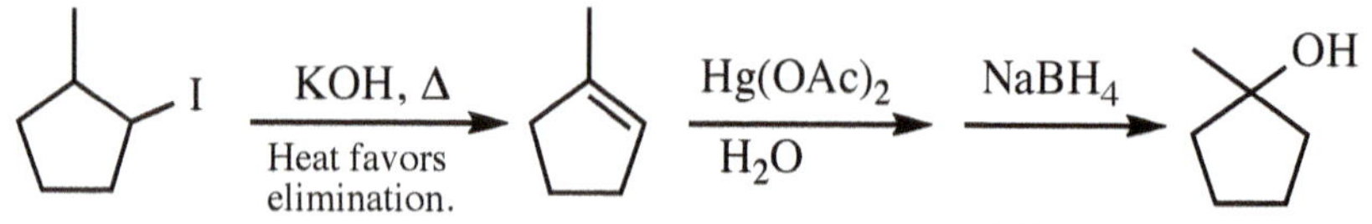

(c) $Hg(OAc)_2$ / H_2O → $NaBH_4$ → OH

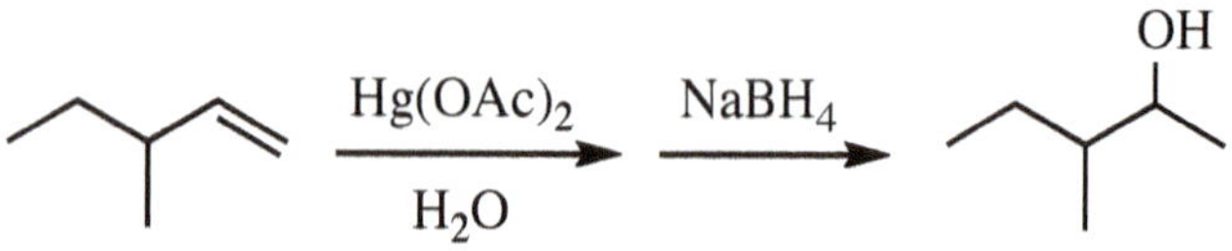

Using an acid-catalyzed hydration in part (c) would initially form a 2° carbocation that would quickly rearrange to 3° on carbon-3. The desired product would not be synthesized.

8-10

(a), (b) $BH_3 \bullet THF$ → H, BH_2 — H_2O_2 / HO^- → OH

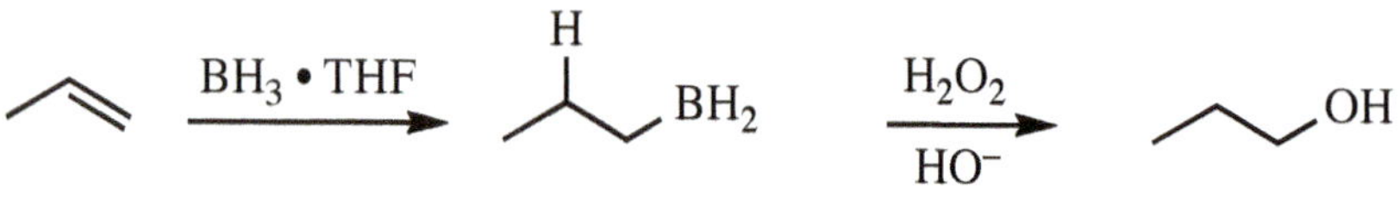

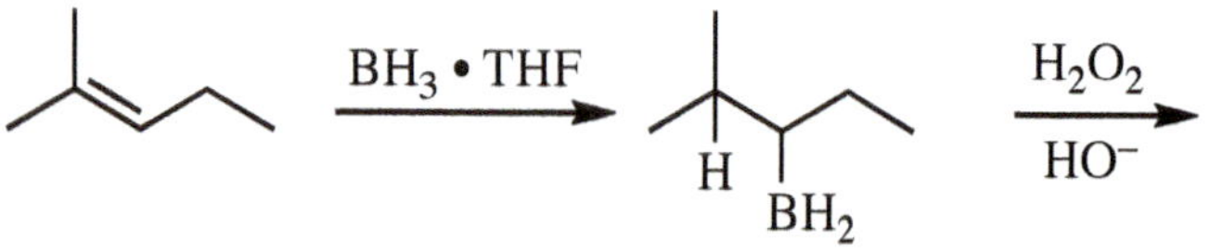

(c), (d) $BH_3 \bullet THF$ → H, BH_2 — H_2O_2 / HO^- → OH racemic mixture

(e), (f) $BH_3 \bullet THF$ → CH_3, H, BH_2, H — H_2O_2 / HO^- → CH_3, H, OH, H plus the enantiomer

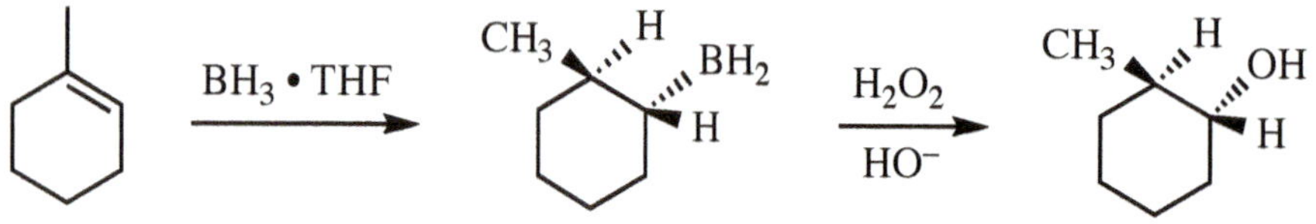

8-11

(a) $BH_3 \bullet THF$ → H_2O_2 / HO^- → HO

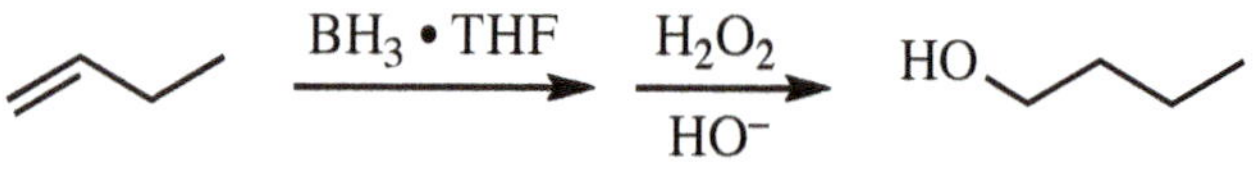

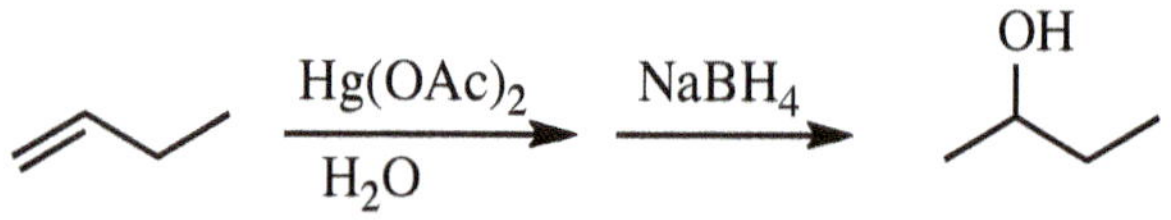

(b) $Hg(OAc)_2$ / H_2O → $NaBH_4$ → OH racemic mixture

(c) Br — KOH, Δ → — $BH_3 \bullet THF$ → H_2O_2 / HO^- → OH

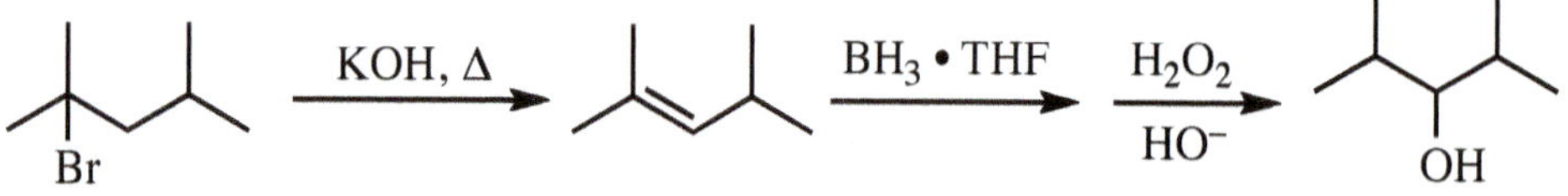

8-12 The attack of borane on 1-methylcyclopentene is equally likely from the top face or the bottom face, leading to a racemic mixture of the *trans* isomer.

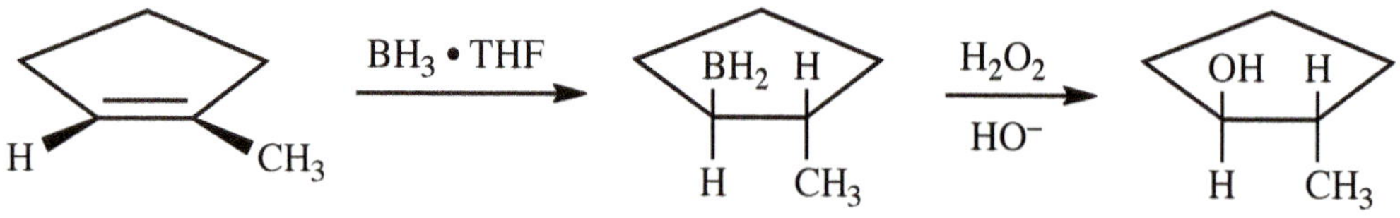

H, CH_3 — $BH_3 \bullet THF$ → BH_2 H, H CH_3 — H_2O_2 / HO^- → OH H, H CH_3 racemic mixture

8-13 The products are racemic.

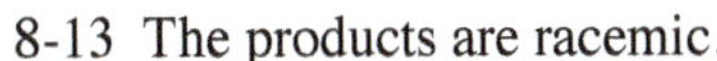

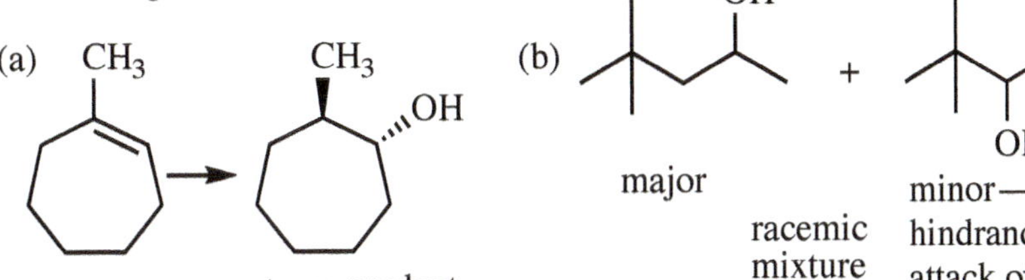

(b) OH + OH

major

racemic mixture

minor—steric hindrance to attack of BH_3

(c) OH, H, H, CH_3

BH_3 goes on the less hindered top (exo) face of the C=C.

8-14

(a) Et, Et, Me, H — *Z* — $BH_3 \cdot THF$ → H_2O_2, HO^- →

R R: Et, Et, Me, H, HO, H

S S: Et, Et, H, Me, H, OH

enantiomers

(b) Et, H, Me, Et — *E* — $BH_3 \cdot THF$ → H_2O_2, HO^- →

R S: Et, H, Me, H, HO, Et

S R: Et, H, H, Me, Et, OH

enantiomers

The enantiomeric pair produced from the *Z*-alkene is diastereomeric with the other enantiomeric pair produced from the *E*-alkene. Hydroboration-oxidation is stereospecific, that is, each alkene gives a specific set of stereoisomers, not a random mixture.

8-15

(a) $Hg(OAc)_2$, H_2O → $NaBH_4$ → OH

(b) $BH_3 \cdot THF$ → H_2O_2, HO^- → H, OH plus the enantiomer

(c) H_3C, OH → H_2SO_4, Δ → CH_3 → $BH_3 \cdot THF$ → H_2O_2, HO^- → CH_3, OH plus the enantiomer

8-16 (a)

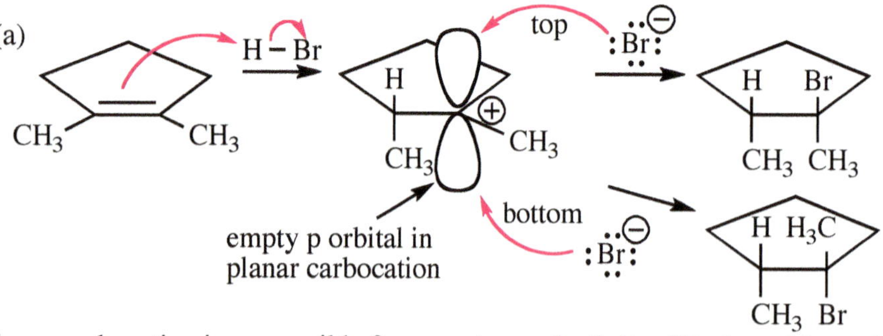

The *planar* carbocation is responsible for non-stereoselectivity. The bromide nucleophile can attack from the top or bottom, leading to a mixture of stereoisomers. The addition is therefore a mixture of syn and anti addition.

(b)

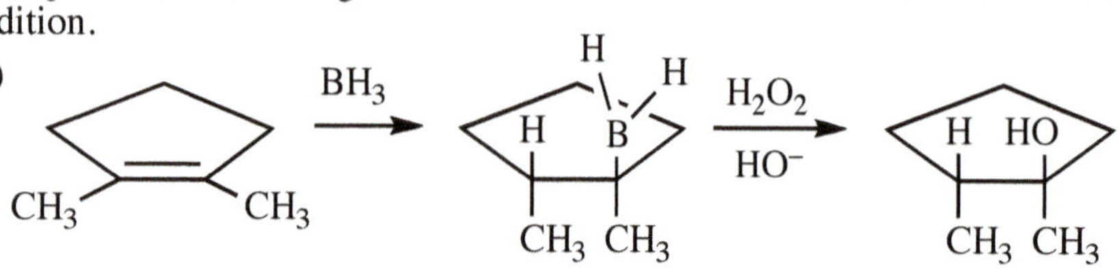

Two methyl groups are always *cis* to each other.

In contrast to part (a), hydroboration has no planar intermediate. Borane adds in a molecular addition with *syn* stereochemistry, and replacement of B with O proceeds with retention of stereochemistry. All of the steps in the process are stereospecific, so the product will be one diastereomer (although a racemic mixture).

8-17 During bromine addition to either the *cis*- or *trans*-alkene, two new chiral centers are being formed. Neither alkene (nor bromine) is optically active, so the product cannot be optically active.

The *cis*-but-2-ene gives two chiral products, a racemic mixture. However, *trans*-but-2-ene (shown on the next page), because of its symmetry, gives only one *meso* product, that can never be chiral. The "optical *in*activity" is built into this symmetric molecule.

This can be seen by following what happens to the configuration of the chiral centers from the intermediates to products, below. (The key lies in the symmetry of the intermediate and *inversion* of configuration when bromide attacks.)

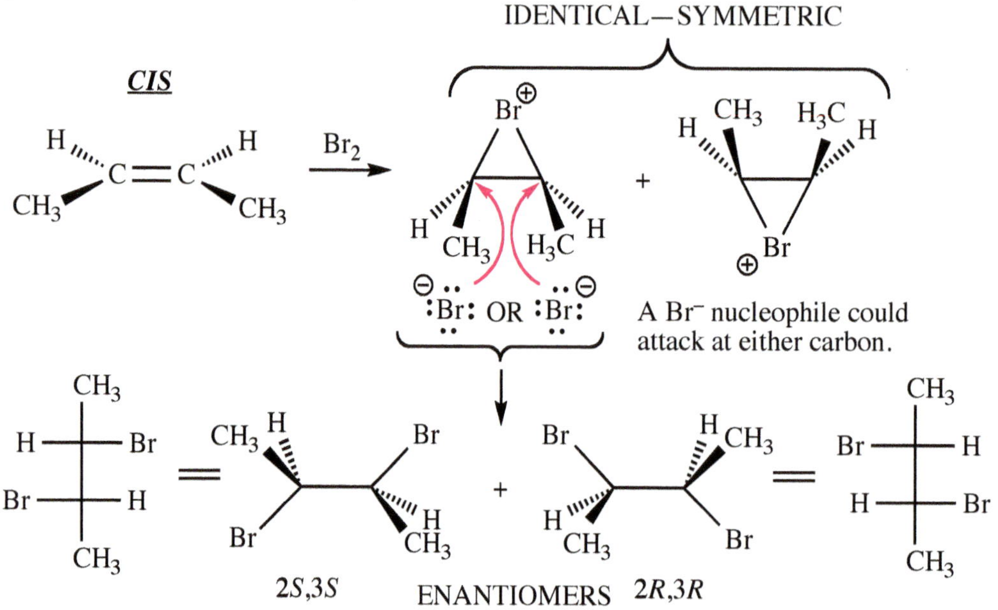

IDENTICAL—SYMMETRIC

CIS

A Br^- nucleophile could attack at either carbon.

2*S*,3*S* ENANTIOMERS 2*R*,3*R*

See the ***TRANS*** case on the next page.

8-17 continued

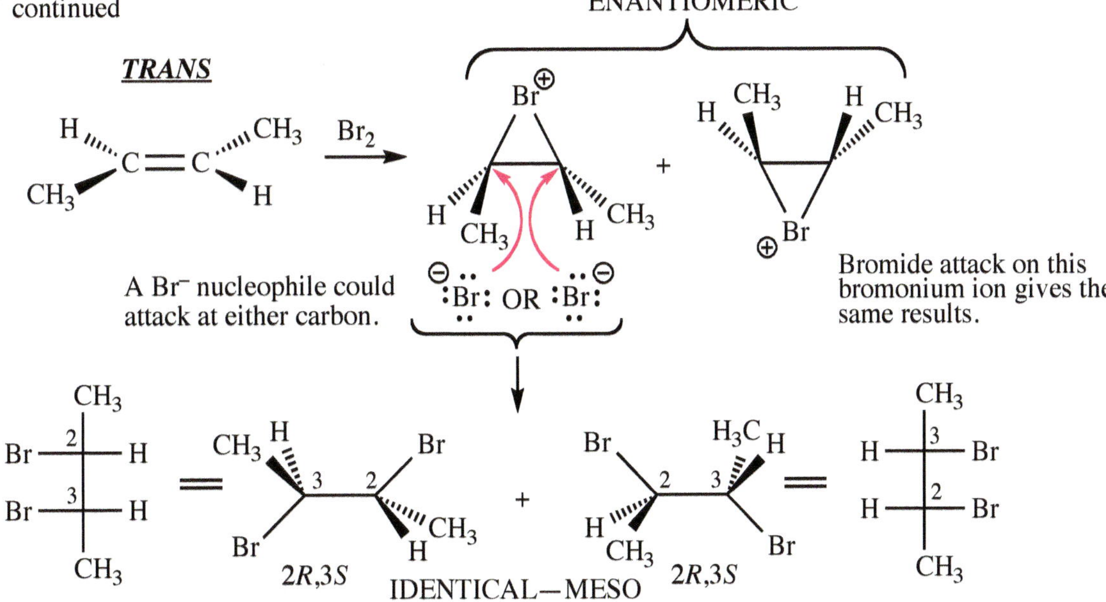

ENANTIOMERIC
TRANS
Br2
A Br⁻ nucleophile could attack at either carbon.
:Br: OR :Br:
Bromide attack on this bromonium ion gives the same results.
2R,3S
IDENTICAL—MESO
2R,3S

CONCLUSION: Anti addition of a symmetric reagent to a symmetric *cis*-alkene gives racemic product, while anti addition to a *trans*-alkene gives meso product. (We will see shortly that syn addition to a *cis*-alkene gives meso product, and syn addition to a *trans*-alkene gives racemic product. Stay tuned.)

8-18 Enantiomers of chiral products are also produced but not shown.

(a) + Br—Br ⟶ + :Br: ⟶ + the enantiomer

(b) Three new asymmetric carbons are produced in this reaction. All stereoisomers will be produced with the restriction that the two adjacent chlorines on the ring must be *trans*.

(c) The bromonium ion could form on either face of the C=C, and bromide will attack the other carbon of the bromonium ion as well. Either leads to equal amounts of the two enantiomers.

plus the enantiomer

8-18 continued

(d)

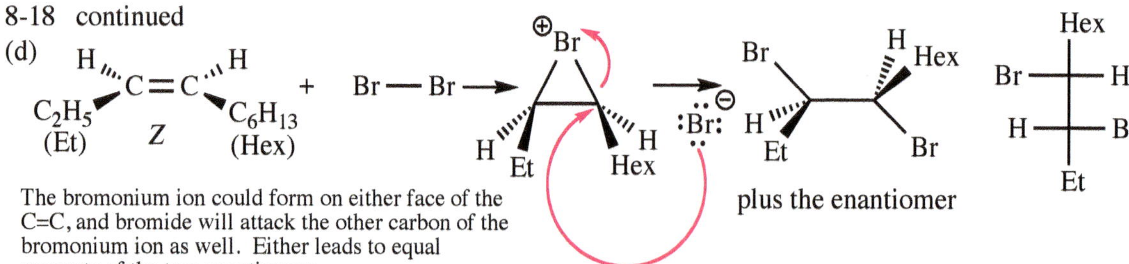

The bromonium ion could form on either face of the C=C, and bromide will attack the other carbon of the bromonium ion as well. Either leads to equal amounts of the two enantiomers.

plus the enantiomer

8-19 The *trans* product results from water attacking the bromonium ion from the face opposite the bromine. Equal amounts of the two enantiomers result from the equal probability that water will attack either C-1 or C-2.

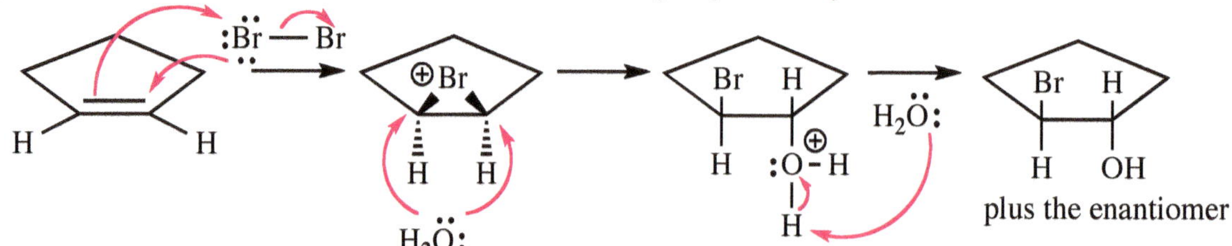

H_2O:

Water will do nucleophilic attack at either carbon.

plus the enantiomer

equal amounts of enantiomers = racemic mixture

8-20

from Solved Problem 8-5

The bromonium ion shown here is the enantiomer of the one shown in Solved Problem 8-5.

enantiomer of product in Solved Problem 8-5

from Solved Problem 8-6: The bromonium ion shown is meso as it has a plane of symmetry; attack by the nucleophile at the other carbon from what is shown in the text will create the enantiomer.

Show products from both nucleophiles attacking at this carbon from the "top" face.

These are the enantiomers of the structures shown in the text.

8-21 The chiral products shown here will be racemic mixtures.

(a) plus the enantiomer

(b) plus the enantiomer

(c) plus the enantiomer

(d) plus the enantiomer

(e) plus the enantiomers

8-22 (a) $\xrightarrow[H_2O]{Cl_2}$ Cl, OH

(b) Cl $\xrightarrow[\Delta]{KOH}$ $\xrightarrow[H_2O]{Cl_2}$ Cl, OH

plus the enantiomer

(c) CH_3, OH $\xrightarrow[\Delta]{H_2SO_4}$ CH_3 $\xrightarrow[H_2O]{Cl_2}$ CH_3, OH, Cl plus the enantiomer

8-23

(a) (b) (c) CH_3, H, H, CH_3 *cis* (d)

8-24 Limonene, $C_{10}H_{16}$, has three elements of unsaturation. Upon catalytic hydrogenation, the product, $C_{10}H_{20}$, has one element of unsaturation. Two elements of unsaturation have been removed by hydrogenation—these must have been pi bonds, either two double bonds or one triple bond. The one remaining unsaturation must be a ring. Thus, limonene must have one ring and either two double bonds or one triple bond. (The structure of limonene is shown in the text in Problem 8-23(d), and the hydrogenation product is shown above in the solution to 8-23(d).)

8-25 The BINAP ligand is an example of a conformationally hindered biphenyl as described in text section 5-9A and Figure 5-18. The groups are too large to permit rotation around the single bond connecting the rings, so the molecules are locked into one chiral twist or its mirror image.

This *simplified* three-dimensional drawing of one of the enantiomers shows that the two naphthalene rings are twisted almost perpendicular to each other. Molecular models will help visualize this concept.

Cl, Ru, Cl, P, P

8-26 Stereochemistry of the starting material is retained in the product. Products are racemic.

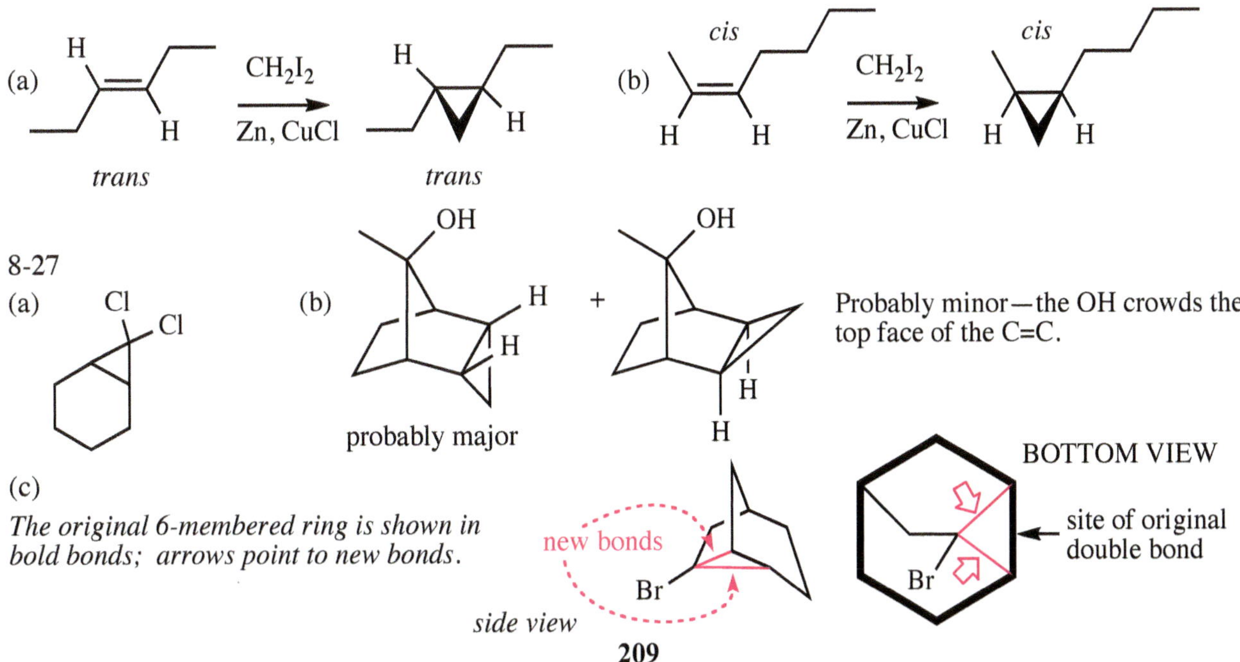

8-28

(a)

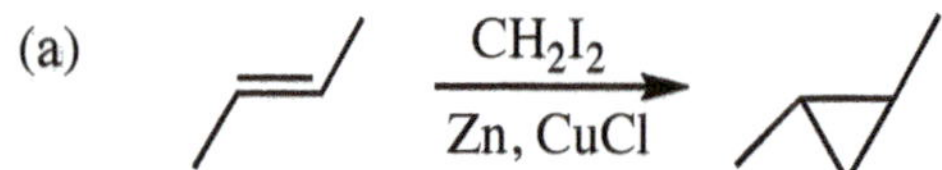

(b)

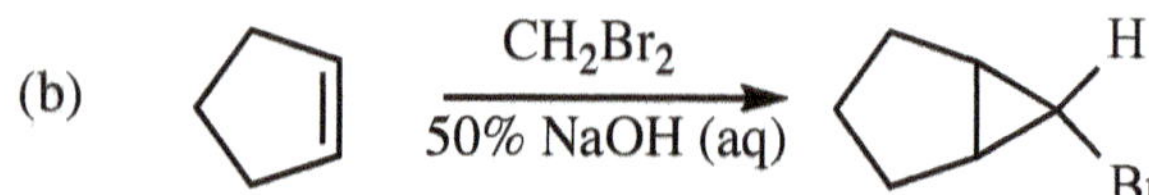

(c)

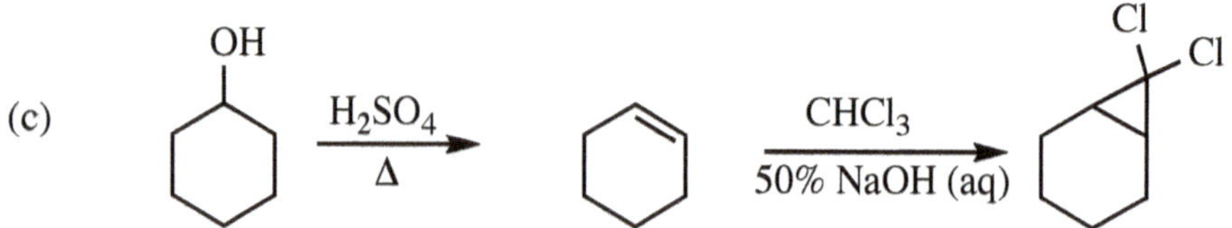

8-29 All chiral products in this problem are racemic mixtures.

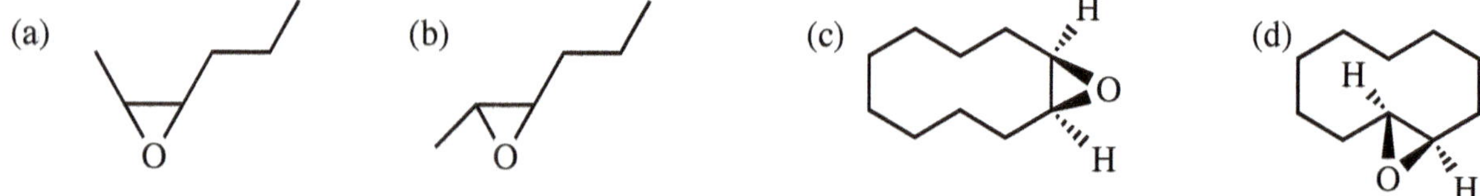

8-30 HA is a generic acid catalyst. It could be RCOOH produced in the epoxidation.

(a)

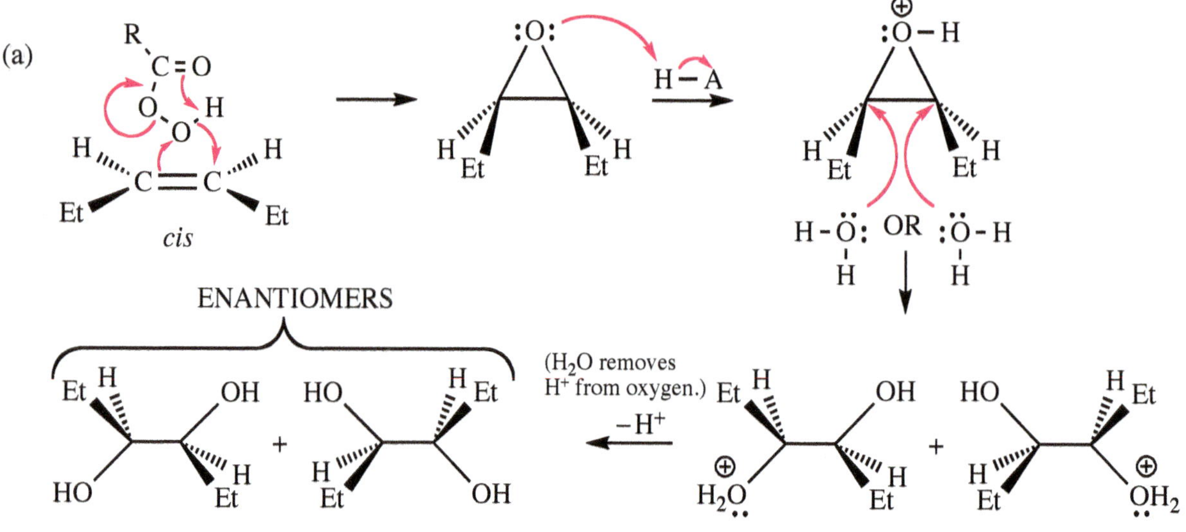

(b)

mechanism continued on next page

8-30 continued IDENTICAL—MESO

from previous page

(H_2O removes H^+ from oxygen.) $- H^+$

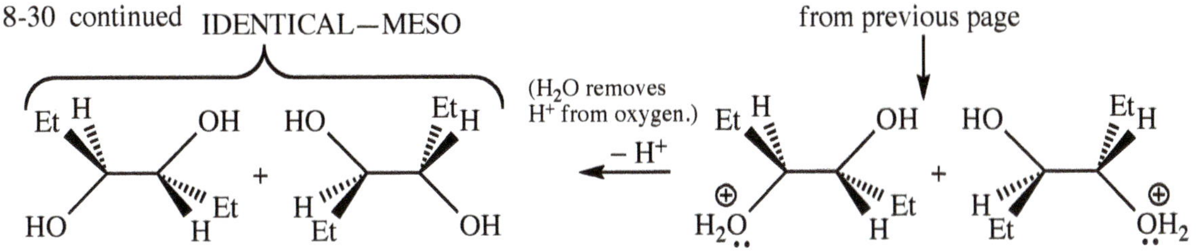

stereochemistry shown in Newman projections:

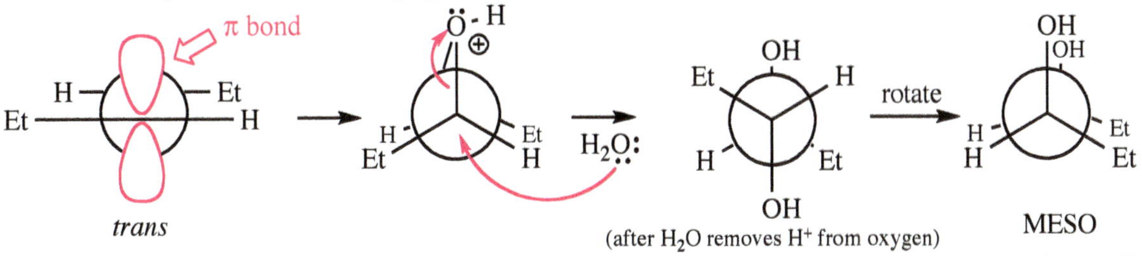

π bond

trans

rotate

(after H_2O removes H^+ from oxygen)

MESO

Remember the lesson from Problem 8-17: anti addition of a symmetric reagent to a symmetric *cis*-alkene gives racemic product, while anti addition to a *trans*-alkene gives meso product. This fits the definition of a *stereospecific* reaction, where different stereoisomers of the starting material (*cis* and *trans*) are converted into different stereoisomers of product (a *dl-pair* and *meso form*).

8-31

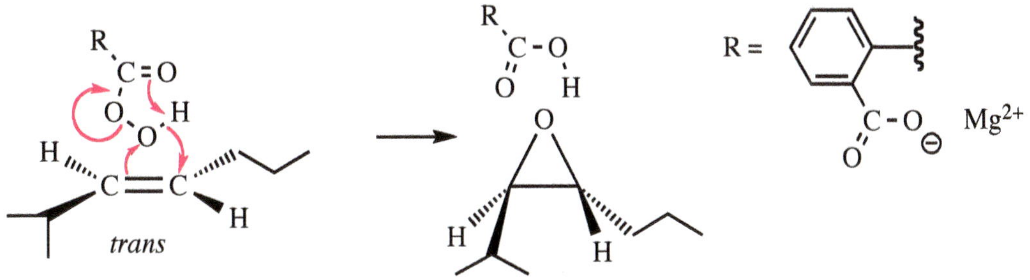

trans

R = ... Mg^{2+}

8-32 All chiral products in this problem are racemic mixtures.

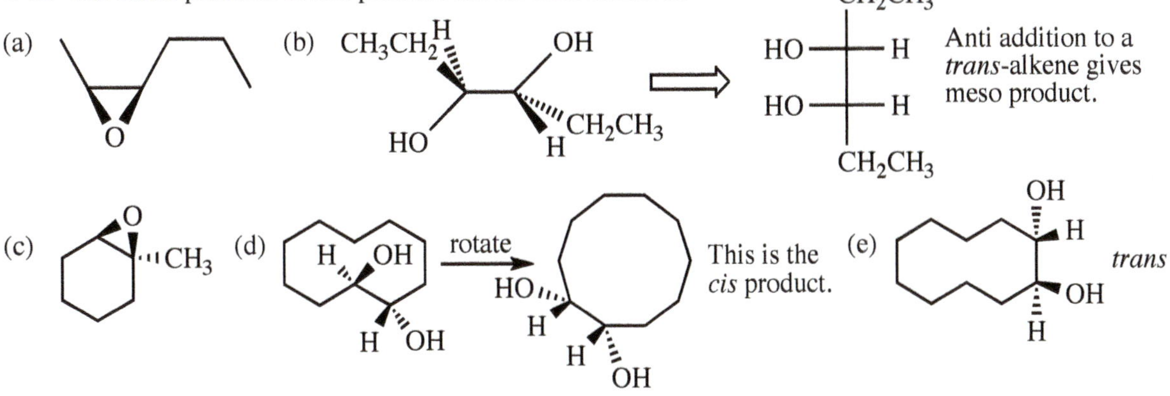

(a)

(b) Anti addition to a *trans*-alkene gives meso product.

(c)

(d) rotate — This is the *cis* product.

(e) *trans*

8-33

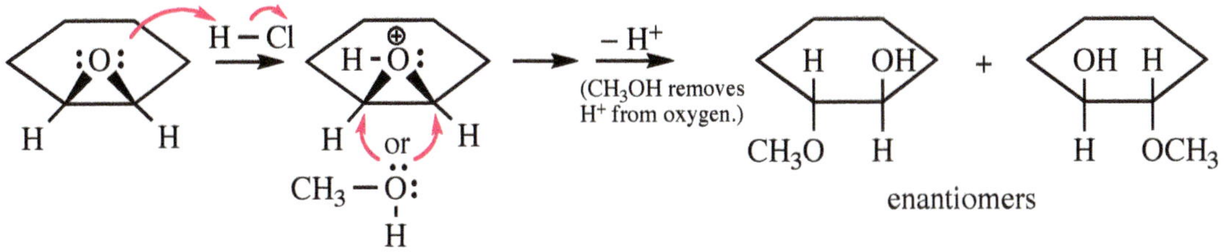

$- H^+$ (CH_3OH removes H^+ from oxygen.)

enantiomers

8-34 All these reactions begin with achiral reagents; therefore, all the chiral products are racemic.

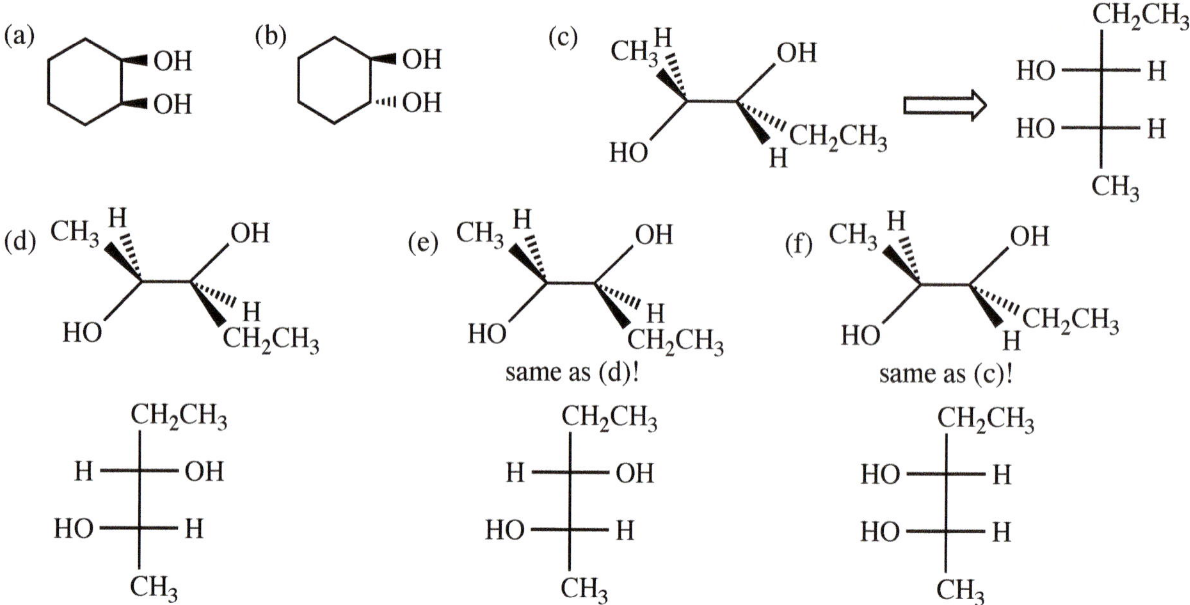

Refer to the observation in the solution to Problem 8-35.

8-35

(a) $\xrightarrow[H_2O_2]{OsO_4}$ meso

(b) $\xrightarrow[H_2O]{CH_3CO_3H}$ racemic (*d*,*l*)

(c) $\xrightarrow[H_2O]{CH_3CO_3H}$ $\xrightarrow{\text{rotate}}$ meso

(d) $\xrightarrow[H_2O_2]{OsO_4}$ $\xrightarrow{\text{rotate}}$ racemic (*d*,*l*)

Have you noticed yet? For symmetric alkenes and symmetric reagents (addition of two identical X groups):

cis-alkene + **syn** addition → meso
cis-alkene + **anti** addition → racemic
trans-alkene + **syn** addition → racemic
trans-alkene + **anti** addition → meso

Assume that *cis*/syn/meso are "same", and *trans*/anti/racemic are "opposite". Then any combination can be predicted, just like math!

$+1 \times +1 = +1$
$+1 \times -1 = -1$
$-1 \times +1 = -1$
$-1 \times -1 = +1$

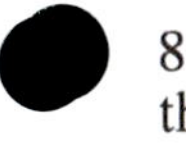

8-36 Solve these ozonolysis problems by working backwards, that is, by "reattaching" the two carbons of the new carbonyl groups into alkenes. Here's a hint. When you cut a circular piece of string, you still have only one piece. When you cut a linear piece of string, you have two pieces. Same with molecules. If ozonolysis forms only one product with two carbonyls, the C=C had to have been in a ring. If ozonolysis gives two molecules, the C=C had to have been in a chain.

(a) Two carbonyls from ozonolysis are in a chain, so alkene had to have been in a ring.

(b) Two carbonyls from ozonolysis are in two different products, so alkene had to have been in a chain, not a ring.

(c) Two carbonyls from ozonolysis are in two different products, so alkene had to have been in a chain, not a ring.

E or *Z* of alkene cannot be determined from products.

8-37

(a) $\xrightarrow[-78\ °C]{O_3}$ $\xrightarrow{Me_2S}$ O + O H

(b) $\xrightarrow[\Delta]{KMnO_4}$ O + O OH

(c) $\xrightarrow[-78\ °C]{O_3}$ $\xrightarrow{Me_2S}$ =O + O=

(d)  $\xrightarrow[-78\ °C]{O_3}$ $\xrightarrow{Me_2S}$ O O H

(e) 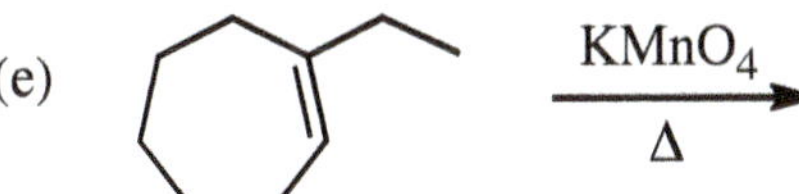 $\xrightarrow[\Delta]{KMnO_4}$ O O OH

(f) 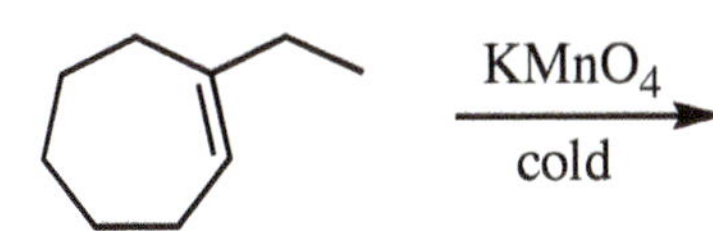 $\xrightarrow[\text{cold}]{KMnO_4}$ OH OH OH groups are *cis* ; product is racemic.

8-38 The representation for a generic acid is H—A, where A is the conjugate base.

(a) H $\xrightarrow{H-A}$ H H ⊕C H → H H ⊕ C H A:⊖ →

8-38 continued

(b) Catalytic BF_3 reacts with trace amounts of water to form the probable catalyst:

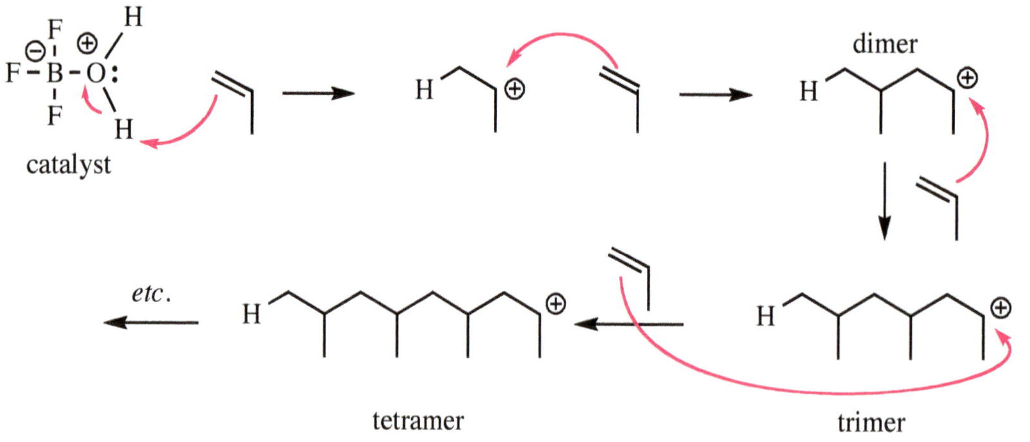

catalyst

dimer

etc.

tetramer

trimer

8-39 H—A symbolizes a generic acid.

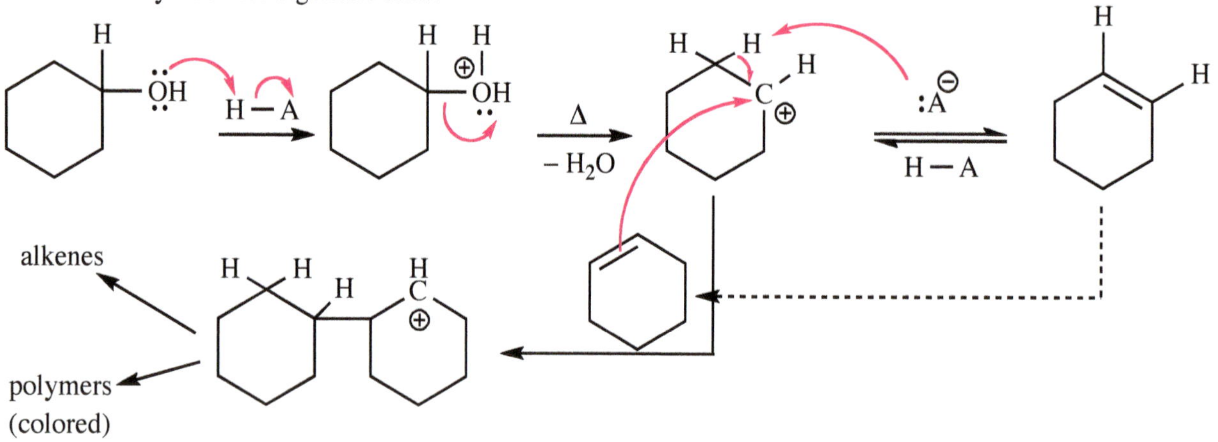

Δ

$- H_2O$

H—A

alkenes

polymers (colored)

Note: The wavy bond symbol is used in the following solutions to indicate the continuation of a polymer chain.

8-40

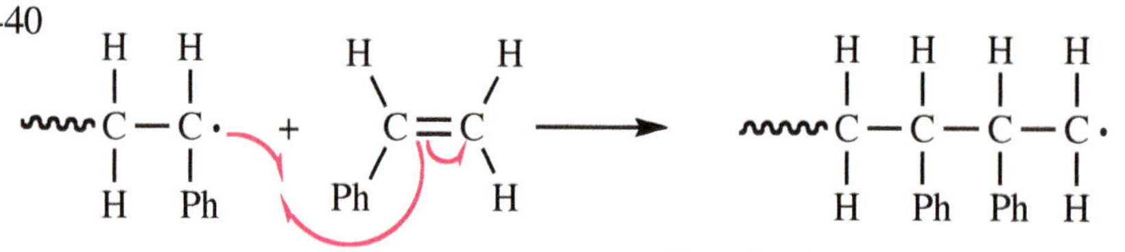

1° radical, and *not* resonance-stabilized—
this orientation is not observed

Orientation of addition always generates the more stable intermediate; the energy difference between a 1° radical (shown above) and a benzylic radical is huge. The phenyl substituents must necessarily be on alternating carbons because the orientation of attack is always the same—not a random process.

8-41 Each monomer has two carbons in the backbone, so the substituents on the monomer will repeat every two carbons in the polymer. Wavy bonds indicate continuation of the polymer chain.

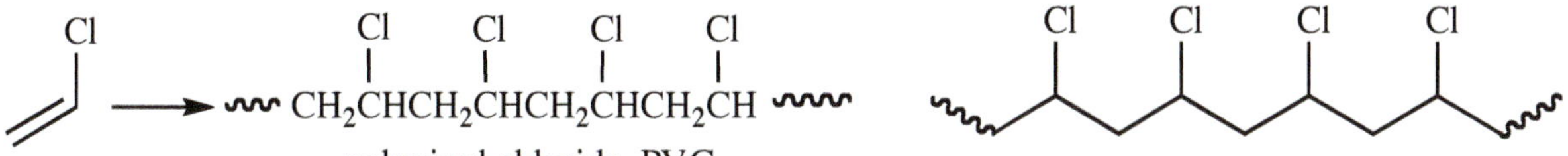

$CH_2{=}CHCl \longrightarrow$ ~ $CH_2CH(Cl)CH_2CH(Cl)CH_2CH(Cl)CH_2CH(Cl)$ ~

polyvinyl chloride, PVC

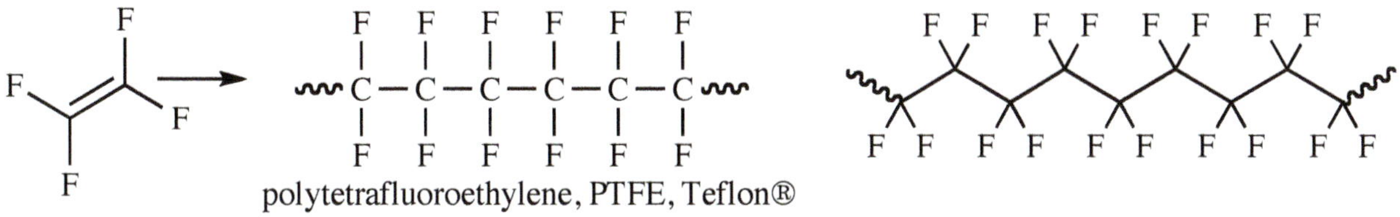

$CF_2{=}CF_2 \longrightarrow$ ~ $CF_2{-}CF_2{-}CF_2{-}CF_2{-}CF_2{-}CF_2$ ~

polytetrafluoroethylene, PTFE, Teflon®

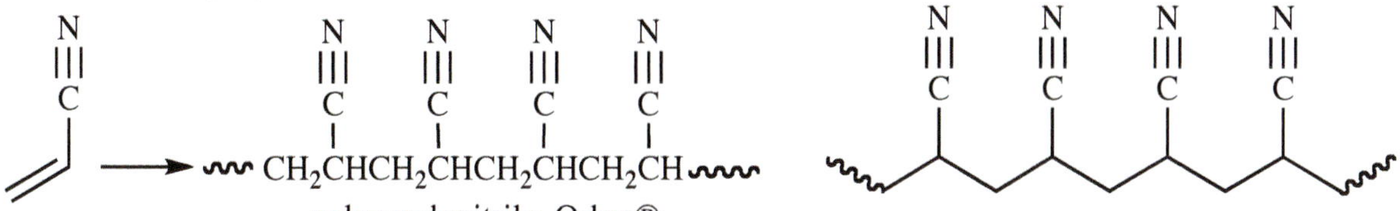

$CH_2{=}CHC{\equiv}N \longrightarrow$ ~ $CH_2CH(C{\equiv}N)CH_2CH(C{\equiv}N)CH_2CH(C{\equiv}N)CH_2CH(C{\equiv}N)$ ~

polyacrylonitrile, Orlon®

8-42

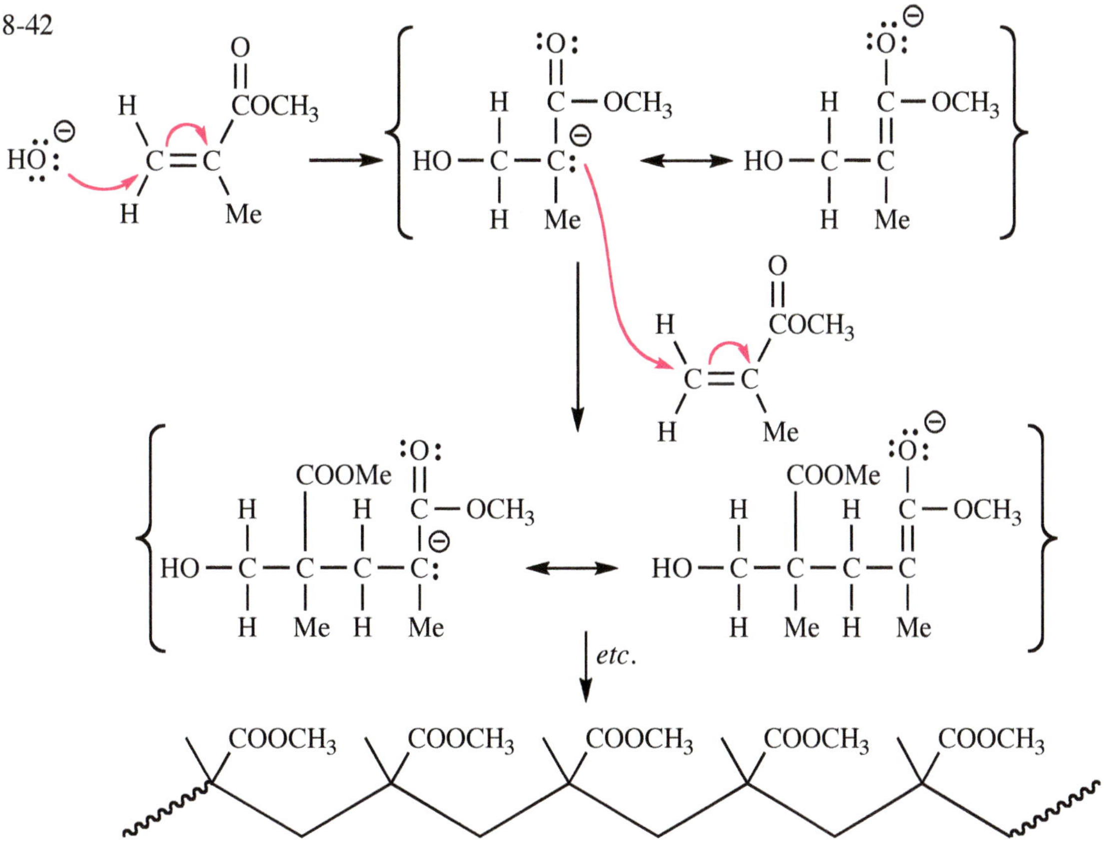

$HO:^{\ominus}$ + $H_2C{=}C(Me)COCH_3$ (C=O) $\longrightarrow$ { $HO{-}CH_2{-}\overset{\ominus}{C}(Me){-}C(=O){-}OCH_3 \longleftrightarrow HO{-}CH_2{-}C(Me){=}C(O^{\ominus}){-}OCH_3$ }

+ $H_2C{=}C(Me)COOCH_3$ $\longrightarrow$

{ $HO{-}CH_2{-}C(Me)(COOMe){-}CH_2{-}\overset{\ominus}{C}(Me){-}C(=O){-}OCH_3 \longleftrightarrow HO{-}CH_2{-}C(Me)(COOMe){-}CH_2{-}C(Me){=}C(O^{\ominus}){-}OCH_3$ }

$\downarrow$ *etc.*

~ $[CH_2{-}C(CH_3)(COOCH_3)]_n$ ~

Wavy bonds mean that the chain continues.

Plexiglas®

8-43 The accepted mechanism of olefin metathesis includes an intermediate with a four-membered ring where one of the atoms in the ring is the metal catalyst, abbreviated [M]. All steps are equilibria.

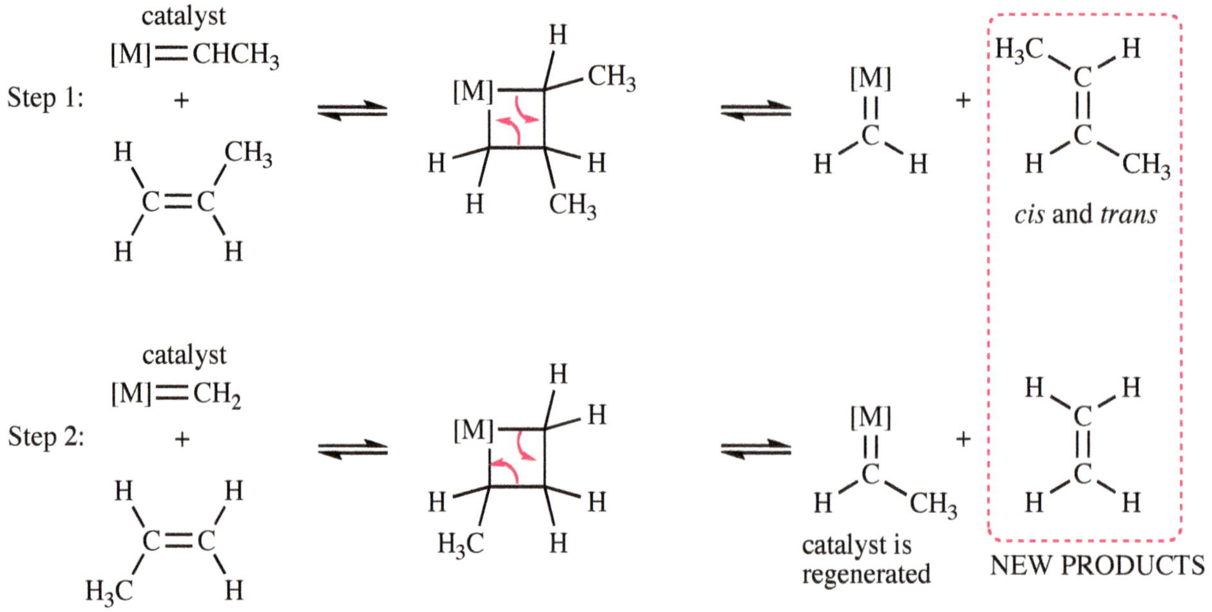

Step 1: catalyst $[M]{=}CHCH_3$ + $CH_3CH{=}CH_2$ ⇌ [metallacyclobutane intermediate] ⇌ $[M]{=}CH_2$ + $CH_3CH{=}CHCH_3$ (*cis* and *trans*)

Step 2: catalyst $[M]{=}CH_2$ + $CH_3CH{=}CH_2$ ⇌ [metallacyclobutane intermediate] ⇌ $[M]{=}CHCH_3$ (catalyst is regenerated) + $CH_2{=}CH_2$

NEW PRODUCTS

8-44 An excellent discussion of the olefin metathesis reaction is available at the Nobel Institute website:
http://www.nobelprize.org/nobel_prizes/chemistry/laureates/2005/ then click on Advanced Information

The catalyst needs to have the short end of the molecule attached to the metal:

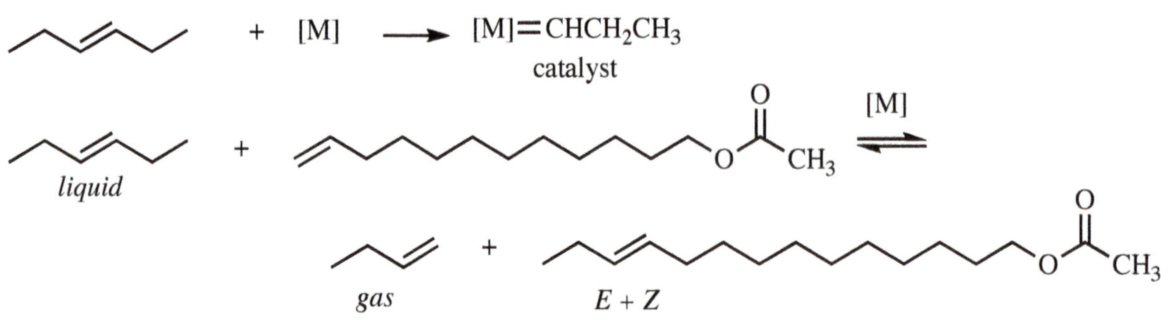

3-hexene + [M] → $[M]{=}CHCH_2CH_3$ catalyst

3-hexene (*liquid*) + undec-10-enyl acetate $\xrightleftharpoons{[M]}$ propene (*gas*) + tridec-10-enyl acetate (*E* + *Z*)

Unnumbered problem in the Problem-Solving Strategy section:

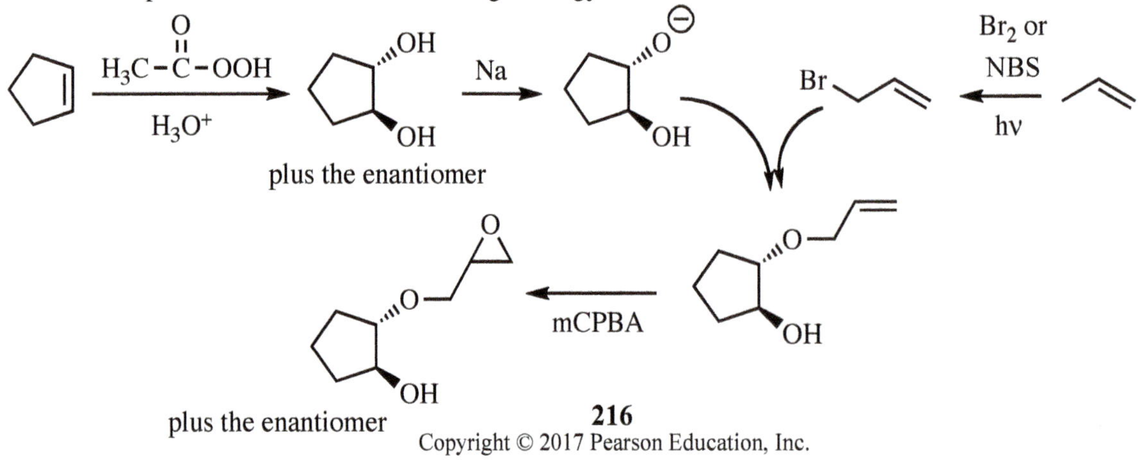

cyclopentene $\xrightarrow[H_3O^+]{H_3C{-}C(=O){-}OOH}$ *trans*-cyclopentane-1,2-diol (plus the enantiomer) $\xrightarrow{Na}$ alkoxide; propene $\xrightarrow[h\nu]{Br_2 \text{ or NBS}}$ Br–$CH_2CH{=}CH_2$; alkoxide + allyl bromide → 2-(allyloxy)cyclopentanol $\xrightarrow{mCPBA}$ 2-(glycidyloxy)cyclopentanol (plus the enantiomer)

8-45 In the spirit of this problem, all starting materials will have six carbons or fewer and only one carbon-carbon double bond. Reagents may have other atoms.

(a)

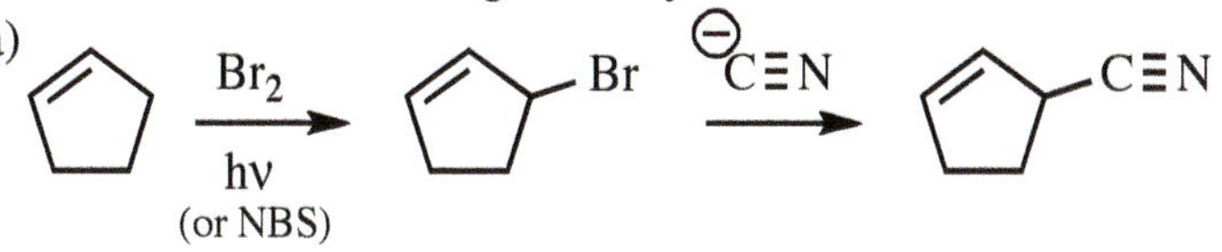

Br_2, hν (or NBS) → Br; $^{\ominus}C\equiv N$ → $C\equiv N$

(b)

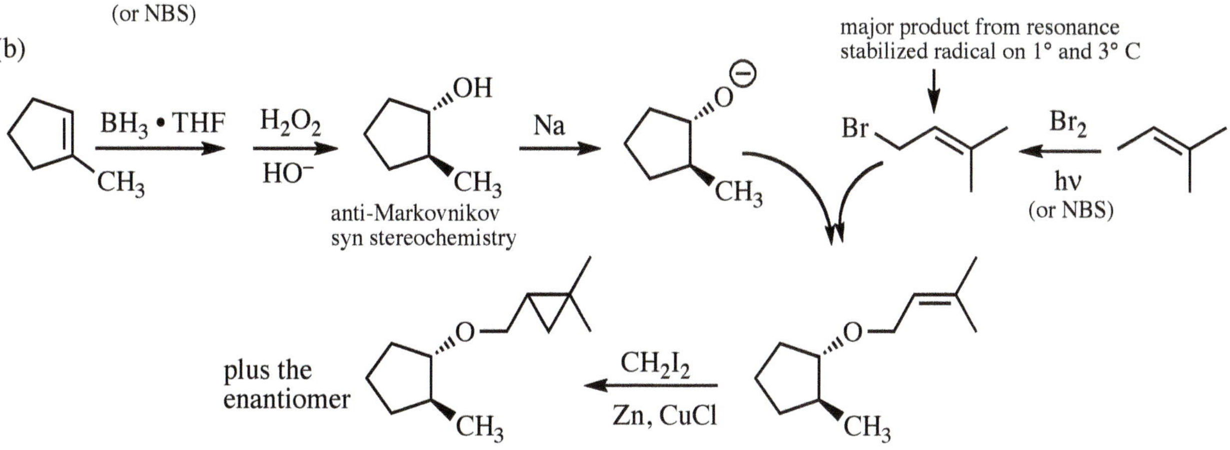

$BH_3 \bullet THF$; H_2O_2, HO^- → OH, CH_3 — anti-Markovnikov syn stereochemistry; Na → $O^{\ominus}$, CH_3

major product from resonance stabilized radical on 1° and 3° C

Br ← Br_2, hν (or NBS)

plus the enantiomer — O, CH_3 ← CH_2I_2, Zn, CuCl — O, CH_3

(c) Br_2 → Br, Br; KOH, Δ →; Br_2, hν (or NBS) → Br; KOH → OH — Only substitution is possible.

Na → $O^{\ominus}$

mCPBA → O

OH, O ← H_2O

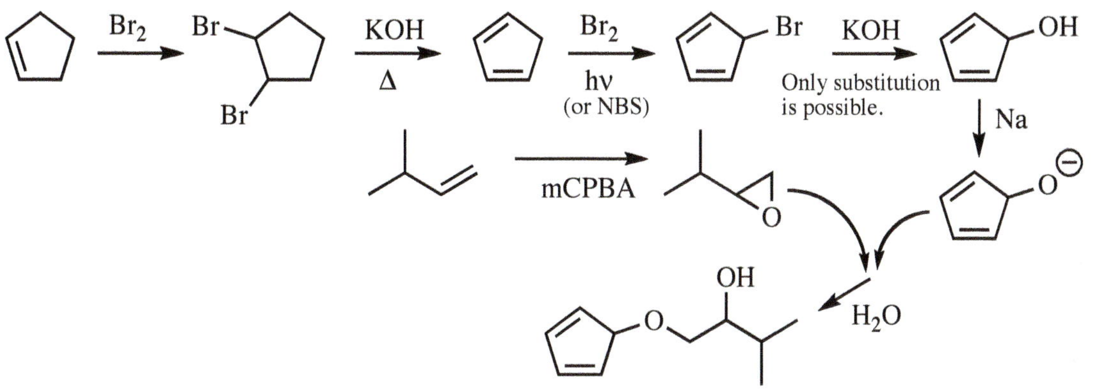

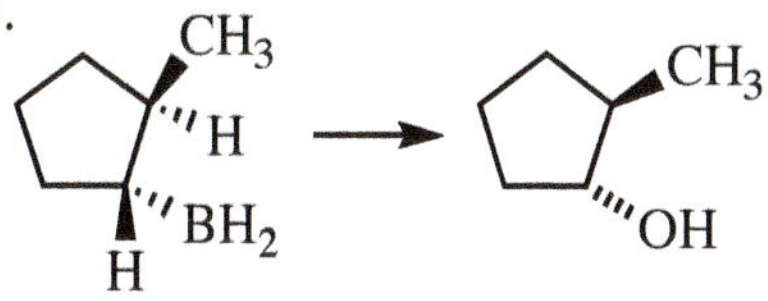

8-46 All chiral products in this problem are racemic mixtures.

(a) Cl, H

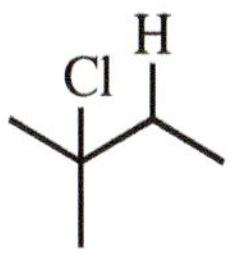

(b) CH_3, Br, Br

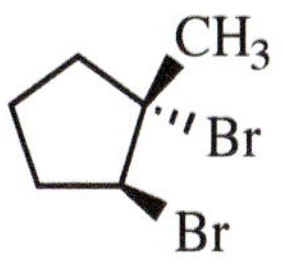

(c) CH_3, H, BH_2, H — intermediate → CH_3, OH

(d) H, O + O

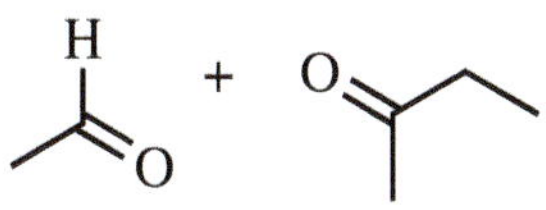

(e) Br

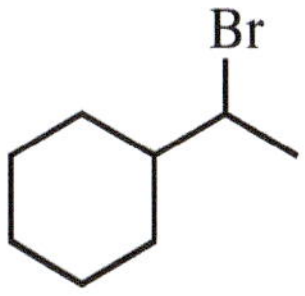

(f) Cl — Peroxides do not affect HCl addition.

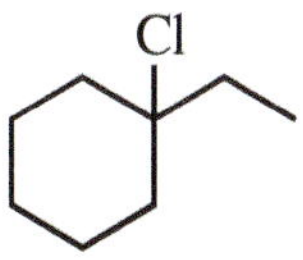

(g) O — *trans*

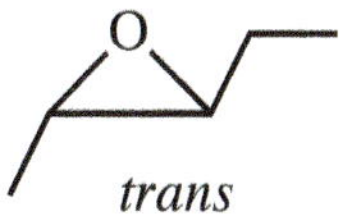

(h) CH_3, OH, OH

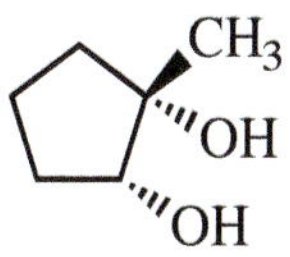

(i) CH_3, OH, OH

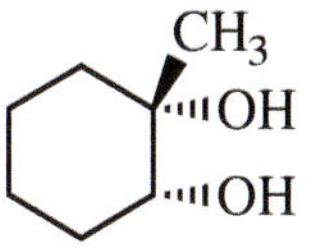

(j) CH_3, OH, OH

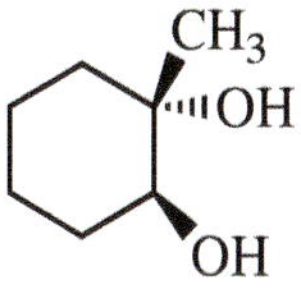

(k) O, O, OH

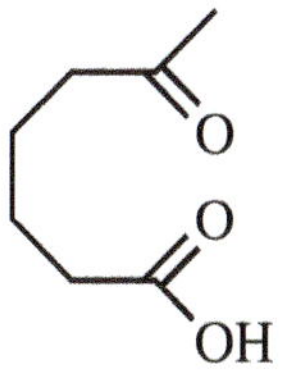

8-46 continued

(l) 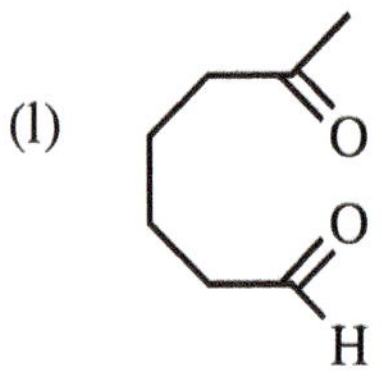

(m) (n) (o) $+ CH_2{=}CH_2$

(p) $\xrightarrow{NaBH_4}$

intermediate

(q)

8-47

(a) (b) (c) $+ CH_2{=}O$

(d) (e) $+ CO_2$ (f)

(g) (h) (i)

(j) (k) (l)

(m)

8-48 Recall these facts about ozonolysis: each alkene cleaved by ozone produces two carbonyl groups; an alkene in a chain produces two separate products; an alkene in a ring produces one product in which the two carbonyls are connected.

(a) $+ CH_2{=}O$ (b) +

8-48 continued

(c)

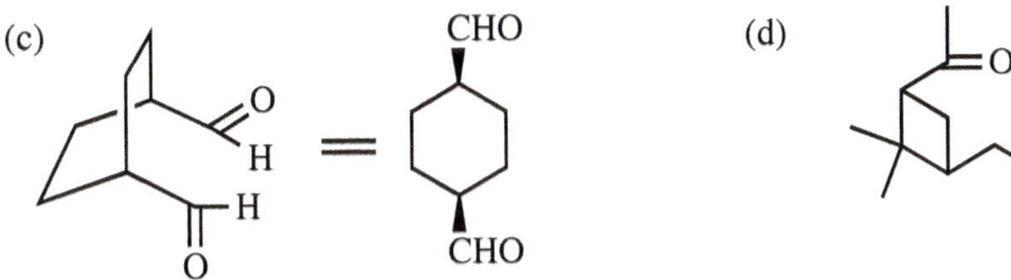

(d) O H O

8-49 Chiral products in this problem are racemic.

(a)

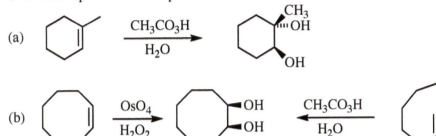

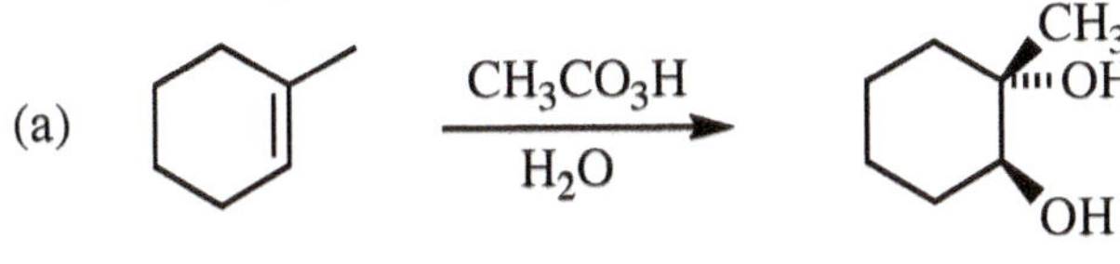

(b) OsO_4 / H_2O_2 (or cold, dilute $KMnO_4$) → OH OH ← CH_3CO_3H / H_2O

cis + **syn**
(*cis* double bond plus
syn stereochemistry of addition)

trans + **anti**
(*trans* double bond plus
anti stereochemistry of addition)

(c)

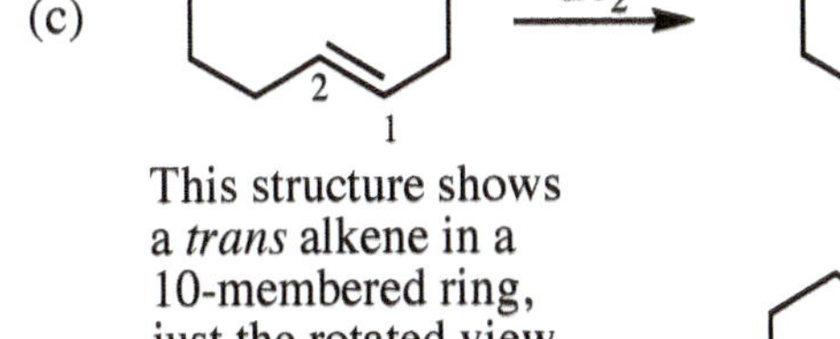

Br_2 → 1 Br 2 Br

Anti addition of Br_2 *requires trans* alkene to give meso product.

This structure shows a *trans* alkene in a 10-membered ring, just the rotated view of the structure to the right.

rotate around C-2

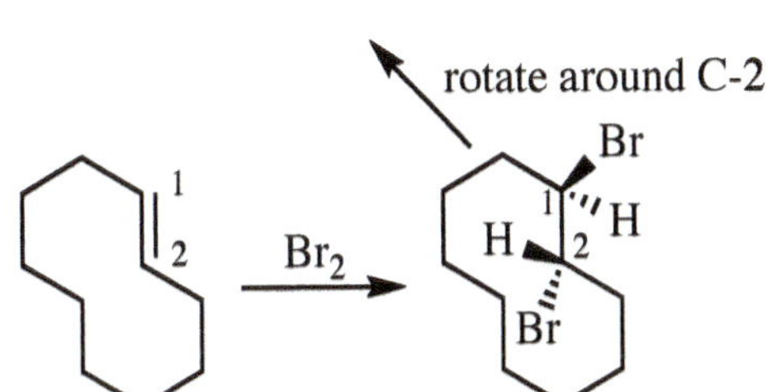

trans-cyclodecene

Br H 1 H 2 Br

(d) Cl_2 / H_2O → Cl OH

(e) $BH_3 \bullet THF$ → H_2O_2 / HO^- → H OH

(f) or or → $Hg(OAc)_2$ / CH_3OH → $NaBH_4$ → OCH_3

8-50

(a) Br_2 , hν or NBS → Br

Even though this product comes from an allylic radical intermediate with the possibility of two products, the major product is this one, with the double bond in the more stable tetrasubstituted position.

8-50 continued

(b) 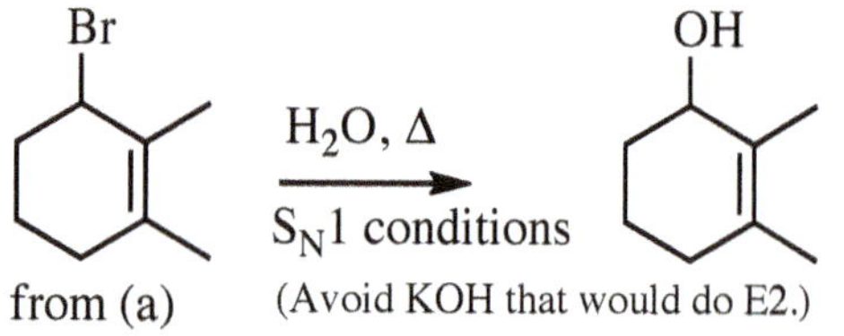 from (a) — Br — H_2O, Δ / S_N1 conditions → OH product
(Avoid KOH that would do E2.)

(c) 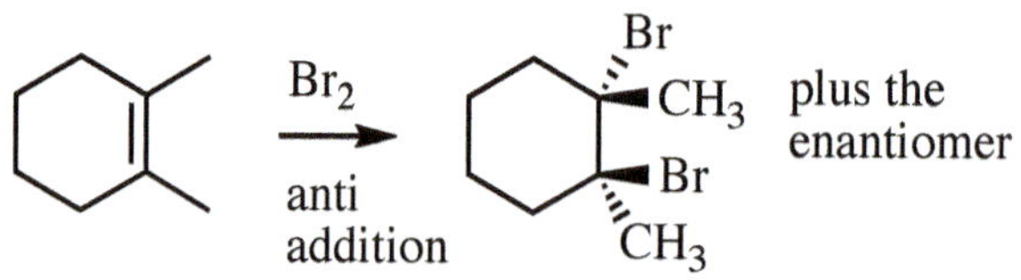 Br_2 / anti addition → Br, CH_3, Br, CH_3 plus the enantiomer

(d) 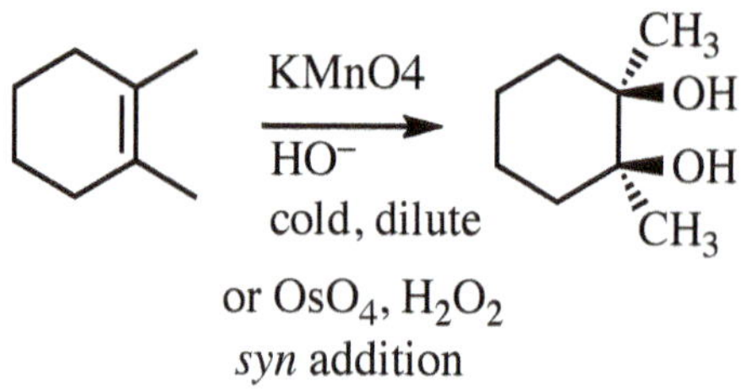 KMnO4 / HO^- / cold, dilute → CH_3, OH, OH, CH_3
or OsO_4, H_2O_2
syn addition

(e) 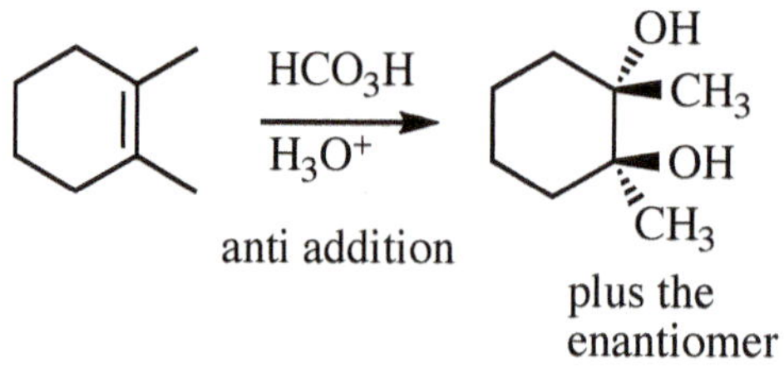 HCO_3H / H_3O^+ → OH, CH_3, OH, CH_3
anti addition
plus the enantiomer

(f) 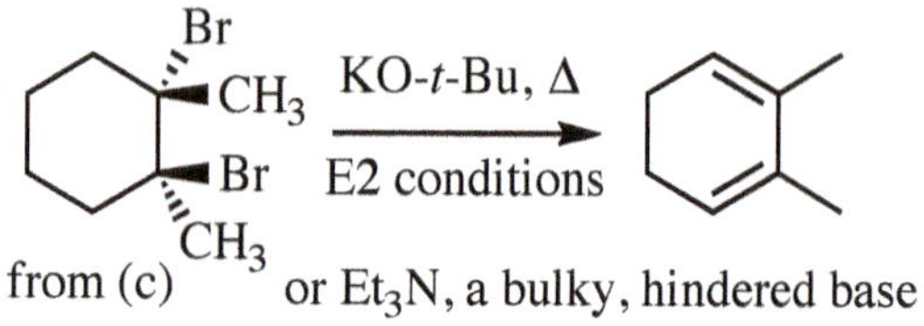 Br, CH_3, Br, CH_3 — KO-*t*-Bu, Δ / E2 conditions →
from (c)
or Et_3N, a bulky, hindered base

(g) 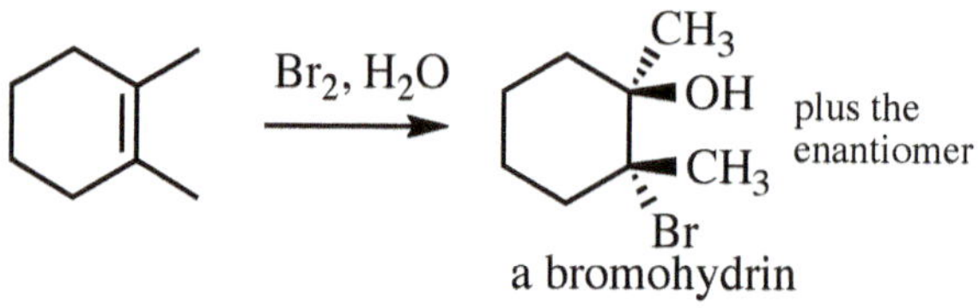 Br_2, H_2O → CH_3, OH, CH_3, Br plus the enantiomer
a bromohydrin

(h) CH_3, OH, CH_3, Br — KOH, Δ / E2 conditions → CH_3, OH
from (g)

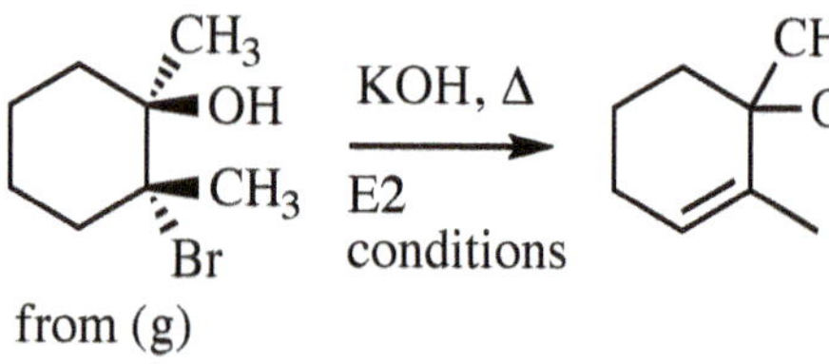

(i) 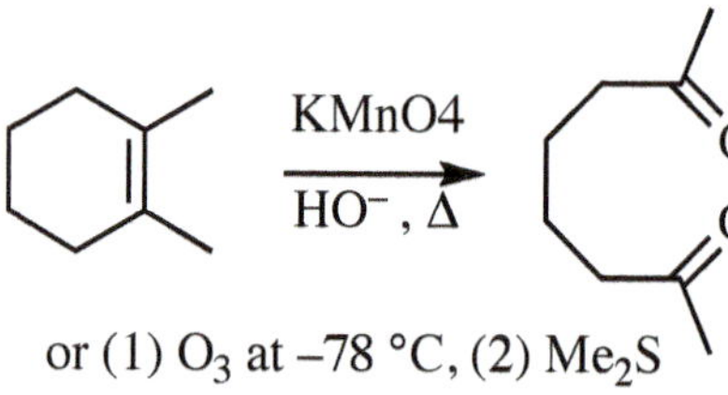 KMn04 / HO^-, Δ → O, O
or (1) O_3 at –78 °C, (2) Me_2S

8-51

(a) 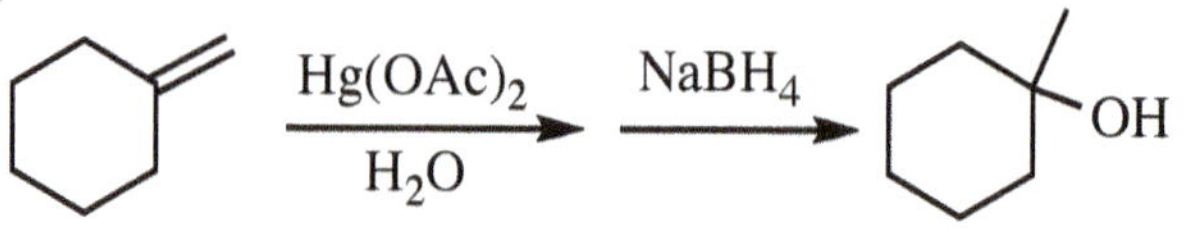 $Hg(OAc)_2$ / H_2O → $NaBH_4$ → OH

(b)  HBr / ROOR → Br

(c) 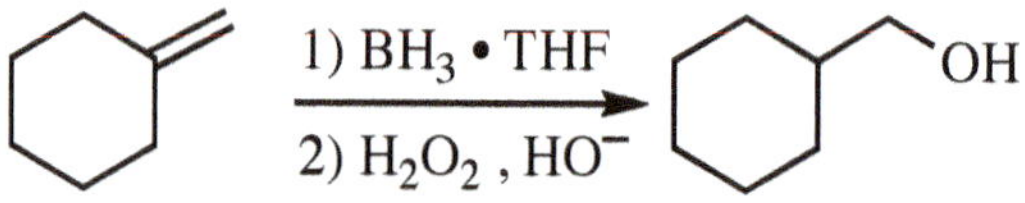 1) BH_3 • THF / 2) H_2O_2 , HO^- → OH

(d) 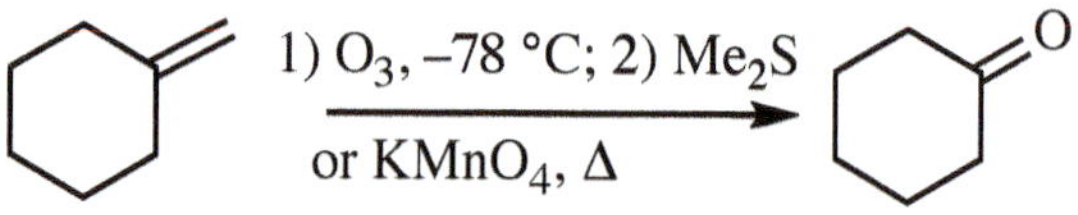 1) O_3, –78 °C; 2) Me_2S / or $KMnO_4$, Δ → O

(e) 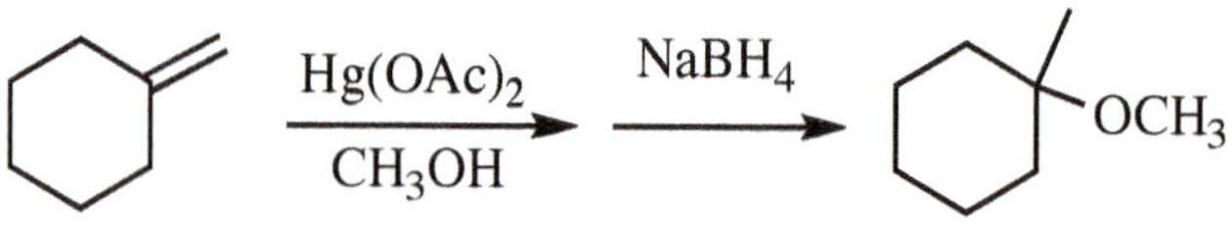 $Hg(OAc)_2$ / CH_3OH → $NaBH_4$ → OCH_3

(f) 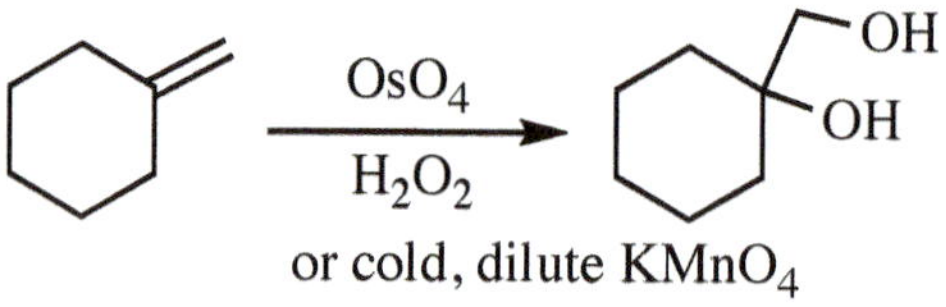 OsO_4 / H_2O_2 → OH, OH
or cold, dilute $KMnO_4$

(g) 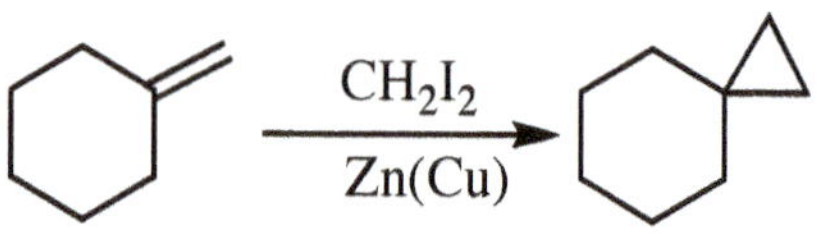 CH_2I_2 / Zn(Cu) →

(h) Cl_2 / H_2O → Cl, OH

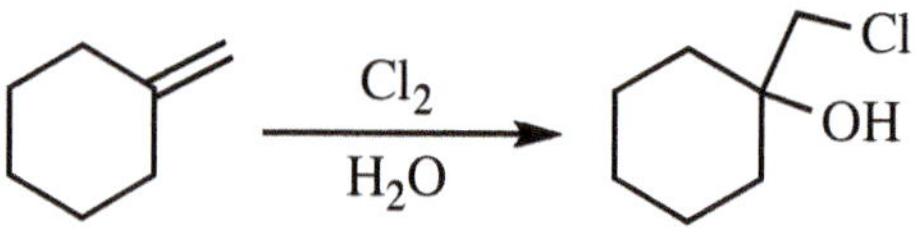

(i) 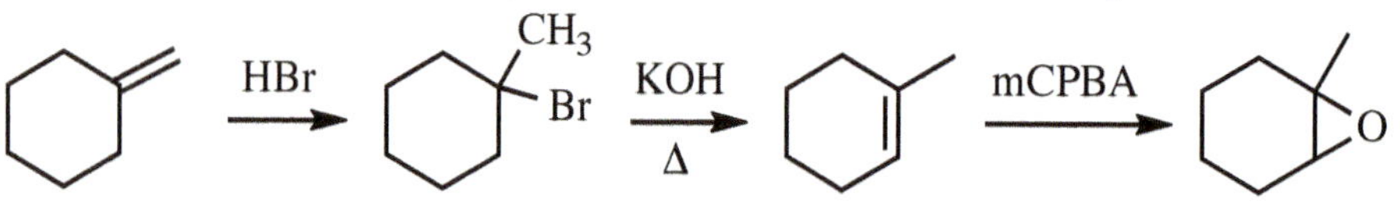 HBr → CH_3, Br → KOH / Δ → mCPBA → O

Alternatively, hydration followed by dehydration to methylcyclohexene would give the same product.

8-52

(a)

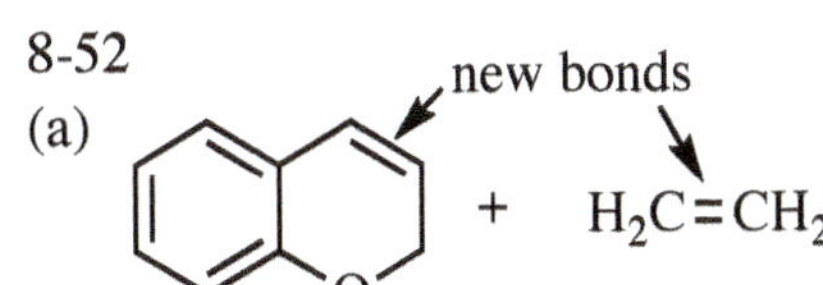

(b) HO, OCH_3, OH; new bonds; + $H_2C{=}CHCH_2OH$

(c) new bonds; + $H_2C{=}CH_2$

8-53

(a) 2 —[M]=CHR→ Z + E; + $H_2C{=}CH_2$

(b) + —[M]=CHR→ Z + E, Z + E

Ring-opening metathesis: the 5 carbons from cyclopentene become the middle 5 carbons of the product.

8-54

(a) O, O, H

(b) *t*-Bu, O; + $O{=}CH_2$

(c) H, $(CH_2)_7COCH_3$, O, O; + O, H, $(CH_2)_7CH_3$

(d) O, O; + $O{=}CH_2$

8-55 Chiral products are racemic mixtures.

(a) Br_2 / H_2O → **A** bromohydrin (Br, OH) — excess NH_3 / S_N2 → **B** (NH_2, OH)

(b) $Hg(OAc)_2$ / H_2O, $NaBH_4$ → **C** (OH) — H_2SO_4 / Δ → **D** — OsO_4 / H_2O_2 → **E** (OH, OH)

8-55 continued

(c)

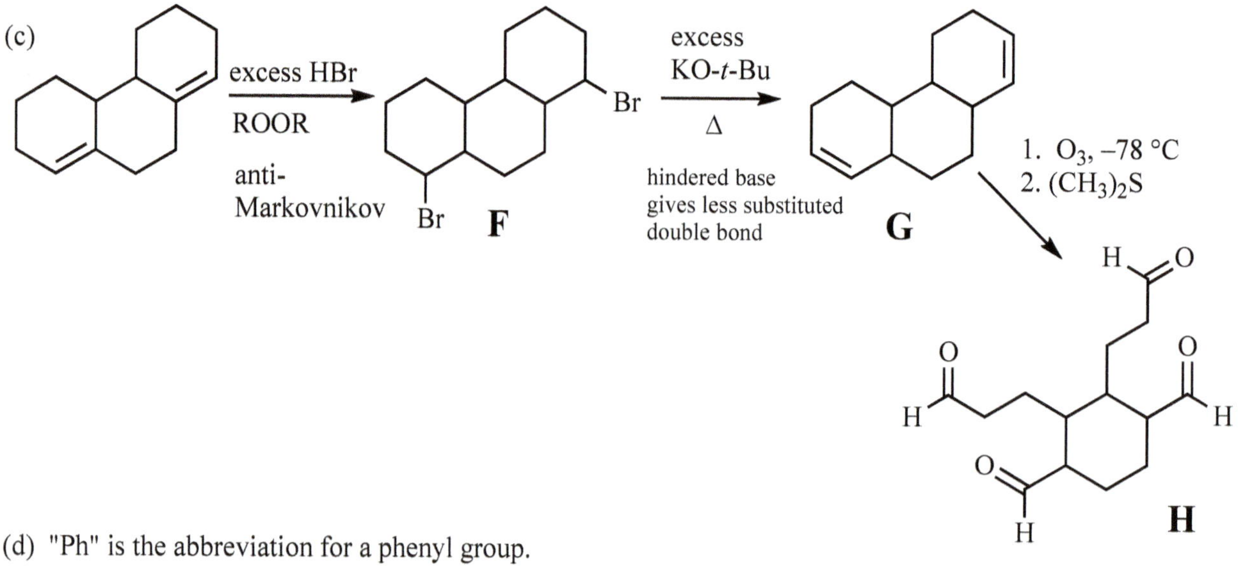
excess HBr
ROOR
anti-
Markovnikov
Br
Br
F
excess
KO-*t*-Bu
Δ
hindered base
gives less substituted
double bond
G
1. O3, –78 °C
2. (CH3)2S
H
O
O
O
O
H
H
H
H

(d) "Ph" is the abbreviation for a phenyl group.

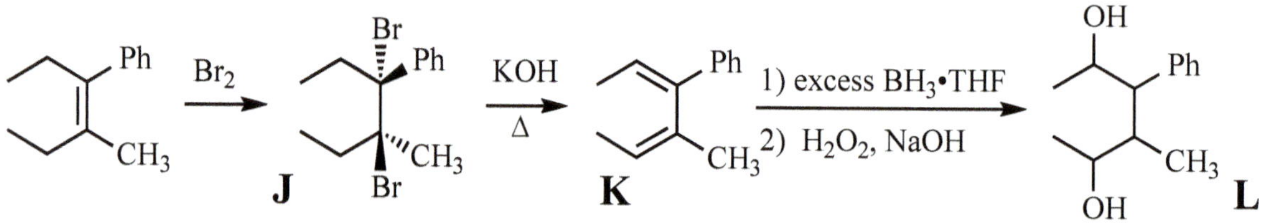
Ph
CH3
Br2
Br
Ph
CH3
Br
J
KOH
Δ
Ph
CH3
K
1) excess BH3•THF
2) H2O2, NaOH
OH
Ph
CH3
OH
L

8-56 (a)

initiation

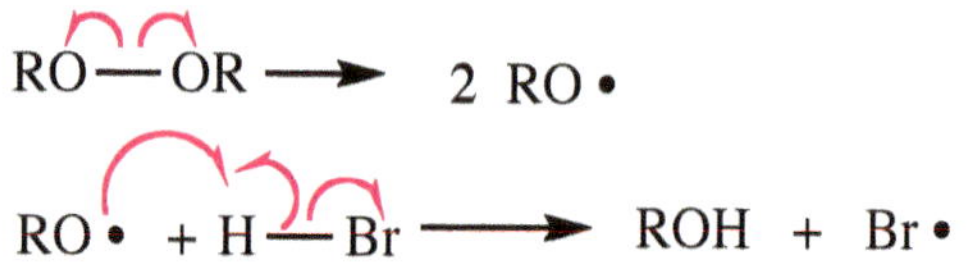
RO—OR
2 RO •
RO • + H—Br
ROH + Br •

propagation

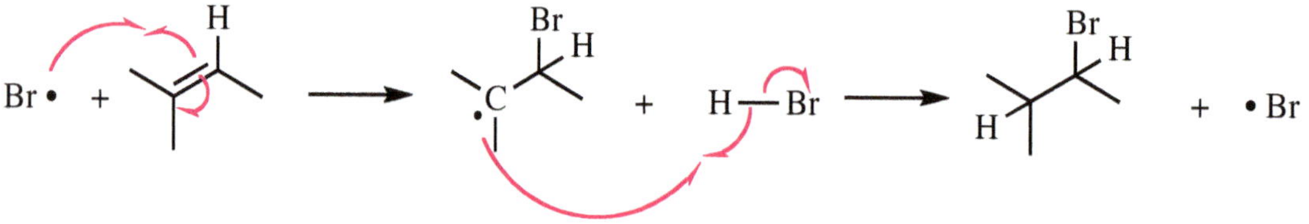
Br • +
H
Br
H
C
+
H—Br
Br
H
H
+ • Br

(b)

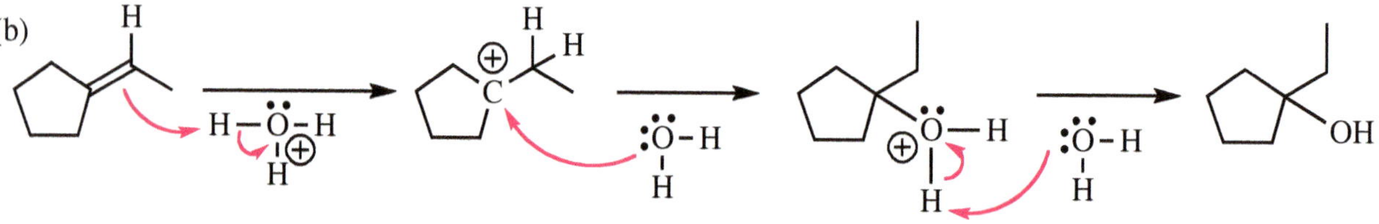
H
H–O–H
H
H
H
C
O–H
H
O–H
H
O–H
H
OH

8-56 continued

(c)

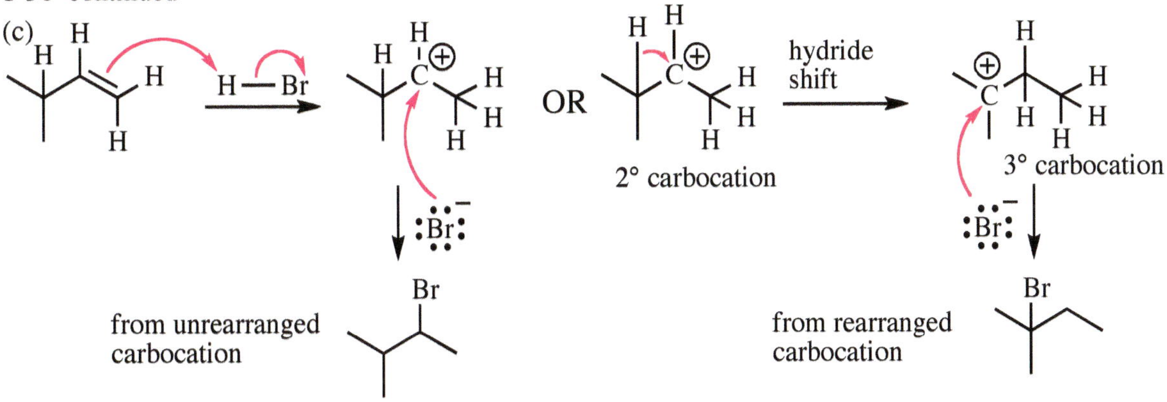

H—Br

OR

hydride shift

2° carbocation

3° carbocation

from unrearranged carbocation

from rearranged carbocation

(d)

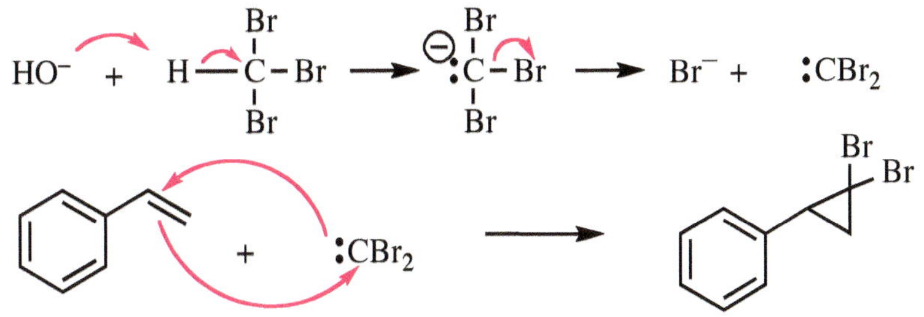

HO⁻ + H—C—Br → C—Br → Br⁻ + :CBr₂

+ :CBr₂ →

(e)

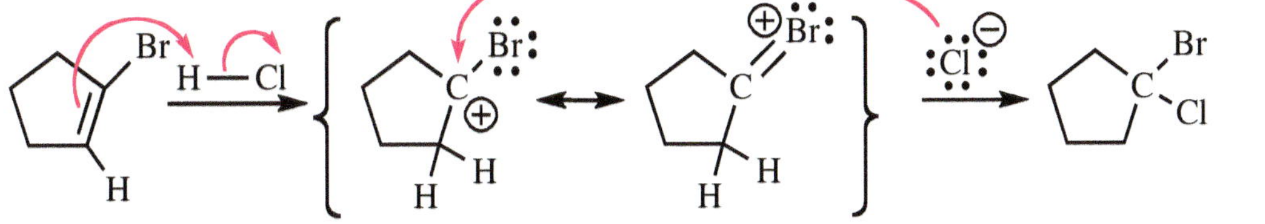

H—Cl

OR

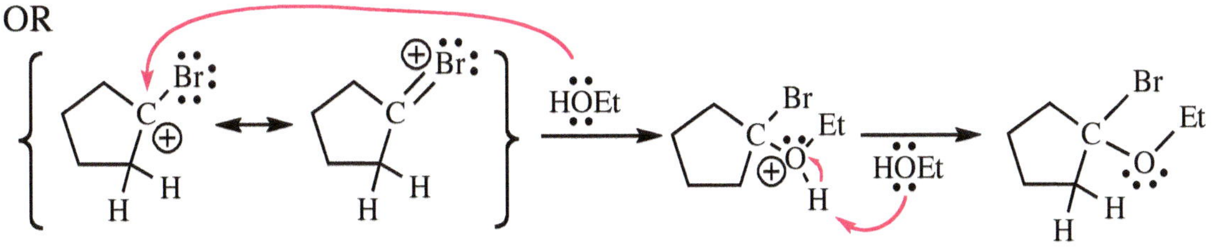

HOEt

HOEt

(f)

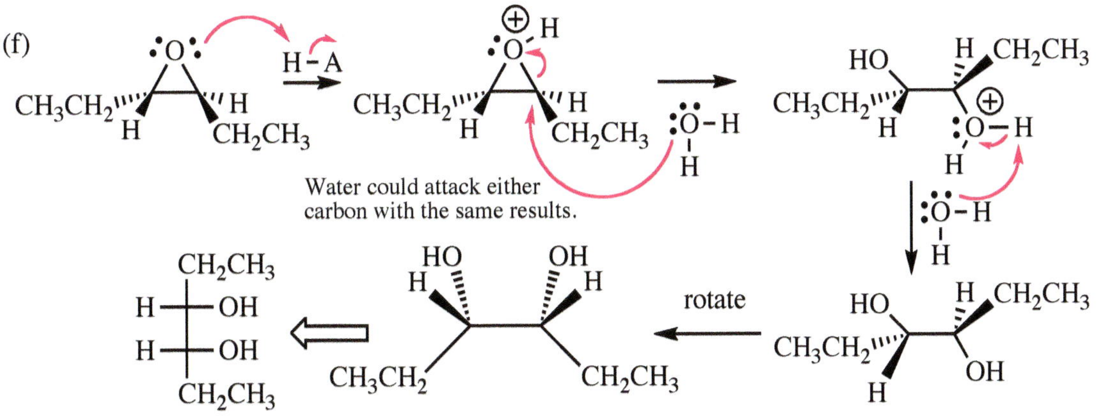

H–A

Water could attack either carbon with the same results.

rotate

8-56 continued

(g)

Methanol could attack either carbon of the epoxide but it must attack from the opposite side of the ring.

racemic mixture

(h)

Recall that "Ph" is the abbreviation for phenyl.

8-57

The slow step is the first step, formation of the carbocation, in electrophilic addition. The major product is the one whose intermediate carbocation is formed faster, as it has the lower activation energy. Assume that the energy of the products will be roughly the same, and both pathways begin with the same reactants, so the only difference occurs in the transition states and the intermediates.

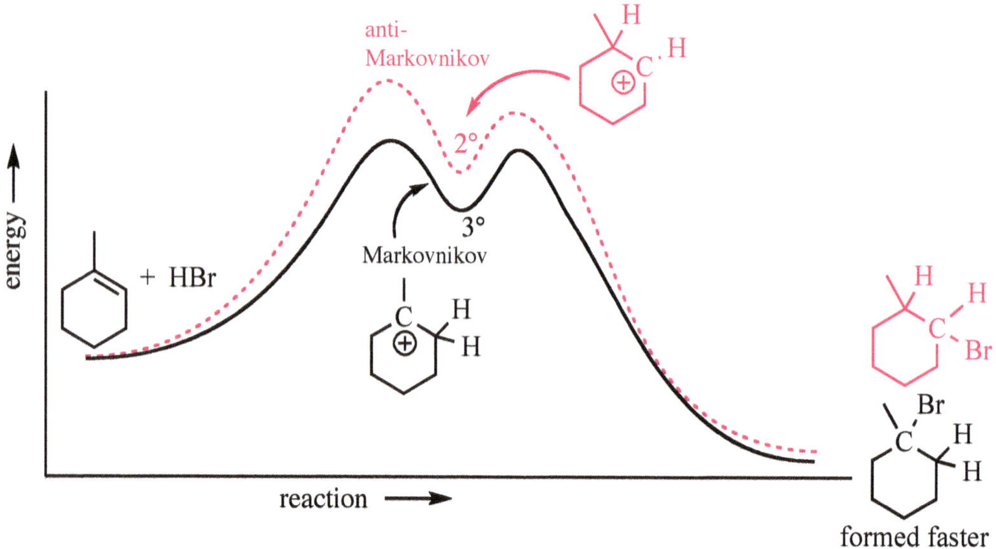

8-58 Once the bromonium ion is formed, it can be attacked by either nucleophile, bromide or chloride, leading to the mixture of products.

8-59 Two orientations of attack of bromine radical are possible:

(A) anti-Markovnikov

(alkene) + •Br ⟶ (3° radical, $\dot{C}$–CH_2Br) $\xrightarrow{HBr}$ (H, Br product) + •Br

3° radical

(B) Markovnikov

(alkene) + •Br ⟶ (Br, $\dot{C}H_2$) $\xrightarrow{HBr}$ (Br, H product) + •Br

1° radical

The first step in the mechanism is endothermic and rate determining. The 3° radical produced in anti-Markovnikov attack (A) of bromine radical is several kJ/mole more stable than the 1° radical generated by Markovnikov attack (B). Hammond's Postulate tells us that it is reasonable to assume that the activation energy for anti-Markovnikov addition is lower than for Markovnikov addition. This defines the first half of the energy diagram.

The relative stabilities of the final products are somewhat difficult to predict. (Remember that stability of final products does not necessarily reflect relative stabilities of intermediates; this is why a thermodynamic product can be different from a kinetic product.) From bond dissociation energies (kJ/mole) in Table 4-2:

anti-Markovnikov		Markovnikov	
H to 3° C	403	H to 1° C	423
Br to 1° C	303	Br to 3° C	304
	706 kJ/mole		727 kJ/mole

If it takes more energy to break bonds in the Markovnikov product, it must be lower in energy, therefore, more stable—OPPOSITE OF STABILITY OF THE INTERMEDIATES! Now we are ready to construct the energy diagram.

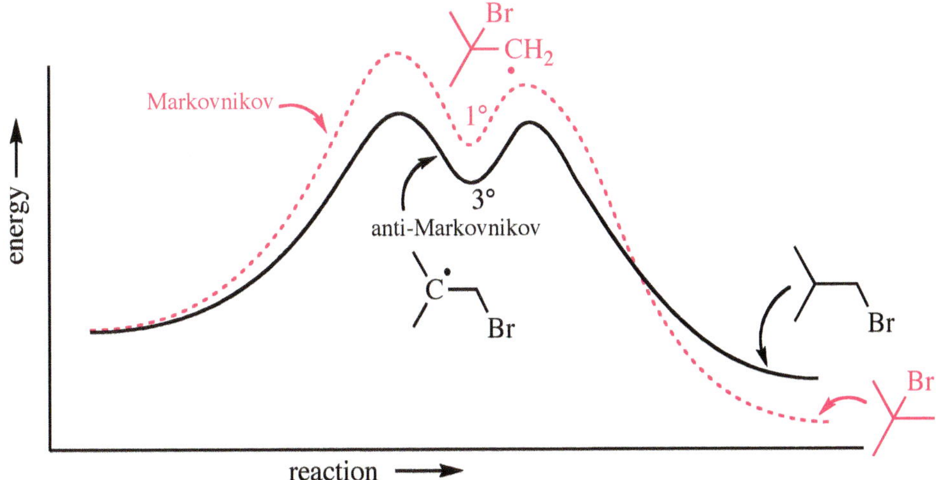

It is the anti-Markovnikov product that is the kinetic product, not the thermodynamic product; the anti-Markovnikov product is obtained since its rate-determining step has the lower activation energy.

8-60

A) Unknown X, C_5H_9Br, has one element of unsaturation. **X** reacts with neither bromine nor $KMnO_4$, so the unsaturation in **X** cannot be an alkene; it must be a ring.

B) Upon treatment with strong base (*tert*-butoxide), **X** loses H and Br to give **Y**, C_5H_8, which does react with bromine and $KMnO_4$; it must have an alkene and a ring. Only one isomer is formed.

C) Catalytic hydrogenation of **Y** gives methylcyclobutane. This is a BIG clue because it gives the carbon skeleton of the unknown. **Y** must have a double bond in the methylcyclobutane skeleton, and **X** must have a Br on the methylcyclobutane skeleton.

D) Ozonolysis of **Y** gives a dialdehyde **Z**, $C_5H_8O_2$, which contains all the original carbons, so the alkene cleaved in the ozonolysis had to be in the ring.

Let's consider the possible answers for **X** and see if each fits the information.

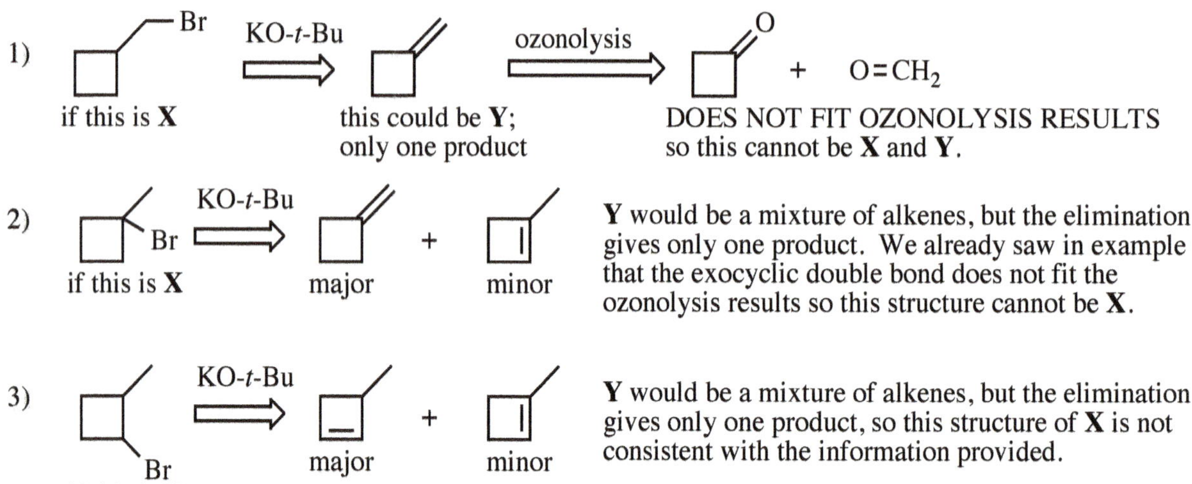

1)
Br
KO-*t*-Bu
ozonolysis
O
+ O=CH$_2$
if this is **X**
this could be **Y**; only one product
DOES NOT FIT OZONOLYSIS RESULTS so this cannot be **X** and **Y**.
2)
KO-*t*-Bu
Br
+
if this is **X**
major
minor
Y would be a mixture of alkenes, but the elimination gives only one product. We already saw in example 1 that the exocyclic double bond does not fit the ozonolysis results so this structure cannot be **X**.
3)
KO-*t*-Bu
+
Br
if this is **X**
major
minor
Y would be a mixture of alkenes, but the elimination gives only one product, so this structure of **X** is not consistent with the information provided.

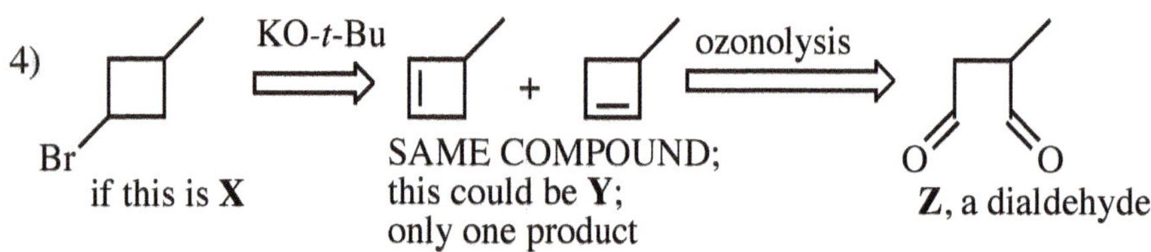

The correct structures for **X**, **Y**, and **Z** are given in the fourth possibility. The only structural feature of **X** that remains undetermined is whether it is the *cis* or *trans* isomer.

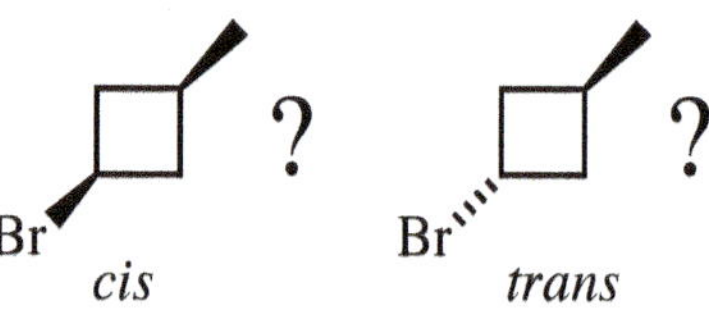

8-61 The clue to the structure of α-pinene is the ozonolysis. Working backwards shows the alkene position.

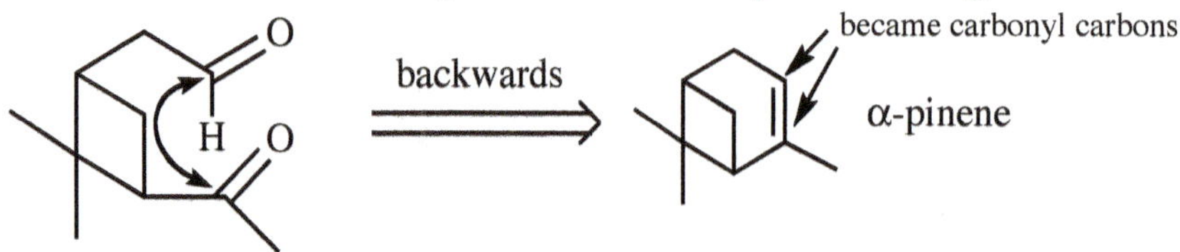

continued on next page

8-61 continued

After ozonolysis, the two carbonyls are still connected; the alkene must have been in a ring, so reconnect the two carbonyl carbons with a double bond.

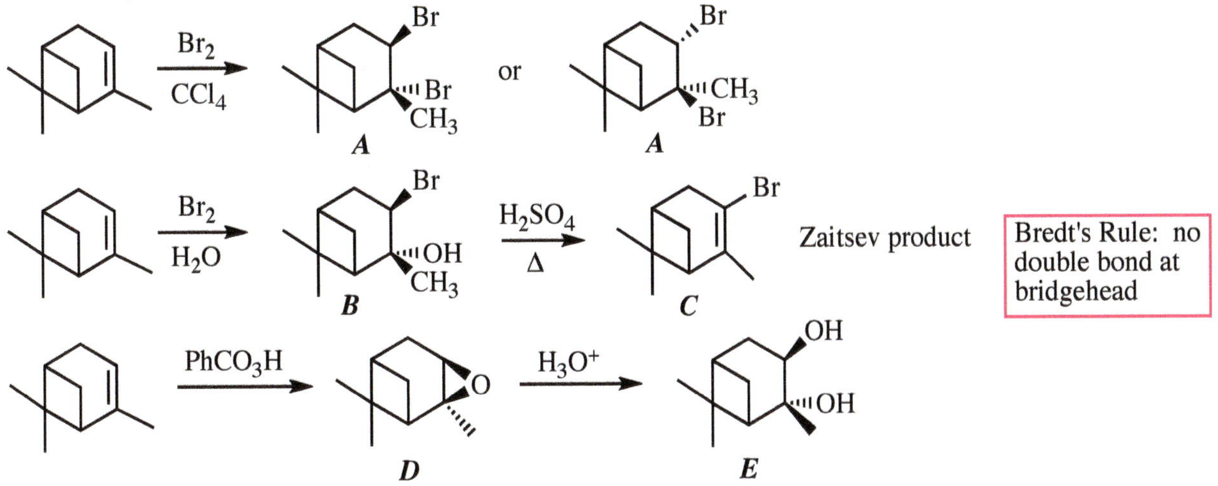

Br_2 / CCl_4 → *A* (Br, Br, CH_3) or *A* (Br, CH_3, Br)

Br_2 / H_2O → *B* (Br, OH, CH_3) — H_2SO_4 / Δ → *C* (Br) Zaitsev product

Bredt's Rule: no double bond at bridgehead

$PhCO_3H$ → *D* (O) — H_3O^+ → *E* (OH, OH)

8-62 The two products from permanganate oxidation must have been connected by a double bond at the carbonyl carbons. Whether the alkene was *E* or *Z* cannot be determined by this experiment.

$CH_3(CH_2)_{12}CH{=}CH(CH_2)_7CH_3$

shown here as Z, the naturally occurring isomer

8-63

Unknown X $\xrightarrow[Pt]{3\ H_2}$ ⟹ must have three alkenes in this skeleton

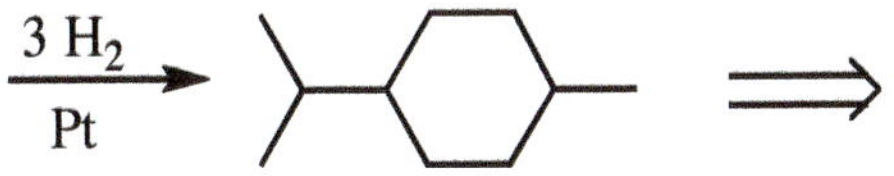

Unknown X $\xrightarrow{O_3}$ $\xrightarrow{Me_2S}$ $CH_2{=}O$ + (H, O, O, O) + (O, O, H)

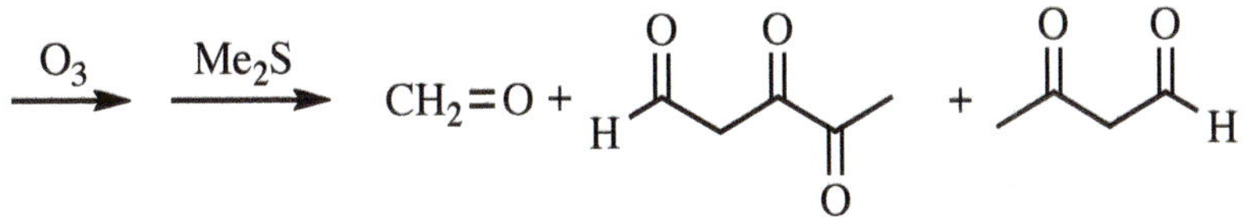

There are several ways to attack a problem like this. One is the trial-and-error method, that is, put double bonds in all possible positions until the ozonolysis products match. There are times when the trial-and-error method is useful (as in simple problems where the number of possibilities is few), but this is not one of them.

Let's try logic. Analyze the ozonolysis products carefully—what do you see? There are only two methyl groups, so one of the three terminal carbons in the skeleton (C-8, C-9, or C-10) has to be a $=CH_2$. Do we know which terminal carbon has the double bond? Yes, we can deduce that. If C-10 were double-bonded to C-4, then after ozonolysis, C-8 and C-9 must still be attached to C-7. However, in the ozonolysis products, there is no branched chain, that is, no combination of C-8 + C-9 + C-7 + C-1. What if C-7 had a double bond to C-1? Then we would have acetone, CH_3COCH_3, as an ozonolysis product—we don't. Thus, we can't have a double bond from C-4 to C-10. One of the other terminal carbons (C-8 or C-9) must have a double bond to C-7.

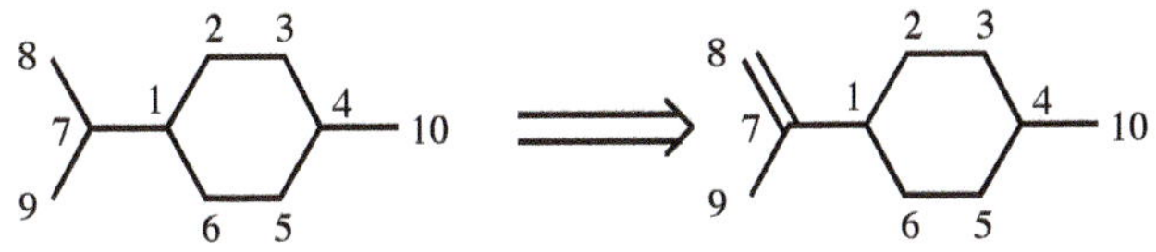

8, 2, 3, 1, 4, 7, 10, 9, 6, 5 ⟹ 8, 2, 3, 1, 4, 7, 10, 9, 6, 5

continued on next page

8-63 continued

The other two double bonds have to be in the ring, but where? The products do not have branched chains, so double bonds must appear at both C-1 and C-4. There are only two possibilities for this requirement.

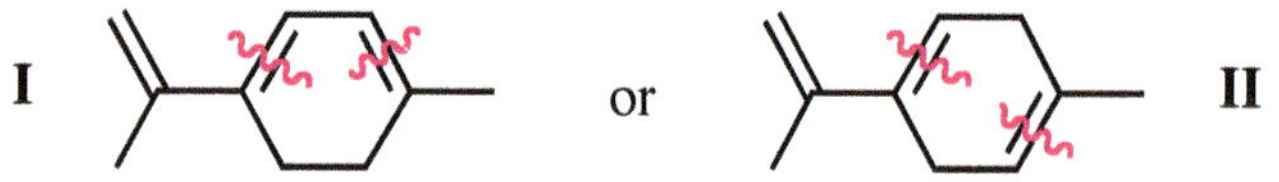

I
or
II

Ozonolysis of **I** would give fragments containing one carbon, two carbons, and seven carbons. Ozonolysis of **II** would give fragments containing one carbon, four carbons, and five carbons. Aha! Our mystery structure must be **II**.

> Note to the student: Science is more than a collection of facts. The application of observation and logic to solve problems by *deduction* and *inference* is a critical scientific skill, one that distinguishes humans from algae.

8-64 In this type of problem, begin by determining which bonds are broken and which are formed. These will always give clues as to what is happening.

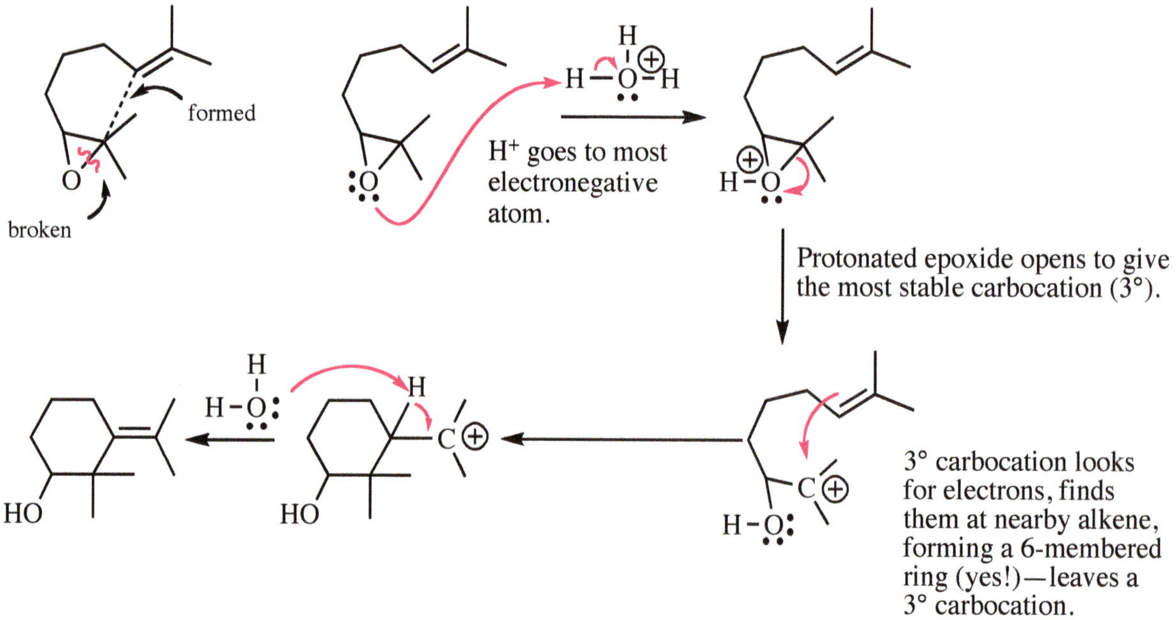

formed
broken
H⁺ goes to most electronegative atom.
Protonated epoxide opens to give the most stable carbocation (3°).
3° carbocation looks for electrons, finds them at nearby alkene, forming a 6-membered ring (yes!)—leaves a 3° carbocation.
HO

8-65 See the solution to Problem 8-35 for simplified examples of these reactions.

(a)

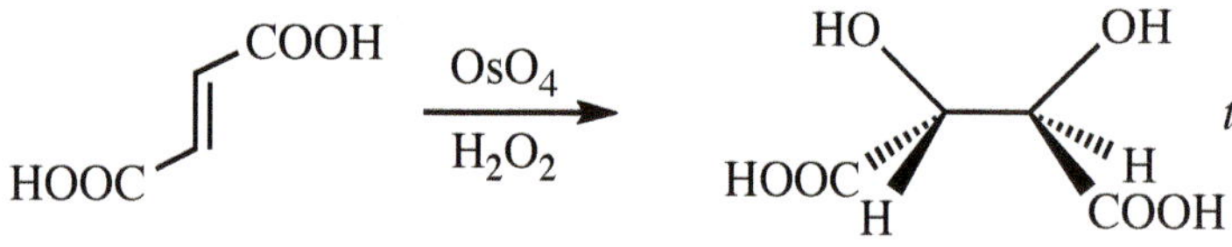

COOH
HOOC
OsO4
H2O2
HO
OH
HOOC
H
H
COOH

trans + **syn** → racemic

(b)

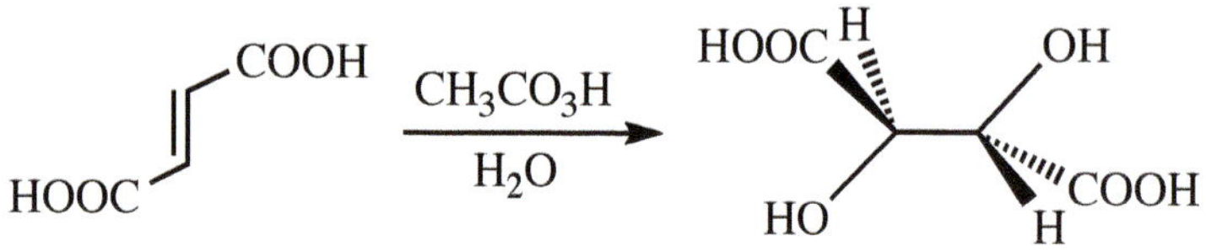

COOH
HOOC
CH3CO3H
H2O
HOOC
H
OH
HO
COOH
H

trans + **anti** → meso

8-65 continued

(c) cis-$HOOC-CH=CH-COOH$ $\xrightarrow[H_2O]{CH_3CO_3H}$ (racemic tartaric acid) *cis* + **anti** → racemic

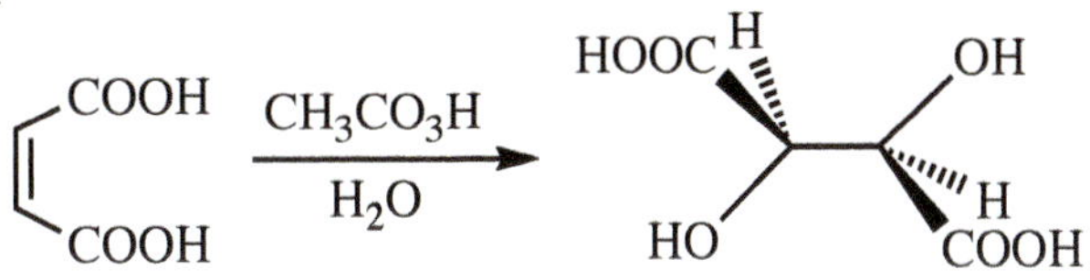

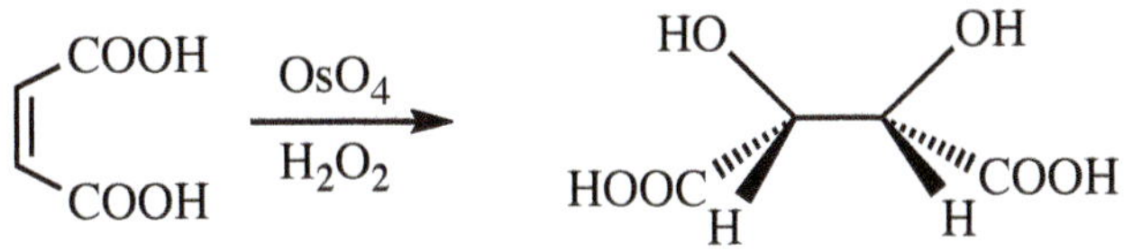

(d) cis-$HOOC-CH=CH-COOH$ $\xrightarrow[H_2O_2]{OsO_4}$ (meso tartaric acid) *cis* + **syn** → meso

8-66 1-methylcyclohexene $\xrightarrow{BD_3 \cdot THF}$ $\xrightarrow[HO^-]{H_2O_2}$ (product with CH_3, D, OH, H)

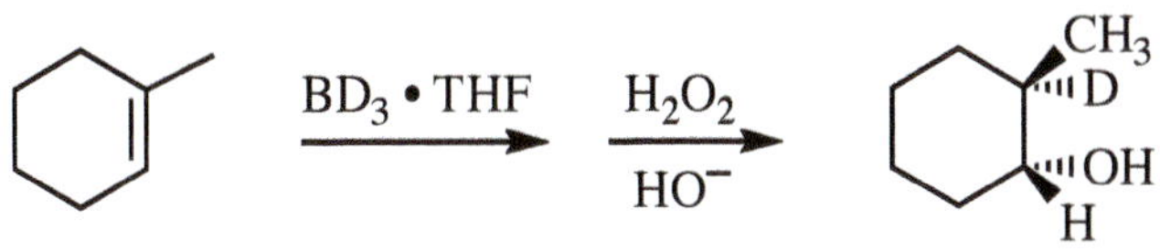

8-67 By now, these rearrangements should not be so "unexpected".

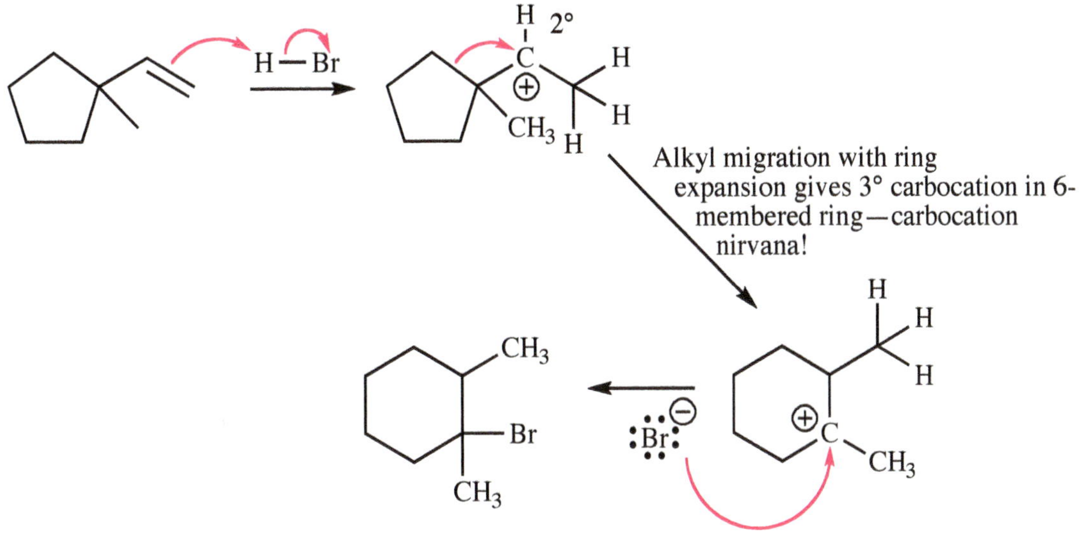

Alkyl migration with ring expansion gives 3° carbocation in 6-membered ring—carbocation nirvana!

You must be asking yourself, "Why didn't the methyl group migrate?" To which you answered by drawing the carbocation that would have been formed:

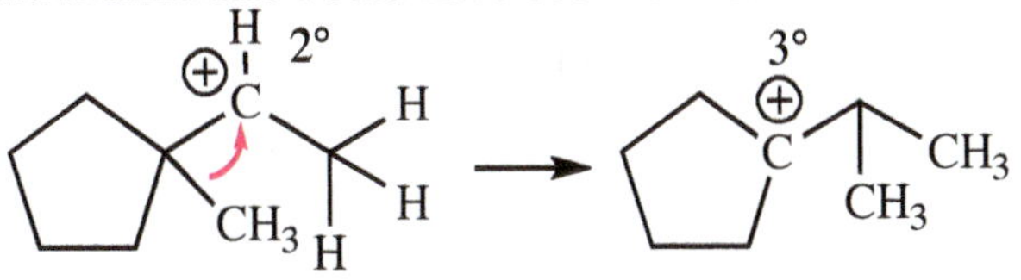

The new carbocation is indeed 3°, but it is only in a 5-membered ring, not quite as stable as in a 6-membered ring. In all probability, some of the product from methyl migration would be formed, but the 6-membered ring would likely be the major product.

8-68 Each alkene will produce two carbonyls upon ozonolysis or permanganate oxidation. Oxidation of the unknown generated four carbonyls, so the unknown must have had two alkenes. There is only one possibility for their positions.

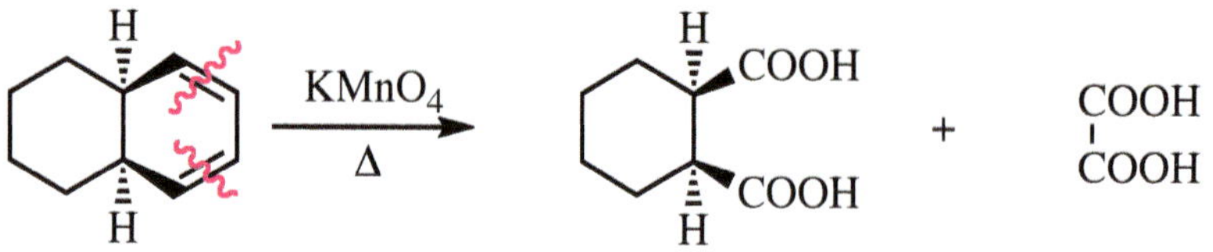

H H COOH KMnO$_4$ Δ COOH + COOH COOH H H the unknown

8-69

(a) Fumarase catalyzes the addition of H and OH, a hydration reaction.
(b) Fumaric acid is planar and cannot be chiral. Malic acid does have a chiral center and is chiral. The enzyme-catalyzed reaction produces only the *S* enantiomer, so the product must be optically active.
(c) One of the fundamental rules of stereochemistry is that optically inactive starting materials produce optically inactive products. Sulfuric-acid-catalyzed hydration would produce a racemic mixture of malic acid, that is, equal amounts of *R* and *S*.
(d) If the product is optically active, then either the starting materials or the catalyst were chiral. We know that water and fumaric acid are not chiral, so we must infer that fumarase is chiral.
(e) The D and the OD are on the "same side" of the Fischer projection (sometimes called the "erythro" stereoisomer). These are produced from either: (1) syn addition to *cis* alkenes, or (2) anti addition to *trans* alkenes. We know that fumaric acid is *trans*, so the addition of D and OD must necessarily be anti.
(f) Hydroboration is a syn addition.

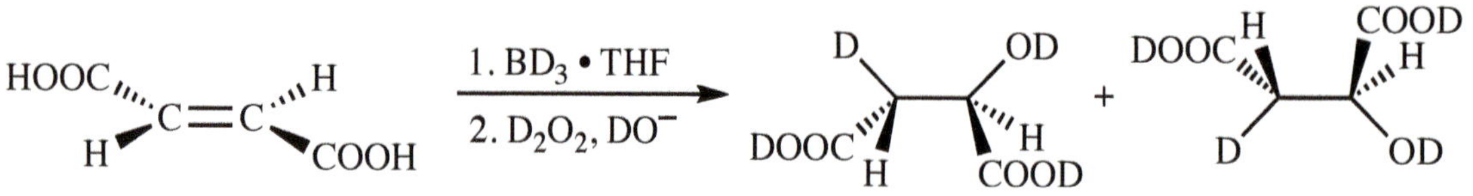

HOOC H C=C H COOH 1. BD_3 • THF 2. D_2O_2, DO^- D OD DOOC H H COOD + DOOC H COOD H D OD

(Note that OH exchanges with D in DO^-.)

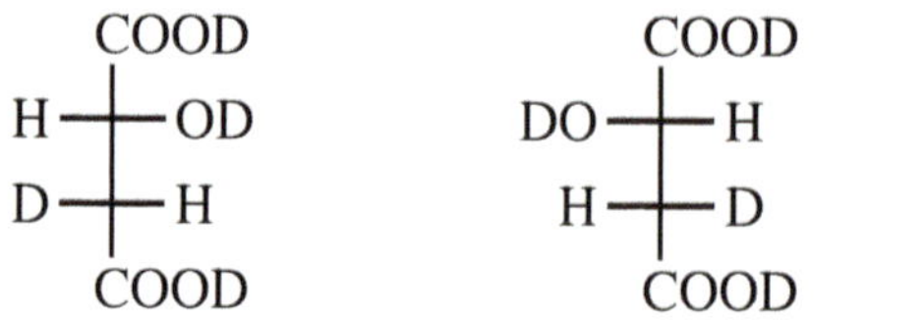

COOD H OD D H COOD COOD DO H H D COOD

As expected, *trans* alkene plus syn addition puts the two groups on the "opposite" side of the Fischer projection (sometimes called "threo").

8-70

(a) OH Hg(OAc) HgOAc OH mercurinium ion Hg(OAc) O H AcO Hg(OAc) O NaBH$_4$ O

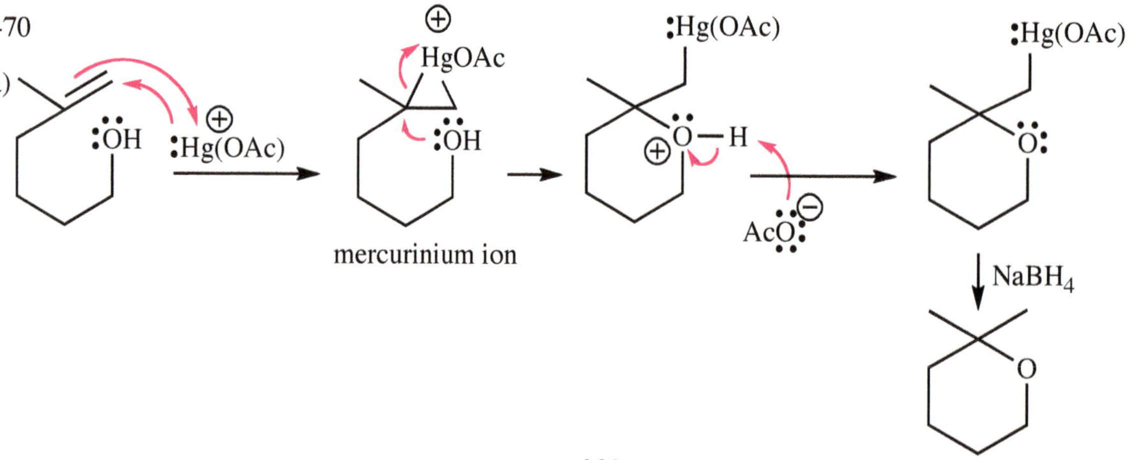

8-70 continued
(b)

bromonium ion
$C_7H_{13}BrO$

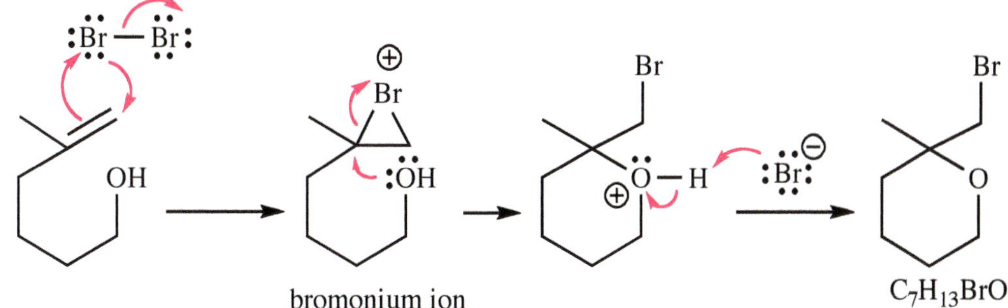

8-71
This mechanism is very similar to formation of a halohydrin (text Mechanism 8-8); the difference is that the nucleophile comes from the carboxylic acid on the other side of the molecule, making a new ring in the process.
Step 1: formation of the iodonium ion;
Step 2: internal nucleophilic attack (the O of the C=O is more nucleophilic than the OH because of resonance stabilization of the intermediate);
Step 3: removal of the proton by H_2O.

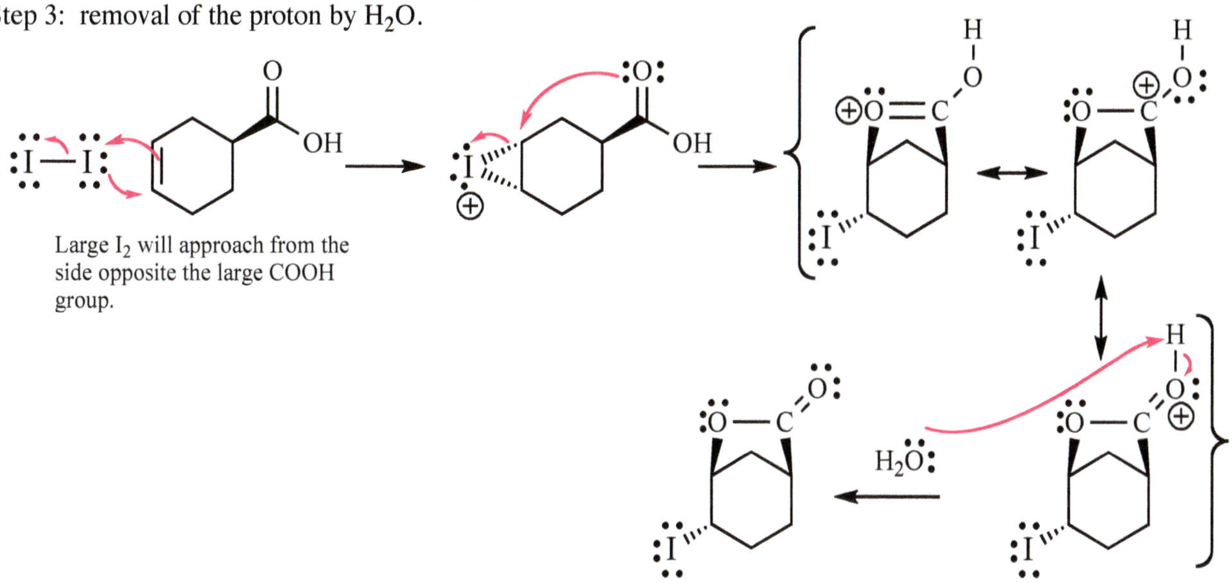

Large I_2 will approach from the side opposite the large COOH group.

8-72 For clarity, the new bonds formed in this mechanism are shown in bold.

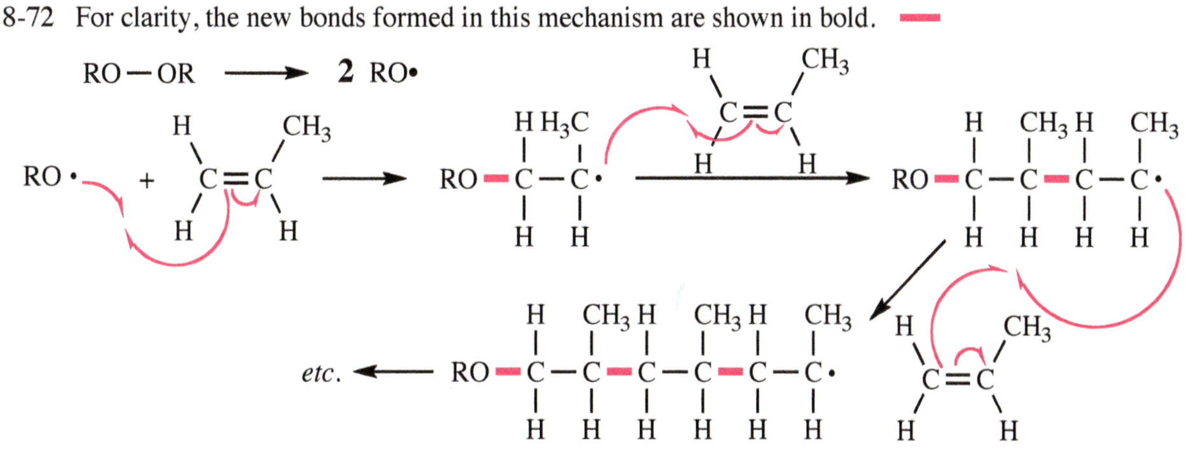

RO—OR → 2 RO•
etc.

8-73 Without divinylbenzene, individual chains of polystyrene are able to slide past one another. Addition of divinylbenzene during polymerization forms bridges, or *crosslinks*, between chains, adding strength and rigidity to the polymer. Divinylbenzene and similar molecules are called crosslinking agents.

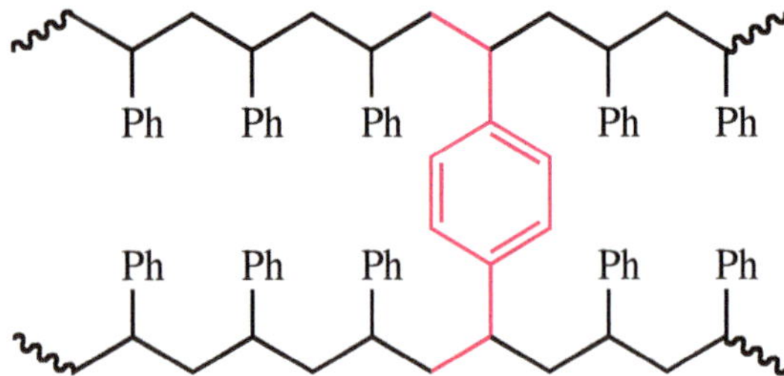

Two polystyrene chains crosslinked by a divinylbenzene monomer shown in red.

8-74 A peroxy radical is shown as the initiator. Newly formed bonds are shown in bold.

RO• + (H)(H)C=C(CH_3)(CH_3) ⟶ RO–C(H)(H)–C•(CH_3)(CH_3) ⟶ RO–C–C–C–C• ⟶ RO–C–C–C–C–C–C• ⟶ *etc.*

8-75

tetrasubstituted, hindered

H_3CO ... OCH_3

1) **2** O_3 at –78 °C

2) $(CH_3)_2S$

esters—not typical

H_3CO ... OCH_3

O O

O O

H H

aldehydes—expected

8-76
The stability of the intermediate formed in the first step determines the orientation.

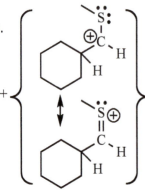

H^+ ⟶

3° carbocation is more stable than secondary; explains Markovnikov orientation.

S

H^+ ⟶

3° carbocation is still good

+

RESONANCE STABILIZED!

8-77 The addition of BH_3 to an alkene is reversible. Given heat and time, the borane will eventually "walk" its way to the end of the chain through a series of addition-elimination cycles. The most stable alkylborane has the boron on the end carbon; eventually, the series of equilibria leads to the ultimate borane product that is oxidized to the primary alcohol.

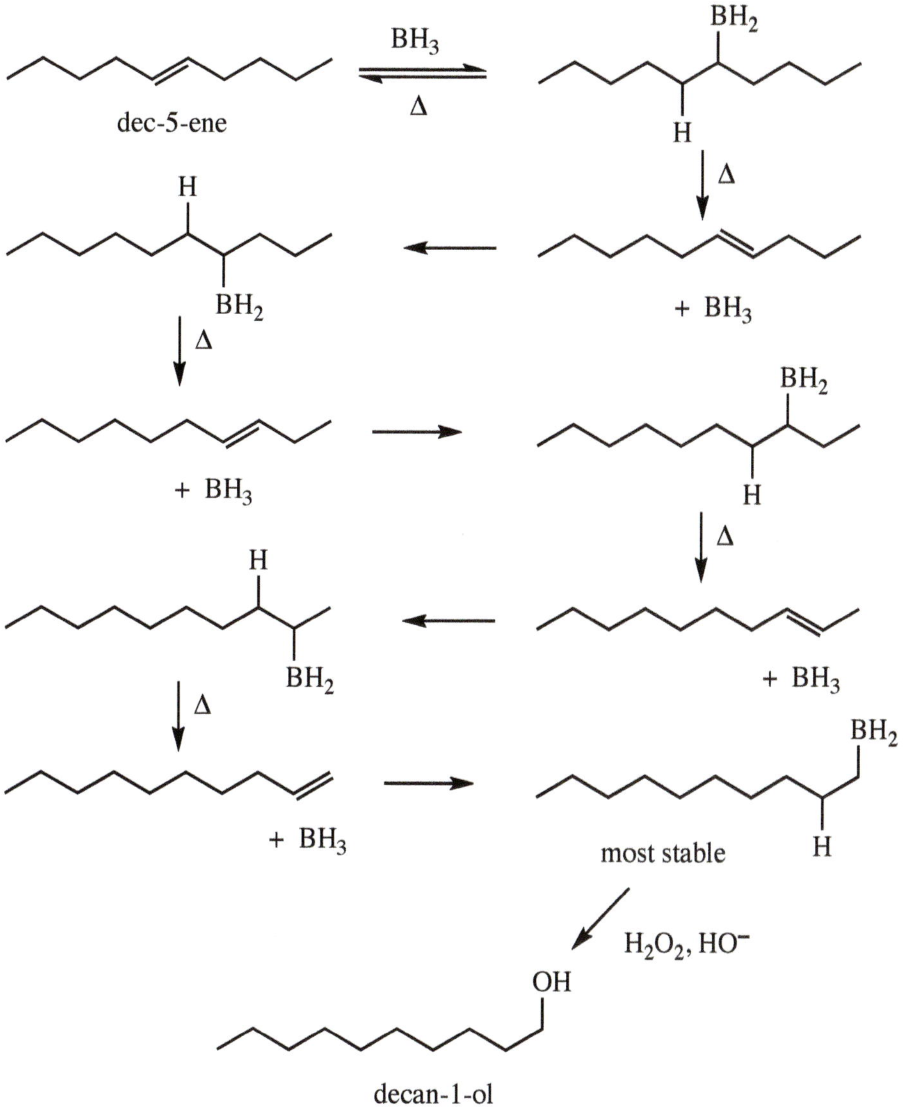

dec-5-ene $\xrightleftharpoons[\Delta]{BH_3}$ (5-boryldecane: BH_2, H)

$\downarrow \Delta$

dec-4-ene + BH_3 → (4-boryldecane: H, BH_2)

$\downarrow \Delta$

dec-3-ene + BH_3 → (3-boryldecane: BH_2, H)

$\downarrow \Delta$

dec-2-ene + BH_3 → (2-boryldecane: H, BH_2)

$\downarrow \Delta$

dec-1-ene + BH_3 → (1-boryldecane: BH_2, H) most stable

H_2O_2, HO^- ↙

OH

decan-1-ol

8-78 First, we explain *how* the mixture of stereoisomers results, then *why*.

We have seen many times that the bridged halonium ion permits attack of the nucleophile only from the opposite side.

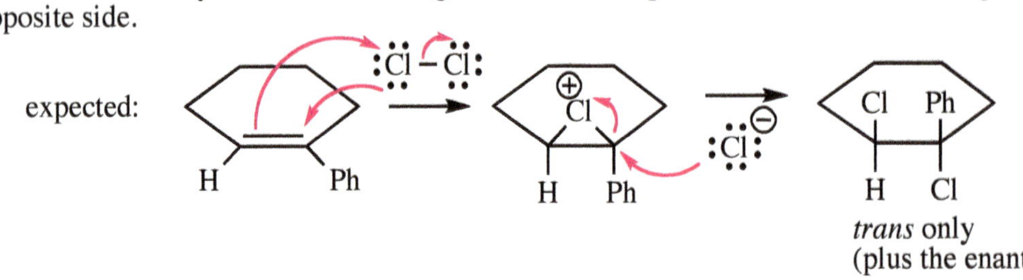

expected:

H Ph → Cl–Cl → Cl⊕ H Ph :Cl:⊖ → Cl Ph H Cl

trans only
(plus the enantiomer)

A mixture of *cis* and *trans* could result only if attack of chloride were possible from both top and bottom, something possible only if a *carbocation* existed at this carbon.

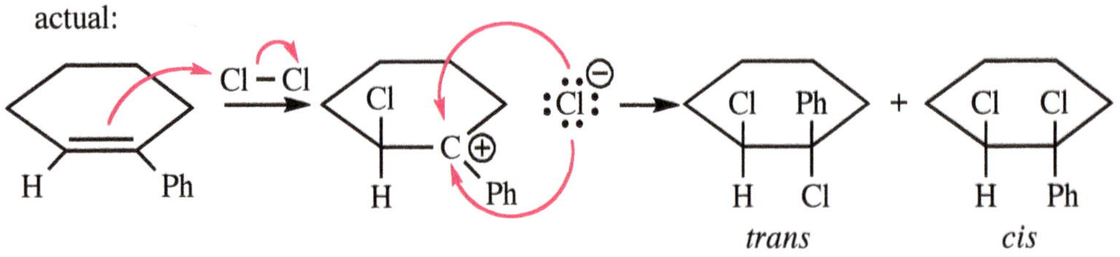

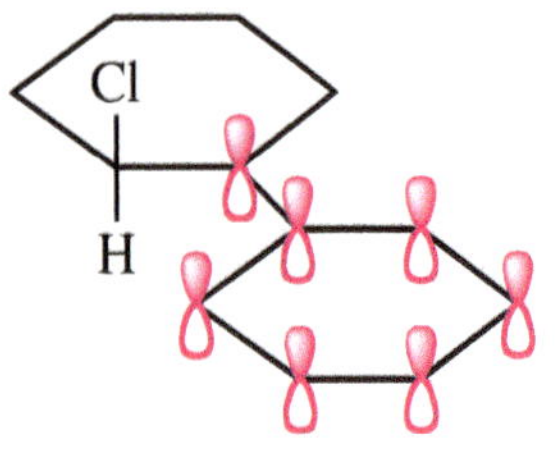

actual:

H Ph → Cl–Cl → Cl H C⊕ Ph :Cl:⊖ → Cl Ph H Cl + Cl Cl H Ph

trans *cis*

Cl H

This picture of the p orbitals of benzene shows resonance overlap with the p orbital of the carbocation. The chloride nucleophile can form a bond to the positive carbon from either the top or the bottom.

Why does a carbocation exist here? Not only is it 3°, *it is also next to a benzene ring (benzylic) and therefore resonance-stabilized*. This resonance stabilization would be forfeited in a halonium ion intermediate.

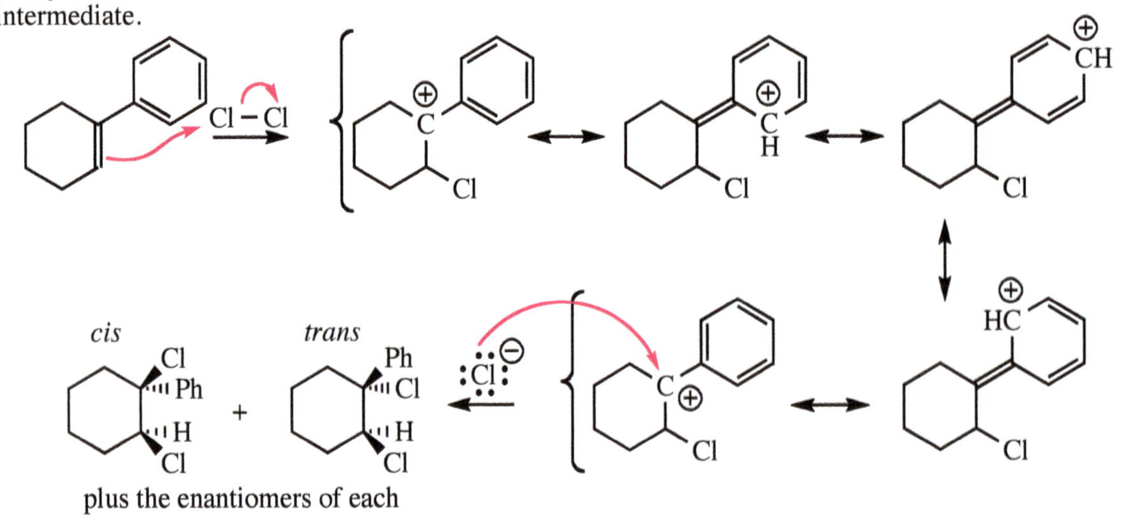

Cl–Cl → C⊕ Cl ↔ ⊕CH Cl ↔ ⊕CH Cl ↕ ⊕HC Cl ↔ C⊕ Cl

:Cl:⊖ ←

cis Cl Ph H Cl + *trans* Ph Cl H Cl

plus the enantiomers of each

8-79

(a) The three bonds to boron, indicated by the squiggly line, will be replaced by an OH.

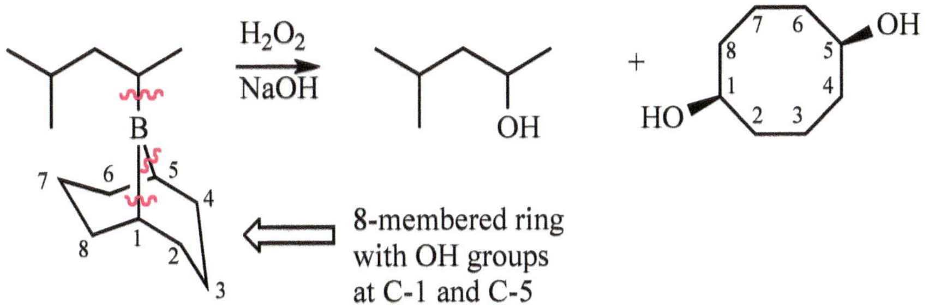

H_2O_2 / NaOH

B

OH

+ HO ... OH

7 6 5 4 3 2 1 8

8-membered ring with OH groups at C-1 and C-5

(b)

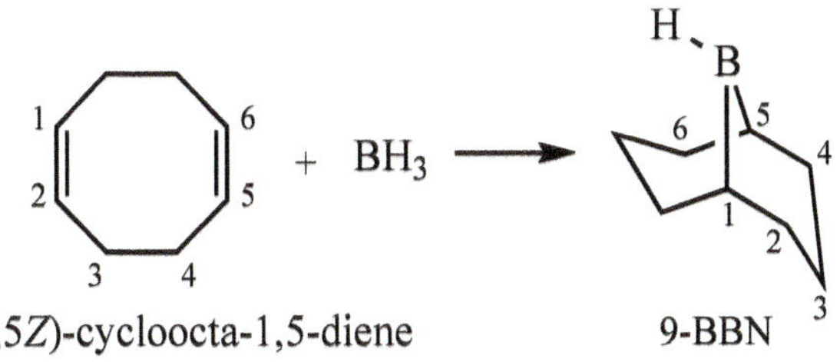

1 2 3 4 5 6

+ BH_3 ⟶

H B 6 5 4 1 2 3

(1*Z*,5*Z*)-cycloocta-1,5-diene

9-BBN

Note to the student: Chapter 8 introduces a type of graphical reaction summary called a "starburst" diagram. These are visually helpful in organizing the reactions, and very useful in reviewing before quizzes and exams. Your instructor may not cover all of the reactions in the chapter, so it would benefit you to customize the reaction diagrams to optimize their utility for your course.

CHAPTER 9—ALKYNES

9-1 Each pi bond = 1 element of unsaturation; each ring = 1 e.u.
(a) parsalmide: 4 double bonds + 1 triple bond + 1 ring = **7** unsaturations
ethynyl estradiol: 3 double bonds + 1 triple bond + 4 rings = **9** unsaturations
dynemicin A: 11 double bonds + 2 triple bonds + 7 rings = **22** unsaturations

(b) Many other structures are possible in each case; the number of unsaturations must be consistent.

(1) C_6H_{10} ⇨ **2**

$CH_3CH_2C{\equiv}CCH_2CH_3$

$CH_3CH_2CH_2CH_2C{\equiv}CH$

(2) C_8H_{12} ⇨ **3**

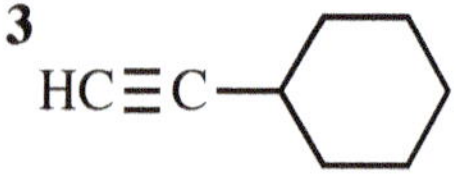

HC≡C–(cyclohexyl)

$HC{\equiv}C{-}CH_2CH{=}CHCH_2CH_2CH_3$

(3) C_7H_8 ⇨ **4**

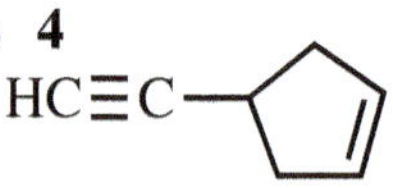

HC≡C–(cyclopentenyl)

$HC{\equiv}C{-}C{\equiv}C{-}CH_2CH_2CH_3$

9-2 New IUPAC names are given. The asterisk (*) denotes acetylenic hydrogens of terminal alkynes.

(a) $CH_3CH_2CH_2{-}C{\equiv}C{-}\overset{*}{H}$ pent-1-yne; $CH_3CH_2{-}C{\equiv}C{-}CH_3$ pent-2-yne; $(CH_3)_2CH{-}C{\equiv}C{-}\overset{*}{H}$ 3-methylbut-1-yne

(b) Do these isomer problems systematically: draw the 6-carbon chains first, then the 5-carbon chains, etc.

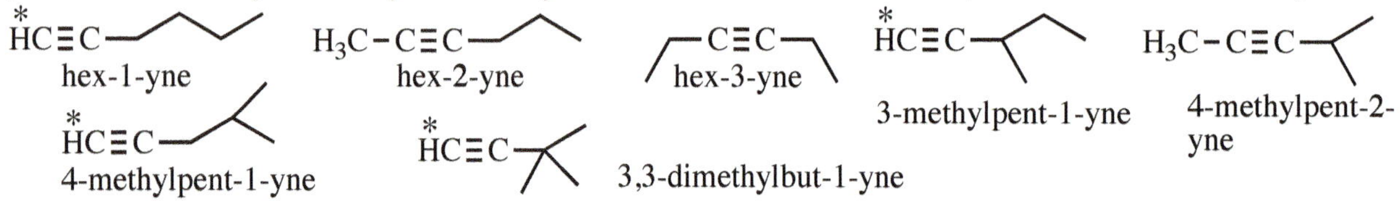

$H\overset{*}{}C{\equiv}C$– hex-1-yne; $H_3C{-}C{\equiv}C$– hex-2-yne; –C≡C– hex-3-yne; $H\overset{*}{}C{\equiv}C$– 3-methylpent-1-yne; $H_3C{-}C{\equiv}C$– 4-methylpent-2-yne

$H\overset{*}{}C{\equiv}C$– 4-methylpent-1-yne; $H\overset{*}{}C{\equiv}C$– 3,3-dimethylbut-1-yne

9-3 Acetylene would likely decompose into its elements. The decomposition reaction below is exothermic ($\Delta H°$ = – 234 kJ/mole) as well as having an increase in entropy. Thermodynamically, at 1500°C, an increase in entropy will have a large effect on ΔG (remember $\Delta G = \Delta H - T\Delta S$). Kinetically, almost any activation energy barrier will be overcome at 1500°C.

$$HC{\equiv}CH \xrightarrow{1500\ °C} 2\ C + H_2$$

9-4 Adding sodium amide to the mixture will produce the sodium salt of hex-1-yne, leaving hex-1-ene untouched. Distillation will remove the hex-1-ene, leaving the non-volatile acetylide salt behind.

$$\left.\begin{matrix} CH_2{=}CHCH_2CH_2CH_2CH_3 \\ H{-}C{\equiv}C{-}CH_2CH_2CH_2CH_3 \end{matrix}\right\} \xrightarrow{NaNH_2} \left\{\begin{matrix} CH_2{=}CHCH_2CH_2CH_2CH_3 \\ Na^+\ {:}\overset{\ominus}{C}{\equiv}C{-}CH_2CH_2CH_2CH_3 \end{matrix}\right.$$ non-volatile salt

9-5 The key to this problem is to understand that *a proton donor will react only with the conjugate base of a **weaker** acid.*

(a) $H{-}C{\equiv}C{-}H + NaNH_2 \longrightarrow H{-}C{\equiv}\overset{\ominus}{C}{:}\ Na^+ + NH_3$

(b) $H{-}C{\equiv}C{-}H + CH_3Li \longrightarrow H{-}C{\equiv}\overset{\ominus}{C}{:}\ Li^+ + CH_4$

(c) no reaction: $NaOCH_3$ is not a strong enough base

(d) no reaction: NaOH is not a strong enough base

(e) $H{-}C{\equiv}\overset{\ominus}{C}{:}\ Na^+ + CH_3OH \longrightarrow H{-}C{\equiv}C{-}H + NaOCH_3$ (opposite of (c))

(f) $H{-}C{\equiv}\overset{\ominus}{C}{:}\ Na^+ + H_2O \longrightarrow H{-}C{\equiv}C{-}H + NaOH$ (opposite of (d))

(g) no reaction : $H{-}C{\equiv}\overset{\ominus}{C}{:}\ Na^+$ is not a strong enough base

(h) no reaction: $NaNH_2$ is not a strong enough base

(i) $CH_3OH + NaNH_2 \longrightarrow NaOCH_3 + NH_3$

9-6

$$H-C\equiv C-H \xrightarrow{NaNH_2} H-C\equiv C:^{\ominus}\ Na^+ \xrightarrow{CH_3CH_2Br} H-C\equiv C-CH_2CH_3$$

$$H-C\equiv C-CH_2CH_3 \xrightarrow{NaNH_2} Na^+\ {}^{\ominus}:C\equiv C-CH_2CH_3 \xrightarrow{CH_3(CH_2)_5Br} CH_3(CH_2)_5-C\equiv C-CH_2CH_3$$

9-7 To be a feasible synthesis, the desired product must have the sp carbons from acetylene bonded to CH_2 groups that came from 1° halides in the precursor.

(a) $$H-C\equiv C-H \xrightarrow[2)\ CH_3CH_2CH_2CH_2Br]{1)\ NaNH_2} H-C\equiv C-CH_2CH_2CH_2CH_3$$

(b) $$H-C\equiv C-H \xrightarrow[2)\ CH_3CH_2CH_2Br]{1)\ NaNH_2} H-C\equiv C-CH_2CH_2CH_3 \xrightarrow[2)\ CH_3I]{1)\ NaNH_2} CH_3-C\equiv C-CH_2CH_2CH_3$$

(c) $$H-C\equiv C-H \xrightarrow[2)\ CH_3CH_2Br]{1)\ NaNH_2} H-C\equiv C-CH_2CH_3 \xrightarrow[2)\ CH_3CH_2Br]{1)\ NaNH_2} CH_3CH_2-C\equiv C-CH_2CH_3$$

(d) The desired product cannot be synthesized by an S_N2 reaction—would require attack of a strong base on a 2° alkyl halide.

$$CH_3-C\equiv C:^{\ominus}\ Na^+ + CH_3CH_2CH(Br)CH_3 \not\longrightarrow CH_3-C\equiv C-CH(CH_3)CH_2CH_3$$

strong nucleophile; must be second order (Br on 2° carbon)

low yields; not practical

(e) $$H-C\equiv C-H \xrightarrow[2)\ CH_3I]{1)\ NaNH_2} CH_3-C\equiv C-H \xrightarrow[2)\ BrCH_2CH(CH_3)CH_3]{1)\ NaNH_2} CH_3-C\equiv C-CH_2CH(CH_3)CH_3$$

(f) $$H-C\equiv C-H \xrightarrow[2)\ Br(CH_2)_8Br]{1)\ NaNH_2} H-C\equiv C-CH_2(CH_2)_7Br$$

$$\xrightarrow{NaNH_2} Na^+\ {}^{\ominus}:C\equiv C-CH_2(CH_2)_7Br \longrightarrow \text{cyclic } H_2C-C\equiv C-CH_2 \text{ (ring)}$$

Intramolecular cyclization of large rings must be carried out in dilute solution so the last S_N2 displacement will be *intra*molecular (within one molecule) and not *inter*molecular (between two molecules).

9-8

(a) $$H-C\equiv C-H \xrightarrow[2)\ H_2C=O]{1)\ NaNH_2} \xrightarrow{H_3O^+} H-C\equiv C-CH_2OH$$

(b) $$H-C\equiv C-H \xrightarrow[2)\ CH_3I]{1)\ NaNH_2} CH_3-C\equiv C-H \xrightarrow[2)\ HC(=O)CH_2CH_2CH_3\quad 3)\ H_3O^+]{1)\ NaNH_2} CH_3-C\equiv C-CH(OH)CH_2CH_2CH_3$$

9-8 continued

(c) $H-C\equiv C-H \xrightarrow[\text{2) Ph(C=O)CH}_3]{\text{1) NaNH}_2} \xrightarrow{H_3O^+} H_3C-C(OH)(Ph)-C\equiv C-H$

(d) $H-C\equiv C-H \xrightarrow[\text{2) CH}_3\text{I}]{\text{1) NaNH}_2} CH_3-C\equiv C-H \xrightarrow[\text{2) butan-2-one}]{\text{1) NaNH}_2} \xrightarrow{H_3O^+} CH_3-C\equiv C-C(OH)(CH_3)CH_2CH_3$

9-9 In a synthesis problem, draw the target to see what new bonds need to be made during the synthesis.

$H-C\equiv C-CH_2CH_3 \xrightarrow[\text{2) Ph(C=O)CH}_3]{\text{1) NaNH}_2} \xrightarrow{H_3O^+} H_3C-C(OH)(Ph)-C\equiv C-CH_2CH_3$ 2-phenylhex-3-yn-2-ol

(new C—C bond between C2 and C3; carbons numbered 1 2 3 4 5 6)

9-10 The mechanism is likely to be two sequential E2 reactions. KOH usually makes internal alkynes.

$PhCH_2-CBr_2-CH_3 \xrightarrow[\Delta]{^-:OH} PhCH=CBr-CH_3 \xrightarrow[\Delta]{^-:OH} Ph-C\equiv C-CH_3$

1-phenylprop-1-yne

9-11 The mechanism is likely to be two sequential E2 reactions. Where possible, $NaNH_2$ makes terminal alkynes because the terminal proton is removed by $NaNH_2$ giving a stable acetylide ion as the initial product. (This is called a *thermodynamic sink*, or a *potential energy well*, that is, a valley on the energy diagram.)

$PhCH_2-CBr_2-CH_3 \xrightarrow[\Delta]{^-:NH_2} PhCH_2-CBr=CH_2 \xrightarrow[\Delta]{^-:NH_2} PhCH_2-C\equiv C-H \xrightarrow{H_2N:^-} PhCH_2-C\equiv C:^- \xrightarrow{H_2O} PhCH_2-C\equiv CH$

3-phenylprop-1-yne

This acetylide ion remains in the reaction mixture until a proton source is added.

9-12

(a) $CH_3CH_2-C\equiv C-CH_2CH_2CH_2CH_3 \xrightarrow[\text{Lindlar's catalyst}]{H_2}$ cis-$CH_3CH_2(H)C=C(H)CH_2CH_2CH_2CH_3$

(b) $H_3C-C\equiv C-CH_2CH_3 \xrightarrow[NH_3]{Na}$ trans-$H_3C(H)C=C(H)CH_2CH_3$

9-12 continued

(c)

[Reaction scheme: *cis*-cyclodecene $\xrightarrow{Br_2,\ CCl_4}$ trans-1,2-dibromocyclodecane (Br, H / H, Br) $\xrightarrow{NaNH_2,\ 150°}$ cyclodecyne ($-C\equiv C-$) $\xrightarrow{Na,\ NH_3}$ *trans*-cyclodecene]

cis *trans*

One of the useful "tricks" of organic chemistry is the ability to convert one stereoisomer into another. Having a pair of reactions like the two reductions shown in parts (a) and (b) that give opposite stereochemistry is very useful, because a common intermediate (the alkyne) can be transformed into either stereoisomer.

(d) $CH_3CH_2-C\equiv C-H \xrightarrow[2)\ CH_3CH_2Br]{1)\ NaNH_2} CH_3CH_2-C\equiv C-CH_2CH_3 \xrightarrow[\text{Lindlar's catalyst}]{H_2}$ (cis) $CH_3CH_2(H)C=C(H)CH_2CH_3$

9-13

$CH_3CH_2CH_2-C\equiv C-H \xrightarrow[2)\ PhCHO;\ 3)\ H_3O^+]{1)\ NaNH_2} CH_3CH_2CH_2-C\equiv C-CH(OH)Ph \xrightarrow[\text{Lindlar's catalyst}]{H_2}$ (cis) $CH_3CH_2CH_2(H)C=C(H)-CH(OH)Ph$

(*Z*)-1-phenylhex-2-en-1-ol

9-14 The goal is to add only one equivalent of bromine, always avoiding an excess of bromine, because two molecules of bromine could add to the triple bond if bromine were in excess. If the alkyne is added to the bromine, the first drops of alkyne will encounter a large excess of bromine. Instead, adding bromine to the alkyne will always ensure an excess of alkyne and should give a good yield of dibromo product.

9-15

$CH_3CH_2CH_2-C\equiv C-H \xrightarrow{H-Br} CH_3CH_2CH_2-\overset{\oplus}{C}=CH-H$ (2° better than 1° carbocation) $\xrightarrow{:Br:^{\ominus}} CH_3CH_2CH_2-C(Br)=CH-H$

$CH_3CH_2CH_2-C(Br)=CH_2 \xrightarrow{H-Br} \{CH_3CH_2CH_2-\overset{\oplus}{C}(Br)-CH_2-H \longleftrightarrow CH_3CH_2CH_2-C(=Br^{\oplus})-CH_2-H\}$

2° carbocation and **resonance-stabilized**

$\xrightarrow{:Br:^{\ominus}} CH_3CH_2CH_2-CBr_2-CH_2-H$

9-16

(a) $C_6H_5-C\equiv CH$ + 2 HBr → $C_6H_5-CBr_2-CH_3$

(b) $CH_3CH_2CH_2CH_2-C\equiv CH$ + 2 HCl → $CH_3CH_2CH_2CH_2-CCl_2-CH_3$

Electrophilic addition to terminal alkynes follows Markovnikov orientation.

(c) cyclooctyne + 2 HBr → 1,1-dibromocyclooctane

(d) $CH_3CH_2CH_2-\overset{3}{C}\equiv\overset{2}{C}-\overset{1}{C}H_3$ + 2 HCl → 3,3-dichloro product + 2,2-dichloro product

Electrophilic addition to unsymmetric internal alkynes will give mixtures of isomers.

9-17

initiation:

$RO-OR \xrightarrow{h\nu} 2\ RO\bullet$

$RO\bullet + H-Br \longrightarrow RO-H + Br\bullet$

propagation:

$Br\bullet + H-C\equiv C-CH_2CH_2CH_3 \longrightarrow H-C(Br)=\dot{C}-CH_2CH_2CH_3$

$H-C(Br)=\dot{C}-CH_2CHCH_3 + H-Br \longrightarrow H-C(Br)=C(H)-CH_2CH_2CH_3 + Br\bullet$

The 2° radical is more stable than 1°. The anti-Markovnikov orientation occurs because the bromine radical attacks first to make the most stable radical, in contrast to electrophilic addition where the H^+ bonds first (see the solution to 9-15).

9-18

(a) $H-C\equiv C-(CH_2)_3CH_3 \xrightarrow{1\text{ eq. } Cl_2} H-C(Cl)=C(Cl)-(CH_2)_3CH_3$ *E* + *Z*

(b) $H-C\equiv C-(CH_2)_3CH_3 \xrightarrow[ROOR]{HBr} H-C(Br)=C(H)-(CH_2)_3CH_3$ *E* + *Z*

(c) $H-C\equiv C-(CH_2)_3CH_3 \xrightarrow{HBr} H-C(H)=C(Br)-(CH_2)_3CH_3$

(d) $H-C\equiv C-(CH_2)_3CH_3 \xrightarrow[CCl_4]{2\ Br_2} H-CBr_2-CBr_2-(CH_2)_3CH_3$

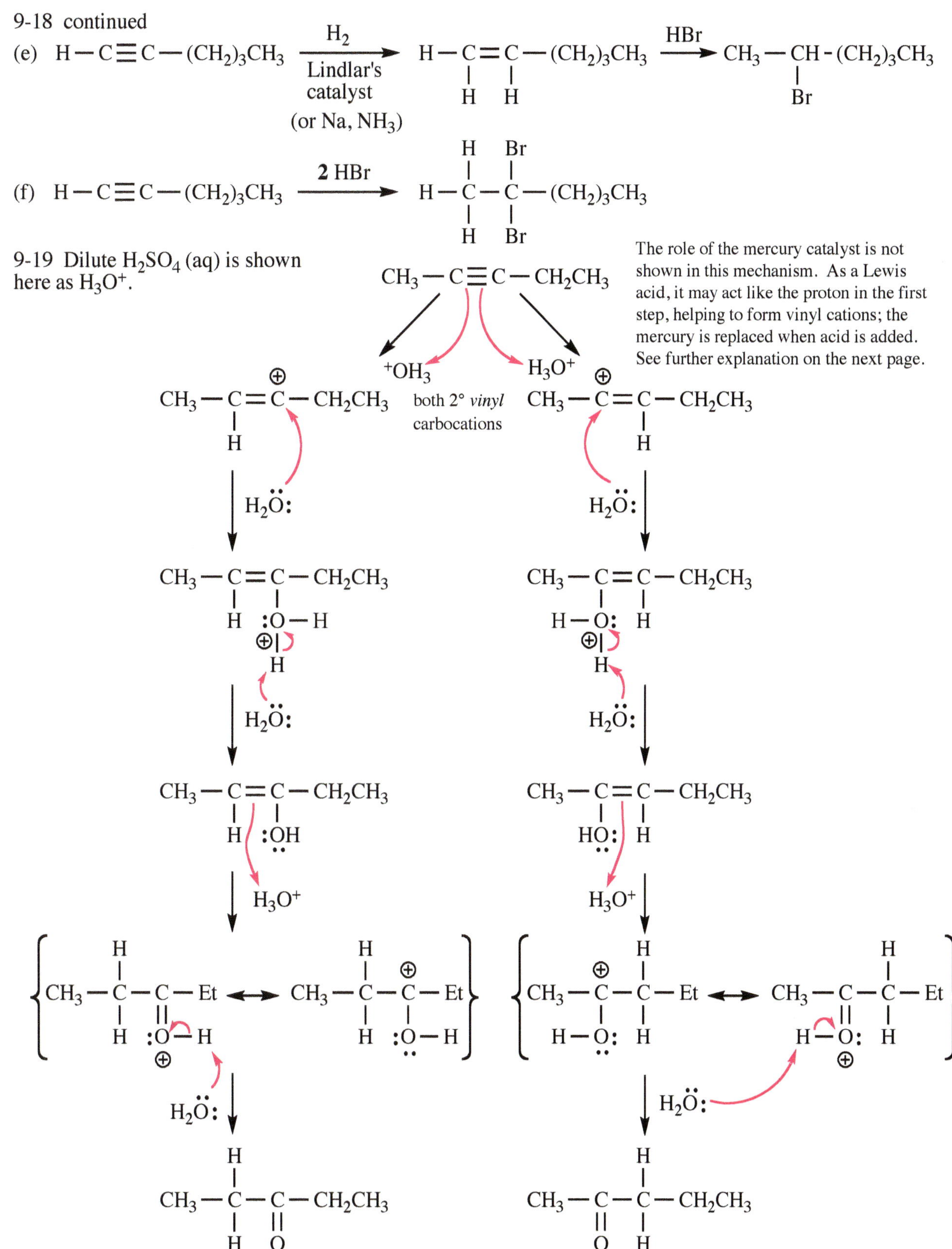
9-18 continued
(e)
H_2
Lindlar's catalyst
(or Na, NH_3)
HBr
(f)
2 HBr
9-19 Dilute H_2SO_4 (aq) is shown here as H_3O^+.
The role of the mercury catalyst is not shown in this mechanism. As a Lewis acid, it may act like the proton in the first step, helping to form vinyl cations; the mercury is replaced when acid is added. See further explanation on the next page.
both 2° *vinyl* carbocations

pentan-3-one

pentan-2-one

continued on next page

9-19 continued

The role of the mercury catalyst is not shown in this mechanism. As a Lewis acid, it may act like the proton in the first step, helping to form vinyl cations; the mercury is replaced when acid is added.

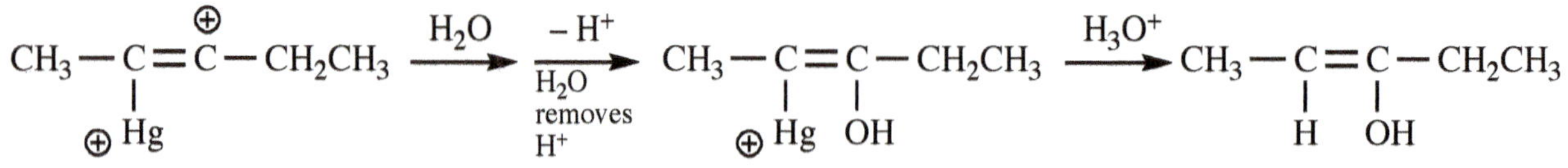

$$CH_3-\overset{}{\underset{\oplus Hg}{C}}=\overset{\oplus}{C}-CH_2CH_3 \xrightarrow{H_2O} \xrightarrow[\text{H}_2\text{O removes H}^+]{-H^+} CH_3-\underset{\oplus Hg}{C}=\underset{OH}{C}-CH_2CH_3 \xrightarrow{H_3O^+} CH_3-\underset{H}{C}=\underset{OH}{C}-CH_2CH_3$$

9-20

(a) But-2-yne is symmetric. Either orientation produces the same product.

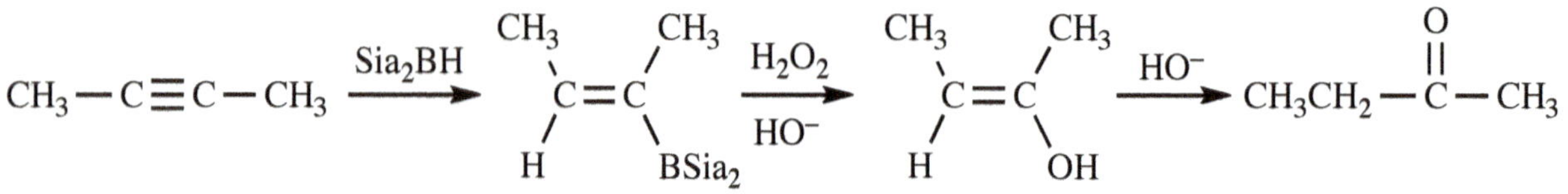

$$CH_3-C\equiv C-CH_3 \xrightarrow{Sia_2BH} (CH_3)(H)C=C(CH_3)(BSia_2) \xrightarrow[HO^-]{H_2O_2} (CH_3)(H)C=C(CH_3)(OH) \xrightarrow{HO^-} CH_3CH_2-\overset{O}{\overset{\|}{C}}-CH_3$$

(b) Pent-2-yne is not symmetric. Different orientations of attack will lead to different products on any unsymmetrical internal alkyne.

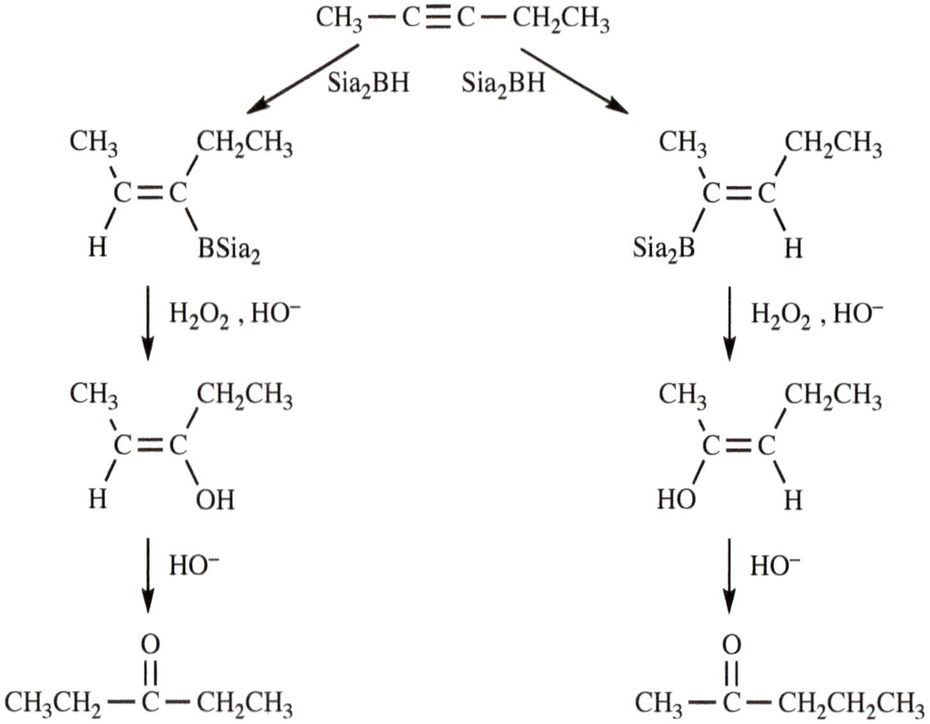

$$CH_3-C\equiv C-CH_2CH_3$$

Left path: $\xrightarrow{Sia_2BH}$ $(CH_3)(H)C=C(CH_2CH_3)(BSia_2)$ $\xrightarrow{H_2O_2,\ HO^-}$ $(CH_3)(H)C=C(CH_2CH_3)(OH)$ $\xrightarrow{HO^-}$ $CH_3CH_2-\overset{O}{\overset{\|}{C}}-CH_2CH_3$

Right path: $\xrightarrow{Sia_2BH}$ $(CH_3)(Sia_2B)C=C(CH_2CH_3)(H)$ $\xrightarrow{H_2O_2,\ HO^-}$ $(CH_3)(HO)C=C(CH_2CH_3)(H)$ $\xrightarrow{HO^-}$ $CH_3-\overset{O}{\overset{\|}{C}}-CH_2CH_2CH_3$

9-21

(a) (1) 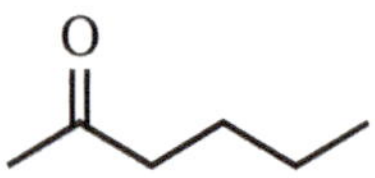 [hexan-2-one structure] (2) H [hexanal structure]

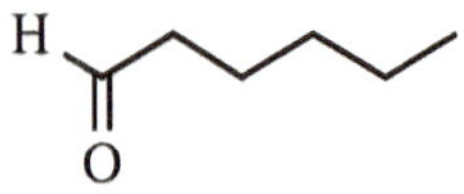

(b) (1) 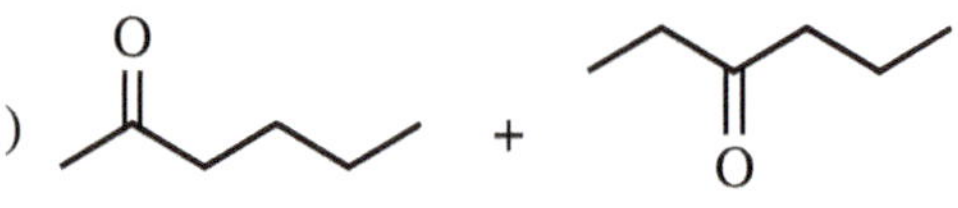 [hexan-2-one structure] + [hexan-3-one structure]

(2) same mixture as in (b) (1)

(c) (1) 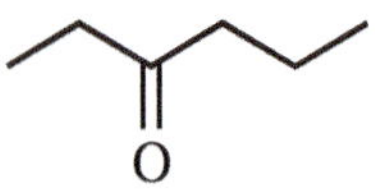 [hexan-3-one structure]

(2) same as in (c) (1)

(d) (1) 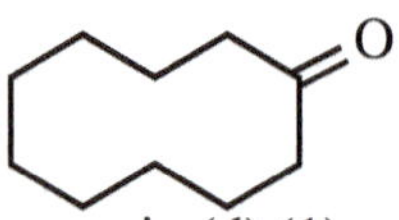 [cyclodecanone structure]

(2) same as in (d) (1)

9-22

(a) $(H_3C)_2C=CH(CH_3)$ + BH_3 ⟶ $H-C(CH_3)_2-C(H)(CH_3)-BH_2$ + $(H_3C)_2C=CH(CH_3)$

↓

$H-C(CH_3)_2-C(H)(CH_3)-B(H)-C(H)(CH_3)-C(CH_3)_2-H$

disiamylborane, Sia_2BH

(b) There is too much steric hindrance in Sia_2BH for the third B—H to add across another alkene. The reagent can add to alkynes because alkynes are linear and attack is not hindered by bulky substituents.

9-23

(a) (1) $HO-C(=O)-C(=O)-(CH_2)_3CH_3$

Oxidation of a terminal alkyne with neutral $KMnO_4$ produces the ketone and carboxylic acid without cleaving the carbon-carbon bond.

(2) CO_2 + $HO-C(=O)-(CH_2)_3CH_3$

Oxidation of a terminal alkyne with warm, basic $KMnO_4$ cleaves the carbon-carbon bond, producing the carboxylic acid after acid work-up, and carbon dioxide.

(b) (1) $CH_3-C(=O)-C(=O)-CH_2CH_2CH_3$

(2) $CH_3-C(=O)-OH$ + $HO-C(=O)-CH_2CH_2CH_3$

(c) (1) $CH_3CH_2-C(=O)-C(=O)-CH_2CH_3$

(2) $CH_3CH_2-C(=O)-OH$ (2 equivalents)

(d) (1) $CH_3CH(CH_3)-C(=O)-C(=O)-CH_2CH_3$

(2) $CH_3CH(CH_3)-C(=O)-OH$ + $HO-C(=O)-CH_2CH_3$

(e) (1) [cyclic 1,2-dione: ring with two adjacent C=O]

(2) HO–C(=O)–(CH_2)_n–C(=O)–OH [dicarboxylic acid; O, HO, OH, O labels]

9-24

Work backwards on ozonolysis problems like this: ozonolysis of trip bonds produces RCOOH, so reconnect the fragments at the COOH groups, creating a C≡C.

(a) $CH_3-C\equiv C-(CH_2)_4-C\equiv C-CH_3$

(b) [cyclic alkyne: ring containing –C≡C–, with substituent $C\equiv C-CH_2CH_3$]

9-25 When proposing syntheses, begin by analyzing the target molecule, looking for smaller pieces that can be combined to make the desired compound. This is especially true for targets that have more carbons than the starting materials; immediately, you will know that a carbon-carbon bond forming reaction will be necessary.

People who succeed at synthesis *know the reactions*—there is no shortcut. Practice the reactions for each functional group until they become automatic. Use the graphical reaction summaries ("starburst diagrams"), or customize your own—these are very effective learning tools.

(a) analysis of target

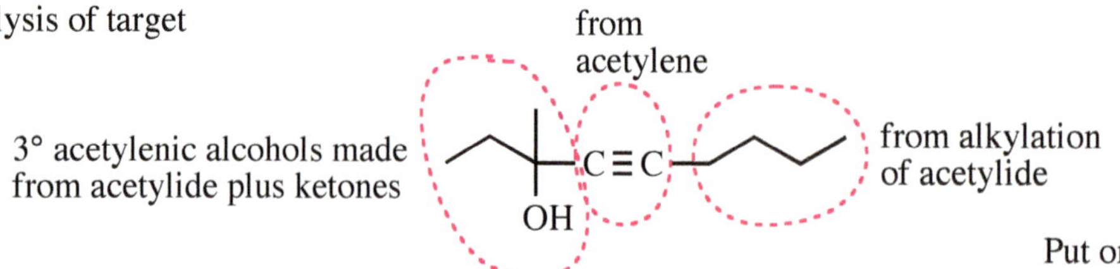

from acetylene
3° acetylenic alcohols made from acetylide plus ketones
C≡C
OH
from alkylation of acetylide

forward direction:

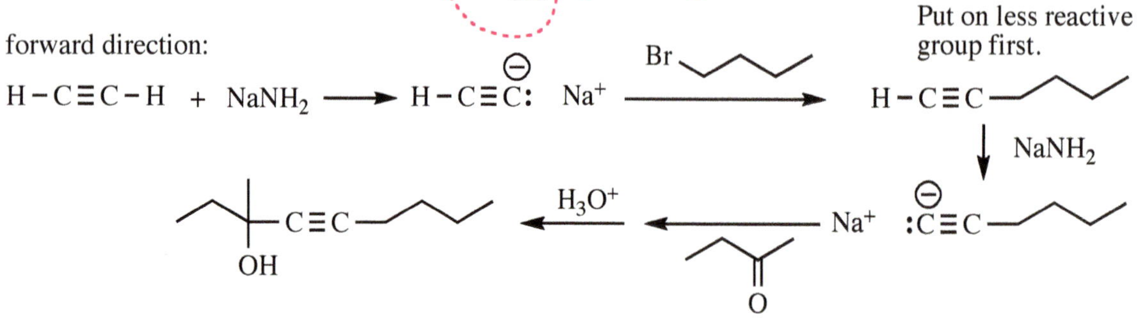

H−C≡C−H + NaNH₂
H−C≡C: Na⁺
Br
Put on less reactive group first.
H−C≡C
NaNH₂
Na⁺ :C≡C
O
H₃O⁺
C≡C
OH

(b)

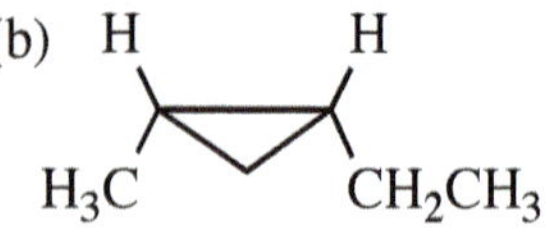

H
H
H₃C
CH₂CH₃

Analysis of target: cyclopropanes are made by carbene insertion into alkenes. To get *cis* substitution around cyclopropane, stereochemistry of the alkene precursor must be *cis*. *Cis* alkenes come from catalytic hydrogenation of an alkyne using Lindlar's catalyst.

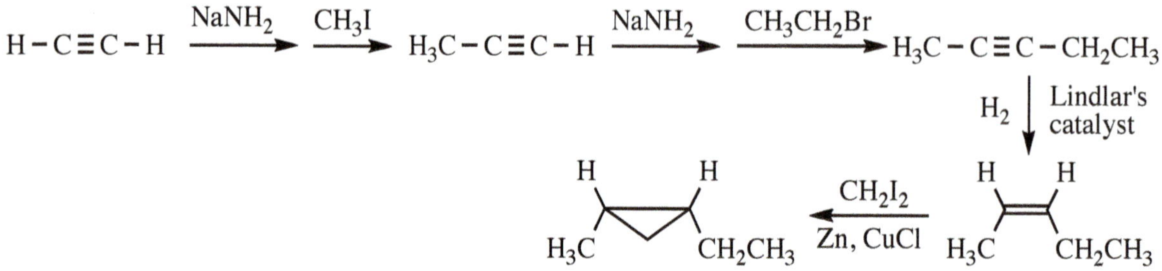

H−C≡C−H
NaNH₂
CH₃I
H₃C−C≡C−H
NaNH₂
CH₃CH₂Br
H₃C−C≡C−CH₂CH₃
H₂
Lindlar's catalyst
H H
H₃C CH₂CH₃
CH₂I₂
Zn, CuCl
H H
H₃C CH₂CH₃

(c)

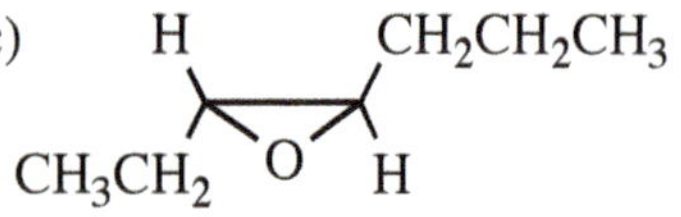

H
CH₂CH₂CH₃
CH₃CH₂
O
H

Analysis of target: epoxides are made by direct epoxidation of alkenes. To get *trans* substitution around the epoxide, stereochemistry of the alkene precursor must be *trans*. *Trans* alkenes come from sodium/ammonia reduction of an alkyne.

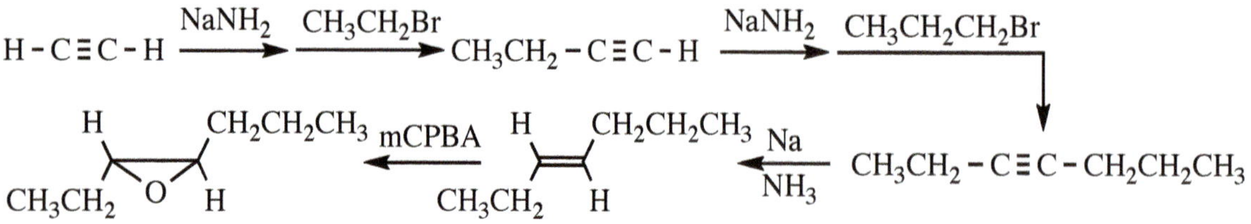

H−C≡C−H
NaNH₂
CH₃CH₂Br
CH₃CH₂−C≡C−H
NaNH₂
CH₃CH₂CH₂Br
CH₃CH₂−C≡C−CH₂CH₂CH₃
Na
NH₃
H CH₂CH₂CH₃
CH₃CH₂ H
mCPBA
H CH₂CH₂CH₃
CH₃CH₂ O H

9-25 continued

(d) *meso*-hexane-3,4-diol

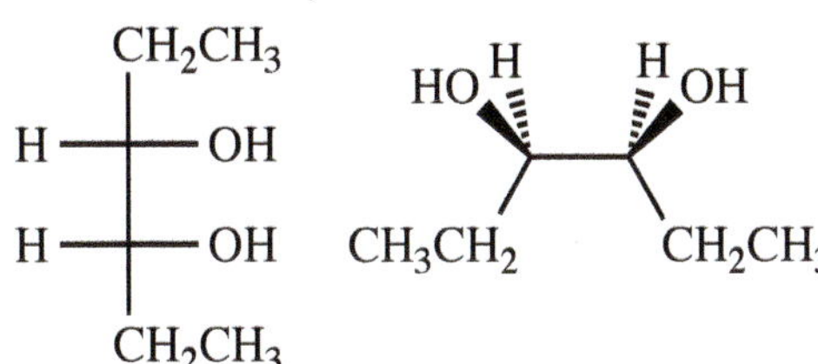

Analysis of target: diols are prepared from the C=C by two methods: cold, dilute $KMnO_4$ (or OsO_4), a syn addition; or with a peroxy acid that makes the epoxide, followed by hydrolysis, an anti addition of the two OH groups. Recall from the solution to 8-35 that in order to produce a meso structure, a syn addition requires a *cis* alkene, and an anti addition requires a *trans* alkene.

Both syntheses begin with the preparation of hex-3-yne:

$$H-C\equiv C-H \xrightarrow{NaNH_2} \xrightarrow{CH_3CH_2Br} CH_3CH_2-C\equiv CH \xrightarrow{NaNH_2} \xrightarrow{CH_3CH_2Br} CH_3CH_2-C\equiv C-CH_2CH_3$$

Route A: via an epoxide on a *trans* alkene

$CH_3CH_2-C\equiv C-CH_2CH_3 \xrightarrow[NH_3]{Na}$ (*trans*-hex-3-ene: H, CH_2CH_3 / CH_3CH_2, H) $\xrightarrow{mCPBA}$ (epoxide: H, CH_2CH_3 / CH_3CH_2, O, H) $\xrightarrow[H_3O^+]{}$ meso product

Route B: via *syn* hydroxylation on a *cis* alkene

$CH_3CH_2-C\equiv C-CH_2CH_3 \xrightarrow[\text{Lindlar's catalyst}]{H_2}$ (*cis*-hex-3-ene: H, H / CH_3CH_2, CH_2CH_3) $\xrightarrow[\text{or } OsO_4\text{ , } H_2O_2]{\text{cold, dilute } KMnO_4}$ meso product

9-26

(a) $CH_3-C\equiv C-(CH_2)_4CH_3$ (b) $CH_3CH_2-C\equiv C-CH_2CH_2CH(CH_3)_2$ (c) (phenyl)$-C\equiv C-H$

(d) (cyclohexyl)$-C\equiv C-H$ (e) $CH_3CH_2-C\equiv C-CHCH_2CH_2CH_3$ with CH_3 on the CH

(f)

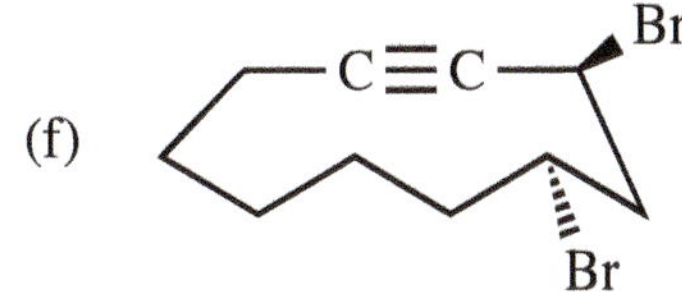

(g) Br Br

(h) C≡C

(i) $H-C\equiv C-CH_2-C\equiv C-CH_2CH_3$ (j) $H-C\equiv C-CH=CH_2$ (k) $H-C\equiv C-C(CH_3)(H)-CH=CH_2$

9-27

(a) ethylmethylacetylene (b) phenylacetylene (c) *sec*-butylpropylacetylene
(d) *sec*-butyl-*tert*-butylacetylene

9-28

(a) 4-phenylpent-2-yne (b) (*E*)-3-methylhept-2-en-4-yne (c) 2,2,5-trimethylhept-3-yne
(d) 4,4-dibromopent-2-yne (e) 3-methylhex-4-yn-3-ol (f) 1-cyclopentylbut-2-yne

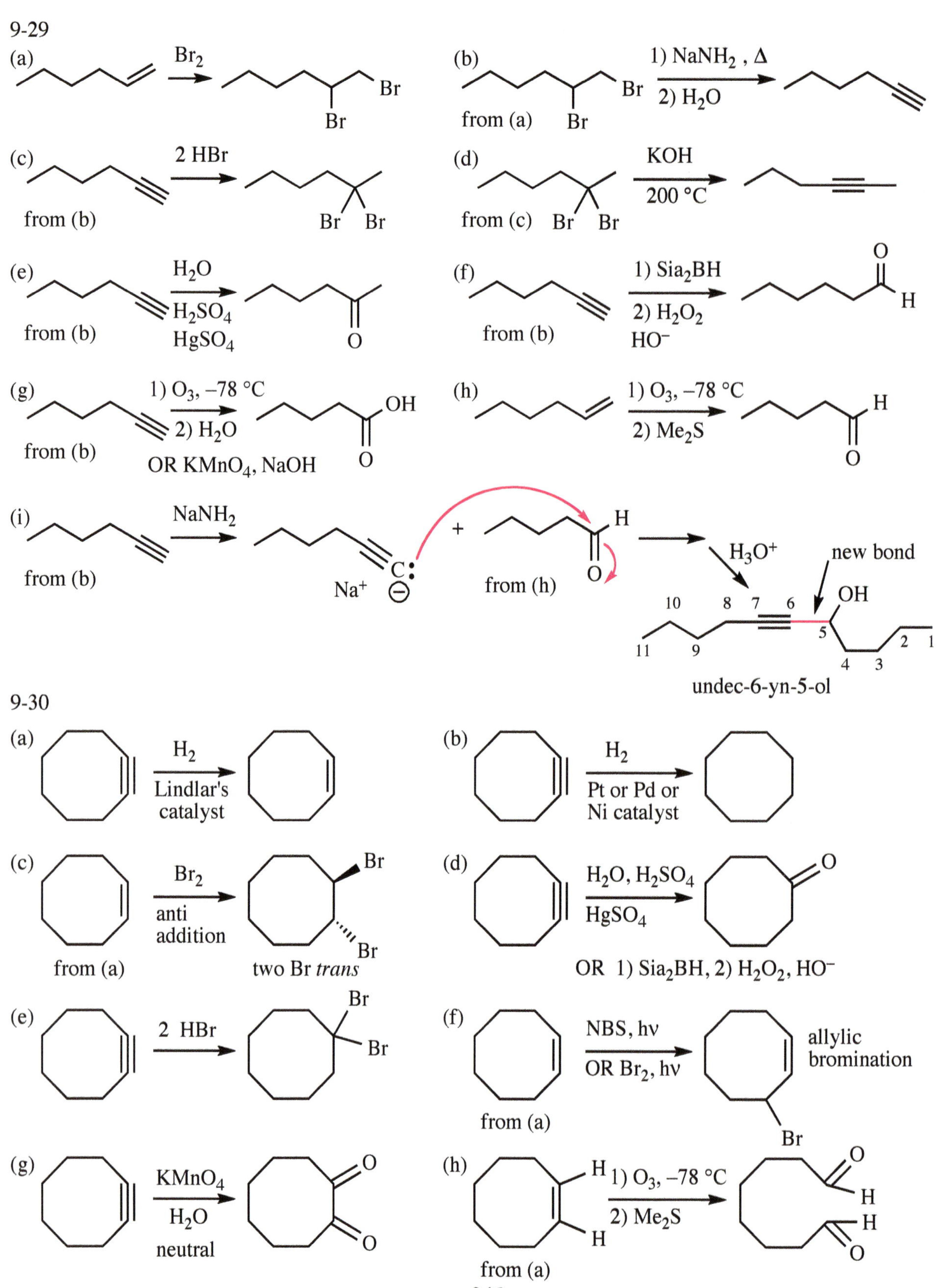
9-29
(a)
Br_2
Br
Br
(b)
1) $NaNH_2$, Δ
2) H_2O
Br
from (a)
Br
(c)
2 HBr
from (b)
Br
Br
(d)
KOH
200 °C
from (c)
Br
Br
(e)
H_2O
H_2SO_4
$HgSO_4$
from (b)
O
(f)
1) Sia_2BH
2) H_2O_2
HO^-
from (b)
O
H
(g)
1) O_3, –78 °C
2) H_2O
OR $KMnO_4$, NaOH
from (b)
OH
O
(h)
1) O_3, –78 °C
2) Me_2S
H
O
(i)
$NaNH_2$
from (b)
Na^+
+
from (h)
H
O
H_3O^+
new bond
OH
10
8
7
6
5
11
9
4
3
2
1
undec-6-yn-5-ol
9-30
(a)
H_2
Lindlar's catalyst
(b)
H_2
Pt or Pd or Ni catalyst
(c)
Br_2
anti addition
from (a)
Br
Br
two Br *trans*
(d)
H_2O, H_2SO_4
$HgSO_4$
O
OR 1) Sia_2BH, 2) H_2O_2, HO^-
(e)
2 HBr
Br
Br
(f)
NBS, hν
OR Br_2, hν
from (a)
allylic bromination
Br
(g)
$KMnO_4$
H_2O
neutral
O
O
(h)
H
H
1) O_3, –78 °C
2) Me_2S
from (a)
O
H
H
O

9-30 continued

(i) cyclooctyne $\xrightarrow[\text{2) } H_3O^+]{\text{1) } KMnO_4, NaOH}$ OR $\xrightarrow[\text{2) } H_3O^+]{\text{1) } O_3, -78\ °C}$ HOOC–$(CH_2)_6$–COOH

9-31

$CH_3CH_2-C(=O)-CH=CHCl$ + $:\overset{\ominus}{C}\equiv CH$ (from $HC\equiv CH$ + $NaNH_2$) $\longrightarrow \xrightarrow{H_3O^+}$ $CH_3CH_2-C(OH)(CH=CHCl)-C\equiv CH$ ethchlorvynol

9-32

$H-C\equiv C-H \xrightarrow{NaNH_2} \xrightarrow{CH_3(CH_2)_7Br} CH_3(CH_2)_7-C\equiv C-H \xrightarrow{NaNH_2} \xrightarrow{CH_3(CH_2)_{12}Br} CH_3(CH_2)_7-C\equiv C-(CH_2)_{12}CH_3$

$\xrightarrow[\text{Lindlar's catalyst}]{H_2}$ cis-$CH_3(CH_2)_7CH=CH(CH_2)_{12}CH_3$ muscalure

9-33

(a) $CH_2=CClCH_2CH_2CH_3$ (b) $H_3C-CCl_2-CH_2CH_2CH_3$ (c) $CH_3CH_2CH_2CH_2CH_3$

(d) $CH_2=CHCH_2CH_2CH_3$ (e) $(H)(Br)C=C(Br)(CH_2CH_2CH_3)$ (f) $H-CBr_2-CBr_2-CH_2CH_2CH_3$

(g) $HO-C(=O)-C(=O)-CH_2CH_2CH_3$ (h) CO_2 + $HO-C(=O)-CH_2CH_2CH_3$ (i) $H_2C=CHCH_2CH_2CH_3$

(j) Na^+ $:\overset{\ominus}{C}\equiv C-CH_2CH_2CH_3$ (k) $H_3C-C(=O)-CH_2CH_2CH_3$ (l) $H-C(=O)-CH_2CH_2CH_2CH_3$

9-34

(a) $H_3C-CBr_2-CH_2CH_3 \xrightarrow[150°]{NaNH_2} \xrightarrow{H_2O} H-C\equiv C-CH_2CH_3$

(b) $H_3C-CBr_2-CH_2CH_3 \xrightarrow[200°]{KOH} H_3C-C\equiv C-CH_3$

(c) $CH_3CH_2-C\equiv C-H \xrightarrow{NaNH_2} CH_3CH_2-C\equiv \overset{\ominus}{C}:\ Na^+ \xrightarrow{CH_3CH_2CH_2CH_2Br} CH_3CH_2-C\equiv C-CH_2CH_2CH_2CH_3$

9-34 continued

(d) $\underset{H}{\overset{H_3C}{}}C{=}C\underset{CH_2CH_2CH_3}{\overset{H}{}} \xrightarrow[CCl_4]{Br_2} CH_3\underset{H}{\overset{Br}{C}}-\underset{H}{\overset{Br}{C}}CH_2CH_2CH_3 \xrightarrow[200°]{KOH} H_3C-C{\equiv}C-CH_2CH_2CH_3$

(e) $H_3C-\underset{Br}{\overset{Br}{C}}-CH_2CH_2CH_2CH_3 \xrightarrow[2)\ H_2O]{1)\ NaNH_2,\ 150°} H-C{\equiv}C-CH_2CH_2CH_2CH_3$

(f) cyclodecyne $\xrightarrow[\text{Lindlar's catalyst}]{H_2}$ cyclodecene — cis

(g) cyclodecyne $\xrightarrow[NH_3]{Na}$ cyclodecene — trans

(h) $HC{\equiv}C-CH_2CH_2CH_2CH_3 \xrightarrow[H_2SO_4,\ HgSO_4]{H_2O} \left[CH_2{=}\overset{OH}{C}CH_2CH_2CH_2CH_3\right] \longrightarrow H_3C-\overset{O}{\overset{\|}{C}}CH_2CH_2CH_2CH_3$

unstable enol

(i) $HC{\equiv}C-CH_2CH_2CH_2CH_3 \xrightarrow[2)\ H_2O_2,\ HO^-]{1)\ Sia_2BH} \left[\overset{OH}{CH}{=}CHCH_2CH_2CH_2CH_3\right] \longrightarrow H-\overset{O}{\overset{\|}{C}}CH_2CH_2CH_2CH_2CH_3$

unstable enol

(j) $\underset{H}{\overset{H_3C}{}}C{=}C\underset{CH_2CH_2CH_3}{\overset{H}{}} \xrightarrow[CCl_4]{Br_2} H_3C-\underset{H}{\overset{Br}{C}}-\underset{H}{\overset{Br}{C}}CH_2CH_2CH_3 \xrightarrow[200°]{KOH} H_3C-C{\equiv}C-CH_2CH_2CH_3$ (major)

$\xrightarrow{H_2,\ \text{Lindlar's catalyst}}$ $\underset{H}{\overset{H_3C}{}}C{=}C\underset{H}{\overset{CH_2CH_2CH_3}{}}$

9-35 All four syntheses in this problem begin with the same reaction of benzyl bromide with acetylide ion:

$HC{\equiv}CH \xrightarrow{NaNH_2} HC{\equiv}C{:}^{\ominus} + CH_2Br$ (benzyl bromide, $C_6H_5CH_2Br$) $\longrightarrow PhCH_2-C{\equiv}CH \xrightarrow{NaNH_2} PhCH_2-C{\equiv}C{:}^{\ominus}$

use this below

9-35 continued

(a) $PhCH_2-C\equiv C:^{\ominus}$ + Br (allyl bromide) ⟶ $PhCH_2-C\equiv C-CH_2CH=CH_2$ 6-phenylhex-1-en-4-yne

(b) $PhCH_2-C\equiv C:^{\ominus}$ + Br (ethyl bromide) ⟶ $PhCH_2-C\equiv C-CH_2CH_3$ $\xrightarrow[\text{Pd/BaSO}_4\text{, quinoline}]{H_2}$ *cis*-1-phenylpent-2-ene

Lindlar's catalyst

(c) $PhCH_2-C\equiv C:^{\ominus}$ + Br (ethyl bromide) ⟶ $PhCH_2-C\equiv C-CH_2CH_3$ $\xrightarrow[NH_3]{Na}$ *trans*-1-phenylpent-2-ene

(d) The diol with the two OH groups on the same side in the Fischer projection is the equivalent of a meso structure, although this one is not meso because the top and bottom group are different. Still, it gives a clue as to its synthesis. The "meso" diol can be formed by either a syn addition to a *cis* double bond, or by an anti addition to a *trans* double bond. We saw the same thing in the solution to 9-25 (d).

cis product from part (b) $\xrightarrow[H_2O_2]{OsO_4}$ (syn addition) diol $\xleftarrow[H_3O^+]{HCO_3H}$ (anti addition) *trans* product from part (c)

Fischer projection: CH_2Ph (top); H—C—OH; H—C—OH; CH_2CH_3 (bottom)

racemic

9-36

(a) $CH_3CH_2-C\equiv C-CH_2CH_3$

(b) $CH_3CH_2-C\equiv C-H$ + $CH_2=C(CH_3)_2$ elimination on 3° halide

(c) $CH_3CH_2-C\equiv C-CH_2OH$ (after H_2O workup)

(d) $CH_3CH_2-C\equiv C-$(1-hydroxycyclohexyl) (after H_2O workup)

(e) $CH_3CH_2-C\equiv C-CH(OH)CH_2CH_2CH_3$ (after H_2O workup)

(f) $CH_3CH_2-C\equiv C-H$ + Na^+ $^{\ominus}O-$cyclohexyl

(g) $CH_3CH_2-C\equiv C-C(OH)(CH_3)-CH_2CH_3$ (after H_2O workup)

9-37

(a) $HC\equiv C-H \xrightarrow{NaNH_2} HC\equiv C:^{\ominus}\ Na^+ \xrightarrow{CH_3CH_2CH_2CH_2Br} HC\equiv C-CH_2CH_2CH_2CH_3$

9-37 continued

(b) $HC{\equiv}C{-}H \xrightarrow{NaNH_2} HC{\equiv}\overset{\ominus}{C}:\ Na^+ \xrightarrow{CH_3CH_2CH_2Br} HC{\equiv}C{-}CH_2CH_2CH_3 \xrightarrow[CH_3I]{NaNH_2} H_3C{-}C{\equiv}C{-}CH_2CH_2CH_3$

(c) $H_3C{-}C{\equiv}C{-}CH_2CH_2CH_3 \xrightarrow[\text{Lindlar's catalyst}]{H_2}$ cis-alkene (H_3C and $CH_2CH_2CH_3$ on the same side; H and H on the same side)

synthesized in part (b)

(d) $H_3C{-}C{\equiv}C{-}CH_2CH_2CH_3 \xrightarrow[NH_3]{Na}$ trans-alkene (H_3C and H on top; H and $CH_2CH_2CH_3$ on bottom)

synthesized in part (b)

(e) $HC{\equiv}C{-}CH_2CH_2CH_2CH_3 \xrightarrow[HOOH]{\text{2 equiv. } HBr} Br_2HC{-}CH_2CH_2CH_2CH_2CH_3$ anti-Markovnikov orientation

synthesized in part (a)

(f) $HC{\equiv}C{-}CH_2CH_2CH_2CH_3 \xrightarrow{\text{2 HBr}} H_3C{-}CBr_2{-}CH_2CH_2CH_2CH_3$ Markovnikov orientation

synthesized in part (a)

(g) $H{-}C{\equiv}C{-}CH_2CH_2CH_3 \xrightarrow[\text{2) } H_2O_2,\ HO^-]{\text{1) } Sia_2BH} H{-}\overset{O}{\overset{\|}{C}}CH_2CH_2CH_2CH_3$

from (b)

(h) $H{-}C{\equiv}C{-}CH_2CH_2CH_3 \xrightarrow[H_2SO_4,\ HgSO_4]{H_2O} H_3C{-}\overset{O}{\overset{\|}{C}}CH_2CH_2CH_3$

from (b)

(i) $HC{\equiv}C{-}H \xrightarrow{NaNH_2} HC{\equiv}\overset{\ominus}{C}:\ Na^+ \xrightarrow{CH_3CH_2Br} HC{\equiv}C{-}CH_2CH_3 \xrightarrow{NaNH_2} \xrightarrow{CH_3CH_2Br} CH_3CH_2{-}C{\equiv}C{-}CH_2CH_3 \xrightarrow[\text{Lindlar's catalyst}]{H_2}$ cis-alkene (CH_3CH_2 and CH_2CH_3 on the same side; H and H) $\xrightarrow{Br_2}$ $CH_3CH_2CHBr{-}CHBrCH_2CH_3$ (H, Br wedged/hashed as drawn)

Alkene must be *cis* to produce the (±) product from anti addition.

Review the stereochemistry in the solution to Problem 8-35 of this Solutions Manual.

9-37 continued

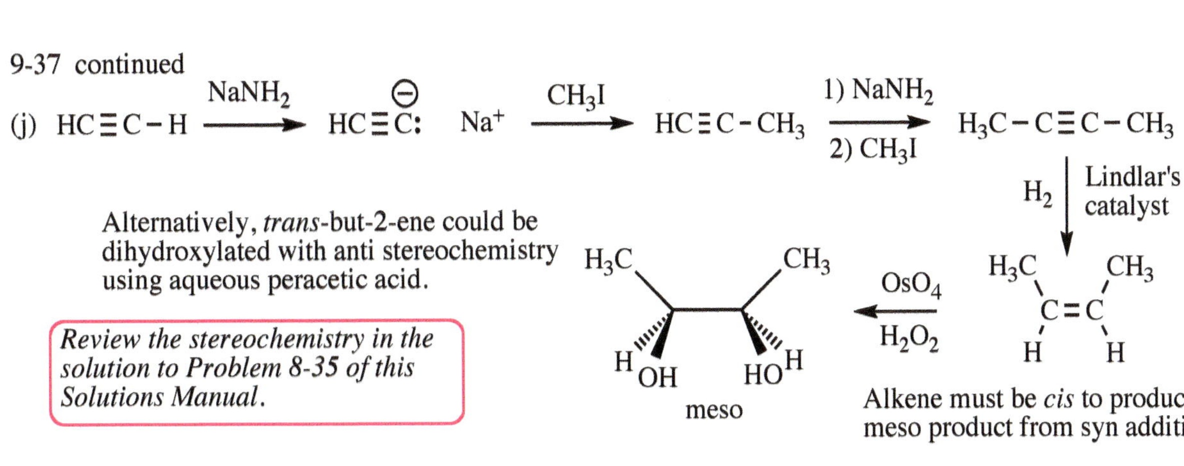

(j) HC≡C−H
NaNH2
HC≡C:⊖ Na+
CH3I
HC≡C−CH3
1) NaNH2
2) CH3I
H3C−C≡C−CH3
H2
Lindlar's catalyst
OsO4
H2O2
meso
Alternatively, trans-but-2-ene could be dihydroxylated with anti stereochemistry using aqueous peracetic acid.
Review the stereochemistry in the solution to Problem 8-35 of this Solutions Manual.
Alkene must be cis to produce the meso product from syn addition.

(k) $HC{\equiv}C{-}H \xrightarrow{NaNH_2} HC{\equiv}C{:}^{\ominus}\ Na^+ \xrightarrow{CH_3CH_2Br} HC{\equiv}C{-}CH_2CH_3$

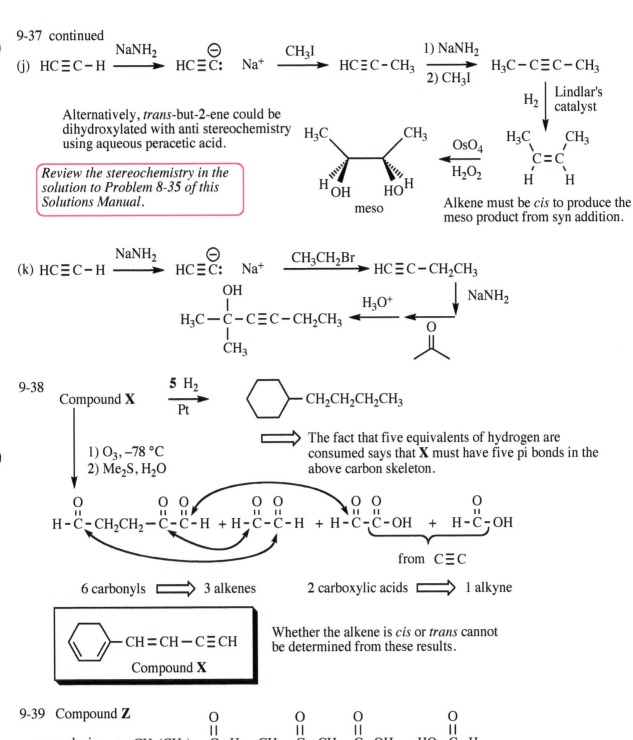

$\xrightarrow{NaNH_2}$ then acetone ($(CH_3)_2C{=}O$), then H_3O^+ → $H_3C{-}C(OH)(CH_3){-}C{\equiv}C{-}CH_2CH_3$

9-38 Compound **X** $\xrightarrow[Pt]{5\ H_2}$ cyclohexyl$-CH_2CH_2CH_2CH_3$

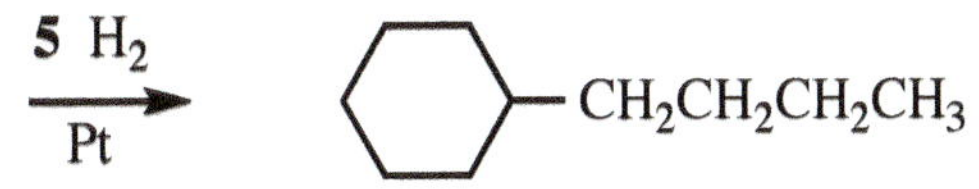

⟹ The fact that five equivalents of hydrogen are consumed says that **X** must have five pi bonds in the above carbon skeleton.

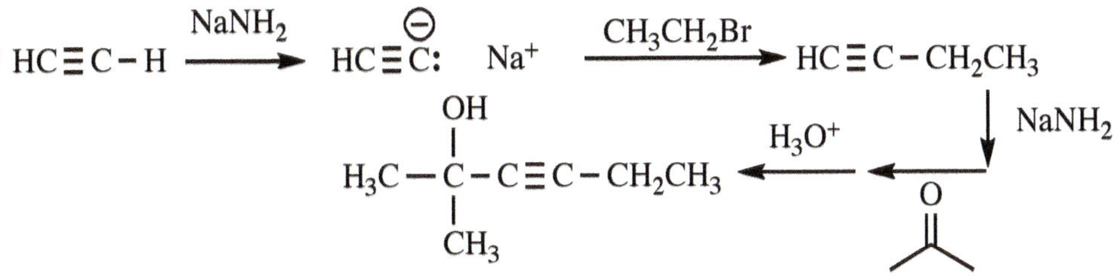

Compound **X** → 1) O_3, −78 °C 2) Me_2S, H_2O →

$H{-}C(=O){-}CH_2CH_2{-}C(=O){-}C(=O){-}H$ + $H{-}C(=O){-}C(=O){-}H$ + $H{-}C(=O){-}C(=O){-}OH$ + $H{-}C(=O){-}OH$

from C≡C

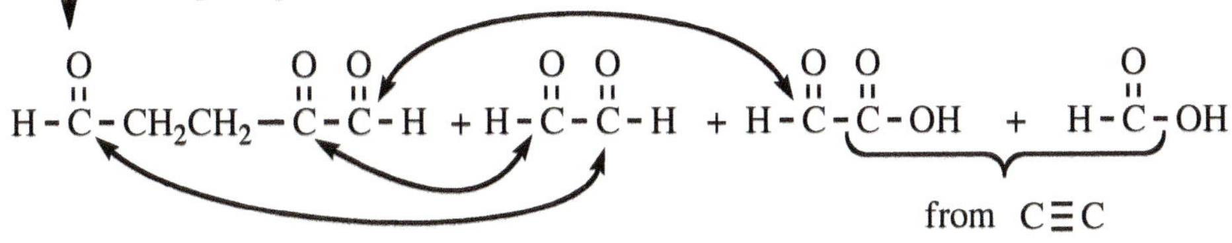

6 carbonyls ⟹ 3 alkenes 2 carboxylic acids ⟹ 1 alkyne

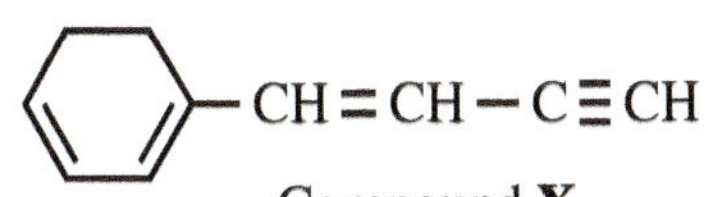

cyclohexadienyl$-CH{=}CH{-}C{\equiv}CH$ Compound **X**

Whether the alkene is *cis* or *trans* cannot be determined from these results.

9-39 Compound **Z**

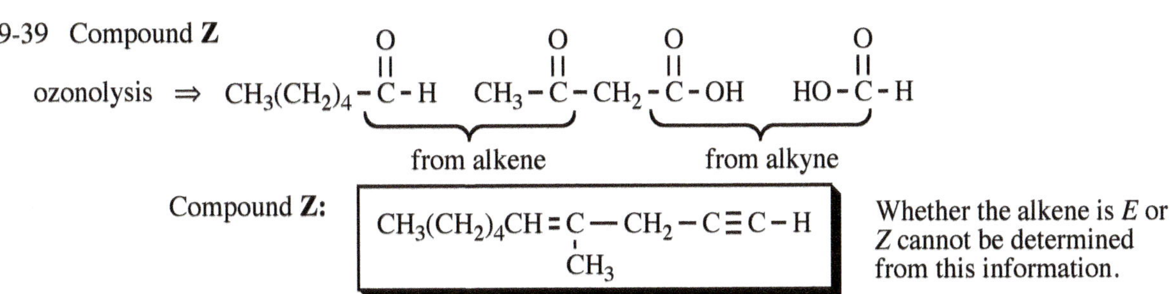

ozonolysis ⇒ CH3(CH2)4−C(=O)−H CH3−C(=O)−CH2−C(=O)−OH HO−C(=O)−H
from alkene
from alkyne
Compound Z:
CH3(CH2)4CH=C(CH3)−CH2−C≡C−H
Whether the alkene is E or Z cannot be determined from this information.

9-40 This synthesis begins the same as the solution to problem 9-35:

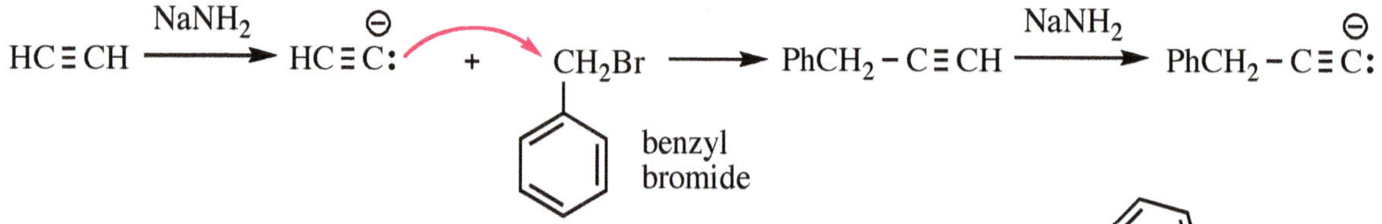

$$HC\equiv CH \xrightarrow{NaNH_2} HC\equiv C:^{\ominus} + CH_2Br \text{ (benzyl bromide)} \longrightarrow PhCH_2-C\equiv CH \xrightarrow{NaNH_2} PhCH_2-C\equiv C:^{\ominus}$$

benzyl
bromide

The anion will add across the carbonyl group of the aldehyde:

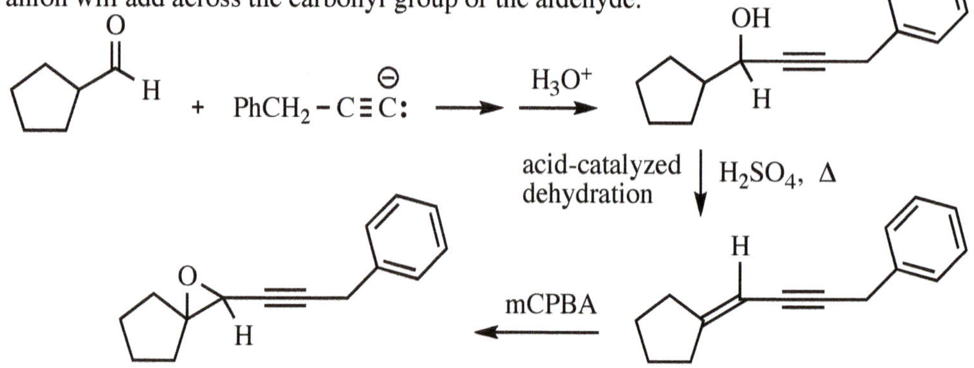

$$\text{cyclopentanecarbaldehyde} + PhCH_2-C\equiv C:^{\ominus} \longrightarrow \xrightarrow{H_3O^+} \text{propargylic alcohol (OH, H)}$$

acid-catalyzed dehydration ↓ H_2SO_4, Δ

mCPBA

9-41

(a) $CH_3CH_2-C\equiv C-H \xrightarrow[2)\ H_2O_2,\ HO^-]{1)\ Sia_2BH} CH_3CH_2CH_2-\overset{O}{\overset{\|}{C}}-H$

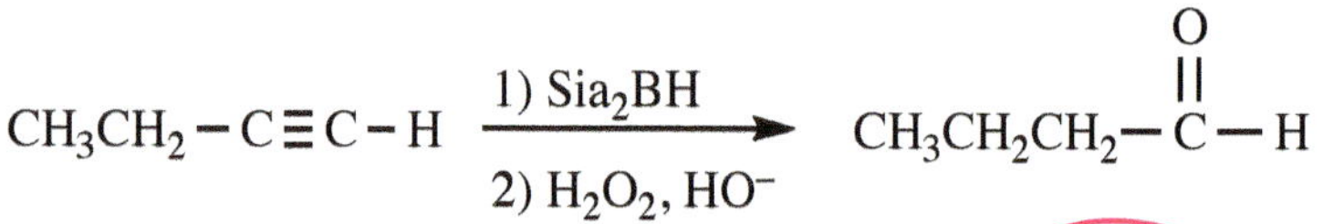

(b) $R-C\equiv C-H + {}^{\ominus}:\ddot{O}-Et \xrightarrow{\text{reaction 1}} R-\overset{..}{C}^{\ominus}=CH(:\ddot{O}-Et)$ + H—O—Et

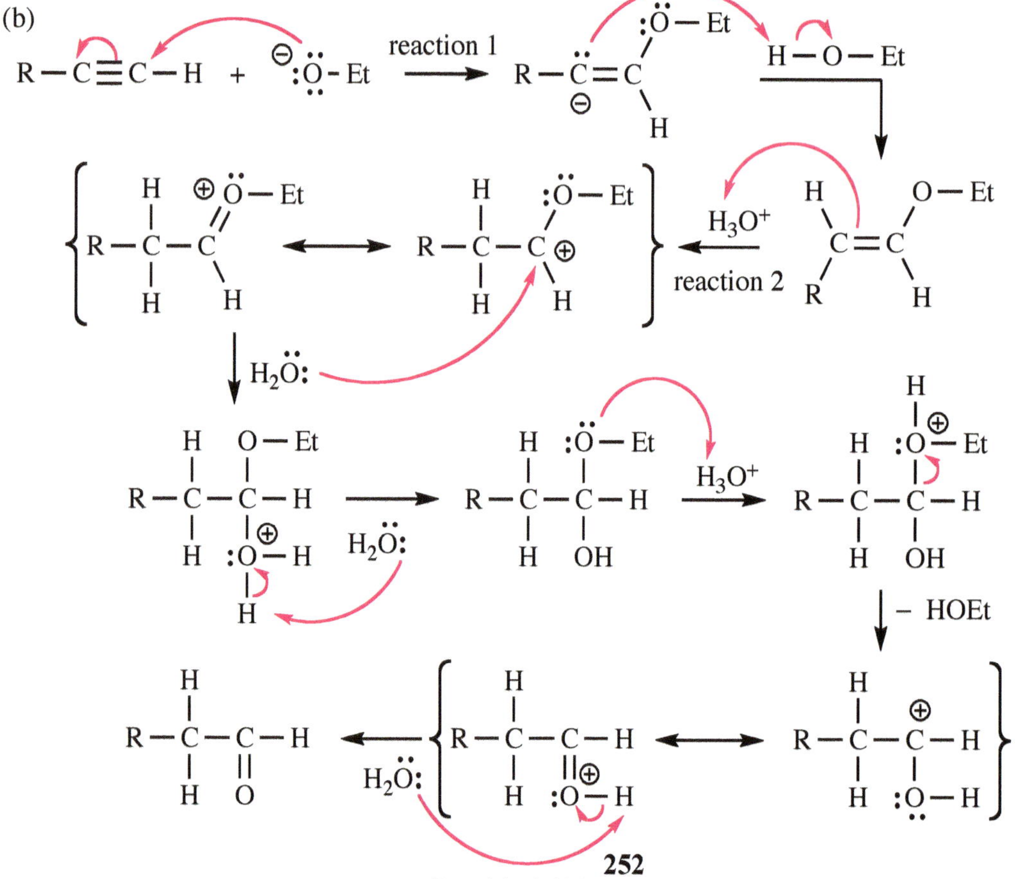

$\xrightarrow{H_3O^+,\ \text{reaction 2}}$ {R—CH₂—CH=O⁺Et ⟷ R—CH₂—C⁺H(:Ö—Et)}

$H_2\ddot{O}:$

reaction 1

reaction 2

H_3O^+

– HOEt

$H_2\ddot{O}:$

9-41 continued

(c) alkyne

$$R-C\equiv CH \xrightarrow{RO^-} R-\overset{\ominus\,\ddot{}}{C}(=CH)-OR$$

sp^2 carbanion

alkene

$$R-C(H)=CH_2 \xrightarrow{RO^-} R-\overset{\ominus\,\ddot{}}{C}(H)-CH_2-OR$$

sp^3 carbanion

The closer that electrons are to the nucleus, the more stable. An s orbital is closer to a nucleus than a p orbital is, as p orbitals are elongated away from the nucleus. An sp^2 carbanion is more stable than an sp^3 carbanion because the sp^2 carbanion has 33% s character and the electron pair is closer to the positive nucleus than in an sp^3 carbanion which is only 25% s character. The sp^2 carbanion is easier to form because of its relative stability.

9-42 Diols are made by two reactions from Chapter 8: either *syn*-dihydroxylation with OsO_4 or cold $KMnO_4$, or *anti*-dihydroxylation via an epoxide using a peroxyacid and water. As this problem says to use inorganic reagents, the solution shown here will use OsO_4 rather than a peroxyacid which is organic.

Recall the stereochemical requirements of syn addition as outlined in this Solutions Manual, Problem 8-35:

cis-alkene + **syn** addition → meso	*trans*-alkene + **syn** addition → racemic (±)
cis-alkene + **anti** addition → racemic (±)	*trans*-alkene + **anti** addition → meso

Part (a) asks for the synthesis of the meso isomer, so syn addition will have to occur on the *cis*-alkene. Part (b) will require syn addition to the *trans*-alkene to give the (±) product.

(a)

$$HC\equiv C-H \xrightarrow{NaNH_2} HC\equiv C{:}^{\ominus}\ Na^+ \xrightarrow{\text{Br}} HC\equiv C- \xrightarrow[\text{Br}]{NaNH_2} -C\equiv C- \xrightarrow[\text{Lindlar's catalyst}]{H_2} \text{C=C (H, H cis)} \xrightarrow[H_2O_2]{OsO_4} \text{HO, H ... H, OH}$$

Alkene must be *cis* to produce the meso product from syn addition, so reduction is done with Lindlar's catalyst to produce the *cis* alkene.

(b)

$$HC\equiv C-H \xrightarrow{NaNH_2} HC\equiv C{:}^{\ominus}\ Na^+ \xrightarrow{\text{Br}} HC\equiv C- \xrightarrow[\text{Br}]{NaNH_2} -C\equiv C- \xrightarrow[NH_3]{Na} \text{C=C (H, H trans)} \xrightarrow[H_2O_2]{OsO_4} \text{H, OH ... H, OH}$$

Alkene must be *trans* to produce the (±) product from syn addition, so reduction is done with Na/NH_3.

9-43 Each unknown has the molecular formula C_8H_{12} with 3 elements of unsaturation.

(a)

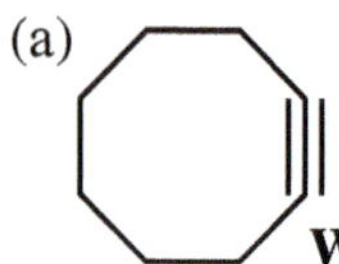

Addition of H_2 to the triple bond gives cyclooctane. Cleavage of the triple bond with ozone will produce an 8-carbon chain with a carboxylic acid at each end, as shown in the solution to 9-30(i).

(b)

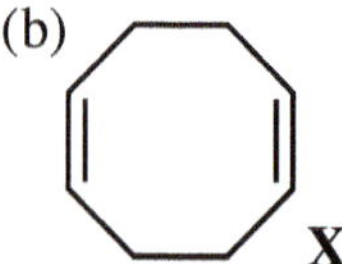

Addition of H_2 to the double bonds gives cyclooctane. Cleavage of the double bonds with ozone will produce two 4-carbon chains with an aldehyde at each end.

(c)

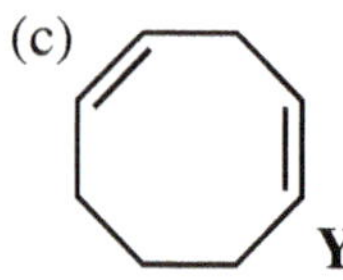

Addition of H_2 to the double bonds gives cyclooctane. Cleavage of the double bonds with ozone will produce one 3-carbon chain and one 5-carbon chain with an aldehyde at each end.

(d)

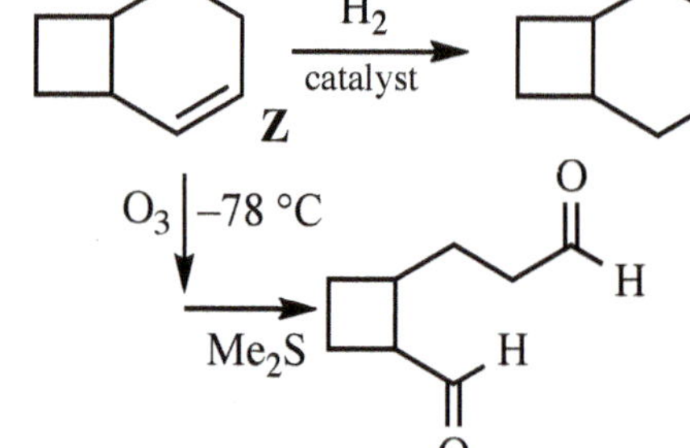

Addition of H_2 to the double bond gives bicyclo[4.2.0]octane. Cleavage of the double bond with ozone opens the larger ring but retains the cyclobutane with 1-carbon and 3-carbon chains with an aldehyde at each end.

CHAPTER 10—STRUCTURE AND SYNTHESIS OF ALCOHOLS

10-1 The 1993 IUPAC recommendations put the position number before the group it describes.

(a) 2-phenylbutan-2-ol
(b) (*E*)-5-bromohept-3-en-2-ol
(c) 4-methylcyclohex-3-en-1-ol ("1" is optional)
(d) *trans*-2-methylcyclohexan-1-ol ("1" is optional) or (1*R*,2*R*)-2-methylcyclohexan-1-ol
(e) (*E*)-2-chloro-3-methylpent-2-en-1-ol
(f) (2*R*,3*S*)-2-bromohexan-3-ol

10-2 IUPAC name first, then common name.

(a) butan-2-ol; *sec*-butyl alcohol
(b) cyclopropanol; cyclopropyl alcohol
(c) 1-cyclobutylpropan-2-ol; no common name
(d) 3-methylbutan-1-ol; isopentyl alcohol (also isoamyl alcohol)

10-3 Only constitutional isomers are requested, not stereoisomers, and only structures with an alcohol group.

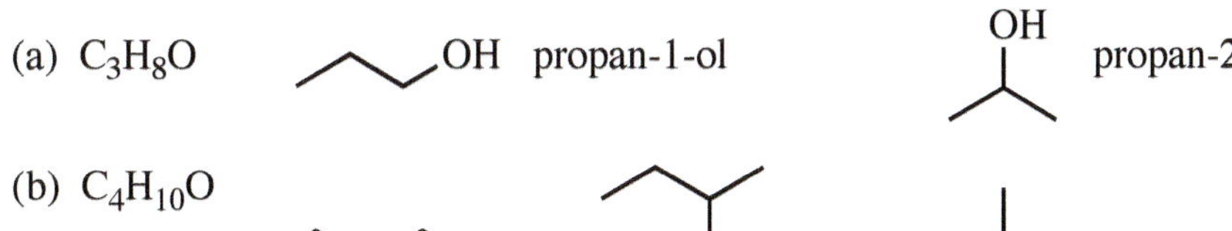

(b) $C_4H_{10}O$

OH — butan-1-ol
OH — butan-2-ol
OH — 2-methylpropan-1-ol
OH — 2-methylpropan-2-ol

(c) C_3H_6O has one element of unsaturation, either a double bond or a ring.

OH — cyclopropanol
OH — prop-2-en-1-ol
OH * — prop-1-en-2-ol
HO * — prop-1-en-1-ol (*E* or *Z*)

(d) C_3H_4O has two elements of unsaturation, so each structure must have either a triple bond, or two double bonds, or a three-membered ring and a double bond. All structures must contain an OH. (In the name, the "e" is dropped from "yne" because it is followed by a vowel in "ol".)

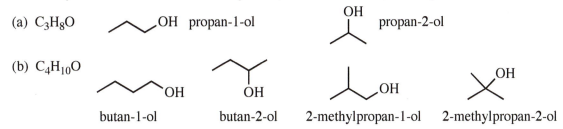

$HO-C\equiv C-CH_3$ * — prop-1-yn-1-ol
$HC\equiv C-CH_2-OH$ — prop-2-yn-1-ol
$H_2C=C=CH-OH$ * — propa-1,2-dien-1-ol
OH * — cycloprop-1-en-1-ol
OH — cycloprop-2-en-1-ol

*The structures with the OH bonded directly to the carbon-carbon double bond are called *enols* or *vinyl alcohols*. The structure with OH on a carbon-carbon triple bond is called an *ynol*. These compounds are unstable, although the structures are legitimate.

10-4 (a) 8,8-dimethylnonane-2,7-diol
(b) octane-1,8-diol
(c) *cis*-cyclohex-2-ene-1,4-diol
(d) 3-cyclopentylheptane-2,4-diol
(e) *trans*-cyclobutane-1,3-diol

10-5 There are four structural features to consider when determining solubility in water: 1) molecules with fewer carbons will be more soluble in water (assuming other things being equal); 2) branched or otherwise compact structures are more soluble than linear structures; 3) more hydrogen-bonding groups will increase solubility; 4) an ionic form of a compound will be more soluble in water than the nonionic form.

(a) Cyclohexanol is more soluble because its alkyl group is more compact than in hexan-1-ol.
(b) 4-Methylphenol is more soluble because its hydrocarbon portion is more compact than in heptan-1-ol, and phenols form particularly strong hydrogen bonds with water.
(c) 3-Ethylhexan-3-ol is more soluble because its alkyl portion is more spherical than in octan-2-ol.
(d) Cyclooctane-1,4-diol is more soluble because it has two OH groups which can hydrogen bond with water, whereas hexan-2-ol has only one OH group. (The ratio of carbons to OH is 4 to 1 in the former compound and 6 to 1 in the latter; the smaller this ratio, the more soluble.)
(e) These are enantiomers and will have identical solubility.

10-6 Dimethylamine molecules can hydrogen bond among themselves so it takes more energy (higher temperature) to separate them from each other. Trimethylamine has no N-H and cannot hydrogen bond, so it takes less energy to separate these molecules from each other, despite its higher molecular weight.

10-7
(a) Methanol is more acidic than *tert*-butyl alcohol. The greater the substitution, the lower the acidity.
(b) 2-Chloropropan-1-ol is more acidic because the electron-withdrawing chlorine atom is closer to the OH group than in 3-chloropropan-1-ol.
(c) 2,2-Dichloroethanol is more acidic because two electron-withdrawing chlorine atoms increase acidity more than just the one chlorine in 2-chloroethanol.
(d) 2,2-Difluoropropan-1-ol is more acidic because fluorine is more electronegative than chlorine; the stronger the electron-withdrawing group, the more acidic the alcohol.

10-8

sulfuric acid >> acetic acid > 2-chloroethanol > water > ethanol >
H_2SO_4 (*most acidic*) CH_3COOH $ClCH_2CH_2OH$ H_2O CH_3CH_2OH

tert-butyl alcohol > hex-1-yne > ammonia > hexane (*least acidic*)
$(CH_3)_3COH$ $C_4H_9-C\equiv C-H$ NH_3 C_6H_{14}

Sulfuric acid is one of the strongest acids known, and acetic acid is the strongest type of organic acid. On the other extreme, alkanes like hexane are the least acidic compounds. Alkynes are more acidic than ammonia. The N–H bond in ammonia is less acidic than any O–H bond. Among the four compounds with O–H bonds, the tertiary alcohols are the least acidic. Water is more acidic than most alcohols including ethanol. However, if a strong electron-withdrawing substituent like chlorine is near the alcohol group, the acidity increases enough so that it is more acidic than water. (Determining exactly where water appears in this list is the most difficult part.)

10-9 Resonance forms of phenoxide anion show the negative charge delocalized onto the ring only at carbons 2, 4, and 6:

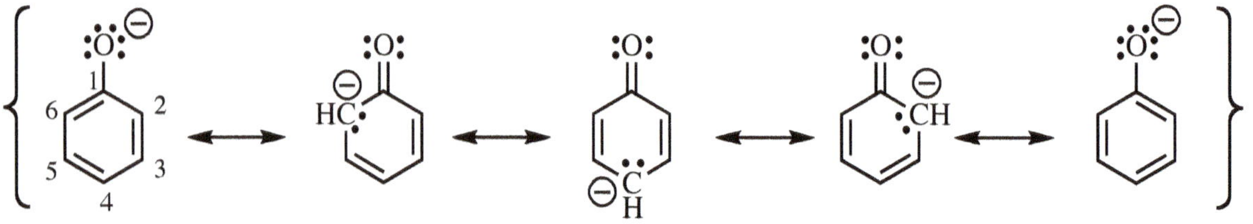

Nitro group at position 2

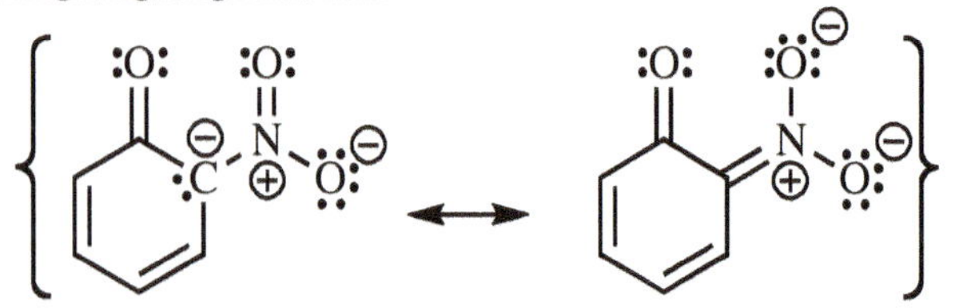

Nitro at position 2 delocalizes negative charge.

Nitro group at position 3

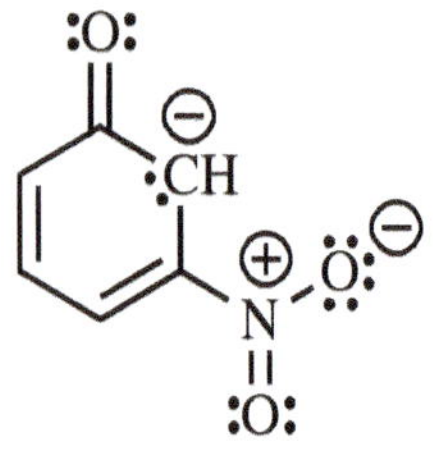

Nitro at position 3 cannot delocalize negative charge at position 2 or 4—no resonance stabilization.

continued on next page

10-9 continued

Nitro group at position 4

Nitro at position 4 delocalizes negative charge.

Only when the nitro group is at one of the negative carbons will the nitro have a stabilizing effect (via resonance). Thus, 2-nitrophenol and 4-nitrophenol are substantially more acidic than phenol itself, but 3-nitrophenol is only slightly more acidic than phenol (due to the inductive effect).

10-10

A   B

(a) Structure **A** is a phenol because the OH is bonded to a benzene ring. As a phenol, it will be acidic enough to react with sodium hydroxide to generate a phenoxide ion that will be fairly soluble in water. Structure **B** is a 2° benzylic alcohol, not a phenol, not acidic enough to react with NaOH.

(b) Both of these organic compounds will be soluble in an organic solvent like dichloromethane. Shaking this organic solution with aqueous sodium hydroxide will ionize the phenol **A**, making it more polar and water soluble; it will be extracted from the organic layer into the water layer, while the alcohol will remain in the organic solvent. Separating these immiscible solvents will separate the original compounds. The alcohol can be retrieved by evaporating the organic solvent. The phenol can be isolated by acidifying the basic aqueous solution and filtering if the phenol is a solid, or separating the layers if the phenol is a liquid.

10-11 The Grignard reaction needs a solvent containing an ether functional group: (b), (f), (g), and (h) are possible solvents. Dimethyl ether, (b), is a gas at room temperature, however, so it would have to be liquefied at low temperature for it to be a useful solvent.

10-12 (a) CH_3CH_2MgBr (b) Li + LiI (c) F—⬡—MgBr (d) Li + LiCl

10-13 Any of three halides—chloride, bromide, iodide, but not fluoride—can be used. Ether is the typical solvent for Grignard reactions. The new C—C bond is shown in bold.

(a) ⬡—MgCl + $H_2C{=}O$ $\xrightarrow{\text{ether}}$ $\xrightarrow{H_3O^+}$ ⬡—CH_2OH

(b) MgBr + $H_2C{=}O$ $\xrightarrow{\text{ether}}$ $\xrightarrow{H_3O^+}$ OH

(c) MgI + $H_2C{=}O$ $\xrightarrow{\text{ether}}$ $\xrightarrow{H_3O^+}$ CH_2OH

In this solution, the portion of the product that came from the Grignard reagent is shown in red.

Note: the alternative arrow symbolism could also be used, where the two steps are numbered around one arrow:

NO! My BAD! This means that water is present with ether during the Grignard reaction!

10-14 Any of three halides—chloride, bromide, iodide, but not fluoride—can be used. Grignard reactions are always performed in ether solvent; ether is not shown here.

(a) two methods

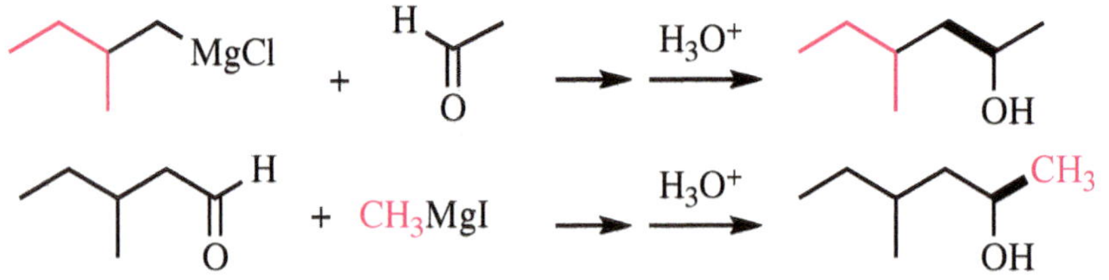

MgCl + H, O → H_3O^+ → OH

The newly-formed C—C bond is shown in bold.

H, O + CH_3MgI → H_3O^+ → CH_3, OH

In this solution, the portion of the product that came from the Grignard reagent is shown in red.

(b) two methods

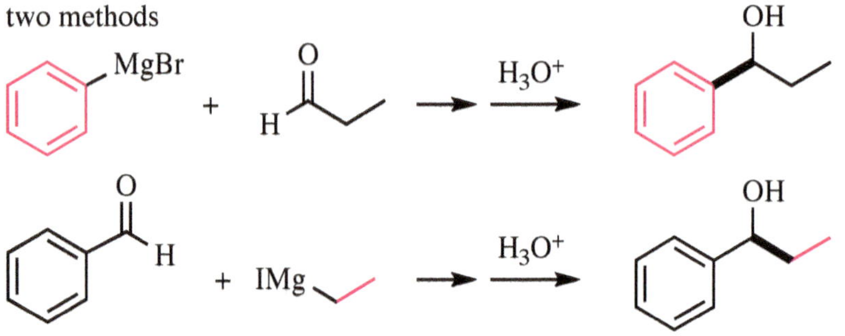

MgBr + O, H → H_3O^+ → OH

O, H + IMg → H_3O^+ → OH

(c) two methods

MgCl + O, H → H_3O^+ → OH

MgCl + O, H → H_3O^+ → OH

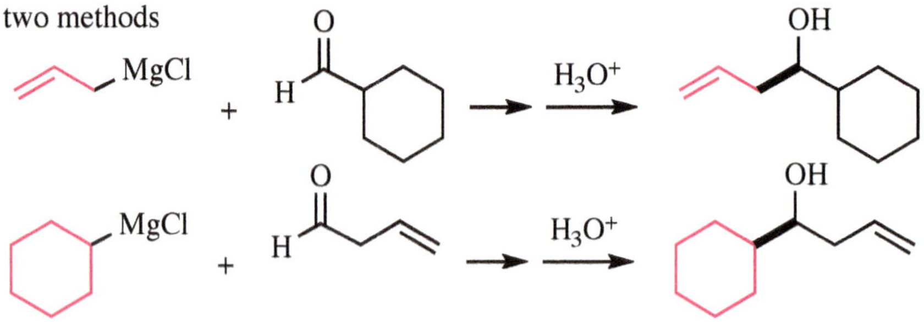

10-15 Grignard reactions are always performed in ether. Here, the ether is not shown.

(a) Any of the three bonds shown in bold can be formed by adding a Grignard reagent across a ketone, followed by aqueous acid workup.

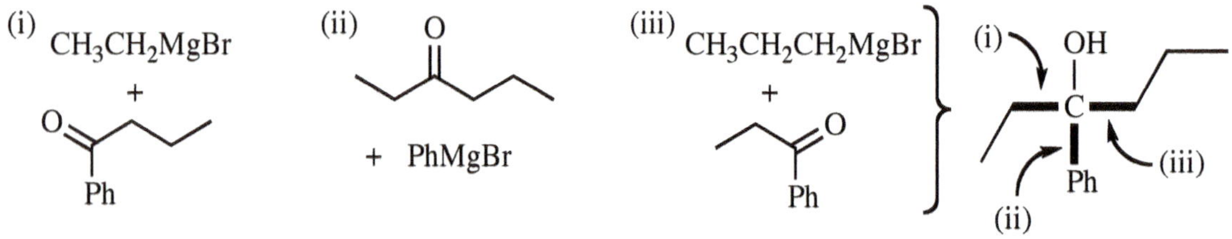

(i) CH_3CH_2MgBr + O, Ph

(ii) O + PhMgBr

(iii) $CH_3CH_2CH_2MgBr$ + O, Ph

(i) OH, C, Ph, (iii), (ii)

(b) Ph−MgBr + Ph, C=O, Ph → H_3O^+ → Ph, Ph−C−OH, Ph

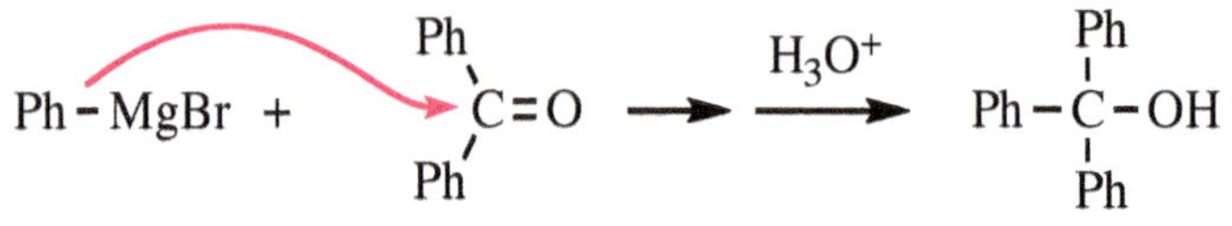

The newly-formed C—C bond is shown in bold.

(c) CH_3CH_2−MgI + O → H_3O^+ → HO, CH_2CH_3

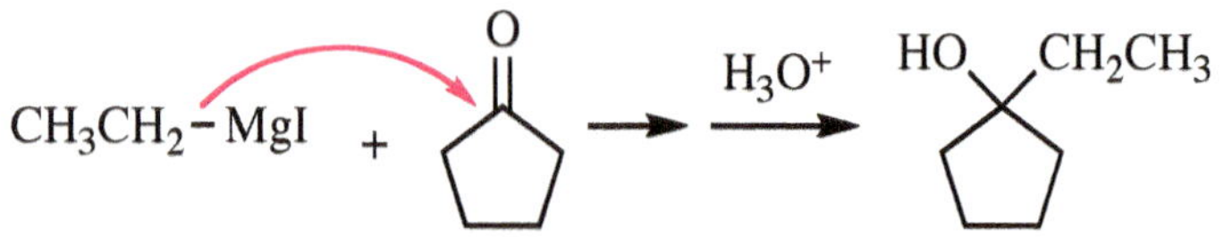

10-15 continued

(d) three methods, all with H_3O^+ workup

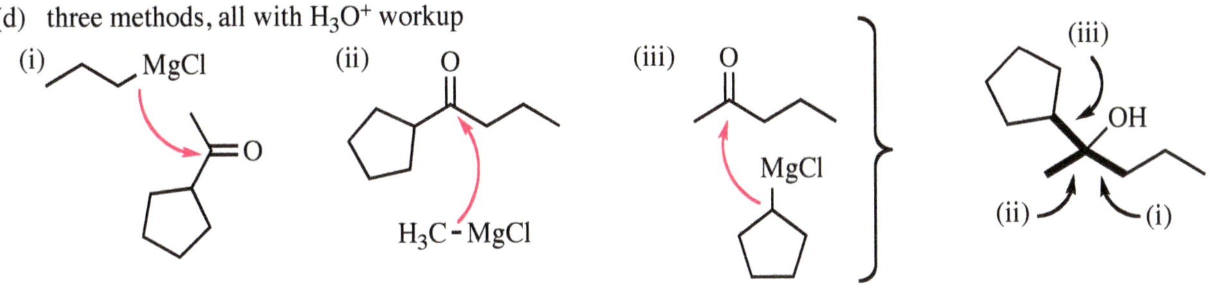

(i)
MgCl
O
(ii)
O
H3C-MgCl
(iii)
O
MgCl
(iii)
OH
(ii)
(i)

10-16

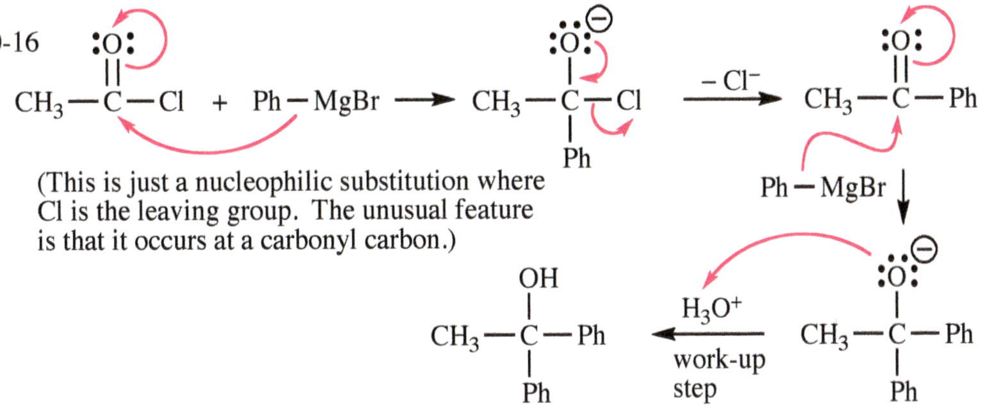

CH3—C—Cl + Ph—MgBr
CH3—C—Cl
Ph
– Cl–
CH3—C—Ph
ketone intermediate
Ph—MgBr
CH3—C—Ph
Ph
H3O+
work-up step
OH
CH3—C—Ph
Ph

(This is just a nucleophilic substitution where Cl is the leaving group. The unusual feature is that it occurs at a carbonyl carbon.)

10-17 Acid chlorides or esters will work as starting materials in these reactions. The typical solvent for Grignard reactions is ether; it is not shown here.

(a) 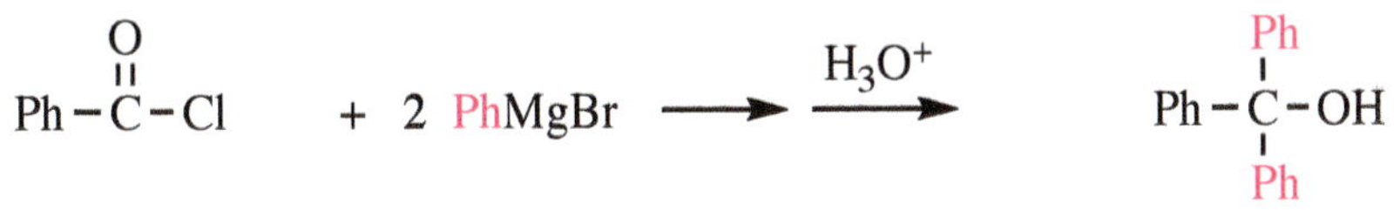

Ph–C–Cl + 2 PhMgBr
H3O+
Ph
Ph–C–OH
Ph

(b) 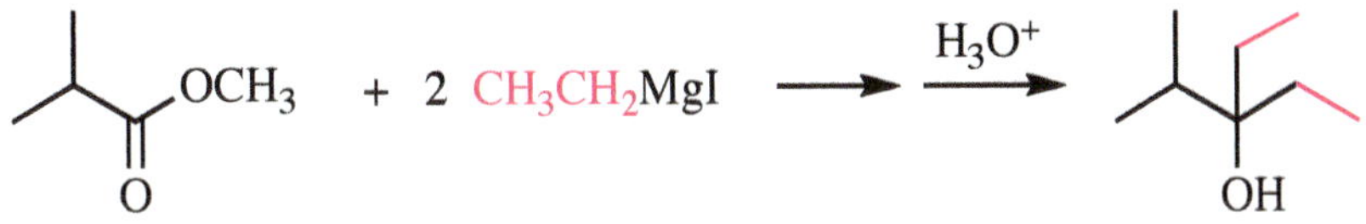

OCH3
O
+ 2 CH3CH2MgI
H3O+
OH

(c) 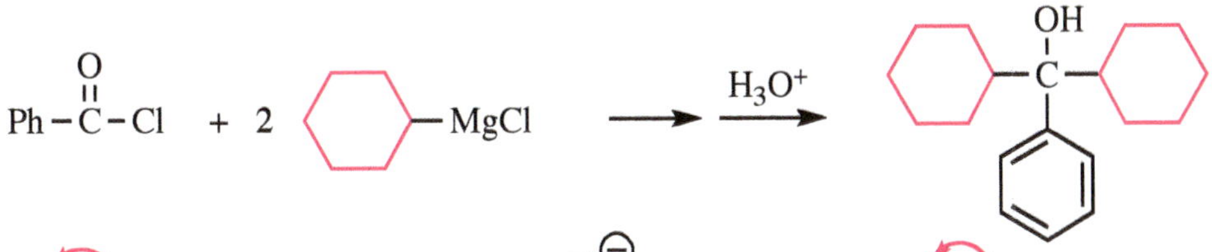

Ph–C–Cl + 2
MgCl
H3O+
OH
C

In this solution, the portion of the product that came from the Grignard reagent is shown in red.

10-18

(a) 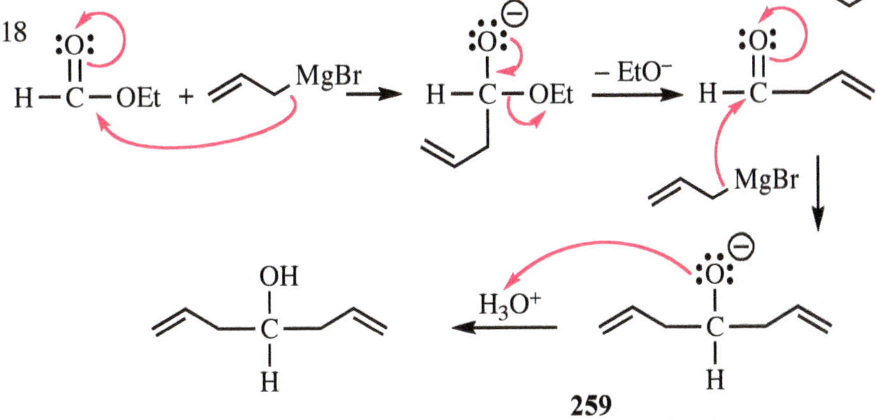

H—C—OEt +
MgBr
H—C—OEt
– EtO–
H—C
aldehyde intermediate
MgBr
H3O+
OH
C
H

10-18 continued

(b) (i) $HC(=O)-OEt + 2\ CH_3CH_2MgBr \longrightarrow \xrightarrow{H_3O^+}$ (product, OH)

In this solution, the portion of the product that came from the Grignard reagent is shown in red.

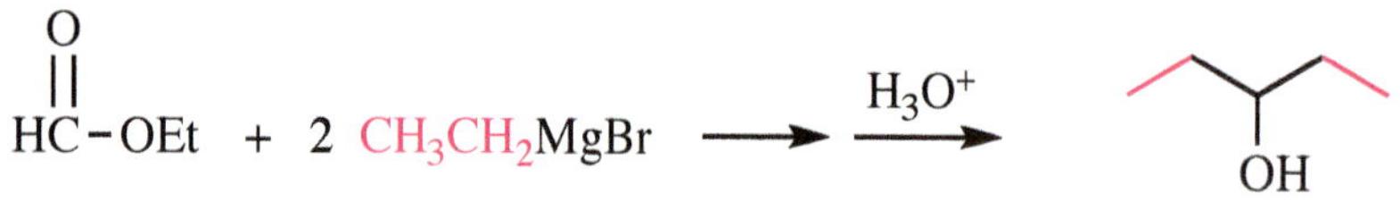

(ii) 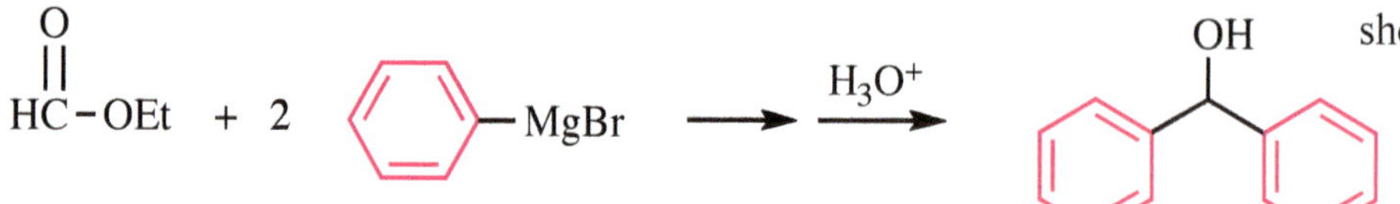 $HC(=O)-OEt + 2$ PhMgBr $\longrightarrow \xrightarrow{H_3O^+}$ (product, OH)

(iii) 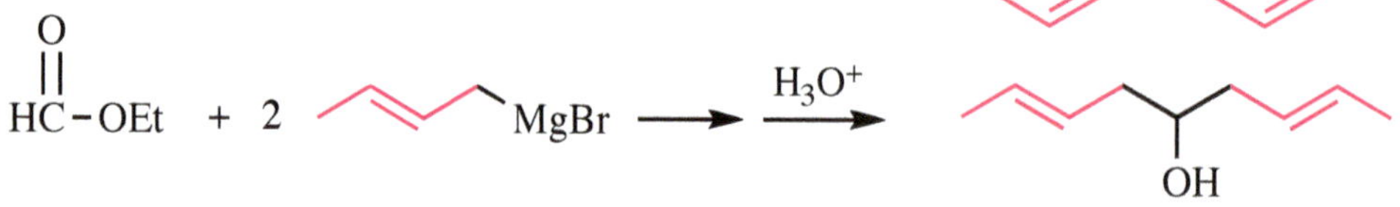 $HC(=O)-OEt + 2$ MgBr $\longrightarrow \xrightarrow{H_3O^+}$ (product, OH)

10-19 Ether is the typical solvent in Grignard reactions. The new C—C bond is shown in bold.

(a) 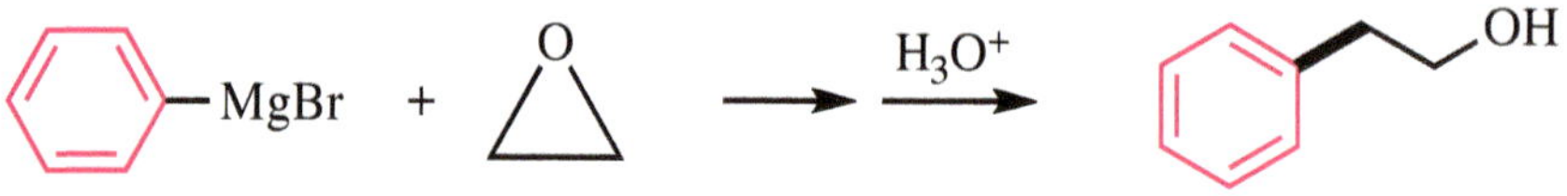 PhMgBr + ethylene oxide (O) $\longrightarrow \xrightarrow{H_3O^+}$ (product, OH)

(b)  MgCl + ethylene oxide (O) $\longrightarrow \xrightarrow{H_3O^+}$ (product, OH)

(c)  MgI + ethylene oxide (O) $\longrightarrow \xrightarrow{H_3O^+}$ (product, OH)

10-20

(a) 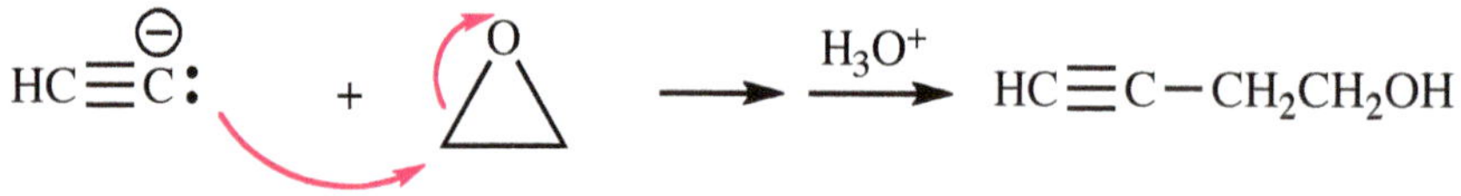 $HC\equiv C:^{\ominus}$ + ethylene oxide (O) $\longrightarrow \xrightarrow{H_3O^+}$ $HC\equiv C-CH_2CH_2OH$

(b) 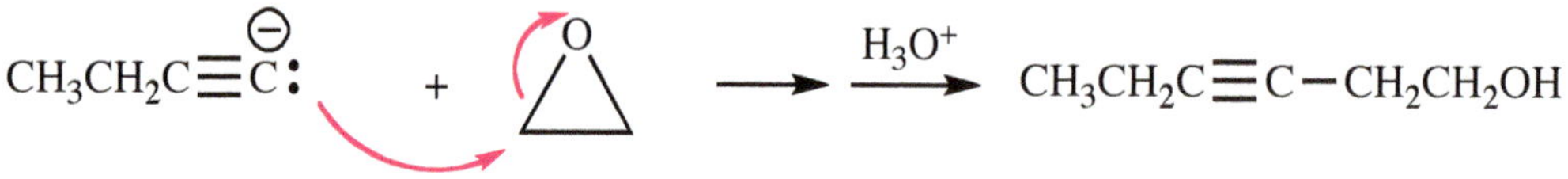 $CH_3CH_2C\equiv C:^{\ominus}$ + ethylene oxide (O) $\longrightarrow \xrightarrow{H_3O^+}$ $CH_3CH_2C\equiv C-CH_2CH_2OH$

10-21 Often, there are several synthetic routes to each structure; the ones shown here are representative. The new bonds formed are shown here in bold. Your answers may be different and still be correct.

(a)  Br $\xrightarrow{Li} \xrightarrow{CuI}$ $(\ )_2CuLi$ $\xrightarrow{\text{Br}}$ (product)

(b) 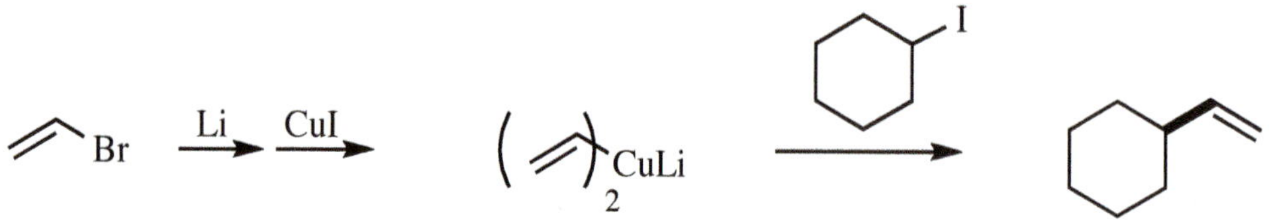 Br $\xrightarrow{Li} \xrightarrow{CuI}$ $(\ )_2CuLi$ $\xrightarrow{\text{cyclohexyl–I}}$ (product)

10-21 continued

(c) 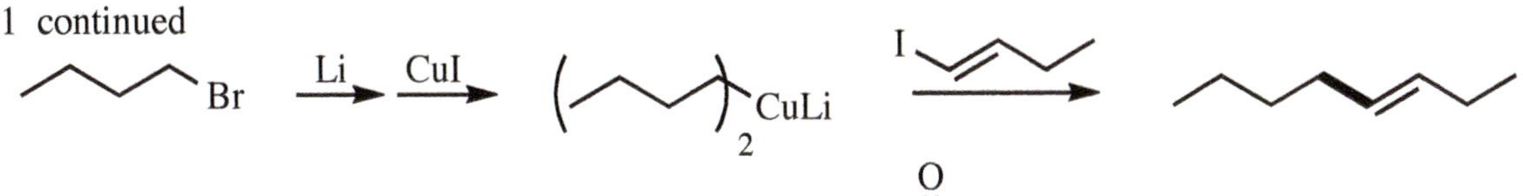 [reaction scheme: 1-bromopropane, Li, CuI → lithium dipropylcuprate $(\text{propyl})_2CuLi$; with (E)-1-iodo-1-butene → (E)-3-heptene]

(d) 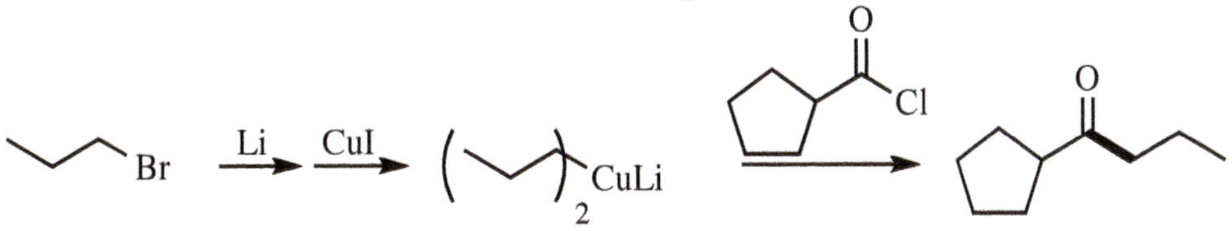 [reaction scheme: ethyl bromide, Li, CuI → $(\text{ethyl})_2CuLi$; with cyclopentanecarbonyl chloride → cyclopentyl ethyl ketone]

10-22 These reactions are acid-base reactions in which an acidic proton (or deuteron) is transferred to a basic carbon in either a Grignard reagent or an alkyllithium.

(a) 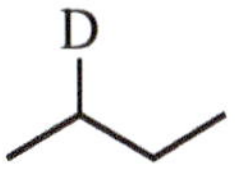 [2-deuterobutane] + Mg(OD)I (b) $CH_3CH_2CH_2CH_3$ + $LiOCH_2CH_3$

(c) 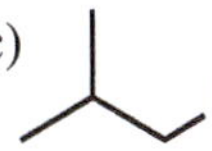 [isobutane] H + $BrMg^{\oplus}$ $:^{\ominus}C{\equiv}C-CH_2CH_3$ (d) 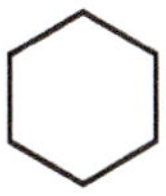 [cyclohexane] + $H_3C-C(=O)-O^-\ Li^+$ (e) 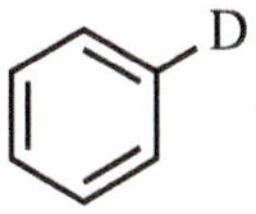 [deuterobenzene] + Mg(OD)Br

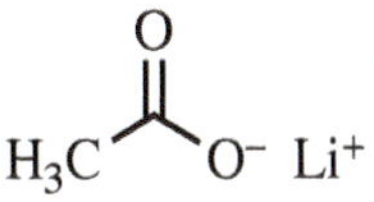

10-23 Grignard reagents are incompatible with acidic hydrogens and with electrophilic, polarized multiple bonds like C=O, NO_2, etc.

(a) As the Grignard reagent is formed, it would instantaneously be protonated by the N—H present in other molecules of the same substance.

(b) As the Grignard reagent is formed, it would immediately attack the ester functional group present in other molecules of the same substance.

(c) Care must be taken in how reagents are written above and below arrows. If reagents are numbered "1. ... 2. ... etc.", it means they are added in separate steps, the same as writing reagents over separate arrows. If reagents written around an arrow are not numbered, it means they are added all at once in the same mixture. In this problem, the ketone is added in the presence of aqueous acid. The acid will immediately protonate and destroy the Grignard reagent before reaction with the ketone can occur.

(d) The ethyl Grignard reagent will be immediately protonated and consumed by the OH. This reaction *could* be made to work, however, by adding two equivalents of ethyl Grignard reagent: the first to consume the OH proton, the second to add across the ketone. Aqueous acid will then protonate both oxygens.

10-24 Sodium borohydride does not reduce carboxylic acids or esters.

(a) $CH_3(CH_2)_8CH_2OH$ (b) no reaction (c) no reaction ($PhCOO^-$ before acid work-up) (d) 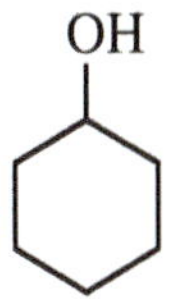 [cyclohexanol, OH]

(e) 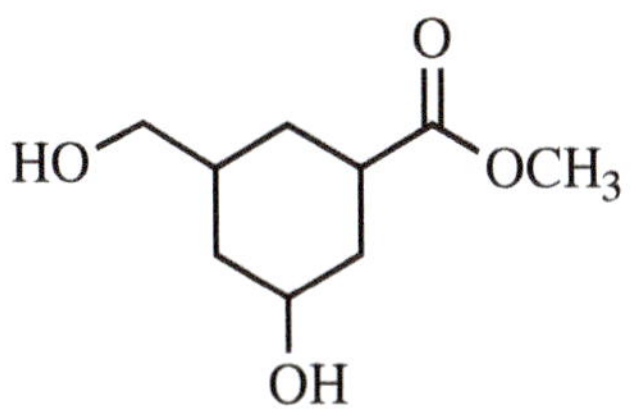 [HO–CH_2–cyclohexane ring bearing OH and $C(=O)OCH_3$] (f) ester { [lactone ring (O, O=) bearing CH_2OH] }

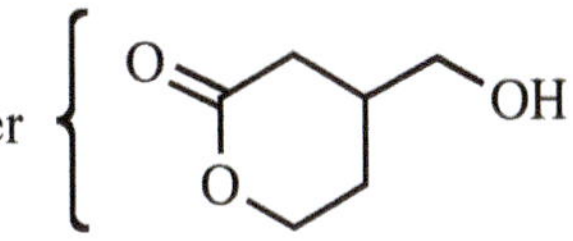

10-25 Lithium aluminum hydride reduces carboxylic acids and esters as well as other carbonyl groups.

(a) $CH_3(CH_2)_8CH_2OH$ (b) $CH_3CH_2CH_2OH$ + $HOCH_3$ (c) $PhCH_2OH$ (d) 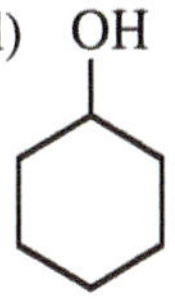 [cyclohexanol, OH]

(e) 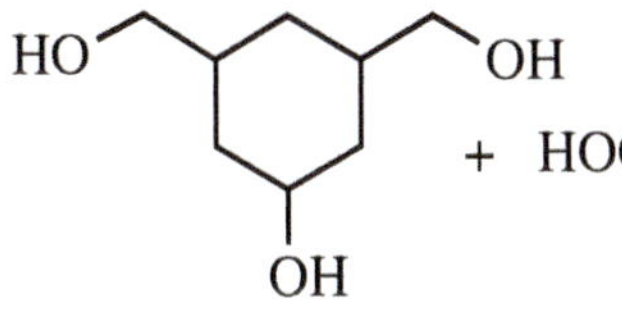 [HO–CH_2–cyclohexane ring bearing OH and CH_2OH] + $HOCH_3$ (f) 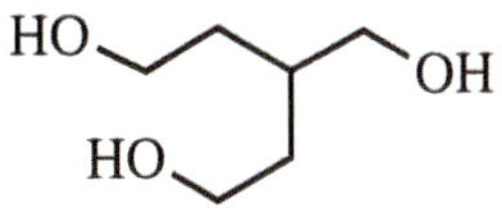 [triol: HO, HO, OH]

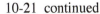

10-26

(a) (hexanal) $\xrightarrow[CH_3OH]{NaBH_4}$ (hexan-1-ol) OR $\xrightarrow[2)\ H_3O^+]{1)\ LiAlH_4}$

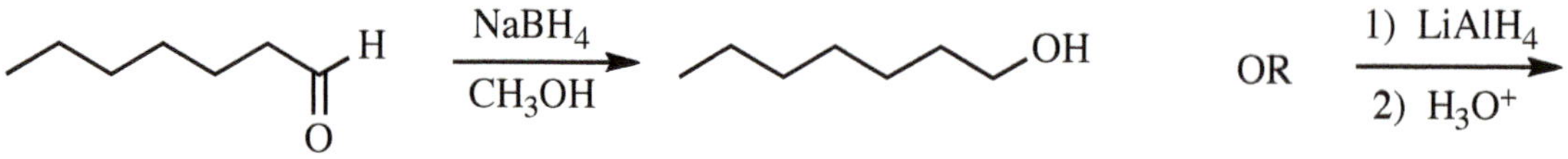

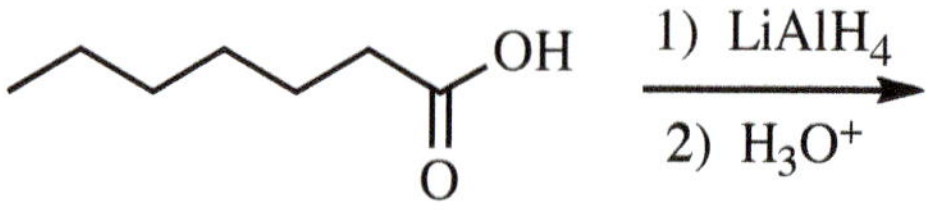

OR

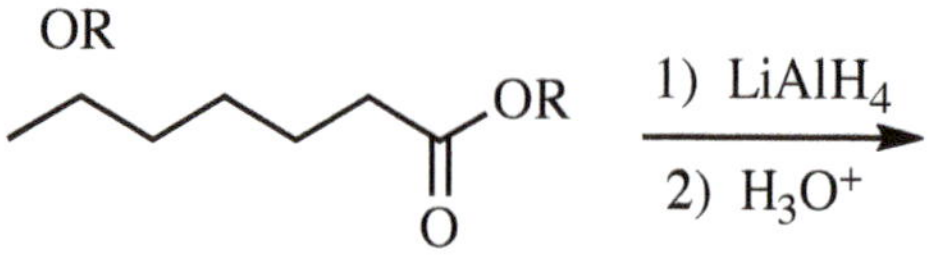

(hexanoic acid) $\xrightarrow[2)\ H_3O^+]{1)\ LiAlH_4}$

OR

(hexanoate ester, OR) $\xrightarrow[2)\ H_3O^+]{1)\ LiAlH_4}$

(b) (hexan-2-one) $\xrightarrow[CH_3OH]{NaBH_4}$ (hexan-2-ol) OR $\xrightarrow[2)\ H_3O^+]{1)\ LiAlH_4}$

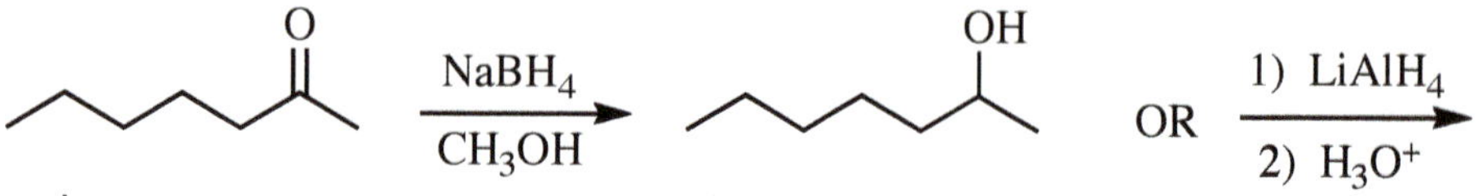

(c) (2-methylhexan-3-one) $\xrightarrow[CH_3OH]{NaBH_4}$ (2-methylhexan-3-ol) OR $\xrightarrow[2)\ H_3O^+]{1)\ LiAlH_4}$

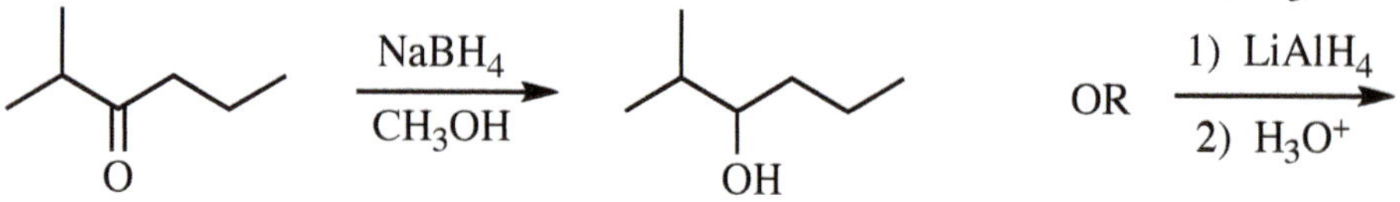

(d) (keto lactone) $\xrightarrow[CH_3OH]{NaBH_4}$ (hydroxy lactone)

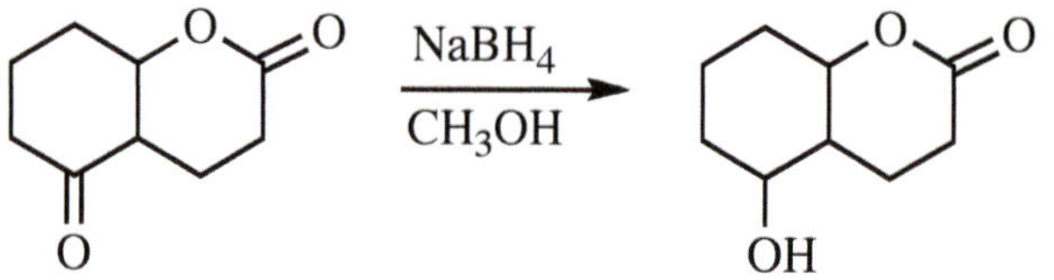

$LiAlH_4$ will NOT give the desired product. $LiAlH_4$ will also reduce the ester in addition to the ketone.

10-27 Approximate pKa values are shown below each compound. Refer to Appendix 4 at the back of the text.

CH_3SO_3H	>	CH_3COOH	>	CH_3SH	>	CH_3OH	>	$CH_3C{\equiv}CH$	>	CH_3NH_2	>	CH_3CH_3
< 0		4.74		≈ 10.5		15.5		25		40		50

most acidic — least acidic

10-28

(a) 4-methylpentane-2-thiol

(b) (*Z*)-2,3-dimethylpent-2-ene-1-thiol

(c) cyclohex-2-ene-1-thiol ("1" is optional)

10-29

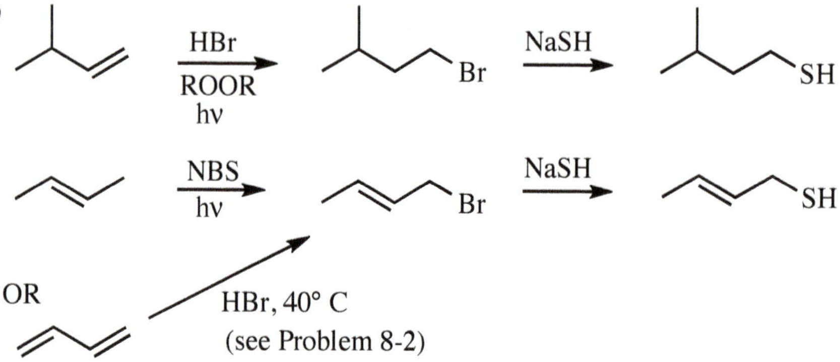

(3-methylbut-1-ene) $\xrightarrow[ROOR,\ h\nu]{HBr}$ (1-bromo-3-methylbutane, Br) $\xrightarrow{NaSH}$ (SH) 3-methylbutane-1-thiol

(but-2-ene) $\xrightarrow[h\nu]{NBS}$ (1-bromobut-2-ene, Br) $\xrightarrow{NaSH}$ (SH) but-2-ene-1-thiol

OR

(buta-1,3-diene) $\xrightarrow{}$ HBr, 40° C (see Problem 8-2)

10-30
(a) 5-methyl-4-propylheptan-2-ol; 2°
(b) 4-(1-bromoethyl)heptan-3-ol; 2°
(c) 6-chloro-3-phenyloctan-3-ol; 3°
(d) 3-bromocyclohex-3-en-1-ol; 2° ("1" is optional)
(e) *cis*-4-chlorocyclohex-2-en-1-ol; 2° ("1" is optional) also possible is (1*R*,4*S*)-
(f) (*E*)-4,5-dimethylhex-3-en-1-ol; 1°
(g) (1-cyclopentenyl)methanol; 1°

10-31
(a) 4-chloro-1-phenylhexane-1,5-diol
(b) *trans*-cyclohexane-1,2-diol
(c) 3-nitrophenol
(d) 4-bromo-2-chlorophenol

10-32
(a) 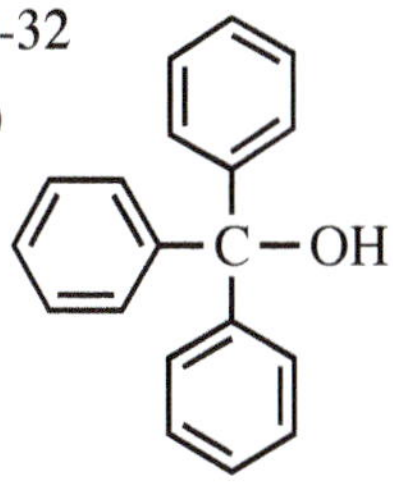 C—OH
(b) Cl OH
(c) OH
(d) 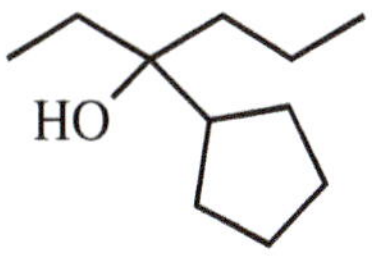 HO
(e) 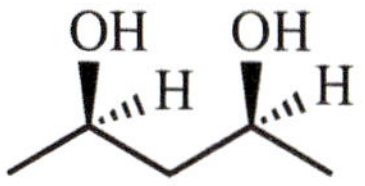 OH OH H H
(f) OH OH
(g) OH I
(h) HO H H OH
(i) SH
(j) CH_3S-SCH_3
(k) 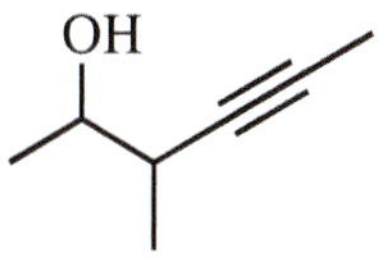 OH

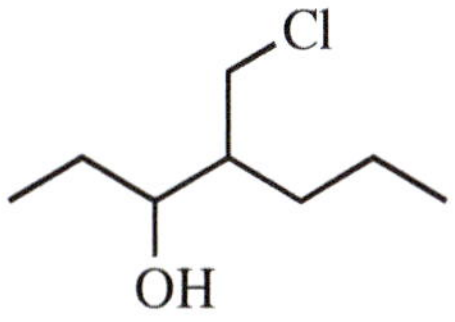

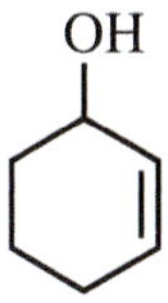

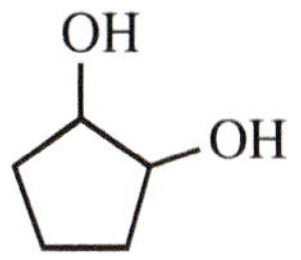

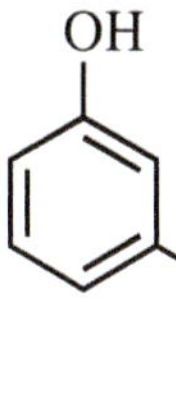

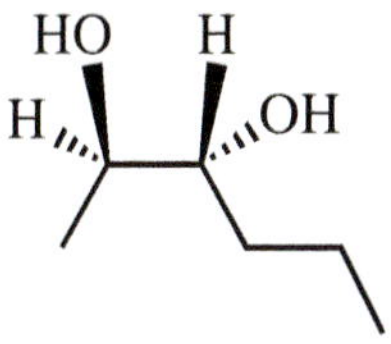

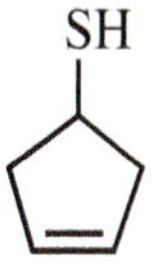

10-33
(a) Hexan-1-ol will boil at a higher temperature as it is less branched than 3,3-dimethylbutan-1-ol.
(b) Hexan-2-ol will boil at a higher temperature because its molecules hydrogen bond with each other, whereas molecules of hexan-2-one have no intermolecular hydrogen bonding.
(c) Hexane-1,5-diol will boil at a higher temperature as it has two OH groups for hydrogen bonding. Hexan-2-ol has only one group for hydrogen bonding.
(d) Hexan-2-ol will boil at a higher temperature because it has a higher molecular weight than pentan-2-ol. All other structural features of the two molecules are the same, so they should have the same intermolecular forces.

10-34 Refer to Table 10-4 to compare acidities of different functional groups. The strength of an acid is determined by the stability of its conjugate base.
(a) 3-Chlorophenol is more acidic than cyclopentanol. In general, phenols are many orders of magnitude more acidic than alcohols because phenoxide anions are stabilized by resonance.
(b) Cyclohexanethiol is more acidic than cyclohexanol. S is beneath O on the periodic table, and acidity increases down the periodic table. Larger anions are more stable because a negative charge on a larger atom is distributed over a larger volume, with lower electron density and greater delocalization of the negative charge.
(c) Cyclohexanecarboxylic acid is more acidic than cyclohexanol. In general, carboxylic acids are many orders of magnitude more acidic than alcohols because carboxylate anions are stabilized by resonance.
(d) 2,2-Dichlorobutan-1-ol is more acidic than butan-1-ol because of the two electron-withdrawing substituents near the acidic functional group.

10-35
(a) Propan-2-ol is the most soluble in water as it has the fewest carbons and the most branching.
(b) Cyclohexane-1,2-diol is the most soluble as it has two OH groups for hydrogen bonding. Cyclohexanol has only one OH group; chlorocyclohexane cannot hydrogen bond and is the least soluble.
(c) Phenol is the most soluble as it can hydrogen bond better than cyclohexanol. Phenol is a stronger acid and will be ionized to a greater extent than cyclohexanol. Cyclohexanol has low solubility, and 4-methylcyclohexanol has the added hydrophobic methyl group, decreasing its water solubility.

10-36 Products after aqueous acid workup:

(a) 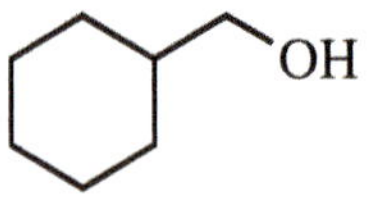 OH (b) 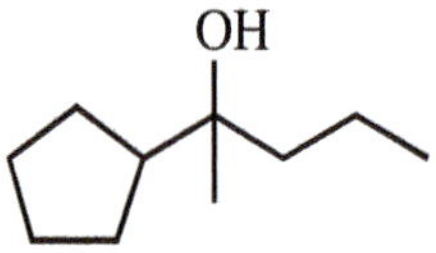 OH (c) 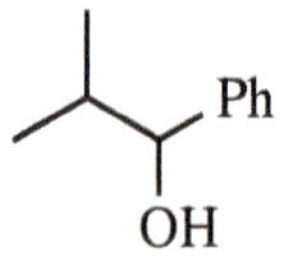 Ph OH

(d) This problem confuses a lot of students. When a Grignard reagent is added to a compound that has an OH group, the first thing that happens is that the Grignard reacts by removing the H^+ to give O^-.

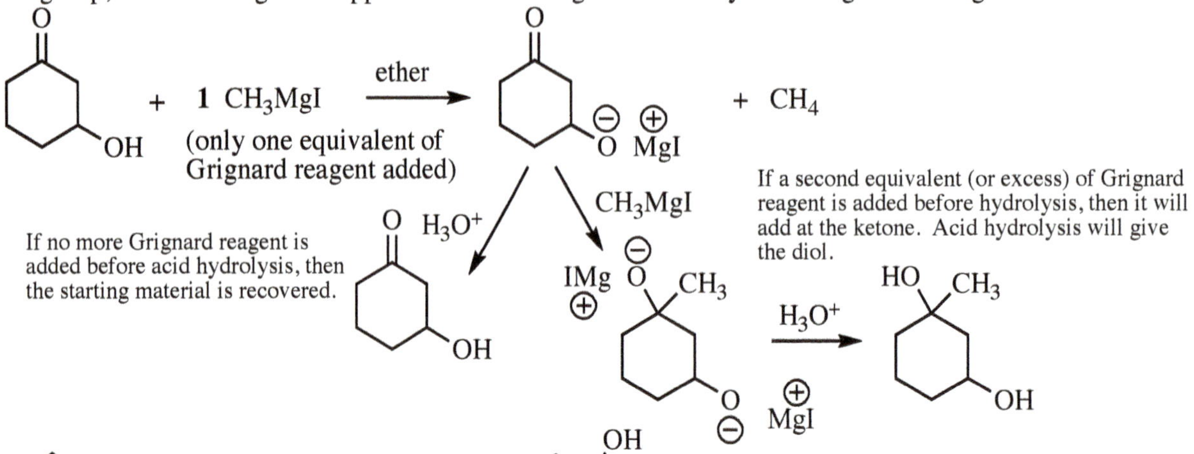

O OH + **1** CH_3MgI (only one equivalent of Grignard reagent added) —ether→ O $O^{\ominus}$ $^{\oplus}MgI$ + CH_4

If no more Grignard reagent is added before acid hydrolysis, then the starting material is recovered.

H_3O^+ → O OH

CH_3MgI → $IMg^{\oplus}$ $O^{\ominus}$ CH_3 $O^{\ominus}$ $^{\oplus}MgI$ —H_3O^+→ HO CH_3 OH

If a second equivalent (or excess) of Grignard reagent is added before hydrolysis, then it will add at the ketone. Acid hydrolysis will give the diol.

(e) 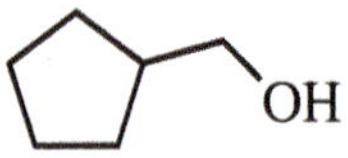 OH (f) 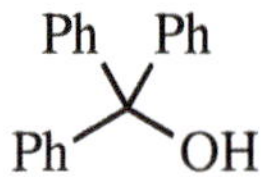 Ph Ph Ph OH (g) 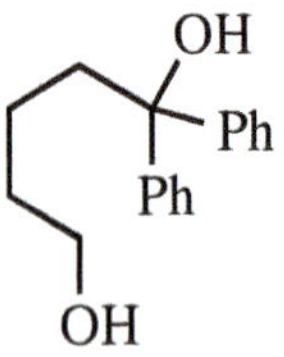 OH Ph Ph OH (h) Ph OH

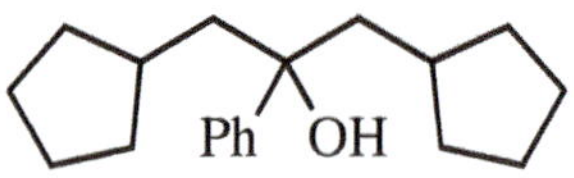

(i) 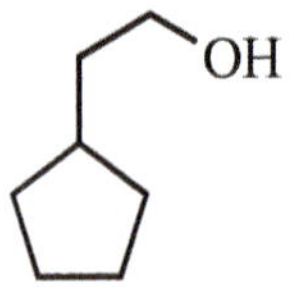 OH (j) 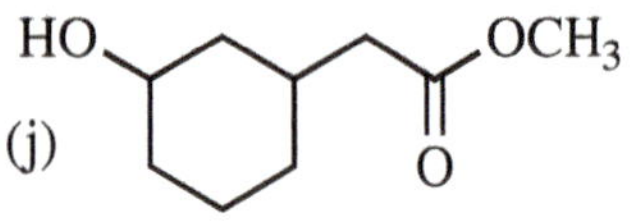 HO OCH_3 O (k) 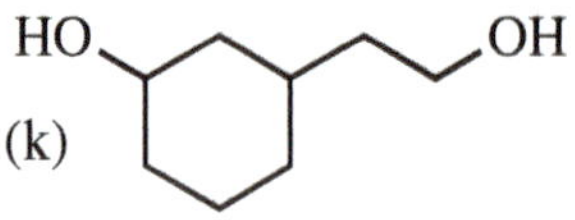 HO OH (l) 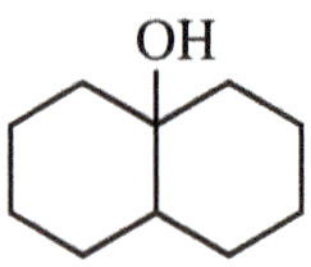 OH

(m) 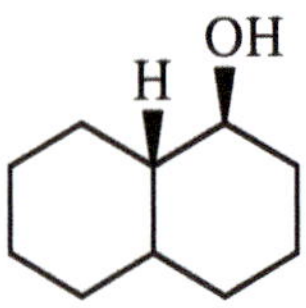 OH H plus the enantiomer (n) 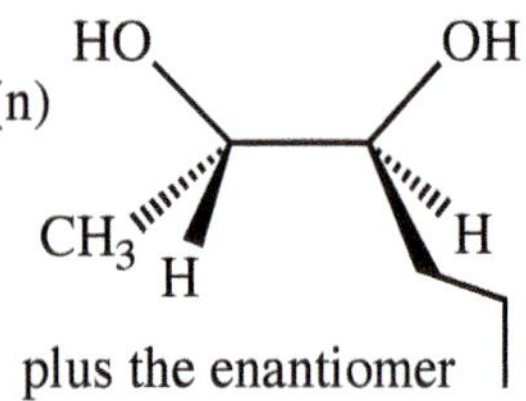 HO OH CH_3 H H plus the enantiomer (o) 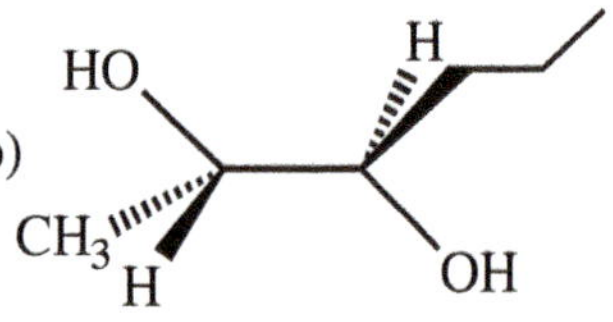 HO H CH_3 H OH plus the enantiomer (p)

10-37 Bromobenzene is abbreviated PhBr. The phenyl Grignard reagent is abbreviated PhMgBr.

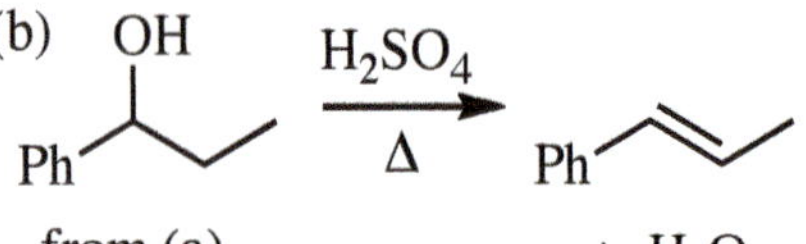

(a) 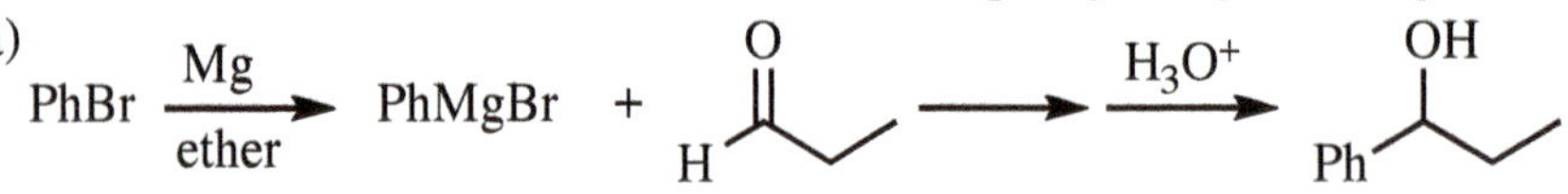 PhBr —Mg, ether→ PhMgBr + O H → —H_3O^+→ OH Ph

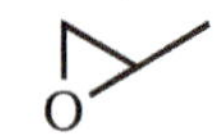

OR

PhMgBr from (a) + O with H_3O^+ workup

(b) OH Ph from (a) —H_2SO_4, Δ→ Ph + H_2O

(c) two methods

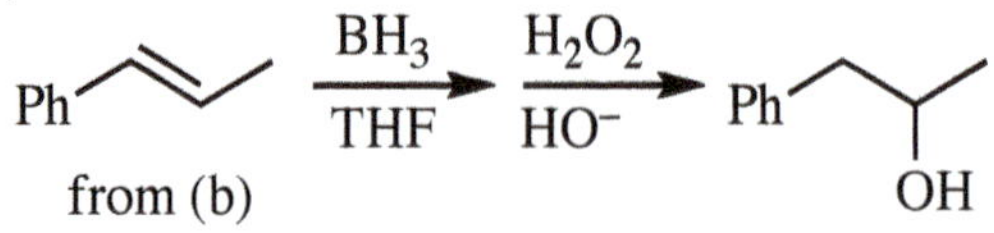

Ph from (b) —BH_3, THF→ —H_2O_2, HO^-→ Ph OH

10-37 continued

(d) Ph–CH=CH–CH$_3$ (from (b)) $\xrightarrow[h\nu]{NBS}$ Ph–CH=CH–CH$_2$Br $\xrightarrow[\Delta]{H_2O}$ Ph–CH=CH–CH$_2$OH *Trans* is more stable. Either S_N1 or S_N2 with NaOH will work.

(e) PhMgBr (from (a)) + 2-butanone $\longrightarrow \xrightarrow{H_3O^+}$ 2-phenyl-2-butanol

(f) 2-phenyl-2-butanol $\xrightarrow[\Delta]{H_2SO_4}$ 2-phenyl-2-butene (major)

10-38

(a) 1-butene $\xrightarrow[H_2O]{Hg(OAc)_2} \xrightarrow{NaBH_4}$ 2-butanol (OH)

(b) 1-methylcyclohexene $\xrightarrow[2)\ H_2O_2,\ HO^-]{1)\ BH_3 \bullet THF}$ trans-2-methylcyclohexanol (CH_3, H, OH)

(c) butylidenecyclopentane $\xrightarrow{BH_3 \bullet THF} \xrightarrow[HO^-]{H_2O_2}$ 1-cyclopentyl-1-butanol (OH)

(d) butylidenecyclopentane $\xrightarrow[H_2O]{Hg(OAc)_2} \xrightarrow{NaBH_4}$ 1-butylcyclopentanol (OH)

10-39 All Grignard reactions are run in ether solvent. Two arrows are shown indicating that the Grignard reaction is allowed to proceed, and then in a second step, dilute aqueous acid is added. The new C—C bonds are shown in bold. ▬

In this solution, the portion of the product that came from the Grignard reagent is shown in red.

(a) hexanal (O, H) + BrMg–CH$_2$CH$_3$ $\longrightarrow \xrightarrow{H_3O^+}$ 3-octanol (OH)

(b) 1-bromoheptane (Br) $\xrightarrow[ether]{Mg} \xrightarrow{CH_2O} \xrightarrow{H_3O^+}$ 1-octanol (OH)

(c) acetaldehyde (O, H) + BrMg–cyclohexyl $\longrightarrow \xrightarrow{H_3O^+}$ 1-cyclohexylethanol (OH)

(d) bromocyclohexane (Br) $\xrightarrow[ether]{Mg}$ $\xrightarrow{\text{ethylene oxide (O)}}$ $\xrightarrow{H_3O^+}$ 2-cyclohexylethanol (OH)

(e) bromobenzene (Br) $\xrightarrow[ether]{Mg} \xrightarrow{CH_2O} \xrightarrow{H_3O^+}$ benzyl alcohol (OH)

(f) cyclohexyl–C(=O)OCH$_2$CH$_3$ + 2 CH$_3$MgI $\longrightarrow \xrightarrow{H_3O^+}$ cyclohexyl–C(CH$_3$)(CH$_3$)–OH

(g) benzaldehyde (O, H) + BrMg–cyclopentyl $\longrightarrow \xrightarrow{H_3O^+}$ cyclopentyl(phenyl)methanol (OH)

10-39 continued

(h) Br $\xrightarrow[\text{ether}]{Mg}$ (ethylene oxide) $\xrightarrow{}$ $\xrightarrow{H_3O^+}$ OH

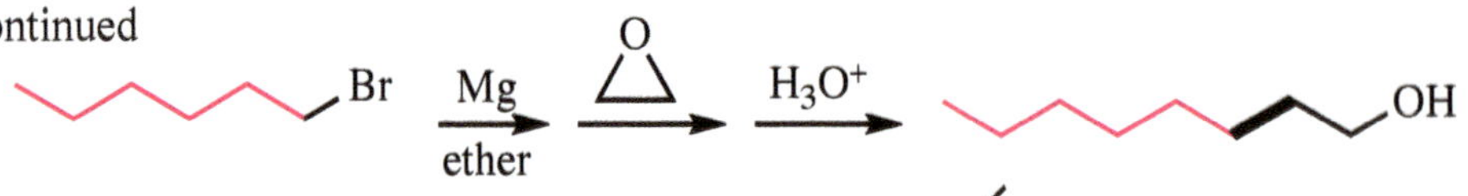

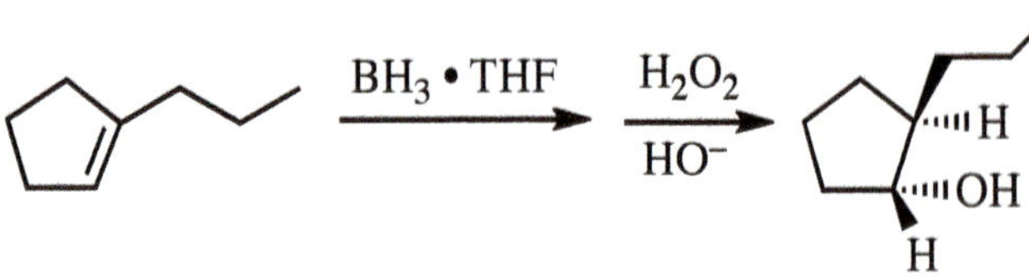

10-40

(a) $\xrightarrow{BH_3 \bullet THF}$ $\xrightarrow[HO^-]{H_2O_2}$ H, OH, H plus the enantiomer

(b) SH $\xrightarrow[\Delta]{KMnO_4 \text{ or } HNO_3 \text{ or } NaOCl}$ SO_3H Many strong oxidizing agents will transform thiols to sulfonic acids.

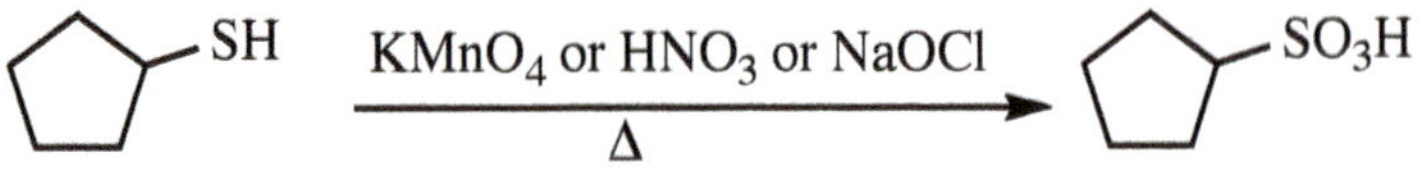

(c) O $\xrightarrow[CH_3OH]{NaBH_4}$ OH

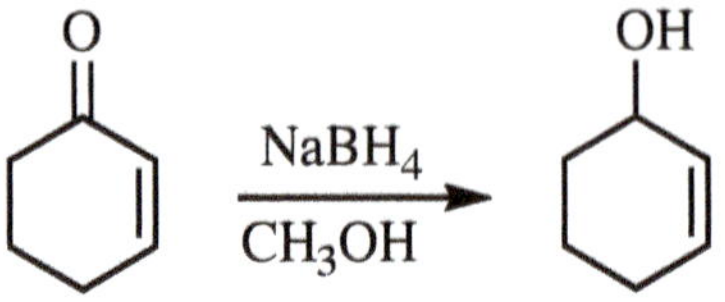

(d) O $\xrightarrow[\text{Pt catalyst}]{1 \text{ eq. } H_2}$ O

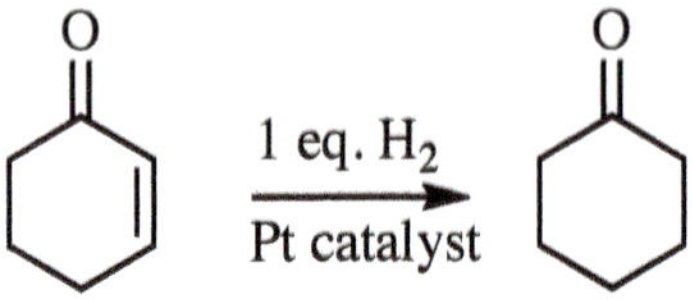

H_2/Pt reduces C=C faster than C=O

(e) O, OEt, O $\xrightarrow[CH_3OH]{NaBH_4}$ OH, OEt, O

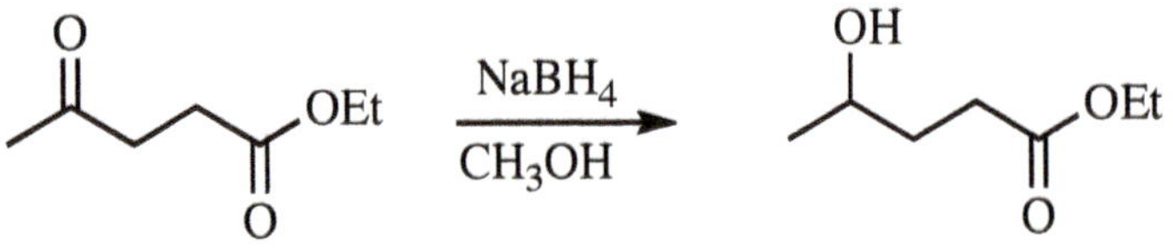

(f) O, OEt, O $\xrightarrow{LiAlH_4}$ $\xrightarrow{H_3O^+}$ OH, OH

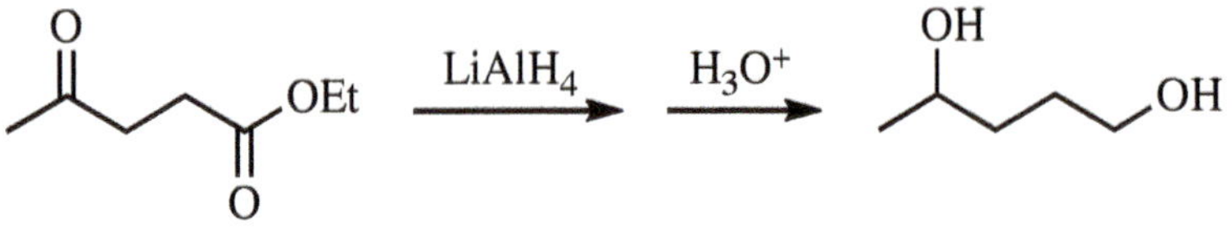

10-41

(a) MgBr + CH_2O $\xrightarrow{\text{ether}}$ $\xrightarrow{H_3O^+}$ OH

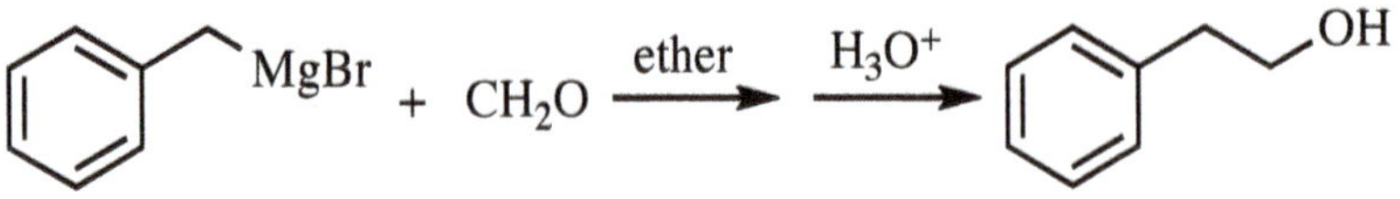

(b) $\xrightarrow{BH_3 \bullet THF}$ $\xrightarrow[HO^-]{H_2O_2}$ OH

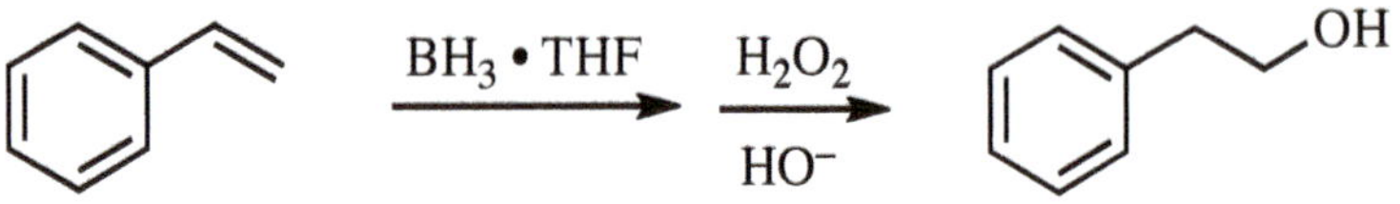

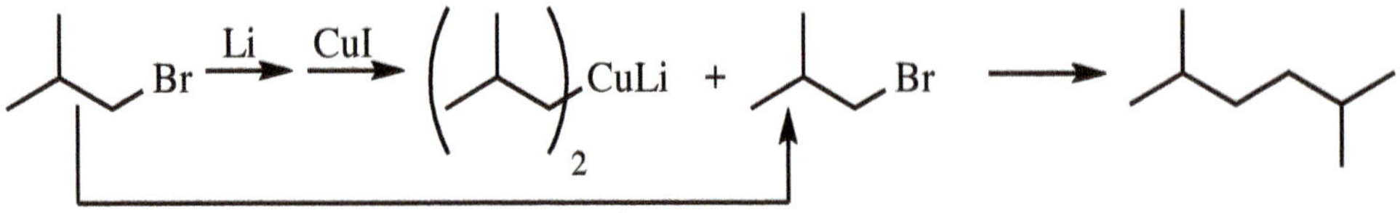

(c) Br $\xrightarrow{NaOH}$ OH

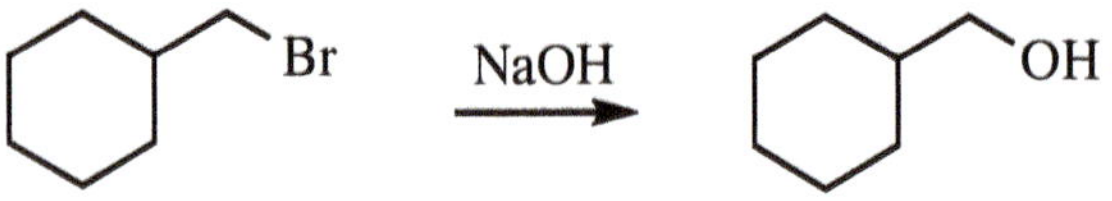

(d) Br $\xrightarrow[\text{ether}]{Mg}$ (ethylene oxide) $\xrightarrow{}$ $\xrightarrow{H_3O^+}$ OH

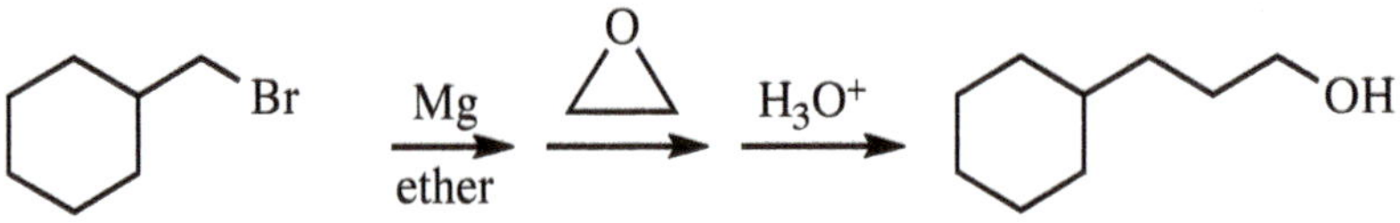

(e) Br + NaSH $\longrightarrow$ SH

(f) Br $\xrightarrow{Li}$ $\xrightarrow{CuI}$ $(\ )_2$CuLi + Br $\longrightarrow$

10-42 The position of the equilibrium can be determined by the strength of the acids or the bases. The stronger acid and stronger base will always react to give the weaker acid and base, so the side of the equation with the weaker acid and base will be favored at equilibrium.

(a) $CH_3CH_2O^{-}$ + C_6H_5–OH ⇌ CH_3CH_2OH + C_6H_5–O^{-}

stronger base | stronger acid | weaker acid | weaker base

products favored

(b) KOH + 2,4-dichlorophenol (Cl–C_6H_3(Cl)–OH) ⇌ H_2O + Cl–C_6H_3(Cl)–O^{-} K^+

stronger base | stronger acid | weaker acid | weaker base

products favored

(c) 1-naphthol (OH) + CH_3O^{-} ⇌ 1-naphthoxide (O^{-}) + CH_3OH

stronger acid | stronger base | weaker base | weaker acid

products favored

(d) cyclopentanol (OH) + KOH ⇌ H_2O + cyclopentoxide (O^{-} K^+)

weaker acid | weaker base | stronger acid | stronger base

reactants favored

(e) $(CH_3)_3CO^{-}$ + CH_3CH_2OH ⇌ $(CH_3)_3COH$ + $CH_3CH_2O^{-}$

stronger base | stronger acid | weaker acid | weaker base

products favored

(f) $(CH_3)_3CO^{-}$ + H_2O ⇌ $(CH_3)_3COH$ + HO^{-}

stronger base | stronger acid | weaker acid | weaker base

products favored

(g) KOH + CH_3CH_2OH ⇌ H_2O + $CH_3CH_2O^{-}$ K^+

weaker base | weaker acid | stronger acid | stronger base

reactants favored

10-43

(a) H, O $\xrightarrow[CH_3OH]{NaBH_4}$ OH OR $\xrightarrow[2)\ H_3O^+]{1)\ LiAlH_4}$

OR

OH, O $\xrightarrow[2)\ H_3O^+]{1)\ LiAlH_4}$

OR

OR, O $\xrightarrow[2)\ H_3O^+]{1)\ LiAlH_4}$

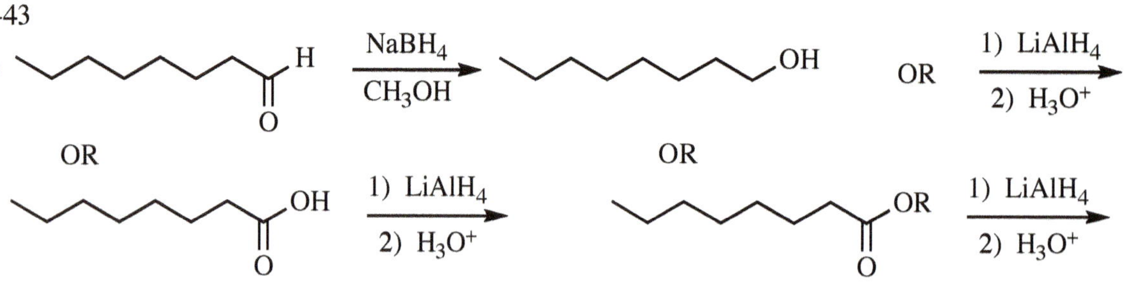

(b) O $\xrightarrow[CH_3OH]{NaBH_4}$ OH OR $\xrightarrow[2)\ H_3O^+]{1)\ LiAlH_4}$

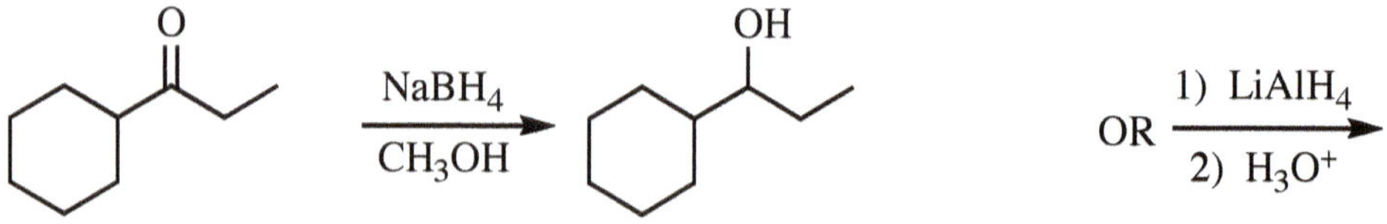

(c) O $\xrightarrow[CH_3OH]{NaBH_4}$ OH OR $\xrightarrow[2)\ H_3O^+]{1)\ LiAlH_4}$

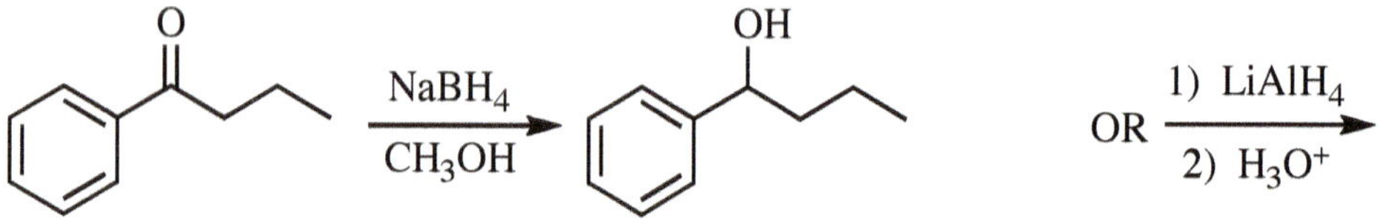

(d) O $\xrightarrow[CH_3OH]{NaBH_4}$ OH OR $\xrightarrow[2)\ H_3O^+]{1)\ LiAlH_4}$

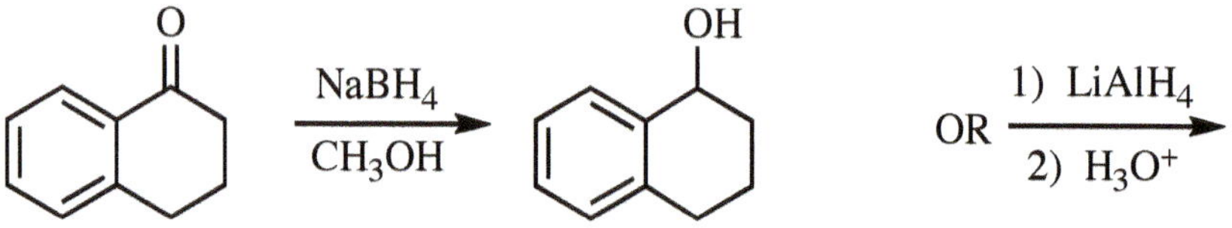

(e) O, H, O $\xrightarrow[CH_3OH]{NaBH_4}$ OH, HO OR $\xrightarrow[2)\ H_3O^+]{1)\ LiAlH_4}$

OR

O, OH (or ester), O $\xrightarrow[2)\ H_3O^+]{1)\ LiAlH_4}$ OH, HO

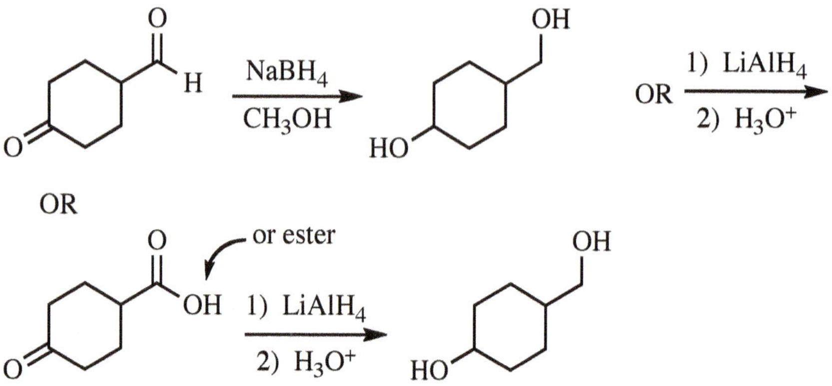

(f) O, OCH_3, O $\xrightarrow[CH_3OH]{NaBH_4}$ O, OCH_3, HO

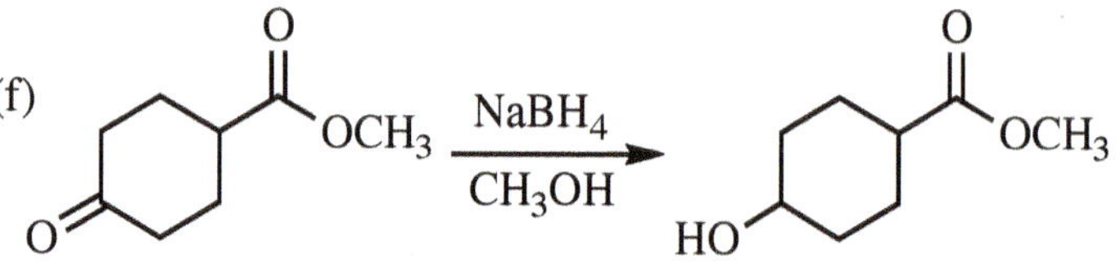

The milder $NaBH_4$ reduces aldehydes and ketones but not carboxylic acids or esters; the stronger $LiAlH_4$ is needed to reduce acids and esters.

10-44 (a) The goal is to synthesize the target compound (boxed) from starting materials of six carbons or fewer. The product has 11 carbons, so the logical "disconnection" in working backwards is one six-carbon fragment and the cyclopentane ring that could be joined in a Grignard reaction. The tetrasubstituted C=C will come from dehydration of a tertiary alcohol produced in two possible Grignard reactions.

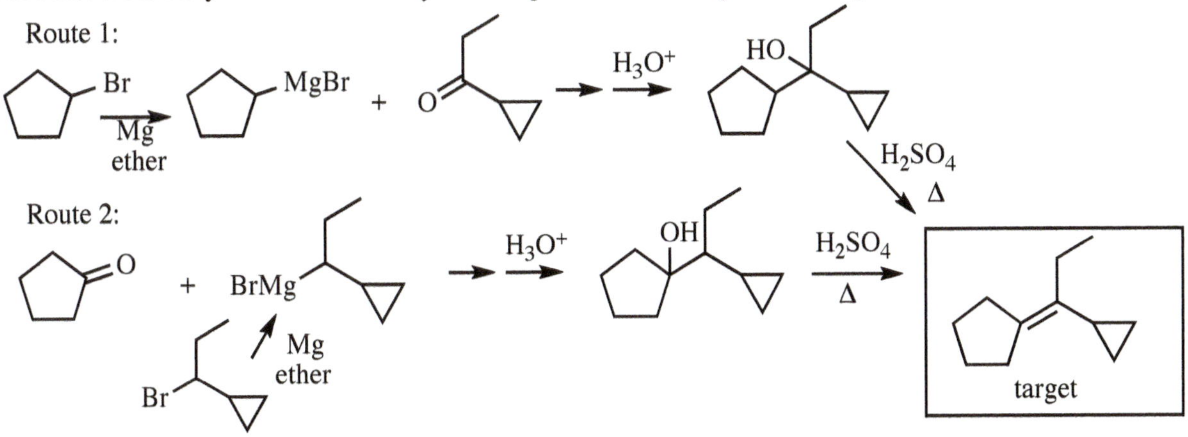

Route 1:

Br → (Mg, ether) MgBr + O → H_3O^+ → HO → (H_2SO_4, Δ)

Route 2:

O + BrMg ← (Mg, ether) Br → H_3O^+ → OH → (H_2SO_4, Δ)

target

(b) The goal is to synthesize the target compound (boxed) from starting materials of six carbons or fewer. The product has 12 carbons, so the logical "disconnection" in working backwards is two six-carbon fragments which could be joined in a Grignard reaction. The best way to make epoxides is from the double bond, and double bonds are made from alcohols which are the products of Grignard reactions.

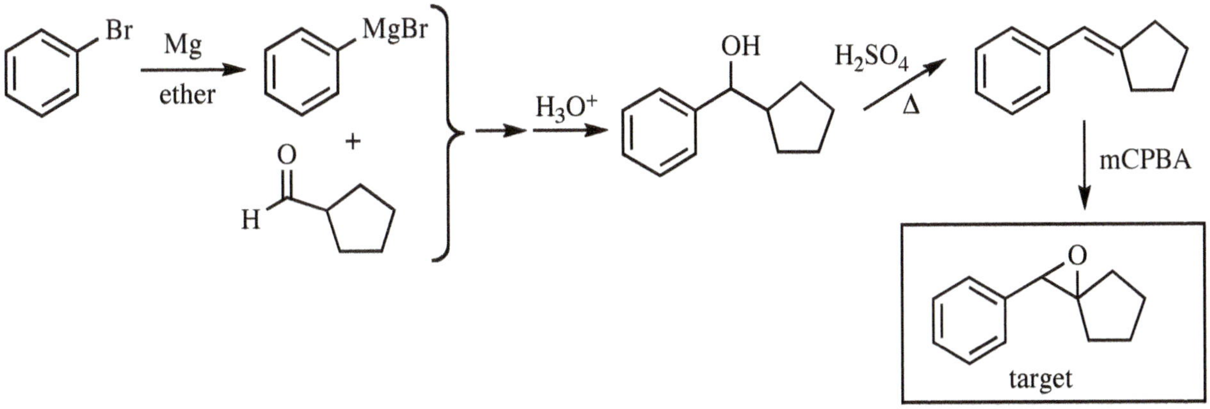

Br → (Mg, ether) MgBr + H, O → H_3O^+ → OH → (H_2SO_4, Δ) → (mCPBA) → O

target

(c) The goal is to synthesize the target compound (boxed) from starting materials of six carbons or fewer. The product has 14 carbons, so the logical "disconnection" in working backwards is two six-carbon fragments and the ethyl group put on with a Williamson ether synthesis at the end.

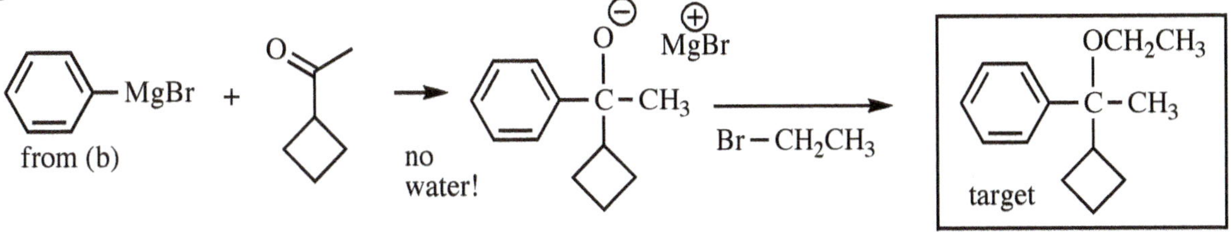

MgBr (from (b)) + O → (no water!) O^- ^+MgBr, C–CH_3 → ($Br-CH_2CH_3$) → OCH_2CH_3, C–CH_3

target

10-45 All steps are reversible.

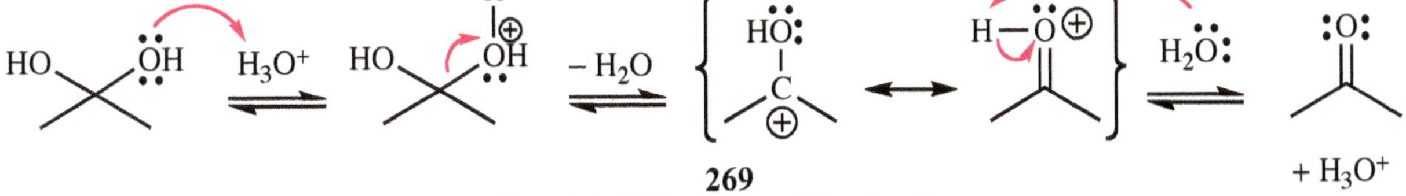

HO, OH ⇌ (H_3O^+) HO, H, OH⊕ ⇌ ($-H_2O$) { HO:, C⊕ ⟷ H–O⊕ } ⇌ (H_2O:) :O: $+ H_3O^+$

10-46 The symbol H—A represents a generic acid, where A^- is the conjugate base.

(a)

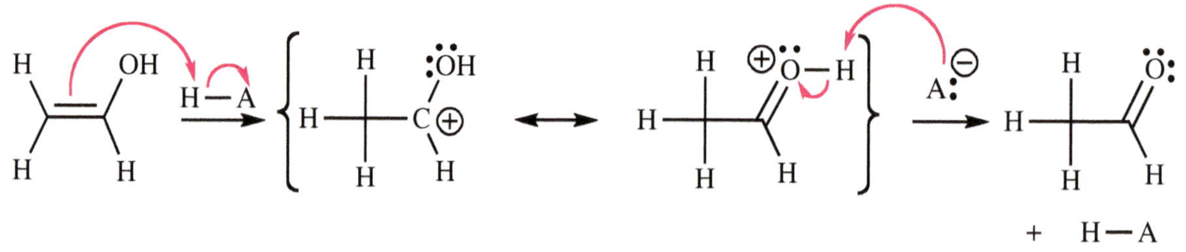

(b)

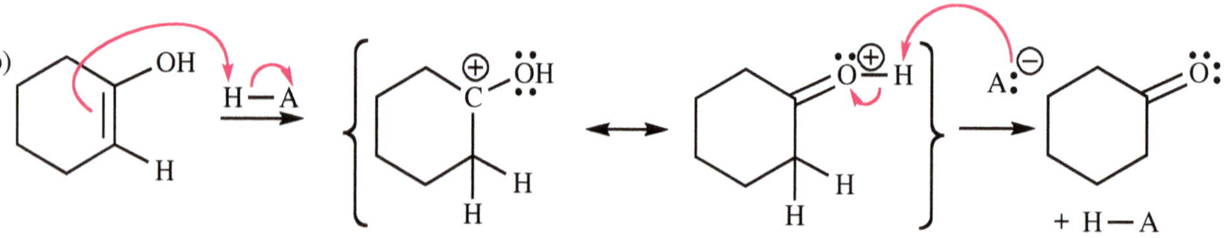

(c)

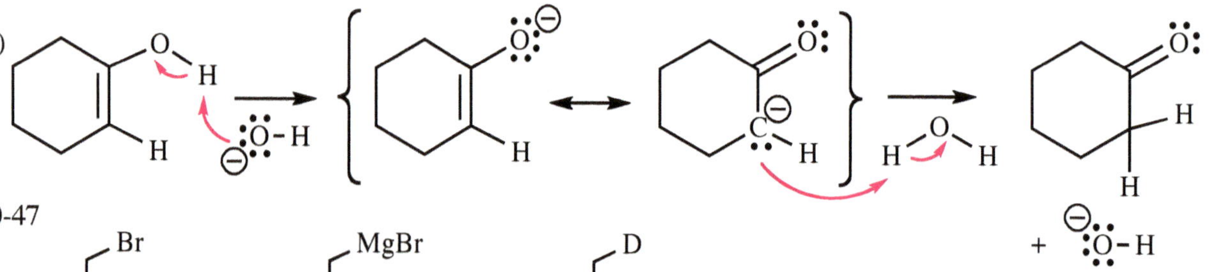

10-47

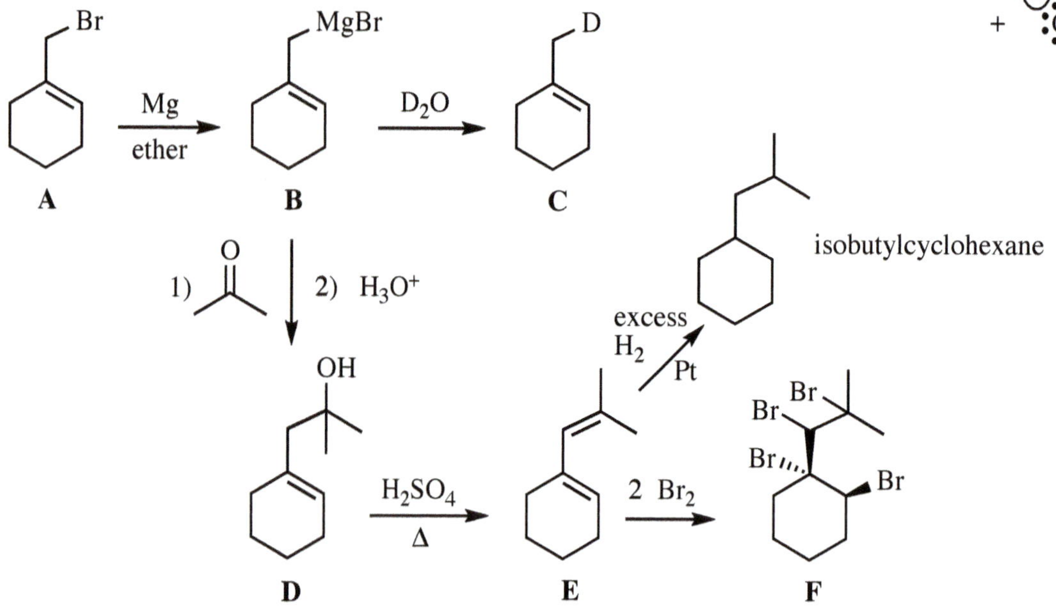

A

B

C

1) 2) H_3O^+

isobutylcyclohexane

excess H_2 Pt

Mg ether

D_2O

H_2SO_4 Δ

2 Br_2

D

E

F

10-48 This mechanism is similar to cleavage of the epoxide in ethylene oxide by Grignard reagents. The driving force for the reaction is relief of ring strain in the 4-membered cyclic ether, which is why it will undergo a Grignard reaction whereas most other ethers will not.

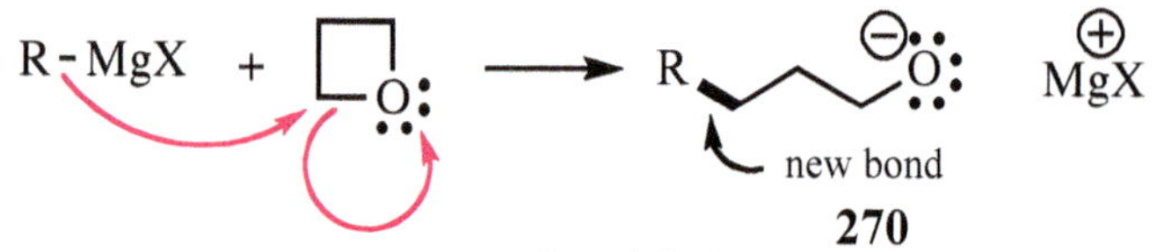

R–MgX + → R ... $\ddot{O}^{\ominus}$ $\overset{\oplus}{MgX}$

new bond

10-49 When mixtures of isomers can result, only the major product is shown.

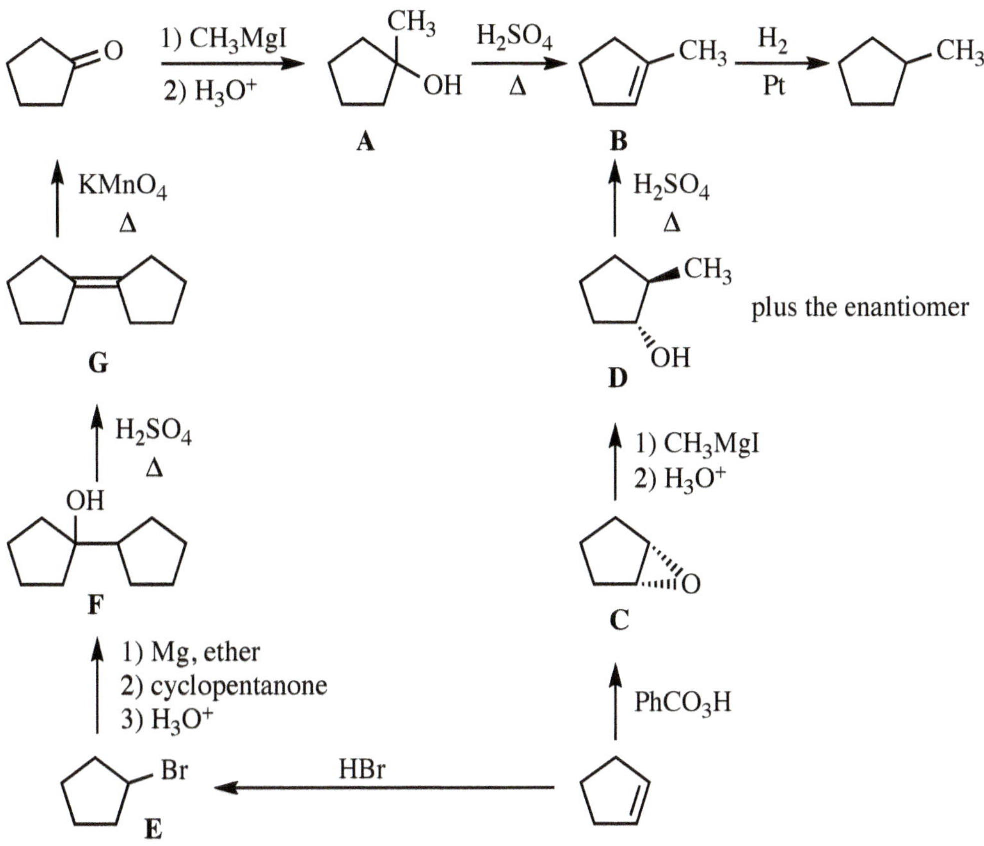

10-50 The most important component in the deskunking mixture is hydrogen peroxide. Thiols are oxidized to structures having one, two, or three oxygens on the sulfur; all of these functional groups are acidic, less volatile so they don't reach the nose, and less stinky. The sodium bicarbonate is basic enough to ionize these acids, making them water soluble where the soap can wash them away.

sulfenic acids — sulfinic acids — sulfonic acids

SH $\xrightarrow{H_2O_2}$ S–OH $\xrightarrow{H_2O_2}$ S(=O)OH $\xrightarrow{H_2O_2}$ S(=O)(=O)–OH

3-methylbutane-1-thiol

SH $\xrightarrow{H_2O_2}$ S–OH $\xrightarrow{H_2O_2}$ S(=O)OH $\xrightarrow{H_2O_2}$ S(=O)(=O)–OH

2-butene-1-thiol
(but-2-ene-1-thiol)

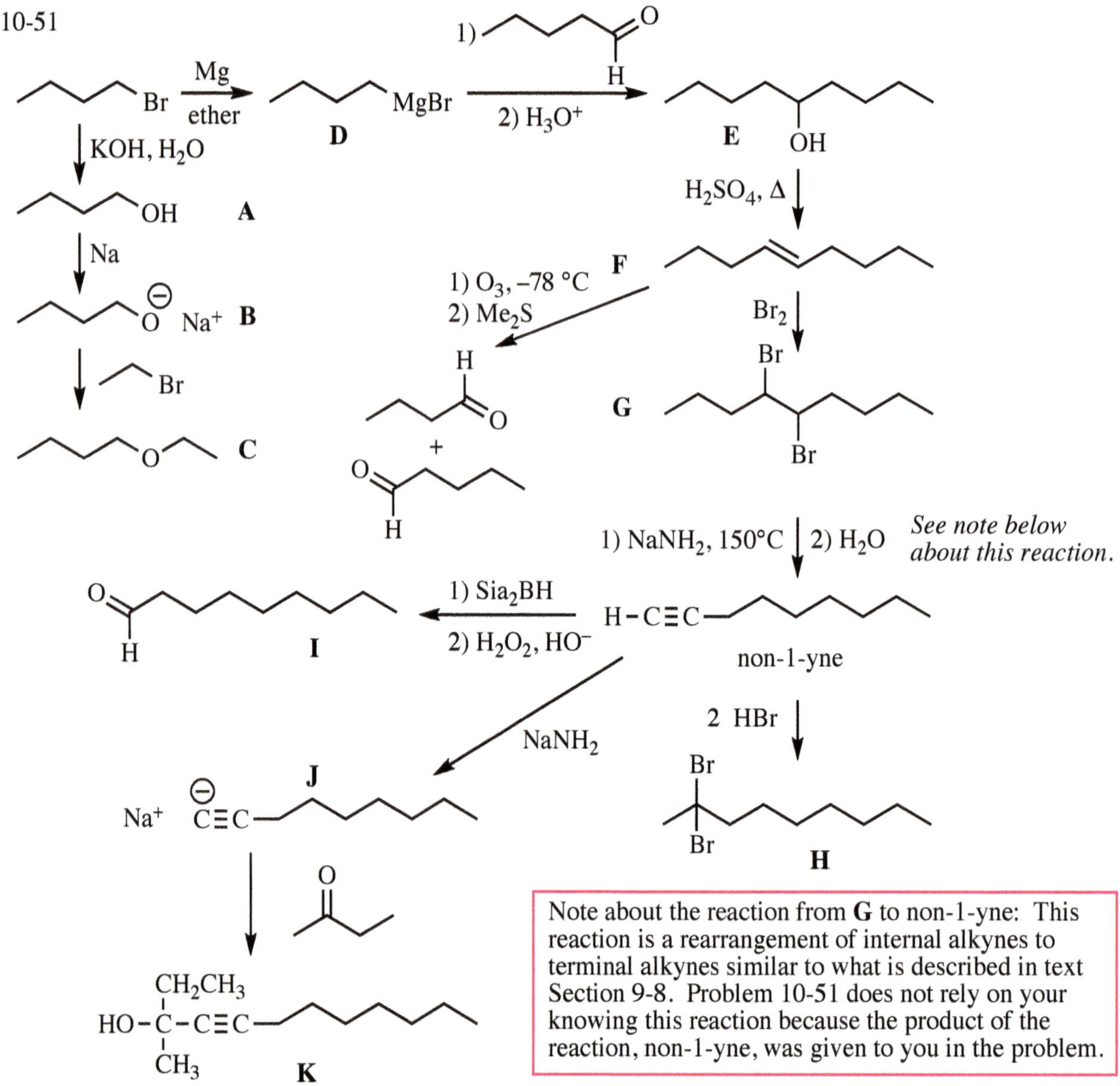

10-52 This problem is reminiscent of problem 8-51.

(a) $Hg(OAc)_2$, H_2O → $NaBH_4$ → OH

(b) 1) BH_3 • THF 2) H_2O_2, HO^- → OH

(c) OsO_4, H_2O_2 or cold, dilute $KMnO_4$ → OH, OH

catalytic H_2SO_4

HBr

Br

KOH, Δ

Parts (d) and (e) require moving the double bond into the ring, to form methylcyclohexene. Adding a catalytic amount of strong acid would do this, and addition-elimination will accomplish the same thing.

Alternatively, the tertiary alcohol product in part (a) could be dehydrated to give methylcyclohexene as the major product.

10-52 continued—See the note at the bottom of the previous page regarding (d) and (e).

(d) 1-methylcyclohexene —[1) $BH_3 \cdot THF$; 2) H_2O_2, HO^-]→ trans-2-methylcyclohexanol (CH_3 wedge, OH hashed) syn addition of H and OH with anti-Markovnikov orientation

plus the enantiomer

(e) 1-methylcyclohexene —[Cl_2, H_2O]→ 2-chloro-1-methylcyclohexanol (OH, CH_3, Cl) chlorohydrin formation with anti addition of Cl and OH

plus the enantiomer

(f) methylenecyclohexane —[mCPBA]→ epoxide (O) —[PhMgBr, ether]→ —[H_3O^+]→ 1-benzylcyclohexanol (Ph, OH)

New C—C bond requires Grignard; OH is not on same carbon where new C—C is formed suggesting the epoxide intermediate.

10-53

I	II
$(H_3C)_2CHOH$	$(F_3C)_2CHOH$
mol. wt. 60 g/mole	mol. wt. 168 g/mole
boiling point 82 °C	boiling point 58 °C
dipole moment 1.66 D	dipole moment 0.32 D
pK_a 16.5	pK_a 9.3

(a) Boiling point is a rough indicator of the strength of intermolecular forces, of which three common ones apply to organic compounds: van der Waal's forces (or London forces), the weakest; dipole-dipole interactions for molecules with a permanent dipole moment; and hydrogen-bonding in compounds with OH or NH bonds, the strongest. Both **I** and **II** have the alcohol functional group and are likely to form hydrogen bonds with their neighbors, although an argument could be made that the slightly larger F substituents with the greater electron clouds would make hydrogen bonding more difficult in **II**. Rather than stretch a possibility, let's focus on something obvious: there is a huge difference in the dipole moments of these two compounds. Even though the molecular weight of **II** is much greater, thereby increasing van der Waal's forces, the dipole moment of **II** is very small. We must conclude that the significantly decreased dipole-dipole interaction of **II** is more important in boiling point than the increased van der Waal's interaction, a weaker force.

(b) **I** has a large dipole moment because of the bond polarizations due to the electronegative oxygen. In **II**, however, the bond polarization in the alcohol is counteracted by six F atoms; recall that F is the most electronegative element in the periodic table. The oxygen pulling in one direction is balanced against the six F atoms pulling partially away from the oxygen. (If you have studied vectors in physics, consider each polarized C—F bond as a vector with a "down" component and a "left" or "right" component; you will see that some portion of the left CF_3 group cancels the same portion of the right CF_3, leaving only a portion of the C—F polarization to cancel the C—O polarization. Make a model!)

(c) **The strength of an acid is determined by the stability of its conjugate base**. The anion of **I** has no particular stabilization. The anion of **II**, however, has six F atoms pulling electron density by the inductive effect; that is, the negative charge on the oxygen is partially shared by the six electronegative F atoms pulling electron density through sigma bonds. The anion of **II** is much more stable than the anion of **I**, making **II** the stronger acid.

10-54

When synthesizing isotopically-labeled compounds, it is critical to keep track of what new atoms come from what source. If the D is wanted on a carbon, it has to come from $LiAlD_4$ but if the D is on an oxygen, then it must come from D_2O.

(a) 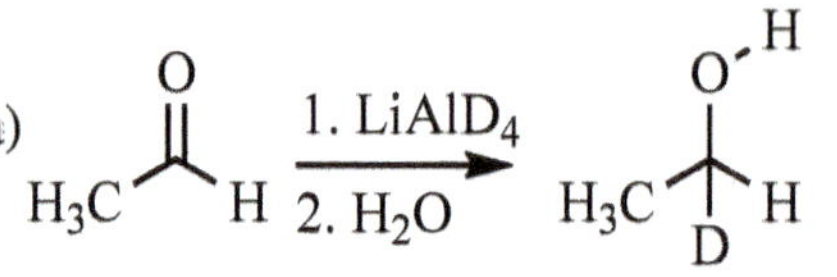 H_3C–CHO (O, H) → 1. $LiAlD_4$ 2. H_2O → H_3C–C(D)(H)–O–H

(b) 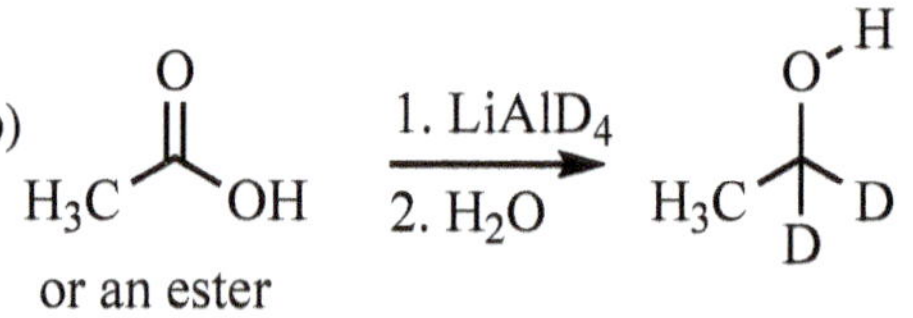 H_3C–COOH (O, OH) or an ester → 1. $LiAlD_4$ 2. H_2O → H_3C–C(D)(D)–O–H

(c) H_3C–COOH (O, OH) or an ester → 1. $LiAlD_4$ 2. D_2O → H_3C–C(D)(D)–O–D

10-55 Grignard reactions are typically performed in ether solvent. Ether is not shown here.

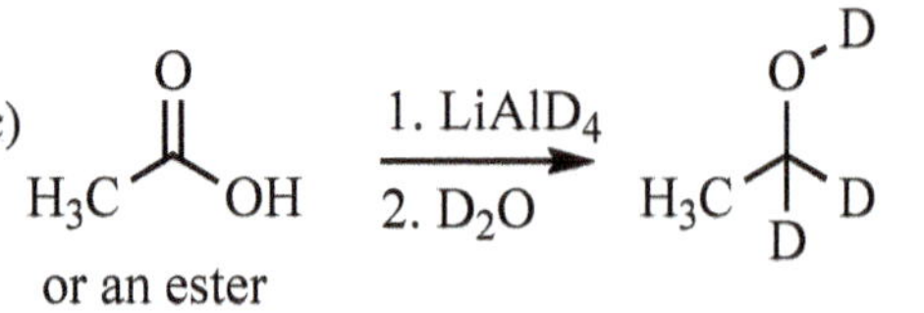

(a)  H_3C–CHO (O, H) → 1. CD_3MgBr 2. D_2O → H_3C–C(H)(CD_3)–O–D

(b) 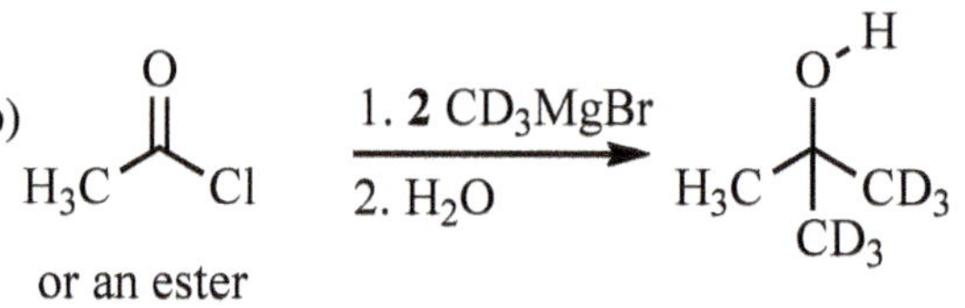 H_3C–COCl (O, Cl) or an ester → 1. **2** CD_3MgBr 2. H_2O → H_3C–C(CD_3)(CD_3)–O–H

(c) epoxide (O) → 1. CD_3MgBr 2. H_2O → D_3C–CH_2–CH_2–O–H

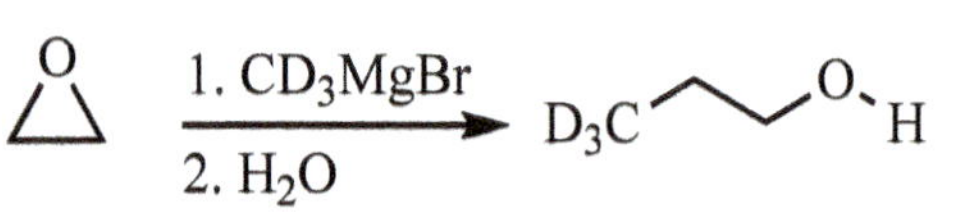

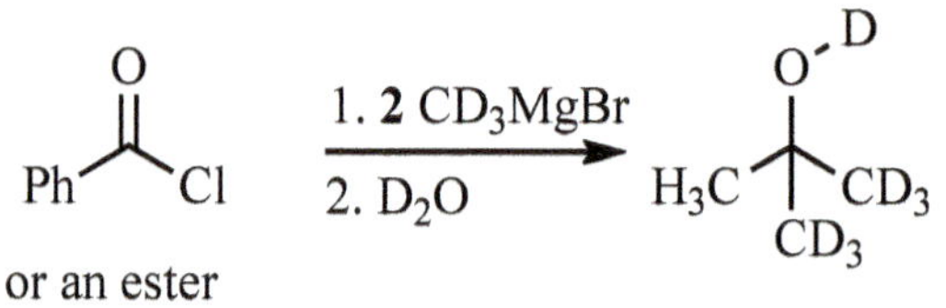

(d) Ph–COCl (O, Cl) or an ester → 1. **2** CD_3MgBr 2. D_2O → H_3C–C(CD_3)(CD_3)–O–D

10-56 For simplicity, hydrolysis steps after Grignards are not shown.

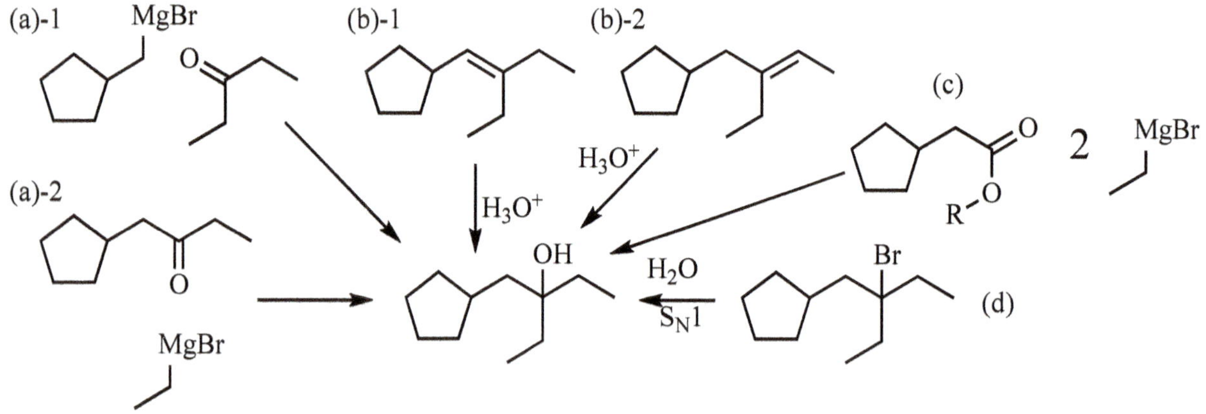

(a)-1 MgBr, O

(b)-1

(b)-2

(c) O, R–O, 2 MgBr

(a)-2 O, MgBr

H_3O^+

H_3O^+

OH

H_2O

S_N1

Br

(d)

10-57 For simplicity, hydrolysis steps after Grignards are not shown.

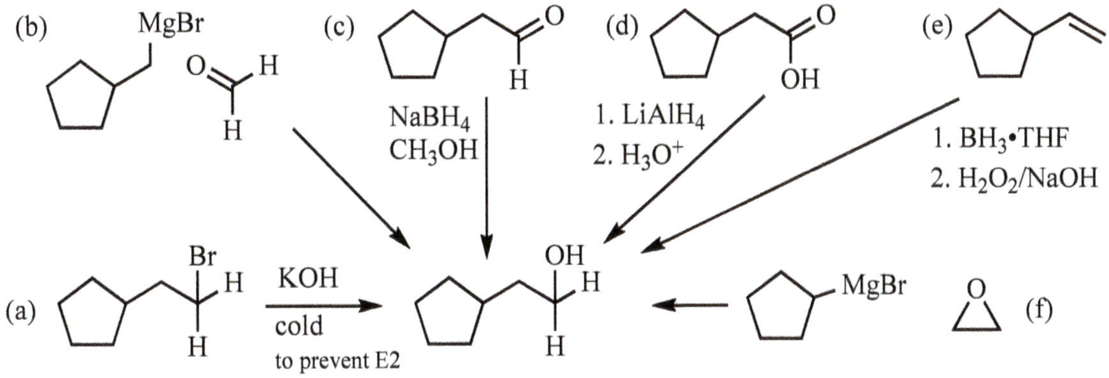

(b) MgBr O H H

(c) O H — $NaBH_4$, CH_3OH

(d) O OH — 1. $LiAlH_4$ 2. H_3O^+

(e) — 1. BH_3•THF 2. H_2O_2/NaOH

(a) Br H H — KOH, cold, to prevent E2 → OH H H ← MgBr O (f)

10-58

(a) Br — KOH, EtOH, Δ → **A** — 1. O_3, −78 °C 2. Me_2S → **B** O H O H — MgBr, ether — H_3O^+ → **C** OH OH

(b) Br — KOH, EtOH, Δ → **A** — $KMnO_4$, NaOH, Δ → **E** O OH O OH — $LiAlH_4$, ether — H_3O^+ → **F** OH OH

(c) Br — KOH, EtOH, Δ → **A** — mCPBA → **H** O — PhMgBr, ether — H_3O^+ → **I** OH Ph — H_2SO_4, Δ → **J** Ph

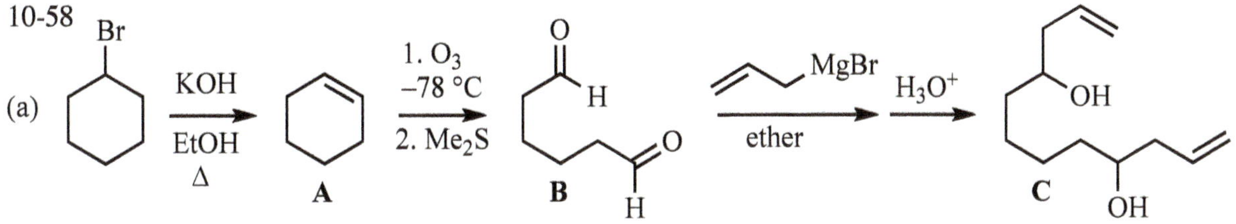

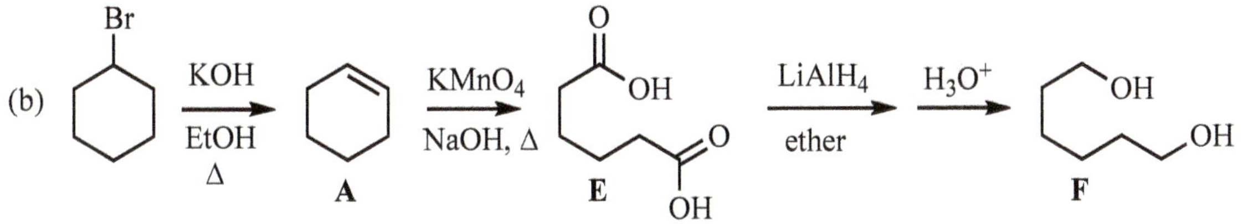

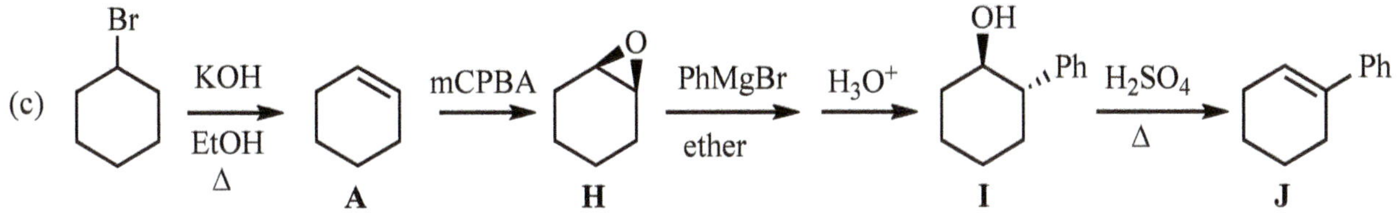

10-59

Approach of a nucleophile, including a Grignard reagent, on an epoxide must be from the back side, as if it were an S_N2 attack. In this example, the epoxide must have been epoxycyclohexane, and allyl Grignard must have come in from the side of the ring opposite the epoxide. An epoxide from a disubstituted double bond can be sluggish to react; in this example, the rigidity of the ring helps.

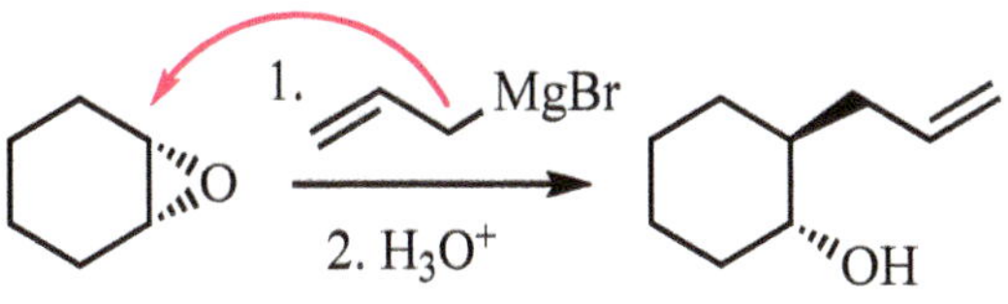

O — 1. MgBr 2. H_3O^+ → OH

10-60

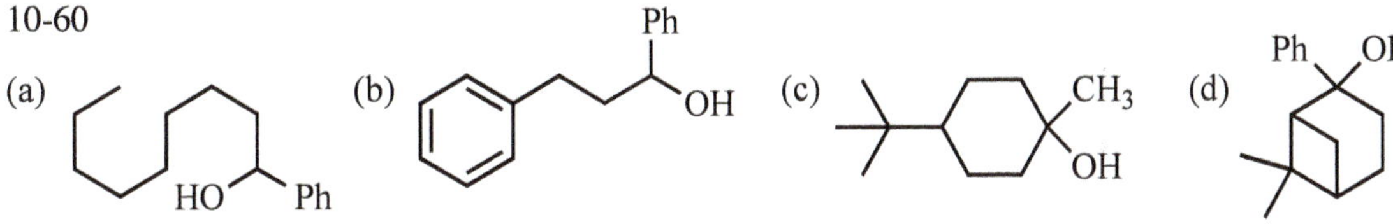

(a) HO Ph

(b) Ph OH

(c) CH_3 OH

(d) Ph OH

Students: use this space for notes or to solve problems.

CHAPTER 11—REACTIONS OF ALCOHOLS

11-1

(a) Both reactions are oxidations.
(b) oxidation, oxidation, reduction, oxidation
(c) One carbon is oxidized and one carbon is reduced—no net change (elimination of H and OH).
(d) reduction: C—O is replaced by C—H
(e) oxidation (addition of X_2)
(f) Neither oxidation nor reduction—the C still has two bonds to O.
(g) neither oxidation nor reduction (addition of HX)
(h) first step: neither oxidation nor reduction (elimination of H_2O); second step: reduction (addition of H_2)
(i) oxidation: adding an O to each carbon of the double bond
(j) The first reaction is oxidation as a new C—O bond is formed to each carbon of the alkene; the second reaction is neither oxidation nor reduction, as H_2O is added to the epoxide, and each carbon still has one bond to oxygen.
(k) oxidation: adding a Cl to one carbon and an O to the other
(l) Neither oxidation nor reduction: overall, only H and OH are added, so there is no net oxidation nor reduction.

11-2

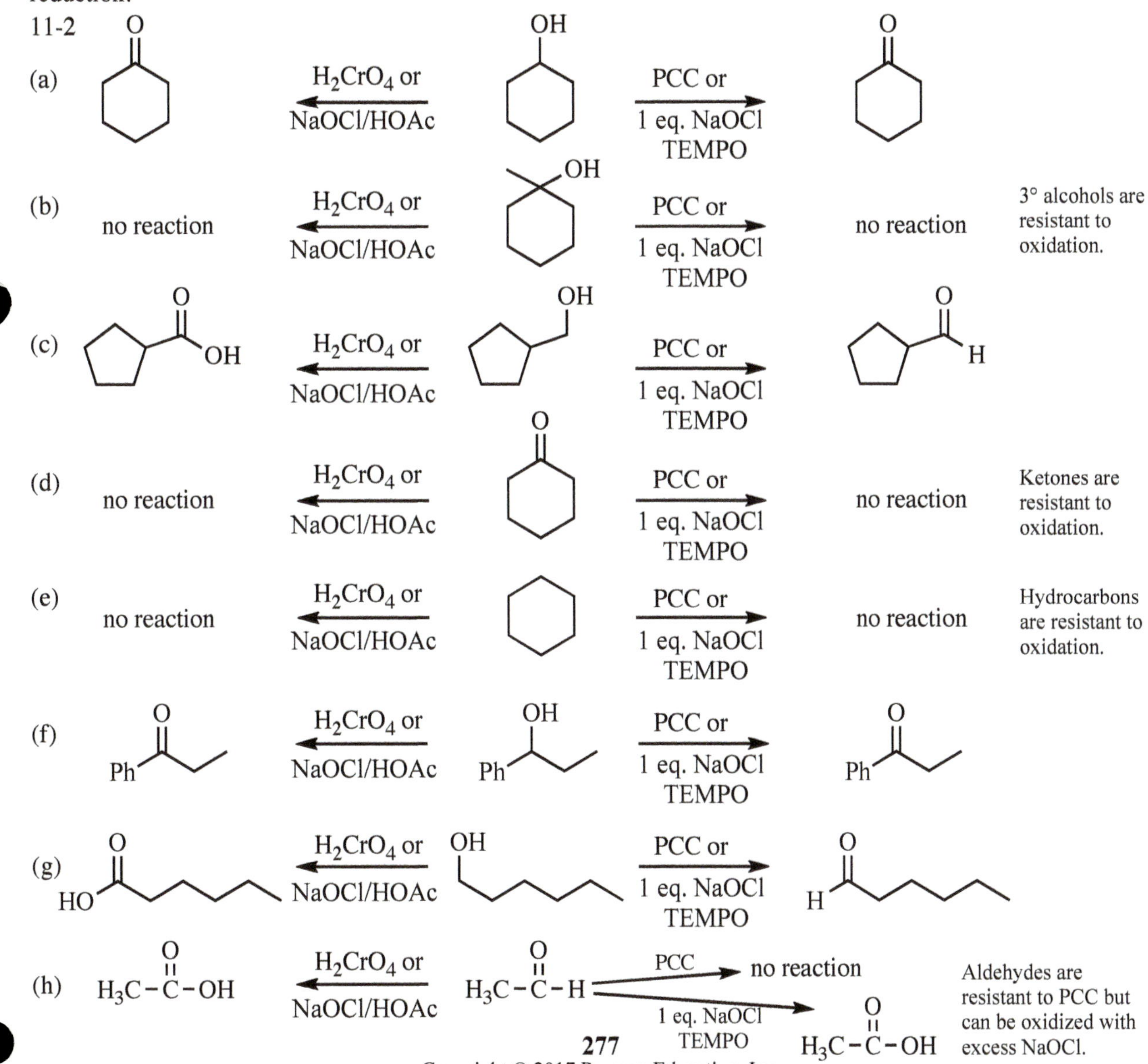

11-3

(a)  $HO-Cr(=O)_2-OH \longrightarrow HO-Cr(=O)-OH$

Cr begins with bonds to four oxygen atoms and ends with bonds to three oxygen atoms. Whether the bonds between the metals and the oxygen are single or double is not important—note this is NOT true of carbon! What matters here is the number of oxygen atoms bonded to the oxidizing atom.

(b)  $^{\ominus}O-Cl \longrightarrow Cl^{\ominus}$ Cl begins with one bond to oxygen and ends with no bonds to oxygen.

(c)  $H_3C-S(=O)-CH_3 \longrightarrow H_3C-S-CH_3$ S begins bonded to one oxygen and ends with no bonds to oxygen.

(d) 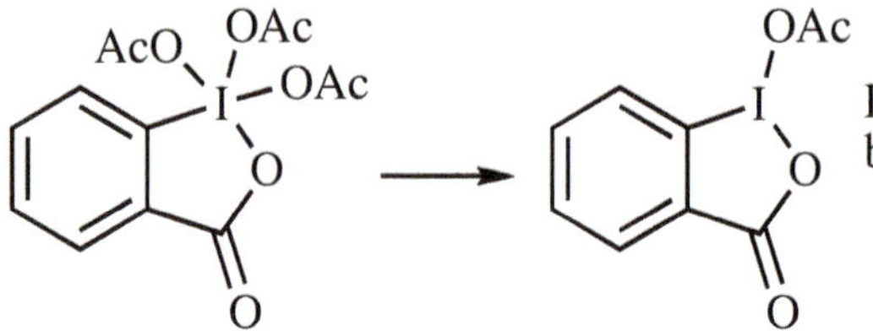 Iodine begins with four bonds to oxygen and ends with two bonds to oxygen.

(e)  C begins with one bond to O and (at least) one bond to H. It ends with two bonds to O and one fewer bond to H. This is one definition of oxidation: replacing a C—H bond with a C—O bond.

11-4 Note that PCC, DMP, Swern oxidation, and 1 equiv. NaOCl/TEMPO stop at the aldehyde when oxidizing a primary alcohol. Chromic acid and excess NaOCl take a primary alcohol to the carboxylic acid.

(a) 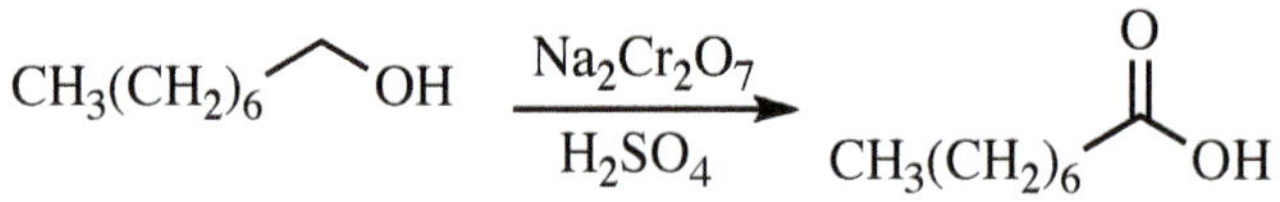 $CH_3(CH_2)_6CH_2OH \xrightarrow[H_2SO_4]{Na_2Cr_2O_7} CH_3(CH_2)_6COOH$

$CH_3(CH_2)_6CH_2OH$ PCC AND DMP AND 1 equiv. NaOCl / TEMPO $\longrightarrow CH_3(CH_2)_6CHO$

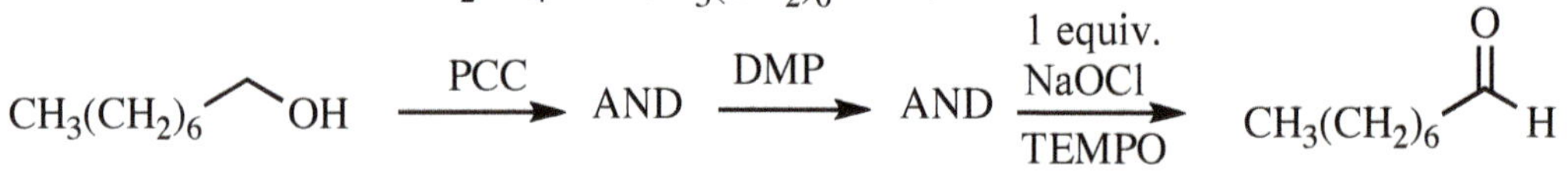

(b) All four reagents give the same ketone product with a secondary alcohol.

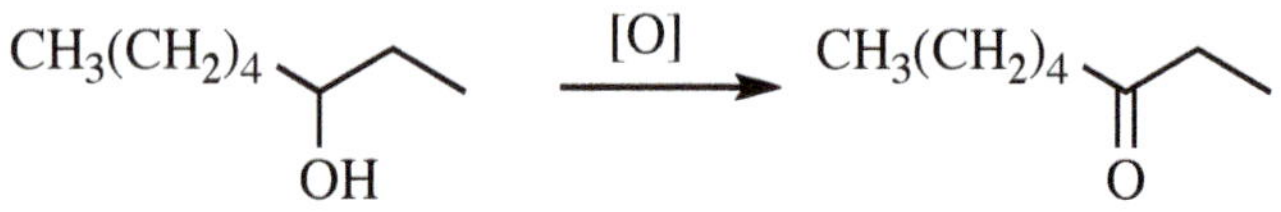

$CH_3(CH_2)_4CH(OH)CH_2CH_3 \xrightarrow{[O]} CH_3(CH_2)_4C(=O)CH_2CH_3$

[O] is the general abbreviation for an oxidizing agent.

(c) 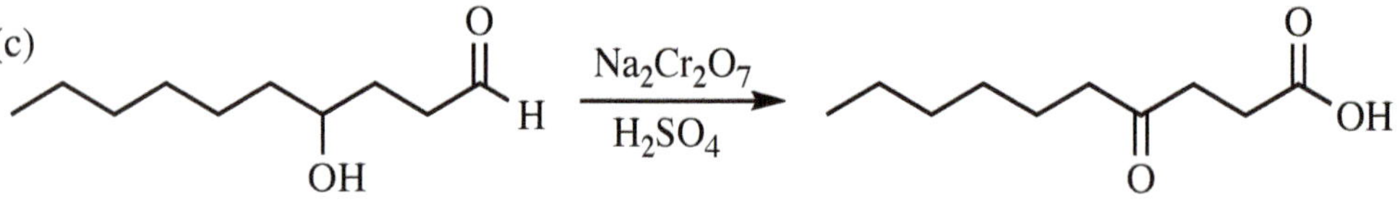 $\xrightarrow[H_2SO_4]{Na_2Cr_2O_7}$

1 equiv. NaOCl / TEMPO

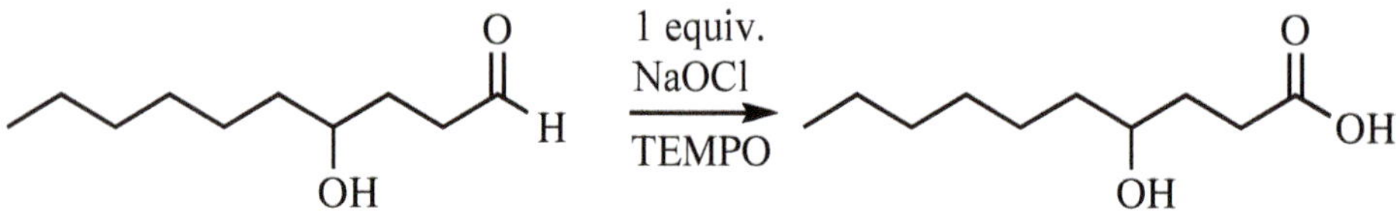

NaOCl oxidizes aldehydes faster than 2° alcohols. If you assumed the opposite, the correct answer would have been the structure below.

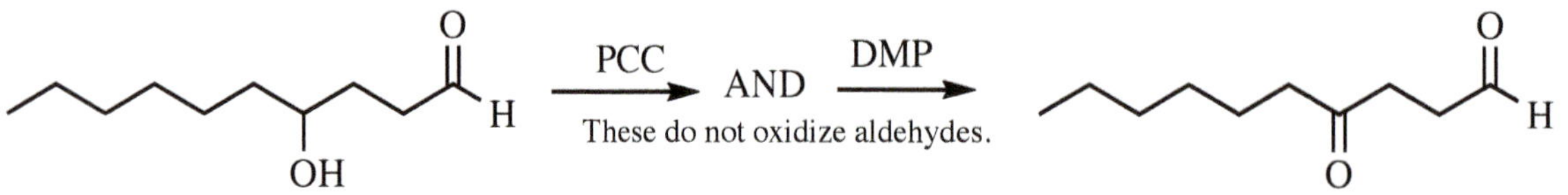

PCC AND DMP

These do not oxidize aldehydes.

11-4 continued

(d) 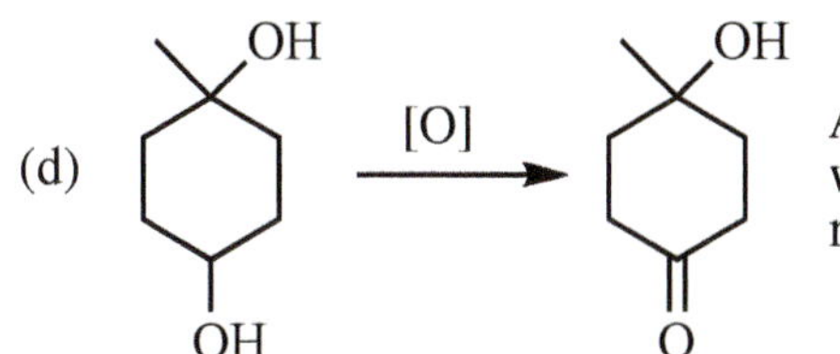

All four reagents give the same ketone product with a secondary alcohol. Tertiary alcohols are resistant to oxidation.

To the student: For consistency, this Solutions Manual will use these laboratory methods of oxidation:

—1 eq. NaOCl/TEMPO to oxidize 1° alcohols to aldehydes;
—excess NaOCl/TEMPO to oxidize 1° alcohols to carboxylic acids;
—NaOCl/HOAc to oxidize 2° alcohols to ketones.

Understand that other choices are legitimate and you should follow the guidelines given by your instructor; for example, DMP and Swern oxidation and PCC work as well as 1 eq. NaOCl in the preparation of aldehydes, and chromic acid will oxidize a 1° alcohol to a carboxylic acid as well as excess NaOCl does. All of these five oxidizing agents will convert a 2° alcohol to a ketone. If you have a question about the appropriateness of a reagent you choose, consult the table in the text before Problem 11-3.

11-5 None of the five oxidation reagents affects the 3° alcohol. All five oxidize the 2° alcohol to a ketone. Chromic acid and excess NaOCl oxidize the 1° alcohol to COOH, whereas PCC, DMSO/oxalyl chloride (Swern), and DMP oxidize the 1° alcohol to an aldehyde.

HO, 2°, OH, 1°, CH_3, 3° OH

(b) PCC OR
(d) DMSO/ClCOCOCl OR
(e) DMP

O, O, CH_3, H, OH

(a) H_2CrO_4 OR
(c) NaOCl

O, O, CH_3, OH, OH

11-6 NaOCl reagents shown over the arrow; Cr reagent shown beneath the arrow. Other non-Cr reagents added in text after the reaction.

(a) OH — 1 eq. NaOCl, TEMPO / PCC → O, H — other non-Cr: DMP or DMSO/ClCOCOCl (Swern)

(b) OH — excess NaOCl, TEMPO / H_2CrO_4 → O, OH — (*Z* or *E* not specified) other non-Cr: none

(c) OH — NaOCl, HOAc / H_2CrO_4 or PCC → O — other non-Cr: DMP or DMSO/ClCOCOCl (Swern)

11-6 continued

(d) cyclopentanol $\xrightarrow[\text{H}_2\text{CrO}_4 \text{ or PCC}]{\text{NaOCl, HOAc}}$ cyclopentanone $\xrightarrow[\text{ether}]{\text{EtMgBr}}$ $\xrightarrow{H_3O^+}$ 1-ethylcyclopentanol (CH_2CH_3, OH)

other non-Cr: DMP or DMSO/ClCOCOCl (Swern)

(e) cyclopentylmethanol (OH) $\xrightarrow[\text{PCC}]{\text{1 eq. NaOCl, TEMPO}}$ cyclopentanecarbaldehyde (O, H) $\xrightarrow[\text{ether}]{\text{EtMgBr}}$ $\xrightarrow{H_3O^+}$ 1-cyclopentylpropan-1-ol (OH, CH_2CH_3)

other non-Cr: DMP or DMSO/ClCOCOCl (Swern)

(f) 1-methylcyclohexanol (OH) $\xrightarrow[\Delta,\ -H_2O]{H_2SO_4}$ 1-methylcyclohexene $\xrightarrow[\text{2) } H_2O_2,\ HO^-]{\text{1) } BH_3 \bullet \text{THF}}$ 2-methylcyclohexanol (OH) $\xrightarrow[\text{H}_2\text{CrO}_4 \text{ or PCC}]{\text{NaOCl, HOAc}}$ 2-methylcyclohexanone (O)

other non-Cr: DMP or DMSO/ClCOCOCl (Swern)

11-7 A chronic alcoholic has induced more ADH enzyme to be present to handle large amounts of imbibed ethanol, so requires more ethanol "antidote" molecules to act as a competitive inhibitor to "tie up" the extra enzyme molecules.

11-8 $CH_3-CH(OH)-CH_2(OH)$ $\xrightarrow{[\,O\,]}$ $CH_3-C(=O)-CH(=O)$ $\xrightarrow{[\,O\,]}$ $CH_3-C(=O)-C(=O)OH$

pyruvaldehyde — pyruvic acid

Pyruvic acid is a normal metabolite in the breakdown of glucose ("blood sugar").

11-9 From this problem on, "Ts" will refer to the "tosyl" or "*p*-toluenesulfonyl" group:

Ts $\Rightarrow$ $-S(=O)_2-C_6H_4-CH_3$

(a) CH_3CH_2-OTs + $KO-C(CH_3)_2-CH_3$ $\longrightarrow$ $CH_3CH_2O-C(CH_3)_2-CH_3$ + KOTs

Lower temperature favors substitution. Higher temperature favors elimination.

(E2 is possible with this hindered base; the product would be ethylene, $CH_2{=}CH_2$.)

(b) isobutyl–OTs + NaI $\longrightarrow$ isobutyl–I + NaOTs

(c) (*R*)-2-pentyl tosylate (TsO, H) + NaCN $\longrightarrow$ (*S*) nitrile (H, CN) + NaOTs inversion—S_N2

11-9 continued

(d) [cyclohexyl-CH_2-OTs] $\xrightarrow{NH_3}$ [cyclohexyl-CH_2-$\overset{\oplus}{N}H_3$ $\overset{\ominus}{O}Ts$] $\xrightarrow{\text{excess } NH_3}$ [cyclohexyl-CH_2-NH_2] + $\overset{\oplus}{N}H_4$ $\overset{\ominus}{O}Ts$

(e) $CH_3CH_2CH_2CH_2$-OTs + Na^+ $:\overset{\ominus}{C}\equiv CH$ $\longrightarrow$ $CH_3CH_2CH_2CH_2$-$C\equiv CH$ + NaOTs

11-10 All parts begin with forming the tosylate.

$CH_3CH_2CH_2$-OH $\xrightarrow[\text{pyridine}]{TsCl}$ $CH_3CH_2CH_2$-OTs

- NaBr → $CH_3CH_2CH_2$-Br (a)
- excess NH_3 → $CH_3CH_2CH_2$-NH_2 (b)
- $NaOCH_2CH_3$ → $CH_3CH_2CH_2$-O-CH_2CH_3 (c)
- KCN → $CH_3CH_2CH_2$-CN (d)

11-11

(a) [cyclohexyl]-CH_2OH $\xrightarrow[\text{pyridine}]{TsCl}$ [cyclohexyl]-CH_2OTs OR [cyclohexyl]-CH_2O-S(=O)(=O)-[benzene ring]-CH_3

(b) [cyclohexyl]-CH_2OTs $\xrightarrow{LiAlH_4}$ [cyclohexyl]-CH_3

(c) [1-methylcyclohexanol, HO / CH_3] $\xrightarrow[\Delta]{H_2SO_4}$ [1-methylcyclohexene, CH_3] (major) + [methylenecyclohexane, CH_2] (minor)

(d) $\xrightarrow[Pt]{H_2}$ [methylcyclohexane, CH_3]

11-12

(a) S_N1 on 3° alcohol

[1-methylcyclohexanol, $\ddot{O}$-H] + H—Br $\longrightarrow$ [protonated alcohol, $\overset{\oplus}{O}H_2$] $\xrightarrow{-H_2O}$ [$\overset{\oplus}{C}$ carbocation] + $:\ddot{Br}:^{\ominus}$ $\longrightarrow$ [1-bromo-1-methylcyclohexane, Br]

carbocation intermediate

(b) S_N2 on 1° alcohol

[cyclohexyl-CH_2CH_2-$\ddot{O}H$] + H—Br $\longrightarrow$ [cyclohexyl-CH_2CH_2-$\overset{\oplus}{O}H_2$] + $:\ddot{Br}:^{\ominus}$ $\longrightarrow$ [cyclohexyl-CH_2CH_2-Br] + H_2O

11-13 $H_3C-\underset{CH_3}{\overset{CH_3}{C}}-\ddot{O}H$ + H—Cl $\longrightarrow$ $H_3C-\underset{CH_3}{\overset{CH_3}{C}}-\overset{\oplus}{O}H_2$ $\xrightarrow{-H_2O}$ $H_3C-\underset{CH_3}{\overset{CH_3}{\overset{\oplus}{C}}}$ + $:\ddot{Cl}:^{\ominus}$ $\longrightarrow$ $H_3C-\underset{CH_3}{\overset{CH_3}{C}}-Cl$

11-14 The two standard qualitative tests are:

1) chromic acid—distinguishes 3° alcohol from either 1° or 2°

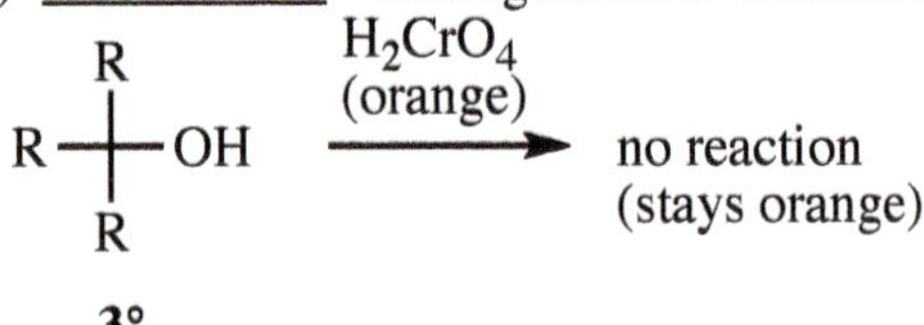

R(or H)—C(R)(H)—OH $\xrightarrow[\text{(orange)}]{H_2CrO_4}$ R—C(R(or OH))=O + Cr^{3+} blue-green

1°, 2°

2) Lucas test—distinguishes 1° from 2° from 3° alcohol by the rate of reaction

3° R_3C—OH (soluble) + HCl $\xrightarrow{ZnCl_2}$ R_3C—Cl + H_2O insoluble—"cloudy" in < 1 minute

2° R_2CH—OH (soluble) + HCl $\xrightarrow{ZnCl_2}$ R_2CH—Cl + H_2O insoluble—"cloudy" in 1-5 minutes

1° RCH_2—OH (soluble) + HCl $\xrightarrow{ZnCl_2}$ RCH_2—Cl + H_2O insoluble—"cloudy" in > 6 minutes (No observable reaction at room temp.)

(a) OH (2-propanol) | +OH (tert-butyl alcohol)

Lucas: cloudy in 1-5 min. | cloudy in < 1 min.
H_2CrO_4: immediate blue-green | no reaction—stays orange

(b) OH (2-propanol) | O (2-butanone)

Lucas: cloudy in 1-5 min. | no reaction
H_2CrO_4: immediate blue-green | no reaction—stays orange

(c) OH (1-hexanol) | OH (cyclohexanol)

Lucas: no reaction | cloudy in 1-5 min.
H_2CrO_4: DOES NOT DISTINGUISH—immediate blue-green for both

(d) OH (allyl alcohol) | OH (1-propanol)

Lucas: cloudy in < 1 min. ** | no reaction
H_2CrO_4: DOES NOT DISTINGUISH—immediate blue-green for both

(**Remember that allylic cations are resonance-stabilized and are about as stable as 3° cations. Thus, they will react as fast as 3° in the Lucas test, even though they may be 1°. Be careful to notice subtle but important structural features!)

(e) O (2-butanone) | +OH (tert-butyl alcohol)

Lucas: no reaction | cloudy in < 1 min.
H_2CrO_4: DOES NOT DISTINGUISH—stays orange for both

11-15

[Scheme: neopentyl alcohol + H—Br → protonated alcohol (OH_2^+) → $-H_2O$ → [1° carbocation] → methyl shift → 3° carbocation; $:Br:^-$ attacks → product with Br]

Even though 1°, the neopentyl carbon is hindered to backside attack, so S_N2 cannot occur easily. Instead, an S_N1 mechanism occurs, with rearrangement.

11-16

[Scheme: cyclohexanol with CH_3 groups + H—Cl → protonated alcohol → $-H_2O$ → 3° carbocation; approach from above / approach from below by Cl^- → methyls *trans* and methyls *cis* products]

This 3° carbocation is planar at the C^+ so that the Cl^- can approach from the top or bottom giving both the *cis* and *trans* isomers.

approach from above

approach from below

methyls *trans*

methyls *cis*

11-17

[Scheme: 2-methylcyclohexanol + $ZnCl_2$ → complex → 2° carbocation → hydride shift *rearrangement* → 3° carbocation; $:Cl:^-$ (from HCl) → 1-chloro-1-methylcyclohexane]

2° hydride shift *rearrangement*

3° (from HCl)

11-18

3 $(CH_3)_3CCH_2OH$ + PBr_3 ⟶ **3** $(CH_3)_3CCH_2Br$ + $P(OH)_3$

6 $CH_3(CH_2)_{14}CH_2OH$ + **2** P + **3** I_2 ⟶ **6** $CH_3(CH_2)_{14}CH_2I$ + **2** $P(OH)_3$

3 cyclopentanol + PBr_3 ⟶ **3** bromocyclopentane + $P(OH)_3$

11-19

(a) [*cis*-4-methylcyclohexanol (OH, CH_3)] $\xrightarrow{SOCl_2}$ [4-methylcyclohexyl chloride (Cl, CH_3)] *retention*

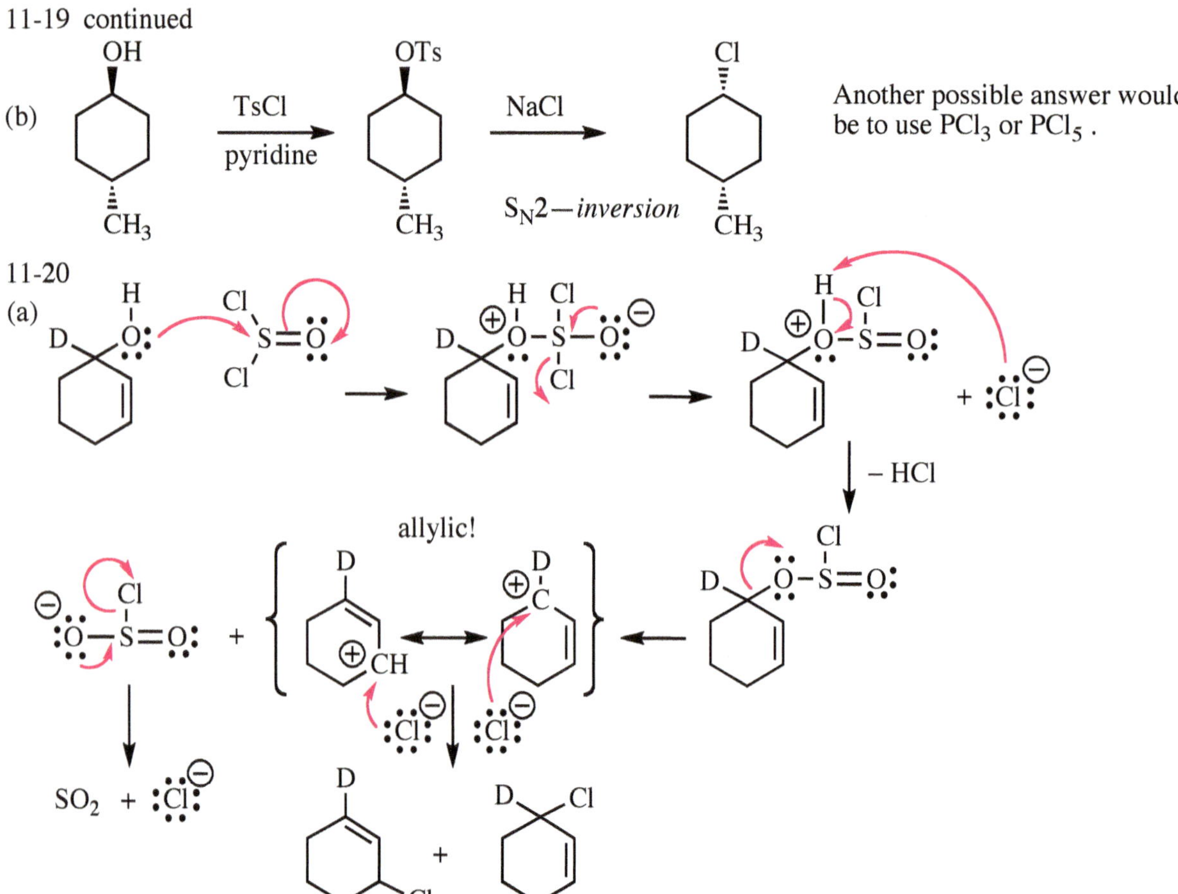

(b) The key is that the intermediate carbocation is allylic, very stable, and relatively long-lived. It can therefore escape the ion pair and become a "free" carbocation. The nucleophilic chloride can attack any carbon with positive charge, not just the one closest. Since two carbons have partial positive charge, two products result.

11-21

(a) butanol (OH)
- $\xrightarrow{HCl,\ ZnCl_2}$ no reaction unless heated, then 1-chlorobutane (Cl)
- $\xrightarrow{HBr}$ Br
- $\xrightarrow{PBr_3}$ Br
- $\xrightarrow{P,\ I_2}$ I
- $\xrightarrow{SOCl_2}$ Cl

(b) 3° alcohol (OH)
- $\xrightarrow{HCl,\ ZnCl_2}$ Cl
- $\xrightarrow{HBr}$ Br

On 3° alcohols, these reagents often give more elimination than substitution.
- $\xrightarrow{PBr_3}$ Br (poor reaction on 3°)
- $\xrightarrow{P,\ I_2}$ I (poor reaction on 3°)
- $\xrightarrow{SOCl_2}$ Cl (poor reaction on 3°)

11-21 continued

(c) 1°, neopentyl

$\xrightarrow[ZnCl_2]{HCl}$ no reaction unless heated, then S_N1—rearrangement

Cl

$\xrightarrow{HBr}$ Br + Br

S_N2—minor (hindered)

S_N1—rearrangement

$\xrightarrow{PBr_3}$ Br

$\xrightarrow[I_2]{P}$ I

$\xrightarrow{SOCl_2}$ Cl

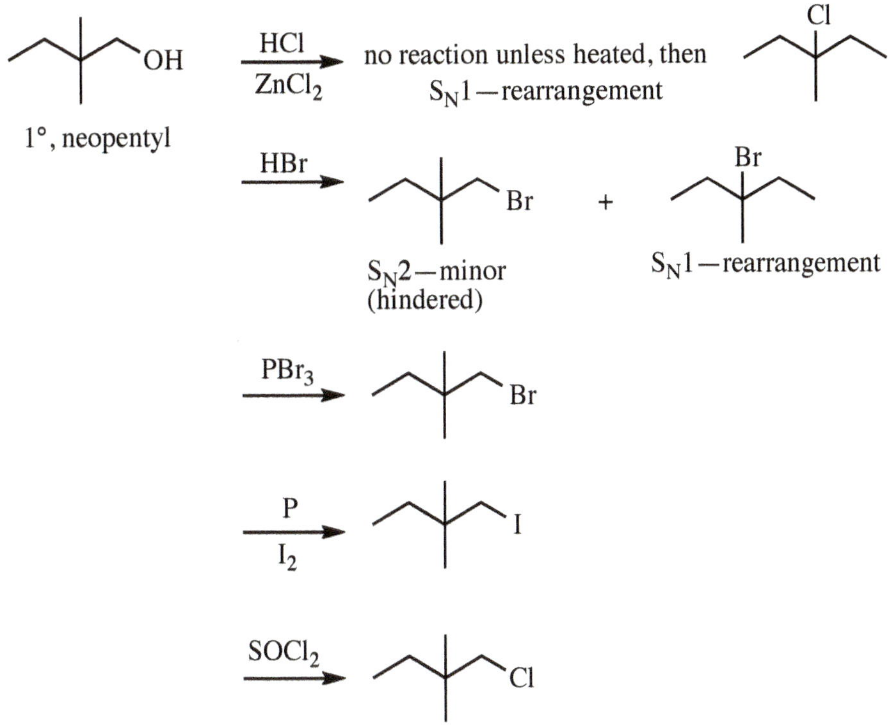

(d) OH

$\xrightarrow[ZnCl_2]{HCl}$ Cl + Cl

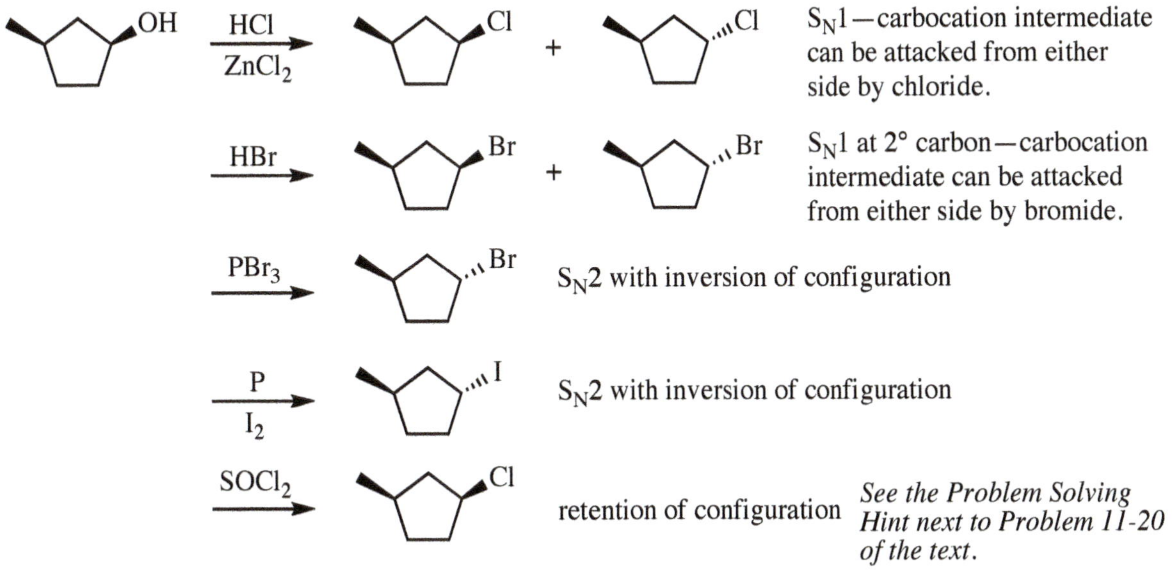

S_N1—carbocation intermediate can be attacked from either side by chloride.

$\xrightarrow{HBr}$ Br + Br

S_N1 at 2° carbon—carbocation intermediate can be attacked from either side by bromide.

$\xrightarrow{PBr_3}$ Br

S_N2 with inversion of configuration

$\xrightarrow[I_2]{P}$ I

S_N2 with inversion of configuration

$\xrightarrow{SOCl_2}$ Cl

retention of configuration

See the Problem Solving Hint next to Problem 11-20 of the text.

11-22 Water is also produced in each of these dehydration reactions.

(a) $\xrightarrow{H_2SO_4,\ \Delta}$ major + minor

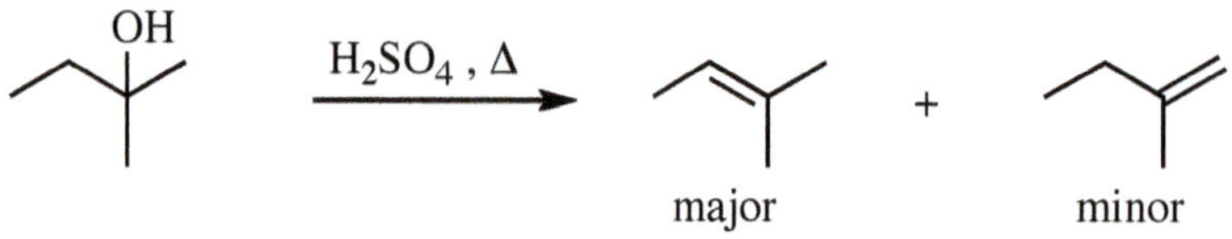

(b) OH $\xrightarrow{H_2SO_4,\ \Delta}$ major (*cis* + *trans*) + minor

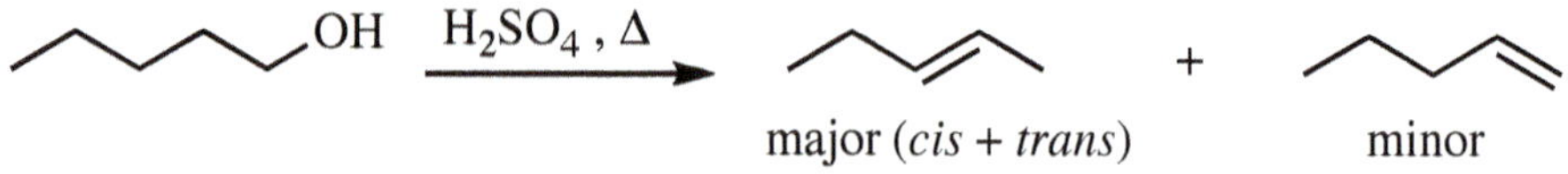

Rearrangement is likely.

(c) OH $\xrightarrow{H_2SO_4,\ \Delta}$ major (*cis* + *trans*) + minor

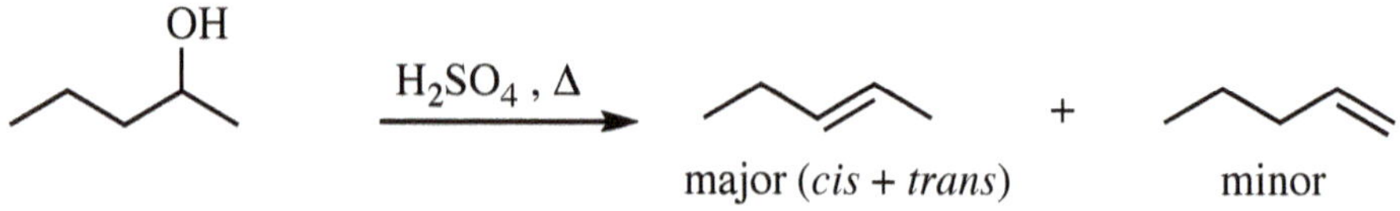

(d) OH $\xrightarrow{H_2SO_4,\ \Delta}$ major + minor

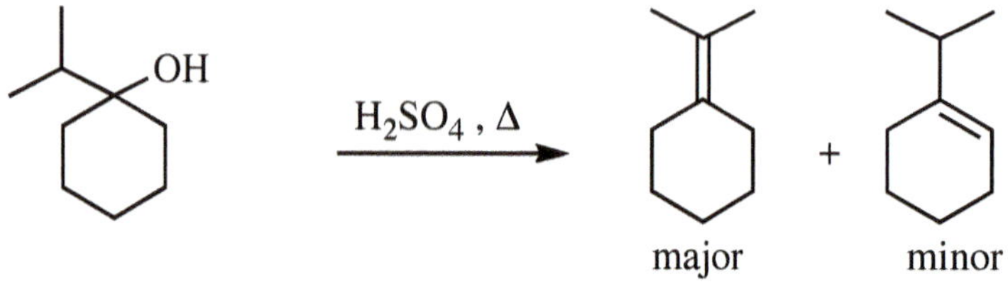

(e) OH $\xrightarrow{H_2SO_4,\ \Delta}$ major + minor + trace

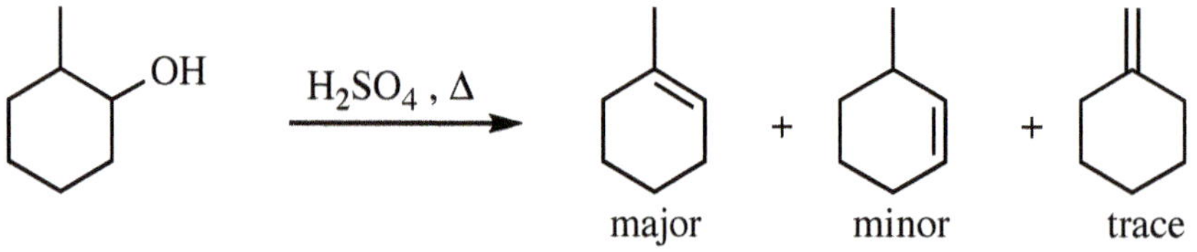

Rearrangement is likely.

11-23

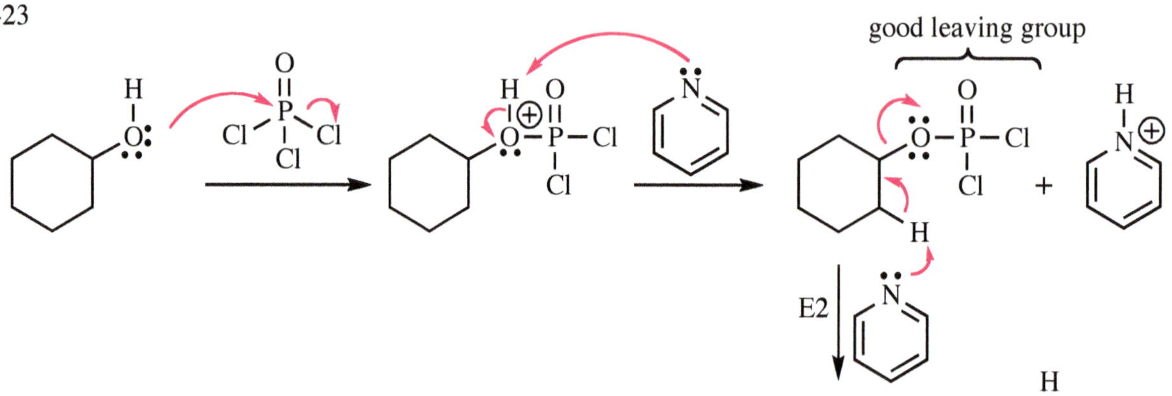

H O: + O=P(Cl)(Cl)Cl → H O⊕ O P Cl Cl → (pyridine N:) → good leaving group O P Cl Cl + H N⊕

E2 (N: removes H)

Cyclohexene was formed without a carbocation intermediate.

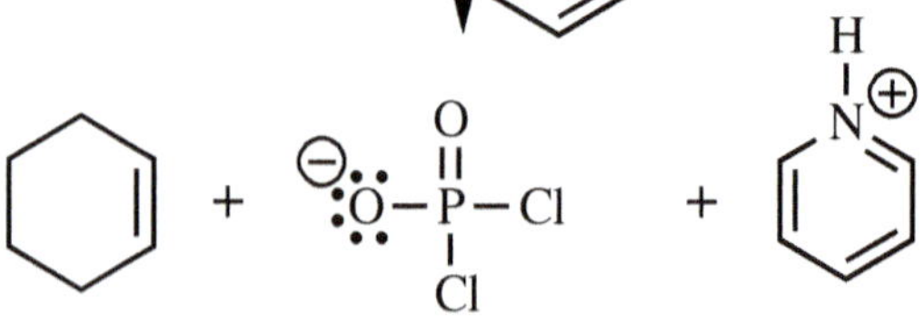

+ ⊖ :O–P–Cl (O, Cl) + H N⊕

This product can react with two more alcohols to become leaving groups in the E2 elimination.

11-24
Both mechanisms begin with protonation of the oxygen.

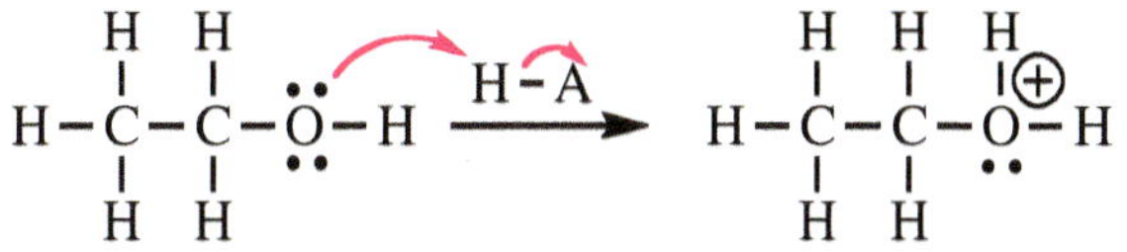

$$CH_3CH_2\ddot{O}H + H{-}A \longrightarrow CH_3CH_2\overset{\oplus}{O}H_2$$

One mechanism involves another molecule of ethanol acting as a base, giving elimination.

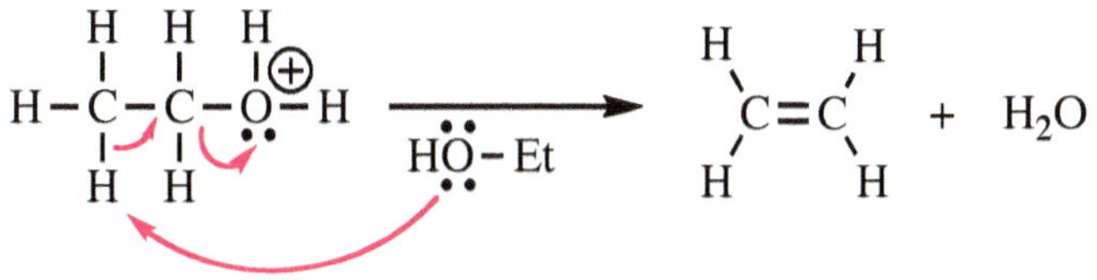

$$CH_3CH_2\overset{\oplus}{O}H_2 \xrightarrow{HO{-}Et} H_2C{=}CH_2 + H_2O$$

The other mechanism involves another molecule of ethanol acting as a nucleophile, giving substitution.

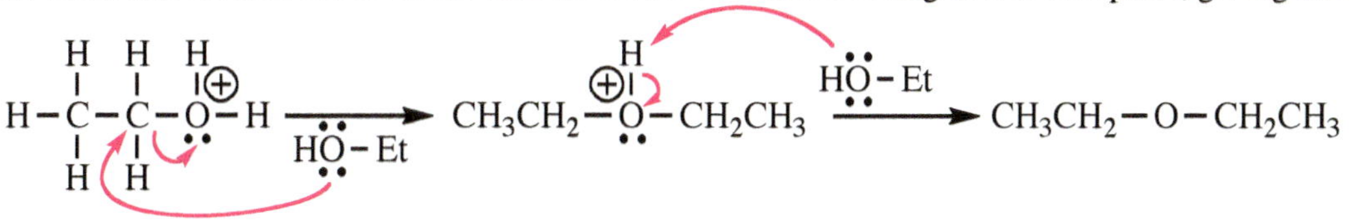

$$CH_3CH_2\overset{\oplus}{O}H_2 \xrightarrow{HO{-}Et} CH_3CH_2{-}\overset{\oplus}{O}H{-}CH_2CH_3 \xrightarrow{HO{-}Et} CH_3CH_2{-}O{-}CH_2CH_3$$

11-25 An equimolar mixture of methanol and ethanol would produce all three possible ethers. The difficulty in separating these compounds would preclude this method from being a practical route to any one of them. This method is practical only for symmetric ethers, that is, where both alkyl groups are identical.

$$CH_3CH_2OH + HOCH_3 \xrightarrow[\Delta]{H_2SO_4} H_2O + CH_3CH_2OCH_3 + CH_3OCH_3 + CH_3CH_2OCH_2CH_3$$

11-26

(a) 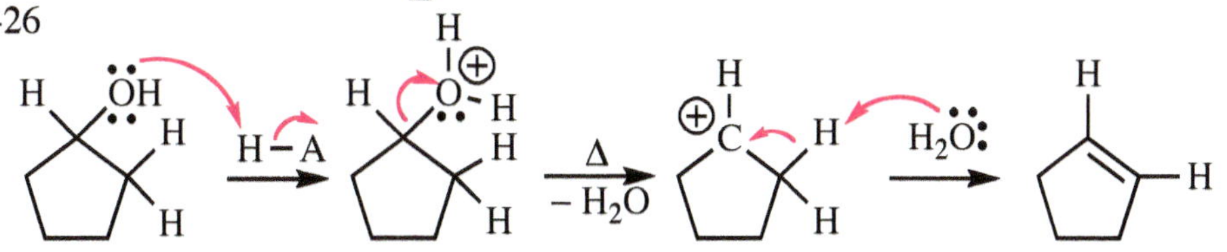 H, OH, H, H; H−A; Δ, $-H_2O$; $H_2\ddot{O}$:; H

(b) 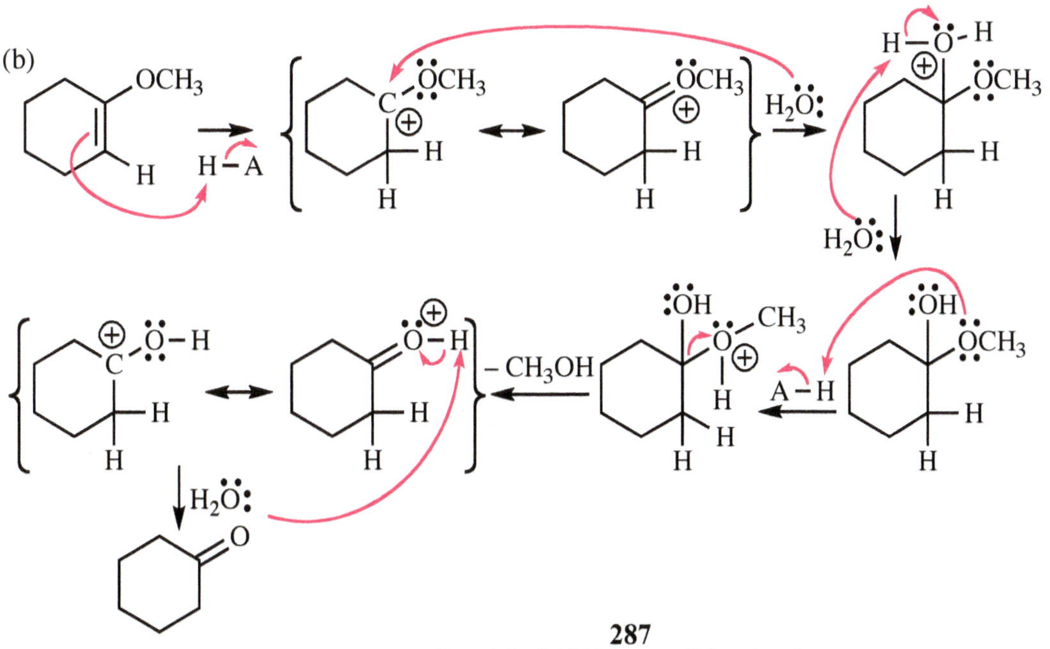 OCH_3, H; H−A; $H_2\ddot{O}$:; A−H; $-CH_3OH$; $H_2\ddot{O}$:; O

11-26 continued

(c) $CH_2\ddot{O}H$ H — H–A → CH_2–Ö–H ⊕ H — Δ, $-H_2O$ → [1° ⊕ CH_2 H] — hydride shift → H–C–H ⊕C H H — $H_2\ddot{O}$: → CH_2 + CH_3

This 1° carbocation would be very unstable.

ring expansion

⊕C H H H — $H_2\ddot{O}$: → H H

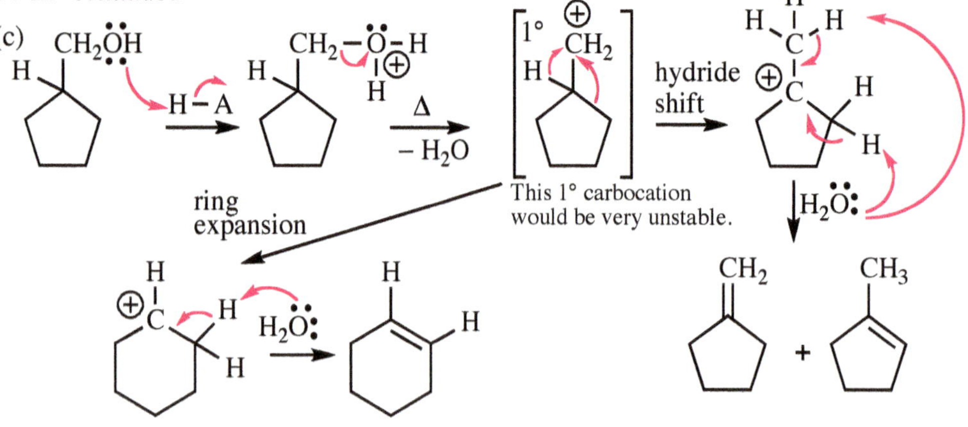

(d) :Ö — H–A → HÖ ⊕ → HO ⊕C → HO C ⊕ — H–ÖEt → HO H :ÖEt ⊕ — H–ÖEt → HO OEt

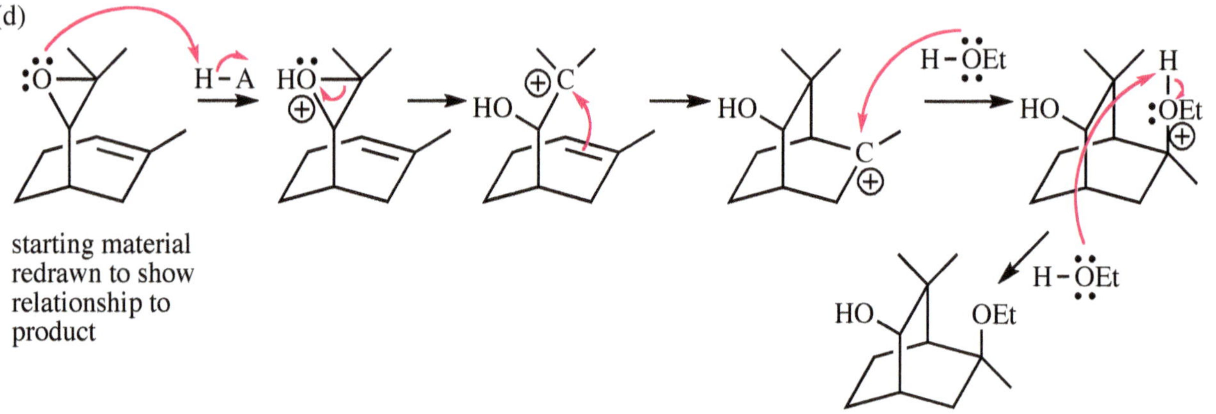

starting material redrawn to show relationship to product

11-27

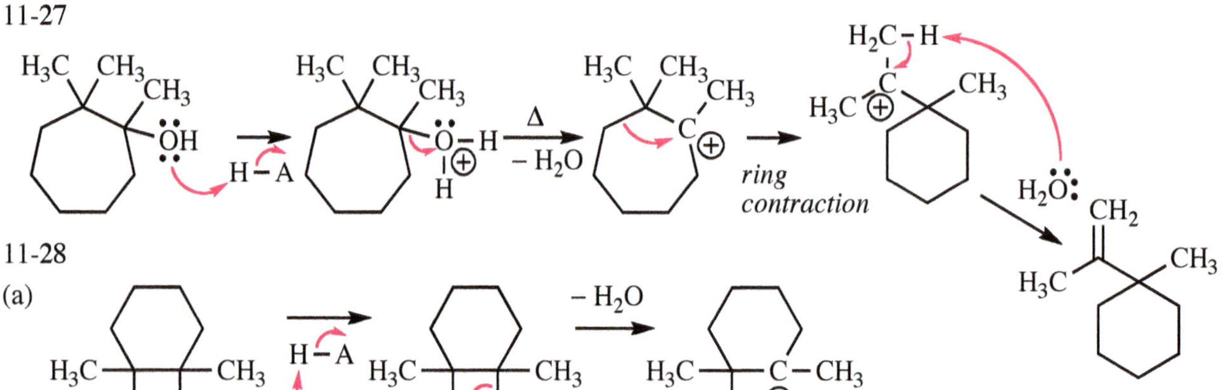

H_3C CH_3 CH_3 ÖH — H–A → H_3C CH_3 CH_3 Ö–H ⊕ H — Δ, $-H_2O$ → H_3C CH_3 CH_3 C ⊕ → *ring contraction* — H_2C–H H_3C C⊕ CH_3 — $H_2\ddot{O}$: → CH_2 H_3C CH_3

11-28

(a) H_3C CH_3 HO :ÖH — H–A → H_3C CH_3 HO :Ö–H ⊕ H — $-H_2O$ → H_3C C–CH_3 ⊕ HO

Methyl shift

H_3C C–CH_3 ⊕ HO → { ⊕ C CH_3 H–Ö: CH_3 ↔ CH_3 H–Ö⊕ CH_3 } → :O CH_3 CH_3

$H_2\ddot{O}$:

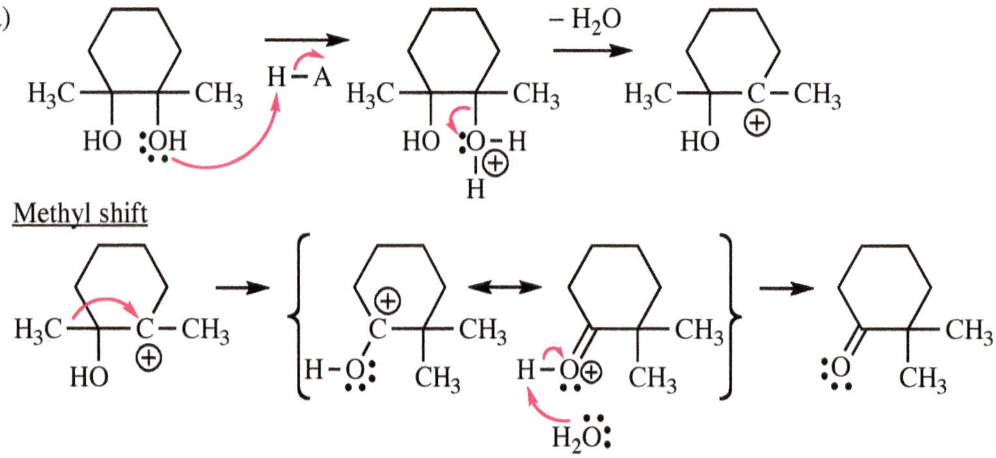

11-28(a) continued on next page

11-28(a) continued

Alkyl shift—ring *contraction*

H_3C C–CH_3 (+) HO → { H_3C (+) C CH_3 H–O: ↔ H_3C CH_3 H–O(+) } H_2O: → H_3C CH_3 :O:

(b)

OH, :OH, Ph, Ph, H–A → OH H, O–H (+), Ph, Ph – H_2O → OH, C(+), Ph, Ph *3° and doubly benzylic carbocation*

ring expansion

O, Ph, Ph ← H_2O: { (+)O–H, Ph, Ph ↔ O–H, C, (+), Ph, Ph }

11-29

OH, H, C=C, H, H H_2SO_4 → OH, H, C, C–C–H, H (+), H ring expansion { (+) O–H, C, C–CH_3, H ↔ (+)O–H, C, C–CH_3, H }

↓ B: O, CH_3

Another oxygen in the reaction mixture is the likely base, removing this proton.

Similar to the pinacol rearrangement, this mechanism involves a carbocation next to an alcohol, with rearrangement to a protonated carbonyl. Relief of some ring strain in the cyclopropane is an added advantage of the rearrangement.

11-30

(a) **2** H_3C–C(=O)–H (b) cyclopentanone, O=CH_2 (c) acetophenone + H–C(=O)–CH_2CH_3 (d) O, H, H, O

11-31 (a) $CH_3CH_2CH_2-\overset{O}{\overset{||}{C}}-Cl$ + $HOCH_2CH_2CH_3$ (b) $CH_3(CH_2)_3OH$ + $Cl-\overset{O}{\overset{||}{C}}CH_2CH_3$

(c) H_3C–C₆H₄–OH + $Cl-\overset{O}{\overset{||}{C}}CH(CH_3)_2$ (d) cyclopropyl–OH + $Cl-\overset{O}{\overset{||}{C}}-C_6H_5$

11-32 *The strength of an acid is determined by the stability of its conjugate base*. The more stable the conjugate base, the stronger the acid.

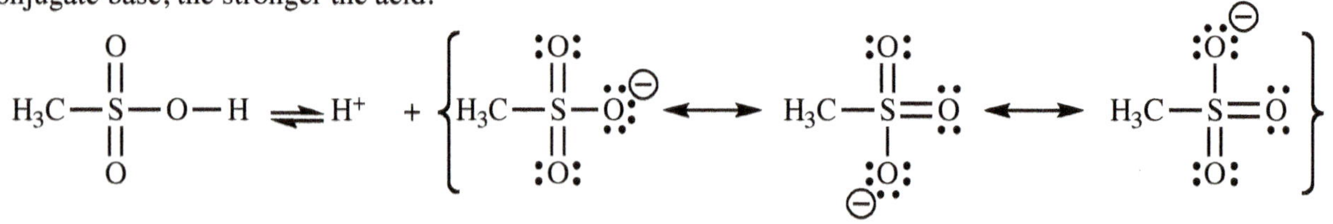

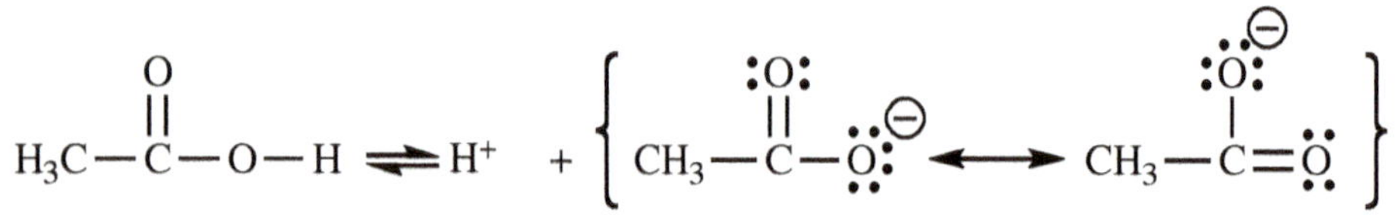

$$H_3C-SO_2-O-H \rightleftharpoons H^+ + \left\{ H_3C-SO_2-O^- \longleftrightarrow H_3C-S(=O)(O^-)=O \longleftrightarrow H_3C-S(O^-)(=O)=O \right\}$$

$$H_3C-C(=O)-O-H \rightleftharpoons H^+ + \left\{ CH_3-C(=O)-O^- \longleftrightarrow CH_3-C(O^-)=O \right\}$$

The methanesulfonate anion is stabilized by resonance and by induction: It has three equivalent resonance forms, plus the sulfur atom is more electronegative than carbon and plays a small role in stabilizing the negative charge on oxygen. The acetate ion has two equivalent resonance forms, but no inductive effect to stabilize the anion. Acetate is good, but the methanesulfonate ion is even better.

Question in Key Mechanism 11-6

In many, but not all, cases of the Williamson ether synthesis, there will be two possible pathways to make new bonds to oxygen. Because this is an S_N2 reaction, always choose the pathway in which an alkoxide attacks a 1° carbon, or if that is not possible, a 2°carbon. S_N2 reactions cannot take place at 3° carbons, nor at sp^2 carbons.

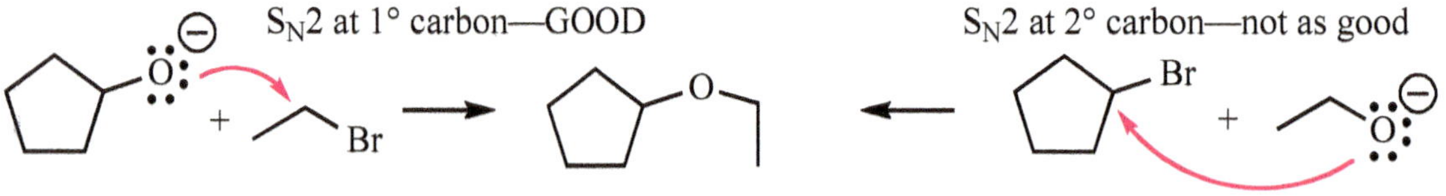

S_N2 at 1° carbon—GOOD

S_N2 at 2° carbon—not as good

Br

Br

O

11-33 Proton transfer (acid-base) reactions are much faster than almost any other reaction. Methoxide will act as a base and remove a proton from the oxygen much faster than methoxide will act as a nucleophile and displace water.

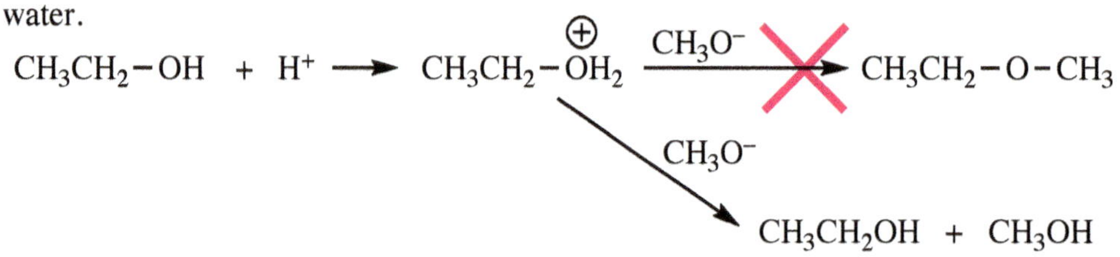

$$CH_3CH_2-OH + H^+ \longrightarrow CH_3CH_2-\overset{\oplus}{O}H_2 \xrightarrow{CH_3O^-} \times \; CH_3CH_2-O-CH_3$$

$$CH_3CH_2-\overset{\oplus}{O}H_2 \xrightarrow{CH_3O^-} CH_3CH_2OH + CH_3OH$$

11-34

(a) $CH_3CH_2OH \xrightarrow[\text{pyridine}]{TsCl} CH_3CH_2OTs$

cyclohexanol (OH) $\xrightarrow{Na}$ cyclohexyl–O^- Na^+

$\longrightarrow$ cyclohexyl–OCH_2CH_3

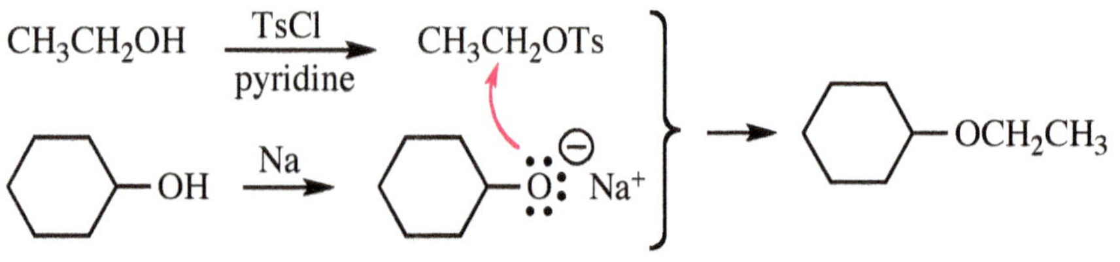

(b) There are two problems with this attempted bimolecular dehydration. First, all three possible ether combinations of cyclohexanol and ethanol would be produced. Second, heat and sulfuric acid are the conditions for dehydrating secondary alcohols like cyclohexanol, so elimination would compete with substitution.

11-35

(a) What the student did:

Na⁺ $O^{\ominus}$ H, H_3C CH_2CH_3 + $TsO-CH_2CH_3$ → CH_3CH_2-O H, H_3C CH_2CH_3

sodium (*S*)-but-2-oxide — (*S*)-2-ethoxybutane

The product also has the *S* configuration, not the *R*. Why? The substitution is indeed an S_N2 reaction, but the *substitution did not take place at the chiral center*, so the configuration of the starting material is retained, not inverted.

(b) There are two ways to make (*R*)-2-ethoxybutane. Start with (*R*)-butan-2-ol, make the anion, and substitute on ethyl tosylate similar to part (a), or do an S_N2 inversion at the chiral center of (*S*)-butan-2-ol. S_N2 works better at 1° carbons so the former method would be preferred to the latter.

(c) This is not the optimum method because it requires S_N2 at a 2° carbon, as discussed in part (b).

(*S*)-butan-2-ol —TsCl, pyridine→ (*S*)-2-butyl tosylate (no inversion yet) —$Na^+\ ^-OCH_2CH_3$, low temperature (high temp. favors elimination)→ (*R*)-2-ethoxybutane INVERSION!

11-36

Phenol (Ph–OH) —NaOH→ $Ph-\ddot{O}:^{\ominus}\ Na^+$ + $CH_3-O-SO_3CH_3$ → $Ph-OCH_3$ + $^{\ominus}:\ddot{O}-SO_3CH_3$

11-37

(a) $CH_3CH_2CH_2OH$ —1 equiv. NaOCl, TEMPO→ $CH_3CH_2-\overset{O}{\overset{\|}{C}}-H$ —1) CH_3CH_2MgBr 2) H_3O^+→ pentan-3-ol —NaOCl, HOAc→ pentan-3-one

(b) $CH_3CH_2-\overset{O}{\overset{\|}{C}}-Cl$ —1) **2** CH_3CH_2MgBr 2) H_3O^+→ $CH_3CH_2-C(OH)(CH_2CH_3)-CH_2CH_3$ —H_2SO_4→ $CH_3CH=C(CH_2CH_3)-CH_2CH_3$ —1) $BH_3 \bullet THF$ 2) H_2O_2, HO^-→ $CH_3CH(OH)CH(CH_2CH_3)CH_2CH_3$ —NaOCl, HOAc→ $CH_3C(=O)CH(CH_2CH_3)CH_2CH_3$

11-38

(a) Cyclopentanol —H_2SO_4, Δ→ cyclopentene —HCO_3H, H_3O^+→ *trans*-cyclopentane-1,2-diol (OH, OH)

Any peroxy acid can be used to form the epoxide, which is cleaved to the *trans*-diol in aqueous acid.

(b) Cyclopentanol —NaOCl, HOAc→ cyclopentanone —CH_3CH_2MgBr→ —H_3O^+→ 1-ethylcyclopentanol (OH) —HCl→ 1-chloro-1-ethylcyclopentane (Cl)

CH_3CH_2OH —PBr_3→ CH_3CH_2Br —Mg, ether→ CH_3CH_2MgBr

11-38 continued

(c)

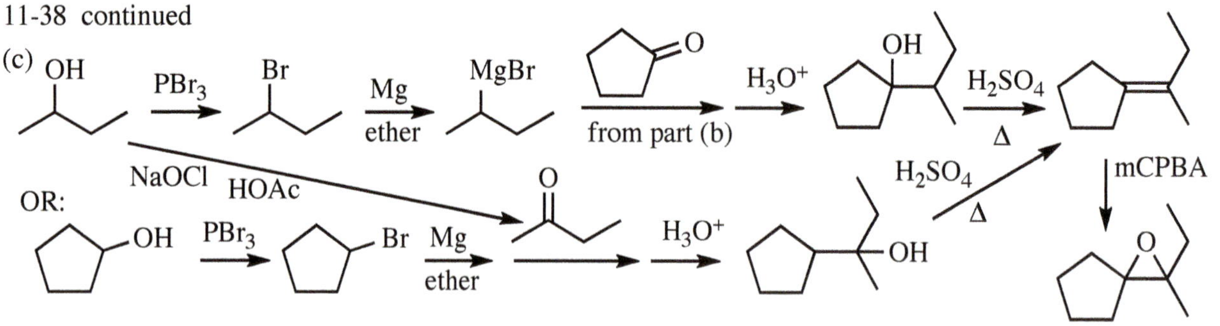

$CH_3CH(OH)CH_2CH_3$ —PBr_3→ 2-bromobutane —Mg, ether→ $CH_3CH(MgBr)CH_2CH_3$ —cyclopentanone (from part (b))→ —H_3O^+→ 1-(sec-butyl)cyclopentanol —H_2SO_4, Δ→ alkene —mCPBA→ epoxide

OR:

$CH_3CH(OH)CH_2CH_3$ —NaOCl, HOAc→ 2-butanone

cyclopentanol —PBr_3→ bromocyclopentane —Mg, ether→ —2-butanone→ —H_3O^+→ 2-cyclopentyl-2-butanol —H_2SO_4, Δ→ alkene

(d)

cyclopentanol —PBr_3→ bromocyclopentane —Mg, ether→ —2-butanone (from (c))→ alkoxide $O^{\ominus}$ $^{\oplus}MgBr$ —CH_3I→ 2-cyclopentyl-2-methoxybutane

CH_3OH —P, I_2→ CH_3I

alternative: alkoxide —H_3O^+→ 2-cyclopentyl-2-butanol —S_N1, CH_3OH, H_2SO_4→ 2-cyclopentyl-2-methoxybutane

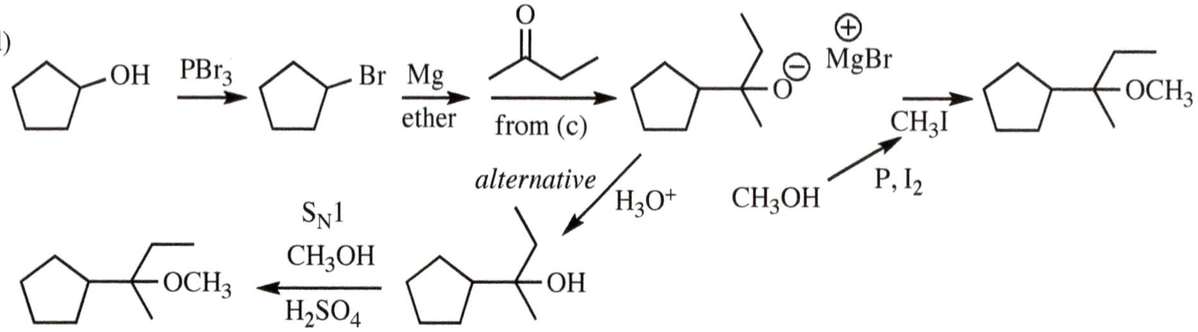

(e) There are several possible combinations of Grignard reactions on aldehydes or ketones. This is one example. Your example may be different and still be correct. Compare with others in your study group.

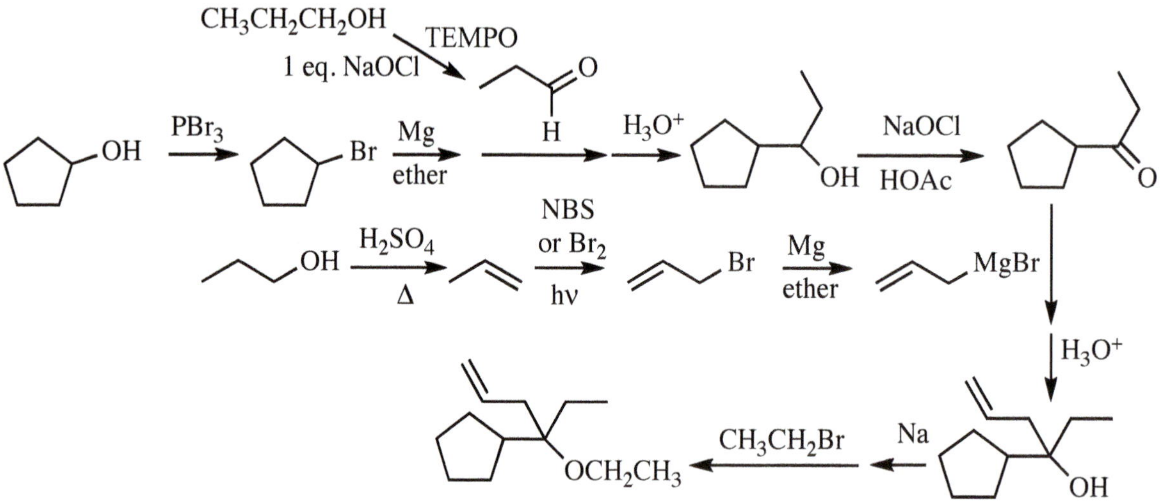

$CH_3CH_2CH_2OH$ —TEMPO, 1 eq. NaOCl→ propanal (H, O)

cyclopentanol —PBr_3→ bromocyclopentane —Mg, ether→ —propanal→ —H_3O^+→ 1-cyclopentyl-1-propanol (OH) —NaOCl, HOAc→ 1-cyclopentyl-1-propanone (O)

1-propanol (OH) —H_2SO_4, Δ→ propene —NBS or Br_2, hν→ allyl bromide (Br) —Mg, ether→ allyl MgBr

ketone + allyl MgBr —H_3O^+→ tertiary alcohol (OH) —Na→ —CH_3CH_2Br→ ether (OCH_2CH_3)

(f)

OCH_3

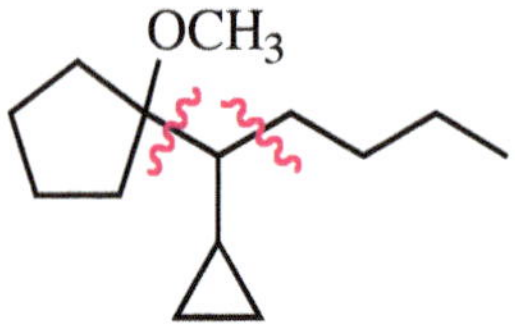

In a complex synthesis, it is worth the time to analyze what pieces need to be put together to create the carbon skeleton; this is called "retrosynthesis" (reverse synthesis), or the "disconnection" approach. This target has a cyclopentane ring and two four-carbon fragments, so the new C—C bonds should be made using Grignard reactions.

11-38(f) continued on next page

11-38(f) continued

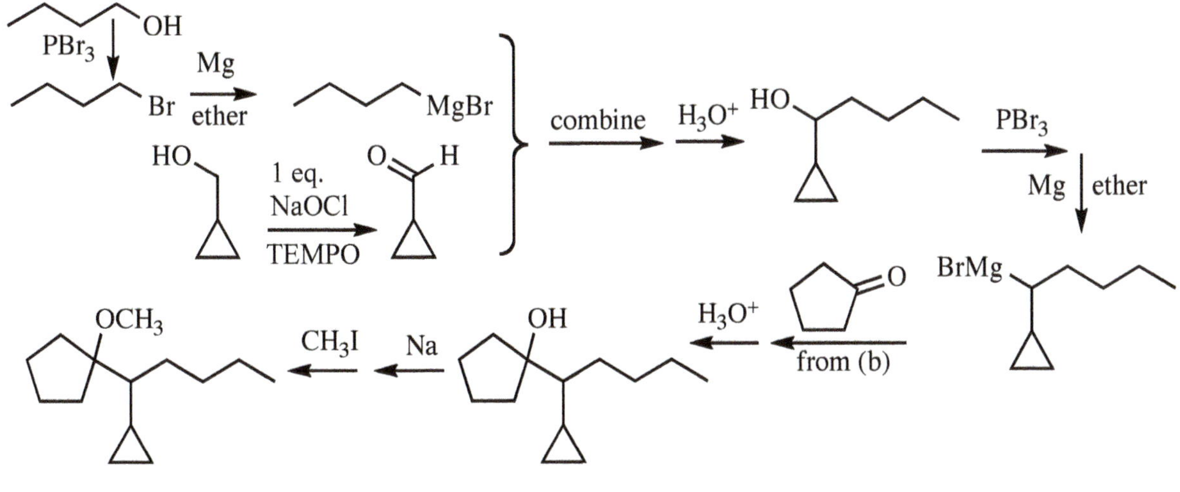

11-39

(a) H, OTs — *R*

(b) H, Br — *R* (from inversion)

(c) O

(d) O, H

(e) COOH

(f) Cl

(g) Br

(h) CH_2OMgBr + CH_3CH_3

(i) OCH_3

(j) + CH_3OH

(k)

(l) H, O, O, H

(m) O

(n) major + minor + EtOH

(o) O, H

(p) O, H

11-40 Stereochemistry is not specified in this problem.

(a) OH $\xrightarrow[\Delta]{H_2SO_4}$

(b) OH $\xrightarrow[\text{pyridine}]{TsCl}$ OTs

(c) OH $\xrightarrow[HOAc]{NaOCl}$ O

or H_2CrO_4 or PCC or DMP or Swern

(d) from (a) $\xrightarrow[H_2SO_4]{H_2O}$ OH

11-40 continued

(e) [2-methylcyclopentanone, from (c)] $\xrightarrow[\text{ether}]{CH_3MgI}$ $\xrightarrow{H_3O^+}$ [1,2-dimethylcyclopentanol] OH

(f) [2-methylcyclopentanol] OH $\xrightarrow{PBr_3}$ [1-bromo-2-methylcyclopentane] Br

OR

[2-methylcyclopentyl tosylate, from (b)] OTs $\xrightarrow[\Delta]{KBr}$ [1-bromo-2-methylcyclopentane] Br

(g) [2-methylcyclopentanol] OH + acetyl chloride (Cl–C(=O)–CH₃) $\longrightarrow$ [2-methylcyclopentyl acetate] O, O

(h) [1-methylcyclopentene, from (a)] $\xrightarrow{HBr}$ [1-bromo-1-methylcyclopentane] Br $\xleftarrow{HBr}$ [1-methylcyclopentanol, from (d)] OH

11-41

(a) [1-butanol] OH 1° $\xrightarrow{SOCl_2}$ [1-chlorobutane] Cl

$\xrightarrow{PBr_3}$ [1-bromobutane] Br

$\xrightarrow[I_2]{P}$ [1-iodobutane] I

(b) [cyclopentanol] 2° OH $\xrightarrow{SOCl_2}$ [chlorocyclopentane] Cl

$\xrightarrow{PBr_3}$ [bromocyclopentane] Br

$\xrightarrow[I_2]{P}$ [iodocyclopentane] I

(c) [1-methylcyclohexanol] 3° OH $\xrightarrow{HCl}$ [1-chloro-1-methylcyclohexane] Cl

$\xrightarrow{HBr}$ [1-bromo-1-methylcyclohexane] Br

$\xrightarrow{HI}$ [1-iodo-1-methylcyclohexane] I

(d) [2-methylcyclohexanol] 2° OH $\xrightarrow{SOCl_2}$ [1-chloro-2-methylcyclohexane] Cl

$\xrightarrow{PBr_3}$ [1-bromo-2-methylcyclohexane] Br

$\xrightarrow[I_2]{P}$ [1-iodo-2-methylcyclohexane] I

11-42

(a) [cyclohexanol] OH $\xrightarrow{Na}$ [cyclohexoxide] O^- Na^+ $\xrightarrow{CH_3CH_2Br}$ [ethoxycyclohexane] OCH_2CH_3 Williamson ether synthesis

11-42 continued

(b) Br (benzyl bromide) —NaOH→ benzyl alcohol (OH) —1 eq. NaOCl, TEMPO→ benzaldehyde (O, H) + benzylmagnesium bromide (MgBr) ←Mg, ether— benzyl bromide (Br)

↓ H_3O^+

1,2-diphenylethanol (OH) —H_2SO_4, Δ→ stilbene

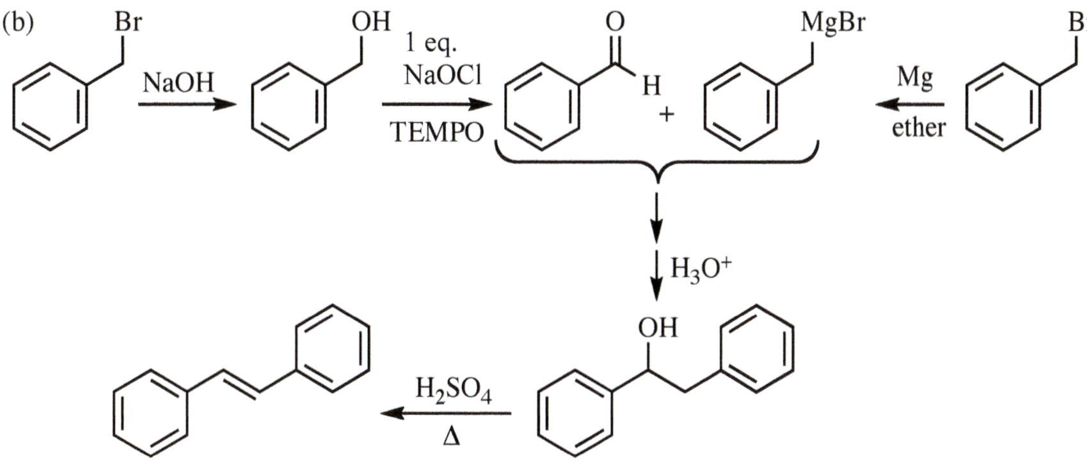

(c) bromocyclohexane (Br) —Mg, ether→ cyclohexylmagnesium bromide (MgBr) —1) H(C=O)CH₂CH₂CH₃, 2) H_3O^+→ 1-cyclohexylbutan-1-ol (OH) —H_2CrO_4→ 1-cyclohexylbutan-1-one (O)

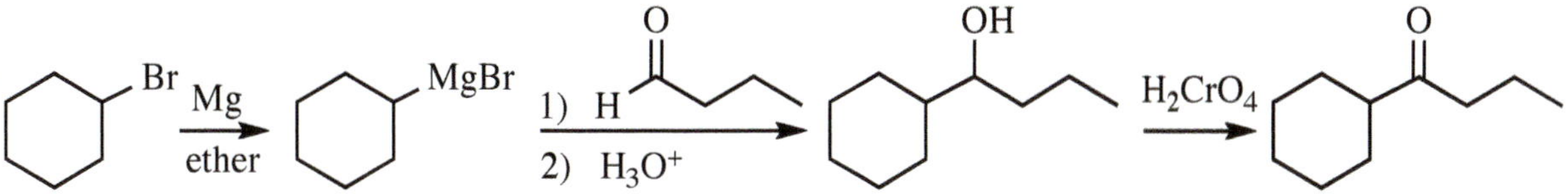

(d) cyclohexylmethanol (OH) —1 eq. NaOCl, TEMPO→ cyclohexanecarbaldehyde (H, O) —1) CH_3CH_2MgBr, 2) H_3O^+→ 1-cyclohexylpropan-1-ol (OH)

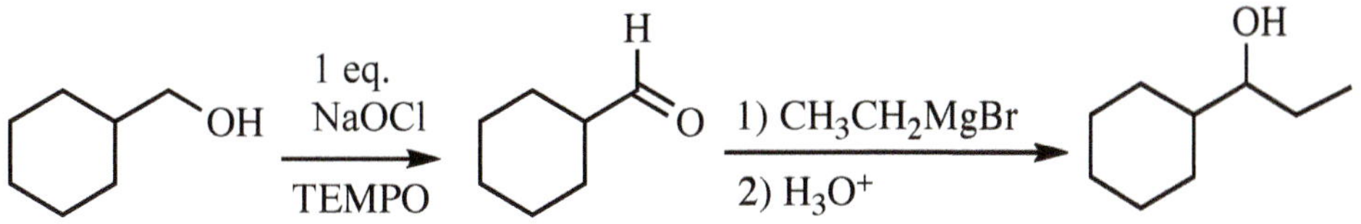

11-43 Major product for each reaction is shown.

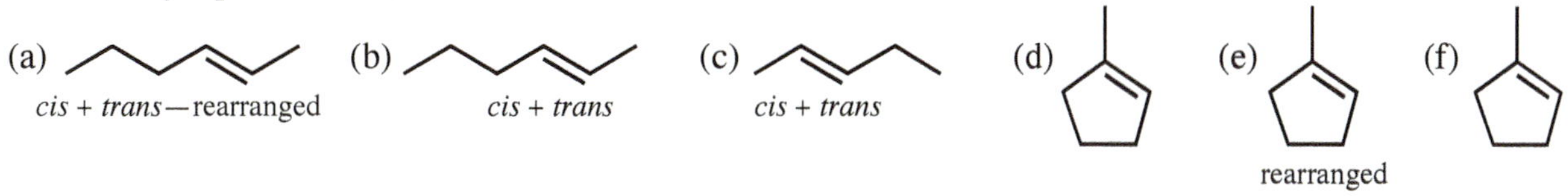

(a) *cis* + *trans*—rearranged (b) *cis* + *trans* (c) *cis* + *trans* (d) (e) rearranged (f)

Note that (d), (e), and (f) produce the same alkene.

11-44

(a) 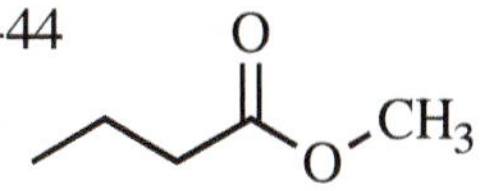 $CH_3CH_2CH_2C(=O)OCH_3$ (b) CH_3ONO_2 (c) $CH_3CH_2O-\overset{O}{\overset{\|}{P}}-OH$ with OCH_2CH_3 (d) 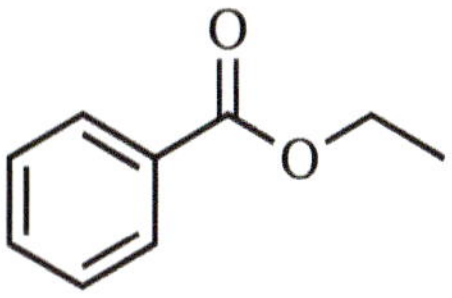 ethyl benzoate (O, O)

(e) 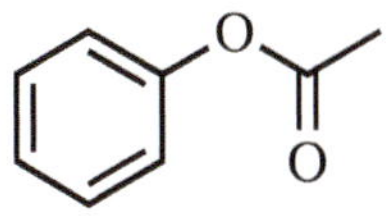 phenyl acetate (O, O)

11-45 *cis* 2-methylcyclohexanol (OH) —H_2SO_4, Δ→ *Zaitsev product* ←H_2SO_4, Δ— *trans* (OH)

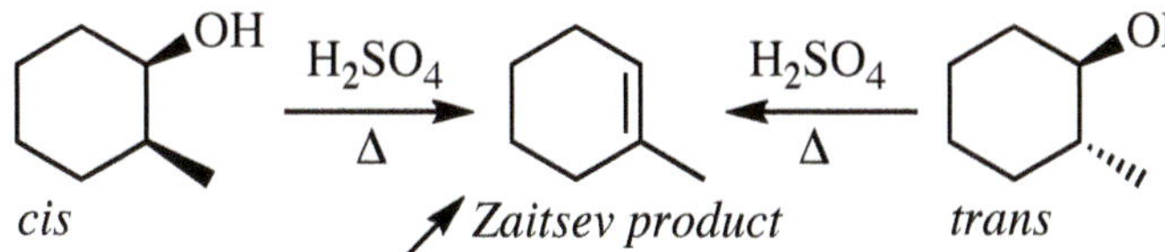

These two dehydrations follow the E1 mechanism with a common carbocation intermediate. The stereochemistry plays no role in the E1 mechanism.

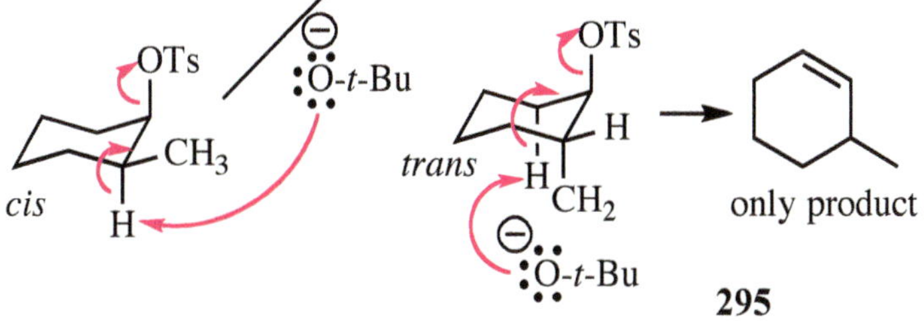

cis: OTs, CH_3, H; $^{\ominus}$:Ö-*t*-Bu → *Zaitsev product*

trans: OTs, H, H, CH_2; $^{\ominus}$:Ö-*t*-Bu → only product

Elimination of the tosylate with base follows the E2 mechanism with the stereochemical requirement that the H and the OTs must be anti-coplanar; see text section 7-14.

11-46

(a)  $SOCl_2$ — S → S—retention

(b) 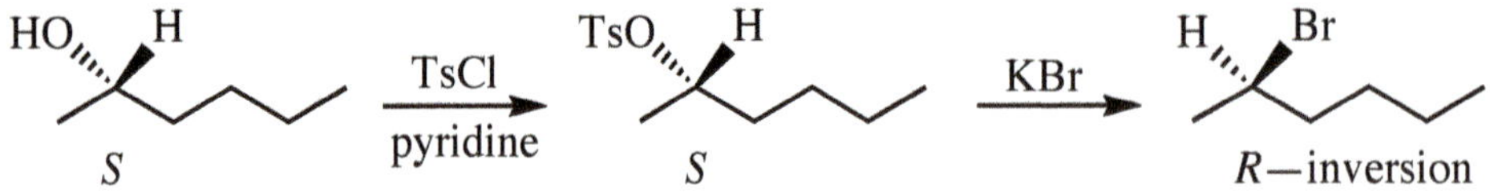 TsCl, pyridine; KBr — S → S → R—inversion

Alternatively, PBr_3 could be used.

(c) 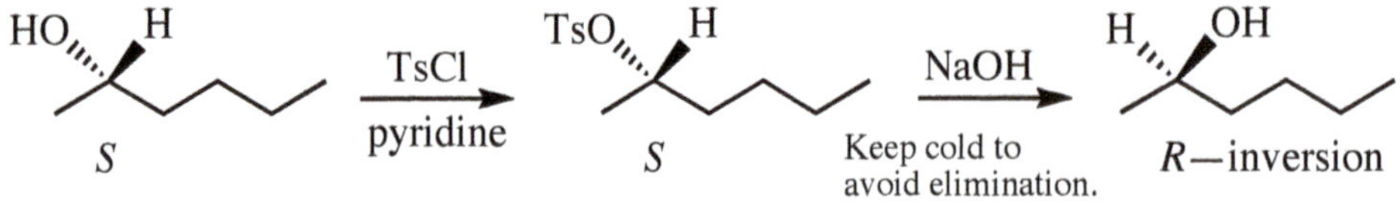 TsCl, pyridine; NaOH — S → S → R—inversion

Keep cold to avoid elimination.

11-47

(a) 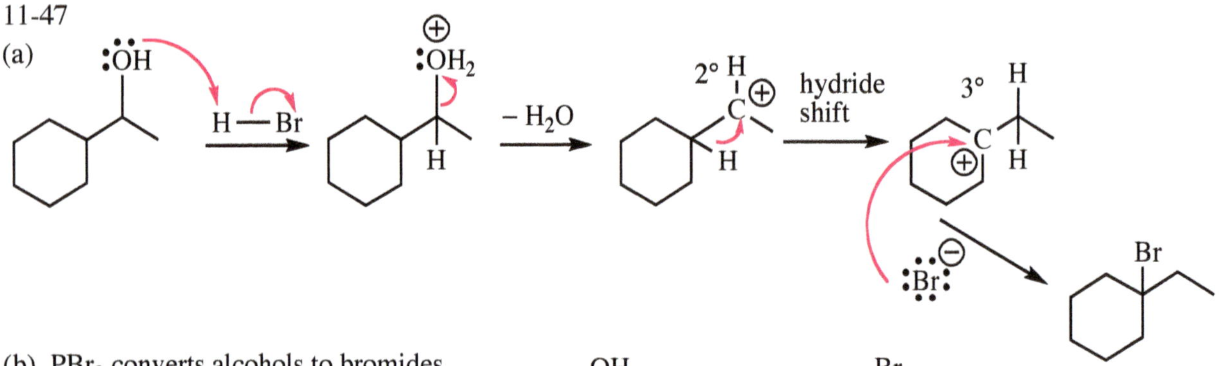 H—Br; $-H_2O$; 2° → hydride shift → 3°; Br

(b) PBr_3 converts alcohols to bromides without rearrangement because no carbocation intermediate is produced. Alternatively, making the tosylate and displacing with bromide would also work.

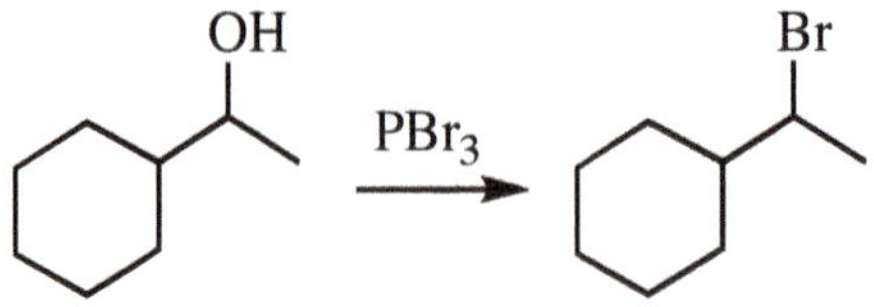

OH —PBr_3→ Br

11-48

(a) 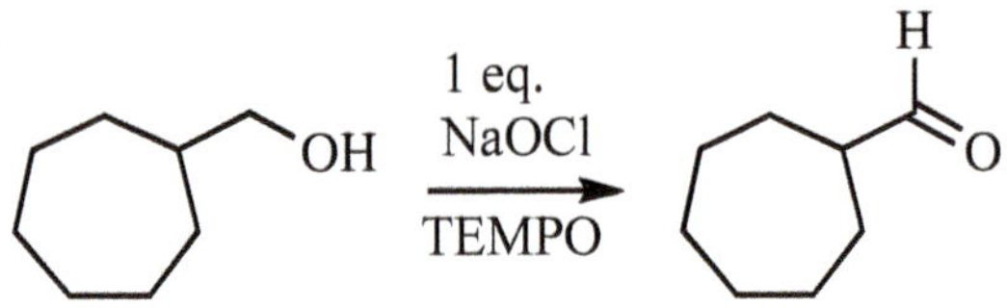 OH — 1 eq. NaOCl, TEMPO → H, O

(b) 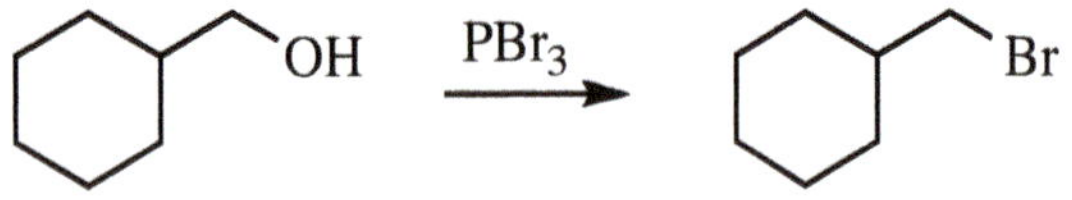 OH — PBr_3 → Br

(c) 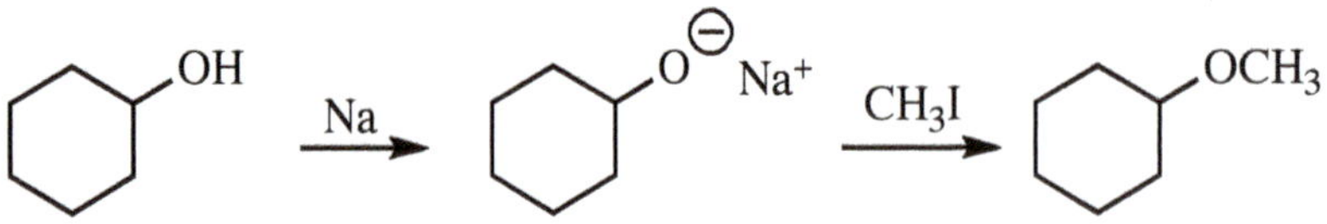 OH — Na → O^- Na^+ — CH_3I → OCH_3

(d) 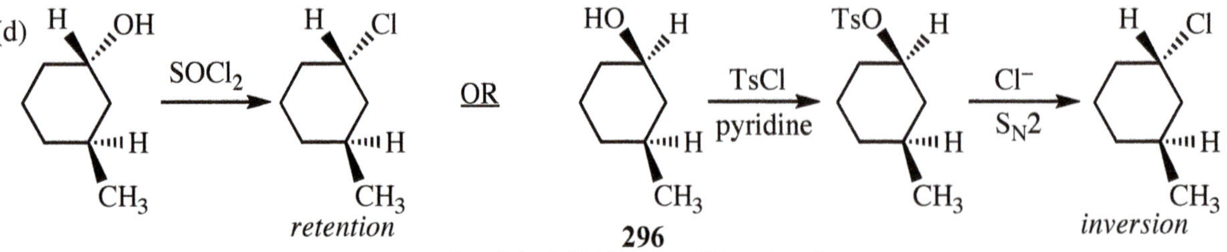 H, OH, H, CH_3 — $SOCl_2$ → H, Cl, H, CH_3 *retention*

OR

HO, H, H, CH_3 — TsCl, pyridine → TsO, H, H, CH_3 — Cl^-, S_N2 → H, Cl, H, CH_3 *inversion*

11-48 continued

(e) OH → H_2SO_4, Δ → 1) O_3, –78 °C; 2) Me_2S → O, O ← HIO_4 ← OH, OH

(f) CH_2OH → excess NaOCl, TEMPO → COOH

(g) OH → NaOCl, HOAc → O

(h) H, H, CH_3, OH → TsCl, pyridine → H, H, CH_3, OTs

cis *cis*

11-49 (a) Br ← PBr_3 ← OH → 1) TsCl, pyridine; 2) NaBr → Br (e)

inversion *inversion*, S_N2

(b) $SOCl_2$ → Cl, *retention*

(c) HCl, $ZnCl_2$ → Cl, S_N1, *cis* and *trans*

(d) HBr → Br, S_N1, *cis* and *trans*

11-50

(a) OH OH

Lucas: no reaction cloudy in 1-5 min.

(b) OH OH

Lucas: cloudy in 1-5 min. cloudy in < 1 min.
H_2CrO_4: immediate blue-green no reaction—stays orange

(c) OH

Lucas: cloudy in 1-5 min. no reaction
H_2CrO_4: immediate blue green no reaction—stays orange

(d) OH O

Lucas: cloudy in 1-5 min. no reaction
H_2CrO_4: immediate blue green no reaction—stays orange

(e) O OH

Lucas: no reaction cloudy in < 1 min.

11-51

The first equivalent of NaH will remove the most acidic proton from the COOH; the second equivalent will remove the phenolic proton; the final equivalent will remove the proton from the 1° alcohol.

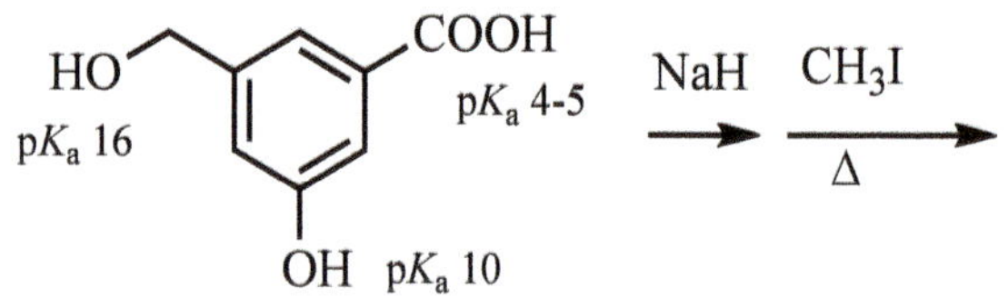

HO, COOH, pK_a 16, pK_a 4-5, OH pK_a 10; NaH; CH_3I, Δ

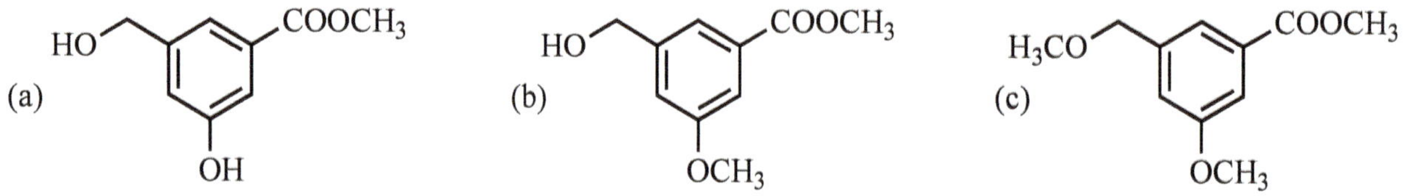

(a) HO, $COOCH_3$, OH (b) HO, $COOCH_3$, OCH_3 (c) H_3CO, $COOCH_3$, OCH_3

11-52 A OH; PBr_3; C Br; Mg, ether; D MgBr; + B O; NaOCl/HOAc; E OMgBr; H_3O^+; OH

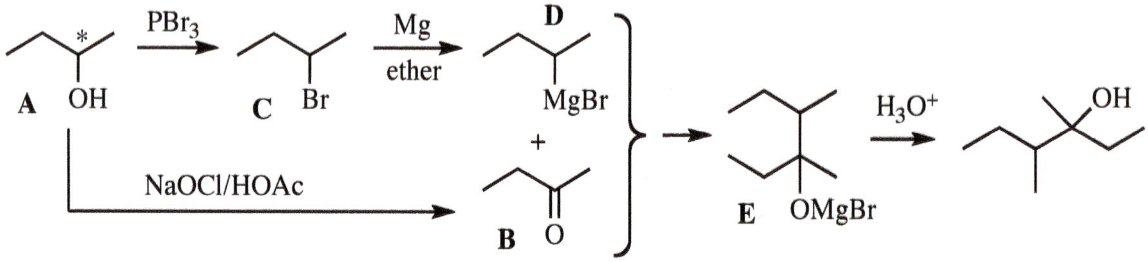

11-53 OH; PBr_3; W Br; Mg, ether; X MgBr; O; H_3O^+; HO; Y

H_2CrO_4; V O

Alternatively, the last step could dehydrate the 3° alcohol, making this alkene **Z**.

$CH_3\overset{O}{\overset{\|}{C}}$–O; $CH_3\overset{O}{\overset{\|}{C}}-Cl$; Z

Keep cold to avoid elimination.

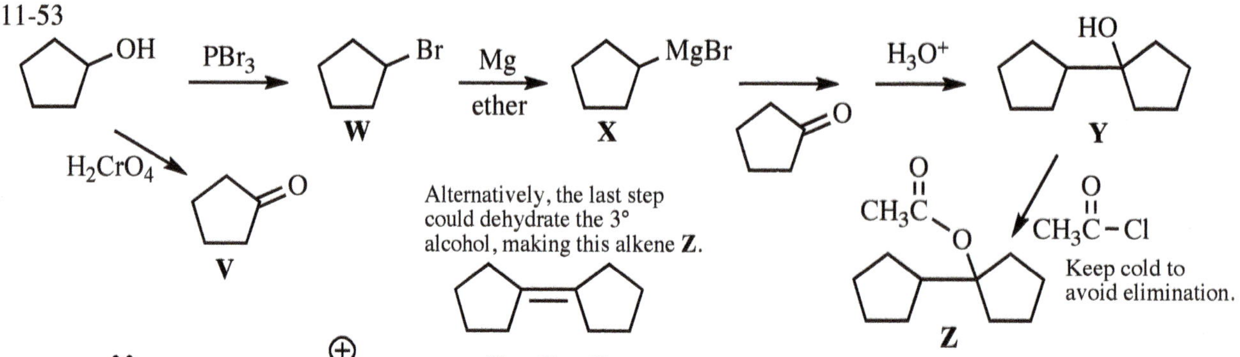

11-54 OH; H–A; $\overset{\oplus}{O}H_2$; $-H_2O$; $\overset{\oplus}{C}H_2$, H; O

alkyl shift—ring expansion

Recall that 1° carbocations probably do not exist; this could be considered a transition state.

ring expansion from 1° carbocation to 2°, *resonance-stabilized* carbocation

$H_2\ddot{O}$:

NOTE:

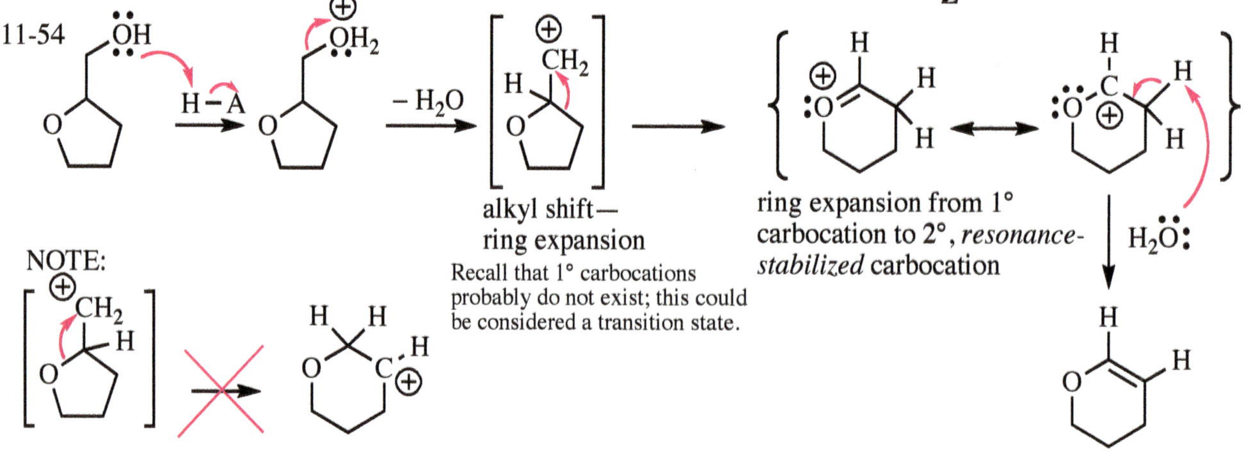

The migration directly above does NOT occur as the cation produced is not resonance-stabilized.

continued on next page

An alternative mechanism could be proposed: protonate the ring oxygen, open the ring to a 2° carbocation followed by a hydride shift to a resonance-stabilized cation, ring closure, and dehydration.

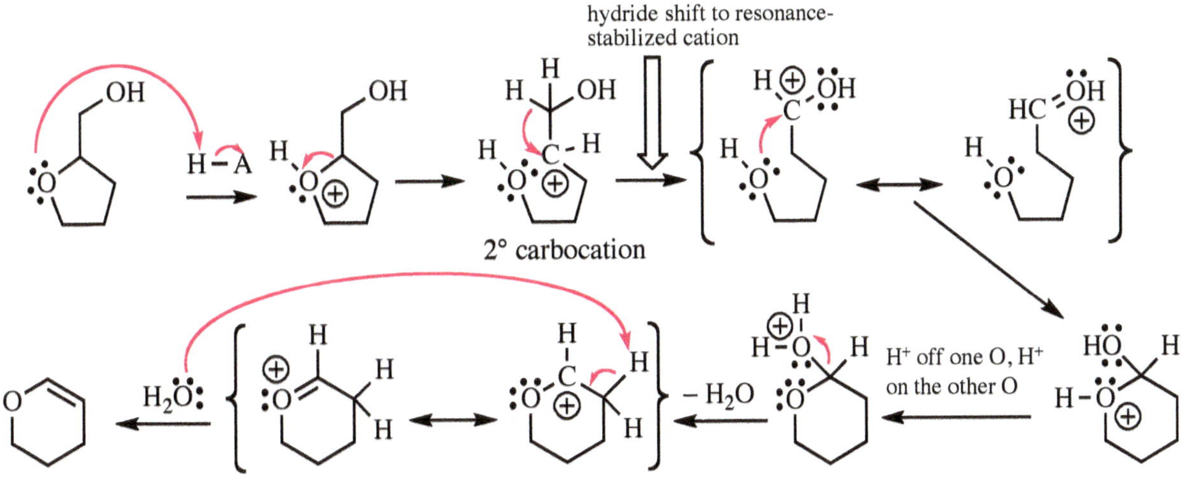

hydride shift to resonance-stabilized cation

2° carbocation

H⁺ off one O, H⁺ on the other O

$- H_2O$

$H_2\ddot{O}$:

11-55

(a)

$ZnCl_2$

$\ominus$ $O-ZnCl_2$

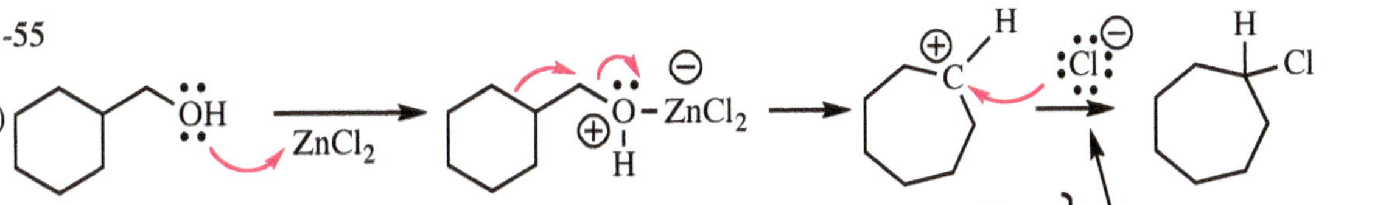

In this presentation of the mechanism, the rearrangement is shown concurrently with cleavage of the C-O bond with no 1° carbocation intermediate.

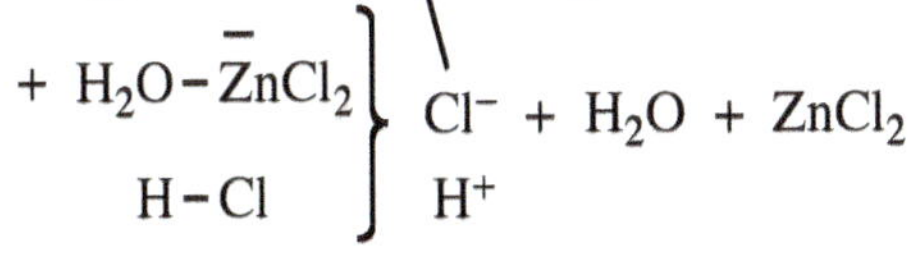

$+ H_2O-\bar{Z}nCl_2$ → $Cl^- + H_2O + ZnCl_2$

$H-Cl$ → H^+

(b)

$H-OSO_3H$

Δ

$+ H_2O$

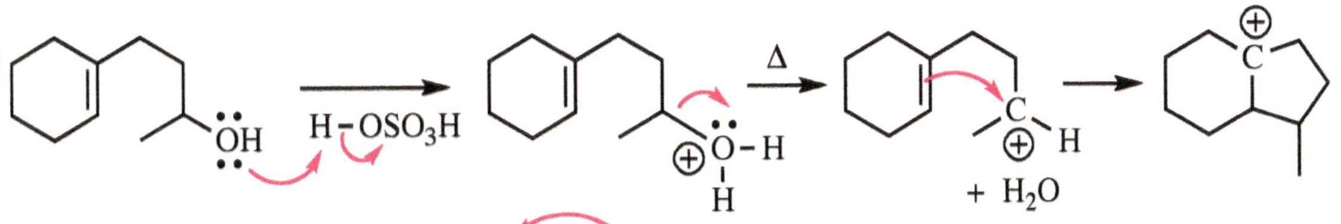

Once this 3° carbocation is formed, removal of adjacent protons produces the compounds shown.

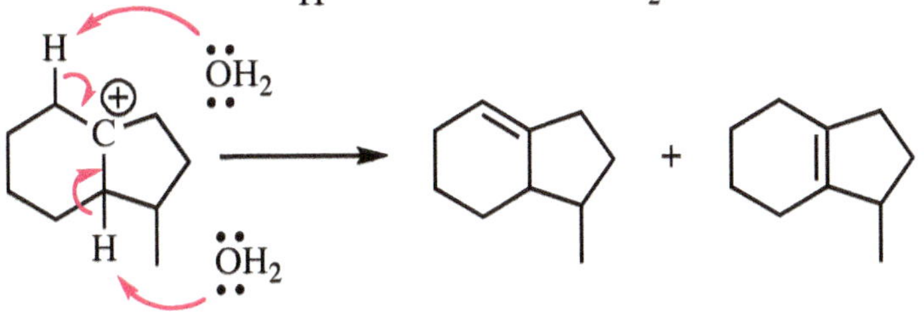

$\ddot{O}H_2$

11-55 continued

(c) All three of the products go through a common carbocation intermediate.

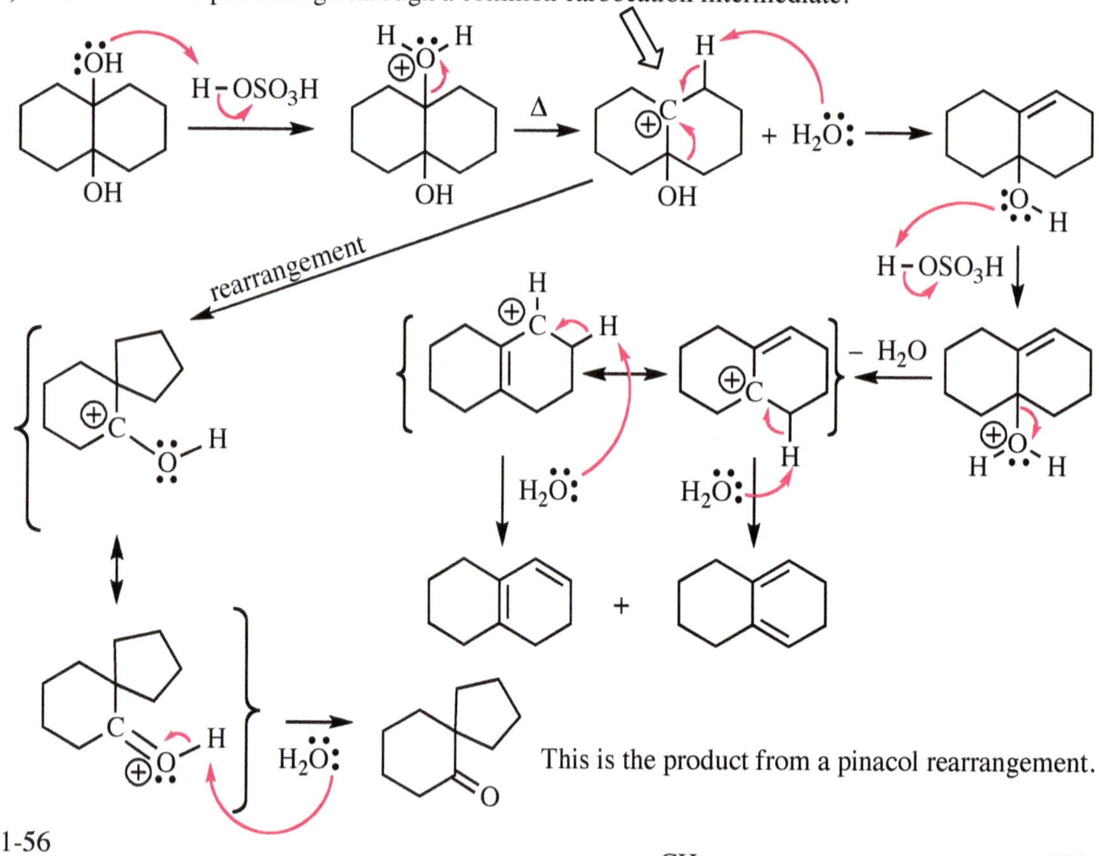

rearrangement

H_2O

$- H_2O$

$H-OSO_3H$

This is the product from a pinacol rearrangement.

11-56

(a) NaOCl, HOAc; CH_3MgBr; Mg, ether; PBr_3; $CH_3OH \rightarrow CH_3Br$; MgBr ⊕; Br; PBr_3; HO

OR: alternative ending—not as good as above as more side products are possible

after hydrolysis of Grignard product

H_2SO_4

S_N1

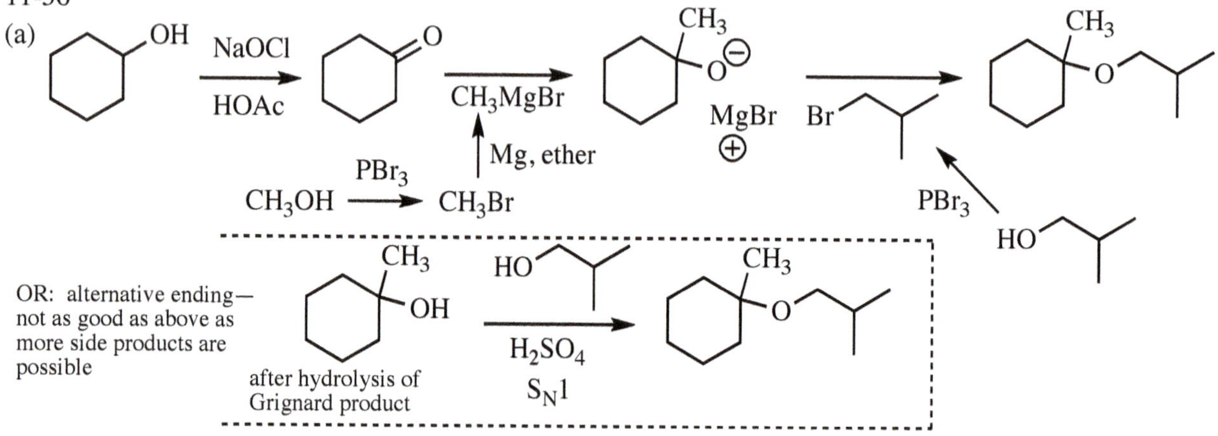

(b) OH; PBr_3; Br; Mg, ether; MgBr; H_3O^+; OH; NaOCl, HOAc

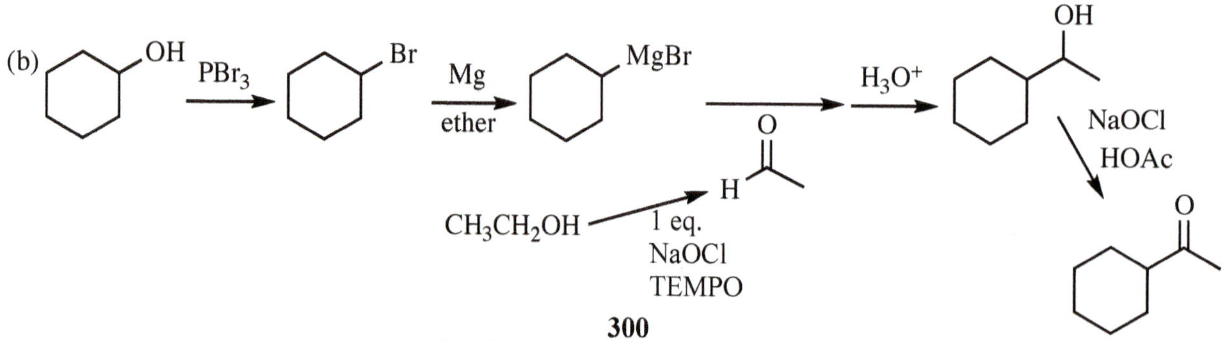

CH_3CH_2OH — 1 eq. NaOCl, TEMPO

11-56 continued

(c) OH → NaOCl / HOAc → O → (MgBr, from (b)) → $O^{\ominus}$ $^{\oplus}MgBr$ → CH_3I → OCH_3

(d) 2 OH → PBr_3 → 2 Br → Mg / ether → 2 MgBr

OH → excess NaOCl / TEMPO → O, C, OH → EtOH, H^+ → O, C, OEt → H_3O^+ → OH

(e) MgBr, from (b) → 1) O (epoxide) 2) H_3O^+ → OH

(f) HO → PBr_3 → Mg / ether → BrMg, from (d)

OH → 1 eq. NaOCl / TEMPO → H, O

→ H_3O^+ → OH → PBr_3 → Br

(g) HO, H → TsCl / pyridine → TsO, H → KCN → H, NC

(h) OH → H_2SO_4 / Δ → NBS / hν → Br → Mg / ether → MgBr + H, O ← 1 eq. NaOCl / TEMPO ← HO

→ $BrMg^{\oplus}$ $O^{\ominus}$ → CH_3CH_2Br (from CH_3CH_2OH, PBr_3) → O → OsO_4 / H_2O_2 → O, OH, OH

11-57 For a complicated synthesis like this, begin by working backwards. Try to figure out where the carbon framework came from; in this problem we are restricted to alcohols containing five or fewer carbons. The dashed boxes show the fragments that must be assembled. The most practical way of forming carbon-carbon bonds is by Grignard reactions. The epoxide must be formed from an alkene, and the alkene must have come from dehydration of an alcohol produced in a Grignard reaction.

fragment A *fragment B* *fragment C* O

1 eq. NaOCl TEMPO *fragment A* OH H O

BrMg *fragment B* Mg ether Br HO PBr_3

mCPBA H_3O^+ HO NaOCl/HOAc

major isomer 1. TsCl, py 2. KOH, Δ

Avoid carbocation conditions to prevent rearrangement.

OH H_3O^+ 1. PBr_3 2. Mg ether OH *fragment C* MgBr O

11-58

(a) Both of these pseudo-syntheses suffer from the misconception that incompatible reagents or conditions can co-exist. In the first example, the S_N1 conditions of ionization cannot exist with the S_N2 conditions of sodium methoxide. The tertiary carbocation in the first step would not wait around long enough for the sodium methoxide to be added in the second step. (The irony is that the first step by itself, the solvolysis of *tert*-butyl bromide in methanol, would give the desired product without the sodium methoxide.)

In the second reaction, the acidic conditions of the first step in which the alcohol is protonated are incompatible with the basic conditions of the second step. If basic sodium methoxide were added to the sulfuric acid solution, the instantaneous acid-base neutralization would give methanol, sodium sulfate, and the starting alcohol. No reaction on the alcohol would occur.

(b)

Br $\xrightarrow[\text{warm}]{CH_3OH}$ OCH_3 S_N1 solvolysis conditions

Several synthetic sequences are possible for the second synthesis.

OH $\xrightarrow{Na}$ $O^-\ Na^+$ $\xrightarrow{CH_3I}$ O

OR OH $\xrightarrow[\text{pyridine}]{TsCl}$ OTs $\xrightarrow{NaOCH_3}$ O

OR OH $\xrightarrow{PBr_3}$ Br $\xrightarrow{NaOCH_3}$ O

11-59

Compound X : —must be a 1° or 2° alcohol with an alkene; no reaction with Lucas leads to a 1° alcohol; can't be allylic as this would give a positive Lucas test.

OH

Compound Y : —must be a cyclic ether, not an alcohol and not an alkene; other isomers of cyclic ethers are possible.

O

11-60

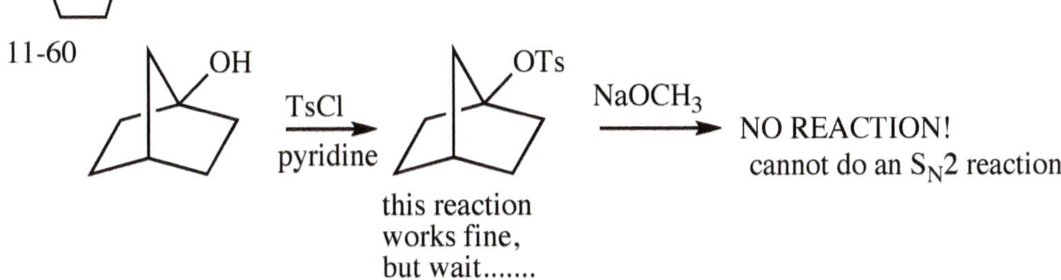

The Williamson ether synthesis is an S_N2 displacement of a leaving group by an alkoxide ion. In addition to being a 3° substrate, this tosylate cannot undergo an S_N2 reaction for two reasons. First, backside attack cannot occur because the back side of the bridgehead carbon is blocked by the other bridgehead. Second, the bridgehead carbon cannot undergo inversion because of the constraints of the bridged ring system.

Backside attack is blocked. H— —OTs side view

H— —OTs side view This carbon cannot invert which is required in the S_N2 mechanism.

alternative synthesis: $R{-}OH \xrightarrow{Na} R{-}O^- Na^+ \xrightarrow{CH_3I} R{-}OCH_3$

11-61 Let's begin by considering the facts.

The axial alcohol is oxidized ten times as fast as the equatorial alcohol. (In the olden days, this observation was used as evidence suggesting the stereochemistry of a ring alcohol.)

CH_3 OH H $\xrightarrow[\text{faster}]{H_2CrO_4}$ CH_3 O H $\xleftarrow[\text{slower}]{H_2CrO_4}$ CH_3 HO H

Second, it is known that the oxidation occurs in two steps: 1) formation of the chromate ester; and 2) loss of H and chromate to form the C=O. Let's look at each mechanism.

AXIAL CH_3 OH H $\xrightarrow[\text{Step 1}]{H_2CrO_4}$ CH_3 H O CrO_3H H Base (H_2O) $\xrightarrow{\text{Step 2}}$ CH_3 O H FASTER

continued on next page

11-61 continued

EQUATORIAL

CH_3 HO ... H $\xrightarrow[\text{Step 1}]{H_2CrO_4}$ CrO_3H O ... CH_3 H H Base (H_2O) $\xrightarrow{\text{Step 2}}$ CH_3 O ... H SLOWER

So what do we know about these systems? We know that substituents are more stable in the equatorial position than in the axial position because any group at the axial position has 1,3-diaxial interactions. So what if Step 1 were the rate-limiting step? We would expect that the equatorial chromate ester would form faster than the axial chromate ester; since this is contrary to what the data show, Step 1 is not likely to be rate-limiting. How about Step 2? If the elimination is rate limiting, we would expect the approach of the base (probably water) to the equatorial hydrogen (axial chromate ester) would be faster than the approach of the base to the axial hydrogen (equatorial chromate ester). Moreover, the axial ester is more motivated to leave due to steric congestion associated with such a large group. This is consistent with the relative rates of reaction from experiment. Thus, it is reasonable to conclude that the second step of the mechanism is rate-limiting.

11-62

(a)

H :OH — start with *R* — H—Br → H :O—H (+) H

Protonation of the OH makes a good leaving group. There are two reaction paths possible, S_N1 and S_N2.

:Br:⊖ ↙ ↘ $- H_2O$

Br H — S_N2 would give 100% *S*.

H C (+) planar — :Br:⊖ →

S_N1 would give a racemic mixture.

Br H — 50% *S* + H Br — 50% *R*

The data show that the product has "racemization with excess inversion", that is, more *S* than *R*. The best explanation is that a mixture of S_N1 and S_N2 is happening. This is not surprising as secondary halides are on the fence between the two mechanisms.

There is another explanation: when the water leaves, it might not leave all the way so it blocks the incoming bromide ion from giving the product with retention of configuration, and the S_N2 type backside attack is the dominant pathway.

(b)

Br H H :OH — *chiral* — H—Br → :Br: H H H_2O^+—H $\xrightarrow{- H_2O}$ bromonium ion! symmetric! (+)Br H H — :Br:⊖ → racemic mixture! 50:50 Br H H Br + H Br Br H

This mechanism is the reverse of bromohydrin formation, passing through the same bromonium ion intermediate, leading to a racemic mixture of *trans*-1,2-dibromocyclopentane. The participation of the Br is called *neighboring group assistance* and explains the difference in results between (a) and (b).

11-63

(a)

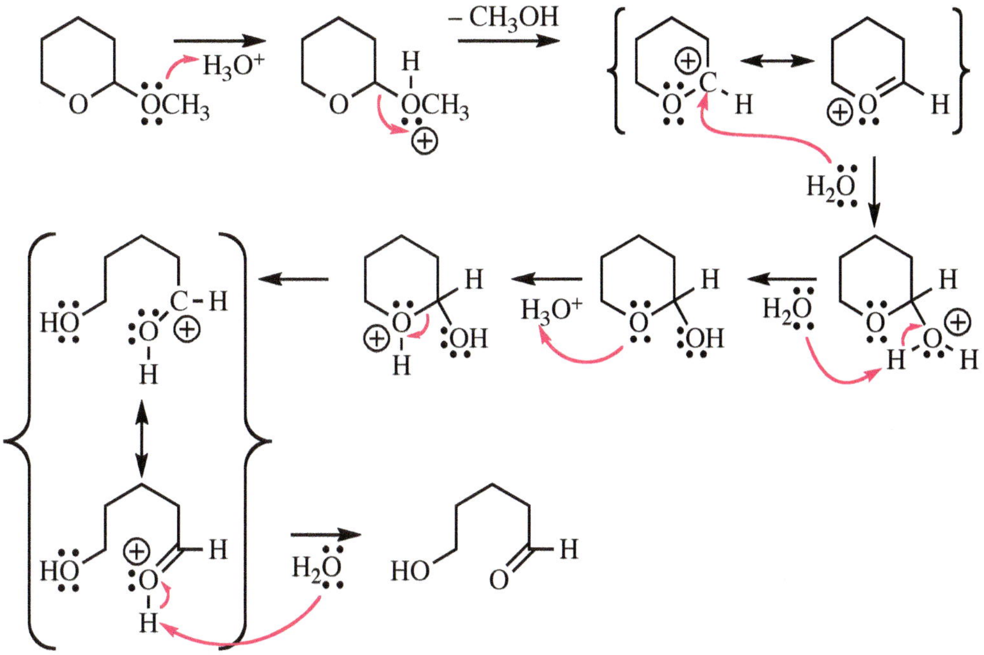

It is equally likely for protonation to occur first on the ring oxygen, followed by ring opening, then replacement of OCH_3 by water.

My colleague Dr. Kantorowski suggests this alternative. He and I will arm wrestle to determine which mechanism is correct.

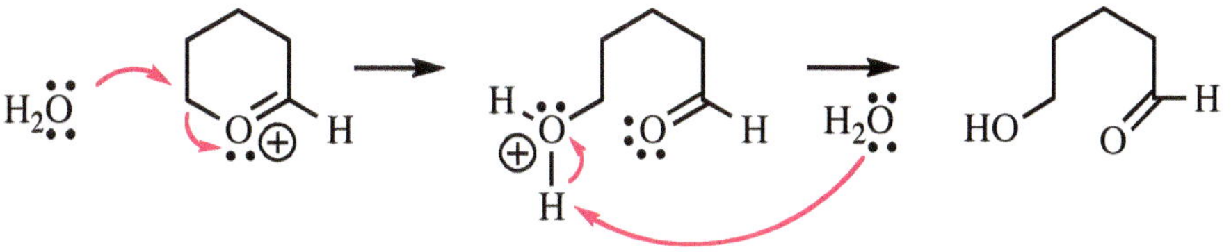

(b)

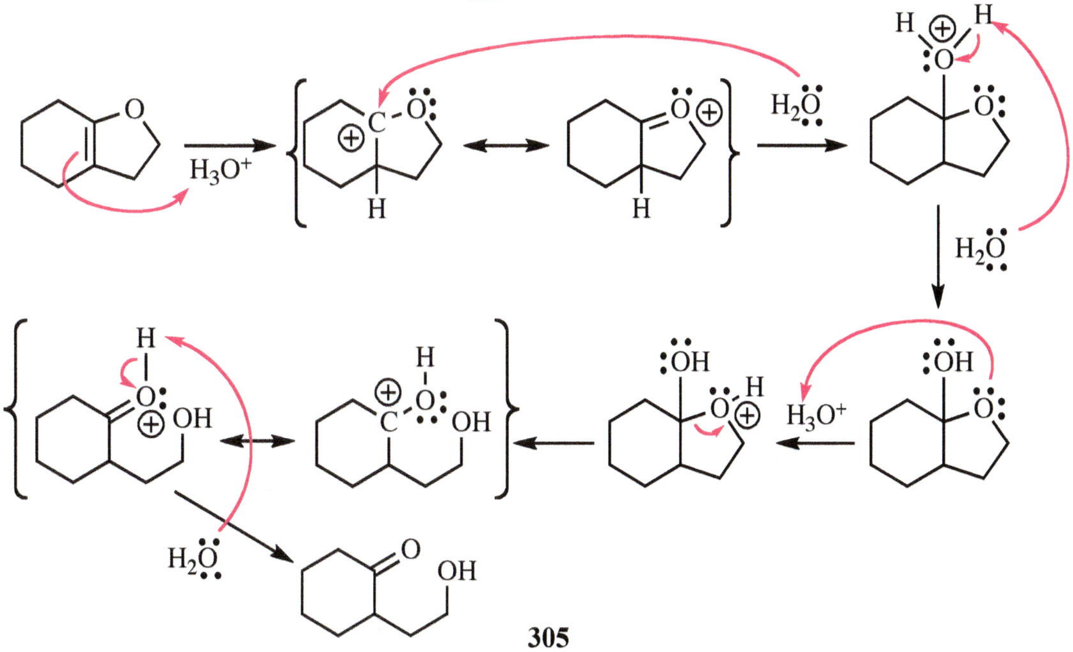

11-64 One of the keys to solving these "roadmap" or "structure proof" problems is making inferences from each piece of information given. Ask the question: "What is this fact telling me?"

Q has molecular formula $C_6H_{12}O$ ⟹ **Q** has one element of unsaturation ⟹ **Q** has a ring OR C=C or C=O

Q cannot be separated into enantiomers ⟹ **Q** does not have an asymmetric carbon atom

Q does not react with Br_2, $KMnO_4$, H_2 ⟹ **Q** has no C=C

Q reacts with H_2SO_4 and loses H_2O ⟹ **Q** is an alcohol, not a C=O ⟹ **Q** has a ring

R has C=C; ozonolysis gives one acyclic product, **S** ⟹ **R** has a C=C in its ring

R has enantiomers ⟹ the dehydration that produced the C=C also created an asymmetric carbon atom

S is a ketoaldehyde ⟹ one C of the C=C has an H, the other has an R

Putting all this together gives a 4-membered ring: 3 is too small to fit the optical activity results, 5 is too large.

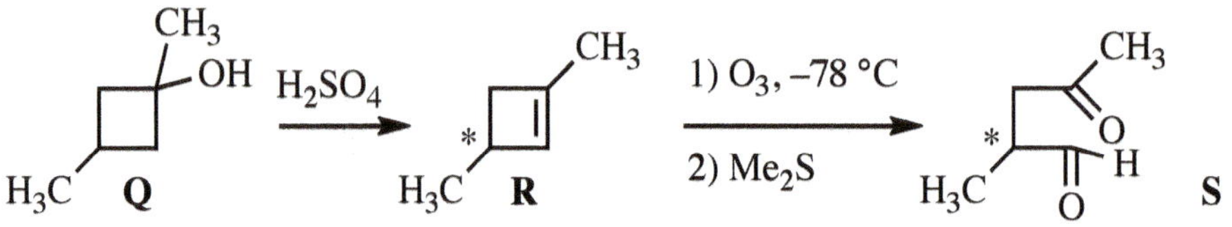

Q → (H_2SO_4) → R → (1) O_3, −78 °C; 2) Me_2S) → S

11-65

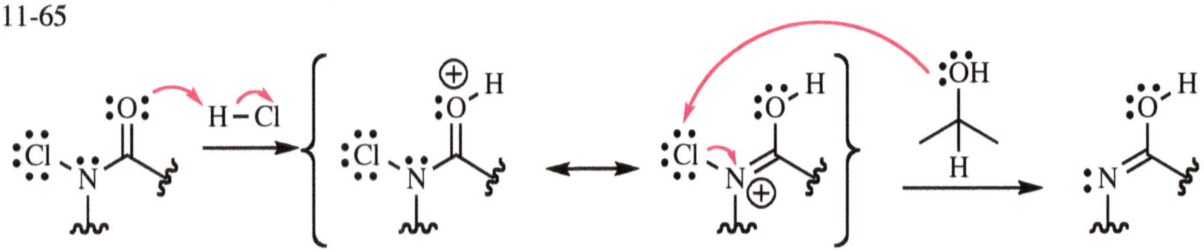

Protonation of TCICA increases the polarization of the N-Cl bond.

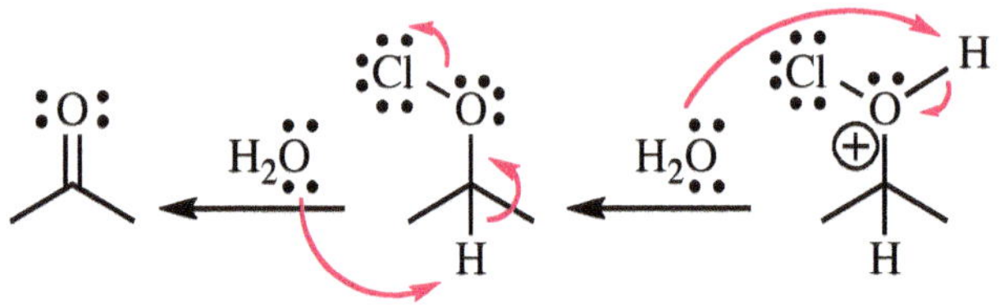

11-66

(a) (b) 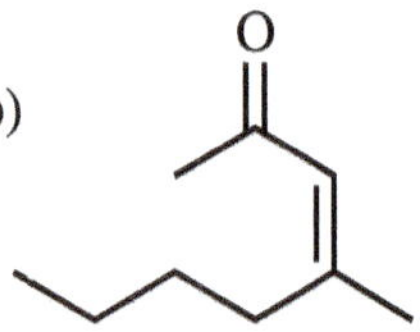 (c) 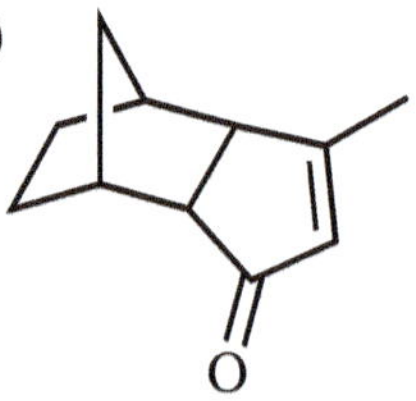 (d)

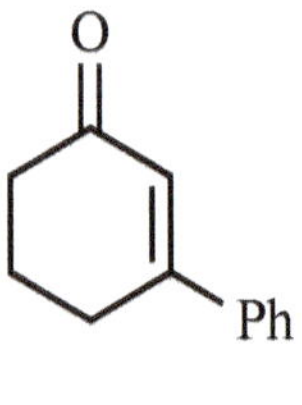

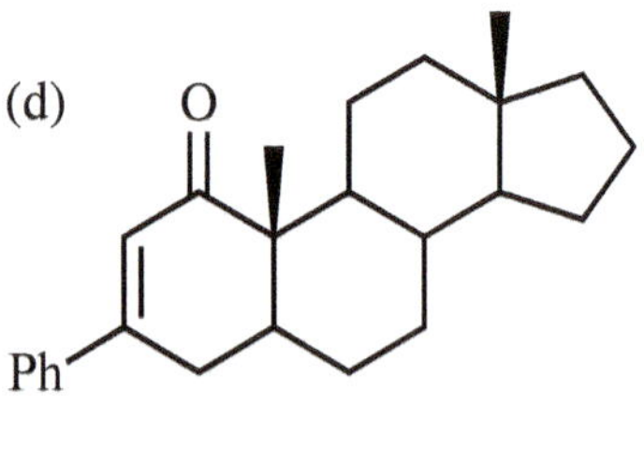

CHAPTER 12—INFRARED SPECTROSCOPY AND MASS SPECTROMETRY

See p. 320 in this Manual for some useful web sites with infrared and mass spectra.

12-1 The table is completed by recognizing that: $(\bar{\nu})(\lambda) = 10{,}000$

$\bar{\nu}$ (cm^{-1})	4000	**3300**	**3003**	**2198**	1700	1640	1600	400
λ (μm)	2.50	3.03	3.33	4.55	**5.88**	**6.10**	**6.25**	25.0

12-2 In general, only bonds with dipole moments will have an IR absorption. Bonds in a completely symmetric environment will have at most a weak absorption.

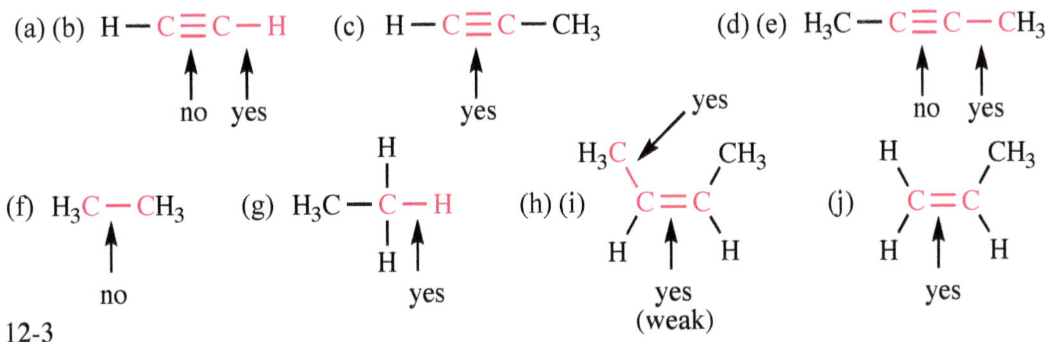

12-3

(a) Alkene: C=C at 1640 cm^{-1}, =C—H at 3080 cm^{-1}, saturated C—H below 3000 cm^{-1}; 1820 is overtone of 910 cm^{-1}.

(b) Alkane: no peaks indicating sp or sp^2 carbons present; only saturated C—H below 3000 cm^{-1}.

(c) This IR shows more than one group. There is a terminal alkyne shown by: C≡C at 2100 cm^{-1}, and ≡C—H at 3300 cm^{-1}. These signals indicate an aromatic hydrocarbon as well: =C—H at 3050 cm^{-1}, and C=C at ≈1600 cm^{-1}. It is unusual to see no saturated C—H below 3000 cm^{-1}.

12-4

(a) 2° amine, R—NH—R: one peak at 3300 cm^{-1} indicates an N—H bond; this spectrum also shows a C=C at 1640 cm^{-1} , unsaturated =C—H at 3070 cm^{-1}, saturated C—H below 3000 cm^{-1} ; 1840 cm^{-1} is overtone of 920 cm^{-1}.

(b) Carboxylic acid: the extremely broad absorption in the 2500-3500 cm^{-1} range, with a "shoulder" around 2500-2700 cm^{-1}, and a C=O at 1710 cm^{-1}, are compelling evidence for a carboxylic acid; saturated C—H below 3000 cm^{-1}.

(c) Alcohol: strong, broad O—H at 3330 cm^{-1}; saturated C—H below 3000 cm^{-1}.

12-5

(a) Conjugated ketone: the small peak at 3030 cm^{-1} suggests =C—H, and the strong peak at 1685 cm^{-1} is consistent with a ketone conjugated with the alkene. The C=C is indicated by a very small peak around 1620 cm^{-1} consistent with conjugation; saturated C—H below 3000 cm^{-1}; 3440 is overtone of 1685 cm^{-1}.

(b) Ester: the C=O absorption at 1738 cm^{-1} (higher than the ketone's 1710 cm^{-1}), in conjunction with the strong C—O at 1200 cm^{-1}, points to an ester; saturated C—H below 3000 cm^{-1}.

(c) Amide: the two peaks at 3160-3360 cm^{-1} are likely to be an NH_2 group; the strong peak at 1640 cm^{-1} is too strong for an alkene, so it must be a different type of C=X, in this case a C=O, so low because it is part of an amide; saturated C—H below 3000 cm^{-1}.

12-6

(a) The moderate peak at 1642 cm^{-1} indicates a C=C, consistent with the =C—H at 3080 cm^{-1}. This appears to be a simple alkene. Saturated C—H below 3000 cm^{-1}; 1825 cm^{-1} is overtone of peak at 915 cm^{-1}.

12-6 continued

(b) The strong absorption at 1691 cm^{-1} is unmistakably a conjugated C=O. The smaller peak at 1626 cm^{-1} indicates a C=C, probably conjugated with the C=O. The two peaks at 2712 cm^{-1} and at 2814 cm^{-1} represent H—C=O confirming that this is an aldehyde. Unsaturated =C—H above 3000 cm^{-1}; saturated C—H below 3000 cm^{-1} .

(c) The strong peak at 1650 cm^{-1} is C=C, probably conjugated with C=O as it is unusually strong. The 1703 cm^{-1} peak appears to be a conjugated C=O, undeniably a carboxylic acid because of the strong, broad O—H absorption from 2400-3400 cm^{-1}. The unsaturated =C—H is obscured by the strong O—H; saturated C—H below 3000 cm^{-1}.

(d) The C=O absorption at 1742 cm^{-1} coupled with C—O at 1220 cm^{-1} suggest an ester but 1742 cm^{-1} is too high to be conjugated. The small peak at 1604 cm^{-1}, peaks above 3000 cm^{-1}, and peaks in the 600-800 cm^{-1} region indicate a benzene ring. It has both unsaturated C—H above 3000 cm^{-1} and saturated C—H below 3000 cm^{-1}.

12-7

(a) The M and M+2 peaks of equal intensity identify the presence of bromine. The mass of M (156) minus the weight of the lighter isotope of bromine (79) gives the mass of the rest of the molecule: $156 - 79 = 77$. The C_6H_5 (phenyl) group weighs 77; this compound is bromobenzene, C_6H_5Br .

(b) The m/z 127 peak shows that iodine is present. The molecular ion minus iodine gives the remainder of the molecule: $156 - 127 = 29$. The C_2H_5 (ethyl) group weighs 29; this compound is iodoethane, C_2H_5I .

(c) The M and M+2 peaks have relative intensities of about 3:1, a sure sign of chlorine. The mass of M minus the mass of the lighter isotope of chlorine gives the mass of the remainder of the molecule: $90 - 35 = 55$. A fragment of mass 55 is not one of the common alkyl groups (15, 29, 43, 57, etc., increasing in increments of 14 mass units (CH_2)), so the presence of unsaturation and/or an atom like oxygen must be considered. In addition to the chlorine atom, mass 55 could be C_4H_7 or C_3H_3O. Possible molecular formulas are C_4H_7Cl or C_3H_3ClO.

(d) The odd-mass molecular ion indicates the presence of an odd number of nitrogen atoms (always begin by assuming *one* nitrogen). The rest of the molecule must be: $115 - 14 = 101$; this is most likely C_7H_{17}. $C_7H_{17}N$ is the correct formula of a molecule with no elements of unsaturation. The seven carbons probably include alkyl groups like ethyl or propyl or isopropyl. A hint about spectra with odd-mass molecular ions: Look for fragment peaks with even masses to confirm that the odd-mass peak is the parent and not just a fragment from an invisible molecular ion.

12-8 Recall that radicals are not detected in mass spectrometry; only positively charged ions are detected.

$$\left[CH_3-\underset{\displaystyle CH_3}{\overset{|}{C}}H-CH_2 \,\big|^{57}\, CH_2-CH_3 \right]^{+\bullet} \longrightarrow \underset{m/z\ 57}{CH_3-\underset{}{\overset{CH_3}{\overset{|}{C}H}}-\overset{\oplus}{C}H_2} + \underset{\text{mass 29 radicals not detected}}{\dot{C}H_2-CH_3}$$

The fragment giving m/z 57 is a primary carbocation, less stable than the more abundant secondary carbocations of m/z 43 and 71.

12-9 $\left[CH_3-CH(CH_3)-CH_2-CH(CH_3)-CH_3\right]^{+\cdot}$ (cleavage labels: 43, 57, 85)

m/z 100

$\longrightarrow CH_3-\overset{+}{C}H-CH_3$ (m/z 43) $+$ $\dot{C}H_2-CH(CH_3)-CH_3$ (mass 57)

$\longrightarrow CH_3-\dot{C}H-CH_3$ (mass 43) $+$ $\overset{+}{C}H_2-CH(CH_3)-CH_3$ (m/z 57)

$\longrightarrow CH_3-CH(CH_3)-CH_2-\overset{+}{C}H-CH_3$ (m/z 85) $+$ $\dot{C}H_3$ (mass 15)

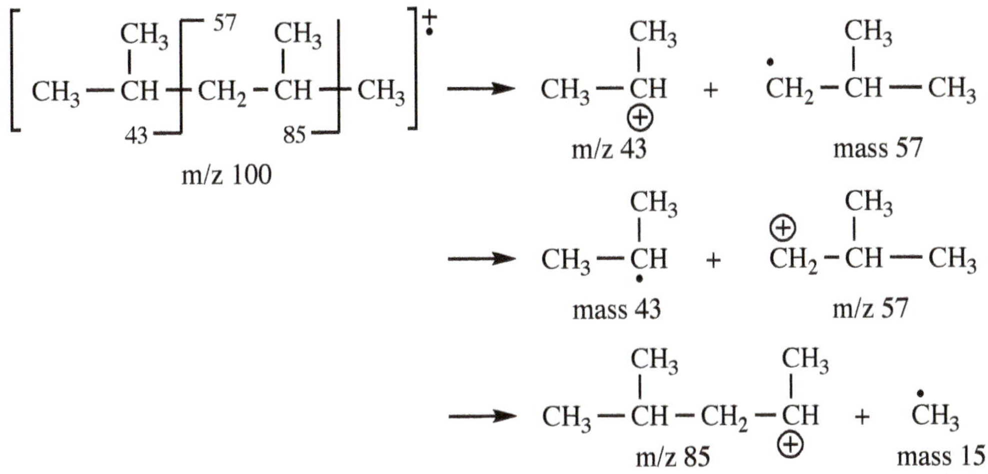

12-10 The molecular weight of each isomer is 116 g/mole, so the molecular ion appears at m/z 116. The left half of each structure is the same; loss of a three-carbon radical gives a stabilized cation, each with m/z 73:

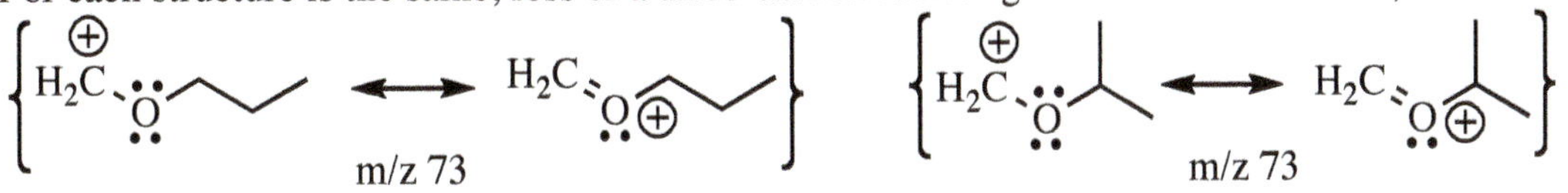

m/z 73 m/z 73

Where the two structures differ is in the alpha-cleavage on the right side of the oxygen. Alpha-cleavage on the left structure loses two carbons, whereas alpha-cleavage on the right structure loses only one carbon.

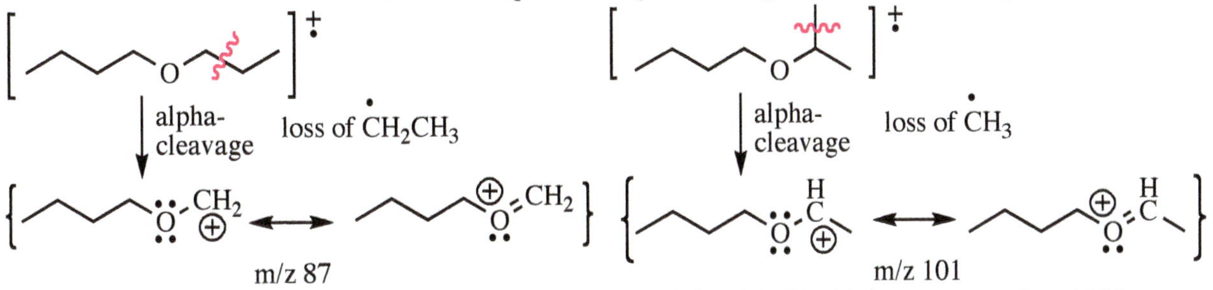

alpha-cleavage loss of $\dot{C}H_2CH_3$

alpha-cleavage loss of $\dot{C}H_3$

m/z 87 m/z 101

12-11 2,6-Dimethylheptan-4-ol, $C_9H_{20}O$, has molecular weight 144. The highest mass peak at 126 is *not* the molecular ion, but rather is the loss of water (18) from the molecular ion.

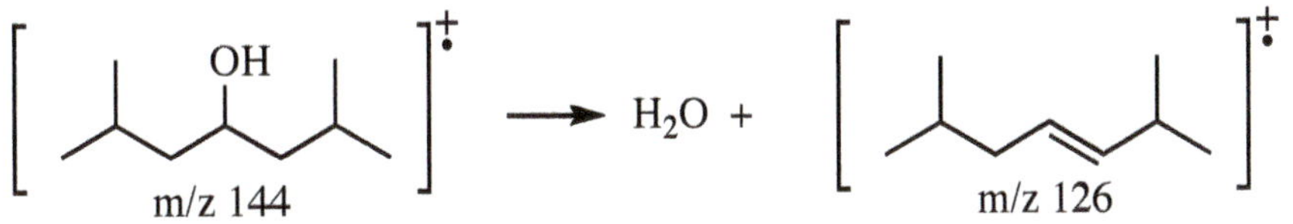

m/z 144 $\longrightarrow H_2O +$ m/z 126

The peak at m/z 111 is loss of another 15 (CH_3) from the fragment of m/z 126. This is called allylic cleavage; it generates a 2°, allylic, resonance-stabilized carbocation.

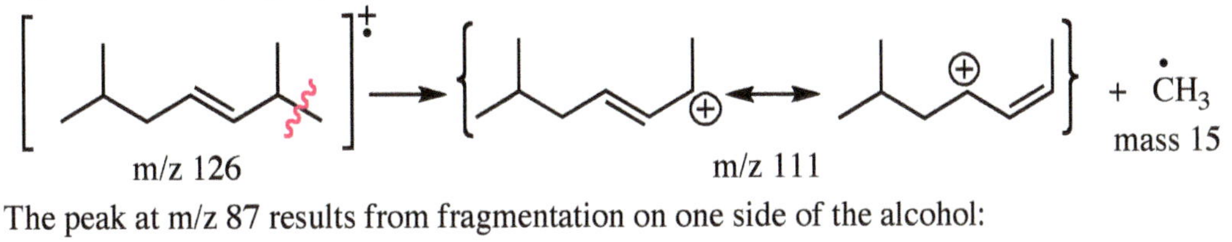

m/z 126 $\longrightarrow$ m/z 111 $+ \dot{C}H_3$ mass 15

The peak at m/z 87 results from fragmentation on one side of the alcohol:

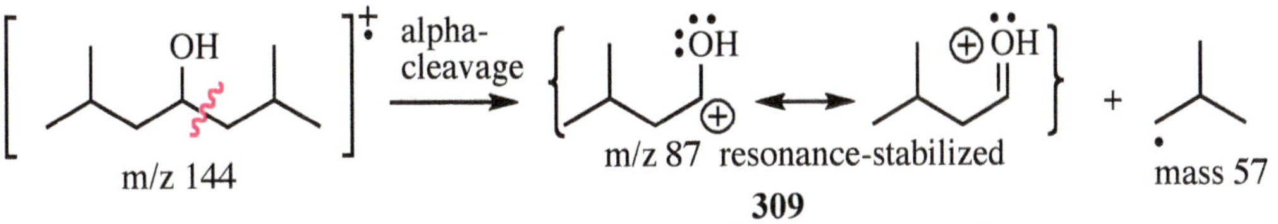

m/z 144 alpha-cleavage $\longrightarrow$ m/z 87 resonance-stabilized + mass 57

12-12 Table 12-2 and text Appendix 2 are particularly helpful in this problem. All values have units of cm^{-1}.
(a) C=C stretch between 1640 and 1680, weak or absent because it is symmetrically substituted, so there is essentially no dipole change; the =C—H stretch appears just above 3000.
(b) broad, strong O—H stretch centered around 3300
(c) strong C=O stretch about 1710
(d) C≡C stretch below 2200; ≡C—H stretch , sharp, around 3300
(e) broad N—H stretch of medium strength centered around 3300, with one spike
(f) broad, strong O—H stretch centered around 3000, covering the C—H stretch, often with a shoulder around 2700; strong C=O stretch around 1710
(g) sharp, strong C≡N stretch around 2200
(h) strong C=O stretch about 1735; often a strong C—O stretch around 1200
(i) strong C=O stretch around 1650—sometimes amides show two peaks; strong N—H stretch around 3300 with two spikes

12-13 Divide the numbers into 10,000 to arrive at the answer.

(a) 1603 cm^{-1} (b) 2959 cm^{-1} (c) 1709 cm^{-1} (d) 1739 cm^{-1} (e) 2212 cm^{-1} (f) 3300 cm^{-1}

12-14

(a) $H_2C=CHCH_2CH_3$ (C=C) 1660 cm^{-1} or O=CH–CH_2CH_3 (C=O) 1725 cm^{-1} **stronger absorption**—larger dipole

(b)

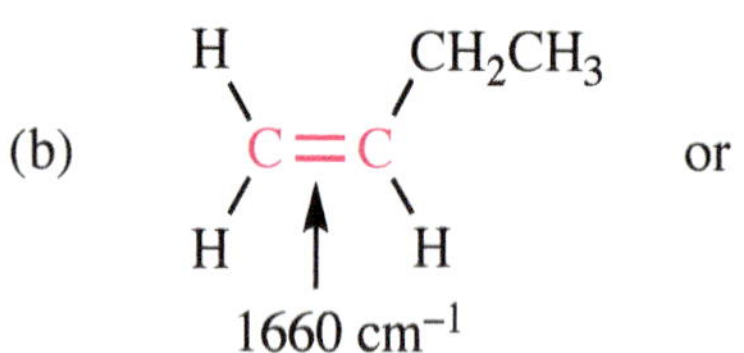

$H_2C=CHCH_2CH_3$ 1660 cm^{-1} or $H_2C=CHOCH_2CH_3$ 1640 cm^{-1} **stronger absorption**—larger dipole

(c)

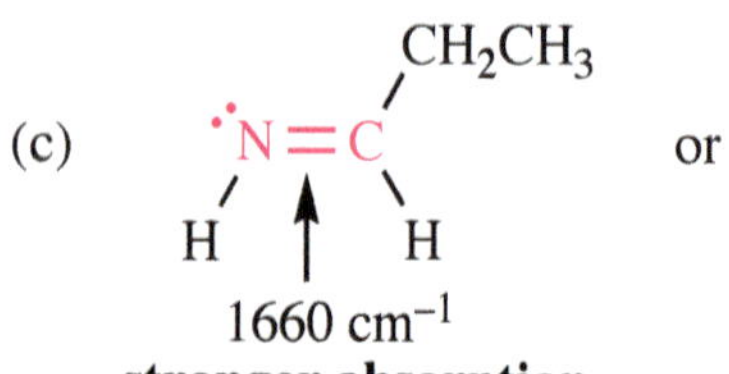

HN=CH–CH_2CH_3 1660 cm^{-1} **stronger absorption**—larger dipole or $H_2C=CHCH_2CH_3$ 1660 cm^{-1}

(d) H_3C–CH=CH–CH_3 1660 cm^{-1} (no dipole moment) or $H_2C=CHCH_2CH_3$ 1660 cm^{-1} **stronger absorption**—larger dipole

12-15

(a) $(H_3C)_2C=C(CH_3)_2$ 1660 cm^{-1} weak or non-existent and $H_2C=C(CH_3)CH(CH_3)_2$ 3000-3100 cm^{-1} (=C–H); 1660 cm^{-1} moderate intensity

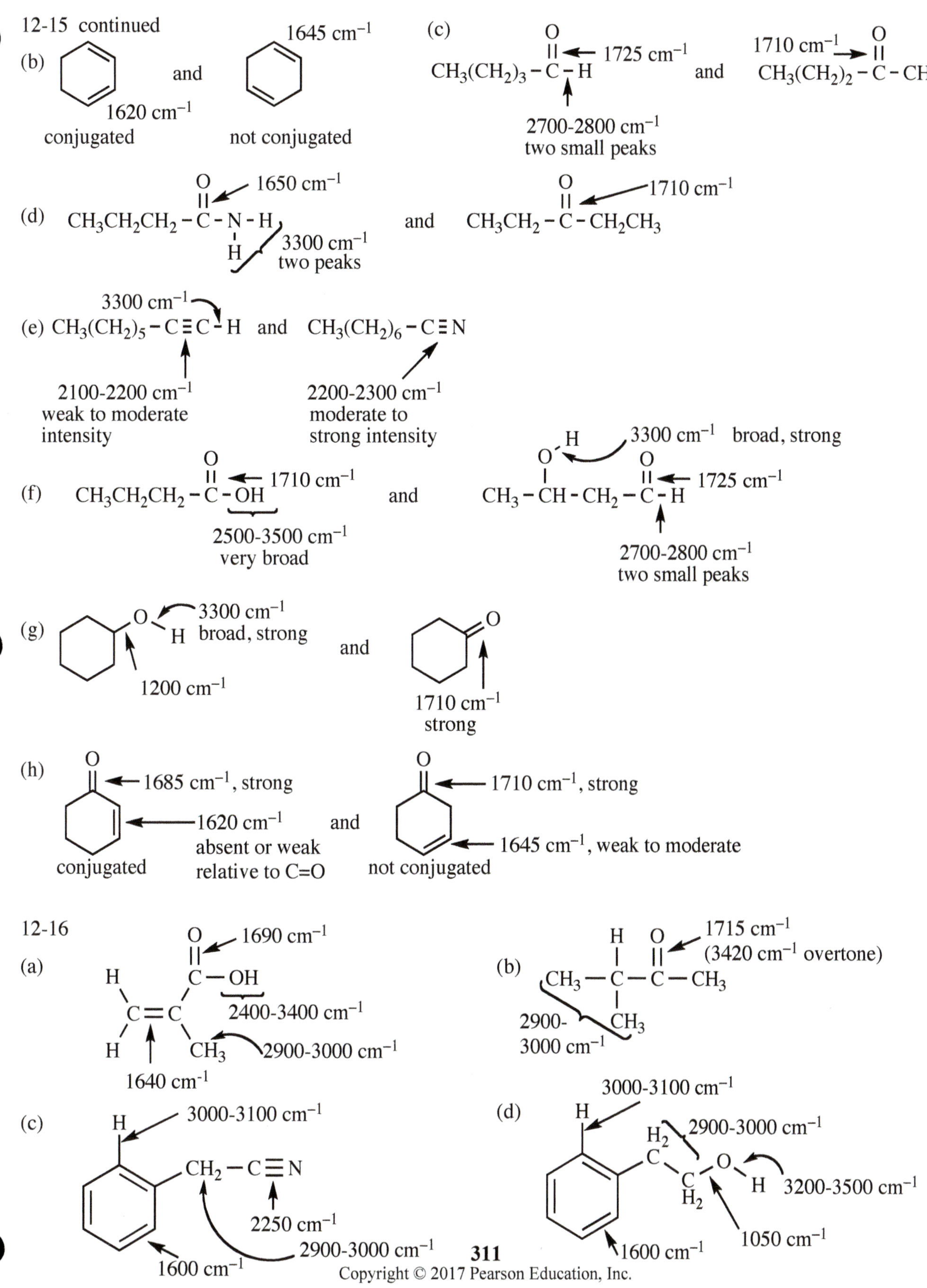
12-15 continued
(b)
1645 cm−1
and
1620 cm−1
conjugated
not conjugated
(c)
O
1725 cm−1
CH3(CH2)3 − C − H
2700-2800 cm−1
two small peaks
and
1710 cm−1
O
CH3(CH2)2 − C − CH3
(d)
O
1650 cm−1
CH3CH2CH2 − C − N − H
H
3300 cm−1
two peaks
and
O
1710 cm−1
CH3CH2 − C − CH2CH3
3300 cm−1
(e) CH3(CH2)5 − C ≡ C − H and CH3(CH2)6 − C ≡ N
2100-2200 cm−1
weak to moderate
intensity
2200-2300 cm−1
moderate to
strong intensity
(f)
O
1710 cm−1
CH3CH2CH2 − C − OH
2500-3500 cm−1
very broad
and
H
3300 cm−1 broad, strong
O
O
1725 cm−1
CH3 − CH − CH2 − C − H
2700-2800 cm−1
two small peaks
(g)
O
3300 cm−1
H broad, strong
1200 cm−1
and
O
1710 cm−1
strong
(h)
O
1685 cm−1, strong
1620 cm−1
absent or weak
relative to C=O
conjugated
and
O
1710 cm−1, strong
1645 cm−1, weak to moderate
not conjugated
12-16
(a)
O
1690 cm−1
H
C — OH
C = C
2400-3400 cm−1
H
CH3
2900-3000 cm−1
1640 cm−1
(b)
H O
1715 cm−1
(3420 cm−1 overtone)
CH3 — C — C — CH3
2900-
3000 cm−1
CH3
(c)
H
3000-3100 cm−1
CH2 — C ≡ N
2250 cm−1
2900-3000 cm−1
1600 cm−1
(d)
3000-3100 cm−1
H
H2
2900-3000 cm−1
C
C
O
H
3200-3500 cm−1
H2
1050 cm−1
1600 cm−1

12-17

(a) $\left[CH_3 - CH(CH_3) - CH_2CH_2CH_3\right]^{+\cdot}$ (cleavage labels 71, 43) m/z 86 $\longrightarrow$ $\oplus CH(CH_3) - CH_2CH_2CH_3$ m/z 71

$\longrightarrow$ $CH_3 - \overset{\oplus}{C}H - CH_3$ m/z 43

(b) $\left[CH_3\text{-}CH{=}C(CH_3)\text{-}CH_2 \text{-} CH_2CH_3\right]^{+\cdot}$ (cleavage label 69) m/z 98—allylic cleavage $\longrightarrow$ $\left\{CH_3\text{-}CH{=}C(CH_3)-\overset{\oplus}{C}H_2 \longleftrightarrow CH_3-\overset{\oplus}{C}H-C(CH_3){=}CH_2\right\}$ m/z 69

(c) $\left[CH_3 - CH(OH) - CH_2\text{-}CH(CH_3)\text{-}CH_3\right]^{+\cdot}$ (cleavage labels 87, 45) m/z 102

alpha-cleavage $\longrightarrow$ $\left\{\overset{\oplus}{C}H(\ddot{O}H)-CH_2\text{-}CH(CH_3)\text{-}CH_3 \longleftrightarrow CH({=}\overset{\oplus}{\ddot{O}}H)-CH_2\text{-}CH(CH_3)\text{-}CH_3\right\}$ m/z 87

alpha-cleavage $\longrightarrow$ $\left\{CH_3-\overset{\oplus}{C}H(\ddot{O}H) \longleftrightarrow CH_3-CH{=}\overset{\oplus}{\ddot{O}}H\right\}$ m/z 45

$\downarrow - H_2O$

$\left[CH_3\text{-}CH{=}CH-CH(CH_3)\text{-}CH_3\right]^{+\cdot}$ m/z 84

(d) [phenylalkane radical cation, cleavage labels 77, 57, 91, 43]$^{+\cdot}$ m/z 134 $\longrightarrow$ benzyl cation ($C_6H_5\overset{\oplus}{C}H_2$) $\xrightarrow{\text{rearrange}}$ tropylium ion m/z 91

The *tropylium ion* is a characteristic fragment from phenylalkanes. It has six more equivalent resonance forms—particularly stable.

$\longrightarrow$ phenyl cation ($C_6H_5^{\oplus}$) m/z 77

$\longrightarrow$ $H\overset{\oplus}{C}(CH_3)-CH_3$ m/z 43

$\longrightarrow$ $H_2\overset{\oplus}{C}-CH(CH_3)_2$ $\xrightarrow{\text{rearrange}}$ $H_3C-\overset{\oplus}{C}(CH_3)-CH_3$ m/z 57

(e) [cyclohexyl isopropyl ether radical cation, cleavage labels 83, 43, 127]$^{+\cdot}$ m/z 142 $\longrightarrow$ cyclohexyl cation ($\oplus CH$) m/z 83

$\longrightarrow$ $H\overset{\oplus}{C}(CH_3)-CH_3$ m/z 43

$\downarrow$ alpha-cleavage

$\left\{C_6H_{11}-\ddot{O}-\overset{\oplus}{C}H(CH_3) \longleftrightarrow C_6H_{11}-\overset{\oplus}{\ddot{O}}{=}CH(CH_3)\right\}$ m/z 127

12-17 continued

(f) $[CH_3CH_2-CH_2-NH-C(CH_3)_2CH_3]^{+\bullet}$ (cleavage labels: 57, 86, 100)

m/z 115

$\rightarrow \{H_2C^{+}-\ddot{N}H-C(CH_3)_2-CH_3 \leftrightarrow H_2C{=}N^{+}H-C(CH_3)_2-CH_3\}$

m/z 86

$^{+}C(CH_3)_2-CH_3$ m/z 57

$\{CH_3CH_2-CH_2-\ddot{N}H-C^{+}(CH_3)_2 \leftrightarrow CH_3CH_2-CH_2-N^{+}H{=}C(CH_3)_2\}$ m/z 100

(g) $[C_6H_5-C(=O)-CH_3]^{+\bullet}$ (cleavage labels: 77, 105, 43)

m/z 120

$\rightarrow \{C_6H_5-C^{+}{=}\ddot{O}: \leftrightarrow C_6H_5-C{\equiv}O^{+}: \leftrightarrow$ plus other resonance forms with (+) on ring$\}$

m/z 105

$\{:\ddot{O}{=}C^{+}-CH_3 \leftrightarrow {}^{+}O{\equiv}C-CH_3\}$ m/z 43

These are three alpha-cleavages.

$C_6H_5^{+}$ m/z 77

(h) $[(CH_3)_2CH-CH(Br)-CH_2-CH_3]^{+\bullet}$ (cleavage labels: 121,123; 43; 85)

m/z 164, 166

$\rightarrow \{HC^{+}(-\ddot{Br}:)-CH_2-CH_3 \leftrightarrow HC({=}Br^{+}:)-CH_2-CH_3\}$

m/z 121, 123

$(CH_3)_2CH^{+}$ m/z 43

$(CH_3)_2CH-C^{+}H-CH_2-CH_3$ m/z 85

Any fragment containing Br will reflect the two isotopes of Br with approximately 50:50 natural abundance.

12-18

(a) $[CH_3(CH_2)_6CH_3]^{+\bullet}$ (cleavage labels: 71, 57, 85)

m/z 114

$\rightarrow$ $CH_3(CH_2)_4CH_2^{+}$ m/z 85 + $H_2\dot{C}-CH_3$ mass 29

$CH_3(CH_2)_3CH_2^{+}$ m/z 71 + $H_2\dot{C}CH_2CH_3$ mass 43

$CH_3(CH_2)_2CH_2^{+}$ m/z 57 + $H_2\dot{C}CH_2CH_2CH_3$ mass 57

12-18 continued

(b)

$$\left[\text{methylcyclohexane}\right]^{+\cdot} \text{ m/z 98} \longrightarrow \text{cyclohexyl cation (m/z 83)} + \cdot CH_3 \text{ (mass 15)}$$

(c)

$$\left[CH_3-\overset{CH_3}{\overset{|}{C}}=CH-CH_2 \mid CH_3\right]^{+\cdot} \text{ m/z 84 (cleavage at 69)} \longrightarrow \left\{CH_3-\overset{CH_3}{\overset{|}{C}}=CH-\overset{\oplus}{C}H_2 \longleftrightarrow CH_3-\overset{CH_3}{\overset{|}{\underset{\oplus}{C}}}-CH=CH_2\right\} \text{ m/z 69} + \cdot CH_3 \text{ mass 15}$$

(d)

$$\left[\overset{OH}{\overset{|}{CH_2}} \mid CH_2-CH_2CH_2CH_3\right]^{+\cdot} \text{ m/z 88 (cleavage at 31)} \longrightarrow \left\{\overset{:\ddot{O}H}{\overset{|}{\underset{\oplus}{CH_2}}} \longleftrightarrow \overset{\oplus\ddot{O}H}{\overset{||}{CH_2}}\right\} \text{ m/z 31} + \cdot CH_2CH_2CH_2CH_3 \text{ mass 57}$$

$\downarrow -H_2O$

$$\left[CH_2=CH-CH_2 \mid CH_2 \mid CH_3\right]^{+\cdot} \text{ m/z 70 (cleavages at 41 and 55)} \longrightarrow CH_2=CH-CH_2-\overset{\oplus}{C}H_2 \text{ m/z 55} + \cdot CH_3 \text{ mass 15}$$

$\downarrow$

$$\left\{CH_2=CH-\overset{\oplus}{C}H_2 \longleftrightarrow \overset{\oplus}{C}H_2-CH=CH_2\right\} \text{ m/z 41} + \cdot CH_2CH_3 \text{ mass 29}$$

(e)

$$\left[C_6H_5 \mid NH-CH_2 \mid CH_3\right]^{+\cdot} \text{ m/z 121 (cleavages at 77 and 106)} \longrightarrow C_6H_5^{\oplus} \text{ m/z 77} + \cdot NHCH_2CH_3 \text{ mass 44}$$

alpha-cleavage

$$\left\{C_6H_5-\ddot{N}H-\overset{\oplus}{C}H_2 \longleftrightarrow C_6H_5-\overset{\oplus}{N}H=CH_2\right\} \text{ m/z 106} + \cdot CH_3 \text{ mass 15}$$

12-18 continued

(f) [Br–CH₂–CH(CH₃)–CH₂–CH₃]⁺• (cleavages labeled 121,123 and 71), m/z 150, 152 → $Br-CH_2-\overset{+}{C}H-CH_3$ m/z 121, 123 + $\bullet CH_2CH_3$ mass 29

Any fragment containing Br will reflect the two isotopes of Br with approximately 50:50 natural abundance.

→ $\overset{+}{C}H_2-CH(CH_3)-CH_2-CH_3$ m/z 71 + $\bullet Br$ mass 79, 81

12-19

(a) The characteristic frequencies of the OH absorption and the C=C absorption will indicate the presence or absence of the groups. A spectrum with an absorption around 3300 cm^{-1} will have some cyclohexanol in it; if that same spectrum also has a peak at 1645 cm^{-1}, then the sample will also contain some cyclohexene. Pure samples will have peaks representative of only one of the compounds and not the other. Note that *quantitation* of the two compounds would be very difficult by IR because the strength of absorptions are very different. Usually, other methods are used in preference to IR for quantitative measurements.

Cyclohexanol: O–H 3300 cm^{-1} broad, strong; C–O 1200 cm^{-1}

Cyclohexene: C=C 1645 cm^{-1} moderate

(b) Mass spectrometry can be misleading with alcohols. Usually, alcohols dehydrate in the inlet system of a mass spectrometer, and the characteristic peaks observed in the mass spectrum are those of the alkene, not of the parent alcohol. For this particular analysis, mass spectrometry would be unreliable and perhaps misleading.

12-20

(a) The "student prep" compound must be 1-bromobutane. The most obvious feature of the mass spectrum is the pair of peaks at M and M+2 of approximately equal heights, characteristic of a bromine atom. Loss of bromine (79) from the molecular ion at 136 gives a mass of 57, C_4H_9, a butyl group. Which of the four possible butyl groups is it? The peaks at 107 (loss of 29, C_2H_5) and 93 (loss of 43, C_3H_7) are consistent with a linear chain, not a branched chain. A branched chain is more likely to lose CH_3 (loss of 15).

(b) The base peak at 57 is so strong because the carbon-halogen bond is the weakest in the molecule. Typically, loss of a halogen is the dominant fragmentation in alkyl halides.

[1-bromobutane]⁺• (cleavages labeled 93, 57, 107), m/z 136 → $\overset{+}{C}H_2CH_2Br$ m/z 107 + $H_2\dot{C}-CH_3$ mass 29

{$\overset{+}{C}H_2-\ddot{Br}:$ ↔ $H_2C=\overset{+}{Br}:$} m/z 93 + $H_2\dot{C}CH_2CH_3$ mass 43

$CH_3CH_2CH_2\overset{+}{C}H_2$ m/z 57 + $\bullet Br$ mass 79

12-21

(a) Deuterium has twice the mass of hydrogen, but similar spring constant, k. Compare the frequency of C—D vibration to C—H vibration by setting up a ratio, changing only the mass (substitute 2m for m).

$$\frac{\nu_D}{\nu_H} = \frac{\sqrt{k/2m}}{\sqrt{k/m}} = \frac{\sqrt{1/2}\ \sqrt{k/m}}{\sqrt{k/m}} = \sqrt{1/2} = 0.707$$

$\nu_D = 0.707\ \nu_H = 0.707\ (3000\ cm^{-1}) \approx$ **2100 cm^{-1}**

(b) The functional group most likely to be confused with a C—D stretch is the alkyne (carbon-carbon triple bond), which appears in the same region and is often very weak.

(If you have had physics, you may recognize this version of Hooke's Law that describes the motion of weights on a spring.)

12-22

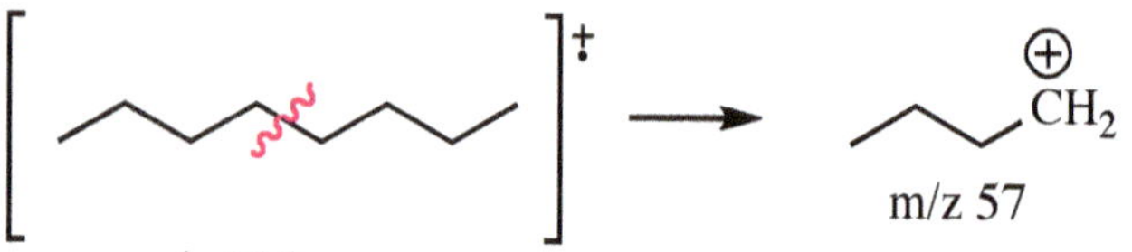

m/z 114 → $CH_2^{\oplus}$ m/z 57 a 1° carbocation—not favorable

m/z 114 → $CH^{\oplus}$ m/z 57 a 2° carbocation—reasonable

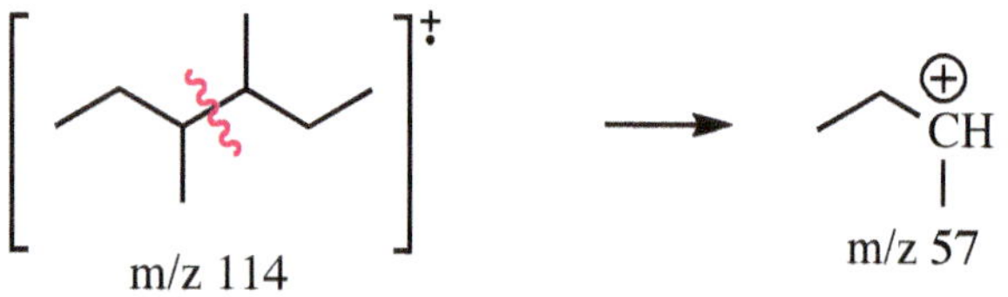

m/z 114 → $C^{\oplus}$ m/z 57 a 3° carbocation—the best

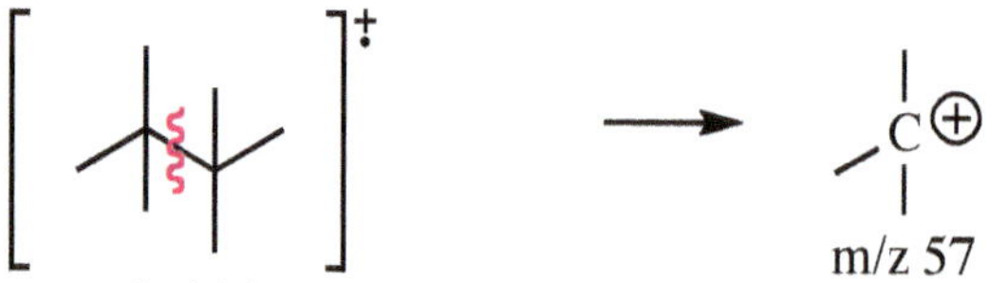

The most likely fragmentation of 2,2,3,3-tetramethylbutane will give a 3° carbocation, the most stable of the common alkyl cations. The molecular ion should be small or non-existent while m/z 57 is likely to be the base peak, whereas the molecular ion peaks will be more prominent for octane and for 3,4-dimethylhexane.

12-23

(a) The information that this mystery compound is a hydrocarbon makes interpreting the mass spectrum much easier. (It is relatively simple to tell if a compound has chlorine, bromine, or nitrogen by a mass spectrum, but oxygen is difficult to determine by mass spectrometry alone.) A hydrocarbon with a molecular ion of 110 can have only 8 carbons (8 x 12 = 96) and 14 hydrogens. The formula C_8H_{14} has two elements of unsaturation.

(b) The IR will be useful in determining what the elements of unsaturation are. Cycloalkanes are generally not distinguishable in the IR. An alkene should have an absorption around 1600-1650 cm^{-1}; none is present in this IR. An alkyne should have a small, sharp peak around 2200 cm^{-1}—PRESENT AT 2120 cm^{-1}! Also, a sharp peak around 3300 cm^{-1} indicates a hydrogen on an alkyne, so the alkyne is at one end of the molecule. Both elements of unsaturation are accounted for by the alkyne. (Also, alkynes smell bad, perhaps because of that linear geometry poking the nose like a sharp stick!)

(c) The only question is how are the other carbons arranged. The mass spectrum shows a progression of peaks from the molecular ion at 110 to 95 (loss of CH_3), to 81 (loss of C_2H_5), to 67 (loss of C_3H_7). The mass spectrum suggests it is a linear chain. The extra evidence that hydrogenation of the mystery compound gives *n*-octane verifies that the chain is linear. The original compound must be oct-1-yne.

12-23 continued

(d) The m/z 39 peak is so strong because the ion produced is stabilized by resonance.

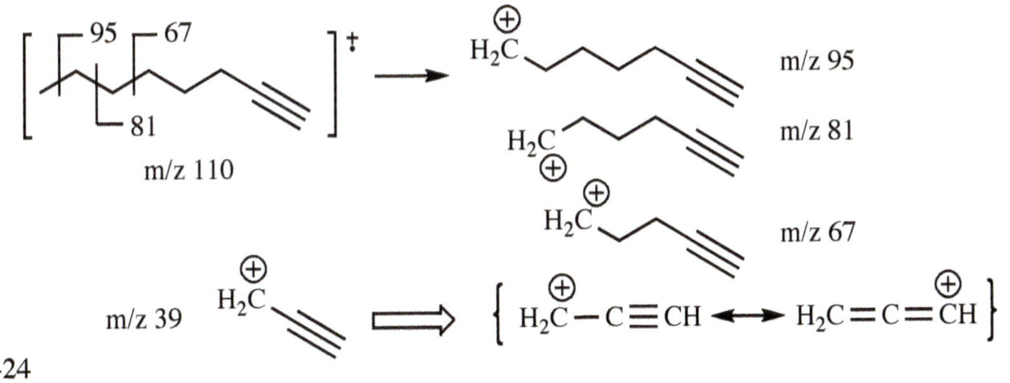

95 67 81

m/z 110

m/z 95

m/z 81

m/z 67

m/z 39

$\{ H_2\overset{\oplus}{C}-C{\equiv}CH \longleftrightarrow H_2C{=}C{=}\overset{\oplus}{C}H \}$

12-24

(a) and (b) The mass spec is consistent with the formula of the alkyne, C_8H_{14}, mass 110. The IR is not consistent with the alkyne, however. Often, symmetrically substituted alkynes have a miniscule C≡C peak, so the fact that the IR does not show this peak does not prove that the alkyne is absent. The important evidence in the IR is the significant peak at 1620 cm^{-1} and the =C—H absorption above 3000 cm^{-1}; this absorption is characteristic of a conjugated diene. Instead of the alkyne being formed, the reaction must have been a double elimination to the diene.

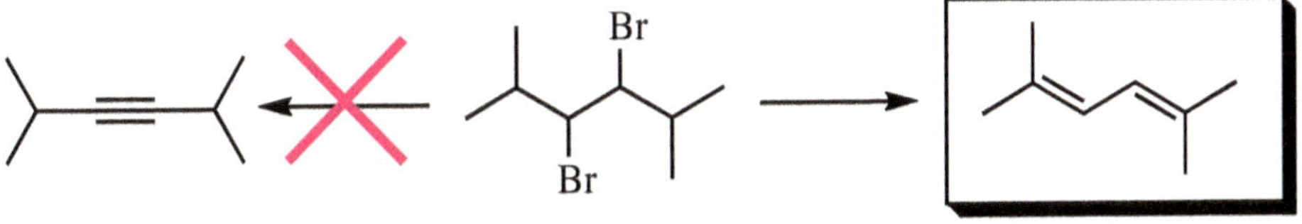

Br

Br

12-25

(a) $H_3C-C(CH_3)_2-C{\equiv}C-H$

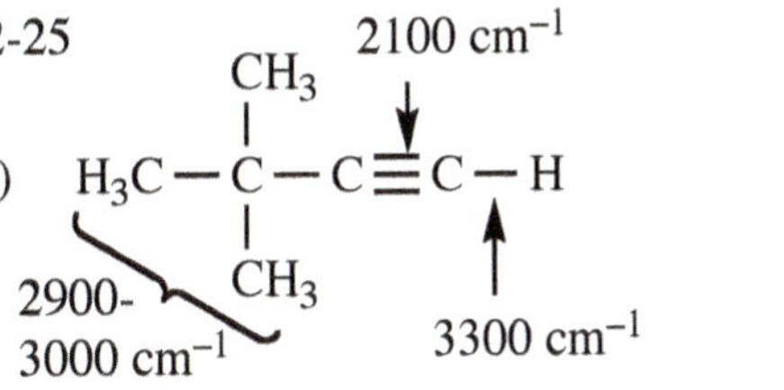

2100 cm^{-1} (C≡C); 3300 cm^{-1} (≡C—H); 2900-3000 cm^{-1} (C—H)

(b) $C_6H_5-CH_2-CHO$

1720 cm^{-1} (overtone at 3440 cm^{-1}); 2710, 2820 cm^{-1}; 1600 cm^{-1}

(c) $C_6H_5-O-C(=O)-CH_3$

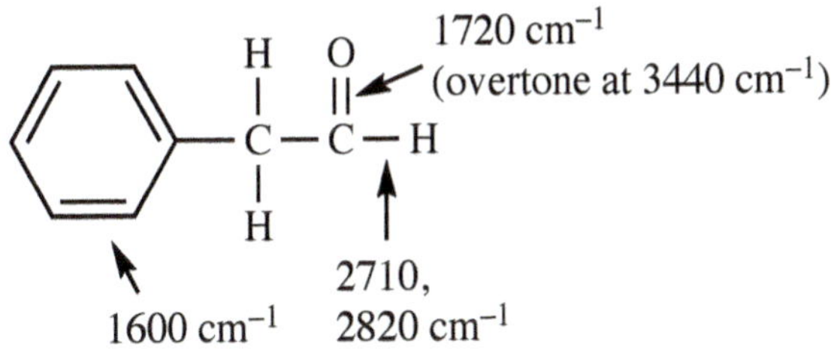

1760 cm^{-1}; 1200 cm^{-1}; 1590 cm^{-1}

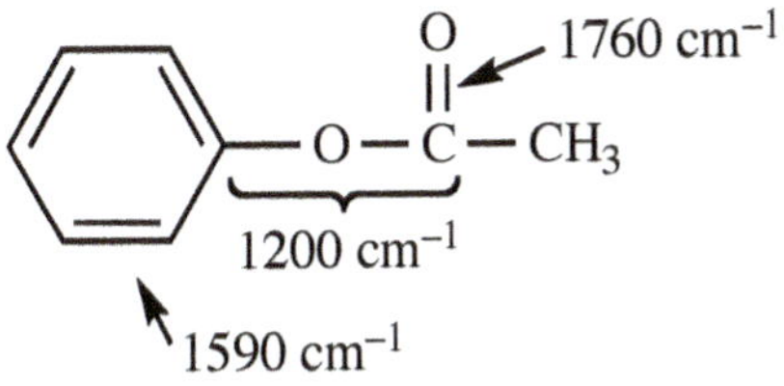

12-26

(a)

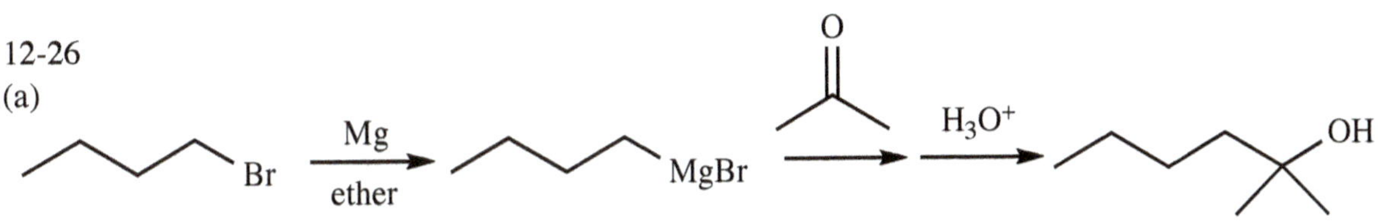

Br $\xrightarrow[\text{ether}]{\text{Mg}}$ MgBr $\xrightarrow{\text{acetone}}$ $\xrightarrow{H_3O^+}$ OH

2-methylhexan-2-ol

$C_7H_{16}O$ mol. wt. 116

12-26 continued

(b) The molecular ion is not visible in the spectrum. Alcohols typically dehydrate in the hot inlet system of the mass spectrometer, especially true for 3° alcohols that are the easiest type to dehydrate. The two fragmentations that produce a resonance-stabilized carbocation give the major peaks in the spectrum at m/z 59 and 101.

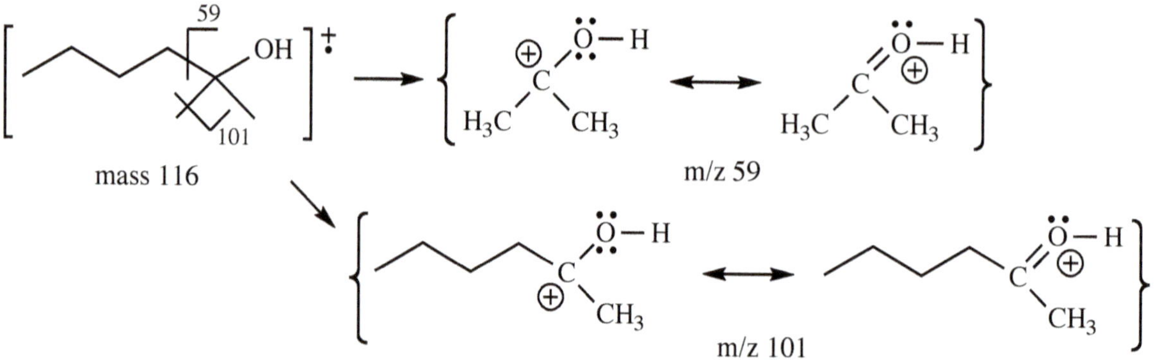

59

OH

101

mass 116

H_3C C CH_3 ↔ H_3C C CH_3

m/z 59

CH_3 ↔ CH_3

m/z 101

This is a good example of how the peak at m/z 101 was probably NOT the molecular ion, because there are no significant even mass peaks that come from it. The peak at 101 must have been a fragment peak.

12-27 The unknown compound has peaks in the mass spec at m/z 198, 155, 127, 71, and 43. It is helpful that the masses of two of the fragments, 155 and 43, sum to 198, as do the other two fragment masses, 127 and 71. We can say with certainty that the molecular ion is at m/z 198, and that the unknown is a relatively simple molecule with two main fragmentations.

This is a fairly high mass for a simple compound; some heavy group must be present. What is NOT present is N because of the even molecular ion mass, nor Cl nor Br because of the lack of isotope peaks, nor a phenyl group because of the absence of a peak at 77. The progression of alkyl group masses: 15, 29, **43**, 57, **71**, 85, 99—includes two of the peaks, so it appears that the unknown contains a propyl group and a pentyl group (the propyl could be part of the pentyl group). The 127 fragment is key; the fragment C_9H_{19} has this mass, but we would expect much more fragmentation from a nine carbon piece. There must be some other explanation for this 127 peak.

And there is! There is one piece—more specifically, one atom—that has mass 127: iodine! In all probability, the iodine atom is attached to a fragment of mass 71 which is C_5H_{11}, a pentyl group. We cannot tell with certainty what isomer it is, so unless there is some other evidence, let's propose a branched chain isomer, 1-iodo-2-methylbutane.

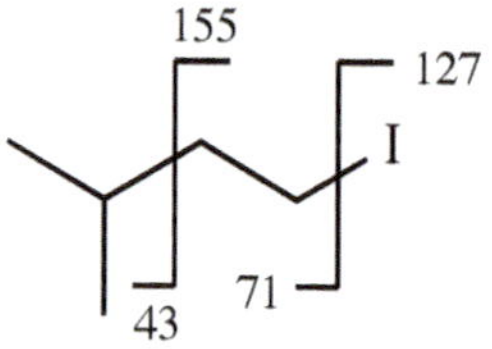

155 127 I 71 43

The 155 fragment probably has this bridged structure because iodine is so big:

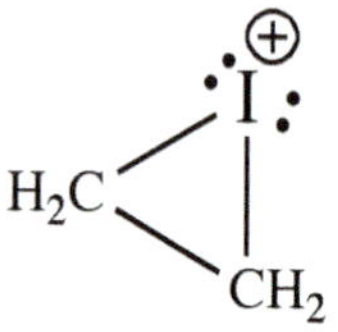

H_2C I CH_2

12-28

(a) $H_3C-C(CH_3)(OH)-C(CH_3)(OH)-CH_3$

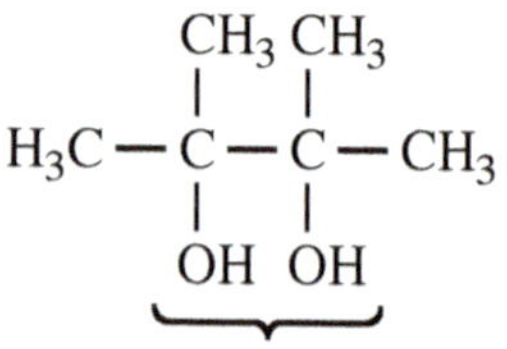

broad, strong OH peak around 3300 cm^{-1}

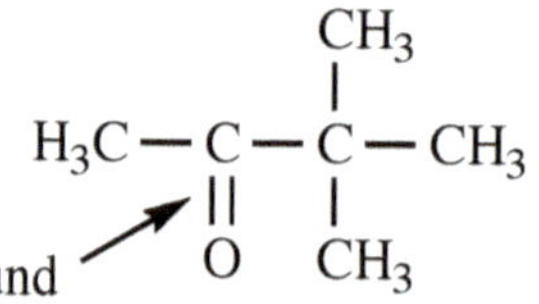

$H_3C-C(=O)-C(CH_3)_2-CH_3$

strong peak around 1715 cm^{-1}

12-28 continued

(b)

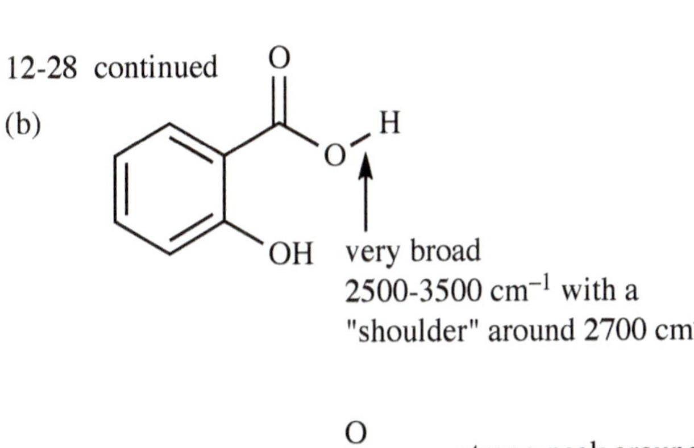

O
CH$_3$
O
OH
broad COOH band missing
still has a strong OH at 3300 cm^{-1}
but different from COOH

(c)

O
1620 cm^{-1}
strong peak around
1690 cm^{-1}
H
two peaks at
2700 and 2800 cm^{-1}
1600 cm^{-1}

OH
1640 cm^{-1}
broad, strong OH peak
around 3300 cm^{-1}
1610-1620 cm^{-1} —less conjugation

12-29

Use the MS to determine how large a molecule the unknown is. With the highest peak at m/z 121, an odd molecular ion suggests the presence of N. If one N, then 121 – 14 = 107 for C and H: the only plausible formula is $C_8H_{11}N$, unless oxygen or some other element is present. If $C_8H_{11}N$, this formula has 4 elements of unsaturation, the minimum needed for a benzene ring. The mass spectrum shows a peak at 77, the phenyl group, so a monosubstituted benzene ring is part of the structure. The other significant peak in the MS is at 106, M – 15, so a methyl group is present.

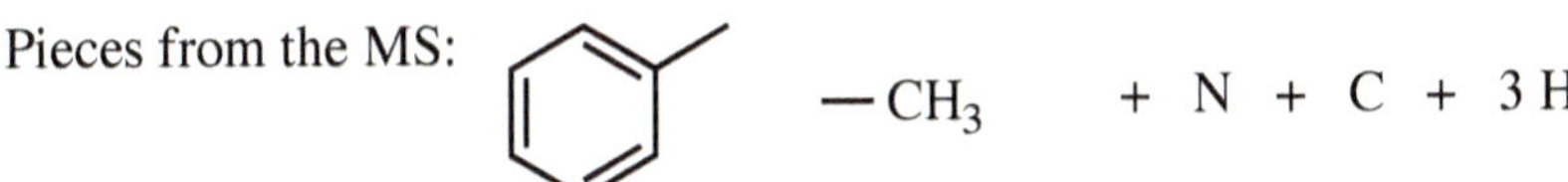

From the IR: The most important peak is the single peak at 3400 cm^{-1} showing the presence of N—H. The only two possible structures are:

H
N
CH$_2$CH$_3$

OR

H$_2$
C
N
H
CH$_3$

This structure would have lost CH$_3$ as its major fragment, predicting that m/z 106 should be the base peak—it is! This must be the correct structure.

This structure would have a m/z 91 for the benzyl group much higher than the m/z 106. This structure does not fit the data.

12-30

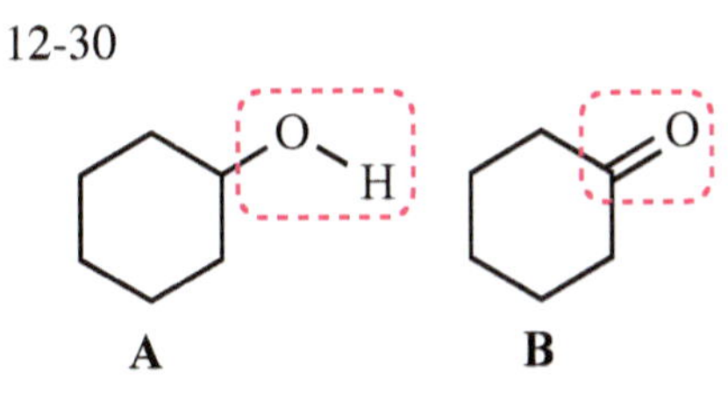

H
N
H
C

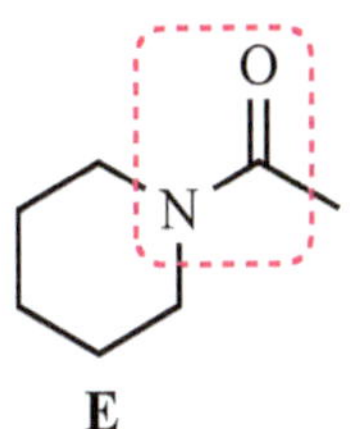

O
N
E

A: (2) very broad, centered about 3330 cm^{-1}

B: (5) strong, sharp 1717 cm^{-1}

C: (4) broad with spikes at 3367 and 3292 cm^{-1}

D: (1) sharp, 2254 cm^{-1}

E: (3) strong, slightly broadened, 1645 cm^{-1}

12-31 Resonance forms of fragments are not shown.

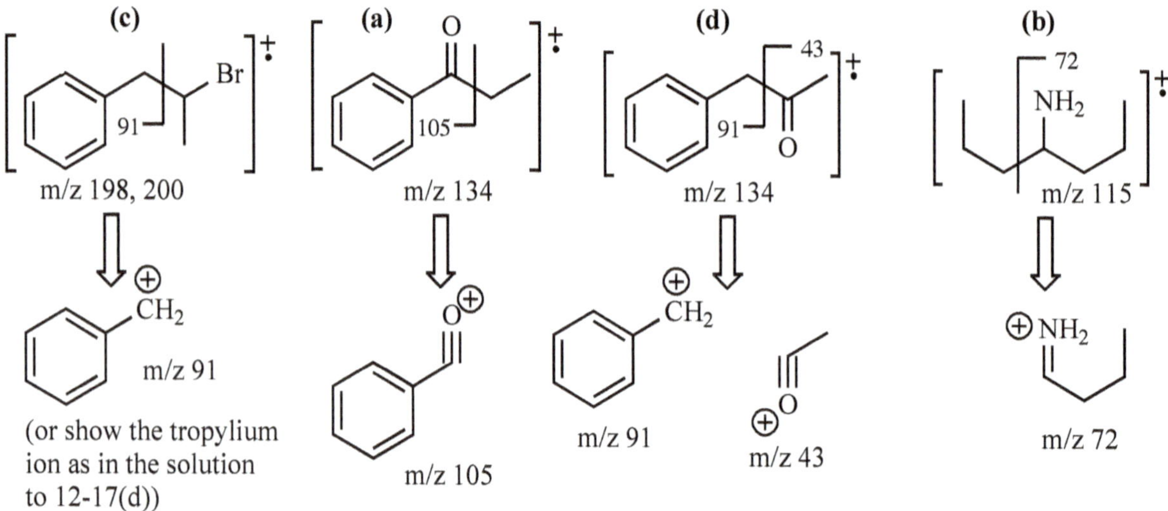

If you wish to find IR spectra and mass spectra of common compounds, there are two websites that are very helpful. Entering a name or molecular formula will give isomers from which to choose the desired structure and the IR spectrum or MS if available in their database.

http://webbook.nist.gov/
"NIST" is the U.S. National Institute of Standards and Technology.

http://sdbs.db.aist.go.jp/sdbs/cgi-bin/cre_index.cgi?lang=eng
This is from the National Institute of Advanced Industrial Science and Technology of Japan.

CHAPTER 13—NUCLEAR MAGNETIC RESONANCE SPECTROSCOPY

A benzene ring can be written with three alternating double bonds or with a circle in the ring. All of the carbons and hydrogens in an unsubstituted benzene ring are equivalent, regardless of the symbolism used.

 equivalent to 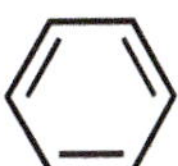

The Japanese website listed at the bottom of p. 320 also gives proton and carbon NMR spectra.

Reminder: The word "spectrum" is singular; the word "spectra" is plural.

13-1

(a) $\frac{650\ \text{Hz}}{300 \times 10^6\ \text{Hz}} = 2.17 \times 10^{-6} = 2.17$ ppm downfield from TMS

(b) The chemical shift does not change with field strength: δ 2.17 at both 60 MHz and 300 MHz.

(c) (2.17 ppm) x (60 MHz) = $(2.17 \times 10^{-6}) \times (60 \times 10^6\ \text{Hz}) = 130$ Hz

13-2 Numbers are chemical shift values, in ppm, derived from Table 13-3 and Appendix 1 in the text. *Your predictions should be in the given range, or within 0.5-1.0 ppm of the given value.*

(a) $(CH_3)_3C$–CH=CH–$C(CH_3)_3$ (CH₃ groups labeled b; vinyl H labeled a)

a = δ 5-6
b = δ 0.9

(b)

a = δ 7.2
b = δ 2.3

(c) CH_3O–C_6H_4–OCH_3

a = δ 7.2
b = δ 3.6

(d) CH_3–C(CH_3)(OH)–C≡C–H (CH_3 labeled c; OH labeled a; ≡C–H labeled b)

a = δ 2-5
b = δ 2.5
c = δ 1-2

(e) C_6H_5–CH_2–CH_2–C(=O)–O–H

a = δ 10-12
b ≈ δ 2.5 (CH_3 next to C=O is around 2.1)
c ≈ δ 2.7 (CH_3 next to benzene is around 2.3)
} a CH_2 is about 0.4 greater than CH_3
d = δ 7.2

(The hydrogens labeled "d" are not equivalent. They appear at roughly the same chemical shift because the substituent is neither strongly electron-donating nor withdrawing.)

(f) CH_3–C(CH_3)(Br)–CH_2Br (CH_3 labeled b; CH_2 labeled a)

a = δ 3-4
b = δ 1-2

13-3

(a) $\overset{c}{CH_3}\overset{b}{CH_2}\overset{a}{CH_2}Cl$

three types of H

(b) $\overset{b}{CH_3}\overset{a}{CH}\overset{b}{CH_3}$ with Cl on the central carbon

two types of H

(c) $CH_3-C(CH_3)_2-CH_2CH_3$ (the three CH_3 groups on the quaternary carbon = a; CH_2 = b; terminal CH_3 = c)

three types of H

(d) $(H_3C)_2CH-CH(CH_3)_2$ (CH_3 = a; CH = b)

two types of H

(e) 4-bromotoluene: aromatic H ortho to Br = a; aromatic H ortho to CH_3 = b; CH_3 = c

three types of H

(f)

five types of H

13-4

(a) toluene: CH_3 = a; ortho H = b; meta H = c; para H = d

four types of H

(b) The three types of aromatic hydrogens appear in a relatively small space around δ 7.2. The signal is complex because all the peaks from the three types of hydrogens overlap.

Note: NMR spectra drawn in this Solutions Manual will represent peaks as single lines. These lines may not look like "real" peaks, but this avoids the problem of variation among spectrometers and printers. Individual spectra may look different from the ones presented here, but all of the important information will be contained in these representational spectra.

13-5

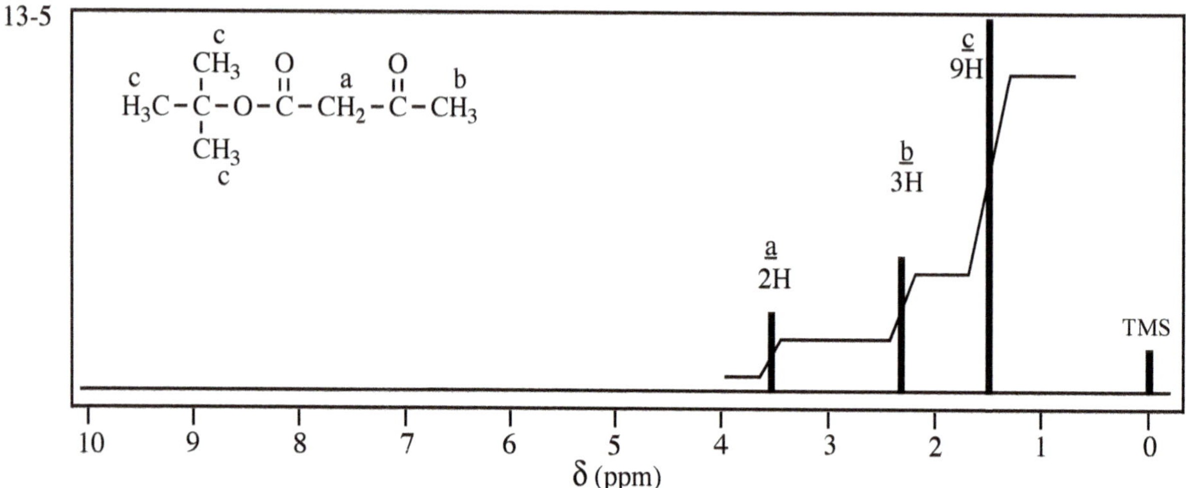

13-6 The three spectra are identified with their structures. Data are given as chemical shift values, with the integration ratios of each peak given in parentheses.

Spectrum (a)

$(CH_3)_2C(OH)-C\equiv C-H$ (C≡C–H = a; OH = b; CH_3 = c)

a = δ 2.4 (1) (1H)
b = δ 2.6 (1) (1H)
c = δ 1.5 (6) (6H)

Spectrum (b)

1,4-dimethoxybenzene: $CH_3O-C_6H_4-OCH_3$ (aromatic H = a; OCH_3 = b)

a = δ 6.8 (2) (4H)
b = δ 3.7 (3) (6H)

13-6 continued

Spectrum (c)

$$\begin{array}{c} \text{b} \\ \text{CH}_3 \\ | \\ \overset{\text{b}}{\text{CH}_3}-\text{C}-\overset{\text{a}}{\text{CH}_2}\text{Br} \\ | \\ \text{Br} \end{array}$$

a = δ 3.9 (1) (2H)
b = δ 1.9 (3) (6H)

(This is the compound in Problem 13-2(f). You may wish to check your answer to that question against the spectrum.)

13-7 Chemical shift values are approximate and may vary slightly from yours. The splitting and integration values should match exactly, however.

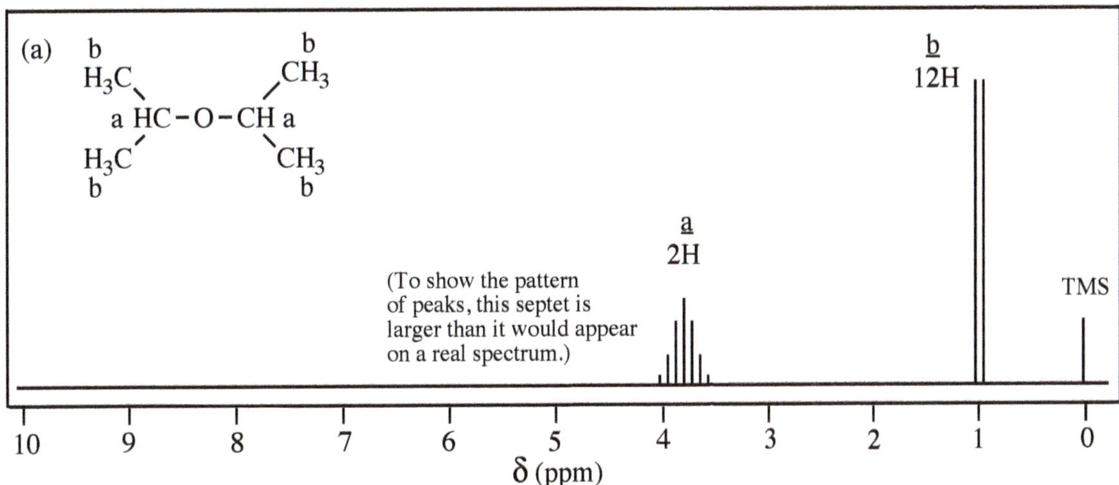

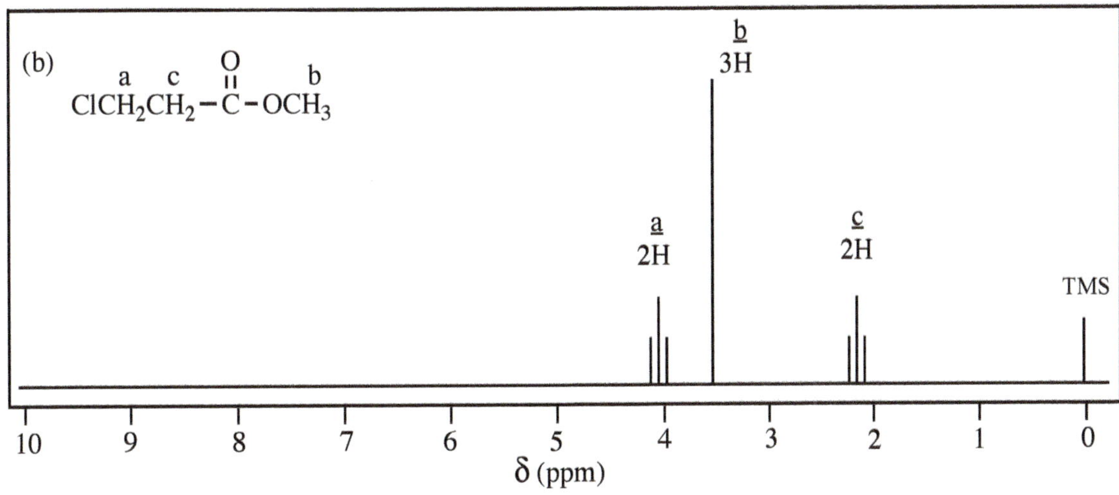

13-7 continued

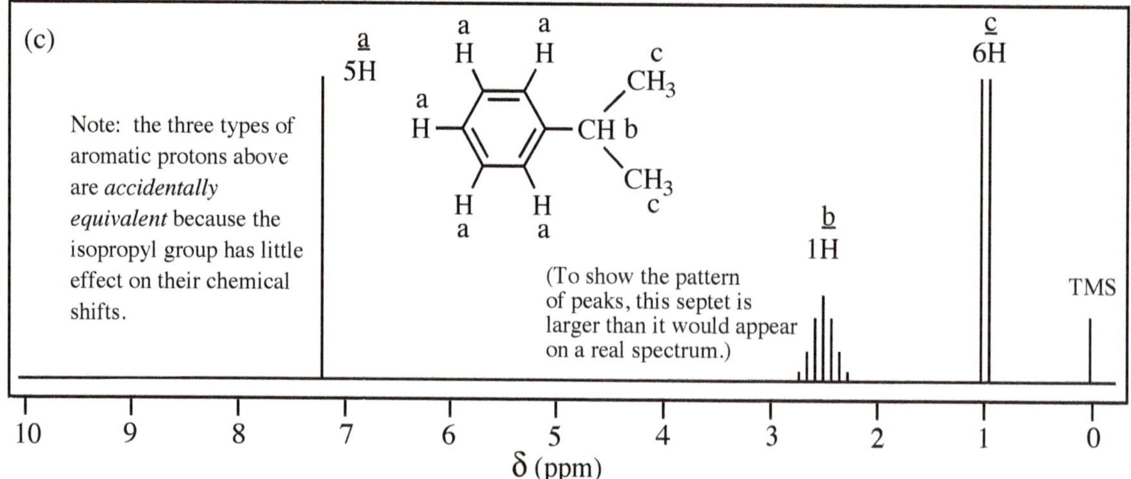

(c)
a
5H
Note: the three types of aromatic protons above are *accidentally equivalent* because the isopropyl group has little effect on their chemical shifts.
c
CH3
CH b
CH3
c
b
1H
(To show the pattern of peaks, this septet is larger than it would appear on a real spectrum.)
c
6H
TMS
10 9 8 7 6 5 4 3 2 1 0
δ (ppm)

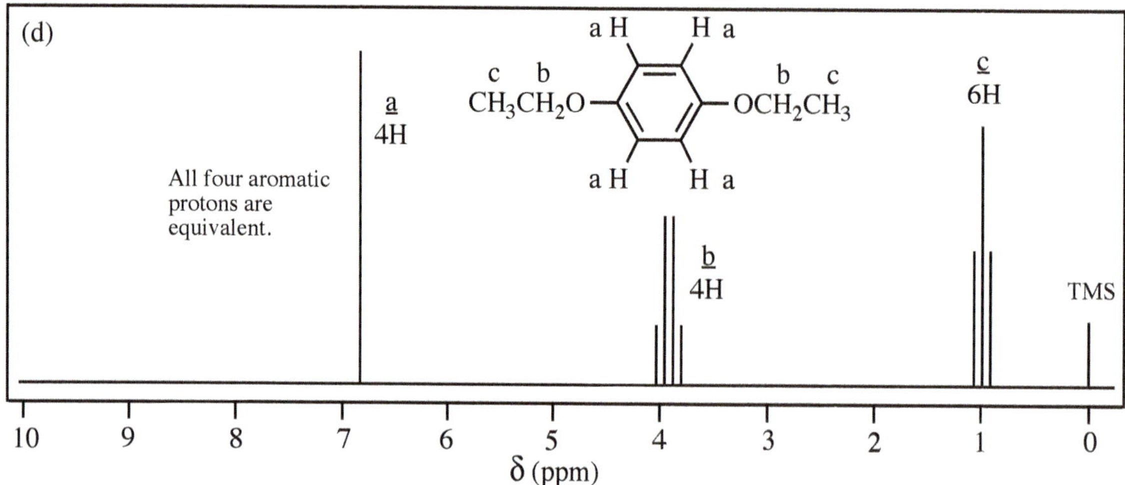

(d)
c b
CH3CH2O
b c
OCH2CH3
a
4H
All four aromatic protons are equivalent.
b
4H
c
6H
TMS
10 9 8 7 6 5 4 3 2 1 0
δ (ppm)

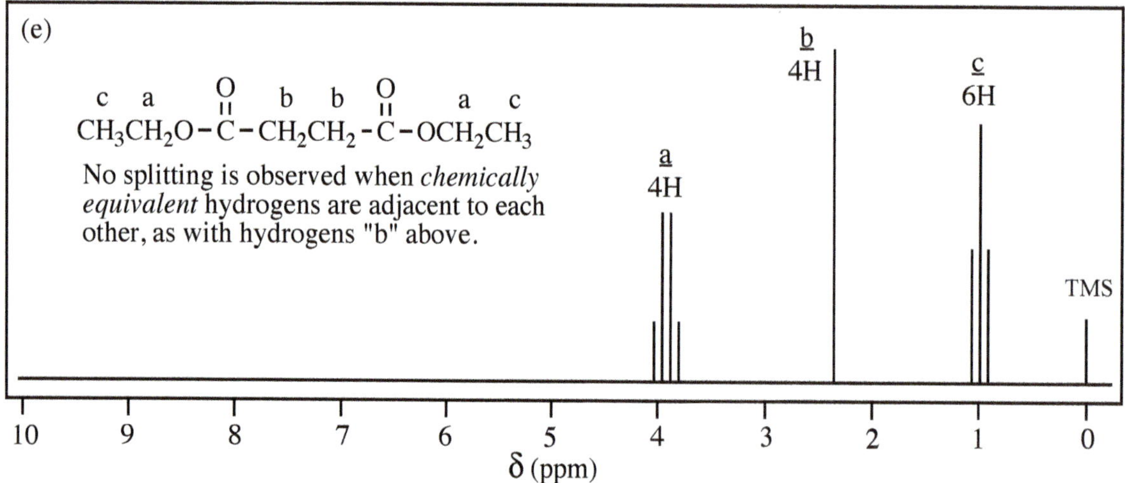

(e)
c a b b a c
CH3CH2O-C-CH2CH2-C-OCH2CH3
No splitting is observed when *chemically equivalent* hydrogens are adjacent to each other, as with hydrogens "b" above.
a
4H
b
4H
c
6H
TMS
10 9 8 7 6 5 4 3 2 1 0
δ (ppm)

13-8

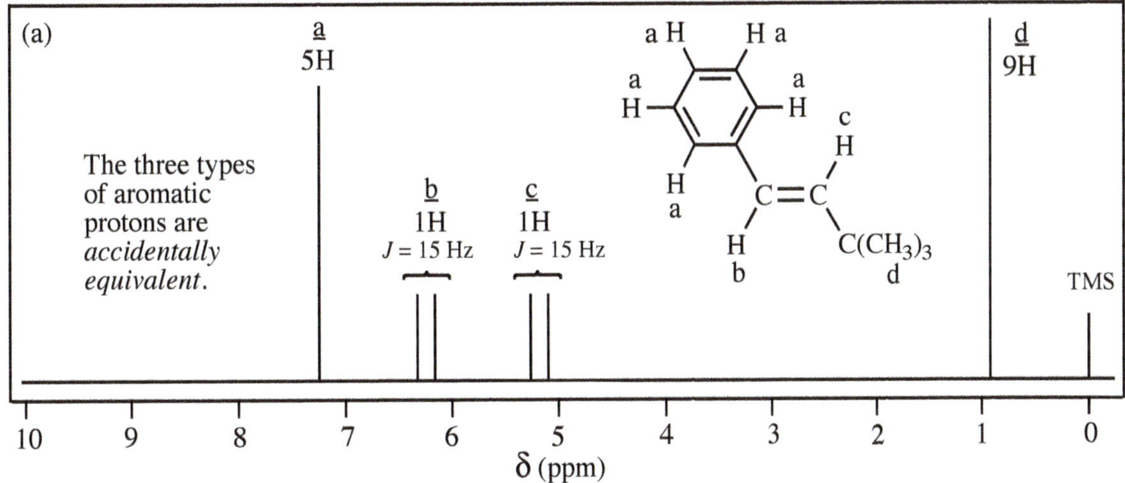

(a)
a
5H
The three types of aromatic protons are *accidentally equivalent*.
b
1H
J = 15 Hz
c
1H
J = 15 Hz
a H
H a
a
H
a
H
c
H
H
a
C=C
H
b
C(CH3)3
d
d
9H
TMS
10 9 8 7 6 5 4 3 2 1 0
δ (ppm)

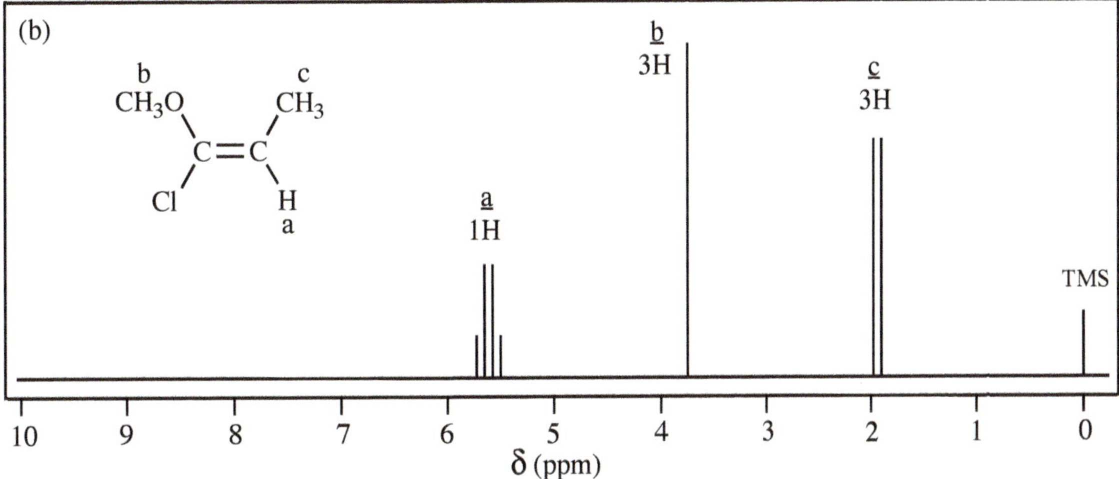

(b)
b
CH3O
c
CH3
C=C
Cl
H
a
a
1H
b
3H
c
3H
TMS
10 9 8 7 6 5 4 3 2 1 0
δ (ppm)

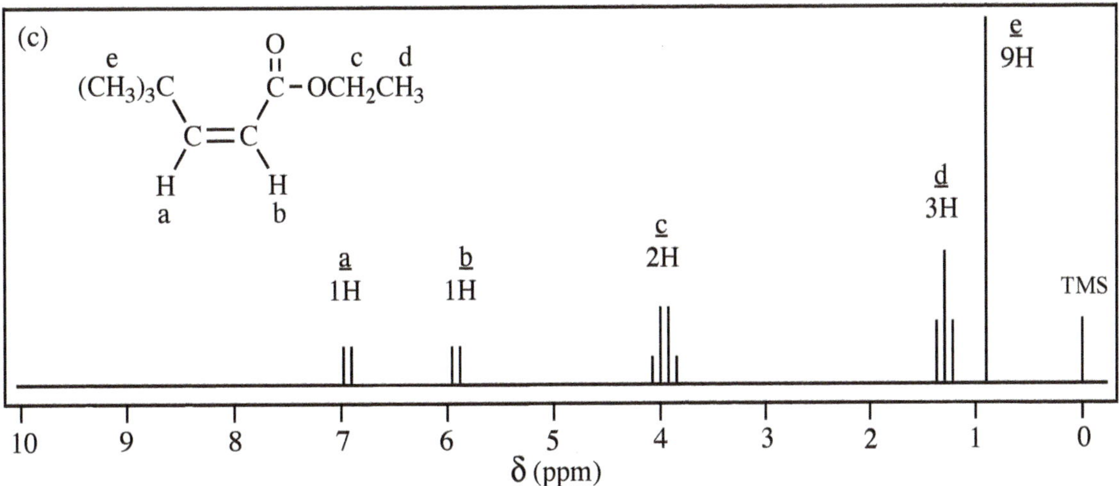

(c)
e
(CH3)3C
O
c d
C–OCH2CH3
C=C
H
a
H
b
a
1H
b
1H
c
2H
d
3H
e
9H
TMS
10 9 8 7 6 5 4 3 2 1 0
δ (ppm)

13-8 continued

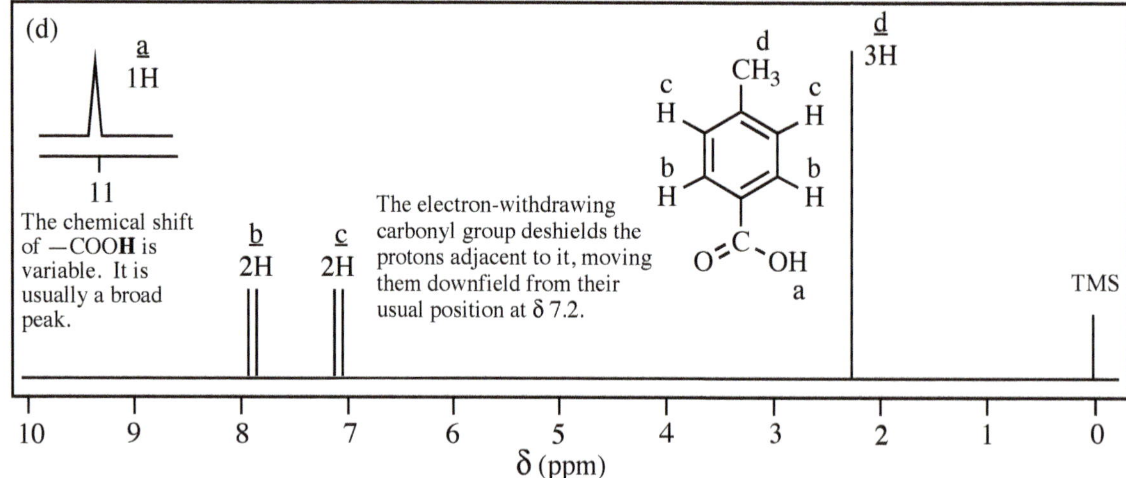

13-9

(a)

O

δ 2.4 H_2C — H δ 5.8

δ 1.8 H_2C — H δ 6.7

H_3C CH_3

δ 1.2

The signals at 1.8 and 2.4 are triplets because each set of protons has two neighboring hydrogens. The N+1 rule correctly predicts each to be a triplet.

(b)

δ 1.1

H_3C CH_3

δ 7.3 H

O

δ 1.5 H_2C

δ 1.65 H_2C

CH_3 δ 2.3

H δ 6.2

C H_2 δ 2.1 allylic

CH_3 δ 1.7 allylic

The three CH_2 groups can be distinguished by chemical shift and by splitting. The allylic CH_2 will be the farthest downfield of the three, at 2.1. The signal at 1.5 is a triplet, so that must be the one with only two neighboring protons. The CH_2 showing the multiplet at 1.65 must be the one between the other two CH_2 groups.

13-10 The formula C_3H_2NCl has three elements of unsaturation. The IR peak at 1650 cm^{-1} indicates an alkene, while the absorption at 2200 cm^{-1} must be from a nitrile (not enough carbons left for an alkyne). These two groups account for the three elements of unsaturation. So far, we have:

C=C—C≡N + 2 H + Cl

The NMR gives the coupling constant for the two protons as 14 Hz. This large J value shows the two protons as *trans* (*cis*, J = 10 Hz; geminal, J = 2 Hz). The structure must be the one in the box.

H, C≡N / C=C / Cl, H

13-11

(a) C_3H_7Cl—no elements of unsaturation; three types of protons in the ratio of 2 : 2 : 3 .

Cl—$\overset{a}{CH_2}\overset{b}{CH_2}\overset{c}{CH_3}$ a = δ 3.55 (triplet, 2H); b = δ 1.8 (multiplet, 2H); c = δ 1.05 (triplet, 3H)

(The only other isomer is 2-chloropropane with two types of protons in the ratio of 6 : 1. This is why you did all of those isomer problems early in the course!)

13-11 continued

(b) $C_9H_{10}O_2$—five elements of unsaturation; four protons in the aromatic region of the NMR indicate a disubstituted benzene; the pair of doublets with $J = 8$ Hz indicate the substituents are on opposite sides of the ring (*para*).

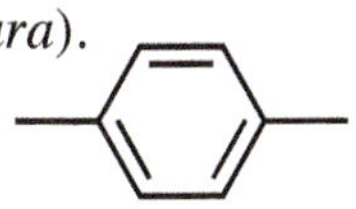

+ 3 C + 6 H + 2 O + 1 element of unsaturation

The other NMR signals are two 3H singlets, two CH_3 groups. One at δ 3.9 must be a CH_3O group. The other at δ 2.4 is most likely a CH_3 group on the benzene ring.

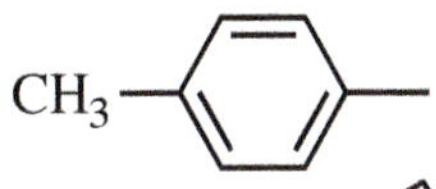

CH_3— + OCH_3 + C + O + 1 element of unsaturation

One way to assemble these pieces consistent with the NMR is:

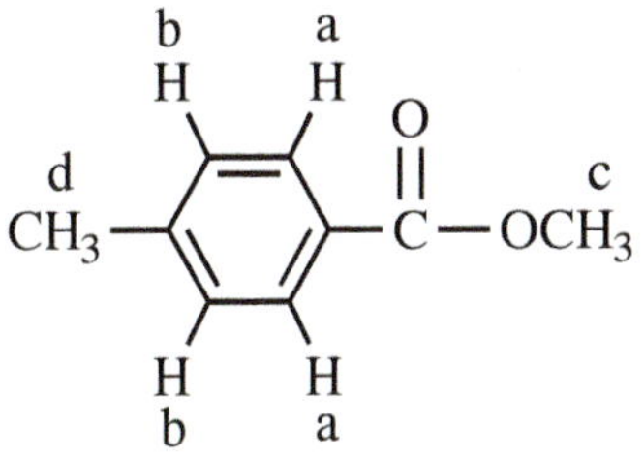

a = δ 7.9 (doublet, 2H)
b = δ 7.2 (doublet, 2H)
c = δ 3.9 (singlet, 3H)
d = δ 2.4 (singlet, 3H)

(Another plausible structure is to have the methoxy group directly on the ring and to put the carbonyl between the ring and the methyl. This does not fit the chemical shift values quite as well as the above structure, as the methyl would appear around δ 2.1 or 2.2 instead of 2.4. It would also put the OCH_3 on the ring, moving the ring H upfield to about 6.8.)

13-12 H_c, δ 5.16

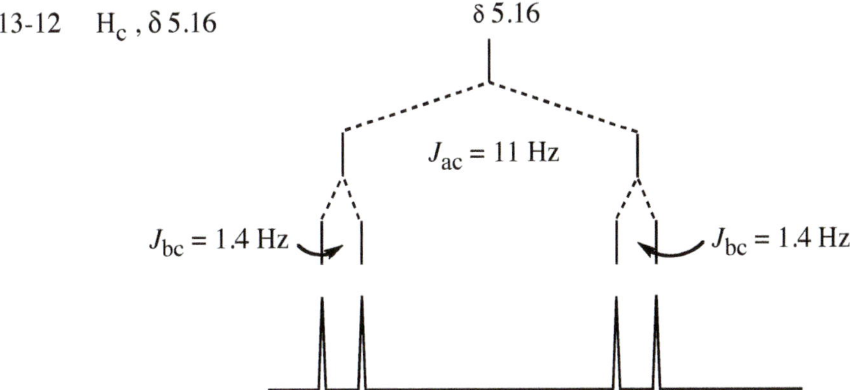

13-13

(a)

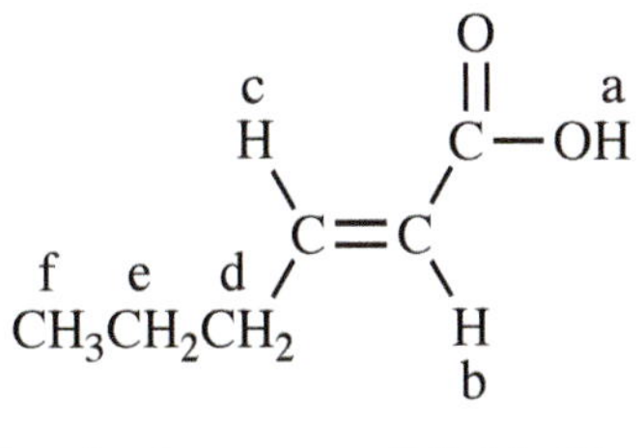

a = δ 12.1 (broad singlet, 1H)
b = δ 5.8 (doublet, 1H)
c = δ 7.1 (multiplet, 1H)
d = δ 2.2 (quartet, 2H)
e = δ 1.5 (sextet, 2H)
f = δ 0.9 (triplet, 3H)

(b) The vinyl proton at δ 7.1 is H_c; it is coupled with H_b and H_d, with two different coupling constants, J_{bc} and J_{cd}, respectively. The value of J_{bc} can be measured most precisely from the signal for H_b at δ 5.8; the two peaks are separated by about 15 Hz, corresponding to 0.05 ppm in a 300 MHz spectrum. The value of J_{cd} appears to be about the standard value 8 Hz, judging from the signal at δ 7.1. The splitting tree would appear as shown on the next page.

13-13(b) continued

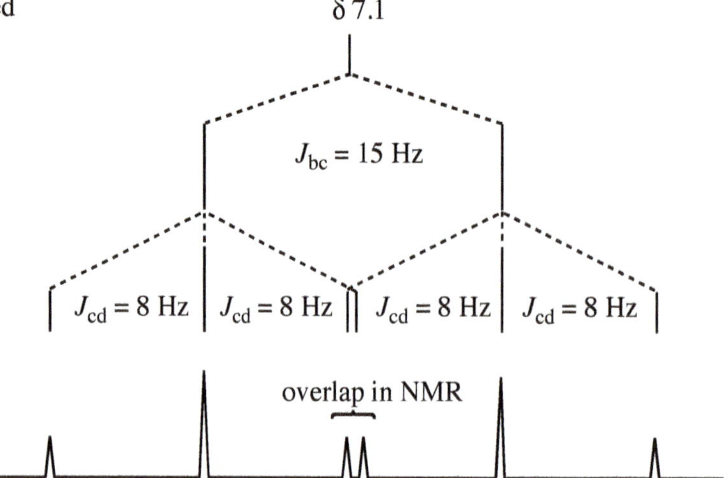

13-14

(a)

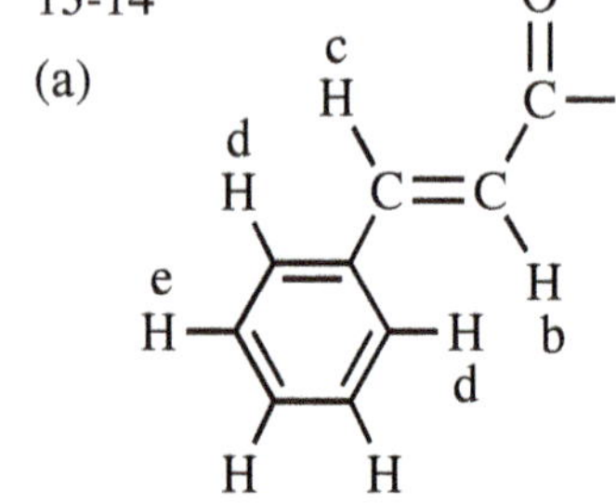

a = δ 9.7 (doublet, 1H)
b = δ 6.7 (doublet of doublets, 1H)
c = δ 7.5 (doublet, 1H)
d = δ 7.5 (doublet, 2H ortho to substituent)
e = δ 7.4 (broad singlet, 3H)

The doublet for H_c at δ 7.4 OVERLAPS the 5 benzene protons.

(b) J_{ab} can be determined most accurately from H_a at δ 9.7: $J_{ab} \approx 8$ Hz, about the same as "normal" alkyl coupling.

J_{bc} can be measured from H_b at δ 6.7, as the distance between either the first and third peaks or the second and fourth peaks (see diagram to the right): $J_{bc} \approx 18$ Hz, characteristic of *trans* coupling, about double the "normal" alkyl coupling.

(c)

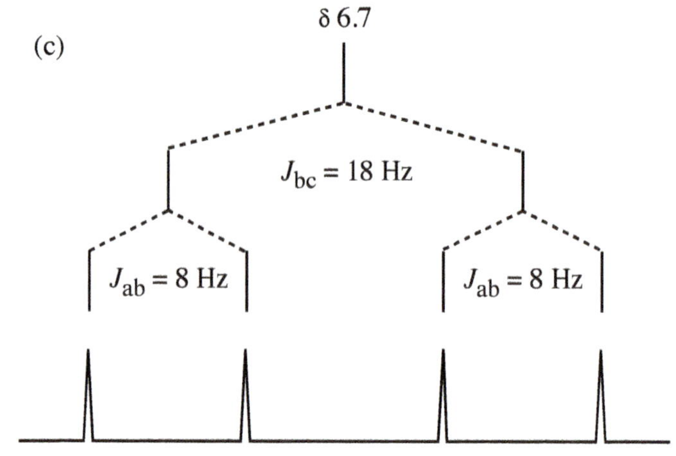

13-15

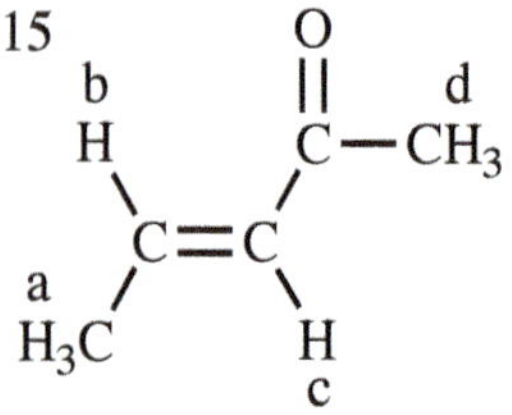

(a) a = δ 1.7
b ≈ δ 6.8
c = δ 5-6
d = δ 2.1

(b) a = doublet
b = multiplet (two overlapping quartets—see part (c))
c = doublet
d = singlet

13-15 continued

(c) using J_{bc} = 15 Hz and J_{ab} = 7 Hz:

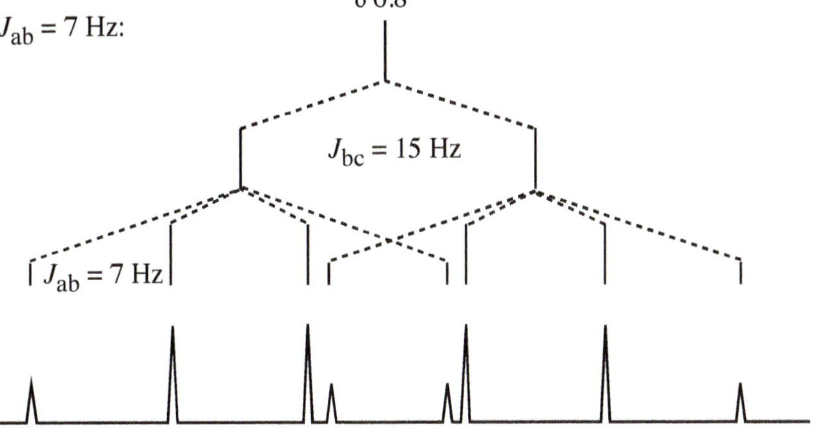

13-16

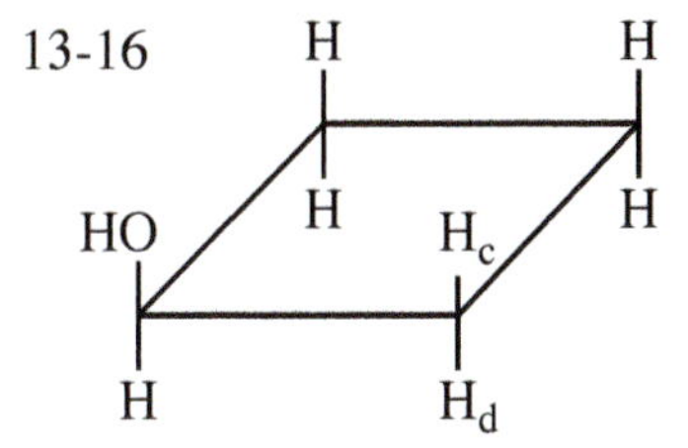

replace H_c →

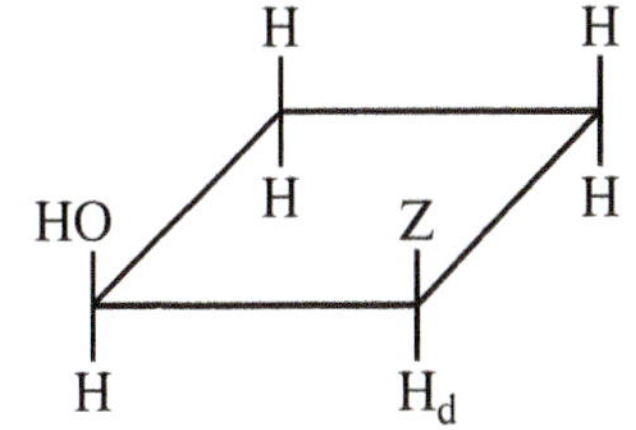

cis diastereomer

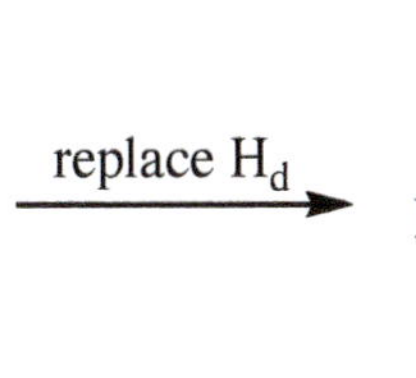

HO, H, H, H, H_c, H, H, Z

trans diastereomer

13-17

(a)

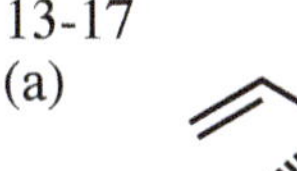

replace H_a → (Z, H_b, Br)

replace H_b → (H_a, Z, Br)

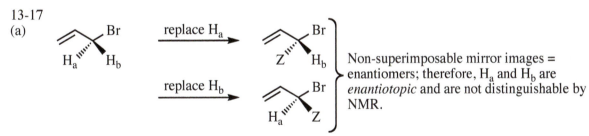

Non-superimposable mirror images = enantiomers; therefore, H_a and H_b are *enantiotopic* and are not distinguishable by NMR.

(b)

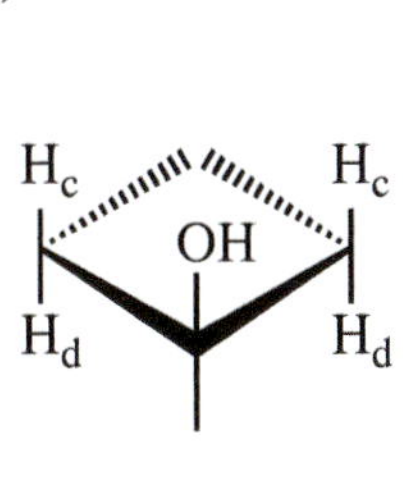

replace H_c on the left →

replace H_c on the right →

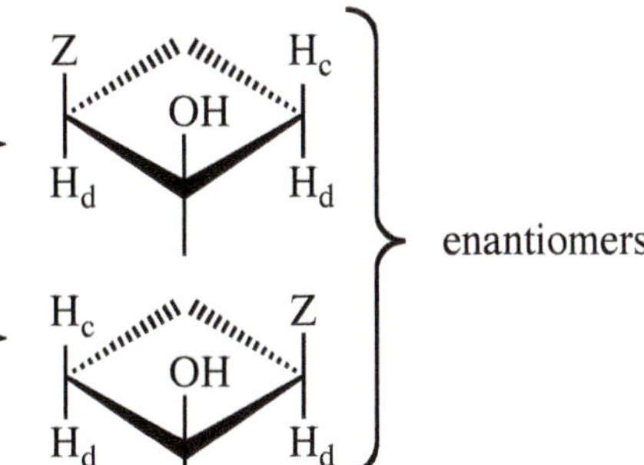

enantiomers

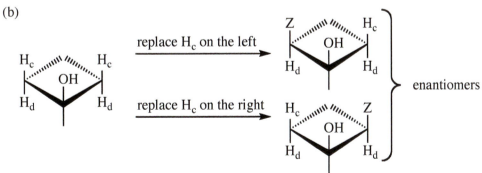

(c) The H_d protons are also enantiotopic.

13-18

(a)

This compound has five types of protons. H_c and H_d are diastereotopic. a = δ 1.5; b = δ 3.6; c,d = δ 1.7; e = δ 1.0

(b)

This compound has six types of protons. H_c and H_d are diastereotopic, as are H_e and H_f. a = δ 2-5; b = δ 3.9; c,d = δ 1.6; e,f = δ 1.3

(c)

This compound has six types of protons. H_e and H_f are diastereotopic. a,b,c = δ 7.2; d ≈ δ 5.0; e,f = δ 3.6

(d)

This compound has three types of protons. H_a and H_b are diastereotopic. a,b = δ 5-6; c = δ 7-8

13-19 (a)

$CH_3CH_2-\ddot{O}-H + H-A \rightarrow CH_3CH_2-\overset{+}{O}(H)-H + :A^- \rightarrow CH_3CH_2-\ddot{O}(H): + H-A$

(b)

$CH_3CH_2-\ddot{O}-H + :A^- \rightarrow CH_3CH_2-\ddot{O}:^- + H-A \rightarrow CH_3CH_2-\ddot{O}(H): + :A^-$

13-20 The protons from the OH in ethanol exchange with the deuteriums in D_2O. Thus, the OH in ethanol is replaced with OD, which does not absorb in the NMR. What happens to the H? It becomes HOD, which can usually be seen as a broad singlet around δ 5.25. (If the solvent is $CDCl_3$, the immiscible HOD will float on top of the solvent, out of the spectrometer beam, and its signal will be missing.)

$$ROH + D_2O \longrightarrow ROD + HOD$$

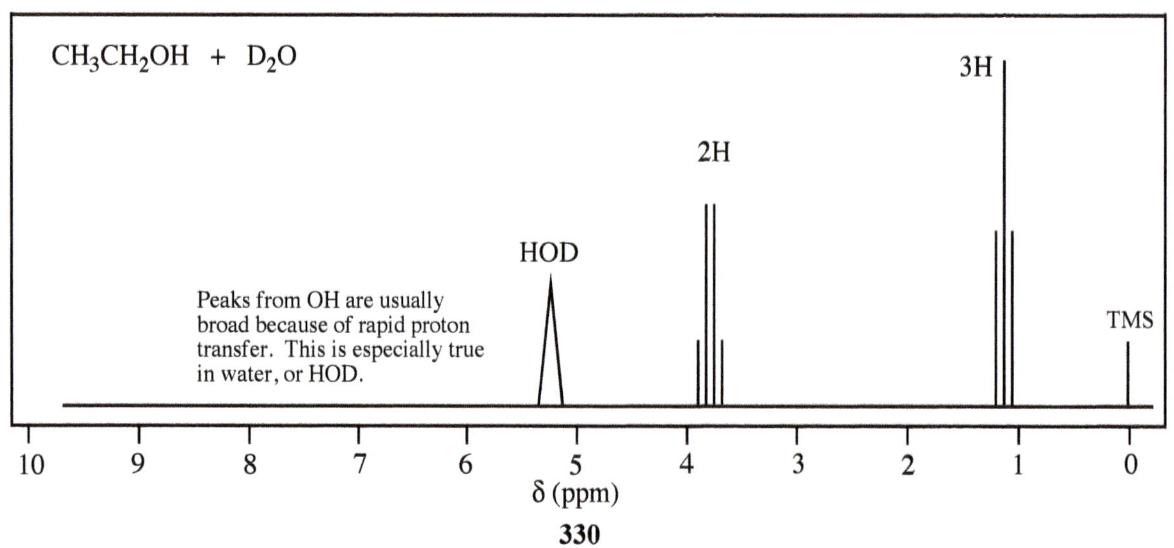

13-21

(a) The formula $C_4H_{10}O_2$ has no elements of unsaturation, so the oxygens must be alcohol or ether functional groups. The doublet at δ 1.2 represents 3H and must be a CH_3 next to a CH. The peaks centered at δ 1.65 integrating to 2H appear to be an uneven quartet and signify a CH_2 between two sets of non-equivalent protons; apparently the coupling constants between the non-equivalent protons are not equal, leading to a complicated pattern of overlapping peaks. The remaining five hydrogens appear in four groups of 1H, 1H, 1H, and 2H, between δ 3.7 and 4.2. The 2H multiplet at δ 3.78 is a CH_2 next to O, with complex splitting (doublet of doublets) due to diastereotopic neighbors. The 1H multiplet at δ 4.0 is a CH between many neighbors. The broad singlet integrating to 2H at δ 4.12 appears to be two OH peaks.

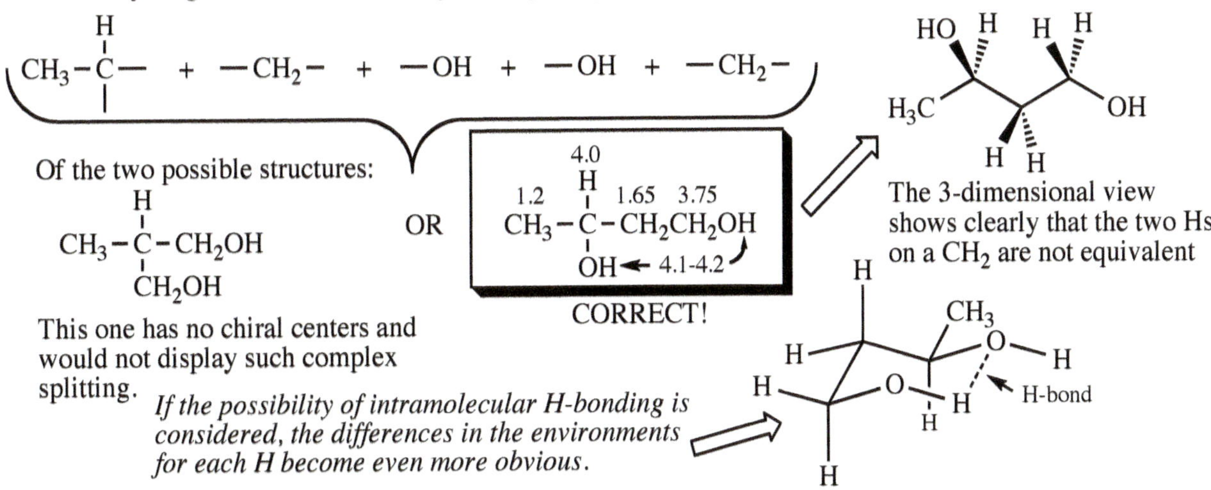

(b) The formula C_2H_7NO has no elements of unsaturation. The N must be an amine and the O must be an alcohol or ether. Two triplets, each with 2H, are certain to be $-CH_2CH_2-$. Since there are no carbons left, the N and O, with enough hydrogens to fill their valences, must go on the ends of this chain. The rapidly exchanging OH and NH_2 protons appear as a very broad, 3H hump at δ 2.8. The broad "peak" is hard to detect; the gradually increasing integration is one clue that something is there.

$HOCH_2CH_2NH_2$

δ 3.6 (CH_2 next to O)   δ 2.8 (CH_2 next to N)

13-22

a: CH_3O; b: CH_2; c: CH_3 — $CH_3O-C(=O)-CH_2CH_3$

a 3H

b 2H

c 3H

TMS

10 9 8 7 6 5 4 3 2 1 0

δ (ppm)

(s) $CH_3-C(=O)-OCH_2CH_3$ (q)

δ 2.05   δ 4.1

ethyl acetate

The key is what protons are adjacent to the oxygen. The CH_3O in methyl propionate (IUPAC: methyl propanoate), above, absorbs at δ 3.9 as a singlet, whereas the CH_2 next to the carbonyl absorbs at δ 2.2 as a quartet. This is in contrast to ethyl acetate, at the left.

13-23

H_a = δ 2.4 (singlet, 1H)
H_b = δ 3.4 (doublet, 2H)
H_c = δ 1.8 (multiplet, 1H)
H_d = δ 0.9 (doublet, 6H)

$$\underset{a}{H}-O-\underset{b}{CH_2}-\overset{\overset{d}{CH_3}}{\underset{\underset{d}{CH_3}}{C}}-\underset{c}{H}$$

13-24

(a) The formula $C_4H_8O_2$ has one element of unsaturation. The 1H singlet at δ 12.1 indicates a carboxylic acid. The 1H multiplet and the 6H doublet scream isopropyl group.

$$(H_3C)_2CH-\overset{O}{\overset{\|}{C}}-OH$$

(b) The formula $C_9H_{10}O$ has five elements of unsaturation. The 5H pattern between δ 7.2 and 7.4 indicates monosubstituted benzene. The peak at δ 9.85 is unmistakably an aldehyde, trying to be a triplet because it is weakly coupled to an adjacent CH_2. The two triplets at δ 2.7-3.0 are adjacent CH_2 groups.

$$C_6H_5-\overset{H_2}{C}-\underset{H_2}{C}-\overset{O}{\overset{\|}{C}}-H$$

(c) The formula $C_5H_8O_2$ has two elements of unsaturation. A 3H singlet at δ 2.3 is probably a CH_3 next to carbonyl. The 2H quartet and the 3H triplet are certain to be ethyl; with the CH_2 at δ 2.7, this also appears to be next to a carbonyl.

$$CH_3CH_2-\overset{O}{\overset{\|}{C}}-\overset{O}{\overset{\|}{C}}-CH_3$$

(d) The formula C_4H_8O has one element of unsaturation, and the signals from δ 5.0 to 6.0 indicate a vinyl pattern ($CH_2{=}CH-$). The complex quartet for 1H at δ 4.3 is a CH bonded to an alcohol, next to CH_3. The OH appears as a 1H singlet at δ 2.5, and the CH_3 next to CH is a doublet at δ 1.3. Put together:

$$CH_2{=}CH-\overset{OH}{\overset{|}{C}H}-CH_3 \qquad \text{but-3-en-2-ol}$$

(e) The formula $C_7H_{16}O$ is saturated; the oxygen must be an alcohol or an ether, and the broad 1H peak at δ 1.2 is probably an OH. Let's analyze the spectrum from left to right. The expansion of the 1H multiplet at δ 1.7 shows seven peaks, a septet—an isopropyl group! Six of the nine protons in the pattern at δ 0.9 must be the doublet from the two methyls from the isopropyl. The 2H quartet at δ 1.5 must be part of an ethyl pattern, from which the methyl triplet must be the other 3H of the pattern at δ 0.9. This leaves only the 3H singlet at δ 1.1 which must be a CH_3 with no neighboring Hs.

$$-OH \; + \; -CH_2CH_3 \; + \; CH_3-\overset{H}{\underset{CH_3}{C}}- \; + \; -CH_3 \; + \; 1\,C$$

These pieces can be assembled in only one way:

$$HO-\overset{CH_3}{\underset{CH_2CH_3}{C}}-CH(CH_3)_2 \qquad \text{2,3-dimethylpentan-3-ol}$$

13-25 Chemical shift values are estimates from Figure 13-41 and from text Appendix 1C, except in (c) and (d), where the values are exact.

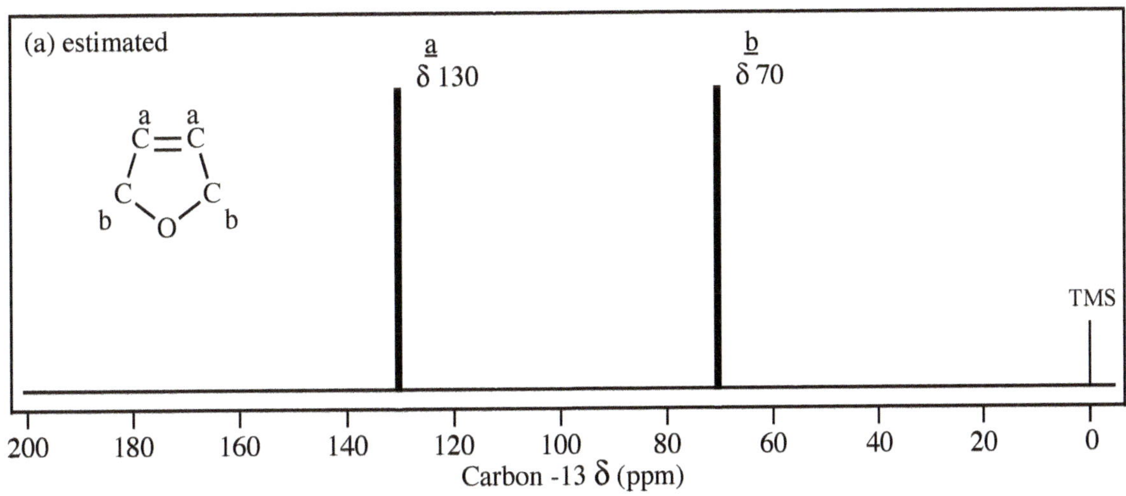

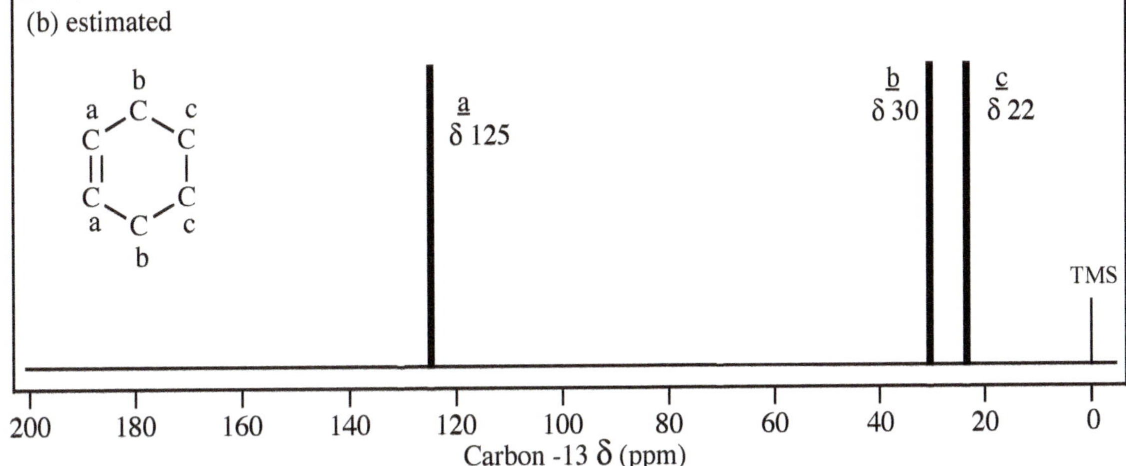

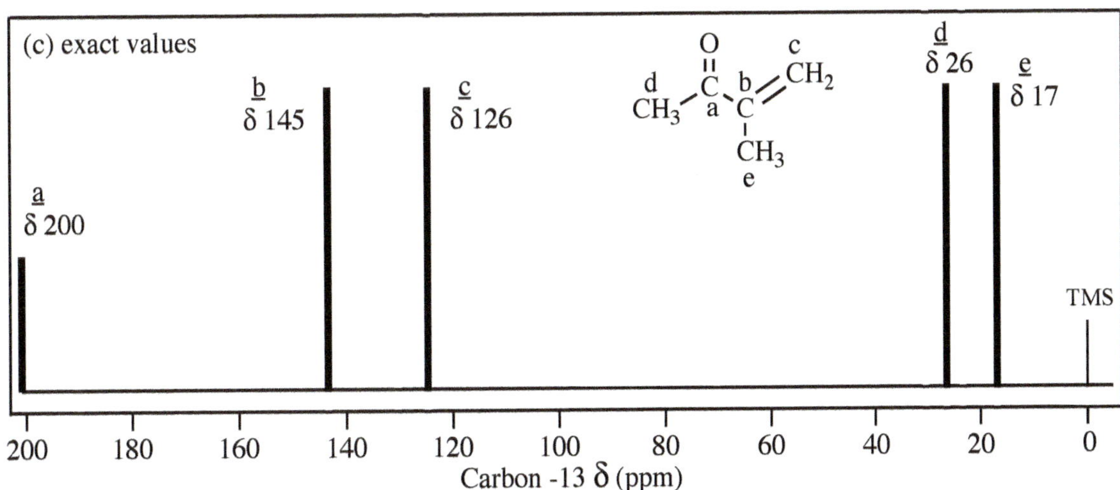

13-25 continued

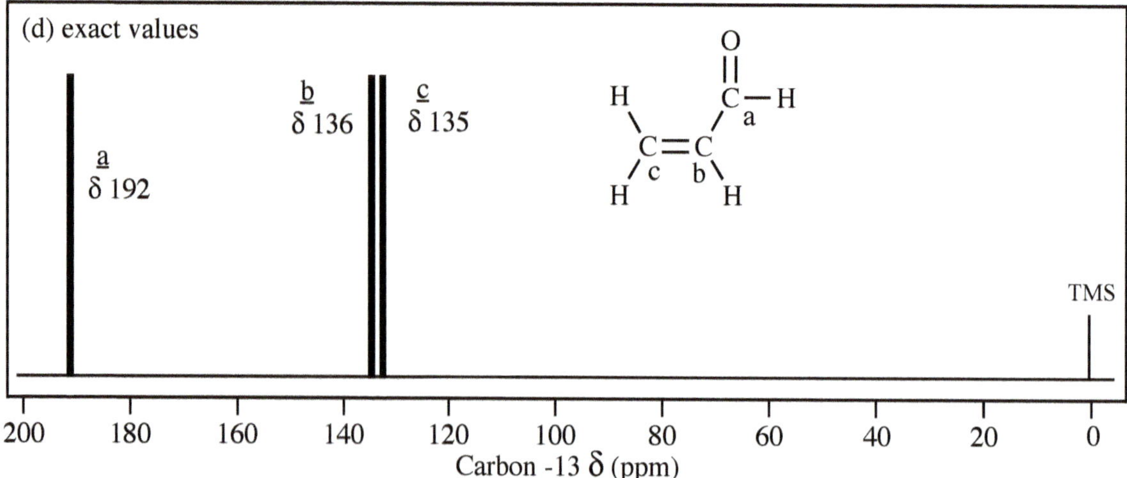

13-26

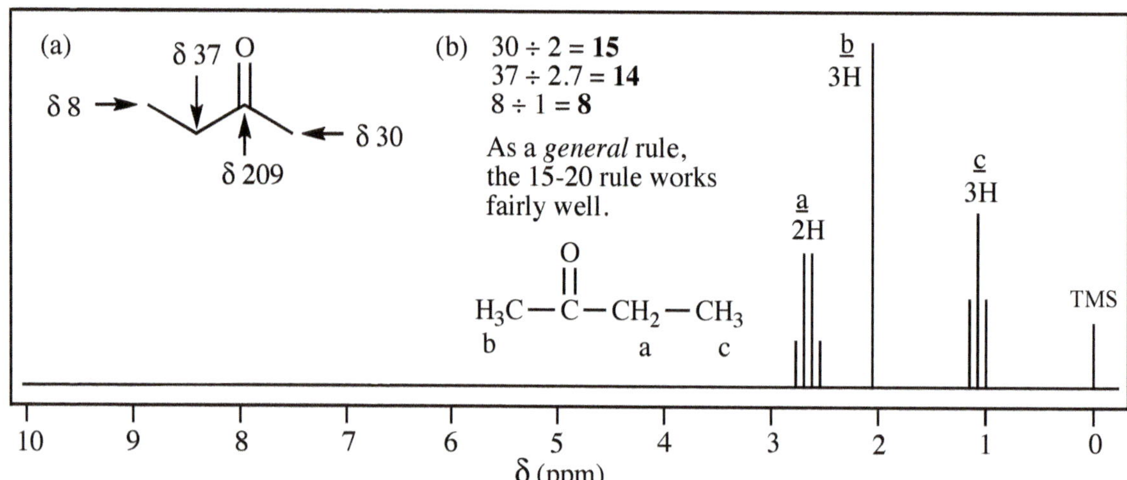

13-27

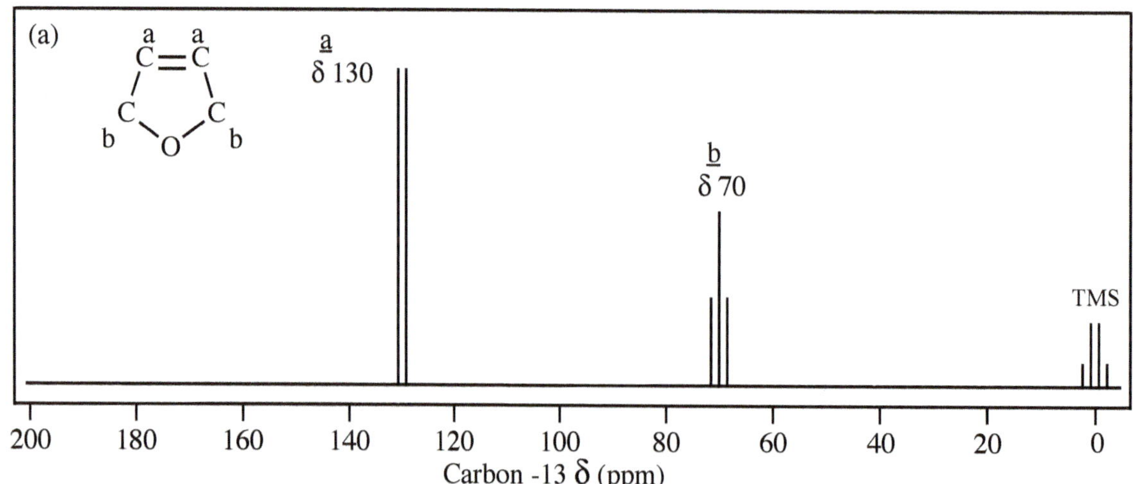

13-27 continued

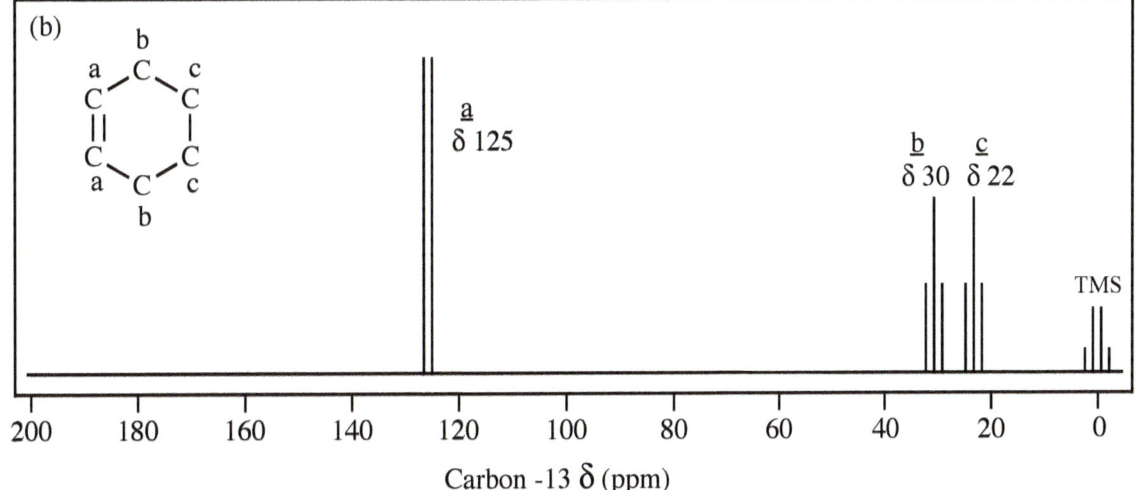

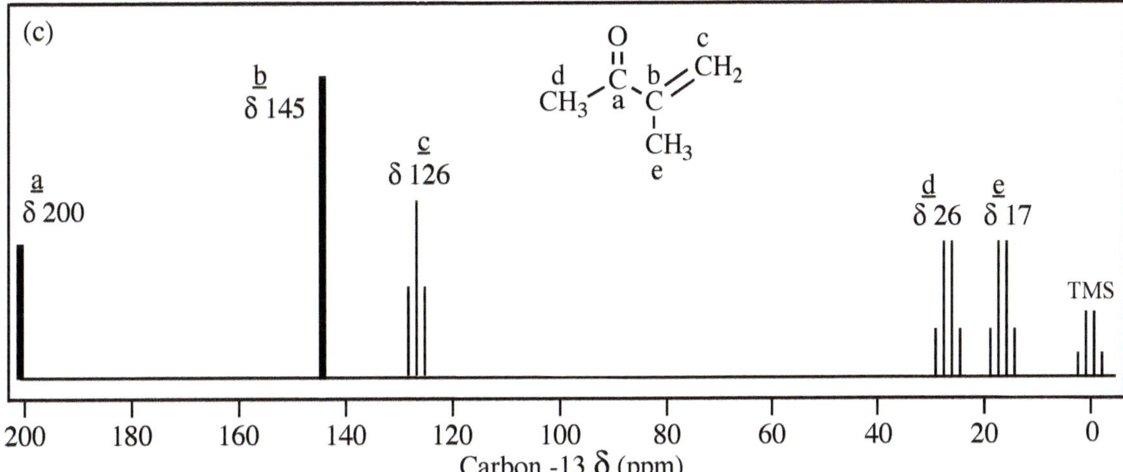

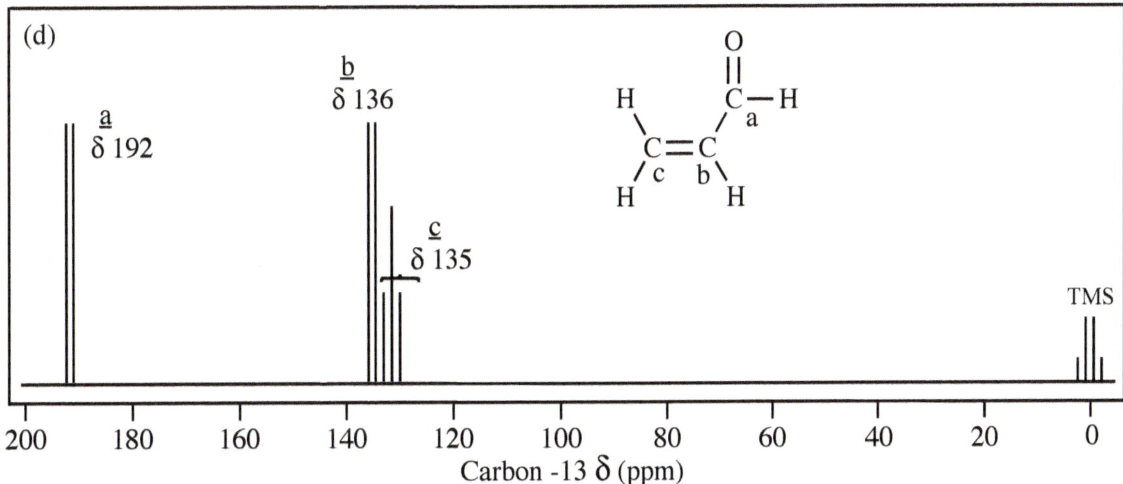

13-28 The full carbon spectrum of phenyl propanoate is presented below.

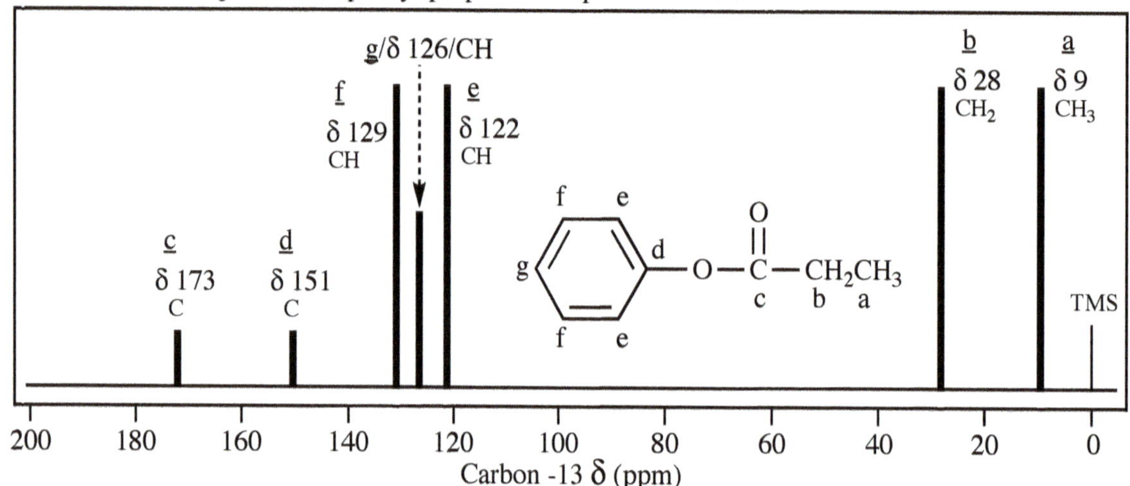

The DEPT-90 will show only the methine carbons, i.e., CH. All other peaks disappear.

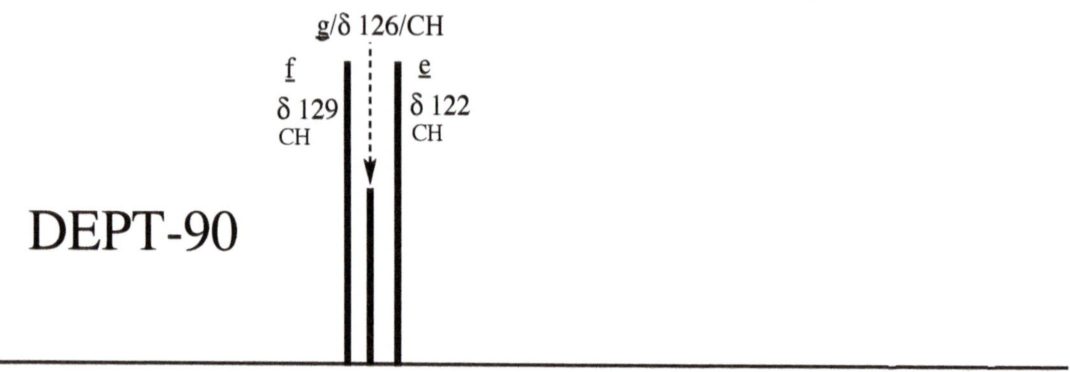

The DEPT-135 will show the methyl, CH_3, and methine, CH, peaks pointed up, and the methylene, CH_2, peaks pointed down.

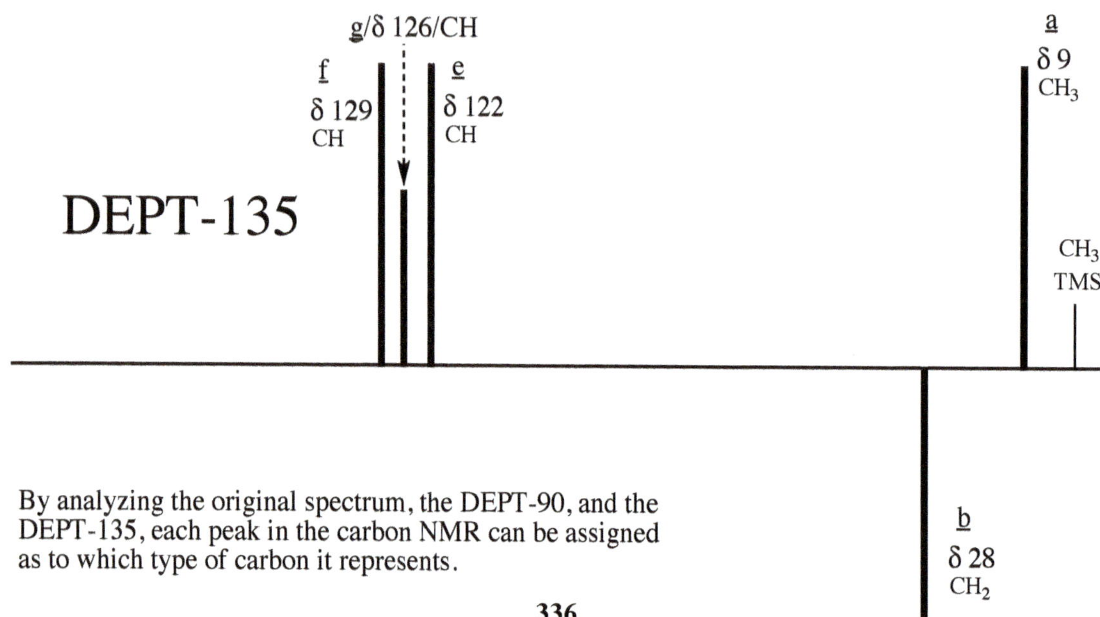

By analyzing the original spectrum, the DEPT-90, and the DEPT-135, each peak in the carbon NMR can be assigned as to which type of carbon it represents.

13-29

(a) and (b) Since allyl bromide was the starting material, it is reasonable to expect the allyl group to be present in the impurity: the peak at 115 is a $=CH_2$, the peak at 138 is =CH– , and the peak at 63 is a deshielded aliphatic CH_2 ; assembling the pieces forms an allyl group. The formula has changed from C_3H_5Br to C_3H_6O, so OH has replaced the Br.

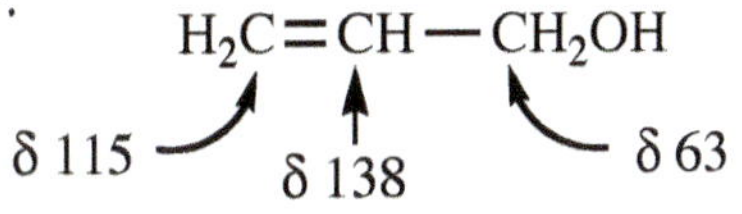

$H_2C{=}CH{-}CH_2OH$ (δ 115 at $H_2C=$, δ 138 at $=CH-$, δ 63 at CH_2OH)

(c) Allyl bromide is easily hydrolyzed by water, probably an S_N1 process.

$$H_2C{=}CH{-}CH_2Br \xrightarrow{H_2O} H_2C{=}CH{-}CH_2OH + HBr$$

13-30 (a) and (b)

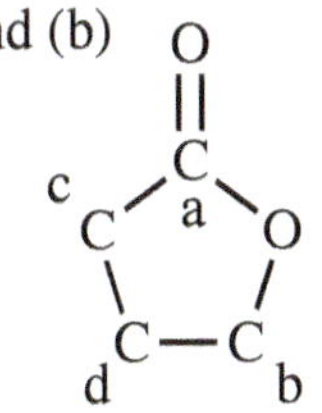

a = δ 180
b = δ 70
c = δ 28
d = δ 22

Two elements of unsaturation in $C_4H_6O_2$, one of which is a carbonyl, and no evidence of a C=C, prove that a ring must be present. This product is formed from 4-hydroxybutanoic acid with loss of a molecule of water, as you will see in Chapter 21.

13-31 (a) and (b)

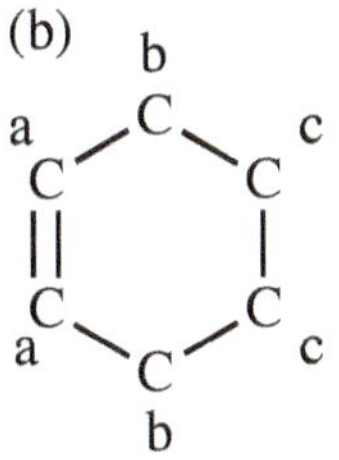

a = δ 128
b = δ 25
c = δ 23

Two elements of unsaturation in C_6H_{10} must be a C=C and a ring. Only three peaks indicates symmetry.

(c) Using PBr_3 instead of H_2SO_4/NaBr would give a higher yield of bromocyclohexane.

13-32 Compound 2

Mass spectrum: the molecular ion at m/z 136 shows a peak at 138 of about equal height, indicating a bromine atom is present: 136 – 79 = 57. The fragment at m/z 57 is the base peak; this fragment is most likely a butyl group, C_4H_9, so a likely molecular formula is C_4H_9Br.
Infrared spectrum: Notable for the absence of functional groups: no O—H, no N—H, no =C—H, no C=C, no C=O ⇒ most likely an alkyl bromide.
NMR spectrum: The 6H doublet at δ 1.0 suggests two CH_3 groups split by an adjacent H—an isopropyl group. The 2H doublet at δ 3.2 is a CH_2 between a CH and the Br.

Putting the pieces together gives isobutyl bromide.

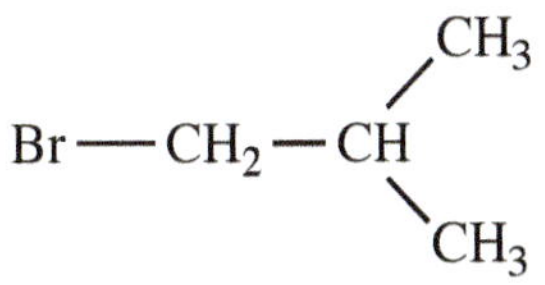

$Br{-}CH_2{-}CH(CH_3)_2$

13-33

The formula $C_9H_{11}Br$ indicates four elements of unsaturation, just enough for a benzene ring.

Here is the most accurate method for determining the number of protons per signal from integration values *when the total number of protons is known*. Add the integration heights: 44 mm + 130 mm + 67 mm = 241 mm. Divide by the total number of hydrogens: 241 mm ÷ 11H = 22 mm/H. Each 22 mm of integration height = 1H, so the ratio of hydrogens is 2 : 6 : 3.

The 2H singlet at δ 7.1 means that only two hydrogens remain on the benzene ring, that is, it has four substituents. The 6H singlet at δ 2.3 must be two CH_3 groups on the benzene ring in identical environments. The 3H singlet at δ 2.2 is another CH_3 in a slightly different environment from the first two. Substitution of the three CH_3 groups and the Br in the most symmetric way leads to the structures on the next page.

13-33 continued

a = δ 7.1 (singlet, 2H)
b = δ 2.3 (singlet, 6H)
c = δ 2.2 (singlet, 3H)

A second structure is also possible although it is less likely because the Br would probably deshield the Hs labeled "a" to about 7.3–7.4.

13-34 The numbers in italics indicate the number of peaks in each signal. Approximate chemical shift values are in bold.

(a) $CH_3—CH_2—CCl_2—CH_3$ — **1.0** (*3*), **2.2** (*4*), **1.7** (*1*)

(b) $CH_3—CH(CH_3)—OH$ — CH **4.0** (*7*), OH **2-5** (*1*), CH_3 **1.2** (*2*)
(assume OH exchanging rapidly—no splitting)

(c) $CH_3—CH(CH_3)—CH_3$ — CH **1.4** (*10*), CH_3 **0.9** (*2*)

(d) Toluene: ring H **7.2** (*1*), CH_3 **2.3** (*1*)

(All of these benzene H atoms are accidentally equivalent and do not split each other.)

(e) $CH_3—CH_2—C(=O)-O-CH_2-CH_3$ — **1.0** (*3*), **2.2** (*4*), **4.0** (*4*), **1.2** (*3*)

(f) 2-methyltetrahydropyran: H **4.2** (*6*), CH_3 **1.2** (*2*)

13-35 Consult Appendix 1 in the text for chemical shift values. *Your predictions should be in the given range, or within 0.5-1.0 ppm of the given value.*

(a) Benzene H — all at δ 7.2

(b) Cyclohexane H — all at δ 1.3

(c) $CH_3-O-CH_2-CH_2-CH_2Cl$ — δ 3.4, δ 3.8, δ 1.6, δ 3.2

(d) $CH_3-CH_2-C≡C-H$ — δ 1.2, δ 2.2, δ 2.5

(e) $CH_3—CH_2—C(=O)—CH_3$ — δ 1.0, δ 2.5, δ 2.0

(f) $(CH_3)_2CH—O-CH_2—CH_2-OH$ — CH_3 δ 0.9, CH δ 4.3, δ 3.8, δ 3.8, OH δ 2-5

(g) Benzaldehyde: C(=O)−H δ 9-10; ring H δ 8.0 (ortho), δ 7.5 (meta), δ 8.0 (para)

The C=O has its strongest deshielding effect at the adjacent H (ortho) and across the ring (para). The remaining H (meta) is less deshielded.

(h) $CH_3—CH=CH—C(=O)—H$ — δ 1.7, δ 6-7, δ 5-6, δ 9-10

13-35 continued

(i)

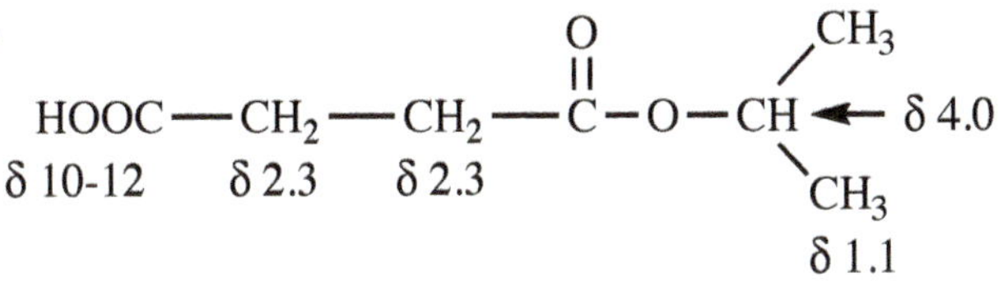

(j)

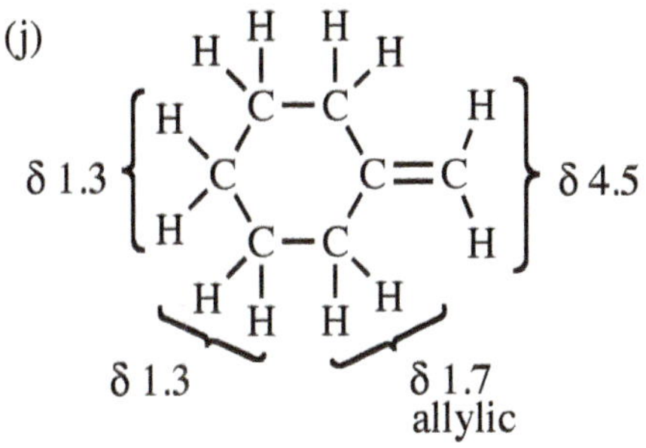

(k)

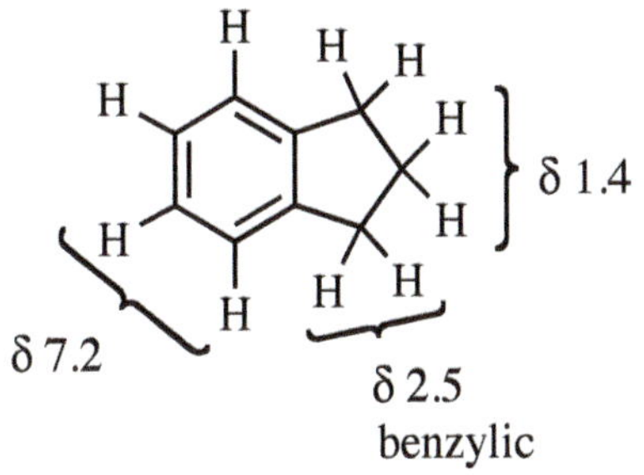

(l)

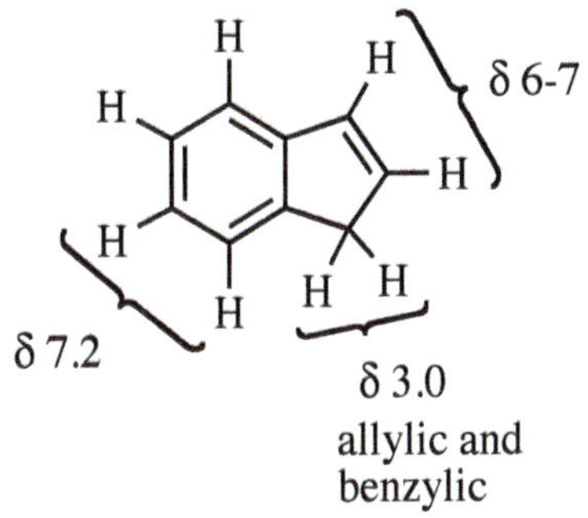

13-36

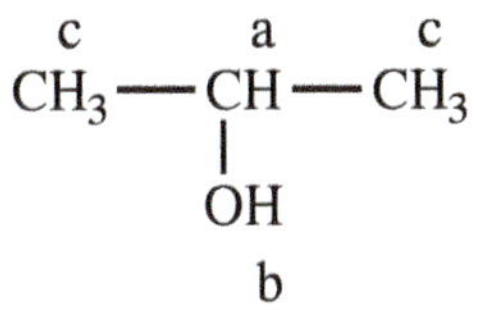

a = δ 4.0 (septet, 1H)
b = δ 2.5 (broad singlet, 1H) (rapidly exchanging)
c = δ 1.2 (doublet, 6H)

13-37

(a) The chemical shift *in ppm* would not change: δ 4.00.

(b) Coupling constants do not change with field strength: $J = 7$ Hz, regardless of field strength.

(c) At 60 MHz, δ 4.00 = 4.00 ppm = $(4.00 \times 10^{-6}) \times (60 \times 10^{6}\text{ Hz}) = 240$ Hz

The signal is 240 Hz downfield from TMS in a 60 MHz spectrum.

At 300 MHz, $(4.00 \times 10^{-6}) \times (300 \times 10^{6}\text{ Hz}) = 1200$ Hz

The signal is 1200 Hz downfield from TMS in a 300 MHz spectrum. Necessarily, 1200 Hz is exactly 5 times 240 Hz because 300 MHz is exactly 5 times 60 MHz. They are directly proportional.

13-38

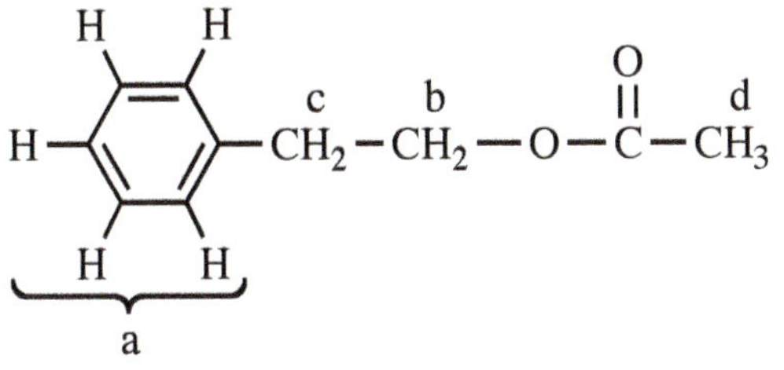

a = δ 7.2-7.3 (multiplet, 5H)
b = δ 4.3 (triplet, 2H)
c = δ 2.9 (triplet, 2H)
d = δ 2.0 (singlet, 3H)

13-39

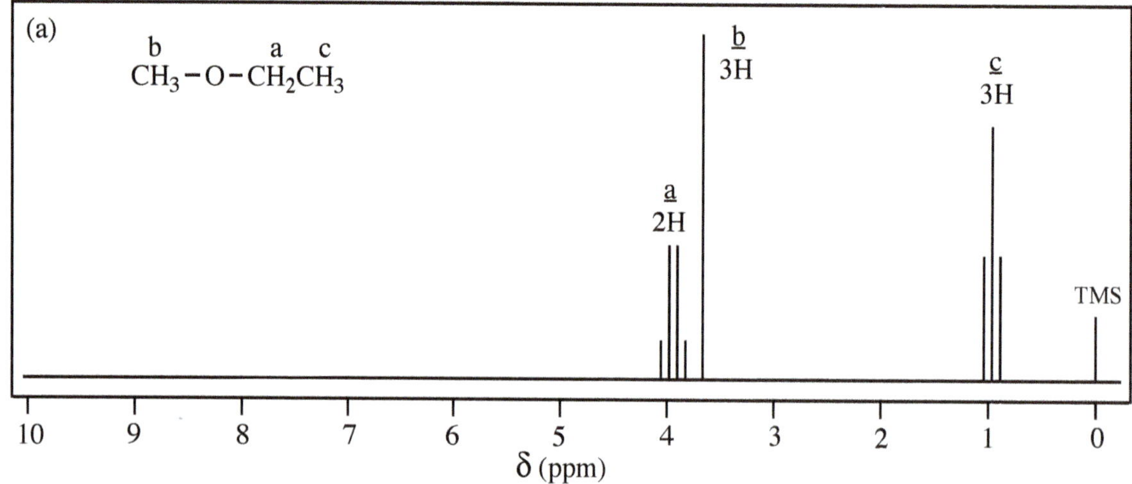

(a)
b a c
$CH_3-O-CH_2CH_3$
b
3H
c
3H
a
2H
TMS
10 9 8 7 6 5 4 3 2 1 0
δ (ppm)

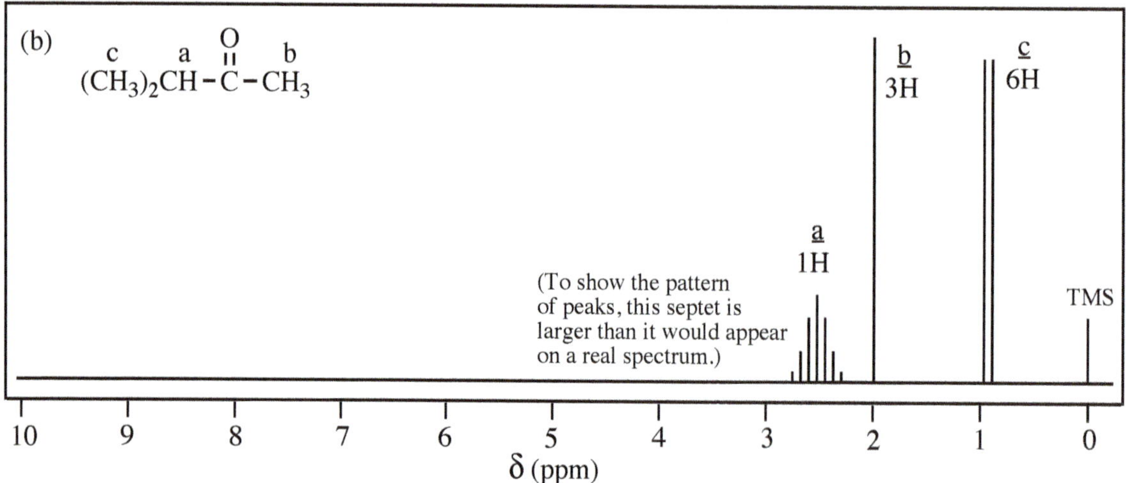

(b)
c a O b
$(CH_3)_2CH-\overset{O}{\overset{\|}{C}}-CH_3$
b
3H
c
6H
a
1H
(To show the pattern of peaks, this septet is larger than it would appear on a real spectrum.)
TMS
10 9 8 7 6 5 4 3 2 1 0
δ (ppm)

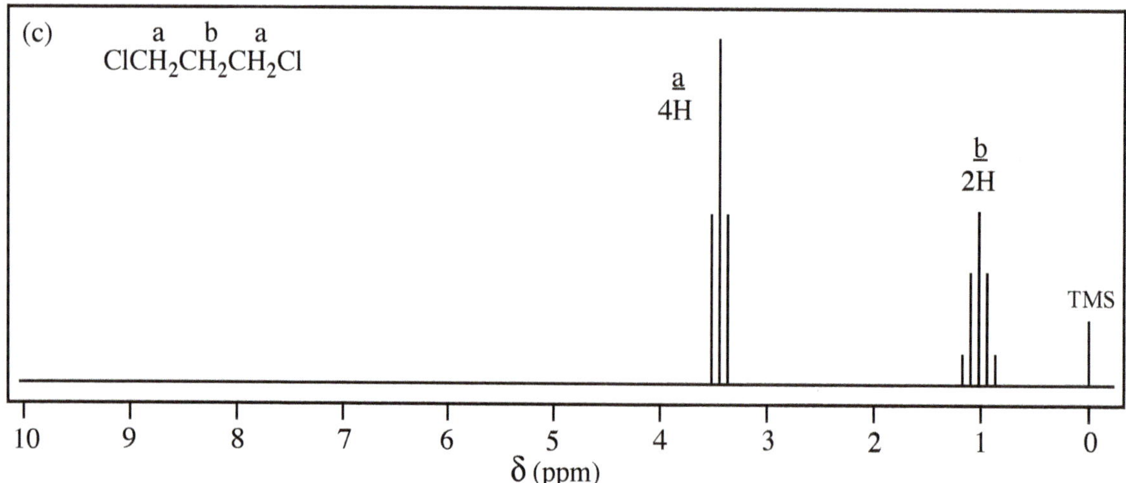

(c)
a b a
$ClCH_2CH_2CH_2Cl$
a
4H
b
2H
TMS
10 9 8 7 6 5 4 3 2 1 0
δ (ppm)

13-39 continued

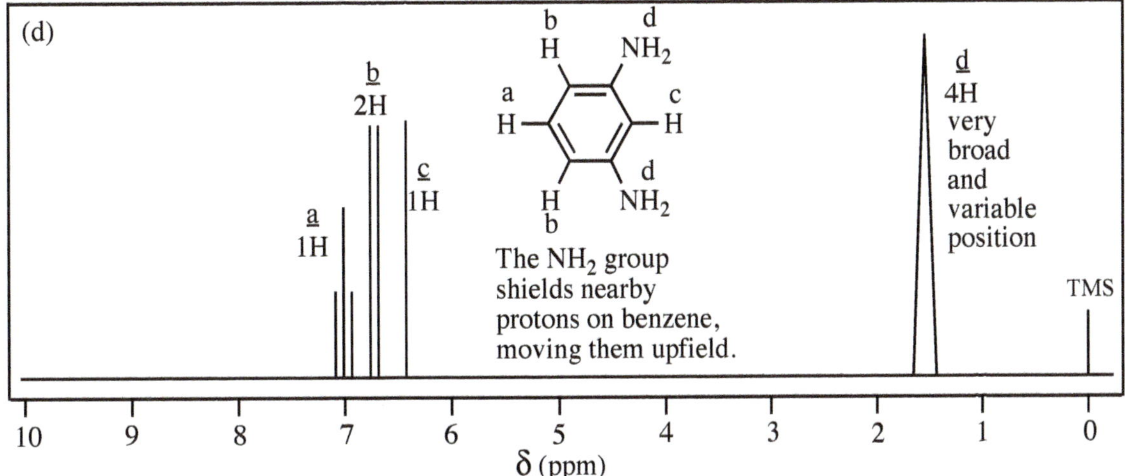

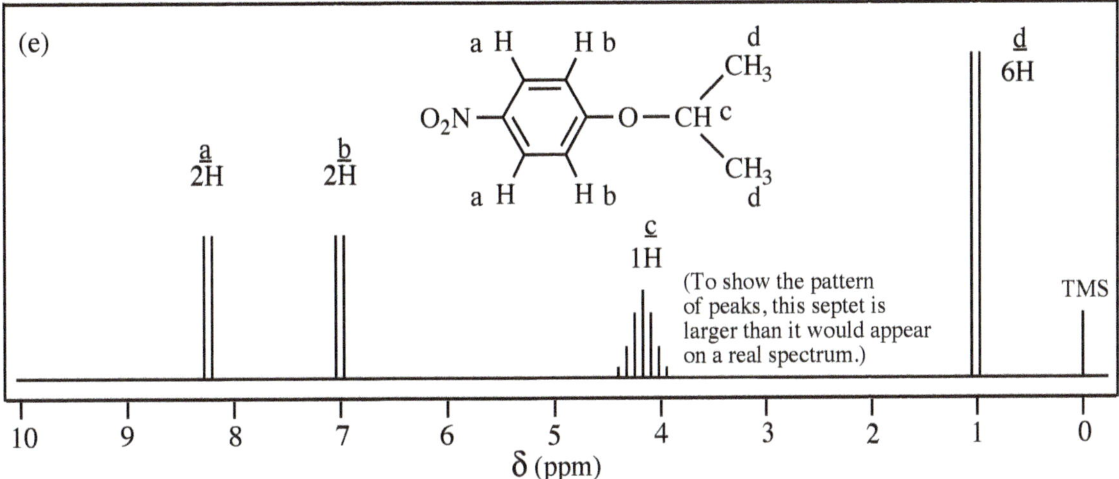

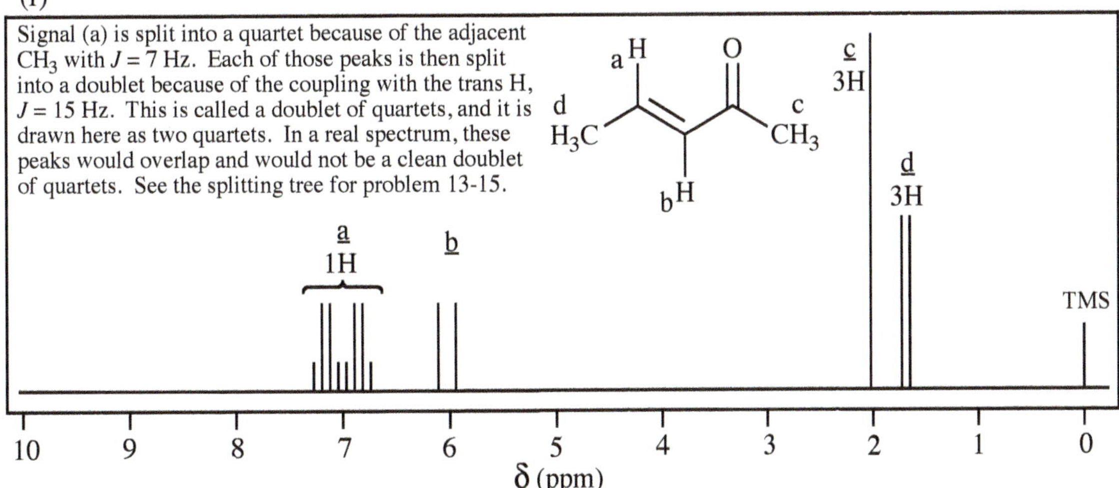

13-40

(a) The NMR of 1-bromopropane would have three sets of signals, whereas the NMR of 2-bromopropane would have only two sets (a septet and a doublet, the typical isopropyl pattern).

(b) Each spectrum would have a methyl singlet at δ 2. The left structure would show an ethyl pattern (a 2H quartet and a 3H triplet), whereas the right structure would exhibit an isopropyl pattern (a 1H septet and a 6H doublet).

(c) The most obvious difference is the chemical shift of the CH_3 singlet. In the compound on the left, the CH_3 singlet would appear at δ 2.1, while the compound on the right would show the CH_3 singlet at δ 3.8. Refer to the solution of 13-22 for the spectrum of the second compound.

(d) The splitting and integration for the peaks in these two compounds would be identical, so the chemical shift must make the difference. As described in text section 13-5B, the alkyne is not nearly as deshielding as a carbonyl, so the protons in pent-2-yne would be farther upfield than the protons in butan-2-one, by about 0.5 ppm. For example, the methyl on the carbonyl would appear near δ 2.1 while the methyl on the alkyne would appear about δ 1.7.

13-41 The multiplicity in the off-resonance decoupled spectrum is given below each chemical shift: s = singlet; d = doublet; t = triplet; q = quartet. It is often difficult to predict exact chemical shift values; your predictions should be in the right vicinity. There should be no question about the multiplicity and DEPT spectra, however.

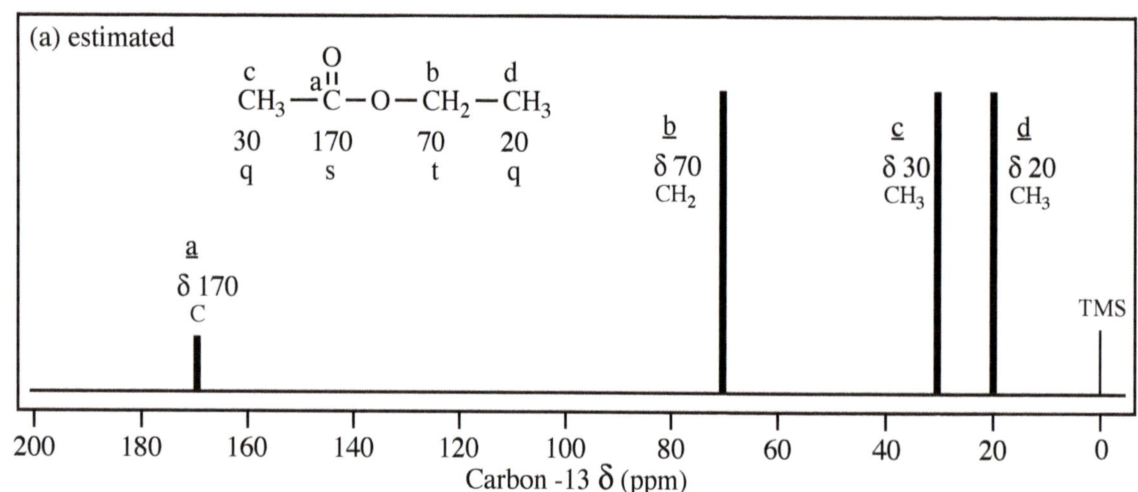

The DEPT-90 spectrum for ethyl acetate would have no peaks because there are no CH groups.

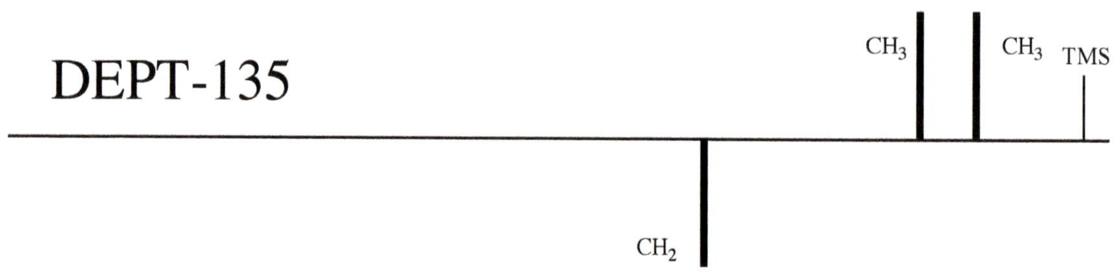

13-41 continued

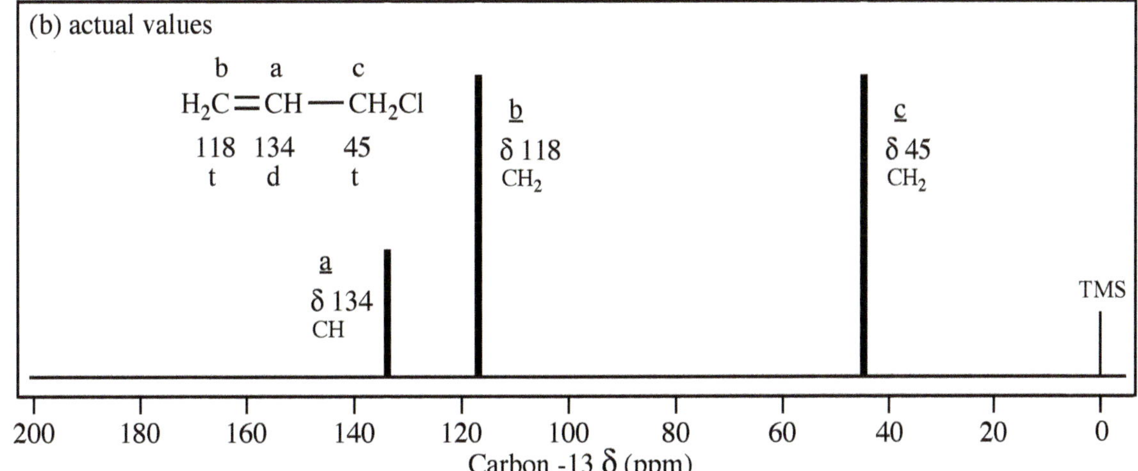

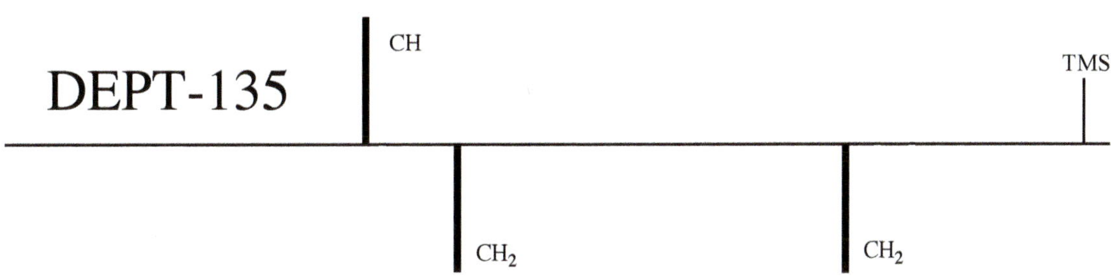

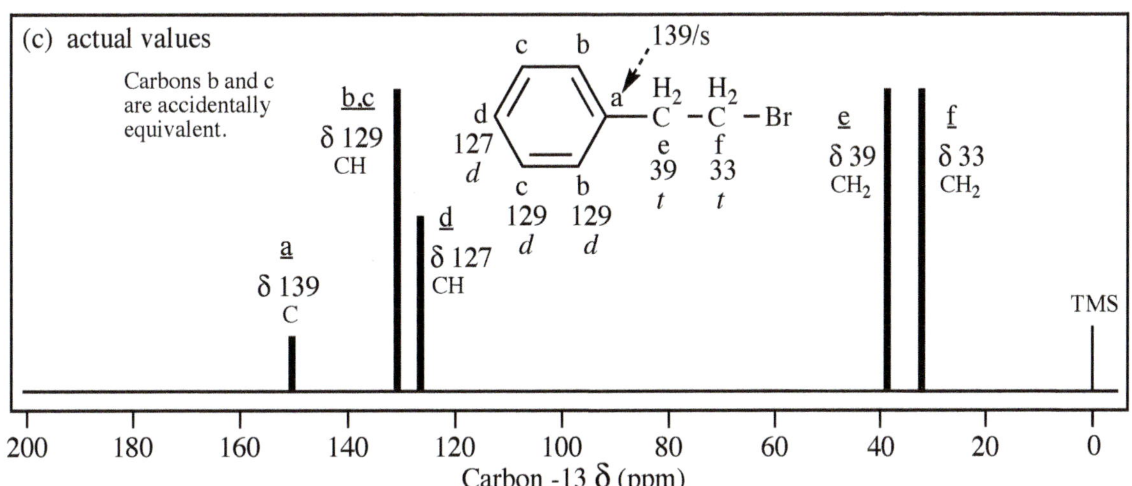

DEPT-90 and DEPT-135 on next page

13-41(c) continued

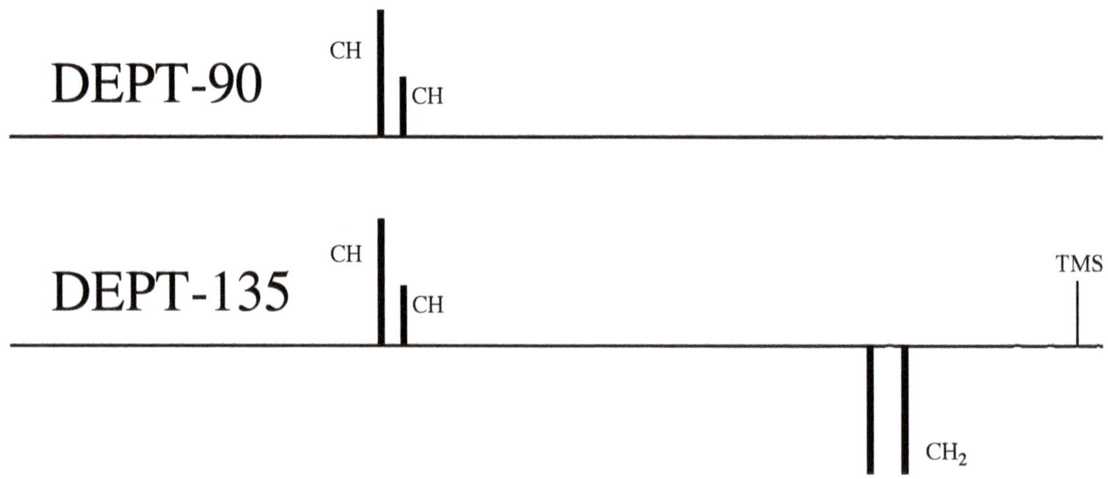

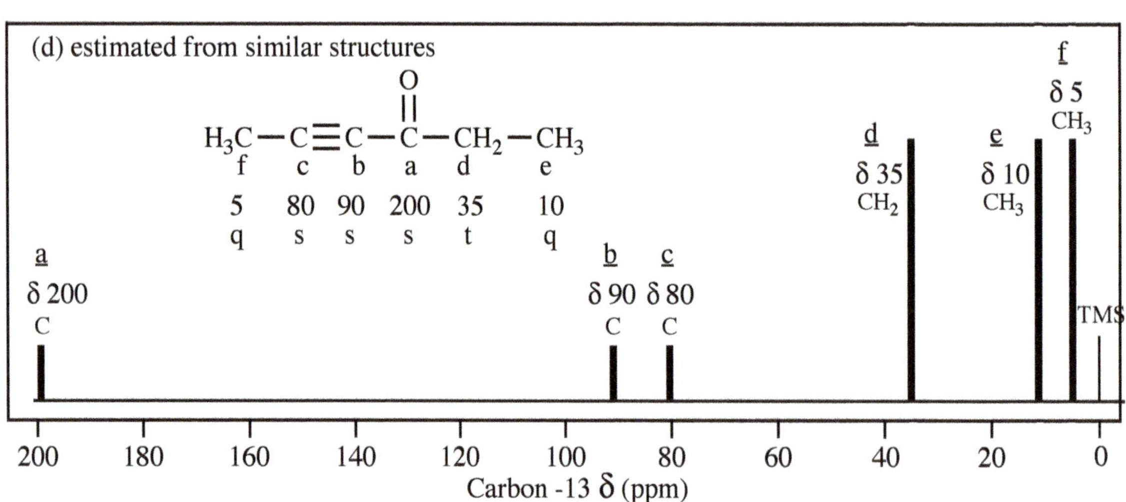

The DEPT-90 spectrum would have no peaks because there are no CH groups.

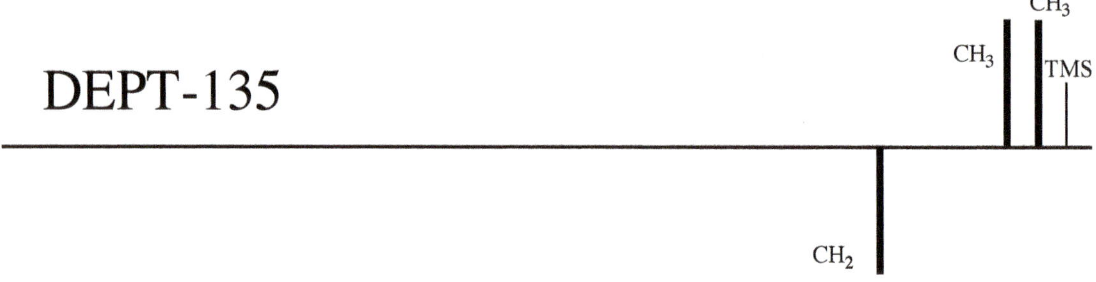

13-42 The multiplicity of the peaks in this off-resonance-decoupled spectrum show two different CHs and a CH_3 . There is only one way to assemble these pieces with three chlorines. (The multiplet at 0 is TMS.)

CH_3—CH(Cl)—CHCl_2

δ 20 q (CH_3); δ 60 d (CHCl); δ 75 d ($CHCl_2$)

13-43 There is no evidence for vinyl hydrogens, so the double bond is gone. Integration gives eight hydrogens, so the formula must be $C_4H_8Br_2$, and the four carbons must be in a straight chain because the starting material was but-2-ene. From the integration, the four carbons must be present as one CH_3, one CH, and two CH_2 groups. The methyl is split into a doublet, so it must be adjacent to the CH. The two CH_2 groups must follow in succession, with two bromine atoms filling the remaining valences. (The spectrum is complex because the asymmetric carbon atom causes the neighboring protons to be diastereotopic.)

CH_3(d)—CHBr(a)—CH_2(c)—CH_2Br(b)

a = δ 4.3 (sextet, 1H)
b = δ 3.6 (triplet, 2H)
c = δ 2.3 (multiplet, 2H)
d = δ 1.7 (doublet, 3H)

13-44 There is no evidence for vinyl hydrogens, so the compound must be a small, saturated, oxygen-containing molecule. Starting upfield (toward TMS), the first signal is a 3H triplet; this must be a CH_3 next to a CH_2. The CH_2 is the signal at δ 1.5, but it has six peaks: it must have five neighboring hydrogens, a CH_3 on one side and a CH_2 on the other side. The third carbon must therefore be a CH_2; its signal is a quartet at δ 3.6, split by a CH_2 and an OH. To be so far downfield, the final CH_2 must be bonded to oxygen. The remaining 1H signal must be from an OH. The compound must be propan-1-ol.

b a c d
$HOCH_2CH_2CH_3$

a = δ 3.6 (quartet, 2H)
b = δ 3.2 (triplet, 1H)
c = δ 1.5 (6 peaks, 2H)
d = δ 0.9 (triplet, 3H)

13-45

CH_3—C(CH_3)(Cl)—CH_2CH_3 —base→ $(CH_3)_2C$=CHCH_3 (Isomer A) + H_2C=C(CH_3)CH_2CH_3 (Isomer B)

(a)

Isomer A: $(CH_3)_2C$=CHCH_3 — b and c are the two CH_3 groups on one carbon, a is the vinyl H, d is the CH_3 on the other carbon.

Isomer B: H_2C=C(CH_3)CH_2CH_3 — d is the two vinyl H, f is the CH_3, e and g are the CH_2 and CH_3 of the ethyl.

Isomer A
a = δ 5.2 (quartet, 1H)
b = δ 1.7 (singlet, 3H)
c = δ 1.6 (singlet, 3H)
d = δ 1.5 (doublet, 3H)

Isomer B
d = δ 4.7 (singlet, 2H)
e = δ 2.0 (quartet, 2H)
f = δ 1.7 (singlet, 3H)
g = δ 1.0 (triplet, 3H)

(b) With NaOH as base, the more highly substituted alkene, Isomer A, would be expected to predominate—the Zaitsev Rule. With KO-*t*-Bu as a hindered, bulky base, the less substituted alkene, Isomer B, would predominate (the Hofmann product).

13-46 "Nuclear waste" is composed of radioactive products from either nuclear reactions—for example, from electrical generating stations powered by nuclear reactors—or residue from medical or scientific studies using radioactive nuclides as therapeutic agents (like iodine for thyroid treatment) or as molecular tracers (carbon-14, tritium H-3, phosphorus-32, nitrogen-15, and many others). The physical technique of *nuclear* magnetic resonance neither uses nor generates any radioactive elements, and does not generate "nuclear waste". (Some people assume that the medical application of NMR, **m**agnetic **r**esonance **i**maging or MRI, purposely dropped the word "nuclear" from the technique to avoid the confusion between *nuclear* and *radioactive*.)

13-47

Mass spectrum: The molecular ion of m/z 117 suggests the presence of an odd number of nitrogens.
Infrared spectrum: No NH or OH appears. Hydrogens bonded to both sp^2 and sp^3 carbon are indicated around 3000 cm^{-1}. The characteristic C≡N peak appears at 2250 cm^{-1} and aromatic C=C is suggested by the peak at 1600 cm^{-1}.
NMR spectrum: Five aromatic protons are shown in the NMR at δ 7.3. A CH_2 singlet appears at δ 3.7.
Assemble the pieces:

C_6H_5– (mass 77) + CH_2 (mass 14) + C≡N (mass 26) → C_6H_5–CH_2–C≡N (mass 117)

13-48 This is a challenging problem, despite the molecule being relatively small.
Mass spectrum: The molecular ion at 96 suggests no Cl, Br, or N. The molecule must have seven carbons or fewer.
Infrared spectrum: The dominant functional group peak is at 1685 cm^{-1}, a carbonyl that is conjugated with C=C (lower wave number than normal, very intense peak). The presence of an oxygen and a molecular ion of 96 lead to a formula of C_6H_8O, with three elements of unsaturation, a C=O and one or two C=C.
Carbon NMR spectrum: The six peaks show, by chemical shift, one carbonyl carbon (196), two alkene carbons (129, 151), and three aliphatic carbons (23, 26, 36). From the DEPT information at the top of the spectrum (blue letters), the groups are: three CH_2 groups, two alkene CH groups, and carbonyl.

Since the structure has one carbonyl and only two alkene carbons, the third element of unsaturation must be a ring.

HC=CH–C=O CH_2 + CH_2 + CH_2 + 1 ring

Since the structure has no methyl group, and no H_2C=, all of the carbons must be included in the ring. The only way these pieces can fit together is in cyclohex-2-enone. Notice that the proton NMR was unnecessary to determine the structure, fortunately, since the HNMR was not easily interpreted except for the two alkene hydrogens; the hydrogen on carbon-2 appears as the doublet at 6.0.

cyclohex-2-enone ⟹ [CH_2=CH–CH=C=O]$^{+\cdot}$ m/z 68 + H_2C=CH_2 loss of mass 28

The mystery mass spec peak at m/z 68 comes from a fragmentation that will be discussed later; it is called a retro-Diels-Alder fragmentation.

13-49 The key to the carbon NMR lies in the symmetry of these structures.

ortho-xylene   *meta*-xylene   *para*-xylene

(a) In each molecule, the methyl carbons are equivalent, giving one signal in the CNMR. Considering the ring carbons, the symmetry of the structures shows that *ortho*-xylene would have 3 carbon signals from the ring (total of 4 peaks), *meta*-xylene would have 4 carbon signals from the ring (total of 5 peaks), and *para*-xylene would have only 2 carbon signals from the ring (total of 3 peaks). These compounds would be instantly identifiable simply by the number of peaks in the carbon NMR.

(b) The proton NMR would be a completely different problem. Unless the substituent on the benzene ring is moderately electron-withdrawing or donating, the ring protons absorb at roughly the same position. A methyl group has essentially no electronic effect on the ring hydrogens, so while the *para* isomer would give a clean singlet because all its ring protons are equivalent, the *ortho* and *meta* isomers would have only slightly broadened singlets for their proton signals. (Only a very high field NMR, 500 MHz or higher, would be able to distinguish these isomers in the proton NMR.)

13-50 (a), (b) and (c) The six isomers are drawn here. Below each structure is the number of proton signals and the number of carbon signals. (Note that splitting patterns would give even more clues in the proton NMR.)

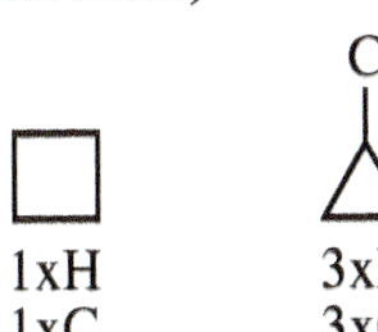

1xH	3xH*	2xH	2xH	2xH	5xH
1xC	3xC	3xC	2xC	2xC	4xC

*Rings always present challenges in stereochemistry. When viewed in three dimensions, it becomes apparent that the two hydrogens on a CH_2 are not equivalent: on each CH_2, one H is *cis* to the methyl and one H is *trans* to the methyl. These are diastereotopic protons. A more correct answer to part (b) would be four types of protons; whether all four could be distinguished in the NMR is a harder question to answer. For the purpose of this problem, whether it is 3 or 4 types of H does not matter because either one, in combination with three types of carbon, will distinguish it from the other 5 structures.

(d) Two types of H and three types of C can be only one isomer: 2-methylpropene (isobutylene). (The only isomers that would not be distinguished from each other would be *cis*- and *trans*-but-2-ene.)

13-51

	peaks in CNMR	peaks in DEPT-90	peaks in DEPT-135
	6 peaks	1 peak (c)	up: 1 CH (c), 1 CH_3 (a) down: 4 CH_2 (b,d,e,f)
	5 peaks	2 peaks (b,c) of equal height	up: 2 CH (b,c), 1 CH_3 (a) down: 2 CH_2 (d,e)
	5 peaks	1 peak (c)	up: 1 CH (c), 1 CH_3 (a) down: 2 CH_2 (d,e)
	5 peaks	2 peaks (b,c) of unequal height	up: 2 CH (b,c), 2 CH_3 (a,e) down: 1 CH_2 (d)

This example clearly shows the utility of DEPT NMR. These four compounds cannot be distinguished easily by regular CNMR. The DEPT-90 in combination with the regular CNMR could distinguish them. However, the easiest method that would give an unambiguous assignment for each structure would be the DEPT-135 where each structure has a unique pattern of peaks.

13-52 Symmetry and chemical shift of the CH_3 signals in the CNMR can distinguish these 5 compounds.

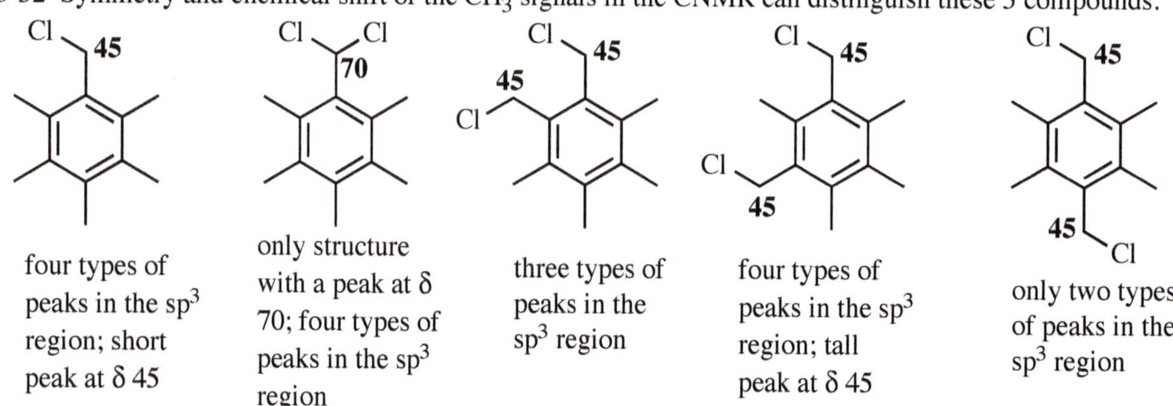

Cl
45
Cl
Cl
70
Cl
45
Cl
45
Cl
45
Cl
45
Cl
45
45
Cl
four types of peaks in the sp³ region; short peak at δ 45
only structure with a peak at δ 70; four types of peaks in the sp³ region
three types of peaks in the sp³ region
four types of peaks in the sp³ region; tall peak at δ 45
only two types of peaks in the sp³ region

13-53

Protons in the description are shown in bold.

Structure 1	Structure 2	Structure 3	Structure 4
(d) broad 1H singlet at δ 11.5 and 2H triplet at δ 2.3	(c) sharp 3H singlet at δ 3.7 and 2H quartet at δ 2.3	(b) sharp 3H singlet at δ 2.0 and 2H quartet at δ 4.1	(a) sharp 1H singlet at δ 8.0 and 2H triplet at δ 4.0

13-54 Note that chemical shift values are not needed to differentiate these four compounds. Symmetry is especially important in CNMR spectra.

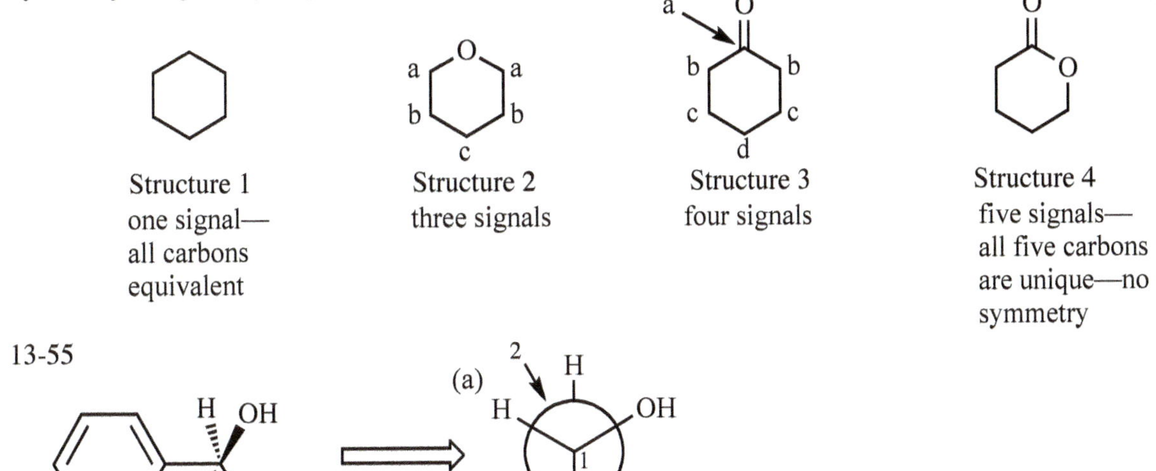

Structure 1
one signal—
all carbons
equivalent

Structure 2
three signals

Structure 3
four signals

Structure 4
five signals—
all five carbons
are unique—no
symmetry

13-55

(a)

(b) In the starting aldehyde, the two methyl groups are equivalent. In the product of the Grignard reaction, however, a new chiral center is created, and the two methyl groups are no longer equivalent. As the Newman projection demonstrates, rotating the back carbon cannot ever put the two methyls into identical environments. This is reflected in the NMR as separate signals with different chemical shifts. These methyl groups are called *diastereotopic*.

13-56

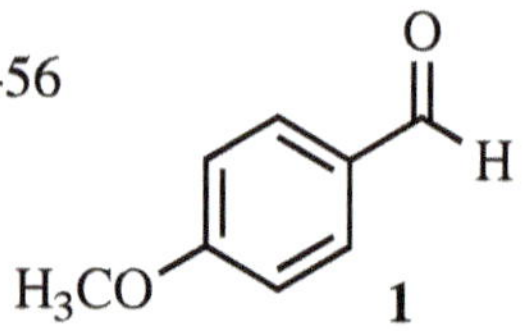

(a) IR only: all have benzene and C=O; **1** has aldehyde C—H at 2710 and 2810; **2** has OH at 3100-3500; **3** has characteristic COOH from 2500-3500 with shoulder around 2700 cm^{-1}

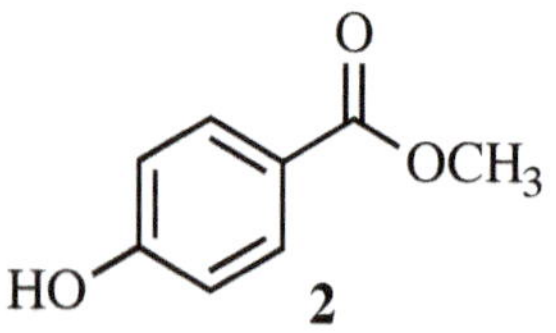

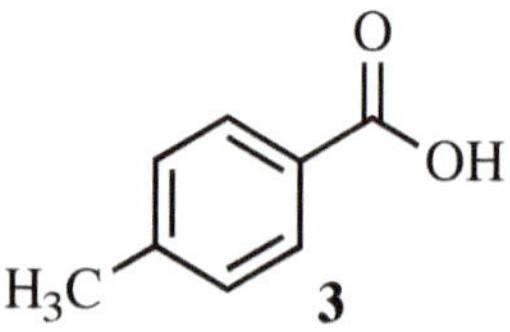

(b) HNMR only: all have benzene para pattern (two doublets): **1** and **2** have peaks at δ 6.8, **3** does not; **1** has sharp singlet from δ 9-10; **3** has broad singlet from δ 10-14; the CH_3 in **3** is at δ 2.5, while in **1** and **2** the OCH_3 is at δ 3.7.

(c) CNMR only: **1** has the C=O around δ 190; **2** and **3** have C=O around δ 170; **2** has a benzene peak at δ 160 from C—O; all of the benzene signals in **3** are from δ 120-140. Also, the CH_3 peaks in **1** and **2** are from δ 50-60 whereas **3** has CH_3 signal around δ 20.

CHAPTER 14—ETHERS, EPOXIDES AND THIOETHERS

14-1
The four solvents decrease in polarity in this order: water, ethanol, ethyl ether, and dichloromethane. The three solutes decrease in polarity in this order: sodium acetate (ionic, most polar), 2-naphthol, and naphthalene (no electronegative atoms). The guiding principle to determine solubility is, "Like dissolves like." Compounds of similar polarity will dissolve (in) each other. Thus, sodium acetate will dissolve in water, will dissolve only slightly in ethanol, and will be virtually insoluble in ethyl ether and dichloromethane. 2-Naphthol will be insoluble in water, somewhat soluble in ethanol, and soluble in ether and dichloromethane. Naphthalene will be insoluble in water, partially soluble in ethanol, and soluble in ethyl ether and dichloromethane. (Actual solubilities are difficult to predict, but you should be able to predict *trends*.)

14-2

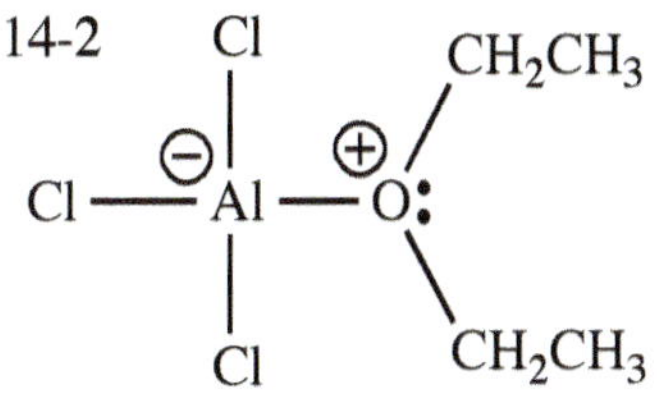

Cl, Cl, Cl, Al ⊖, O ⊕, CH_2CH_3, CH_2CH_3

Oxygen shares one of its electron pairs with aluminum; oxygen is the Lewis base, and aluminum is the Lewis acid. An oxygen atom with three bonds and one unshared pair has a positive formal charge. An aluminum atom with four bonds has a negative formal charge.

14-3 The crown ether has two effects on $KMnO_4$: first, it makes $KMnO_4$ much more soluble in benzene; second, it holds the potassium ion tightly, making the permanganate more available for reaction. Chemists call this a "naked anion" because it is not complexed with solvent molecules.

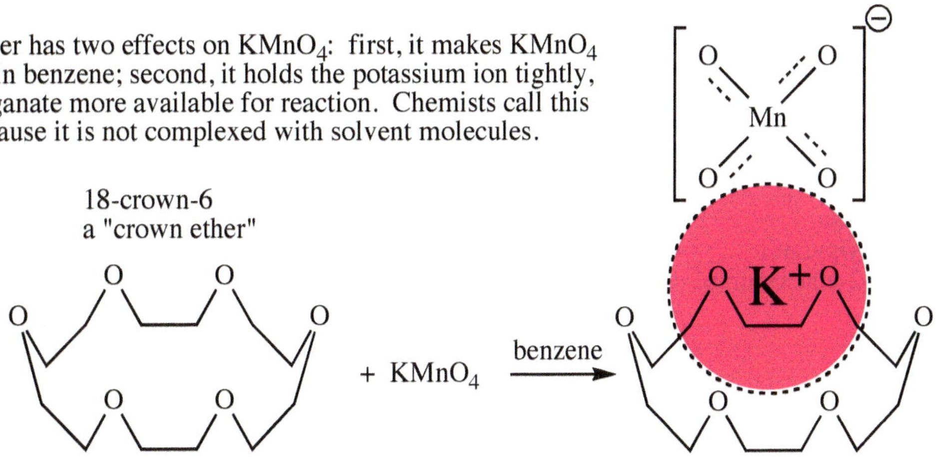

18-crown-6
a "crown ether"

+ $KMnO_4$ —benzene→ K^+ complex; $[MnO_4]^-$

14-4 IUPAC name first; then common name (see Appendix 5 in the text for a summary of IUPAC nomenclature). *Current IUPAC recommendations place the position number immediately before the group that it modifies.*

(a) methoxyethene; methyl vinyl ether
(b) 2-ethoxypropane; ethyl isopropyl ether
(c) 1-chloro-2-methoxyethane; 2-chloroethyl methyl ether
(d) 2-ethoxy-2,3-dimethylpentane; no common name
(e) 1,1-dimethoxycyclopentane; no common name
(f) *trans*-2-methoxycyclohexan-1-ol; also possible: (1*R*,2*R*)-2-methoxycyclohexan-1-ol; no common name
(g) methoxycyclopropane; cyclopropyl methyl ether
(h) 1-methoxybut-2-yne; no common name
(i) (*Z*)-2-methoxypent-2-ene; no common name

14-5

(a) HO–CH₂CH₂–OH + HO–CH₂CH₂–OH —H^+→ 1,4-dioxane + **2** H_2O

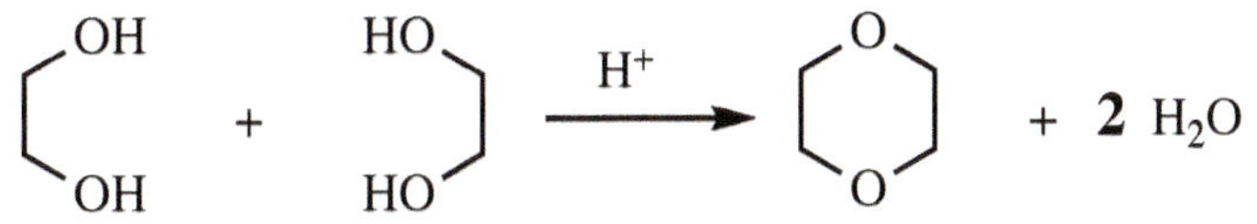

The alcohol is ethane-1,2-diol; the common name is ethylene glycol.

14-5 continued

(b)

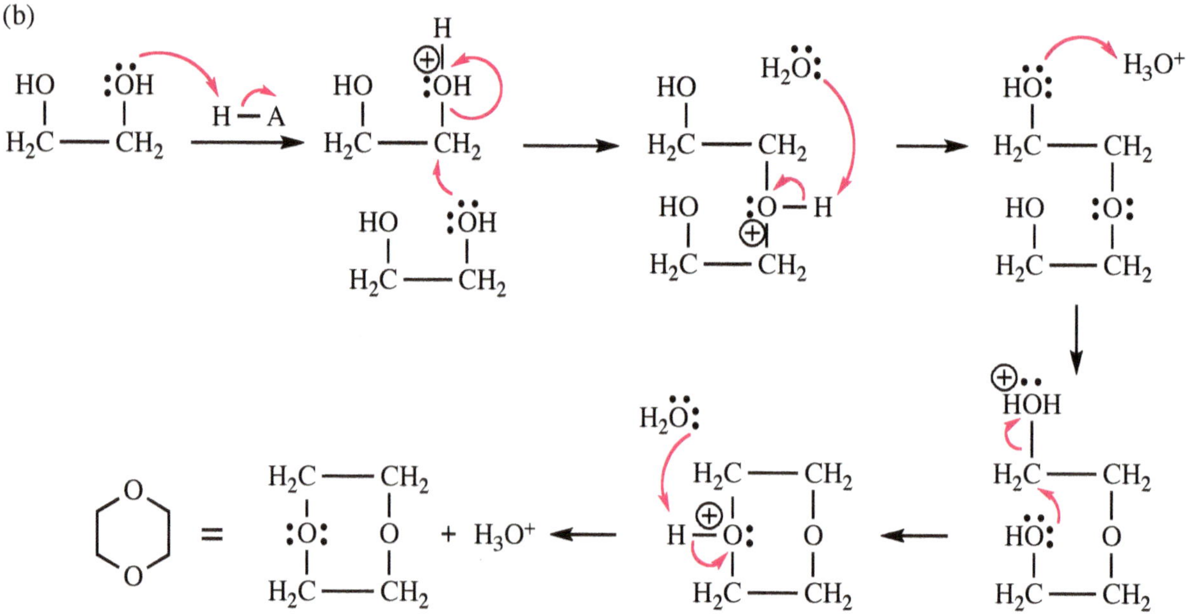

HO
:ÖH
H₂C
CH₂
H—A
H
⊕
H₂Ö:
H₃O⁺
HÖ:
:O—H
:O:
⊕..
HOH
HÖ:
O
H—Ö:
H₂C—CH₂
:O:
+ H₃O⁺

The mechanism shows that the acid catalyst is regenerated at the end of the reaction.

14-6

(a) dihydropyran
(b) 2-chloro-1,4-dioxane
(c) 3-isopropylpyran
(d) *trans*-2,3-diethyloxirane; *trans*-3,4-epoxyhexane; *trans*-hex-3-ene oxide
(e) 3-bromo-2-ethoxyfuran
(f) 3-bromo-2,2-dimethyloxetane

14-7

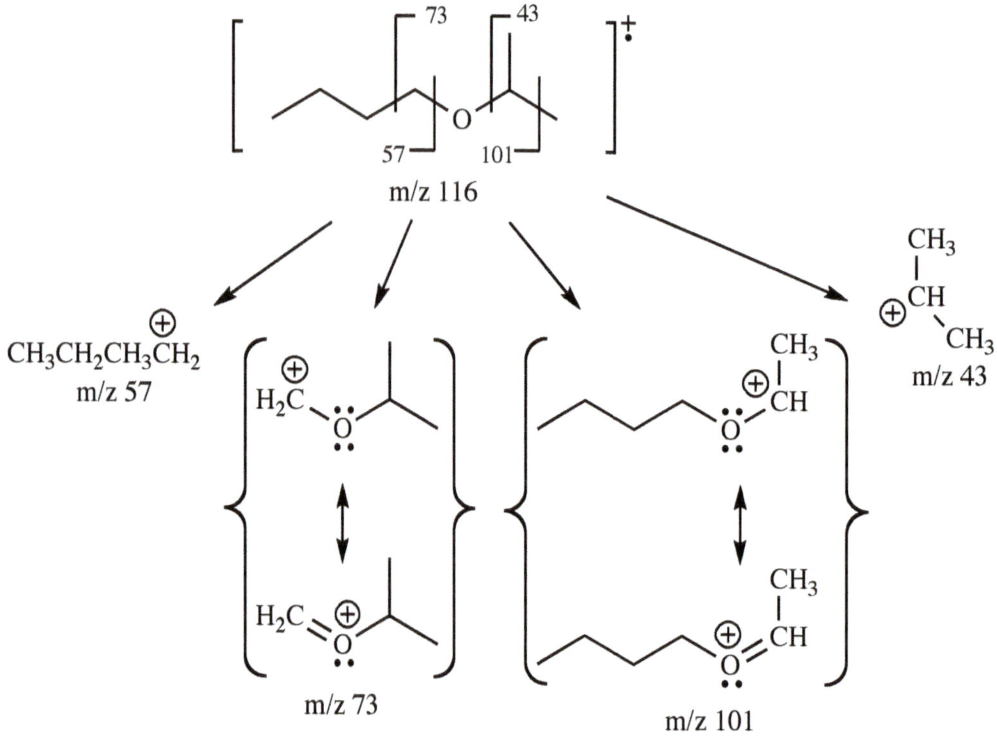

73
43
57
101
O
m/z 116
CH₃CH₂CH₃CH₂
m/z 57
H₂C
Ö
m/z 73
CH₃
CH
m/z 101
CH₃
CH
CH₃
m/z 43

14-8 S_N2 reactions, including the Williamson ether synthesis, work best when the nucleophile attacks a 1° or methyl carbon. Instead of attempting to form the bond from oxygen to the 2° carbon on the ring, form the bond from oxygen to the 1° carbon of the butyl group.

The OH must first be transformed into a good leaving group: either a tosylate, or one of the halides (not fluoride).

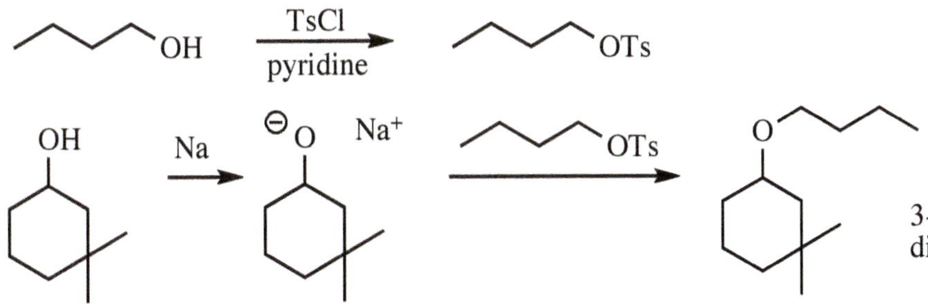

OH —(TsCl, pyridine)→ OTs

OH —(Na)→ $^{\ominus}$O Na^+ —(OTs)→ O

3-butoxy-1,1-dimethylcyclohexane

14-9 Always put the leaving group on the less substituted carbon to maximize substitution: S_N2 favors methyl > 1° >> 2°.

(a) 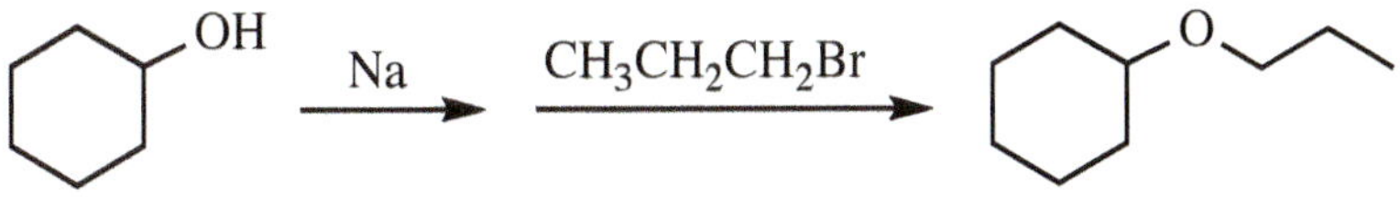 OH —(Na)→ —($CH_3CH_2CH_2Br$)→ O

(b) 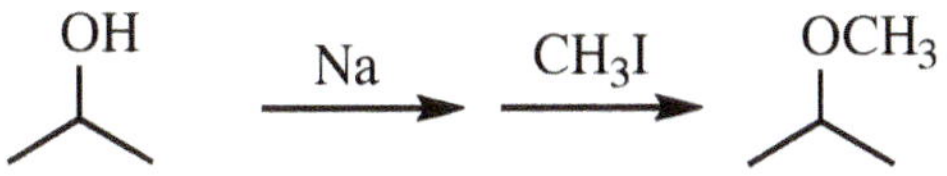 OH —(Na)→ —(CH_3I)→ OCH_3

(c) 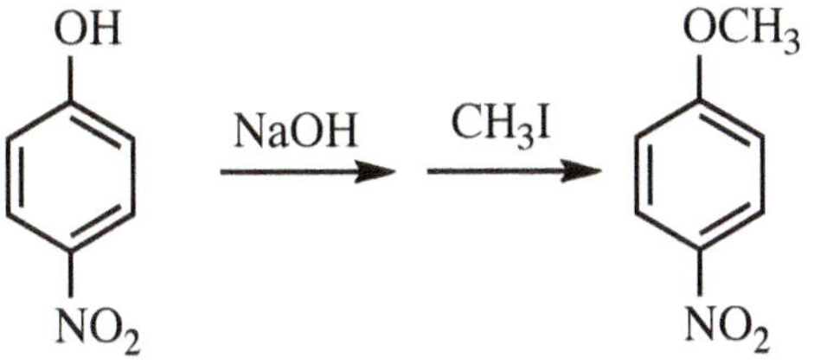 OH, NO_2 —(NaOH)→ —(CH_3I)→ OCH_3, NO_2

(d) $CH_3CH_2CH_2OH$ —(Na)→ —(CH_3CH_2Br)→ $CH_3CH_2CH_2OCH_2CH_3$

CH_3CH_2OH —(Na)→ —($CH_3CH_2CH_2Br$)→ $CH_3CH_2CH_2OCH_2CH_3$

Both carbons bonded to oxygen are 1°, so either method will work.

(e) 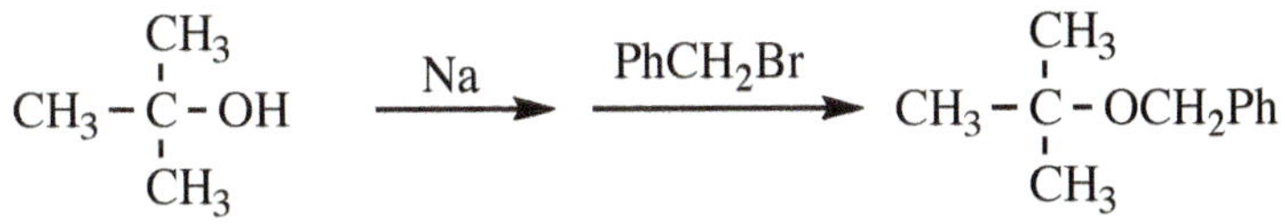 $CH_3-\underset{CH_3}{\overset{CH_3}{C}}-OH$ —(Na)→ —($PhCH_2Br$)→ $CH_3-\underset{CH_3}{\overset{CH_3}{C}}-OCH_2Ph$

14-10

(a) (1) 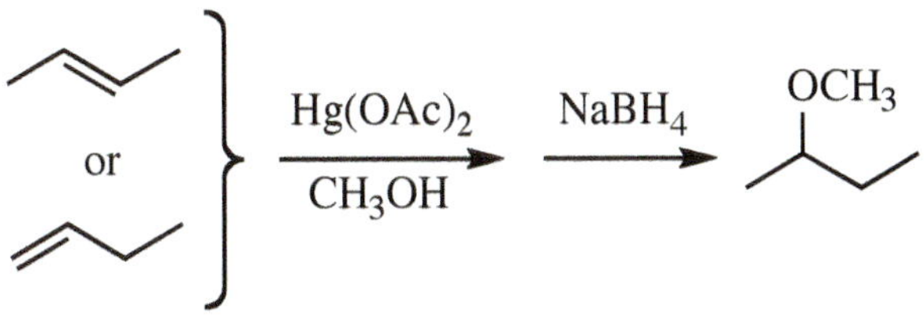 [alkene] or [alkene] —($Hg(OAc)_2$, CH_3OH)→ —($NaBH_4$)→ OCH_3

(2)  OH —(Na)→ —(CH_3I)→ OCH_3

14-10 continued

(b) (1) cyclohexene $\xrightarrow[CH_3CH_2OH]{Hg(OAc)_2}$ $\xrightarrow{NaBH_4}$ cyclohexyl–OCH_2CH_3

(2) cyclohexanol (OH) $\xrightarrow{Na}$ $\xrightarrow{CH_3CH_2Br}$ cyclohexyl–OCH_2CH_3

(c) (1) Alkoxymercuration is not practical here; the product does not have Markovnikov orientation.

(2) 2-methylcyclopentanol (OH) $\xrightarrow{Na}$ $\xrightarrow{CH_3I}$ OCH_3

(d) (1) 1-methylcyclopentene $\xrightarrow[CH_3OH]{Hg(OAc)_2}$ $\xrightarrow{NaBH_4}$ 1-methyl-1-OCH_3-cyclopentane (methylenecyclopentane also possible as a starting material)

(2) 1-methylcyclopentanol (OH) $\xrightarrow{Na}$ $\xrightarrow{CH_3I}$ OCH_3

(e) (1) 1-methylcyclopentene $\xrightarrow[\text{(CH}_3)_2\text{CHOH}]{Hg(OAc)_2}$ $\xrightarrow{NaBH_4}$ O–isopropyl (methylenecyclopentane also possible as a starting material)

OH

(2) Williamson ether synthesis would give a poor yield of product as the halide is on a 2° carbon.

OH $\xrightarrow{Na}$ $\xrightarrow{\text{2-bromopropane (Br)}}$ O (2°)

(f) (1) isobutylene $\xrightarrow[\text{phenol (OH)}]{Hg(OAc)_2}$ $\xrightarrow{NaBH_4}$ phenyl–O–tert-butyl

(2) Williamson ether synthesis is not feasible here. S_N2 does not work on either a benzene or a 3° halide.

14-11 An important principle of synthesis is to avoid mixtures of isomers wherever possible; minimizing separations increases recovery of products. Bimolecular condensation is a random process, assuming similar structures for the two alkyl groups on the ether. Heating a mixture of ethanol and methanol with acid will produce all possible combinations: dimethyl ether, ethyl methyl ether, and diethyl ether. This mixture would be troublesome to separate.

14-12

Ether formation

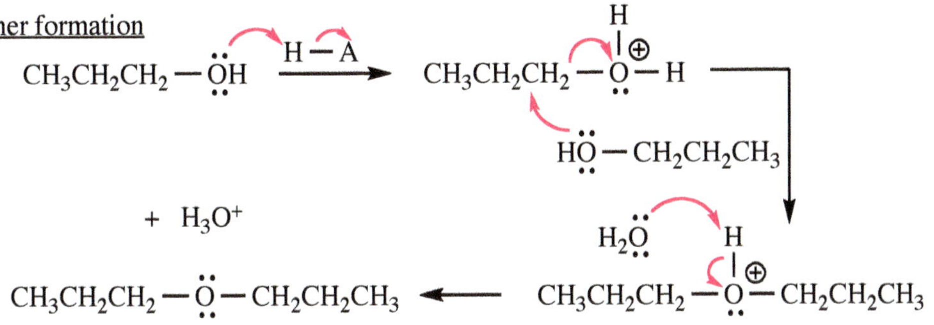

$CH_3CH_2CH_2—\ddot{O}H \xrightarrow{H—A} CH_3CH_2CH_2—\overset{+}{O}H_2$; with $H\ddot{O}—CH_2CH_2CH_3$ → $CH_3CH_2CH_2—\overset{+}{O}(H)—CH_2CH_2CH_3$; with $H_2\ddot{O}$ → $CH_3CH_2CH_2—\ddot{O}—CH_2CH_2CH_3$ + H_3O^+

Dehydration

$CH_3CH_2CH_2—\ddot{O}H \xrightarrow{H—A} CH_3CH(H)CH_2—\overset{+}{O}H_2$; with $H_2\ddot{O}$ → $CH_3CH=CH_2$ + H_2O

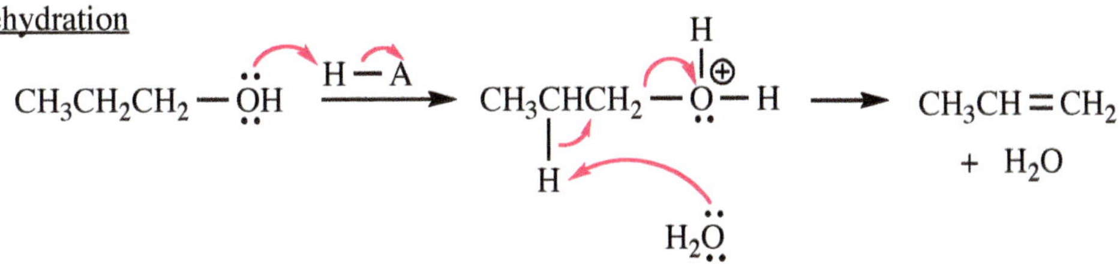

Remember $\Delta G = \Delta H - T\Delta S$? Thermodynamics of a reaction depend on the sign and magnitude of ΔG. As temperature increases, the entropy term grows in importance. In ether formation, the ΔS is small because two molecules of alcohol give one molecule of ether plus one molecule of water—no net change in the number of molecules. In dehydration, however, one molecule of alcohol generates one molecule of alkene plus one molecule of water—a large increase in entropy. So $T\Delta S$ is more important for dehydration than for ether formation. As temperature increases, the competition will shift toward more dehydration (elimination).

14-13

(a) This symmetrical ether at 1° carbons could be produced in good yield by bimolecular condensation.

(b) This unsymmetrical ether could not be produced in high yield by bimolecular condensation. Williamson synthesis would be preferred.

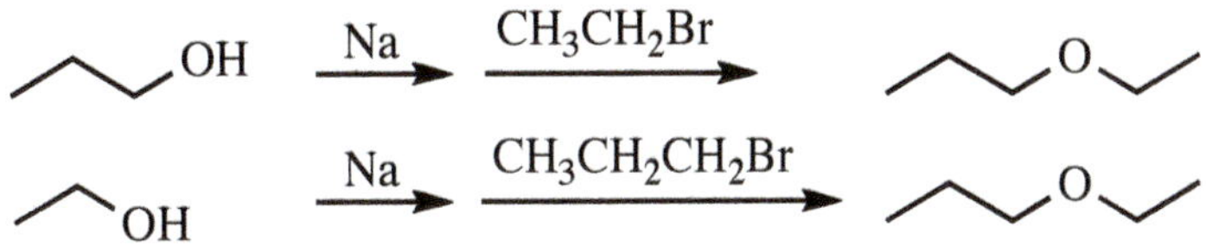

propanol (OH) $\xrightarrow{Na}$ $\xrightarrow{CH_3CH_2Br}$ ethyl propyl ether (O)

ethanol (OH) $\xrightarrow{Na}$ $\xrightarrow{CH_3CH_2CH_2Br}$ ethyl propyl ether (O)

(c) Even though this ether is symmetrical, both carbons are 2°, so bimolecular condensation would give low yields. Unimolecular dehydration to give alkenes would be the dominant pathway. Alkoxymercuration-demercuration is the preferred route.

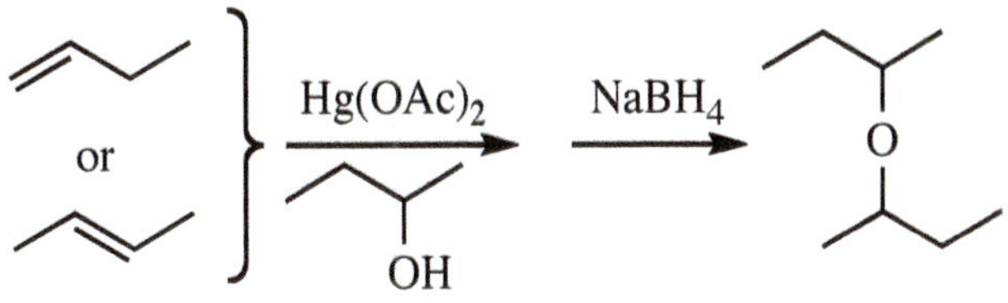

1-butene or 2-butene, with 2-butanol (OH) $\xrightarrow{Hg(OAc)_2}$ $\xrightarrow{NaBH_4}$ di-sec-butyl ether (O)

14-14

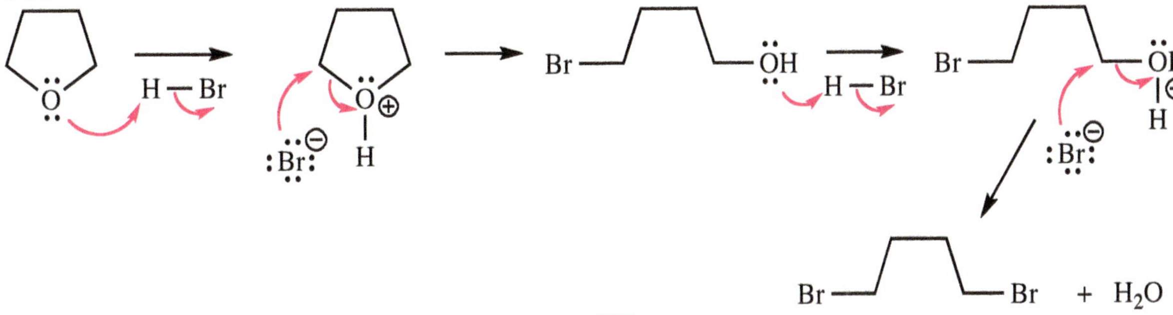

tetrahydrofuran ($\ddot{O}$) + H—Br → protonated THF ($O^{\oplus}$—H) + $:\ddot{Br}:^{\ominus}$ → Br—$(CH_2)_4$—$\ddot{O}H$ + H—Br → Br—$(CH_2)_4$—$\overset{\oplus}{O}H_2$ + $:\ddot{Br}:^{\ominus}$ → Br—$(CH_2)_4$—Br + H_2O

14-15

(a) cyclohexyl–OCH_2CH_3 $\xrightarrow{HBr}$ cyclohexyl–Br + $Br-CH_2CH_3$ + H_2O

(b) tetrahydropyran $\xrightarrow{HI}$ I–$(CH_2)_5$–I + H_2O

(c) Ph–OCH_3 $\xrightarrow{HBr}$ Ph–OH + CH_3Br

no S_N at sp^2 carbon

(d) $\xrightarrow{HI}$ OH, I

no S_N at sp^2 carbon

(e) Ph–$O-CH_2CH_2-CH(CH_3)-CH_2-O-CH_2CH_3$ $\xrightarrow{HBr}$

Ph–OH + $BrCH_2CH_2-CH(CH_3)-CH_2-Br$ + $BrCH_2CH_3$ + H_2O

no S_N at sp^2 carbon

14-16

BBr_3; H_3C; $Br^{\ominus}$; CH_3Br; $H_2\ddot{O}$:; Bu

Two proton transfers happen here: one H^+ comes off of oxygen, another H^+ goes on the other oxygen; several different scenarios of how this happens could be proposed.

Nucleophilic attack on CH_3 is faster than on 1° carbon.

BuOH + $HOBBr_2$

$\xrightarrow{2\ H_2O}$

$B(OH)_3$ + **2** HBr

14-17 Begin by transforming the alcohols into good leaving groups like halides or tosylates:

OH $\xrightarrow{PBr_3}$ Br $\xrightarrow{NaSH}$ SH $\xrightarrow{NaOH}$ $S^{\ominus}$ Na^+

OH $\xrightarrow[\text{pyridine}]{TsCl}$ OTs

S

14-18 The sulfur at the center of mustard gas is an excellent nucleophile, and chloride is a decent leaving group. Sulfur can do in *internal* nucleophilic subsitution to make a reactive sulfonium salt and the sulfur equivalent of an epoxide.

(a)

Cl, S, Cl $\xrightarrow{S_Ni}$ S⊕, Cl

very reactive alkylating agent

enzyme–$\ddot{N}H_2$ →

enzyme–HN, S, Cl

inactivated enzyme

14-18 continued

(b) NaOCl is a powerful oxidizing agent. It oxidizes sulfur to a sulfoxide or more likely a sulfone, either of which is no longer nucleophilic, preventing formation of the cyclic sulfonium salt.

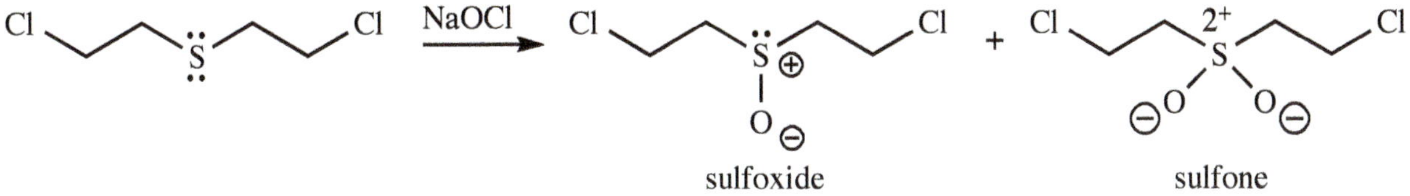

NaOCl

sulfoxide

sulfone

14-19

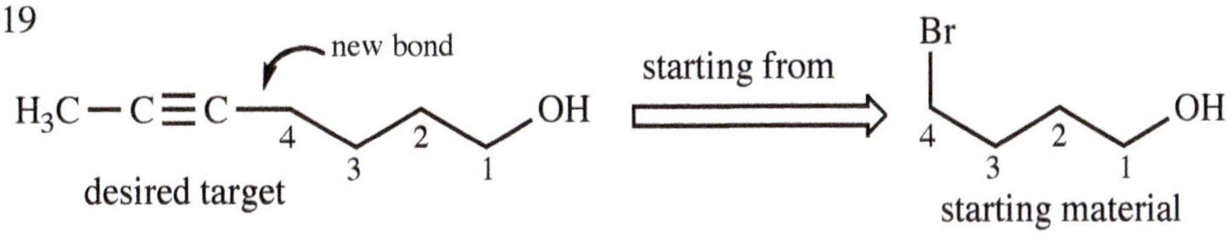

new bond

$H_3C-C\equiv C-$ ... OH

desired target

starting from

starting material

Analysis of the desired target structure shows that the new bond can be made from an acetylide ion plus the given starting material, 4-bromobutan-1-ol. However, an acetylide ion is a strong base and will remove H^+ from the alcohol, making an alkyne that is no longer nucleophilic. The alcohol must be protected before attempting the nucleophilic substitution.

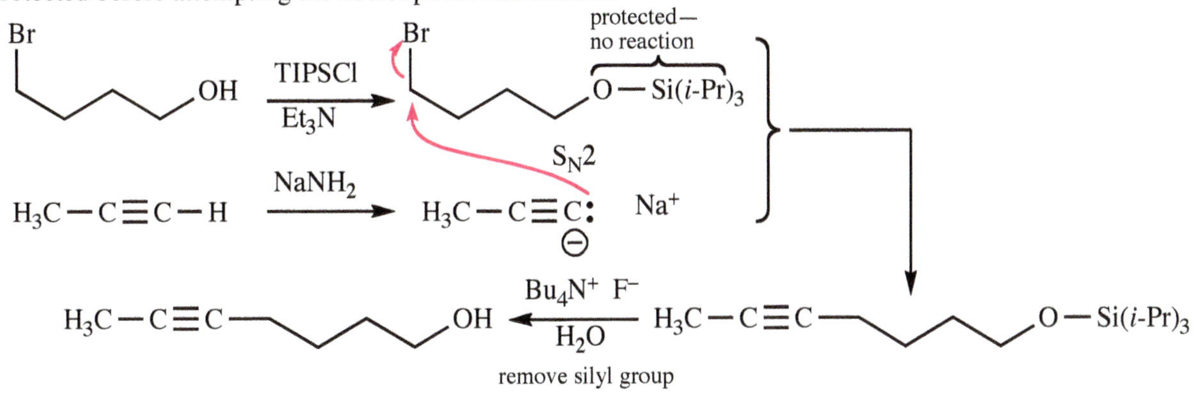

TIPSCl / Et_3N

protected—no reaction

$O-Si(i\text{-}Pr)_3$

S_N2

$H_3C-C\equiv C-H$ $\xrightarrow{NaNH_2}$ $H_3C-C\equiv C:^{-}$ Na^+

$Bu_4N^+ F^-$ / H_2O

remove silyl group

14-20

Generally, chemists prefer the peroxyacid method of epoxide formation to the halohydrin method. Reactions (a) and (b) show the peroxyacid method, but the halohydrin method could also be used.

(a)  mCPBA / CH_2Cl_2

(b) 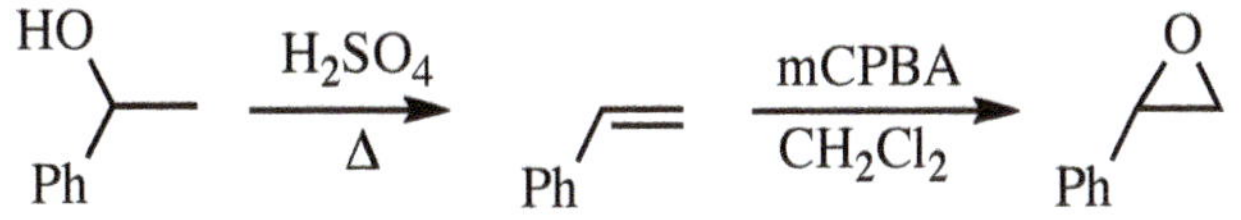 H_2SO_4 / Δ; mCPBA / CH_2Cl_2

(c) 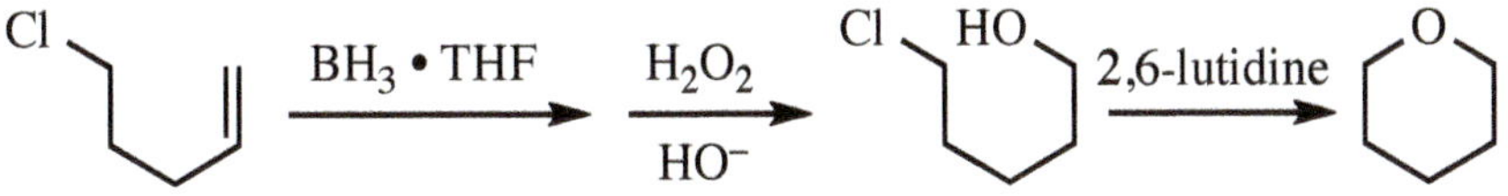 $BH_3 \bullet THF$; H_2O_2 / HO^-; 2,6-lutidine

2,6-Lutidine is a non-nucleophilic base.

(d) 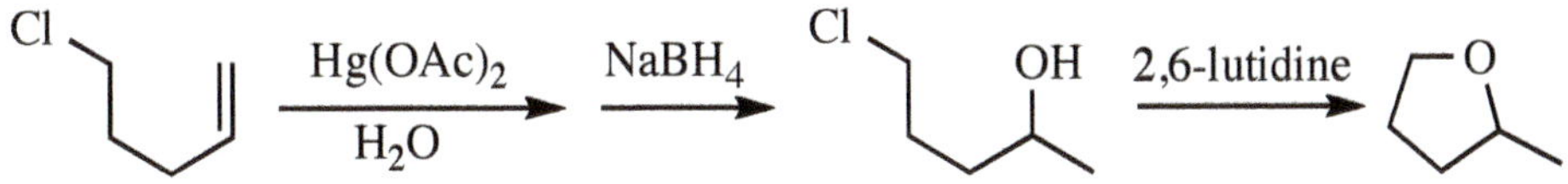 $Hg(OAc)_2$ / H_2O; $NaBH_4$; 2,6-lutidine

(e) 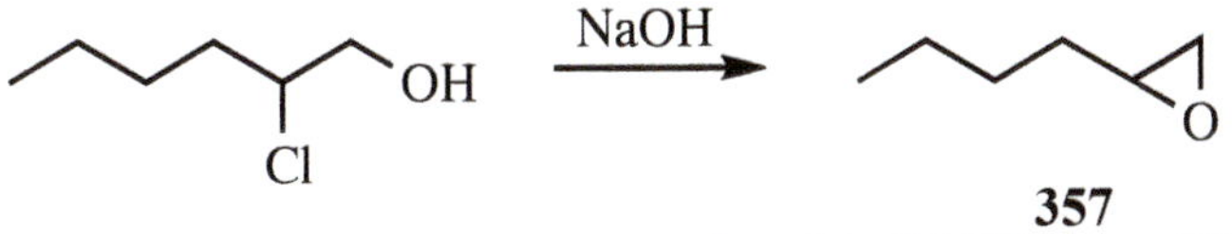 NaOH

NaOH can also do unwanted S_N2 and E2 with the chloride.

14-21

(a) 1) *tert*-Butyl hydroperoxide is the oxidizing agent. The $(CH_3)_3COOH$ contains the O—O bond just like a peroxyacid. 2) Diethyl tartrate has two asymmetric carbons and is the source of asymmetry; its function is to create a chiral transition state that is of lowest energy, leading to only one enantiomer of product. This process is called *chirality transfer*. 3) The function of the titanium(IV) isopropoxide is to act as the glue that holds all of the reagents together. The titanium holds an oxygen from each reactant—geraniol, *t*-BuOOH, and diethyl tartrate—and tethers them so that they react together, rather than just having them in solution and hoping that they will eventually collide.

(b) All three reactants are required to make Sharpless epoxidation work, but the key to *enantioselective* epoxidation is the chiral molecule, diethyl tartrate. When it complexes (or *chelates*) with titanium, it forms a large structure that is also chiral. As the *t*-BuOOH and geraniol approach the complex, the steric requirements of the complex allow the approach in one preferred orientation. When the reaction between the alkene and *t*-BuOOH occurs, it occurs preferentially from one face of the alkene, leading to one major stereoisomer of the epoxide. Without the chiral diethyl tartrate in the complex, the alkene could approach from one side just as easily as the other, and a racemic mixture would be formed.

(c) Using the enantiomer of diethyl L-tartrate, called diethyl D-tartrate, would give exactly the opposite stereochemical results, *i.e.*, the enantiomer of the first product.

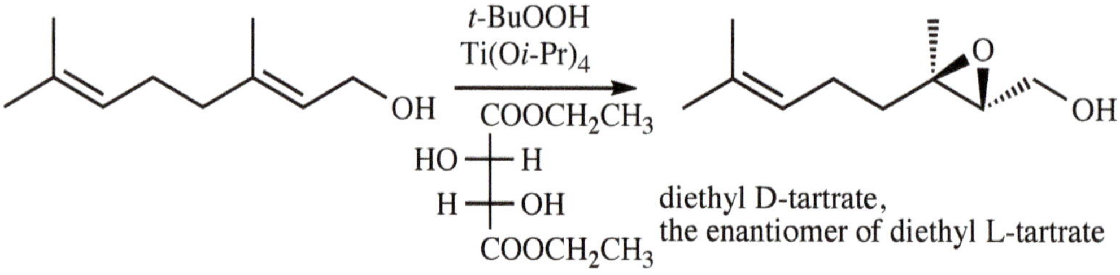

OH

t-BuOOH

Ti(O*i*-Pr)$_4$

$COOCH_2CH_3$

HO—H

H—OH

$COOCH_2CH_3$

O

OH

diethyl D-tartrate,
the enantiomer of diethyl L-tartrate

14-22

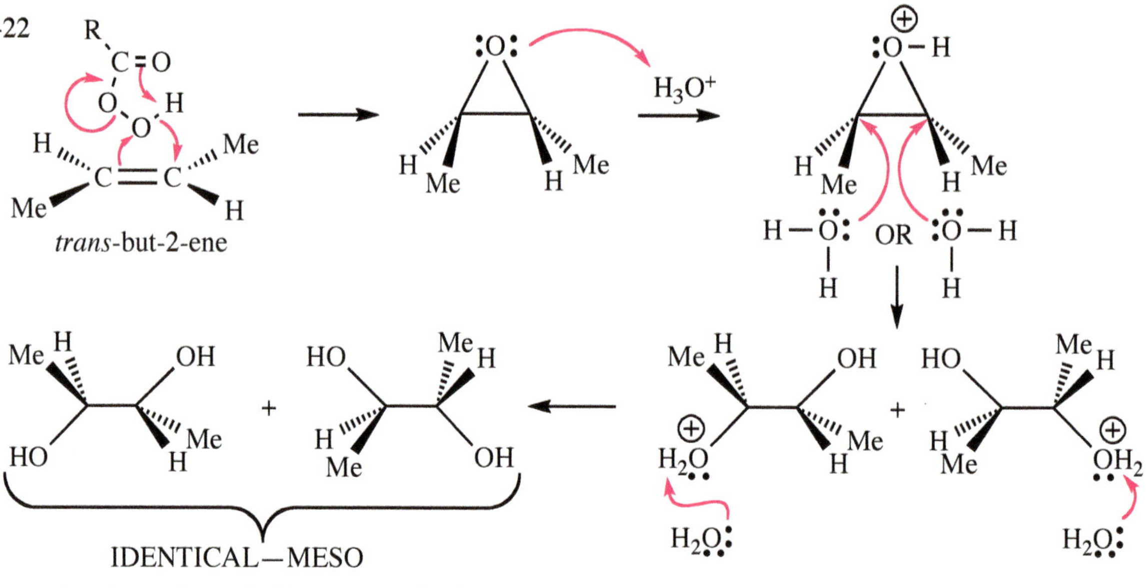

R, C=O, O, H, O, H, Me, C=C, Me, H

trans-but-2-ene

:O:

H, Me, H, Me

H_3O^+

:O—H (⊕)

H, Me, H, Me

H—Ö: OR :Ö—H

H H

Me H OH HO Me H

⊕ H_2O Me H + H Me ⊕ OH_2

H_2O: H_2O:

Me H OH HO Me H

HO Me H + H Me OH

IDENTICAL—MESO

stereochemistry shown in Newman projections:

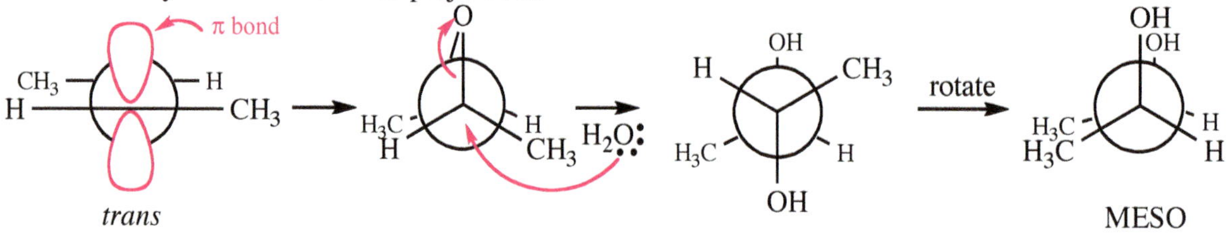

π bond

CH_3 H H CH_3

trans

O

H_3C H H CH_3 H_2O:

OH H CH_3 H_3C H OH

rotate

OH OH H_3C H_3C H H

MESO

See cis-but-2-ene on next page.

14-22 continued

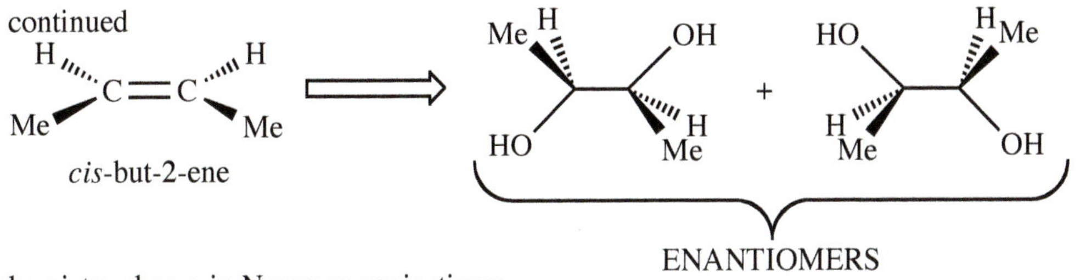

H_3C–CH=CH–CH_3 (cis-but-2-ene) ⟹ enantiomers

cis-but-2-ene

ENANTIOMERS

stereochemistry shown in Newman projections:

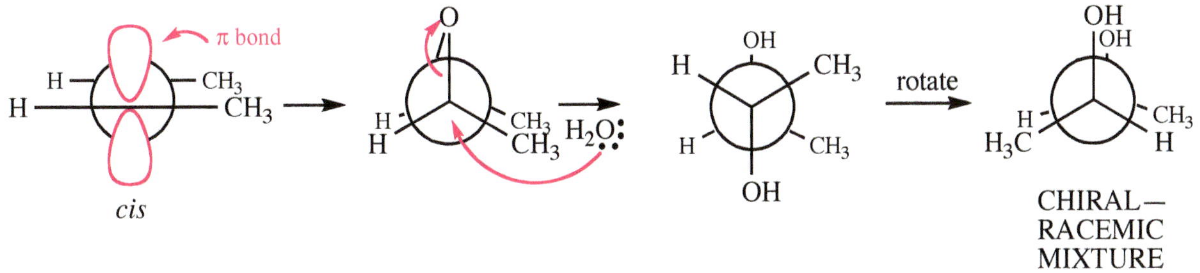

π bond

cis

rotate

CHIRAL—RACEMIC MIXTURE

14-23

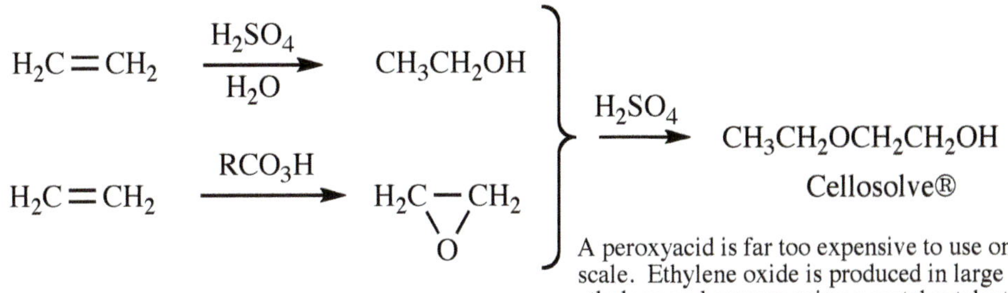

$H_2C{=}CH_2 \xrightarrow[H_2O]{H_2SO_4} CH_3CH_2OH$

$H_2C{=}CH_2 \xrightarrow{RCO_3H}$ H_2C-CH_2 (epoxide, O)

$\xrightarrow{H_2SO_4} CH_3CH_2OCH_2CH_2OH$

Cellosolve®

A peroxyacid is far too expensive to use on the industrial scale. Ethylene oxide is produced in large quantity from ethylene and oxygen using a metal catalyst.

14-24 The cyclization of squalene via the epoxide is an excellent (and extraordinary) example of how Nature uses organic chemistry to its advantage. In one enzymatic step, Nature forms four rings and eight chiral centers! Out of 256 possible stereoisomers, only one is formed!

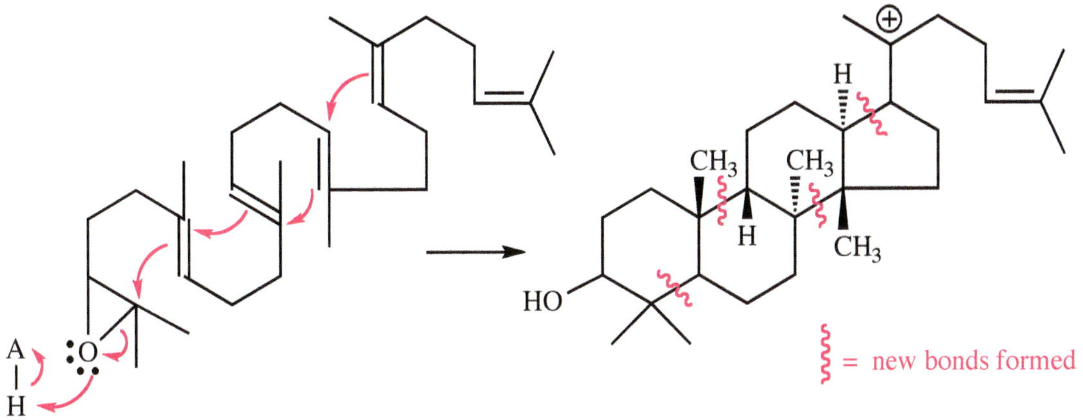

= new bonds formed

14-25

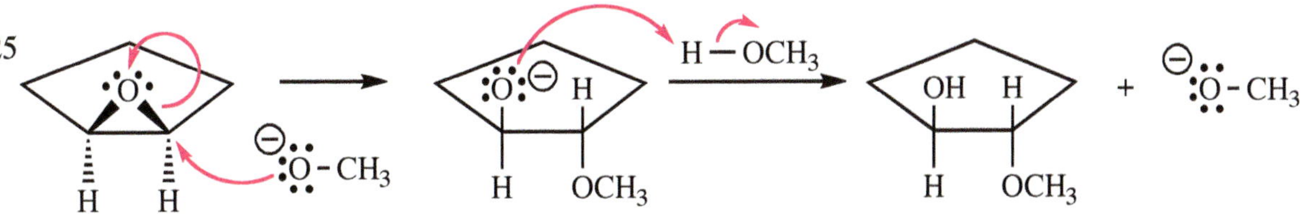

$H-OCH_3$

$^{\ominus}\!:\!\ddot{O}-CH_3$

plus the other enantiomer

14-26

(a)

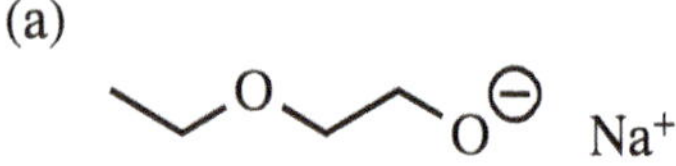

(b) (c) (d) (e) (f)

14-27

(a) (b) (c)

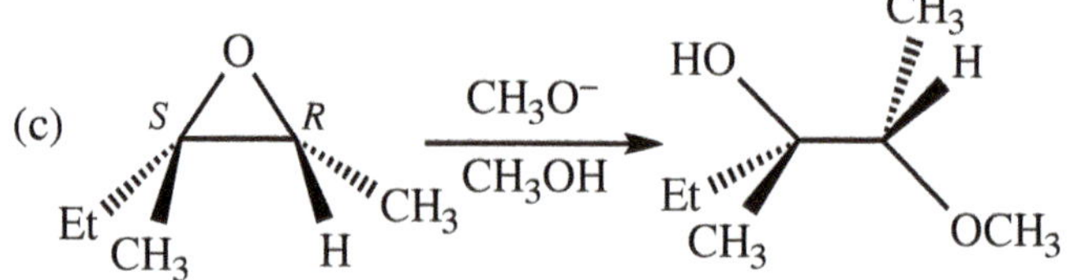

Basic nucleophiles react at the less substituted carbon.

(d)

In acid, nucleophiles react at the more substituted carbon.

14-28 Newly formed bonds are shown in bold.

(a) (b) (c)

14-29

(a) (b) (c) (d) (e) (f)

(g)

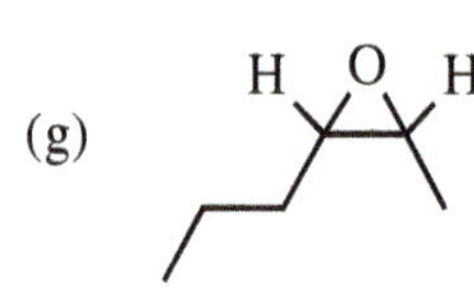

(h) (i)

14-30

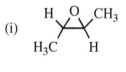

(a) *sec*-butyl isopropyl ether
(b) *tert*-butyl isobutyl ether
(c) ethyl phenyl ether
(d) chloromethyl propyl ether
(e) methyl *trans*-2-hydroxycyclohexyl ether
(f) cyclopentyl methyl ether
(g) propylene oxide
(h) cyclopropyl vinyl ether
(i) cyclopentene oxide
(j) 2-methyltetrahydrofuran

14-31

(a) 2-methoxypropan-1-ol
(b) ethoxybenzene or phenoxyethane
(c) methoxycyclopentane
(d) 2,2-dimethoxycyclopentan-1-ol
(e) *trans*-1-methoxy-2-methylcyclohexane
(f) *trans*-3-chloro-1,2-epoxycycloheptane
(g) *trans*-1,2-epoxy-1-methoxybutane; or, *trans*-2-ethyl-3-methoxyoxirane
(h) 3-bromooxetane
(i) 1,3-dioxane

14-32 As is often true when explaining the properties of molecules, hydrogen bonding is the key.

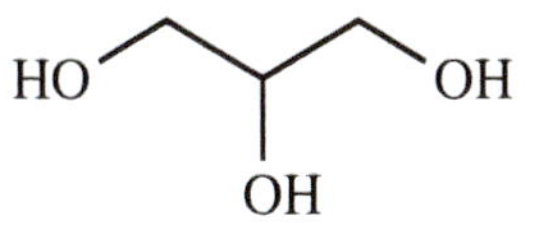

HO OH OH

glycerol
mol. wt. 92 g/mol
b.p. 290 °C
d 1.24 g/mL

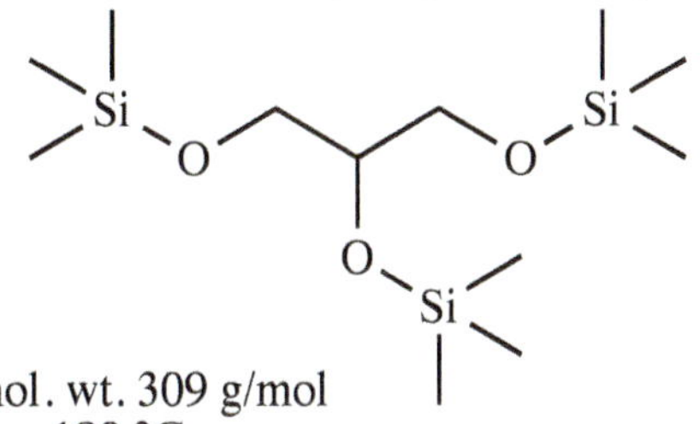

Si O O Si O Si

mol. wt. 309 g/mol
b.p. 180 °C
d 0.88 g/mL

Glycerol has extremely strong intermolecular hydrogen bonding because of the three OH groups per molecule. Overcoming these intermolecular forces requires a lot of energy: thus, glycerol has a high boiling point despite its fairly low molecular weight, and it flows slowly because hydrogen bonding must be overcome in order for molecules to slide past each other. The density is high because these molecules pack together tightly to maximize hydrogen bonding.

In contrast, the TMS (trimethylsilyl) ether of glycerol not only has no hydrogen bonding, but on each end of the molecule, there is a nonpolar and essentially spherical group—this is like putting on boxing gloves and trying to pick up a dime, or anything! So in spite of the high molecular weight, these molecules tend to stay far apart, explaining the ease of flow, the low density, and the relatively low boiling point.

For a boiling point comparison, look up the structure of "isocetane," which is a highly branched alkane of molecular weight 226 g/mol, with boiling point 240 °C. The van der Waals forces in the TMS ether above must be even lower, despite it having a higher molecular weight than isocetane.

14-33

(a) Br + Br + H_2O

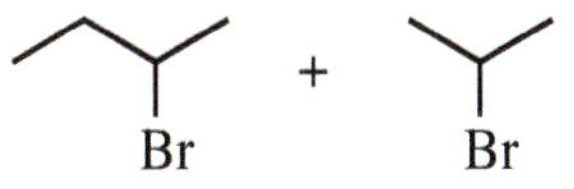

(b) Br + Br + H_2O

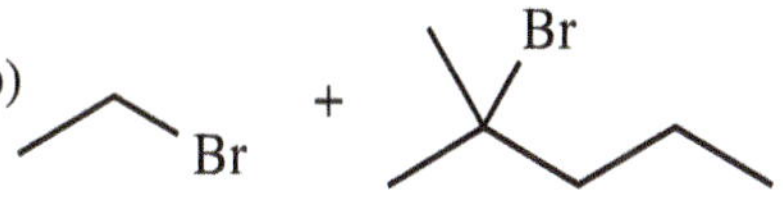

(c) and (d) no reaction—ethers are cleaved only under acid conditions

(e) OH + CH_3CH_2I
no S_N at sp^2 carbon

(f) OH OCH_3

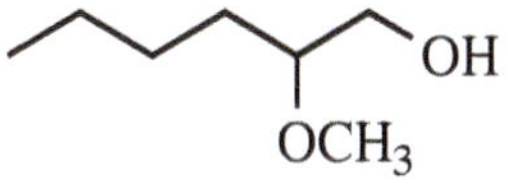

(g) H OH HO H plus the enantiomer

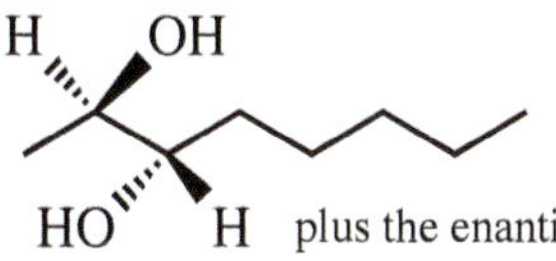

(h) OH $NHCH_3$

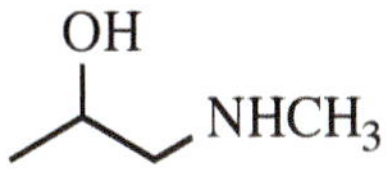

(i) O

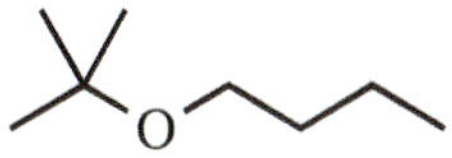

Keep at low temperature to minimize the competing E2 reaction.

(j) HO

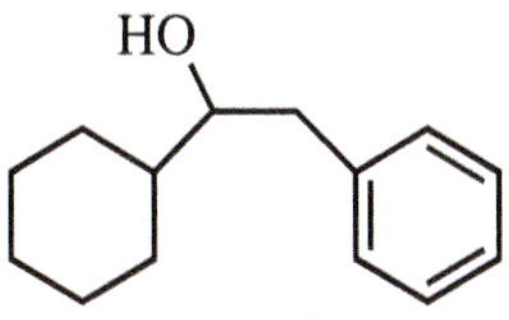

(k) O

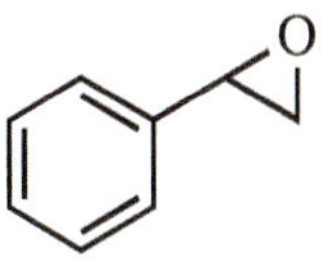

(l) Br OH

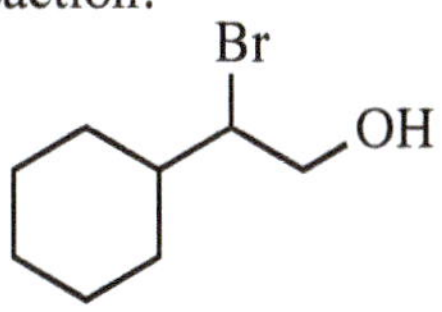

(m) CH_3 OH H OCH_3 plus the enantiomer

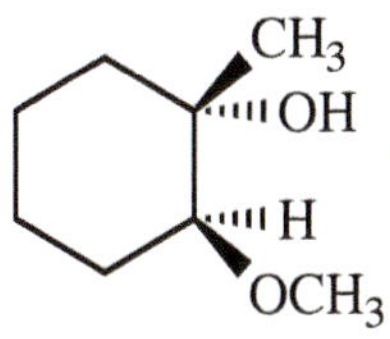

(n) OCH_3 CH_3 OH H plus the enantiomer

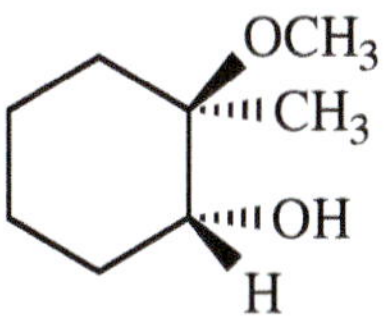

14-34

(a) at least three possible methods:

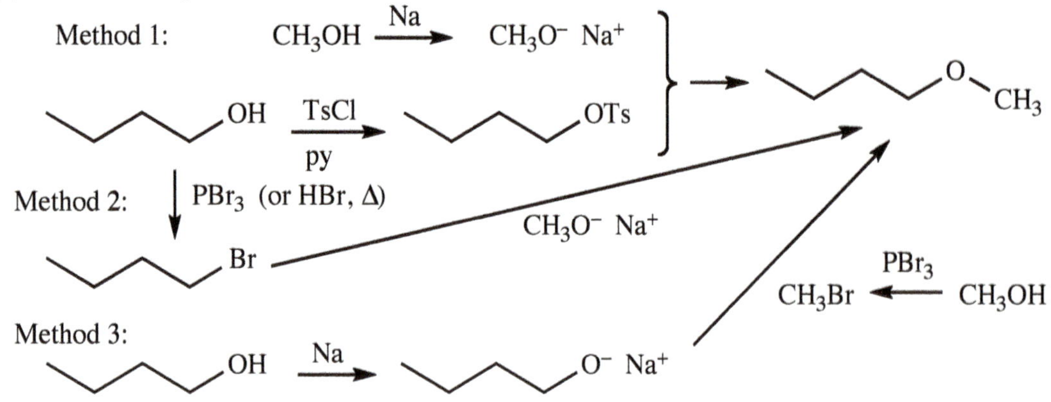

(b) more practical way:

+ $HOCH_2CH_3$ $\xrightarrow[S_N1]{H_2SO_4}$ OCH_2CH_3

OH

3°—forms carbocation

Bimolecular condensation works well on 3° carbons.

less practical method by S_N2, requiring more steps:

$O^-\ Na^+$ + LG

LG = leaving group like Br or OTs

Recall that ethoxide plus *tert*-butyl bromide gives only E2! See Solved Problem 14-1 in the text.

(c) more practical way:

OH + HO $\xrightarrow[S_N1]{H_2SO_4}$ O

benzylic—forms carbocation

less practical method by S_N2, requiring more steps:

LG + $Na^+\ ^-O$

LG = leaving group like Br or OTs

Also possible is alkoxymercuration-demercuration:

OH + $\xrightarrow{Hg(OAc)_2}$ $\xrightarrow{NaBH_4}$ O

(d) OH $\xrightarrow[\Delta]{H_2SO_4}$ $\xrightarrow{mCPBA}$ O $\xrightarrow[H_2SO_4]{CH_3OH}$ OH, OCH_3

plus the enantiomer

See the solution to 14-25.

(e) OH, OCH_3 $\xrightarrow[Et_3N]{TIPSCl}$ O–$Si(i\text{-}Pr)_3$, OCH_3

14-34 continued

(f) Alkoxymercuration-demercuration is the more practical method. Williamson ether synthesis on 2° carbons would not give good yields.

OH + (cyclopentene) 1) $Hg(OAc)_2$ 2) $NaBH_4$ OR (cyclohexene) + HO (cyclopentanol) 1) $Hg(OAc)_2$ 2) $NaBH_4$ → O

14-35

(a) On long-term exposure to air, ethers form peroxides. Peroxides are explosive when concentrated or heated. (For exactly this reason, ethers should *never* be distilled to dryness.)

(b) Peroxide formation can be prevented by excluding oxygen. Ethers can be checked for the presence of peroxides, and peroxides can be destroyed safely by treatment with reducing agents.

14-36

(a) In each case, the new bond will be shown in bold.

(i)

Ph MgBr + O ⇒ Ph OH ⇐ Ph O

first method

second method

BrMg — CH_3

(ii)

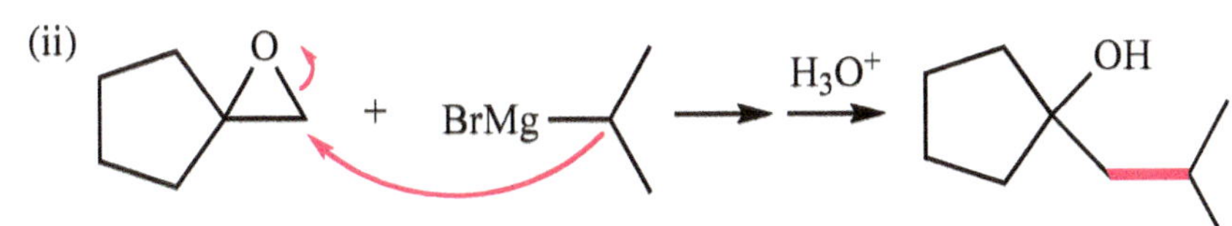

(b) In each case, the epoxide is derived from a tetrasubstituted double bond, far too hindered for the Grignard reagent to approach.

(i)

MgBr + O

(ii)

O + BrMg

14-37

(a) Beginning with *(R)*-butan-2-ol and producing the *(R)*-sulfide requires two inversions of configuration.

HO H, *R* — PBr_3 → H Br, *S* — $Na^+\ ^-SCH_3$ → H_3C-S H, *R*

An alternative approach would be to make the tosylate, displace with chloride or bromide (S_N2 with inversion), then do a second inversion with $NaSCH_3$.

(b) Synthesis of the *(S)* isomer directly requires only one inversion.

HO H, *R* — TsCl, pyridine → TsO H, *R* — $Na^+\ ^-SCH_3$ → H $S-CH_3$, *S*

14-38

(a)

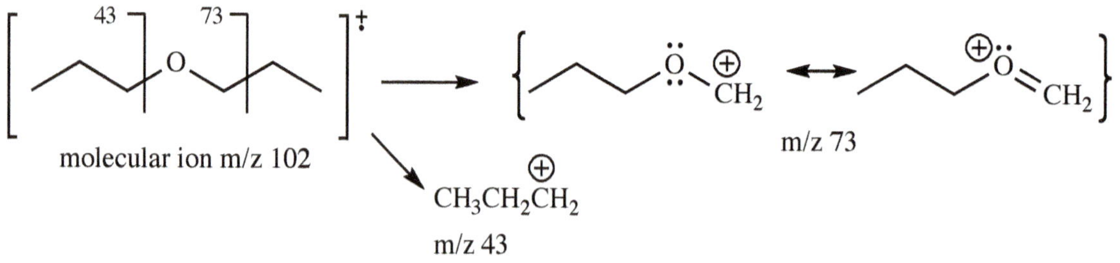

molecular ion m/z 102

m/z 73

$CH_3CH_2CH_2^+$

m/z 43

(b)

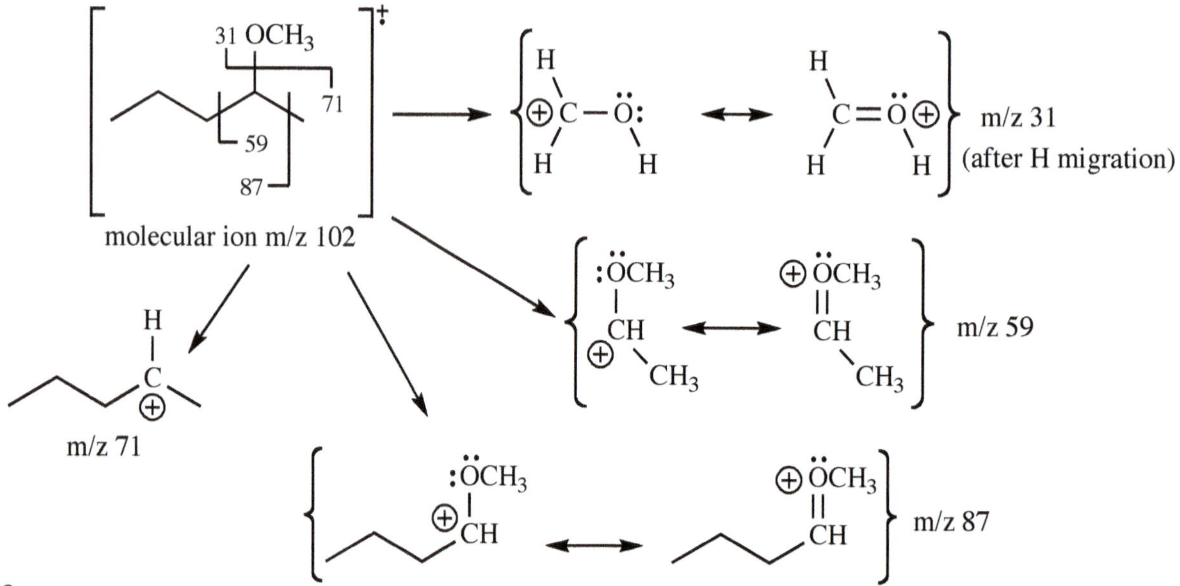

molecular ion m/z 102

m/z 31 (after H migration)

m/z 59

m/z 71

m/z 87

14-39

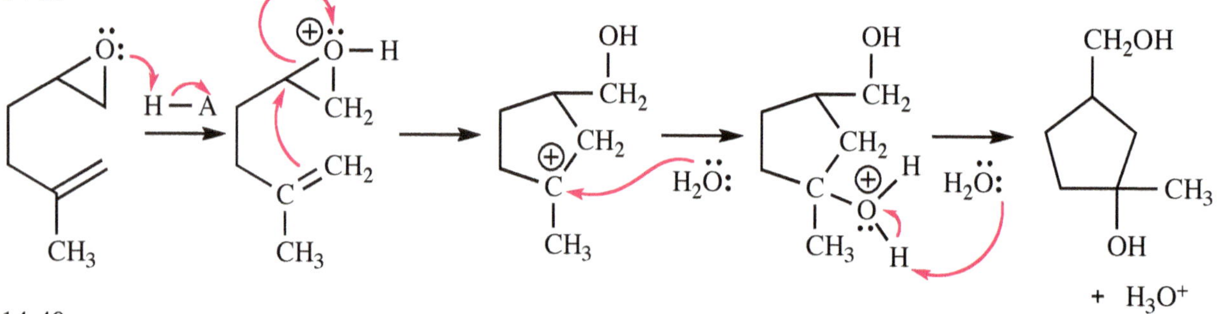

$+ H_3O^+$

14-40

(a) $Hg(OAc)_2$, CH_3OH; $NaBH_4$

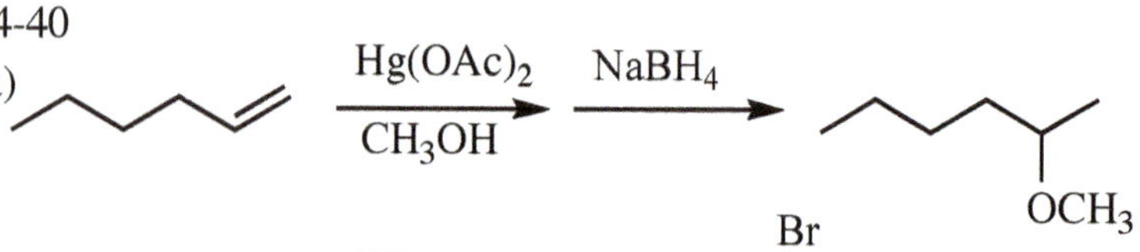

You could also make hexan-2-ol, then Na followed by CH_3I.

(b) HBr, HOOH; $NaOCH_3$, CH_3OH

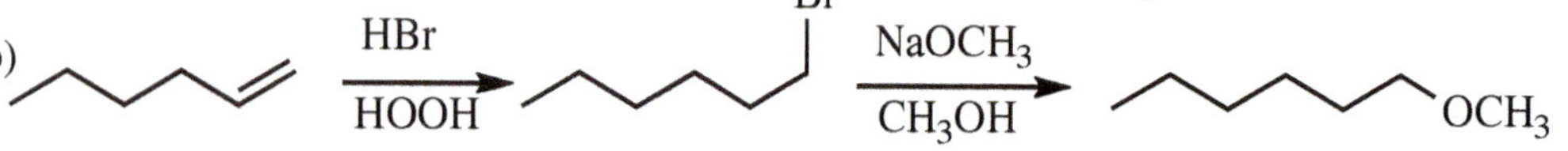

You could also make hexan-1-ol, then Na followed by CH_3I, similar to the solution to 14-34(a).

(c) mCPBA, CH_2Cl_2; $NaOCH_3$, CH_3OH

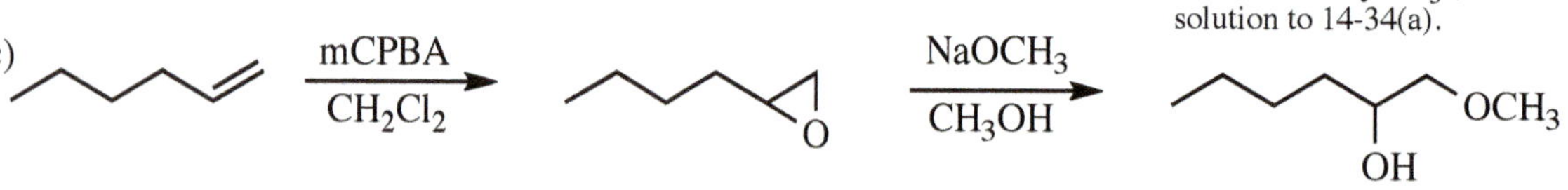

14-40 continued

(d) mCPBA, CH_2Cl_2 → epoxide; H^+, CH_3OH → product (OH, OCH_3)

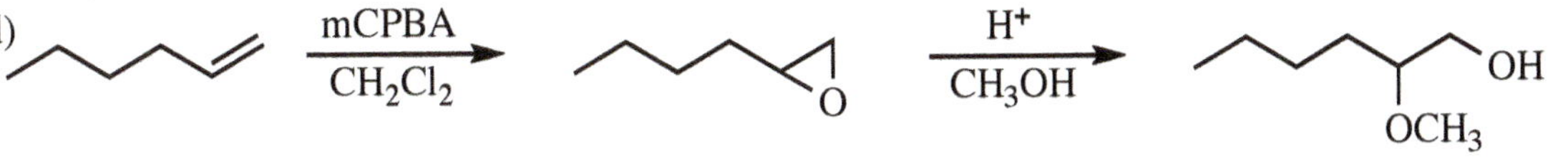

(e) mCPBA, CH_2Cl_2 → epoxide; 1) PhMgBr 2) H_2O → product (Ph, OH)

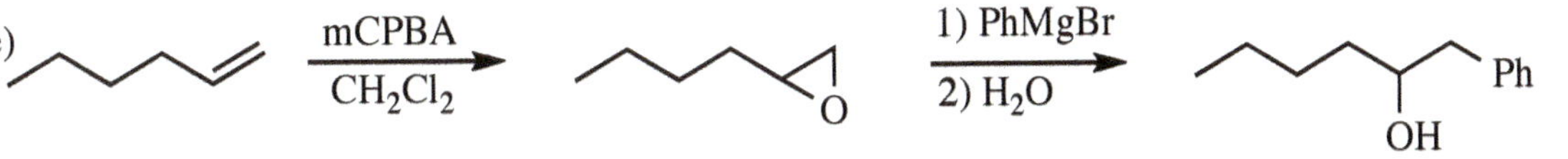

(e) mCPBA, CH_2Cl_2 → epoxide; PhMgBr, *do not perform aqueous workup* → product (Ph, O^- $MgBr^+$); CH_3I → product (Ph, OCH_3)

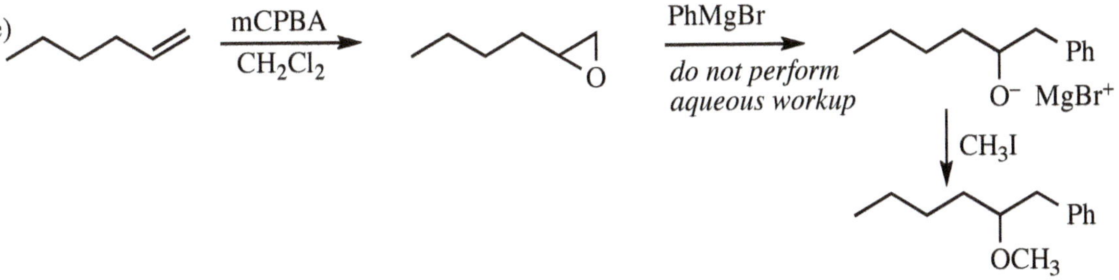

14-41

Two hints for this problem: 1) the last step in each sequence is a Williamson ether synthesis which follows an S_N2 mechanism, so always try to do this substitution on a 1° carbon; 2) as you learn more reactions, you will see that there are often multiple options of how to perform a synthesis. In general, syntheses with fewer steps are better.

In this problem, the abbreviation "LG" will indicate a "leaving group" = a halide or a tosylate.

(a) ketone (O) + $LiAlH_4$ → alkoxide (O^- Li^+) + benzyl–LG → ether (O)

OR

aldehyde (O, H) + BrMg–allyl → alkoxide (O^- ^+MgBr) + benzyl–LG → ether

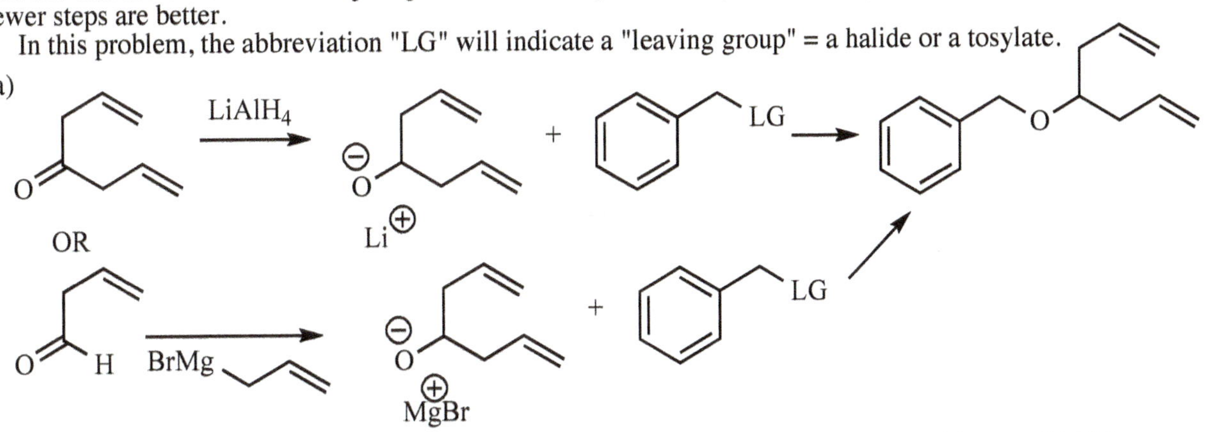

(b) The intermediate alkoxide can be made by three different Grignard reactions. In this case, $LiAlH_4$ is not an option because there is no H on the carbinol carbon.

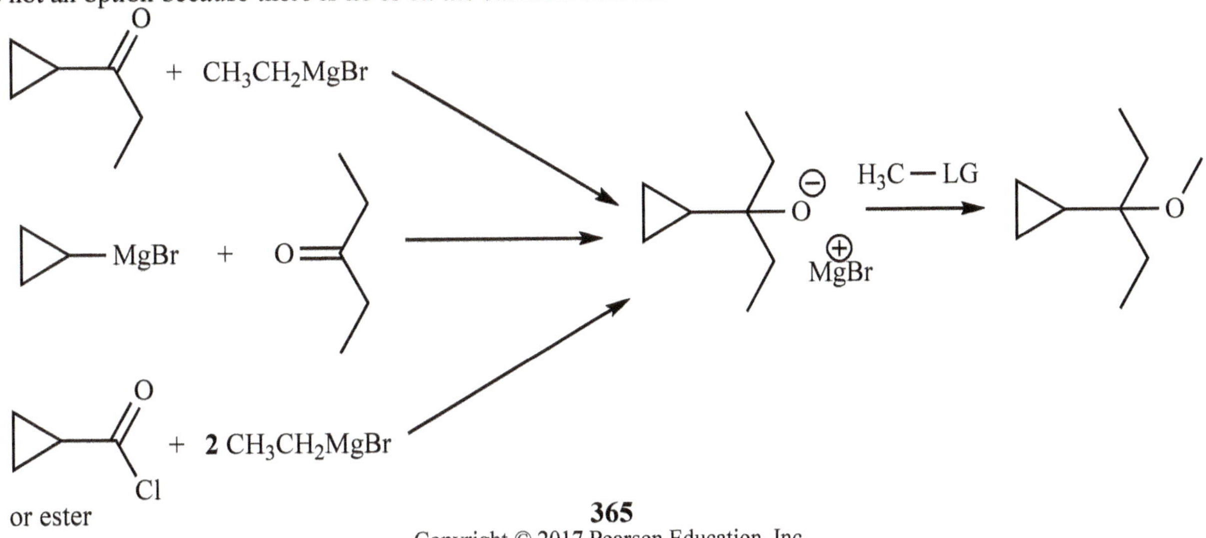

cyclopropyl ketone (O) + CH_3CH_2MgBr →

cyclopropyl–MgBr + O= (3-pentanone) →

cyclopropyl acid chloride (O, Cl) + **2** CH_3CH_2MgBr →

or ester

→ alkoxide (O^- ^+MgBr); $H_3C—LG$ → ether (O)

14-41 continued

(c)

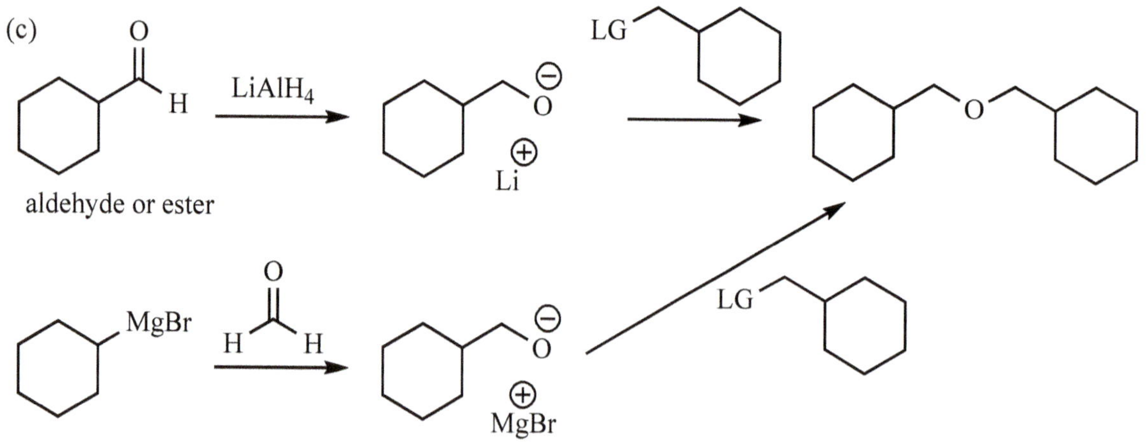

$\xrightarrow{LiAlH_4}$ ... O^- Li^+ → (LG) ...

aldehyde or ester

MgBr $\xrightarrow{H_2C=O}$ ... O^- ^+MgBr → (LG) ...

14-42 All chiral products in this problem are racemic.

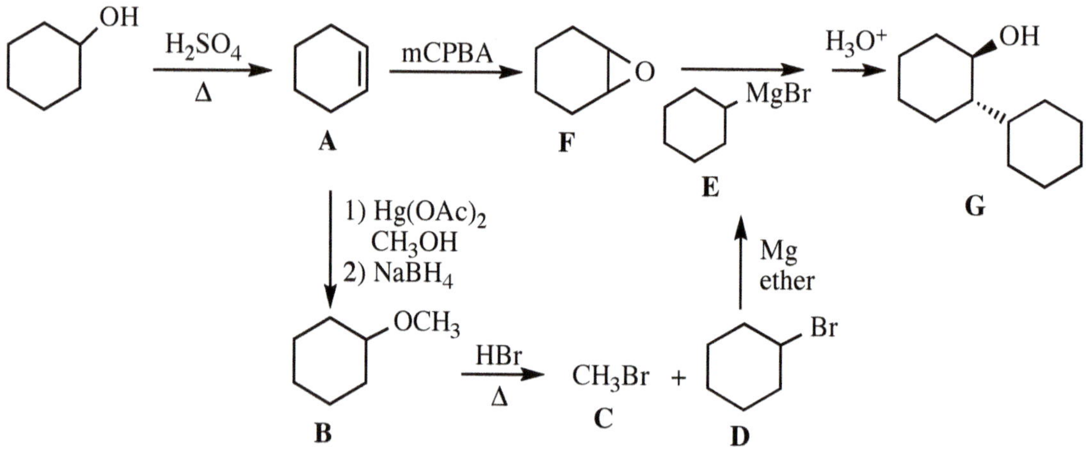

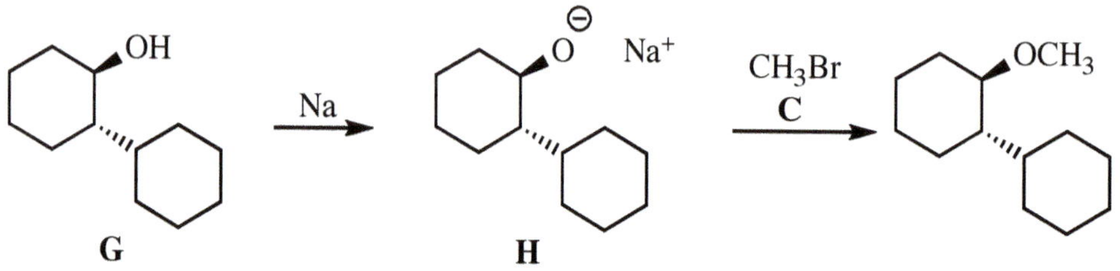

OH $\xrightarrow[\Delta]{H_2SO_4}$ **A** $\xrightarrow{mCPBA}$ **F** (O) → MgBr (**E**) $\xrightarrow{H_3O^+}$ OH **G**

A ↓ 1) $Hg(OAc)_2$, CH_3OH 2) $NaBH_4$

OCH_3 **B** $\xrightarrow[\Delta]{HBr}$ CH_3Br (**C**) + Br (**D**)

D ↑ Mg, ether → **E**

OH **G** $\xrightarrow{Na}$ O^- Na^+ **H** $\xrightarrow{CH_3Br\ \mathbf{C}}$ OCH_3

14-43

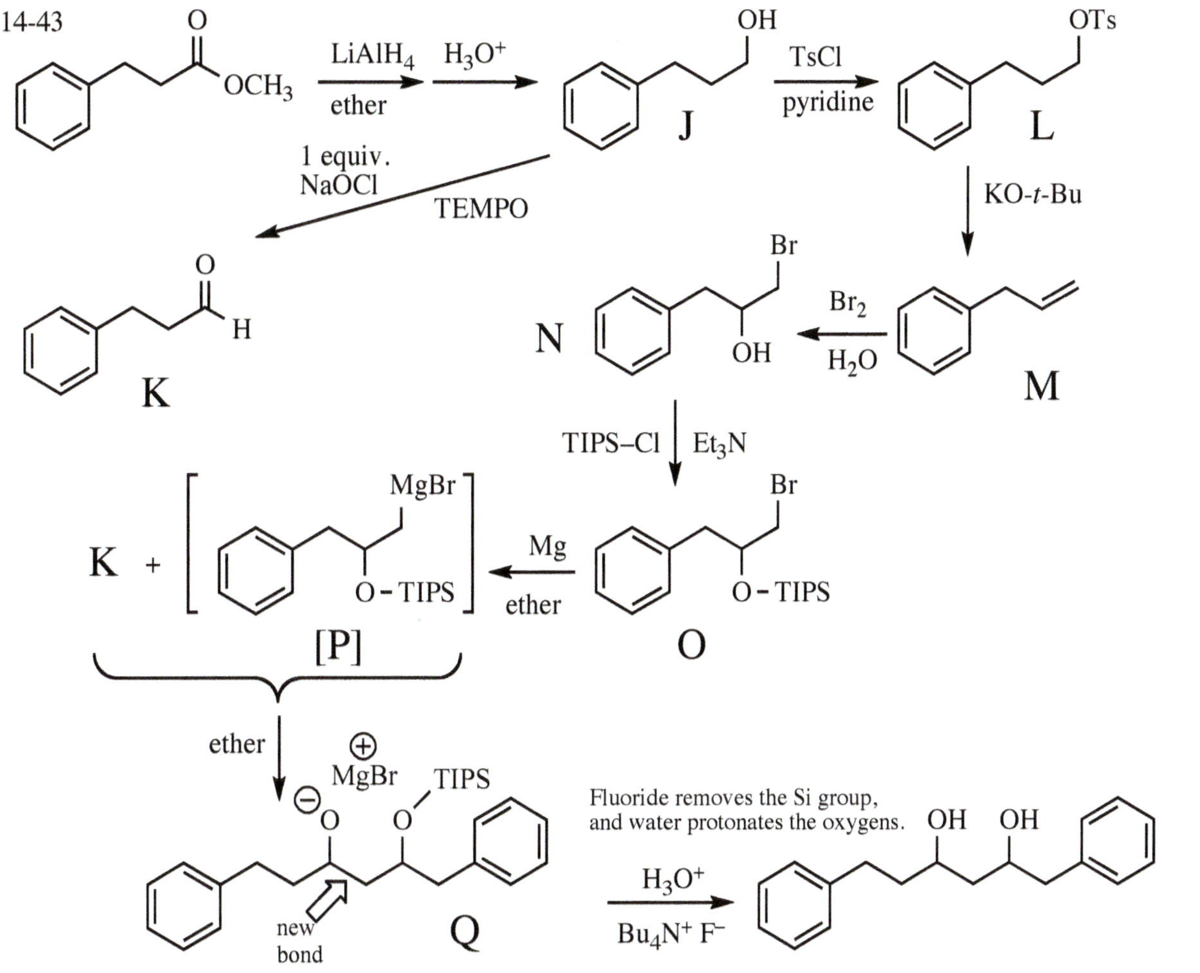

J, L, K, N, M, O, [P], Q

1 equiv. NaOCl, TEMPO; LiAlH₄ ether; H_3O^+; TsCl pyridine; KO-*t*-Bu; Br_2, H_2O; TIPS–Cl, Et_3N; Mg ether; ether

Fluoride removes the Si group, and water protonates the oxygens.

H_3O^+

Bu_4N^+ F^-

new bond

14-44 The student turned in the wrong product! Three pieces of information are consistent with the desired product: molecular formula $C_4H_{10}O$; O—H stretch in the IR at 3300 cm^{-1} (although it should be strong, not weak); and mass spectrum fragment at m/z 59 (loss of CH_3). The NMR of the product should have a 9H singlet at δ 1.0 and a 1H singlet between δ 2 and δ 5. Instead, the NMR shows CH_3CH_2 bonded to oxygen. The student isolated diethyl ether, *the typical **solvent** used in Grignard reactions*.

Predicted product	Isolated product
$(CH_3)_3C—OH$ (59)	$H_3C—CH_2—O—CH_2—CH_3$ (59)
$C_4H_{10}O$	$C_4H_{10}O$
strong O—H at 3300 cm^{-1}	weak O—H at 3300 cm^{-1} due to water contamination

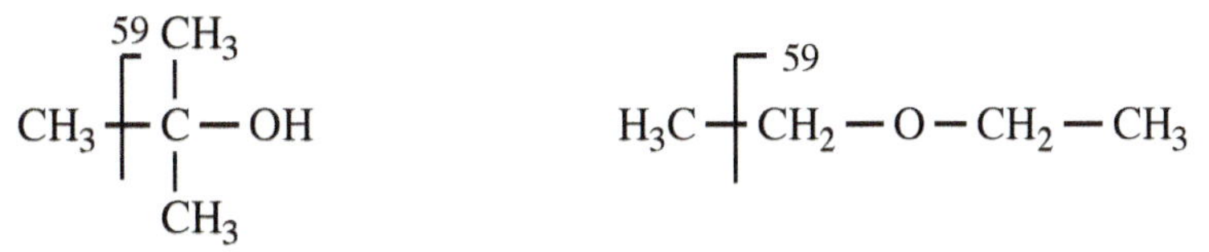

14-45

(a) cyclopentanol —Na→ cyclopentoxide Na^+; propanol —PBr_3→ propyl bromide; → cyclopentyl propyl ether

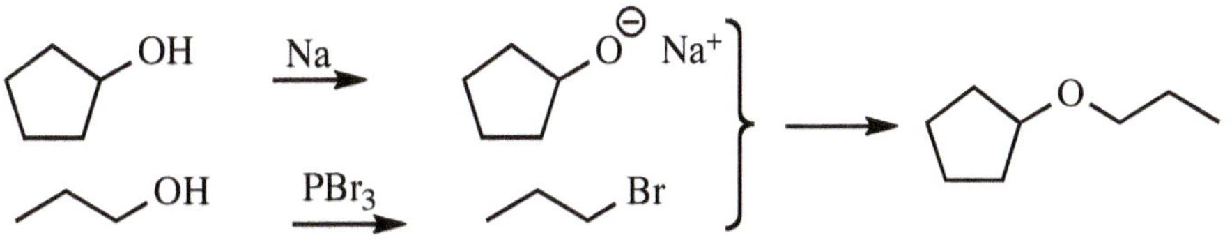

14-45 continued

(b) [reaction scheme: phenol + NaOH → sodium phenoxide; 1-butanol + TsCl, pyridine → butyl tosylate; together → butyl phenyl ether]

NaOH

$O^{\ominus}$ Na^+

TsCl / pyridine

OH → OTs

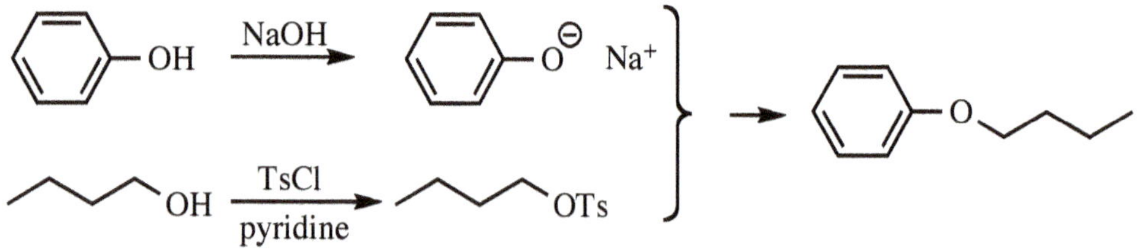

(c) Hg(OAc)$_2$ / CH_3CH_2OH, then $NaBH_4$ → OCH_2CH_3

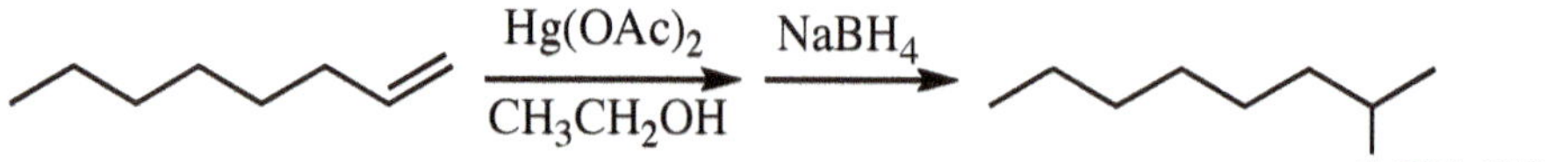

(d) 1) $BH_3 \cdot THF$ 2) H_2O_2, HO^- → OH

HBr, ROOR → Br

1) Na 2) CH_3I

$NaOCH_3$ → OCH_3

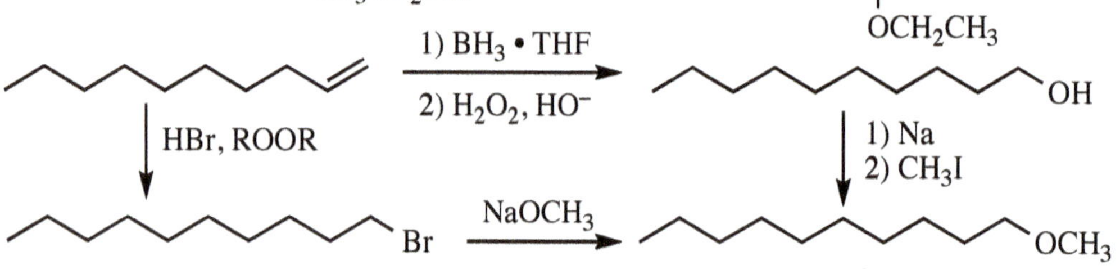

(e) OH → H_2SO_4, Δ → 1) Hg(OAc)$_2$, CH_3CH_2OH 2) $NaBH_4$ → OCH_2CH_3

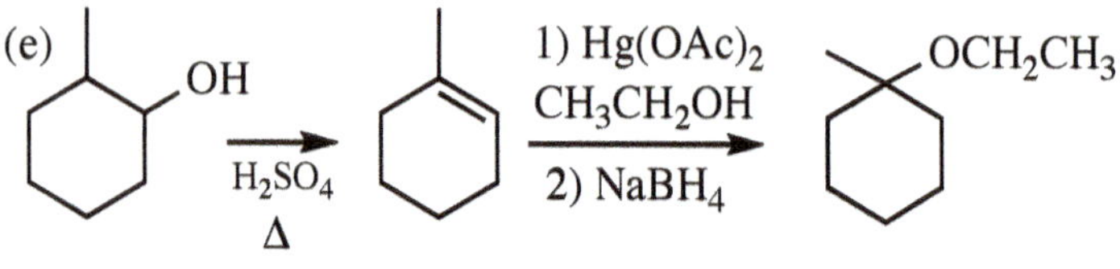

(f) OH → H_2SO_4, Δ → *trans* + *cis* → mCPBA, CH_2Cl_2 → O

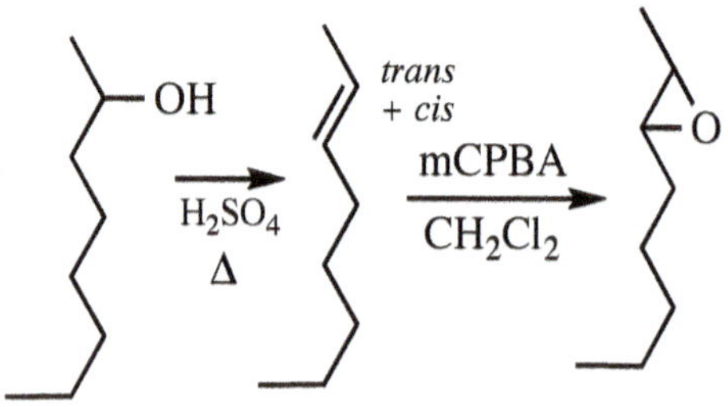

The *trans* isomer would be the major product from the dehydration because *trans* is more stable than *cis*. Epoxidation with mCPBA is stereospecific and will retain the stereochemistry of the starting double bond. The major product will be *trans*, with some *cis* as a contaminant.

14-46 Approaching a good synthesis problem begins with comparing the product to the starting material. If new carbons appear in the product, then the synthesis must include a carbon-carbon bond-forming reaction, of which there are very few.

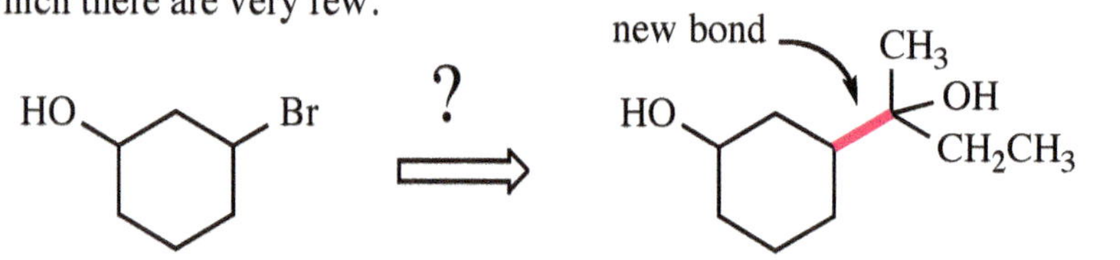

HO, Br ? ⟹ HO, new bond, CH_3, OH, CH_2CH_3

The new bond is shown in bold.

continued on next page

14-46 continued

At first glance, this appears to require a simple Grignard reaction, but then we recall that a Grignard reagent cannot coexist with an OH group in the same molecule. Aha! The OH group needs to be protected before the Grignard can proceed.

HO, Br; TIPSCl, Et_3N → $(i\text{-Pr})_3Si-O$, Br; Mg, ether → $(i\text{-Pr})_3Si-O$, MgBr

This could be isolated.

Grignard reagents are stable only in solution; they cannot be isolated.

Fluoride removes the Si group, and water protonates the oxygens.

HO, CH_3, OH, CH_2CH_3 ← Bu_4N^+ F^-, H_2O ← $(i\text{-Pr})_3Si-O$, CH_3, O^-, ^+MgBr, CH_2CH_3

14-47 In the first sequence, no bond is broken to the chiral center, so the configuration of the product is the same as the configuration of the starting material.

H, OH; Na → H, O^- Na^+; CH_3CH_2I → H, OCH_2CH_3

$[\alpha]_D = -8.24°$

(Assume the enantiomer shown is levorotatory.)

$[\alpha]_D = -15.6°$

In the second reaction sequence, however, a bond to the chiral carbon is broken once, so the stereochemistry of the process will be a net inversion.

H, OH, Cl—S(=O)(=O)— → H, ⊕O—S, H; :N

H, OTs ← H, O—S

Undoubtedly, some E2 products will also form in this reaction.

$CH_3CH_2\ddot{O}:^-$ Na^+

S_N2—INVERSION

CH_3CH_2O, H

RETENTION OF CONFIGURATION

The second sequence involves *retention* followed by *inversion*, thereby producing the *enantiomer* of the 2-ethoxyoctane generated by the first sequence. The optical rotation of the final product will have equal magnitude but opposite sign, $[\alpha]_D = +15.6°$.

14-48

(a)

anhydrous HBr—only Br^- nucleophiles present

$$H_2C\text{—}CH_2 \text{ (epoxide, :O:)} \xrightarrow{H-Br} H_2C\text{—}CH_2 \text{ (protonated epoxide, } O^{\oplus}\text{—H)} + :\ddot{Br}:^{\ominus} \longrightarrow CH_2CH_2Br\text{—}\ddot{O}\text{—H} \xrightarrow{H-Br} CH_2CH_2Br\text{—}\overset{\oplus}{O}H_2 + :\ddot{Br}:^{\ominus} \longrightarrow BrCH_2CH_2Br + H_2O$$

aqueous HBr—many more H_2O nucleophiles than Br^- nucleophiles

$$H_2C\text{—}CH_2 \text{ (epoxide, :O:)} \xrightarrow{H-Br} H_2C\text{—}CH_2 \text{ (protonated epoxide, } O^{\oplus}\text{—H)} + :\ddot{O}\text{—H (H)} \longrightarrow H_2C(HO)\text{—}CH_2\overset{\oplus}{O}H_2 + H_2\ddot{O} \longrightarrow H_2C(HO)\text{—}CH_2(OH) + H_3O^+$$

(b)

$$H\ddot{O}:^{\ominus} + H_2C\text{—}CH(CH_3) \text{ (epoxide)} \longrightarrow HO\text{-}CH_2\text{-}CH(CH_3)\text{-}\ddot{O}:^{\ominus} + H_2C\text{—}CH(CH_3) \text{ (epoxide)} \longrightarrow HO\text{-}CH_2\text{-}C(CH_3)(H)\text{—}O\text{-}CH_2\text{-}CH(CH_3)\text{-}\ddot{O}:^{\ominus}$$

$$+ H_2C\text{—}CH(CH_3) \text{ (epoxide)} \longrightarrow HO\text{-}CH_2\text{-}CH(CH_3)\text{-}O\text{-}CH_2\text{-}CH(CH_3)\text{-}O\text{-}CH_2\text{-}CH(CH_3)\text{-}\ddot{O}:^{\ominus}$$

$$\xrightarrow{HO-H} HO\text{-}CH_2\text{-}CH(CH_3)\text{-}O\text{-}CH_2\text{-}CH(CH_3)\text{-}O\text{-}CH_2\text{-}CH(CH_3)\text{-}OH + HO^-$$

14-49 methyl cellosolve® $CH_3OCH_2CH_2OH$

To begin, what can be said about methyl cellosolve®? Its molecular weight is 76; its IR would show C—O in the 1000–1200 cm^{-1} region and a strong O—H around 3300 cm^{-1}; and its NMR would show four sets of signals in the ratio of 3:2:2:1.

The unknown has molecular weight 134; this is double the weight of methyl cellosolve®, minus 18 (water). The IR shows no OH, only ether C—O. The NMR shows no OH, only H—C—O in the ratio of 3:2:2. Apparently, two molecules of methyl cellosolve® have combined in an acid-catalyzed, bimolecular condensation.

$$CHOCH_2CH_2\ddot{O}H \xrightarrow{H-A} CH_3OCH_2CH_2\text{—}\overset{\oplus}{O}H_2 \xrightarrow[-\,H_2O]{CH_3OCH_2CH_2\ddot{O}H} CH_3OCH_2CH_2\text{—}\overset{\oplus}{\ddot{O}}(H)\text{—}CH_2CH_2OCH_3 \xrightarrow{H_2\ddot{O}:} \boxed{CH_3OCH_2CH_2\text{—}O\text{—}CH_2CH_2OCH_3}$$

(This compound is called *diethylene glycol dimethyl ether*, or a shortened, common name is *diglyme*.)

14-50

(a)

S　　　　　R

(b)

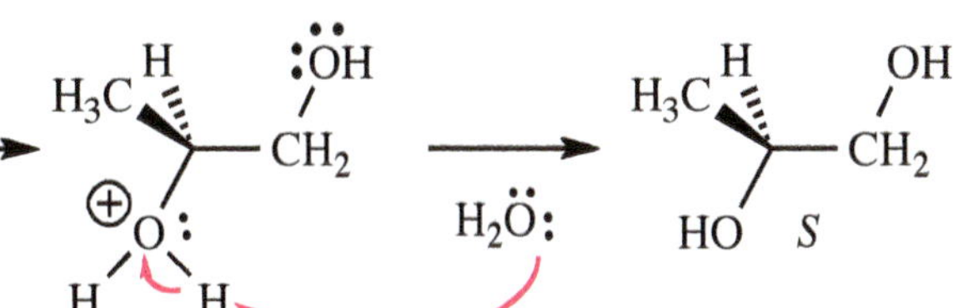

Attack of water gave *inversion* of configuration at the chiral center; *R* became *S*.

(c)

No bond to the chiral center was broken. Configuration is retained; *R* stays as *R*.

(d) The difference in these mechanisms lies in where the nucleophile attacks. Attack at the chiral carbon gives inversion; attack at the achiral carbon retains the configuration at the chiral carbon. These products are enantiomers and must necessarily have optical rotations of opposite sign. The lower enantiomeric excess in the acid catalyzed mechanism probably comes from some opening of the epoxide ring to form a secondary carbocation that can be attacked from either top or bottom by the water nucleophile, producing a mixture of *R* and *S* configuration.

14-51

The formula C_8H_8O has five elements of unsaturation (enough (4) for a benzene ring). The IR is useful for what it does *not* show. There is neither OH nor C=O, so the oxygen must be an ether functional group.

The HNMR shows a 5H signal at δ 7.2, a monosubstituted benzene. No peaks in the δ 4.5-6.0 range indicate the absence of an alkene, so the remaining element of unsaturation must be a ring. The three protons are non-equivalent, with complex splitting.

+ 2 C + O + 3 H + ring
ether

These pieces can be assembled in only one manner consistent with the data.

(Note that the CH_2 hydrogens are not equivalent (one is *cis* and one is *trans* to the phenyl) and therefore have distinct chemical shifts.)

14-52

(a) The product has a new pi bond so this must be an elimination reaction. The strong base *tert*-butoxide points to E2 as the probable mechanism.

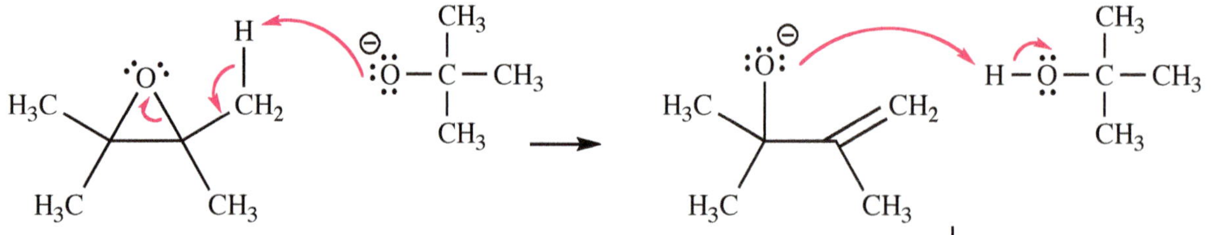

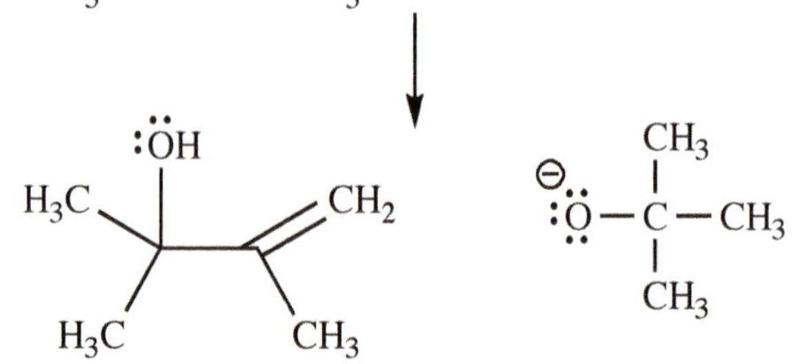

The epoxide oxygen is the leaving group in this elimination. Ordinarily, ether oxygens are not leaving groups; in epoxides, however, the severe ring strain makes the ring open fairly readily.

(b) This result does not occur with dimethyloxirane in acidic conditions; something must be special about the two phenyl groups.

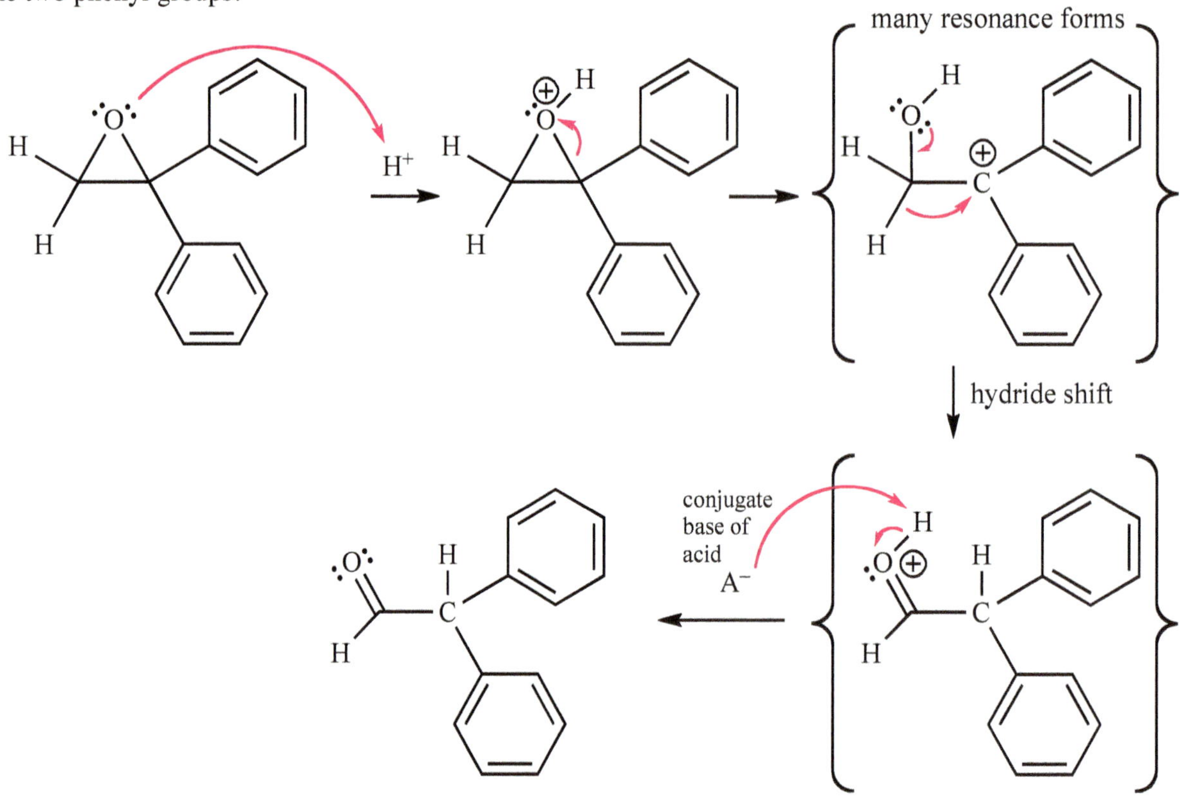

many resonance forms

hydride shift

conjugate base of acid A^-

What's different because of the phenyl groups is that a free carbocation is present because it is doubly benzylic. Once the free carbocation is present, it can undergo a hydride shift to form a protonated carbonyl.

14-53 HA is an acid; A^- is the conjugate base.

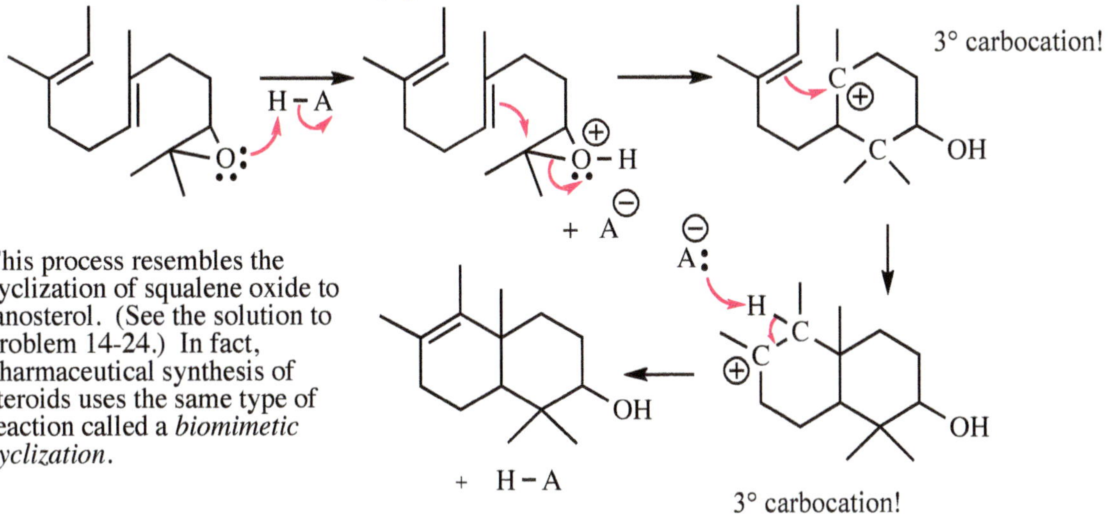

3° carbocation!

This process resembles the cyclization of squalene oxide to lanosterol. (See the solution to problem 14-24.) In fact, pharmaceutical synthesis of steroids uses the same type of reaction called a *biomimetic cyclization*.

+ H − A

3° carbocation!

14-54

(a) There is only one true ether in Byrostatin 1, although there are two groups called a hemiacetal that include an ether portion, so the answer that you might have given is "three".
(b) The other oxygen-containing functional groups are indicated in *italics*: five esters and two alcohols, although you likely included the two OH groups in the hemiacetals as alcohols.
(c) The minimum number of atoms making a continuous path around the large ring is 20 (numbered).
(d) Chiral centers around the main ring are at carbons 4, 6, 8, 10, 12, 16, and 18, plus three more at atoms outside of the main ring (indicated by a large dot •): total of 10 chiral centers.
(e) From section 5-12: "A compound with n asymmetric carbon atoms might have as many as 2^n stereoisomers." For 10 chiral centers, $2^{10} = 1024$ possible stereoisomers! Nature makes only one!

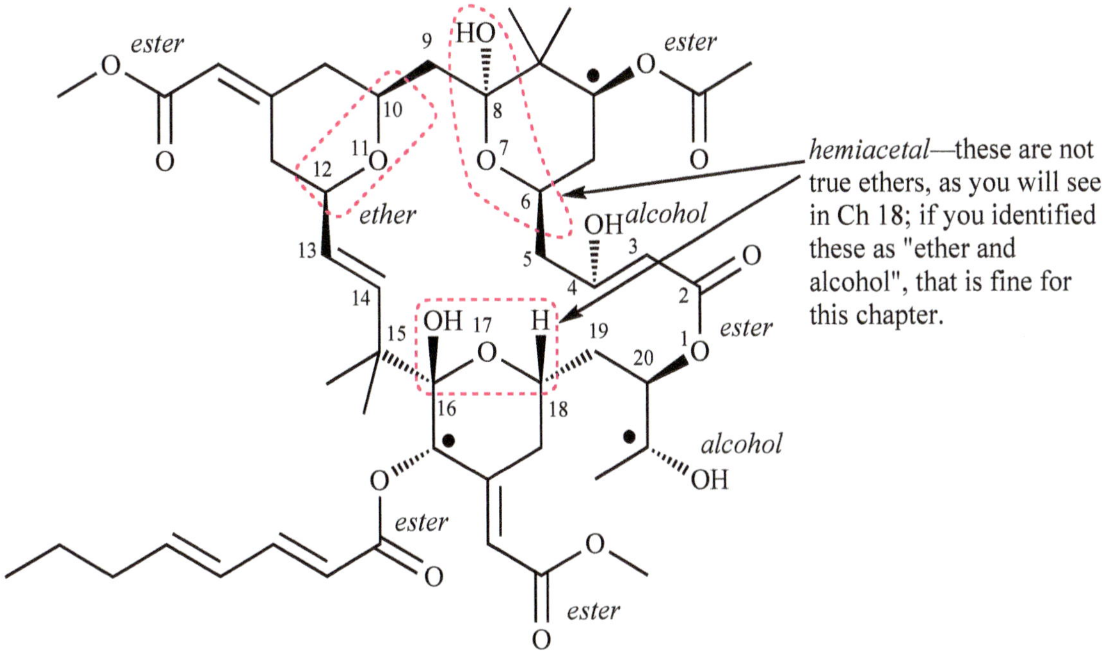

hemiacetal—these are not true ethers, as you will see in Ch 18; if you identified these as "ether and alcohol", that is fine for this chapter.

14-55

(a) The ether groups are marked with a large dot •. The count is 24.
(b) The alcohols are indicated as OH. There are thirteen 2° alcohols and no 1° or 3° alcohols.
(c) There is one aldehyde.

This record-setting structure containing 17 fused (contiguous) rings, plus 7 other rings! Remember that this structure is made by an algae.

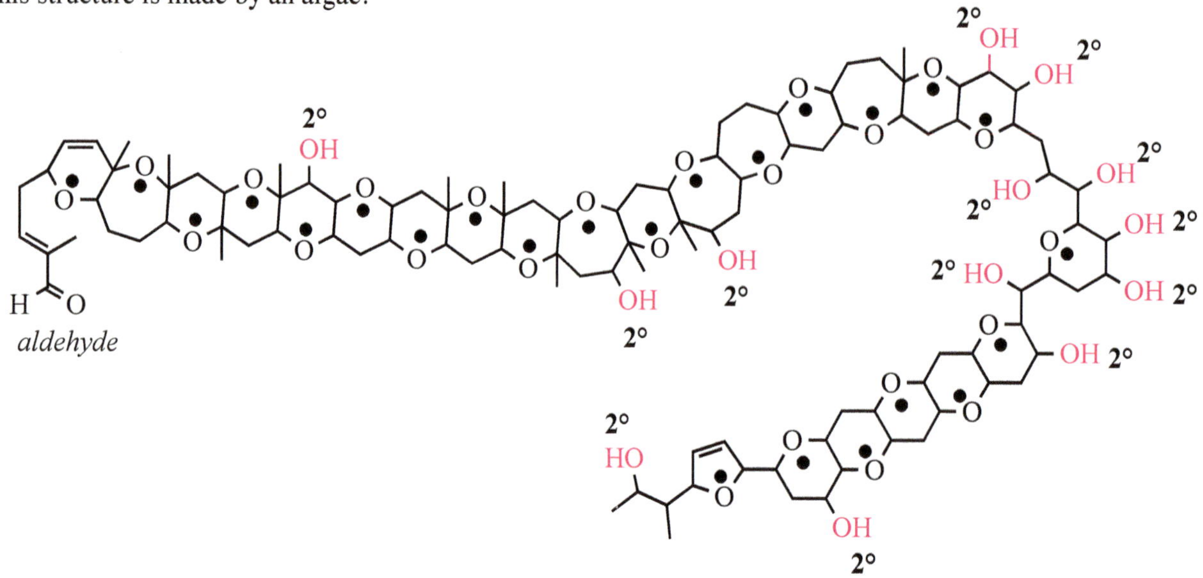

aldehyde

CHAPTER 15—CONJUGATED SYSTEMS, ORBITAL SYMMETRY, AND ULTRAVIOLET SPECTROSCOPY

15-1 Look for: 1) the number of double bonds to be hydrogenated—the fewer C=C, the smaller the ΔH; 2) conjugation—the more conjugated, the more stable, the lower the ΔH; 3) degree of substitution of the alkenes—the more substituted, the more stable, the lower the ΔH.

(a)

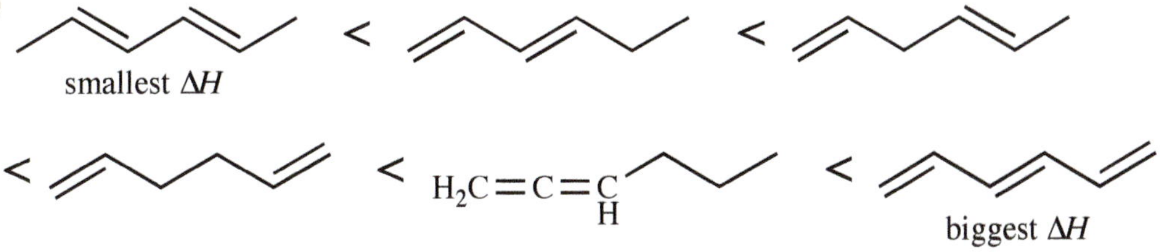

< <

smallest ΔH

< < $H_2C{=}C{=}\underset{H}{C}$ <

biggest ΔH

(b)

< < < <

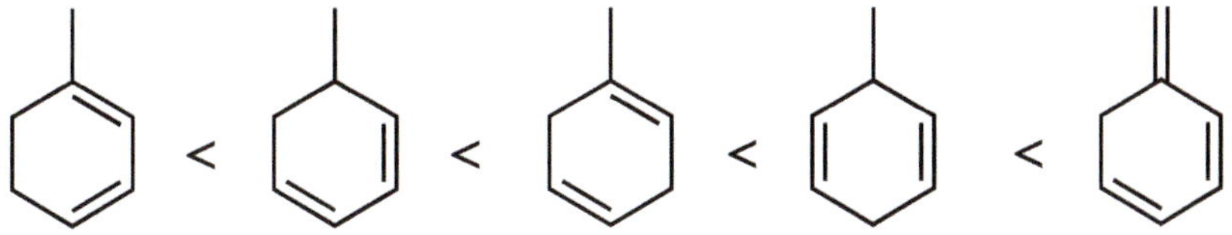

smallest ΔH
most stable
of the dienes

triene
biggest ΔH

15-2 Reminder: H—A is used to symbolize the general form for an acid, that is, a protonated base; A^- is the conjugate base.

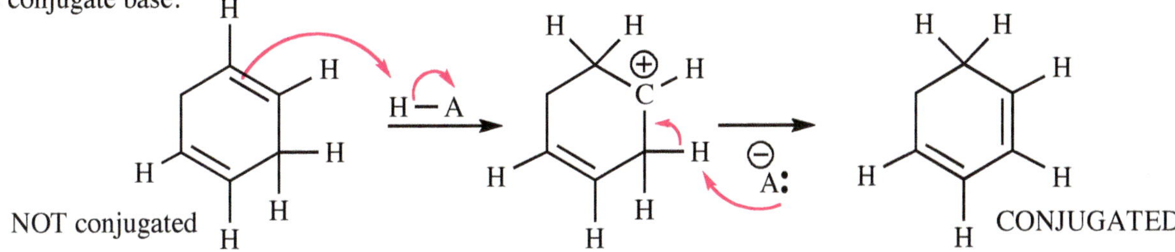

15-3 (You may wish to refer to Problem 1-20.)

(a)

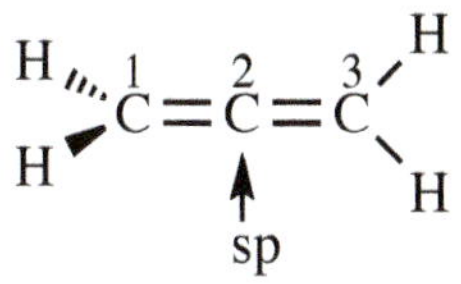

H, H (C1) = C2 = C3 (H, H); C2 is sp

orbital picture

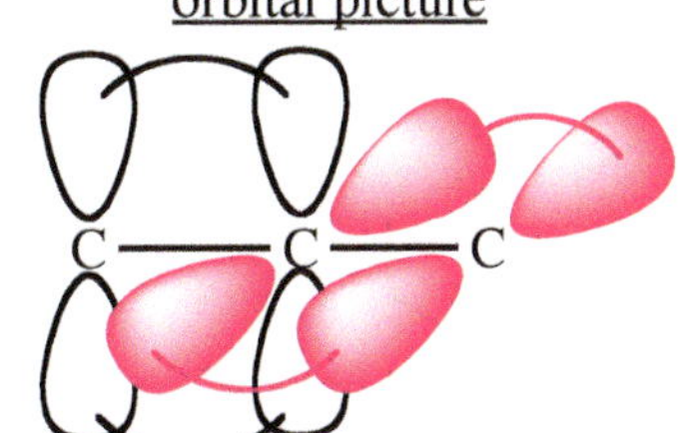

The shaded orbitals are perpendicular to the unshaded orbitals.

The central carbon atom makes two π bonds with two p orbitals. These p orbitals must necessarily be perpendicular to each other, thereby forcing the groups on the ends of the allene system perpendicular.

(b)

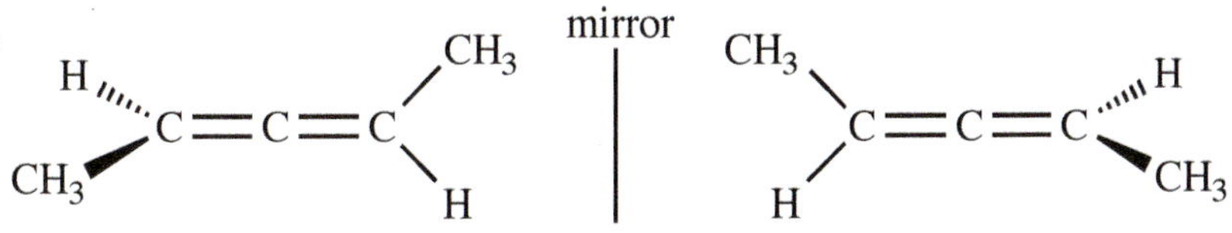

H, CH_3 C=C=C CH_3, H | mirror | CH_3, H C=C=C H, CH_3

non-superimposable mirror images = enantiomers

15-4 Carbocation stability depends first on conjugation (benzylic, allylic), then on degree of carbon carrying the positive charge.

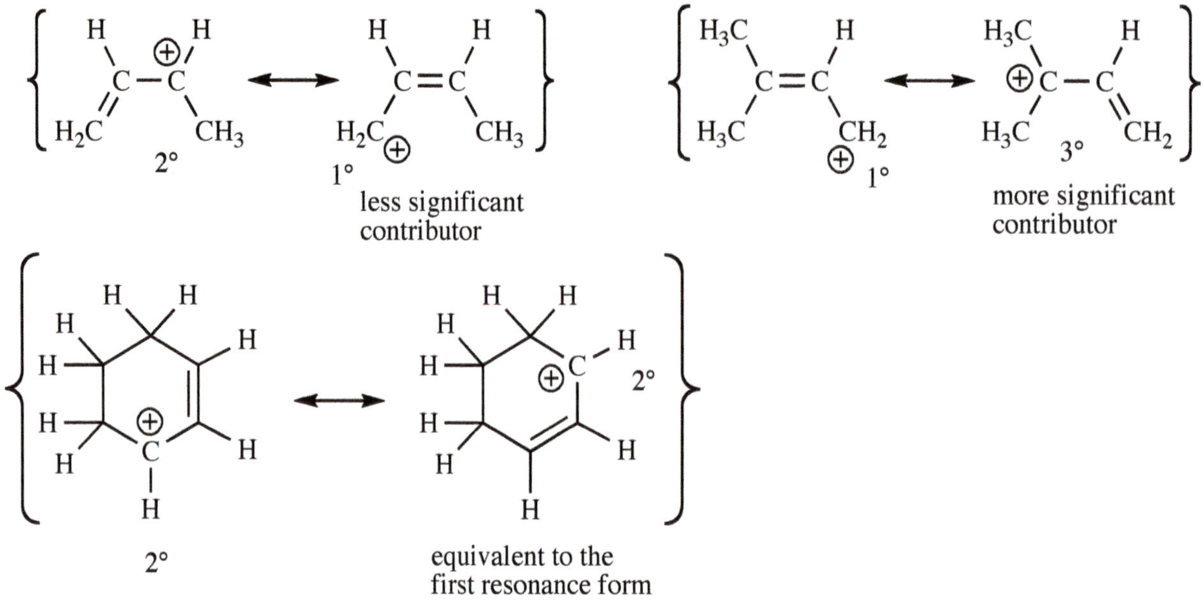

2°

1°

less significant contributor

1°

3°

more significant contributor

2°

2°

equivalent to the first resonance form

15-5

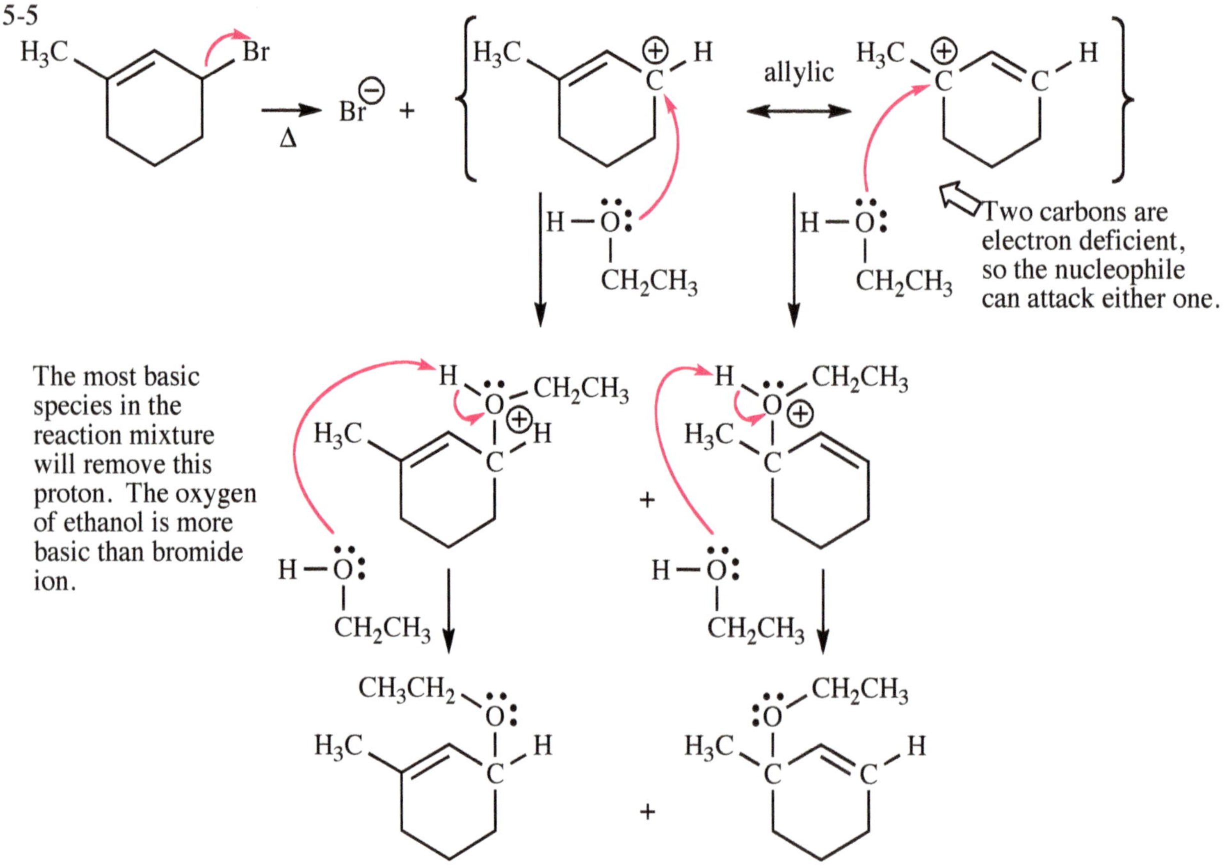

Δ

$Br^{\ominus}$ +

allylic

Two carbons are electron deficient, so the nucleophile can attack either one.

The most basic species in the reaction mixture will remove this proton. The oxygen of ethanol is more basic than bromide ion.

15-6

Two carbons are electron deficient, so the nucleophile can attack either one.

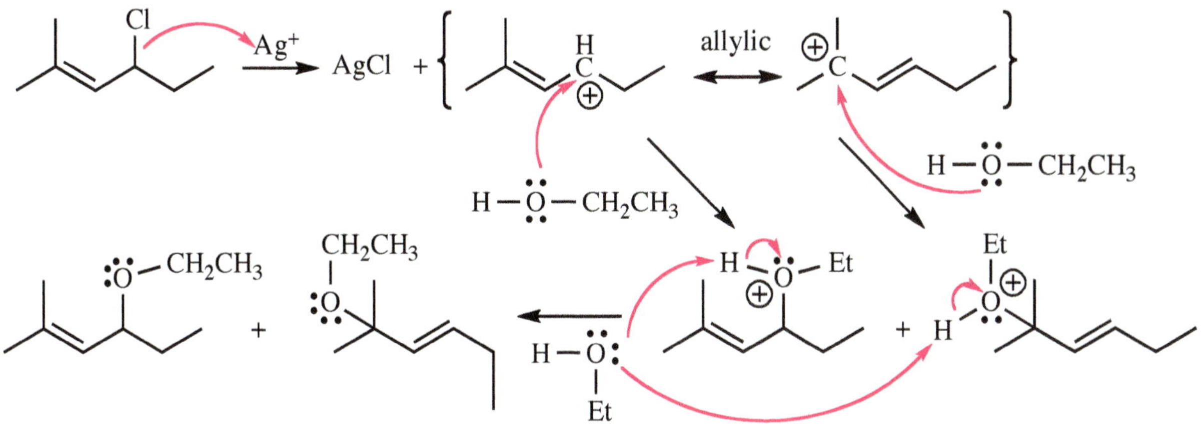

Cl

Ag^+

AgCl +

allylic

$H-\ddot{O}-CH_2CH_3$

$H-\ddot{O}-CH_2CH_3$

$\ddot{O}-CH_2CH_3$ + CH_2CH_3

H–O–Et

H–O–Et + H

Et

15-7

(a) $(H_3C)_2C{=}CH-CH_2-OH$ + H–Br → $(H_3C)_2C{=}CH-CH_2-OH_2^+$ → (– H_2O) {$(H_3C)_2C{=}CH-CH_2^+$ ⟷ allylic ⟷ $(H_3C)_2C^+-CH{=}CH_2$}

$:\ddot{Br}:^-$

Two carbons are electron deficient, so the nucleophile can attack either one.

$(H_3C)_2C{=}CH-CH_2-Br$ + $Br-C(CH_3)_2-CH{=}CH_2$

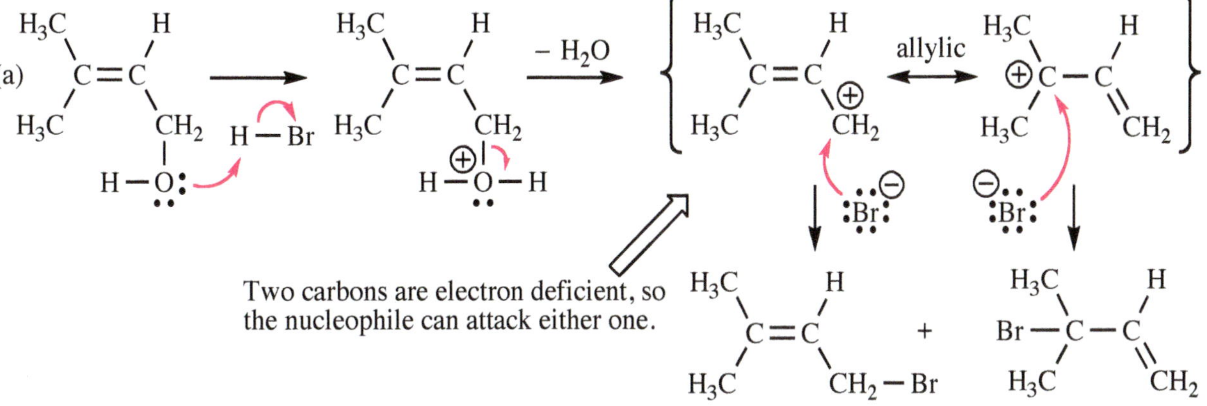

(b) $H-\ddot{O}-C(CH_3)_2-CH{=}CH_2$ + H–Br → $H_2O^+-C(CH_3)_2-CH{=}CH_2$

– H_2O

same carbocation as in (a)

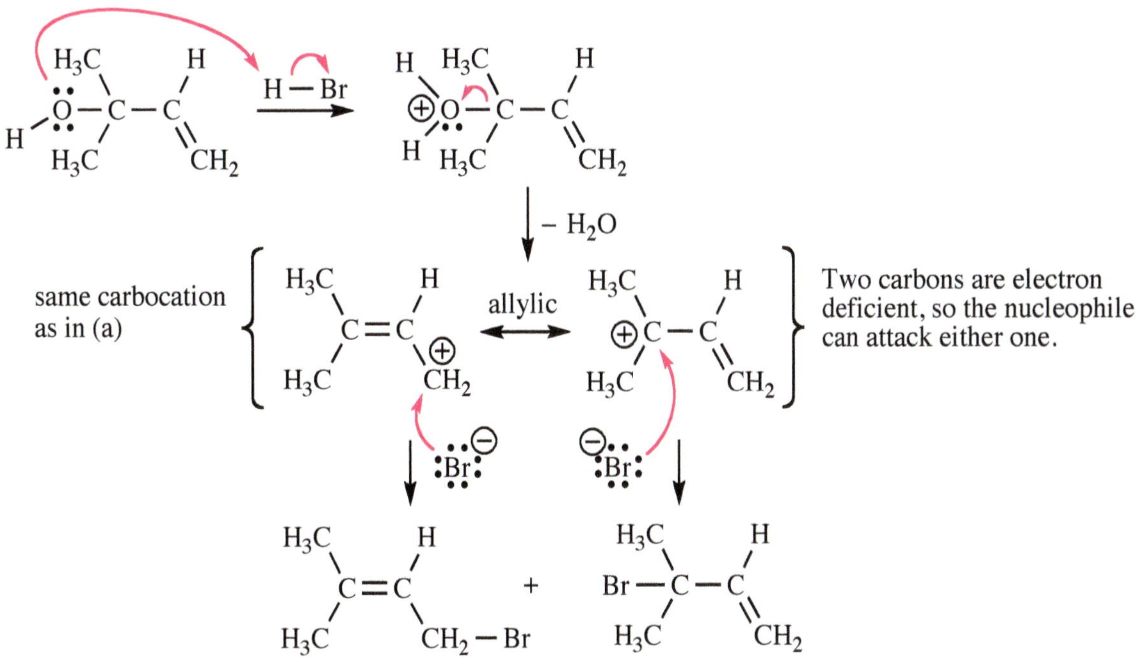

{$(H_3C)_2C{=}CH-CH_2^+$ ⟷ allylic ⟷ $(H_3C)_2C^+-CH{=}CH_2$}

Two carbons are electron deficient, so the nucleophile can attack either one.

$:\ddot{Br}:^-$

$(H_3C)_2C{=}CH-CH_2-Br$ + $Br-C(CH_3)_2-CH{=}CH_2$

15-7 continued

(c)

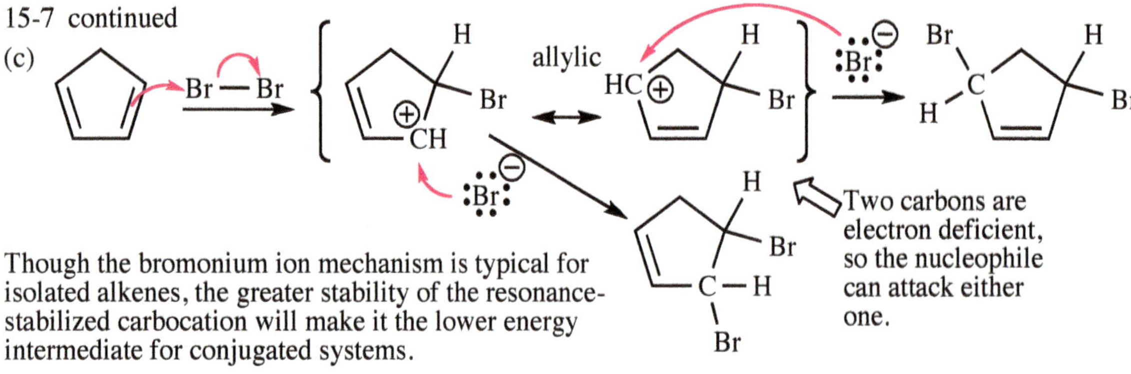

allylic
Two carbons are electron deficient, so the nucleophile can attack either one.

Though the bromonium ion mechanism is typical for isolated alkenes, the greater stability of the resonance-stabilized carbocation will make it the lower energy intermediate for conjugated systems.

(d)

$$H_3C-CH=CH-CH_2Cl \xrightarrow[-\,AgCl]{Ag^+} \left\{ H_3C-CH=CH-\overset{\oplus}{C}H_2 \xleftrightarrow{\text{allylic}} H_3C-\overset{\oplus}{C}H-CH=CH_2 \right\}$$

Two carbons are electron deficient, so the nucleophile can attack either one.

$$\xrightarrow{H_2O} H_3C-CH=CH-CH_2-\overset{\oplus}{O}H_2 \;+\; H_3C-CH(\overset{\oplus}{O}H_2)-CH=CH_2$$

$$\xrightarrow{H_2O} H_3C-CH=CH-CH_2OH \;+\; H_3C-CH(OH)-CH=CH_2$$

(e)

$$H_3C-CHCl-CH=CH_2 \xrightarrow[-\,AgCl]{Ag^+} \left\{ H_3C-\overset{\oplus}{C}H-CH=CH_2 \xleftrightarrow{\text{allylic}} H_3C-CH=CH-\overset{\oplus}{C}H_2 \right\}$$

same carbocation as in (d)

Two carbons are electron deficient, so the nucleophile can attack either one.

$$\xrightarrow{H_2O} H_3C-CH(\overset{\oplus}{O}H_2)-CH=CH_2 \;+\; H_3C-CH=CH-CH_2-\overset{\oplus}{O}H_2$$

$$\xrightarrow{H_2O} H_3C-CH(OH)-CH=CH_2 \;+\; H_3C-CH=CH-CH_2OH$$

15-8

(a) $H_2C(Br)-CH(Br)-CH=CH_2$ (A) + $H_2C(Br)-CH=CH-CH_2(Br)$ (B)

(b) $H_2C=CH-CH=CH_2 \xrightarrow{Br-Br} \left\{ H_2C(Br)-\overset{\oplus\,2°}{C}H-CH=CH_2 \;(A^+) \xleftrightarrow{\text{allylic}} H_2C(Br)-CH=CH-\overset{\oplus\,1°}{C}H_2 \;(B^+) \right\}$

Two carbons are electron deficient, so the nucleophile can attack either one.

$\downarrow :\ddot{Br}:^{\ominus}$

$H_2C(Br)-CH(Br)-CH=CH_2$ (A) + $H_2C(Br)-CH=CH-CH_2(Br)$ (B)

(c) The resonance form A^+, which eventually leads to product **A**, has positive charge on a 2° carbon and is a more significant resonance contributor than structure B^+. With greater positive charge on the 2° carbon than on the 1° carbon, we would expect bromide ion attack on the 2° carbon to have lower activation energy. Therefore, **A** must be the *kinetic* product. At higher temperature, however, the last step becomes reversible, and the stability of the products becomes the dominant factor in determining product ratios. As **B** has a disubstituted alkene whereas **A** is only monosubstituted, it is reasonable that **B** is the major, *thermodynamic* product at 60° C.

(d) At 60° C, ionization of **A** would lead to the same allylic carbocation as shown in (b), which would give the same product ratio as formation of **A** and **B** from butadiene.

$H_2C(Br)-CH(Br)-CH=CH_2 \;(A) \xrightarrow{-\,Br^-} \left\{ H_2C(Br)-\overset{\oplus\,2°}{C}H-CH=CH_2 \xleftrightarrow{\text{allylic}} H_2C(Br)-CH=CH-\overset{\oplus\,1°}{C}H_2 \right\}$

$\downarrow :\ddot{Br}:^{\ominus}$

Two carbons are electron deficient, so the nucleophile can attack either one.

$H_2C(Br)-CH(Br)-CH=CH_2$ (A) 10% + 90% $H_2C(Br)-CH=CH-CH_2(Br)$ (B)

15-9

(a) 2-bromo-1-methylenecyclohexane (Br) + 1-(bromomethyl)cyclohexene (Br)

15-9 continued

(b) Br—Br $\xrightarrow{h\nu}$ **2** Br•

Br• + [methylenecyclohexane] → HBr + { [allylic radical] ↔ [allylic radical] } allylic

Br—Br ↓

Br• + [1-bromo-2-methylenecyclohexane] + [1-(bromomethyl)cyclohexene]

recycles in chain mechanism

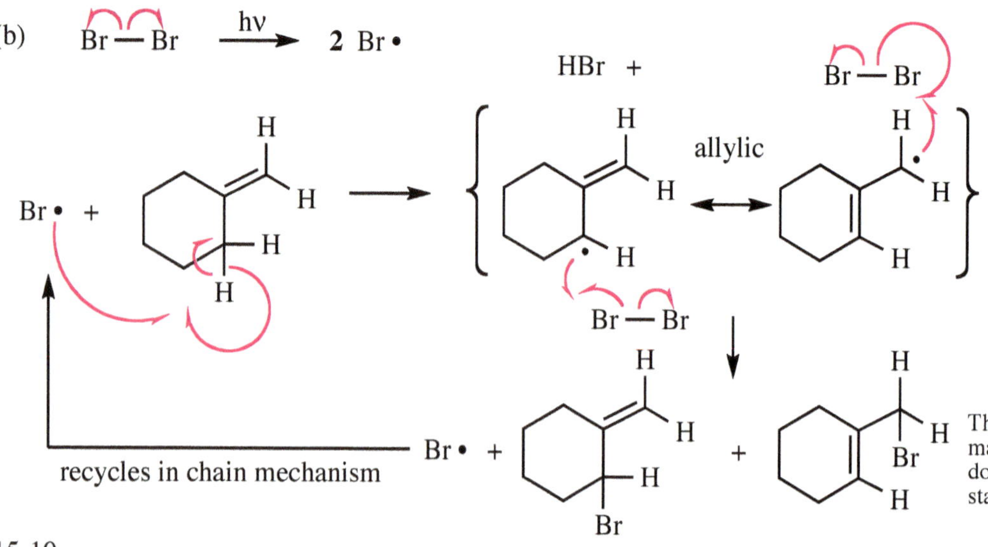

This would probably be the major product because the double bond is in a more stable position.

15-10

(a) [3-bromohex-1-ene] + [1-bromohex-2-ene] *trans* + [1-bromohex-2-ene] *cis*

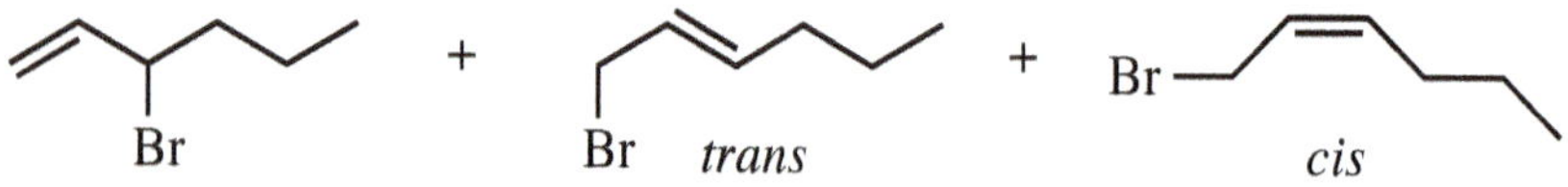

(b) ("Pr" is the abbreviation for propyl, used below.)

NBS + HBr generate a low concentration of Br_2

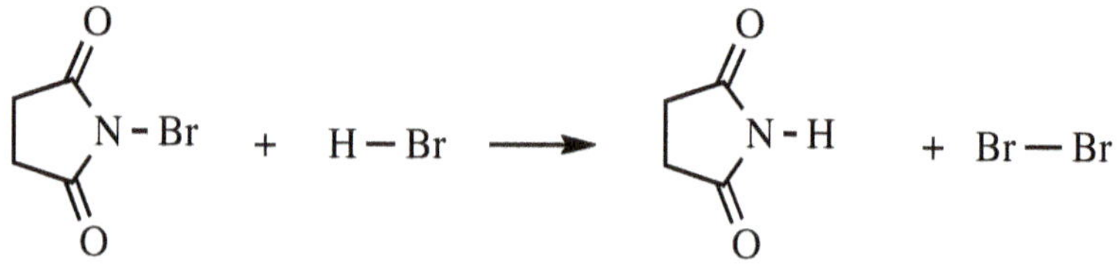

NBS + H—Br → succinimide (N—H) + Br—Br

initiation Br—Br $\xrightarrow{h\nu}$ **2** Br•

propagation

The first step is abstraction of an allylic hydrogen to generate the allylic radical.

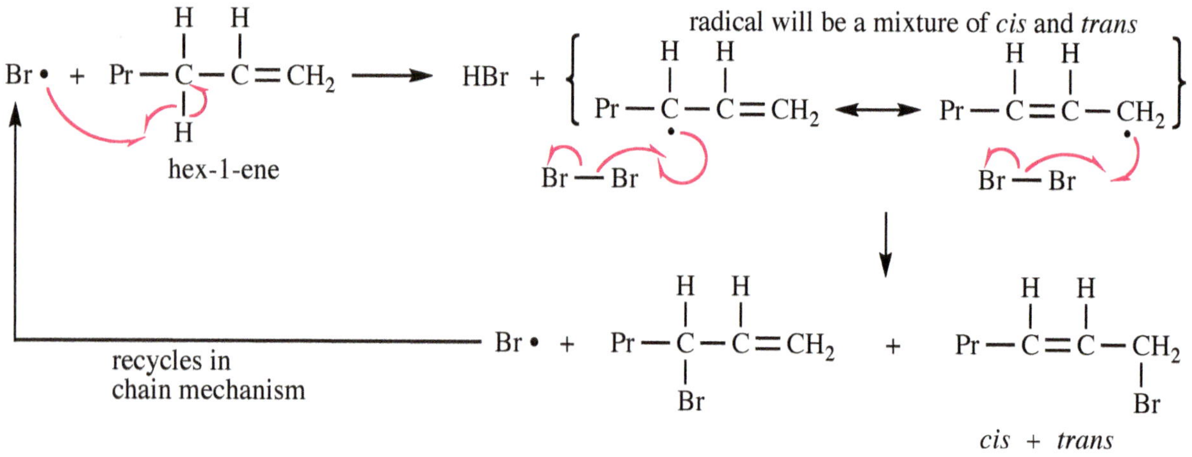

Br• + Pr—CH(H)—CH=CH_2 (hex-1-ene) → HBr + { Pr—ĊH—CH=CH_2 ↔ Pr—CH=CH—ĊH$_2$ }

radical will be a mixture of *cis* and *trans*

Br—Br ↓ Br—Br

Br• + Pr—CH(Br)—CH=CH_2 + Pr—CH=CH—CH_2Br

cis + *trans*

recycles in chain mechanism

15-11

(a) 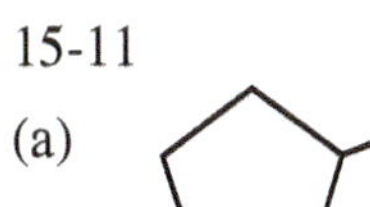

(b) Br + Br

These are the major products from abstraction of a 1° allylic H.

(c) CH_2Br

Benzylic radicals are even more stable than allylic.

15-12 Both halides generate the same allylic carbanion.

$$H_3C-CH=CH-CH_2Br \xrightarrow{Mg} \left\{ H_3C-CH=CH-\ddot{C}H_2^{\ominus} \longleftrightarrow H_3C-\ddot{C}H^{\ominus}-CH=CH_2 \right\} \overset{\oplus}{MgBr} \xrightarrow{H_2O} H_3C-CH=CH-CH_2H \; + \; H_3C-CHH-CH=CH_2$$

$$H_3C-CHBr-CH=CH_2 \xrightarrow{Mg}$$

15-13

(a) Br $\xrightarrow[\text{ether}]{Mg}$ MgBr $\xrightarrow{H_2C=CHCH_2Br}$

(b) $CH_3CHBrCH_3 \xrightarrow[\text{ether}]{Mg} CH_3CH(MgBr)CH_3 + CH_3CH=CHCH_2Br \longrightarrow CH_3CH=CHCH_2-CH(CH_3)CH_3$

(c) $CH_3CH_2CH_2Br \xrightarrow[\text{ether}]{Mg} CH_3CH_2CH_2MgBr$

\+ H_2C-Br / $Br-CH_2$ *add one-half equivalent* $\longrightarrow$ $H_2C-CH_2CH_2CH_3$ / $CH_3CH_2CH_2-CH_2$ dec-5-ene

This synthesis could also be performed sequentially.

15-14 Chiral products from achiral reactants produce racemic mixtures. New bonds are in bold.

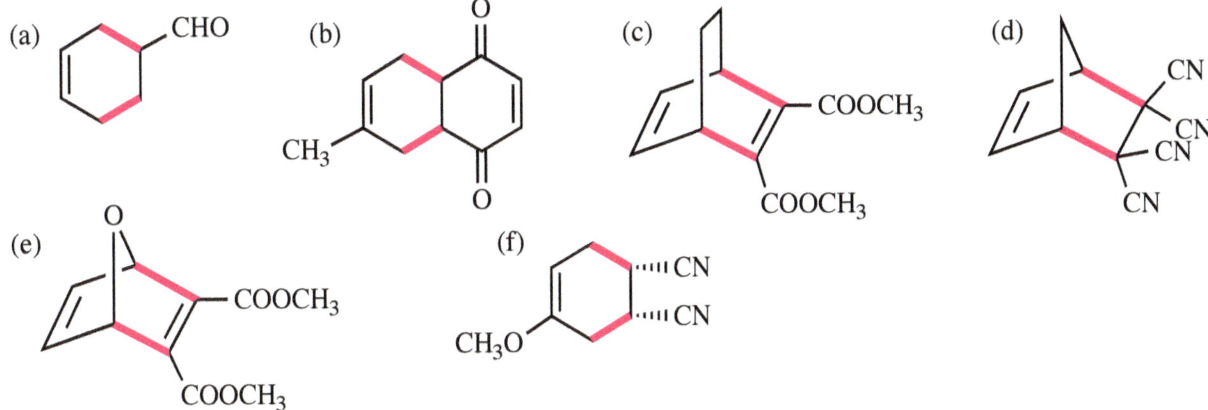

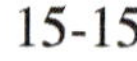

15-15

(a)

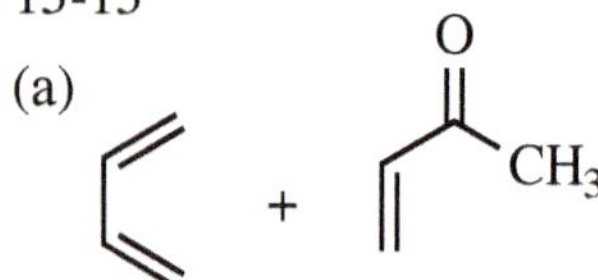

(b) (c)

(d) (e) (f)

15-16 Chiral products from achiral reactants produce racemic mixtures.

(a)

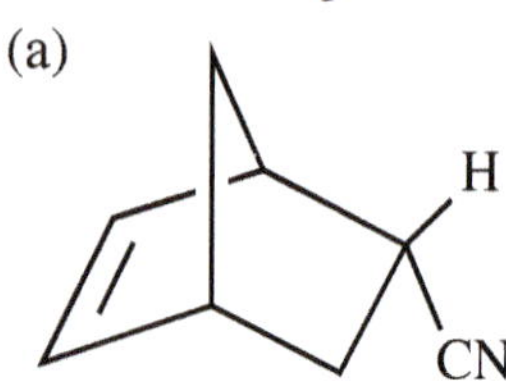

(b) (c) all *cis*

15-17 These structures show the alignment of diene and dienophile in the Diels-Alder transition state, leading to 1,4-orientation in (a) and 1,2-orientation in (b).

(a)

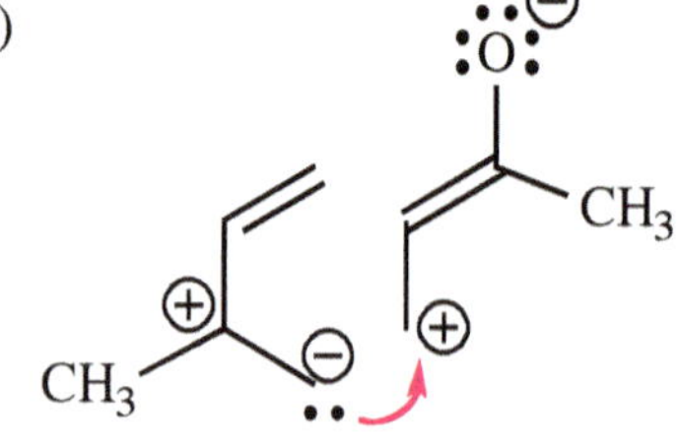

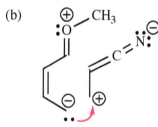

This left structure is a VERY minor resonance contributor; however, it explains the orientation for the diene as carbon-1 is more negative and carbon-2, a 3° carbon, is slightly more positive because of methyl group stabilization.

(b)

15-18 For clarity, the sigma bonds formed in the Diels-Alder reaction are shown in bold. Chiral products from achiral reactants produce racemic mixtures.

(a)

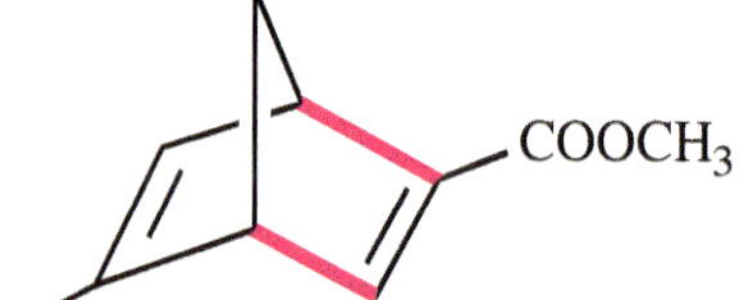

(b)

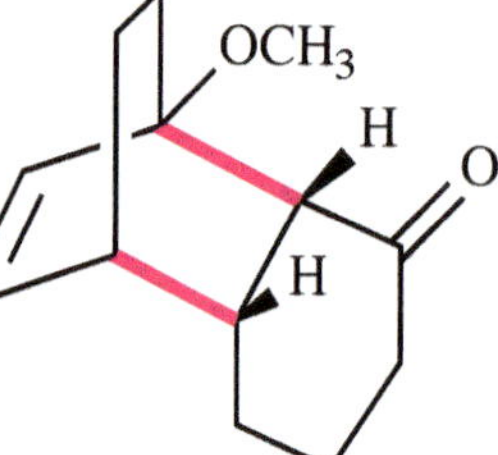

(c)

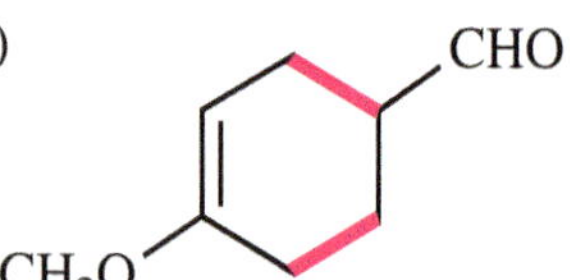

(d)

15-19 For a photochemically *allowed* process, one molecule must use an excited state in which an electron has been promoted to the first antibonding orbital. All orbital interactions between the excited molecule's HOMO* and the other molecule's LUMO must be bonding for the interaction to be allowed; otherwise, it is a forbidden process.

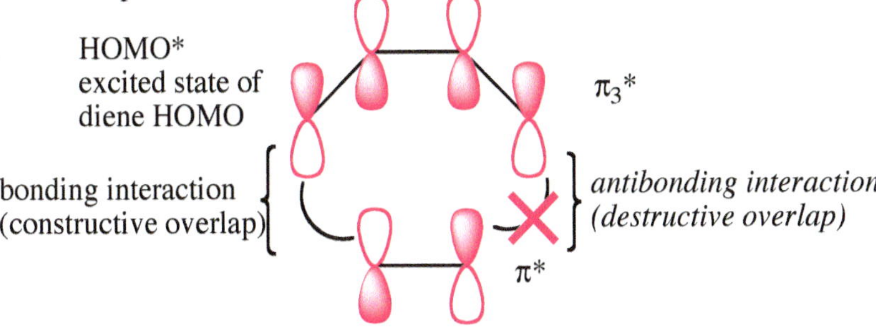

In the Diels-Alder cycloaddition, the LUMO of the dienophile and the excited state of the HOMO of the diene (labeled HOMO*) produce one bonding interaction and one antibonding interaction. Thus, this is a **photochemically forbidden** process.

15-20 For a [4 + 4] cycloaddition:

(a)

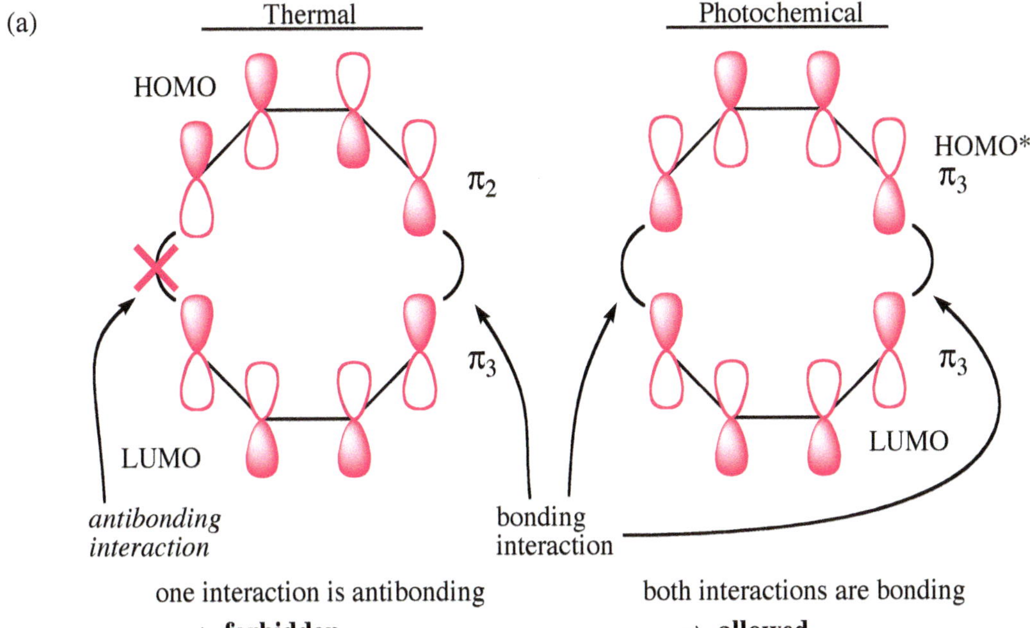

one interaction is antibonding ⇒ **forbidden**

both interactions are bonding ⇒ **allowed**

(b) A [4 + 4] cycloaddition is not thermally allowed, but a [4 + 2] (Diels-Alder) is!

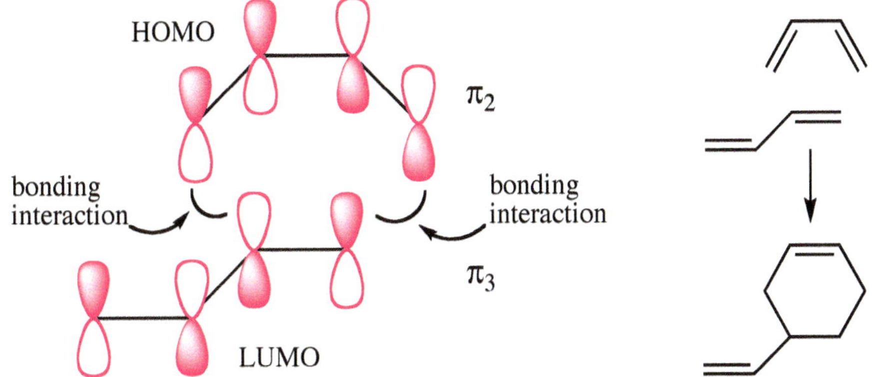

15-21 $A = \varepsilon c\,l \qquad \varepsilon = \dfrac{A}{c\,l} \qquad l = 1 \text{ cm} \qquad A = 0.50$

convert mass to moles: $1 \text{ mg} \times \dfrac{1 \text{ g}}{1000 \text{ mg}} \times \dfrac{1 \text{ mole}}{160 \text{ g}} = 6 \times 10^{-6} \text{ moles}$

$$c = \frac{6 \times 10^{-6} \text{ moles}}{10 \text{ mL}} \times \frac{1000 \text{ mL}}{1 \text{ L}} = 6 \times 10^{-4} \text{ M}$$

$$\varepsilon = \frac{0.50}{(6 \times 10^{-4})(1)} = 833 \approx \boxed{800}$$

15-22

(a) 353 nm: a conjugated tetraene—must have highest absorption maximum among these compounds;
(b) 313 nm: closest to the bicyclic conjugated triene in Table 15-2; the diene is in a more substituted ring, so it is not surprising for the maximum to be slightly higher than 304 nm;
(c) 232 nm: similar to 3-methylenecyclohexene in Table 15-2;
(d) 292 nm: like cyclohexa-1,3-diene (256 nm) + 1 additional conjugated double bond (30 nm) + 1 alkyl group (5 nm) = predicted value of 291 nm

15-23

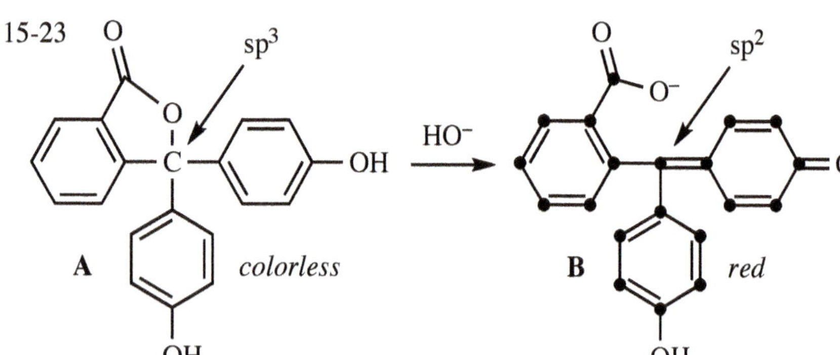

Extended conjugation is required for an organic compound to be colored. The acid form of phenolphthalein **A** has an sp^3 carbon disrupting the conjugation, whereas the basic form **B** has a central carbon with sp^2 hybridization, permitting conjugation over three rings; the conjugated sp^2 carbons are indicated with a large dot in this picture.

15-24

(a) isolated

1, 4, 5

(b) conjugated

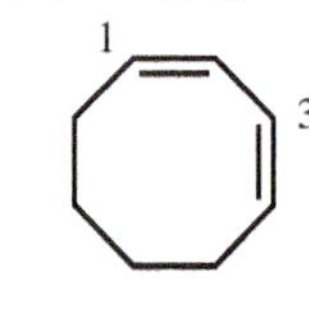

(c) cumulated

C

The sp carbon would require linear geometry.

(d) conjugated

1, 3, 5, 7

(e) conjugated

(f) cumulated (1,2) and conjugated (2,4)

$H_2C{=}C{=}CH{-}CH{=}CH_2$ (carbons numbered 1, 2, 3, 4, 5)

15-25

(a)

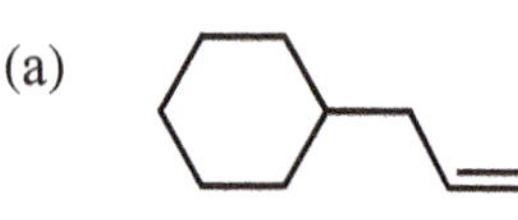

(b) Cl

one equivalent of HCl

(c)

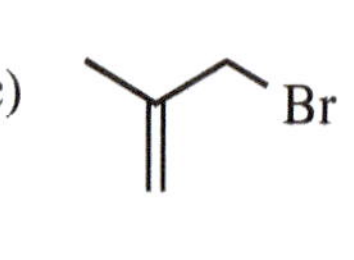

(d)

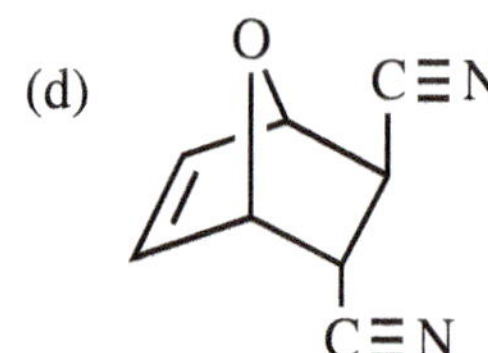

(e) Br, OH

+ Br, OH

(f) Br, Br + Br, Br minor—not conjugated + Br, Br

15-25 continued

(g) OCH_3 ... + ... OCH_3 + ... + ...

substitution elimination

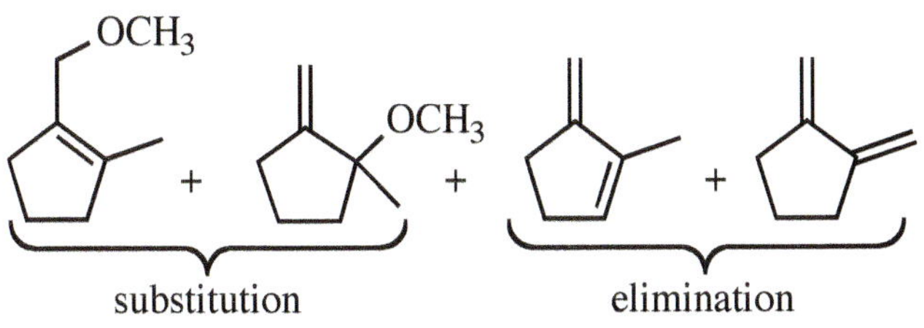

(h) $COOCH_3$

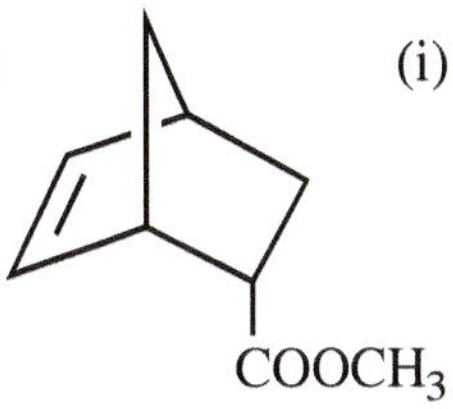

(i) O O

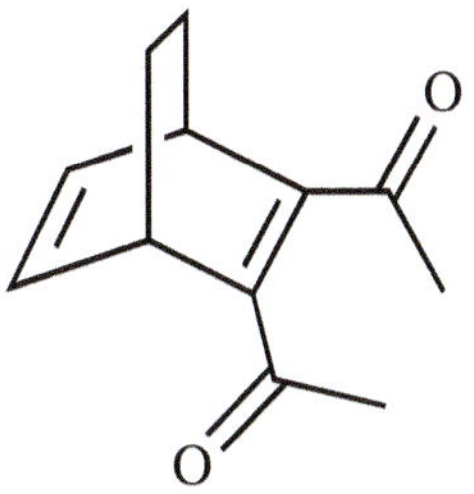

15-26 Grignard reactions are performed in ether solvent. The bonds formed are shown in bold.

(a) Br + BrMg ⟶

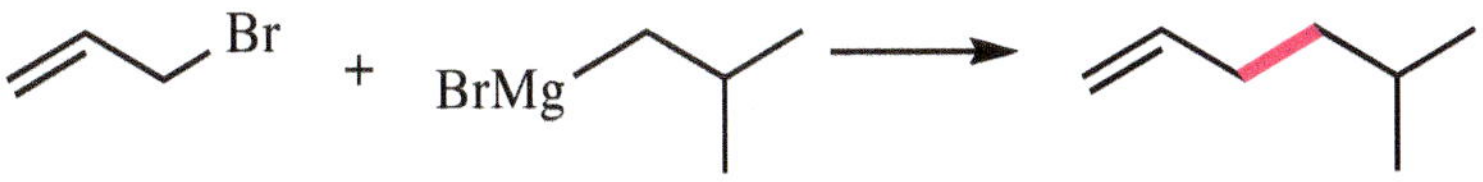

(b) Br + BrMg ⟶

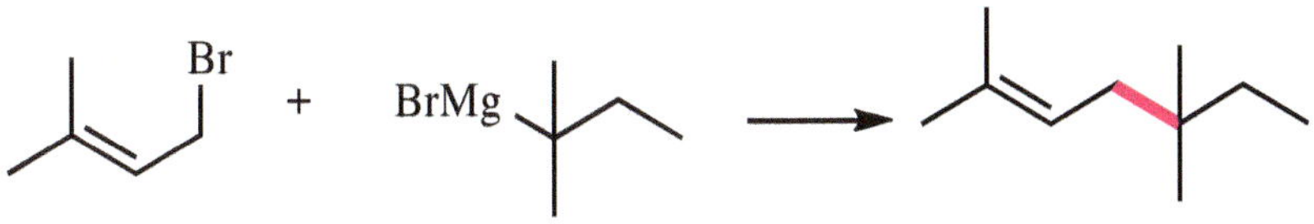

(c) Br + BrMg ⟶ 5 3 1 4 2

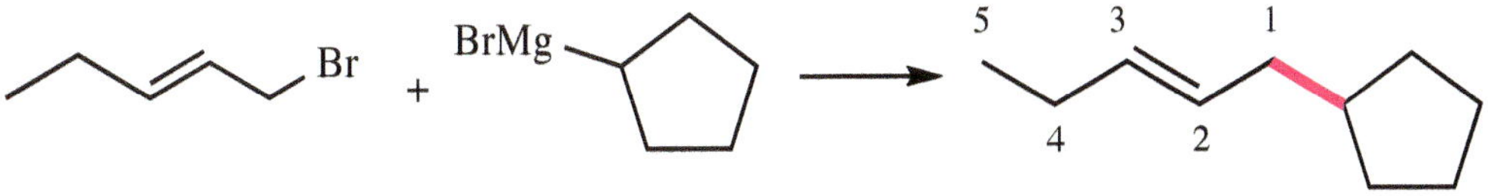

15-27

(a) CH_2 CH_2 H ⟷ CH_2 C H CH_2 ⟷ CH_2 CH_2 H

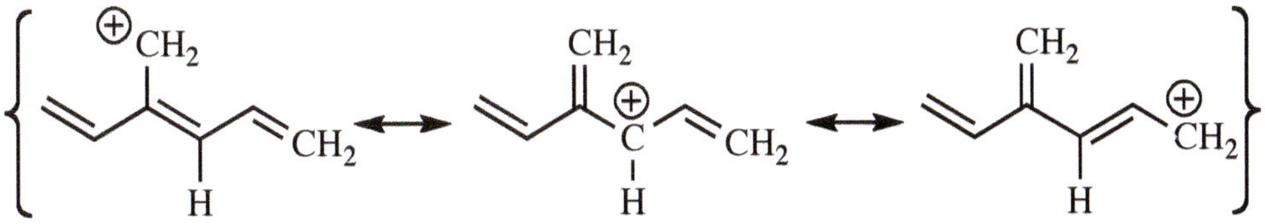

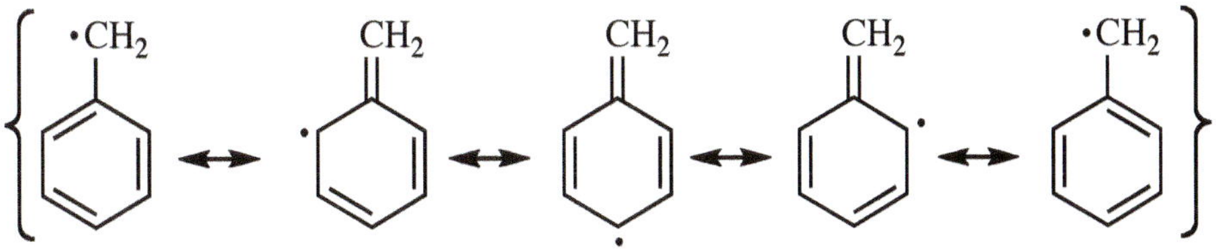

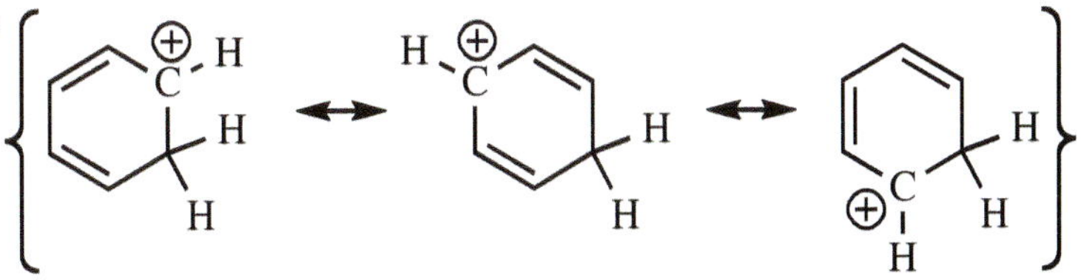

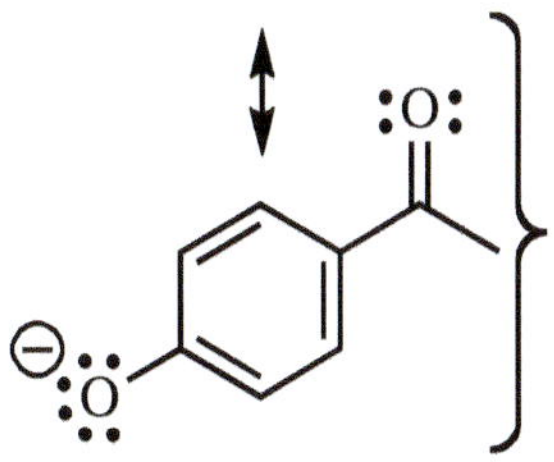

(b) $\cdot CH_2$ ⟷ CH_2 ⟷ CH_2 ⟷ CH_2 ⟷ $\cdot CH_2$

(c) :O: :O: ⟷ HC :O: ⟷ C :O: ⟷ :O: ⟷ :O: C H ⟷ :O:

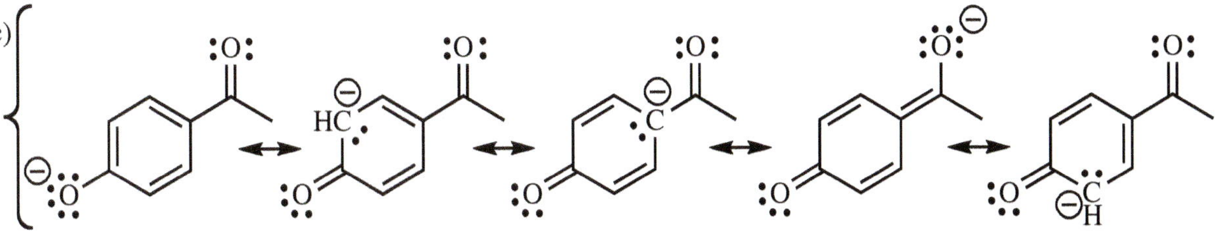

(d) C H H H ⟷ H C H H ⟷ H C H H

15-27 continued

(e)

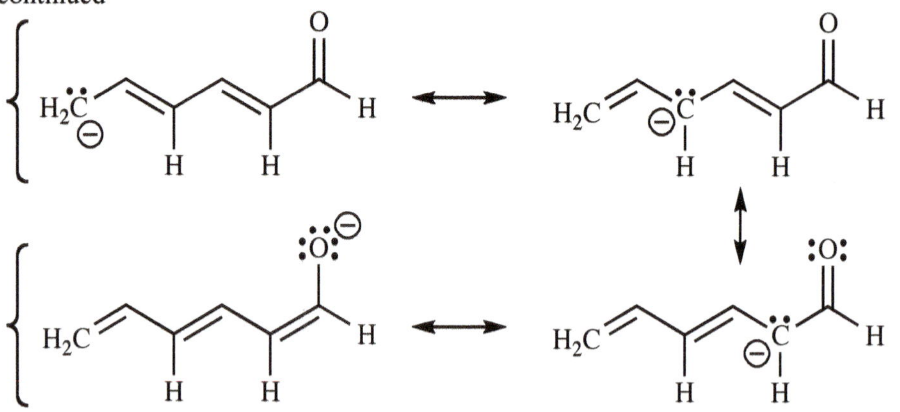

most significant contributor—
negative charge on more electronegative atom

(f)

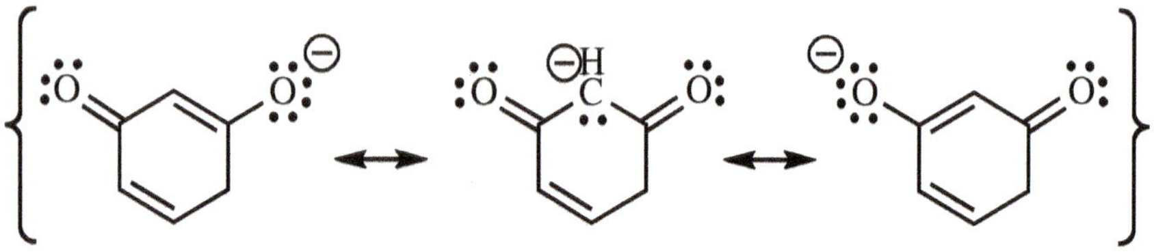

(g)

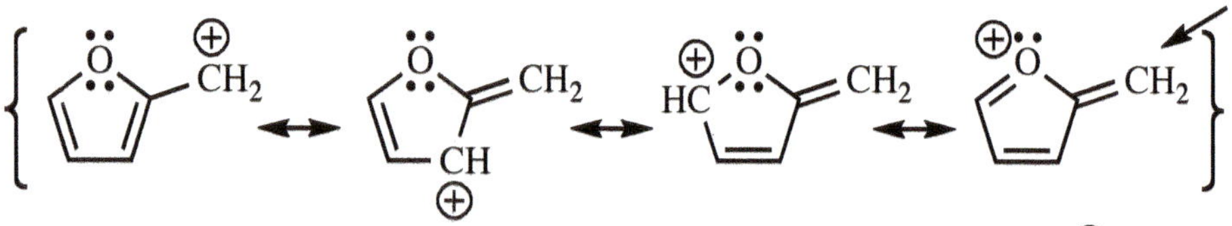

Most significant contributor—all atoms have full octets.

(h)

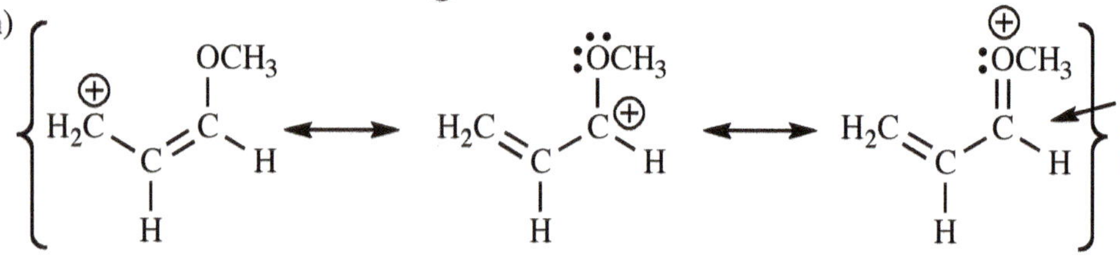

Most significant contributor—all atoms have full octets.

15-28

(a) $A = \varepsilon c\, l$ $\qquad \varepsilon = \dfrac{A}{c\, l}$ $\qquad l = 1$ cm $\qquad A = 0.74$

convert mass to moles: $0.0010\text{ g} \times \dfrac{1\text{ mole}}{255\text{ g}} = 3.9 \times 10^{-6}$ moles

$$c = \frac{3.9 \times 10^{-6}\text{ moles}}{100\text{ mL}} \times \frac{1000\text{ mL}}{1\text{ L}} = 3.9 \times 10^{-5}\text{ M}$$

$$\varepsilon = \frac{0.74}{(3.9 \times 10^{-5})(1)} \approx \boxed{19{,}000}$$

(b) This large value of ε could only come from a conjugated system, eliminating the first structure. The absorption maximum at 235 nm is most likely a diene rather than a triene. The most reasonable structure is:

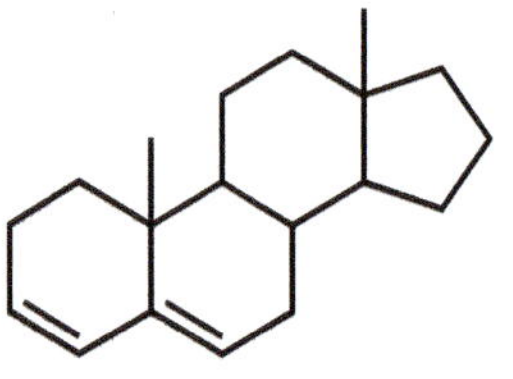

compare with dienes:

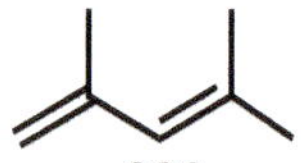

λ_{max} = 232 nm
Solved Problem 15-3

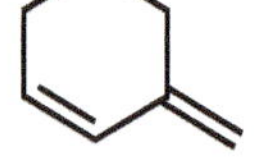

λ_{max} = 232 nm
Table 15-2

15-29

(a) align δ+ and δ−

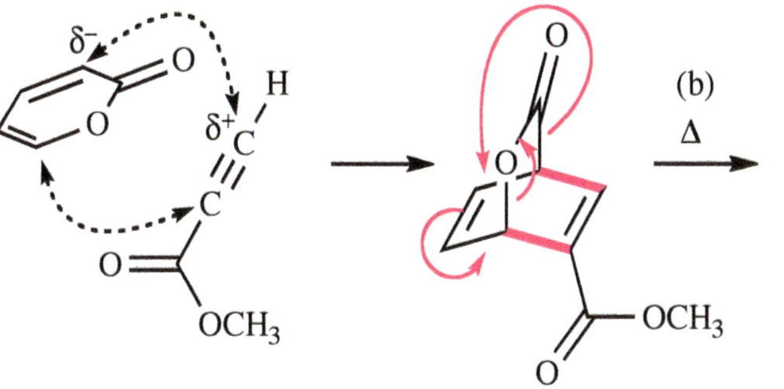

new sigma bonds shown in bold

O=C=O

OCH3

O

methyl benzoate

(b) The second reaction is called a retro (reverse) Diels-Alder reaction. It is also an electrocyclic rearrangement of electrons, in this case breaking sigma bonds, forming new pi bonds, and creating an *aromatic* ring in the process. The stability of the aromatic ring accounts for the thermodynamic preference for the benzene product.

15-30 For clarity, the sigma bonds formed in the Diels-Alder reaction are shown in bold.
Chiral products from achiral reactants produce racemic mixtures.

(a)

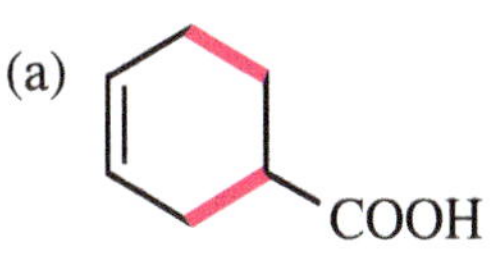

(b)

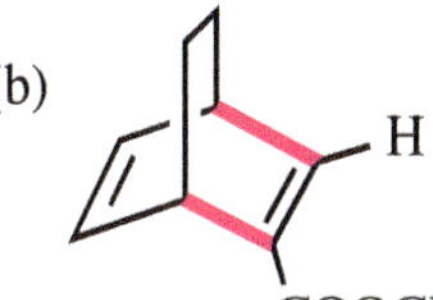

(c)

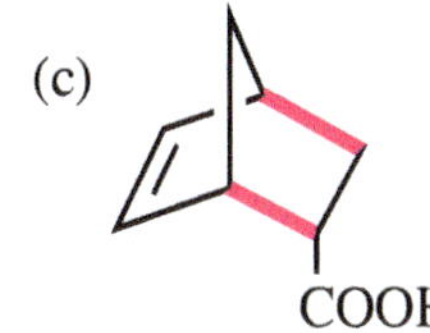

Endo Rule

(d)

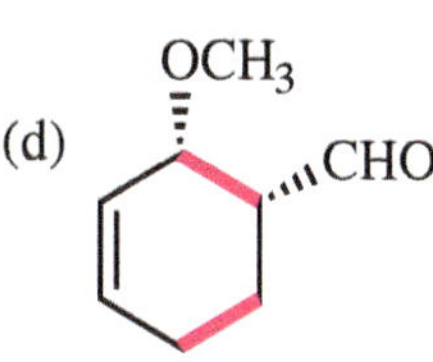

(e)

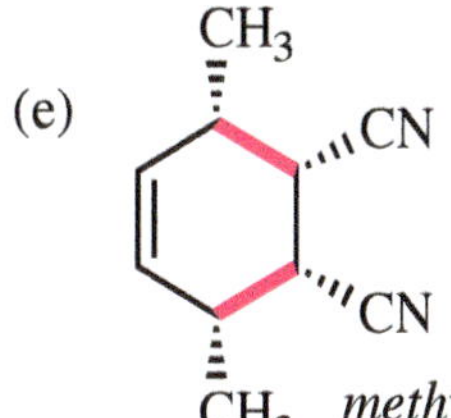

methyls cis

(f)

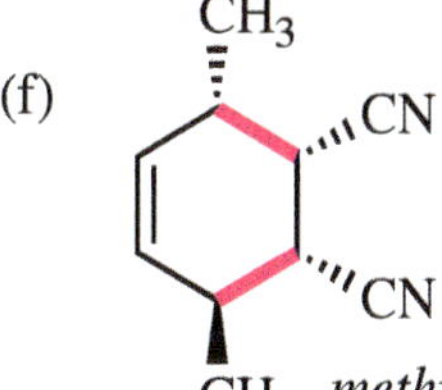

methyls trans

15-31 For clarity, the sigma bonds formed in the Diels-Alder reaction are shown in bold.
Chiral products from achiral reactants produce racemic mixtures.

(a)

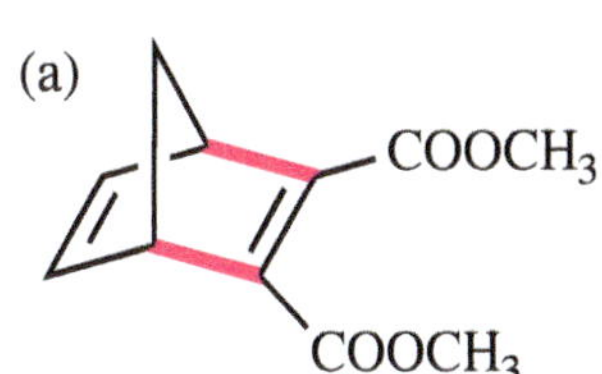

(b)

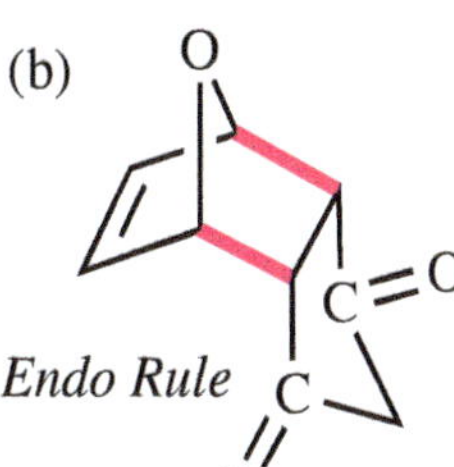

Endo Rule

(c)

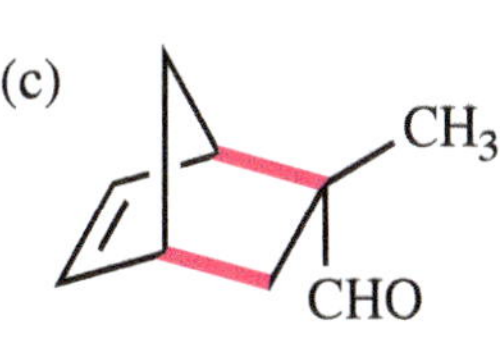

Endo Rule

(d)

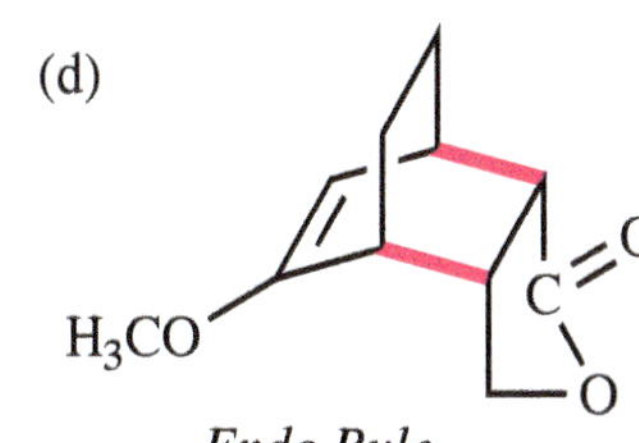

Endo Rule

(e) N(CH3)2

C=O

Endo Rule OEt

(f) OCH3 O

Ring juncture is cis.

15-32

(a) The absorption at 1630 cm^{-1} suggests a conjugated alkene. The higher temperature allowed for migration of the double bond.

(b)

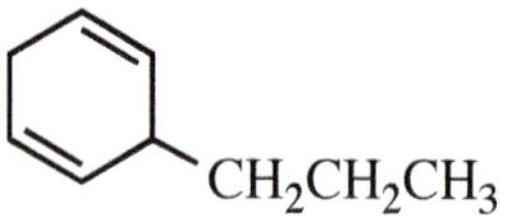

$CH_2CH_2CH_3$

desired

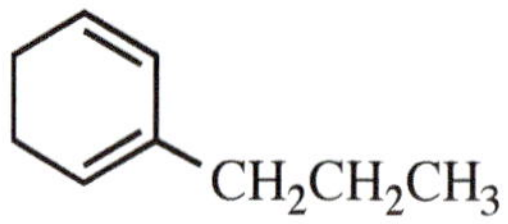

$CH_2CH_2CH_3$

actual

The UV λ_{max} at 261 nm verifies that the two double bonds have become conjugated. Isolated double bonds absorb at < 200 nm.

(c)
expected:

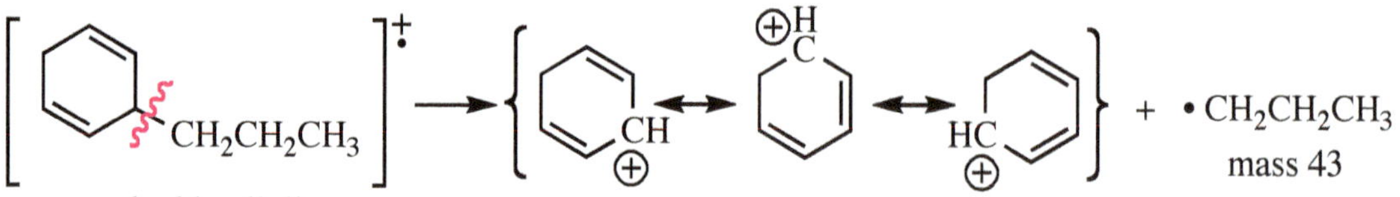

$CH_2CH_2CH_3$

doubly allylic
m/z 122

CH

H C

HC

m/z 79

+ • $CH_2CH_2CH_3$
mass 43

actual:

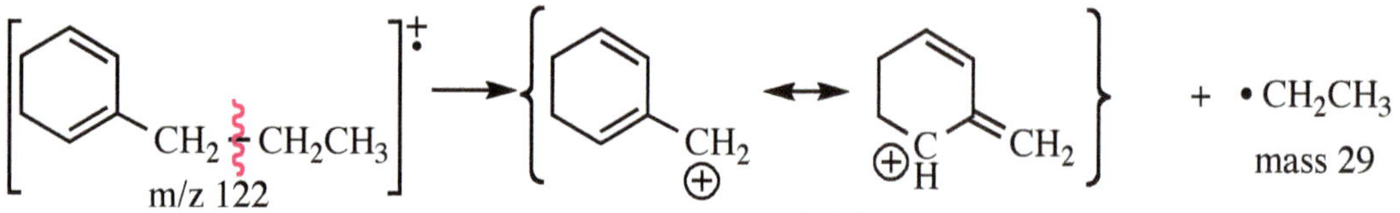

CH_2 CH_2CH_3
m/z 122

CH_2

C H CH_2

m/z 93

+ • CH_2CH_3
mass 29

15-33

(a) + H O

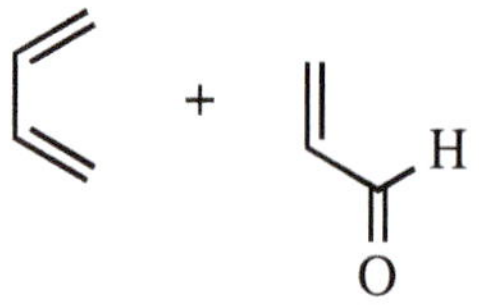

(b) + Cl Cl

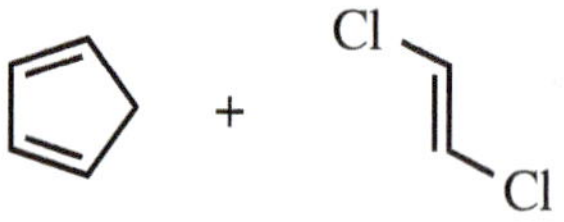

(c) CH_3 CH_3 + $COOCH_3$ C CH

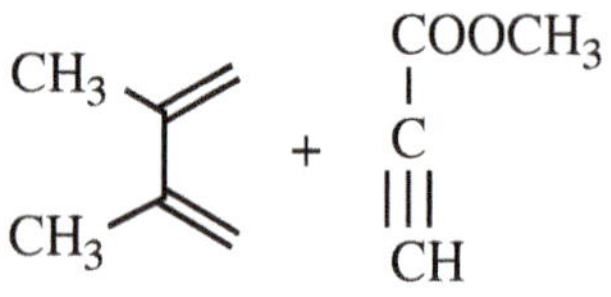

(d) + O O

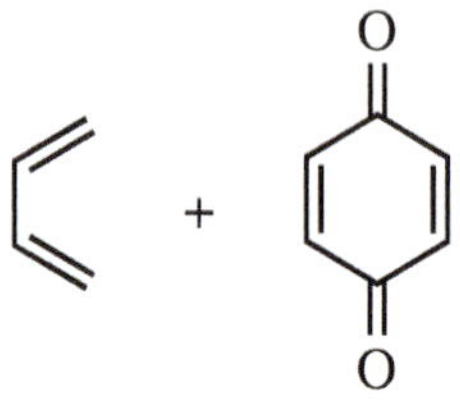

(e) + O O O

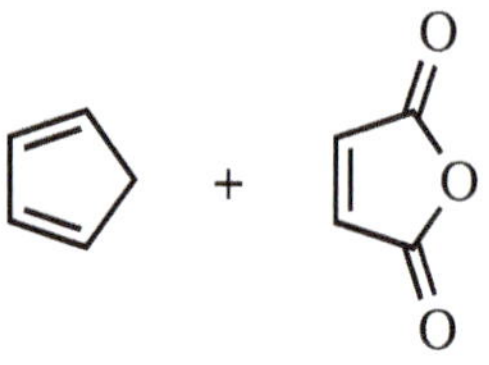

(f) O + CN CN

(g) Cl Cl Cl Cl Cl Cl + Cl Cl

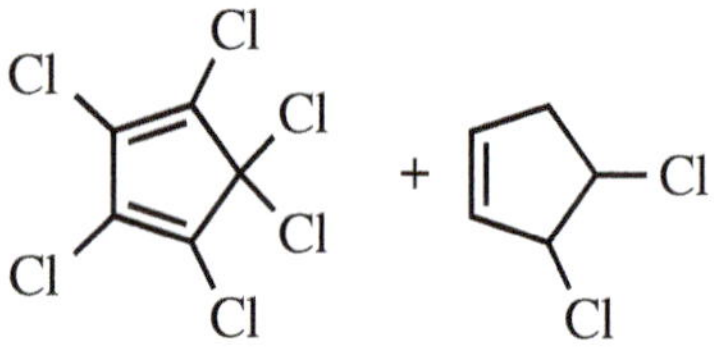

(h) Cl Cl Cl Cl Cl Cl +

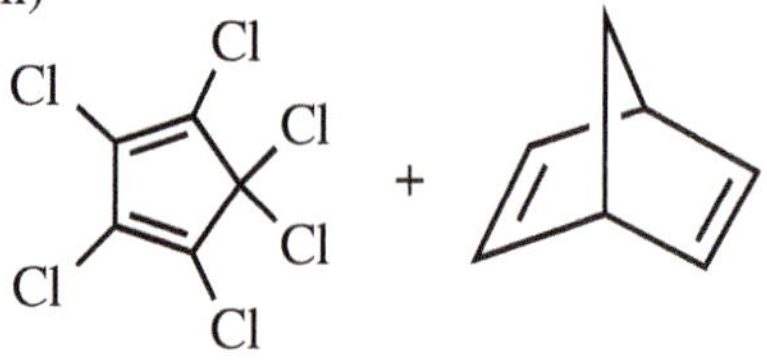

(i) O S O + CN NC

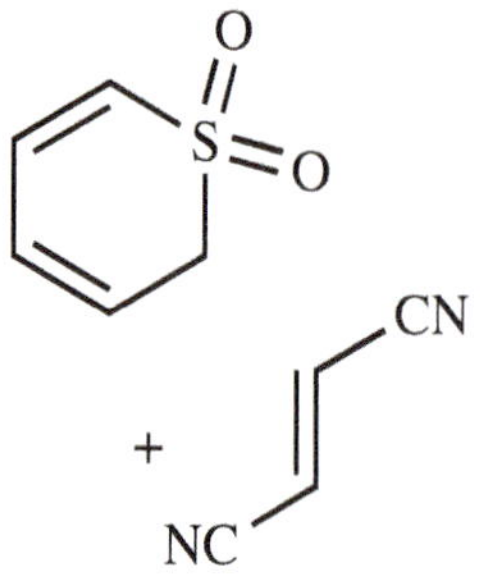

15-34

PART 1

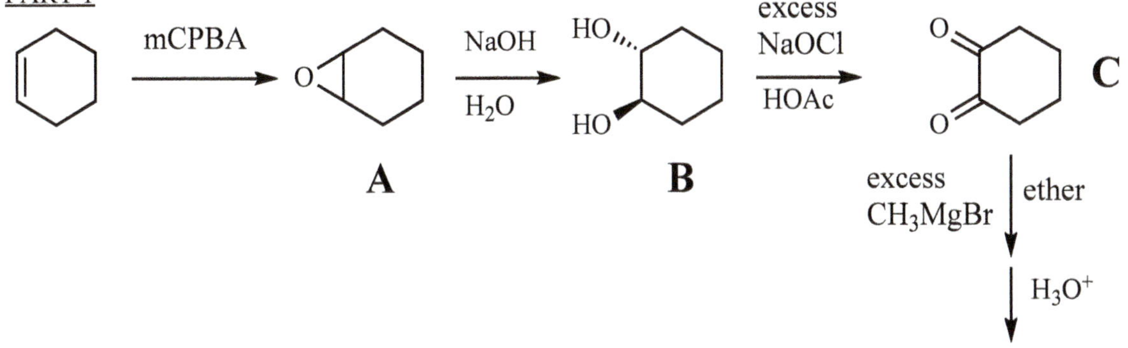

cyclohexene —mCPBA→ **A** —NaOH, H_2O→ **B** —excess NaOCl, HOAc→ **C**

C —excess CH_3MgBr, ether; H_3O^+→ **D** (mixture of diastereomers)

D —H_2SO_4, Δ→ **E** C_8H_{12}

This conjugated diene is the most stable product, and therefore the major product in this E1 reaction.

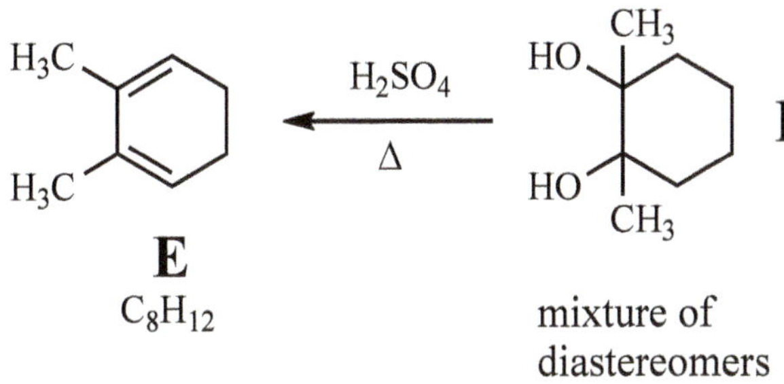

PART 2

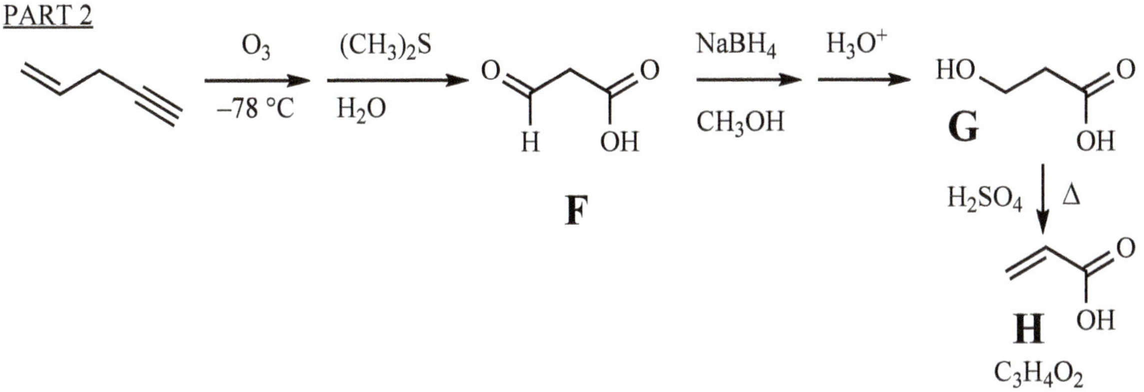

—O_3, −78 °C; $(CH_3)_2S$, H_2O→ **F** —$NaBH_4$, CH_3OH; H_3O^+→ **G**

G —H_2SO_4, Δ→ **H** $C_3H_4O_2$

PART 3

E + **H** —Δ→ **I** $C_{11}H_{16}O_2$

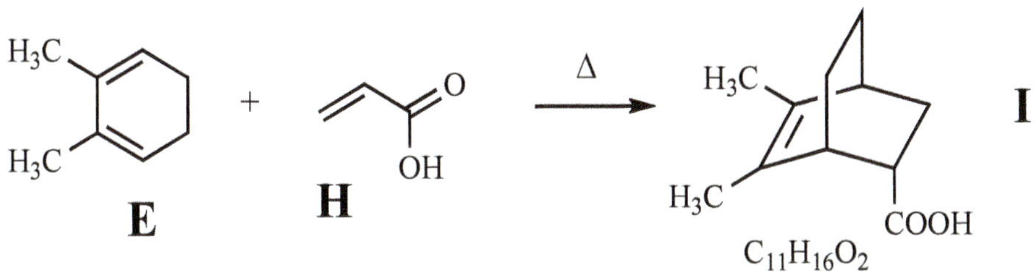

All of the atoms in **E** and **H** appear in *I*—good evidence that this is a Diels-Alder reaction.

15-35

(a)

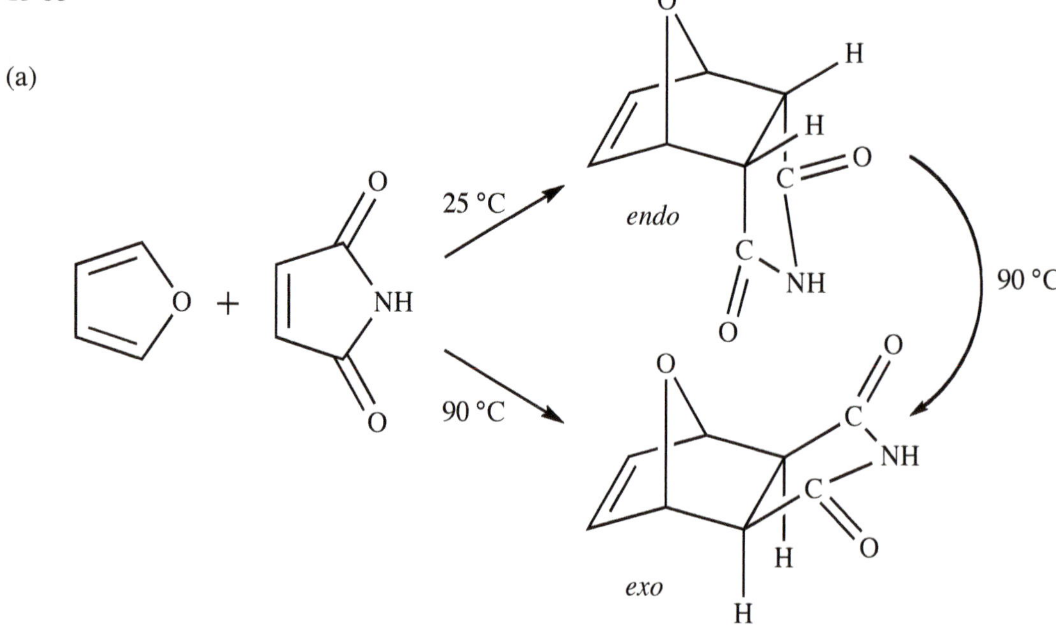

(b) The *endo* isomer is usually preferred because of secondary p orbital overlap of C=O with the diene in the transition state.

(c) The reasoning in (b) applies to stabilization of the transition state of the reaction, not the stability of the product. Arguments based on transition state stability apply to the rate of reaction, inferring that the *endo* product is the kinetic product.

(d) At 25 °C, the reaction cannot easily reverse, or at least not very rapidly. The *endo* product is formed faster and is the major product because its transition state is lower in energy—the reaction is under kinetic control. At 90 °C, the reverse reaction is not as slow and equilibrium is achieved. The *exo* product is less crowded and therefore more stable—equilibrium control gives the *exo* as the major product.

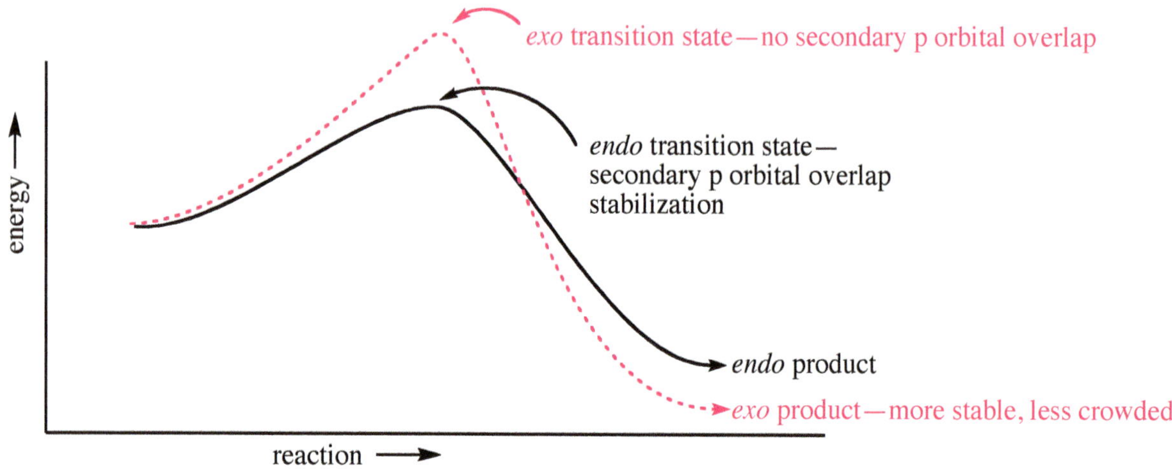

15-36 Nodes are represented by dashed lines.

(a)

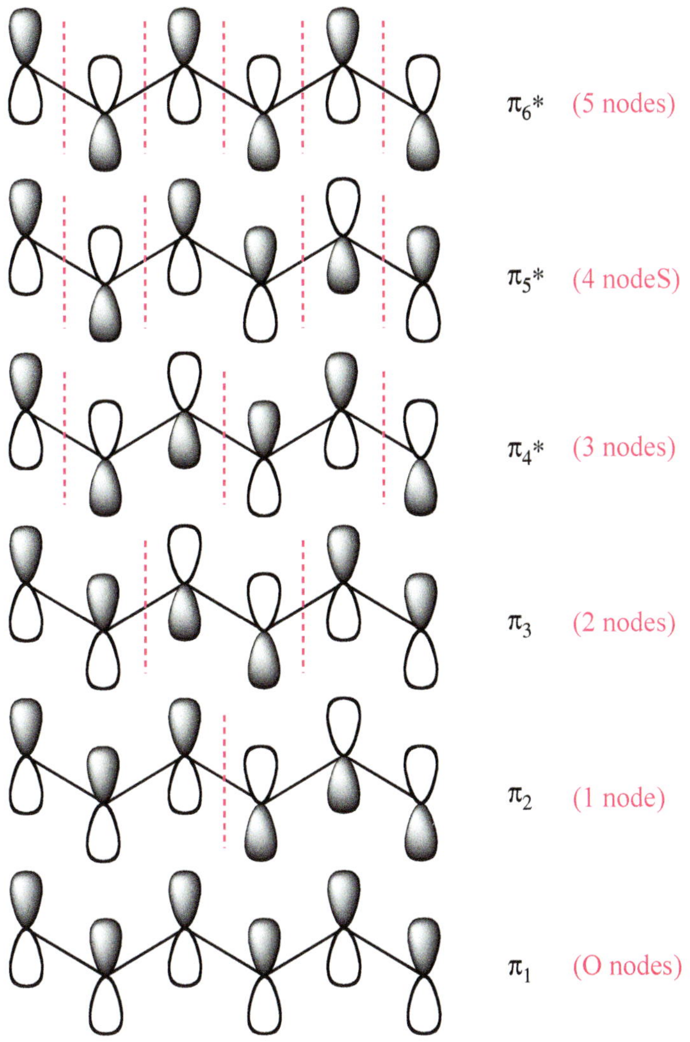

π_6* (5 nodes)

π_5* (4 nodeS)

π_4* (3 nodes)

π_3 (2 nodes)

π_2 (1 node)

π_1 (O nodes)

(b)

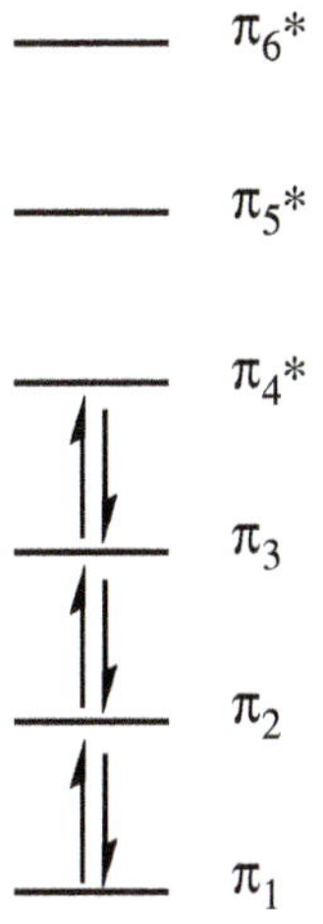

π_6*

π_5*

π_4*

π_3

π_2

π_1

(c)

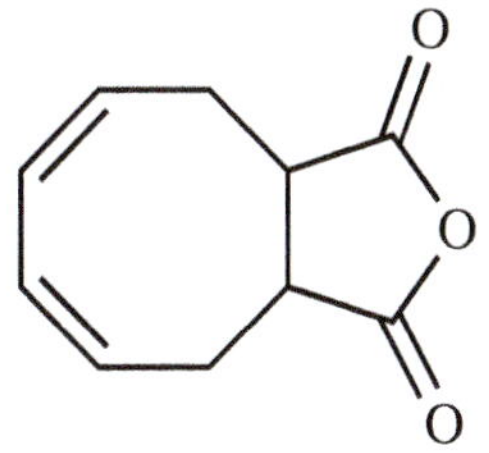

O

O

O

(d) Whether the triene is the HOMO and the alkene is the LUMO, or *vice versa*, the answer will be the same.

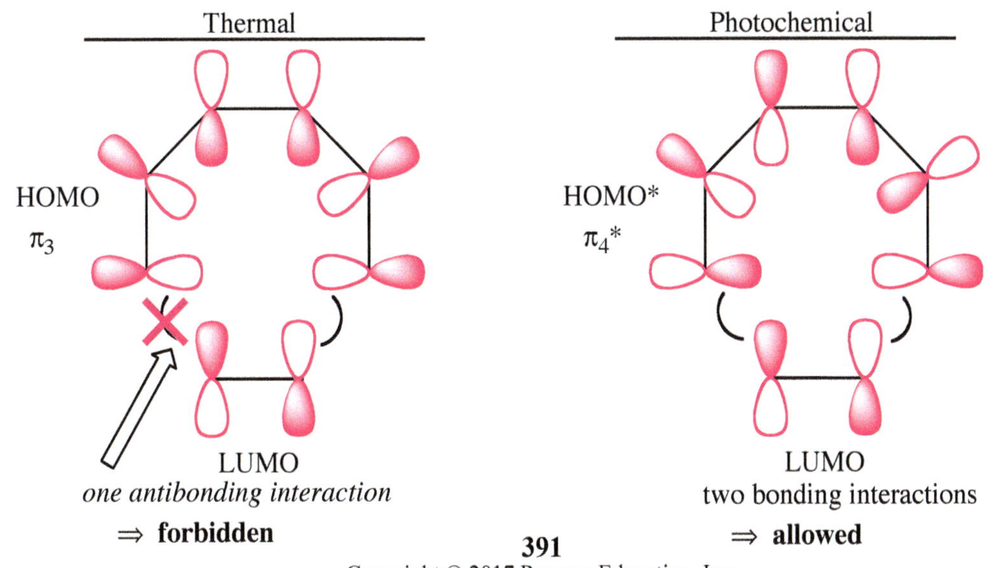

15-36 continued

(e)

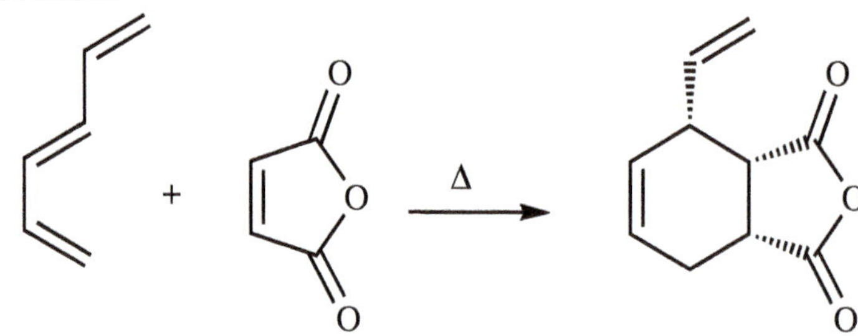

15-37

(a) $\left\{ H_2C{=}\underset{\substack{|\\H}}{C}{-}\underset{\substack{|\\H}}{C}{=}\underset{\substack{|\\H}}{C}{-}\dot{C}H_2 \longleftrightarrow H_2C{=}\underset{\substack{|\\H}}{C}{-}\underset{\substack{|\\H}}{\dot{C}}{-}\underset{\substack{|\\H}}{C}{=}CH_2 \longleftrightarrow H_2\dot{C}{-}\underset{\substack{|\\H}}{C}{=}\underset{\substack{|\\H}}{C}{-}\underset{\substack{|\\H}}{C}{=}CH_2 \right\}$

(b) Five p atomic orbitals will generate five pi molecular orbitals.

(c) The lowest energy molecular orbital has no nodes. Each higher molecular orbital will have one more node, so the fifth molecular orbital will have four nodes.

(d) Nodes are represented by dashed lines.

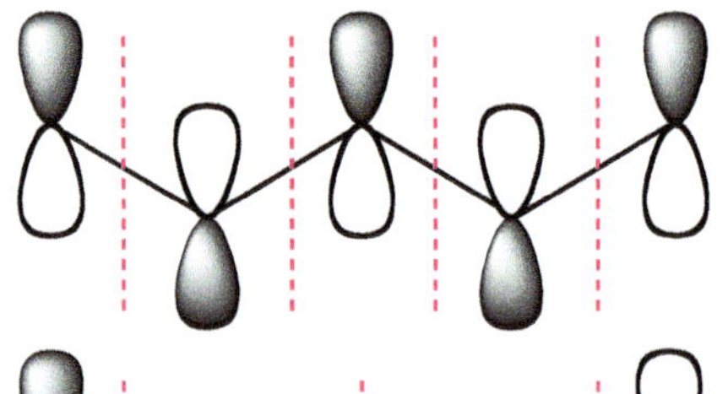

π_5*

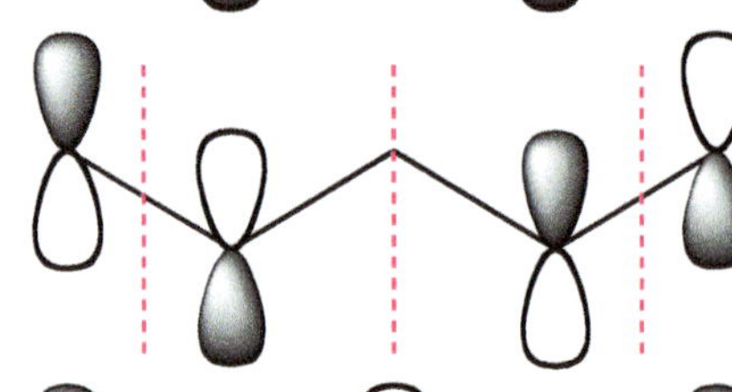

π_4*

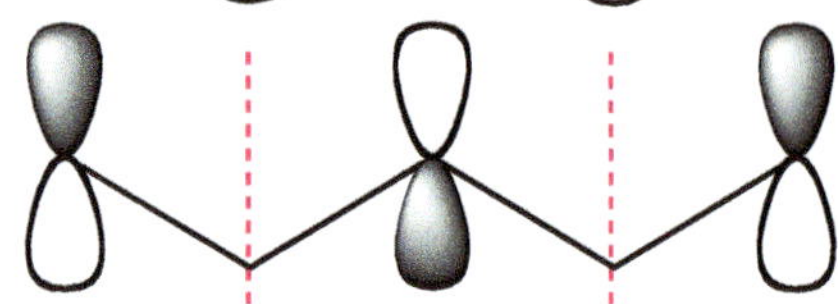

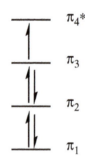

π_3 (nonbonding)

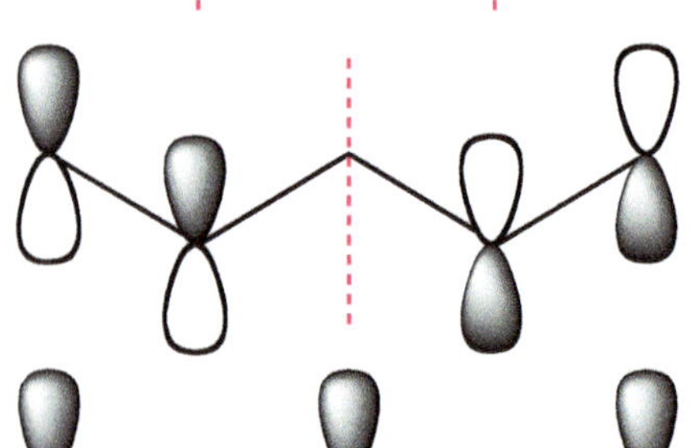

π_2

π_1

(e)

π_5*

π_4*

π_3 ↑

π_2 ↑↓

π_1 ↑↓

(f) The HOMO, π_3, contains an unpaired electron giving this species its radical character. The HOMO is a nonbonding orbital with lobes only on carbons 1, 3, and 5, consistent with the resonance picture.

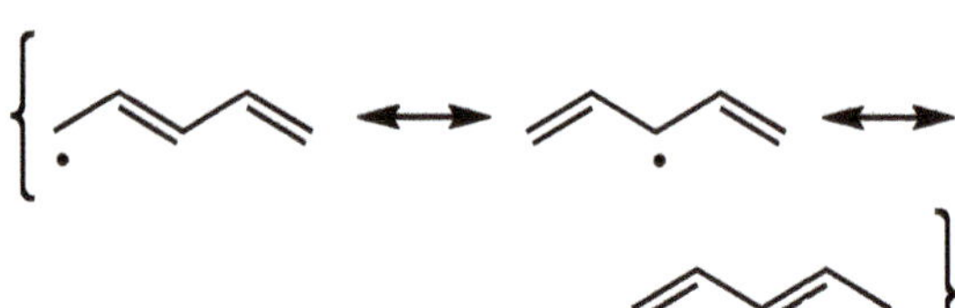

15-37 continued

(g)

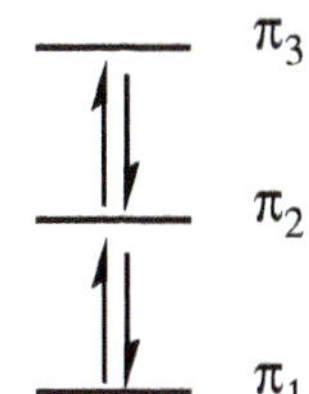

Again, it is π_3 that determines the character of this species. When the single electron in π_3 of the neutral radical is removed, positive charge appears only in the position(s) which that electron occupied. That is, the positive charge depends on the now *empty* π_3, with *empty* lobes (positive charge) on carbons 1, 3, and 5, consistent with the resonance description.

(h)

Again, it is π_3 that determines the character of this species. The negative charge depends on the *filled* π_3, with lobes (negative charge) on carbons 1, 3, and 5, consistent with the resonance description.

15-38

(a)

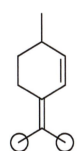

(b) Use text Section 15-13D to predict λ_{max} values. Alkyl substituents are circled.

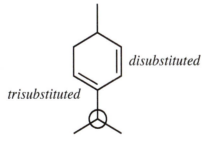

transoid cyclic diene with one exocyclic double bond = 232 nm
2 extra alkyl groups = 10 nm
TOTAL = 242 nm

desired product—**not consistent with UV data**

cisoid cyclic diene = 256 nm
1 extra alkyl group = 5 nm
TOTAL = 261 nm

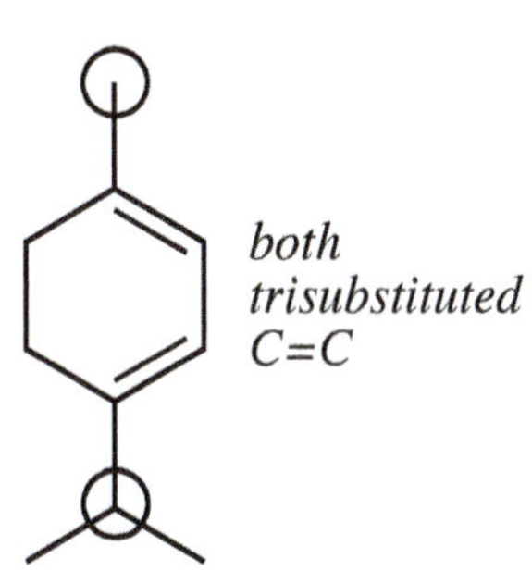

cisoid cyclic diene = 256 nm
2 extra alkyl groups = 10 nm
TOTAL = 266 nm—AHA!—closest to observed value

actual product **B**

15-38 continued

(c)

allylic!

hydride shift

$- H_2O$

H_2SO_4

3°

B is produced in preference to the other *cisoid* diene above because **B**'s diene system is more highly substituted and therefore more stable.

B

both trisubstituted C=C

15-39 It is stunningly clever reactions like this that earned E. J. Corey his Nobel Prize.

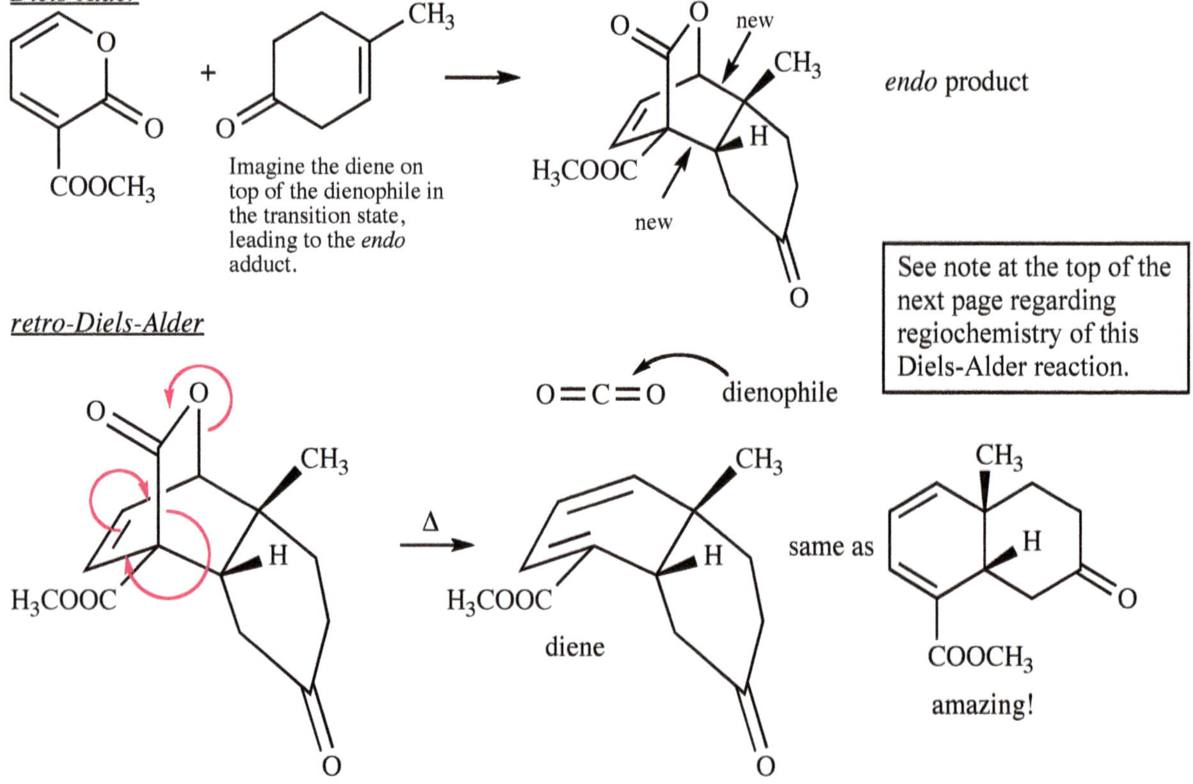

15-39 continued

The regiochemistry of this Diels-Alder reaction is not obvious, and it merited significant discussion in the original publication: E. J. Corey and David S. Watt, *Journal of the American Chemical Society*, **1973**, *95(7)*, 2303. That discussion is beyond the scope of this solution. Suffice it to say that if you chose the opposite orientation of the dienophile and concluded with the structure at the right, that is just as good as the actual one shown on the previous page.

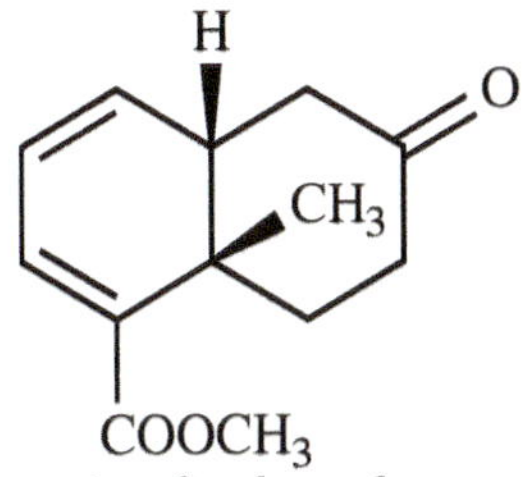

H O CH_3 $COOCH_3$

alternative final product

15-40 Look for *extended conjugation*, meaning an unbroken string of sp^2 hybridized atoms, noted here with a large dot on carbon atoms. Other sp^2 atoms, like O and N, are designated with a *.

(a) Basic Blue 6, aka Meldola Blue, a biological stain

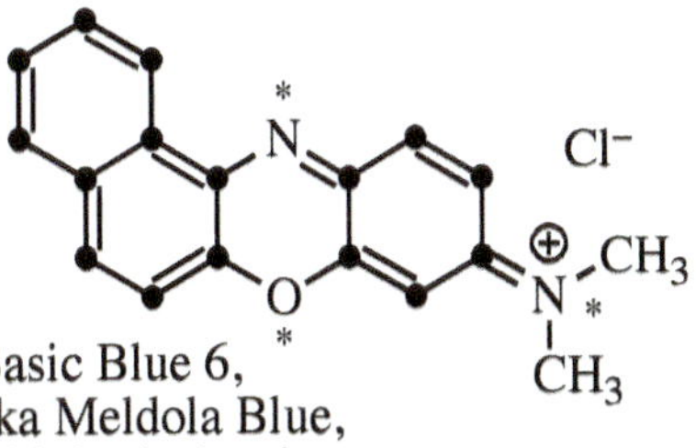

(b) *NOT COLORED— no extended conjugation; all pi bonds are isolated*

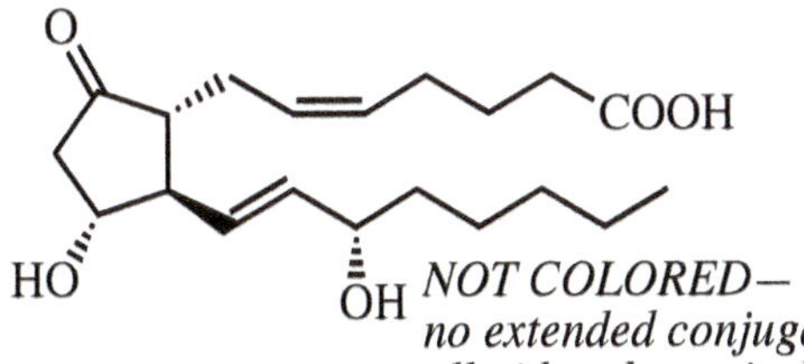

(c) kermesic acid, a red dye from insects

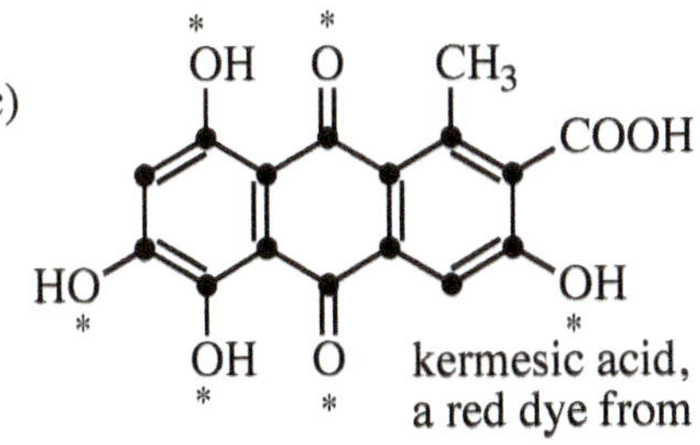

(d) Acid Alizarin Violet N, a violet dye

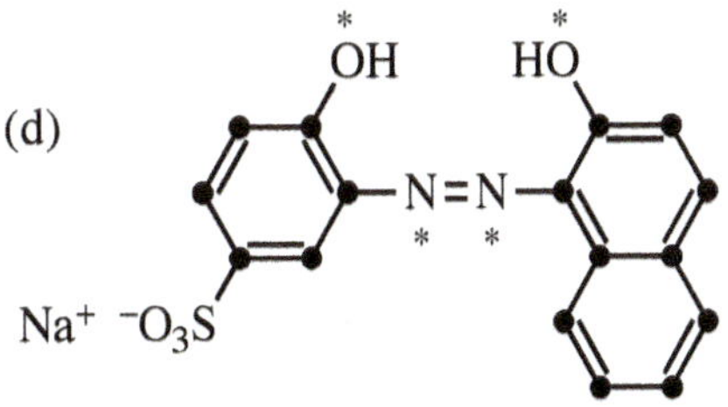

(e) *NOT COLORED— no extended conjugation; only two pi bonds are conjugated, not enough to be colored*

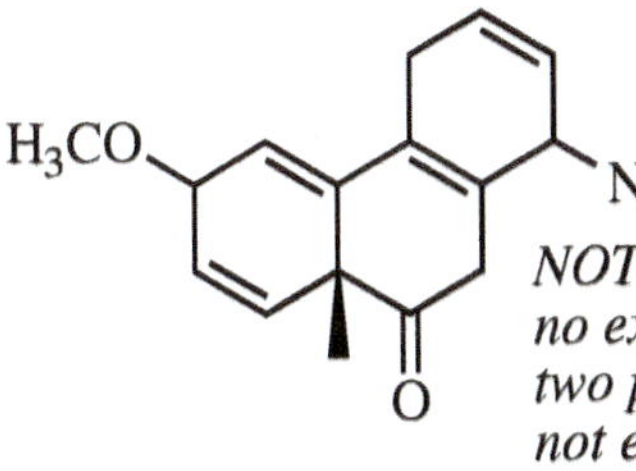

(f) p-naphtholbenzein, a red indicator

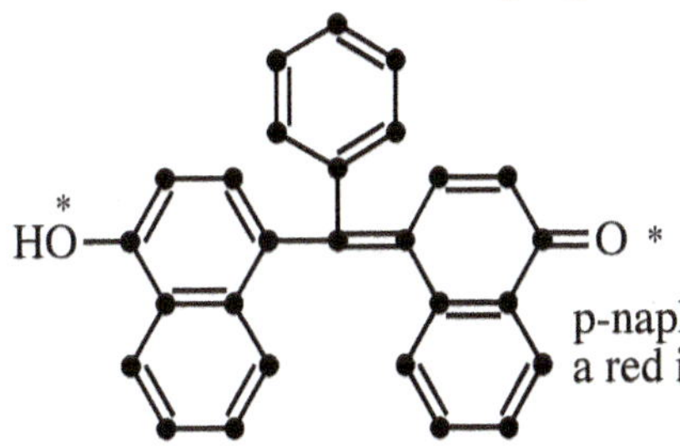

(g) food dye: FD&C Red #3

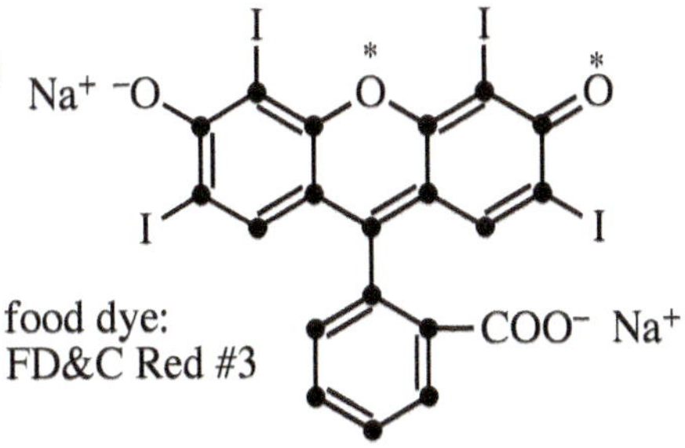

(h) *NOT COLORED— no extended conjugation; all pi bonds are isolated*

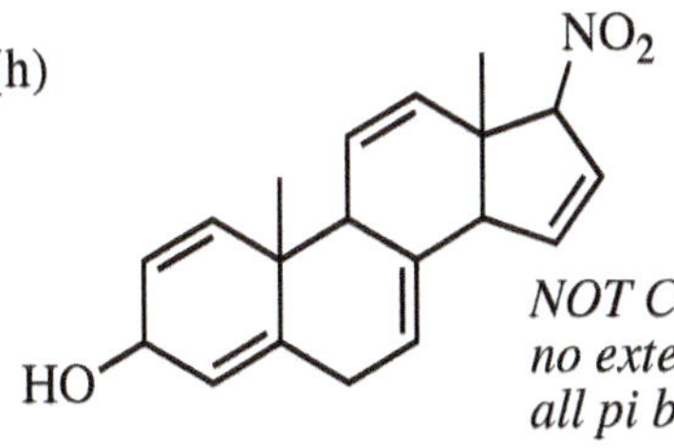

(i) rhodoxanthin, a purple dye found in plants and bird feathers

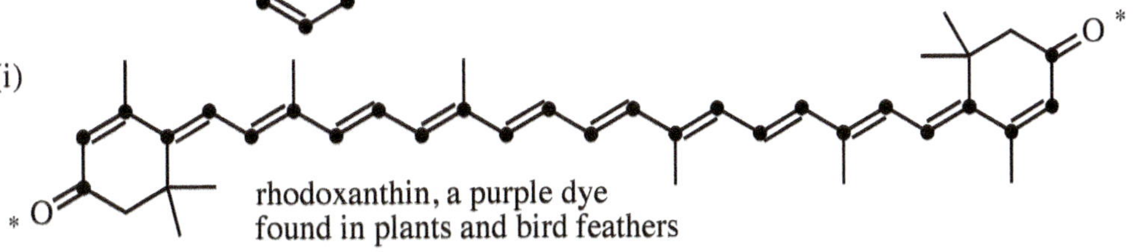

15-40 continued

(j) food dye: FD&C Yellow #5

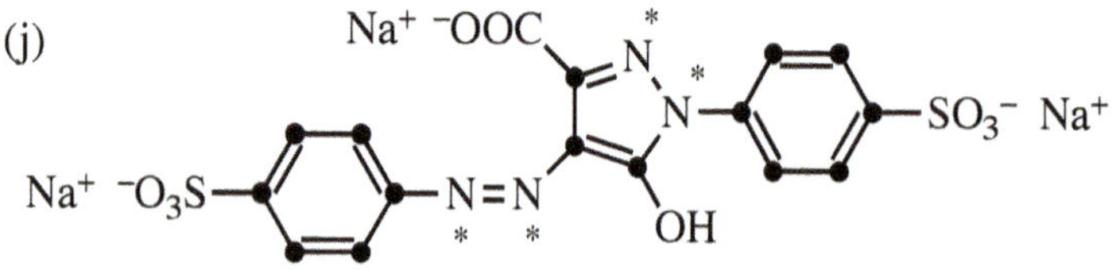

Na^+ ^-OOC, Na^+ ^-O_3S, N=N, N, OH, SO_3^- Na^+

All 4 N and COO⁻ are also sp².

(k) Cl, Cl, Cl, Cl, OH, NH_2, O

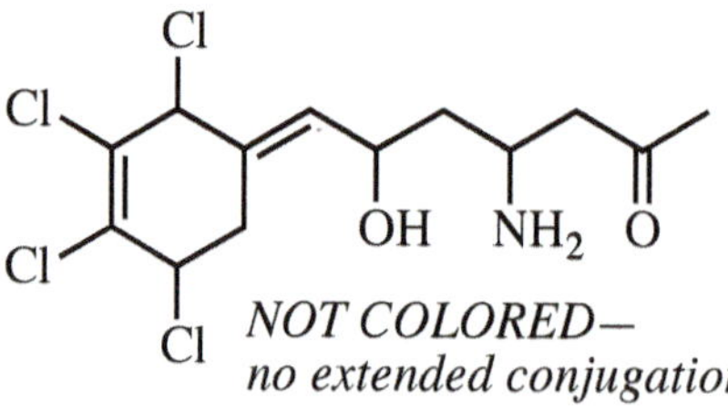

NOT COLORED—
no extended conjugation

15-41 In some cases, the stereochemistry is not obvious without making a model. New bonds are indicated with an arrow.

(a) Δ → H, H, C≡N

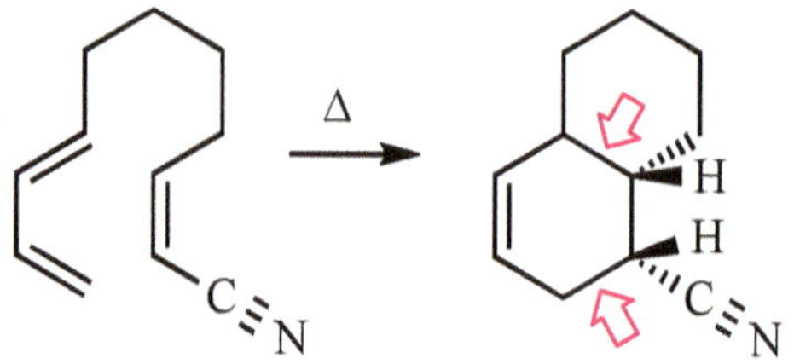

(b) O, $OSiR_3$, OR^2 Δ → O, $OSiR_3$, OR^2

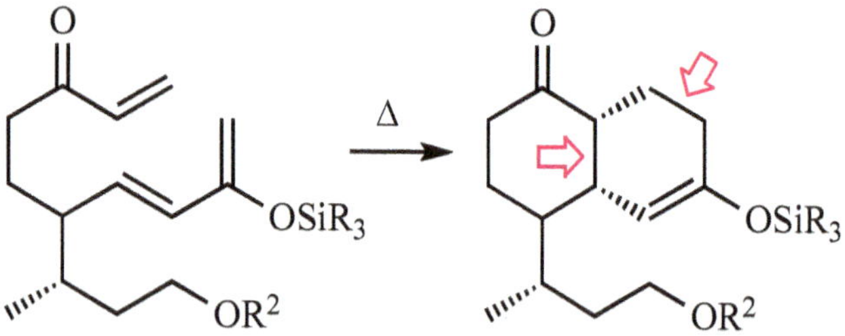

(c) $SiMe_3$, O, O, O, O Δ → $SiMe_3$, O, O, O, O

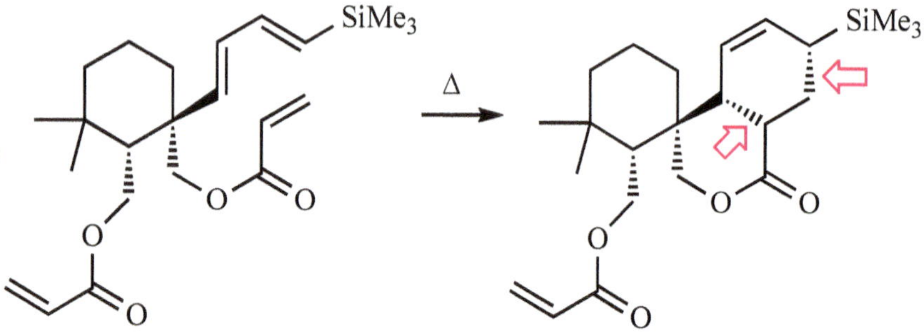

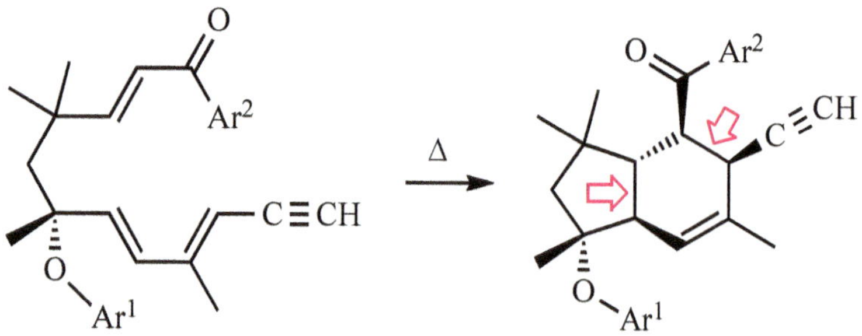

(d) O, Ar^2, C≡CH, O, Ar^1 Δ → O, Ar^2, C≡CH, O, Ar^1

Note: The representation of benzene with a circle to represent the π system is fine for questions of nomenclature, properties, isomers, and reactions. For questions of mechanism or reactivity, however, the representation with three alternating double bonds (the Kekulé picture) is more informative. For clarity and consistency, this Solutions Manual will use the Kekulé form exclusively.

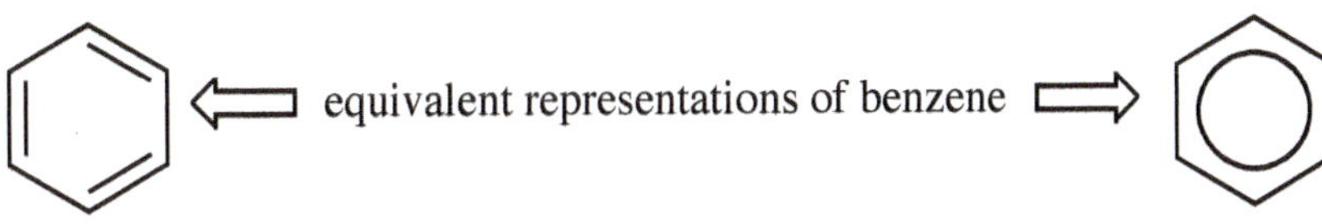

Kekulé form used in the Solutions Manual

16-1

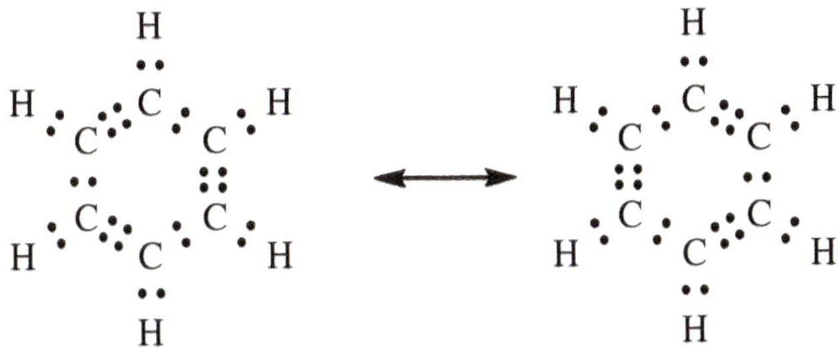

16-2 All values are per mole.

(a)	benzene	−208 kJ	− 49.8 kcal
	− cyclohexa-1,4-diene	−240 kJ	− 57.4 kcal
	ΔH =	+ 32 kJ	+ 7.6 kcal
(b)	benzene	−208 kJ	− 49.8 kcal
	− cyclohexene	−120 kJ	− 28.6 kcal
	ΔH =	− 88 kJ	− 21.2 kcal
(c)	cyclohexa-1,3-diene	−232 kJ	− 55.4 kcal
	− cyclohexene	−120 kJ	− 28.6 kcal
	ΔH =	− 112 kJ	−26.8 kcal

16-3

(a)

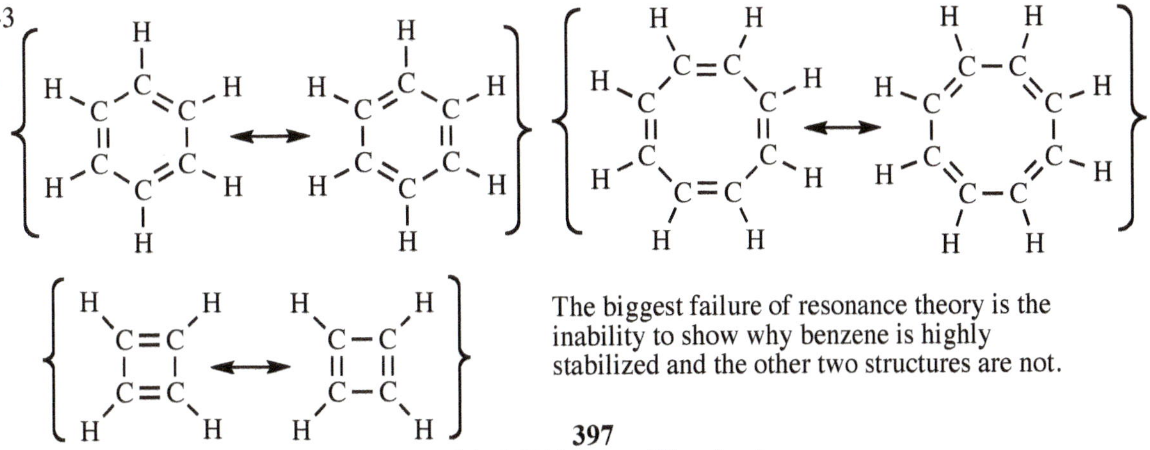

The biggest failure of resonance theory is the inability to show why benzene is highly stabilized and the other two structures are not.

16-3 continued

(b)

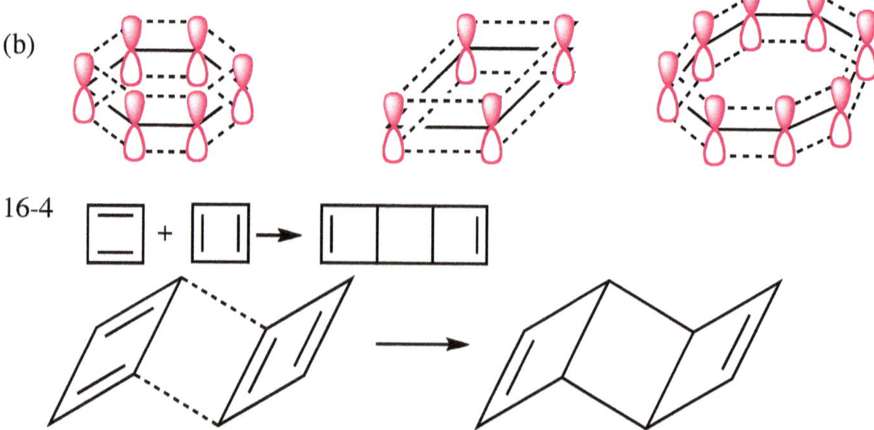

16-4

16-5 Figure 16-8 shows that the first three pairs of electrons are in three bonding molecular orbitals of cyclooctatetraene. Electrons 7 and 8, however, are located in two different nonbonding orbitals. As in cyclobutadiene, a planar cyclooctatetraene is predicted to be a diradical, a particularly unstable electron configuration.

16-6

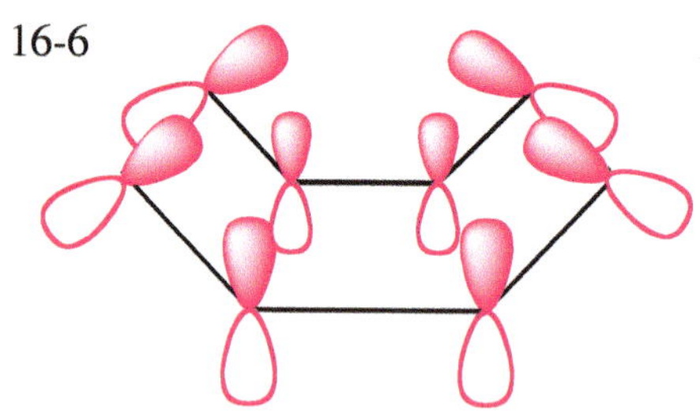

Models show that the angles between p orbitals on adjacent π bonds approach 90°.

16-7 Planarity of a real structure is difficult to predict just by looking at a planar drawing.
(a) nonaromatic: internal hydrogens prevent planarity
(b) nonaromatic: not all atoms in the ring have a p orbital, as one carbon is sp^3 hybridized
(c) aromatic if planar: [14]annulene
(d) aromatic: also a [14]annulene in the outer ring: the internal alkene is not part of the aromatic system

16-8 Azulene satisfies all the criteria for aromaticity, and it has a Huckel number of π electrons: 10. Both heptalene (12 π electrons) and pentalene (8 π electrons) are antiaromatic.

16-9

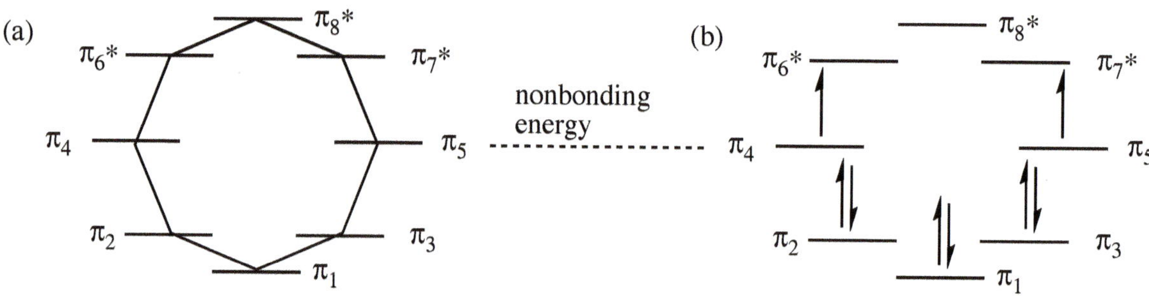

This electronic configuration is antiaromatic. It would be aromatic if it lost 2 electrons to make the double-positive ion.

16-9 continued

(c)

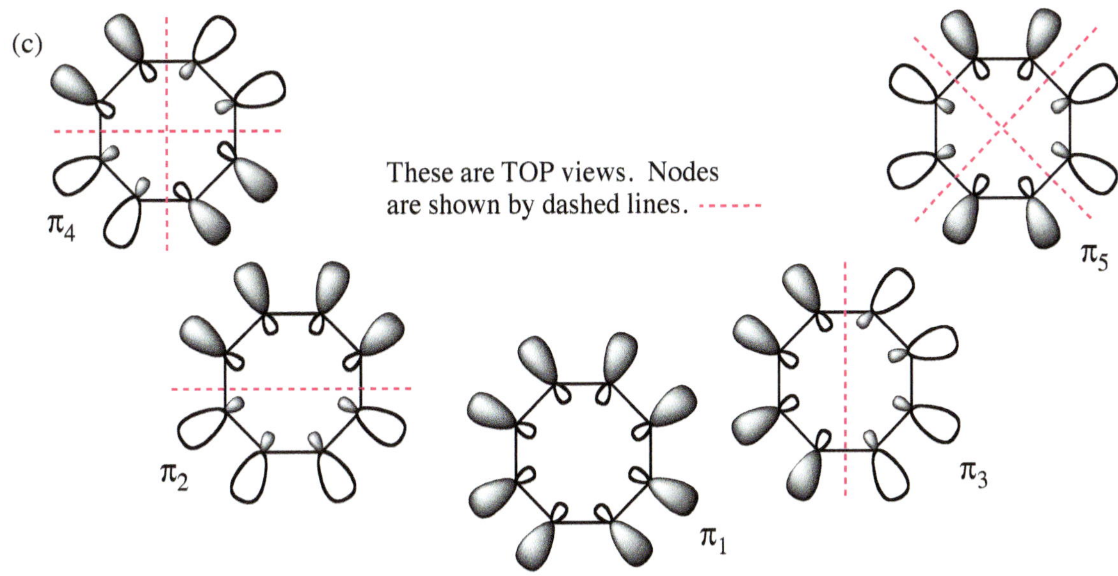

16-10 Nodes in these orbital pictures are indicated by dashed lines.

(a)

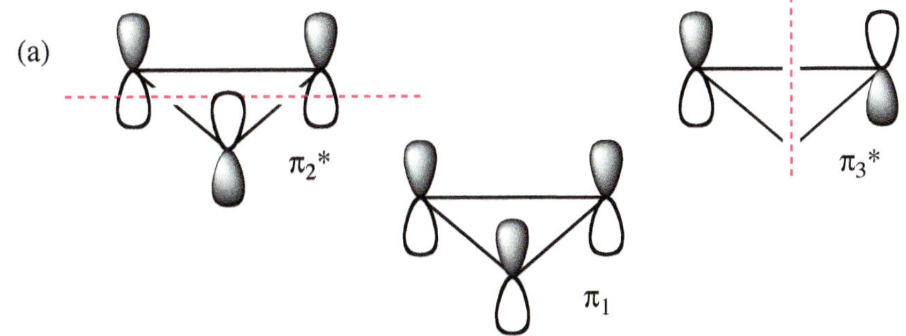

(b)

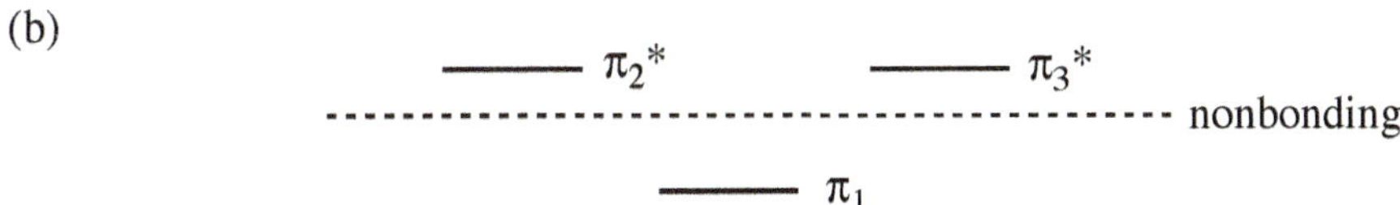

π_1 is bonding; π_2* and π_3* are antibonding.

(c)

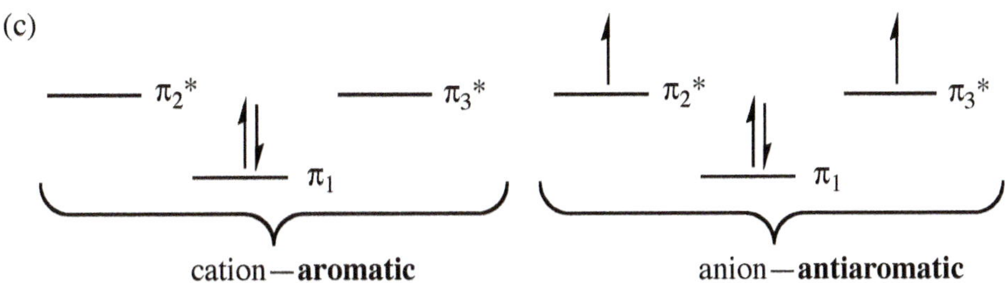

16-11 Nodes in these orbital pictures are indicated by dashed lines.

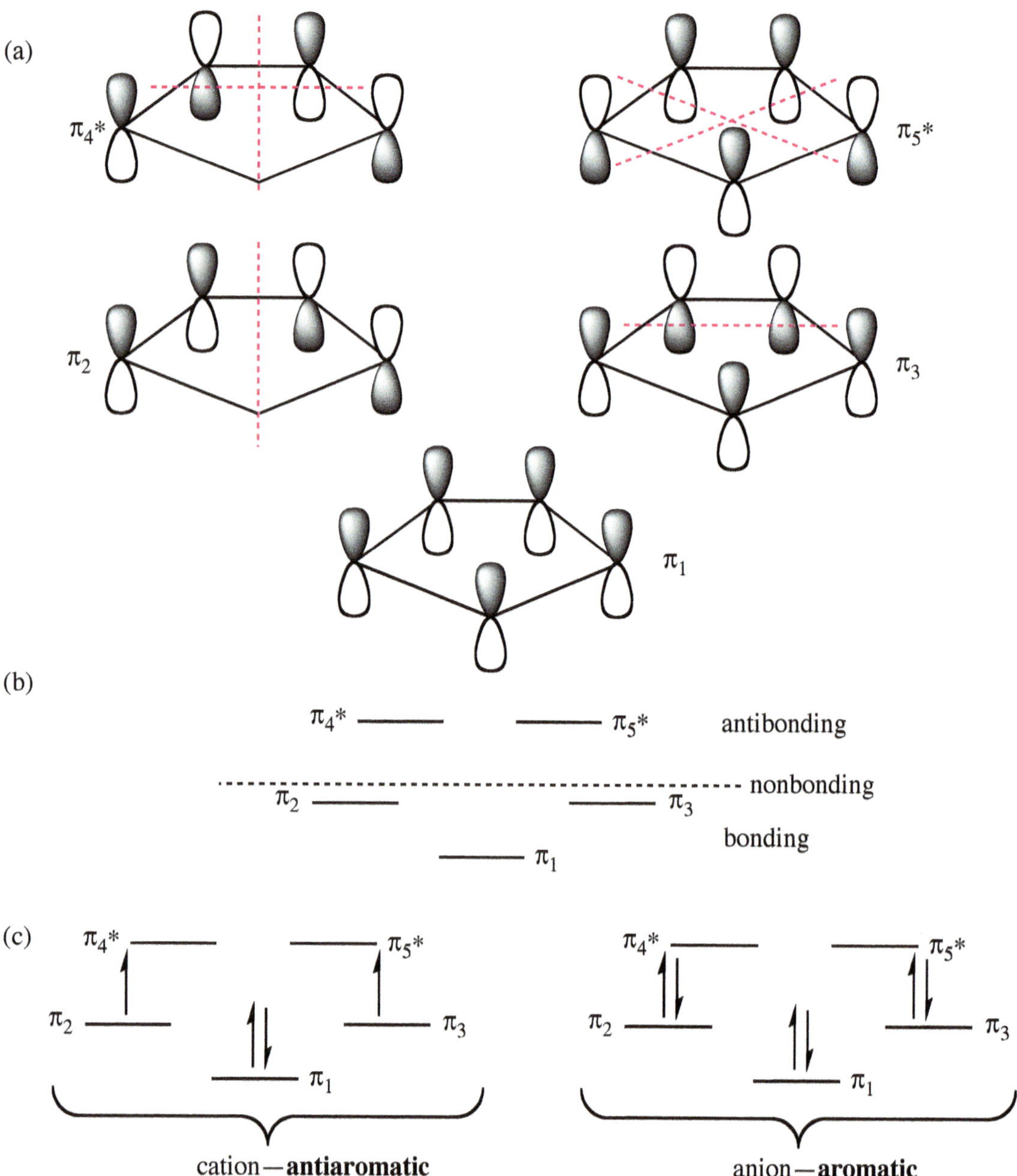

16-12 Assume planarity in these ring systems for the purpose of determining aromaticity.
(a) antiaromatic: 8 π electrons (4n), not a Huckel number
(b) aromatic: 10 π electrons (4n + 2), a Huckel number
(c) aromatic if planar: 18 π electrons (4n + 2), a Huckel number
(d) antiaromatic: 20 π electrons (4n), not a Huckel number
(e) nonaromatic: no cyclic π system
(f) aromatic if planar: 20 – 2 = 18 π electrons (4n + 2), a Huckel number

16-13 The reason for the dipole can be seen in a resonance form distributing the electrons to give each ring 6 π electrons. This resonance picture gives one ring a negative charge and the other ring a positive charge.

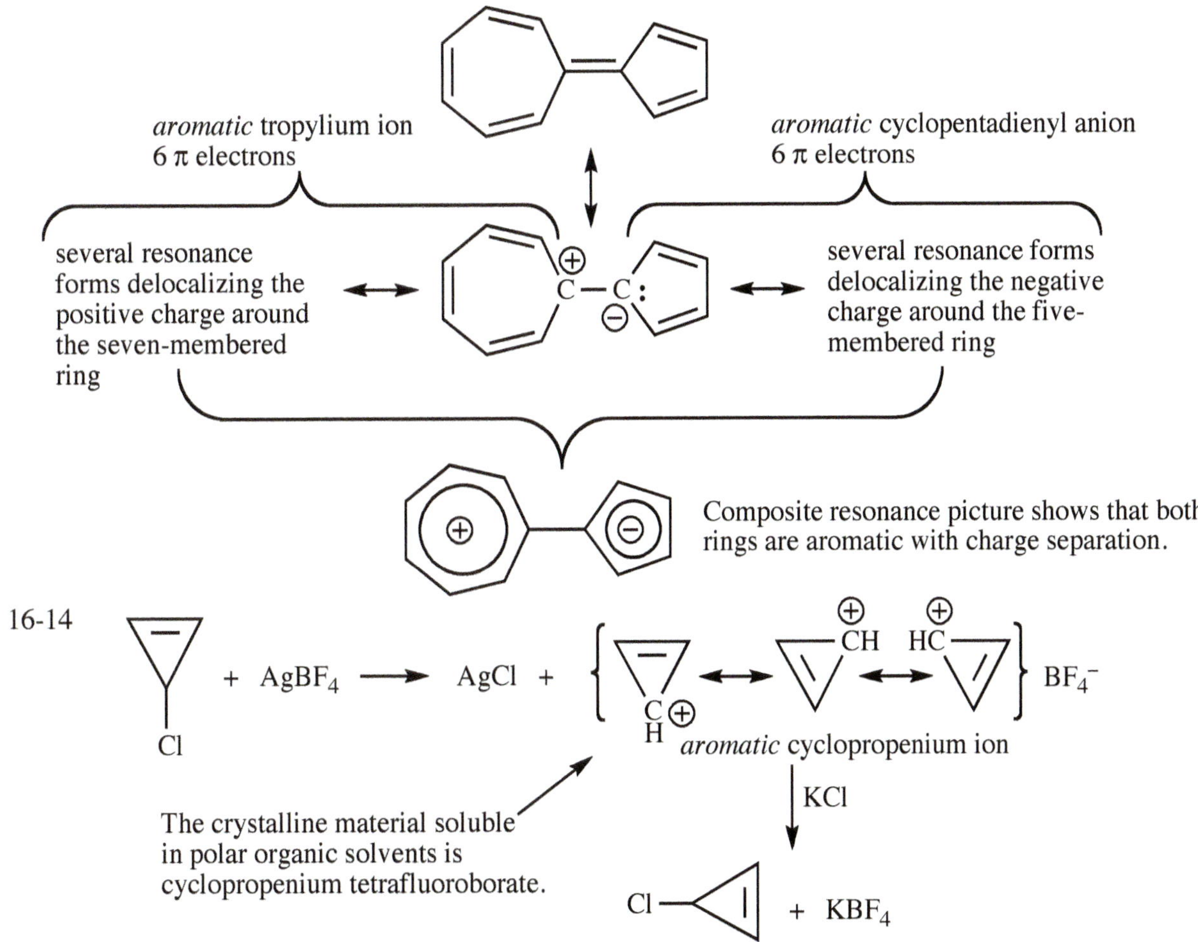

aromatic tropylium ion
6 π electrons
aromatic cyclopentadienyl anion
6 π electrons
several resonance forms delocalizing the positive charge around the seven-membered ring
several resonance forms delocalizing the negative charge around the five-membered ring
Composite resonance picture shows that both rings are aromatic with charge separation.
16-14
Cl
+ $AgBF_4$ ⟶ AgCl +
CH HC
BF_4^-
aromatic cyclopropenium ion
KCl
The crystalline material soluble in polar organic solvents is cyclopropenium tetrafluoroborate.
Cl
+ KBF_4

16-15 Draw resonance forms showing the carbonyl polarization, leaving a positive charge on the carbonyl carbon.

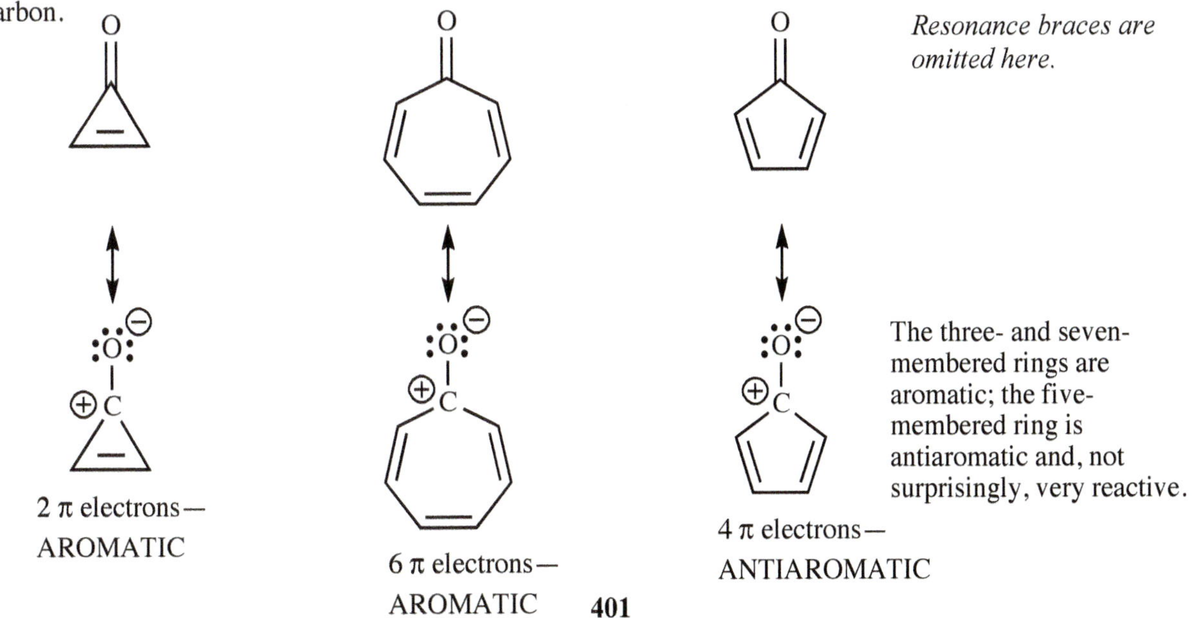

O
Resonance braces are omitted here.
2 π electrons—
AROMATIC
6 π electrons—
AROMATIC
4 π electrons—
ANTIAROMATIC
The three- and seven-membered rings are aromatic; the five-membered ring is antiaromatic and, not surprisingly, very reactive.

16-16

(a)

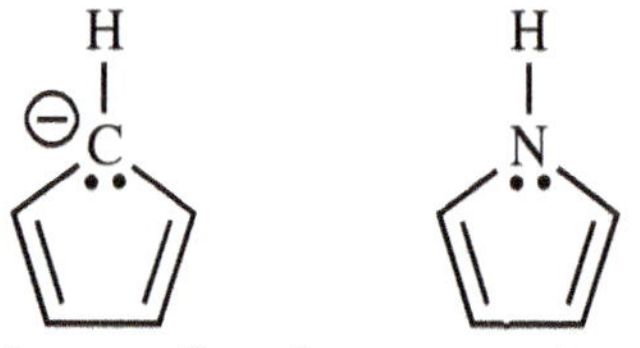

cyclopentadienyl anion   pyrrole

Both pyrrole and cyclopentadienyl anion are aromatic systems with 6 pi electrons, two electrons coming from either the carbanion carbon of cyclopentadienyl anion or the nitrogen atom in pyrrole.

(b) Since the two structures are isoelectronic (same number of electrons), the only difference is the identity of the "fifth" atom of the ring, C or N, and the difference between those is one proton: N has one more proton than C, which is why pyrrole does not have a negative charge. It is likely that pyrrole has two extra neutrons as ^{14}N is the most abundant isotope of nitrogen and ^{12}C is the most abundant isotope of C, but this could vary depending on the particular isotopes present. The extra proton in pyrrole makes the biggest difference.

(c)

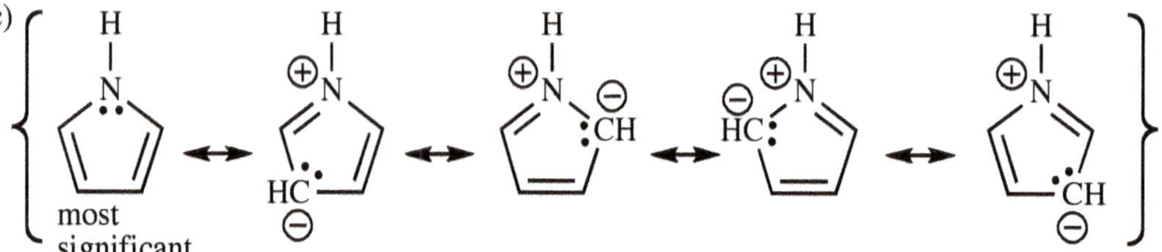

most significant contributor

Note that all of the minor resonance contributors with charge separation have (+) charge on the nitrogen atom.

16-17

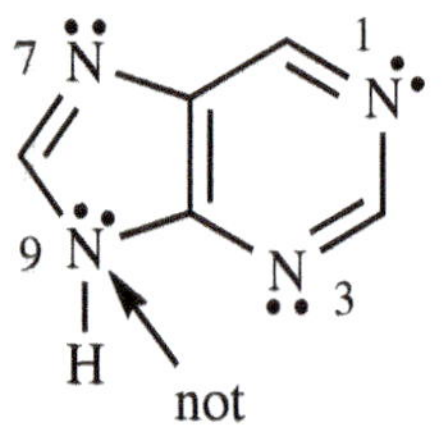

not basic

The structure of purine shows two types of nitrogens. One type (N-1, N-3, and N-7) has an electronic structure like the nitrogen in pyridine; the pair of electrons is in an sp^2 orbital planar with the ring. These electrons are available for bonding, and these three nitrogens are basic. The other type of nitrogen at N-9 has an electronic structure like the nitrogen of pyrrole; its electron pair is in a p orbital, perpendicular to the ring system, and more importantly, an essential part of the aromatic pi system. With this pair of electrons, the pi system is aromatic and has 10 electrons, a Huckel number, so the electron pair is not available for bonding and N-9 is not a basic nitrogen.

16-18

(a) The proton NMR of benzene shows a single peak at δ 7.2; alkene hydrogens absorb at δ 4.5–6. The chemical shifts of 2-pyridone are more similar to benzene's absorptions than they are to alkenes. It would be correct to infer that 2-pyridone is aromatic.

(b)

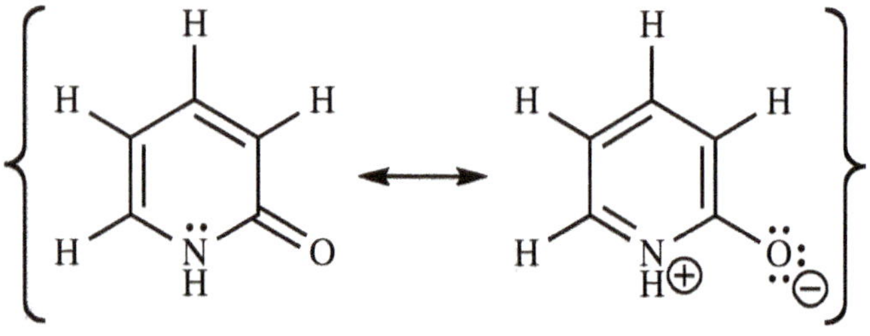

The lone pair of electrons on the nitrogen in the first resonance form is part of the cyclic pi system. The second resonance form shows three alternating double bonds with 6 electrons in the cyclic pi system, consistent with an aromatic electronic system.

continued on next page

16-18(b) continued

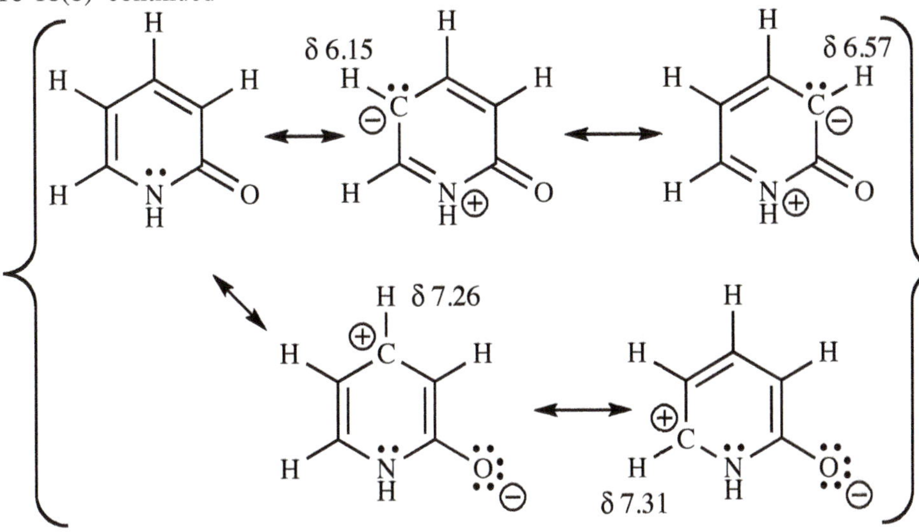

δ 6.15   δ 6.57

These resonance forms show that the hydrogens at positions with greater electron density are shielded, decreasing chemical shift.

δ 7.26   δ 7.31

These resonance forms show that the hydrogens at positions with greater positive charge are deshielded, increasing chemical shift.

(c)

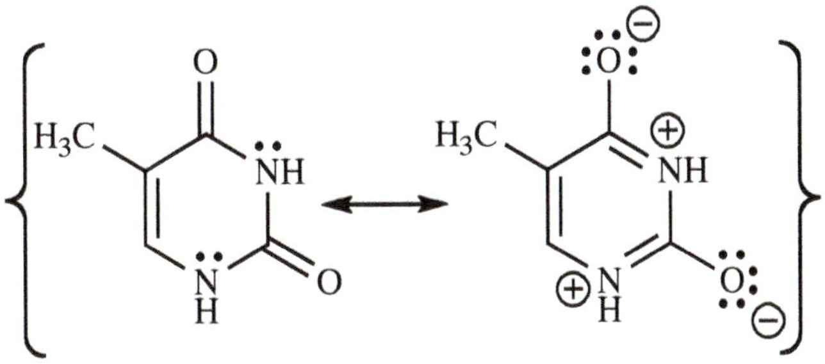

This resonance form of thymine shows a cyclic pi system with 6 pi electrons, consistent with an aromatic system. Four of these electrons came from the two lone pairs of electrons on nitrogens in the first resonance form.

(d)

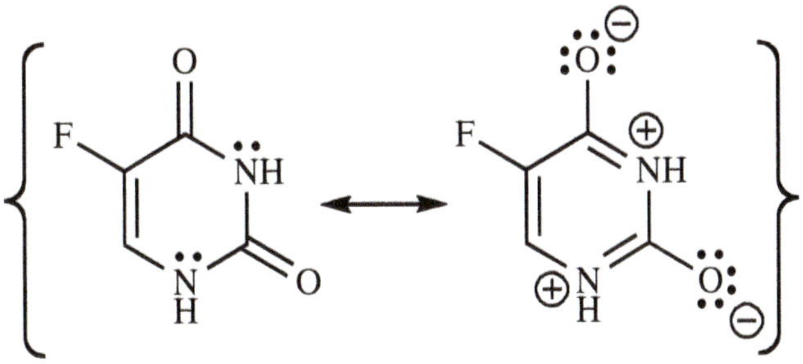

The ring portion of 5-fluorouracil is identical to the ring structure of thymine in part (c). Both structures have the aromatic ring in common.

16-19

(a) aromatic: 4 π electrons in double bonds plus one pair from oxygen

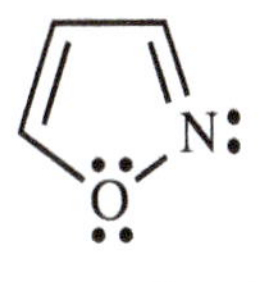

(b) aromatic: 4 π electrons in double bonds plus one pair from sulfur

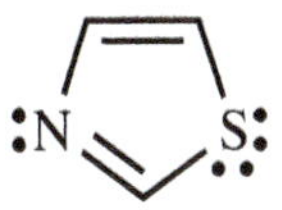

(c) sp^3 not aromatic: no cyclic π system because of sp^3 carbon

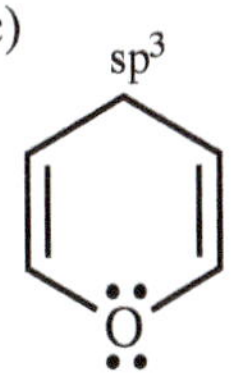

(d) aromatic: Cation on carbon-4 indicates an empty p orbital; two π bonds plus a pair of electrons from oxygen makes a 6 π electron system.

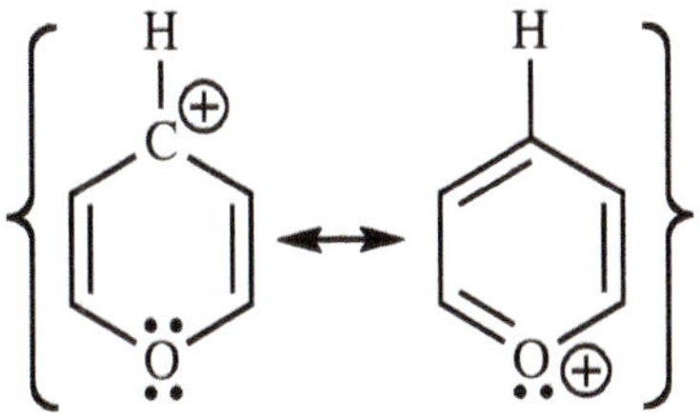

16-19 continued

(e)

aromatic: Resonance form shows "push-pull" of electrons from one O to the other, making a cyclic π system with 6 electrons.

(f)

not aromatic: no cyclic π system because of sp^3 carbon

(g)

aromatic: Resonance form shows electron pair from N making a cyclic π system with 6 electrons.

(h)

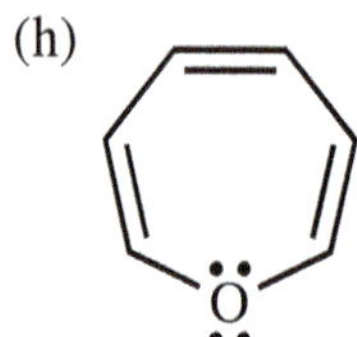

not aromatic: If 2 π electrons from oxygen are included, then 8 π electrons; if oxygen does not contribute electrons, then oxygen is sp^3 and there is no cyclic π system.

16-20

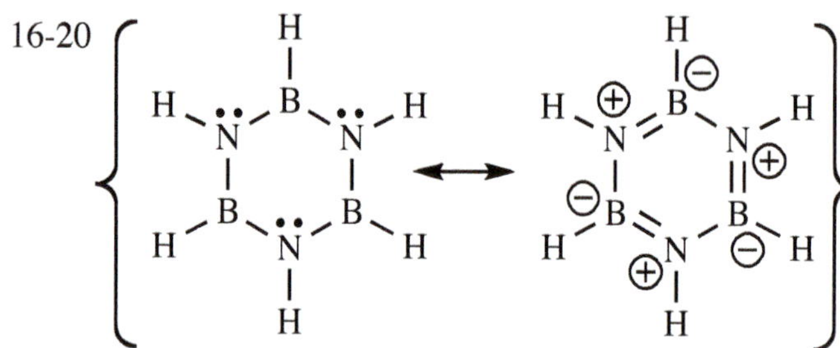

Borazole is a non-carbon equivalent of benzene. Each boron is hybridized in its normal sp^2. Each nitrogen is also sp^2 with its pair of electrons in its p orbital. The system has 6 π electrons in 6 p orbitals—aromatic! Although borazole is isoelectronic with benzene, benzene boils at 80 °C whereas borazole boils at 161 °C because of its polar nature.

16-21

(a) resonance braces omitted

anthracene

phenanthrene

(b) anthracene

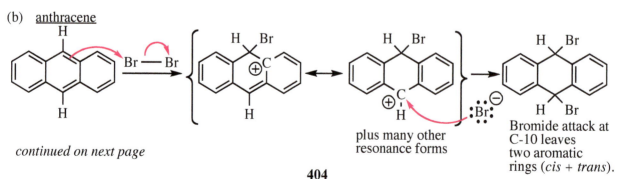

plus many other resonance forms

Bromide attack at C-10 leaves two aromatic rings (*cis* + *trans*).

continued on next page

16-21(b) continued

phenanthrene

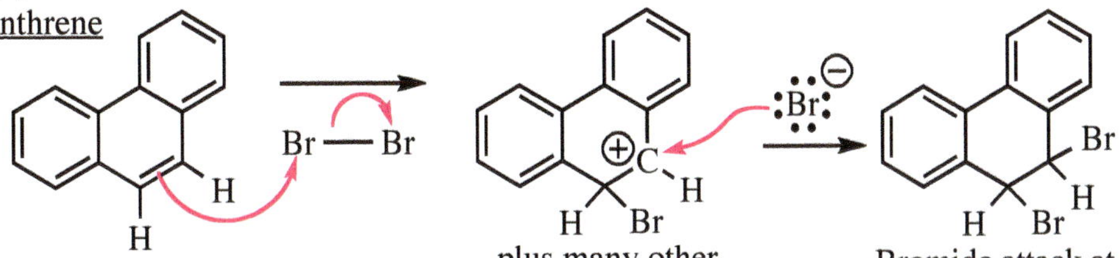

H, H, Br — Br

H, Br, H, :Br:⊖

plus many other resonance forms

Br, H, H, Br *cis* and *trans*

Bromide attack at C-9 leaves two aromatic rings.

(c) A typical addition of bromine occurs with a bromonium ion intermediate, which can give only anti addition. Addition of bromine to phenanthrene, however, generates a free carbocation because the carbocation is benzylic, stabilized by resonance over two rings. In the second step of the mechanism, bromide nucleophile can attack either side of the carbocation, giving a mixture of *cis* and *trans* products.

(d)

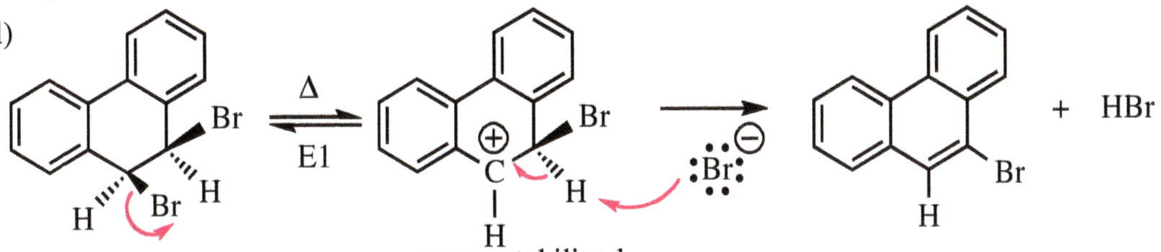

Br, H, Br, H ⇌ (Δ, E1) Br, H, H

resonance-stabilized

:Br:⊖ → Br, H + HBr

16-22

(a)

F, O, COOH, HN, N, N ⟷ F, :O:⊖, COOH, HN, N, ⊕N

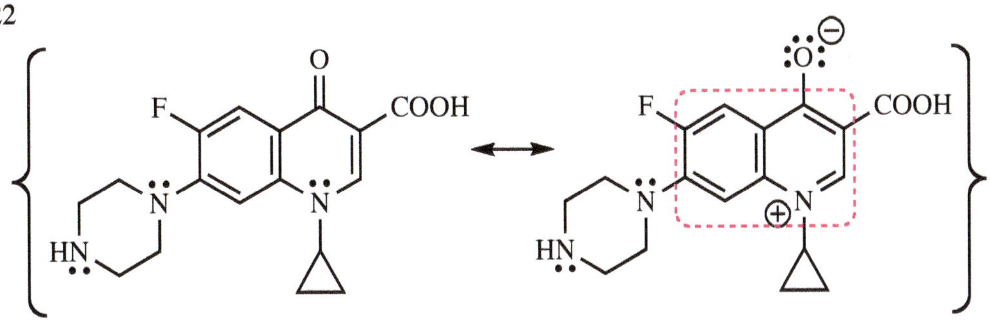

The second resonance form shows that the two rings in the dashed box form a 10 electron pi system, isoelectronic with naphthalene. Cipro is aromatic.

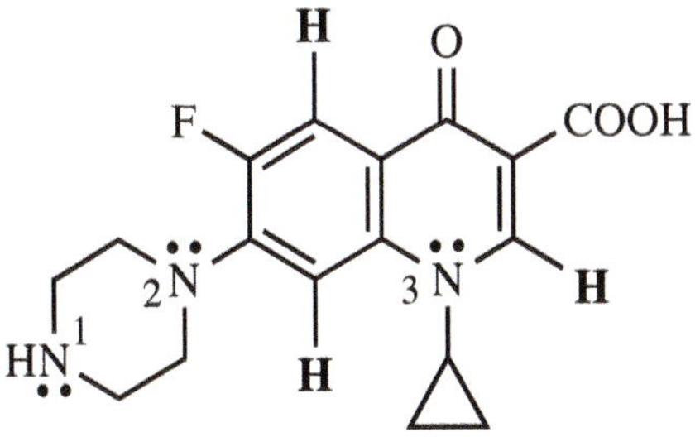

(b) For a nitrogen (or any atom) to be basic, it must have an electron pair to share by forming a sigma bond with a proton. The nitrogen atoms labeled 1 and 2 are basic because they have an unshared electron pair, but nitrogen 3 does not have an unshared pair; its electron pair is delocalized in the aromatic pi system.

(c) The three H atoms in bold are on the aromatic ring and will appear in the HNMR between δ 6–8.

H, O, F, COOH, 2N, 3N, **H**, HN 1, **H**

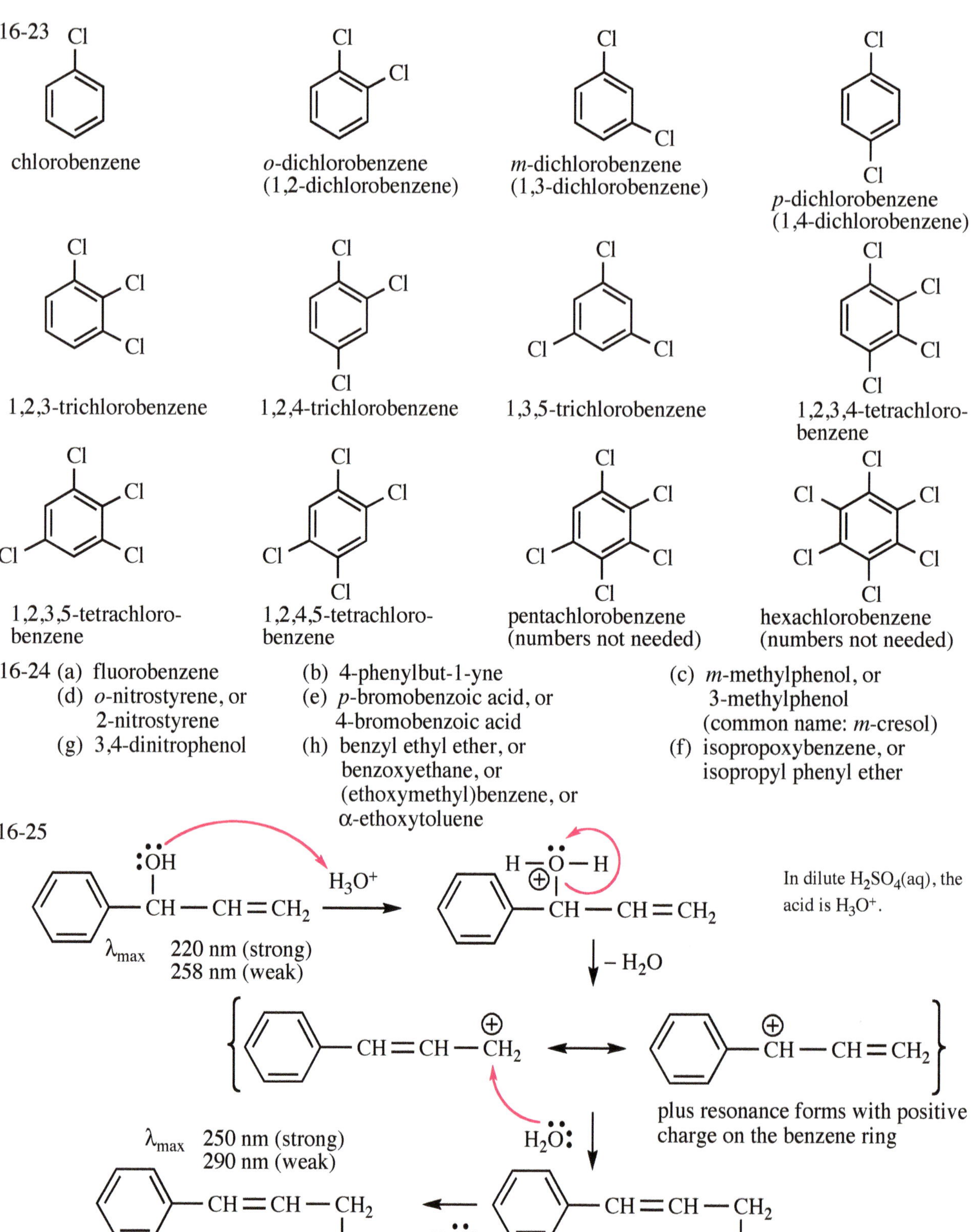

16-24 (a) fluorobenzene
(b) 4-phenylbut-1-yne
(c) *m*-methylphenol, or 3-methylphenol (common name: *m*-cresol)
(d) *o*-nitrostyrene, or 2-nitrostyrene
(e) *p*-bromobenzoic acid, or 4-bromobenzoic acid
(f) isopropoxybenzene, or isopropyl phenyl ether
(g) 3,4-dinitrophenol
(h) benzyl ethyl ether, or benzoxyethane, or (ethoxymethyl)benzene, or α-ethoxytoluene

16-26

(a) OCH_3 NO_2

(b) OH OCH_3 OCH_3

(c) COOH NH_2

(d) NH_2 NO_2

(e) CH_3 Cl

(f) $HC=CH_2$ $HC=CH_2$

(g) $HC=CH_2$ Br

(h) CHO CH_3O OCH_3

(i) H $\oplus$C Cl^-

(j) H $\ominus$C Na^+

(k) HO

(l) CH_2OCH_3

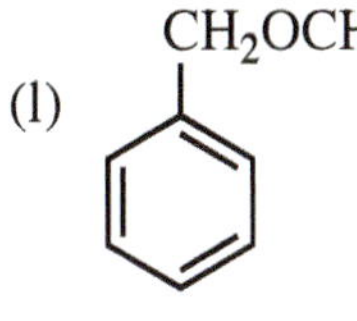

(m) SO_3H CH_3

(n) CH_3 CH_3

(o) N

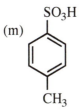

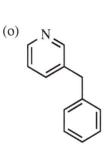

16-27 The IUPAC system recommends using position numbers for substituted benzenes. The terms *ortho*, *meta*, and *para* for disubstituted benzenes are commonly used; they are presented in parentheses.

(a) 1,2-dichlorobenzene (*ortho*)
(b) 4-nitroanisole (*para*)
(c) 2,3-dibromobenzoic acid
(d) 2,7-dimethoxynaphthalene
(e) 3-chlorobenzoic acid (*meta*)
(f) 2,4,6-trichlorophenol
(g) 2-*sec*-butylbenzaldehyde (*ortho*)
(h) cyclopropenium tetrafluoroborate

16-28 CH_3 toluene

CH_3 CH_3 *o*-xylene

CH_3 CH_3 *m*-xylene

CH_3 CH_3 *p*-xylene

CH_3 CH_3 CH_3 1,2,3-trimethylbenzene

CH_3 CH_3 CH_3 1,2,4-trimethylbenzene

CH_3 CH_3 CH_3 1,3,5-trimethylbenzene (common name: mesitylene)

16-29 The key concept in parts (a)–(c) is that an aromatic product is created.

(a)

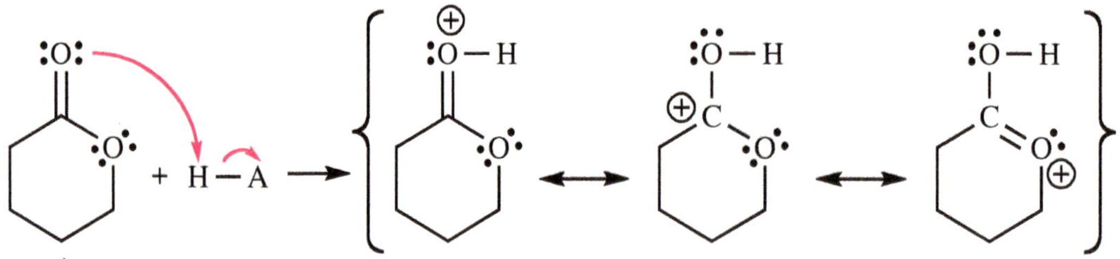

A

The protonated carbonyl gives a resonance-stabilized cation. Protonation of the singly bonded oxygen does not generate a resonance-stabilized product.

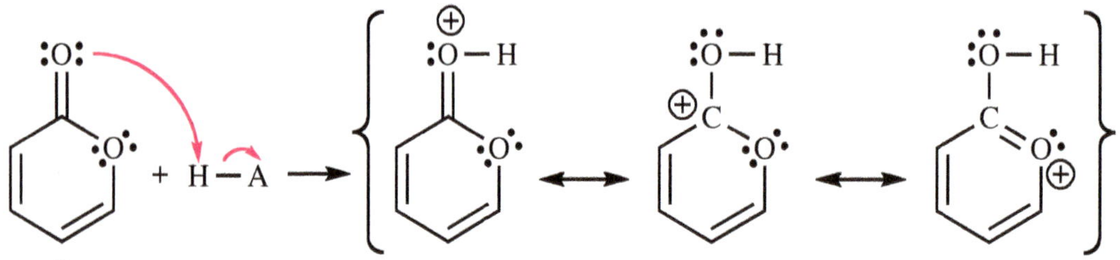

B

AROMATIC

The last resonance form shows that the cation produced is aromatic and therefore more stable than the corresponding nonaromatic ion. In this second reaction, the products are more favored than in the first reaction, which is interpreted as the reactant **B** being more basic than **A**.

(b)

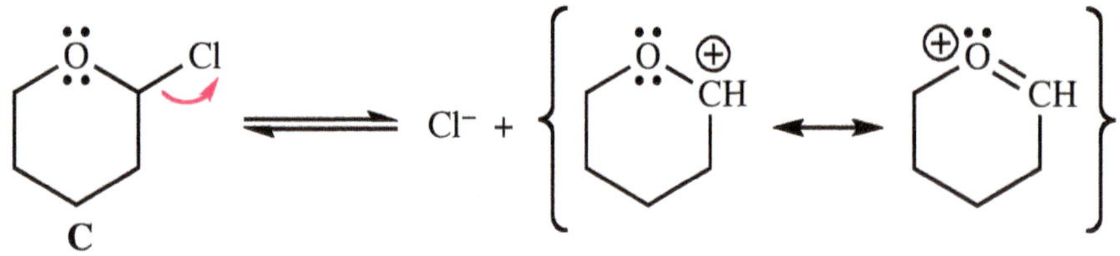

C

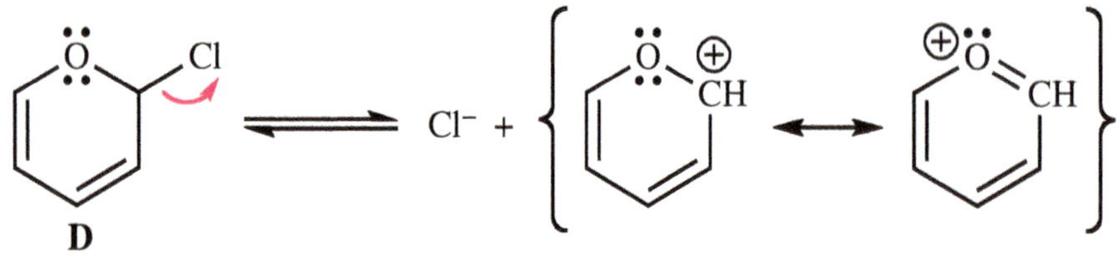

D

AROMATIC

The product from ionization of **C** is stabilized by resonance. The ionization product of **D** is not only resonance-stabilized but is also aromatic and therefore more stable. A reaction that produces a more stable product will usually happen faster under milder conditions because the transition state leading to that product will be stabilized, leading to a lower activation energy.

16-29 continued

(c)

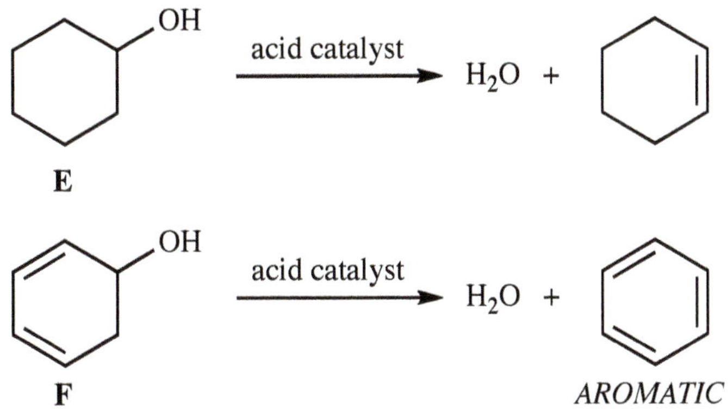

E

F

acid catalyst → H_2O +

acid catalyst → H_2O +

AROMATIC

Dehydration of **F** produces an aromatic product that is more stable than the product from **E**. A reaction that produces a more stable product will usually happen faster under milder conditions because the transition state leading to that product will be stabilized, leading to a lower activation energy.

(d)

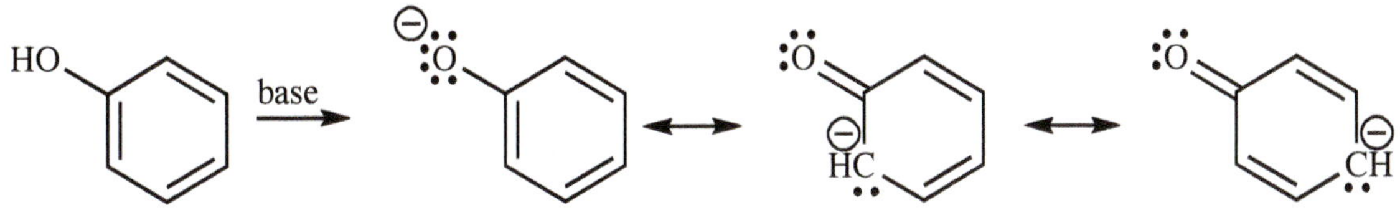

phenol

base →

Resonance stabilization of the phenoxide anion shows the negative charge distributed over the one para and two ortho carbons.

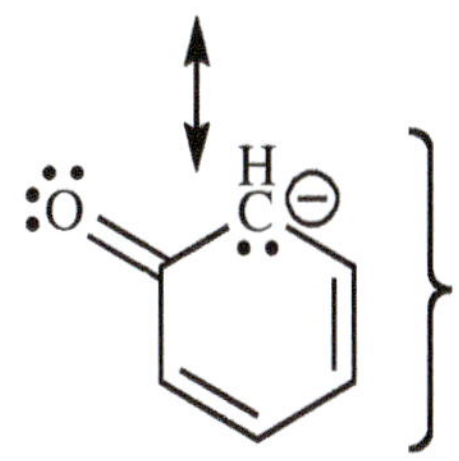

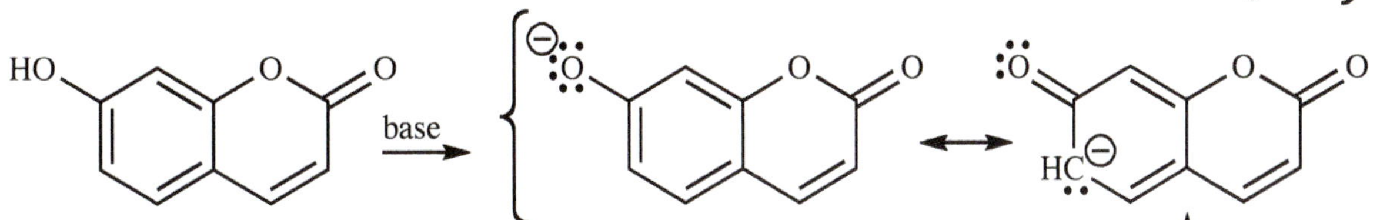

umbelliferone

base →

(Umbelliferone is one of about 1000 coumarins isolated from plants, primarily the families of Angiosperms: Fabaceae, Asteraceae, Apiaceae, and Rutaceae.)

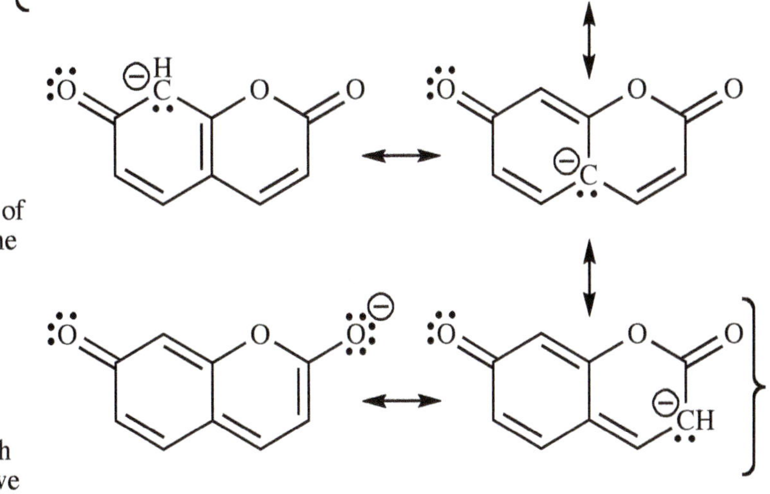

Resonance stabilization of the anion of umbelliferone gives not only the same three forms as the phenoxide anion, but in addition, gives an extra resonance form with (–) charge on a carbon, and the most significant resonance contributor, another form with the (–) charge on the other carbonyl oxygen. This anion is much more stable than phenoxide, which we interpret as enhanced acidity of the starting material, umbelliferone. In fact, the pK_a of umbelliferone is 7.7 whereas phenol is about 10.

16-30 Aromaticity is one of the strongest stabilizing forces in organic molecules. The cyclopentadienyl system is stabilized in the anion form where it has 6 π electrons, a Huckel number. The question then becomes: which of the four structures can lose a proton to become aromatic?

While the first, third, and fourth structures can lose protons from sp^3 carbons to give resonance-stabilized anions, only the second structure can make a cyclopentadienide anion. It will lose a proton most easily of these four structures which, by definition, means it is the strongest acid.

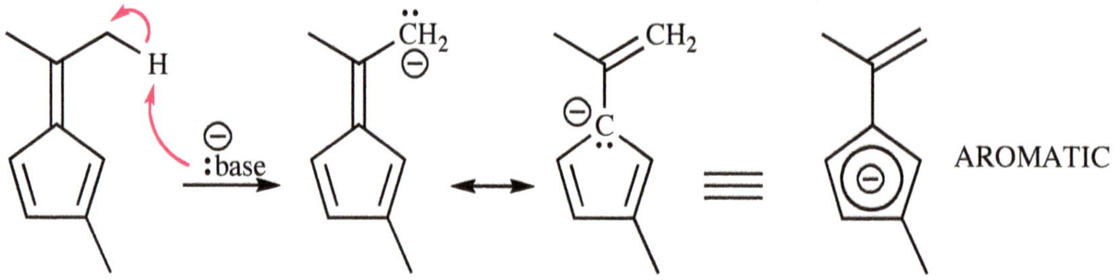

H :base CH_2 CH_2 C AROMATIC

16-31

(a) H H H H H H H H H H H H H H H H H H H H H H H H H H H H H H

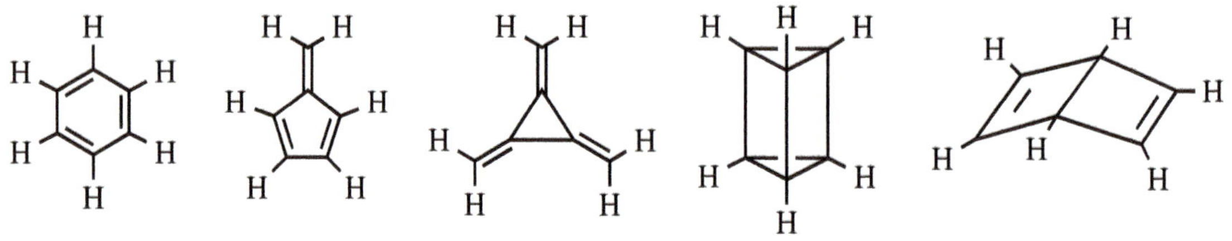

(b) Br Br + Br + Br Br Br Br + Br

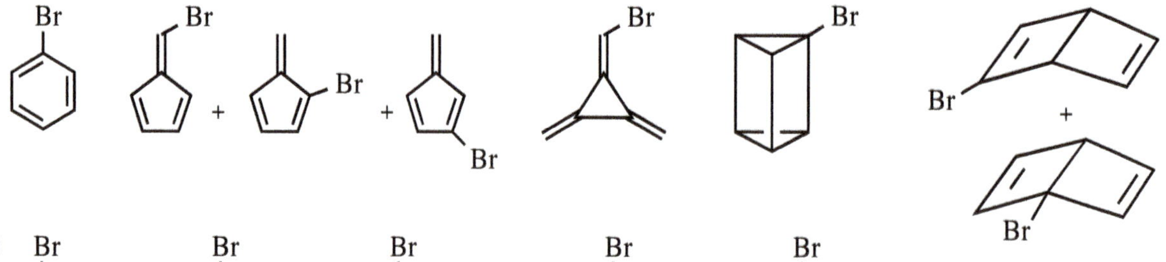

(c) Br Br + Br + Br Br Br Br + Br Br

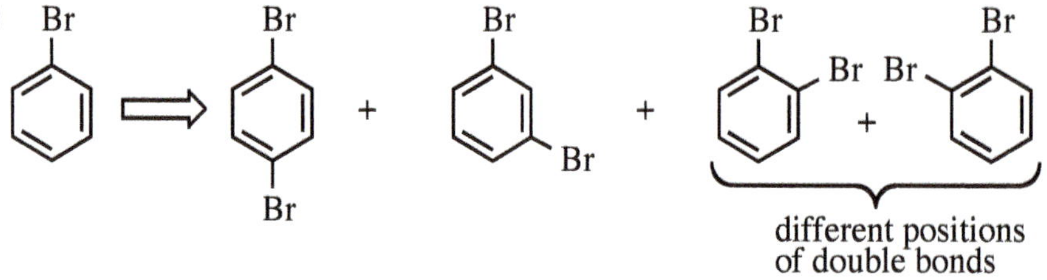

different positions of double bonds

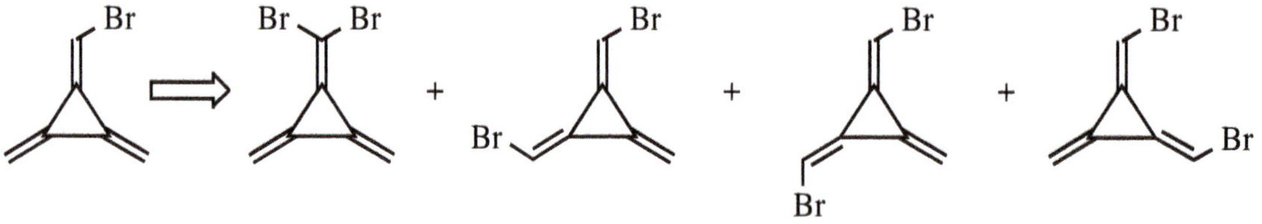

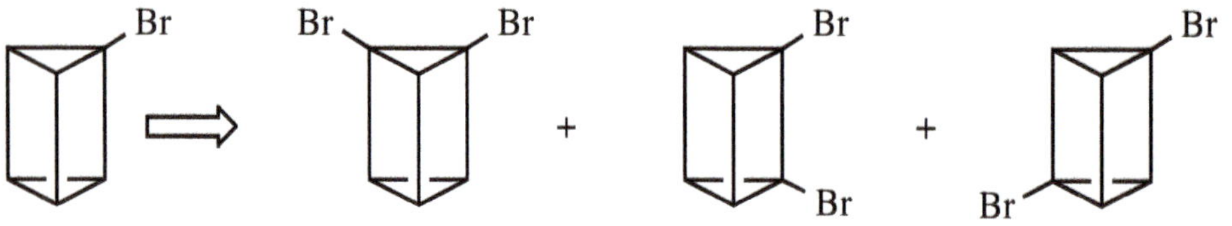

Br Br Br + Br + Br + Br Br Br Br

Br Br Br + Br + Br Br Br

(ignoring enantiomers)

16-31 continued

(d) The only structure consistent with three isomers of dibromobenzene is the prism structure, called Ladenburg benzene. It also gives a negative test for alkenes, consistent with the behavior of benzene. (Kekulé defended his structure by claiming that the "two" structures of *ortho*-dibromobenzene were rapidly interconverted, equilibrating so quickly that they could never be separated.)

(e) We now know that three- and four-membered rings are the least stable, but this fact was unknown to chemists during the mid-1800s when the benzene controversy was raging. Ladenburg benzene has two three-membered rings and three four-membered rings (of which only four of the rings are independent), which we would predict to be unstable. (In fact, the structure has been synthesized. Called *prismane*, it is NOT aromatic, but rather, is very reactive toward addition reactions.)

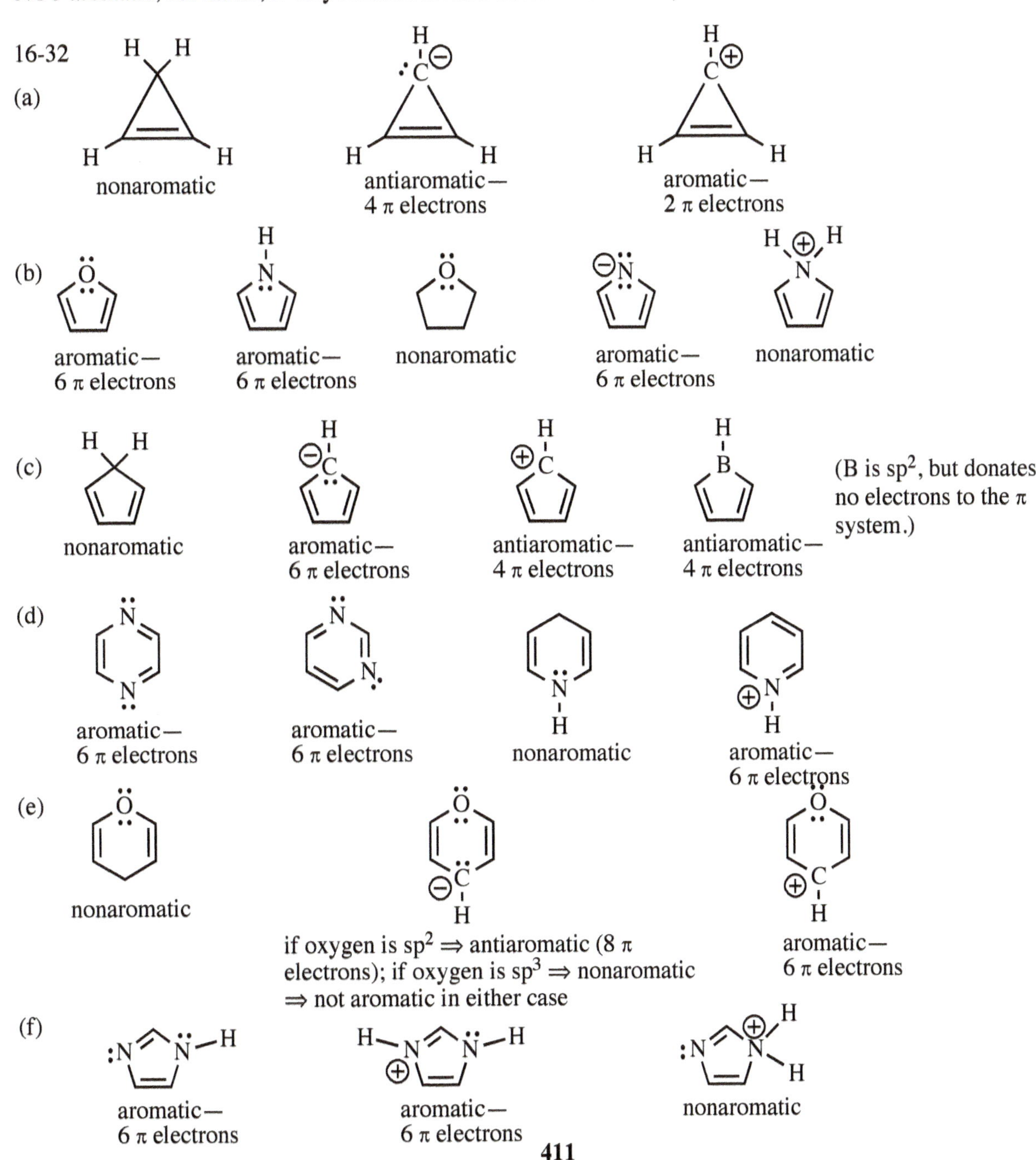

16-32 continued

(g)

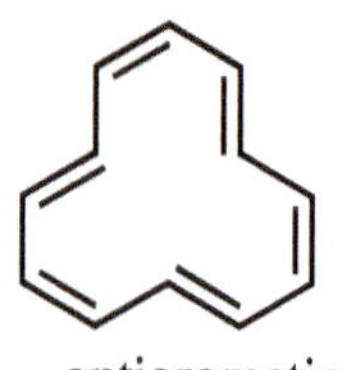

antiaromatic—
12 π electrons

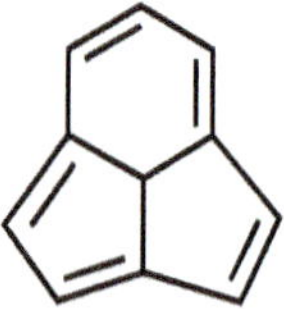

aromatic—
10 π electrons
Experiment shows the outer carbons to be close to planar but not exactly planar.

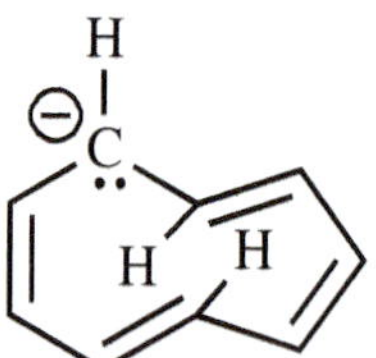

This is a tough call—it has 10 π electrons so it could be aromatic, but internal H's might force it out of planarity.

(h)

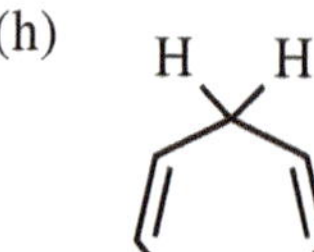

nonaromatic

aromatic—
6 π electrons

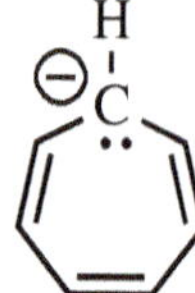

antiaromatic—
8 π electrons

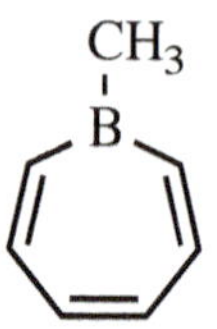

aromatic—
6 π electrons

(B is sp^2, but donates no electrons to the π system.)

16-33 The clue to azulene is recognition of the five- and seven-membered rings. To attain aromaticity, a seven-membered carbon ring must have a positive charge; a five-membered carbon ring must have a negative charge. Drawing a resonance form of azulene shows this:

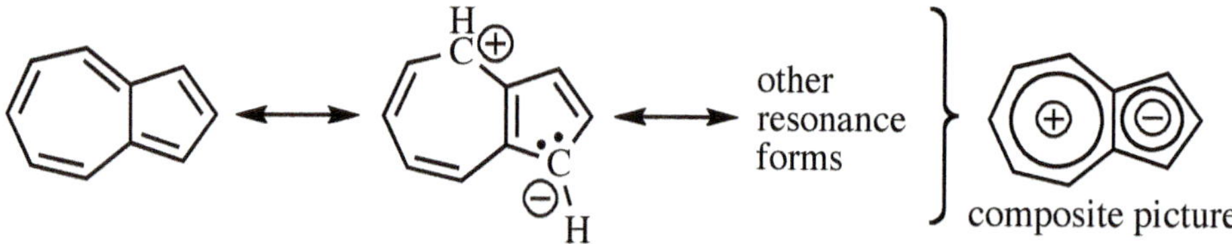

The composite picture shows that the negative charge is concentrated in the five-membered ring, giving rise to the dipole.

16-34 Whether a nitrogen is strongly basic or weakly basic depends on the location of its electron pair. If the electron pair is needed for an aromatic π system, the nitrogen will not be basic (shown here as "weak base"). If the electron pair is in either an sp^2 or sp^3 orbital, it is available for bonding, and the nitrogen is a "strong base".

(a)

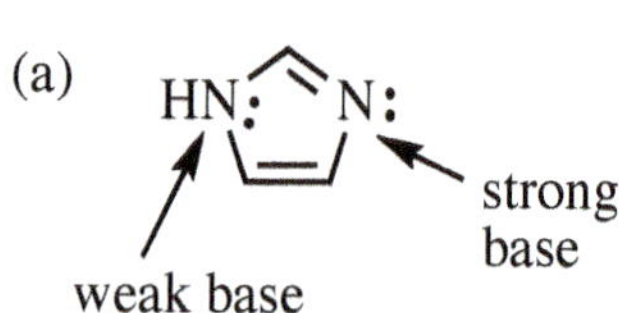

(b)

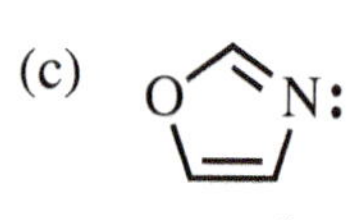

strong base

(c)

strong base

(d)

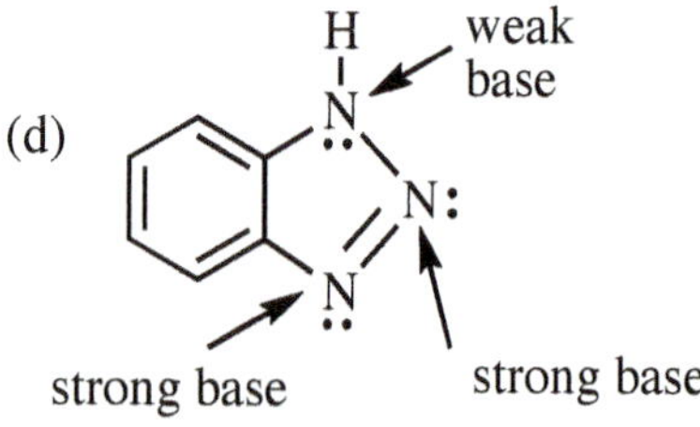

(e)

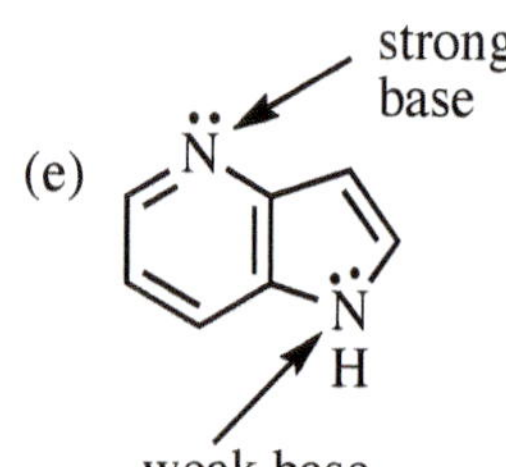

16-35 Where a resonance form demonstrates aromaticity, the resonance form is shown.

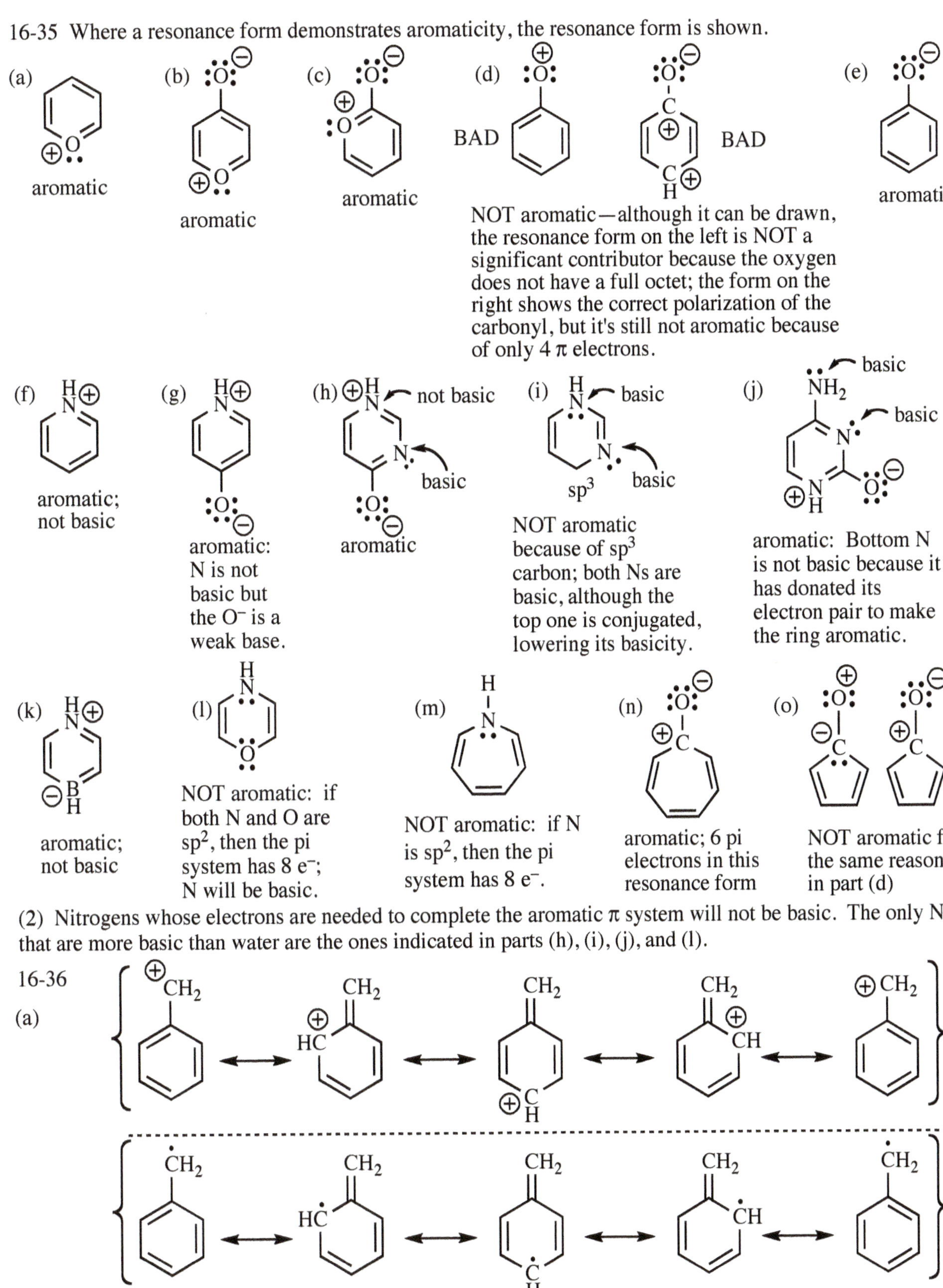

(d) NOT aromatic—although it can be drawn, the resonance form on the left is NOT a significant contributor because the oxygen does not have a full octet; the form on the right shows the correct polarization of the carbonyl, but it's still not aromatic because of only 4 π electrons.

(i) NOT aromatic because of sp^3 carbon; both Ns are basic, although the top one is conjugated, lowering its basicity.

(j) aromatic: Bottom N is not basic because it has donated its electron pair to make the ring aromatic.

(l) NOT aromatic: if both N and O are sp^2, then the pi system has 8 e^-; N will be basic.

(m) NOT aromatic: if N is sp^2, then the pi system has 8 e^-.

(o) NOT aromatic for the same reason as in part (d)

(2) Nitrogens whose electrons are needed to complete the aromatic π system will not be basic. The only N that are more basic than water are the ones indicated in parts (h), (i), (j), and (l).

16-36

(a)

continued on next page

16-36(a) continued

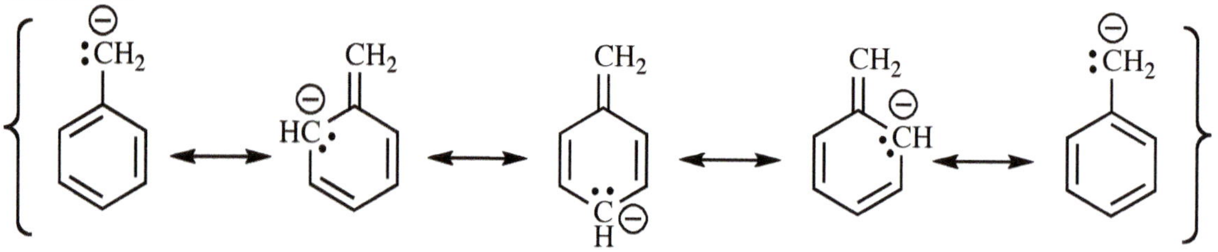

:$CH_2^{\ominus}$ ⟷ HC:$^{\ominus}$ / CH_2 ⟷ CH_2 / :C$^{\ominus}$ H ⟷ CH_2 / :CH$^{\ominus}$ ⟷ :$CH_2^{\ominus}$

(b) initiation

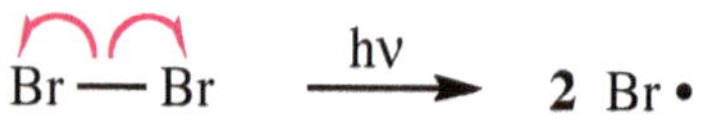

Br — Br $\xrightarrow{h\nu}$ **2** Br •

propagation Br • + H-CH_2 (benzyl) ⟶ HBr + •CH_2 (benzyl) resonance-stabilized

Br — Br + •CH_2 (benzyl) ⟶ Br • + Br − CH_2 (benzyl)

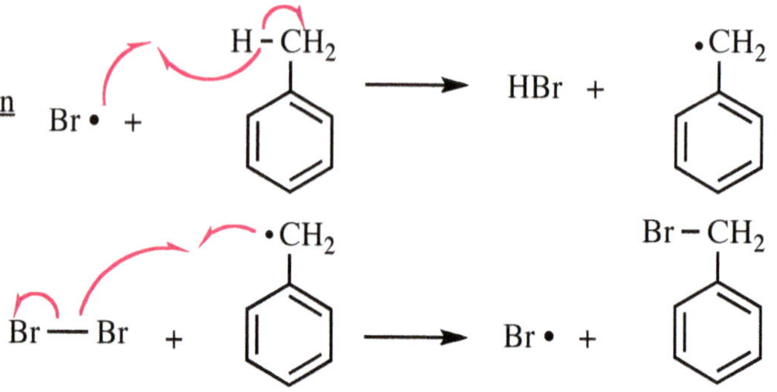

(c) Both reactions are S_N2 on primary carbons, but the one at the benzylic carbon occurs faster. In the transition state of S_N2, as the nucleophile is approaching the carbon and the leaving group is departing, the electron density resembles that of a p orbital. As such, it can be stabilized through overlap with the π system of the benzene ring.

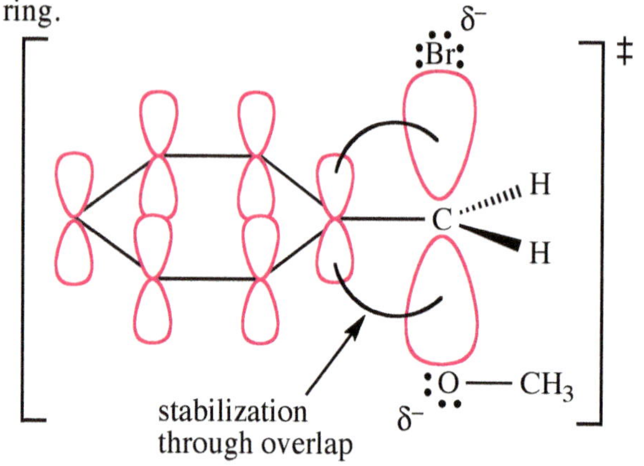

δ− :Br: ‡ H C H :O — CH_3 δ−

stabilization through overlap

16-37

(a) CH_3 / CH_3 ⟶ CH_3 / NO_2 / CH_3 + CH_3 / CH_3 / NO_2 + CH_3 / O_2N / CH_3

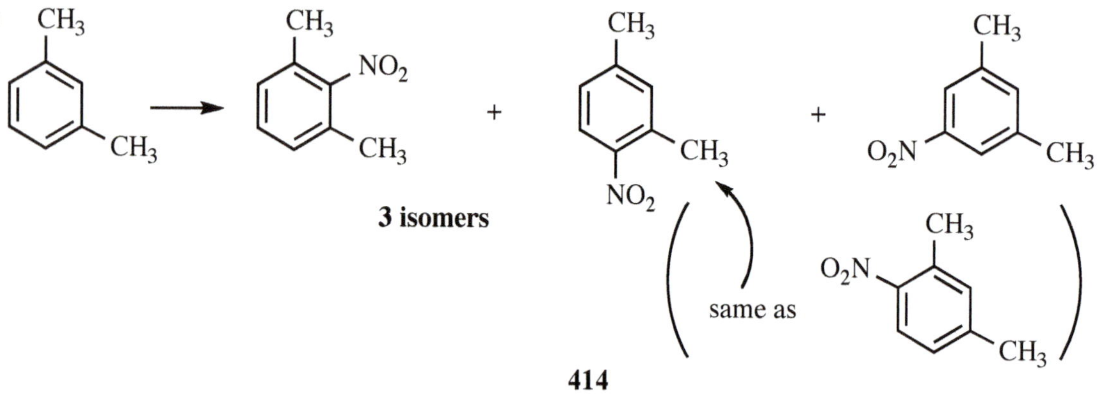

3 isomers

(same as CH_3 / O_2N / CH_3)

16-37 continued

(b) CH_3 ... CH_3 → CH_3, NO_2, CH_3 **only 1 isomer**

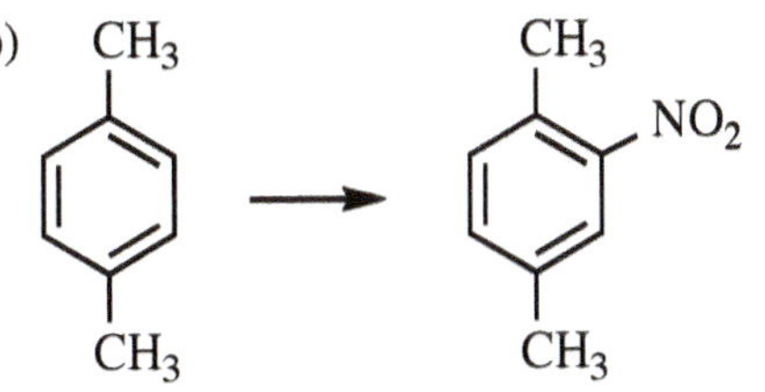

(c) The original compound had to have been *meta*-dibromobenzene as this is the only dibromo isomer that gives three mononitrated products.

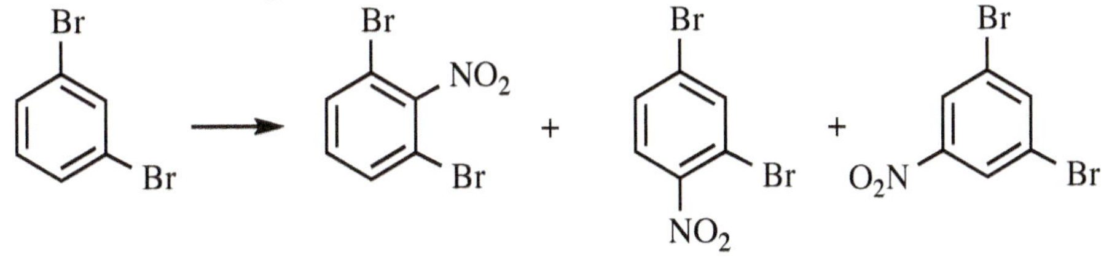

Br, Br → Br, NO_2, Br + Br, Br, NO_2 + Br, O_2N, Br

16-38

(a) The formula C_8H_7OCl has five elements of unsaturation, probably a benzene ring (4) plus either a double bond or a ring. The IR suggests a conjugated carbonyl at 1690 cm^{-1} and an aromatic ring at 1602 cm^{-1}. The NMR shows a total of five aromatic protons, indicating a monosubstituted benzene. A 2H singlet at δ 4.7 is a deshielded methylene.

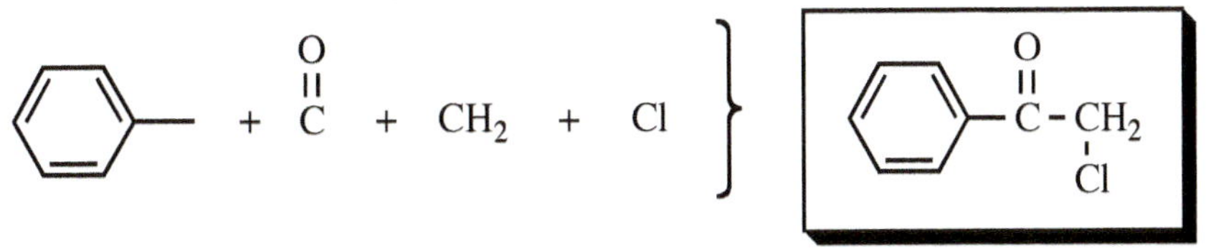

+ C=O + CH_2 + Cl → Ph–C(=O)–CH_2–Cl

(b) The mass spectral evidence of molecular ion peaks of 1 : 1 intensity at 184 and 186 shows the presence of a bromine atom. The m/z 184 minus 79 for bromine gives a mass of 105 for the rest of the molecule, which is about a benzene ring plus two carbons and a few hydrogens. The NMR shows four aromatic hydrogens in a typical *para* pattern (two doublets), indicating a *para*-disubstituted benzene. The 2H quartet and 3H triplet are characteristic of an ethyl group.

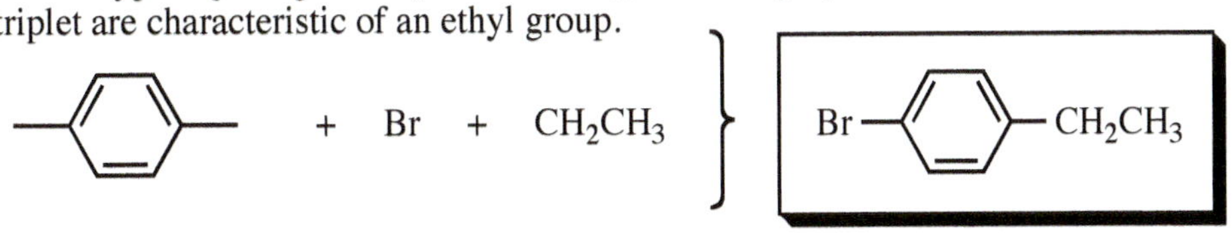

+ Br + CH_2CH_3 → Br–C_6H_4–CH_2CH_3

16-39

(a)

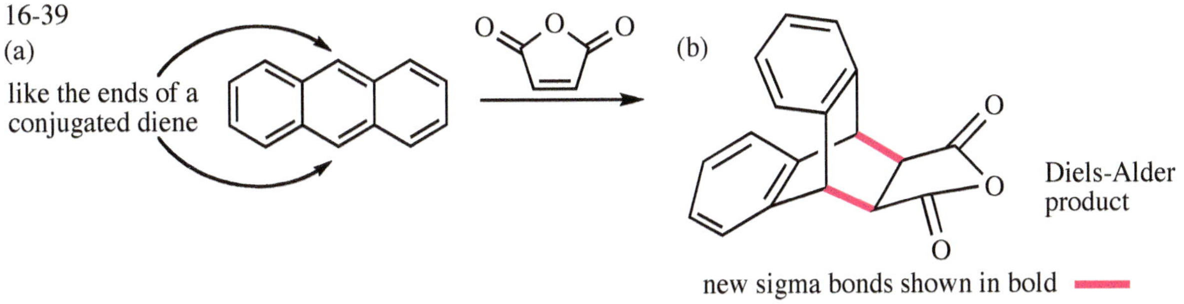

like the ends of a conjugated diene

(b)

Diels-Alder product

new sigma bonds shown in bold

16-40

(a) No, biphenyl is not fused. The rings must share two atoms to be labeled "fused".
(b) There are 12 π electrons in biphenyl compared with 10 for naphthalene.
(c) Biphenyl has 6 "double bonds". An isolated alkene releases 120 kJ/mole upon hydrogenation.

predicted:	6 x 120 kJ/mole (28.6 kcal/mole) $\approx$	720 kJ/mole (172 kcal/mole)
observed:		418 kJ/mole (100 kcal/mole)
	resonance energy:	302 kJ/mole (72 kcal/mole)

(d) On a "per ring" basis, biphenyl is 302 ÷ 2 = 151 kJ/mole, the same as the value for benzene. Naphthalene's resonance energy is 252 kJ/mole (60 kcal/mole); on a "per ring" basis, naphthalene has only 126 kJ/mole of stabilization per ring. This is consistent with the greater reactivity of naphthalene compared with benzene. In fact, the more fused rings, the lower the resonance energy per ring, and the more reactive the compound. (Refer to Problem 16-21.)

16-41 Two protons are removed from sp^3 carbons to make sp^2 carbons and to generate a π system with 10 π electrons.

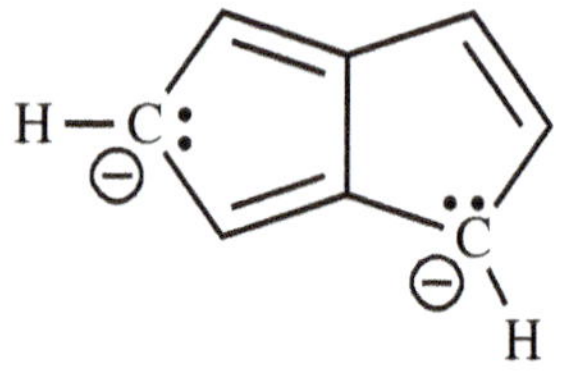

H—C:⊖ ⊖C:—H

16-42

(a) 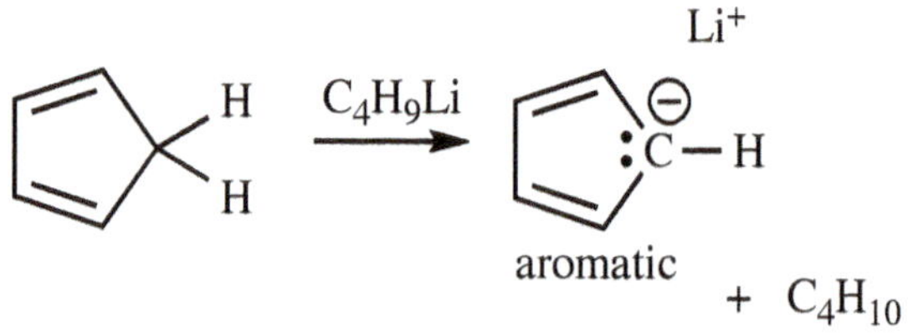 cyclopentadiene (H, H) $\xrightarrow{C_4H_9Li}$ cyclopentadienyl anion :C⊖—H Li^+ aromatic + C_4H_{10}

(b) cycloheptatrienol —OH $\xrightarrow{H_2SO_4}$ ⊕CH HSO_4^- aromatic + H_2O

(c) cyclopropenyl —Cl $\xrightarrow{Ag^+}$ ⊕CH + AgCl aromatic

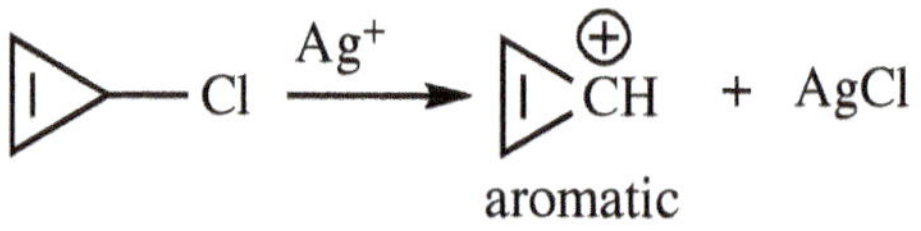

(d) 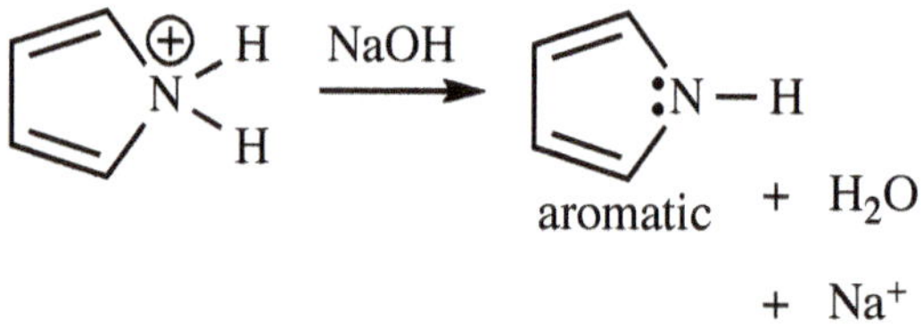 ⊕N(H)(H) $\xrightarrow{NaOH}$:N—H aromatic + H_2O + Na^+

(e) 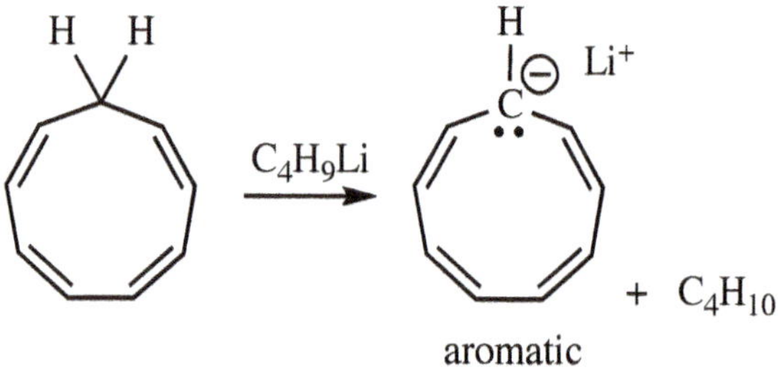 H H cyclononatetraene $\xrightarrow{C_4H_9Li}$ H—C:⊖ Li^+ + C_4H_{10} aromatic

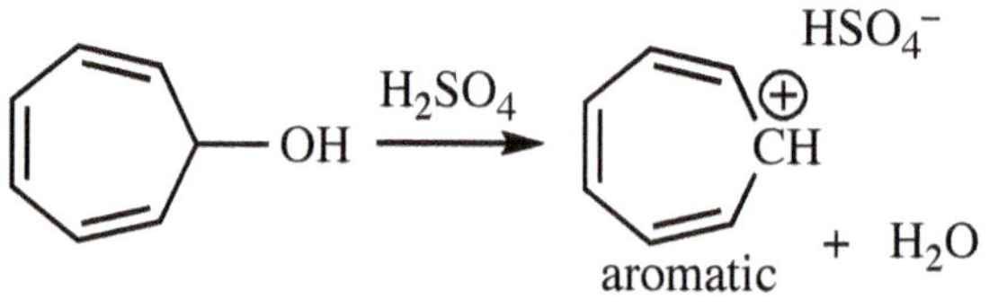

(f) 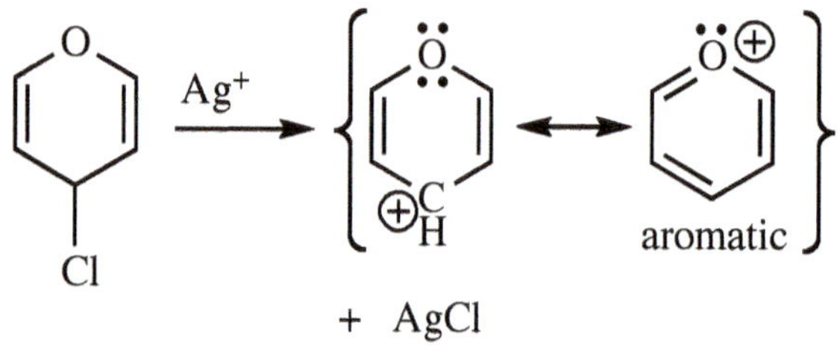 O ... Cl $\xrightarrow{Ag^+}$ { Ö ... ⊕CH ⟷ Ö⊕ aromatic } + AgCl

16-43 These four bases can be aromatic, partially aromatic, or aromatic in a tautomeric form. In other words, aromaticity plays an important role in the chemistry of all four structures. (Only electron pairs involved in the important resonance are shown.) To answer part (b), nitrogens that are basic are denoted by ***B***. Those that are not basic are shown as ***NB***. Note that some nitrogens change depending on the tautomeric form.

(a) and (c)

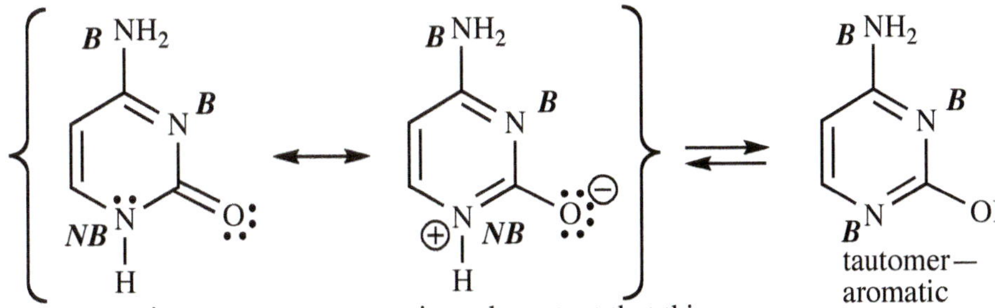

cytosine

aromatic to the extent that this resonance form contributes

tautomer—aromatic

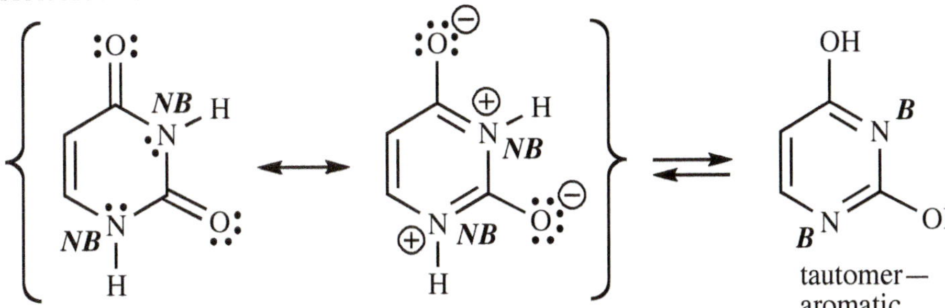

uracil

aromatic to the extent that this resonance form contributes

tautomer—aromatic

---

aromatic

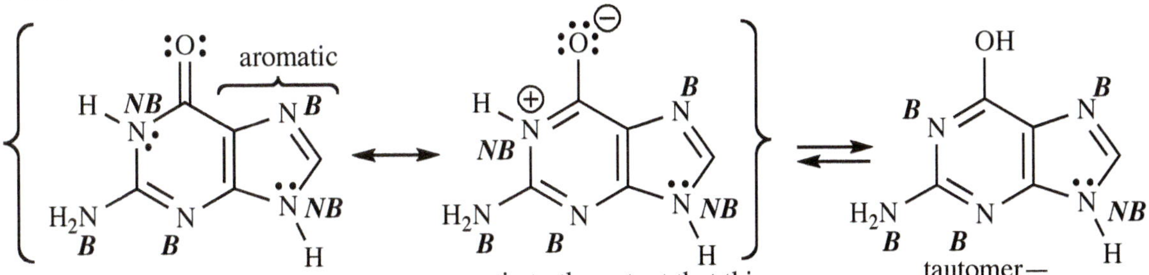

guanine

aromatic to the extent that this resonance form contributes

tautomer—fully aromatic

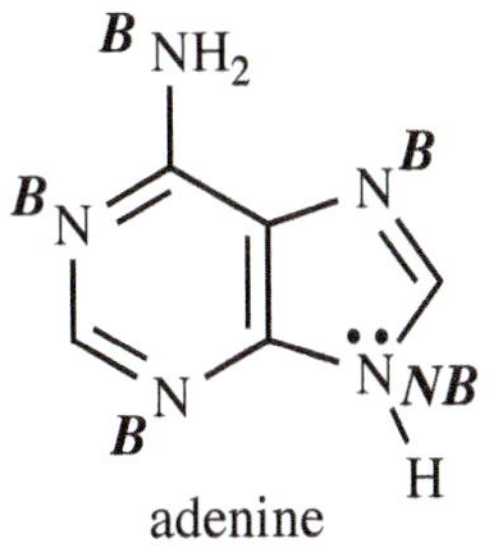

---

fully aromatic

adenine

16-44 (a) Antiaromatic—only 4 π electrons.

(b) This molecule is electronically equivalent to cyclobutadiene. Cyclobutadiene is unstable and undergoes a Diels-Alder reaction with another molecule of itself. The *tert*-butyl groups prevent dimerization by blocking approach of any other molecule.

(c) Yes, the nitrogen should be basic. The pair of electrons on the nitrogen is in an sp^2 orbital parallel to the ring and is perpendicular to the π system. It cannot be part of the π system as that would require two p orbitals occupying the same space.

(d)

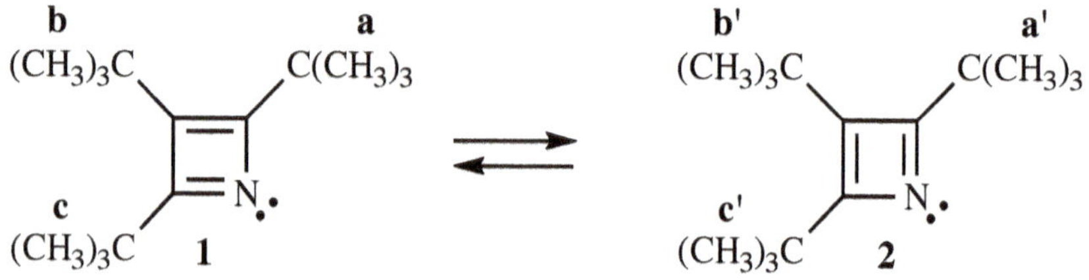

b $(CH_3)_3C$ a $C(CH_3)_3$ c $(CH_3)_3C$ N: **1** ⇌ b' $(CH_3)_3C$ a' $C(CH_3)_3$ c' $(CH_3)_3C$ N: **2**

Analysis of structure **1** shows the three *tert*-butyl groups in unique environments in relation to the nitrogen. We would expect three different signals in the NMR, as is observed at –110° C. Why do signals coalesce as the temperature is increased? Two of the *tert*-butyl groups become equivalent—which two? Most likely, they are **a** and **c** that become equivalent as they are symmetric around the nitrogen. But they are *not* equivalent in structure **1**—what is happening here?

What must happen is an equilibration between structures **1** and **2**, very slow at –110° C, but very fast at room temperature, faster than the NMR can differentiate. So the signal that has coalesced is an average of **a** and **a'** and **c** and **c'**. (This type of low-temperature NMR experiment is also used to differentiate axial and equatorial hydrogens on a cyclohexane.)

The NMR data prove that **1** is not aromatic, and that **1** and **2** are isomers, not resonance forms. If **1** were aromatic, then **a** and **c** would have identical NMR signals at all temperatures.

16-45

Mass spectrum: Molecular ion at 150; base peak at 135, M – 15, is loss of methyl.

Infrared spectrum: The broad peak at 3500 cm^{-1} is OH; thymol must be an alcohol. The peak at 1620 cm^{-1} suggests an aromatic compound.

NMR spectrum: The singlet at δ 4.8 is OH; it disappears upon shaking with D_2O. The 6H doublet at δ 1.2 and the 1H multiplet at δ 3.2 are an isopropyl group, apparently on the benzene ring. A 3H singlet at δ 2.3 is a methyl group, also on the benzene ring.

Analysis of the aromatic protons suggests the substitution pattern. The three aromatic hydrogens confirm that there are three substituents. The singlet at δ 6.6 is a proton between two substituents (no neighboring Hs). The doublets at δ 6.75 and δ 7.1 are ortho hydrogens, splitting each other.

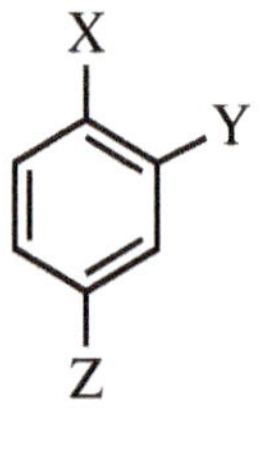

X, Y, Z (benzene ring) + CH_3 + OH + $CH(CH_3)_2$

continued on next page

16-45 continued

Several isomeric combinations are consistent with the spectra (although the single H giving δ 6.6 suggests that either Y or Z is the OH group—an OH on a benzene ring shields hydrogens ortho to it, moving them upfield).

The structure of thymol is:

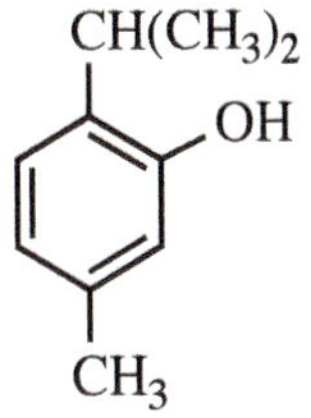

thymol

These structures fit the data equally well.

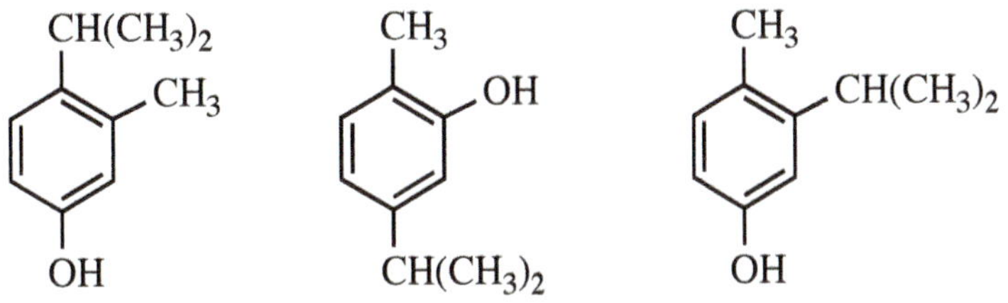

The final question is how the molecule fragments in the mass spectrometer:

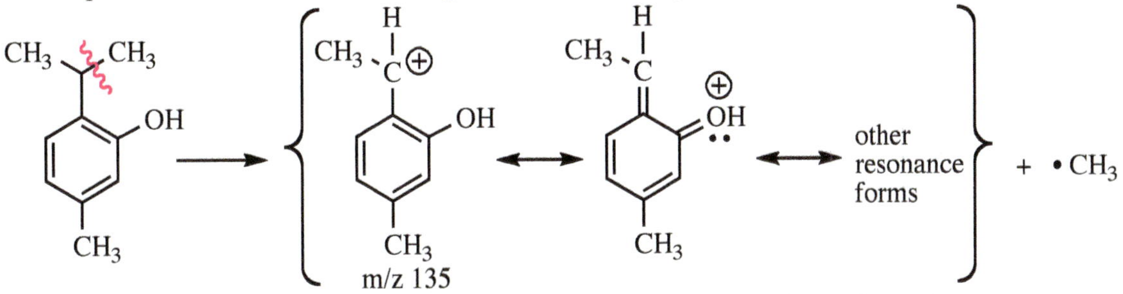

m/z 135

other resonance forms

$+ \bullet CH_3$

Resonance stabilization of this benzylic cation includes forms with positive charge on three ring carbons and on oxygen (shown).

16-46

Mass spectrum: Molecular ion at 170; two prominent peaks are M – 15 (loss of methyl) and M – 43 (as we shall see, most likely the loss of acetyl, CH_3CO).

Infrared spectrum: The two most significant peaks are at 1680 cm^{-1} (conjugated carbonyl) and 1600 cm^{-1} (aromatic C=C).

NMR spectrum: A 3H singlet at δ 2.7 is methyl next to a carbonyl, shifted slightly downfield by an aromatic ring. The other signals are seven aromatic protons. The 1H at δ 8.7 is a deshielded proton next to a carbonyl. Since there is only one, the carbonyl can have only one neighboring hydrogen.

Conclusions:

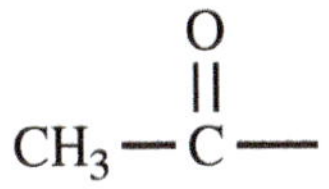

$CH_3-C(=O)-$ + m/z 127 including 7H ⇒ mass 120 for carbons ⇒ 10 C

The fragment $C_{10}H_7$ is almost certainly a naphthalene. The correct isomer (box) is indicated by the NMR.

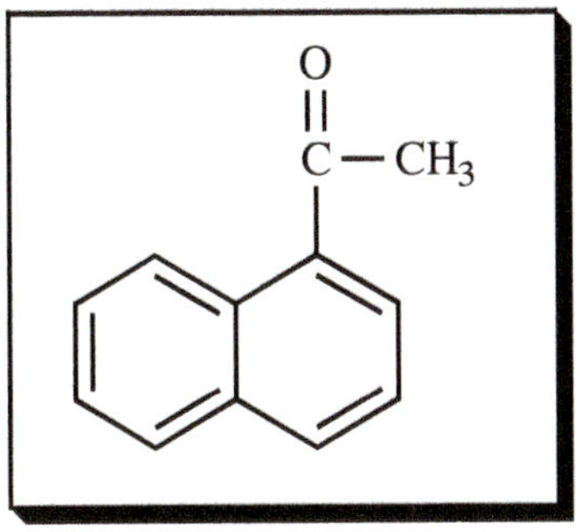

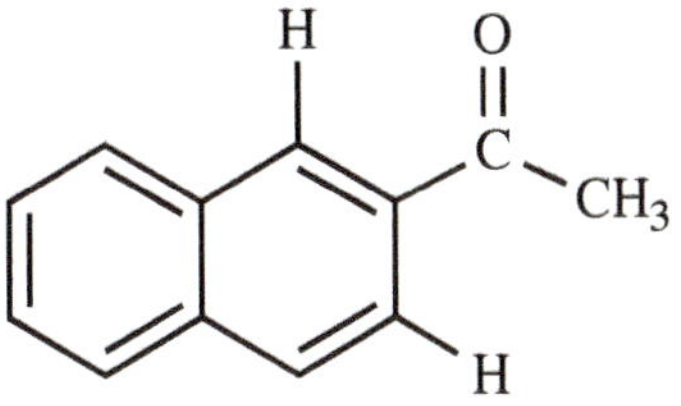

This isomer is a less good answer as it would have two deshielded protons in the NMR, one of which would be a singlet (it's a doublet in the spectrum).

16-47

Although all carbons in hexahelicene are sp^2, the molecule is not flat. Because of the curvature of the ring system, one end of the molecule has to sit on top of the other end—the carbons and the hydrogens would bump into each other if they tried to occupy the same plane. In other words, the molecule is the beginning of a spiral. An "upward" spiral is the nonsuperimposable mirror image of a "downward" spiral, so the molecule is chiral and therefore optically active.

The magnitude of the optical rotation is extraordinary: it is one of the largest rotations ever recorded. In general, alkanes have small rotations and aromatic compounds have large rotations, so it is reasonable to expect that it is the interaction of plane-polarized light (electromagnetic radiation) with the electrons in the twisted pi system (which can also be considered as having wave properties) that causes this enormous rotation.

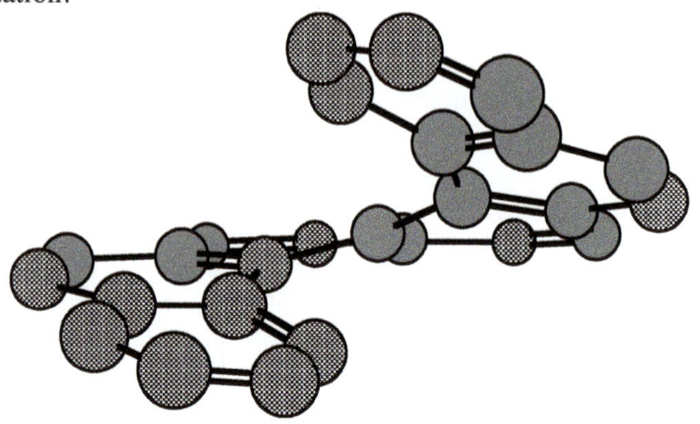

This three-dimensional picture of hexahelicene shows the twist or spiral in the system of six rings.

16-48 The cycloheptatrienyl cation has six pi electrons that just fill the bonding molecular orbitals, making this an aromatic system.

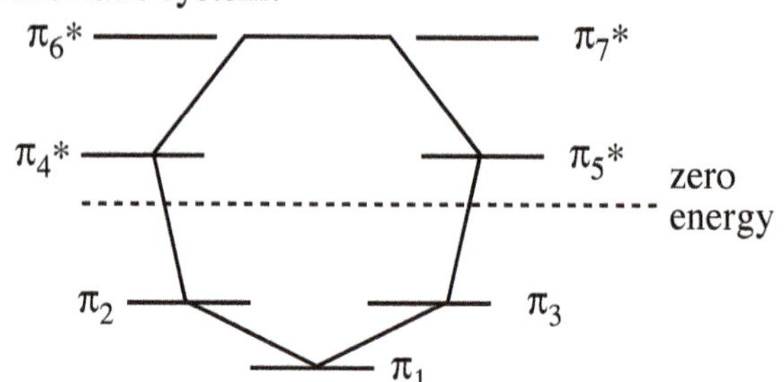

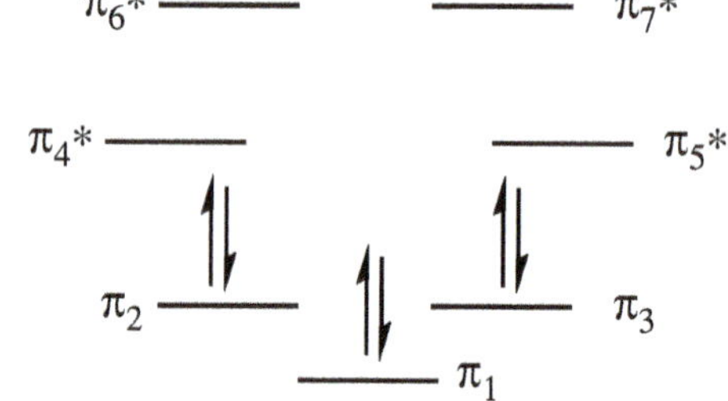

This electronic configuration is aromatic. Electrons fill the bonding molecular orbitals.

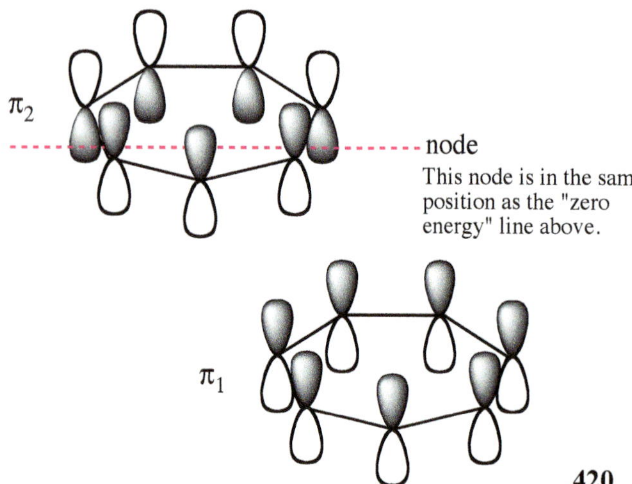

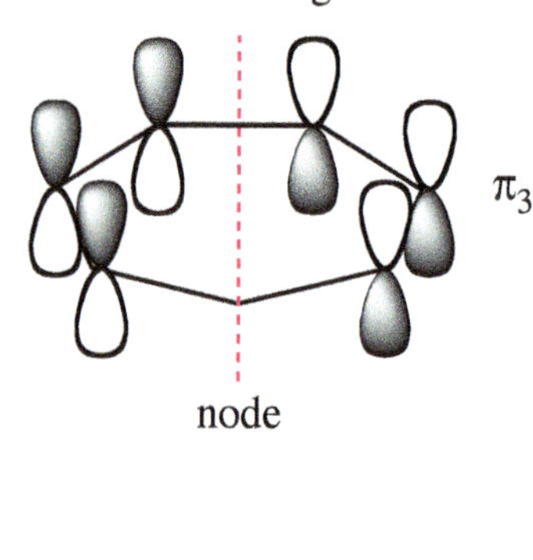

16-49 Are you familiar with the concept "tough love", that is, sometimes you have to be stern with someone you love for their own good? The concept is that sometimes to demonstrate one emotion, the behavior has to appear exactly the opposite. Granted, this is a stretch to apply it to atoms, but the point is that sometimes atoms like oxygen and nitrogen can have conflicting effects: they can withdraw electron density by their strong electronegativity (an inductive effect) but at the same time, they can donate electron density through their resonance effect. This phenomenon will prove important in the reactivity of substituted benzenes described in Chapter 17.

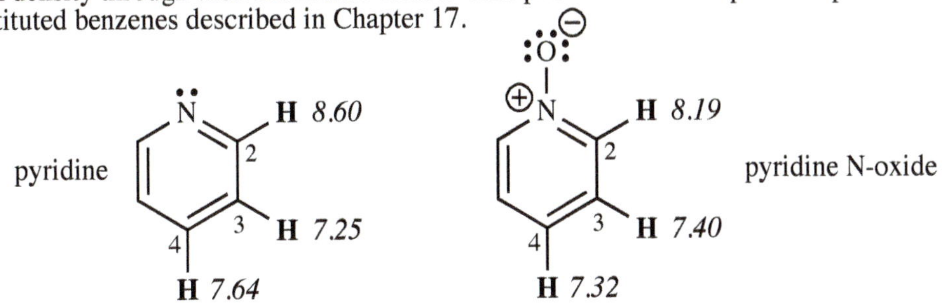

pyridine: H-2 *8.60*, H-3 *7.25*, H-4 *7.64*

pyridine N-oxide: H-2 *8.19*, H-3 *7.40*, H-4 *7.32*

Nitrogen is electronegative, so it exerts a deshielding effect on H-2 in pyridine. The effect diminishes with distance as expected with an inductive effect, although it is harder to explain why H-4 is deshielded more than H-3.

In pyridine N-oxide, with an even more electronegative oxygen attached to the N, it would be reasonable to expect that the hydrogens would be deshielded. This is the case with H-3; we can infer that the effect on H-3 is purely an inductive effect.

Even more interesting is the *shielding* effect on H-2 and H-4; these chemical shifts are shifted *upfield*. This must reflect the other side of oxygen's personality, the donation of electron density through a resonance effect. Drawing the resonance forms clearly shows that the electron density at H-2 and H-4 (and presumably H-6) increases through this donation by resonance, in perfect agreement with the NMR results. As we will see in Chapter 17, in most cases, the resonance effect trumps the inductive effect.

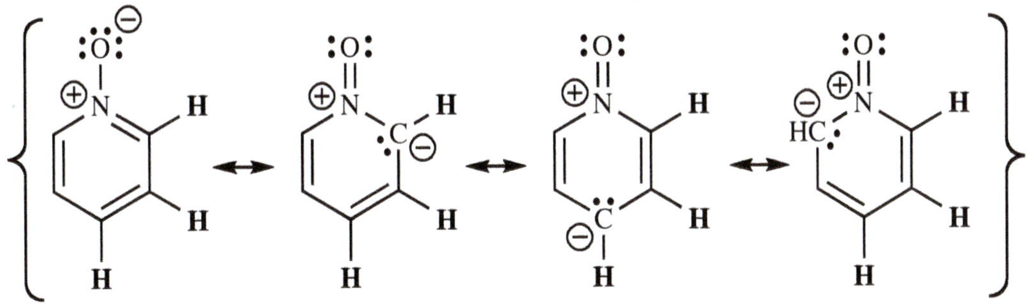

Note: The practice of applying human emotions to inanimate objects is called "anthropomorphizing". For example, we say that an atom is "happy" when it has a full octet of electrons. In casual conversation, this gets a point across, but it is not appropriate in rigorous scientific terms, for example, on exams. Under more formal conditions, we are expected to use the specific terms of science because they are well defined and do not permit sloppy or fuzzy concepts.

As our equipment technician says: "Don't anthropomorphize computers. They hate that."

16-50

It will be easier in this cyclic pi system to put a dot • on each atom in the aromatic system (including two N atoms), rather than to circle the double bonds. Eighteen atoms are connected by 9 double bonds, a Hückel system of 18 pi electrons.

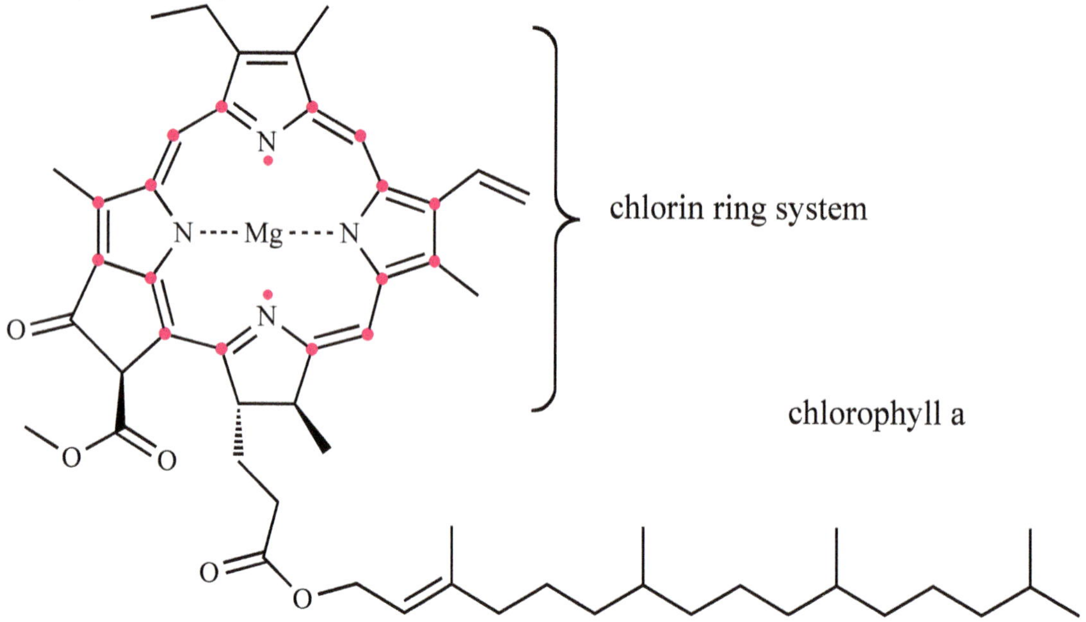

N
N----Mg----N
N
O
O
O
O
O
chlorin ring system
chlorophyll a

16-51

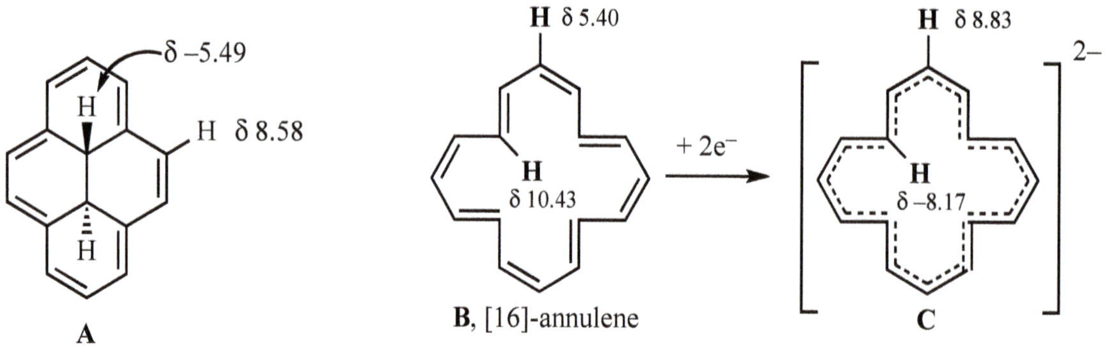

δ –5.49
H
H δ 8.58
H
A
H δ 5.40
H
δ 10.43
B, [16]-annulene
+ 2e⁻
H δ 8.83
2–
H
δ –8.17
C

Structure **A**, a [14]-annulene, demonstrates the observation that protons inside, above, or below an aromatic system are in a "shielding" region that gives them a negative chemical shift, that is, to the right of TMS. Note that the H outside of the pi system has a typical chemical shift for an aromatic proton δ 7-9.

In contrast, structure **B**, not aromatic, shows an external H with a chemical shift typical of a vinyl hydrogen, that is, an H on an sp^2 carbon. The internal H is highly deshielded, about the same deshielding as what an aldehyde hydrogen experiences.

Adding two electrons to **B** gives **C** with 18 pi electrons. The external H moved from a δ value typical of a vinyl hydrogen to a value characteristic of an aromatic system, very similar to the external H in **A**. The internal H shows an enormous difference: it has moved farther than 8 ppm to the right of TMS, a difference of >18 ppm from structure **B**. Just by adding two electrons, the electronic system went from anti-aromatic to aromatic, and the chemical shifts demonstrate how significant a difference this is. Aromatic versus nonaromatic is like two different universes of organic molecules.

CHAPTER 17—REACTIONS OF AROMATIC COMPOUNDS

The representation of benzene with a circle to represent the π system is fine for questions of nomenclature, properties, isomers, and reactions. For questions of mechanism or reactivity, however, the representation with three alternating double bonds (the Kekulé picture) is more informative. For clarity and consistency, this Solutions Manual will use the Kekulé form exclusively.

17-1

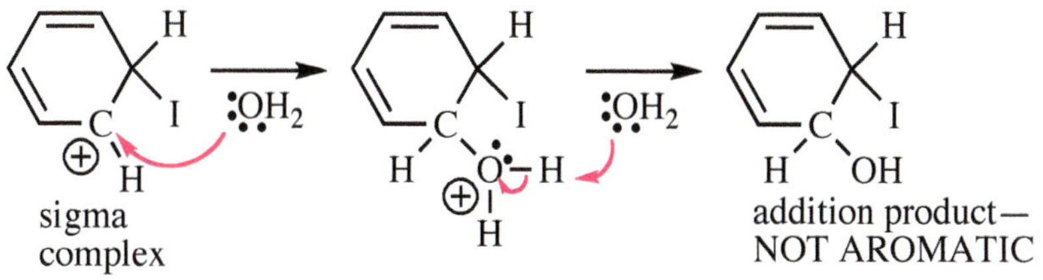

sigma complex

addition product—
NOT AROMATIC

Though the addition of water to the sigma complex can be shown in a reasonable mechanism, the product is not aromatic. Thus, it has lost the 151 kJ/mol (36 kcal/mole) of resonance stabilization energy. The addition reaction is not favorable energetically, and substitution prevails.

17-2

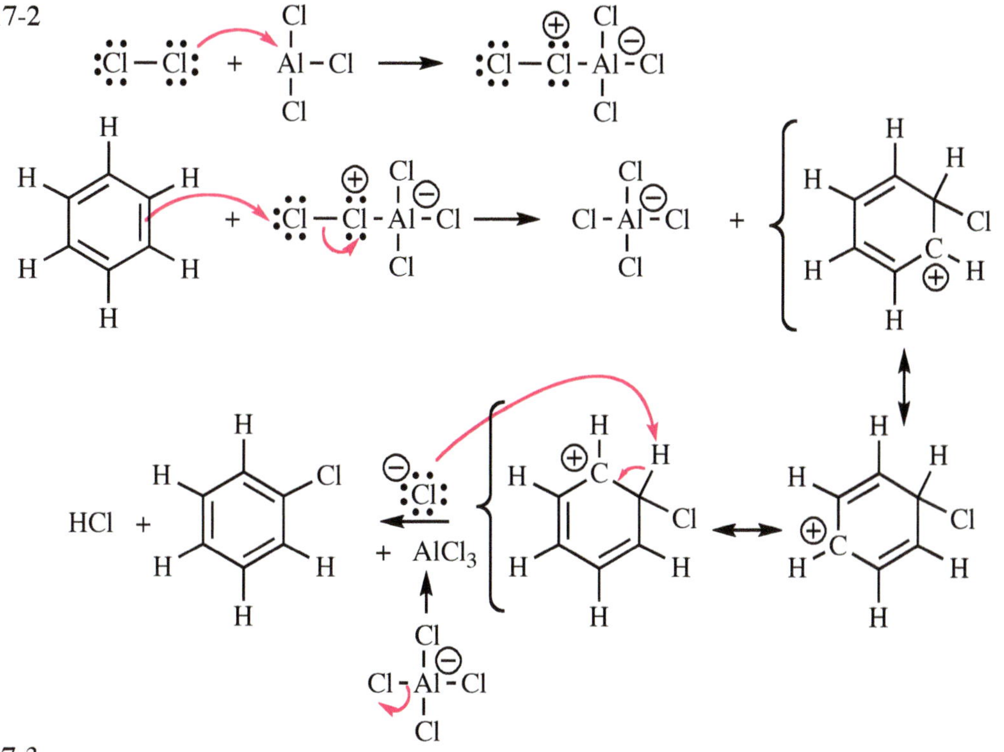

HCl + + $AlCl_3$

17-3

2° 2° 3°

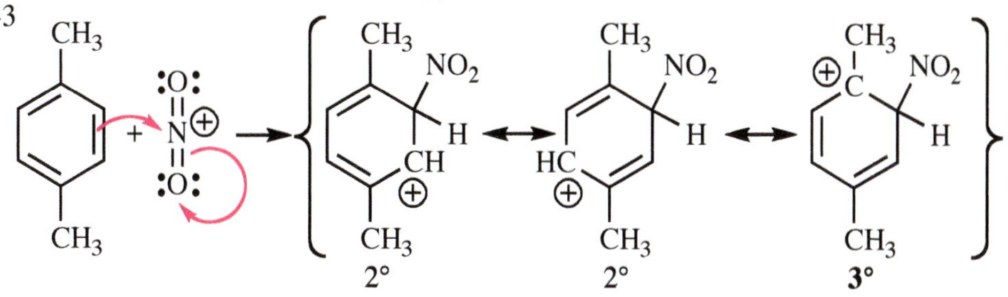

Benzene's sigma complex has positive charge on three 2° carbons. The sigma complex above shows positive charge in one resonance form on a 3° carbon, lending greater stabilization to this sigma complex. The more stable the intermediate, the lower the activation energy required to reach it, and the faster the reaction will be.

17-4 delocalization of the positive charge on the ring

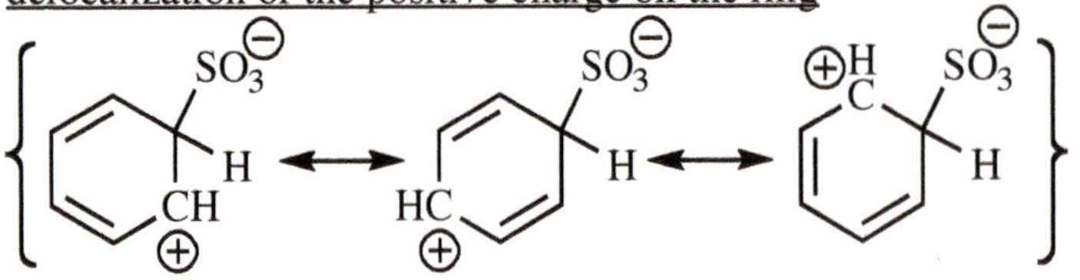

delocalization of the negative charge on the sulfonate group

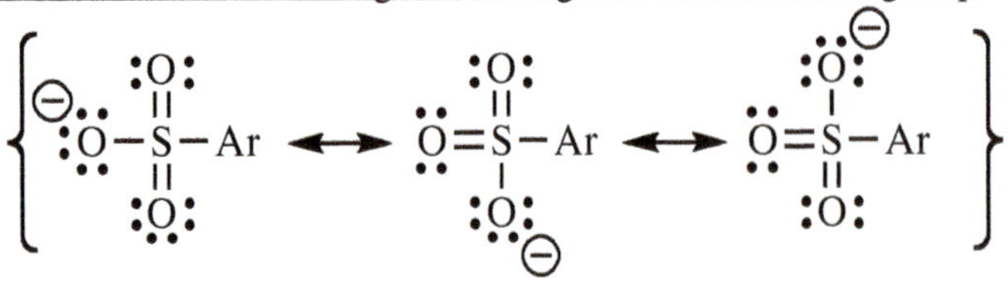

("Ar" is the general abbreviation for an *aromatic* or *aryl* group, in this case, benzene; "R" is the general abbreviation for an *aliphatic* or *alkyl* group. In cases where the identity of the R group does not matter, it has been used to represent alkyl or aryl groups.)

17-5

(a) The key to electrophilic aromatic substitution lies in the stability of the sigma complex. When the electrophile bonds at ortho or para positions of ethylbenzene, the positive charge is shared by the 3° carbon with the ethyl group. Bonding of the electrophile at the meta position lends no particular advantage because the positive charge in the sigma complex is never adjacent to, and therefore never stabilized by, the ethyl group. (Unshared electron pairs on halogen not shown here.)

ortho CH_2CH_3 Br—Br—$FeBr_3$ → 3°—*good*

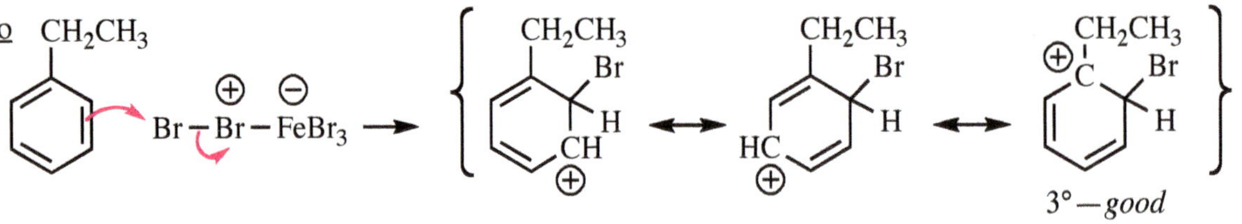

meta CH_2CH_3 Br—Br—$FeBr_3$ →

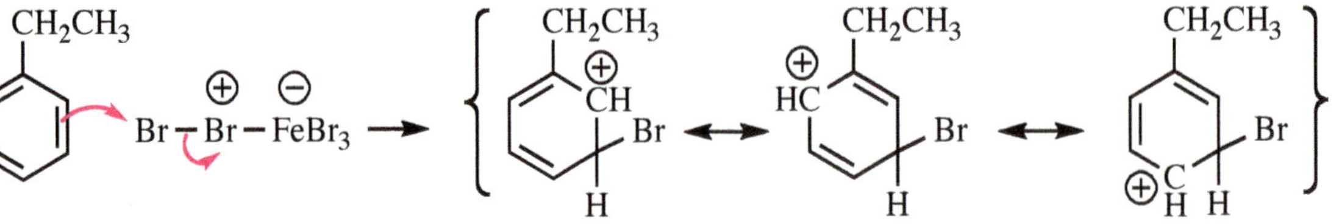

para CH_2CH_3 Br—Br—$FeBr_3$ → 3°—*good*

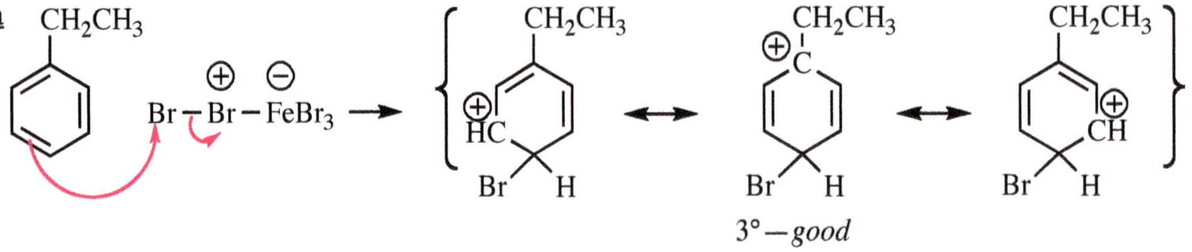

(b) Electrophilic attack on *p*-xylene gives an intermediate in which only one of the three resonance forms is stabilized by a substituent (see the solution to Problem 17-3). *m*-Xylene, however, is stabilized in two of its three resonance forms. A more stable intermediate gives a faster reaction.

m-xylene CH_3 CH_3 $^{+}NO_2$ → 3°—*good* 3°—*good*

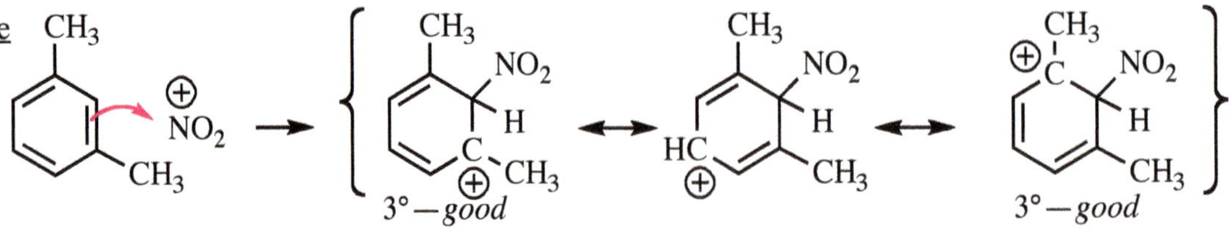

17-6 For ortho and para attack, the positive charge in the sigma complex can be shared by resonance with the vinyl group. This cannot happen with meta attack because the positive charge is never adjacent to the vinyl group. (Ortho attack is shown; para attack gives an intermediate with positive charge on the same carbons.)

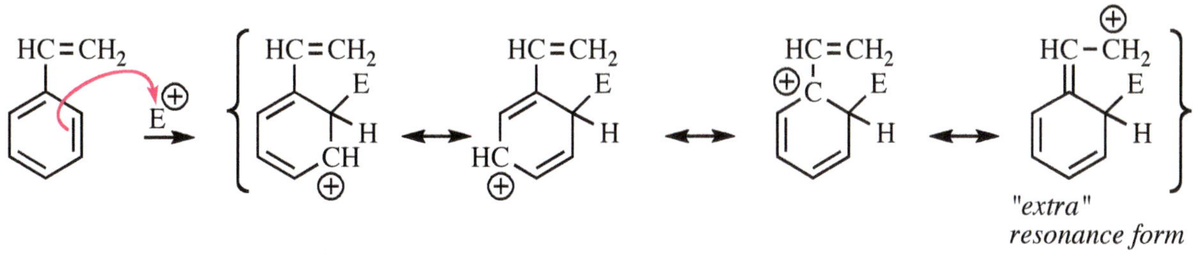

"extra" resonance form

17-7 Attack at only ortho and para positions (not meta) places the positive charge on the carbon with the ethoxy group, where the ethoxy group can stabilize the positive charge by resonance donation of a lone pair of electrons. (Ortho attack is shown; para attack gives a similar intermediate.) (Unshared electron pairs on halogen not shown here.)

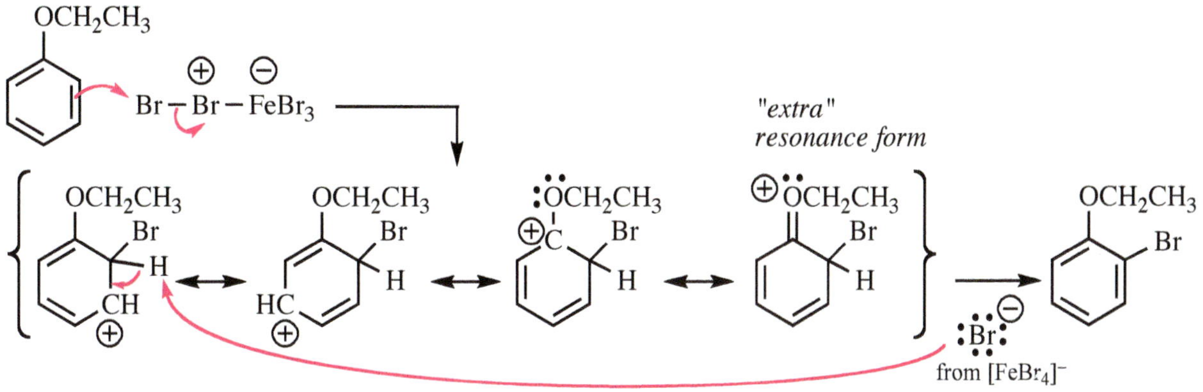

"extra" resonance form

from $[FeBr_4]^-$

17-8

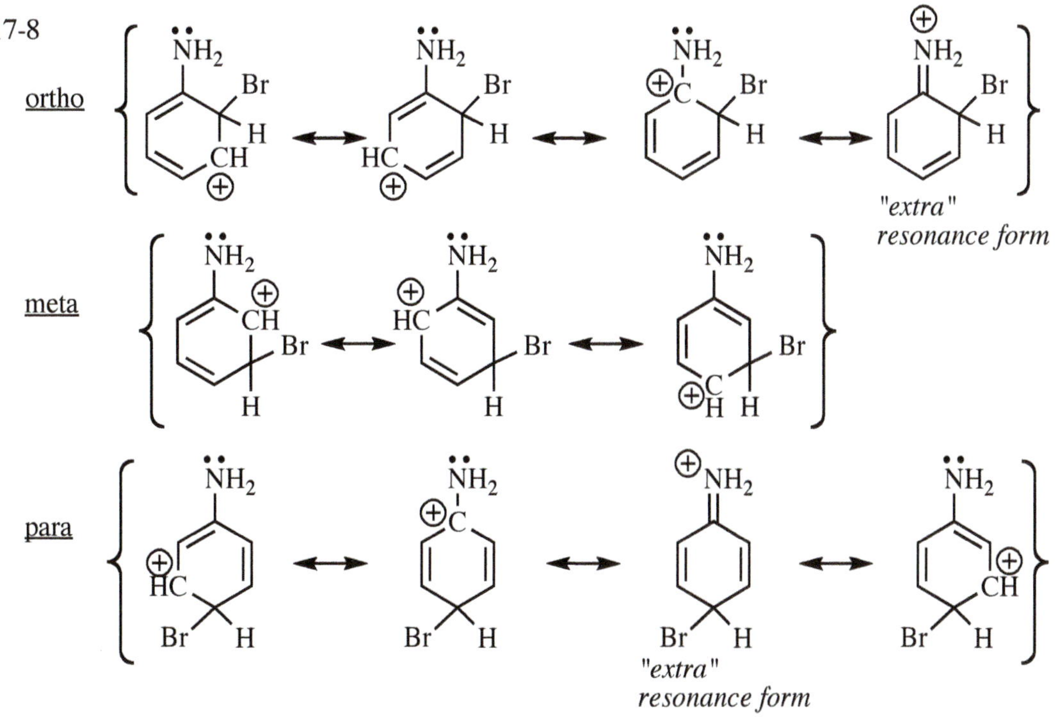

ortho

"extra" resonance form

meta

para

"extra" resonance form

17-9

Substitution: $OCH(CH_3)_2$-benzene + Br_2 → $OCH(CH_3)_2$-bromobenzene + **HBr** (g)

Addition: cyclohexene + Br_2 → trans-1,2-dibromocyclohexane

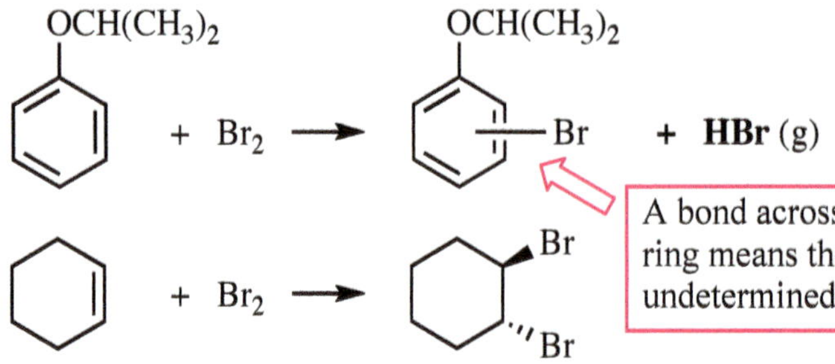

A bond across the middle of a benzene ring means the substituent is at an undetermined (or unimportant) location.

Substitution generates HBr whereas the addition does not. If the reaction is performed in an organic solvent, bubbles of HBr can be observed, and HBr gas escaping into moist air will generate a cloud. If the reaction is performed in water, adding moist litmus paper to test for acid will differentiate the results of the two compounds.

17-10

(a) Nitration is performed with nitric acid and a sulfuric acid catalyst. In strong acid, amines in general, including aniline, are protonated.

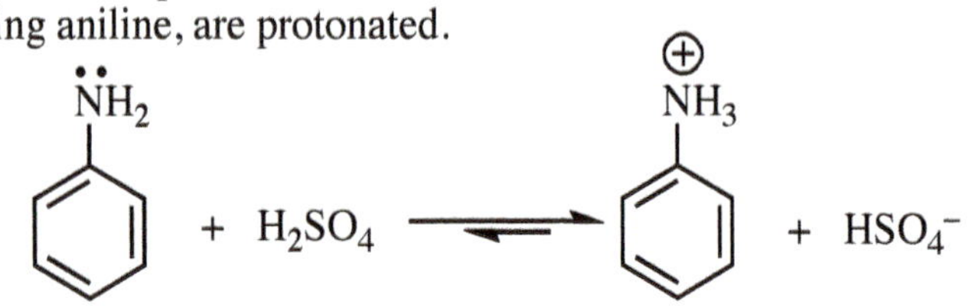

$$C_6H_5NH_2 + H_2SO_4 \rightleftharpoons C_6H_5NH_3^+ + HSO_4^-$$

(b) The NH_2 group is a strongly activating ortho,para-director. In acid, however, it exists as the protonated ammonium ion—a strongly **deactivating meta-director**. The strongly acidic nitrating mixture itself forces the reaction to be slower.

(c) The acetyl group removes some of the electron density from the nitrogen, making it much less basic; the nitrogen of this amide is not protonated under the reaction conditions. The N retains enough electron density to share with the benzene ring, so the $NHCOCH_3$ group is still an activating ortho,para-director, though weaker than NH_2.

$$\left\{ Ph-\ddot{N}H-C(=O)-CH_3 \longleftrightarrow Ph-\overset{\oplus}{N}H=C(-O^{\ominus})-CH_3 \right\}$$

17-11 Nitronium ion attack at the ortho and para positions places positive charge on the carbon adjacent to the bromine, allowing resonance stabilization by an unshared electron pair from the bromine. Meta attack does not give a stabilized intermediate.

<u>ortho</u>

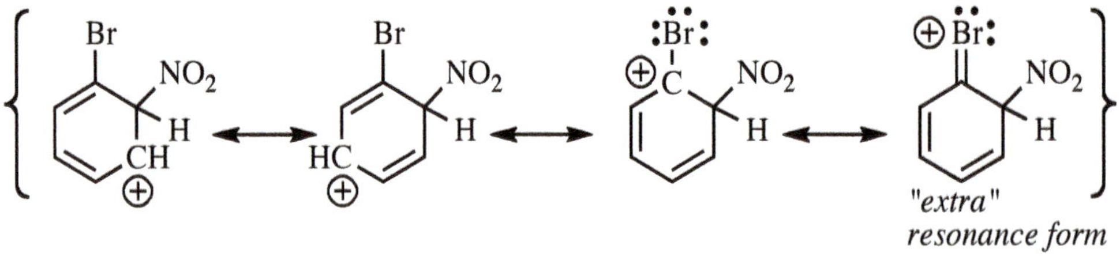

Br, NO_2, H, CH ⊕ ⟷ Br, NO_2, H, HC ⊕ ⟷ :Br:, ⊕C, NO_2, H ⟷ ⊕Br:, NO_2, H

"extra" *resonance form*

<u>meta</u>

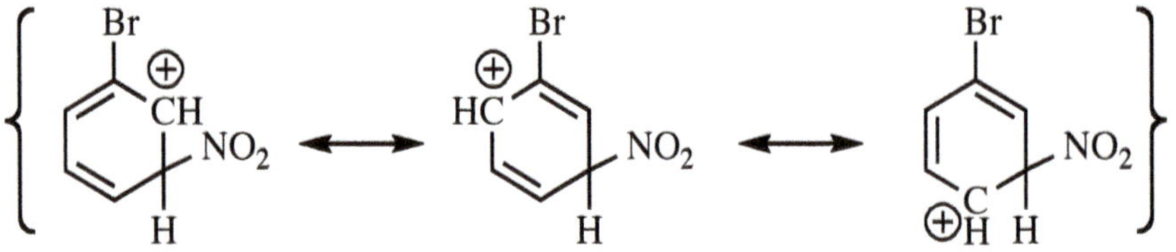

Br, ⊕CH, NO_2, H ⟷ Br, ⊕HC, NO_2, H ⟷ Br, NO_2, ⊕C H H

17-11 continued

para

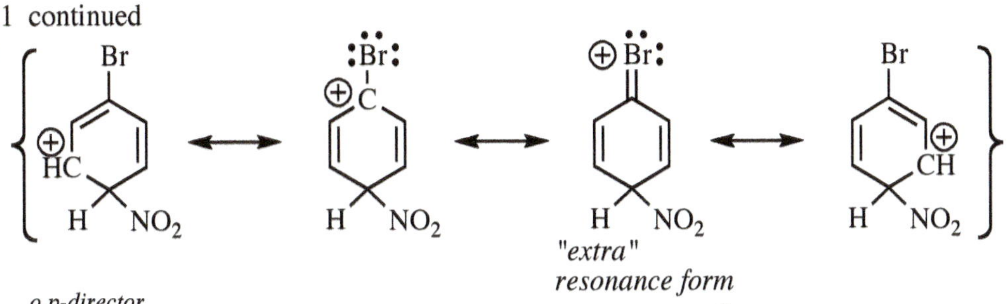

H NO$_2$ — "extra" resonance form

17-12

(a) CH_3 *o,p-director*; NO_2 *m-director* — products: 2,4-dinitrotoluene + 2,6-dinitrotoluene

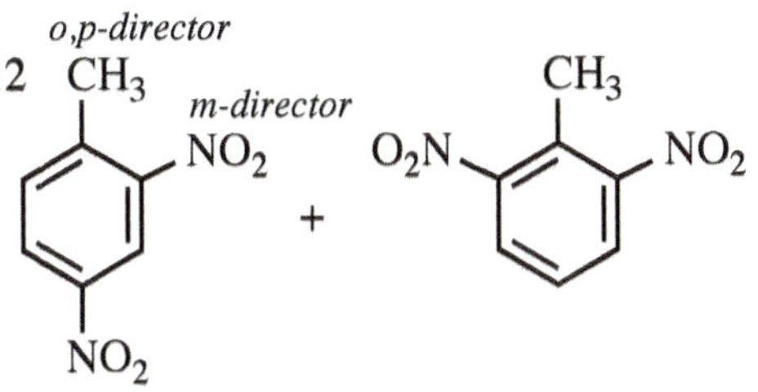

(b) CH_3 *o,p-director*; Cl *o,p-director* — products + trace

trace

Squeezing between two groups is difficult.

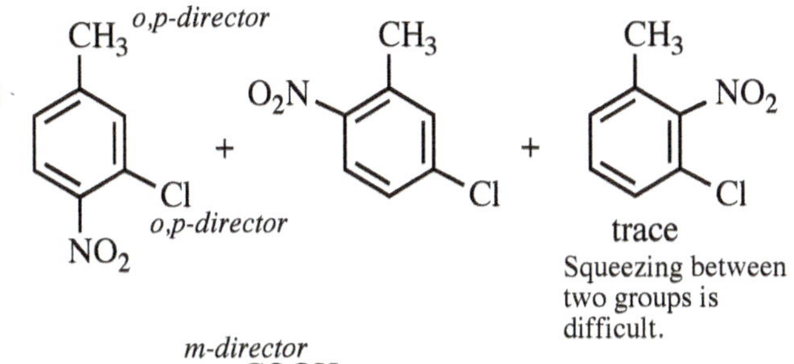

(c) COOH *m-director*; Br *o,p-director*

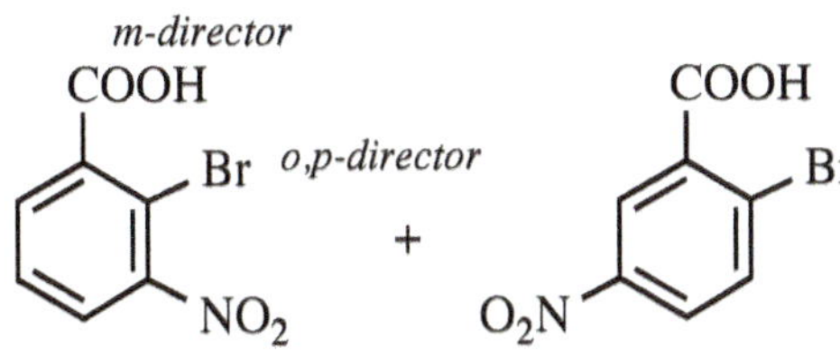

(d) COOH *m-director*; OCH_3 *o,p-director*

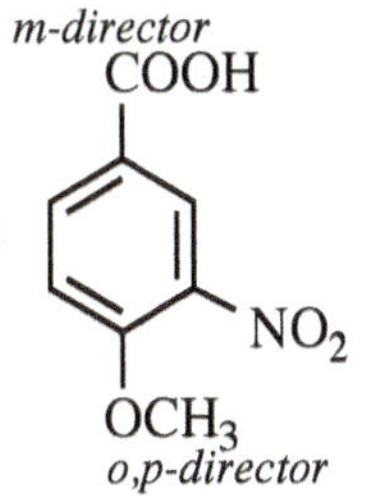

(e) OH *o,p-director*; CH_3 *o,p-director*

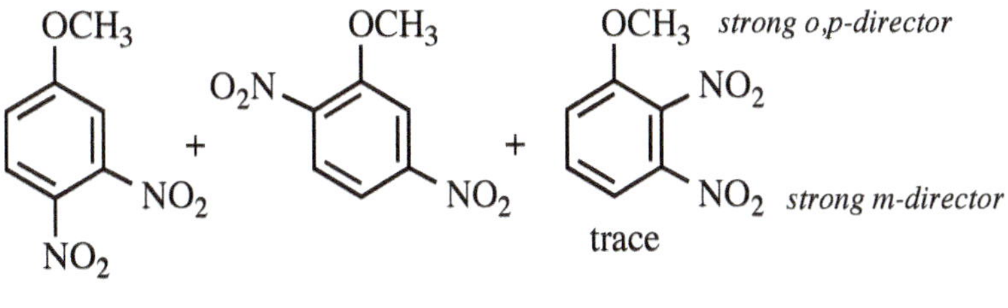

trace

Squeezing between two groups is difficult.

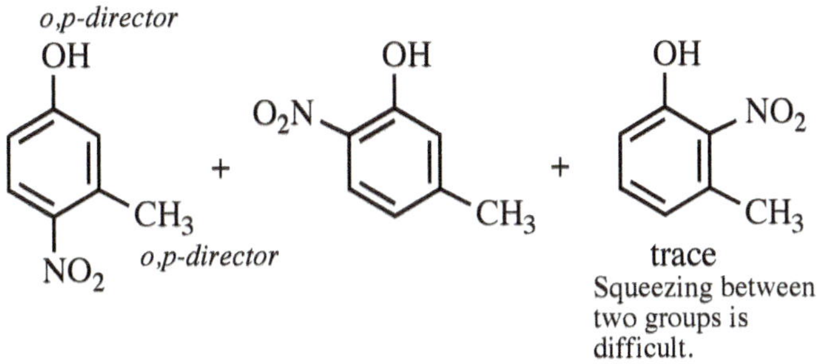

(f) OH *o,p-director*; *m-director*

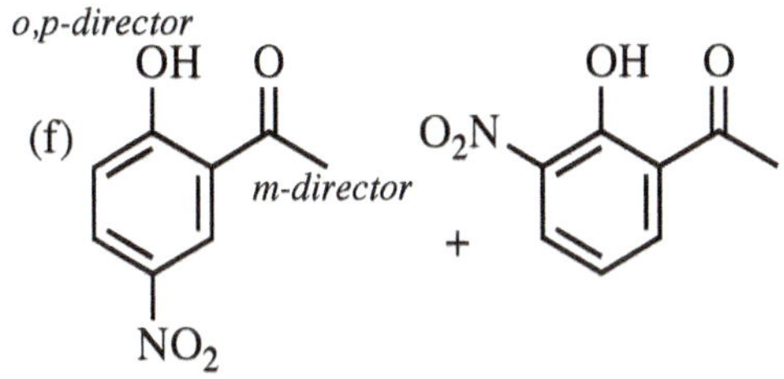

17-13

(a) OCH_3 *strong o,p-director*; CH_3 *weak o,p-director*

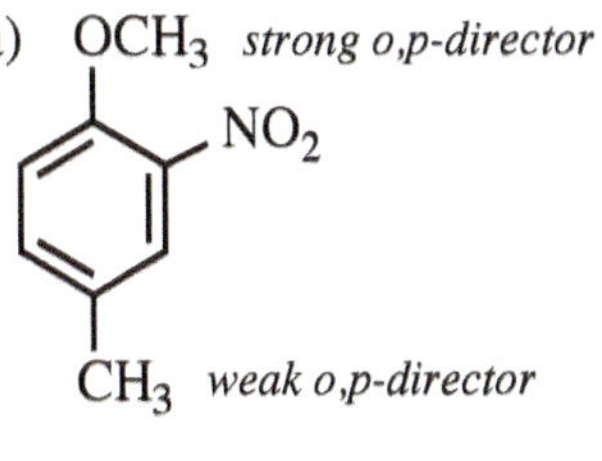

(b) Cl *weak o,p-director*; NO_2 *strong m-director*

trace

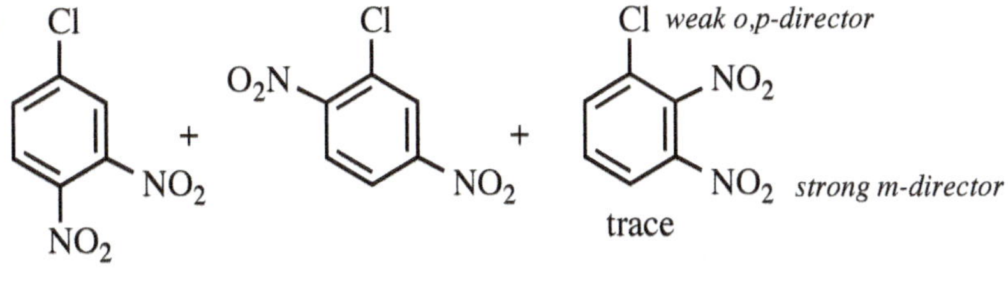

(c) OH *strong o,p-director*; Cl *weak o,p-director*

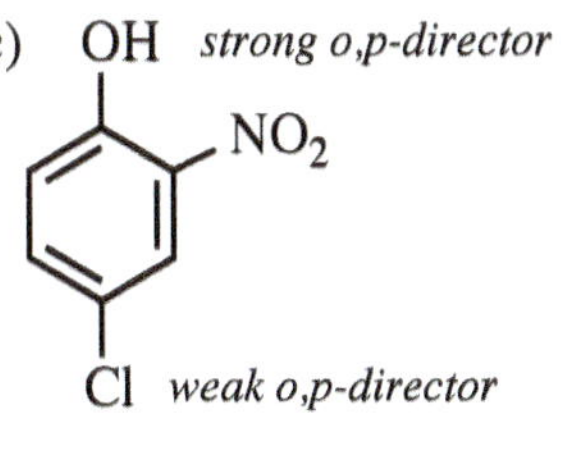

(d) OCH_3 *strong o,p-director*; NO_2 *strong m-director*

trace

17-13 continued

(e) NO_2 ... NH—C(=O)—CH_3 strong o,p-director; CH_3 weak o,p-director + O_2N— ... NH—C(=O)—CH_3, CH_3

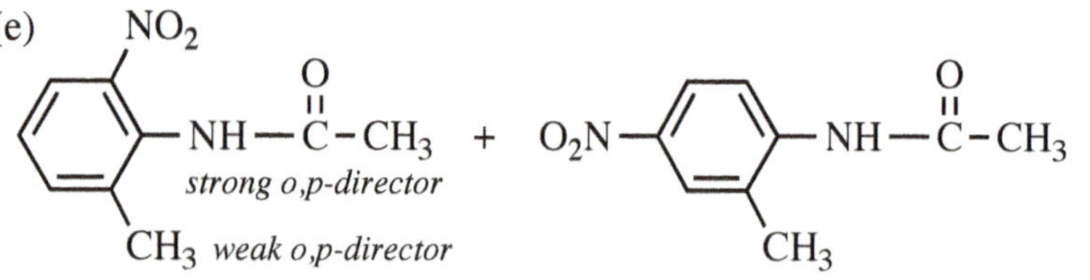

(f) O_2N; CH_3—C(=O)—NH— strong o,p-director, activating; —C(=O)—NH_2 strong m-director, deactivating

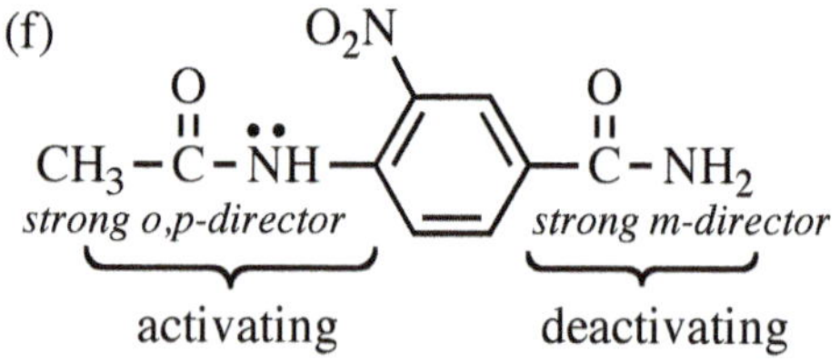

only product

17-14

(a) Sigma complex of ortho attack—the phenyl substituent stabilizes positive charge by resonance (braces omitted):

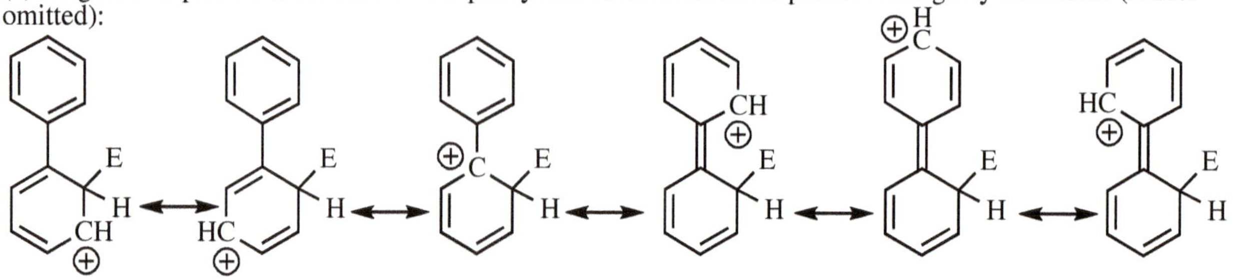

E, H, CH ⊕ ⟷ HC ⊕ ⟷ ⊕C ⟷ CH ⊕ ⟷ ⊕ HC ⟷ HC ⊕

Para attack gives similar stabilization. Meta attack does not permit delocalization of the positive charge on the phenyl substituent.

(b) The nitronium ion electrophile will prefer to attack the more activated ring, or the less deactivated, ring.

(i) —NO_2 + O_2N

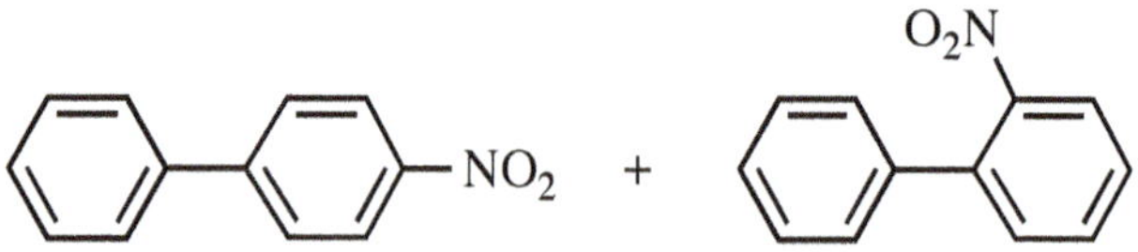

(ii) O_2N OH trace + OH —NO_2 + OH O_2N

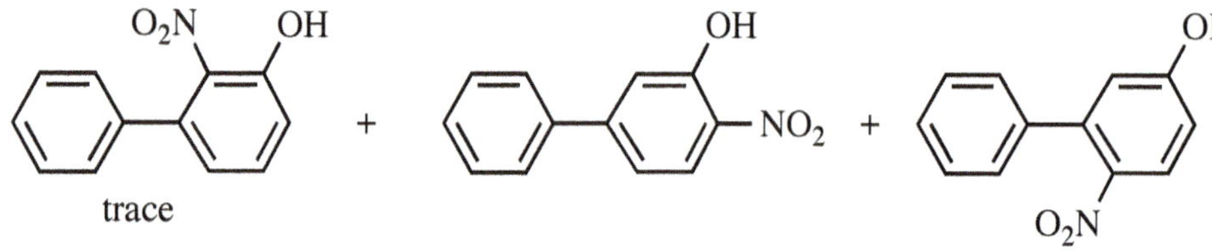

(iii) O_2N— —C(=O)—CH_3 + NO_2 —C(=O)—CH_3

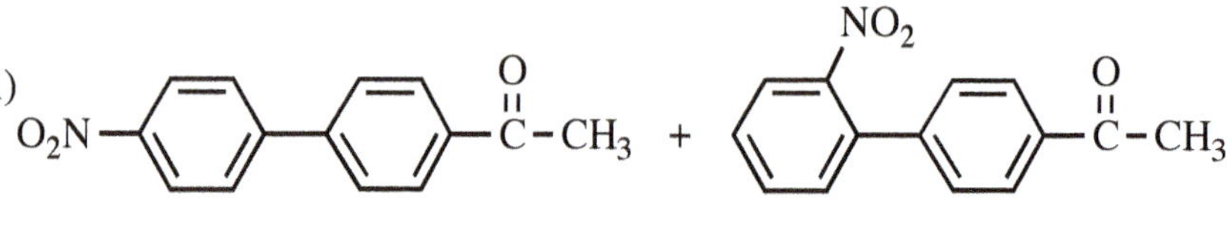

(iv) —NO_2 + NO_2; O_2N NO_2 minor

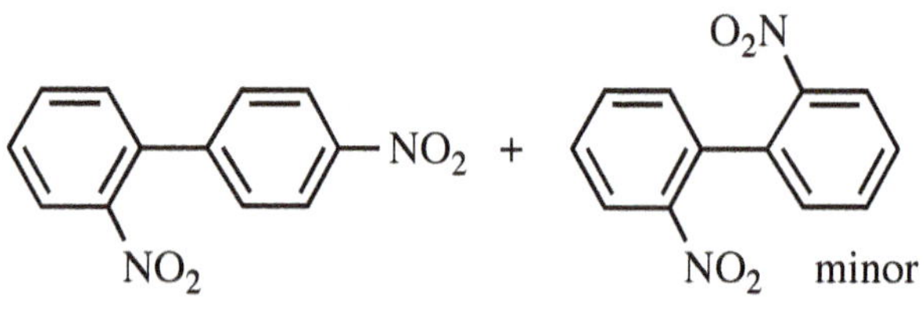

(v) O_2N O O + O_2N O O

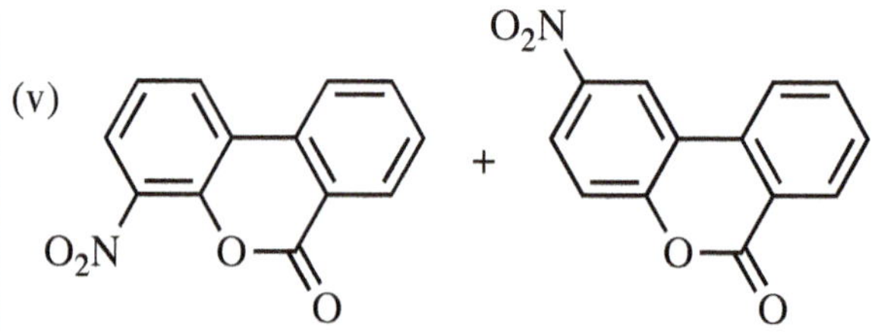

(vi) NO_2

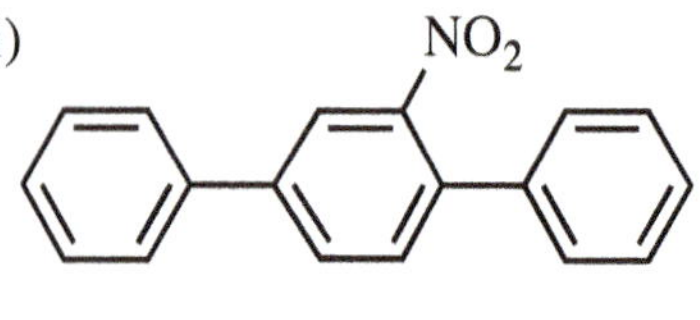

Major product—minor amounts of nitration would occur on the outer rings; the middle ring has two activating substituents while the outer rings have only one.

17-15

(a) + $AlCl_3$ ⇌ $AlCl_4^-$ + C–H

H H H H H H HC CH C H H H

(b) Multiple alkylation can also occur.

CH_3—Cl + $AlCl_3$ ⇌ CH_3—Cl—$AlCl_3$ OCH_3

Para isomer is also formed by similar mechanism.

OCH_3 CH_3 OCH_3 CH_3 H CH OCH_3 CH_3 H HC OCH_3 CH_3 H C OCH_3 CH_3 H

(c) CH_3 Cl CH_3—C—$CHCH_3$ CH_3 $AlCl_3$ → $AlCl_4^-$ CH_3 CH_3—C—$CHCH_3$ CH_3 2° methyl shift → 3° CH_3 CH_3—C–$CHCH_3$ CH_3

+ CH_3 CH_3–C–$CHCH_3$ CH_3 → CH H CH_3–C–$CH(CH_3)_2$ CH_3 C H CH_3–C–$CH(CH_3)_2$ CH_3 HC H CH_3–C–$CH(CH_3)_2$ CH_3

Cl

$CH(CH_3)_2$ C—CH_3 CH_3

Only a small amount of the ortho isomer might be produced, as steric interactions will discourage this approach path of the electrophile.

17-16

(a) cyclohexene + HF → cyclohexyl carbocation ($C^{\oplus}$) —benzene→ cyclohexylbenzene

(b) $CH_3-C(CH_3)_2-OH$ + BF_3 → $CH_3-C^{\oplus}(CH_3)_2$ —benzene→ $CH_3-C(CH_3)_2-C_6H_5$

(c) $CH_2=C(CH_3)_2$ + HF → $CH_3-C^{\oplus}(CH_3)_2$ —$CH_3-C(CH_3)_2-C_6H_5$→ $CH_3-C(CH_3)_2-C_6H_4-C(CH_3)_2-CH_3$

(d) $HO-CH(CH_3)CH_3$ + BF_3 → $\overset{\oplus}{C}HCH_3$ (CH_3) —toluene (CH_3)→ p-isopropyltoluene (CH_3) + o-isopropyltoluene (CH_3)

17-17 In (a), (b), and (d), the electrophile has rearranged.

(a) $CH_3-C(CH_3)_2-C_6H_5$
rearranged

(b) CH_3 (para) + CH_3 (ortho)
rearranged

(c) No reaction: nitrobenzene is too deactivated for the Friedel-Crafts reaction to succeed.

(d) $C_6H_5-C(CH_3)_2-CH(CH_3)_2$ rearranged

17-18

(a) benzene + $CH_3CH_2CH_2CH_2Br$ $\xrightarrow{AlCl_3}$ 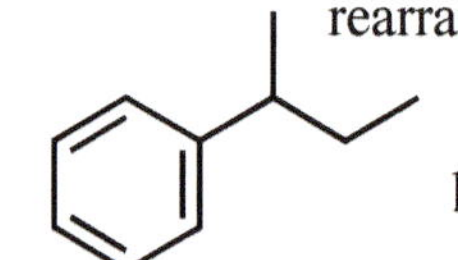 rearranged—not the desired product

plus poly-alkylation products

(b) Gives desired product: activated ring plus 3° carbon in the electrophile gives para as the major product.
(c) Gives desired product plus ortho isomer; use excess bromobenzene to avoid overalkylation.
(d) No reaction; benzamide is too deactivated for a Friedel-Crafts reaction.
(e) Gives desired product: methyl is slightly activating; putting three deactivating nitro groups on a benzene ring requires forcing conditions, and great care is needed to avoid detonation! BOOM!

17-19

(a) Benzene (excess) $\xrightarrow[AlCl_3]{(CH_3)_3CCl}$ $C_6H_5C(CH_3)_3$ $\xrightarrow[H_2SO_4]{HNO_3}$ p-$O_2NC_6H_4C(CH_3)_3$

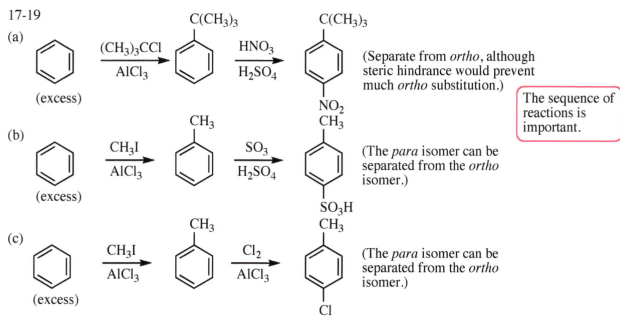

(Separate from *ortho*, although steric hindrance would prevent much *ortho* substitution.)

The sequence of reactions is important.

(b) Benzene (excess) $\xrightarrow[AlCl_3]{CH_3I}$ $C_6H_5CH_3$ $\xrightarrow[H_2SO_4]{SO_3}$ p-$CH_3C_6H_4SO_3H$

(The *para* isomer can be separated from the *ortho* isomer.)

(c) Benzene (excess) $\xrightarrow[AlCl_3]{CH_3I}$ $C_6H_5CH_3$ $\xrightarrow[AlCl_3]{Cl_2}$ p-$CH_3C_6H_4Cl$

(The *para* isomer can be separated from the *ortho* isomer.)

17-20 Friedel-Crafts alkylation is typically followed by a water workup to remove $AlCl_3$ or other catalyst.

(a) Benzene + $Cl-C(=O)-CH_2CH(CH_3)_2$ $\xrightarrow[\text{2) } H_2O \text{ workup}]{\text{1) } AlCl_3}$ $C_6H_5-C(=O)-CH_2CH(CH_3)_2$

(b) Benzene + $Cl-C(=O)-C(CH_3)_3$ $\xrightarrow[\text{2) } H_2O \text{ workup}]{\text{1) } AlCl_3}$ $C_6H_5-C(=O)-C(CH_3)_3$

(c) Benzene + $Cl-C(=O)-C_6H_5$ $\xrightarrow[\text{2) } H_2O \text{ workup}]{\text{1) } AlCl_3}$ $C_6H_5-C(=O)-C_6H_5$

(d) $C_6H_5OCH_3$ $\xrightarrow[AlCl_3/CuCl]{CO, HCl}$ $H_3CO-C_6H_4-CHO$ (para)

Gatterman-Koch

(e) Benzene + $Cl-C(=O)-CH_2CH(CH_3)_2$ $\xrightarrow[\text{2) } H_2O \text{ workup}]{\text{1) } AlCl_3}$ $C_6H_5-C(=O)-CH_2CH(CH_3)_2$ $\xrightarrow[\text{Clemmensen}]{Zn(Hg), HCl}$ $C_6H_5-CH_2CH_2CH(CH_3)_2$

Making a new bond from benzene to a 1° carbon requires Friedel-Crafts acylation followed by reduction.

(f) Benzene + $Cl-C(=O)-C(CH_3)_3$ $\xrightarrow[\text{2) } H_2O \text{ workup}]{\text{1) } AlCl_3}$ $C_6H_5-C(=O)-C(CH_3)_3$ $\xrightarrow[\text{Clemmensen}]{Zn(Hg), HCl}$ $C_6H_5-CH_2C(CH_3)_3$

17-20 continued

(g) Benzene + $Cl-\overset{O}{\overset{\|}{C}}-CH_2CH_2CH_3$ $\xrightarrow[\text{2) } H_2O \text{ workup}]{\text{1) } AlCl_3}$ $C_6H_5-\overset{O}{\overset{\|}{C}}-CH_2CH_2CH_3$ $\xrightarrow[\text{Zn(Hg), HCl, Clemmensen}]{}$ $C_6H_5-CH_2CH_2CH_2CH_3$

(h) Benzene $\xrightarrow[H_2SO_4]{HNO_3}$ $C_6H_5-NO_2$ $\xrightarrow[HCl]{Fe}$ $C_6H_5-NH_2$ $\xrightarrow{\text{"}Ac_2O\text{"}}$ $C_6H_5-NH-C(=O)CH_3$ $\xrightarrow[\text{H}_2\text{O workup}]{CH_3COCl,\ AlCl_3}$ p-$CH_3C(=O)-C_6H_4-NH-C(=O)CH_3$

Reduction of the nitro group is shown first in Section 17-3.

See Solved Problem 17-1 for reaction of the amide group. Another example of this amide reacting by E.A.S. is Problem 17-10(c).

17-21 Another way of asking this question is this: Why is fluoride ion a good leaving group from **A** but not from **B** (either by S_N1 or S_N2)? Nuc = a nucleophile

A (anionic sigma complex: ring carbon bearing F and Nuc, with NO_2 groups ortho and para) **B** (F–C(A)(B)(C) attacked by :Nuc)

Formation of the anionic sigma complex **A** is the rate-determining (slow) step in nucleophilic aromatic substitution. The loss of fluoride ion occurs in a subsequent fast step in which aromaticity is restored, where the nature of the leaving group does not affect the overall reaction rate. In the S_N1 or S_N2 mechanisms, however, the carbon-fluorine bond is breaking in the rate-determining step, so the poor leaving group ability of fluoride does indeed affect the rate.

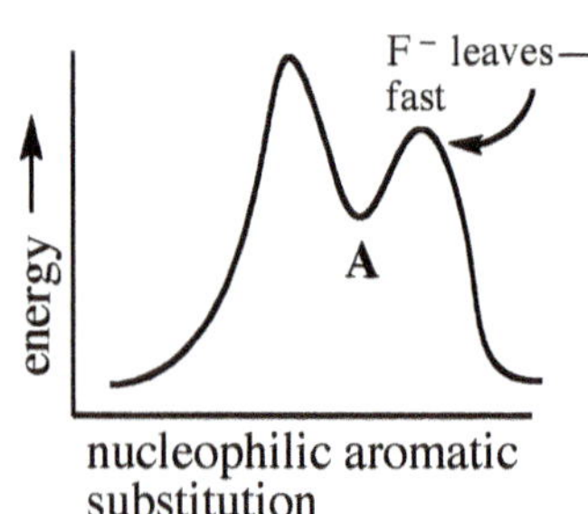

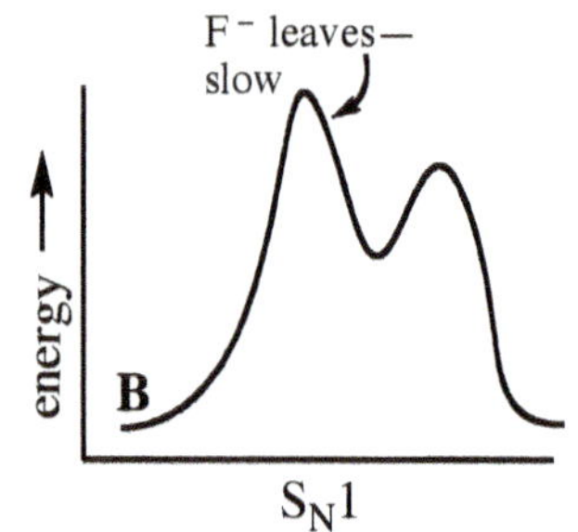

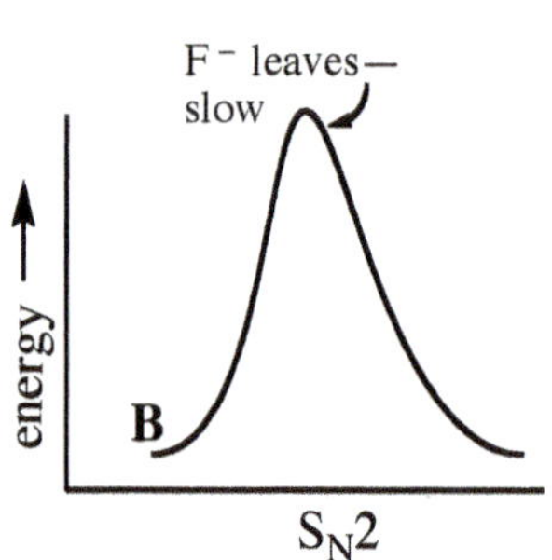

17-22

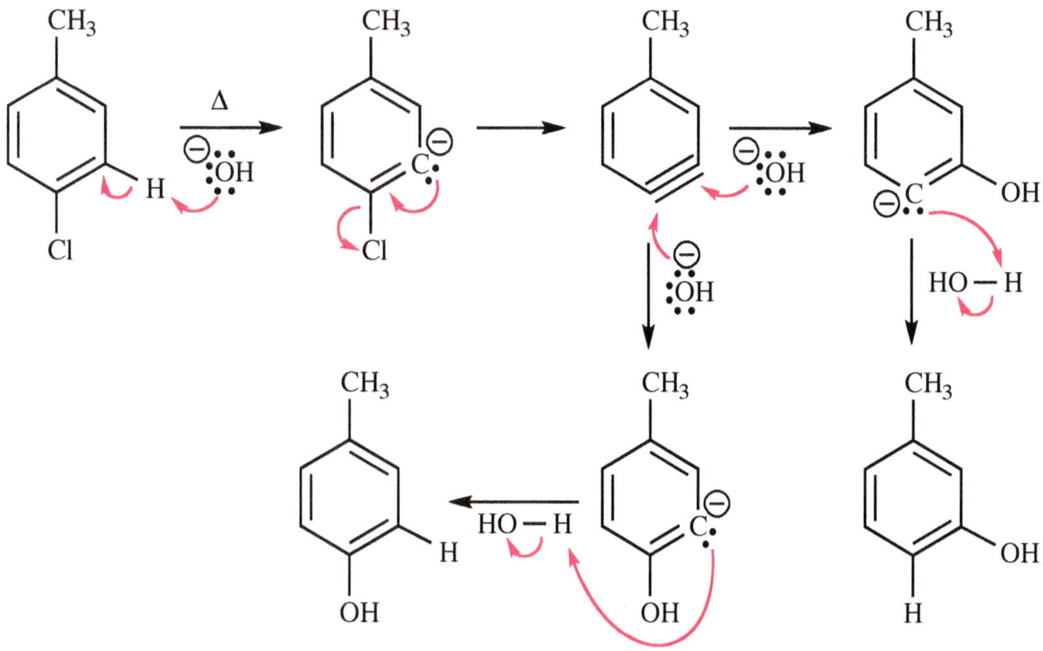

17-23

(a)

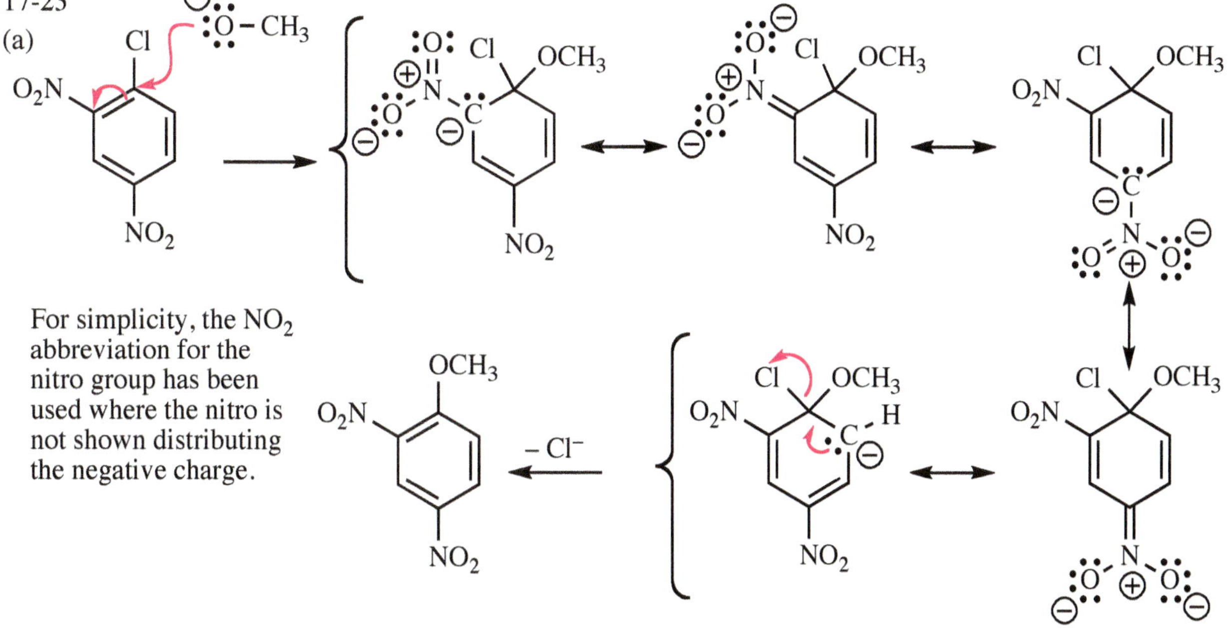

For simplicity, the NO_2 abbreviation for the nitro group has been used where the nitro is not shown distributing the negative charge.

17-23 continued

(b)

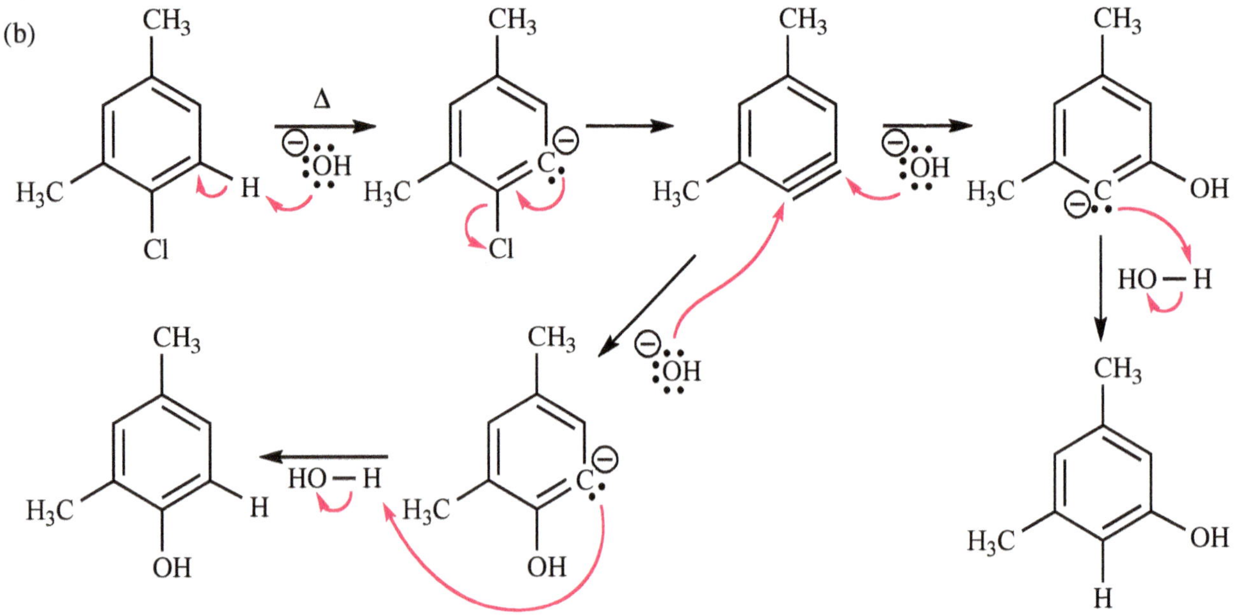

CH3
Δ
OH
H3C
H
Cl
C
HO — H
OH

(c)

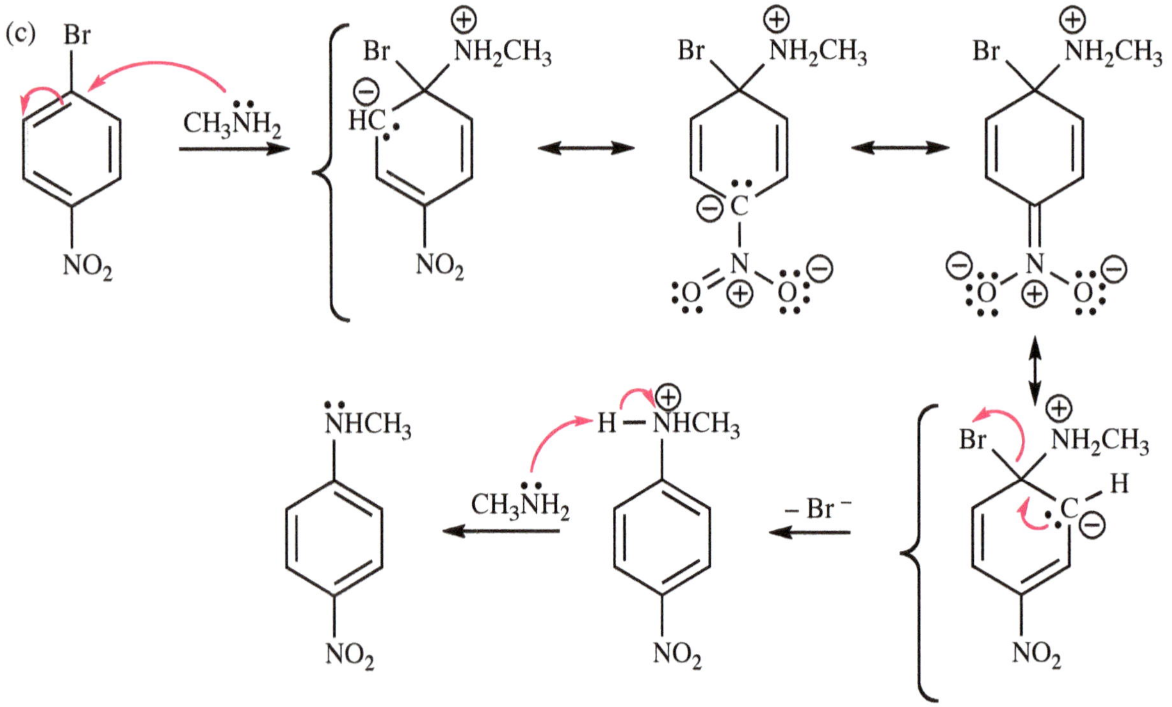

Br
NO2
CH3NH2
NH2CH3
HC
C
N
O
– Br –
H — NHCH3
NHCH3
H

17-23 continued

(d)

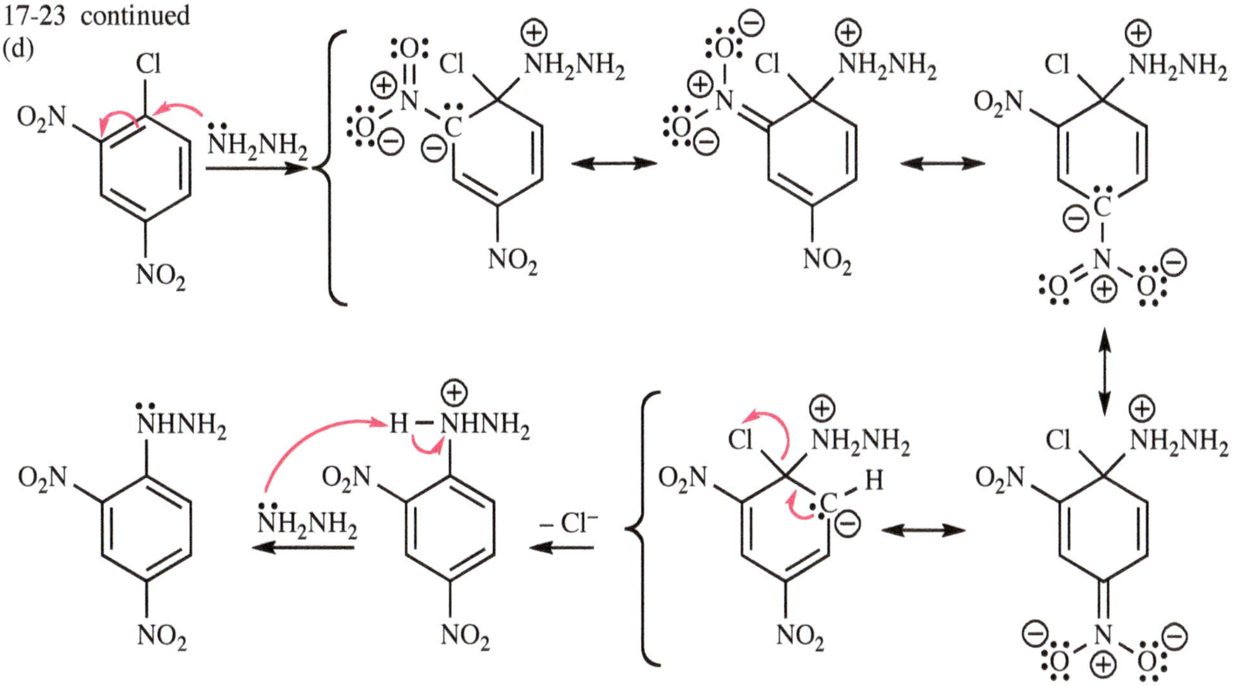

Cl, O_2N, NO_2, $\ddot{N}H_2NH_2$, $\overset{+}{N}H_2NH_2$, O_2N, NO_2, $NHNH_2$, $-Cl^-$, $H-\overset{+}{N}HNH_2$

17-24

from chlorobenzene + NaOH + heat

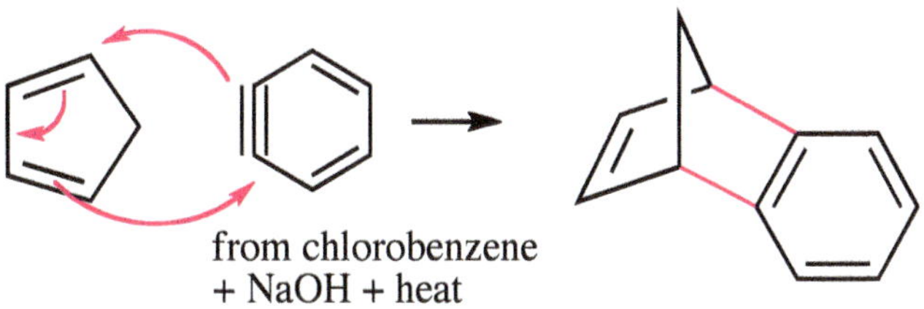

17-25 The new carbon-carbon bonds are shown in bold.

(a) OCH_3, OCH_3

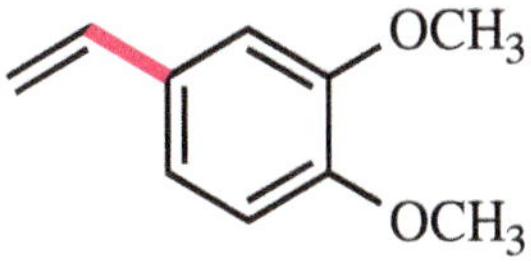

(b) Ph

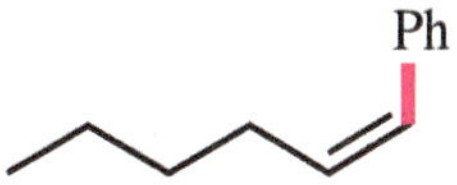

The *cis* stereochemistry of the reactant is retained in the product, a stereospecific reaction.

17-26

(a) $CuLi$

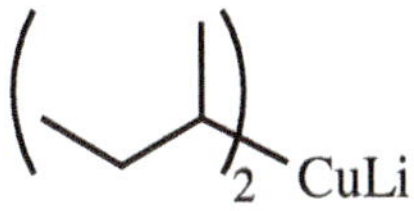

(b) $CuLi$

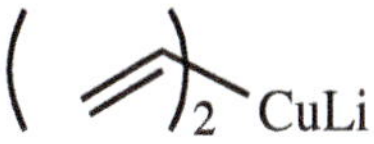

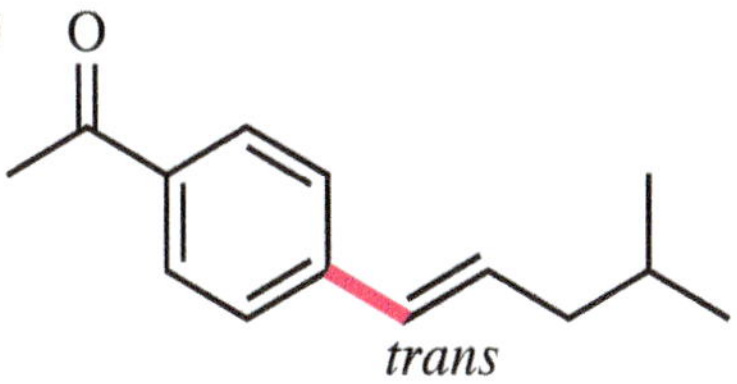

17-27 The new carbon-carbon bonds are shown in bold.

(a) O

trans

The *trans* isomer is the major product in the Heck reaction.

(b) *trans*

OCH_3

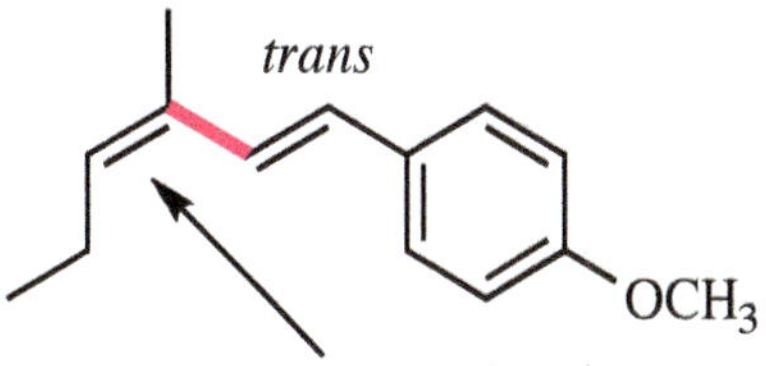

Stereochemistry around this double bond is retained.

17-28 (a) (b) CN

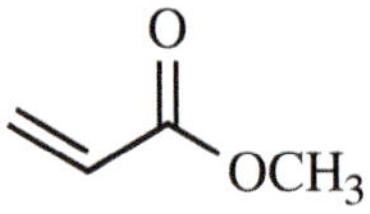

OCH_3

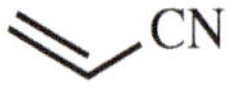

17-29 The new carbon-carbon bonds are shown in bold.

(a) (b)

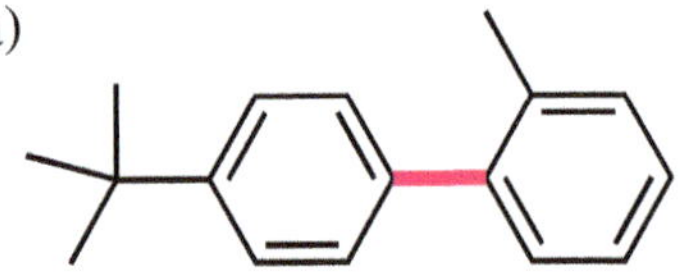

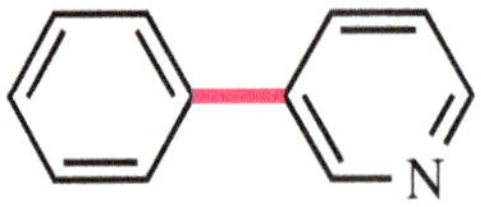

N

17-30 Any palladium catalyst and base can be proposed for these reactions.

(a) OH B OH or ester (b) OH B OH or ester

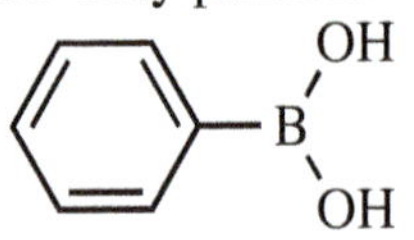

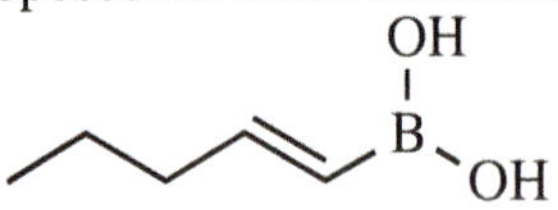

\+ Pd catalyst + a base like NaOH   + Pd catalyst + a base like NaOH

17-31

(a) First, the carboxylic acid proton is neutralized.

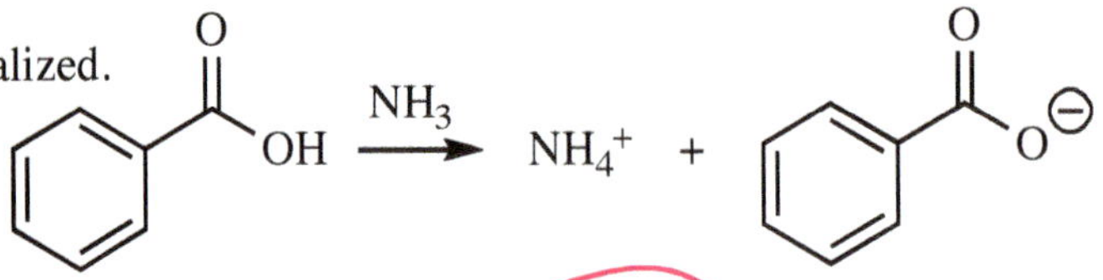

O OH $\xrightarrow{NH_3}$ NH_4^+ + O O

mechanism:

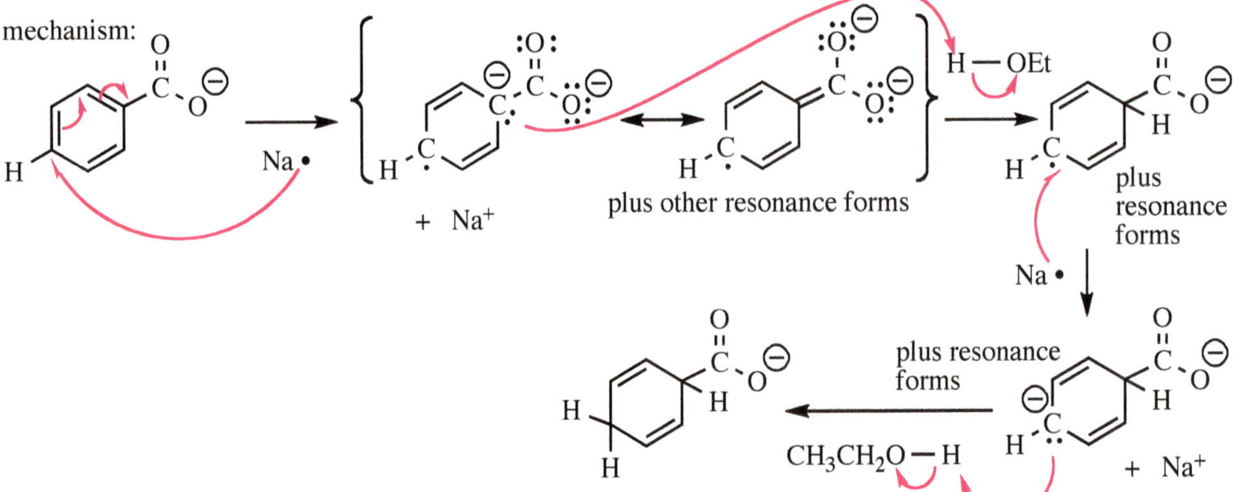

Na •   + Na^+ plus other resonance forms   H—OEt   plus resonance forms

Na •

plus resonance forms $CH_3CH_2O—H$ + Na^+

(b) H OCH$_3$ Li •   + Li^+ H—O-*t*-Bu   Li •   + Li^+ H—O-*t*-Bu

The negative charge avoids the carbon bearing the electron-donating methoxy group.

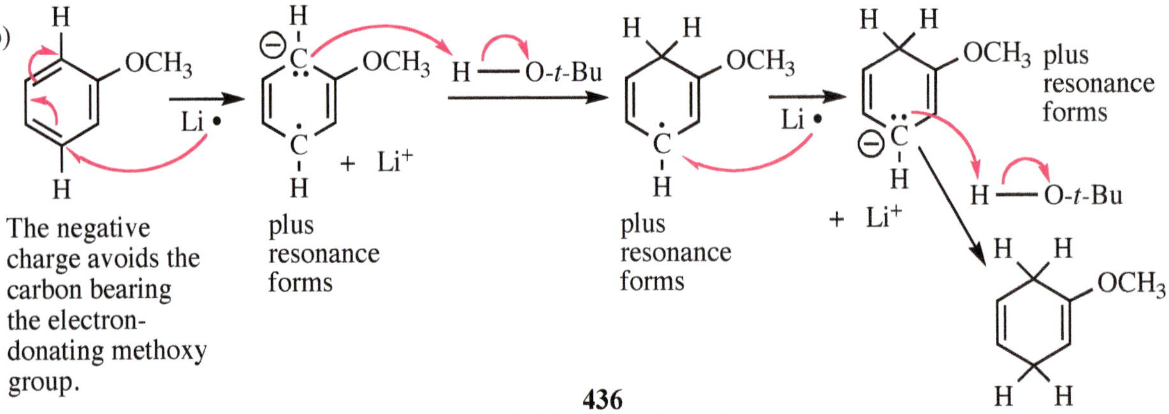

plus resonance forms

plus resonance forms

plus resonance forms

17-32

(a) 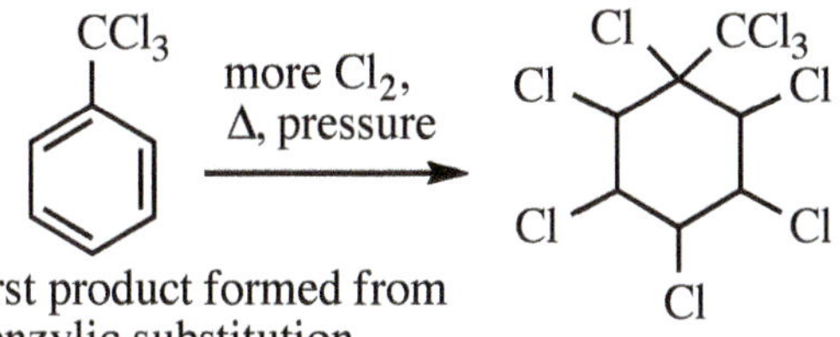 first product formed from benzylic substitution; more Cl_2, Δ, pressure. Addition of chlorine to the π system of the ring gives a mixture of stereoisomers.

(b)

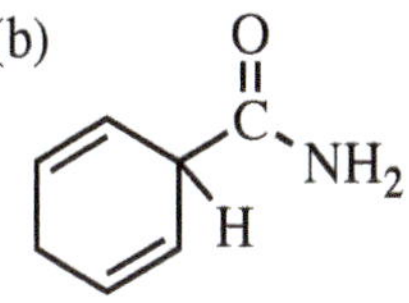

(c) 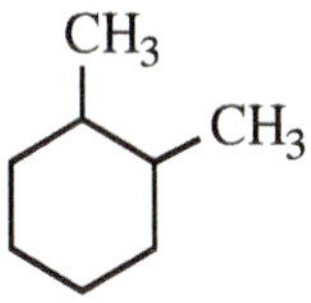 *cis* + *trans*

(d)

(e) 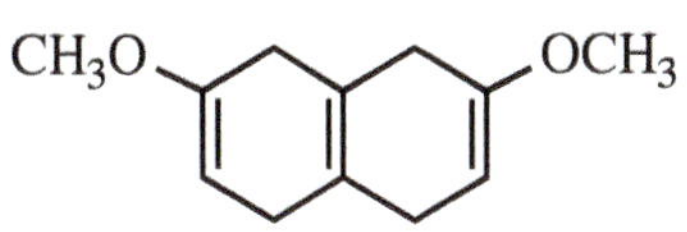 CH_3O … OCH_3

17-33

(a) 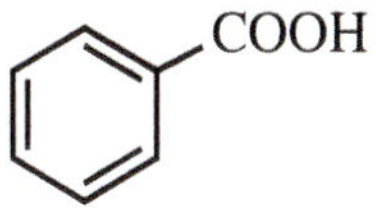 COOH (b) 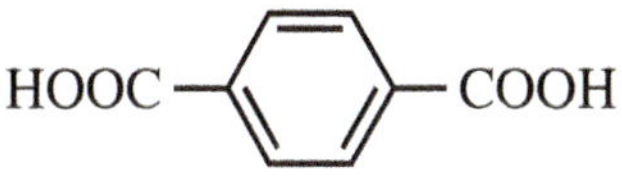 HOOC—⟨benzene⟩—COOH (c) 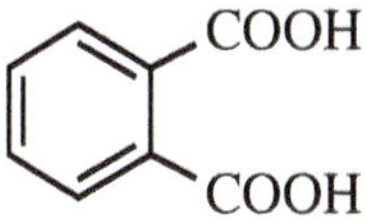 COOH, COOH

17-34 Br—Br $\xrightarrow{h\nu}$ 2 Br• initiation

Br• + ⟨ethylbenzene⟩ → propagation step 1 → resonance-stabilized benzylic radical

Br_2 propagation step 2 → Br• + ⟨1-bromo-1-phenylethane⟩

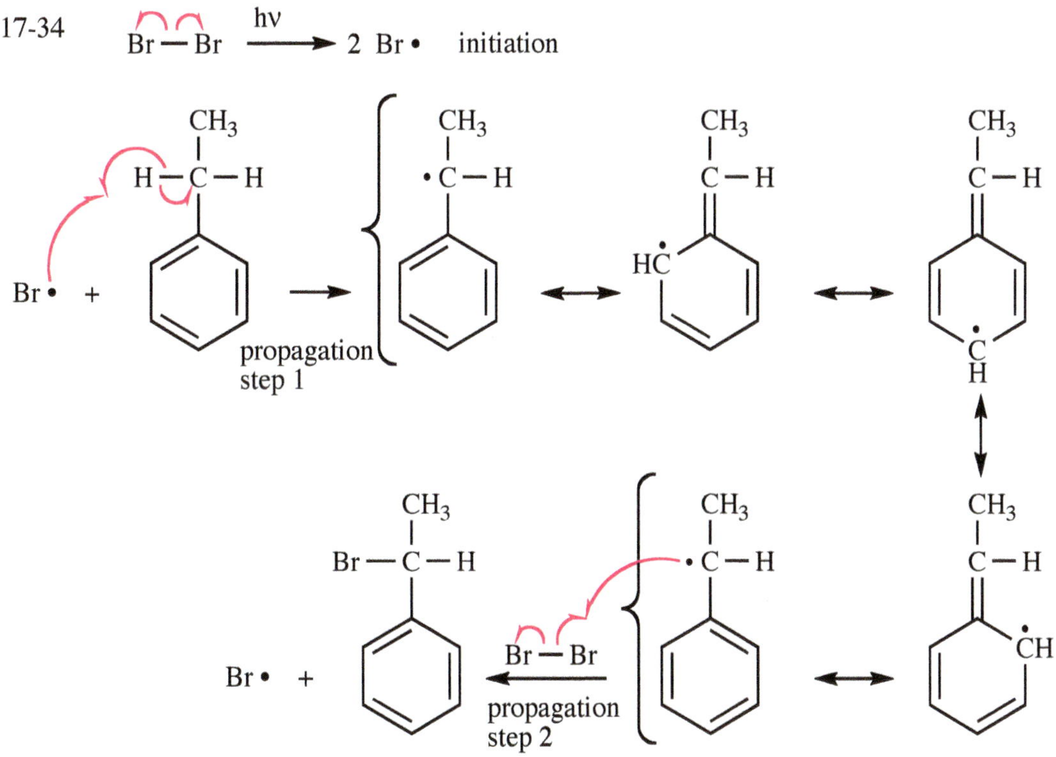

17-35 A statistical mixture would give 2 : 3 or 40% : 60% α to β. To calculate the relative reactivities, the percents must be corrected for the numbers of each type of hydrogen.

α: $\frac{56\%}{2H}$ = 28 relative reactivity β: $\frac{44\%}{3H}$ = 14.7 relative reactivity

The reactivity of α to β is $\frac{28}{14.7}$ = 1.9 to 1

17-36 Replacement of aliphatic hydrogens with bromine can be done under free radical substitution conditions, but reaction at aromatic carbons is unfavorable because of the very high energy of the aryl radical. Benzylic substitution is usually the only product observed.

(a) (1) 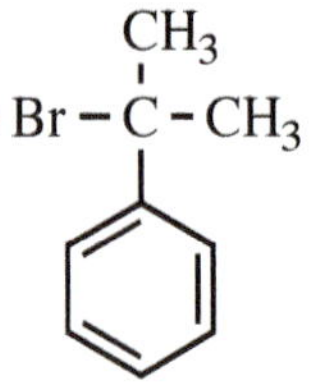 $Br-C(CH_3)_2-C_6H_5$ (2) $Br-C(CBr_3)_2-C_6H_5$

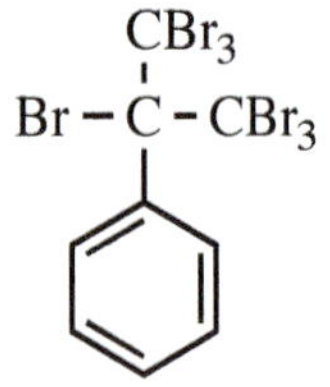

produced under extreme conditions

(b)(1) 1-bromotetralin (2) 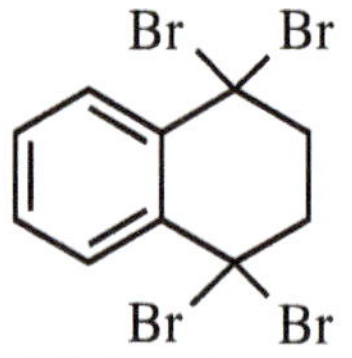 1,1,4,4-tetrabromotetralin

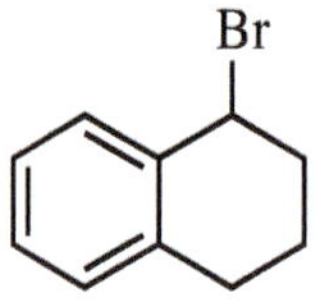

All four benzylic hydrogens are replaced.

17-37

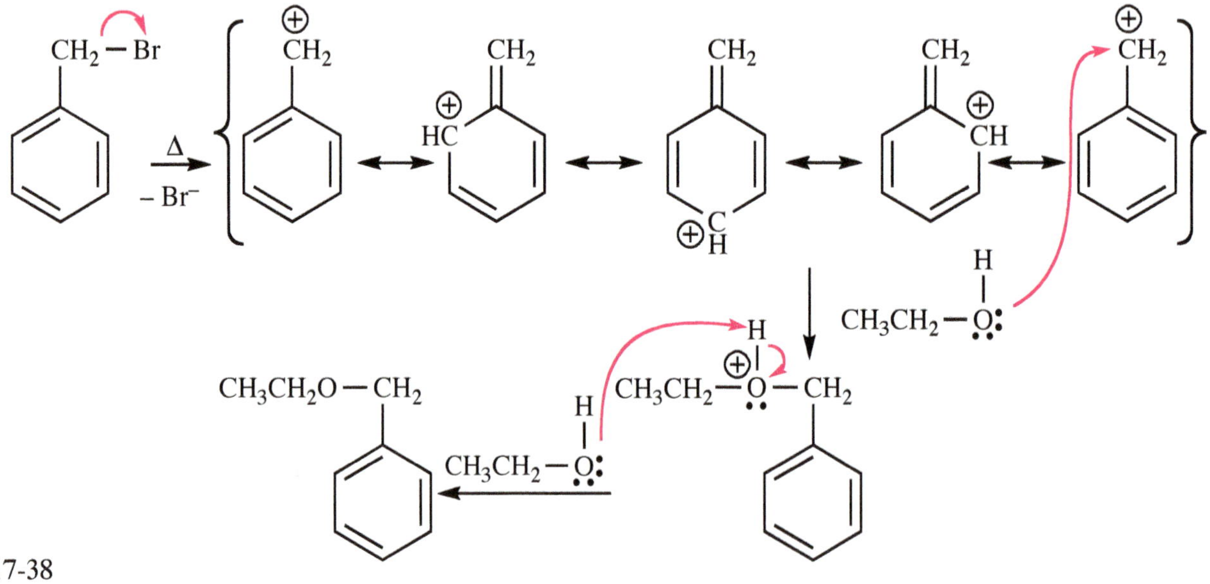

$C_6H_5CH_2-Br \xrightarrow[-\,Br^-]{\Delta}$ {$C_6H_5CH_2^+$ ⟷ ring-cation resonance forms (ortho, para, ortho) ⟷ $C_6H_5CH_2^+$}

$CH_3CH_2-\ddot{O}-H$ attacks → $CH_3CH_2-O^+(H)-CH_2C_6H_5$ → ($CH_3CH_2-\ddot{O}H$ removes H) → $CH_3CH_2O-CH_2C_6H_5$

17-38

(a) Benzylic cations are stabilized by resonance and are much more stable than regular alkyl cations. The product is 1-bromo-1-phenylpropane.

(b)

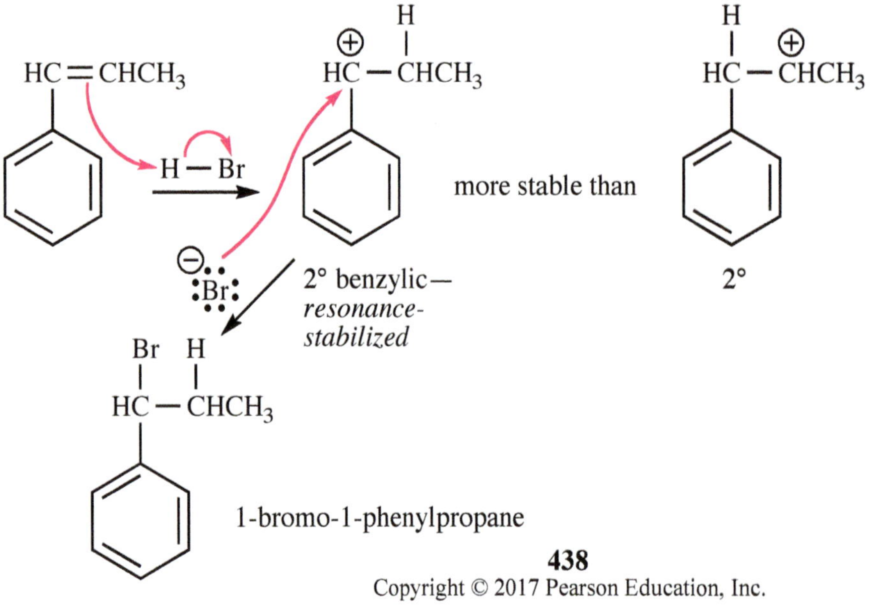

$C_6H_5CH{=}CHCH_3 \xrightarrow{H-Br} C_6H_5\overset{+}{C}H-CH_2CH_3$ more stable than $C_6H_5CH_2-\overset{+}{C}HCH_3$

2° benzylic—*resonance-stabilized* 2°

$:\ddot{Br}:^-$ →

$C_6H_5CH(Br)-CH_2CH_3$

1-bromo-1-phenylpropane

17-38 continued

(c) The combination of HBr with a free-radical initiator generates bromine radicals and leads to anti-Markovnikov orientation. (Recall that whatever species adds *first* to an alkene determines orientation.) The product will be 2-bromo-1-phenylpropane.

(d) Assume the free-radical initiator is a peroxide.

Initiation

RO—OR $\xrightarrow{\Delta}$ **2** RO•

RO• + H—Br ⟶ ROH + Br•

Propagation

HC=CHCH₃ (attached to phenyl) + •Br ⟶ HC•—CH(Br)CH₃ (attached to phenyl)

2° benzylic—*resonance-stabilized*

more stable than HC(Br)—C•HCH₃ (attached to phenyl), 2°

HC•—CH(Br)CH₃ + H—Br ⟶ HC(H)—CH(Br)CH₃ (attached to phenyl) 2-bromo-1-phenylpropane + Br•

(Br• recycles in chain mechanism)

17-39

(a) $C_6H_5CH=CH_2$ $\xrightarrow{HBr}$ $C_6H_5CH(Br)CH_3$ $\xrightarrow[ether]{Mg}$ $C_6H_5CH(MgBr)CH_3$ $\xrightarrow{\text{ethylene oxide}}$ $\xrightarrow{H_3O^+}$ $C_6H_5CH(CH_2CH_2OH)CH_3$

(b) $C_6H_5OCH_3$ $\xrightarrow[AlCl_3]{CH_3CH_2Br}$ p-$CH_3OC_6H_4CH_2CH_3$ + ortho $\xrightarrow[Br_2 \text{ or NBS}]{h\nu}$ p-$CH_3OC_6H_4CH(Br)CH_3$ $\xrightarrow[\Delta]{CH_3OH}$ p-$CH_3OC_6H_4CH(OCH_3)CH_3$

(c) $C_6H_5CH_3$ $\xrightarrow[H_2SO_4]{HNO_3}$ p-$O_2NC_6H_4CH_3$ + ortho $\xrightarrow[Br_2 \text{ or NBS}]{h\nu}$ p-$O_2NC_6H_4CH_2Br$ $\xrightarrow{NaCN}$ p-$O_2NC_6H_4CH_2C\equiv N$

17-40 OH / OH + HO-OH $\xrightarrow{\text{enzyme catalysts}}$ O / O + H-O-H + O-H-H

17-41

(a) OCH_2CH_3, CH_3 (b) O-C(=O)-CH_3, CH_3 (c) OH, Br, CH_3 + OH, Br, CH_3

If excess Br_2 is used, the 2,4,6-tribromo isomer would be produced.

(d) OH, Br, Br, CBr_3, Br (e) O, O, CH_3 (f) OH, $(CH_3)_3C$, CH_3, $C(CH_3)_3$

17-42

(a) O, O (b) H, O, H, O

17-43

(a) SO_3H, Cl

The meta-director went on the ring first; the second substituent had to go to the meta position.

(b) Cl, SO_3H

The ortho,para-director went on the ring first; the second substituent had to go to the ortho or para position; para is usually the major isomer.

The reactions in (a) and (b) were the same; changing the sequence changes the product.

17-44 Two different types of arrows are used in each part to show the directive effects of each group.

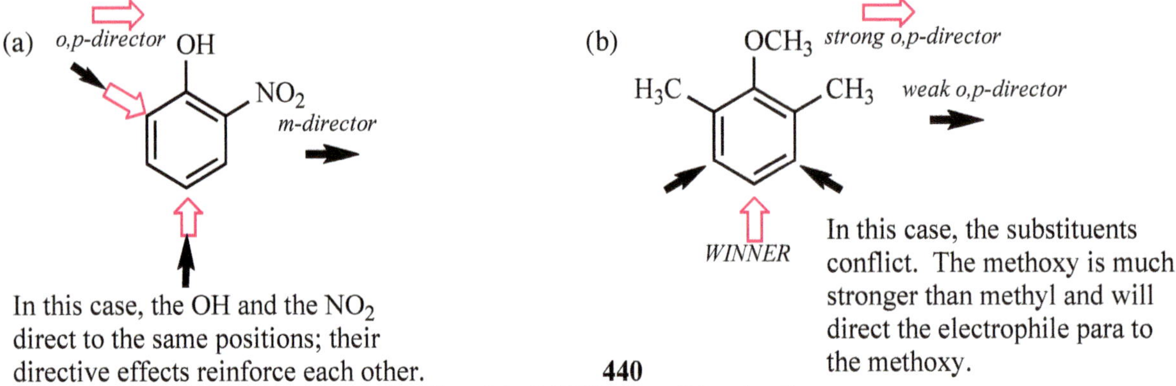

In this case, the OH and the NO_2 direct to the same positions; their directive effects reinforce each other.

In this case, the substituents conflict. The methoxy is much stronger than methyl and will direct the electrophile para to the methoxy.

17-45

(a) 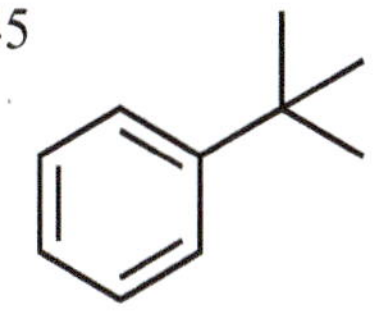 The isobutyl group rearranged to the *tert*-butyl.

(b) 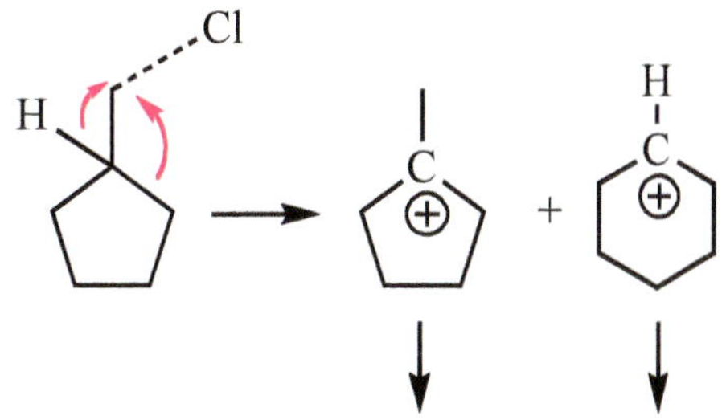 As the Cl is removed by $AlCl_3$, rearrangement occurs by either a hydride shift or a ring expansion. The resulting carbocations will bond to the benzene ring.

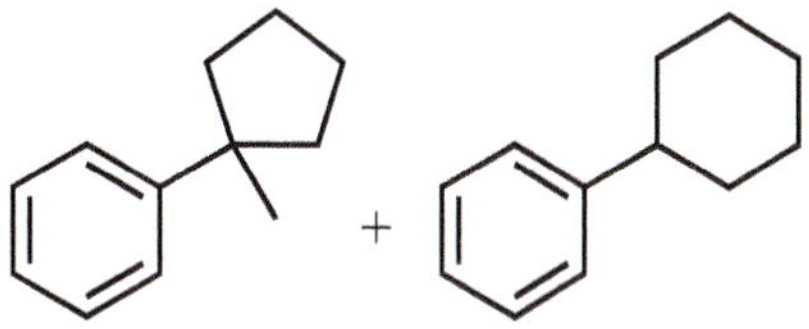

This is probably the major product. Six-membered rings are usually optimum.

17-46

Sadly for the chemist, Friedel-Crafts reactions do not work on deactivated rings, so it is not feasible to add the ethyl group meta to the sulfonic acid group. An alternative has to be used: put an acetyl group (a meta-director) on first, followed by the sulfonic acid, then reduce the acetyl to ethyl. This does not reduce the sulfonic acid (although it would reduce a nitro group to an amine).

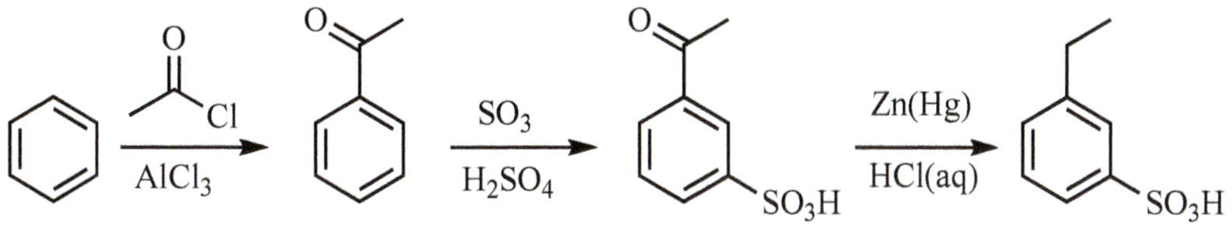

Cl; $AlCl_3$; SO_3, H_2SO_4; SO_3H; Zn(Hg), HCl(aq); SO_3H

17-47

In using the $KMnO_4$ oxidation, timing is important, as is awareness of other substituents: phenols and anilines will be destroyed under these harsh conditions.

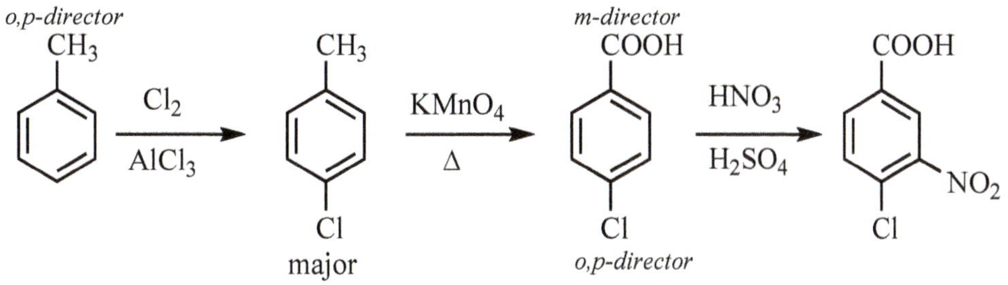

o,p-director CH_3 → Cl_2, $AlCl_3$ → CH_3, Cl, major → $KMnO_4$, Δ → *m-director* COOH, Cl, *o,p-director* → HNO_3, H_2SO_4 → COOH, NO_2, Cl

17-48

CH_3 → SO_3, H_2SO_4 → CH_3, SO_3H → Cl_2, $AlCl_3$ → CH_3, Cl, SO_3H → H_2O, Δ → CH_3, Cl → $KMnO_4$, Δ → COOH, Cl **ORTHO**

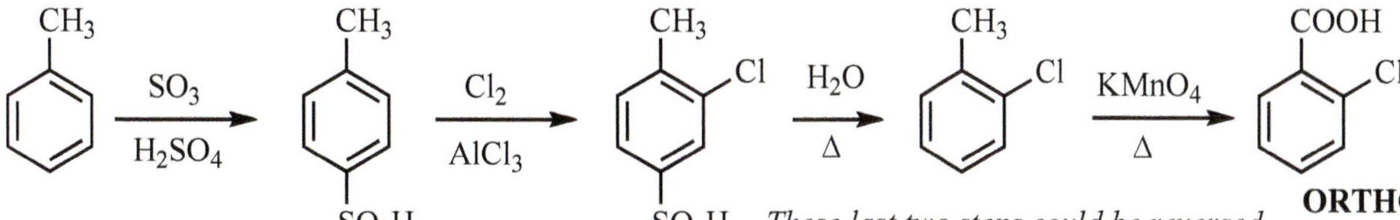

These last two steps could be reversed.

CH_3 → $KMnO_4$, Δ → COOH → Cl_2, $AlCl_3$ → COOH, Cl **META**

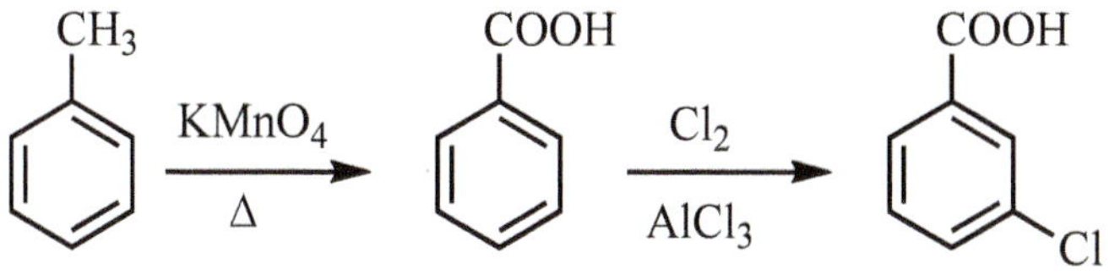

Synthesis of para on the next page.

17-48 continued

CH_3 → (Cl_2, $AlCl_3$) → CH_3 / Cl (major) → ($KMnO_4$, Δ) → COOH / Cl **PARA** Sequence of reactions is critical!

17-49 Analyze the target molecule and devise a strategy.

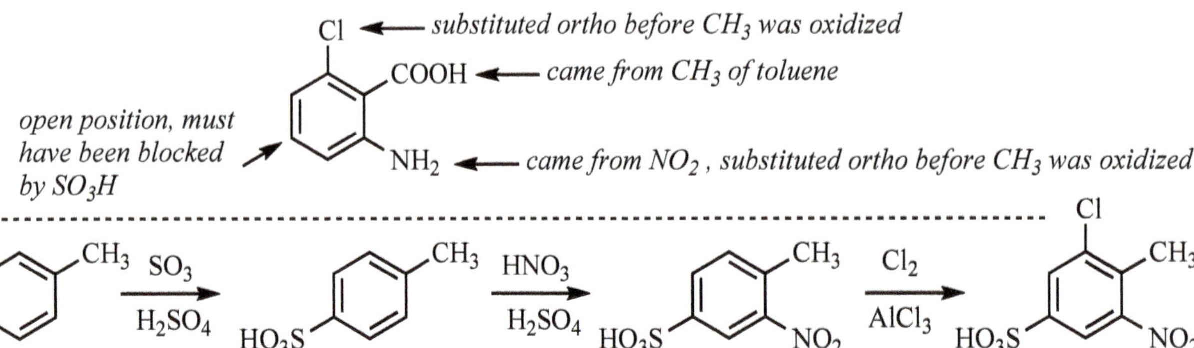

Better to put on NO_2 before Cl so that all groups direct to the correct position.

H_2O | Δ

The NH_2 is the most sensitive group on this molecule; ideally, it should be created last. It would also be OK to keep the SO_3H until the very last step, and remove it after the reduction of the NO_2.

Cl, COOH, NH_2 ← (Fe, HCl(aq)) ← Cl, COOH, NO_2 ← ($KMnO_4$, Δ) ← Cl, CH_3, NO_2

Removal of SO_3H and oxidation steps could be reversed.

17-50

(a) $C(CH_3)_3$ (b) $CH_3-CHCH_2CH_3$ (c) $C(CH_3)_3$ (d) Br (e) $C(CH_3)_3$ (f) SO_3H

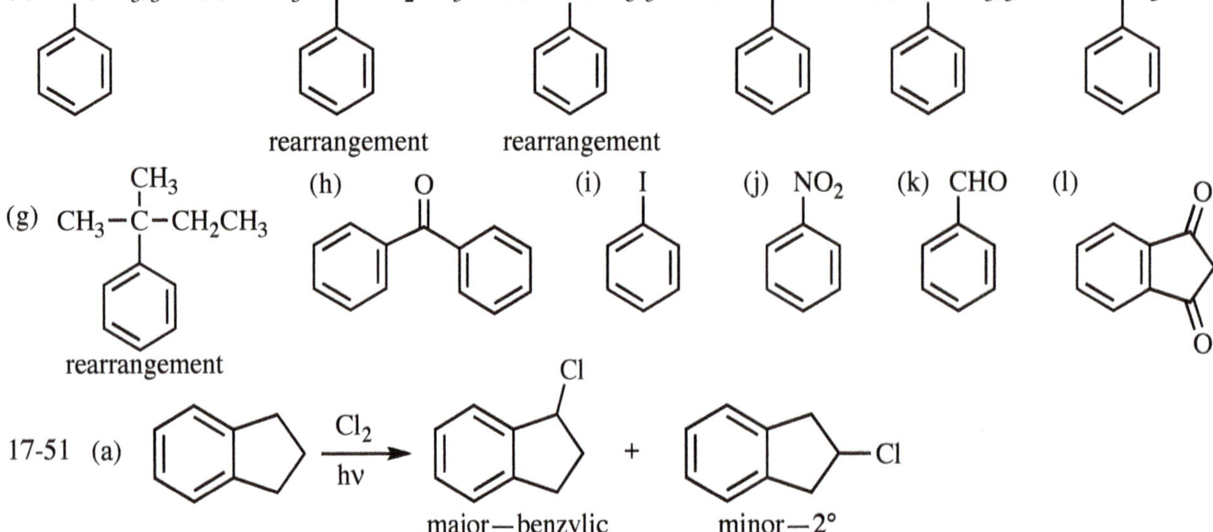

17-51 continued

(b) indane + excess Cl_2, hν → **1** + **2** + **3** + **4**

(c) Gas chromatography will be the optimum technique for separating the six compounds. By itself, GC does not give any structural information, so the more expensive GC instruments have a mass spectrometer as the detector. As the peaks come out of the GC, they go directly into the MS, so a mass spectrum is collected for each GC peak. GC-MS will distinguish the monochloro from the dichloro isomers, but it will not easily identify the specific monochloro or dichloro isomers. (A good GC could also separate *cis-trans* isomers of **1** and **2**.)

(d) The two types of NMR would easily distinguish these four dichloro isomers (ignoring geometric isomers). In HNMR, a Cl—C—H would appear ≈ δ 4 if not benzylic and ≈ δ 5 if benzylic. In CNMR, a C—Cl will appear ≈ δ 65 and a CCl_2 will appear ≈ δ 80; benzylic C appears ≈ δ 40, and nonbenzylic CH_2 comes ≈ δ 20.

structure	HNMR	CNMR
1	2H ≈ δ 5	1 tall peak ≈ δ 65, 1 medium peak ≈ δ 20
2	1H ≈ δ 4, 1H ≈ δ 5	2 peaks ≈ δ 65, 1 medium peak ≈ δ 40
3	2 signals < δ 4	1 peak ≈ δ 80, 1 peak ≈ δ 40, 1 peak ≈ δ 20
4	1 singlet ≈ δ 2.5-3.0	1 peak ≈ δ 80, 1 tall peak ≈ δ 40

17-52

(a) benzene + butanoyl chloride, $AlCl_3$ → butyrophenone; Zn(Hg), HCl → butylbenzene; NBS, hν → 1-bromo-1-phenylbutane

(b) 1-bromo-1-phenylbutane (from (a)); CH_3OH, Δ → 1-methoxy-1-phenylbutane (OCH_3)

(c) benzene + $CH_2{=}CH{-}CH_2Cl$, $AlCl_3$ → allylbenzene; 1) $BH_3 \cdot THF$ 2) H_2O_2, HO^- → 3-phenyl-1-propanol (OH)

OR toluene (CH_3); Br_2, hν or NBS, hν → benzyl bromide (CH_2Br); Mg, ether; ethylene oxide; H_3O^+ → 3-phenyl-1-propanol (OH). New bond in bold.

(d) benzene; Cl_2, $AlCl_3$ → chlorobenzene (Cl); NaOH, 350 °C → phenol (OH); NaOH, CH_3CH_2Br → ethoxybenzene (OCH_2CH_3)

17-52 continued

(e) Benzene $\xrightarrow[AlCl_3]{Cl_2}$ chlorobenzene $\xrightarrow[H_2SO_4]{HNO_3}$ 1-chloro-4-nitrobenzene $\xrightarrow[AlCl_3]{Cl_2}$ 1,2-dichloro-4-nitrobenzene

(f) three methods

Allylbenzene (from (c)) $\xrightarrow[H_2O]{Hg(OAc)_2}$ $\xrightarrow{NaBH_4}$ 1-phenyl-2-propanol (OH)

OR $C_6H_5CH_2MgBr$ (from (c)) + $CH_3-\overset{O}{\overset{||}{C}}H$ $\rightarrow$ $\xrightarrow{H^+}$ 1-phenyl-2-propanol

OR Chlorobenzene (from (d)) $\xrightarrow[ether]{Mg}$ + propylene oxide $\rightarrow$ $\xrightarrow{H_3O^+}$ 1-phenyl-2-propanol

(g) Toluene (CH_3) $\xrightarrow[H_2SO_4]{HNO_3}$ p-nitrotoluene (CH_3, NO_2) $\xrightarrow[\Delta]{KMnO_4}$ p-nitrobenzoic acid ($COOH$, NO_2) $\xrightarrow[HCl(aq)]{\text{Fe or Sn or Zn}}$ p-aminobenzoic acid ($COOH$, NH_2)

(h) two possible methods

Toluene (CH_3) $\xrightarrow[\text{or NBS, hv}]{Br_2\text{, hv}}$ $C_6H_5CH_2Br$ $\xrightarrow[ether]{Mg}$ $C_6H_5CH_2MgBr$ + 2-butanone (O) $\rightarrow$ $\xrightarrow{H_2O}$ 2-methyl-1-phenyl-2-butanol (OH)

New bond in bold.

OR

Benzene $\xrightarrow[AlCl_3]{Br_2}$ bromobenzene (Br) $\xrightarrow[ether]{Mg}$ C_6H_5MgBr + 2-methyl-1,2-epoxybutane (O) $\rightarrow$ $\xrightarrow{H_2O}$ 2-methyl-1-phenyl-2-butanol (OH)

New bond in bold.

(i) Toluene (CH_3) $\xrightarrow[AlCl_3]{Cl_2}$ 4-chlorotoluene (Cl, CH_3; major) $\xrightarrow[H_2SO_4]{HNO_3}$ 4-chloro-2-nitrotoluene (NO_2, CH_3, Cl; major) $\xrightarrow[HCl(aq)]{\text{Fe or Sn or Zn}}$ 5-chloro-2-methylaniline (NH_2, CH_3, Cl)

(j) Toluene (CH_3) $\xrightarrow[AlCl_3]{Br_2}$ p-bromotoluene (CH_3, Br) $\xrightarrow[\Delta]{KMnO_4}$ p-bromobenzoic acid ($COOH$, Br) $\xrightarrow[H_2SO_4]{HNO_3}$ 4-bromo-3-nitrobenzoic acid ($COOH$, NO_2, Br)

17-52 continued

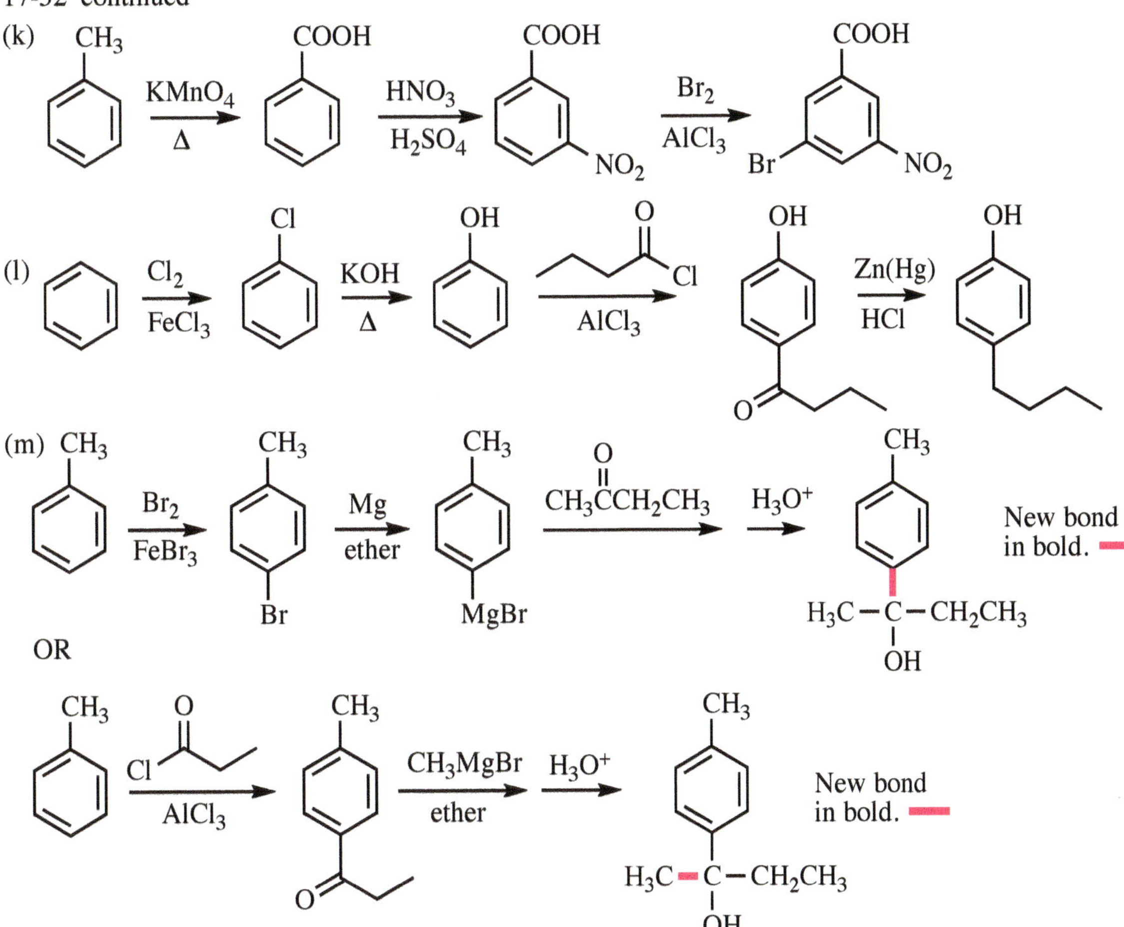

17-53 (a) OCH_3, NO_2, NO_2 (b) OH, $C(CH_3)_3$ (c) NO_2, SO_3H

(d) No reaction: Friedel-Crafts acylation does not occur on rings with strong deactivating groups like NO_2.

(e) CH_3O, $\overset{O}{\overset{||}{C}}$ CH_3, CH_3 (f) OCH_3, CH_2Br or OCH_3, Br, Br, CH_2Br (g) NH_2, Cl, NO_2 (h) CH_3, NH_2 Basic workup isolates free amine.

(i) CH_2CH_3 (j) SO_3H, NO_2, CH_2CH_3 (k) $Ph-\overset{O}{\overset{||}{C}}-\underset{H}{N}-C_6H_4-\overset{O}{\overset{||}{C}}CH_2CH_3$ (l) COOH, COOH

17-54 Major products are shown. Other isomers are possible.

(a) NO_2 activated; Br; deactivated; OCH_3

(b) deactivated; activated; Br; O; O

(c) $COOCH_3$; OCH_3; activated; Br

17-55 The new carbon-carbon bonds are shown in bold.

(a)

(b) O; OH; trans; O

(c) cis—retention of stereochemistry

(d) trans

(e) NO_2; trans

17-56

$$\text{C}_6\text{H}_5\text{-CH}_2\text{CH}_2\text{COOH} \xrightarrow[\Delta]{AlCl_3} \text{product}$$

starting material
molecular weight 150

molecular weight 132 =
loss of 18 = loss of H_2O

IR spectrum: The dominant peak is the carbonyl at 1710 cm^{-1}. No COOH stretch.

NMR spectrum: The splitting is complicated but the integration is helpful. In the region of δ 2.6–3.2, there are two signals, each with an integration value of 2H; these must be the two adjacent methylenes, CH_2CH_2 . The aromatic region from δ 7.3 to 7.8 has integration of 4H, so the ring must be disubstituted.

Carbon NMR: Of the six aromatic signals, four are C—H and two are C indicating a disubstituted benzene. Also indicated are carbons in a carbonyl and two methylenes.

H; lose; O; $CH_2CH_2—C—OH$; lose 1 H; retain ⟹ O; C; CH_2; CH_2; molecular weight 132; or; O; common name: indanone

The product must be the cyclized ketone, formed in an intramolecular Friedel-Crafts acylation.

17-57 O; **A**; O; **B**; NO_2; **C**; COOH; **D**

continued on next page

17-57 continued

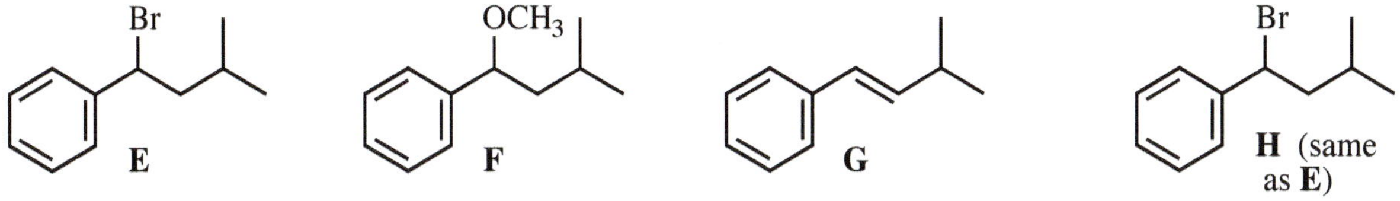

Br OCH$_3$ Br

E F G H (same as E)

17-58 Cl Cl

HO: + H Δ – Cl$^-$

a substituted benzyne

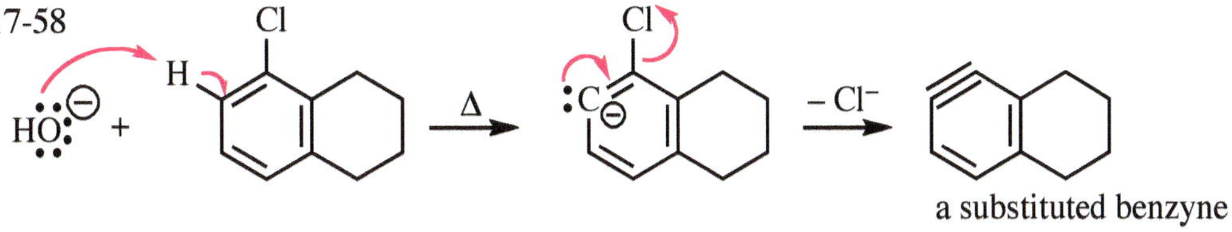

Attack A

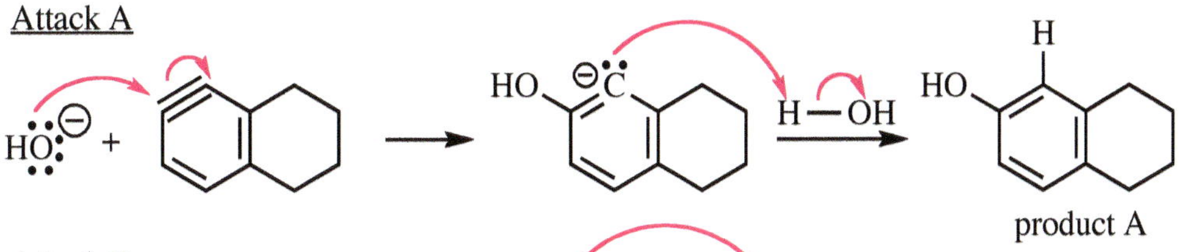

HO: + HO H—OH HO H

product A

Attack B

HO: + OH H—OH H OH

product B

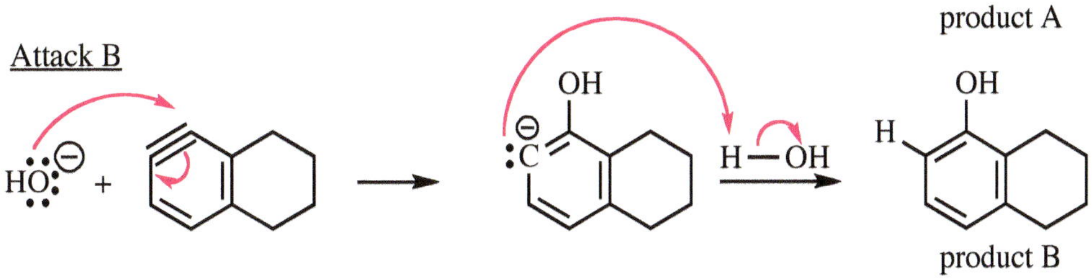

17-59 The electron-withdrawing carbonyl group stabilizes the adjacent negative charge.

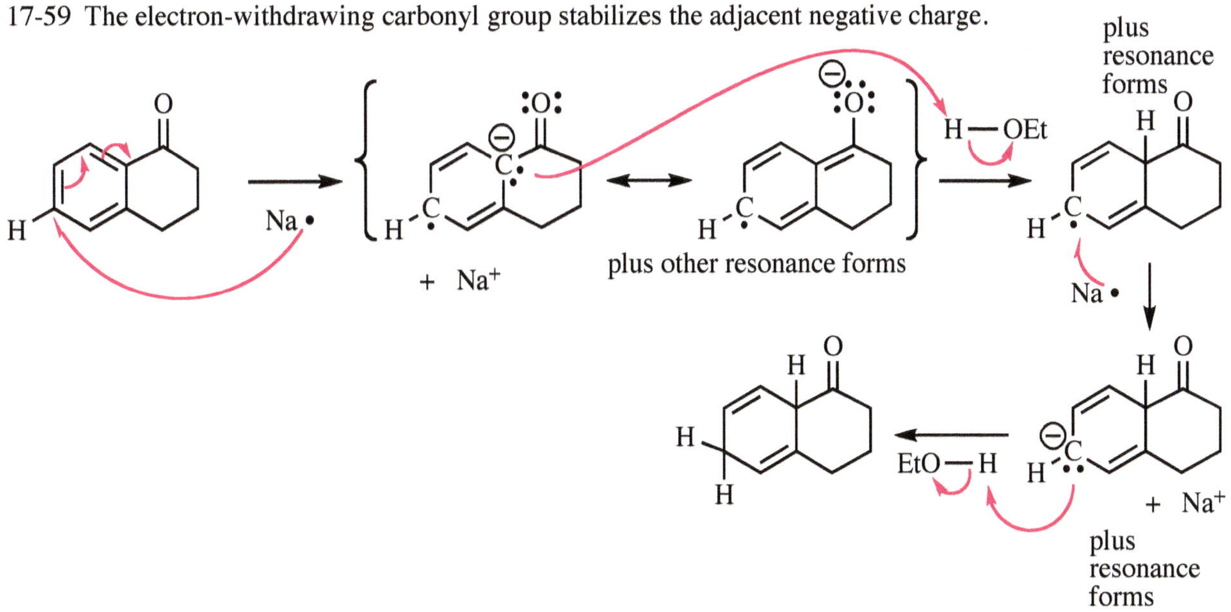

O O O O H

H Na• H + Na$^+$ H plus other resonance forms H—OEt H plus resonance forms

Na•

H O H O

H EtO—H H + Na$^+$

H plus resonance forms

17-60

(a) 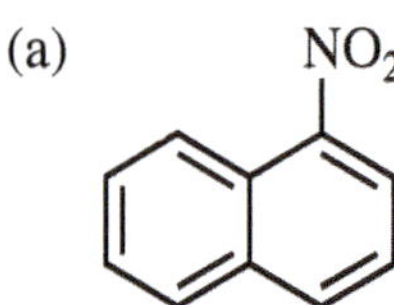

(b) 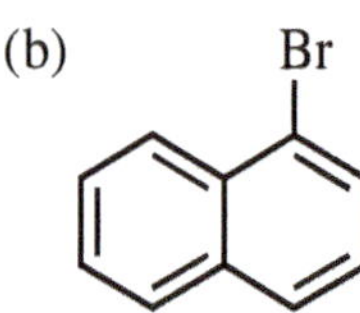

(c) [structure: 1-naphthyl ethyl ketone]

(d) $C(CH_3)_3$

(e)

(f) SO_3H

17-61 Assume an acidic workup to each of these reactions to produce the phenol, not the phenoxide ion.

(a) benzene $\xrightarrow[AlCl_3]{Cl_2}$ chlorobenzene $\xrightarrow[H_2SO_4]{HNO_3}$ p-chloronitrobenzene (+ ortho) $\xrightarrow[\Delta]{NaOH}$ p-nitrophenol — from addition-elimination mechanism; only this isomer

(b) benzene $\xrightarrow[AlCl_3]{Cl_2}$ chlorobenzene $\xrightarrow[350° C]{NaOH}$ phenol (OH strongly activated; no catalyst needed) $\xrightarrow{3\ Br_2}$ 2,4,6-tribromophenol

(c) benzene $\xrightarrow[AlCl_3]{2\ Cl_2}$ p-dichlorobenzene (+ ortho) $\xrightarrow[\Delta]{NaOH}$ p-chlorophenol + m-chlorophenol via benzyne mechanism

(d) toluene $\xrightarrow[AlCl_3]{Cl_2}$ p-chlorotoluene (+ ortho) $\xrightarrow[\Delta]{NaOH}$ m-cresol + p-cresol via benzyne mechanism

17-61 continued

(e)

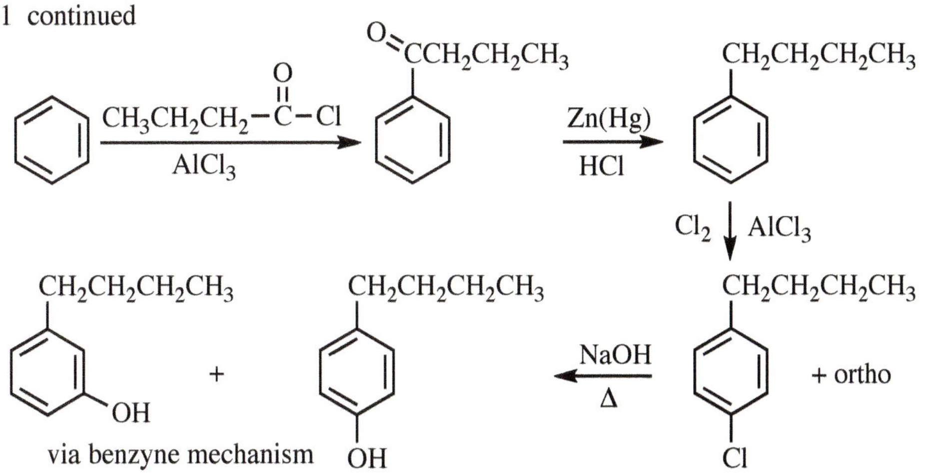

Benzene + $CH_3CH_2CH_2-\overset{O}{\overset{\|}{C}}-Cl$ ($AlCl_3$) → $C_6H_5-C(=O)CH_2CH_2CH_3$ → (Zn(Hg), HCl) → $C_6H_5-CH_2CH_2CH_2CH_3$

(Cl_2, $AlCl_3$) → p-chloro $CH_2CH_2CH_2CH_3$ benzene (Cl) + ortho

(NaOH, Δ) → m-($CH_2CH_2CH_2CH_3$)phenol (OH) + p-($CH_2CH_2CH_2CH_3$)phenol (OH)

via benzyne mechanism

17-62 Professor Suzuki's research group carried out this synthesis of bombykol in 1983.

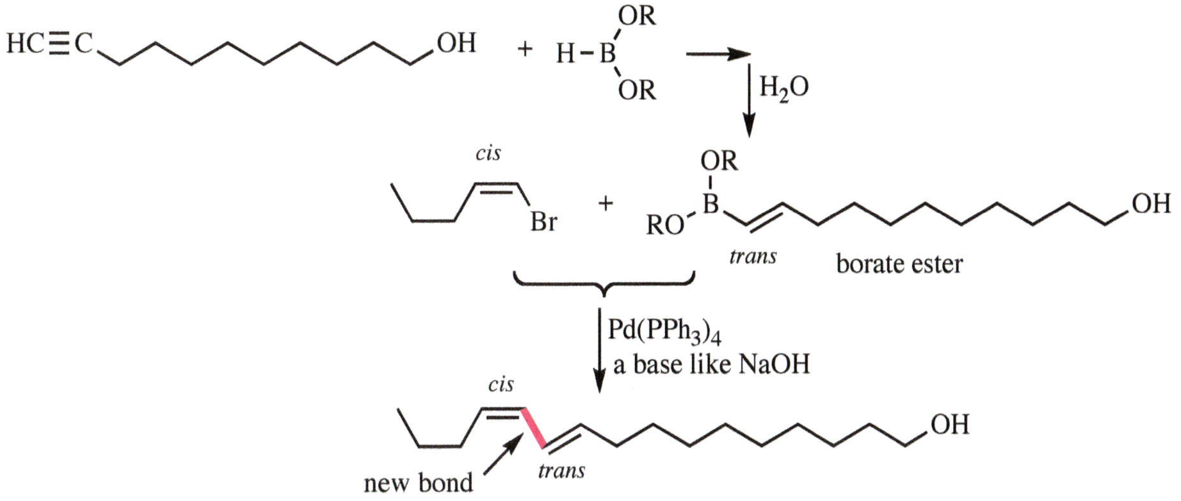

HC≡C–(chain)–OH + H–B(OR)(OR) → (H_2O) → RO–B(OR)–(trans chain)–OH (borate ester)

cis alkenyl bromide (Br) + RO–B(OR)–*trans* borate ester

↓ $Pd(PPh_3)_4$, a base like NaOH

cis, *trans* diene–OH

new bond

17-63

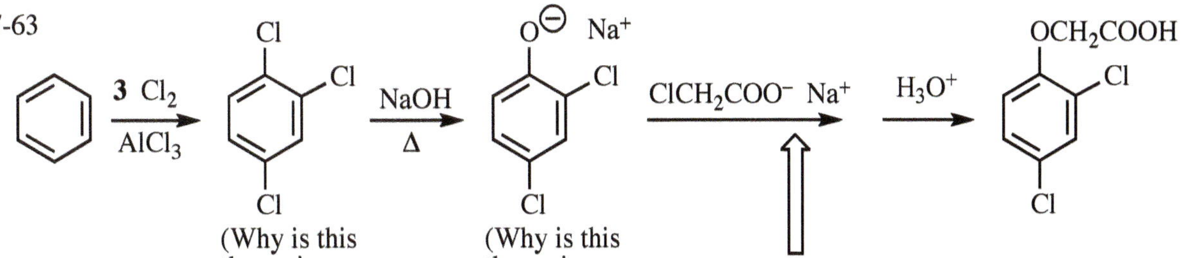

Benzene ($\mathbf{3}$ Cl_2, $AlCl_3$) → 1,2,4-trichlorobenzene (Cl, Cl, Cl)

(Why is this the major isomer?)

(NaOH, Δ) → $O^{\ominus}$ Na^+ 2,4-dichlorophenoxide (Cl, Cl)

(Why is this the major isomer?)

($ClCH_2COO^-$ Na^+) → (H_3O^+) → OCH_2COOH 2,4-dichloro (Cl, Cl)

The sodium salt of the carboxylic acid is typically used in substitutions under basic conditions to prevent the COOH from neutralizing, and thereby deactivating, the phenoxide nucleophile.

17-64

(a)

bromination at C-2

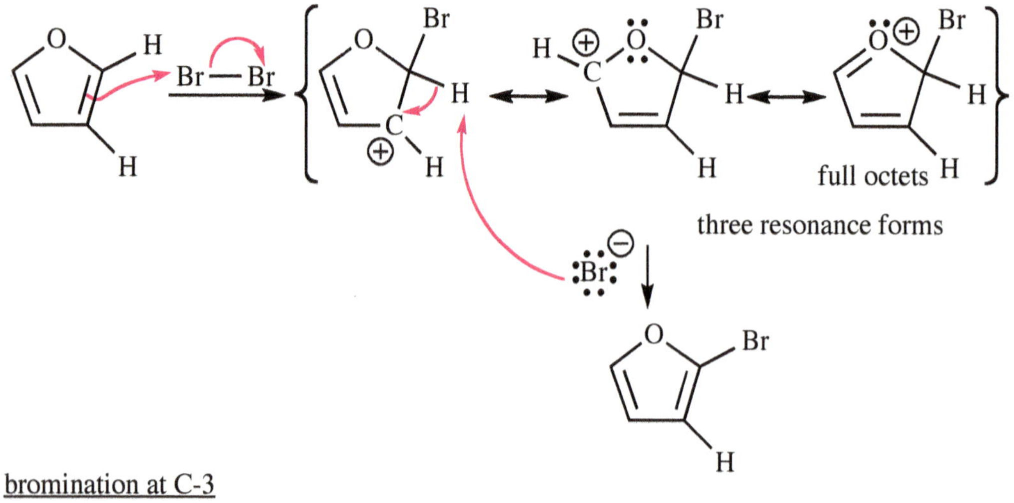

[Reaction scheme: furan + Br—Br → {three resonance forms of the C-2 intermediate} → 2-bromofuran (via :Br:⁻ removing H)]

Br, H, full octets

three resonance forms

bromination at C-3

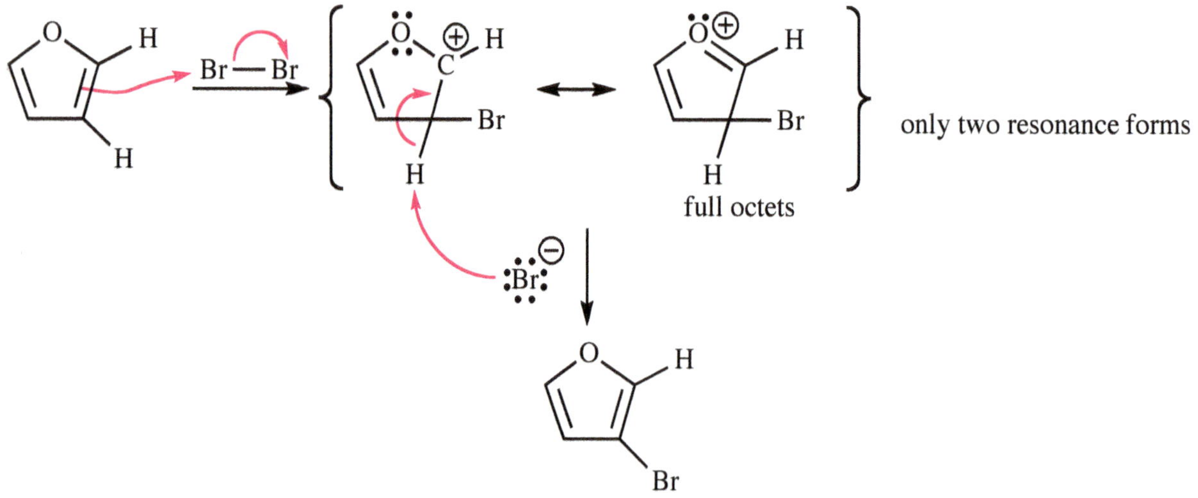

[Reaction scheme: furan + Br—Br → {two resonance forms of the C-3 intermediate} → 3-bromofuran (via :Br:⁻ removing H)]

full octets

only two resonance forms

(b) Attack at C-2 gives an intermediate stabilized by three resonance forms, as opposed to only two resonance forms stabilizing attack at C-3. Bromination at C-2 will occur more readily than at C-3, and both will be faster than benzene.

17-65 Benzenedicarboxylic acid isomers have the common name of phthalic acid isomers.

(a)

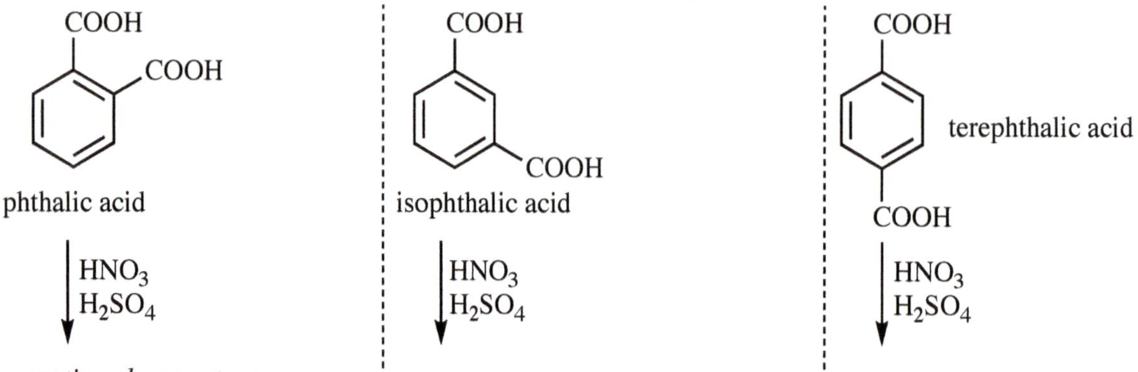

COOH / COOH (ortho)	COOH / COOH (meta)	COOH / COOH (para)
phthalic acid	isophthalic acid	terephthalic acid
↓ HNO_3, H_2SO_4	↓ HNO_3, H_2SO_4	↓ HNO_3, H_2SO_4

continued on next page

17-65 continued

(b)

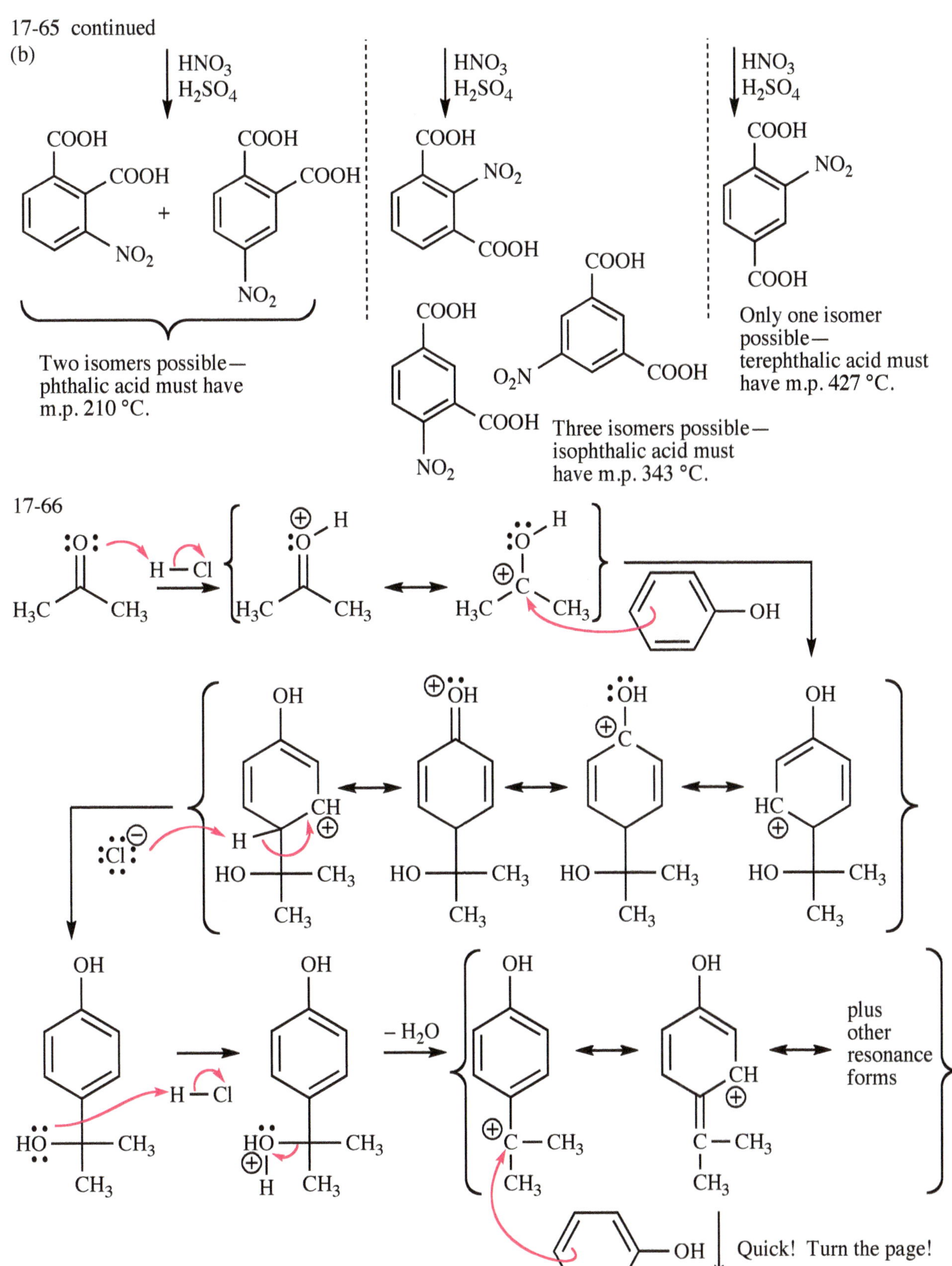
HNO3
H2SO4
COOH
COOH
NO2
+
Two isomers possible—
phthalic acid must have
m.p. 210 °C.
Three isomers possible—
isophthalic acid must
have m.p. 343 °C.
Only one isomer
possible—
terephthalic acid must
have m.p. 427 °C.
17-66
H—Cl
H3C
CH3
OH
HO
CH
HC
Cl
H
– H2O
plus
other
resonance
forms
Quick! Turn the page!

17-66 continued

You may have heard about bisphenol A (BPA) in the news. There is some concern about the presence of BPA in products that are used for babies, like milk bottles, nipples, and pacifiers. Several states have passed legislation to restrict BPA in such products.

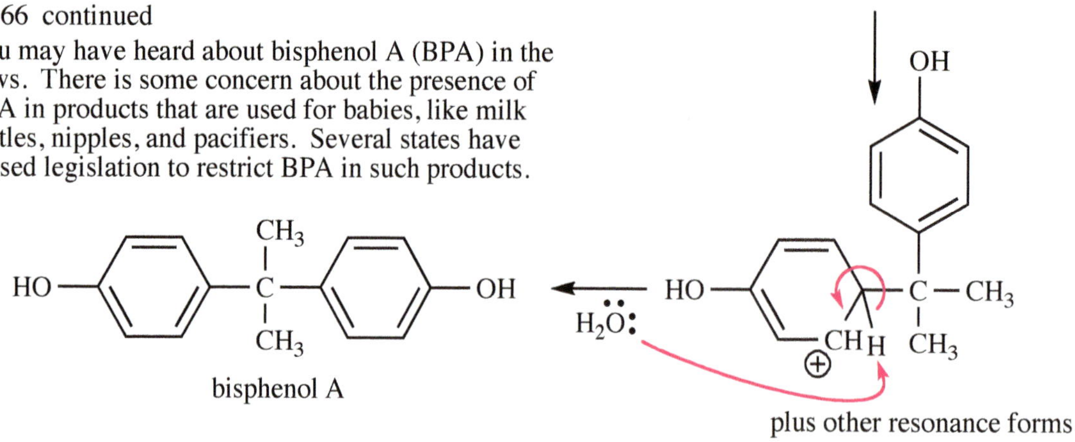

bisphenol A

plus other resonance forms

17-67

(a) This is an example of kinetic versus thermodynamic control of a reaction. At low temperature, the kinetic product predominates: in this case, almost a 1 : 1 mixture of ortho and para. These two isomers must be formed at approximately equal rates at 0 °C. At 100 °C, however, enough energy is provided for the *desulfonation* to occur rapidly; the large excess of the para isomer indicates the para is more stable, even though it is formed initially at the same rate as the ortho.

(b) The product from the 0 °C reaction will equilibrate as it is warmed, and at 100 °C will produce the same ratio of products as the reaction that was run initially at 100 °C.

(c) The principle of a blocking group is to attach it at the position that you want blocked, perform the other reaction, then remove the blocking group. This is accomplished on benzene very effectively with the sulfonic acid group.

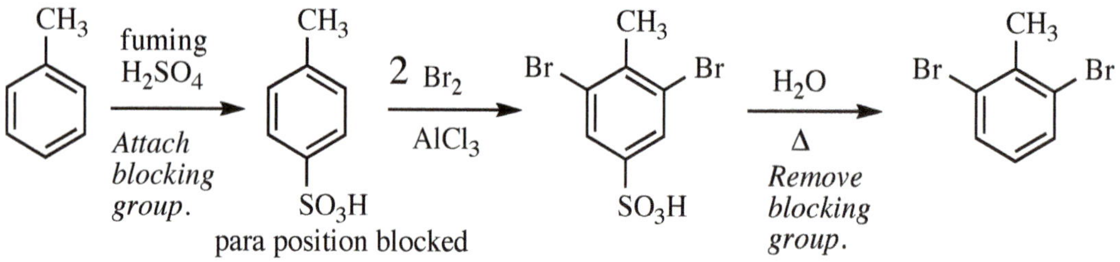

fuming H_2SO_4 — *Attach blocking group.*

para position blocked

2 Br_2, $AlCl_3$

H_2O, Δ — *Remove blocking group.*

17-68 Solve the problem by writing the mechanism for each possible position of attack. (See the solution to problem 17-23(a) for an identical mechanism.)

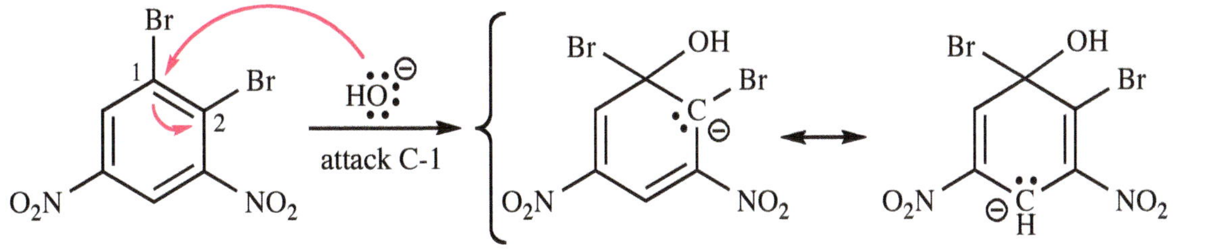

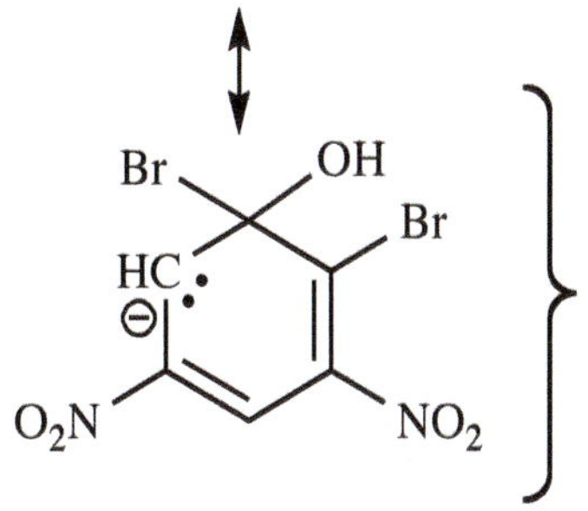

attack C-1

Hydroxide attack on C-1 puts the negative charge on carbons that do not have the NO_2 groups, so these anions are not stabilized.

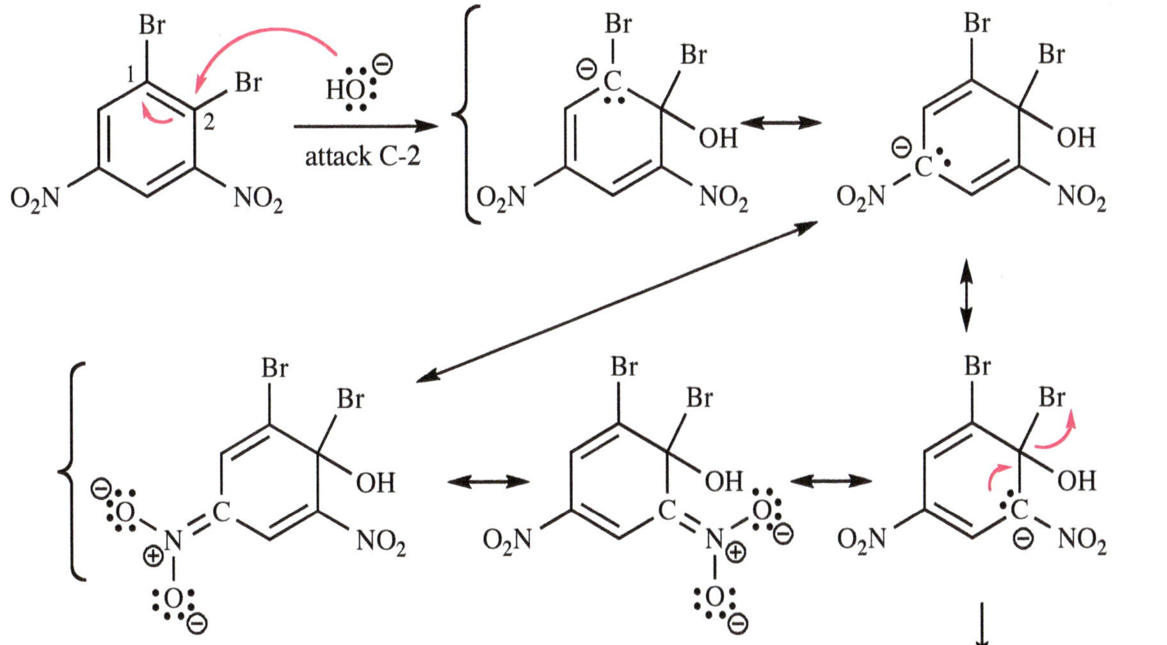

attack C-2

Hydroxide attack on C-2 puts the negative charge on carbons with nitro groups, thereby increasing the stabilization by delocalizing the negative charge. This intermediate is formed preferentially.

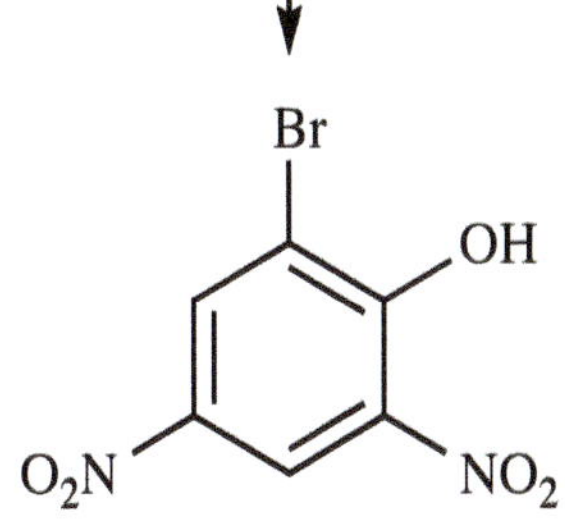

only product formed

17-69

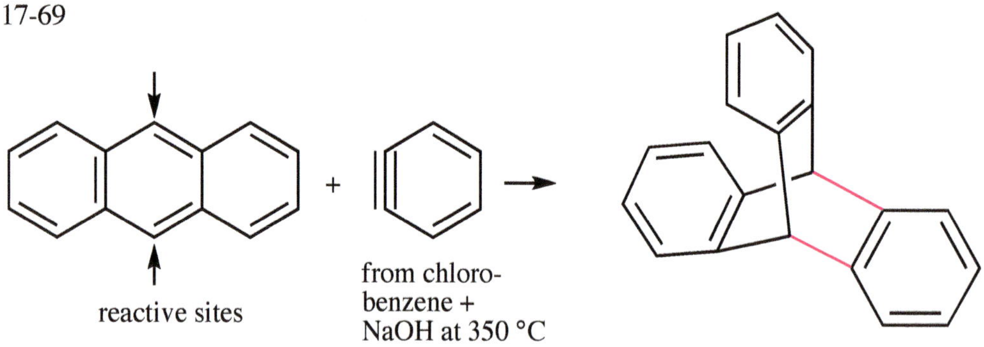

reactive sites

from chloro-
benzene +
NaOH at 350 °C

As we saw in Chapter 16, the carbons of the center ring of anthracene are susceptible to electrophilic addition, leaving two isolated benzene rings on the ends. Benzyne is such a reactive dienophile that the reluctant anthracene is forced into a Diels-Alder reaction.

17-70

(a)

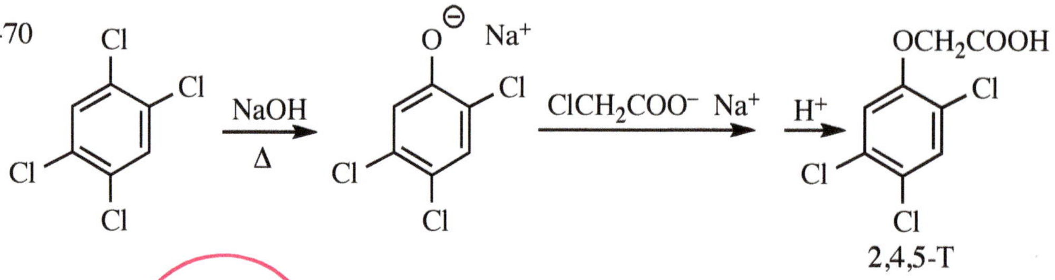

NaOH, Δ

$ClCH_2COO^-$ Na^+

H^+

OCH_2COOH

2,4,5-T

(b)

Δ → 2 Cl^- +

TCDD

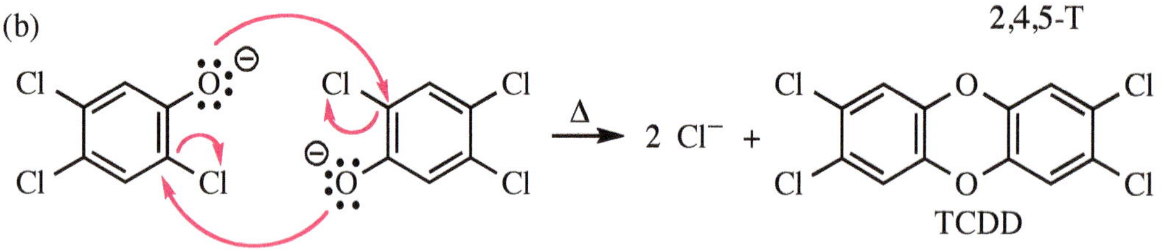

(This is the compound used to poison Ukrainian political leader, Boris Yushchenko, in 2004.)

Two nucleophilic aromatic substitutions form a new six-membered ring. (Though not shown here, this reaction would follow the standard addition-elimination mechanism.)

(c) To minimize formation of TCDD during synthesis: 1) keep the solutions dilute; 2) avoid high temperature; 3) replace chloroacetate with a more reactive molecule like bromoacetate or iodoacetate; 4) add an excess of the haloacetate.

To separate TCDD from 2,4,5-T at the end of the synthesis, take advantage of the acidic properties of 2,4,5-T. The 2,4,5-T will dissolve in an aqueous solution of a weak base like $NaHCO_3$. The TCDD will remain insoluble and can be filtered or extracted into an organic solvent like ether or dichloromethane. The 2,4,5-T can be precipitated from aqueous solution by adding acid.

17-71

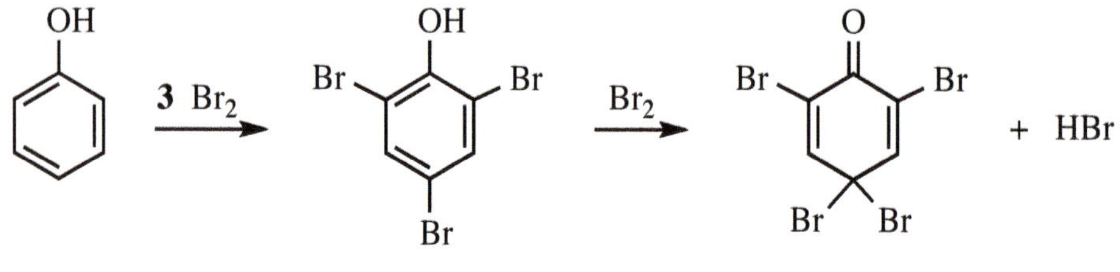

3 Br_2 →

Br_2 →

+ HBr

$C_6H_3OBr_3$

+ **3** HBr

$C_6H_2OBr_4$

17-72

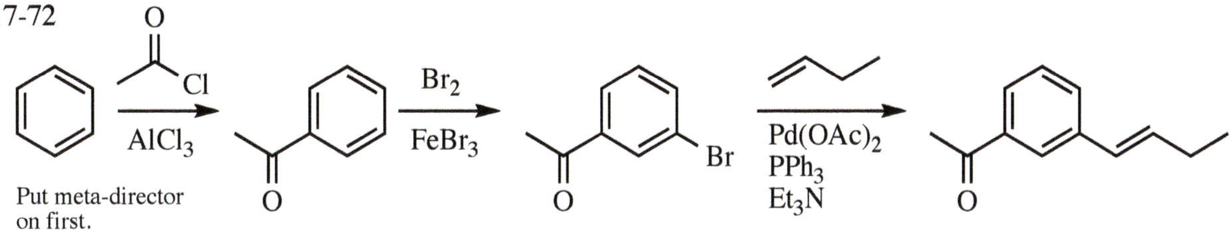

O
Cl
AlCl$_3$
Br$_2$
FeBr$_3$
O
O
Br
Pd(OAc)$_2$
PPh$_3$
Et$_3$N
O

Put meta-director on first.

17-73 A benzyne must have been generated from the Grignard reagent.

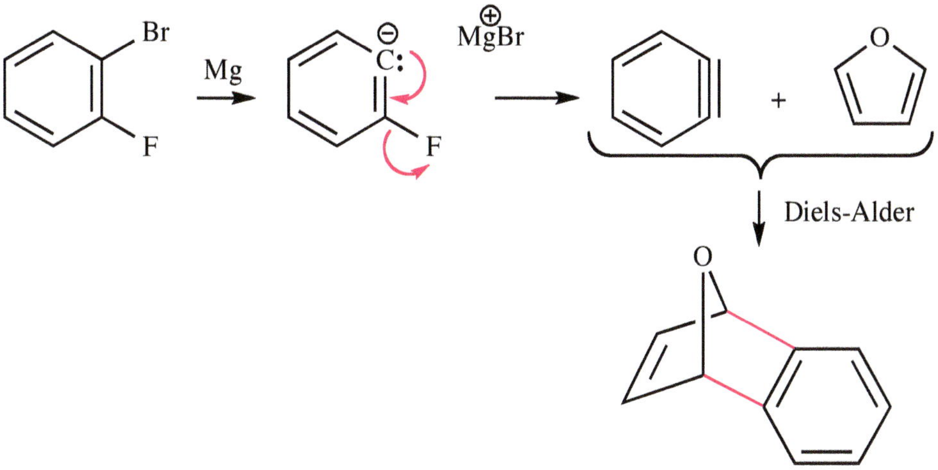

Br
F
Mg
C:
MgBr
F
+
O
Diels-Alder
O

17-74 The intermediate anion forces the loss of hydroxide.

For simplicity, abbreviate

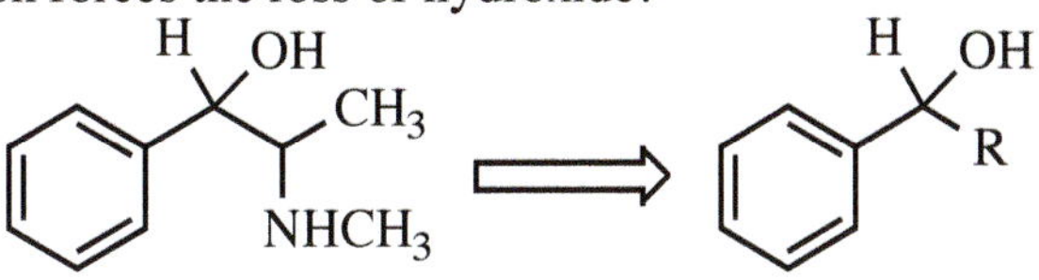

H OH
CH$_3$
NHCH$_3$
H OH
R

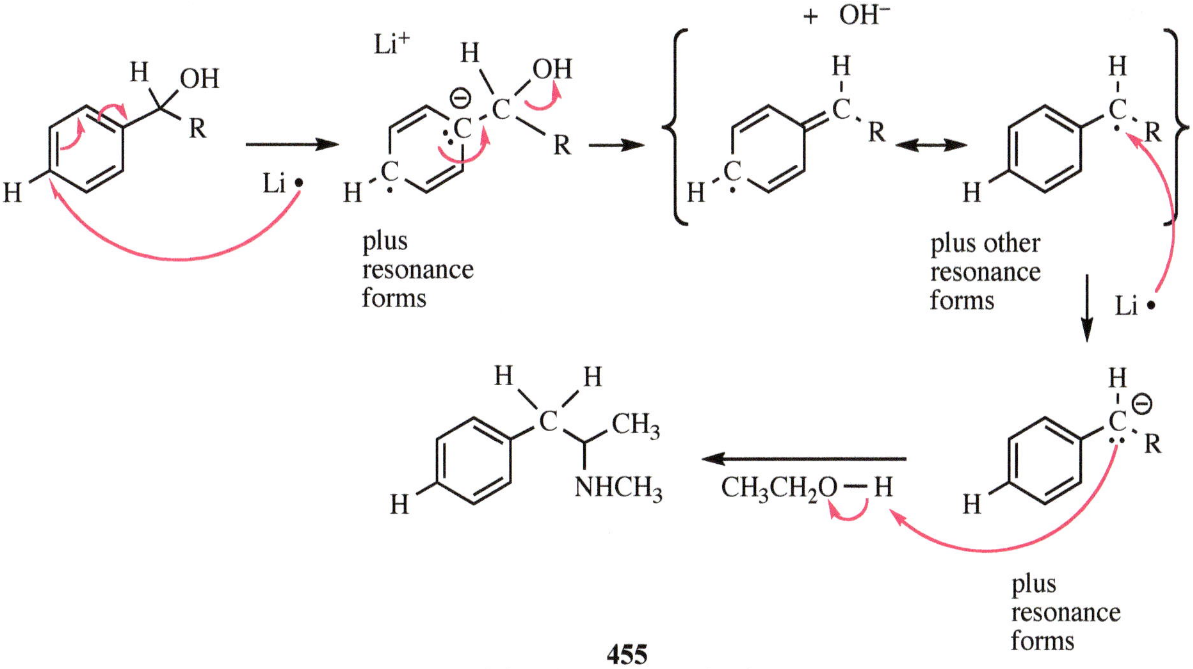

+ OH$^-$

H OH R
H
Li •
Li$^+$
H OH
C C R
H C
plus resonance forms
H C R
H C
H C R
H
plus other resonance forms
Li •

H H
C CH$_3$
H NHCH$_3$
CH$_3$CH$_2$O — H
H C R
H
plus resonance forms

17-75 Make a boronate ester from one of the halides first, then use Suzuki coupling to link them.

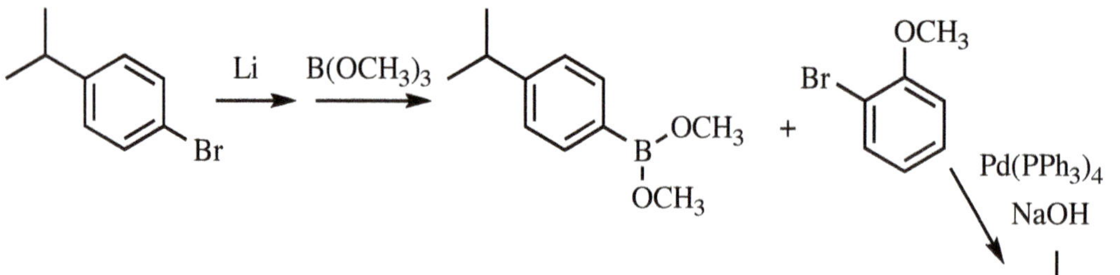

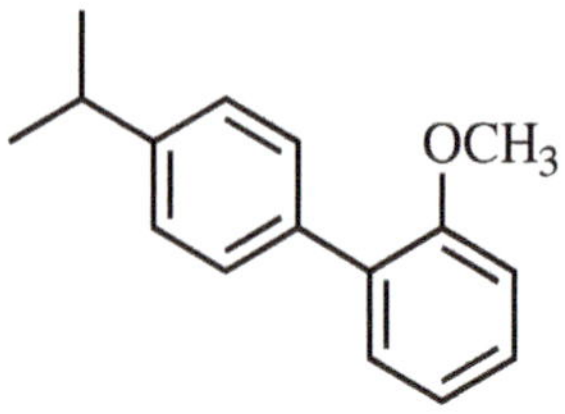

Li; $B(OCH_3)_3$; Br; OCH_3; B; OCH_3; OCH_3; +; Br; OCH_3; $Pd(PPh_3)_4$; NaOH; OCH_3

Alternatively, the other boronate ester could be made to couple with 4-bromoisopropylbenzene.

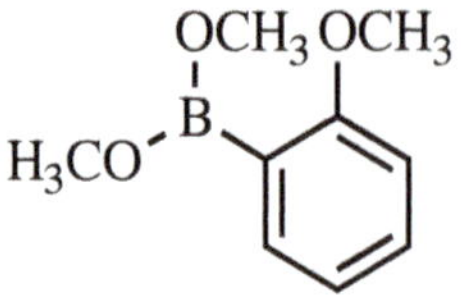

OCH_3 OCH_3; B; H_3CO

17-76 OH; OH; NaOH; 1 equiv. CH_3I; OH; OCH_3; 1 equiv. $Cl-C(CH_3)_3$; $AlCl_3$; OH; $C(CH_3)_3$; OCH_3 BHA, butylated hydroxyanisole

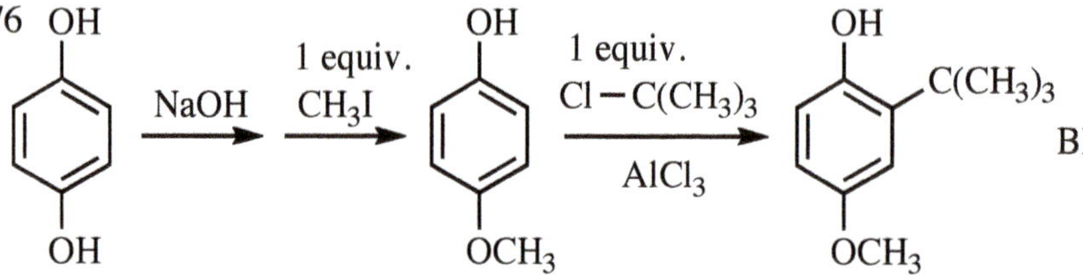

hydroquinone

plus minor isomer with *tert*-butyl *ortho* to OCH_3

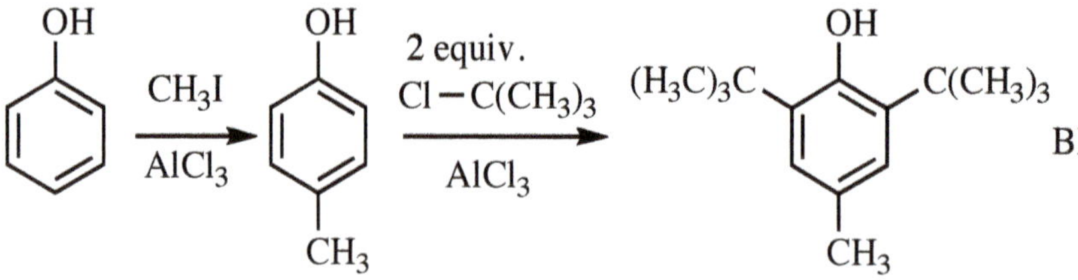

OH; CH_3I; $AlCl_3$; OH; CH_3; 2 equiv. $Cl-C(CH_3)_3$; $AlCl_3$; OH; $(H_3C)_3C$; $C(CH_3)_3$; CH_3 BHT, butylated hydroxytoluene

17-77

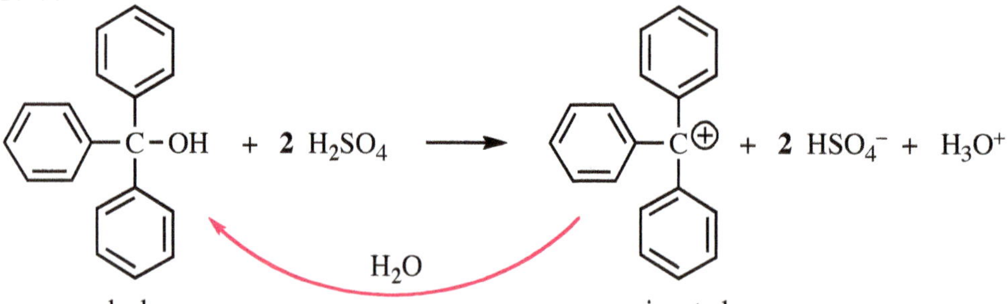

C-OH + **2** H_2SO_4 ⟶ C⊕ + **2** HSO_4^- + H_3O^+

H_2O

colorless

conjugated—yellow

Concentrated sulfuric acid "dehydrates" the alcohol, producing a highly conjugated, colored carbocation, and protonates the water to prevent the reverse reaction. Upon adding more water, however, there are too many water molecules for the acid to protonate, and triphenylmethanol is regenerated.

17-78

(a)

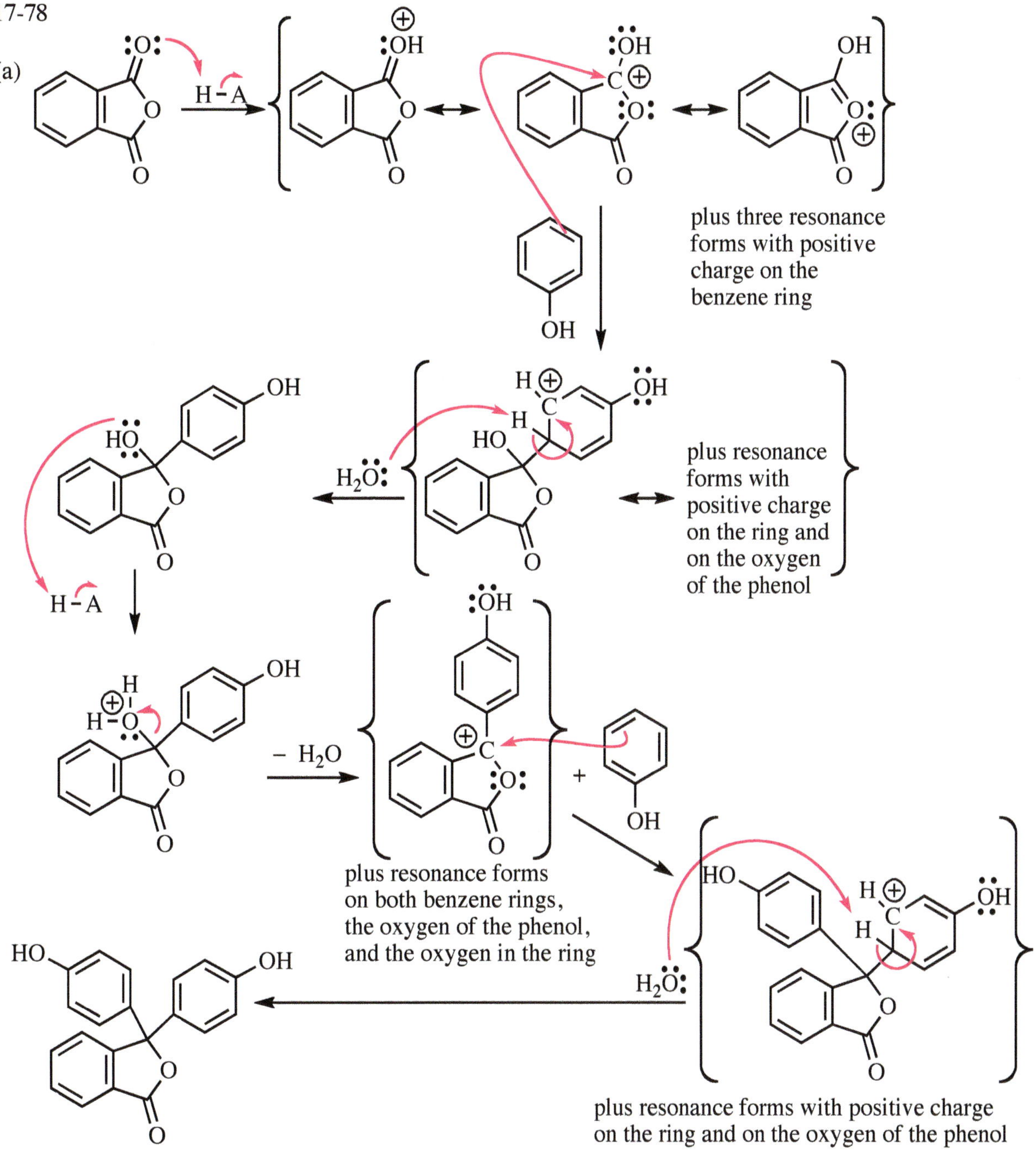

H–A
plus three resonance forms with positive charge on the benzene ring
OH
plus resonance forms with positive charge on the ring and on the oxygen of the phenol
H_2O
H–A
– H_2O
plus resonance forms on both benzene rings, the oxygen of the phenol, and the oxygen in the ring
plus resonance forms with positive charge on the ring and on the oxygen of the phenol

17-78 continued

(b) Color comes from highly conjugated molecules and ions. See text Section 15-14 and Problem 15-23.

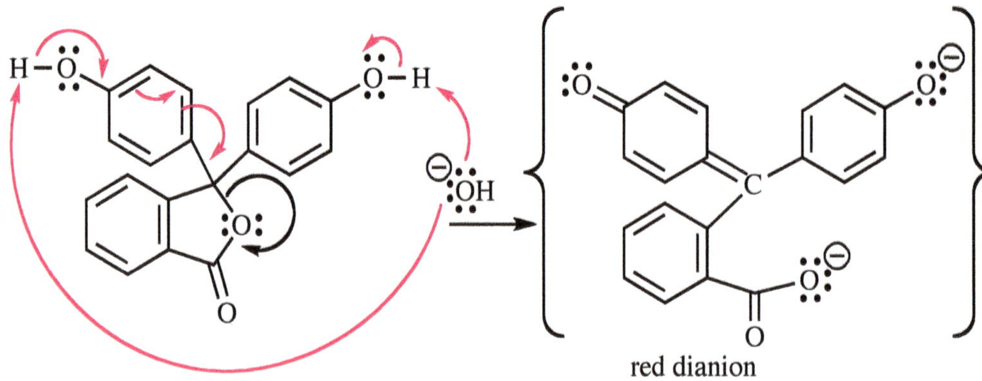

H–O: :OH⁻ → :O: :O:⁻ C :O:⁻ O

numerous resonance forms

red dianion

(c) :O: :O:⁻ C :O:⁻ O ⟷ ⁻:O: :O: C :O:⁻ O

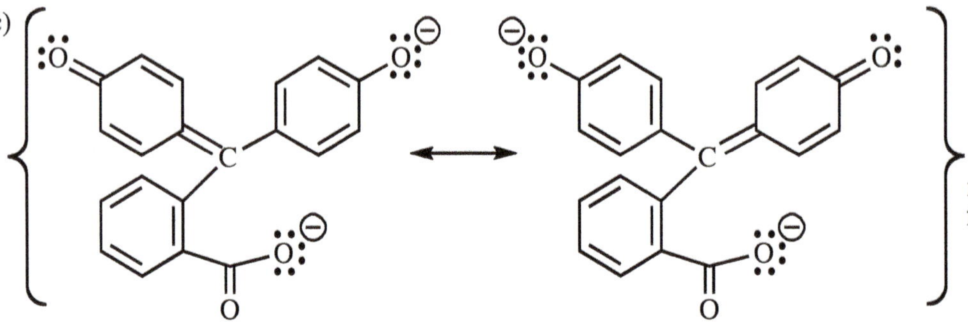

numerous resonance forms

CHAPTER 18—KETONES AND ALDEHYDES

18-1

(a) 5-hydroxyhexan-3-one; ethyl β-hydroxypropyl ketone
(b) 3-phenylbutanal; β-phenylbutyraldehyde
(c) *trans*-2-methoxycyclohexanecarbaldehyde (or (1*R*,2*R*) if you named this enantiomer); no common name
(d) 6,6-dimethylcyclohexa-2,4-dienone; no common name

18-2

(a) $C_9H_{10}O$ ⇒ 5 elements of unsaturation

1H doublet (very small coupling constant) at δ 9.7 ⇒ aldehyde hydrogen, next to CH

5H multiple peaks at δ 7.2-7.4 ⇒ monosubstituted benzene

1H multiplet at δ 3.6 and 3H doublet at δ 1.4 ⇒ $CHCH_3$

C_6H_5- + $-CH(CH_3)-$ + $-CHO$ ⟹ $C_6H_5-CH(CH_3)-C(=O)-H$

The splitting of the hydrogen on carbon-2, next to the aldehyde, is worth examining. In its overall shape, it looks like a quartet due to the splitting from the adjacent CH_3. A closer examination of the peaks shows that each peak of the quartet is split into two peaks: this is due to the splitting from the aldehyde hydrogen. The aldehyde hydrogen and the methyl hydrogens are not equivalent, so it is to be expected that the coupling constants will not be equal. If a hydrogen is coupled to different neighboring hydrogens by different coupling constants, they must be considered separately, just as you would by drawing a splitting tree for each type of adjacent hydrogen.

(b) C_8H_8O ⇒ 5 elements of unsaturation

cluster of 4 peaks at δ 128–145 ⇒ mono- or para-substituted benzene ring

peak at δ 197 ⇒ carbonyl carbon (the small peak height suggests a ketone rather than an aldehyde)

peak at δ 26 ⇒ methyl next to carbonyl or benzene

$C_6H_5-C(=O)-CH_3$

$CH_3-C_6H_4-C(=O)-H$

This structure is also possible from the chemical shift values, but the DEPT information about the type of carbons present proves the monosubstituted benzene.

18-3 A compound has to have a hydrogen on a γ carbon (or other atom) in order for the McLafferty rearrangement to occur. Butan-2-one has no γ-hydrogen.

γ position—no carbon

$CH_3-C(=O)-CH_2-CH_3$ (α, α, β)

18-4 $\left[CH_3 \overset{43}{+} \overset{O}{\overset{||}{C}} \overset{85}{+} CH_2CH_2CH_2CH_2CH_2CH_3 \right]^{\dagger}$ (113 cleavage) m/z 128

$\rightarrow \left\{ :\ddot{O}=C^{\oplus}-CH_3 \longleftrightarrow :O^{\oplus}\equiv C-CH_3 \right\}$ m/z 43

$\rightarrow \overset{\oplus}{C}H_2CH_2CH_2CH_2CH_2CH_3$ m/z 85

$\rightarrow \left\{ :\ddot{O}=C^{\oplus}-(CH_2)_5CH_3 \longleftrightarrow :O^{\oplus}\equiv C-(CH_2)_5CH_3 \right\}$ m/z 113

McLafferty rearrangement

$\left[CH_3-C(=O)-CH_2-CH_2-CH(H)CH_2CH_2CH_3 \right]^{\dagger} \longrightarrow \left[CH_3-C(-OH)=CH_2 \right]^{\dagger} + CH_2=CHCH_2CH_2CH_3$

m/z 58

18-5 Cholesterol has one isolated double bond with π to π* transition at < 200 nm. Therefore, it will show no UV spectrum in the range of 200–400 nm.

Cholest-4-en-3-one has a double bond conjugated with a ketone. The ketone's n to π* transition will be weak, but there will be a large π to π* transition predicted at 240 nm (base value of 210 nm plus 3 alkyl substituents). The actual λ_{max} for cholest-4-en-3-one is 241 nm with $\log_{10}\varepsilon$ of 4.26, a very strong absorption.

REMINDERS ABOUT SYNTHESIS PROBLEMS:
1. There may be more than one legitimate approach to a synthesis, especially as the list of reactions gets longer.
2. Begin your analysis by comparing the target to the starting material. If the product has more carbons than the reactant, you will need to use one of the small number of reactions that form carbon-carbon bonds.
3. Where possible, work backwards from the target back to the starting material.
4. KNOW THE REACTIONS. There is no better test of whether you know the reactions than attempting synthesis problems. The starburst reaction summaries are an excellent technique of organizing reactions for effective studying. Customize your own diagrams based on what your instructor has emphasized.

18-6 All three target molecules in this problem have more than six carbons, so all answers will include carbon-carbon bond-forming reactions. So far, the main types of reactions that form carbon-carbon bonds: at sp^3 carbon: the Grignard reaction, S_N2 substitution by an acetylide ion or cyanide ion, and the Friedel-Crafts alkylation; at sp^2 carbon: Grignard reaction, Friedel-Crafts acylation, olefin metathesis, and organometallic coupling (organocuprate, Heck, Suzuki).

(a) cyclohexyl–Br $\xrightarrow[\text{ether}]{\text{Mg}}$ cyclohexyl–MgBr $\xrightarrow[\text{2) } H_3O^+]{\text{1) } H-C(=O)-CH_2CH_3}$ cyclohexyl–CH(OH)CH₂CH₃ $\xrightarrow[\text{HOAc}]{\text{NaOCl}}$ cyclohexyl–C(=O)CH₂CH₃

18-6 continued

(b)

Br — identical sequence as in part (a) → O (propiophenone)

OR

benzene + $Cl–C(=O)CH_2CH_3$ / $AlCl_3$ → propiophenone

This method using Friedel-Crafts acylation is more efficient as it is only one step.

(c)

cyclopentyl Br — Mg, ether → cyclopentyl MgBr — 1) propylene oxide (O); 2) H_3O^+ → alcohol (OH) — NaOCl, HOAc → ketone

OR

cyclopentylmethyl Br — Na^+ $^{\ominus}C{\equiv}CH$ → cyclopentyl–$CH_2C{\equiv}CH$ — Hg^{2+}, H_2O, H_2SO_4 → ketone (O)

18-7 New sigma bonds are shown in bold.

The first equivalent of R—Li reacts with the H^+ of COOH to produce R—H; the second R—Li adds to the C=O.

(a) benzoic acid (O, OH) — 1) 2 CH_3Li; 2) H_3O^+ → acetophenone (O, CH_3) + CH_4

(b) cyclopentyl Br — 2 Li → cyclopentyl Li + LiBr — $HO–C(=O)–CH_3$, 0.5 equiv. — H_3O^+ → cyclopentyl–C(=O)–CH_3

(c) hexanoic acid (O, OH) — 1) 2 CH_3CH_2Li; 2) H_3O^+ → ketone (O)

CH_3Li, CH_3CH_2Li, PhLi, BuLi, and *t*-BuLi are commercially available.

(d) phenylacetic acid (OH, O) — 1) 2 $(CH_3)_3CLi$; 2) H_3O^+ → ketone (O)

18-8 New sigma bonds are shown in bold.

(a) ketone (O) — same product as 18-7(c)

(b) aldehyde (O, H)

(c) $PhCH_2$—CN (simple S_N2)

(d) $PhCH_2C(=O)$—cyclopentyl

(e) Ph–CH_2–C(=O)–H

18-9

(a) Bromobenzene $\xrightarrow[\text{ether}]{Mg}$ phenylmagnesium bromide (MgBr) $\xrightarrow{CH_3CH_2C\equiv N}$ $\xrightarrow{H_3O^+}$ $C_6H_5\overset{O}{\overset{\|}{C}}CH_2CH_3$

(b) $CH_3CH_2C\equiv N$ + $BrMgCH_2CH_2CH_2CH_3$ $\longrightarrow$ $\xrightarrow{H_3O^+}$ 3-heptanone

(c) Benzoic acid (O, OH) $\xrightarrow{2\ \text{cyclopentyllithium (Li)}}$ $\xrightarrow{H_3O^+}$ phenyl cyclopentyl ketone

(d) Br $\xrightarrow{NaCN}$ $C\equiv N$ $\xrightarrow[2.\ H_3O^+]{1.\ \text{DIBAL-H}}$ $\overset{O}{\overset{\|}{C}}$—H

chain lengthened by 1 C

18-10

(a) $C_6H_5CH_2OH$ (b) $C_6H_5\overset{O}{\overset{\|}{C}}H$ (c) (ketone, O) (d) (OH)

(e) (O, H) (f) (O, O) $\xrightarrow[-78\ °C]{\text{DIBAL-H}}$ $\xrightarrow{H_2O}$ (O, H, OH)

a cyclic ester called a lactone

18-11 Review the reminders on p. 460 of this Manual. There is often more than one correct way to do a synthesis, although a more direct route with fewer steps is usually better.

(a) Benzaldehyde (O, H) $\xrightarrow{PhMgBr}$ $\xrightarrow{H_3O^+}$ diphenylmethanol (OH) $\xrightarrow[\text{HOAc}]{\text{NaOCl}}$ benzophenone (O)

PhBr $\xrightarrow{Mg}$ PhMgBr

(b) Benzophenone (O) $\xrightarrow{PhMgBr}$ $\xrightarrow{H_3O^+}$ triphenylmethanol (OH)

PhBr $\xrightarrow{Mg}$ PhMgBr

(c) Benzophenone (O) $\xrightarrow[\text{(solvent)}]{NaBH_4,\ CH_3OH}$ diphenylmethanol (OH)

also reduces ketones: $\xrightarrow[2)\ H_3O^+]{1)\ LiAlH_4}$

18-11 continued

(d) Benzaldehyde + $Na^+ \; ^{\ominus}C{\equiv}CCH_2CH_3$ (from $HC{\equiv}CCH_2CH_3$ + $NaNH_2$), then H_3O^+ → $PhCH(OH)C{\equiv}CCH_2CH_3$

18-12 The triacetoxyborohydride ion is similar to borohydride, BH_4^-, where three acetoxy groups have replaced three hydrides.

(a) Na^+ $H-\overset{\ominus}{B}(OAc)_3$

$NaBH(OAc)_3$

(b) $(AcO)_3B^{\ominus}-\mathbf{H}$ Na^+ + $RCHO$ → $(AcO)_3B$ + $R\mathbf{H}CH-O^{\ominus}$ Na^+

CH_3COO-H

$R\mathbf{H}CH-O-H$

PROBLEM in Key Mechanism 18-1—Basic Conditions

The Principle of Microscopic Reversibility dictates that the reverse reaction must follow the same steps as the forward reaction, as each must be following a minimum energy path.

$PhCH(OH)C{\equiv}N$ + $^{\ominus}:C{\equiv}N$ → $PhCH(O^{\ominus})C{\equiv}N$ → $PhCH{=}O$ + $^{\ominus}:C{\equiv}N$

PROBLEM in Key Mechanism 18-1—Acidic Conditions

As above, the Principle of Microscopic Reversibility dictates that the reverse reaction must follow the same steps as the forward reaction, as each must be following a minimum energy path.

$PhCH(OH)OCH_3$ + $H-A$ ⇌ $PhCH(OH)\overset{\oplus}{O}HCH_3$ ⇌ { $PhCH^{\oplus}OH$ ↔ $PhCH{=}\overset{\oplus}{O}H$ } $\xrightarrow{CH_3OH}$ $PhCH{=}O$

+ res. forms where the positive charge is delocalized over the benzene ring

18-13

(a) $Cl_3C-C(=O)-H \underset{}{\overset{H-A}{\rightleftharpoons}} \left\{ Cl_3C-C(=\overset{+}{O}H)-H \longleftrightarrow Cl_3C-\overset{+}{C}(OH)-H \right\} \overset{H_2O}{\rightleftharpoons} Cl_3C-C(OH)(\overset{+}{O}H_2)-H \overset{H_2O}{\rightleftharpoons} Cl_3C-C(OH)_2-H$

(b) $CH_3-C(=O)-CH_3 \overset{^-OH}{\rightleftharpoons} CH_3-C(O^-)(OH)-CH_3 \overset{H-OH}{\rightleftharpoons} CH_3-C(OH)_2-CH_3 + HO^-$

18-14 Two general principles apply to hydrate formation. Sterically, less hindered C=O form more hydrate, so aldehydes are more likely to form a hydrate than ketones are. Electronically, electron-withdrawing substituents at the alpha carbon intensify the (+) charge on the C=O, making it more likely to form a hydrate.

cyclohexanone < 2-bromocyclohexanone < cyclopentyl-CH_2-CHO < cyclopentyl-CH(Br)-CHO

least amount of hydrate ... greatest amount of hydrate

18-15

(a) $CH_3CH_2-C(=O)-H \overset{^-:C\equiv N}{\rightleftharpoons} CH_3CH_2-C(O^-)(CN)-H \overset{H-CN}{\rightleftharpoons} CH_3CH_2-C(OH)(CN)-H$

(b) $CH_3CH_2-C(=O)-CH_3 \overset{^-:C\equiv N}{\rightleftharpoons} CH_3CH_2-C(O^-)(CN)-CH_3 \overset{H-CN}{\rightleftharpoons} CH_3CH_2-C(OH)(CN)-CH_3$

(c) $t\text{-Bu}-C(=O)-t\text{-Bu} \overset{^-:C\equiv N}{\rightleftharpoons} t\text{-Bu}-C(O^-)(CN)-t\text{-Bu} \overset{H-CN}{\rightleftharpoons} t\text{-Bu}-C(OH)(CN)-t\text{-Bu}$

18-16

(a) acetophenone $\xrightarrow[HCN]{^-CN}$ $C_6H_5-C(HO)(CN)-CH_3$

18-16 continued

(b) Cyclopentanecarbaldehyde $\xrightarrow[HCN]{^{\ominus}CN}$ cyanohydrin (HO, CN, H) $\xrightarrow{H_3O^+}$ α-hydroxy acid (HO, COOH, H)

(c) Hexan-1-ol $\xrightarrow{PCC}$ hexanal $\xrightarrow[HCN]{^{\ominus}CN}$ cyanohydrin (HO, CN, H) $\xrightarrow[\Delta]{H_3O^+}$ α-hydroxy acid (HO, COOH, H)

Problem in Key Mechanism 18-4

(a) If too much of a strong acid were added—meaning more than 1 equivalent relative to the amount of amine present—the amine nucleophile would be tied up in the protonated form in which it is not nucleophilic. Adding acid past a catalytic amount will slow the reaction, eventually stopping the reaction if no free nucleophile is available to react.

(b) If it were too basic, meaning not enough acid to protonate the carbonyl, the reaction will also slow, because the nucleophile has to attack a free carbonyl. It is possible for this to happen, but the reaction will be significantly slower if no acid is present.

As Goldilocks said in regards to acid catalysis: This amount is *just right*.

18-17 Imine formation has optimum pH 4-5 where protonation of the carbonyl oxygen is the first step. All steps in imine formation are equilibria.

(a) Cyclohexanone + H—A $\xrightarrow{A}$ {protonated carbonyl $\longleftrightarrow$ carbocation} $\xrightarrow[:NH_3]{B}$ $-NH_2^{\oplus}$ adduct $\xrightarrow[H_2O:]{C}$ carbinolamine (O—H, NH_2) $\xrightarrow[D]{H_3O^+}$ protonated OH ($H-O^{\oplus}-H$, NH_2) $\xrightarrow[E]{-H_2O}$ {iminium $\longleftrightarrow$ carbocation} $\xrightarrow[F]{H_2O:}$ imine (N—H)

Imine formation has six steps: four are proton transfers, one is a nucleophilic attack, and one is a leaving group leaving.

A proton on (resonance-stabilized cation)
B nucleophile attacks
C proton off
D proton on
E leaving group leaves (resonance-stabilized cation)
F proton off

18-17 continued

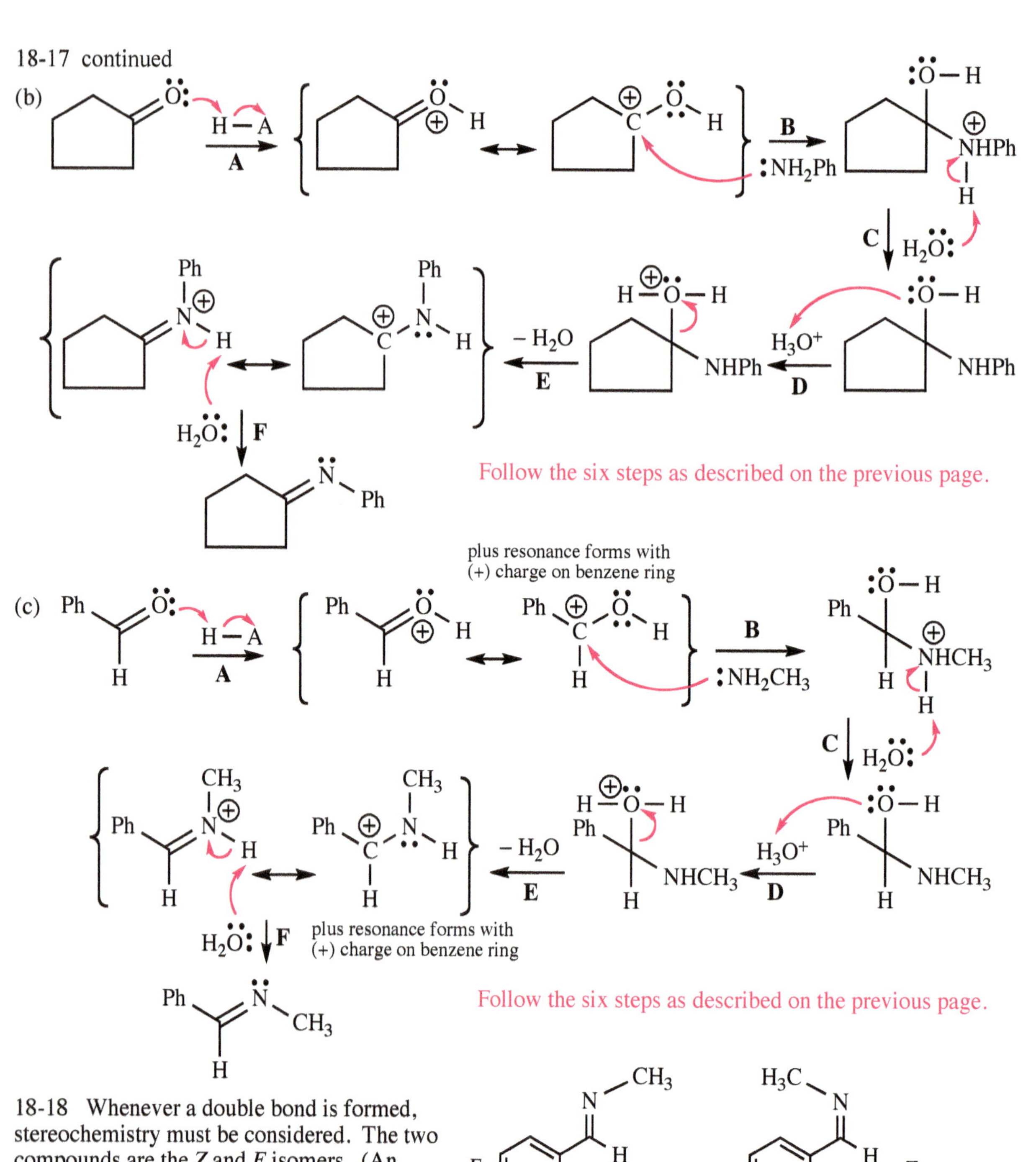

(b)
Follow the six steps as described on the previous page.
(c)
plus resonance forms with (+) charge on benzene ring
Follow the six steps as described on the previous page.

18-18 Whenever a double bond is formed, stereochemistry must be considered. The two compounds are the *Z* and *E* isomers. (An electron pair has lower priority than H in the Cahn-Ingold-Prelog system.)

E *Z*

18-19

(a) + CH_3NH_2

(b) + NH_3

(c) NH_2 +

(d) NH_2 CH_3

(e) NH_2

(f) NH_2

18-20 This mechanism is the reverse of the one shown in 18-17(c) above.

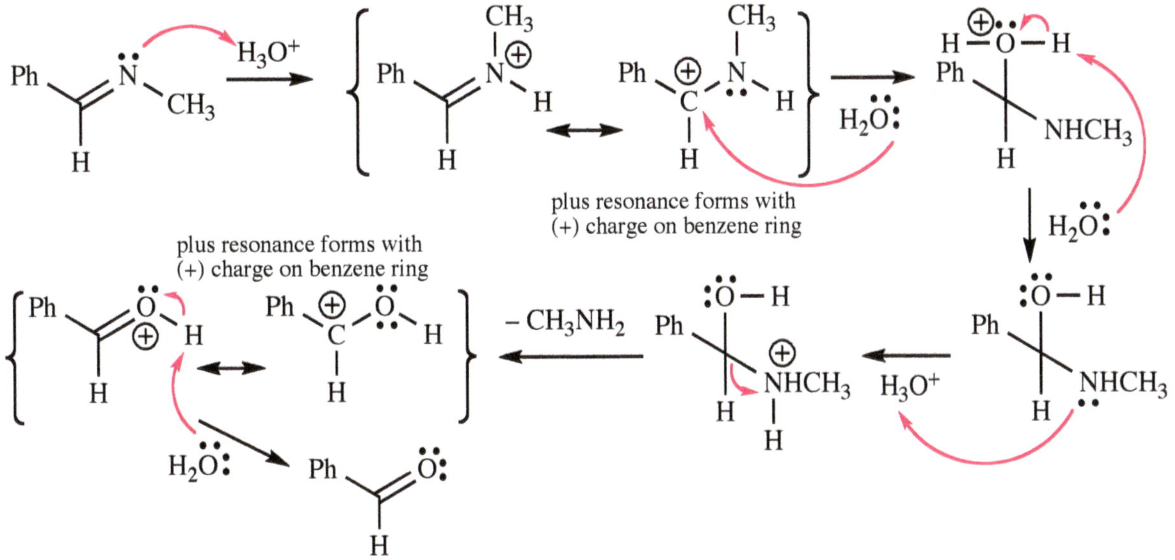

plus resonance forms with (+) charge on benzene ring
plus resonance forms with (+) charge on benzene ring
$- CH_3NH_2$

18-21

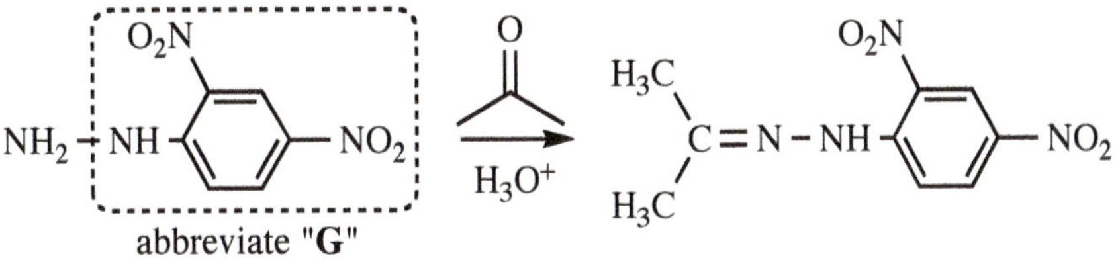

abbreviate "G"

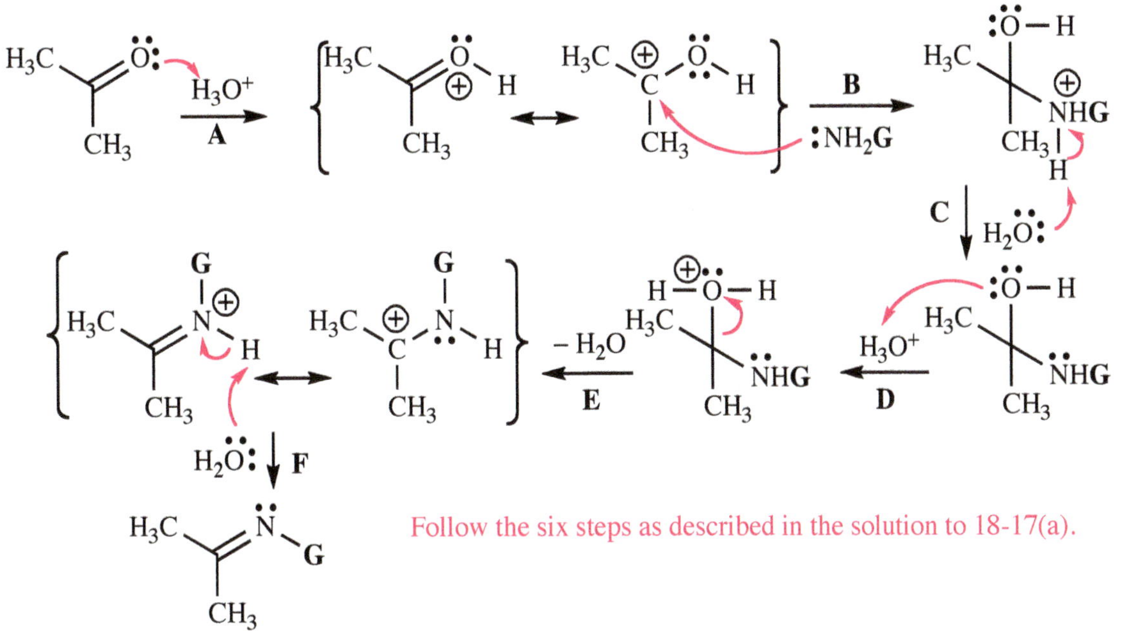

A
B
C
D
E
F

Follow the six steps as described in the solution to 18-17(a).

18-22

(a) 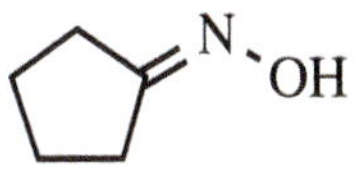

(b) 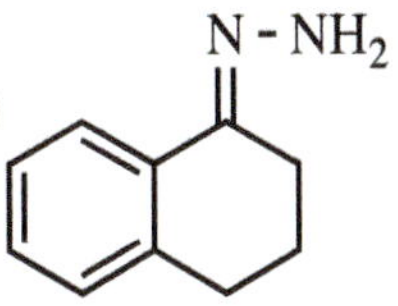

(c) 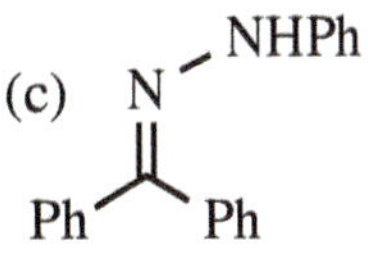

(d) 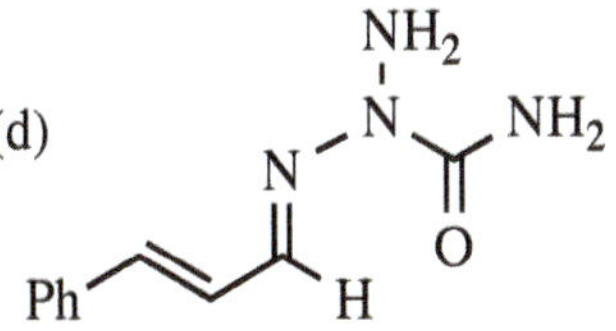

18-23

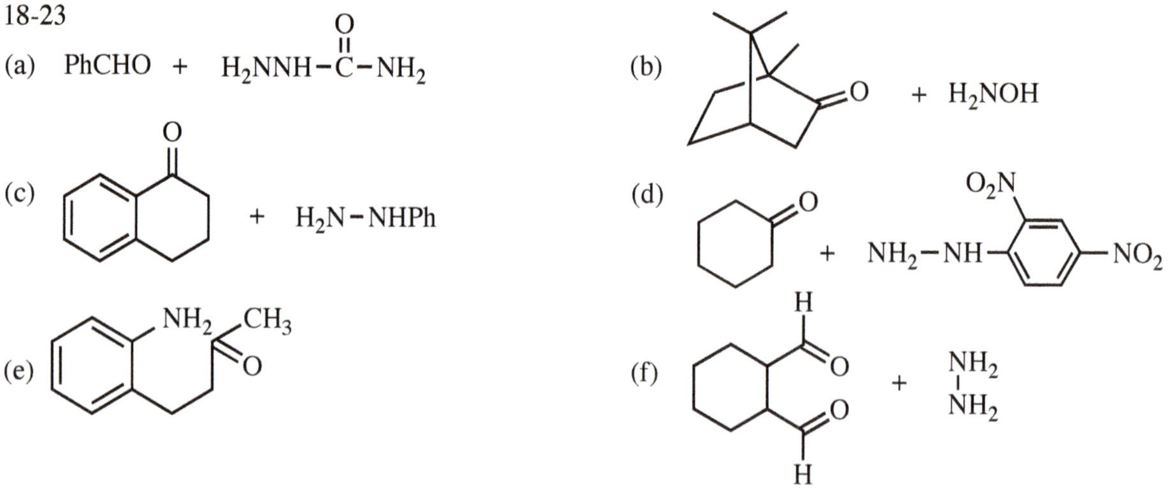

(a) PhCHO + $H_2NNH-C(=O)-NH_2$

(b) [bicyclic ketone] + H_2NOH

(c) [tetralone] + $H_2N-NHPh$

(d) [cyclohexanone] + $NH_2-NH-C_6H_3(NO_2)_2$ (O_2N, NO_2)

(e) NH_2 CH_3 O

(f) H O O H + NH_2-NH_2

18-24

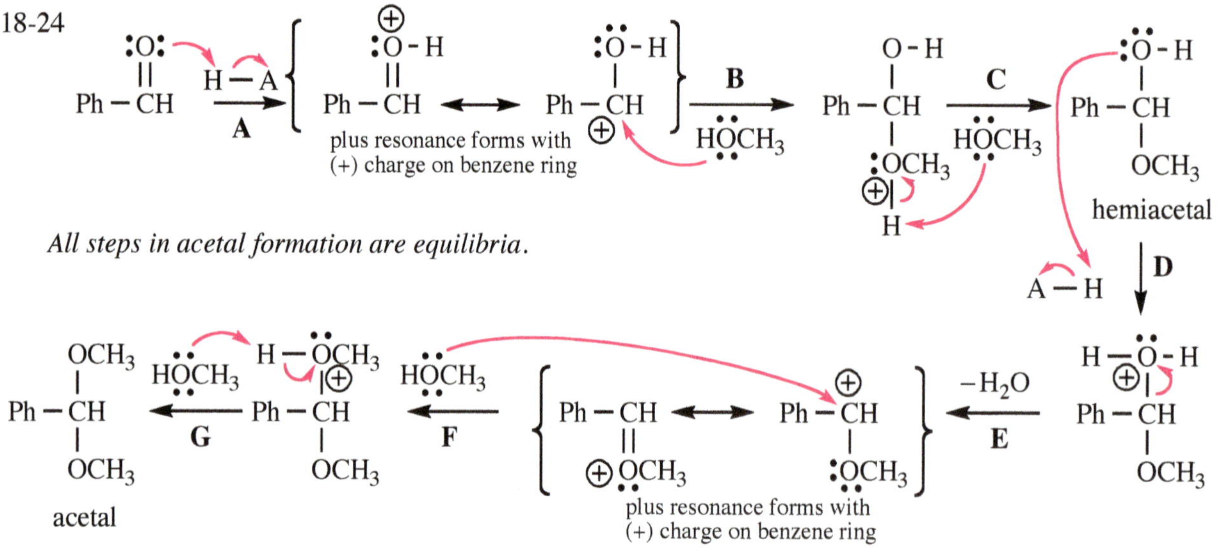

A, B, C, hemiacetal, D, E, F, G, acetal

plus resonance forms with (+) charge on benzene ring

All steps in acetal formation are equilibria.

$-H_2O$

plus resonance forms with (+) charge on benzene ring

Acetal formation has seven steps, the first five of which are similar to imine formation: four are proton transfers, two are nucleophilic attacks, and one is a leaving group leaving.

A proton on (resonance-stabilized cation)
B nucleophile attacks
C proton off
D proton on
E leaving group leaves (resonance-stabilized cation)
F second nucleophile attacks (this is the step that is unique to acetal formation)
G proton off

18-25

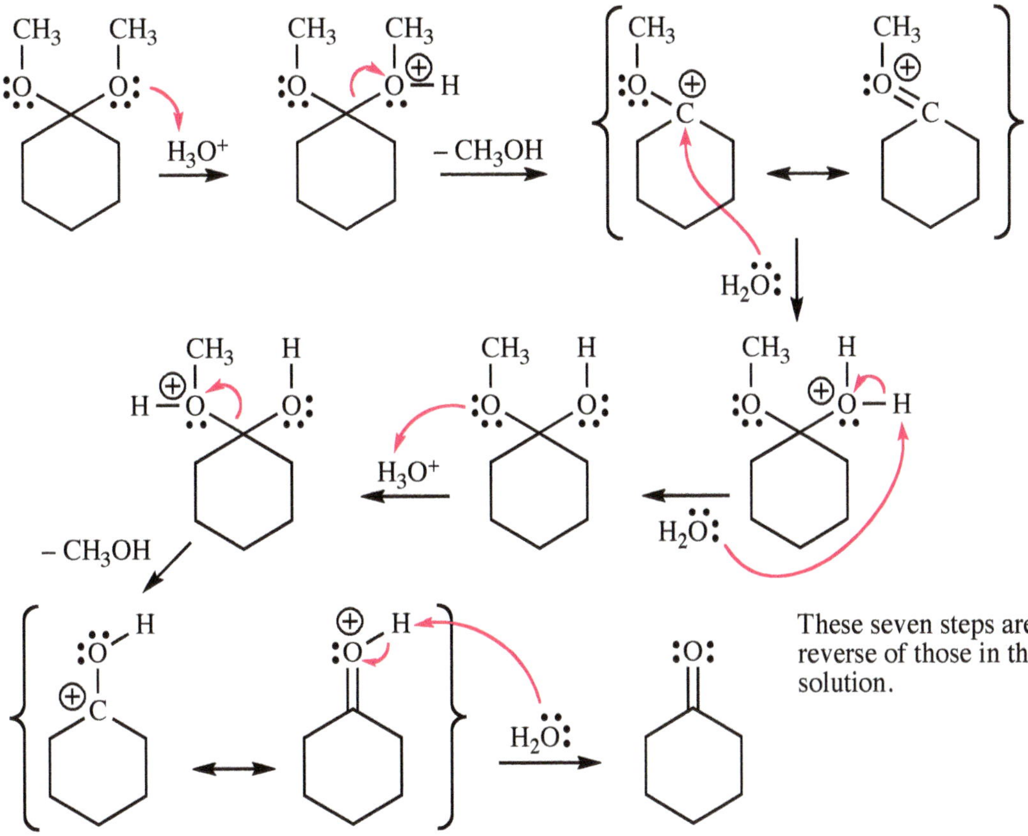

These seven steps are the exact reverse of those in the previous solution.

18-26

(a)

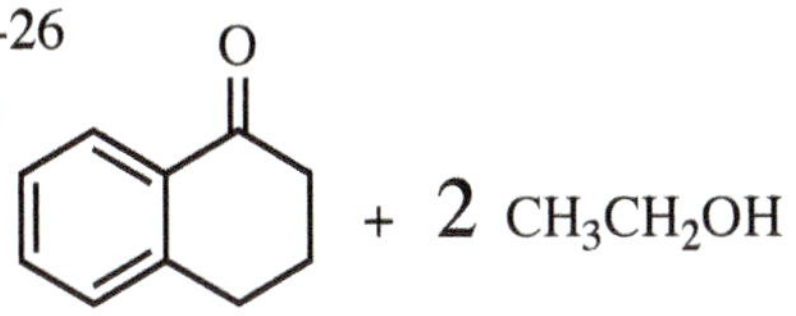

(b)

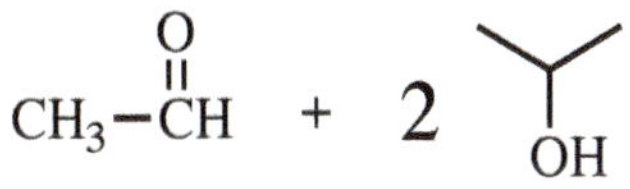

(c)

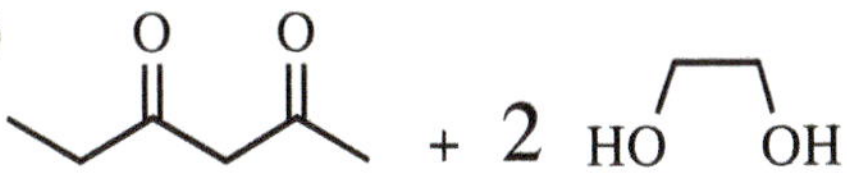

(d)

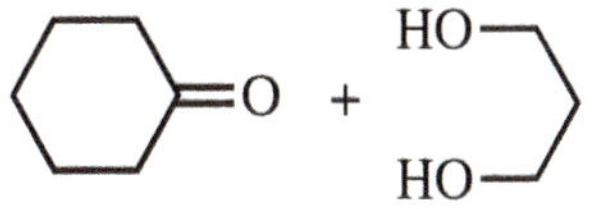

(e)

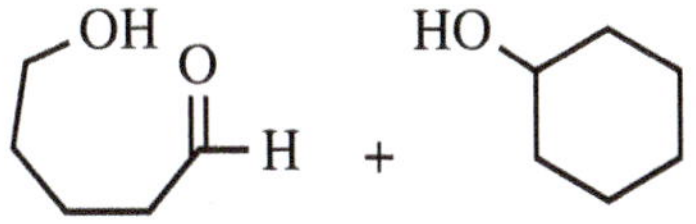

(f)

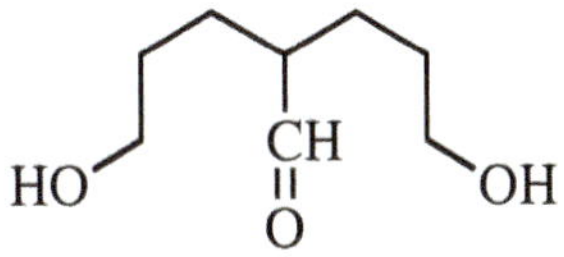

18-27

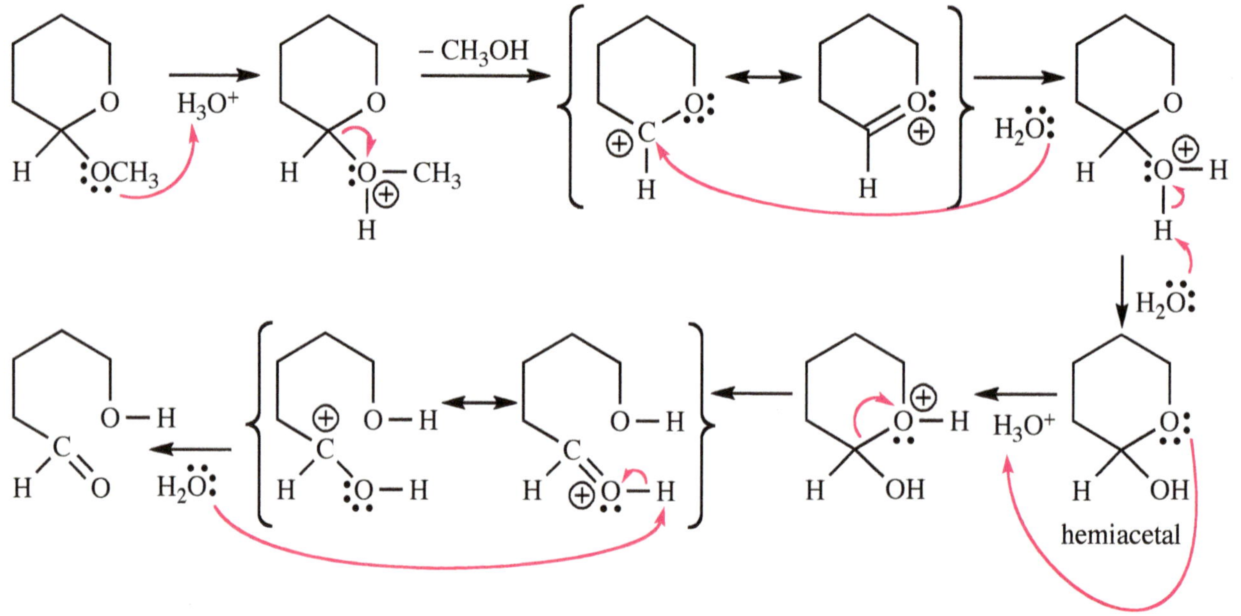

18-28

(a)

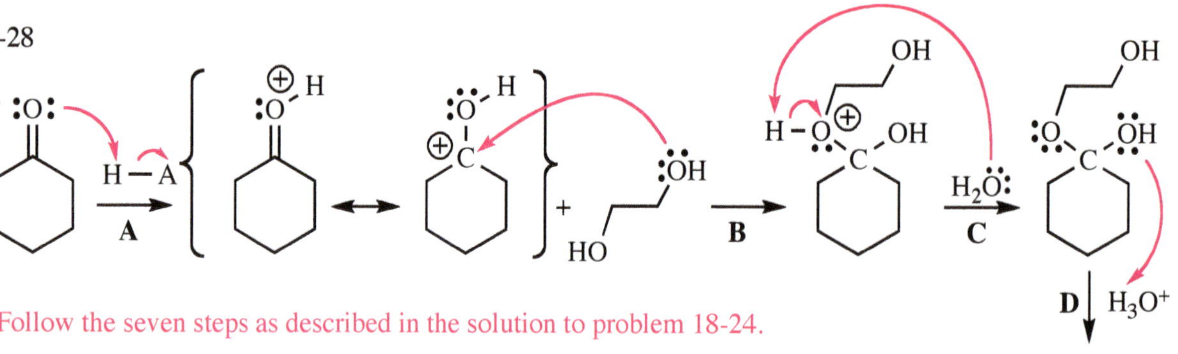

Follow the seven steps as described in the solution to problem 18-24.

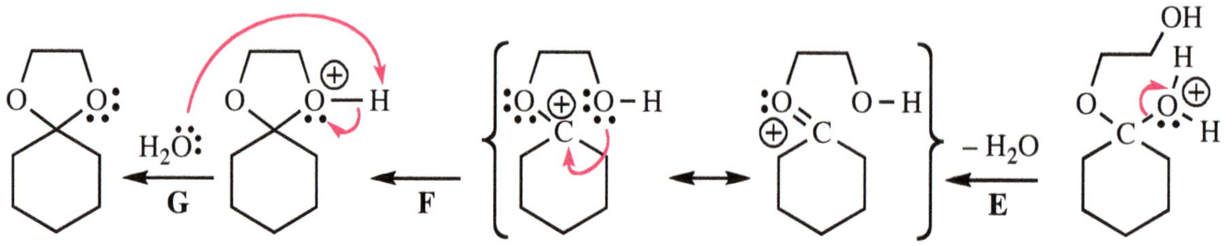

18-28 continued

(b)

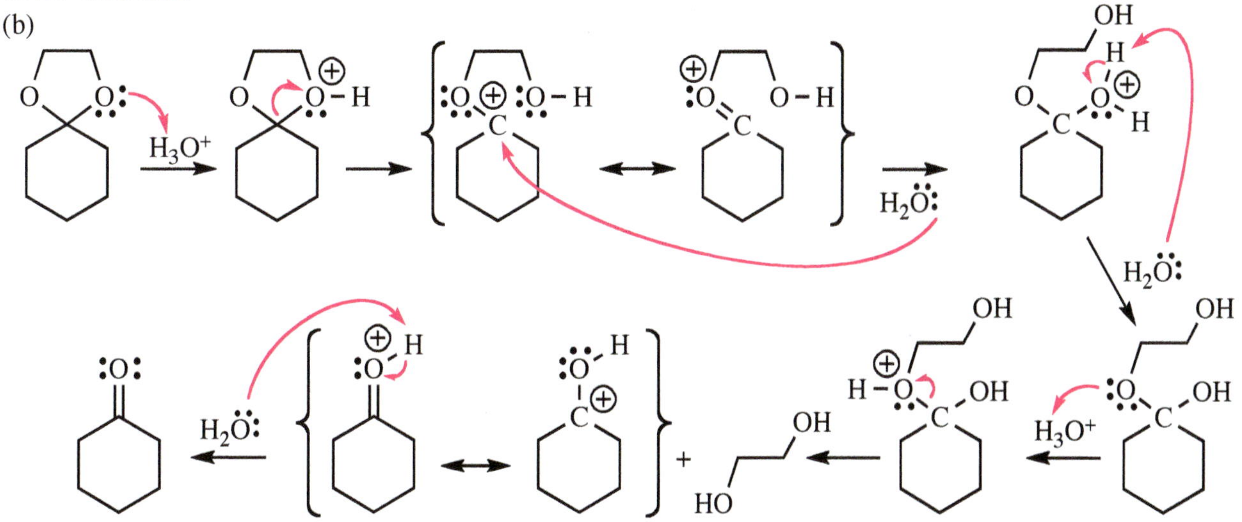

(c) The mechanisms of formation of an acetal and hydrolysis of an acetal are identical, just in reverse order. This has to be true because this process is an equilibrium: if the forward steps follow a minimum energy path, then the reverse steps have to follow the identical minimum energy path. This is the famous Principle of Microscopic Reversibility, text section 8-4A.

(d)

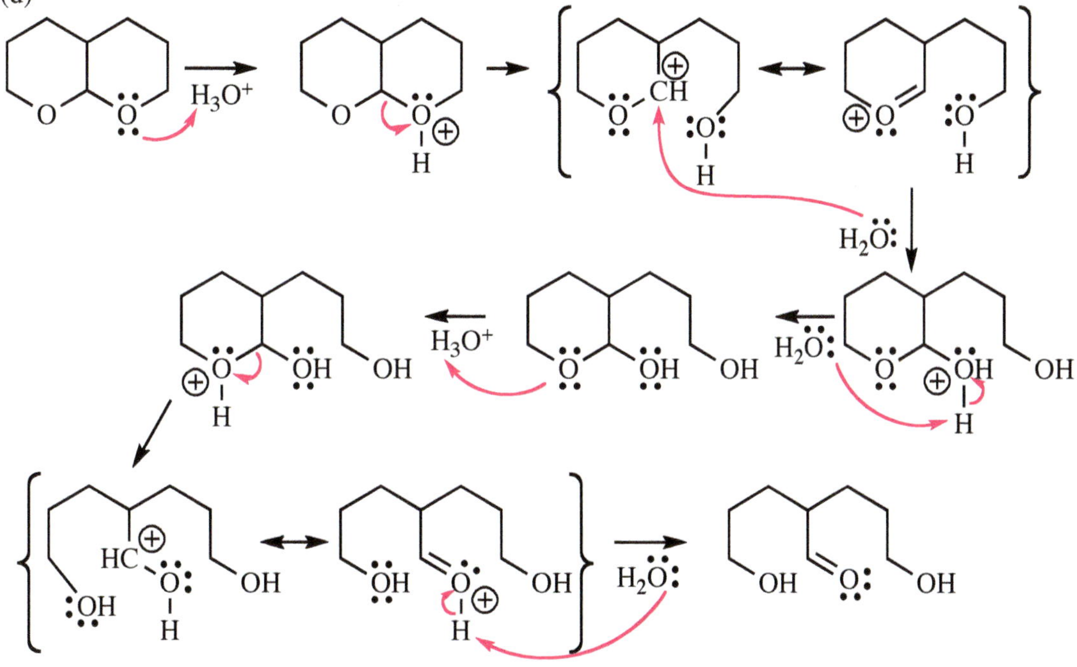

18-29

(a)

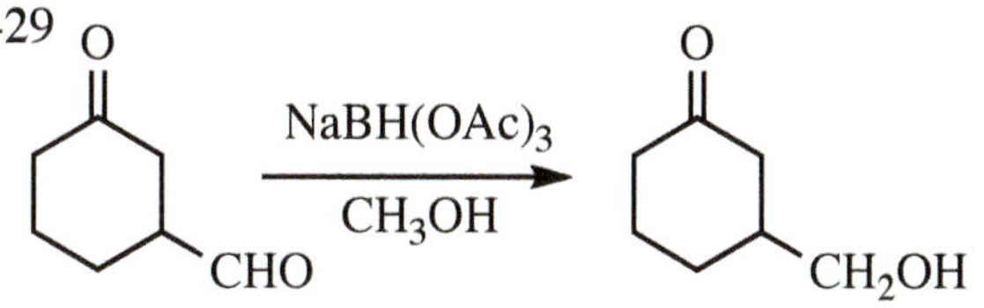

Sodium triacetoxyborohydride is the optimum reagent to reduce aldehydes selectively in the presence of ketones; see the solution to problem 18-12. It does not reduce ketones even with an excess of the reagent. Even with aldehydes, the reaction is slow—slower reactions are more selective.

18-29 continued

(b) [3-formylcyclohexanone] $\xrightarrow[\text{acid catalyst}\ -H_2O]{\text{1 equivalent HO–CH}_2\text{CH}_2\text{–OH}}$ [3-(1,3-dioxolan-2-yl)cyclohexanone] $\xrightarrow{CH_3MgBr}$ $\xrightarrow{H_3O^+}$ [1-methyl-3-formylcyclohexanol (HO, CH_3; CHO)]

Protect the more reactive functional group, then react at the other group.

(c) [2-(bromomethyl)cyclohexanone, CH_2Br] $\xrightarrow[\text{acid catalyst}\ -H_2O]{\text{1 equivalent HO–CH}_2\text{CH}_2\text{–OH}}$ [acetal, CH_2Br] $\xrightarrow[\text{ether}]{Mg}$ [acetal, CH_2MgBr] + [benzaldehyde, H–C(=O)–Ph]

↓

[acetal alkoxide, O^- $MgBr^+$] $\xrightarrow[\Delta]{H_3O^+}$ [2-(2-phenylethenyl)cyclohexanone]

The last step protonates the oxygen, dehydrates the alcohol, and hydrolyzes the acetal.

(d) [2,6-dimethylcyclohexane-1,4-dione, CH_3, CH_3] $\xrightarrow[\text{acid catalyst}\ -H_2O]{\text{1 equivalent HO–CH}_2\text{CH}_2\text{–OH}}$ [monoacetal, CH_3, CH_3] $\xrightarrow[CH_3OH]{NaBH_4}$ [H, OH, CH_3, CH_3 acetal] $\xrightarrow{H_3O^+}$ [4-hydroxy-3,5-dimethylcyclohexanone (H, OH, CH_3, CH_3)]

less-hindered ketone

(e) [cyclopentanone] + [$BrMgCH_2$–C(OCH$_2$CH$_2$O)–Ph] $\longrightarrow$ $\xrightarrow{H_3O^+}$ [1-hydroxycyclopentyl]$-CH_2-\overset{O}{\overset{\|}{C}}-Ph$ (OH)

(f) $BrCH_2CH_2\overset{O}{\overset{\|}{C}}-CH_3$ $\xrightarrow[\text{acid catalyst}\ -H_2O]{\text{HO–CH}_2\text{CH}_2\text{–OH}}$ $BrCH_2CH_2-C(OCH_2CH_2O)-CH_3$ $\xrightarrow{HC\equiv C:^-\ Na^+}$ $HC\equiv C-CH_2-CH_2-C(OCH_2CH_2O)-CH_3$

$\downarrow H_3O^+$

$HC\equiv C-CH_2-CH_2-\overset{O}{\overset{\|}{C}}-CH_3$

18-30 Trimethylphosphine has α-hydrogens that could be removed by butyllithium, generating undesired ylides.

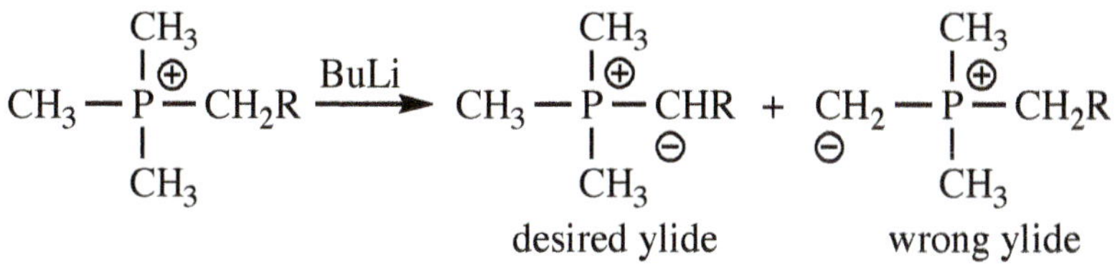
BuLi
desired ylide
wrong ylide

18-31

(a)

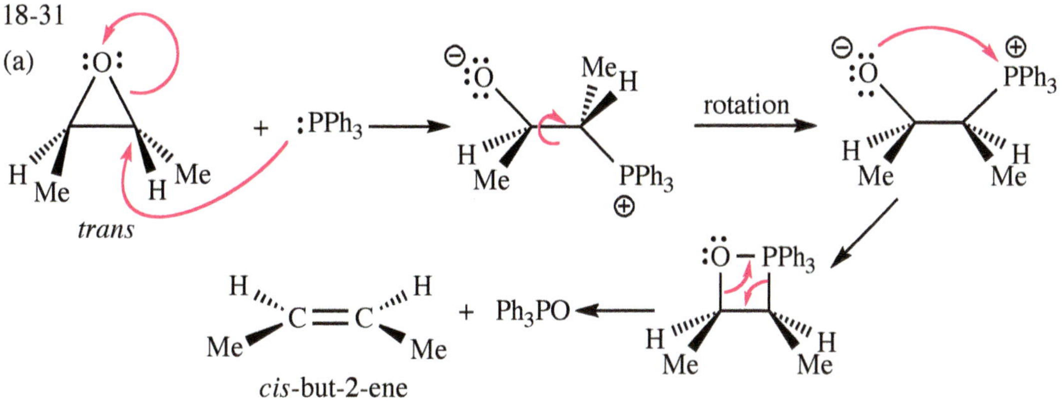
trans
rotation
cis-but-2-ene

The stereochemistry is inverted. The nucleophile triphenylphosphine must attack the epoxide in an *anti* fashion, yet the triphenylphosphine oxide must eliminate with *syn* geometry.

(b)

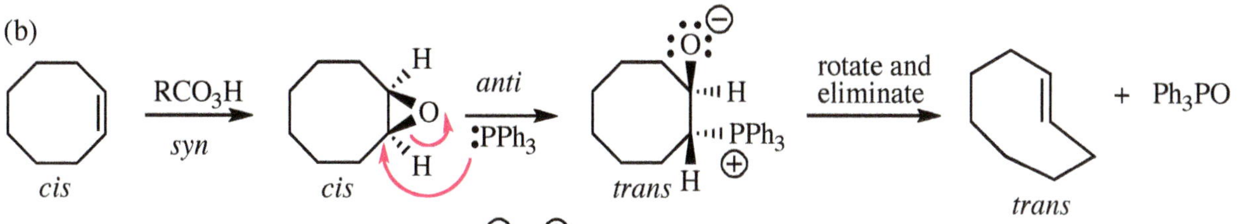
cis
RCO3H
syn
cis
anti
trans
rotate and
eliminate
trans
+ Ph3PO

18-32

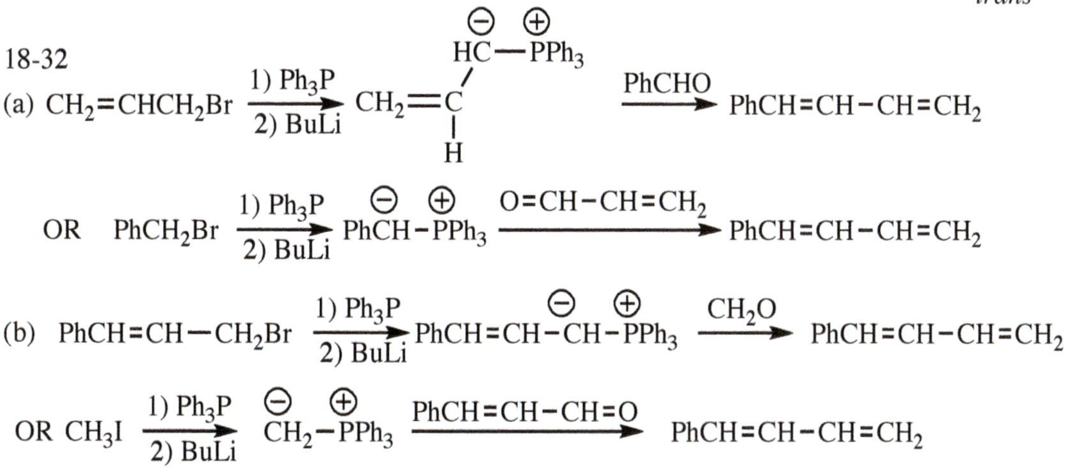
(a) CH2=CHCH2Br
1) Ph3P
2) BuLi
PhCHO
PhCH=CH−CH=CH2
OR PhCH2Br
O=CH−CH=CH2
PhCH=CH−CH=CH2
(b) PhCH=CH−CH2Br
CH2O
PhCH=CH−CH=CH2
OR CH3I
PhCH=CH−CH=O
PhCH=CH−CH=CH2

18-33 Many alkenes can be synthesized by two different Wittig reactions (as in the previous problem). The ones shown here form the phosphonium salt from the less hindered alkyl halide.

(a)

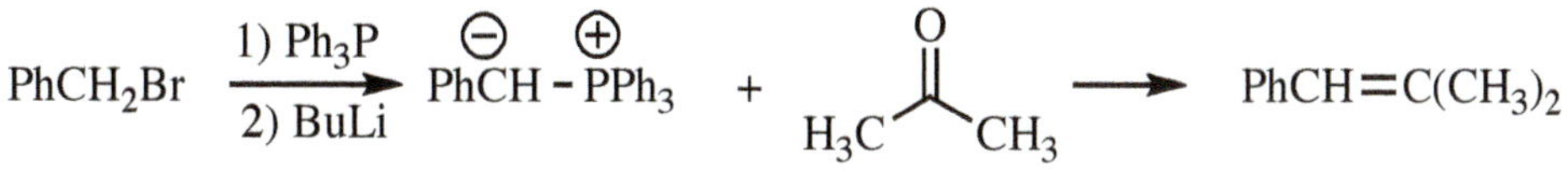
PhCH2Br
1) Ph3P
2) BuLi
PhCH=C(CH3)2

18-33 continued

(b) CH_3I $\xrightarrow[\text{2) BuLi}]{\text{1) }Ph_3P}$ $\overset{\ominus}{C}H_2-\overset{\oplus}{P}Ph_3$ + $PhC(=O)CH_3$ ⟶ $PhC(=CH_2)CH_3$

(c) $PhCH_2Br$ $\xrightarrow[\text{2) BuLi}]{\text{1) }Ph_3P}$ $Ph\overset{\ominus}{C}H-\overset{\oplus}{P}Ph_3$ $\xrightarrow{PhCH=CH-CH=O}$ $PhCH=CH-CH=CHPh$

(d) CH_3CH_2Br $\xrightarrow[\text{2) BuLi}]{\text{1) }Ph_3P}$ $CH_3\overset{\ominus}{C}H-\overset{\oplus}{P}Ph_3$ + cyclohexanone ⟶ cyclohexylidene=$CHCH_3$

18-34

(a) 4-hydroxycyclohexanecarboxylic acid (HO, COOH) + Ag^0, after adding acid

(b) 4-oxocyclohexanecarboxylic acid (O, COOH)

(c) 3-oxocyclohexanecarboxylic acid (O, COOH) + Ag^0, after adding acid

(d) HO, HO, COOH (3,4-dihydroxycyclohexanecarboxylic acid, wedged OH groups)

18-35 Imines can be formed in either acid or base conditions. Here, the hydrazone formation is shown in acid conditions.

Cyclohexanone $\xrightarrow[\textbf{A}]{H-A}$ {protonated carbonyl ($\overset{\oplus}{O}-H$) ⟷ carbocation ($\overset{\oplus}{C}-\ddot{O}-H$)} $\xrightarrow[:NH_2NH_2]{\textbf{B}}$ $O-H$ / $\overset{\oplus}{N}HNH_2$ (H) $\xrightarrow[H_2\ddot{O}:]{\textbf{C}}$ $\ddot{O}-H$ / $NHNH_2$ $\xrightarrow[\textbf{D}]{H_3O^+}$ $H-\overset{\oplus}{O}-H$ / $NHNH_2$ $\xrightarrow[\textbf{E}]{-H_2O}$ {NH_2, $\overset{\oplus}{N}$=C, H ⟷ $\overset{\oplus}{C}-\ddot{N}(NH_2)-H$} $\xrightarrow[\textbf{F}]{H_2\ddot{O}:}$ cyclohexanone hydrazone ($C=\ddot{N}-NH_2$)

Follow the six steps **A–F** of imine formation in the solution to 18-17(a).

Reduction of the hydrazone shown on the next page.

18-35 continued

reduction of the hydrazone

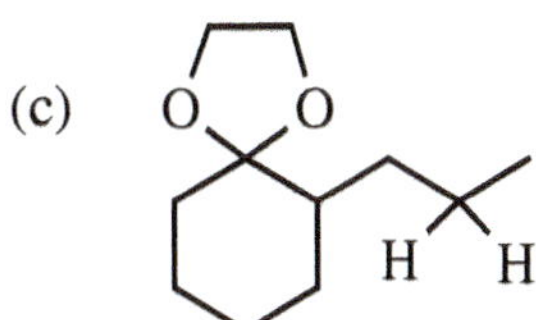

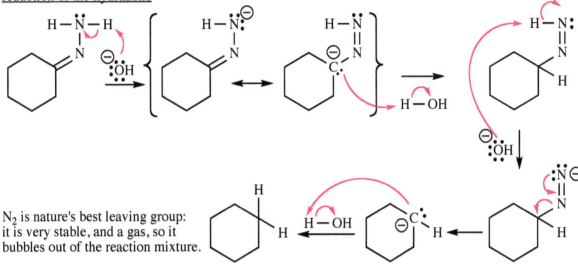

N_2 is nature's best leaving group: it is very stable, and a gas, so it bubbles out of the reaction mixture.

+ :N≡N: (gas)

18-36

(a) common name: indane

(b)

(c)

(d) Aqueous acid first hydrolyzes the acetal; the Zn(Hg)/HCl then reduces both ketones.

18-37

(a)

(b)

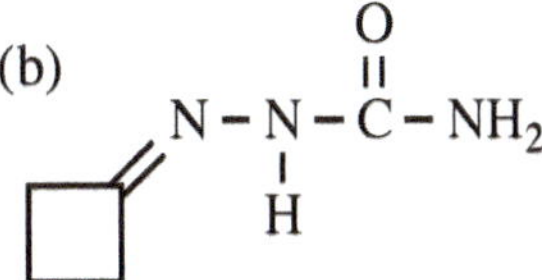

(c)

(d)

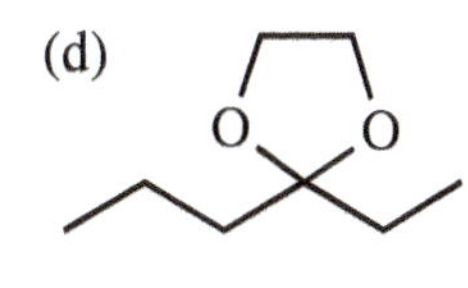

(e)

(f)

(g)

(h) 5-hydroxypentanal → hemiacetal

18-38 IUPAC names first; then common names. Recall that IUPAC recommends placing the position numbers as close as possible to the groups they describe. Parts (i)–(l) have no common names.

(a) heptan-2-one; methyl pentyl ketone
(b) heptan-4-one; dipropyl ketone
(c) heptanal; no simple common name
(d) benzophenone; diphenyl ketone
(e) butanal; butyraldehyde
(f) propanone; acetone (IUPAC accepts "acetone")
(g) 4-bromo-2-methylhexanal; no common name
(h) 3-phenylprop-2-enal; cinnamaldehyde
(i) hexa-2,4-dienal
(j) 2-hydroxycyclohexane-1,3-dione
(k) 3-oxocyclopentanecarbaldehyde
(l) *cis*-2,4-dimethylcyclopentanone

18-39

(a) N–NHPh

(b) O, H

(c) O, H

(d) O

(e) H, O, HO + CH_3OH

(f) H, O

(g) O

(h)

(i) OH, OH + H, O, H

(j) CH_3, N

18-40 In order of increasing equilibrium constant for hydrate formation:

$CH_3-\overset{O}{\overset{\|}{C}}-CH_3$ < $CH_3-\overset{O}{\overset{\|}{C}}-CH_2Cl$ < $CH_3-\overset{O}{\overset{\|}{C}}-H$ < $ClCH_2-\overset{O}{\overset{\|}{C}}-H$ < $H-\overset{O}{\overset{\|}{C}}-H$

least amount of hydration — greatest amount of hydration

Steric hindrance decreases hydrate formation: electron-withdrawing substituents at the alpha carbon increase hydrate formation. Without more information, it would be hard to say if the CH_2Cl group is better or worse than H in formation of a hydrate.

18-41

O, O; O, O; O, O, O, OH

18-42

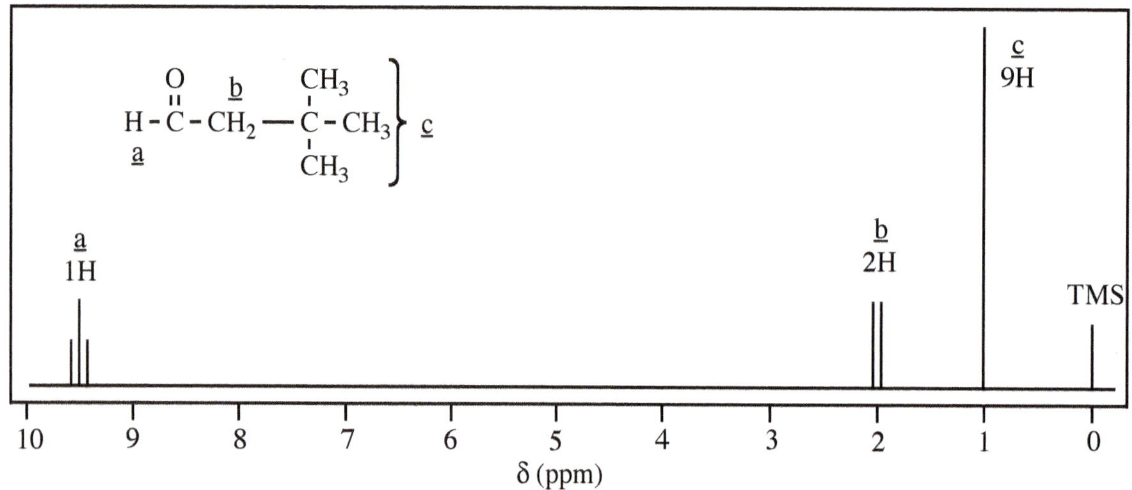

18-43 $C_6H_{10}O_2$ indicates two elements of unsaturation.

The IR absorption at 1708 cm^{-1} suggests a ketone, or possibly two ketones since there are two oxygens and two elements of unsaturation. The NMR singlets in the ratio of 2 : 3 indicate a highly symmetric molecule. The singlet at δ 2.15 is probably methyl next to carbonyl, and the singlet at δ 2.67 integrating to two is likely to be CH_2 on the other side of the carbonyl.

$$H_3C-\overset{O}{\overset{\|}{C}}-CH_2- \Longrightarrow \boxed{H_3C-\overset{O}{\overset{\|}{C}}-CH_2-CH_2-\overset{O}{\overset{\|}{C}}-CH_3}$$

Since the molecular formula is double this fragment, the molecule must be twice the fragment.

Two questions arise. Why is the integration 2 : 3 and not 4 : 6? Integration provides a *ratio*, not absolute numbers, of hydrogens. Why don't the two methylenes show splitting? Adjacent, *identical* hydrogens, with identical chemical shifts, do not split each other; the signals for ethane or cyclohexane appear as singlets.

18-44 The formula $C_{10}H_{12}O$ indicates five elements of unsaturation. A solid 2,4-DNP derivative suggests an aldehyde or a ketone, but a negative Tollens test precludes the possibility of an aldehyde; therefore, the unknown must be a ketone.

The NMR shows the typical ethyl pattern at δ 1.0 (3H, triplet) and δ 2.5 (2H, quartet), and a monosubstituted benzene at δ 7.3 (5H, multiplet). The singlet at δ 3.7 is a CH_2, but quite far downfield, apparently deshielded by two groups. Assemble the pieces:

$$C_6H_5- \; + \; CH_2 \; + \; \overset{O}{\overset{\|}{C}} \; + \; CH_2CH_3 \Longrightarrow \boxed{C_6H_5-CH_2-\overset{O}{\overset{\|}{C}}CH_2CH_3}$$

18-45

(a) $[H-C(=O)-CH_2-CH_2-CH_2CH_3]^{+\bullet}$ (m/z 86) ⟶ $[H-C(OH)=CH_2]^{+\bullet}$ (m/z 44) + $CHCH_3=CH_2$ (mass 42)

(b) $[CH_3-C(=O)-CH(CH_3)-CH_2-CH_2CH_3]^{+\bullet}$ (m/z 114) ⟶ $[CH_3-C(OH)=CH-CH_3]^{+\bullet}$ (m/z 72) + $CH_3CH=CH_2$ (mass 42)

(c) $[H_3C-C(=O)-CH_2-CH(CH_3)-CH_2CH_3]^{+\bullet}$ (m/z 114) ⟶ $[H_3C-C(OH)=CH_2]^{+\bullet}$ (m/z 58) + $CHCH_3=CHCH_3$ (mass 56)

18-46

A molecular ion of m/z 70 means a fairly small molecule. A solid semicarbazone derivative and a negative Tollens test indicate a ketone. The carbonyl (CO) has a mass of 28, so 70 – 28 = 42, enough mass for only three more carbons. The molecular formula is probably C_4H_6O (mass 70); with two elements of unsaturation, we can infer the presence of a double bond or a ring in addition to the carbonyl.

The IR shows a strong peak at 1790 cm^{-1}, indicative of a ketone in a small ring. No peak in the 1600–1650 cm^{-1} region shows the absence of an alkene. The only possibilities for a small ring ketone containing four carbons are these:

A (cyclobutanone, carbons numbered 1–4)   B (2-methylcyclopropanone, CH_3)

The HNMR can distinguish these. No methyl doublet appears in the NMR spectrum, ruling out **B**. The NMR does show a 4H triplet at δ 3.1; this signal comes from the two methylenes (C-2 and C-4) adjacent to the carbonyl, split by the two hydrogens on C-3. The signal for the methylene at C-3 appears at δ 2.0, roughly a quintet because of splitting by four neighboring protons.

The unknown is cyclobutanone, **A**. The symmetry indicated by the carbon NMR rules out structure **B**. The IR absorption of the carbonyl at 1790 cm^{-1} is characteristic of small ring ketones; ring strain strengthens the carbon-oxygen double bond, increasing its frequency of vibration. (See Section 12-9 in the text.)

18-47

(a) cyclopentanone + $Ph_3\overset{+}{P}–\overset{-}{C}H$CH₂CH₂CH₃ —Wittig→ butylidenecyclopentane

Br–CH₂CH₂CH₂CH₃ —PPh_3→ ; —BuLi→ ylide

(b) anisole (OCH_3) —fuming H_2SO_4→ HO_3S-anisole (para position is blocked) —CO, HCl / $AlCl_3$, CuCl (Gatterman-Koch)→ OCH_3, CHO, HO_3S —H_2O, Δ (remove SO_3H)→ OCH_3, CHO

While this Gatterman-Koch formylation reaction (text section 17-11C) would give a decent yield of the ortho product, to avoid reaction at the para position, the SO_3H blocking group could be used. See the solution to problem 17-67(c) for a refresher on how to use the SO_3H as a blocking group.

(c) allylic bromide (Br) —NaCN→ nitrile (C≡N) —1) DIBAL-H; 2) H_3O^+→ aldehyde (C(=O)H)

(d) cyclohexylmethyl Br —1) PPh_3; 2) BuLi→ ylide ($\overset{+}{P}Ph_3$) —cyclopentanone (O), Wittig→ alkene product

18-47 continued

(e) $SOCl_2$; + $(CH_2{=}CH)_2CuLi$ (excess)

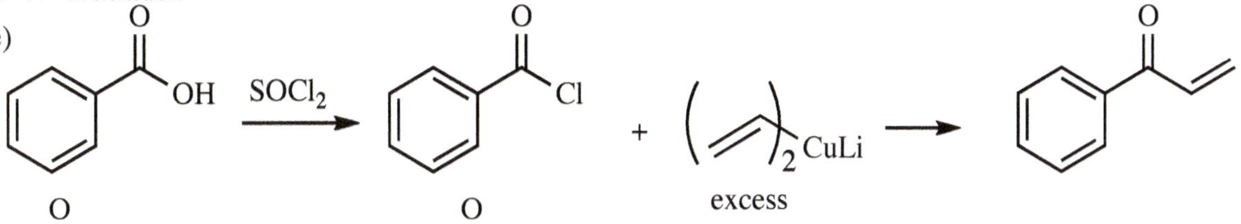

(f) DIBAL-H, −78 °C; H_2O

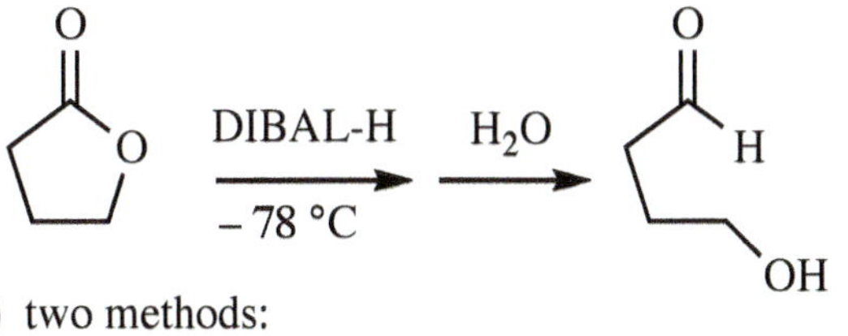

(g) two methods:

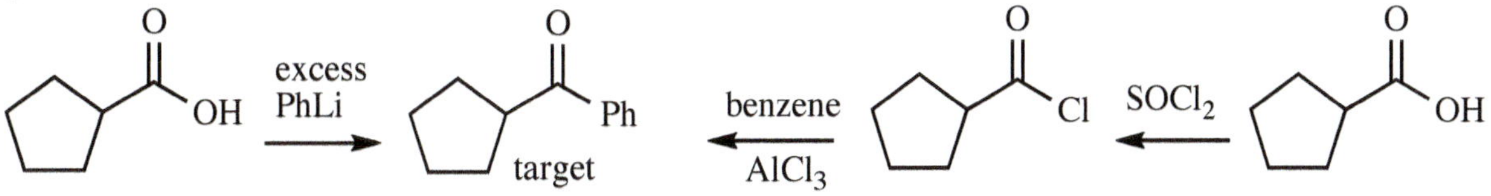

excess PhLi

target

benzene, $AlCl_3$

$SOCl_2$

(h) PCC or Swern or DMP

HO OH, acid catalyst, − H_2O

$Ph_3P^{\oplus}—CH_2^{\ominus}$

H_3O^+

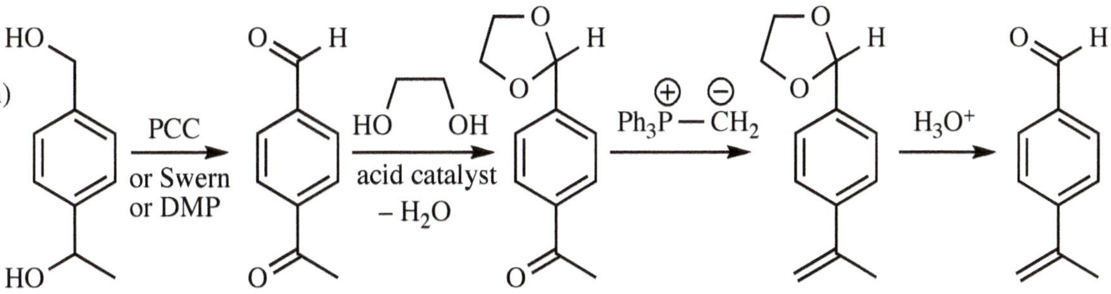

NaOCl would oxidize the 1° alcohol to COOH before oxidizing the 2° alcohol.

18-48 Recall that "dilute acid" means an aqueous solution, and aqueous acid will remove acetals.

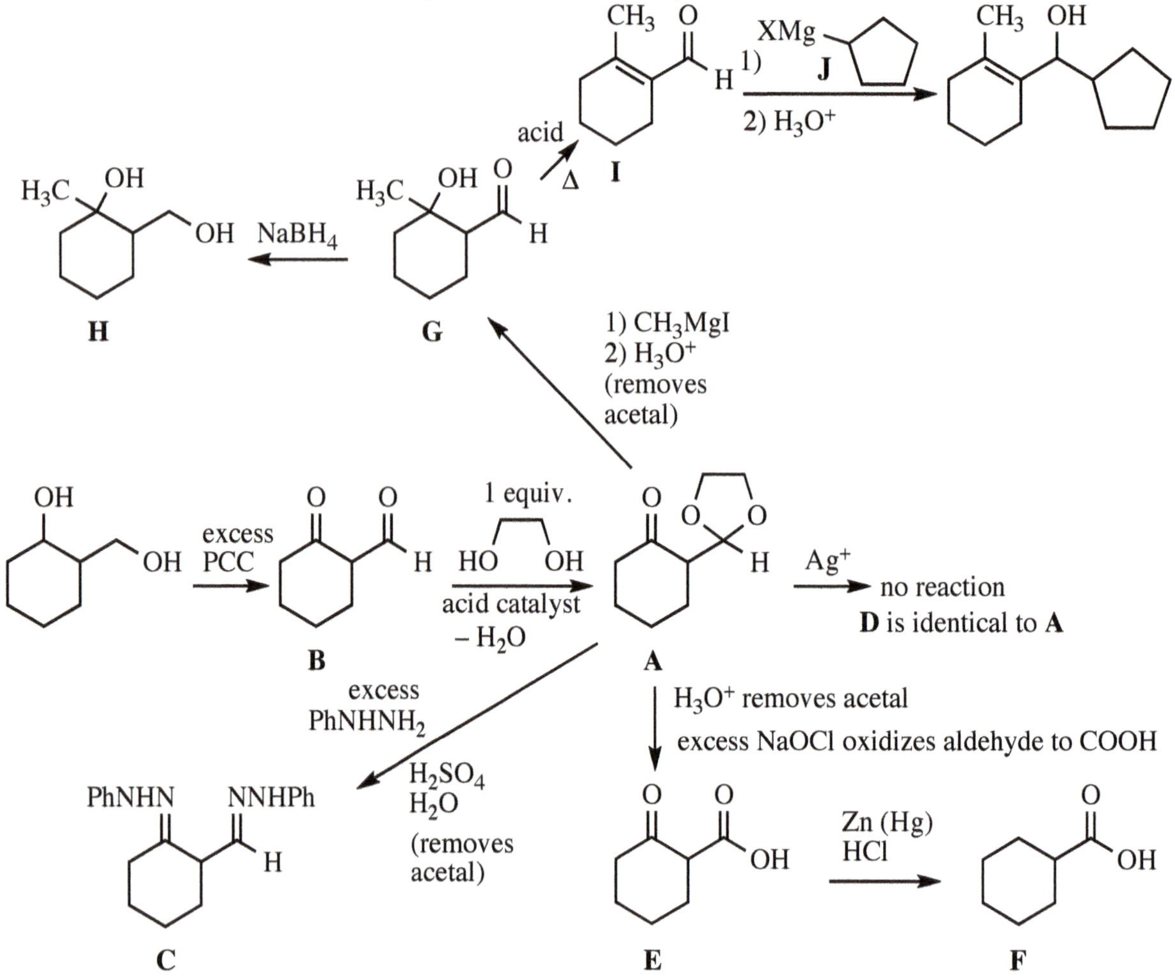

CH_3 O — H; 1) XMg– (J) 2) H_3O^+ → CH_3 OH

acid, Δ — I

H_3C OH — OH (H) ← $NaBH_4$ — H_3C OH O — H (G)

1) CH_3MgI
2) H_3O^+
(removes acetal)

OH — OH (excess PCC) → O O — H (B) → (1 equiv. HO OH, acid catalyst, $-H_2O$) → O, O O — H (A) → Ag^+ → no reaction
D is identical to **A**

excess $PhNHNH_2$, H_2SO_4, H_2O (removes acetal) → PhNHN NNHPh — H (C)

H_3O^+ removes acetal
excess NaOCl oxidizes aldehyde to COOH

O O — OH (E) → Zn (Hg), HCl → O — OH (F)

18-49 Historically, acetals have come from aldehydes and ketals have come from ketones. IUPAC has discouraged but still accepts the *ketal* designation and now recommends just the use of *acetal*.

(a) acetal (old: ketal) $CH_3CH_2CH_2-\overset{O}{\overset{\|}{C}}-CH_3$ + **2** CH_3OH

(b) hemiacetal (old: hemiketal) cyclohexanone + CH_3CH_2OH

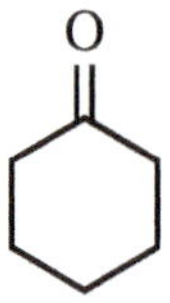

(c) acetal — O H, H, O + HO–cyclopentyl

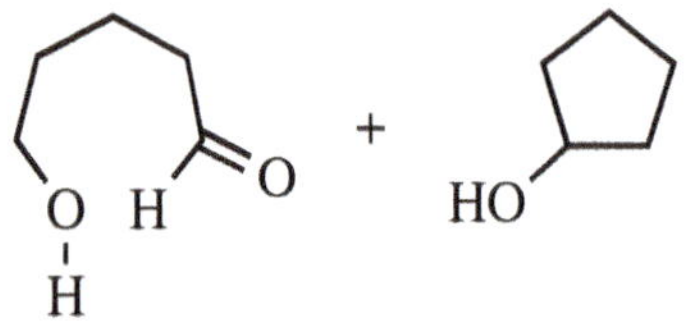

(d) acetal (old: ketal) 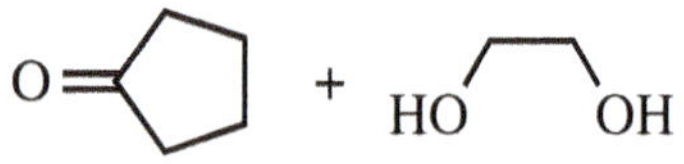 O=cyclopentanone + HO OH

(e) acetal — OH, OH + O=C(H)H

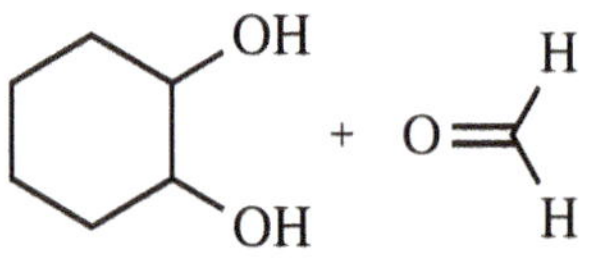

(f) diether, inert to hydrolysis

(g) imine(s) NH_2, NH_2 + O=CH–CH=O

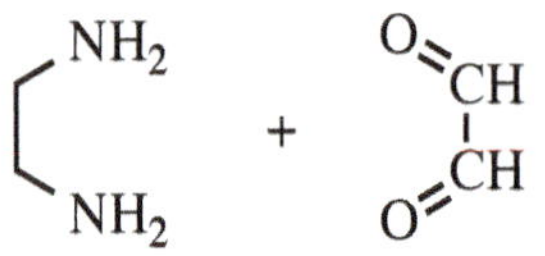

(h) hydrazone (a type of imine) cyclopentanone=O + H_2N-NH_2

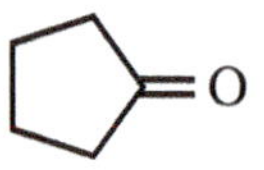

18-50

(a) Ph—CH=O + H—A (A) → {Ph—CH=$\overset{+}{O}$—H ↔ Ph—$\overset{+}{C}$H—OH} (B, $HOCH_3$) → ... (C) → hemiacetal (D, H—A) → ... (E, $-H_2O$) → ... (F, $HOCH_3$) → ... (G) → acetal

A: plus resonance forms that show the (+) charge on the benzene ring

E: plus resonance forms that show the (+) charge on the benzene ring

hemiacetal

acetal

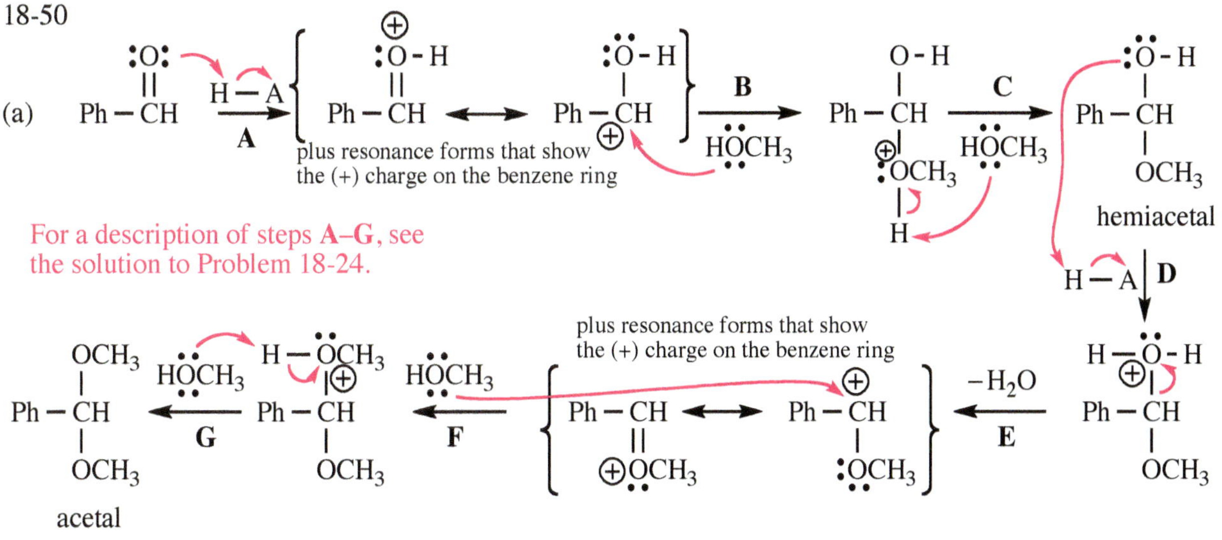

For a description of steps **A–G**, see the solution to Problem 18-24.

(b) Steps A–F: H—A (A); $:NH_2NHPh$ (B); $H_2\ddot{O}:$ (C); H_3O^+ (D); $-H_2O$ (E); $H_2\ddot{O}:$ (F); products with NHNHPh, NHPh

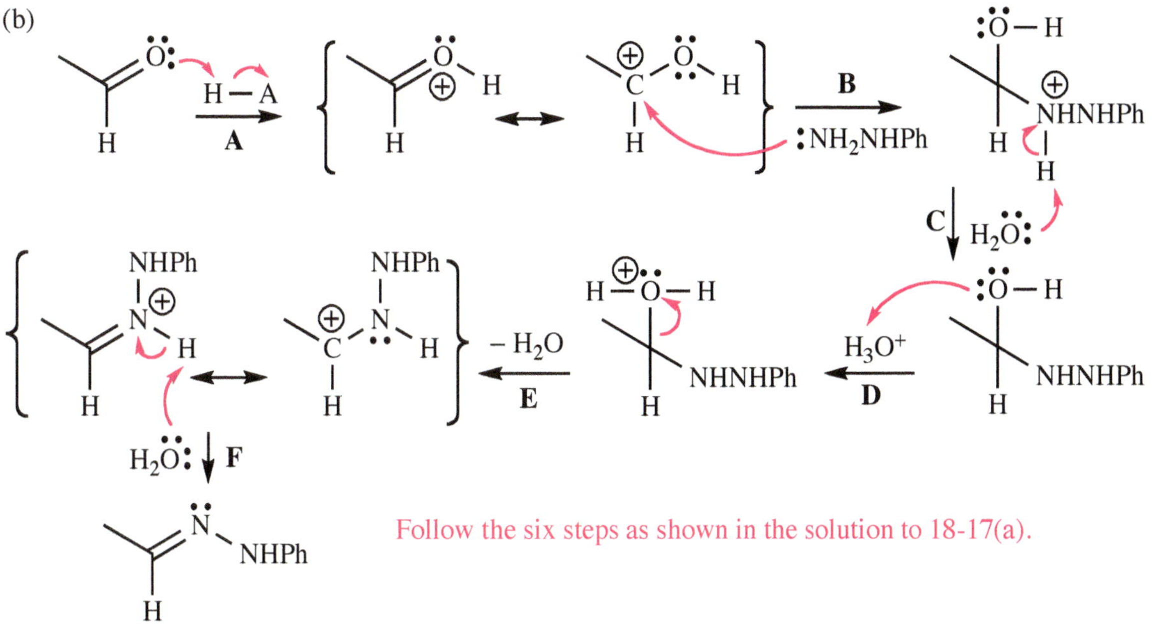

Follow the six steps as shown in the solution to 18-17(a).

(c) Cyclohexanone + $:\overset{\ominus}{C}H_2—\overset{\oplus}{P}Ph_3$ → betaine (O^-, $\overset{+}{P}Ph_3$, CH_2) → oxaphosphetane (O—PPh_3, CH_2) → Ph_3PO + methylenecyclohexane ($=CH_2$)

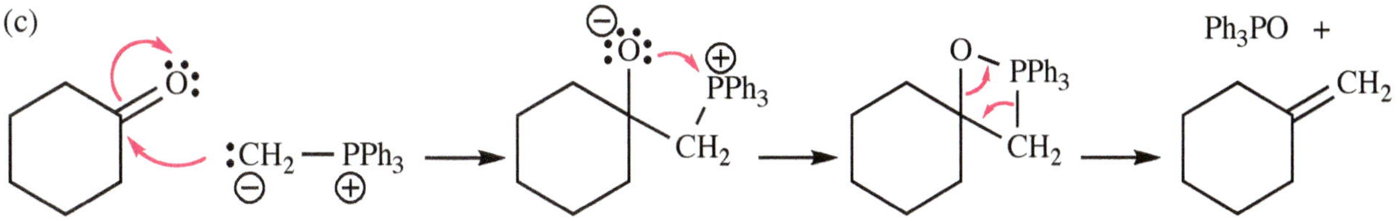

18-50 continued

(d)

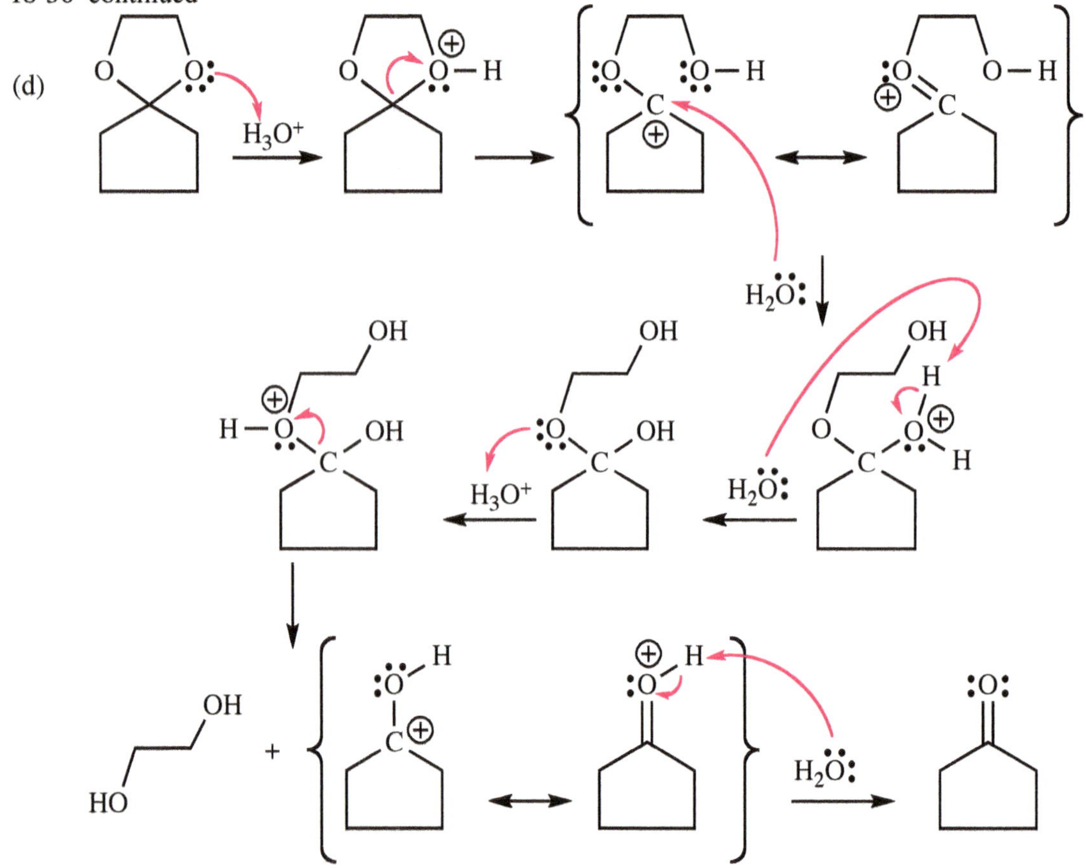

H_3O^+

$H_2\ddot{O}$:

H_3O^+

$H_2\ddot{O}$:

$H_2\ddot{O}$:

(e)

H_3O^+

$H_2\ddot{O}$:

$H_2\ddot{O}$:

H_3O^+

$H_2\ddot{O}$:

H_3O^+

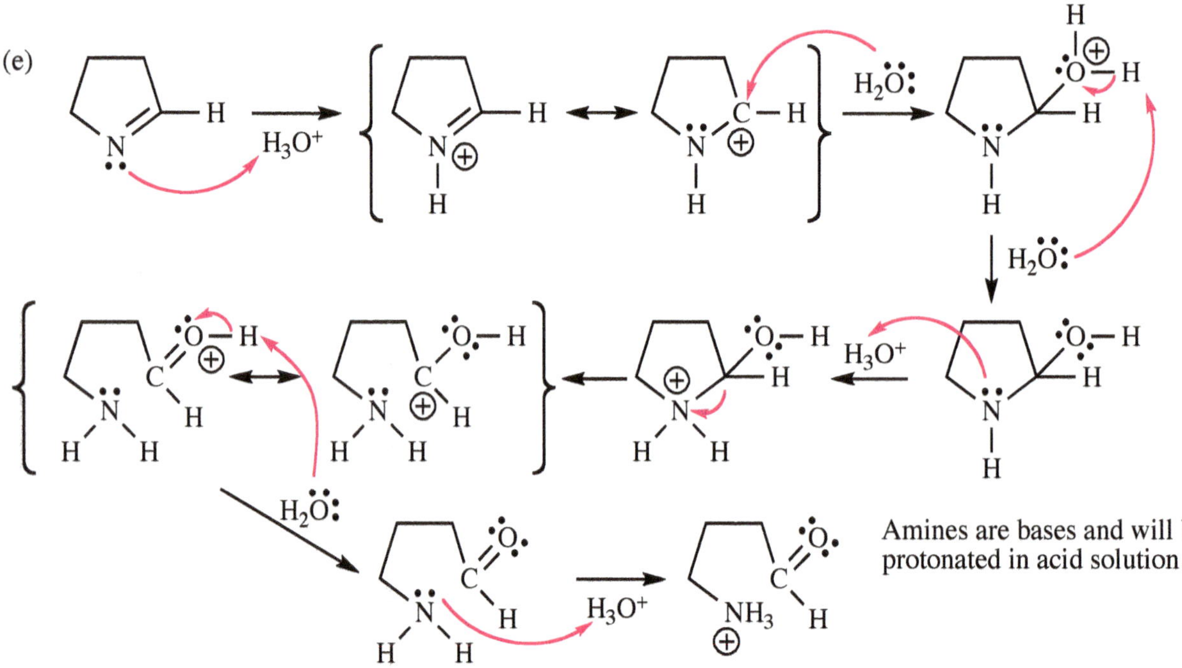

Amines are bases and will be protonated in acid solution.

18-50 continued

(f) For simplicity, $CH_3CH_2NH_2$ will be abbreviated $EtNH_2$. $EtNH_3^+$ serves as the proton source (acid) here.

$EtNH_3^+$; $- CH_3OH$; $EtNH_2$; $EtNH_2$; $EtNH_3^+$; $- CH_3OH$; $EtNH_2$

18-51

(a) $CH_3-\overset{O}{\overset{\|}{C}}H \xrightarrow[HCN]{KCN} CH_3-\overset{OH}{\overset{|}{C}}H-C\equiv N \xrightarrow{H_3O^+} CH_3-\overset{OH}{\overset{|}{C}}H-COOH$

(b) $PhCH_2Br \xrightarrow{Ph_3P} \xrightarrow{BuLi} Ph\overset{\ominus}{C}H-\overset{\oplus}{P}Ph_3 \longrightarrow$ (with cyclopentanone) HC–Ph

(c) $\xrightarrow[CH_3OH]{NaBH(OAc)_3}$ CHO → CH_2OH

Sodium triacetoxyborohydride selectively reduces the aldehyde; see the solutions to problems 18-12 and 18-29(a).

(d) CHO $\xrightarrow[\text{acid catalyst},\ -H_2O]{\text{1 equivalent HO–CH}_2\text{CH}_2\text{–OH}}$ H ... $\xrightarrow[CH_3OH]{NaBH_4}$ H OH ... H $\xrightarrow{H_3O^+}$ H OH ... CHO

(e) CHO $\xrightarrow[\text{acid catalyst},\ -H_2O]{\text{1 equivalent HO–CH}_2\text{CH}_2\text{–OH}}$ H ... $CH_3CH_2CH_2Br \xrightarrow{PPh_3}$ $\xrightarrow{BuLi}$ $CH_3CH_2\overset{\ominus}{C}H-\overset{\oplus}{P}Ph_3$ → $CHCH_2CH_3$ ... H $\xrightarrow{H_3O^+}$ $CHCH_2CH_3$ ... H–C=O

18-51 continued

(f)

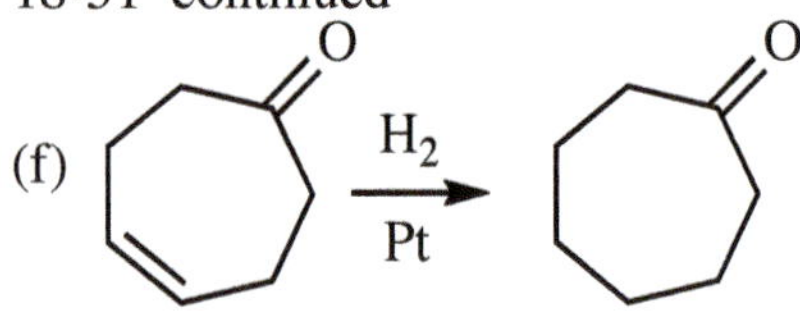

(g) O $\xrightarrow[CH_3OH]{NaBH_4}$ OH

LiAlH$_4$ in ether could also be used—with caution!

18-52 All of these reactions would be acid-catalyzed.

(a) + H_2NOH

(b) CHO + H_2N

(c) NH_2 + O

(d) O + HO OH

(e) NH_2 + O=C(CH_3)(CH_3)

(f) O + **2** CH_3OH

18-53

(a) NCH_3

(b) CH_3O OCH_3

(c) NOH

(d) O O

(e) N−N−Ph, H

(f) Ph OH

(g) no reaction

(h)

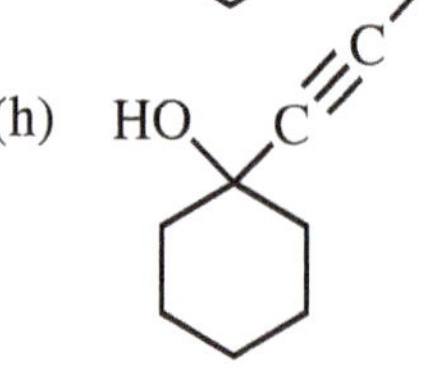

(i)

(j) CH_2

(k) Na^+ O^- C≡N

(l) HO COOH

18-54

(a) OH, C, H, Ph

(b) O, C, O^- + Ag^0

(c) H O, N−N−C−NH_2, C, H

(d) OCH_2CH_3, C, H, OCH_2CH_3

(e) O O, C, H

(f) CH_3

18-55

(a) [aldehyde] $\xrightarrow[\text{ether}]{CH_3MgI}$ $\xrightarrow{H_3O^+}$ [secondary alcohol, OH] $\xrightarrow[\text{HOAc}]{NaOCl}$ [methyl ketone, O]

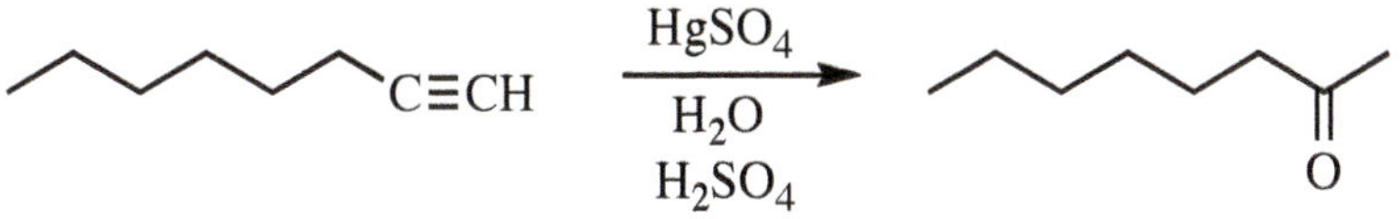

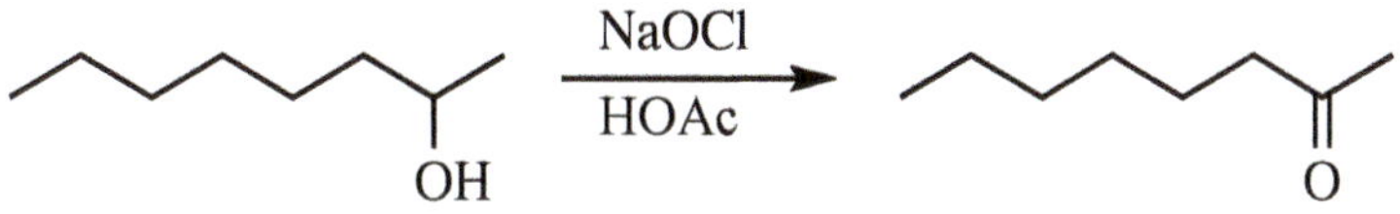

Swern or DMP or PCC or H_2CrO_4 would work too.

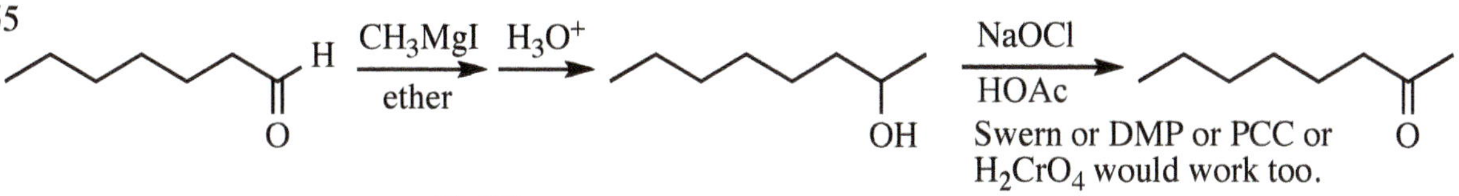

(b) [alkyne, C≡CH] $\xrightarrow[H_2O,\ H_2SO_4]{HgSO_4}$ [methyl ketone, O]

(c) [alkene] $\xrightarrow[\text{2) } Me_2S]{\text{1) } O_3\text{, −78 °C}}$ [ketone, O] + [acetone, O]

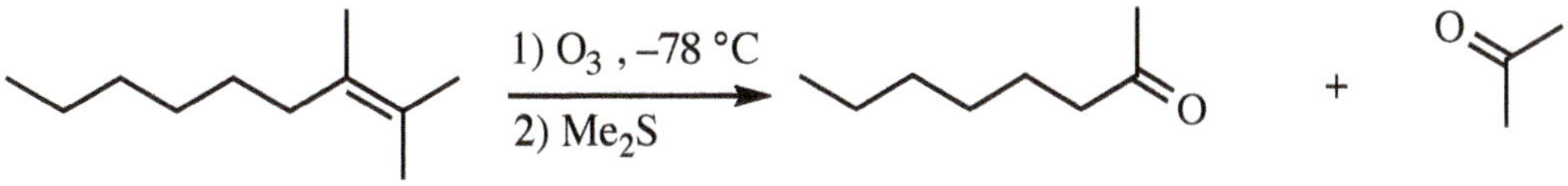

(d) [secondary alcohol, OH] $\xrightarrow[\text{HOAc}]{NaOCl}$ [methyl ketone, O] Swern or DMP or PCC or H_2CrO_4 would work too.

(e) [carboxylic acid, OH, O] $\xrightarrow[\text{2) } H_3O^+]{\text{1) excess } CH_3Li}$ [methyl ketone, O]

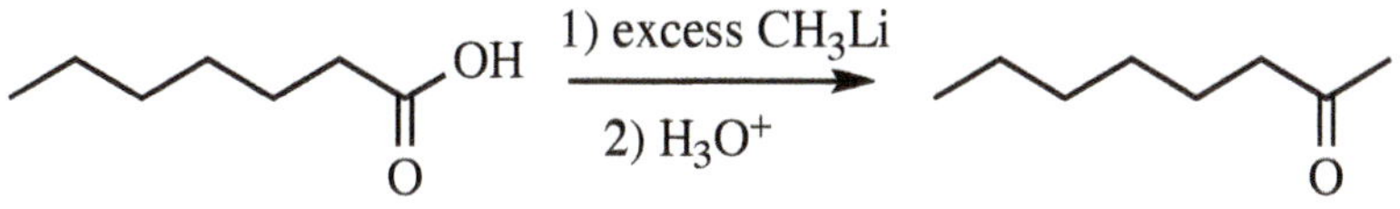

(f) [nitrile, C≡N] $\xrightarrow[\text{ether}]{CH_3MgI}$ $\xrightarrow{H_3O^+}$ [methyl ketone, O]

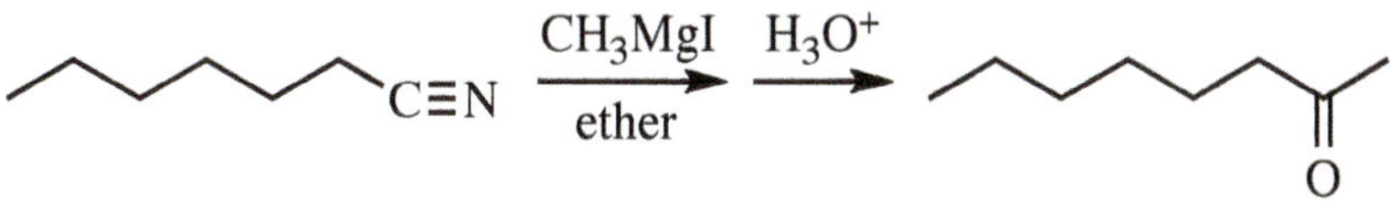

18-56

(a) [primary alcohol, OH] $\xrightarrow[\text{TEMPO}]{NaOCl}$ [aldehyde, O, H] PCC or Swern or DMP are OK too.

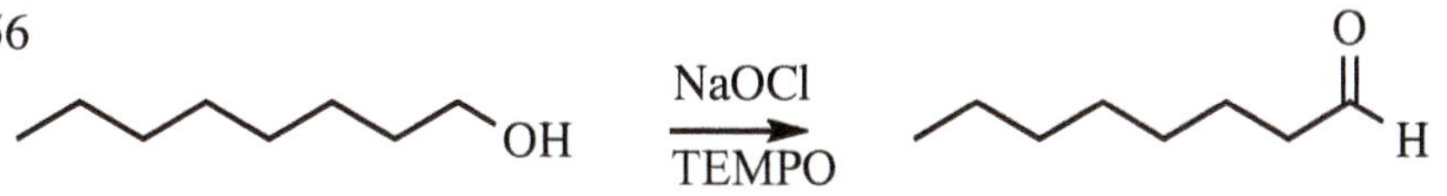

(b) [alkene, CH_2] $\xrightarrow[\text{2) } Me_2S]{\text{1) } O_3\text{, −78 °C}}$ [aldehyde, O, H]

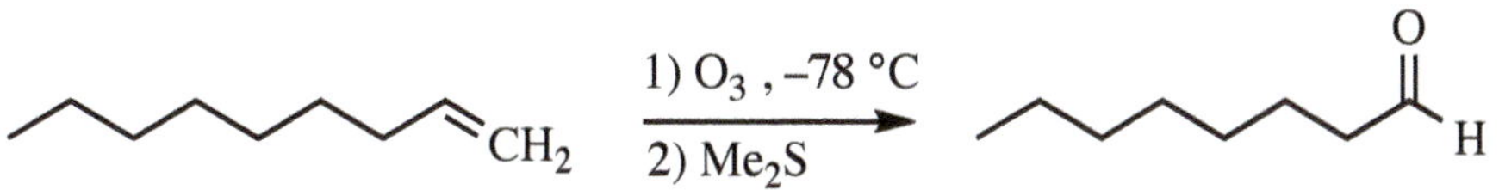

(c) [alkyne, C≡CH] $\xrightarrow[\text{2) } H_2O_2\text{, } HO^-]{\text{1) } Sia_2BH}$ [aldehyde, O, H]

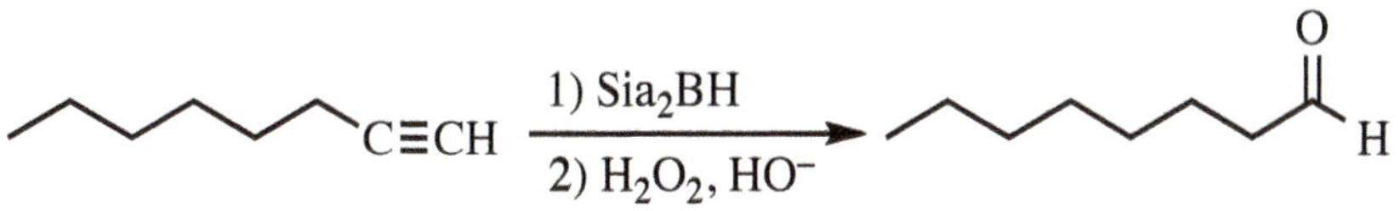

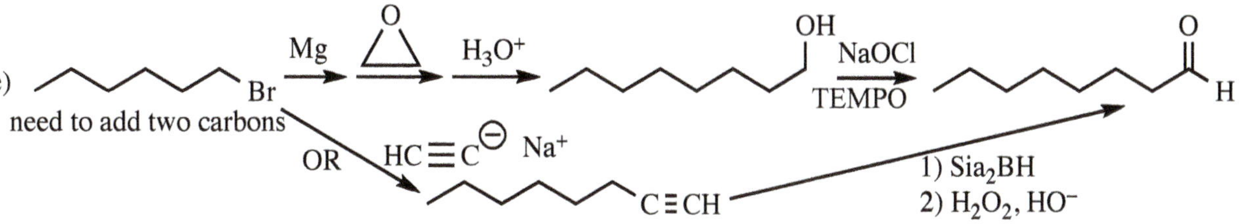

(d) [alkyl bromide, Br] $\xrightarrow{NaCN}$ [nitrile, CN] $\xrightarrow[\text{2) } H_3O^+]{\text{1) DIBAL-H}}$ [aldehyde, O, H]

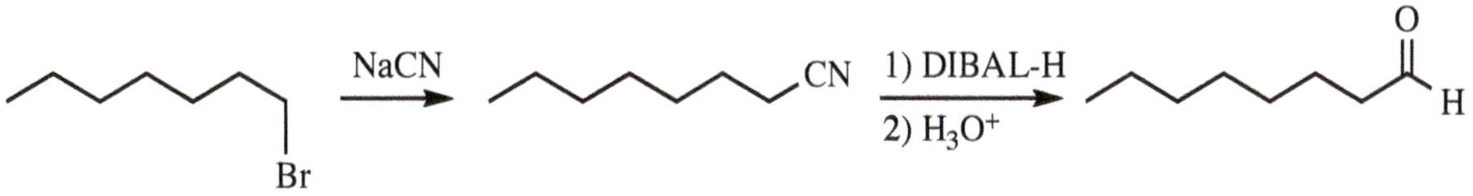

(e) [alkyl bromide, Br] $\xrightarrow{Mg}$ $\xrightarrow{\text{[ethylene oxide, O]}}$ $\xrightarrow{H_3O^+}$ [primary alcohol, OH] $\xrightarrow[\text{TEMPO}]{NaOCl}$ [aldehyde, O, H]

need to add two carbons

OR $HC{\equiv}C^{\ominus}\ Na^+$ → [alkyne, C≡CH] $\xrightarrow[\text{2) } H_2O_2\text{, } HO^-]{\text{1) } Sia_2BH}$ [aldehyde]

18-56 continued

(f) [octanoic acid] $\xrightarrow[2)\ H_3O^+]{1)\ LiAlH_4}$ [1-octanol] $\xrightarrow[TEMPO]{NaOCl}$ [octanal]

OR [octanoic acid] $\xrightarrow{SOCl_2}$ [octanoyl chloride] $\xrightarrow{LiAl(Ot\text{-}Bu)_3H}$ [octanal]

(g) [ethyl octanoate] $\xrightarrow[-78\ °C]{DIBAL\text{-}H}$ $\xrightarrow{H_2O}$ [octanal]

18-57 The new bond to carbon comes from the NaB**H**$_4$ or NaB**D**$_4$, shown in bold below. The new bond to oxygen comes from the protic solvent.

(a) $CH_3-C(=O)-CH_2CH_3 \xrightarrow{NaBD_4} \xrightarrow{H_2O} CH_3-C(OH)(\mathbf{D})-CH_2CH_3$

(b) $CH_3-C(=O)-CH_2CH_3 \xrightarrow{NaBD_4} \xrightarrow{D_2O} CH_3-C(OD)(\mathbf{D})-CH_2CH_3$

(c) $CH_3-C(=O)-CH_2CH_3 \xrightarrow{NaBH_4} \xrightarrow{D_2O} CH_3-C(OD)(\mathbf{H})-CH_2CH_3$

18-58 While hydride is a small group, the actual chemical species supplying it, AlH_4^-, is fairly large, so it prefers to approach from the less hindered side of the molecule, that is, the side opposite the methyl. This forces the oxygen to go to the same side as the methyl, producing the *cis* isomer as the major product.

less hindered face; $H-\overset{\ominus}{Al}H_2-H$ → H_3C ... $O^\ominus$ $\xrightarrow{H_3O^+}$ OH, CH_3

cis major product

18-59

(a) $CH_3Br \xrightarrow{Ph_3P} \xrightarrow{BuLi} \overset{\ominus}{C}H_2-\overset{\oplus}{P}Ph_3 \xrightarrow{\text{cyclohexanone}}$ [cyclohexane]$=CH_2$

methylenecyclohexane

An exocyclic double bond is less stable than an endocyclic double bond.

18-59 continued

(b) The difficulty in synthesizing methylenecyclohexane from cyclohexanone without using the Wittig reaction rests in the stability of the double bond inside the ring (endocyclic) versus outside the ring (exocyclic).

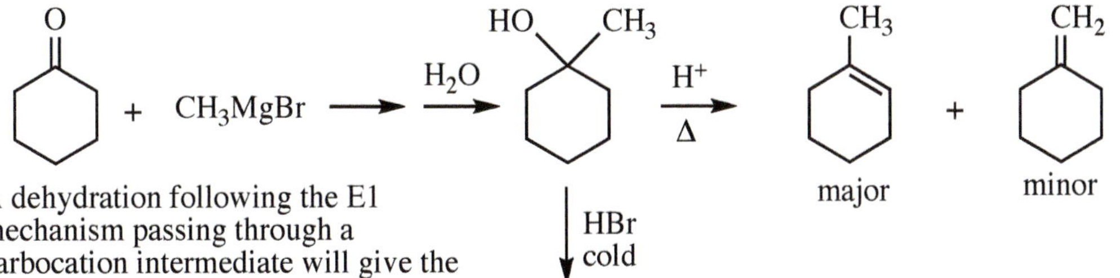

O + CH_3MgBr → H_2O → HO CH$_3$ → H^+, Δ → CH$_3$ (major) + CH$_2$ (minor)

HBr, cold

Br CH$_3$ → K^+ $^-$O-*t*-Bu, E2 → CH$_3$ (minor) + CH$_2$ (major)

A dehydration following the E1 mechanism passing through a carbocation intermediate will give the more substituted, endocyclic double bond as the major product. The only chance of making the exocyclic double bond as the major product is to do an E2 elimination using a bulky base to give Hofmann orientation (Chapter 7).

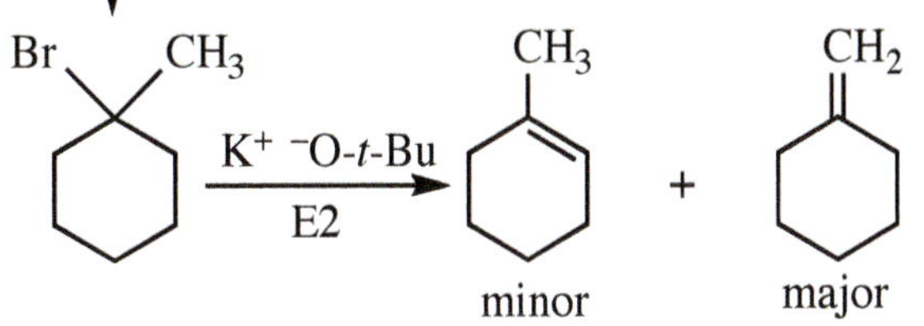

18-60

(a) Cl–C(=O)CH$_2$CH$_2$CH$_3$, $AlCl_3$ → O → Zn(Hg), HCl →

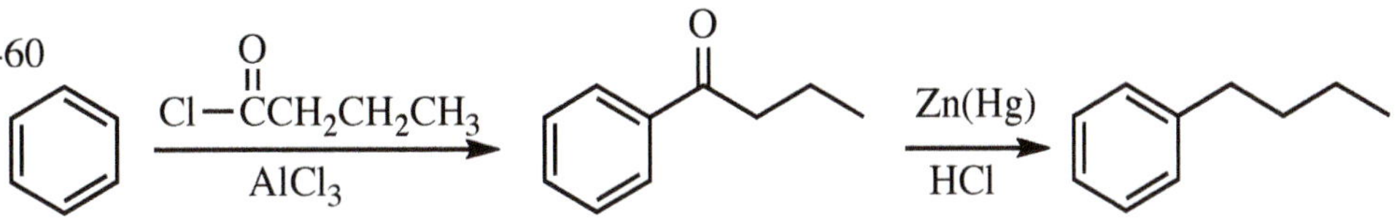

(b) C≡N → CH_3CH_2MgBr, ether → H_3O^+ → C(=O)–CH$_2$CH$_3$

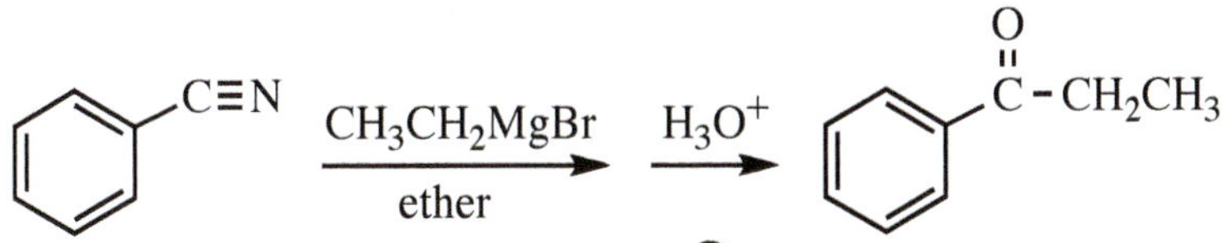

(c) Cl_2, $AlCl_3$ → Cl → NaOH, 350° C → $^{\ominus}$O Na$^+$ → CH_3I → OCH$_3$ → CO, HCl, $AlCl_3$, CuCl → OCH$_3$, CHO

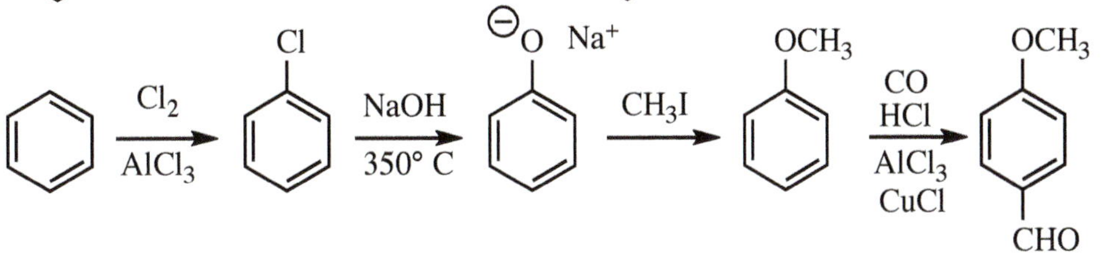

(d) HO → excess NaOCl, TEMPO or H_2CrO_4 → HO, O → conc. H_2SO_4 → O

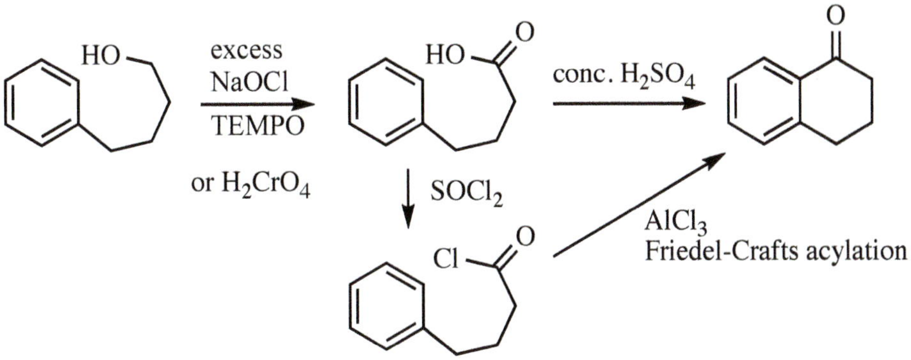

$SOCl_2$

Cl, O → $AlCl_3$ Friedel-Crafts acylation

18-61 The key to this problem is understanding that the relative proximity of the two oxygens can dramatically affect their chemistry.

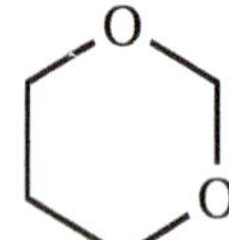

1,2-dioxane	1,3-dioxane	1,4-dioxane
The "second" isomer described: two oxygens connected by a sigma bond are a peroxide. The O—O bond is easily cleaved to give radicals. In the presence of organic compounds, radical reactions can be explosive.	The "third" isomer described: two oxygens bonded to the same sp^3 carbon constitute an acetal, which is hydrolyzed in aqueous acid. See the mechanism below.	The "first" isomer described: an excellent solvent (although toxic), these oxygens are far enough apart to act independently. It is a simple ether.

Mechanism of acetal hydrolysis

This is the acetal carbon. It becomes formaldehyde.

18-62 Building a model will help visualize this problem. Ignore stereochemistry at chiral carbons for this problem.

(a)

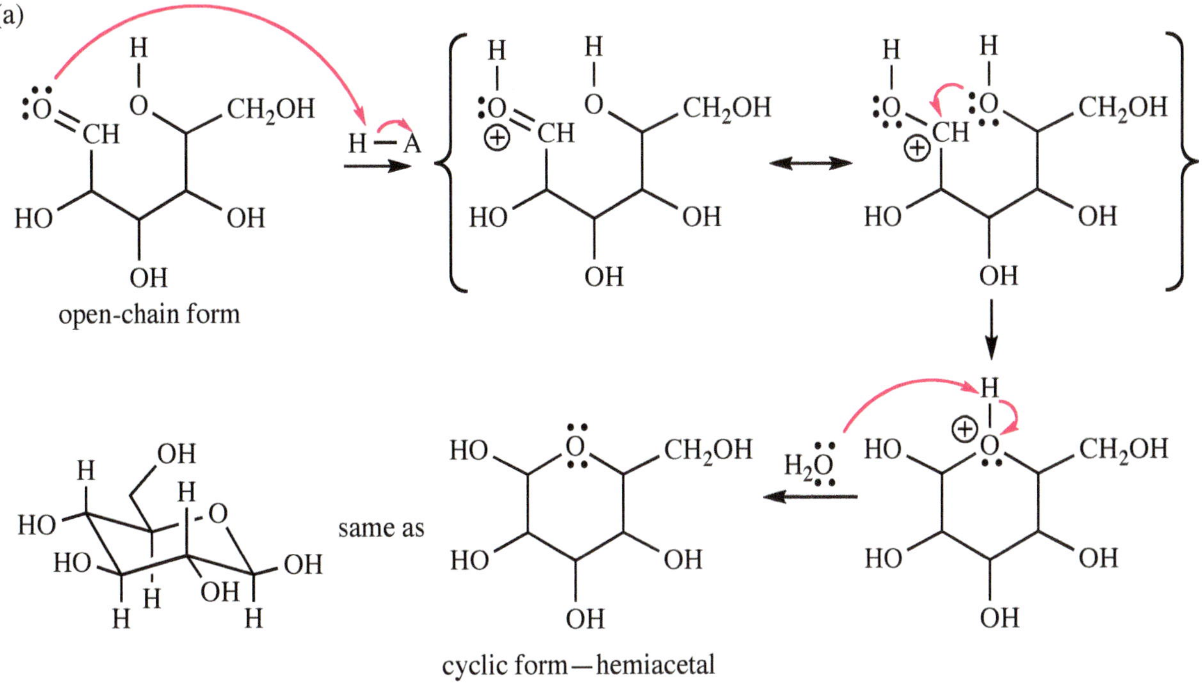

open-chain form

same as

cyclic form—hemiacetal

(b) Yes, the cyclic form of glucose will give a positive Tollens test. In the basic solution of the Tollens test, the hemiacetal is in equilibrium with the open-chain aldehyde with the cyclic form in much larger concentration. However, it is the open-chain aldehyde that reacts with silver ion, so even though there is only a small amount of open-chain form present at any given time, as more of the open-chain form is oxidized by silver ion, more cyclic form will open to replace the consumed open-chain form. Eventually all of the cyclic form will be dragged kicking and screaming through the open-chain form to be oxidized to the carboxylate. Le Châtelier's Principle strikes again!

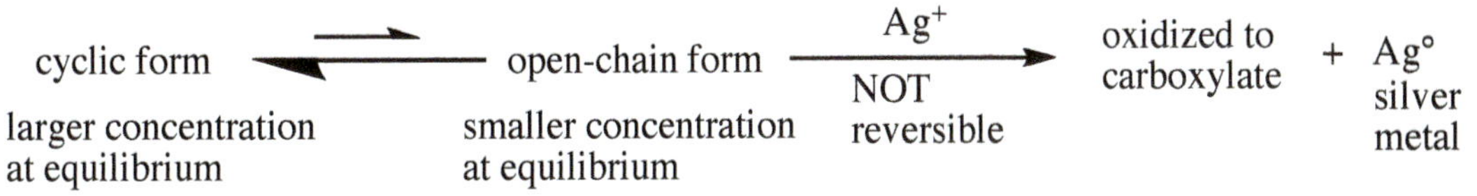

cyclic form ⇌ open-chain form → (Ag^+; NOT reversible) oxidized to carboxylate + Ag° silver metal

cyclic form: larger concentration at equilibrium

open-chain form: smaller concentration at equilibrium

18-63

(a)

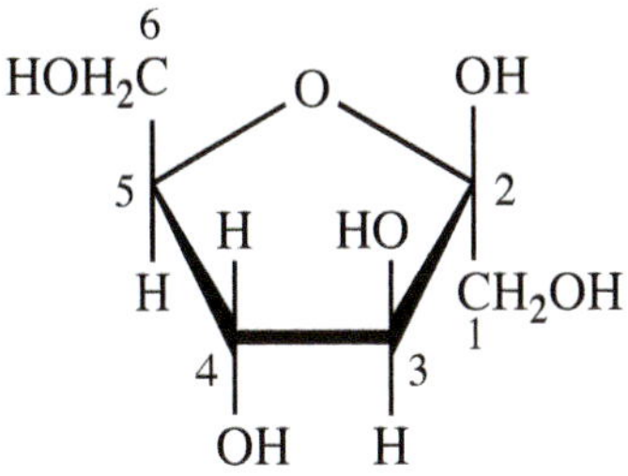

Any carbon with two oxygens bonded to it with single bonds belongs to the acetal family. If one of the oxygen groups is an OH, then the functional group is a hemiacetal. Thus, the functional group at C-2 is a hemiacetal. (The old name for this group is hemiketal as it came from a ketone.)

18-63 continued

(b) Models will help. Ignore stereochemistry for the mechanism.

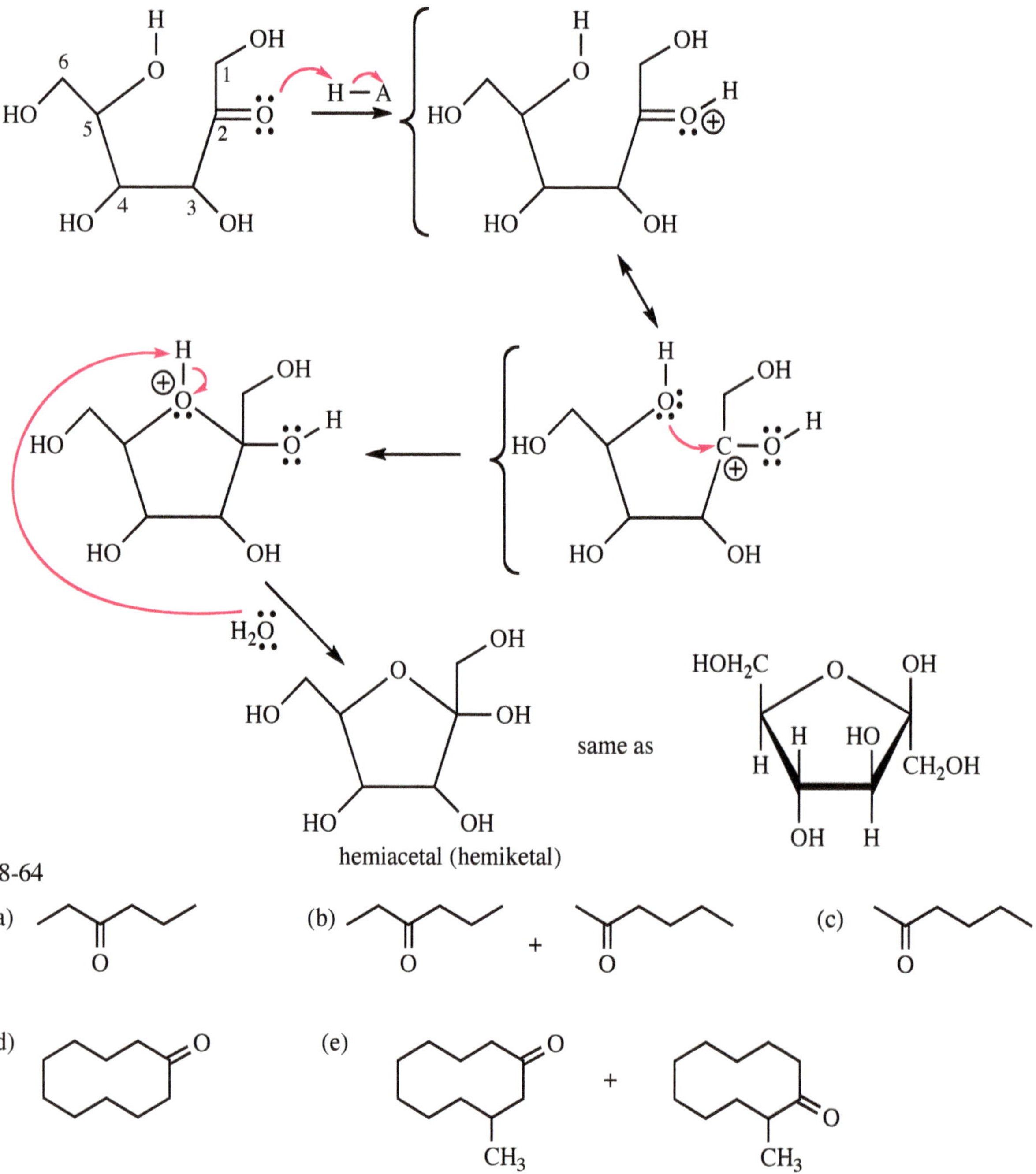

18-65

(a) ketone: no reaction
(b) aldehyde: positive
(c) enol of an aldehyde—tautomerizes to aldehyde in base: positive
(d) hemiacetal of an aldehyde in equilibrium with the aldehyde in base: positive
(e) acetal—stable in base: no reaction
(f) hemiacetal of an aldehyde in equilibrium with the aldehyde in base: positive

18-66 The structure of **A** can be deduced from its reaction with **J** and **K**. What is common to both products of these reactions is the heptan-2-ol part; the reactions must be Grignard reactions with heptan-2-one, so **A** must be heptan-2-one.

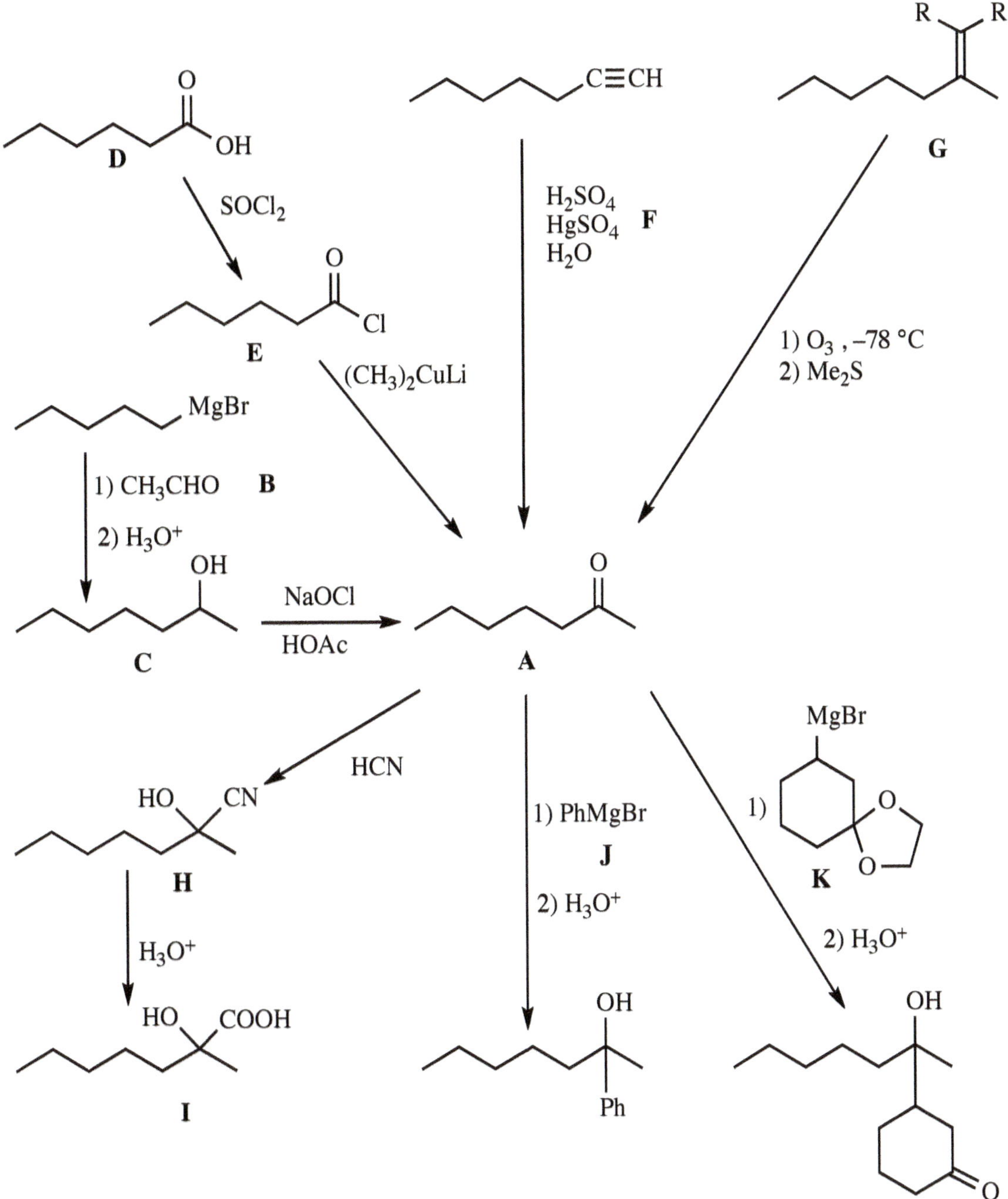

18-67

General trends in reactivity of aldehydes and ketones:
•aldehydes generally react faster than ketones;
•less hindered C=O react faster than more hindered;
•more electron-deficient C=O (because of electron-withdrawing substituents) react faster;
•unconjugated C=O react faster than conjugated.

(a)

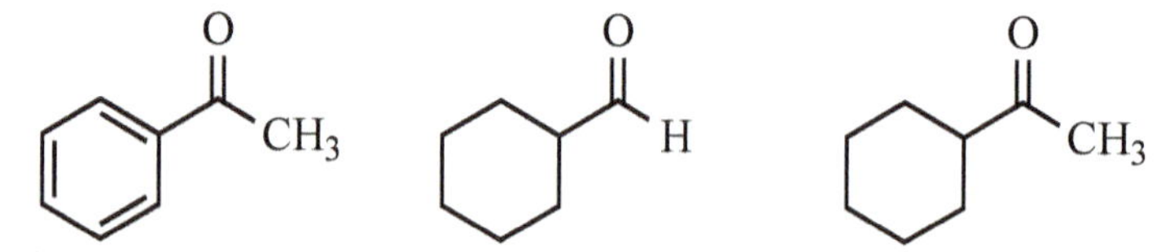

CH_3 — *slowest—conjugated ketone*

H — *fastest—aldehyde*

CH_3

(b)

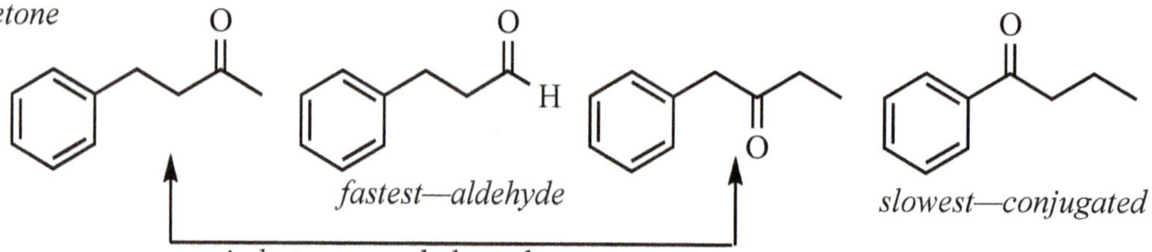

fastest—aldehyde

slowest—conjugated ketone

in between, and about the same

(c)

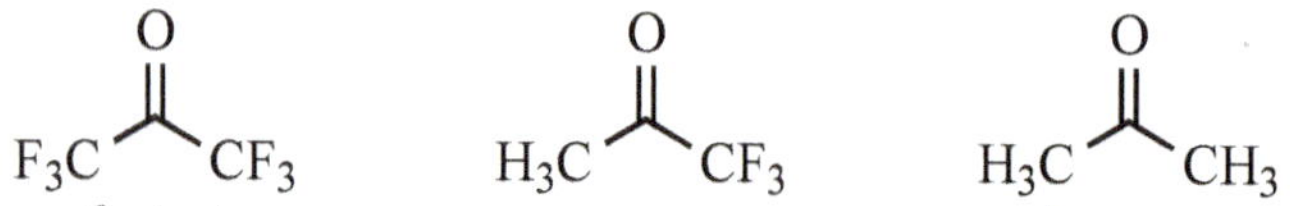

F_3C CF_3 — *fastest—most electron-deficient*

H_3C CF_3

H_3C CH_3 — *slowest—plain old ketone*

18-68

(a) $\xrightarrow{H_3O^+}$ NO REACTION

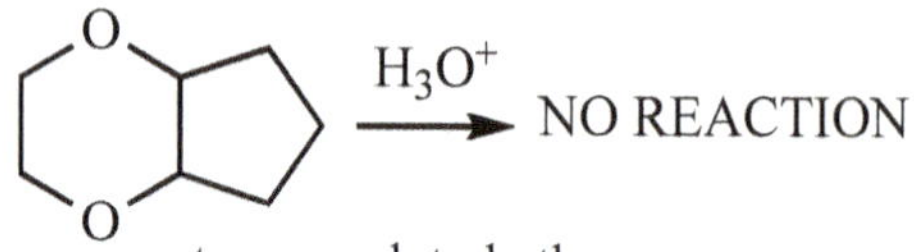

two unrelated ethers

(b) $\xrightarrow{H_3O^+}$ OH, OH $O{=}CH_2$

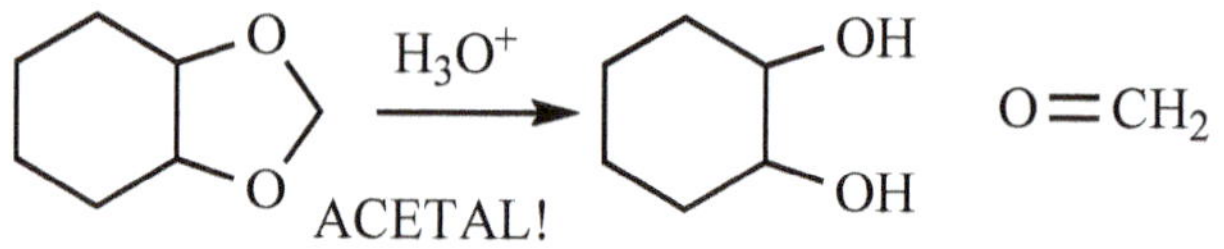

ACETAL!

mechanism

H_3O^+ → CH_2 → CH_2 ↔ CH_2

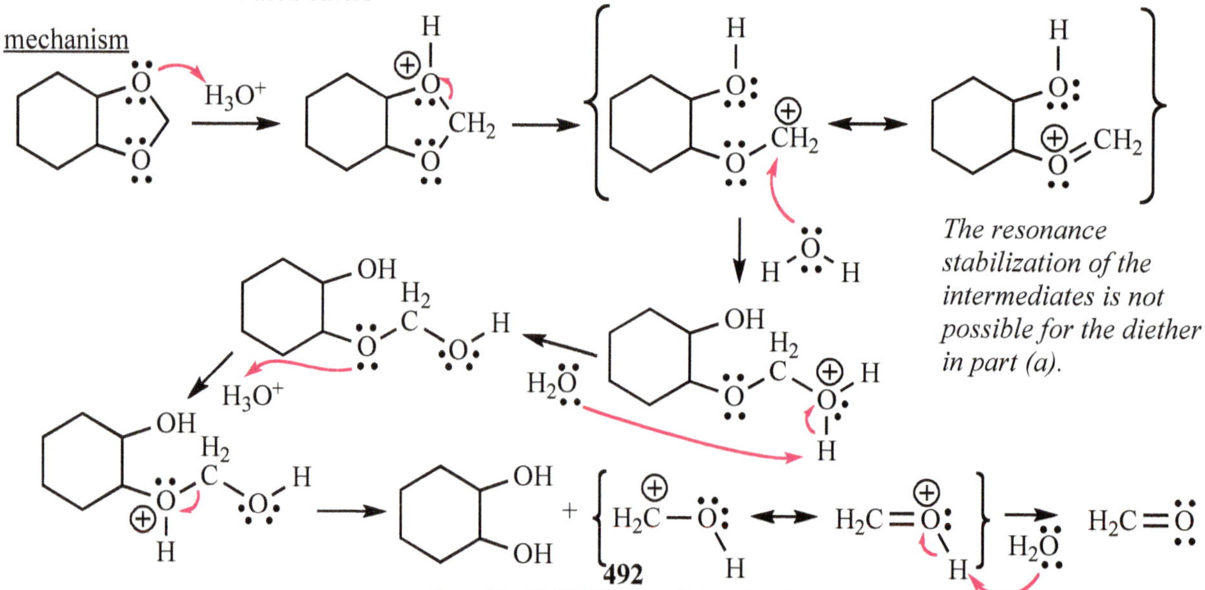

The resonance stabilization of the intermediates is not possible for the diether in part (a).

H_2O, H_3O^+

OH + $H_2C{-}O$ ↔ $H_2C{=}O$ → H_2O $H_2C{=}O$

18-69

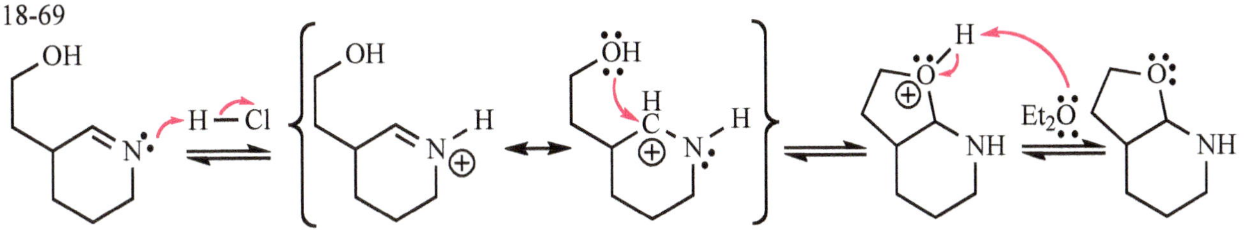

18-70

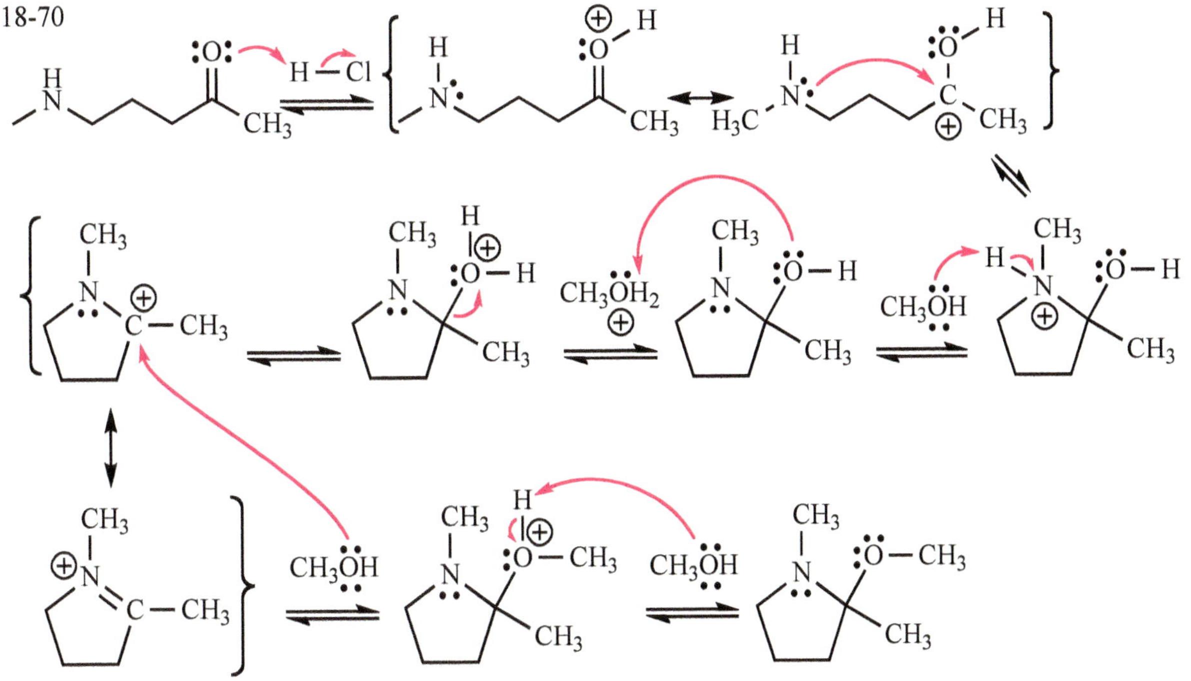

18-71 The very strong π to π* absorption at 225 nm in the UV spectrum suggests a conjugated ketone or aldehyde. The IR confirms this: strong, conjugated carbonyl at 1690 cm^{-1} and small alkene at 1610 cm^{-1}. The absence of peaks at 2700–2800 cm^{-1} shows that the unknown is not an aldehyde, confirmed by the absence of a peak in the proton NMR from δ 9 to δ 10. The peak at 199 in the carbon NMR also suggests a ketone.

The molecular ion at 96 leads to the molecular formula:

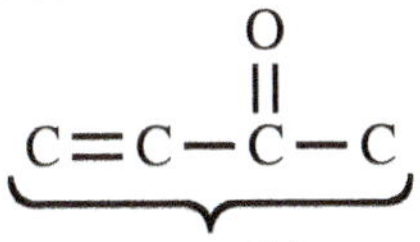

C=C−C(=O)−C, mass 64

$$\begin{array}{r} 96 \\ -\,64 \\ \hline 32 \end{array}$$

32 mass units ⇒ add 2 carbons and 8 hydrogens

molecular formula = C_6H_8O = 3 elements of unsaturation

Two elements of unsaturation are accounted for in the enone. The other one is likely a ring.

The NMR shows two vinyl hydrogens. The doublet at δ 6.0 says that the two hydrogens are on neighboring carbons (two peaks = one neighboring H).

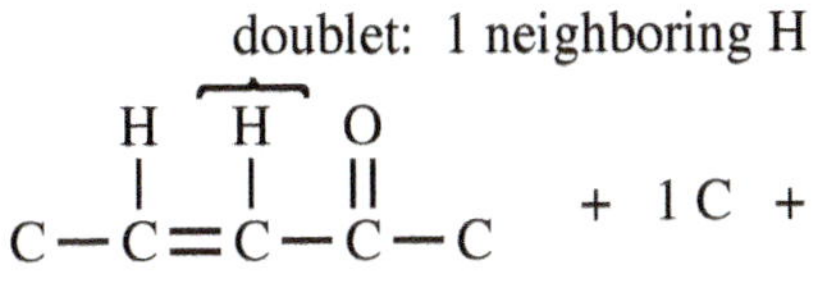

doublet: 1 neighboring H

C−CH=CH−C(=O)−C + 1 C + 6 H + 1 ring

continued on next page

18-71 continued

No methyls are apparent in the NMR, so the 6H group of peaks at δ 2.0–2.4 is most likely 3 CH_2 groups. Combining the pieces:

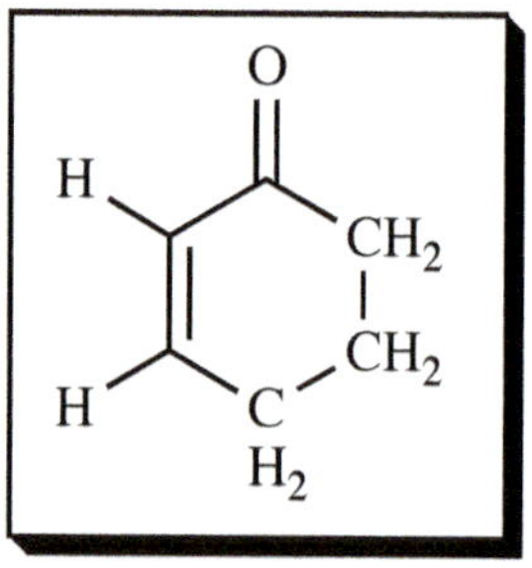

O, H, H, CH_2, CH_2, C, H_2

Carbon NMR values
(actual values)
C1 δ 199
C2 δ 130
C3 δ 151
C4 δ 26
C5 δ 23
C6 δ 38

The mass spectral fragmentation can be explained by a "retro" or reverse Diels-Alder fragmentation:

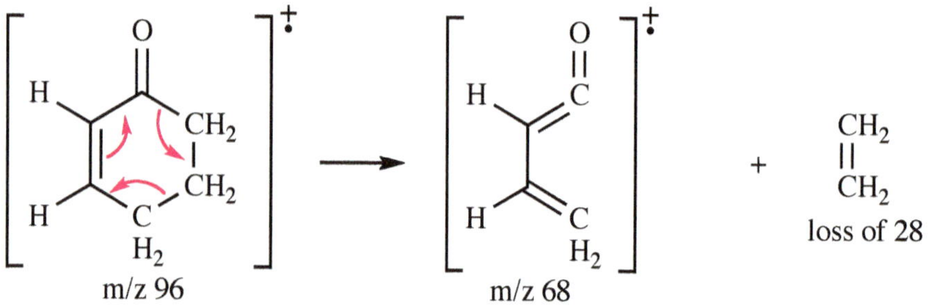

m/z 96 → m/z 68 + $CH_2{=}CH_2$ loss of 28

In the HNMR, one of the vinyl hydrogens appears at δ 7.0. This is typical of an α,β-unsaturated carbonyl because of the resonance form that shows deshielding of the β-hydrogen.

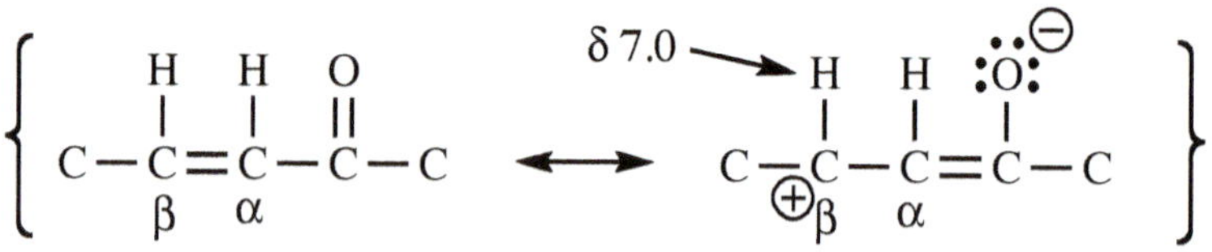

δ 7.0

18-72

(a) H_3O^+, NMe_2, $H_2\ddot{O}$:, H_3O^+, $H_2\ddot{O}$:

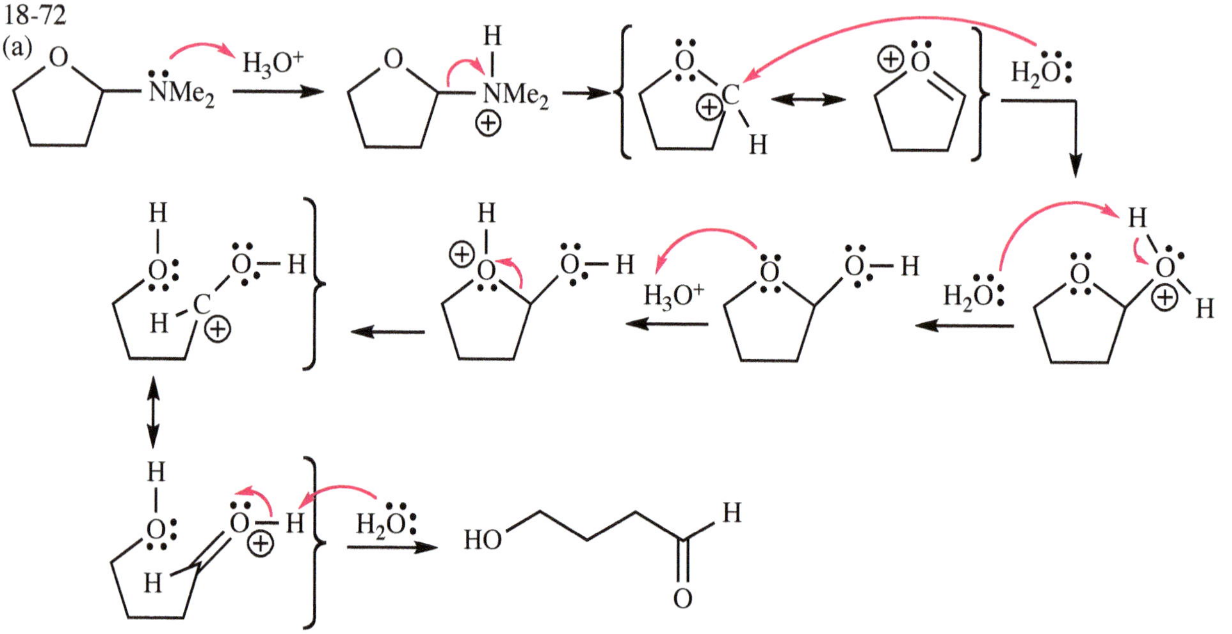

HO–CH₂CH₂CH₂–CHO (H, O)

18-72 continued

(b) Aminoacetal linkages are in the dashed boxes.

NH_2 NH_2 N N N N HO O OH deoxyadenosine

N N O HO O OH deoxycytidine

(c) The first step in the mechanism in part (a) is protonation of the amine's electron pair. The nitrogens of the DNA nucleosides, however, are part of aromatic rings, and the electron pairs are required for the aromaticity of the ring. (See the solution to Problem 16-43 for a description of the aromaticity of these nucleoside bases.) Protonation of the nitrogen will not occur unless the acid is extremely strong; dilute acids will not protonate the N and therefore the nucleoside will be stable.

18-73

First, deduce what functional groups are present in **A** and **B**. The IR of **A** shows no alkene and no carbonyl: the strongest peak is at 1210 cm^{-1}, possibly a C—O bond. After acid hydrolysis of **A**, the IR of **B** shows a carbonyl at 1715 cm^{-1}: a ketone. (If it were an aldehyde, it would have aldehyde C—H around 2700–2800 cm^{-1}, absent in the spectrum of **B**.) What functional group has C—O bonds and is hydrolyzed to a ketone? An acetal (ketal)!

There is only one ketone of formula C_4H_8O: butan-2-one.

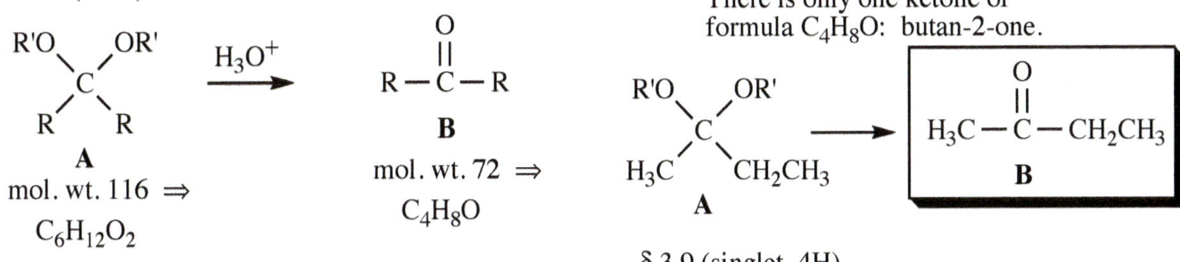

A must have the same alkyl groups as **B**. **A** has one element of unsaturation and is missing only C_2H_4 from the partial structure above. The most likely structure is the ethylene ketal. Is this consistent with the NMR?

δ 3.9 (singlet, 4H)

H_2C—CH_2 O O **A** H_3C CH_2—CH_3

δ 1.3 (singlet, 3H)

δ 1.6 (quartet, 2H)

δ 0.9 (triplet, 3H)

What about the peaks in the MS at m/z 87 and 101? The 87 peak is the loss of 29 from the molecular ion at 116.

101 H_2C—CH_2 87 O O H_3C CH_2—CH_3

m/z 116

H_2C—CH_2 O O C ⊕ H_3C

m/z 87

plus two resonance forms with positive charge on the oxygen atoms

The peak at m/z 101 comes from loss of CH_3 from the molecular ion at 116.

H_2C—CH_2 O O C ⊕ CH_2—CH_3

m/z 101

plus two resonance forms with positive charge on the oxygen atoms

18-74 The strong UV absorption at 220 nm indicates a conjugated aldehyde or ketone. The IR shows a strong carbonyl at 1690 cm^{-1}, alkene at 1625 cm^{-1}, and two peaks at 2720 cm^{-1} and 2810 cm^{-1} —aldehyde!

$$C{=}C{-}\overset{\overset{O}{\|}}{C}{-}H$$

The NMR shows the aldehyde proton at δ 9.5 split into a doublet, so it has one neighboring H. There are only two vinyl protons, so there must be an alkyl group coming off the β carbon:

$$R{-}\overset{H}{\overset{|}{C}}{=}\overset{H}{\overset{|}{C}}{-}\overset{O}{\overset{\|}{C}}{-}H$$

$$CH_3{-}\overset{H}{\overset{|}{C}}{=}\overset{H}{\overset{|}{C}}{-}\overset{O}{\overset{\|}{C}}{-}H$$

"crotonaldehyde"

The only other NMR signal is a 3H doublet: R must be methyl.

18-75

H–C(=O)–C$_6$H$_4$–C(=O)–H $\xrightarrow{H_2O}$ [intermediate] $\longrightarrow$ [cyclic product]

intermediate after the H_2O adds to the C=O; OH is close enough to the second C=O to form a ring with no C=O remaining

OH ← broad singlet at δ 5.0 (2H)

H ← sharp singlet at δ 5.9 (2H)

18-76 Each compound will undergo two α-cleavages and one McLafferty. The location of the D will differentiate the masses of the fragments. All isomers have a starting mass of 73.

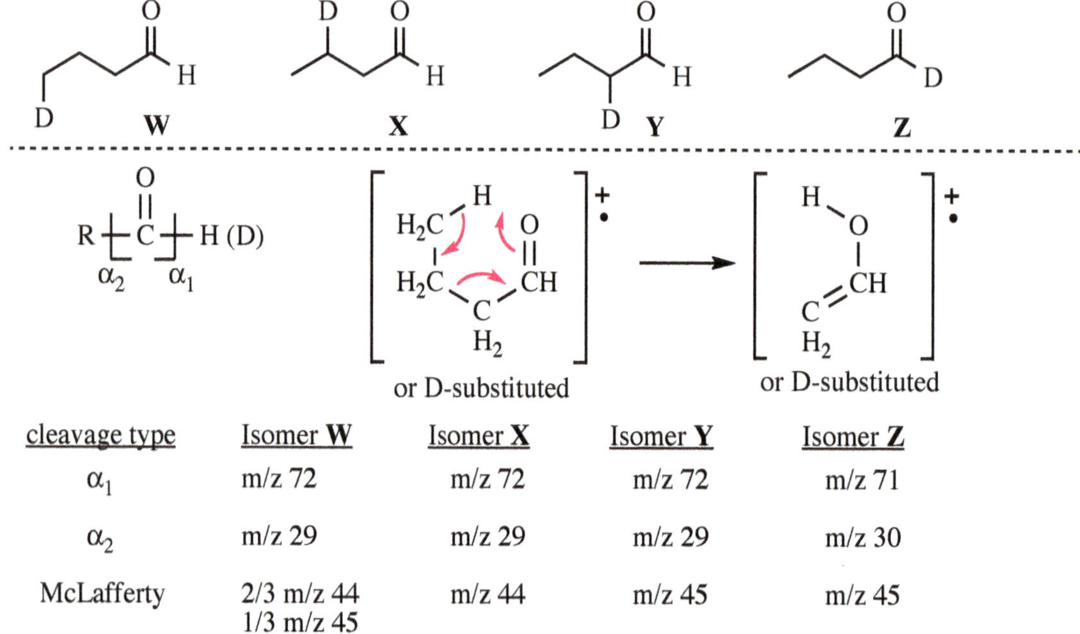

cleavage type	Isomer **W**	Isomer **X**	Isomer **Y**	Isomer **Z**
α_1	m/z 72	m/z 72	m/z 72	m/z 71
α_2	m/z 29	m/z 29	m/z 29	m/z 30
McLafferty	2/3 m/z 44 1/3 m/z 45	m/z 44	m/z 45	m/z 45

Each isomer gives a different set of fragmentation peaks, so that each possible metabolite could be unequivocally detected.

18-77

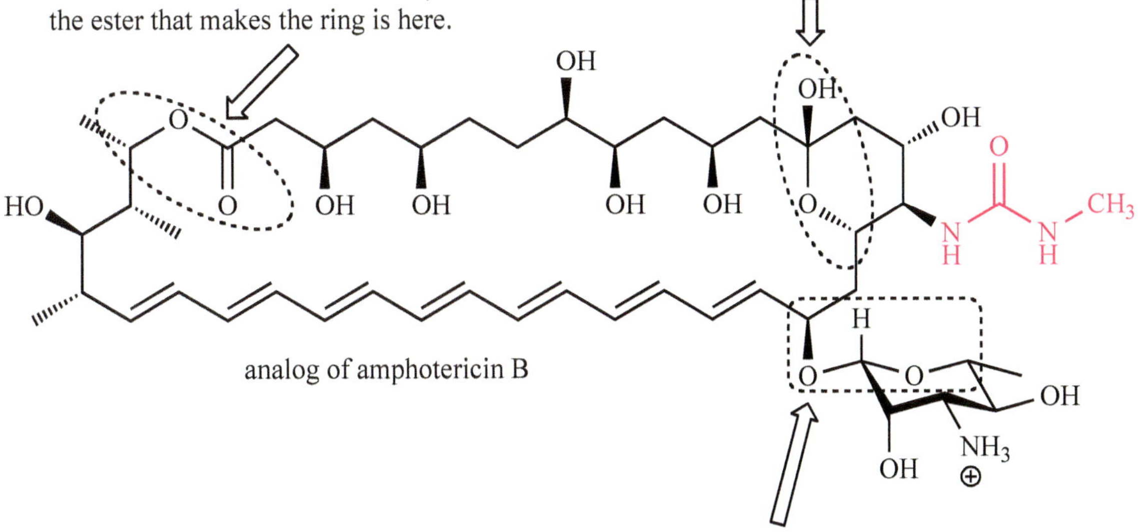
(a) The ester group in a ring is called a lactone. In this molecule, the ester that makes the ring is here.
(b) This group is a hemiketal, although the generic name is hemiacetal. At the center is one carbon bonded to one OH, one O—C, and two other carbons; that defines a hemiketal.
OH
OH
OH
O
O
HO
O
OH
OH
OH
OH
O
O
N
H
N
H
CH3
H
O
O
OH
NH3
OH
⊕
analog of amphotericin B
(b) This group is an acetal. At the center is one carbon bonded to two O—C, one other carbon, and one H; that defines an acetal.

Students: use this space for notes or to solve problems.

CHAPTER 19—AMINES

19-1 These compounds satisfy the criteria for aromaticity (planar, cyclic π system, and the Huckel number of $4n + 2$ π electrons): pyrrole, imidazole, indole, pyridine, 2-methylpyridine, pyrimidine, and purine. The systems with 6 π electrons are: pyrrole, imidazole, pyridine, 2-methylpyridine, and pyrimidine. The systems with 10 π electrons are: indole and purine. The other nitrogen heterocycles shown are not aromatic because they do not have cyclic π systems.

19-2

(a) 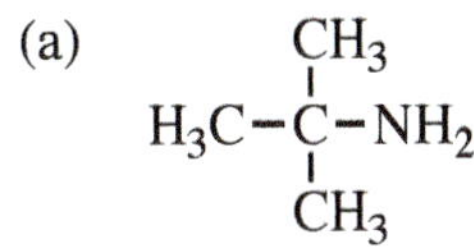

(b) $CH_3-CH(NH_2)CHO$

(c) N; N; H_3C; CH_3

(d) H; N; CH_3

(e) H_3C; N; CH_2CH_3

(f) 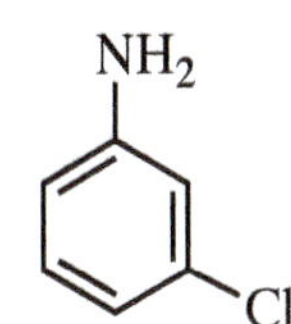

19-3

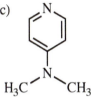

(a) pentan-2-amine (old: 2-pentylamine)
(b) *N*-methylbutan-2-amine
(c) 3-aminophenol (or *meta*-)
(d) 3-methylpyrrole
(e) *trans*-cyclopentane-1,2-diamine (or 1*R*,2*R*)
(f) *cis*-3-aminocyclohexanecarbaldehyde (or 1*S*,3*R*)

19-4
(a) Resolvable: there are two asymmetric carbons; carbon does not invert.
(b) Not resolvable: the nitrogen is free to invert.
(c) Not resolvable: it is symmetric.
(d) Not resolvable: even though the nitrogen is quaternary, one of the groups is a proton which can exchange rapidly, allowing for inversion.
(e) Resolvable: the nitrogen is quaternary and cannot invert when bonded to carbons.

19-5 In order of increasing boiling point (increasing intermolecular hydrogen bonding):
(a) triethylamine and propyl ether have the same b.p. < dipropylamine
(b) dimethyl ether < dimethylamine < ethanol
(c) trimethylamine < diethylamine < diisopropylamine (increased number of carbons)

19-6 Listed in order of increasing basicity.
(a) $PhNH_2 < NH_3 < CH_3NH_2 < NaOH$
(b) *p*-nitroaniline < aniline < *p*-methylaniline (*p*-toluidine)
(c) pyrrole < aniline < pyridine < piperidine
(d) 3-nitropyrrole < pyrrole < imidazole

19-7
(a) secondary amine: one peak in the 3200–3400 cm^{-1} region, indicating NH
(b) primary amine: two peaks in the 3200–3400 cm^{-1} region, indicating NH_2
(c) alcohol: strong, broad peak around 3400 cm^{-1}

19-8 A compound with formula $C_4H_{11}N$ has no elements of unsaturation. The proton NMR shows five types of H, with the NH_2 appearing as a broad peak at δ 1.15, meaning that there are four different groups of hydrogens on the four carbons. The carbon NMR also shows four carbons, so there is no symmetry in this structure; that is, it does not contain a *tert*-butyl group or an isopropyl group.

The multiplet farthest downfield is a CH deshielded by the nitrogen; integration shows it to be one H. There is a 2H multiplet at δ 1.35, the broad NH_2 peak at δ 1.15, a 3H doublet at δ 1.05, and a 3H triplet at δ 0.90. The latter two signals must represent methyl groups next to a CH and a CH_2 respectively. So far:

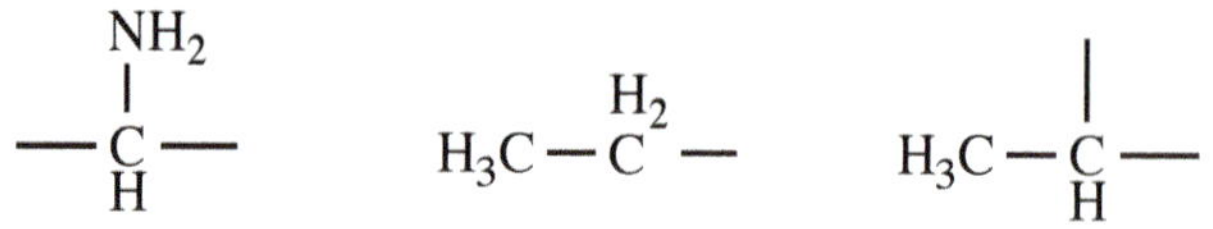

$-CH(NH_2)-$ H_3C-CH_2- $H_3C-CH(-)-$

The pieces shown above have one carbon too many, so there must be one carbon that is duplicated: the only possible one is the CH, and the structure reveals itself.

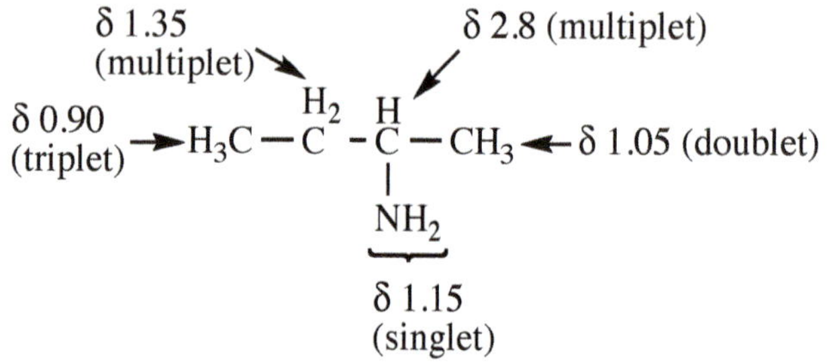

$H_3C-CH_2-CH(NH_2)-CH_3$

δ 0.90 (triplet) → H_3C; δ 1.35 (multiplet) → CH_2; δ 2.8 (multiplet) → CH; δ 1.05 (doublet) → CH_3; δ 1.15 (singlet) → NH_2

19-9

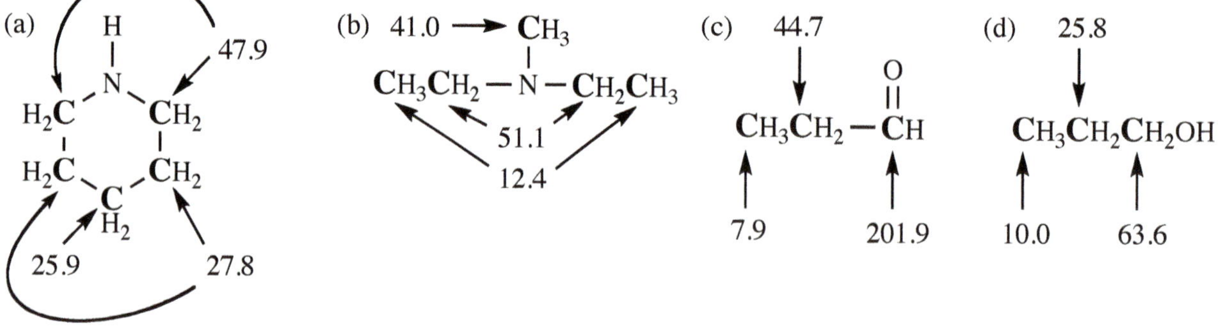

(a) Piperidine: δ 47.9 (CH_2 next to N), 25.9, 27.8 (central CH_2)

(b) $CH_3CH_2-N(CH_3)-CH_2CH_3$: 41.0 ($CH_3$ on N), 51.1 (CH_2), 12.4 (CH_3)

(c) $CH_3CH_2-CH=O$: 44.7 (CH_2), 7.9 (CH_3), 201.9 (CH)

(d) $CH_3CH_2CH_2OH$: 25.8 (middle CH_2), 10.0 (CH_3), 63.6 (CH_2OH)

19-10

(a)

$$[H_3C-CH_2-NH-CH_2-CH_2CH_3]^{+\bullet} \text{ (m/z 87)} \longrightarrow \{CH_3-CH_2-\ddot{N}H-\overset{\oplus}{C}H_2 \longleftrightarrow CH_3-CH_2-\overset{\oplus}{N}H=CH_2\} \text{ (m/z 58)} + CH_3\dot{C}H_2 \text{ (mass 29)}$$

(cleavage labels: 72, 58)

(b)

$$\{H_2\overset{\oplus}{C}-\ddot{N}H-CH_2-CH_2CH_3 \longleftrightarrow CH_2=\overset{\oplus}{N}H-CH_2-CH_2CH_3\} \text{ (m/z 72)} + \bullet CH_3 \text{ (mass 15)}$$

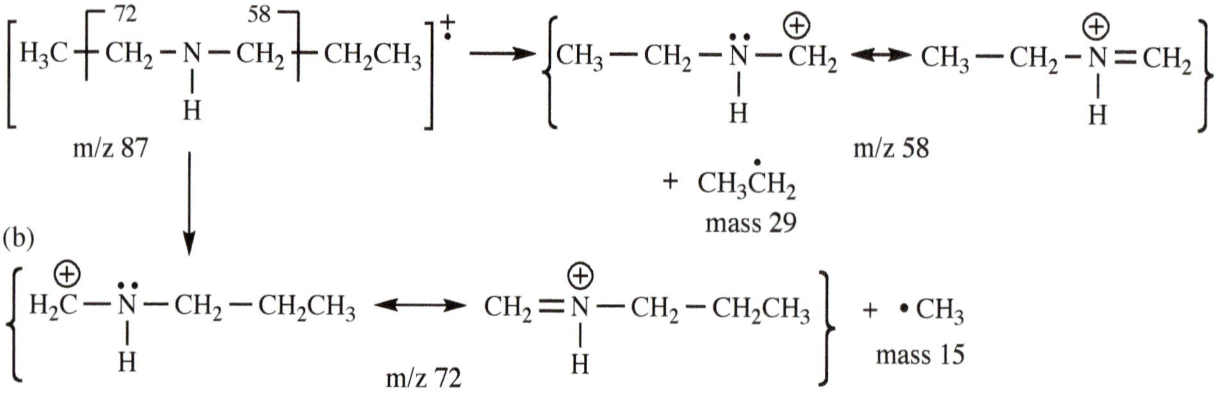

(c) The fragmentation in (a) occurs more often than the one in (b) because of stability of the radicals produced along with the iminium ions. Ethyl radical is much more stable than methyl radical, so pathway (a) is preferred.

19-11 Nitration at the 4-position of pyridine is not observed for the same reason that nitration at the 2-position is not observed: the intermediate puts some positive character on an electron-deficient nitrogen, and electronegative nitrogen hates that. (It is important to distinguish this type of positive nitrogen without a complete octet of electrons, from the quaternary nitrogen, also positively charged but with a full octet. It is the number of electrons around atoms that is most important; the charge itself is less important.)

GOOD: —N⊕— VERY BAD: —N⊕

mechanism

VERY BAD—
N does not have
an octet of electrons.

not produced because of unfavorable intermediate

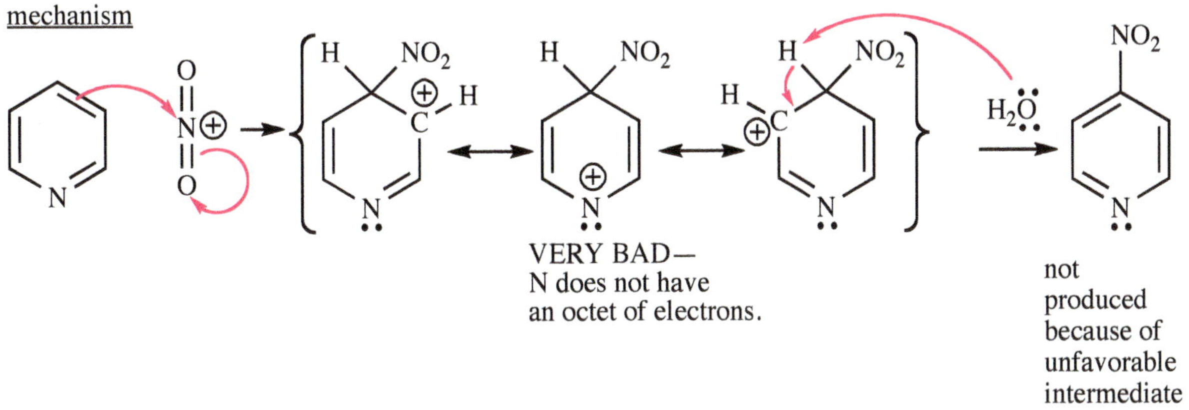

19-12 Any electrophilic attack, including sulfonation, is preferred at the 3-position of pyridine because the intermediate is more stable than the intermediate from attack at either the 2-position or the 4-position. (Resonance forms of the sulfonate group are not shown, but remember that they are important!)

The N of pyridine is basic, and in the strong acid mixture, it will be protonated as shown here. That is part of the reason that pyridine is so sluggish to react: the ring already has a positive charge, so attack of an electrophile is slowed.

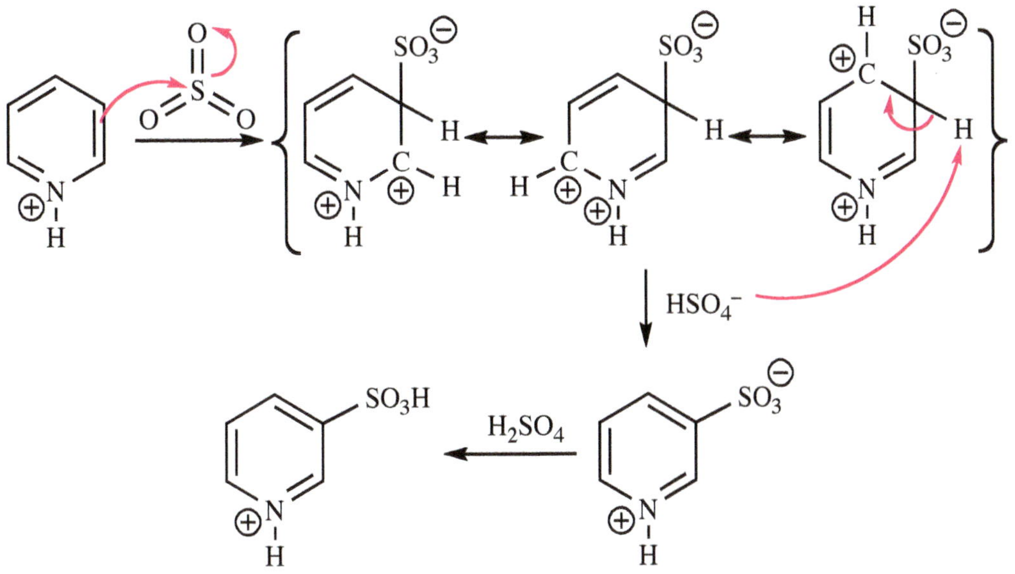

19-13 The middle resonance form demonstrates how the negative charge is distributed onto the more electronegative nitrogen atom.

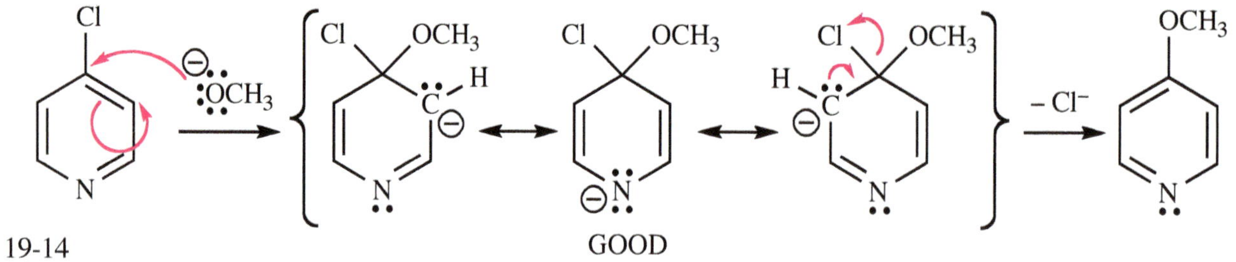

19-14

(a)

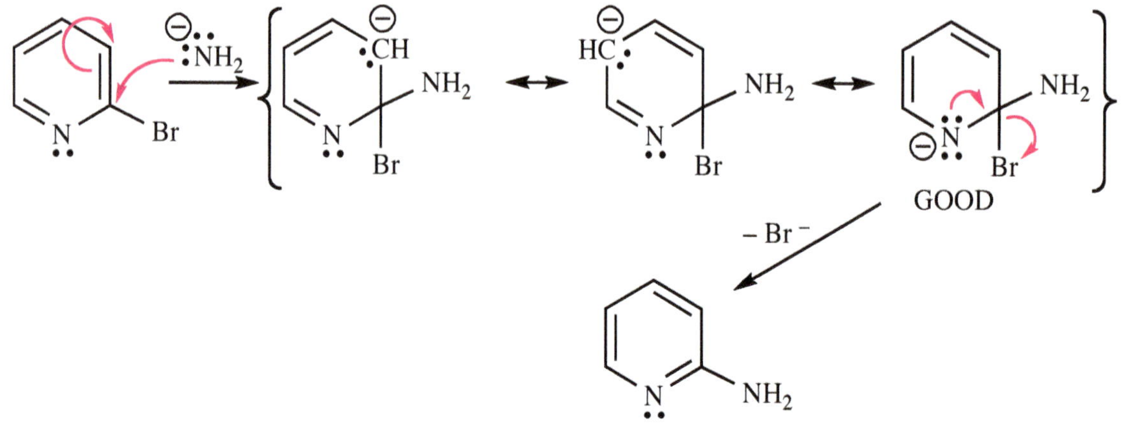

(b)

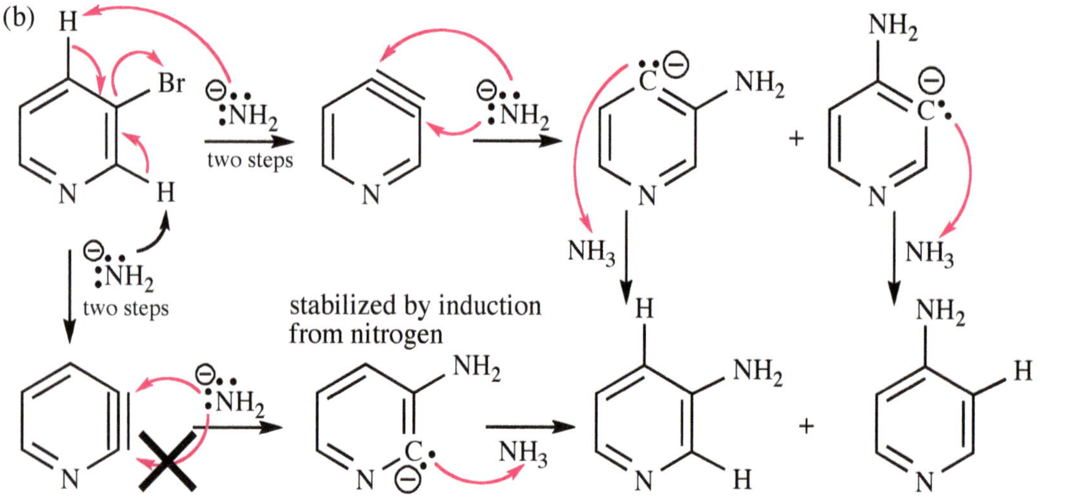

Amide ion will not attack at C-2 because the anion at C-3 is not as stable as the anion at C-2.

This is a benzyne-type mechanism. (For simplicity above, two steps of benzyne generation are shown as one step: first, a proton is abstracted by amide anion, followed by loss of bromide.) Amide ion is a strong enough base to remove a proton from 3-bromopyridine as it does from a halobenzene. Once a benzyne is generated (two possibilities), the amide ion reacts quickly, forming a mixture of products.

Why does the 3-bromo follow this extreme mechanism while the 2-bromo reacts smoothly by the addition-elimination mechanism? Stability of the intermediate! Negative charge on the electronegative nitrogen makes for a more stable intermediate in the 2-bromo substitution. No such stabilization is possible in the 3-bromo case.

19-15

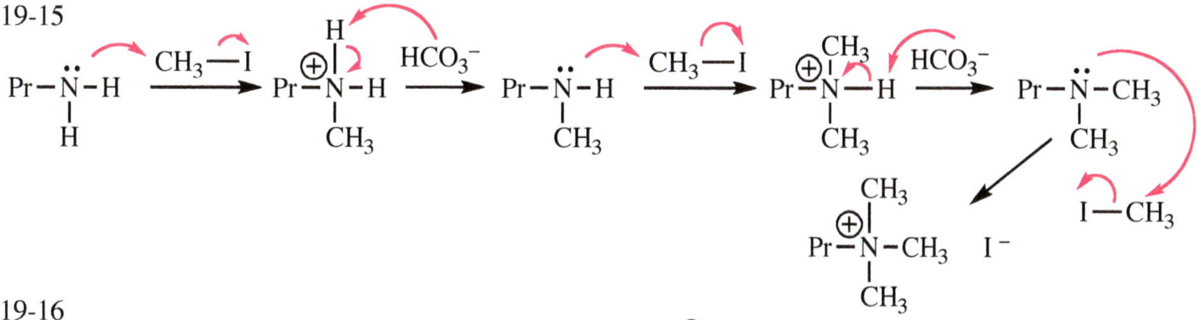

$Pr-\ddot{N}H_2 + CH_3-I \rightarrow Pr-\overset{\oplus}{N}H_2-CH_3 \xrightarrow{HCO_3^-} Pr-\ddot{N}H-CH_3 \xrightarrow{CH_3-I} Pr-\overset{\oplus}{N}H(CH_3)_2 \xrightarrow{HCO_3^-} Pr-\ddot{N}(CH_3)_2 \xrightarrow{I-CH_3} Pr-\overset{\oplus}{N}(CH_3)_3\ I^-$

19-16

(a) $PhCH_2NH_2$ + excess CH_3I $\xrightarrow{NaHCO_3}$ $PhCH_2\overset{\oplus}{N}(CH_3)_3$ I^-

OR $PhCH_2I$ + $N(CH_3)_3$

(b)  excess NH_3 + $Br(CH_2)_5CH_3$ → $H_2N(CH_2)_5CH_3$

(c) excess NH_3 + $PhCH_2Br$ → $PhCH_2NH_2$

19-17

(a) $CH_3C(=O)-NHCH_2CH_3$ (b) 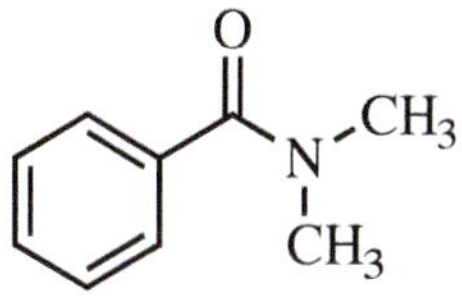 $PhC(=O)N(CH_3)_2$ (c) 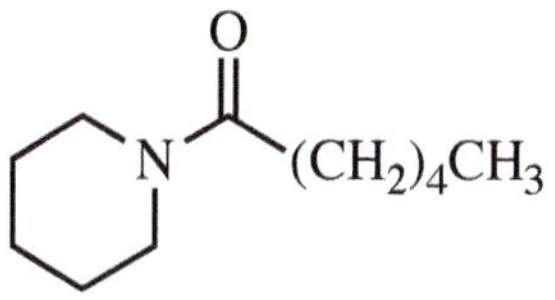 piperidine–$N-C(=O)(CH_2)_4CH_3$

19-18 If the amino group were not protected, it would do a nucleophilic substitution on chlorosulfonic acid. Later in the sequence, this group could not be removed without cleaving the other sulfonamide group.

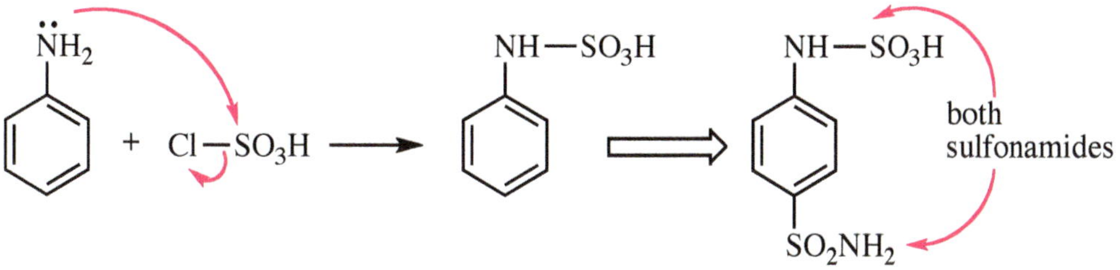

$PhNH_2$ + $Cl-SO_3H$ → $PhNH-SO_3H$ ⇒ p-$(SO_2NH_2)C_6H_4NH-SO_3H$

both sulfonamides

19-19

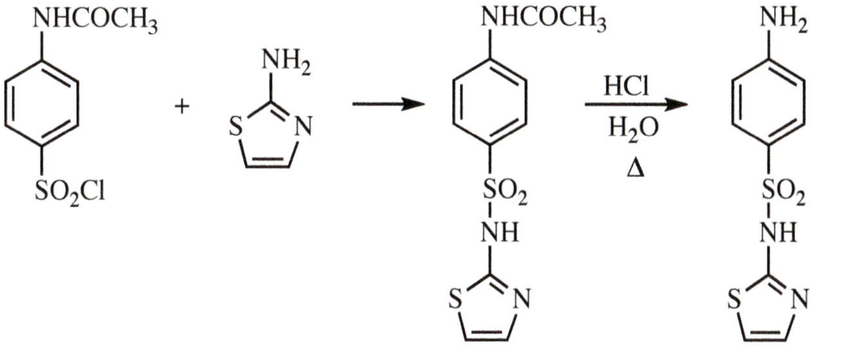

p-$(NHCOCH_3)C_6H_4SO_2Cl$ + 2-aminothiazole → p-$(NHCOCH_3)C_6H_4SO_2-NH$-(2-thiazolyl) $\xrightarrow[\Delta]{HCl,\ H_2O}$ p-$(NH_2)C_6H_4SO_2-NH$-(2-thiazolyl)

sulfathiazole

continued on next page

19-19 continued

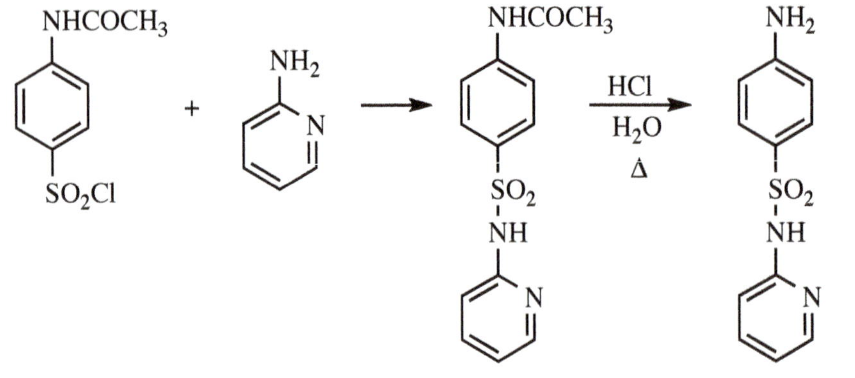

NHCOCH$_3$ / SO$_2$Cl + 2-aminopyridine → (NHCOCH$_3$ / SO$_2$–NH–pyridyl) —HCl, H_2O, Δ→ (NH$_2$ / SO$_2$–NH–pyridyl)

sulfapyridine

19-20

(a)  (b) 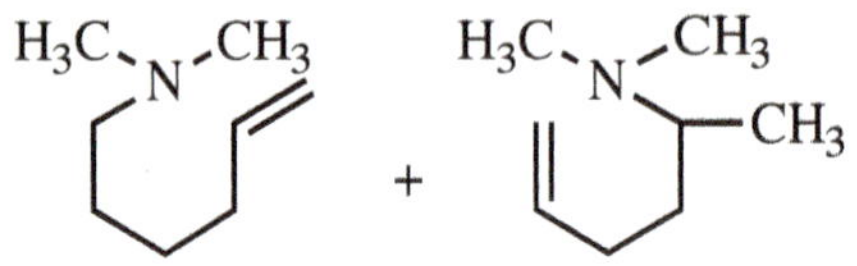 H_3C N CH_3 + H_3C N CH_3 CH_3 (c) 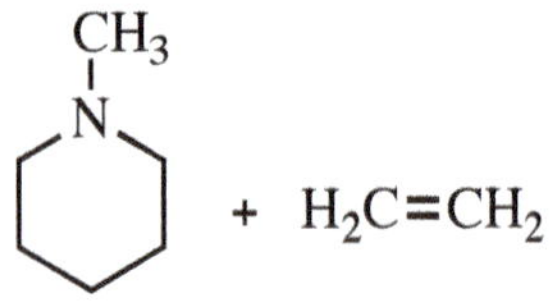 CH_3 N + $H_2C{=}CH_2$

(d) 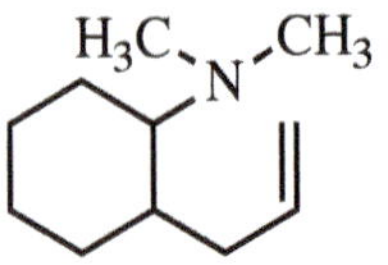 H_3C N CH_3 (e) 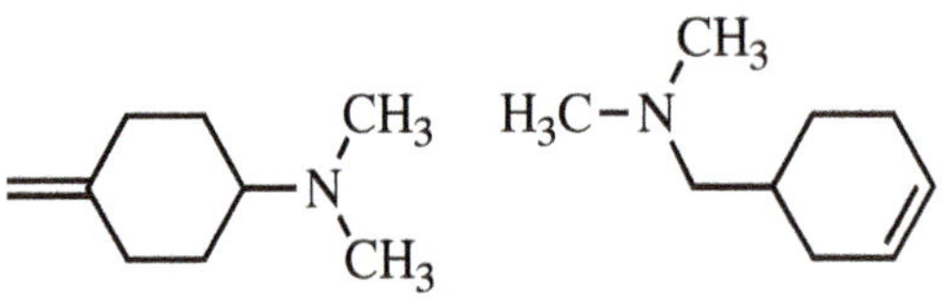 CH_3 N CH_3 CH_3 H_3C-N (f) 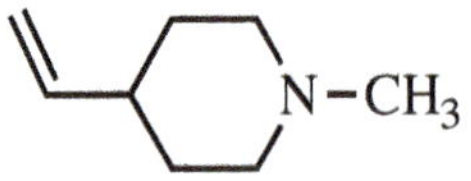 $N-CH_3$

19-21 Orientation of the Cope elimination is similar to Hofmann elimination: the *less* substituted alkene is the major product.

(a)  + $(CH_3)_2NOH$ (b) $H_2C{=}CH_2$ major + HO N + minor + $(CH_3CH_2)_2NOH$

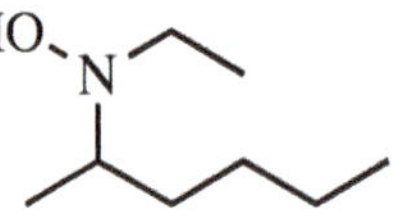

(c)  + $(CH_3)_2NOH$ (d) $H_2C{=}CH_2$ + N OH + N OH minor

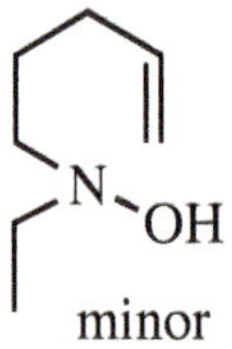

19-22 The key to this problem is to understand that Hofmann elimination occurs via an E2 mechanism requiring *anti* coplanar stereochemistry, whereas Cope elimination requires *syn* coplanar stereochemistry.

(a) 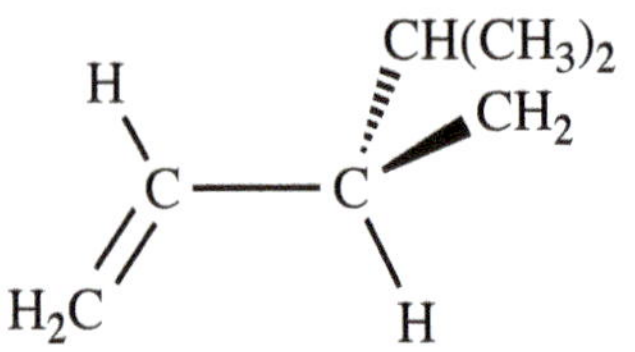 H, H_2C, C—C, $CH(CH_3)_2$, CH_2, H

Hofmann orientation loses a hydrogen from the CH_3 as well as the $N(CH_3)_3$ group to make the less substituted double bond.

19-22 continued

(b) Hofmann elimination

$(CH_3)_3N^{+}$ ... HO^{-} → E (Zaitsev product—more highly substituted)

H, $CH(CH_3)_2$, H_3C, CH_3

Cope elimination

$(CH_3)_2N^{+}-O^{-}$ ... Δ → Z (Zaitsev product—more highly substituted)

H, CH_3, H_3C, $CH(CH_3)_2$

19-23

(a) [cyclohexyl–N_2^{+} Cl^{-}] $\xrightarrow{H_2O}$ cyclohexanol (OH) + cyclohexene

Aliphatic diazonium ions are very unstable, rapidly decomposing to carbocations.

(b) N–NO (c) NO–N piperidine (d) N_2^{+} Cl^{-} benzene

Aryl diazonium ions are relatively stable if kept cold.

19-24 The diazonium ion can do aromatic substitution like any other electrophile.

$\{^{-}O_3S$–C$_6$H$_4$–$\overset{+}{N}\equiv N$ ⟷ $^{-}O_3S$–C$_6$H$_4$–$N=\overset{+}{N}\}$ + C$_6$H$_5$–$N(CH_3)_2$

$\{^{-}O_3S$–C$_6$H$_4$–N=N–(H)C$_6$H$_4$=$\overset{+}{N}(CH_3)_2$ ⟷ $^{-}O_3S$–C$_6$H$_4$–N=N–(H)C$_6$H$_4^{+}$–$N(CH_3)_2\}$

most significant resonance contributor

plus two other resonance forms with positive charge on the ring

$:\ddot{Cl}:^{-}$ (or some other base)

$^{-}O_3S$–C$_6$H$_4$–N=N–C$_6$H$_4$–$N(CH_3)_2$

methyl orange

19-25

(a) NH_2 (aniline) $\xrightarrow[HCl]{NaNO_2}$ $N_2^{\oplus}$ $Cl^{\ominus}$ $\xrightarrow[\Delta]{HBF_4}$ F

(b) $N_2^{\oplus}$ $Cl^{\ominus}$ $\xrightarrow{CuCl}$ Cl

from (a)

(c) NH_2 $\xrightarrow{CH_3\overset{O}{\overset{\|}{C}}-Cl}$ $HN-\overset{O}{\overset{\|}{C}}CH_3$ $\xrightarrow[AlCl_3]{\textbf{3}\ CH_3I}$ CH_3, CH_3, CH_3 (acetanilide ring) $\xrightarrow[\Delta]{H_3O^+}$ NH_2 with CH_3, CH_3, CH_3 $\xrightarrow[HCl]{NaNO_2}$ $Cl^{\ominus}$ $N_2^{\oplus}$ with CH_3, CH_3, CH_3 $\xrightarrow{H_3PO_2}$ H with CH_3, CH_3, CH_3

activating o,p-director

Aniline must be acylated before undergoing Friedel-Crafts reactions.

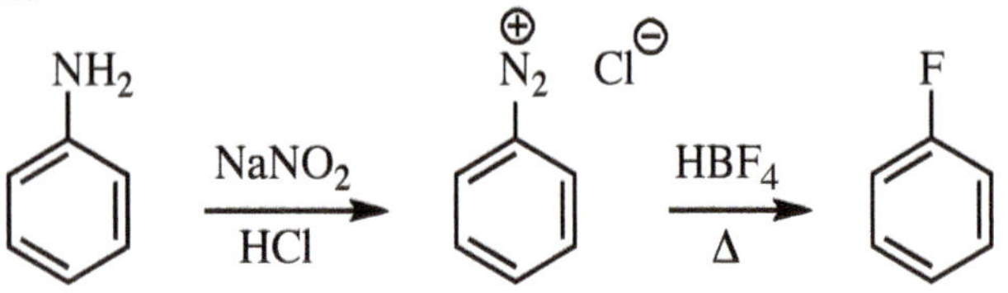

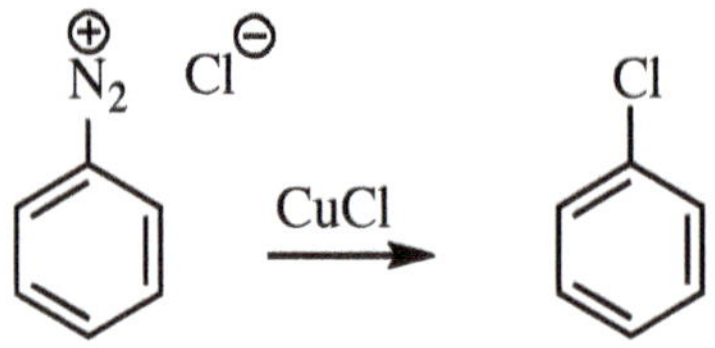

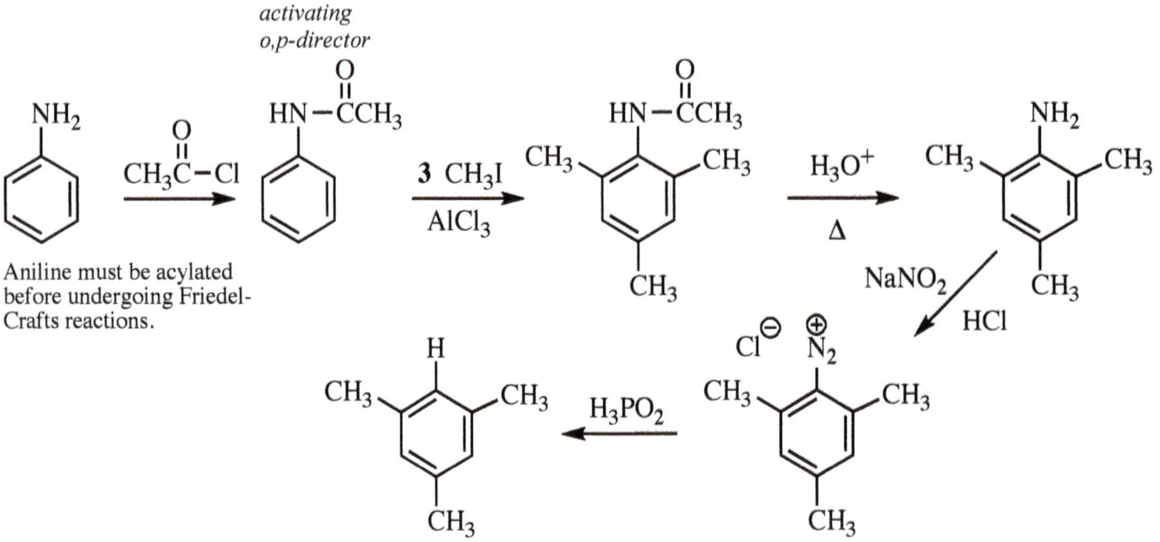

(d) $N_2^{\oplus}$ $Cl^{\ominus}$ $\xrightarrow{CuBr}$ Br

from (a)

(e) $N_2^{\oplus}$ $Cl^{\ominus}$ $\xrightarrow{KI}$ I

(f) $N_2^{\oplus}$ $Cl^{\ominus}$ $\xrightarrow{CuCN}$ CN

(g) $N_2^{\oplus}$ $Cl^{\ominus}$ $\xrightarrow[\Delta]{H_3O^+}$ OH

(h) $N_2^{\oplus}$ $Cl^{\ominus}$ + HO, OH (resorcinol) $\longrightarrow$ $N=N$ —OH, HO

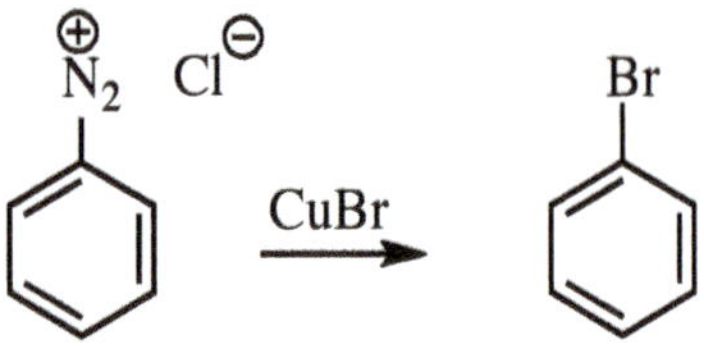

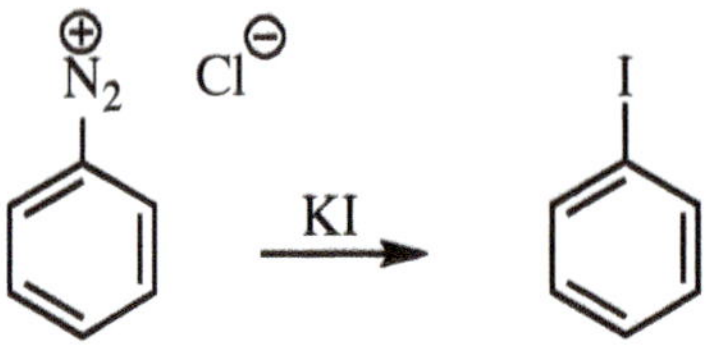

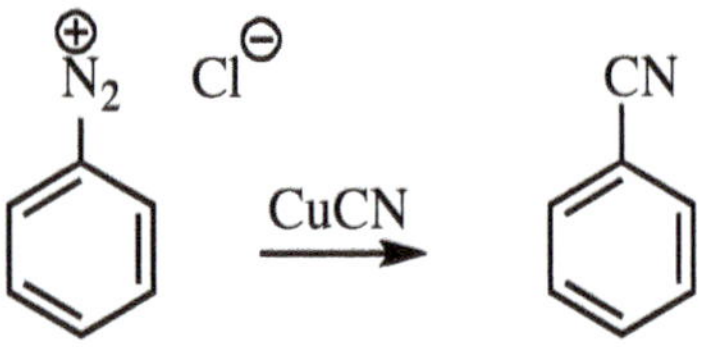

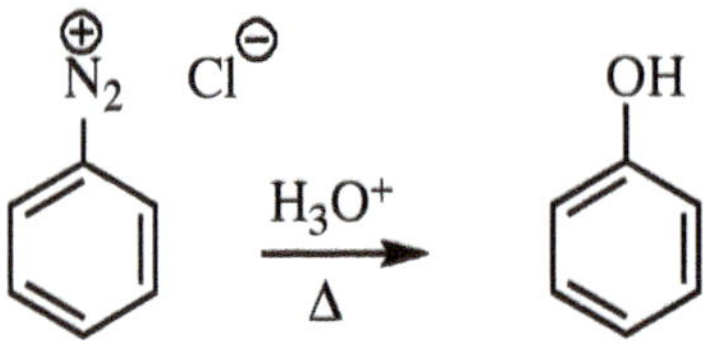

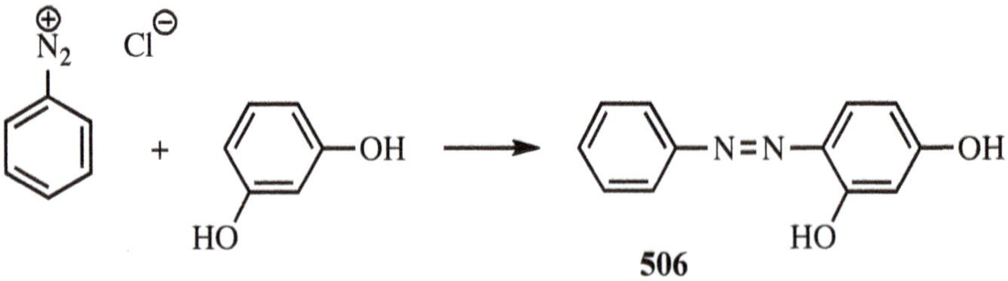

19-26 General guidelines for choice of reagent for reductive amination: use $LiAlH_4$ when the imine or oxime is isolated. Use $NaBH(CH_3COO)_3$, abbreviated $NaBH(OAc)_3$, in solution when the imine or iminium ion is not isolated. Alternatively, catalytic hydrogenation works in most cases.

(a) benzaldehyde (PhCHO) + CH_3NH_2 (1° amine) $\xrightarrow[CH_3COOH]{NaBH(OAc)_3}$ $PhCH_2NHCH_3$ (2° amine)

(b) piperidine + PhCHO $\xrightarrow[CH_3COOH]{NaBH(OAc)_3}$ N-CH_2Ph piperidine

(c) cyclohexanone + aniline ($PhNH_2$) $\xrightarrow{\text{acid catalyst}}$ N-Ph imine of cyclohexanone $\xrightarrow[2)\ H_2O]{1)\ LiAlH_4}$ cyclohexyl-NHPh

(d) cyclohexanone $\xrightarrow[\text{acid catalyst}]{H_2NOH}$ cyclohexanone oxime (NOH) $\xrightarrow[2)\ H_2O]{1)\ LiAlH_4}$ cyclohexyl-NH_2

(e) $PhCH_2-C(=O)-CH_3$ $\xrightarrow[\text{acid catalyst}]{H_2NOH}$ $PhCH_2-C(=NOH)-CH_3$ $\xrightarrow[2)\ H_2O]{1)\ LiAlH_4}$ $PhCH_2-CH(NH_2)-CH_3$

(f) piperidine (NH) + O=cyclopentane $\xrightarrow[CH_3COOH]{NaBH(OAc)_3}$ N-cyclopentylpiperidine

19-27

(a) piperidine + $Cl-C(=O)CH_2CH_2CH_3$ $\rightarrow$ N-butanoylpiperidine $\xrightarrow[2)\ H_2O]{1)\ LiAlH_4}$ N-butylpiperidine

(b) aniline (NH_2) + $Cl-C(=O)Ph$ $\rightarrow$ PhNH-C(=O)Ph $\xrightarrow[2)\ H_2O]{1)\ LiAlH_4}$ $PhNH-CH_2Ph$

19-28 (a)

S_N2 — Br + :NH_3 → Br^- $\oplus$ NH_3^+ ... :NH_3

:N-H ← $\oplus$N (H, H) Br^- ← :NH_2 ... Br

Good nucleophile—alkylation process can continue.

(b) Use a large excess of ammonia to avoid multiple alkylations of each nitrogen.

19-29

(a) phthalimide anion :N: $\xrightarrow{BrCH_2Ph}$ N-CH_2Ph $\xrightarrow[\Delta]{NH_2NH_2}$ H_2NCH_2Ph

(b) phthalimide anion :N: $\xrightarrow{Br(CH_2)_5CH_3}$ N-$(CH_2)_5CH_3$ $\xrightarrow[\Delta]{NH_2NH_2}$ $H_2N(CH_2)_5CH_3$

(c) phthalimide anion :N: $\xrightarrow{Br(CH_2)_3COO^{\ominus}}$ N-$(CH_2)_3COO^{\ominus}$ $\xrightarrow[2)\ H-A]{1)\ NH_2NH_2\ ,\ \Delta}$ $H_2N(CH_2)_3COOH$

Must use anion to avoid protonating the phthalimide anion.

19-30 Assume that $LiAlH_4$ or H_2/catalyst can be used interchangeably to reduce azides or nitriles.

(a) $PhCH_2Br$ + NaN_3 → $PhCH_2N_3$ $\xrightarrow[Pt]{H_2}$ $PhCH_2NH_2$

(b) Br $\xrightarrow{KCN}$ C≡N $\xrightarrow[2)\ H_2O]{1)\ LiAlH_4}$ NH_2

(c) OH (O) $\xrightarrow[2)\ H_2O]{1)\ LiAlH_4}$ OH $\xrightarrow[pyridine]{TsCl}$ OTs $\xrightarrow{NaN_3}$ N_3 $\xrightarrow[2)\ H_2O]{1)\ LiAlH_4}$ NH_2

OR

OH (O) $\xrightarrow{SOCl_2}$ Cl (O) $\xrightarrow{NH_3}$ NH_2 (O) $\xrightarrow{1)\ LiAlH_4\ \ 2)\ H_2O}$ NH_2

19-30 continued

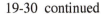

(d) OTs (from (c)) —KCN→ C≡N —1) $LiAlH_4$, 2) H_2O→ NH_2

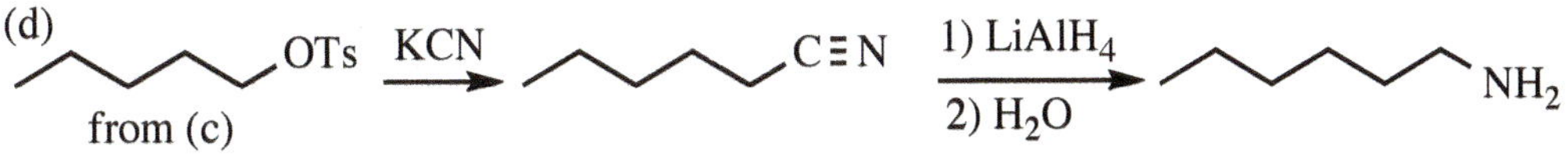

(e) Br, H (R) —NaN_3, S_N2— inversion→ H, N_3 —1) $LiAlH_4$, 2) H_2O→ H, NH_2 (S)

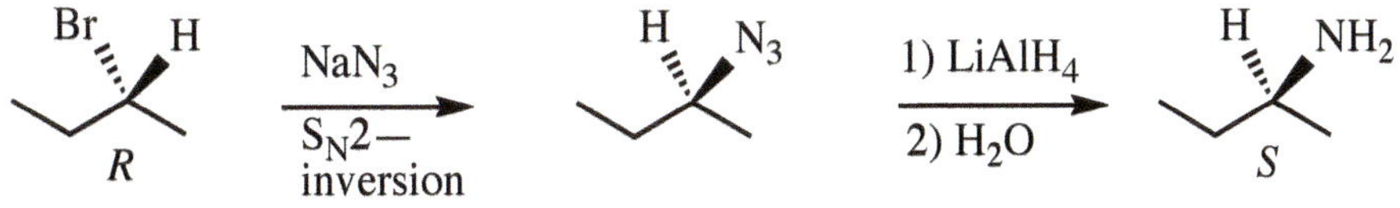

(f) Br, H (R) —NaCN, S_N2— inversion→ H, C≡N —1) $LiAlH_4$, 2) H_2O→ H, CH_2NH_2 (S)

(g) O —KCN, HCN→ HO, CN —1) $LiAlH_4$, 2) H_2O→ HO, NH_2

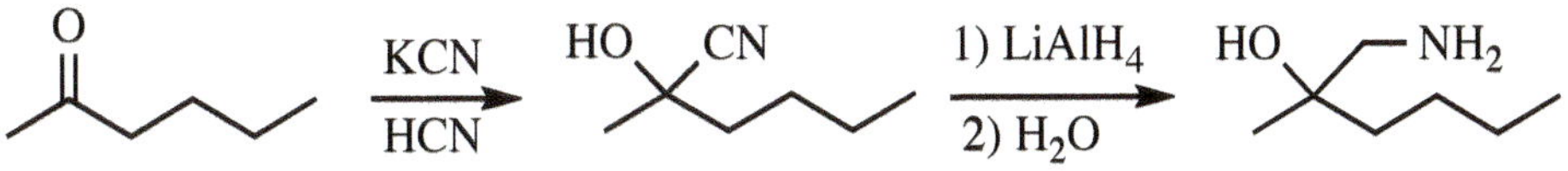

19-31 To reduce nitroaromatics, the reducing reagents (H_2 plus a metal catalyst, or a metal plus HCl) can be used virtually interchangeably. Assume a workup in base to give the free amine final product.

(a) —HNO_3, H_2SO_4→ NO_2 —Sn, HCl→ NH_2

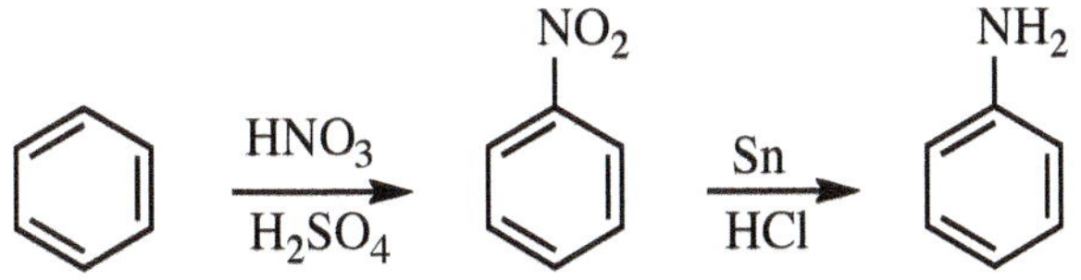

(b) —Br_2, $FeBr_3$→ Br —HNO_3, H_2SO_4→ Br, NO_2 + ortho —Fe, HCl→ Br, NH_2

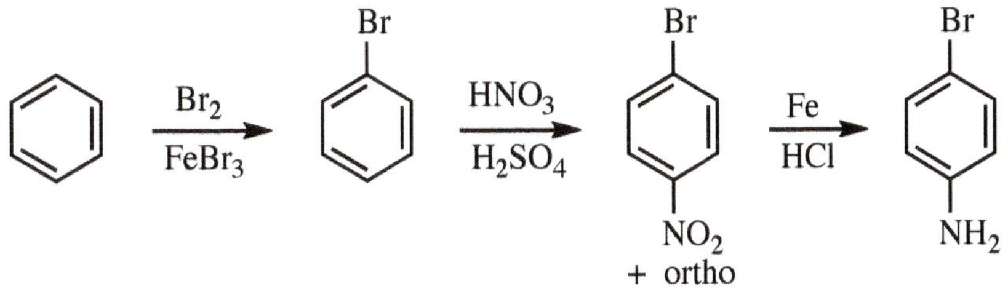

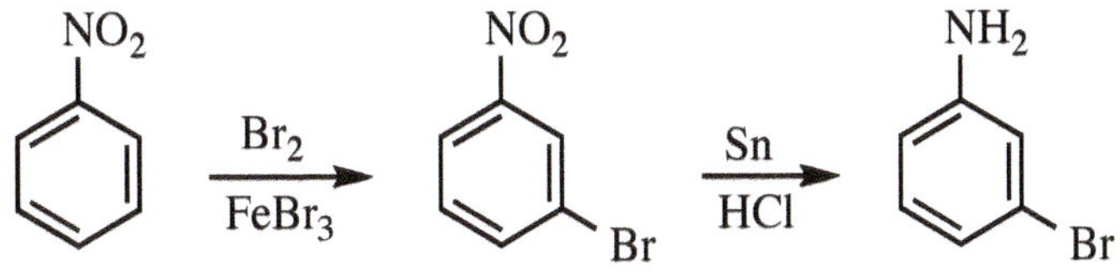

(c) NO_2 (from (a)) —Br_2, $FeBr_3$→ NO_2, Br —Sn, HCl→ NH_2, Br

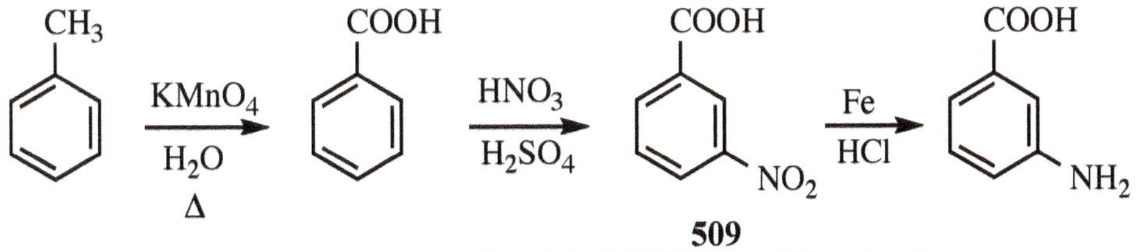

(d) CH_3 —$KMnO_4$, H_2O, Δ→ COOH —HNO_3, H_2SO_4→ COOH, NO_2 —Fe, HCl→ COOH, NH_2

19-32

(a) primary amine; 2,2-dimethylpropan-1-amine, or neopentylamine
(b) secondary amine; N-methylpropan-2-amine, or isopropylmethylamine
(c) tertiary heterocyclic amine and a nitro group; 3-nitropyridine
(d) quaternary heterocyclic ammonium ion; *N*,*N*-dimethylpiperidinium iodide
(e) tertiary aromatic amine oxide; *N*-ethyl-*N*-methylaniline oxide
(f) tertiary aromatic amine; *N*-ethyl-*N*-methylaniline
(g) tertiary heterocyclic ammonium ion; pyridinium chloride
(h) secondary amine; *N*,4-diethylhexan-3-amine

19-33 Shown in order of increasing basicity. In sets a–c, the aliphatic amine is the strongest base.

(a) Ph–N(H)–Ph < $PhNH_2$ < cyclohexyl–NH_2 The aliphatic amine is the strongest base.

(b) pyrrole < pyridine < piperidine The aliphatic amine is the strongest base; pyrrole's aromaticity would be lost if protonated.

(c) pyrrole < imidazole < pyrrolidine The aliphatic amine is the strongest base; pyrrole's aromaticity would be lost if protonated.

(d) O_2N–C_6H_4–NH_2 < C_6H_5–NH_2 < H_3C–C_6H_4–NH_2 Basicity is a measure of the ability to donate the pair of electrons on the amine. Electron-withdrawing groups like NO_2 decrease basicity, while electron-donating groups like CH_3 increase basicity.

(e) C_6H_5–C(=O)–NH_2 < C_6H_5–NH_2 < C_6H_5–CH_2NH_2 The aliphatic amine is the strongest base (amides are not basic).

19-34 Increasing basicity. Begin by identifying the classification of the nitrogen.

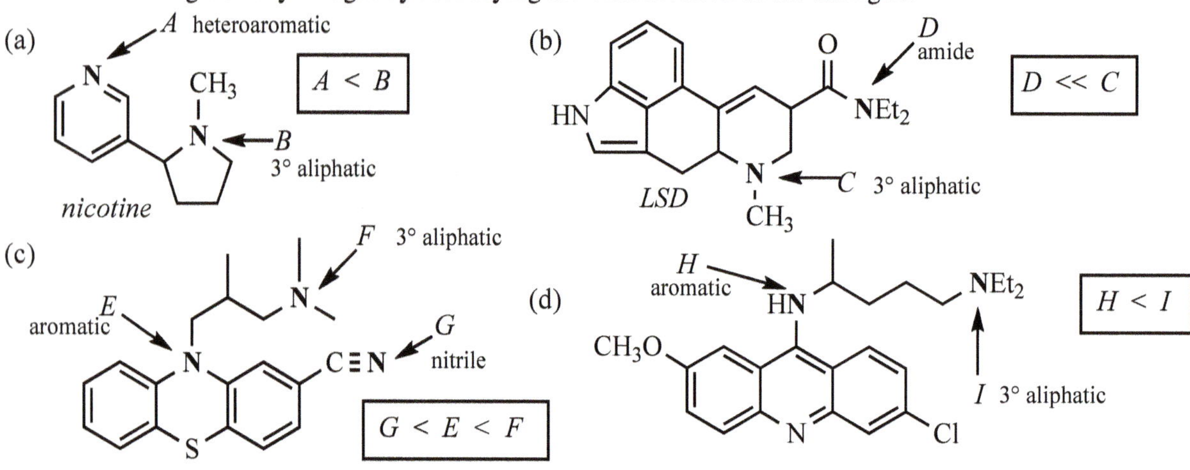

19-34 continued

(e) K amide; J aromatic; L 2° aliphatic

$K < J < L$

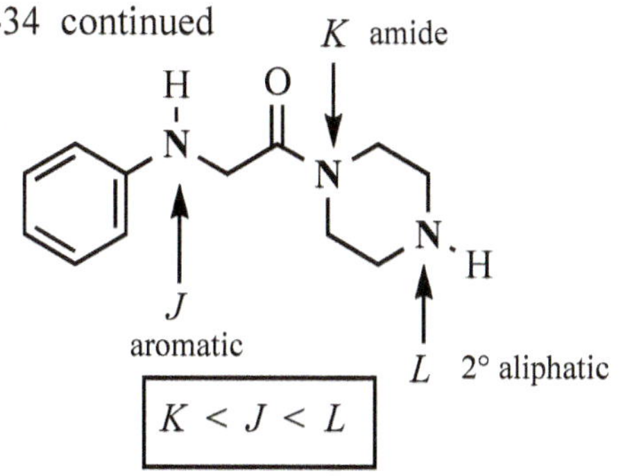

(f) M aromatic; sulfonamide O; 3° aliphatic N

ipsapirone, an anxiolytic drug

$O < M < N$

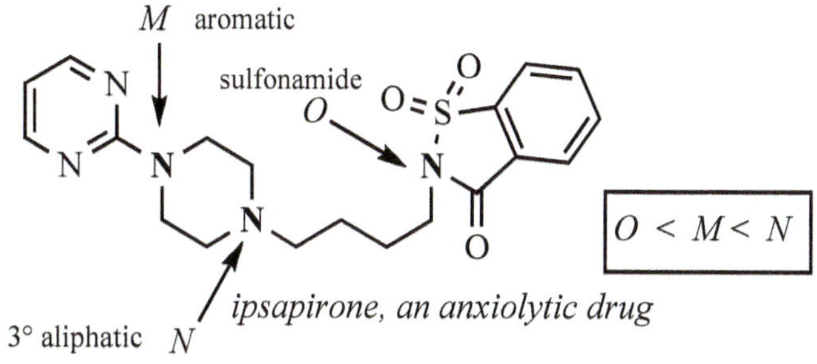

19-35 In each pair of compounds, the stronger base is boxed. Begin by classifying the category of amine, then look for steric and electronic effects.

(a) $HOCH_2CH_2NH_2$ **[$CH_3CH_2NH_2$]**

Both are aliphatic amines, but the first structure has an electron-withdrawing OH substituent, reducing the electron density on the amine.

(b) $PhNH_2$ **[$PhCH_2NH_2$]**

Aliphatic amines are stronger than aromatic amines.

(c) [boxed bicyclic amine] or triethylamine

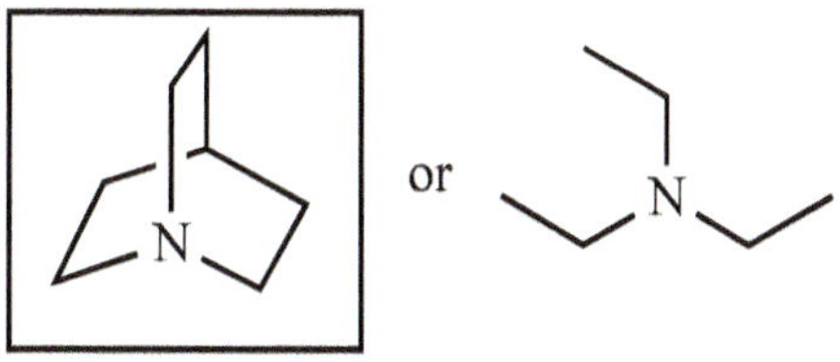

Both are aliphatic amines but the first structure has the R groups tied back. The N: is more exposed, and can be solvated more easily, so this is a steric reason.

*(d) [boxed bicyclic lactam] or NH amide

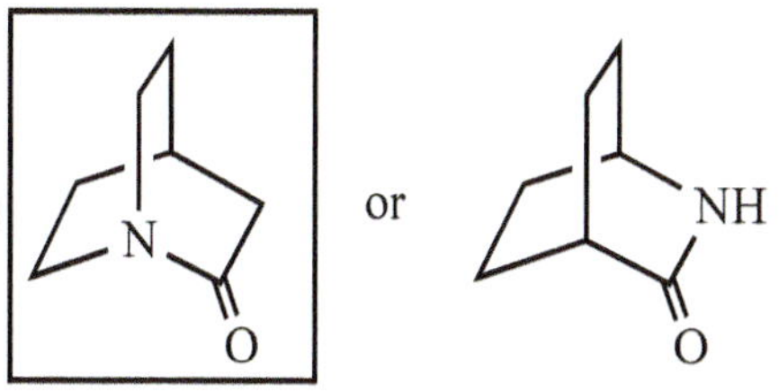

A model would help figure out this one. The structure on the right is a regular amide. The structure on the left is constrained so the the pair of electrons on the N cannot overlap with the C=O, so there is no resonance sharing of the N electrons. That electron pair is available for bonding; the amide's electron pair is not. (This is called Steric Inhibition of Resonance.)

19-36

(a) not resolvable: planar
(b) resolvable: asymmetric carbon
(c) not resolvable: symmetric
(d) resolvable: nitrogen inversion very slow
(e) not resolvable: symmetric
(f) resolvable: asymmetric nitrogen, unable to invert
(g) not resolvable in conditions where the proton on N can exchange
(h) resolvable: asymmetric nitrogen, unable to invert

19-37 The values of pK_b of amines or pK_a of the conjugate acids can be obtained from text Table 19-3 or from text Appendix 4. The side of the reaction with the weaker acid and base will be favored at equilibrium.

(a) pyridine + acetic acid (pK_a 4.74) ⇌ pyridinium (pK_a 8.75) + acetate

Products are favored.

19-37 continued

(b) + OH pK$_a$ 4.74 ⇌ + pK$_a$ ≈ −1 *Reactants are favored.*

(c) pK$_a$ 8.75 + ⇌ + pK$_a$ 11.12 *Products are favored.*

(d) NH_3 + ⇌ NH_2 + pK$_a$ 4.60 pK$_a$ 11.27 *Products are favored.*

19-38 Assume all reductions have aqueous acid workup.

(a) $PhCH_2CH_2CH_2NH_2$ (b) NH_2 (c) CH_3 (d) CH_3 OH

(e) N CH_3 (f) N H_3C CH_3 (g) N NO (h) NH_2 after workup with base

(i) $NHCH_3$ O (j) $CH_2CH_2NHCH_3$ (k) $NHCH_3$ $CH_3(CH_2)_3CHCH_2CH_3$ (l) CH_2NH_2 $PhCH_2CHCH_3$

19-39

(a) N

(b) 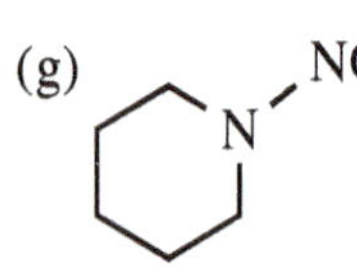 OCH_2CH_3

For a reminder on why fluoride is such a good leaving group in nuc. aromatic subst., see the solution to 17-21 or 19-55(b).

(c) Br NO_2

(d) HO CN *from the first step* → 1) $LiAlH_4$ 2) H_2O → HO CH_2NH_2 (e) H N (f) major product—Hofmann orientation

(g) N Et (h) H N Et

(i) Syn elimination of the amine and D requires the Cope elimination.

H_2O_2 and Δ

(j) Anti elimination of the amine and H requires the Hofmann elimination.

(1) CH_3I
(2) Ag_2O, Δ

19-40

(a) CH_3 … NH_2 $\xrightarrow[HCl]{NaNO_2}$ CH_3 … $\overset{\oplus}{N_2}$ Cl^- $\xrightarrow{CuCN}$ CH_3 … CN

(b) CH_3 … CN (from (a)) $\xrightarrow[2)\ H_2O]{1)\ LiAlH_4}$ CH_3 … CH_2NH_2

(c) CH_3 … $\overset{\oplus}{N_2}$ Cl^- (from (a)) $\xrightarrow{KI}$ CH_3 … I

(d) CH_3 … $\overset{\oplus}{N_2}$ Cl^- (from (a)) $\xrightarrow[\Delta]{H_3O^+}$ CH_3 … OH

(e) CH_3 … NH_2 $\xrightarrow{CH_3\overset{O}{\overset{\|}{C}}-Cl}$ CH_3 … $NHCOCH_3$ $\xrightarrow[H_2SO_4]{HNO_3}$ CH_3 … $NHCOCH_3$, O_2N (+ isomer) $\xrightarrow[\Delta]{H_3O^+}$ after workup with base: CH_3 … NH_2, O_2N

(f) CH_3 … NH_2 + O= (cyclopentanone) $\xrightarrow[CH_3COOH]{NaBH(OAc)_3}$ CH_3 … N(H)–cyclopentyl

19-41 This fragmentation is favorable because the iminium ion produced is stabilized by resonance. Also, there are three possible cleavages that give the same ion. Both factors combine to make the cleavage facile, at the expense of the molecular ion.

$$\left[CH_3 \!\mid\! \underset{CH_3}{\overset{CH_3}{C}} - NH_2\right]^{+\bullet} \longrightarrow \underset{\text{mass 15}}{\bullet CH_3} + \left\{ \oplus\underset{CH_3}{\overset{CH_3}{C}} - \ddot{N}H_2 \longleftrightarrow \underset{CH_3}{\overset{CH_3}{C}} = \overset{\oplus}{N}H_2 \right\}$$

58

m/z 73

m/z 58

19-42

(a) benzene $\xrightarrow[H_2SO_4]{HNO_3}$ nitrobenzene (NO_2) $\xrightarrow[HCl]{Fe}$ aniline (NH_2) $\xrightarrow{Cl-C(=O)-\text{cyclopentyl}}$ $C_6H_5-NH-C(=O)-\text{cyclopentyl}$

(b) aniline (NH_2) + $Cl-SO_2-C_6H_4-CH_3$ → $C_6H_5-N(H)-SO_2-C_6H_4-CH_3$

(c) piperidine (NH) $\xrightarrow{\text{excess } CH_3I}$ I^- N,N-dimethylpiperidinium ($\oplus N(CH_3)_2$) $\xrightarrow[\Delta]{Ag_2O}$ $CH_2=CH-CH_2CH_2CH_2-N(CH_3)_2$

(d) cycloheptanone (O) $\xrightarrow[H-A]{H_2NOH}$ cycloheptanone oxime (NOH) $\xrightarrow[2)\ H_2O]{1)\ LiAlH_4}$ cycloheptylamine (NH_2)

(e) N-methylpyrrolidine (N - CH_3) $\xrightarrow{H_2O_2}$ N-oxide ($\oplus N(-O^\ominus)-CH_3$)

(f) N-methylpyrrolidine (N - CH_3) $\xrightarrow{H_2O_2}$ N-oxide ($\oplus N(-O^\ominus)-CH_3$) $\xrightarrow{\Delta}$ $CH_2=CH-CH_2CH_2-N(OH)-CH_3$

(g) $CH_3-C_6H_4-COOH$ (meta) $\xrightarrow{SOCl_2}$ $CH_3-C_6H_4-C(=O)-Cl$ $\xrightarrow{HN(CH_2CH_3)_2}$ $CH_3-C_6H_4-C(=O)-N(CH_2CH_3)_2$

19-43 The problem restricts the starting materials to six carbons or fewer. Always choose starting materials with as many of the necessary functional groups as possible.

(a)

$HO-C_6H_4-NH_2$ $\xrightarrow[2)\ CH_3CH_2Br]{1)\ NaOH}$ $EtO-C_6H_4-NH_2$ $\xrightarrow{H_3C-C(=O)-Cl}$ $EtO-C_6H_4-N(H)-C(=O)-CH_3$

phenacetin

(b)

C_6H_5-Br $\xrightarrow[\text{ether}]{Mg}$ $\xrightarrow{H_2C\text{—}CHCH_3 \text{ (epoxide, O)}}$ $\xrightarrow{H_3O^+}$ $C_6H_5-CH_2-CH(OH)CH_3$ $\xrightarrow[HOAc]{NaOCl}$ $C_6H_5-CH_2-C(=O)CH_3$

$\xrightarrow[NaBH(OAc)_3,\ CH_3COOH]{CH_3NH_2}$ $C_6H_5-CH_2-CH(NHCH_3)CH_3$

methamphetamine

19-43 continued

(c) HO / HO —[benzene]— $\xrightarrow[AlCl_3]{CH_3I}$ HO / HO —[benzene]— CH_3 $\xrightarrow[h\nu]{NBS}$ HO / HO —[benzene]— CH_2Br $\xrightarrow{KCN}$ HO / HO —[benzene]— CH_2CN $\xrightarrow{H_2,\ Pt}$ HO / HO —[benzene]— $CH_2CH_2NH_2$

dopamine

19-44 This is one possible approach; there may be others as well. As an alternative, make the diazonium ion from p-nitroaniline, then replace with cyanide, and hydrolyze the nitrile to the COOH.

benzene $\xrightarrow[AlCl_3]{CH_3I}$ CH_3–benzene $\xrightarrow[H_2SO_4]{HNO_3}$ O_2N–benzene–CH_3 $\xrightarrow[\Delta]{KMnO_4}$ O_2N–benzene–COOH

Nitro reduction could also be done here.

NH + O (epoxide) → N–OH

H_2SO_4

O_2N–benzene–C(=O)–O–CH₂CH₂–N(Et)₂ $\xrightarrow[HCl]{Fe}$ H_2N–benzene–C(=O)–O–CH₂CH₂–N(Et)₂

Ester formation was first discussed in Section 11-12.

19-45

(a)

benzene $\xrightarrow[AlCl_3]{Cl-C(=O)-CH_3}$ O=C(CH₃)–benzene $\xrightarrow[H_2SO_4]{HNO_3}$ O=C(CH₃)–benzene–NO_2 $\xrightarrow[HCl\ (aq)]{Zn\ (Hg)}$ CH_2CH_3–benzene–NH_2

Clemmensen conditions reduce both the ketone and the nitro group.

1. $NaNO_2$, HCl(aq)
2. CuCN

CH_2CH_3–benzene–C≡N $\xrightarrow[H^+,\ \Delta]{H_2O}$ CH_2CH_3–benzene–COOH

19-45 continued

(b) This synthesis is similar to the one in part (a).

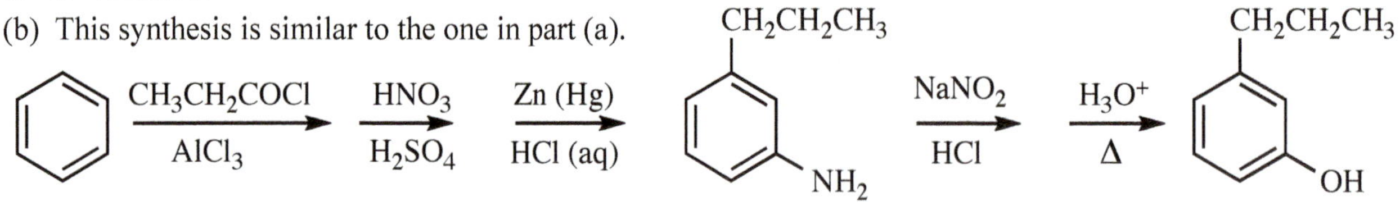

Benzene $\xrightarrow[AlCl_3]{CH_3CH_2COCl}$ $\xrightarrow[H_2SO_4]{HNO_3}$ $\xrightarrow[HCl\ (aq)]{Zn\ (Hg)}$ m-propylaniline ($CH_2CH_2CH_3$, NH_2) $\xrightarrow[HCl]{NaNO_2}$ $\xrightarrow[\Delta]{H_3O^+}$ m-propylphenol ($CH_2CH_2CH_3$, OH)

(c) This long synthesis uses diazonium substitution twice to put on the OH and the COOH.

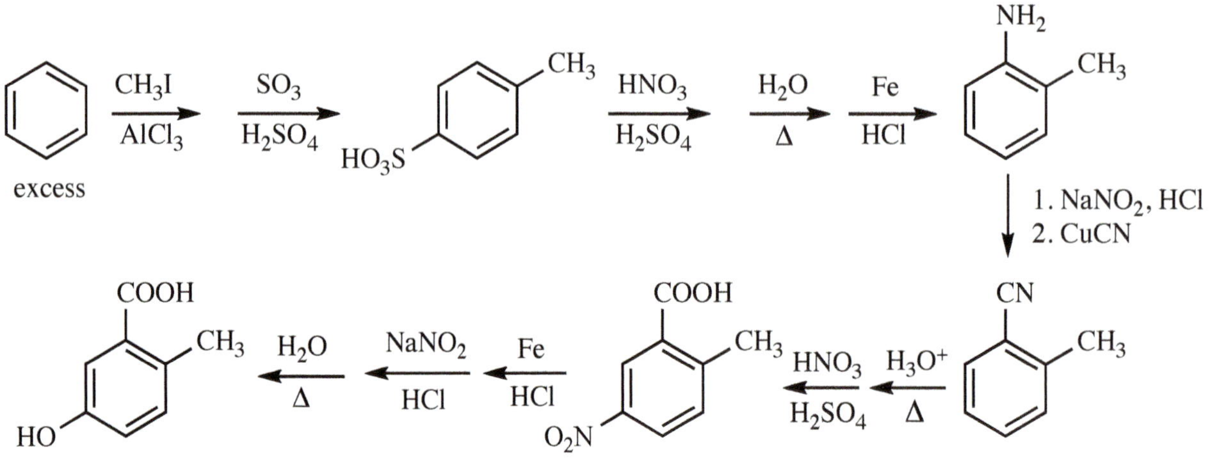

Benzene (excess) $\xrightarrow[AlCl_3]{CH_3I}$ $\xrightarrow[H_2SO_4]{SO_3}$ HO_3S–C6H4–CH_3 $\xrightarrow[H_2SO_4]{HNO_3}$ $\xrightarrow[\Delta]{H_2O}$ $\xrightarrow[HCl]{Fe}$ o-toluidine (NH_2, CH_3) $\xrightarrow{1.\ NaNO_2,\ HCl;\ 2.\ CuCN}$ o-tolunitrile (CN, CH_3) $\xrightarrow[\Delta]{H_3O^+}$ $\xrightarrow[H_2SO_4]{HNO_3}$ COOH, CH_3, O_2N $\xrightarrow[HCl]{Fe}$ $\xrightarrow[HCl]{NaNO_2}$ $\xrightarrow[\Delta]{H_2O}$ COOH, CH_3, HO

(d)

Benzene $\xrightarrow[H_2SO_4]{HNO_3}$ $\xrightarrow[HCl]{Fe}$ aniline (NH_2) $\xrightarrow[HCl]{NaNO_2}$ $\xrightarrow[\Delta]{H_2O}$ phenol (OH) $\xrightarrow[2.\ CH_3I]{1.\ NaOH}$ (Williamson) anisole (OCH_3) $\xrightarrow[H_2SO_4]{HNO_3}$ $\xrightarrow[HCl]{Fe}$ p-anisidine (OCH_3, NH_2)

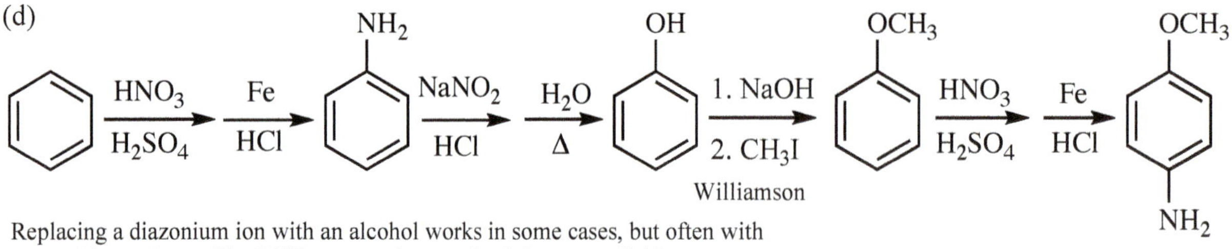

Replacing a diazonium ion with an alcohol works in some cases, but often with many side products. The Williamsom ether synthesis is more reliable.

19-46

(a)

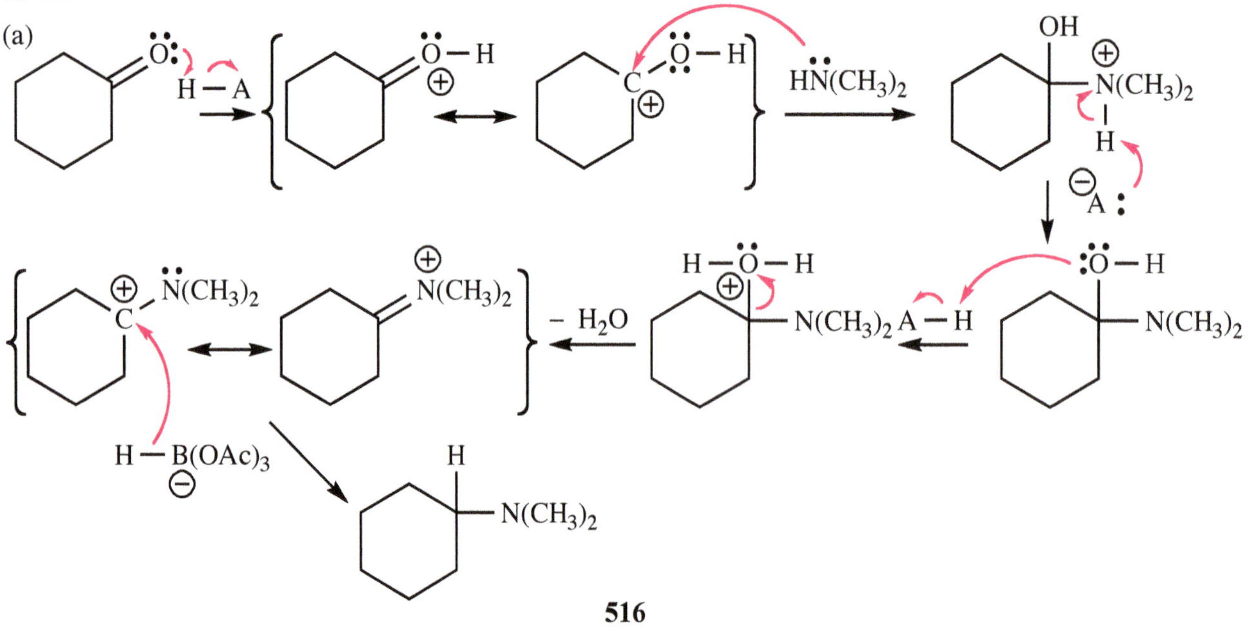

Cyclohexanone (:Ö:) + H—A → [protonated ketone (Ö—H, ⊕) ⟷ carbocation (C⊕, :Ö—H)] + $H\ddot{N}(CH_3)_2$ → carbinolamine intermediate (OH, ⊕$N(CH_3)_2$—H) + $\ominus$A: ↓ :Ö—H, $N(CH_3)_2$ + A—H → H—Ö⊕—H, $N(CH_3)_2$ → − H_2O → [C⊕ $\ddot{N}(CH_3)_2$ ⟷ ⊕$N(CH_3)_2$ (iminium)] + H—B(OAc)$_3$ ⊖ → H, $N(CH_3)_2$ (N,N-dimethylcyclohexylamine)

19-46 continued

(b)

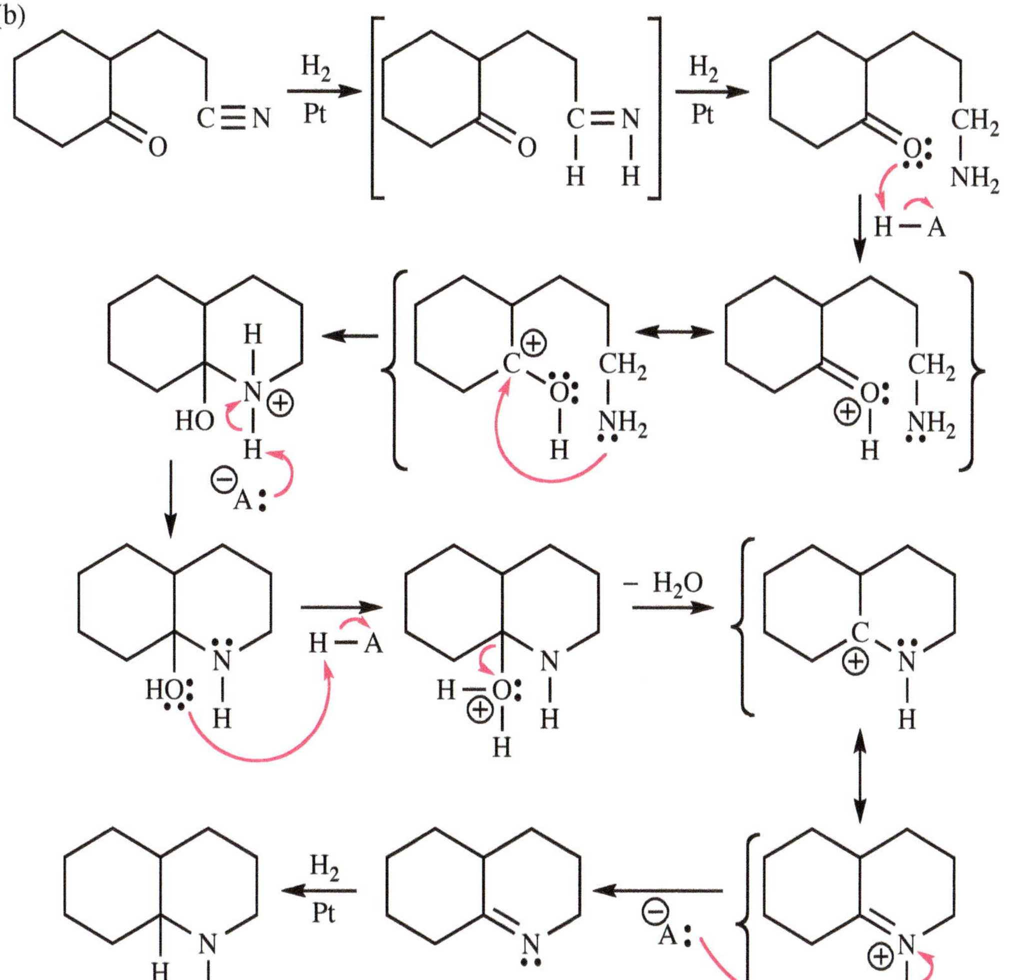

19-47

(a)

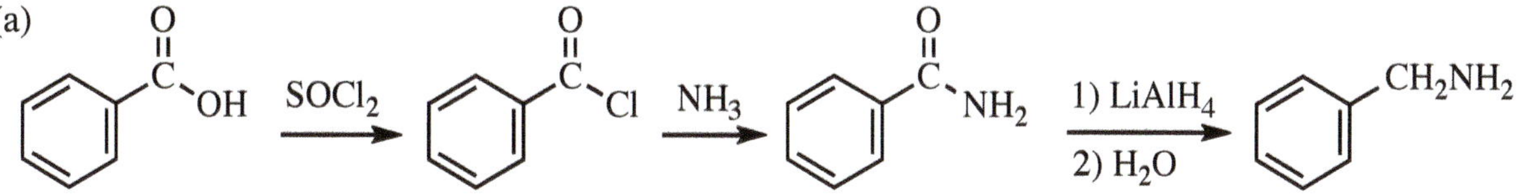

(b)

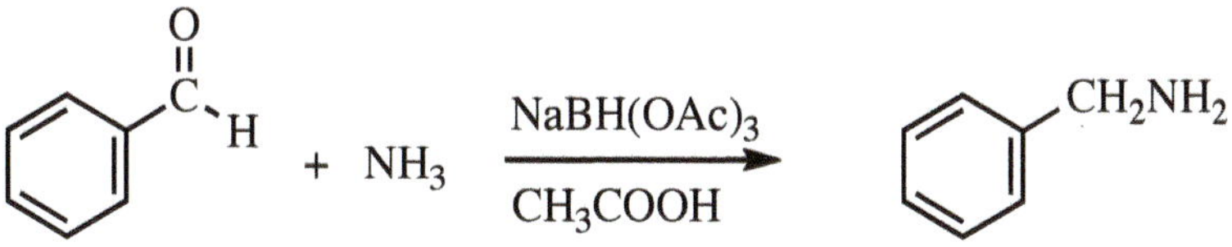

(c)

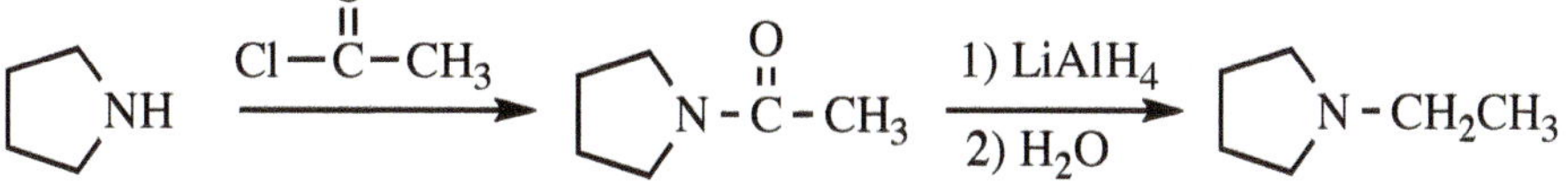

(d)

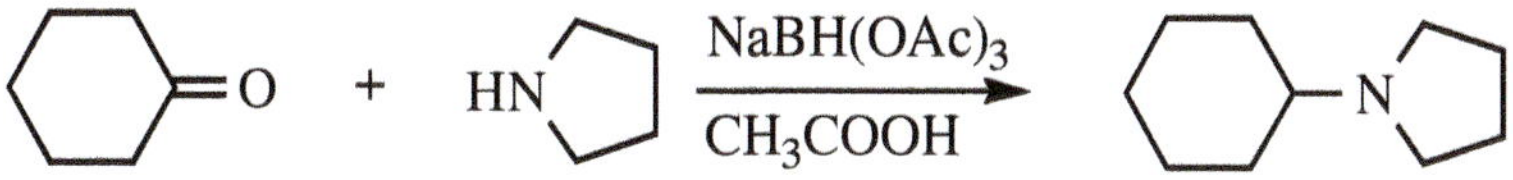

19-47 continued

(e) HO–C(=O)–CH$_2$CH$_2$CH$_2$–C(=O)–OH $\xrightarrow{SOCl_2}$ Cl–C(=O)–CH$_2$CH$_2$CH$_2$–C(=O)–Cl $\xrightarrow{NH_3}$ H_2N–C(=O)–CH$_2$CH$_2$CH$_2$–C(=O)–NH_2 $\xrightarrow{1)\ LiAlH_4\quad 2)\ H_2O}$ H_2N–CH$_2$CH$_2$CH$_2$CH$_2$CH$_2$–NH_2

19-48

(a) allyl Br + phthalimide anion (N$^{\ominus}$) ⟶ N-allylphthalimide $\xrightarrow[\Delta]{NH_2NH_2}$ allyl NH_2

(b) CH_2CH_3 (ethylbenzene) $\xrightarrow[H_2SO_4]{HNO_3}$ CH_2CH_3 / NO_2 (para) $\xrightarrow[HCl]{Sn}$ CH_2CH_3 / NH_2 (para)

(c) Br (Ph) + phthalimide anion (N$^{\ominus}$) ⟶ $\xrightarrow[\Delta]{NH_2NH_2}$ NH_2 (Ph)

(d) Br (Ph) $\xrightarrow{KCN}$ CN (Ph) $\xrightarrow[2)\ H_2O]{1)\ LiAlH_4}$ NH_2 (Ph)

19-49 A neutral, aqueous workup is required in each synthesis.

(a)

Ph–CH$_2$CH$_2$–NH–CH$_2$CH$_2$CH$_3$ + H–C(=O)–CH_3 $\xrightarrow{NaBH(OAc)_3,\ CH_3COOH}$ product

Ph–CH$_2$CH$_2$–NH–CH$_2$CH$_3$ + H–C(=O)–CH_2CH_3 $\xrightarrow[CH_3COOH]{NaBH(OAc)_3}$ product

HN(CH$_2$CH$_3$)(CH$_2$CH$_2$CH$_3$) + Ph–CH$_2$–C(=O)–H $\xrightarrow{NaBH(OAc)_3,\ CH_3COOH}$ product

(b)

Ph–CH$_2$CH$_2$–NH–CH$_2$CH$_2$CH$_3$ + Cl–C(=O)–CH_3 ⟶ $\xrightarrow{LiAlH_4}$ product

Ph–CH$_2$CH$_2$–NH–CH$_2$CH$_3$ + Cl–C(=O)–CH_2CH_3 ⟶ $\xrightarrow{LiAlH_4}$ product

HN(CH$_2$CH$_3$)(CH$_2$CH$_2$CH$_3$) + Ph–CH$_2$–C(=O)–Cl ⟶ $\xrightarrow{LiAlH_4}$ product

Product: Ph–CH$_2$CH$_2$–N(CH$_2$CH$_3$)–CH$_2$CH$_2$CH$_3$

19-50 In this problem, sodium triacetoxyborohydride, $NaBH(OAc)_3$, will be used for reductive alkylation. Other synthetic sequences may be equally effective. (Acetic acid is abbreviated HOAc.)

(a) OH —(TsCl, pyridine)→ OTs —(KCN)→ CN —(1) $LiAlH_4$; 2) H_2O)→ NH_2

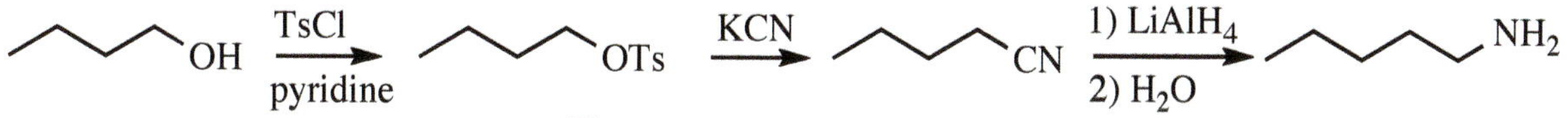

(b) OH —(NaOCl, TEMPO)→ H, O —(CH_3NH_2, $NaBH(OAc)_3$, HOAc)→ $NHCH_3$

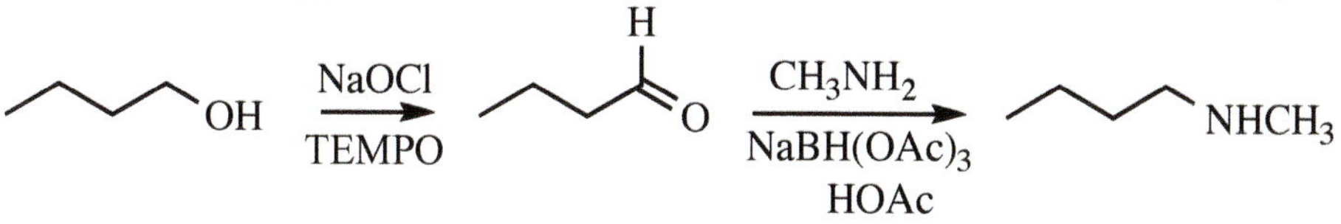

(c) OH —(TsCl, pyridine)→ OTs —(excess NH_3)→ NH_2

OH —(NaOCl, HOAc)→ O + NH_2 —($NaBH(OAc)_3$, HOAc)→ HN

CH_3CH_2OH —(NaOCl, TEMPO)→ $CH_3-\overset{O}{\overset{\|}{C}}H$

—($NaBH(OAc)_3$, HOAc)→ N

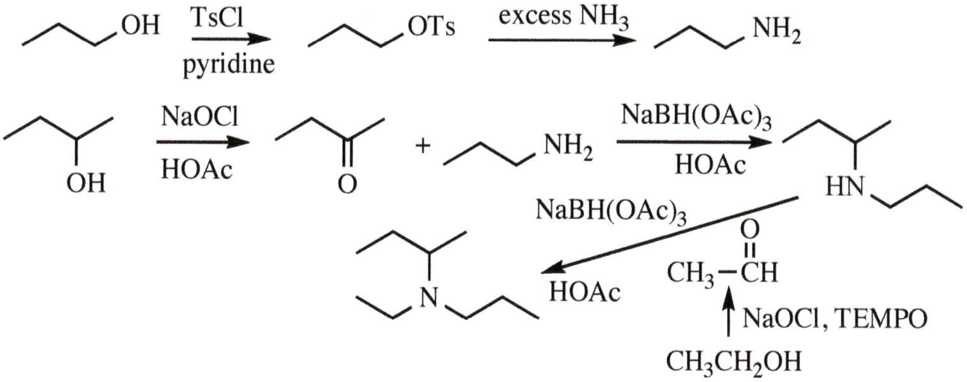

(d) CH_3 —(NBS, hv)→ CH_2Br —(NaOH)→ CH_2OH —(NaOCl, TEMPO)→ $\overset{O}{\overset{\|}{C}}$ H

The Gatterman-Koch reaction would make benzaldehyde from benzene.

—(HOAc; $NaBH(OAc)_3$; NH_2 from (c))→ $CH_2NHCH_2CH_2CH_3$

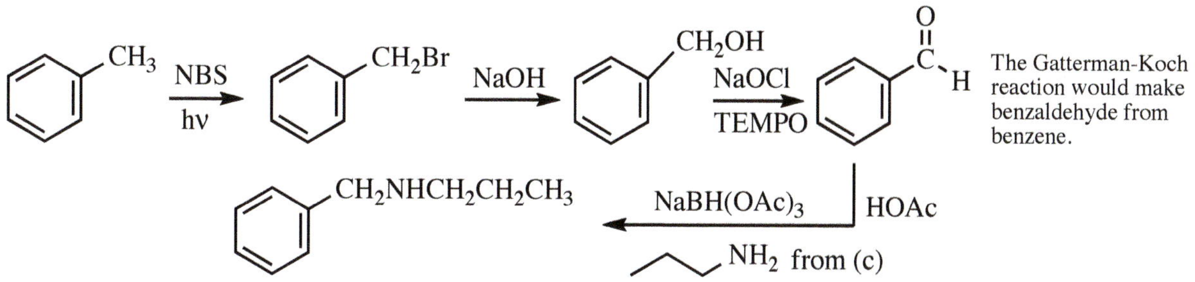

(e) —(HNO_3, H_2SO_4)→ NO_2 —(Fe, HCl)→ NH_2 —($NaNO_2$, HCl)→ $\overset{\oplus}{N_2}$ Cl^- —(H_3O^+)→ OH

Combine phenol and phenyldiazonium ion.

N=N, OH

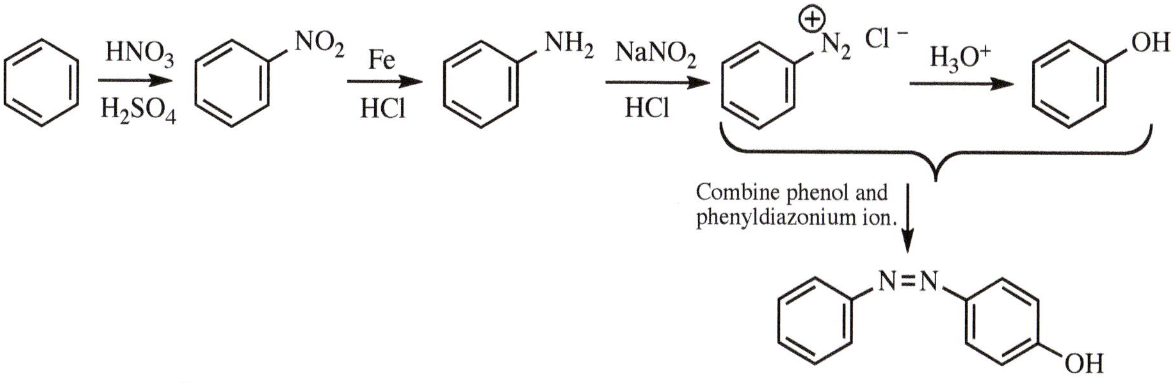

(f) —(O, Cl; $AlCl_3$)→ O —(HNO_3, H_2SO_4)→ O, NO_2 —(Zn(Hg), HCl; Clemmensen reduces both)→ NH_2 after neutral workup

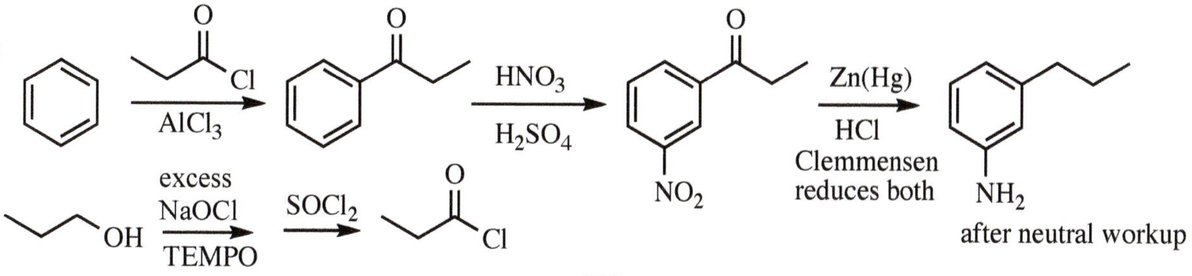

OH —(excess NaOCl, TEMPO)→ —($SOCl_2$)→ O, Cl

19-50 continued

(g)

Benzene + $(CH_3)_2CHCOCl$ ($AlCl_3$) → isobutyrophenone → Zn(Hg), HCl → isobutylbenzene → HNO_3, H_2SO_4; Fe, HCl → H_2N–C_6H_4–$CH_2CH(CH_3)_2$ (para)

after workup with base

$(CH_3)_2CHCH_2OH$ → excess NaOCl, TEMPO → $SOCl_2$ → $(CH_3)_2CHCOCl$

19-51

(a)

Benzene + $Cl-\overset{O}{\overset{\|}{C}}CH_2CH_2CH_3$ ($AlCl_3$) → C_6H_5–$C(=O)CH_2CH_2CH_3$ → Zn(Hg), HCl → C_6H_5–$CH_2CH_2CH_2CH_3$ → HNO_3, H_2SO_4 → p-NO_2–C_6H_4–$CH_2CH_2CH_2CH_3$ → Fe, HCl →

after workup with base H_2N–C_6H_4–$CH_2CH_2CH_2CH_3$

(b) Quinuclidine → CH_3I → N-methylquinuclidinium iodide (N^+-CH_3 I^-) → Ag_2O, Δ → 4-vinyl-1-methylpiperidine (N – CH_3)   Hofmann elimination

(c) Ph–C_6H_5 → HNO_3, H_2SO_4 → Ph–C_6H_4–NO_2 → Fe, HCl → Ph–C_6H_4–NH_2 → $Cl-\overset{O}{\overset{\|}{C}}CH_3$ → Ph–C_6H_4–$NHCOCH_3$ → CH_3CH_2Br, $AlCl_3$ → Ph–$C_6H_3(CH_2CH_3)$–$NHCOCH_3$ → Δ, H_3O^+ → Ph–$C_6H_3(CH_2CH_3)$–NH_2 → 1) $NaNO_2$, HCl 2) CuCN → Ph–$C_6H_3(CH_2CH_3)$–C≡N → 1) $LiAlH_4$ 2) H_2O → Ph–$C_6H_3(CH_2CH_3)$–CH_2NH_2

19-52

coniine (S) → excess CH_3I → N,N-dimethylpiperidinium iodide (CH_3, CH_3, I^-) → Δ, Ag_2O → (S) H, $N(CH_3)_2$

coniine

N,N-dimethyloct-7-ene-4-amine

19-53

(a) unknown **X**

—fishy odor ⇒ amine

—molecular weight 101 ⇒ odd number of nitrogens

⇒ If one nitrogen and no oxygen, the remainder is C_6H_{15} .

Mass spectrum:

—fragment at 86 = M – 15 = loss of methyl ⇒ the compound is likely to have this structural piece: CH_3—C—N (α-cleavage)

IR spectrum:

—no OH, no NH ⇒ must be a 3° amine

—no C=O or C=C or C≡N or NO_2

NMR spectrum:

—only a triplet and quartet, integration about 3 : 2 ⇒ ethyl group(s) only

Assemble the evidence: $C_6H_{15}N$, 3° amine, only ethyl in the NMR:

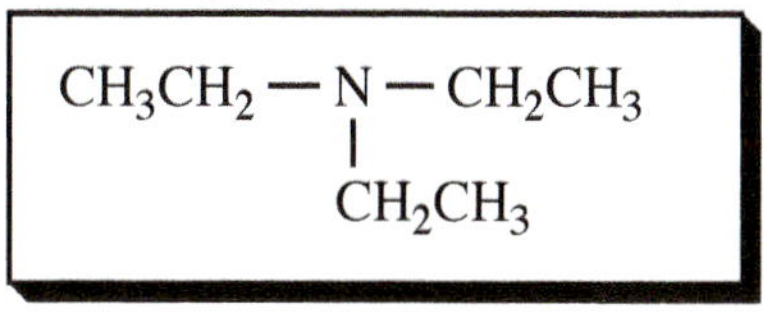

CH_3CH_2—N—CH_2CH_3 (with CH_2CH_3 on N)

(b) React the triethylamine with HCl. The pure salt is solid and odorless.

$$(CH_3CH_2)_3N + HCl \longrightarrow (CH_3CH_2)_3\overset{\oplus}{N}H \quad Cl^-$$

salt

(c) Washing her clothing in dilute acid like vinegar (dilute acetic acid) or dilute HCl would form a water-soluble salt as shown in (b). Normal washing will remove the water-soluble salt.

19-54

(a) $CH_3C\equiv O^{\oplus}$ ⟷ $CH_3\overset{\oplus}{C}=O$ + pyrrole (N–H) ⟶ {resonance forms of the 2-acylated intermediate, H, H, H, N–H} ; $CH_3COO^{\ominus}$ ⟶ 2-acetylpyrrole (N–H)

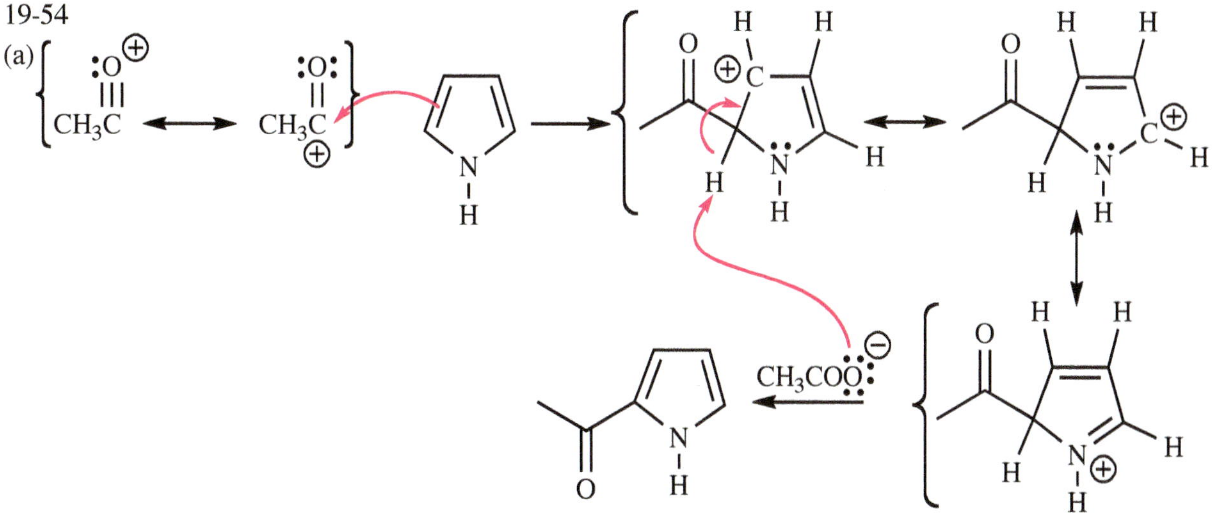

(b) Both answers can be found in the resonance forms of the intermediate, in particular, the resonance form that shows the positive charge on the N. This is the major resonance contributor; what is special about it is that every atom has a full octet, the best of all possible conditions. That does not arise in the benzene intermediate, so it must be easier to form the intermediate from pyrrole than from benzene. Also, acylation at the 3-position puts positive charge at the 2-position and on the N, but never on the other side of the ring, so this substitution has only two resonance forms. The intermediate from acylation at the 3-position is therefore not as stable as the intermediate from acylation at the 2-position.

19-55

(a)

F, O_2N, $\ddot{N}H_2R$, $R = CH_3$, NO_2 → F, $\overset{\oplus}{N}H_2R$, NO_2 ↔ F, $\overset{\oplus}{N}H_2R$, NO_2 ↔ F, $\overset{\oplus}{N}H_2R$, O_2N ↔ F, $\overset{\oplus}{N}H_2R$, O_2N ↔ F, $\overset{\oplus}{N}H_2R$, O_2N, H, NO_2 → $H-\overset{\oplus}{N}HR$, O_2N, NO_2 + $:\ddot{F}:^{\ominus}$ → $H\ddot{N}-CH_3$, O_2N, NO_2

(b) Why is fluoride ion a good leaving group from **A** but not from **B** (either by S_N1 or S_N2)?

A: F, Nu, NO_2, HC, NO_2 **B**: F, A, B, C, :Nu

Formation of the anionic sigma complex **A** is the rate-determining (slow) step in nucleophilic aromatic substitution. The loss of fluoride ion occurs in a subsequent fast step in which aromaticity is restored, where the nature of the leaving group does not affect the overall reaction rate. In the S_N1 or S_N2 mechanisms, however, the carbon-fluorine bond is breaking in the rate-determining step, so the poor leaving group ability of fluoride does indeed affect the rate.

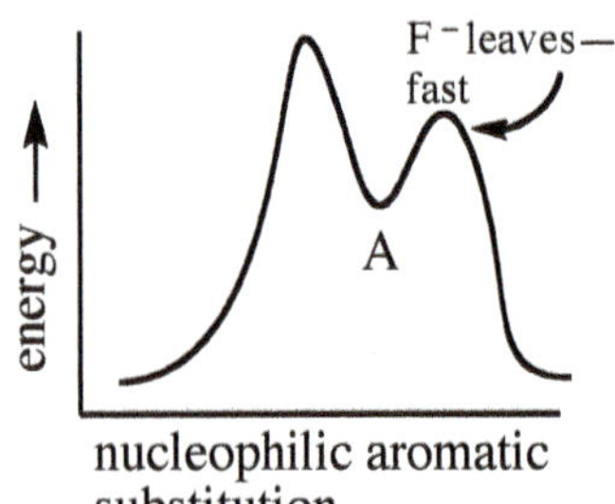

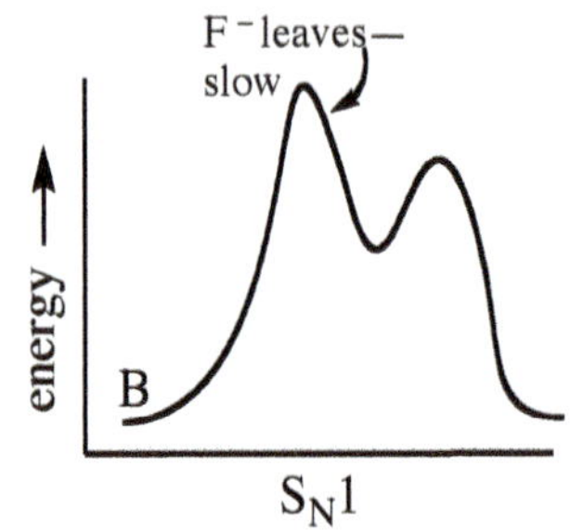

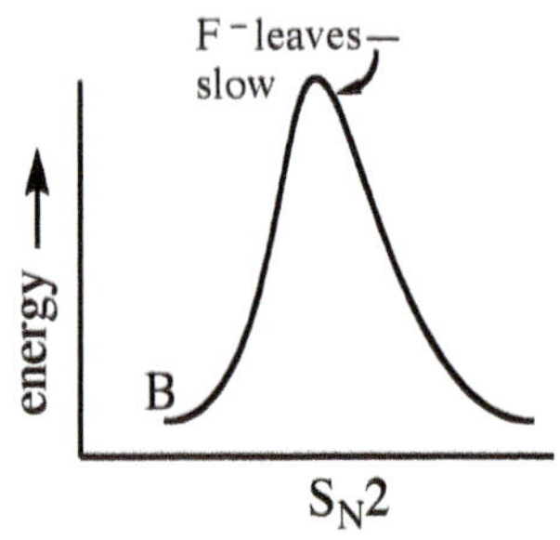

(c) Amines can act as nucleophiles as long as the electron pair on the N is available for bonding. The initial reactant, methylamine, CH_3NH_2, is a very reactive nucleophile. However, once the N is bonded to the benzene ring, the electron pair is delocalized onto the ring, especially with such strong electron-withdrawing groups like NO_2 in the ortho and para positions. The electrons on N are no longer available for bonding so there is no danger of it acting as a nucleophile in another reaction.

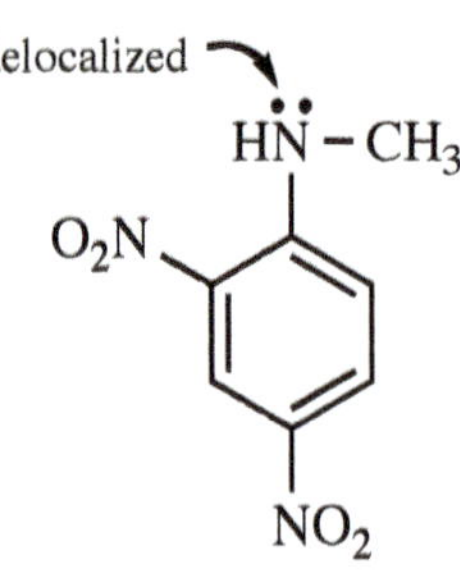

19-56 **Compound A**

Mass spectrum:

—molecular ion at 73 = odd mass = odd number of nitrogens;

if one nitrogen and no oxygen present ⇒ molecular formula $C_4H_{11}N$

19-56 continued

—base peak at 44 is M – 29 ⇒ This fragment must be present:

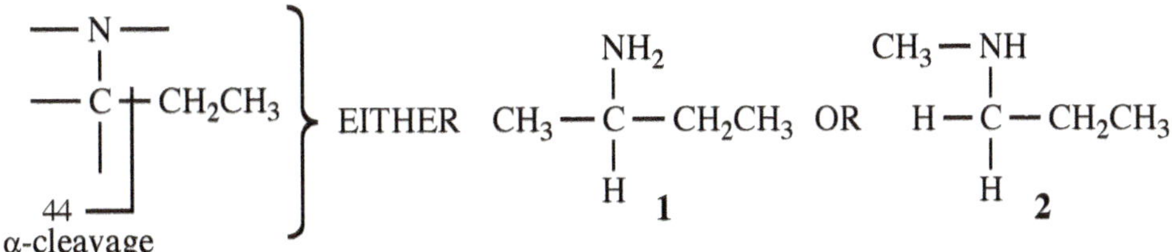

IR spectrum:

—two peaks around 3300 cm^{-1} indicate a 1° amine so **2** is not correct; no indication of oxygen

NMR spectrum:

—two exchangeable protons suggest NH_2 present

—1H multiplet at δ 2.8 means a $CH-NH_2$

A

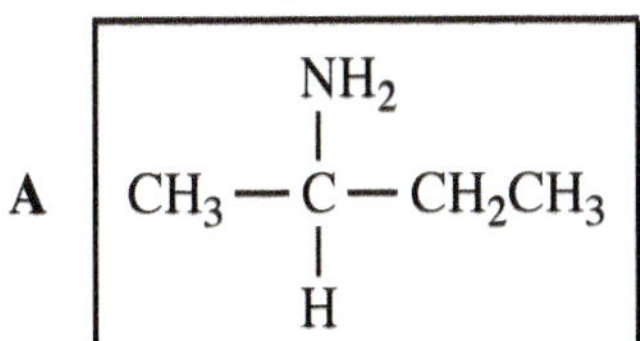

The structure of **A** must be the same as **1** above:

Compound B

an isomer of **A**, so its molecular formula must also be $C_4H_{11}N$

IR spectrum:

—only one peak at 3300 cm^{-1} ⇒ 2° amine

NMR spectrum:

—one exchangeable proton ⇒ NH

—two ethyls present

The structure of **B** must be:

B

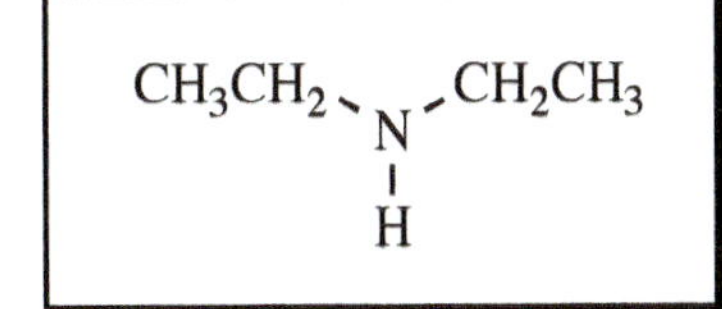

$$\left[CH_3 \overset{58}{+} CH_2-\underset{\underset{H}{|}}{N}CH_2CH_3\right]^{+\cdot} \longrightarrow CH_2=\overset{\oplus}{\underset{\underset{H}{|}}{N}}CH_2CH_3$$

m/z 58 resonance-stabilized

19-57

Mass spectrum:

—molecular ion at 87 = odd mass = odd number of nitrogens present

—if one nitrogen and no oxygens ⇒ molecular formula $C_5H_{13}N$

—base peak at m/z 30 ⇒ structure must include this fragment

$$R + CH_2NH_2 \ (30) \longrightarrow H_2C=NH_2^{\oplus}$$

m/z 30

IR spectrum:

—two peaks in the 3300–3400 cm^{-1} region ⇒ 1° amine

NMR spectrum:

—singlet at δ 0.9 for 9H must be a *t*-butyl group

—2H signal at δ 1.0 exchanges with D_2O ⇒ must be protons on N or O

$$CH_3-\underset{CH_3}{\overset{CH_3}{C}}-CH_2-NH_2$$

δ 0.9 { $CH_3-\underset{CH_3}{\overset{CH_3}{C}}-CH_2-NH_2$ ← δ 1.0

δ 2.4 (CH_2)

19-57 continued

Note that the base peak in the MS arises from cleavage to give these two, relatively stable fragments:

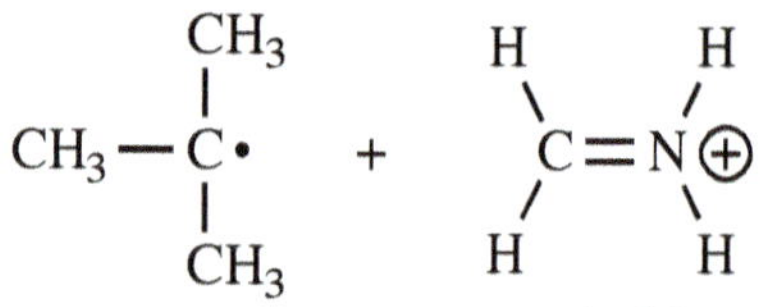

$(CH_3)_3C\bullet$ + $H_2C{=}NH_2^{\oplus}$

m/z 30

19-58 (a tough problem)

Molecular formula $C_{11}H_{16}N_2$ has 5 elements of unsaturation, enough for a benzene ring; no oxygens precludes NO_2 and amide; if C≡N is present, there are not enough elements of unsaturation left for a benzene ring, so benzene and C≡N are mutually exclusive.

IR spectrum:

—one spike around 3300 cm^{-1} suggests a 2° amine

—no C≡N

—CH and C=C regions suggest an aromatic ring

Proton NMR spectrum:

—5H multiplet at δ 7.3 indicates a monosubstituted benzene ring (the fact that all the peaks are huddled around 7.3 precludes N being bonded to the ring)

—1H singlet at δ 2.0 is exchangeable ⇒ NH of secondary amine

—2H singlet at δ 3.5 is CH_2; the fact that it is so strongly deshielded and unsplit suggests that it is between a nitrogen and the benzene ring

fragments so far:

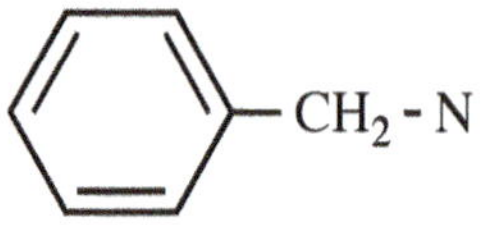

$C_6H_5-CH_2-N$ + NH + 4 C + 8 H + 1 element of unsaturation

Carbon NMR spectrum:

—four signals around δ 125–138 are the aromatic carbons

—the signal at δ 65 is the CH_2 bonded to the benzene

—the other four carbons come as two signals at δ 46 and δ 55; each is a CH_2, so there are two sets of two equivalent CH_2 groups, each bonded to N to shift it downfield

fragments so far:

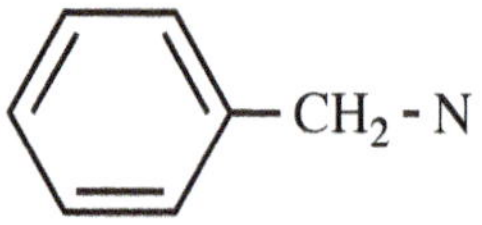

$C_6H_5-CH_2-N$ + NH + $\begin{matrix}CH_2\\CH_2\end{matrix}$ + $\begin{matrix}CH_3\\CH_2\end{matrix}$ + 1 element of unsaturation

There is no evidence for an alkene in any of the spectra, so the remaining element of unsaturation must be a ring. The simplicity of the NMR spectra indicates a fairly symmetric compound.

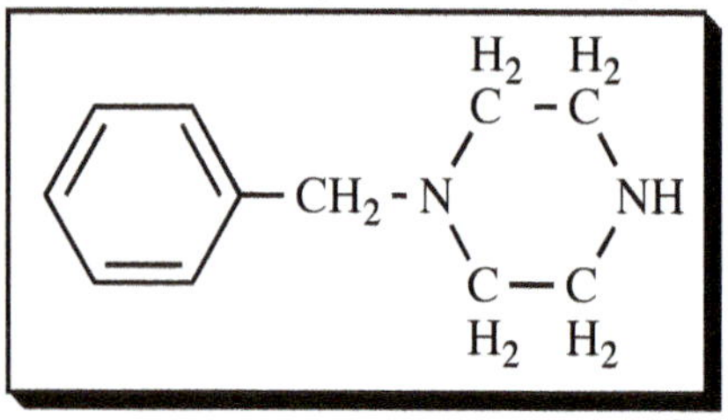

$C_6H_5-CH_2-N(CH_2CH_2)_2NH$ (1-benzylpiperazine)

19-59

(a) The acid-catalyzed condensation of P2P (a controlled substance) with methylamine hydrochloride gives an imine which can be reduced to methamphetamine. The suspect was probably planning to use zinc in muriatic acid (dilute HCl) for the reduction.

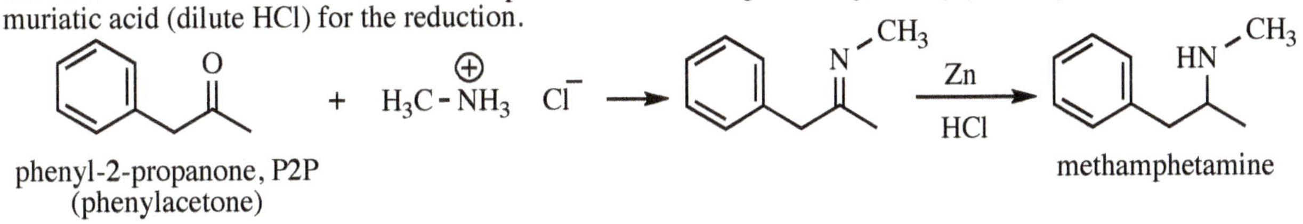

$$\text{P2P} + H_3C\text{-}\overset{\oplus}{N}H_3\ Cl^- \longrightarrow \text{imine} \xrightarrow[HCl]{Zn} \text{methamphetamine}$$

phenyl-2-propanone, P2P
(phenylacetone)

methamphetamine

(b) The jury acquitted the defendant on the charge of attempted manufacture of methamphetamine. There were legal problems with possible entrapment, plus the fact that he had never opened the bottle of the starting material. The defendant was convicted on several possession charges, however, and was awarded four years of institutional time to study organic chemistry.

19-60

(a) When guanidine is protonated, the cation is greatly stabilized by resonance, distributing the positive charge over all atoms (except H):

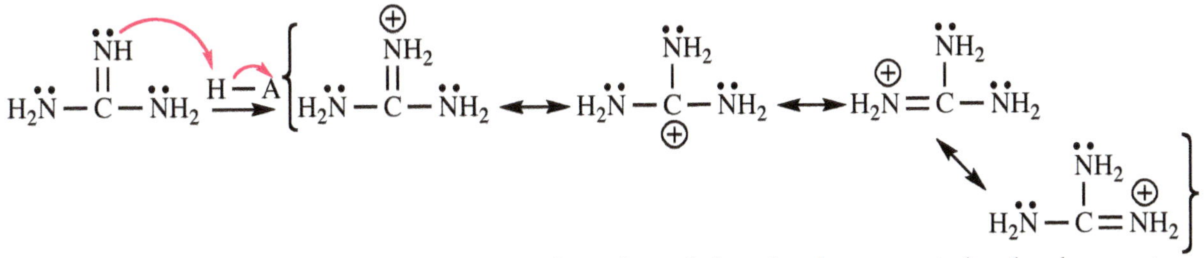

$$H_2N-C(=NH)-NH_2 \xrightarrow{H-A} \left\{ H_2N-C(=\overset{\oplus}{N}H_2)-NH_2 \longleftrightarrow H_2N-\overset{\oplus}{C}(NH_2)-NH_2 \longleftrightarrow H_2\overset{\oplus}{N}=C(NH_2)-NH_2 \longleftrightarrow H_2N-C(NH_2)=\overset{\oplus}{N}H_2 \right\}$$

(b) The unprotonated molecule has a resonance form shown below that the protonated molecule cannot have. Therefore, the unprotonated form is stabilized relative to the protonated form. This greater stabilization of the unprotonated form is reflected in weaker basicity.

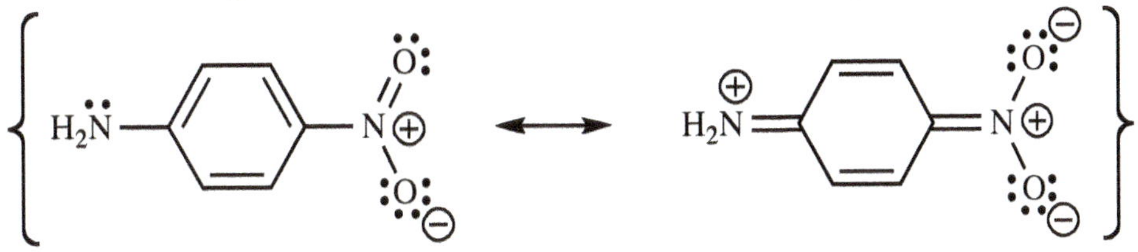

$$\left\{ H_2N-C_6H_4-\overset{\oplus}{N}(=O)-O^{\ominus} \longleftrightarrow H_2\overset{\oplus}{N}=C_6H_4=\overset{\oplus}{N}(O^{\ominus})_2 \right\}$$

19-60 continued

(c) Anilines are weaker bases than aliphatic amines because the electron pair on the nitrogen is shared with the ring, stabilizing the system. There is a steric requirement, however: the p orbital on the N must be parallel with the p orbitals on the benzene ring in order for the electrons on N to be distributed into the π system of the ring.

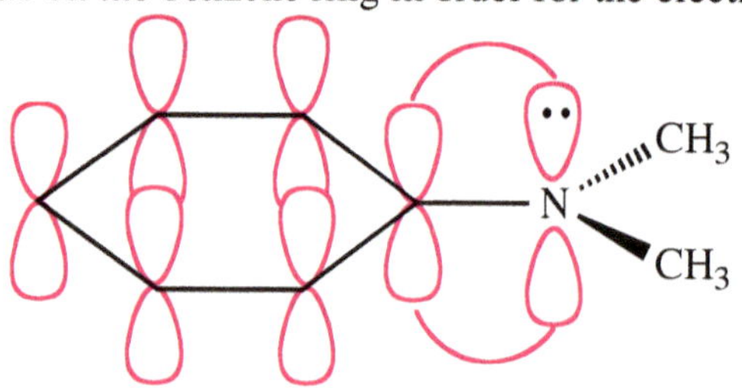

CH3
N
CH3

If the orbital on the nitrogen is forced out of this orientation (by substitution on C-2 and C-6, for example), the electrons are no longer shared with the ring. The nitrogen is hybridized sp^3 (no longer any reason to be sp^2), and the electron pair is readily available for bonding $\Rightarrow$ increased basicity.

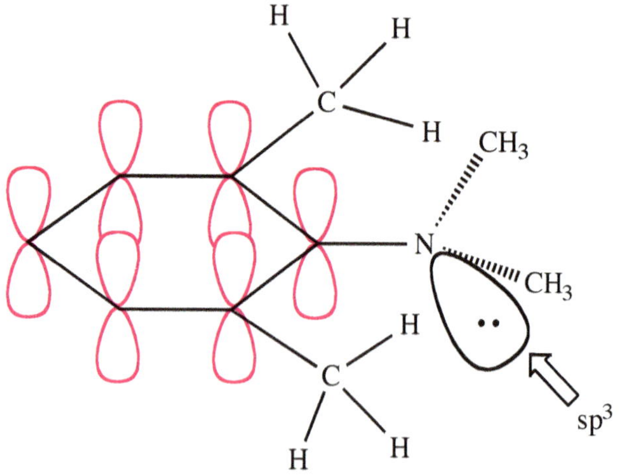

H
H
C
H
CH3
N
CH3
H
C
H
H
sp3

As surprising as it sounds, this aniline is about as basic as a tertiary aliphatic amine, except that the aromatic ring substituent is electron-withdrawing *by induction*, decreasing the basicity slightly. This phenomenon is called *steric inhibition of resonance*. We will see more examples soon.

19-61

(a) Let's label these: **A**, **B**, and **C**. **A** and **C** are not surprising; **B** merits some discussion.

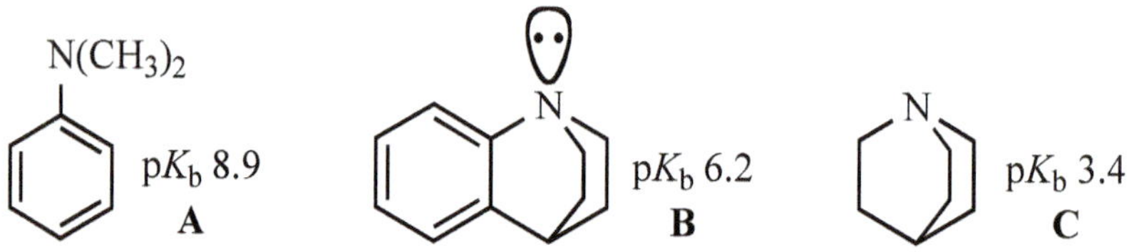

N(CH3)2
pKb 8.9
A
N
pKb 6.2
B
N
pKb 3.4
C

The pK_b value of **A** is typical for an aniline, right around 9. The pair of electrons from N is delocalized around the ring, stabilizing the starting material. In contrast, the pK_b value of **C** is also typical of a 3° aliphatic amine, between 3 and 4. **A** and **C** are right where we would predict.

What is happening with **B**? Isn't it an aniline? Electronically, no. Because of the rigidity of the alkyl groups, the pair of electrons in the sp^3 orbital on N is perpendicular to the aromatic pi system; there is NO electron overlap between N and the benzene ring—another example of *steric inhibition of resonance.*

If the overlap is gone, then why isn't **B** as strong a base as **C**? What electronic effect is occuring other than resonance? *Induction.* The benzene ring is electron-withdrawing by induction so it destabilizes the protonated form, making it less basic.

This is a perfect example of how a solid understanding resonance and induction is critical to solving problems.

19-61 continued

(b) 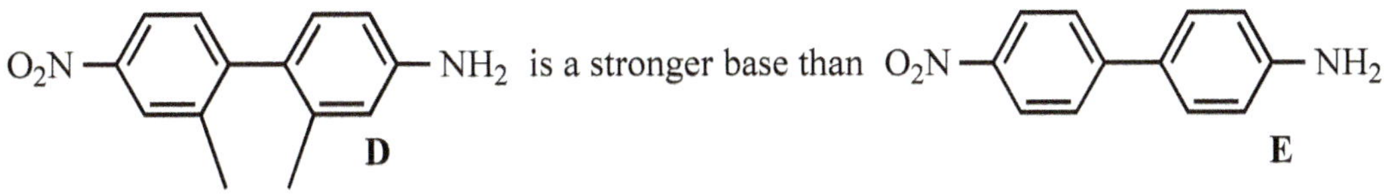 O_2N–(dimethylbiphenyl)–NH_2 **D** is a stronger base than O_2N–(biphenyl)–NH_2 **E**

As we saw in part (a), benzene withdraws electrons from nitrogen by both resonance and induction. In this case, an extra nitro is added to increase the strength of the withdrawal by resonance. So why is **D** a stronger base? Putting the two methyl groups at the ortho positions prevents the two rings from being planar, effectively cutting off any electronic connection between the two rings. The nitrogen's electrons in **D** can still overlap with the first ring, but they would not be delocalized over the second ring, unlike in **E**. **D** would be as strong as a typical aniline, and **E** would hardly be basic at all. This is another example of *steric inhibition of resonance*. Are you seeing a pattern?

(c)

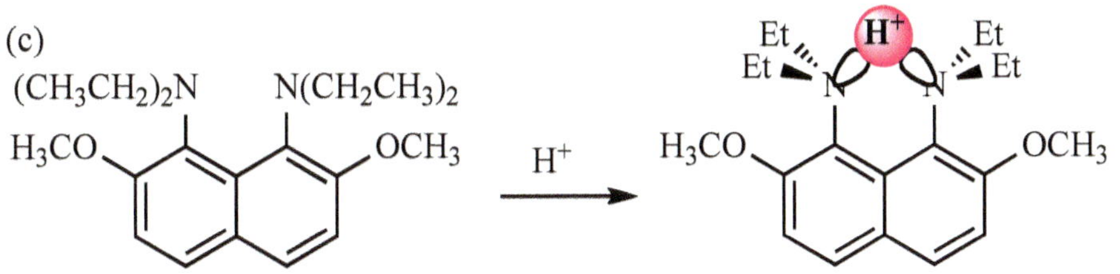

$(CH_3CH_2)_2N$ $N(CH_2CH_3)_2$ H_3CO OCH_3 $\xrightarrow{H^+}$ Et Et H^+ Et Et N N H_3CO OCH_3

OMG! How many factors does this base have going for it? Tertiary. Bonded to aromatic rings should make it weaker, but the presence of the ortho methoxy groups force the amines out of the plane of the rings--although having two close to each other does the same thing--destroying any resonance overlap. The Ns are free to donate their electron pairs, and when they hold onto an H, it makes a perfect 6-membered ring: geometry, plus *steric inhibition of resonance*!

19-62 Not only is substitution at C-2 and C-4 the major products, but substitution occurs under surprisingly mild conditions.

Begin by drawing the resonance forms of pyridine N-oxide:

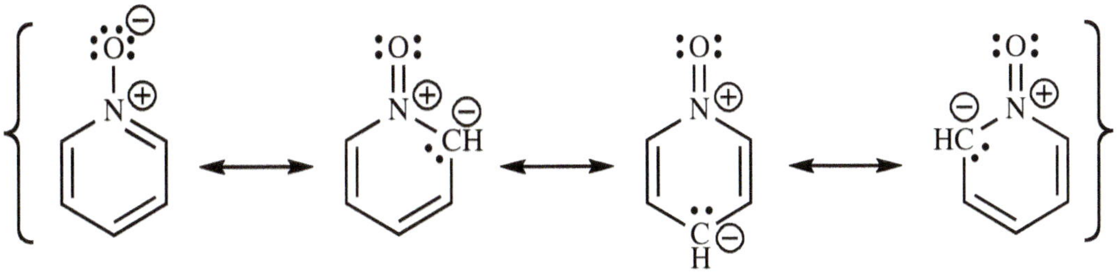

Resonance forms show that the electron density from the oxygen is distributed at C-2 and C-4; these positions would be the likely places for an electrophile to attack.

Resonance forms from electrophilic attack at C-2

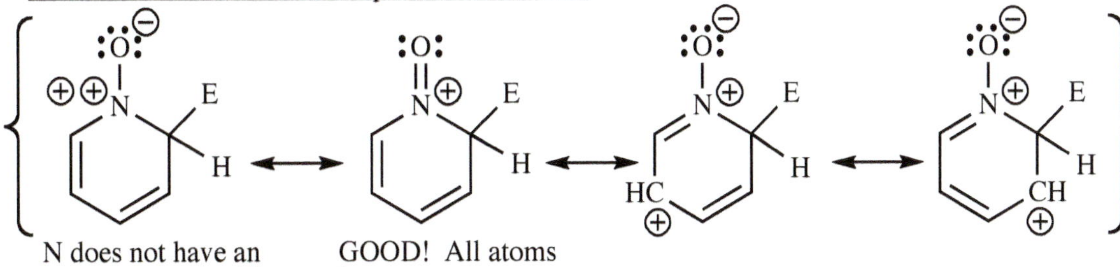

N does not have an octet—not a significant resonance contributor.

GOOD! All atoms have octets.

continued on next page

19-62 continued

Resonance forms from electrophilic attack at C-3

These two resonance forms place two positive charges on adjacent atoms—not good.

Resonance forms from electrophilic attack at C-4

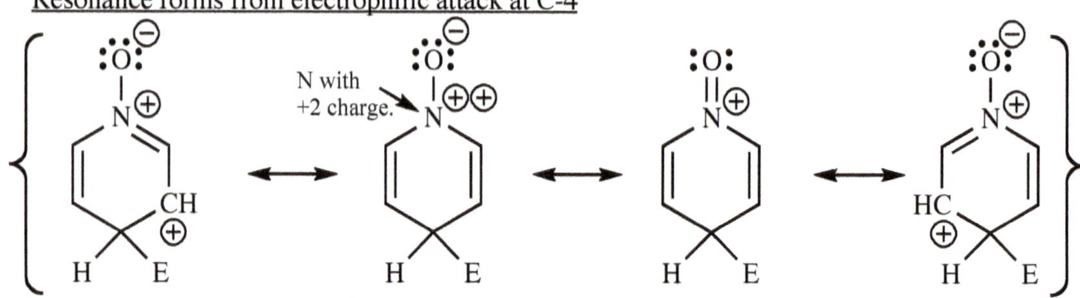

N does not have an octet—not a significant resonance contributor.

GOOD! All atoms have octets.

The resonance forms from electrophilic attack at C-3 are bad; only one of the three is a significant contributor, which means that there is not much resonance stabilization. When the electrophile attacks at C-2 or C-4, however, there are two forms that are good plus one great one that has all atoms with full octets. Clearly, attacks at C-2 and C-4 give the most stable intermediates and will be the preferred sites of attack.

19-63 In each product, the new bond is shown in bold.

(a)

(b)

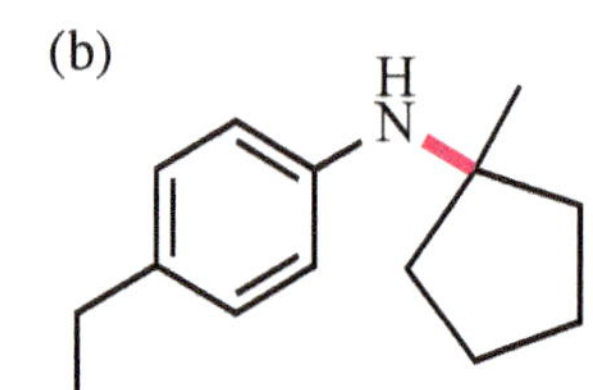

(c)

(d)

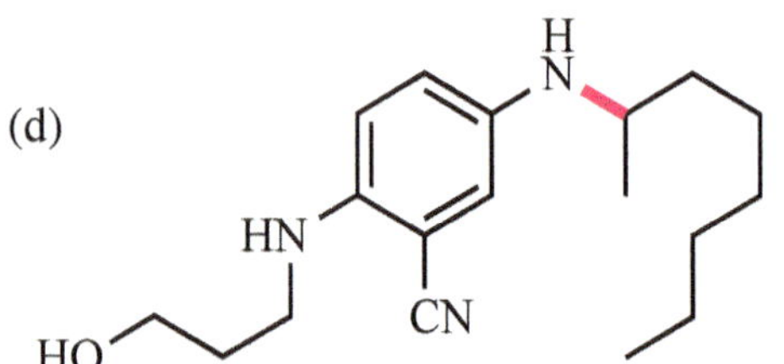

Note to the student: You might have received the impression that everything is known about organic chemistry. That notion is far from the truth. New ways to make bonds and to modify functional groups are being reported in the chemical literature all over the world, and Nobel Prizes acknowledge the most valuable and widely used methods. Problems like this one from the recent literature are intended to show that new methods are being discovered all the time.

19-64 As is typical, show the attack of the electrophile or nucleophile and look for stabilization of the intermediate through a key resonance form.

(a)

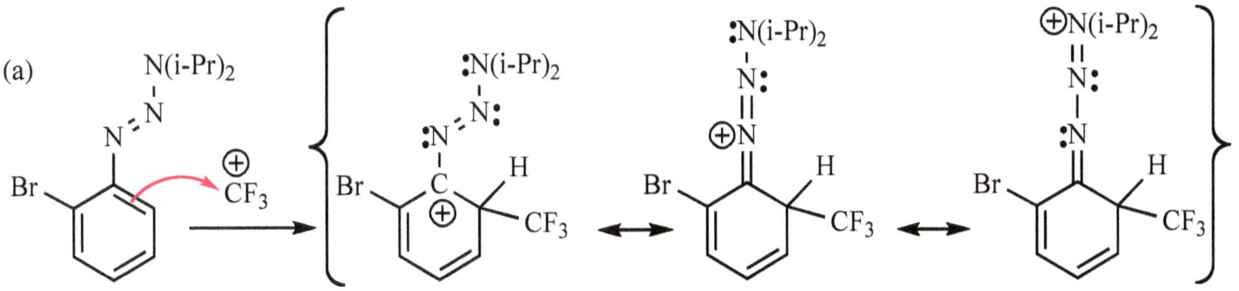

Extensive delocalization of *positive* charge on the substituent stabilizes the intermediate from electrophilic attack.

(b)

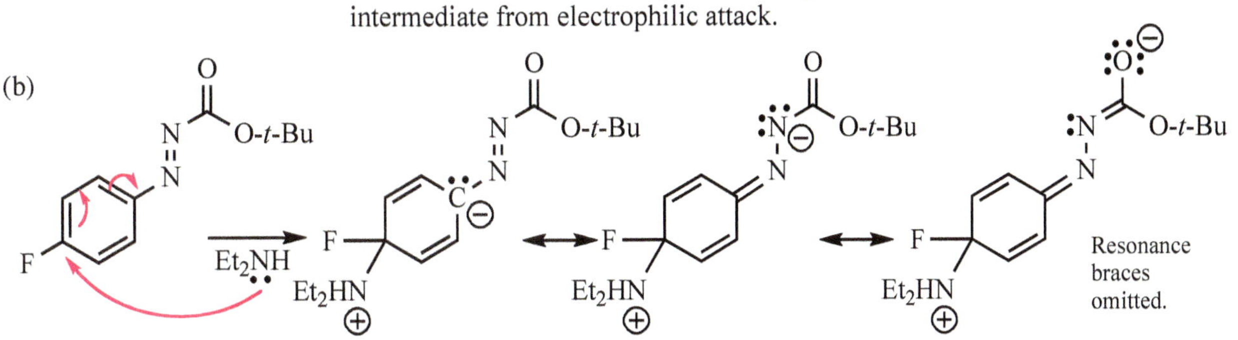

Resonance braces omitted.

Extensive delocalization of *negative* charge on the substituent stabilizes the intermediate from nucleophilic attack.

19-65

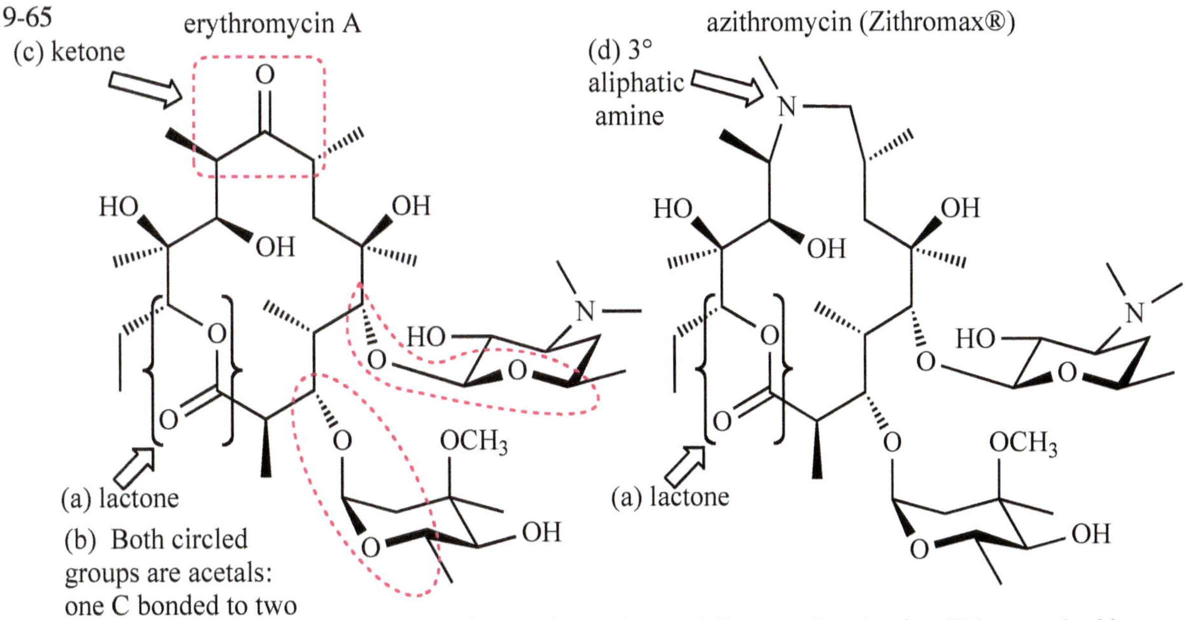

erythromycin A

azithromycin (Zithromax®)

(c) ketone

(d) 3° aliphatic amine

(a) lactone

(a) lactone

(b) Both circled groups are acetals: one C bonded to two O—C groups, one C, and one H.

(e) A ketone is an electrophile, meaning that it will be attacked by a nucleophile. Presumably, bacteria used an enzyme with a nucleophile at its active site to detoxify erythromycin. In contrast, the amine in azithromycin is nucleophilic and will be unaffected by another nucleophile, rendering the detoxifying enzyme useless. Using organic chemistry for better drugs is the foundation of medicinal chemistry.

Students: use this page to stimulate mindfulness and reflection.

CHAPTER 20—CARBOXYLIC ACIDS

20-1

(a) 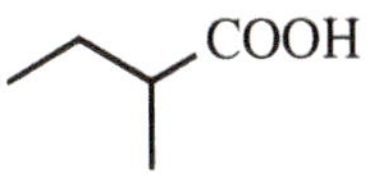 (b) 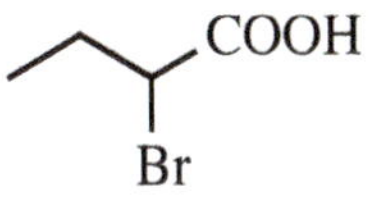 (c) 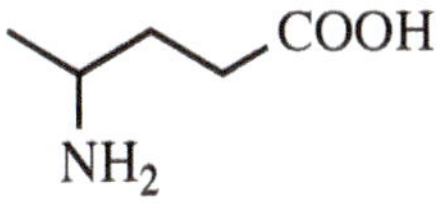 (d) 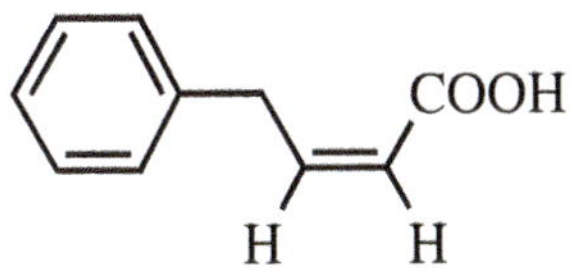 (e) 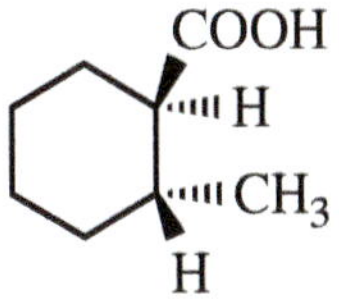 (f) 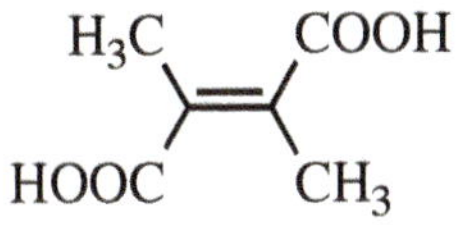 (g) 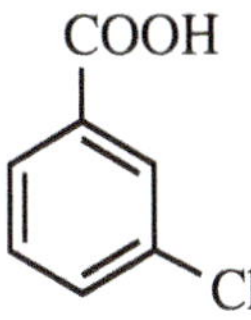 (h) 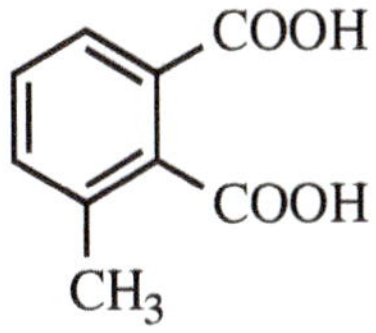 (i) 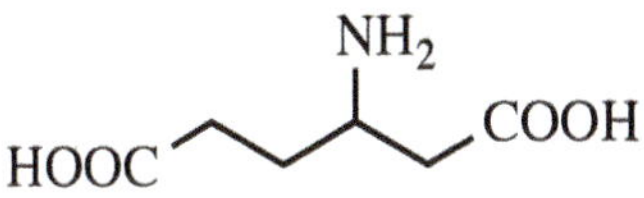 (j) 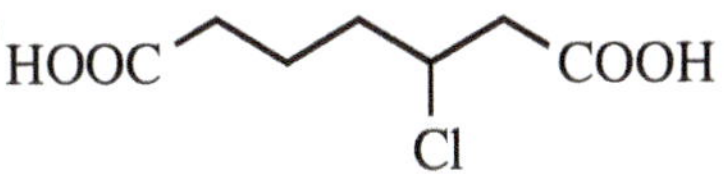 (k) 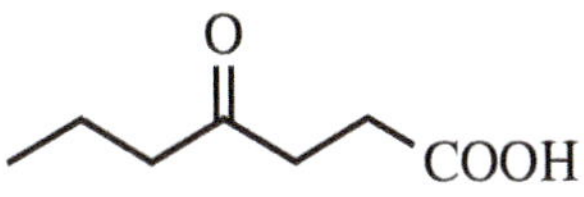 (l)

(b) Br; (c) NH_2; (d) COOH, H, H; (e) COOH, H, CH_3, H — or the enantiomer; (f) H_3C, COOH, HOOC, CH_3; (g) COOH, Cl; (h) COOH, COOH, CH_3; (i) NH_2, HOOC, COOH; (j) HOOC, COOH, Cl; (k) O, COOH; (l) OH, O

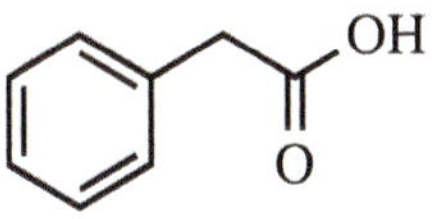

20-2 IUPAC name first; then common name
(a) 2-iodo-3-methylpentanoic acid; α-iodo-β-methylvaleric acid
(b) (*Z*)-3,4-dimethylhex-3-enoic acid; no common name
(c) 2,3-dinitrobenzoic acid; no common name
(d) *trans*-cyclohexane-1,2-dicarboxylic acid (or 1*R*,2*R*); (*trans*-hexahydrophthalic acid)
(e) 2-chlorobenzene-1,4-dicarboxylic acid; 2-chloroterephthalic acid
(f) 3-methylhexanedioic acid; β-methyladipic acid

20-3 Listed in order of increasing acid strength (weakest acid first).

(a) CH_3CH_2COOH (*weakest*) < $CH_3-CH(Br)COOH$ < CH_3-CBr_2COOH (*strongest*)

The greater the number of electron-withdrawing substituents, the greater the stabilization of the carboxylate ion.

(b) $CH_3CH(Br)CH_2CH_2COOH$ (*weakest*) < $CH_3CH_2CH(Br)CH_2COOH$ < $CH_3CH_2CH_2CH(Br)COOH$ (*strongest*)

The closer the electron-withdrawing group, the greater the stabilization of the carboxylate ion.

(c) CH_3CH_2COOH (*weakest*) < $CH_3-CH(Cl)COOH$ < $CH_3-CH(C\equiv N)COOH$ < $CH_3-CH(NO_2)COOH$ (*strongest*)

The stronger the electron-withdrawing effect of the substituent, the greater the stabilization of the carboxylate ion. The nitro group is one of the strongest electron-withdrawing groups known.

20-4

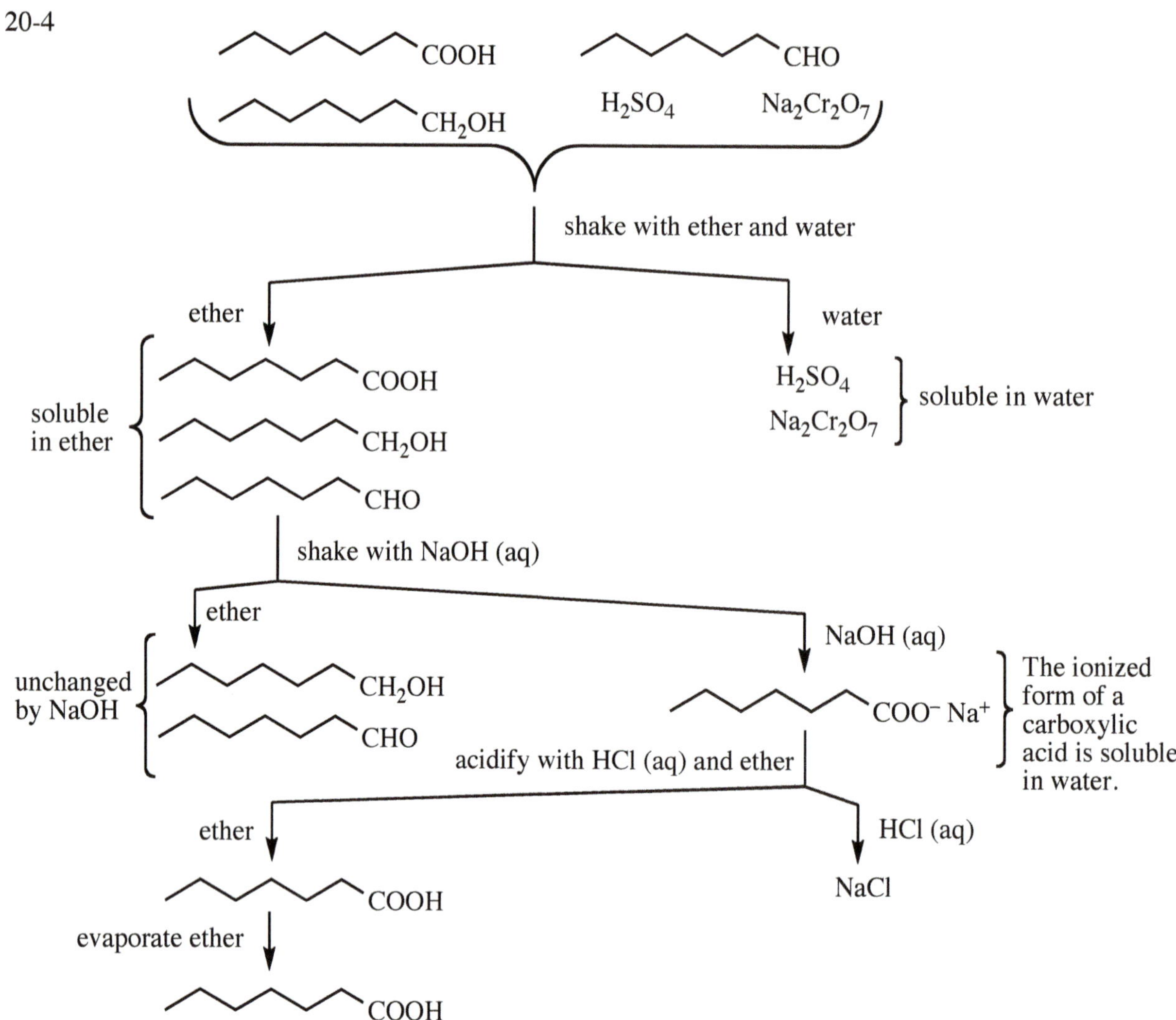
COOH
CHO
CH2OH
H2SO4
Na2Cr2O7
shake with ether and water
ether
water
soluble in ether
COOH
CH2OH
CHO
H2SO4
Na2Cr2O7
soluble in water
shake with NaOH (aq)
ether
NaOH (aq)
unchanged by NaOH
CH2OH
CHO
COO⁻ Na⁺
The ionized form of a carboxylic acid is soluble in water.
acidify with HCl (aq) and ether
ether
HCl (aq)
NaCl
COOH
evaporate ether
COOH

20-5 The principle used to separate a carboxylic acid (a stronger acid) from a phenol (a weaker acid) is to neutralize with a weak base ($NaHCO_3$), a base strong enough to ionize the stronger acid but not strong enough to ionize the weaker acid.

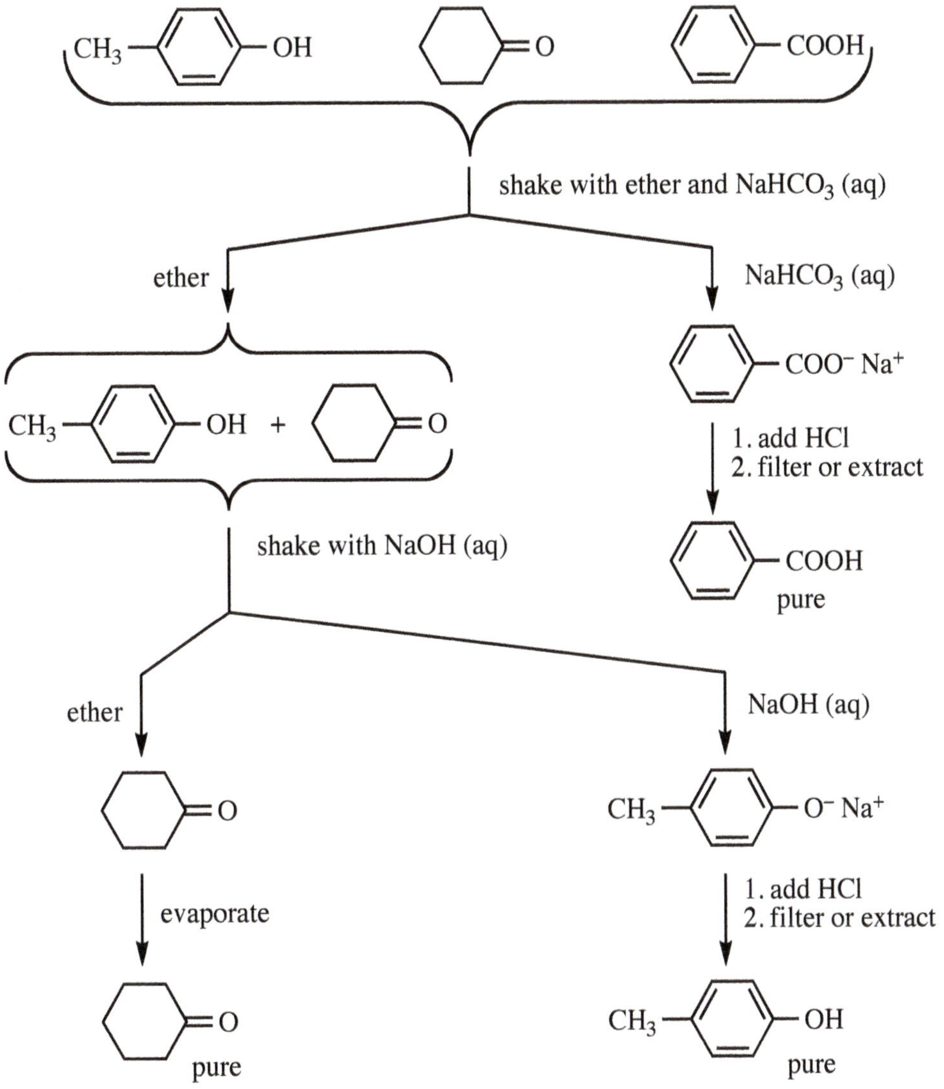

CH_3 OH
O
COOH
shake with ether and $NaHCO_3$ (aq)
ether
$NaHCO_3$ (aq)
$COO^- Na^+$
CH_3 OH + O
1. add HCl
2. filter or extract
shake with NaOH (aq)
COOH
pure
ether
NaOH (aq)
O
CH_3 $O^- Na^+$
evaporate
1. add HCl
2. filter or extract
O
pure
CH_3 OH
pure

20-6 The reaction mixture includes the initial reactant, reagent, desired product, and the overoxidation product—not unusual for an organic reaction mixture.

(a)

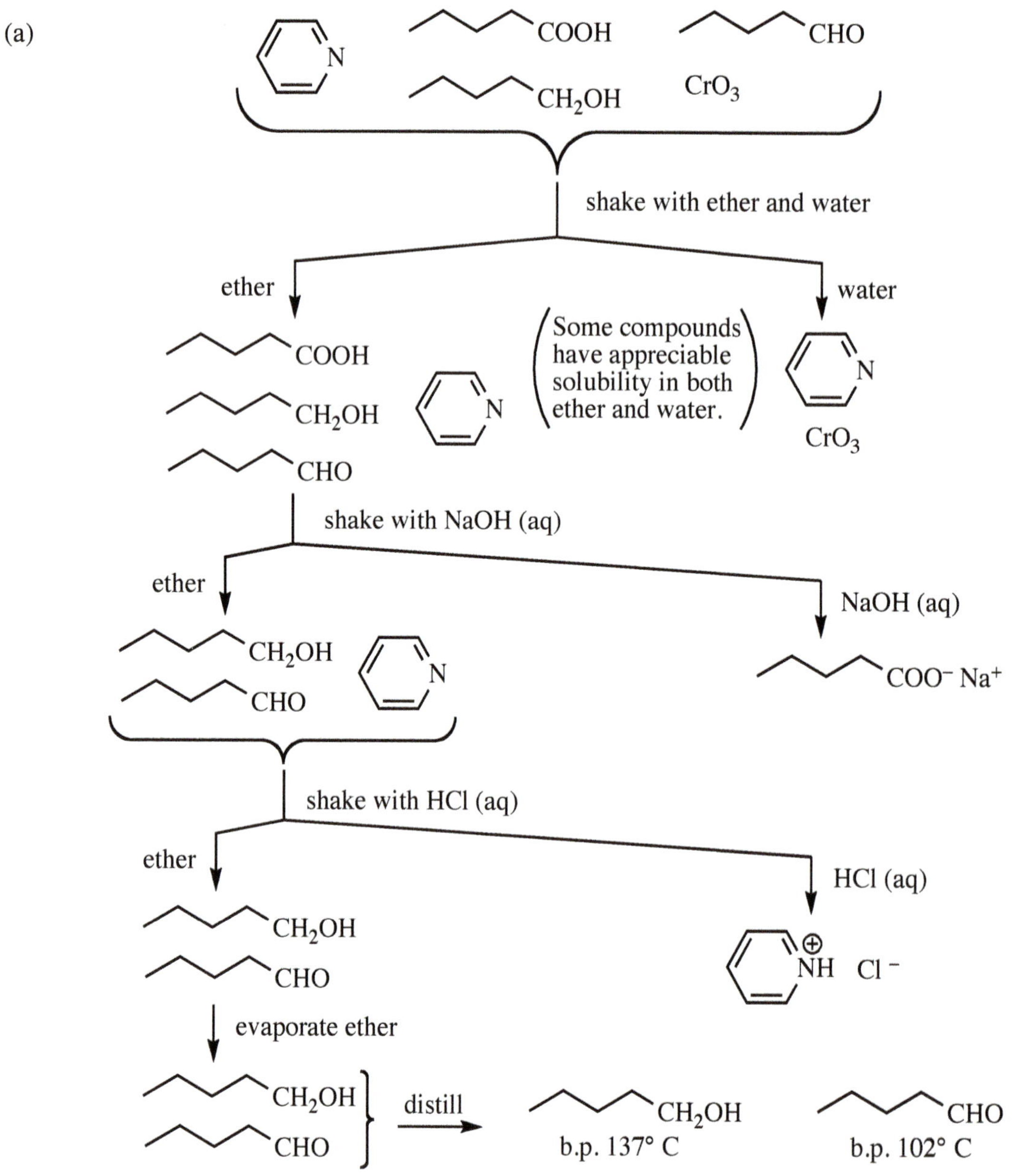

(b) Pentan-1-ol cannot be removed from pentanal by acid-base extraction. These two remaining products can be separated by distillation, the alcohol having the higher boiling point because of hydrogen bonding.

20-7 The COOH has a characteristic IR absorption: a broad peak from 3400 to 2400 cm^{-1}, with a "shoulder" around 2700 cm^{-1}. The carbonyl stretch at 1695 cm^{-1} is a little lower than the standard 1710 cm^{-1}, suggesting conjugation. The strong alkene absorption at 1650 cm^{-1} also suggests it is conjugated.

20-8

(a) The ethyl pattern is obvious: a 3H triplet at δ 1.15 and a 2H quartet at δ 2.4. The only other peak is the COOH at δ 11.9 (a 2.1 δ unit offset added to 9.8).

$$CH_3CH_2—\overset{\overset{\large O}{||}}{C}—OH$$

(b)

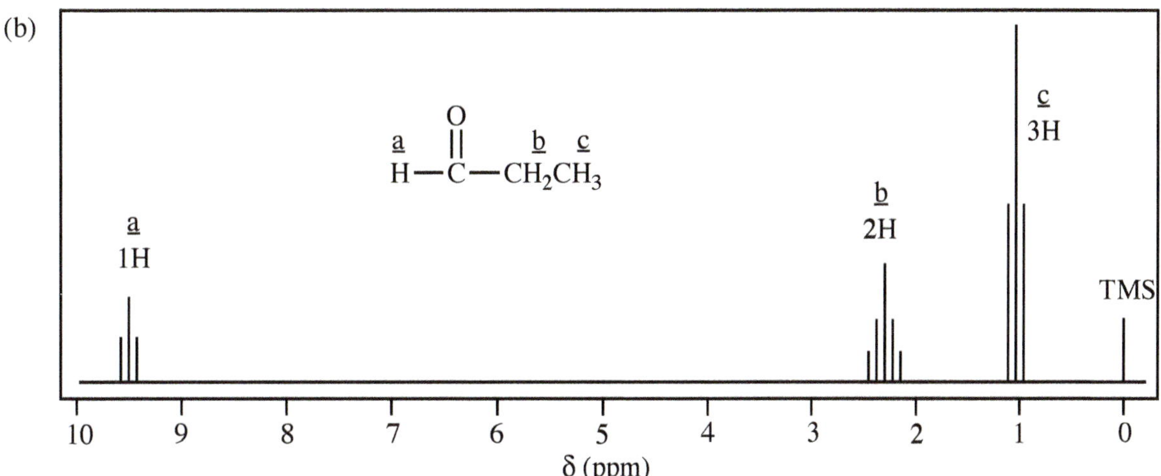

The multiplet between δ 2 and δ 3 is drawn as a pentet as though it were split equally by the aldehyde proton and the CH_2 group. These coupling constants are probably unequal, in which case the actual splitting pattern will be a complex multiplet.

(c) The chemical shift of the aldehyde proton is between δ 9 and 10, not as far downfield as the carboxylic acid proton. Also, the aldehyde proton is split into a triplet by the CH_2, unlike the COOH proton, which always appears as a singlet. Finally, the CH_2 is split by an extra proton, so it will give a multiplet with complex splitting, instead of the quartet shown in the acid.

20-9 { $H_2C{=}CH{-}\overset{\oplus}{C}(OH)(OH)$ ⟷ $\overset{\oplus}{H_2C}{-}CH{=}C(OH)(OH)$ ⟷ $H_2C{=}CH{-}C(=\overset{\oplus}{O}H)(OH)$ ⟷ $H_2C{=}CH{-}C(OH)(=\overset{\oplus}{O}H)$ }

20-10

(a) The linear carboxylic acids typically show a McLafferty peak at m/z 60: $[HO{-}C(OH){=}CH_2]^{+\cdot}$

(b) $[CH_3CH_2{+}CH_2CHCOOH$ (with CH_3 on CH; 87) m/z 116$]^{+\cdot}$ ⟶ $CH_3CH_2\cdot$ (mass 29) + $\overset{\oplus}{H_2C}{-}C(CH_3){=}C(OH)(OH)$ m/z 87

plus resonance forms as shown in 20-9

continued on next page

20-10(b) continued

McLafferty rearrangement

$[H_3C-CH(H)-CH_2-CH(CH_3)-C(=O)OH]^{+\cdot}$ → $H_3C-CH=CH_2$ + $[HC(CH_3)=C(OH)-\ddot{O}H]^{+\cdot}$

m/z 116 — mass 42 — m/z 74

(c) The peak at m/z 60 shows up only if there are no substituents at the alpha or beta position, i.e., in linear carboxylic acids. Any substituent group alpha or beta to the COOH will be present in the McLafferty fragment and will increase the mass of the fragment. The m/z 60 fragment is the lowest possible mass for the McLafferty peak.

20-11

(a) $CH_3CH_2CH_2-C\equiv C-CH_2CH_2CH_3$ —conc. $KMnO_4$ or 1) O_3, 2) H_2O→ $CH_3CH_2CH_2COOH$

(b) cyclodecene —conc. $KMnO_4$, Δ, H_3O^+→ HOOC(CH₂)₈COOH (COOH, COOH)

(c) PhBr —Mg, ether→ PhMgBr —ethylene oxide→ —H_3O^+→ $PhCH_2CH_2OH$ —excess NaOCl, TEMPO→ $PhCH_2COOH$

(d) 2-butanol (OH) —PBr_3→ 2-bromobutane (Br) —Mg, ether→ —CO_2→ —H_3O^+→ 2-methylbutanoic acid (COOH)

(e) p-xylene (CH_3, CH_3) —conc. $KMnO_4$, Δ, H_3O^+→ terephthalic acid (COOH, COOH)

(f) $CH_2=CHCH_2I$ —Mg, ether→ —CO_2→ —H_3O^+→ $CH_2=CHCH_2COOH$

OR $CH_2=CHCH_2I$ —KCN→ $CH_2=CHCH_2CN$ —H_3O^+, Δ→ $CH_2=CHCH_2COOH$

Question in Key Mechanism 20-2

Carboxylic acids are unique in all of the acyl functional groups in that their reactivity changes in basic conditions: a carboxylic acid plus base makes the carboylate anion which is much less susceptible to nucleophilic attack, being negatively charged, as well as having no leaving group. Esterification cannot occur in base.

20-12 (a)

first intermediate: $\left\{ RC(=\overset{+}{O}H)-\ddot{O}H \longleftrightarrow R\overset{+}{C}(-\ddot{O}H)-\ddot{O}H \longleftrightarrow RC(-\ddot{O}H)=\overset{+}{O}H \right\}$

second intermediate: $\left\{ R\overset{+}{C}(-\ddot{O}H)(-\ddot{O}R') \longleftrightarrow RC(-\ddot{O}H)=\overset{+}{O}R' \longleftrightarrow RC(=\overset{+}{O}H)(-\ddot{O}R') \right\}$

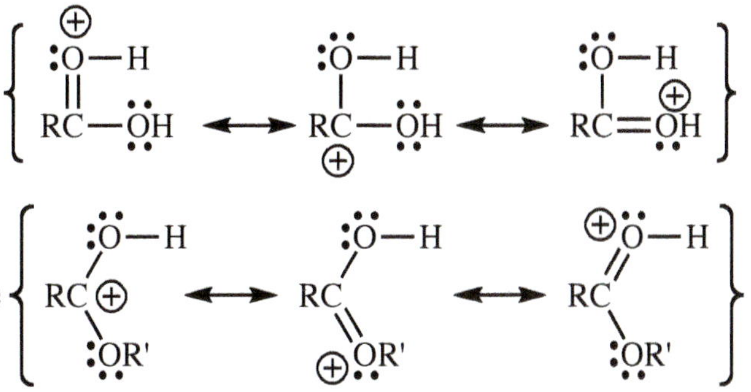

(b) The mechanism of acid-catalyzed nucleophilic acyl substitution may seem daunting, but it is simply a succession of steps that are already very familiar to you.

Typically, these mechanisms have six steps: four proton transfers (two on, two off), a nucleophilic attack, and a leaving group leaving, with a little resonance stabilization thrown in that makes the whole thing work. The six steps are labeled in the mechanism below:

Step A proton on (resonance stabilization)
Step B nucleophile attacks
Step C proton off
Step D proton on
Step E leaving group leaves (resonance stabilization)
Step F proton off

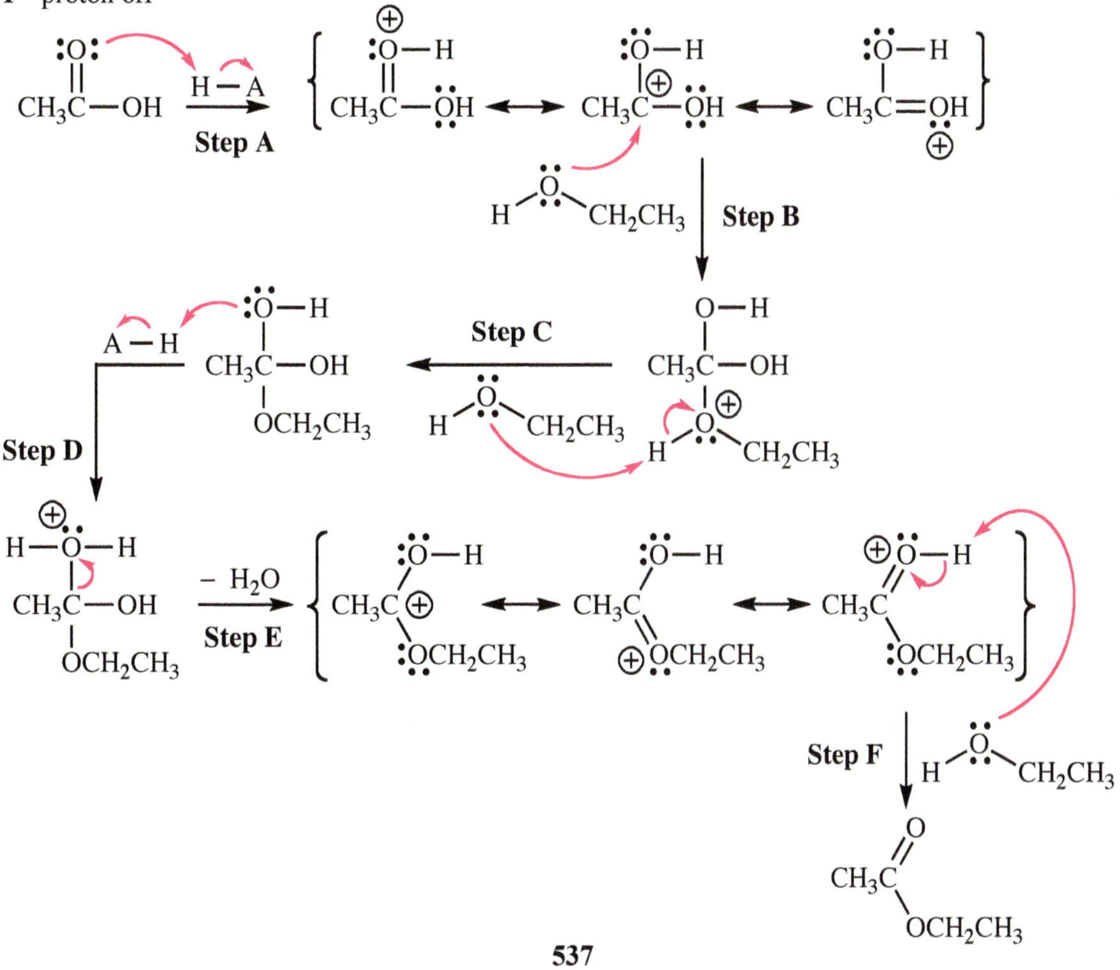

$CH_3C(=O)-OH \xrightarrow[\text{Step A}]{H-A} \left\{ CH_3C(=\overset{+}{O}H)-\ddot{O}H \longleftrightarrow CH_3\overset{+}{C}(-\ddot{O}H)-\ddot{O}H \longleftrightarrow CH_3C(-\ddot{O}H)=\overset{+}{O}H \right\}$

Step B ($H-\ddot{O}-CH_2CH_3$): $CH_3C(-OH)(-OH)(-\overset{+}{O}(H)CH_2CH_3)$

Step C ($H-\ddot{O}-CH_2CH_3$): $CH_3C(-\ddot{O}H)(-OH)(-OCH_2CH_3)$

Step D (A—H): $CH_3C(-\overset{+}{O}H_2)(-OH)(-OCH_2CH_3)$

$\xrightarrow[\text{Step E}]{-\,H_2O} \left\{ CH_3\overset{+}{C}(-\ddot{O}H)(-\ddot{O}CH_2CH_3) \longleftrightarrow CH_3C(-\ddot{O}H)=\overset{+}{O}CH_2CH_3 \longleftrightarrow CH_3C(=\overset{+}{O}H)(-\ddot{O}CH_2CH_3) \right\}$

Step F ($H-\ddot{O}-CH_2CH_3$): $CH_3C(=O)-OCH_2CH_3$

20-12 (c) All steps are reversible, which is the reason the Principle of Microscopic Reversibility applies. Applying the steps as outlined on the previous page (abbreviating OCH_2CH_3 as OEt):

Step A proton on (resonance stabilization)
Step B nucleophile attacks
Step C proton off
Step D proton on
Step E leaving group leaves (resonance stabilization)
Step F proton off

H—A is the acid catalyst

:$A^{\ominus}$ is the conjugate base, although in hydrolysis reactions, water usually removes H^+.

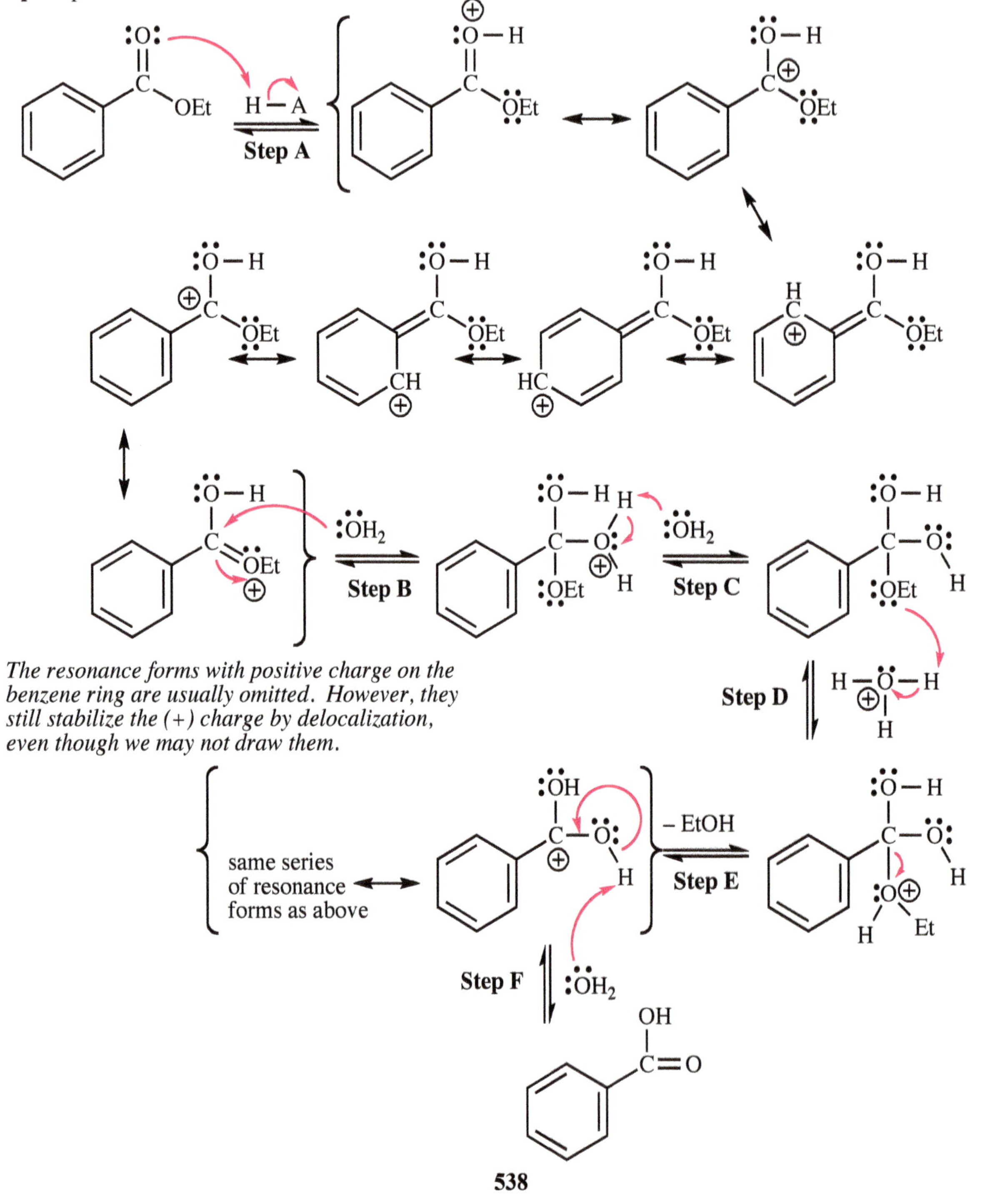

Step A

Step B

Step C

Step D

The resonance forms with positive charge on the benzene ring are usually omitted. However, they still stabilize the (+) charge by delocalization, even though we may not draw them.

same series of resonance forms as above

– EtOH

Step E

Step F

20-13 For the sake of space in this problem, resonance forms will not be drawn, but remember that they are critical! The generic acid HA is represented here by just H^+.

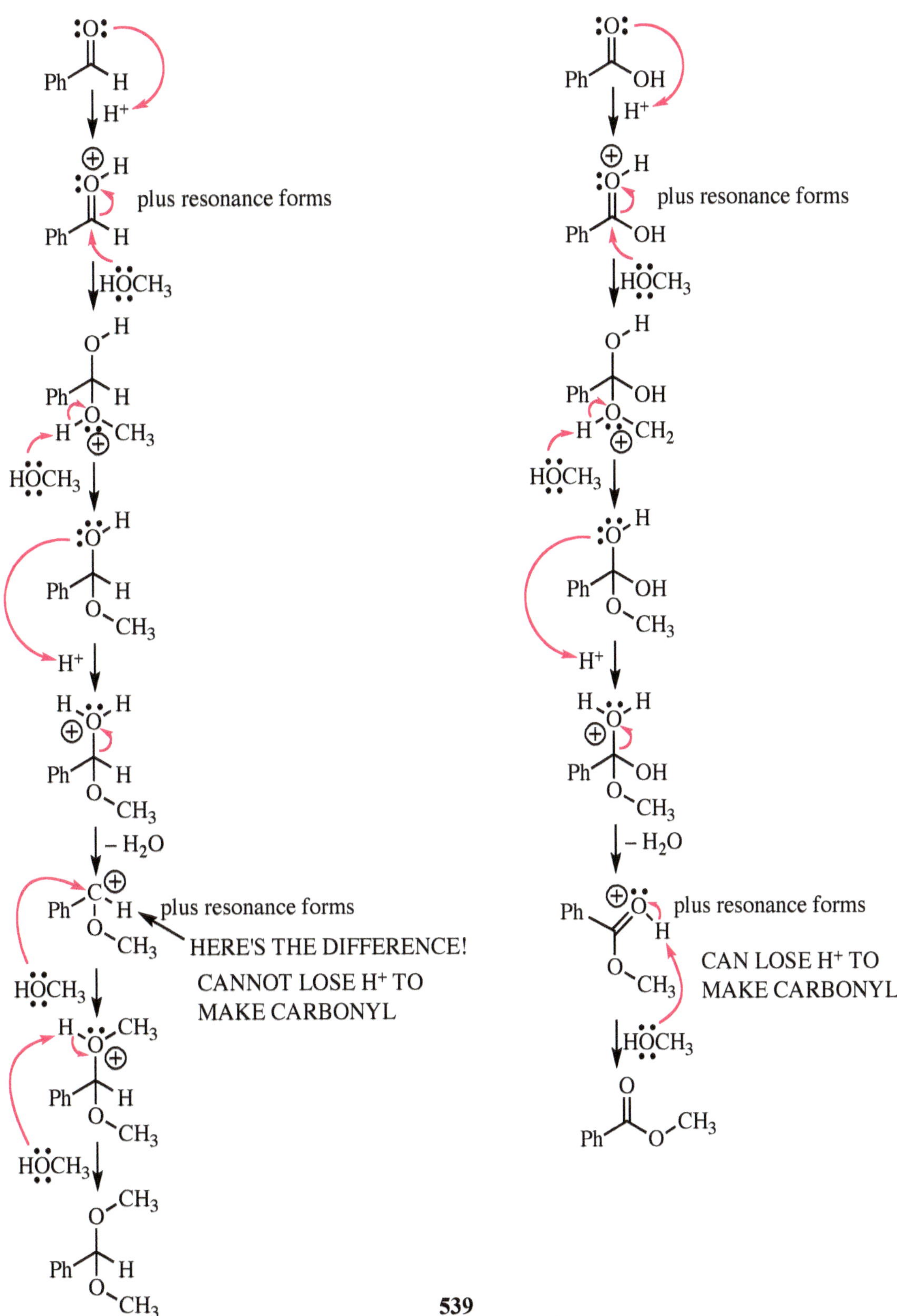

20-14

(a) $R-\overset{O}{\overset{\|}{C}}-\ddot{O}H$ + H–A → { $R-\overset{:O:}{\overset{\|}{C}}-\overset{\oplus}{\ddot{O}}H(-H)$ ↔ $R-\overset{:\ddot{O}:^{\ominus}}{\overset{|}{C}}^{\oplus}-\overset{\oplus}{\ddot{O}}H(-H)$ }

BAD—two adjacent positive charges

(b) Protonation on the OH gives only two resonance forms, one of which is bad because of adjacent positive charges. Protonation on the C=O is good because of three resonance forms distributing the positive charge over three atoms, with no additional charge separation.

$RC(=\ddot{O})-OH$ + H–A → { $RC(=\overset{\oplus}{O}-H)-\ddot{O}H$ ↔ $\overset{\oplus}{R}C(-\ddot{O}-H)-\ddot{O}H$ ↔ $RC(-\ddot{O}-H)=\overset{\oplus}{O}H$ }

(c) The carbonyl oxygen is more "basic" because, by definition, it reacts with a proton more readily. It does so because the intermediate it produces is more stable than the intermediate from protonation of the OH.

20-15

(a) 2-hydroxybenzoic acid (OH, COOH on benzene) + CH_3OH $\underset{}{\overset{H-A}{\rightleftharpoons}}$ methyl 2-hydroxybenzoate (OH, $COOCH_3$ on benzene) + H_2O

Use CH_3OH as solvent.

Remove water with molecular sieves or by distillation.

(b) $HC(=O)-OH$ + CH_3OH $\overset{H-A}{\rightleftharpoons}$ $HC(=O)-OCH_3$ + H_2O

Use CH_3OH as solvent.

Remove by distillation. b.p. 32 °C

(c) $PhCH_2C(=O)OH$ + CH_3CH_2OH $\overset{H-A}{\rightleftharpoons}$ $PhCH_2C(=O)OCH_2CH_3$ + H_2O

Use CH_3CH_2OH as solvent.

Remove water with molecular sieves or by distillation.

20-16 The asterisk ("*") denotes the ^{18}O isotope.

(a) and (b)

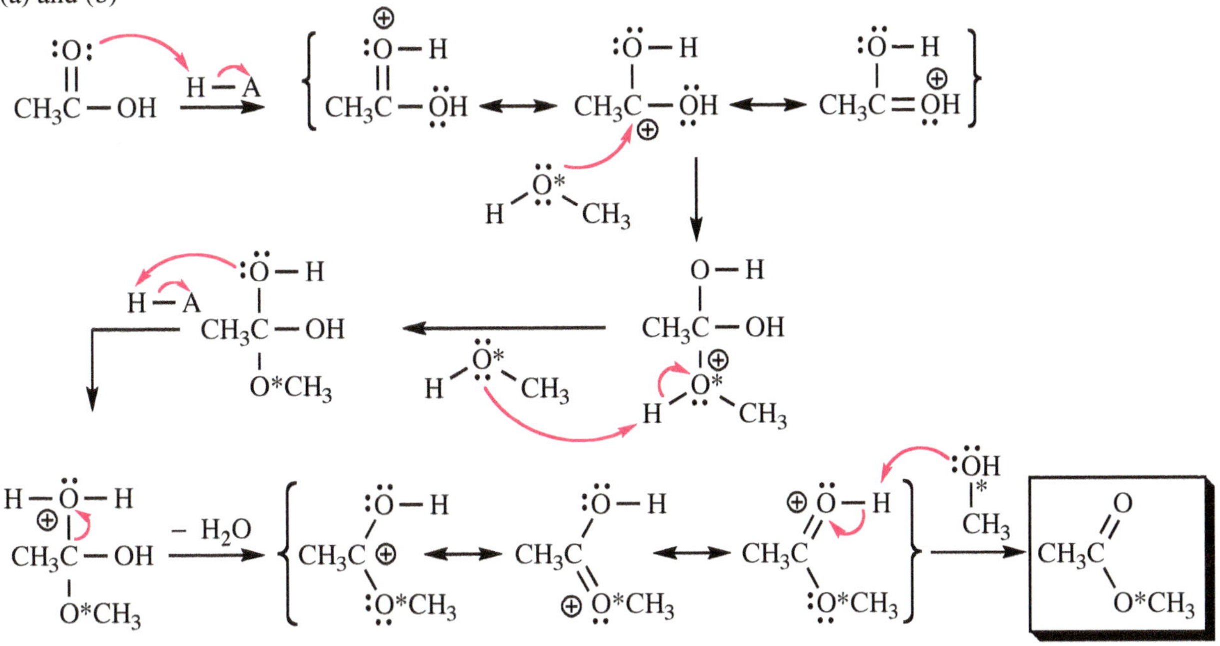

(c) The ^{18}O has two more neutrons, and therefore two more mass units, than ^{16}O. The instrument ideally suited to analyze compounds of different mass is the mass spectrometer.

20-17

(a)

first intermediate:

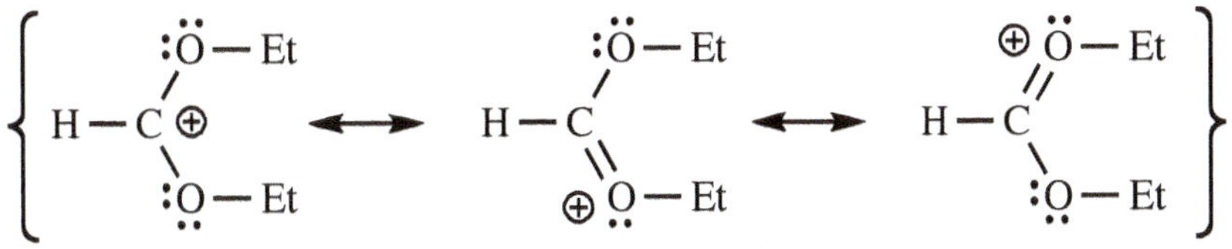

second intermediate:

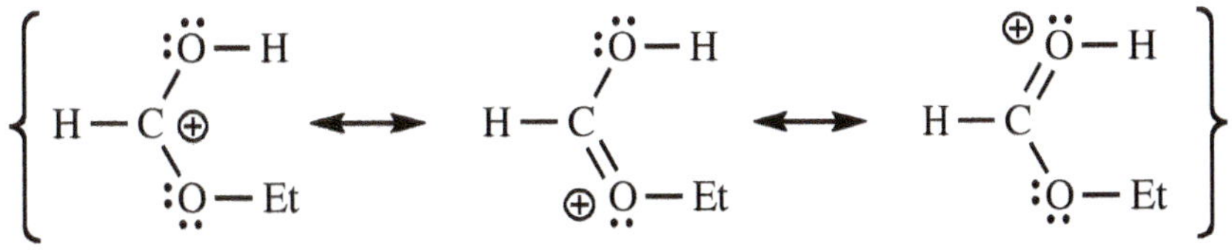

The more resonance forms that can be drawn to represent an intermediate, the more stable the intermediate. The more stable the intermediate, the more easily it can be formed, that is, under milder conditions. These intermediates are highly stabilized due to delocalization of the positive charge over the carbon and both oxygens. A trace of acid is all that is required to initiate this process.

20-17 continued

(b) [Mechanism: protonation of ethyl formate by H–A gives resonance-stabilized cation; water attacks; proton transfer; protonation of OEt; loss of EtOH gives resonance-stabilized cation; deprotonation by water gives formic acid, H–C(=O)OH]

HC—OEt → {HC=O⁺Et / HC⁺—OEt ↔ HC=OH⁺} → HC(OEt)(OH) → H–C(=O)OH; – EtOH

20-18 Amide formation by this industrial method requires high temperature. It is not a convenient method for laboratory synthesis.

(a) m-H_3C–C_6H_4–COOH $\xrightarrow{HN(CH_2CH_3)_2}$ m-H_3C–C_6H_4–COO^- $H_2N^+(CH_2CH_3)_2$ $\xrightarrow{\Delta}$ m-H_3C–C_6H_4–$C(=O)NEt_2$ + H_2O

(b) $C_6H_5NH_2$ $\xrightarrow{HO-C(=O)-CH_3}$ $C_6H_5NH_3^+$ $^-O-C(=O)-CH_3$ $\xrightarrow{\Delta}$ C_6H_5–NH–C(=O)–CH_3 + H_2O

(c) $(CH_3)_2NH$ $\xrightarrow{HO-C(=O)-H}$ $(CH_3)_2NH_2^+$ $^-O-C(=O)-H$ $\xrightarrow{\Delta}$ H_3C–N(CH_3)–C(=O)–H + H_2O

20-19 B_2H_6 and $BH_3 \bullet THF$ can be considered as equivalent reagents.

(a) C_6H_5–CH_2–C(=O)–OH $\xrightarrow{LiAlH_4}$ $\xrightarrow{H_3O^+}$ or $\xrightarrow{B_2H_6}$ $\xrightarrow{H_3O^+}$ C_6H_5–CH_2–CH_2OH

(b) C_6H_5–CH_2–C(=O)–Cl $\xrightarrow{\text{1) LiAlH(O-}t\text{-Bu)}_3\text{; 2) }H_3O^+}$ C_6H_5–CH_2–C(=O)–H

(c) 3-oxocyclopentane-COOH $\xrightarrow{B_2H_6}$ $\xrightarrow{H_3O^+}$ 3-oxocyclopentane-CH_2OH

B_2H_6 (or $BH_3 \bullet THF$) selectively reduces a carboxylic acid in the presence of a ketone. Alternatively, protecting the ketone as an acetal, reducing the COOH, and removing the protecting group would also be possible but longer.

20-20

$$R-C(O^-)(O^-)-R' \;(2\,Li^+) + 2\,H-A \longrightarrow R-C(OH)(OH)-R' \xrightarrow{H-A} R-C(OH)(\overset{+}{O}H_2)-R' \xrightarrow{-\,H_2O} \left\{ R-C(=\overset{+}{O}-H)-R' \longleftrightarrow R-\overset{+}{C}(OH)-R' \right\} \xrightarrow{H_2\ddot{O}:} R-C(=O)-R'$$

a hydrate

20-21

(a)

$$CH_3CH_2-C(=O)-OH + \mathbf{2}\; Li-C_6H_5 \longrightarrow \xrightarrow{H_2O} CH_3CH_2-C(=O)-C_6H_5$$

$$CH_3CH_2-C(=O)-OH \xrightarrow{SOCl_2} CH_3CH_2-C(=O)-Cl + C_6H_6 \xrightarrow{AlCl_3} CH_3CH_2-C(=O)-C_6H_5$$

(b)

$$C_6H_{11}-C(=O)-OH + \mathbf{2}\; CH_3Li \longrightarrow \xrightarrow{H_2O} C_6H_{11}-C(=O)-CH_3$$

20-22

$$PhC(=O)OH + Cl-C(=O)-C(=O)-Cl \longrightarrow \left\{ HO-C(Ph)=\overset{+}{O}-C(O^-)(Cl)-C(=O)Cl \longleftrightarrow HO-\overset{+}{C}(Ph)-O-C(O^-)(Cl)-C(=O)Cl \right\}$$

$$\updownarrow$$

$$H-\overset{+}{O}=C(Ph)-O-C(=O)-C(=O)Cl \;+\; Cl^- \longleftarrow \left\{ \text{plus resonance forms with positive charge on the benzene ring} \longleftrightarrow HO=\overset{+}{C}(Ph)-O-C(O^-)(Cl)-C(=O)Cl \right\}$$

plus resonance forms

$$\longrightarrow O=C(Ph)-O-C(=O)-C(=O)Cl + HCl(g) \xrightarrow{:Cl^-} {}^-O-C(Cl)(Ph)-O-C(=O)-C(=O)Cl \longrightarrow PhC(=O)Cl + O=C=O\ (gas) + {}^-C\equiv O^+\ (gas) + Cl^-$$

20-23

(a) $Ph-\overset{\overset{\large :O:}{\|}}{C}-Cl \quad H-\ddot{O}-CH_2CH_3 \longrightarrow Ph-\overset{\overset{\large :\ddot{O}:^{\ominus}}{|}}{C}(-Cl)-\overset{\oplus}{O}(H)CH_2CH_3 \xrightarrow{-Cl^-} Ph-\overset{\overset{\large :O:}{\|}}{C}-\overset{\oplus}{O}(H)CH_2CH_3 \xrightarrow{Et\ddot{O}H} Ph-\overset{\overset{O}{\|}}{C}-\ddot{O}CH_2CH_3$

(b) $CH_3-\overset{\overset{\large :O:}{\|}}{C}-Cl + H-\ddot{N}(H)-CH_3 \longrightarrow CH_3-\overset{\overset{\large :\ddot{O}:^{\ominus}}{|}}{C}(-Cl)-\overset{\oplus}{N}(H)(H)-CH_3 \xrightarrow{-Cl^-} CH_3-\overset{\overset{\large :O:}{\|}}{C}-\overset{\oplus}{N}(H)(H)-CH_3 \xrightarrow{H_2\ddot{N}-CH_3} CH_3-\overset{\overset{O}{\|}}{C}-\ddot{N}HCH_3$

20-24

(a) benzoic acid (PhCOOH) $\xrightarrow{SOCl_2}$ PhCOCl $\xrightarrow{PhNH_2}$ PhCONHPh

(b) $CH_3CH_2COOH \xrightarrow{SOCl_2} CH_3CH_2COCl \xrightarrow{HO-Ph} CH_3CH_2COOPh$

20-25

(a) pent-2-ynoic acid
(b) 3-amino-2-hydroxybutanoic acid
(c) 3-methylbut-2-enoic acid
(d) *trans*-2-methylcyclopentanecarboxylic acid (relative stereochem.) (or 1*R*,2*R*) (absolute stereochem.)
(e) 2,4,6-trinitrobenzoic acid
(f) 5,5-dimethyl-4-oxohexanoic acid

20-26

(a) 3-phenylpropanoic acid; β-phenylpropionic acid
(b) potassium benzoate; no common name
(c) 2-bromo-3-methylbutanoic acid; α-bromo-β-methylbutyric acid
(d) 2-methylbutane-1,4-dioic acid; α-methylsuccinic acid
(e) sodium 3-methylbutanoate; sodium β-methylbutyrate
(f) 3-aminobutanoic acid; β-aminobutyric acid
(g) 2-bromobenzoic acid; *o*-bromobenzoic acid
(h) magnesium ethanedioate; magnesium oxalate
(i) 4-methoxybenzene-1,2-dicarboxylic acid; 4-methoxyphthalic acid

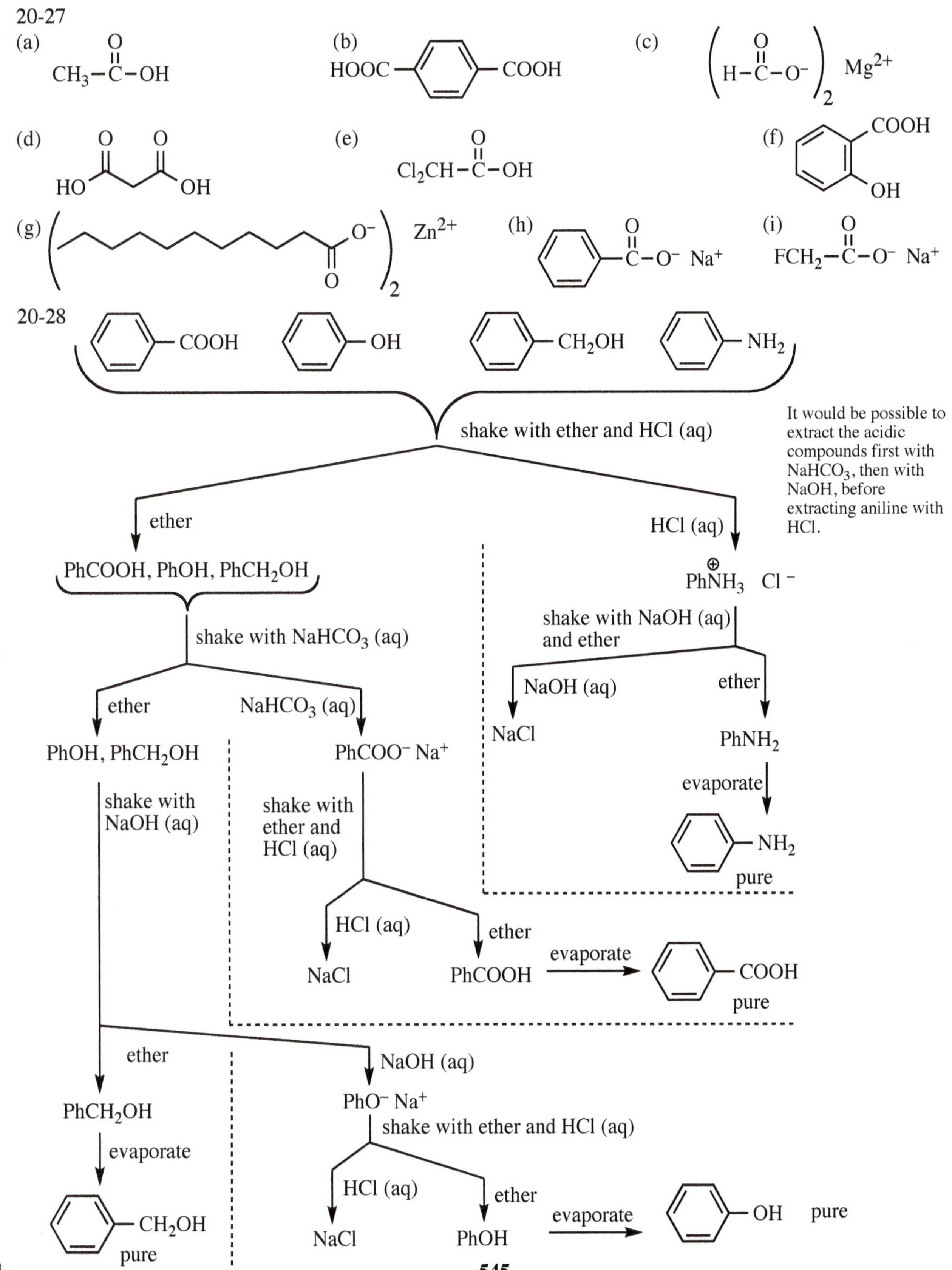
20-27
(a) $CH_3-C(=O)-OH$
(b) HOOC COOH
(c) $(H-C(=O)-O^-)_2$ Mg^{2+}
(d) HO OH
(e) $Cl_2CH-C(=O)-OH$
(f) COOH OH
(g) O^- O 2 Zn^{2+}
(h) $C(=O)-O^-$ Na^+
(i) $FCH_2-C(=O)-O^-$ Na^+
20-28
COOH
OH
CH_2OH
NH_2
shake with ether and HCl (aq)
It would be possible to extract the acidic compounds first with $NaHCO_3$, then with NaOH, before extracting aniline with HCl.
ether
HCl (aq)
PhCOOH, PhOH, $PhCH_2OH$
$PhNH_3^+$ Cl^-
shake with $NaHCO_3$ (aq)
shake with NaOH (aq) and ether
ether
$NaHCO_3$ (aq)
NaOH (aq)
ether
NaCl
$PhNH_2$
PhOH, $PhCH_2OH$
$PhCOO^-$ Na^+
evaporate
shake with NaOH (aq)
shake with ether and HCl (aq)
NH_2
pure
HCl (aq)
ether
NaCl
PhCOOH
evaporate
COOH
pure
ether
NaOH (aq)
$PhCH_2OH$
PhO^- Na^+
shake with ether and HCl (aq)
evaporate
HCl (aq)
ether
NaCl
PhOH
evaporate
OH
pure
CH_2OH
pure

20-29 Weaker base listed first. (Weaker bases come from *stronger* conjugate acids. Some pK_a values of the conjugate acids are listed in italics.)

(a) $ClCH_2COO^-$ < CH_3COO^- < PhO^- (*2.8*, *4.7*, *10.0*)

(b) $CH_3COO^-\ Na^+$ < $HC{\equiv}C^-\ Na^+$ < $NaNH_2$ (*4.7*, *25*, *36*)

(c) $PhCOO^-\ Na^+$ < $PhO^-\ Na^+$ < $CH_3CH_2O^-\ Na^+$ (*4.20*, *10.0*, *15.9*)

(d) $CH_3COO^-\ Na^+$ < pyridine < $CH_3CH_2O^-\ Na^+$ (*4.7*, *5.2*, *15.9*)

20-30

(a) $CH_3COOH + NH_3 \longrightarrow CH_3COO^-\ {}^+NH_4$

(b) phthalic acid (benzene-1,2-(COOH)₂) + 2 NaOH ⟶ benzene-1,2-($COO^-\ Na^+$)₂ + 2 H_2O

(c) CH_3–C₆H₄–COOH + $CF_3COO^-\ K^+$ ⟶ no reaction

(d) $CH_3-CH(Br)COOH + CH_3CH_2COO^-\ Na^+ \longrightarrow CH_3-CH(Br)COO^-\ Na^+ + CH_3CH_2COOH$

(e) C_6H_5–COOH + $Na^+\ {}^-O$–C_6H_5 ⟶ C_6H_5–$COO^-\ Na^+$ + HO–C_6H_5

20-31

$CH_3C(=O)-OCH_2CH_3$ lowest b.p. (77 °C) < $CH_2{=}CH-O-CH_2CH_2-OH$ (b.p. 143 °C) < $CH_3CH_2CH_2-C(=O)-OH$ highest b.p. (162 °C)

The ester cannot form hydrogen bonds and will be the lowest boiling. The alcohol can form hydrogen bonds. The carboxylic acid forms two hydrogen bonds and boils as the dimer, the highest boiling among these three compounds.

20-32 Listed in order of increasing acidity (weakest acid first):

(a) ethanol < phenol < acetic acid
(b) acetic acid < chloroacetic acid < *p*-toluenesulfonic acid
(c) benzoic acid < *m*-nitrobenzoic acid < *o*-nitrobenzoic acid
(d) butyric acid < β-bromobutyric acid < α-bromobutyric acid
(e) 3-bromocyclopentane–COOH < 2-chlorocyclopentane–COOH < 1-fluorocyclopentane–COOH

EWG = electron-withdrawing group

Acidity increases with:
1. closer proximity of EWG
2. greater number of EWG
3. increasing strength (electronegativity) of EWG

20-33 Acetic acid derivatives are often used as a test of electronic effects of a series of substituents: they are fairly easily synthesized (or are commercially available), and pK_a values are easily measured by titration.

Substituents on carbon-2 of acetic acid can express only an inductive effect; no resonance effect is possible because the CH_2 is sp^3 hybridized and no pi overlap is possible.

Two conclusions can be drawn from the given pK_a values. First, all four substituents are electron-withdrawing because all four substituted acids are stronger than acetic acid. Second, the magnitude of the electron-withdrawing effect increases in the order: OH < Cl < CN < NO_2 . (It is always a safe assumption that nitro is the strongest electron-withdrawing group of all the common substituents.)

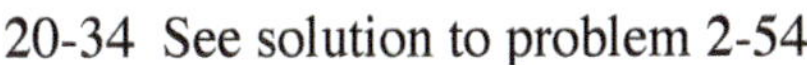

20-34 See solution to problem 2-54.

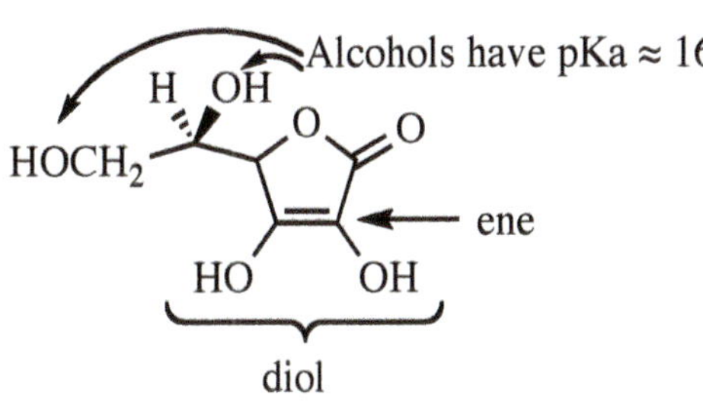

(a) Ascorbic acid is not a carboxylic acid. It is an example of a structure called an ene-diol where one of the OH groups is unusually acidic because of the adjacent carbonyl group. See part (c).

(b) Ascorbic acid, pK_a 4.71, is almost identical to the acidity of acetic acid, pK_a 4.74.

(c) The more acidic H will be the one that, when removed, gives the more stable conjugate base.

start here

$-H^+$ $-H^+$

major

major

major

more acidic

THREE resonance forms, two major with (–) on oxygen

two resonance forms, only one with (–) on oxygen

(d) In the slightly basic pH of physiological fluid, ascorbic acid is present as the conjugate base, ascorbate, whose structure can be represented by any of the three resonance forms on the left of part (c).

20-35

(a) OH

(b) CH_2COOH

(c) O

(d) 2 COOH

(e) COOH

(f) Ph
$CH_3CH_2-CHCH_2OH$

(g) O O

(h) COOH COOH

(i) COOH CHO CH_2CH_3

Aqueous acid also removes the acetal.

(j) O O CH_3

20-36

(a) Br $\xrightarrow[\text{ether}]{Mg}$ $\xrightarrow{CO_2}$ $\xrightarrow{H_3O^+}$ COOH

OR $\xrightarrow{KCN}$ $\xrightarrow{H_3O^+}$

(b) $\xrightarrow[\Delta,\ H_3O^+]{\text{conc. } KMnO_4}$ **2** COOH

(c) O, H $\xrightarrow[NH_3\ (aq)]{Ag^+}$ O, OH

OR $\xrightarrow[\text{TEMPO}]{\text{excess NaOCl}}$

(d) O, OH $\xrightarrow{SOCl_2}$ O, Cl $\xrightarrow{LiAlH(O\text{-}t\text{-}Bu)_3}$ O, H

1) $LiAlH_4$ 2) H_3O^+ → OH $\xrightarrow[\text{TEMPO}]{\text{1 equiv. NaOCl}}$

It is a very common lab practice to reduce RCOOH to the alcohol, then to oxidize to the aldehyde. This avoids making acid chlorides which is sometimes difficult and smells bad.

(e) COOH $\xrightarrow[H{-}A]{CH_3OH}$ $COOCH_3$

OR $\xrightarrow{CH_2N_2}$

OR $\xrightarrow{SOCl_2}$ COCl $\xrightarrow{CH_3OH}$ $COOCH_3$

(f) COOH $\xrightarrow{LiAlH_4}$ $\xrightarrow{H_3O^+}$ CH_2OH

OR $\xrightarrow{B_2H_6}$ $\xrightarrow{H_3O^+}$ B_2H_6 and $BH_3 \bullet THF$ are equivalent reagents.

(g) CH_2COOH $\xrightarrow{SOCl_2}$ $\xrightarrow{CH_3NH_2}$ H, N, CH_3, O

(h) Cl, COOH + → Cl, COOH $\xrightarrow[Pt]{H_2}$ Cl, COOH

Diels-Alder

20-37 Products are boxed.

(a) All steps are reversible in an acid-catalyzed ester hydrolysis (abbreviating OCH_2CH_3 as OEt).

Step A proton on (resonance stabilization)
Step B nucleophile attacks
Step C proton off
Step D proton on
Step E leaving group leaves (resonance stabilization)
Step F proton off

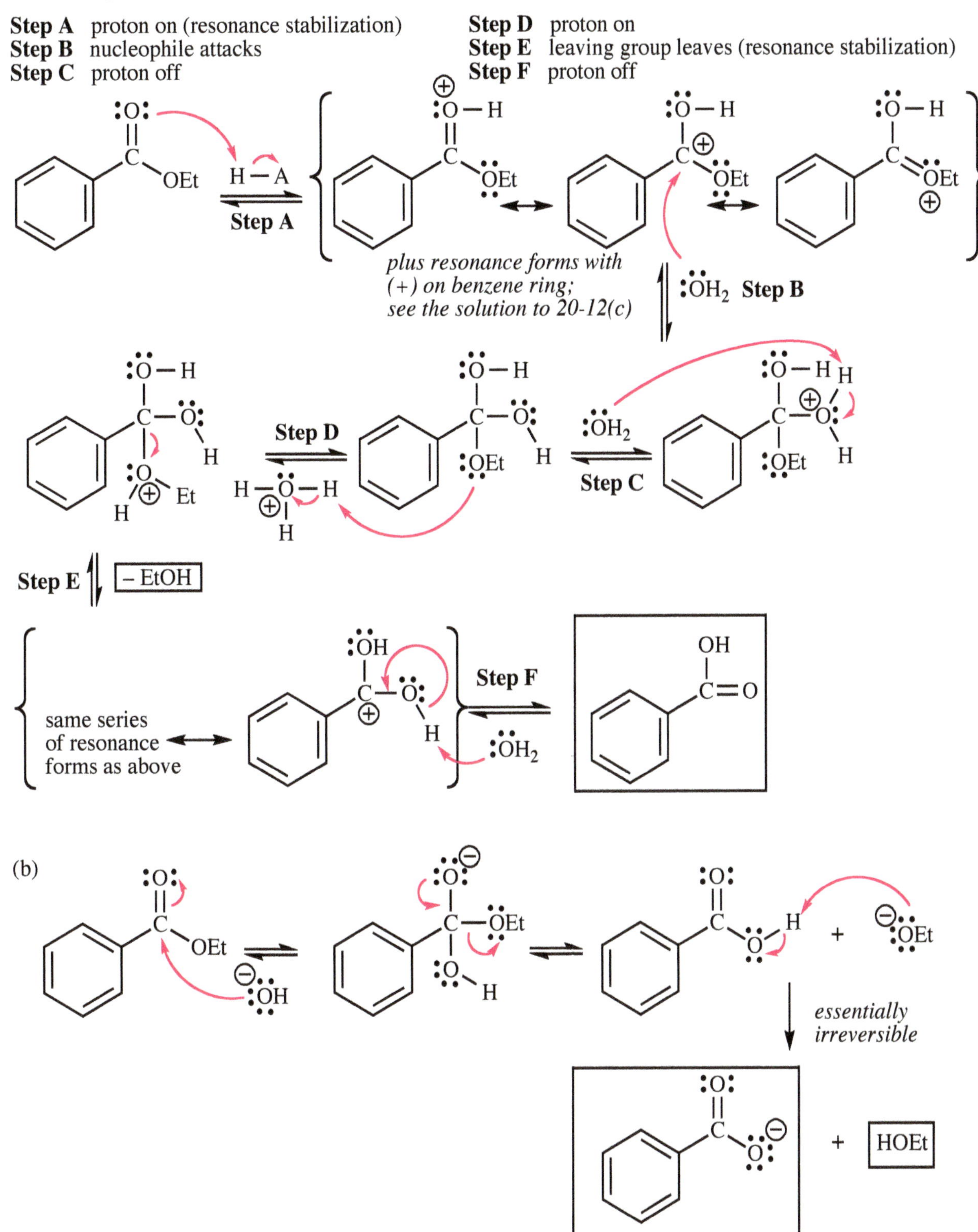

20-37 continued

(c)

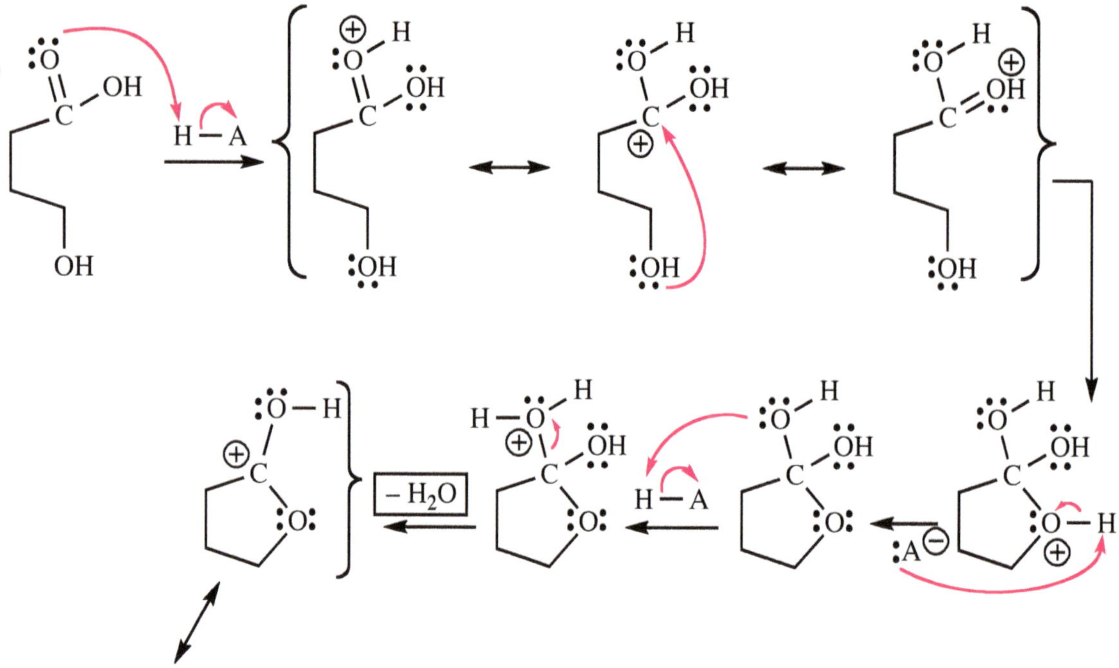

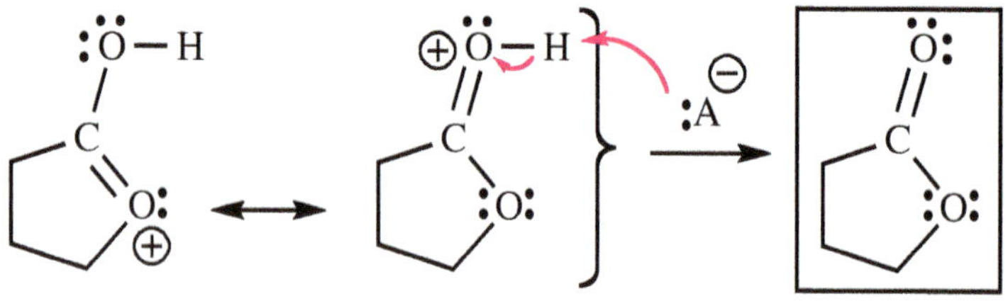

A cyclic ester is called a lactone. Lactones form when the OH nucleophile is just a few carbons away from the carbonyl electrophile, especially if a 5- or 6-membered ring can form.

(d)

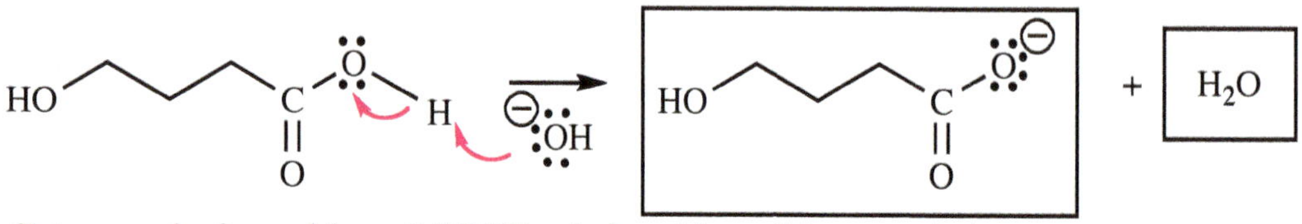

Esters can be formed from RCOOH only in acid conditions, not in base.

20-38

(a)

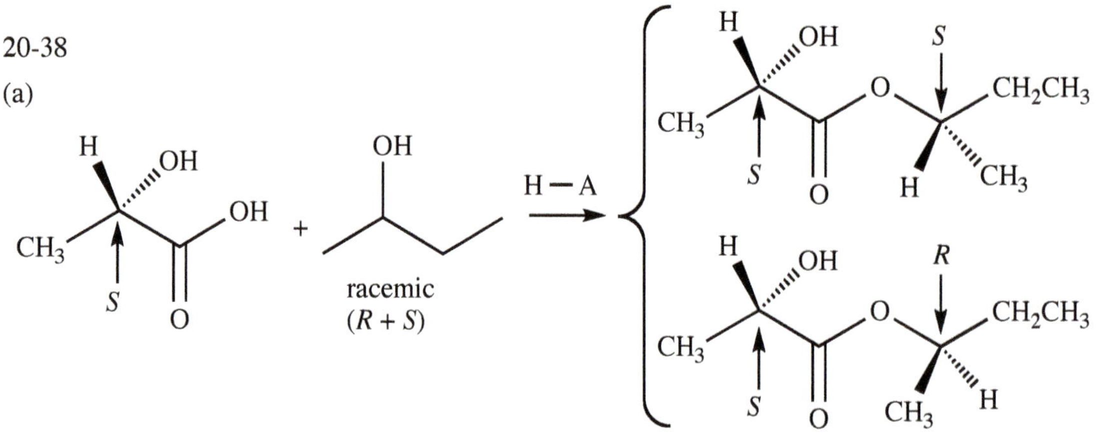

racemic
$(R + S)$

(b) Isomers that are *R*,*S* and *S*,*S* are diastereomers.

20-39

(a) $PhCH_2CH_2OH \xrightarrow[\text{pyridine}]{TsCl} \xrightarrow{KCN} PhCH_2CH_2CN \xrightarrow[\Delta]{H_3O^+} PhCH_2CH_2COOH$

$PhCH_2CH_2OH \xrightarrow{PBr_3} PhCH_2CH_2Br \xrightarrow[\text{ether}]{Mg} \xrightarrow{CO_2} \xrightarrow{H_3O^+} PhCH_2CH_2COOH$

(b) methylenecyclohexane (CH_2) $\xrightarrow{HBr}$ 1-bromo-1-methylcyclohexane (CH_3, Br) $\xrightarrow[\text{ether}]{Mg} \xrightarrow{CO_2} \xrightarrow{H_3O^+}$ 1-methylcyclohexanecarboxylic acid (CH_3, COOH)

(c) methylenecyclohexane (CH_2) $\xrightarrow[\text{ROOR}]{HBr}$ (bromomethyl)cyclohexane (CH_2Br) $\xrightarrow[\text{ether}]{Mg} \xrightarrow{CO_2} \xrightarrow{H_3O^+}$ cyclohexylacetic acid (CH_2COOH)

(d) bromo-tetralone (Br, O) $\xrightarrow[\text{H—A, }\Delta]{\text{HO}\diagup\!\diagdown\text{OH}}$ ethylene ketal (Br, O O) $\xrightarrow[\text{3) } H_3O^+]{\text{1) Mg; 2) } CO_2}$ tetralone carboxylic acid (COOH, O)

(e) cyclohexanecarboxylic acid (COOH) + **2** cyclohexyllithium (Li) $\rightarrow \xrightarrow{H_3O^+}$ dicyclohexyl ketone (O) $\xrightarrow[\text{H—A, }\Delta]{\text{HO}\quad\text{OH}}$ 1,3-dioxane ketal (O O)

(f) cyclopentanecarboxylic acid (O, OH) $\xrightarrow[\text{2) } H_3O^+]{\text{1) } LiAlH_4}$ cyclopentylmethanol (OH) + HO–C(=O)–cyclopentane $\xrightarrow[\Delta]{\text{H—A}}$ cyclopentylmethyl cyclopentanecarboxylate (O, O)

20-40

Spectrum A: $C_9H_{10}O_2$ ⇒ 5 elements of unsaturation
δ 11.8, 1H ⇒ COO**H**
δ 7.3, 5H ⇒ monosubstituted benzene ring
δ 3.8, 1H quartet ⇒ C**H**CH_3
δ 1.5, 3H doublet ⇒ CHC**H**$_3$

Ph–CH(CH$_3$)–COOH

Spectrum B: $C_4H_6O_2$ ⇒ 2 elements of unsaturation
δ 12.1, 1H ⇒ COO**H**
δ 6.2, 1H singlet ⇒ H—C=C
δ 5.7, 1H singlet ⇒ H—C=C
δ 1.9, 3H singlet ⇒ vinyl CH_3 with no H neighbors CH_3–C=C
The carbon NMR does not add more information, other than three sp^2 carbons and one sp^3 carbon.

$H_2C=C(COOH)CH_3$

Spectrum C: $C_6H_{10}O_2$ ⇒ 2 elements of unsaturation
δ 12.15, 1H ⇒ COOH
δ 7.1, 1H multiplet ⇒ H–C=C–COOH
δ 5.8, 1H doublet ⇒ C=C(H)–COOH

δ 2.2-0.9 ⇒ $CH_2CH_2CH_3$

$CH_3CH_2CH_2$(H)C=C(H)COOH (trans)

It must be *trans* due to large coupling constant in doublet at δ 5.8.

20-41

(a) δ-valerolactone

(b) [Mechanism: protonation of the carbonyl oxygen of 5-hydroxypentanoic acid by H–A; resonance forms of the protonated acid; intramolecular attack of the terminal OH on the carbonyl carbon; deprotonation by A^- gives the tetrahedral intermediate; protonation of an OH by H–A; loss of H_2O (– H_2O); resonance forms of the protonated lactone; deprotonation by $H_2\ddot{O}$: gives δ-valerolactone.]

20-42 (a)

$$H_2O_2 \rightleftharpoons H^+ + HOO^- \quad pK_a\ 11.6$$

$$H_2O \rightleftharpoons H^+ + HO^- \quad pK_a\ 15.7$$

Hydrogen peroxide is four pK_a units (10^4 times) stronger acid than water, so the hydroperoxide anion, HOO^-, must be stabilized relative to hydroxide. This is from the *inductive effect* of the electronegative oxygen bonded to the O^-; by induction, the negative charge is distributed over both oxygens. The oxygen in hydroxide has to support the full negative charge with no delocalization.

(b)

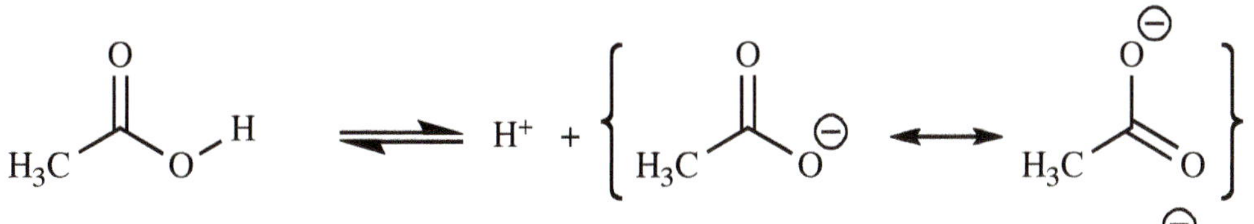

$$CH_3COOH \rightleftharpoons H^+ + \{CH_3COO^- \leftrightarrow CH_3COO^-\} \quad pK_a\ 4.74$$

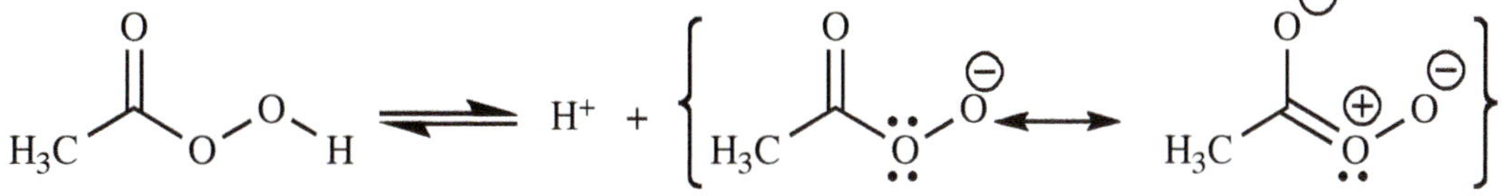

$$CH_3COOOH \rightleftharpoons H^+ + \{CH_3COOO^- \leftrightarrow CH_3C(O^-)=O^+-O^-\} \quad pK_a\ 8.2$$

The reason that carboxylic acids are so acidic (over 10 pK_a units more acidic than alcohols) is because of the resonance stabilization of the carboxylate anion with two equivalent resonance forms in which all atoms have octets and the negative charge is on the more electronegative atom—the best of all possible resonance worlds. The peroxyacetate anion, however, cannot delocalize the negative charge onto the carbonyl oxygen; that negative charge is stuck out on the end oxygen like a wet nose on a frigid morning. There is some delocalization of the electron density onto the carbonyl, but with all the charge separation, this second form is a minor resonance contributor. This resonance does explain, however, why peroxyacetic acid is more acidic than hydrogen peroxide. It does not come close to acetic acid, though.

(c)

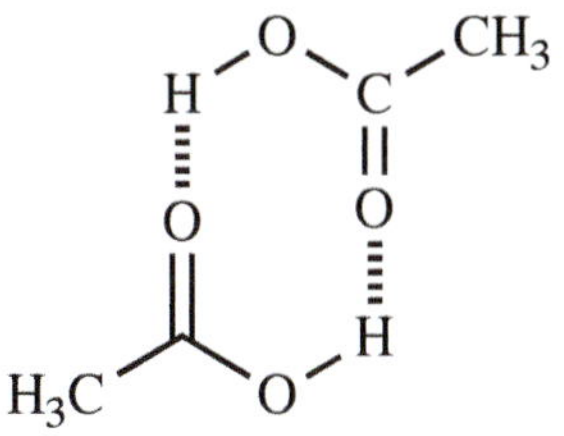

Carboxylic acids boil as the dimer, that is, two molecules are held tightly by hydrogen bonding. The dimer is an 8-membered ring with two hydrogen bonds as shown with dashed lines in the diagram. This works because the carbonyl oxygen has significant negative charge, and the H—O bond is weak because it is a relatively strong acid. The b.p. is 118 °C.

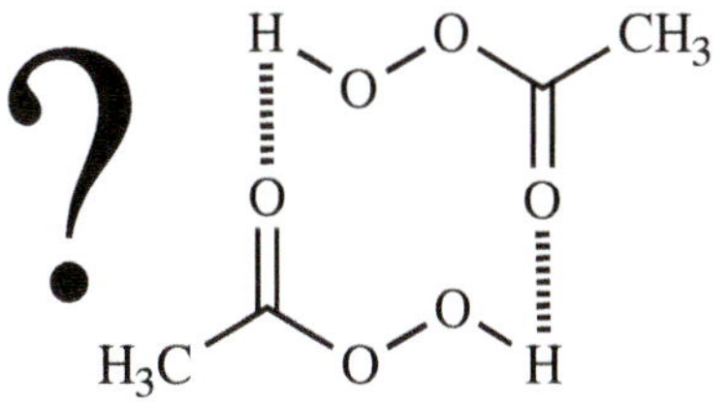

?

Do peroxyacids boil as the dimer? The author does not know, but there are three reasons to suspect that they do not. First, the b.p. is lower (105 °C) instead of higher, suggesting that they do not boil as a couple but rather individually. Second, the dimer shown is a 10-membered ring—still possible but less likely than 8-membered. Third and most important, the electronic nature of the carbonyl group, as implied in the resonance forms in part (b), places less negative charge on the carbonyl oxygen, and the H is less acidic, suggesting that the hydrogren bonding is much less strong.

20-43

(a) Mass spectrum:

—m/z 152 ⇒ molecular ion ⇒ molecular weight 152

—m/z 107 ⇒ M − 45 ⇒ loss of COOH

—m/z 77 ⇒ monosubstituted benzene ring

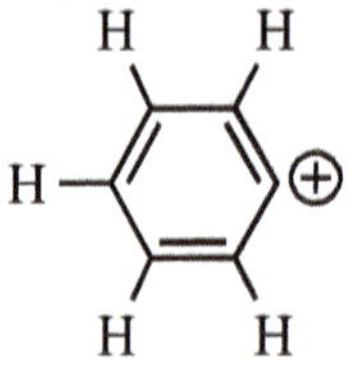

IR spectrum:

—3400–2400 cm^{-1}, broad ⇒ O—H stretch of COOH

—1700 cm^{-1} ⇒ C=O

—1240 cm^{-1} ⇒ C–O

—1600 cm^{-1} ⇒ aromatic C=C

NMR spectrum:

—δ 6.8–7.3, two signals in the ratio of 2H to 3H ⇒ monosubstituted benzene ring

—δ 4.6, 2H singlet ⇒ CH_2, deshielded

Carbon NMR spectrum:

—δ 170, small peak ⇒ carbonyl

—δ 115-157, four peaks ⇒ monosubstituted benzene ring; deshielded peak indicates oxygen substitution on the ring

(b) Fragments indicated in the spectra:

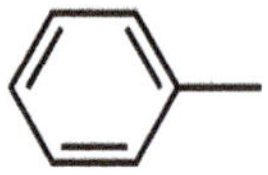

m/z 77	CH_2 m/z 14	COOH m/z 45 and from IR

This appears deceptively simple. The problem is that the mass of these fragments adds to 136, not 152—we are missing 16 mass units ⇒ oxygen! Where can the oxygen be? There are only two possibilities:

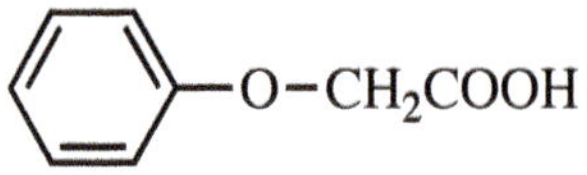

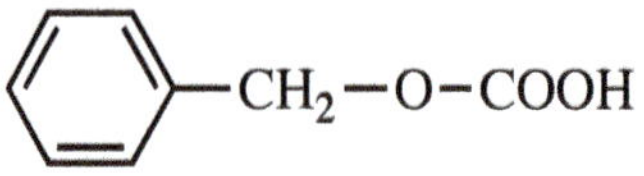

C_6H_5–O–CH_2COOH C_6H_5–CH_2–O–COOH

How can we differentiate? Mass spectrometry!

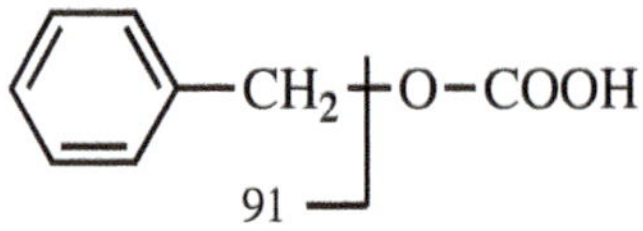

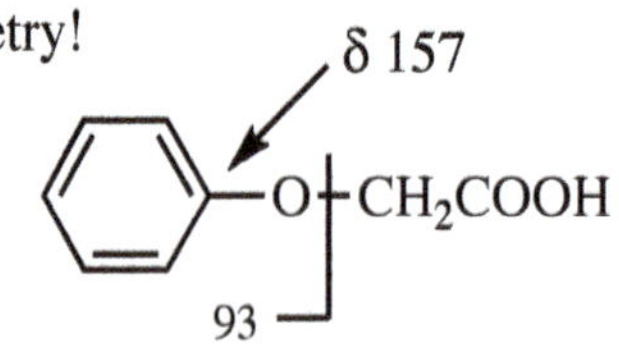

C_6H_5–CH_2 | O–COOH (91)

C_6H_5–O | CH_2COOH (93) — δ 157

phenoxyacetic acid

This structure is consistent with the peak at δ 157 in the carbon NMR.

The m/z 93 peak in the MS confirms the structure is **phenoxyacetic acid**. The CH_2 is so far downfield in the NMR because it is between two electron-withdrawing groups, the O and the COOH.

(c) The COOH proton is missing from the proton NMR. Either it is beyond 10 and the NMR was not scanned (unlikely), or the peak was broadened beyond detection because of hydrogen bonding with DMSO.

20-44 Cyanide substitution is an S_N2 reaction and requires a 1° or 2° carbon with a leaving group. The Grignard reaction is less particular about the type of halide, but is sensitive to, and incompatible with, acidic functional groups and other reactive groups.

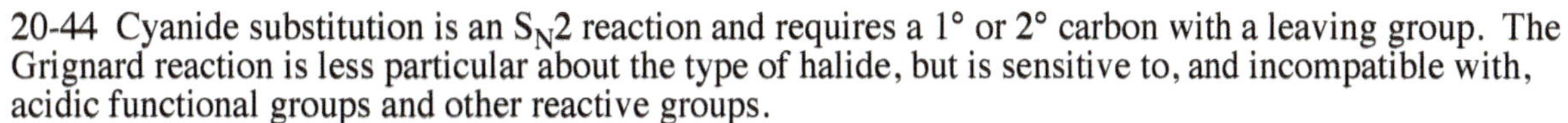

(a) Both methods will work.

$C_6H_5CH_2Br \xrightarrow[\text{ether}]{Mg} \xrightarrow{CO_2} \xrightarrow{H_3O^+}$ **OR** $\xrightarrow{NaCN} \xrightarrow[\Delta]{H_3O^+} C_6H_5CH_2COOH$

(b) Only Grignard will work. The S_N2 reaction does not work on sp^2-hybridized carbons, and nucleophilic aromatic substitution does not work well on unactivated benzene rings.

$C_6H_5Br \xrightarrow[\text{ether}]{Mg} \xrightarrow{CO_2} \xrightarrow{H_3O^+} C_6H_5COOH$

(c) Grignard will fail because of the OH group. The cyanide reaction will work, although an excess of cyanide will need to be added because the first equivalent will deprotonate the phenol.*

$HO-C_6H_4-CH_2Br \xrightarrow{NaCN} \xrightarrow[\Delta]{H_3O^+} HO-C_6H_4-CH_2COOH$

(d) Grignard will fail because of the OH group. The cyanide reaction will fail because S_N2 does not work on unactivated sp^2 carbons. In this case, NEITHER method will work.

$HO-C_6H_4-Br$

The Grignard method can be made to work if the OH of the phenol is protected with an inert group like the TIPS group. Following the Grignard, the TIPS can be removed by fluoride ion.

(e) Both methods will work, although cyanide substitution on secondary C will be accompanied by elimination.

$C_5H_9Br \xrightarrow[\text{ether}]{Mg} \xrightarrow{CO_2} \xrightarrow{H_3O^+}$ **OR** $\xrightarrow{NaCN} \xrightarrow[\Delta]{H_3O^+} C_5H_9COOH$

(f) Grignard will fail because of the OH group. The cyanide reaction will work. Since alcohols are much less acidic than phenols, there is no problem with cyanide deprotonating the alcohol.*

$HO-C_5H_8-CH_2Br \xrightarrow{NaCN} \xrightarrow[\Delta]{H_3O^+} HO-C_5H_8-CH_2COOH$

* The pK_a of HCN is 9.2, and the pK_a of phenol is 10.0. Thus cyanide is strong enough to pull off some of the H from the phenol, although the equilibrium would favor cyanide ion and phenol. The pK_a of secondary alcohols is 16–18, so there is no chance that cyanide would deprotonate an alcohol.

$^{\ominus}CN + HO-C_6H_5 \rightleftharpoons H-CN + {}^{\ominus}O-C_6H_5$

pK_a 10.0 (phenol); pK_a 9.2 (HCN)

The side with the weaker acid is favored.

20-45

(a) stock bottle: $CH_3(CH_2)_5CHO$ — students' samples: $CH_3(CH_2)_5COOH$

(b) The spectrum of the students' samples shows the carboxylic acid present. Contact with oxygen from the air oxidized the sensitive aldehyde group to the acid.

(c) Storing the aldehyde in an inert atmosphere like nitrogen or argon prevents oxidation. Freshly prepared unknowns will avoid the problem.

20-46 Recall that the weaker acid and base (higher pK_a) is the side favored at equilibrium.

(a) pK_a 4.7 (acetic acid) CH_2COOH / CH_2NH_2 ⇌ CH_2COO^- / $CH_2\overset{\oplus}{N}H_3$ pK_a 10.7 (methylammonium ion)

PRODUCT is strongly favored at equilibrium (6 pK_a units--that's a lot)

(b) pK_a 4.7 (acetic acid) CH_2COOH / NH_2 ⇌ CH_2COO^- / $\oplus NH_3$ pK_a 4.6 (anilinium ion)

(The pK_a value of phenylacetic acid of 4.31 suggests that the product is the major form at equilibrum.)

Very close! This might be near a 50:50 mixture. Without more precise pK_a values, we could not say for sure which one is favored.

(c) pK_a 4.2 (benzoic acid) COOH / CH_2NH_2 ⇌ COO^- / $CH_2\overset{\oplus}{N}H_3$ pK_a 10.7 (methylammonium ion)

PRODUCT is strongly favored at equilibrium (6.5 pK_a units--that's a lot)

(d) pK_a 4.9 COOH / NH_2 ⇌ COO^- / $NH_3^{\oplus}$ pK_a 2.3

REACTANT is strongly favored at equilibrium (2.6 pK_a units)

Two notable things: 1. the basicity of the amine is far more variable depending on conjugation of the amine with the benzene ring than the carboxylic acid is; 2. the compound in (d) is a weaker base and a weaker acid than the other structures. What is happening? RESONANCE!

H_2N—C₆H₄—C(=O)—OH ⟷ $H_2\overset{\oplus}{N}$=C₆H₄=C($\ddot{O}:^{\ominus}$)—OH

plus many other resonance forms

The NH_2 donates electrons, becoming a weaker base, and the COOH accepts electrons, becoming a weaker acid.

20-47 Electron-donating groups makes acids weaker; electron-withdrawing groups make acids stronger.

(a) **B** is a stronger acid than **A**, so methoxy in the meta position must be withdrawing.

(b) **C** is a weaker acid than **A**, so methoxy in the para position must be donating.

(c) Any substituent in the meta position exerts only an inductive effect, not a resonance effect. Therefore, methoxy must be electron-withdrawing by induction—not surprising because of the electronegative oxygen. In the para position, methoxy must be electron-donating, and this must be a resonance effect (also explains why methoxy is an ortho,para-director). The resonance effect must be stronger than induction, because methoxy is always withdrawing by induction, even in the para position, but the resonance donation is so strong that it overpowers the inductive effect. (Quick! Go to the fluoro substituent!)

20-47 continued

(d) **E** is a stronger acid than **D**, so fluoro in the meta position must be withdrawing.

(e) **F** is a stronger acid than **D**, although not as strong as **E**, so fluoro in the para position must be slightly withdrawing.

(f) Any substituent in the meta position exerts only an inductive effect, not a resonance effect. Therefore, fluoro must be electron-withdrawing by induction—not surprising because fluorine is the most electronegative element. In the para position where resonance can be expressed, fluoro is still electron-withdrawing, but the net effect is not as strong as the inductive effect expressed at the meta position. So the resonance effect must be electron-donating, but for fluro, the inductive effect must be stronger than resonance.

20-48

(a)

acetal; CH_3; O; O; CH_3; CH_2COOH; carboxylic acid; Compound 1 — H_3O^+ → O; O; CH_3; OH alcohol; ester (An ester in a ring is called a lactone.); Compound 2

(b) Compound 1 has eight carbons, and Compound 2 has six carbons. Two carbons have been lost: the two carbons of the acetal have been cleaved. (This is the best way to figure out reactions and mechanisms: find out which atoms of the reactant have become which atoms of the product, then determine what bonds have been broken and formed.)

(c) Acetals are stable to base, so the acetal must have been cleaved when acid was added.

CH_3; O; O; 5; 3; CH_3; 4; CH_2COOH (2, 1) — H_3O^+ → O; O; 1; 2; 5; 3; CH_3; 4; OH

20-48 continued

(d) The carbons have been numbered in compounds 1 and 2 on the previous page to help you visualize which atoms in the reactant become which atoms in the product. The overall process requires cleavage of the acetal to expose two alcohols. The 3° alcohol at carbon-3 can be found in the product, so it is the primary alcohol at carbon-5 that reacts with the carboxylic acid to form the lactone.

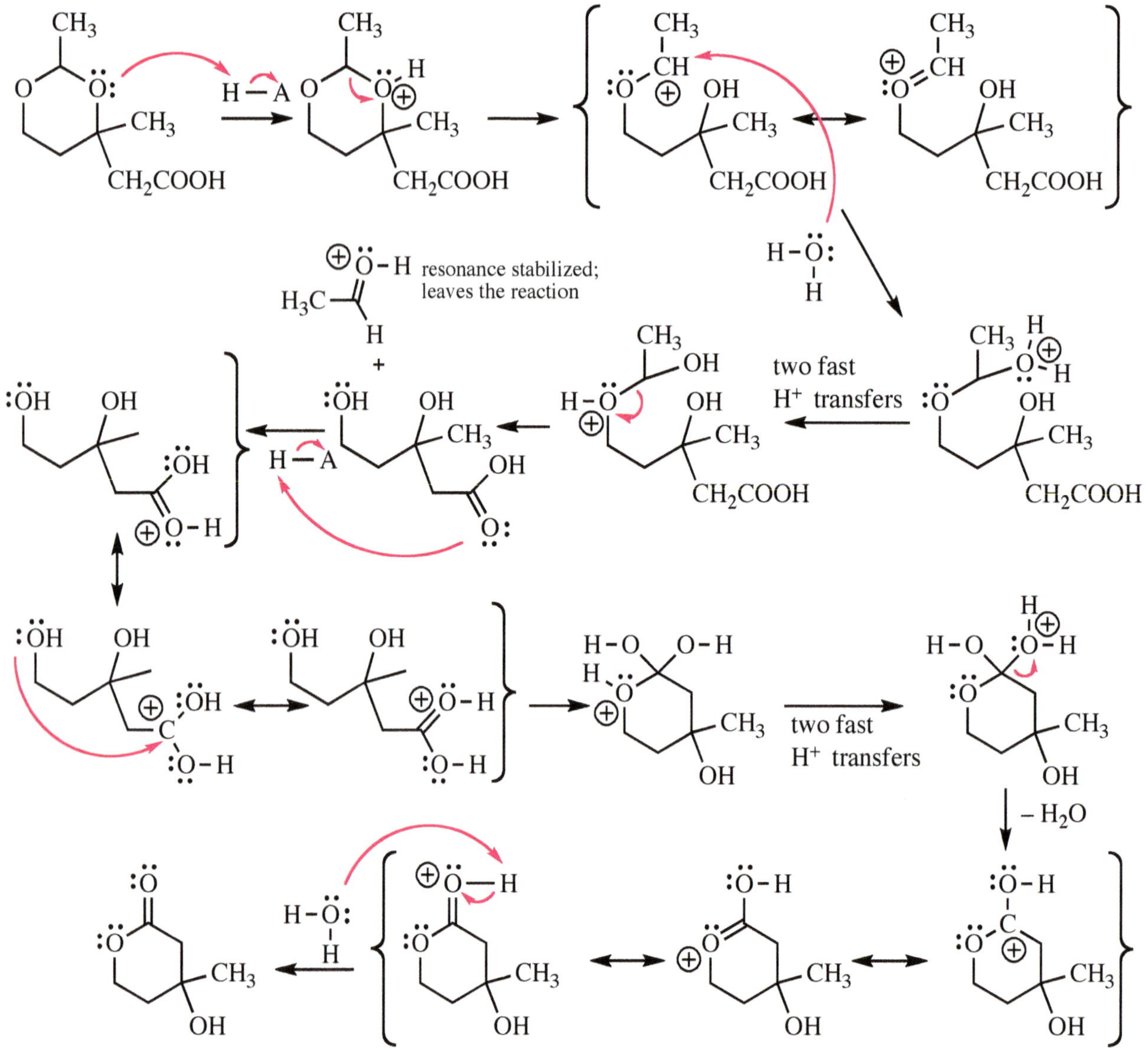

resonance stabilized; leaves the reaction

two fast H^+ transfers

two fast H^+ transfers

$- H_2O$

CHAPTER 21—CARBOXYLIC ACID DERIVATIVES

21-1 IUPAC name first; then common name

(a) isobutyl benzoate (both IUPAC and common)
(b) phenyl methanoate; phenyl formate
(c) methyl 2-phenylpropanoate; methyl α-phenylpropionate
(d) 3-methyl-*N*-phenylbutanamide; β-methylbutyranilide
(e) *N*-benzylethanamide; *N*-benzylacetamide
(f) 3-hydroxybutanenitrile; β-hydroxybutyronitrile
(g) 3-methylbutanoyl bromide; isovaleryl bromide
(h) dichloroethanoyl chloride; dichloroacetyl chloride
(i) methanoic 2-methylpropanoic anhydride; formic isobutyric anhydride
(j) cyclopentyl cyclobutanecarboxylate (both IUPAC and common)
(k) 5-hydroxyhexanoic acid lactone; δ-caprolactone
(l) *N*-cyclopentylbenzamide (both IUPAC and common)
(m) propanedioic anhydride; malonic anhydride
(n) 1-hydroxycyclopentanecarbonitrile; cyclopentanone cyanohydrin
(o) *cis*-4-cyanocyclohexanecarboxylic acid; no common name
(p) 3-bromobenzoyl chloride; *m*-bromobenzoyl chloride
(q) 5-(*N*-methylamino)heptanoic acid lactam; no common name
(r) *N*-ethanoylpiperidine; *N*-acetylpiperidine

21-2 An aldehyde has a C—H absorption (usually 2 peaks) at 2700–2800 cm^{-1}. A carboxylic acid has a strong, broad absorption between 2400 and 3400 cm^{-1}. The spectrum of methyl benzoate has no peaks in this region.

21-3 Sometimes the C=O stretch alone is not definitive in assigning a functional group. Other peaks in the IR spectrum can give evidence that supports or excludes a functional group.

An aldehyde has a C—H absorption (usually 2 peaks) at 2700–2800 cm^{-1}. A carboxylic acid has a strong, broad absorption between 2400–3400 cm^{-1} with a characteristic "shoulder" around 2600 cm^{-1}. An ester has a strong C—O peak around 1000–1200 cm^{-1}; this peak is often as strong as the C=O. A ketone doesn't have any of these extra peaks, so we deduce a ketone *by exclusion*, that is, by eliminating all other possible functional groups.

21-4

(a) acid chloride: single C=O peak at 1800 cm^{-1}; no other carbonyl comes so high;
(b) primary amide: broad C=O at 1650 cm^{-1} and two N—H peaks between 3200–3400 cm^{-1}
(c) anhydride: two C=O absorptions at 1750 and 1820 cm^{-1}

21-5 (a) The formula C_3H_5NO has two elements of unsaturation. The IR spectrum shows two peaks between 3200 and 3400 cm^{-1}, an NH_2 group. The strong peak at 1670 cm^{-1} is a C=O, and the peak at 1610 cm^{-1} is a C=C. This accounts for all of the atoms.

The HNMR corroborates the assignment. The 1H multiplet at δ 5.8 is the vinyl H next to the carbonyl. The 2H multiplet at δ 6.3 is the vinyl hydrogen pair on carbon-3. The 2H singlet at δ 4.8 is the amide hydrogens. The CNMR confirms the structure: two vinyl carbons and a carbonyl.

$$H_2C{=}CH{-}\overset{\overset{\displaystyle O}{\|}}{C}{-}NH_2$$

(b) The formula $C_5H_8O_2$ has two elements of unsaturation. The IR spectrum shows no OH (small peak at 3500 is an overtone of the C=O at 1730), so this compound is neither an alcohol nor a carboxylic acid. The strong peak at 1730 cm^{-1} is likely an ester carbonyl. The C—O appears 1050–1250 cm^{-1}. The IR shows no C=C absorption, so the other element of unsaturation is likely a ring. The carbon NMR spectrum shows the carbonyl carbon at δ 171, the C—O carbon at δ 69, and three more carbons in the aliphatic region, but no carbons in the vinyl region δ 100-150, so there can be no C=C. The proton NMR shows multiplets of 2H at δ 4.3 and 2.5, most likely CH_2 groups next to oxygen and carbonyl respectively.

The only structure with an ester, four CH_2 groups, and a ring, is δ-valerolactone:

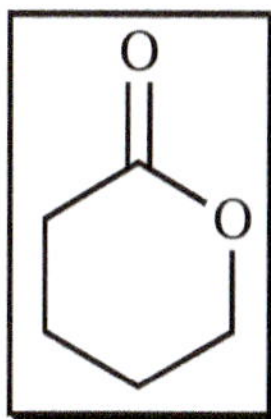

21-6 Chloride ion is shown as the base in these substitutions on acid chlorides, but any other species with an unshared electron pair could also remove a strongly acidic proton.

(a)

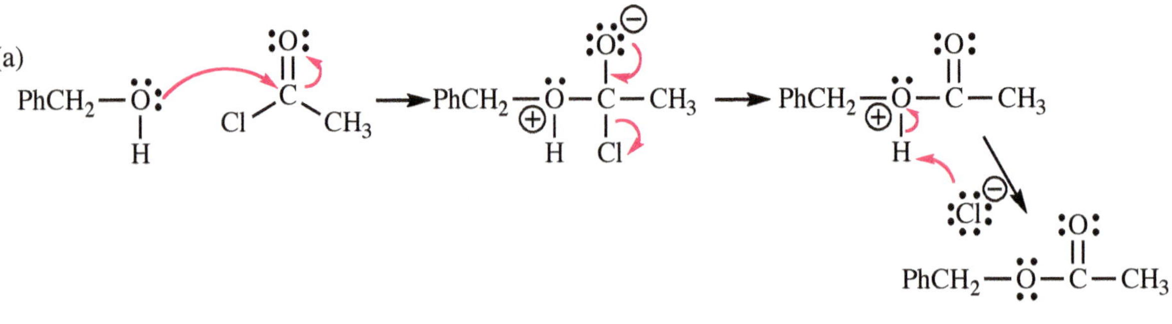

PhCH$_2$—O—C(=O)—CH$_3$

(b)

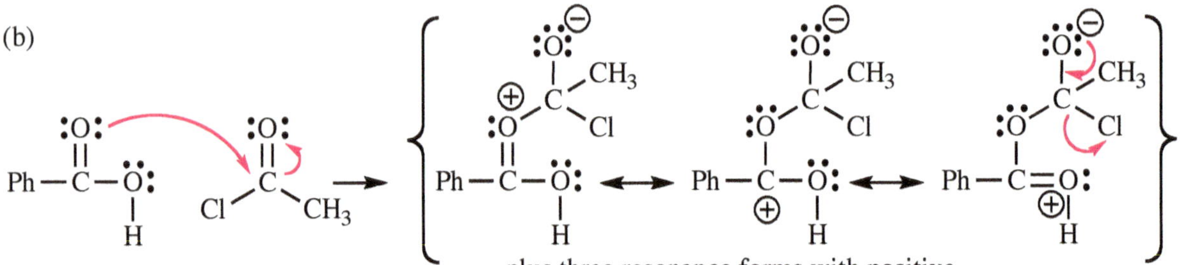

plus three resonance forms with positive charge delocalized on the benzene ring

The carbonyl oxygen is more nucleophilic than the single-bonded oxygen because the product is resonance stabilized.

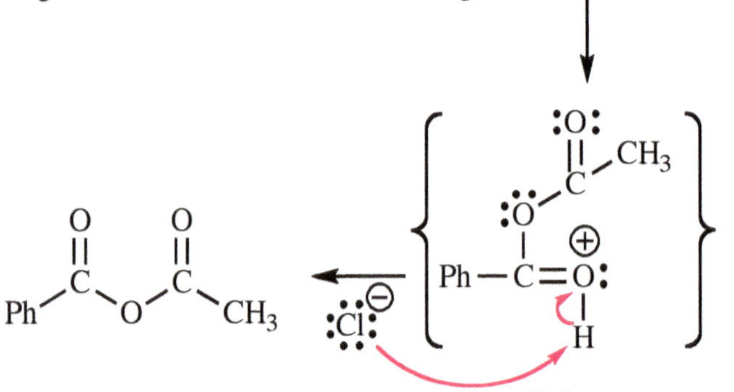

plus all the resonance forms as above

(c)

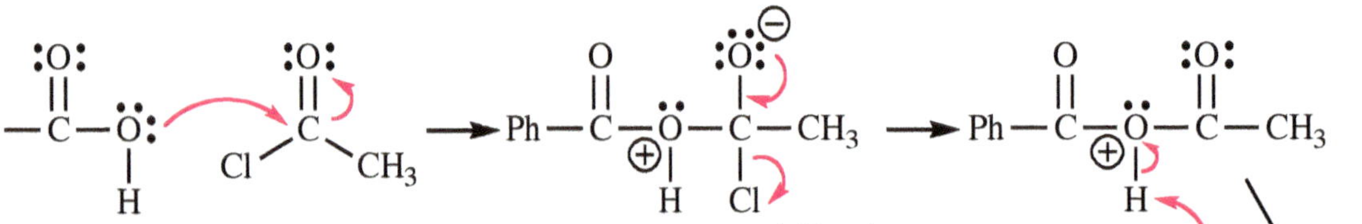

Nucleophilic attack by this oxygen does not generate a resonance-stabilized intermediate and is much less likely than that shown in part (b).

no resonance stabilization

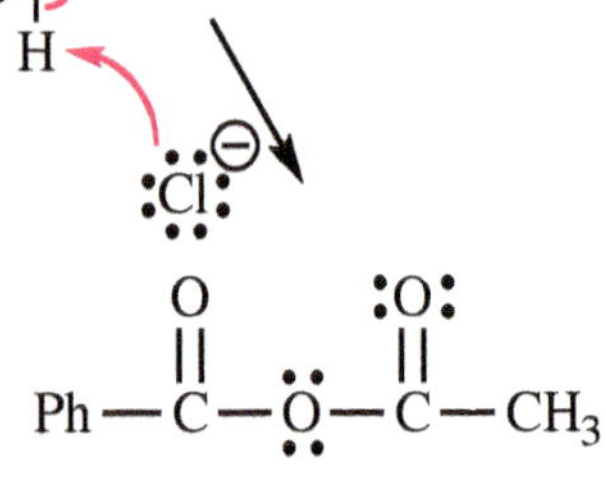

21-6 continued

(d)

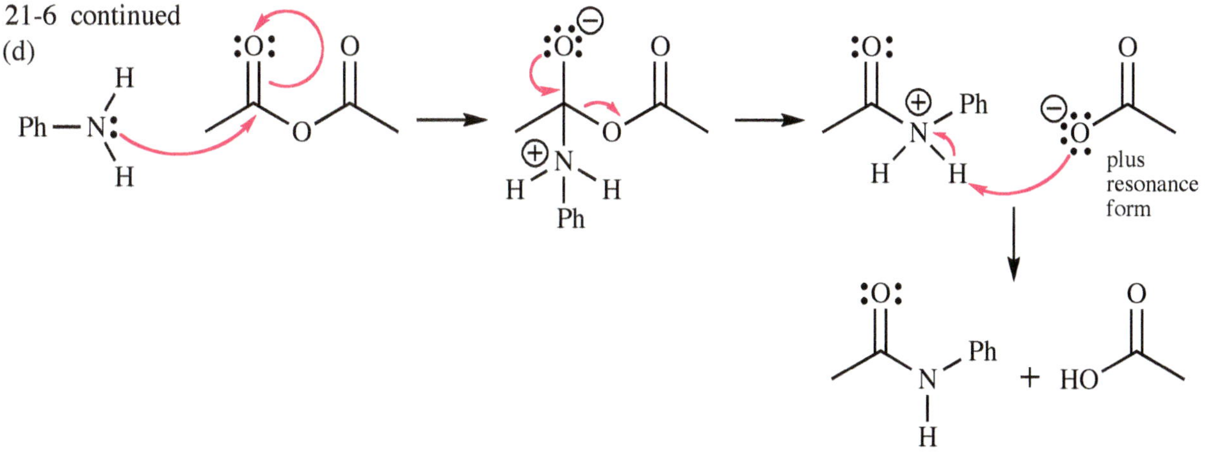

Ph—N(H)(H) + $CH_3C(=O)OC(=O)CH_3$ → tetrahedral intermediate → $CH_3C(=O)N^{+}H_2Ph$ + $^{-}OC(=O)CH_3$ (plus resonance form) → $CH_3C(=O)NHPh$ + $HOC(=O)CH_3$

(e)

Ph—N(H)(H) + $CH_3C(=O)OCH_2CH_3$ → tetrahedral intermediate → $CH_3C(=O)N^{+}H_2Ph$ + $^{-}OCH_2CH_3$ (leaving group) → $CH_3C(=O)NHPh$ + $HOCH_2CH_3$

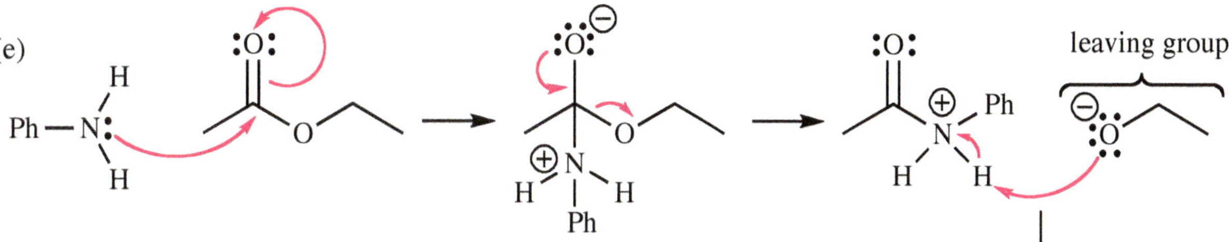

The leaving group is ethoxide ion, $CH_3CH_2O^-$, a very strong base. Ethoxide would never be a leaving group in an S_N2 reaction as it is too strong a base.

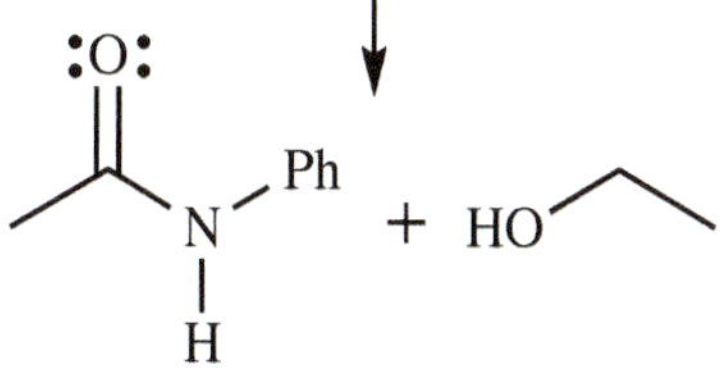

21-7 **Figure 21-9 is critical!** Reactions that go from a more reactive functional group to a less reactive functional group ("downhill reactions") will occur readily.

(a) Amide to acid chloride will NOT occur—it is an "uphill" transformation.
(b) Acid chloride to amide will occur rapidly.
(c) Amide to ester will NOT occur—another "uphill" transformation.
(d) Acid chloride to anhydride will occur rapidly.
(e) Anhydride to amide will occur rapidly.

21-8

(a) $CH_3CH_2-C(=O)-Cl$ + $HOCH_2CH_3$ → $CH_3CH_2-C(=O)-OCH_2CH_3$ + HCl

(b) 3-methylpentanoyl chloride + HO–phenyl → phenyl 3-methylpentanoate + HCl

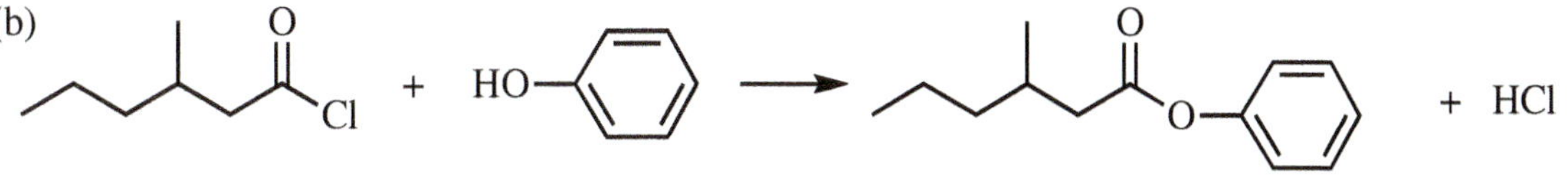

(c) $C_6H_5-C(=O)-Cl$ + $HOCH_2-C_6H_5$ → $C_6H_5-C(=O)-OCH_2-C_6H_5$ + HCl

(d) cyclohexyl$-C(=O)-Cl$ + HO–cyclopropyl → cyclohexyl$-C(=O)-O-$cyclopropyl + HCl

21-8 continued

(e)

$$CH_3-\overset{O}{\overset{\|}{C}}-Cl + HO-C(CH_3)_2-CH_3 \longrightarrow CH_3C(=O)-O-C(CH_3)_2-CH_3 + HCl$$

This reaction would have to be kept cold to avoid elimination. Esters of *tert*-butyl alcohol are hard to make because 3° alcohols eliminate so easily.

(f)

$$ClC(=O)CH_2CH_2C(=O)Cl + 2\ CH_2{=}CHCH_2OH \longrightarrow CH_2{=}CHCH_2O-C(=O)CH_2CH_2C(=O)-OCH_2CH{=}CH_2 + 2\ HCl$$

21-9

(a) $H_3C-\overset{O}{\overset{\|}{C}}-Cl + HN(CH_3)_2 \longrightarrow H_3C-\overset{O}{\overset{\|}{C}}-N(CH_3)_2 + HCl$

(b) $H_3C-\overset{O}{\overset{\|}{C}}-Cl + H_2N-C_6H_5 \longrightarrow H_3C-\overset{O}{\overset{\|}{C}}-NH-C_6H_5 + HCl$

(c) $C_6H_{11}-\overset{O}{\overset{\|}{C}}-Cl + NH_3 \longrightarrow C_6H_{11}-\overset{O}{\overset{\|}{C}}-NH_2 + HCl$

(d) $C_6H_5-\overset{O}{\overset{\|}{C}}-Cl + HN(CH_2)_5 \longrightarrow C_6H_5-\overset{O}{\overset{\|}{C}}-N(CH_2)_5 + HCl$

21-10

(a) (i)

$$H_3C-\overset{O}{\overset{\|}{C}}-O-\overset{O}{\overset{\|}{C}}-CH_3 + HOCH_2-C_6H_5 \longrightarrow H_3C-\overset{O}{\overset{\|}{C}}-OCH_2-C_6H_5 + HO-\overset{O}{\overset{\|}{C}}-CH_3$$

(ii)

$$H_3C-\overset{O}{\overset{\|}{C}}-O-\overset{O}{\overset{\|}{C}}-CH_3 + HN(CH_2CH_3)_2 \longrightarrow H_3C-\overset{O}{\overset{\|}{C}}-N(CH_2CH_3)_2 + HO-\overset{O}{\overset{\|}{C}}-CH_3$$

(b) (i)

$$H_3C-\overset{:O:}{\overset{\|}{C}}-O-\overset{O}{\overset{\|}{C}}CH_3 + H\ddot{O}CH_2Ph \longrightarrow H_3C-C(-\ddot{O}:^{\ominus})(-\overset{\oplus}{O}H-CH_2Ph)-\ddot{O}-\overset{O}{\overset{\|}{C}}CH_3$$

$$\downarrow$$

$$H_3C-\overset{:O:}{\overset{\|}{C}}-\ddot{O}-CH_2Ph + H-\ddot{O}-\overset{O}{\overset{\|}{C}}CH_3 \longleftarrow H_3C-\overset{:O:}{\overset{\|}{C}}-\overset{\oplus}{O}H-CH_2Ph + {}^{\ominus}:\ddot{O}-\overset{O}{\overset{\|}{C}}CH_3$$

plus resonance form

continued on next page

21-10(b) continued

(ii)

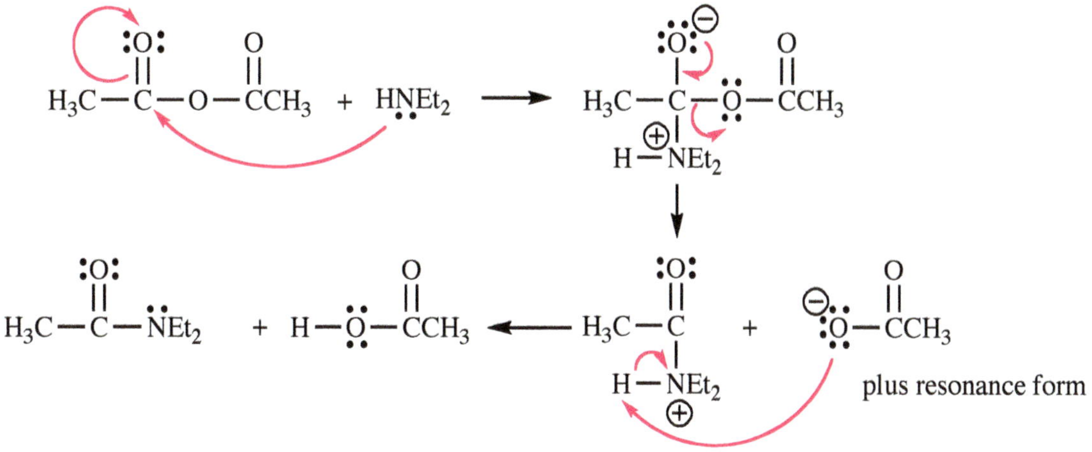

$$H_3C{-}\overset{O}{\overset{\|}{C}}{-}O{-}\overset{O}{\overset{\|}{C}}CH_3 + HNEt_2 \longrightarrow H_3C{-}\underset{H-\overset{+}{N}Et_2}{\overset{O^-}{C}}{-}O{-}\overset{O}{\overset{\|}{C}}CH_3$$

$$\downarrow$$

$$H_3C{-}\overset{O}{\overset{\|}{C}}{-}NEt_2 + H{-}O{-}\overset{O}{\overset{\|}{C}}CH_3 \longleftarrow H_3C{-}\underset{H-\overset{+}{N}Et_2}{\overset{O}{\overset{\|}{C}}} + {}^{-}O{-}\overset{O}{\overset{\|}{C}}CH_3$$

plus resonance form

21-11

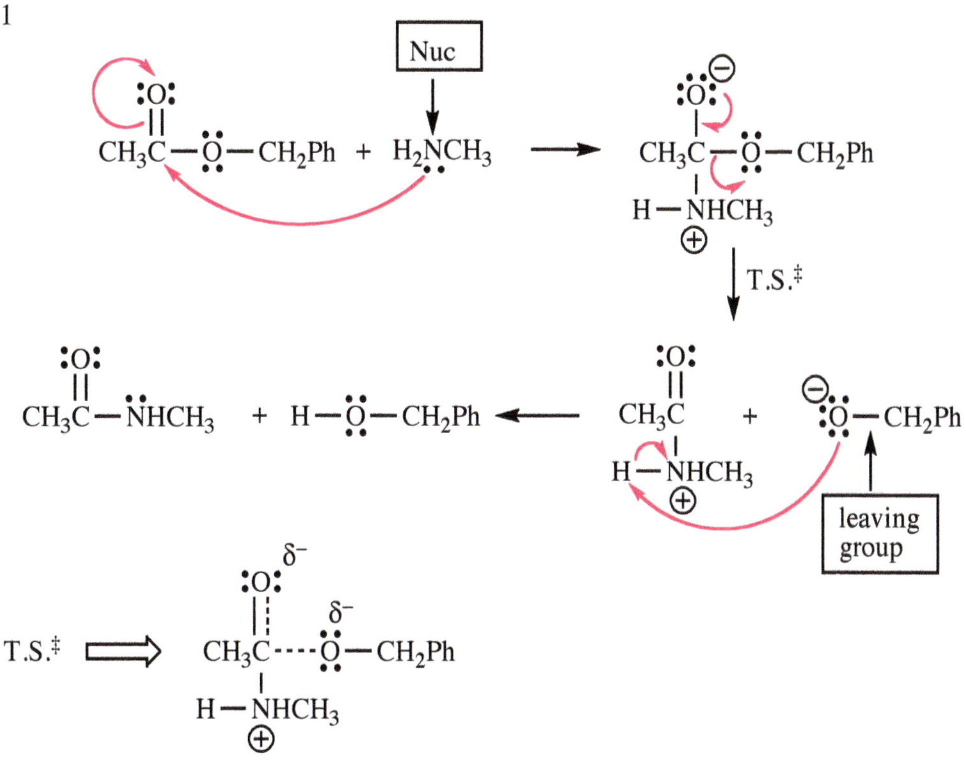

Nuc

$$CH_3\overset{O}{\overset{\|}{C}}{-}O{-}CH_2Ph + H_2NCH_3 \longrightarrow CH_3\underset{H-\overset{+}{N}HCH_3}{\overset{O^-}{C}}{-}O{-}CH_2Ph$$

$$\downarrow \text{T.S.}^{\ddagger}$$

$$CH_3\overset{O}{\overset{\|}{C}}{-}NHCH_3 + H{-}O{-}CH_2Ph \longleftarrow CH_3\underset{H-\overset{+}{N}HCH_3}{\overset{O}{\overset{\|}{C}}} + {}^{-}O{-}CH_2Ph$$

leaving group

$$\text{T.S.}^{\ddagger} \Longrightarrow CH_3\underset{H-\overset{+}{N}HCH_3}{\overset{O^{\delta-}}{\overset{\|}{C}}}\cdots\overset{\delta-}{O}{-}CH_2Ph$$

21-12

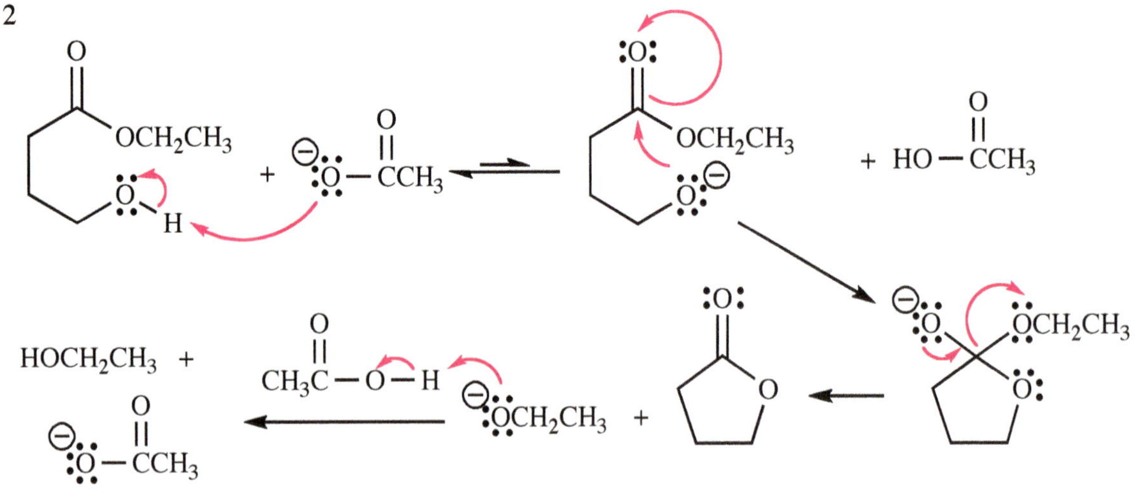

21-13

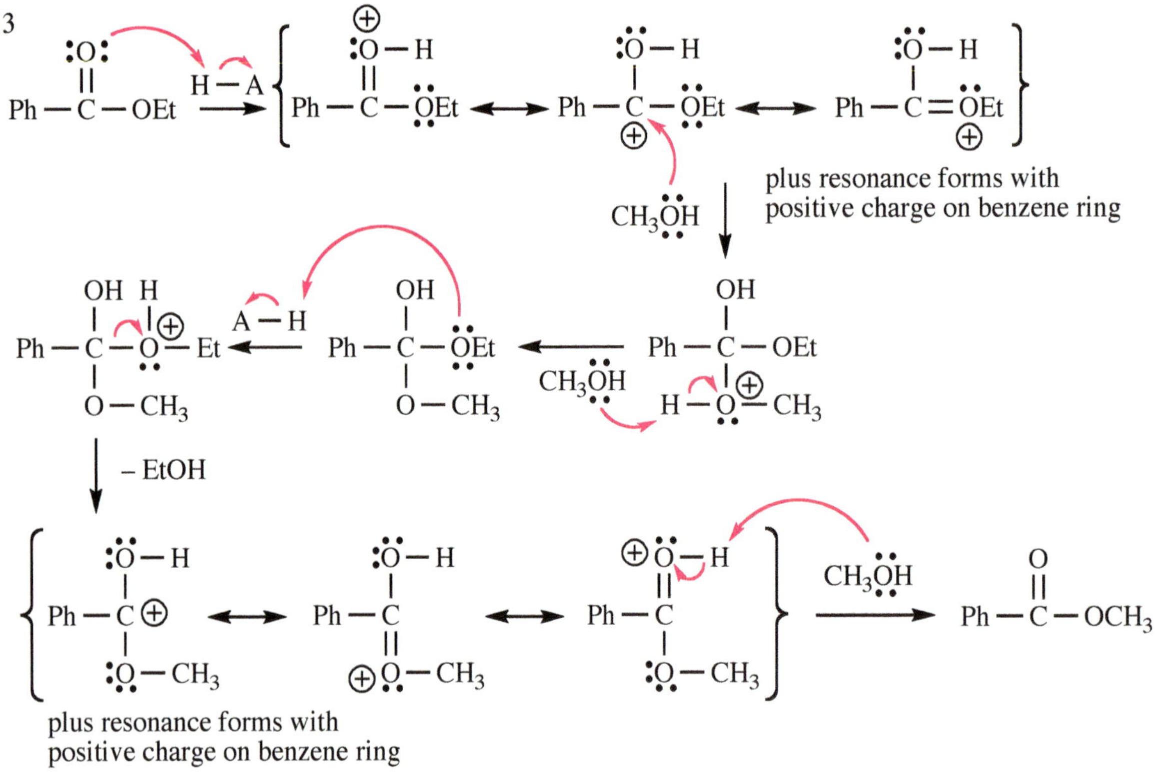
plus resonance forms with positive charge on benzene ring
– EtOH
plus resonance forms with positive charge on benzene ring

21-14

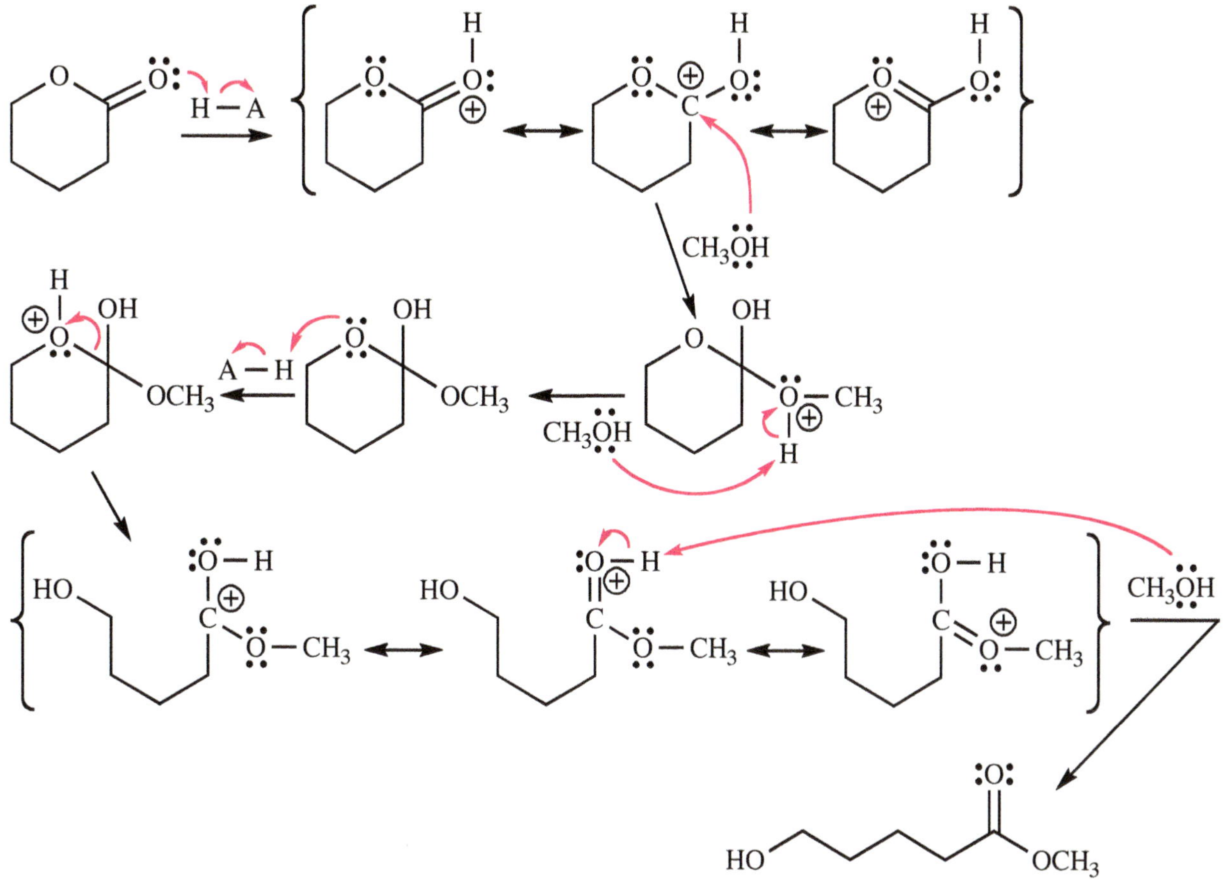

H—A

$CH_3\ddot{O}H$

OH

OCH$_3$

A—H

CH$_3$ÖH

HO

CH$_3$

HO

OCH$_3$

21-15 (a)

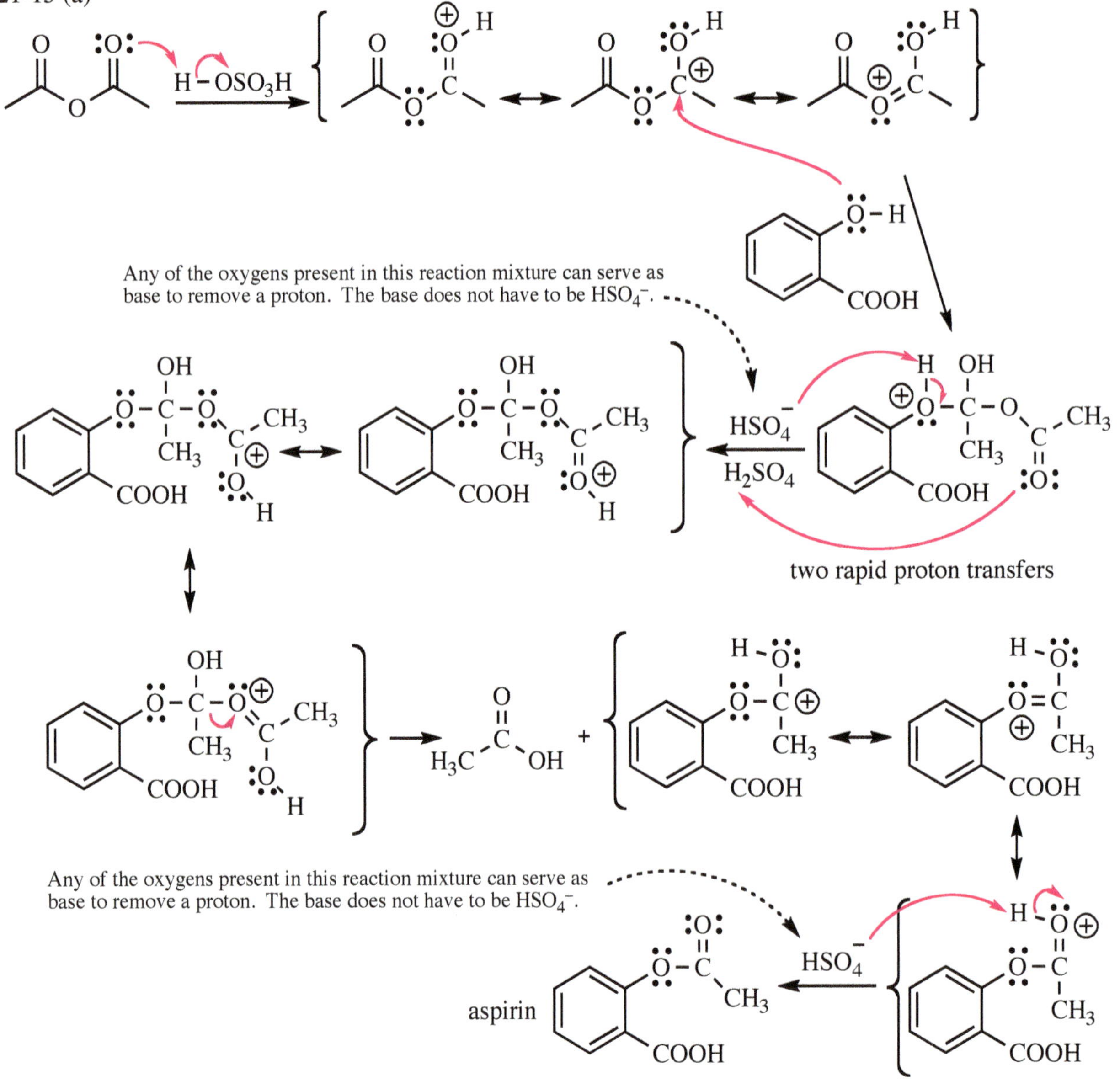

H−OSO$_3$H

COOH

Any of the oxygens present in this reaction mixture can serve as base to remove a proton. The base does not have to be HSO_4^-.

HSO_4^-

H_2SO_4

two rapid proton transfers

$H_3C-C(=O)-OH$ +

Any of the oxygens present in this reaction mixture can serve as base to remove a proton. The base does not have to be HSO_4^-.

aspirin

HSO_4^-

(b) As a true catalyst that speeds the rate of reaction without being consumed, the drop of H_2SO_4 is regenerated in the last step; it is used in the mechanism but not consumed. Its function in the mechanism is to make the carbonyl carbon of the anhydride more positive and therefore more susceptible to nucleophilic attack by the OH of the phenol, which is a weak nucleophile. Without the H_2SO_4 catalyst, this reaction would still occur but it would be much slower, probably by several factors of 10.

21-16 The asterisk (*) will denote ^{18}O.

(a)

$$H_3C-\overset{\overset{\displaystyle :O:}{\|}}{C}-O^*R \xrightarrow{\;^{\ominus}:\ddot{O}H\;} H_3C-\overset{\overset{\displaystyle :\ddot{O}:^{\ominus}}{|}}{\underset{\underset{\displaystyle :\ddot{O}-H}{|}}{C}}-\ddot{O}^*R \longrightarrow H_3C-\overset{\overset{\displaystyle :O:}{\|}}{C}-\underset{\underset{\displaystyle H}{|}}{\ddot{O}:} + {}^{\ominus}:\ddot{O}^*R \longrightarrow H_3C-\overset{\overset{\displaystyle O}{\|}}{C}-O^{\ominus}$$

$$O^*R = {}^{18}O-CH(CH_2CH_3)(CH_3)\ (R)$$

+

$H^{18}O-CH(CH_2CH_3)(CH_3)$ (R)

The alcohol product contains the ^{18}O label, with none in the carboxylate. The bond between ^{18}O and the tetrahedral carbon with (*R*) configuration did not break, so the configuration is retained.

(b) The products are identical regardless of mechanism.

$$H_3C-\overset{\overset{\displaystyle :O:}{\|}}{C}-O^*R \xrightarrow{H-A} \left\{ CH_3\overset{\overset{\displaystyle {}^{\oplus}:O-H}{\|}}{C}-\ddot{O}^*R \longleftrightarrow CH_3\overset{\overset{\displaystyle :\ddot{O}-H}{|}}{C^{\oplus}}-\ddot{O}^*R \longleftrightarrow CH_3\overset{\overset{\displaystyle :\ddot{O}-H}{|}}{C}=\overset{\oplus}{O}^*R \right\}$$

$$\xrightarrow{H_2\ddot{O}:} CH_3\overset{\overset{\displaystyle OH}{|}}{C}-O^*R \ (\text{with } H-\overset{\oplus}{O}-H) \xrightarrow{H_2\ddot{O}:} CH_3\overset{\overset{\displaystyle OH}{|}}{\underset{\underset{\displaystyle O-H}{|}}{C}}-\ddot{O}^*R \xrightarrow{H-A} CH_3\overset{\overset{\displaystyle OH}{|}}{\underset{\underset{\displaystyle O-H}{|}}{C}}-\overset{\overset{\displaystyle H}{|\oplus}}{O}^*R$$

– HO*R

$$\left\{ H_3C-\overset{\overset{\displaystyle :\ddot{O}-H}{|}}{\underset{\underset{\displaystyle :\ddot{O}-H}{|}}{C^{\oplus}}} \longleftrightarrow H_3C-\overset{\overset{\displaystyle :\ddot{O}-H}{|}}{\underset{\underset{\displaystyle \oplus\ddot{O}-H}{\|}}{C}} \longleftrightarrow H_3C-\overset{\overset{\displaystyle \oplus\ddot{O}-H}{\|}}{\underset{\underset{\displaystyle :\ddot{O}-H}{|}}{C}} \right\} \xrightarrow{H_2\ddot{O}:} H_3C-\overset{\overset{\displaystyle :O:}{\|}}{C}-OH$$

acetic acid

(c) The ^{18}O has 2 more neutrons in its nucleus than ^{16}O. Mass spectra of these products would show the molecular ion of acetic acid at its standard value of m/z 60, whereas the molecular ion of butan-2-ol would appear at m/z 76 instead of m/z 74, proving that the heavy isotope of oxygen went with the alcohol. This demonstrates that the bond between oxygen and the carbonyl carbon is broken, not the bond between the oxygen and the alkyl carbon.

To show if the alcohol was chiral or racemic, measuring its optical activity in a polarimeter and comparing to known values would prove its configuration. (The heavy oxygen isotope has a negligible effect on optical rotation.)

21-17

(a) A catalyst is defined as a chemical species that speeds a reaction but is not consumed in the reaction. In the acidic hydrolysis, acid is used in the first and fourth steps of the mechanism but is regenerated in the third and last steps. Acid is not consumed; the final concentration of acid is the same as the initial concentration. In the basic hydrolysis, however, the hydroxide that initially attacks the carbonyl is never regenerated. An alkoxide leaves from the carbonyl, but it quickly neutralizes the carboxylic acid. For every molecule of ester, one molecule of hydroxide is consumed; the base *promotes* the reaction but does not *catalyze* the reaction.

(b) Basic hydrolysis is not reversible. Once an ester molecule is hydrolyzed in base, the carboxylate cannot form an ester. Acid catalysis, however, is an equilibrium: the mixture will always contain some ester, and the yield will never be as high as in basic hydrolysis. Second, long chain fatty acids are not soluble in water until they are ionized; they are soluble only as their sodium salts (soap). Basic hydrolysis is preferred for higher yield and greater solubility of the product.

21-18

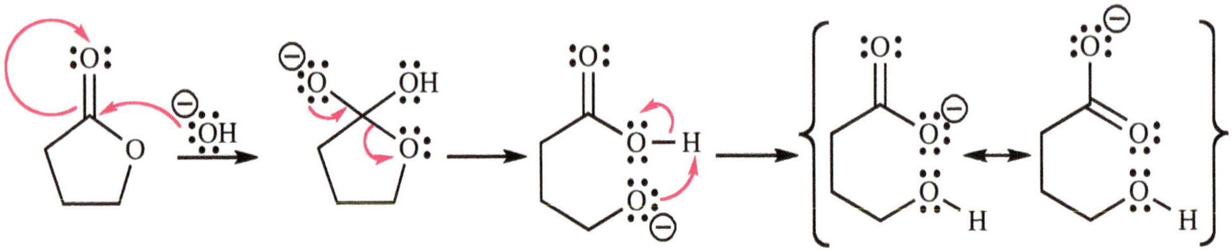

21-19

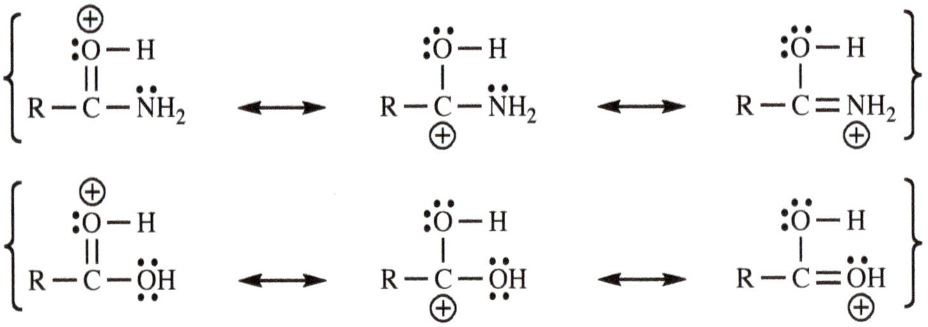

21-20

(a) $H_3C-C(=O)-NMe_2$ + ^-OH → $H_3C-C(O^-)(OH)-NMe_2$ → $H_3C-C(=O)-O-H$ + $^-NMe_2$ → $HN(CH_3)_2$ + $\{H_3C-C(O^-)=O \leftrightarrow H_3C-C(=O)-O^-\}$

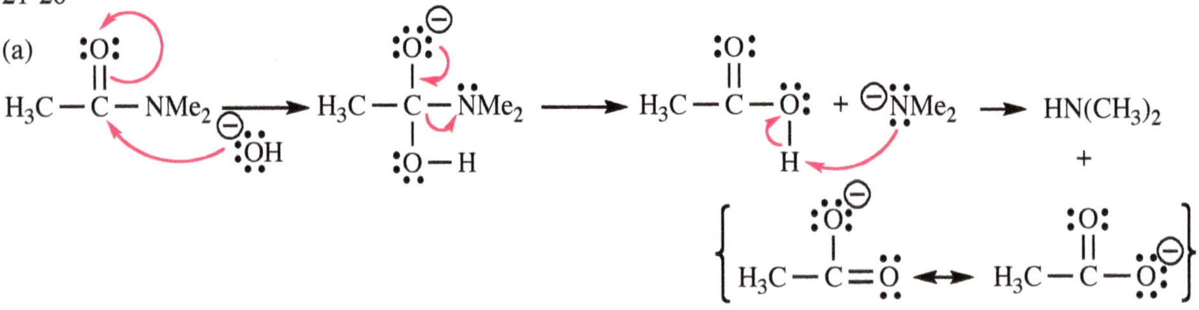

21-20 continued

(b) [Mechanism: acid-catalyzed hydrolysis of $H_3C-C(=O)-NMe_2$; protonation by H–A, resonance forms, attack by H_2O, proton transfers, loss of $HN(CH_3)_2$, final products $H_3C-C(=O)-OH$ + $H_2\overset{\oplus}{N}(CH_3)_2$]

$- HN(CH_3)_2$

$HN(CH_3)_2$

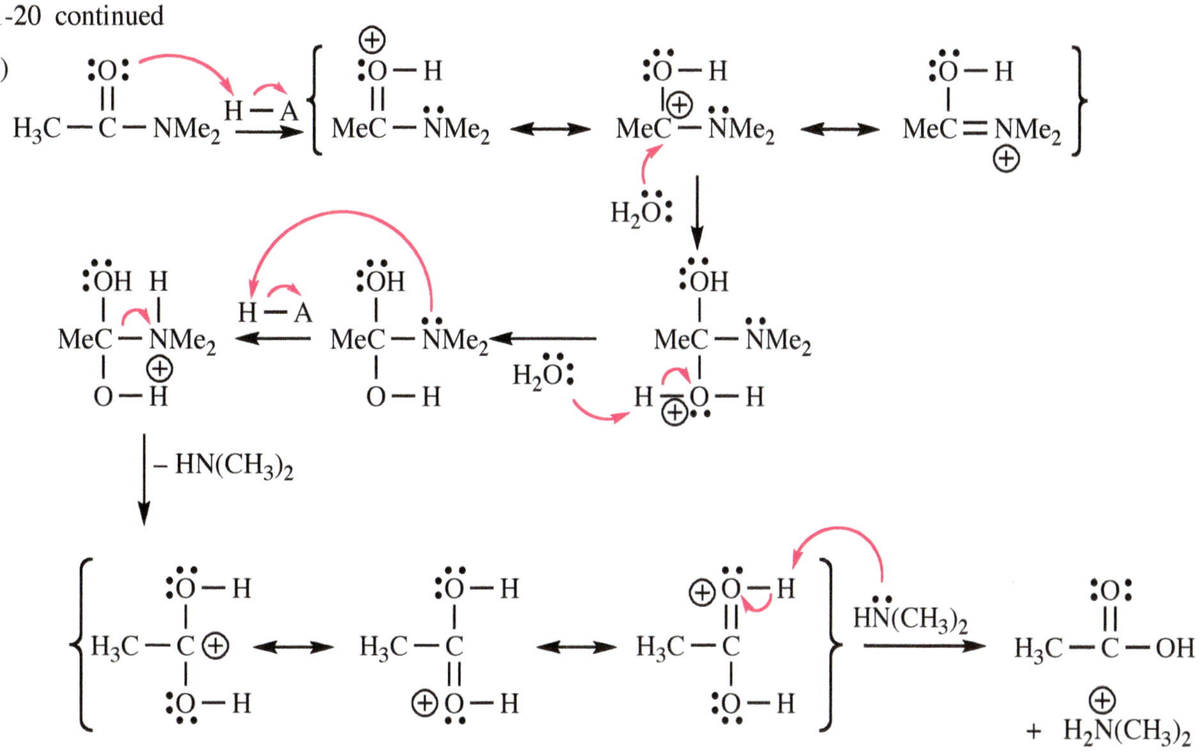

21-21 In the basic hydrolysis (21-20(a)), the step that drives the reaction to completion is the final step, the deprotonation of the carboxylic acid by the amide anion. In the acidic hydrolysis (21-20(b)), protonation of the amine by acid is exothermic and it prevents the reverse reaction by tying up the pair of electrons on the nitrogen so that the amine is no longer nucleophilic.

21-22

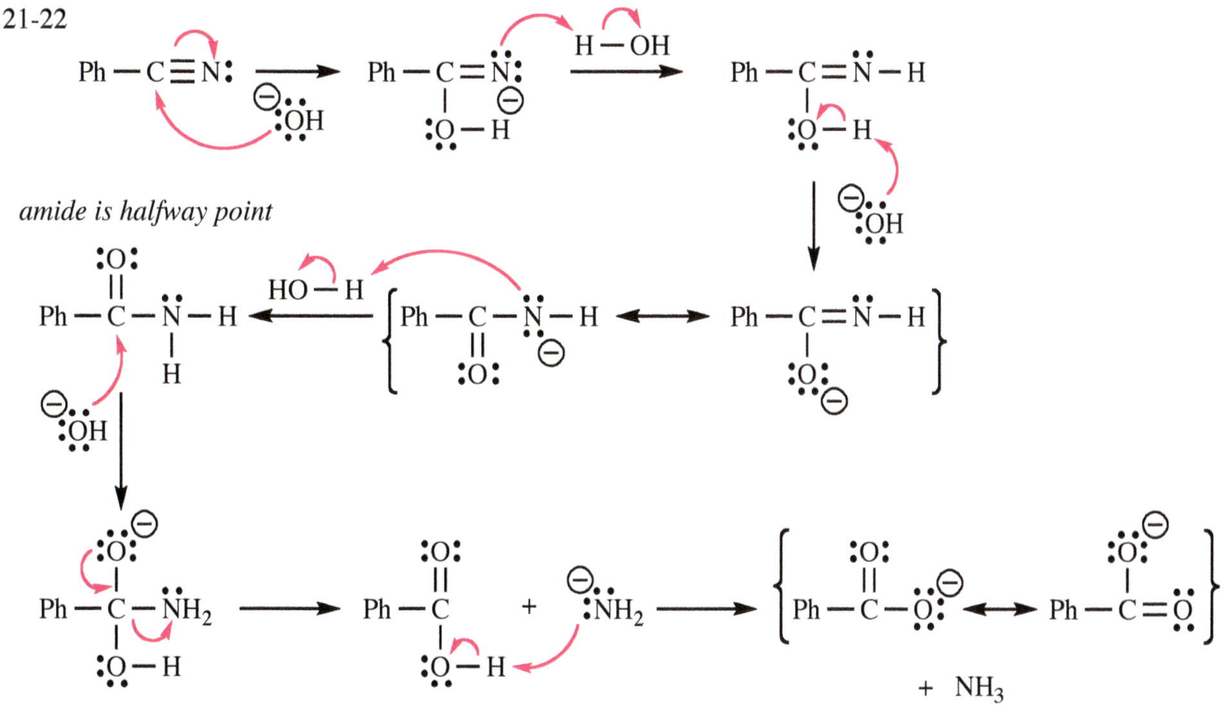

[Mechanism: basic hydrolysis of $Ph-C\equiv N$ by $^{\ominus}OH$ and H–OH to the amide $Ph-C(=O)-NH_2$, then to $Ph-COOH$ + $^{\ominus}NH_2$, giving the carboxylate resonance forms + NH_3]

amide is halfway point

21-23

Ph—C≡N: + H–A → {Ph—C≡N⁺—H ↔ Ph—C⁺=N—H} → (H_2O:) Ph—C=N—H with H—O⁺—H → (H_2O:) Ph—C=N—H with :O—H → (A–H) {Ph—C=N⁺(H)—H with :O—H ↔ Ph—C⁺—N(H)—H with :O—H ↔ Ph—C—N(H)—H with =O⁺—H} → (H_2O:) Ph—C(=O)—NH_2

Note: species with positive charge on carbon adjacent to benzene also have resonance forms (not shown) with the positive charge distributed over the ring.

21-24

(a) Reduction occurs when a new C—H bond is formed. In ester reduction, a new C—H bond is formed in the first step and in the third step. This can also be seen in this mechanism where the steps are similarly labeled.

(b)

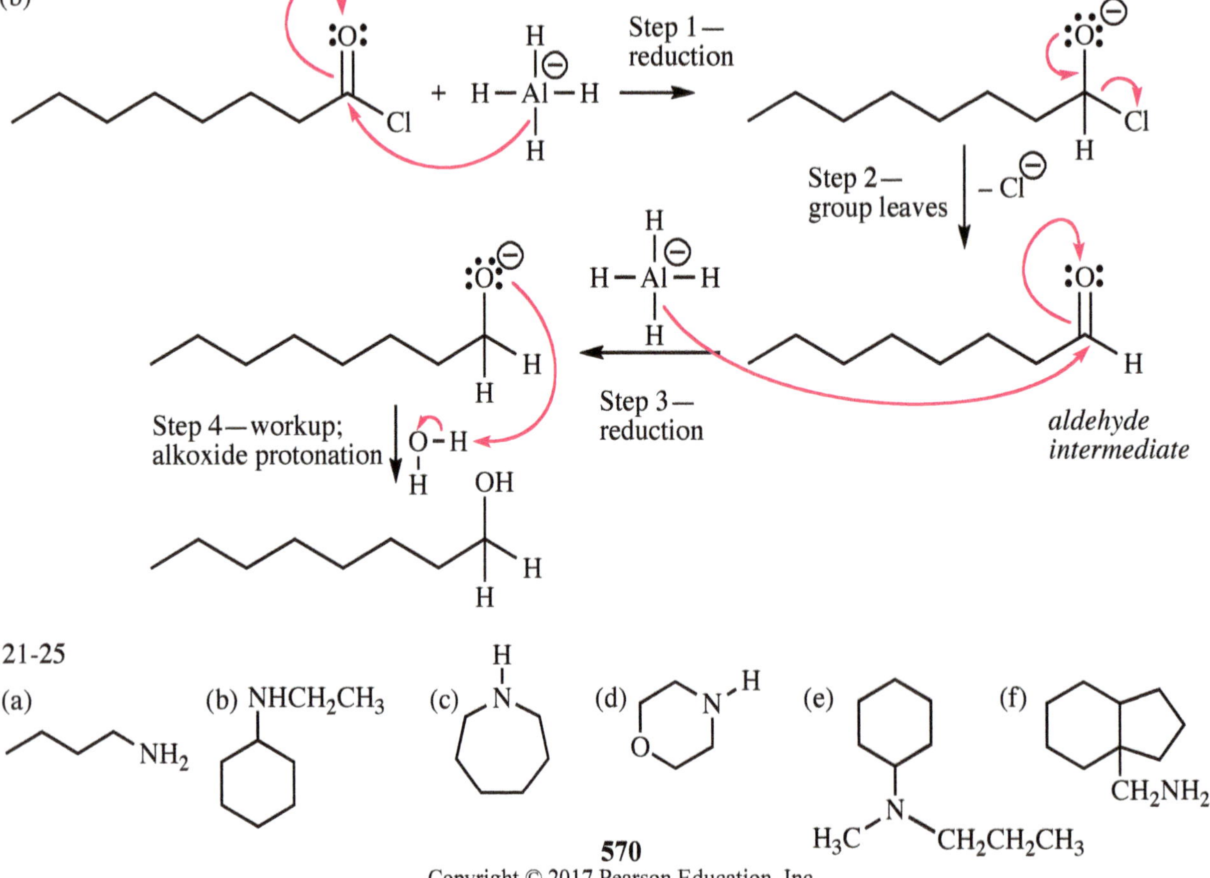

21-26

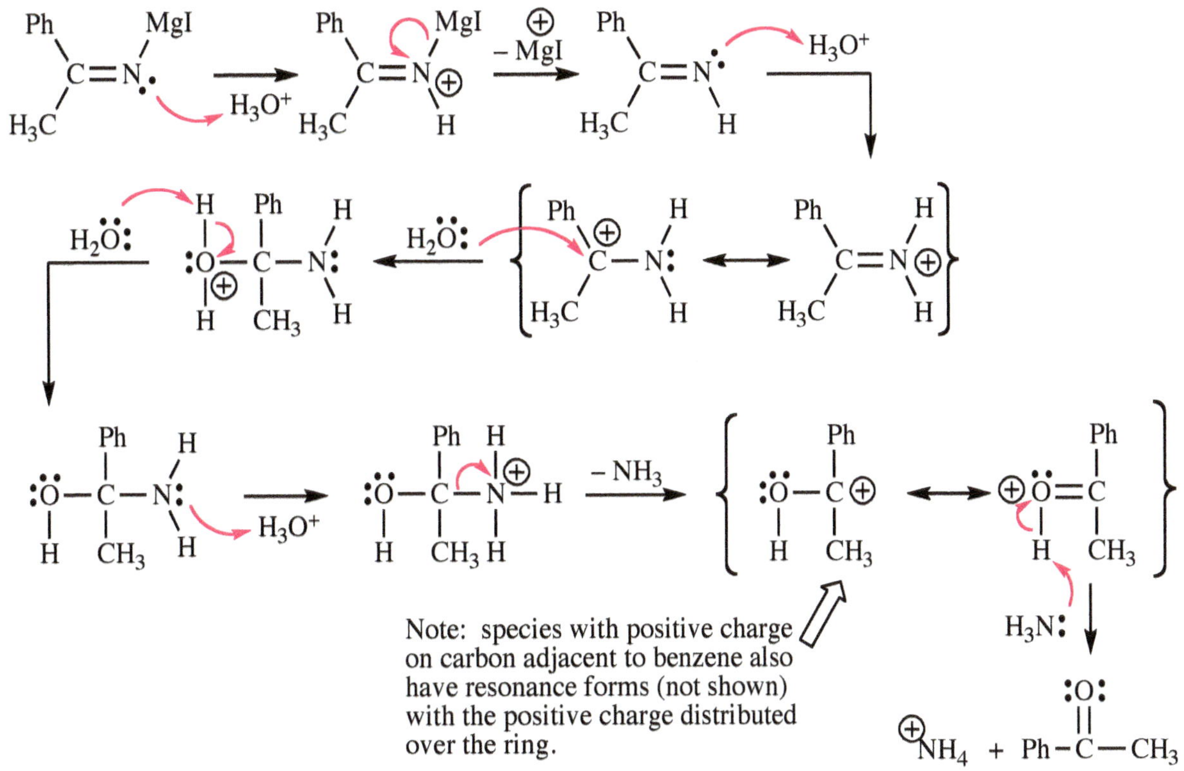

Note: species with positive charge on carbon adjacent to benzene also have resonance forms (not shown) with the positive charge distributed over the ring.

21-27

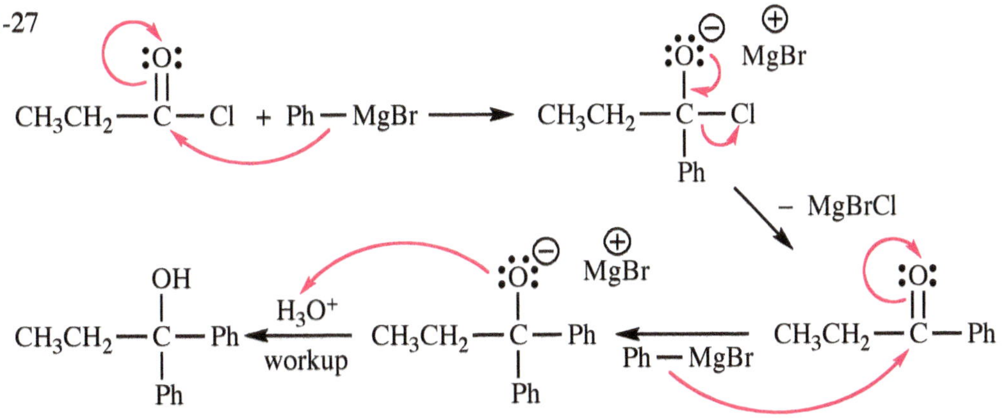

Grignard reagents plus acid chlorides or esters produce tertiary alcohols where two of the R groups are the same, having both come from the Grignard reagent.

21-28 The new carbon-carbon bonds are shown in bold.

(a) $PhCO_2R$ + 2 $CH_3CH_2CH_2MgBr$ → H_3O^+ → Ph, OH (tertiary alcohol)

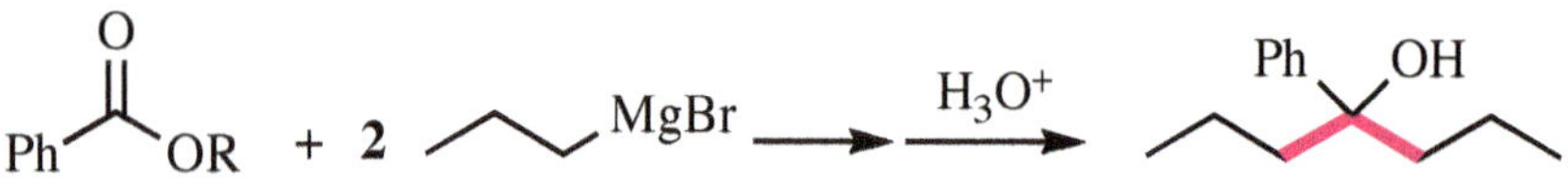

These alcohols can also be synthesized from ketones:

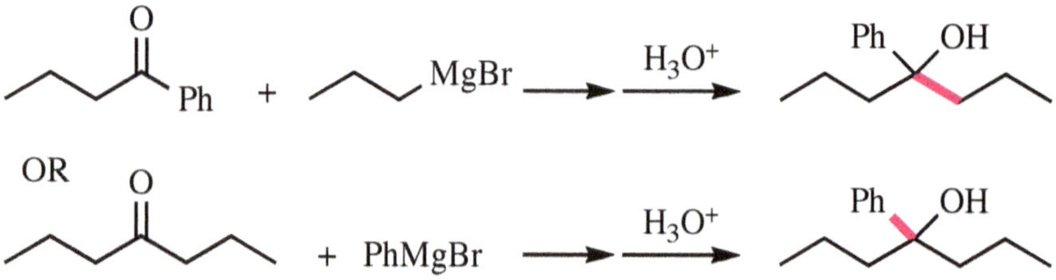

$CH_3CH_2CH_2COPh$ + $CH_3CH_2CH_2MgBr$ → H_3O^+ → Ph, OH

OR

$CH_3CH_2CH_2COCH_2CH_2CH_3$ + PhMgBr → H_3O^+ → Ph, OH

(b) HCOOR + 2 $CH_3CH_2CH_2MgBr$ → H_3O^+ → H, OH

formate ester

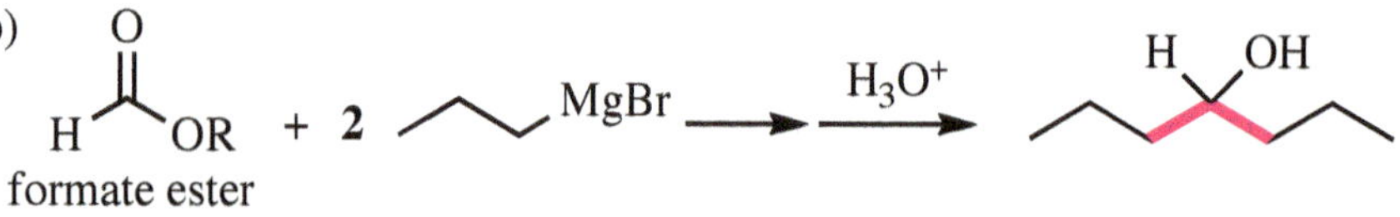

Adding Grignard to a formate ester is a very practical way to make 2° alcohols where both R groups are the same.

(c) $CH_3C{\equiv}N$ + $CH_3CH_2CH_2MgBr$ → H_3O^+ → $CH_3COCH_2CH_2CH_3$

OR

$N{\equiv}CCH_2CH_2CH_3$ + CH_3MgBr → H_3O^+ → $CH_3COCH_2CH_2CH_3$

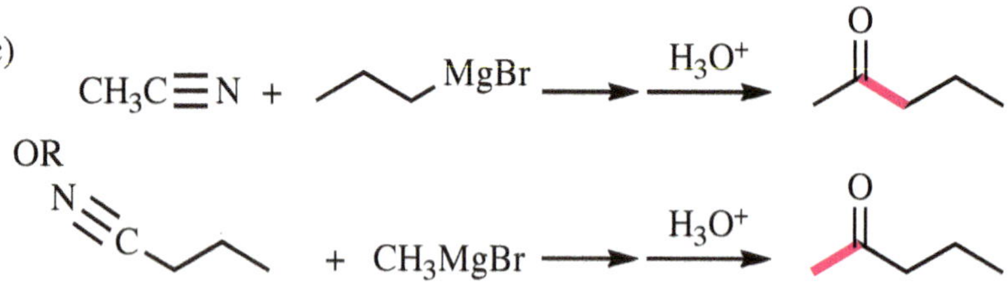

21-29

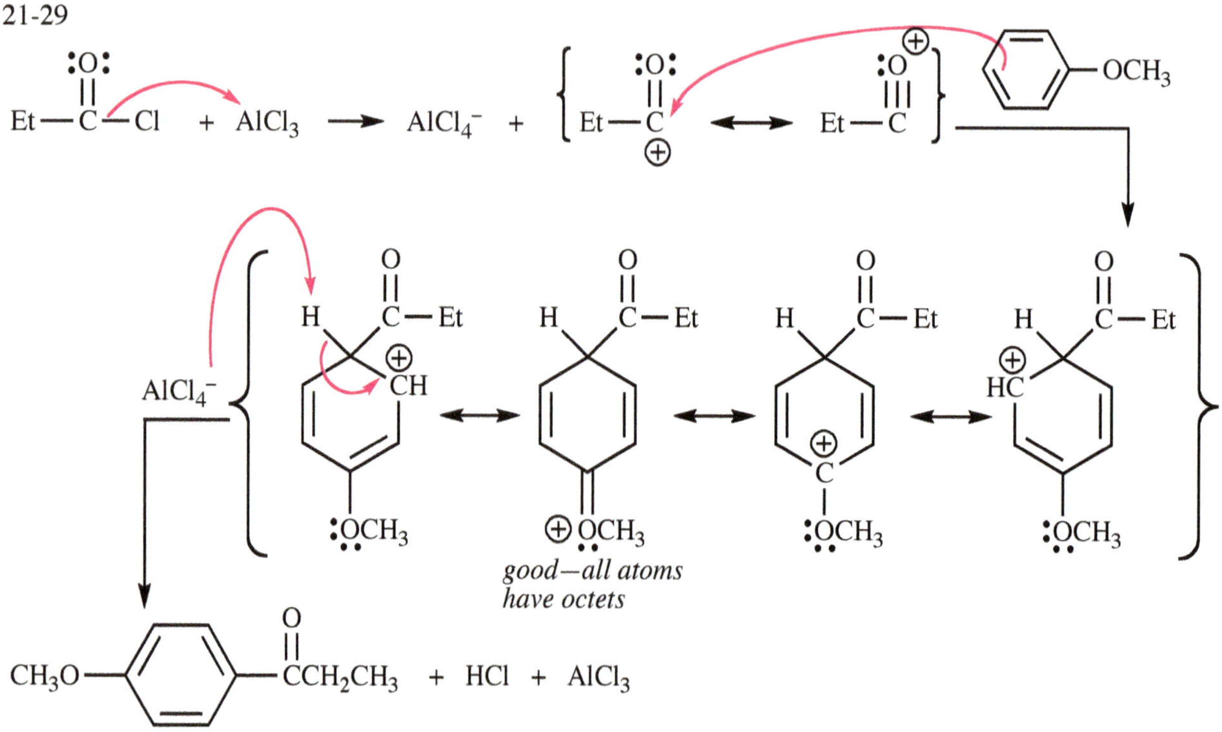

Et—C(=O)—Cl + $AlCl_3$ → $AlCl_4^-$ + {Et—C^+=O ↔ Et—C≡O^+} + $C_6H_5OCH_3$ ↓

$AlCl_4^-$ { resonance forms of sigma complex: H, C(=O)—Et, $\oplus$CH, :OCH_3 ↔ H, C(=O)—Et, $\oplus$$OCH_3$ ↔ H, C(=O)—Et, $\oplus$C, :OCH_3 ↔ H, C(=O)—Et, $\oplus$HC, :OCH_3 }

good—all atoms have octets

↓

CH_3O—C_6H_4—CCH_2CH_3(=O) + HCl + $AlCl_3$

21-30

(a) $C_6H_6 + CH_3COCl \xrightarrow{AlCl_3} C_6H_5COCH_3$

(b) $C_6H_6 + C_6H_5COCl \xrightarrow{AlCl_3} C_6H_5COC_6H_5$

(c) $C_6H_6 + CH_3CH_2CH_2COCl \xrightarrow{AlCl_3} C_6H_5COCH_2CH_2CH_3 \xrightarrow[HCl]{Zn(Hg)} C_6H_5CH_2CH_2CH_2CH_3$

21-31

(a) (i) $HC(=O){-}O{-}C(=O)CH_3 + H_2N{-}C_6H_5 \longrightarrow HC(=O){-}NH{-}C_6H_5 + HO{-}C(=O)CH_3$

(ii) $HC(=O){-}O{-}C(=O)CH_3 + HOCH_2{-}C_6H_5 \longrightarrow HC(=O){-}OCH_2{-}C_6H_5 + HO{-}C(=O)CH_3$

(b) (i)

$H{-}C(=O){-}O{-}C(=O)CH_3 + H_2NPh \longrightarrow H{-}C(O^-)(N^+H_2Ph){-}O{-}C(=O)CH_3 \longrightarrow H{-}C(=O){-}N^+H_2Ph + {}^-O{-}C(=O)CH_3$

$\downarrow$

$H{-}C(=O){-}NHPh + H{-}O{-}C(=O)CH_3$

The nucleophile attacks the less hindered C=O more rapidly.

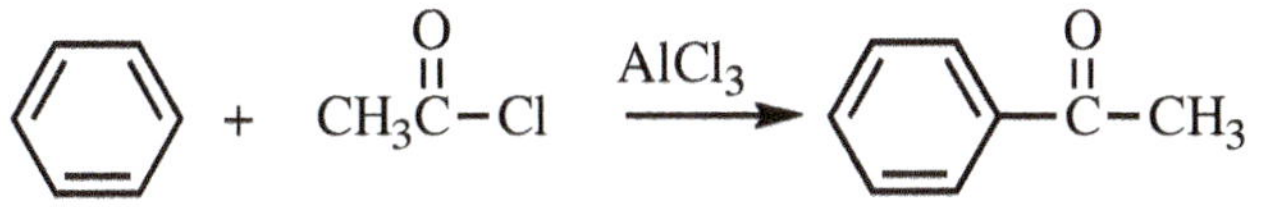

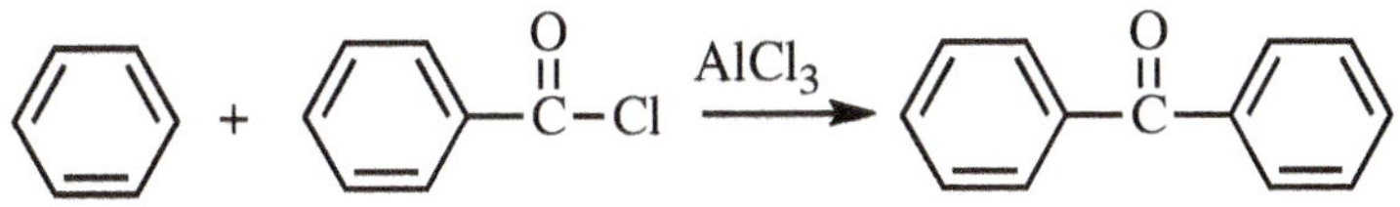

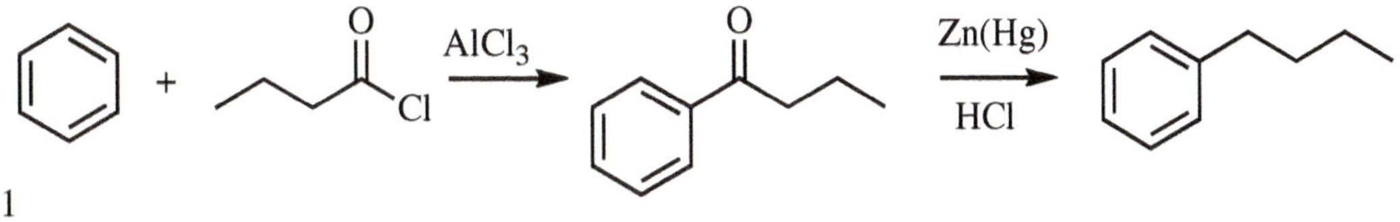

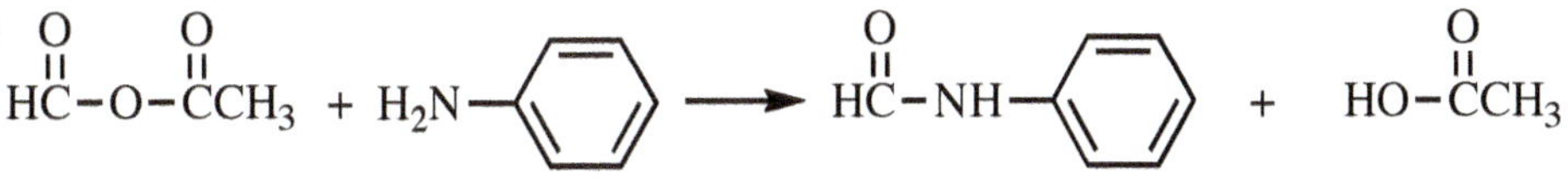

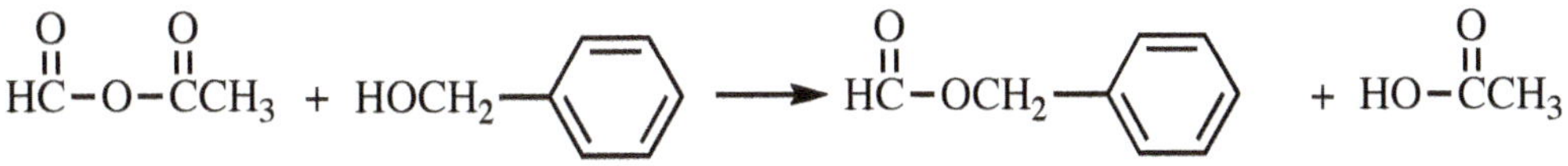

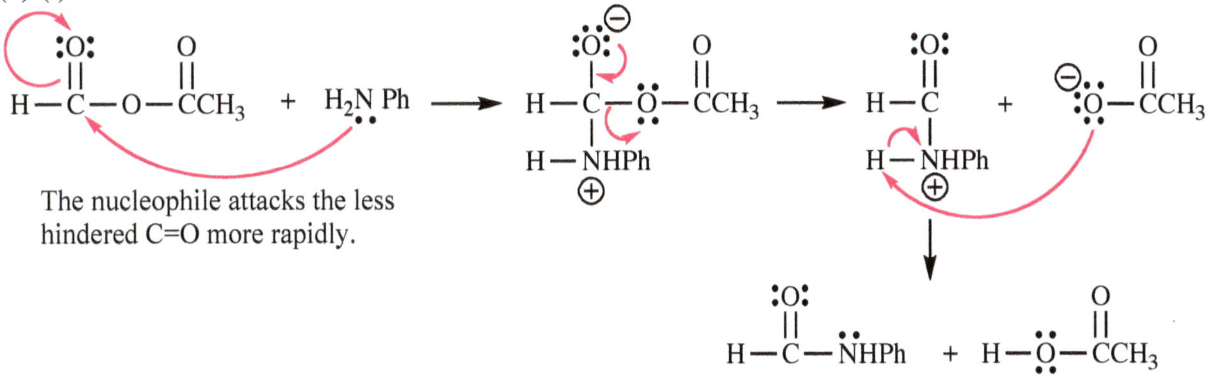

(ii)

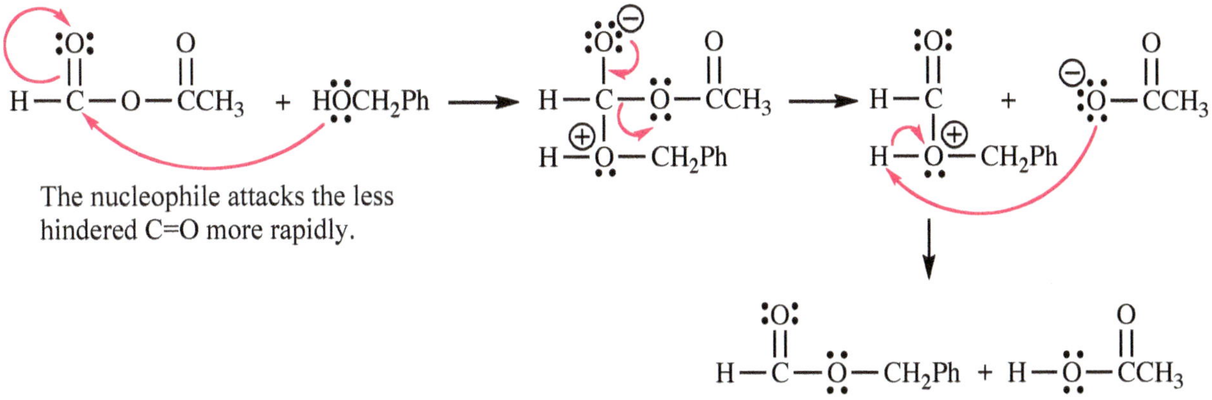

$H{-}C(=O){-}O{-}C(=O)CH_3 + HOCH_2Ph \longrightarrow H{-}C(O^-)(O^+H{-}CH_2Ph){-}O{-}C(=O)CH_3 \longrightarrow H{-}C(=O){-}O^+H{-}CH_2Ph + {}^-O{-}C(=O)CH_3$

$\downarrow$

$H{-}C(=O){-}O{-}CH_2Ph + H{-}O{-}C(=O)CH_3$

The nucleophile attacks the less hindered C=O more rapidly.

21-32

(a)

$$CH_3(CH_2)_7OH + HC(=O)-O-C(=O)CH_3 \longrightarrow CH_3(CH_2)_7-O-C(=O)H$$

$HC(=O)-Cl$ ⇐ "Formyl chloride" is imaginary; it does not exist; it is unstable relative to its decomposition products, CO and HCl. Acetic formic anhydride is the most practical way to formylate the alcohol.

Do not use formyl chloride on exams—it will be marked incorrect!

(b)

$$CH_3(CH_2)_7OH + CH_3C(=O)-O-C(=O)CH_3 \longrightarrow CH_3(CH_2)_7-O-C(=O)CH_3$$

Acetic anhydride is more convenient, easier to handle, less toxic and less expensive than acetyl chloride.

(c)

phthalic anhydride $+ NH_3 \longrightarrow$ 2-(aminocarbonyl)benzoic acid ($C(=O)NH_2$ and $C(=O)OH$)

The acid chloride would tend to react at both carbonyls instead of just one; only the anhydride will give this product. Recall that a carboxylic acid and NH_3 can make an amide only at a very high temperature, so once the COOH is formed, the reaction will stop there.

(d)

succinic anhydride $+ CH_3O^- \longrightarrow \xrightarrow{H_3O^+}$ $CH_3OC(=O)CH_2CH_2C(=O)OH$

The acid chloride would tend to react at both carbonyls instead of just one; only the anhydride will give this product. As in part (c), once the COOH is formed, it will not go onto ester with methoxide under basic conditions.

21-33

OH OH H–A OH :ÖH ÖH OH ⊕ ⊕ÖH OH H₂Ö: O H OH OH :ÖH C ÖH O ÖH H–A O H O H OH – H_2O OH O O O O C H H H H_2O: O O

21-34

(a) (2-hydroxybenzoic acid) + $CH_3\overset{O}{\overset{\|}{C}}-O-\overset{O}{\overset{\|}{C}}CH_3$ $\longrightarrow$ (2-acetoxybenzoic acid, $O-\overset{O}{\overset{\|}{C}}CH_3$, COOH) + CH_3COOH

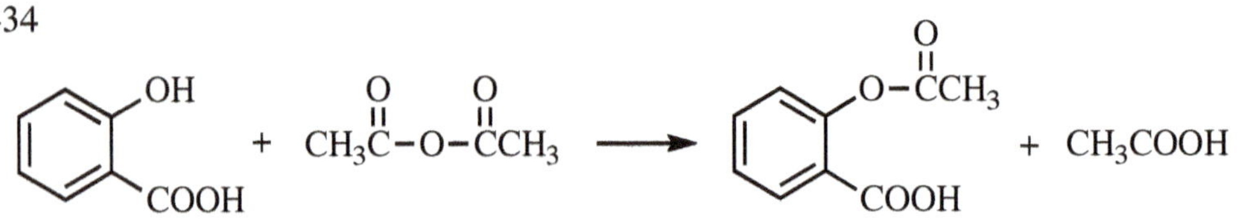

Generally, acetic anhydride is the optimum reagent for the preparation of acetate esters. Acetyl chloride would also react with the carboxylic acid to form a mixed anhydride.

(b) (2-hydroxybenzoic acid; OH, COOH) + CH_3OH $\xrightleftharpoons{H_2SO_4}$ (OH, $COOCH_3$) + H_2O

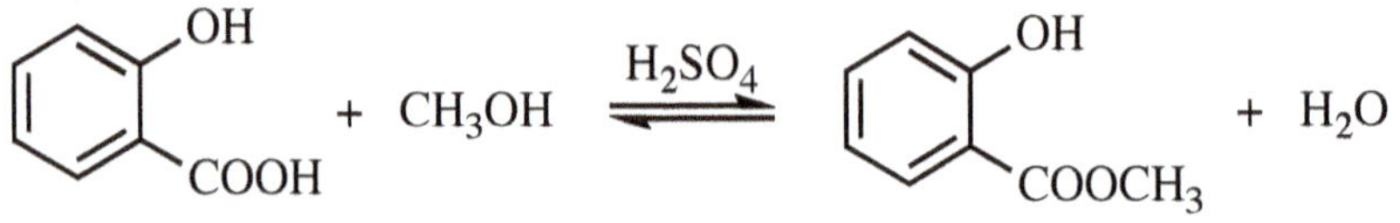

Fischer esterification works well to prepare simple carboxylic esters. The diazomethane method would also react with the phenol, making the phenyl ether.

(c) (CH_3, $COOCH_3$) $\xrightarrow{LiAlH_4}$ $\xrightarrow{H_2O}$ (CH_3, CH_2OH)

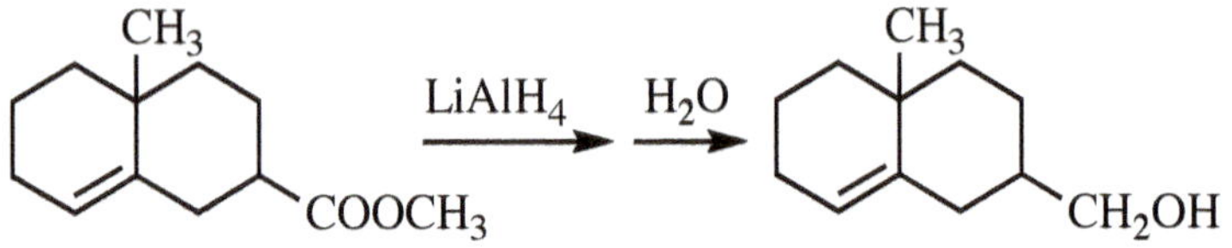

The more powerful reducing agent $LiAlH_4$ is required to reduce esters to primary alcohols.

(d) (indane ester, OEt, O) $\xrightarrow[\Delta]{H_2N-Ph}$ (amide, H N–Ph, O) + EtOH

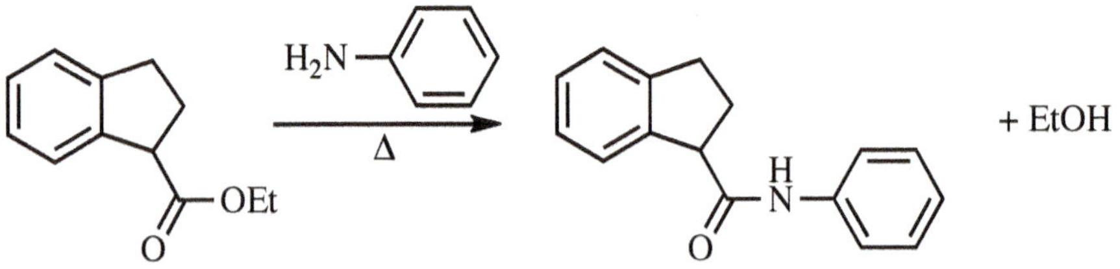

Amides can be made directly from esters plus an amine. Aniline is a poor nucleophile, however, so heating this reaction will be required to complete the reaction in a reasonable time.

21-35 Syntheses may have more than one correct approach.

(a) $Ph-\overset{O}{\overset{\|}{C}}-OCH_3 + \mathbf{2}\ PhMgBr \xrightarrow{ether} \xrightarrow{H_3O^+} Ph-\underset{Ph}{\overset{OH}{C}}-Ph$

(b) $H-\overset{O}{\overset{\|}{C}}-OCH_2CH_3 + \mathbf{2}\ PhCH_2MgBr \xrightarrow{ether} \xrightarrow{H_3O^+} H-\underset{CH_2Ph}{\overset{OH}{C}}-CH_2Ph$

(c) $Ph-\overset{O}{\overset{\|}{C}}-OCH_3 + H_2NCH_2CH_3 \xrightarrow{\Delta} Ph-\overset{O}{\overset{\|}{C}}-NHCH_2CH_3$

(d) $H-\overset{O}{\overset{\|}{C}}-OCH_2CH_3 + \mathbf{2}\ PhMgBr \xrightarrow{ether} \xrightarrow{H_3O^+} H-\underset{Ph}{\overset{OH}{C}}-Ph$

21-35 continued

(e) $Ph-C(=O)-OCH_3 + LiAlH_4 \xrightarrow{ether} \xrightarrow{H_3O^+} PhCH_2OH$

(f) $Ph-C(=O)-OCH_3 \xrightarrow{H_3O^+} Ph-C(=O)-OH + CH_3OH$

(g) $PhCH_2OH$ from (e) $\xrightarrow[2)\ KCN]{1)\ TsCl,\ pyridine} PhCH_2C{\equiv}N \xrightarrow[\Delta]{H_3O^+} PhCH_2-C(=O)-OH \xrightarrow[HO-CH(CH_3)_2]{H_2SO_4} PhCH_2-C(=O)-O-CH(CH_3)_2$

(h) $PhCH_2-C(=O)-OCH(CH_3)_2$ from (g) $+ \mathbf{2}\ CH_3CH_2MgBr \xrightarrow{ether} \xrightarrow{H_3O^+} PhCH_2-C(OH)(CH_2CH_3)-CH_2CH_3$

(i) How to make an 8-carbon diol from an ester that has no more than 8 carbons? Make the ester a lactone!

lactone (ring atoms 1–8) $+ LiAlH_4 \xrightarrow{ether} \xrightarrow{H_3O^+}$ 1,8-octanediol (HO at C1 and C8)

21-36

(a) N-ethylbenzamide $\xrightarrow{LiAlH_4} \xrightarrow{H_2O}$ N-ethylbenzylamine

(b) ethyl benzoate $+ H_2N$-ethyl $\xrightarrow{\Delta}$ N-ethylbenzamide $+$ HO-ethyl

Heating the reaction will form the amide faster.

(c) pyrrolidine (NH) $+$ Cl-C(=O)-CH₃ $\longrightarrow$ N-acetylpyrrolidine $+$ HCl

Acetic anhydride would work as well as acetyl chloride; a base such as pyridine or sodium bicarbonate or triethylamine is usually added to consume the HCl.

(d) gamma-amino-butyric acid (NH_2 at γ; β, α; HO–C=O) $\xrightarrow[-\ H_2O]{300\ °C}$ 2-pyrrolidinone $\xrightarrow{LiAlH_4} \xrightarrow{H_2O}$ pyrrolidine

This high-temperature synthesis of amides is used industrially but not in normal lab reactions. Alternatively, an ester could be produced from the COOH, followed by milder cyclization to the amide.

21-37 There may be other correct approaches to these problems. Consult with your study group.

(a)

$$C_6H_5COOH \xrightarrow{SOCl_2} C_6H_5COCl \xrightarrow{HN(CH_3)_2} C_6H_5CON(CH_3)_2 \xrightarrow[2)\ H_2O]{1)\ LiAlH_4} C_6H_5CH_2N(CH_3)_2$$

(b)

$$\text{pyrrolidine} + CH_3COCl \longrightarrow \text{N-acetylpyrrolidine} \xrightarrow{LiAlH_4} \xrightarrow{H_2O} \text{N-ethylpyrrolidine}$$

See the comment in the solution to 21-36(c) about using acetic anhydride instead of acetyl chloride.

(c)

$$C_5H_9COOH \xrightarrow{SOCl_2} C_5H_9COCl \xrightarrow{NH_3} C_5H_9CONH_2 \xrightarrow{POCl_3} C_5H_9C{\equiv}N$$

21-38

(a)

$$PhCH_2{-}C(=O){-}OH \xrightarrow{SOCl_2} PhCH_2{-}C(=O){-}Cl \xrightarrow{NH_3} PhCH_2{-}C(=O){-}NH_2 \xrightarrow{POCl_3} PhCH_2{-}C{\equiv}N$$

(b)

$$PhCH_2{-}C(=O){-}OH \xrightarrow[2)\ H_2O]{1)\ LiAlH_4} PhCH_2CH_2OH \xrightarrow[\text{pyridine}]{TsCl} PhCH_2CH_2OTs \xrightarrow{NaCN} PhCH_2CH_2CN$$

(c)

$$p\text{-}ClC_6H_4NO_2 \xrightarrow[HCl]{Fe} p\text{-}ClC_6H_4NH_2 \xrightarrow[HCl]{NaNO_2} \xrightarrow{CuCN} p\text{-}ClC_6H_4CN$$

21-39

(a)

$$CH_3(CH_2)_5OH \xrightarrow{PBr_3} CH_3(CH_2)_5Br \xrightarrow{NaCN} CH_3(CH_2)_5CN \xrightarrow[2)\ H_2O]{1)\ LiAlH_4} CH_3(CH_2)_5CH_2NH_2$$

(b)

$$C_6H_{11}C(=O)NH_2 \xrightarrow{POCl_3} C_6H_{11}C{\equiv}N \xrightarrow[2)\ H_3O^+]{1)\ CH_3CH_2MgBr} C_6H_{11}C(=O)CH_2CH_3$$

21-39 continued

(c) PBr$_3$ converts the alcohol (…OH) to the bromide (…Br); or make the tosylate. NaCN gives the nitrile (…CN). 1) CH_3MgI 2) H_3O^+ gives the methyl ketone (…C(=O)CH$_3$).

21-40

$CH_3-N=C=O$ + Ar—OH → { $CH_3-N=C-O^-$ with $H-O^+-Ar$ ↔ $CH_3-N^--C=O$ with $H-O^+-Ar$ } → Ar—OH_2^+ / Ar—OH, two rapid proton transfers → Sevin® (carbaryl)

Ar = (1-naphthyl)

$H_3C-NH-C(=O)-O-$(1-naphthyl)

Sevin®
(carbaryl)

21-41

(a) cyclic carbonate of cyclohexane-1,2-diol $\xrightarrow{H_3O^+}$ cyclohexane-1,2-diol (OH, OH) + CO_2

(i) carbonate ester
(iii) not aromatic

(b) thiolactone $\xrightarrow{H_3O^+}$ HS—CH$_2$CH=CH—C(=O)OH (HO, O)

(i) thiolactone
(iii) not aromatic

(c) cyclic S,S-thiocarbonate $\xrightarrow{H_3O^+}$ SH SH (propane-1,3-dithiol) + CO_2

(i) thiocarbonate ester
(iii) not aromatic

(d) { HN—C(=O)—NH ring ↔ HN—C(—O^-)=NH^+ ring } $\xrightarrow{H_3O^+}$ CO_2 + H_2N—CH=CH—NH_2 → H_2N—CH$_2$—CH=NH

enediamine imine

(i) a substituted urea
(iii) AROMATIC—more easily seen in the resonance form shown

(The enediamine product would not be stable in aqueous acid. It would probably tautomerize to an imine, hydrolyze to ammonia and 2-aminoethanal, then polymerize.)

21-41 continued

(e) At first glance, this AROMATIC compound does not appear to be an acid derivative. Like any enol, however, its tautomer must be considered.

OH ⇌ lactam $\xrightarrow{H_3O^+}$ COOH NH_2 enamine ⟶ COOH imine $\xrightarrow{H_3O^+}$ H COOH + NH_3

(f) $\xrightarrow{H_3O^+}$ CO_2 + HO NH_2 ⟶ polymer

see part (d)

(i) a carbamate or urethane
(iii) AROMATIC—more easily seen in the resonance form

21-42

(a)
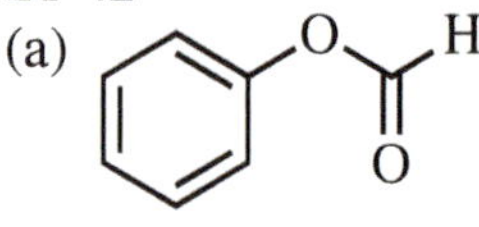

(b)
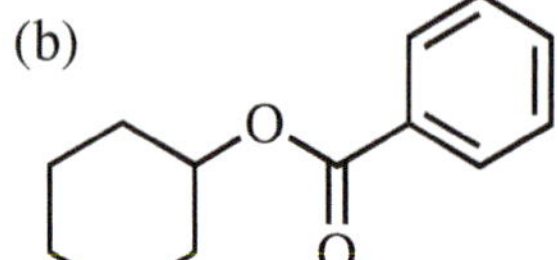

(c)

(d)

(e)

(f)
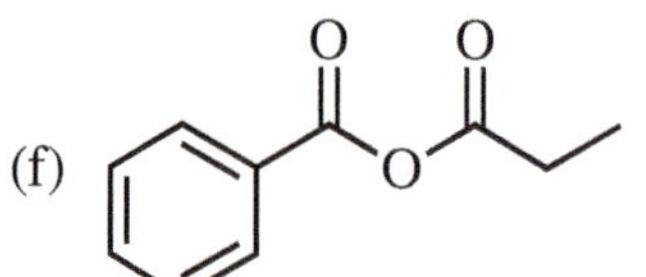

(g)

(h) "Valero" has 5 carbons.

(i)

(j)
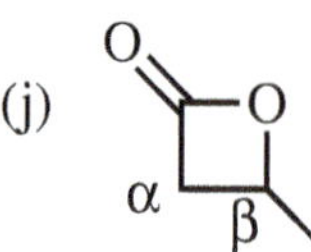
Oxygen is on the beta carbon of the 4-carbon chain.

(k)

(l)
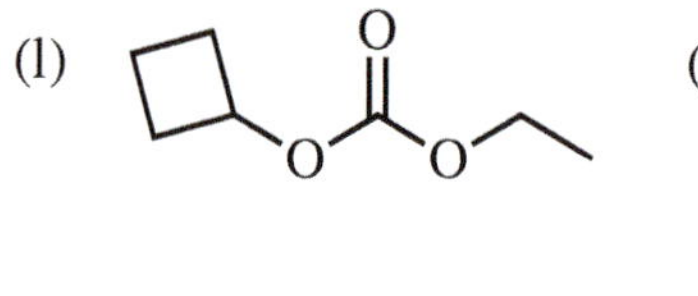

(m)
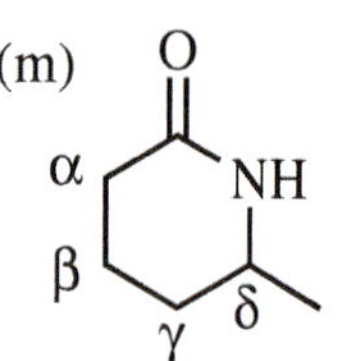
"Capro" has 6 carbons.

(n)
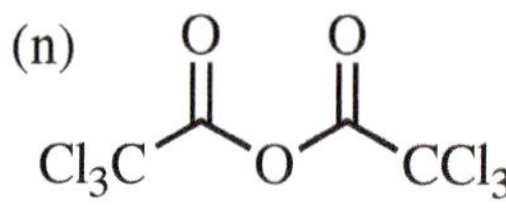

(o)

21-43

(a) 3-methylpentanoyl chloride
(b) benzoic formic anhydride — "Methanoic" is IUPAC, but most chemists use "formic".
(c) acetanilide; *N*-phenylethanamide
(d) *N*-methylbenzamide
(e) phenyl acetate; phenyl ethanoate
(f) methyl benzoate
(g) benzonitrile
(h) 4-phenylbutane nitrile; γ-phenylbutyronitrile
(i) dimethyl isophthalate, or dimethyl benzene-1,3-dicarboxylate
(j) *N,N*-diethyl-3-methylbenzamide
(k) 4-hydroxypentanoic acid lactone; γ-valerolactone
(l) 3-aminopentanoic acid lactam; β-valerolactam

21-44

(a) $PhC(=O)-OCH_2CH_3$ (b) $PhC(=O)-O-C(=O)CH_3$ (c) $PhC(=O)-N(H)-C_6H_5$ (d) $H_3CO-C_6H_4-C(=O)Ph$

(e) $Ph-C(OH)(Ph)-Ph$ (f) $PhC(=O)-H$

21-45

(a) $C_6H_5-O-C(=O)-CH_3$ (b) $C_6H_5-O-C(=O)-H$ (c) $C_6H_4(C(=O)NH-C_6H_5)(C(=O)OH)$ Anhydrides react only once.

(d) $H_3CO-C_6H_4-C(=O)CH_2CH_2C(=O)OH$ (e) $C_6H_5-CH(OH)CH_2-NH-C(=O)CH_3$ Amines are more nucleophilic than alcohols and amides are more stable than esters.

(f) $C_6H_5-CH(O-C(=O)CH_3)CH_2-NH-C(=O)CH_3$

21-46 When a carboxylic acid is treated with a basic reagent, the base removes the acidic proton rather than attacking at the carbonyl (proton transfers are much faster than formation or cleavage of other types of bonds). Once the carboxylate anion is formed, the carbonyl is no longer susceptible to nucleophilic attack: nucleophiles do not attack sites of negative charge. By contrast, in acidic conditions, the protonated carbonyl has a positive charge and is activated to nucleophilic attack.

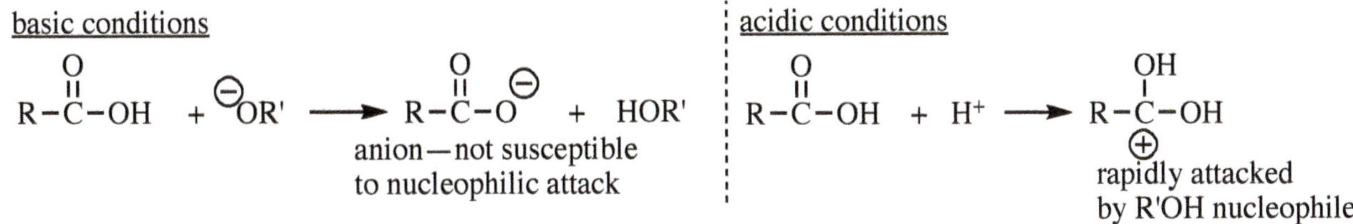

21-47 Products after adding dilute acid in the workup:

(a) $HC(=O)-OH$ + $HO-Ph$ (b) $CH_3CH_2C(=O)OH$ + $HOCH_2CH_3$

(c) $C_6H_4(OH)CH_2CH_2COOH$ (d) $HOCH_2CH_2OH$ + $HOOC-COOH$

21-48

(a) $C_6H_5-NH_2$ + $H-C(=O)-O-C(=O)-CH_3$ → $C_6H_5-NH-C(=O)-H$ + $HO-C(=O)-CH_3$

See the solution to 21-32(a), and text Sec. 21-11, for use of acetic formic anhydride.

21-48 continued

(b) Benzoic acid $\xrightarrow{SOCl_2}$ benzoyl chloride $\xrightarrow{Na^+\ ^-OOCCH_3}$ $C_6H_5C(=O)OC(=O)CH_3$

(c) cis-1,2-cyclohexanediol + $ClC(=O)C(=O)Cl$ $\longrightarrow$ cyclic oxalate ester

(d) Phthalic acid $\xrightarrow[\Delta,\ -H_2O]{HA}$ phthalic anhydride $\xrightarrow[HA]{HOCH(CH_3)_2}$ monoester (C(=O)OCH(CH$_3$)$_2$, C(=O)OH)

(e) 5-hydroxypentanal $\xrightarrow[NH_3\ (aq)]{Ag^+}$ 5-hydroxypentanoate (O^-, OH) $\xrightarrow[\Delta,\ -H_2O]{HA}$ δ-valerolactone

(f) 5-oxopentanoic acid (H, COOH) $\xrightarrow[CH_3OH]{NaBH_4}$ 5-hydroxypentanoic acid (OH, COOH) $\xrightarrow[\Delta,\ -H_2O]{HA}$ δ-valerolactone

(g) 4-oxocyclohexane-1,1-diyl diester (MeOOC, COOMe) $\xrightarrow[HA,\ -H_2O]{HO\quad OH}$ ketal (MeOOC, COOMe) $\xrightarrow[2)\ H_3O^+]{1)\ LiAlH_4}$ ketone (HOH_2C, CH_2OH) $\xrightarrow[pyridine]{Ac_2O}$ ketone ($AcOH_2C$, CH_2OAc)

Any ester where ethylene glycol displaces methanol (by transesterification) will be reduced with $LiAlH_4$.

Aqueous acid workup removes acetal (ketal) protecting group.

(h) cyclopentene $\xrightarrow{mCPBA}$ epoxide $\xrightarrow[H_2O]{KCN}$ trans-2-hydroxycyclopentanecarbonitrile (OH, CN) $\xrightarrow[H_2O,\ \Delta]{KOH}$ trans-2-hydroxycyclopentanecarboxylic acid (OH, COOH) $\xrightarrow[HOAc]{NaOCl}$ 2-oxocyclopentanecarboxylic acid (O, COOH)

Prevent elimination by using base.

OR

cyclopentene $\xrightarrow[H_2O]{Br_2}$ trans-2-bromocyclopentanol (OH, Br) $\xrightarrow[HOAc]{NaOCl}$ 2-bromocyclopentanone (O, Br) $\xrightarrow[HA,\ -H_2O]{HO\quad OH}$ ketal (Br) $\xrightarrow[2.\ CO_2]{1.\ Mg}$ $\xrightarrow{H_3O^+}$ 2-oxocyclopentanecarboxylic acid

21-49

(a)

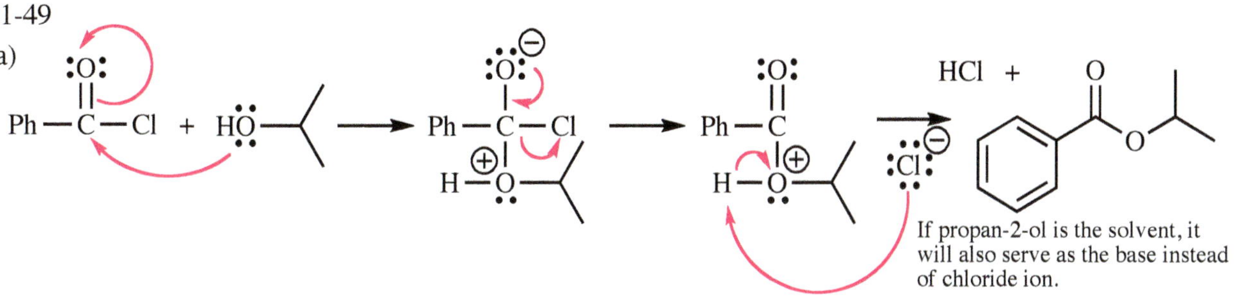

Ph—C(=O)—Cl + HO—CH(CH₃)₂ ⟶ Ph—C(O⁻)(Cl)—O⁺(H)—CH(CH₃)₂ ⟶ Ph—C(=O)—O⁺(H)—CH(CH₃)₂ + Cl⁻ ⟶ HCl + isopropyl benzoate

If propan-2-ol is the solvent, it will also serve as the base instead of chloride ion.

(b)

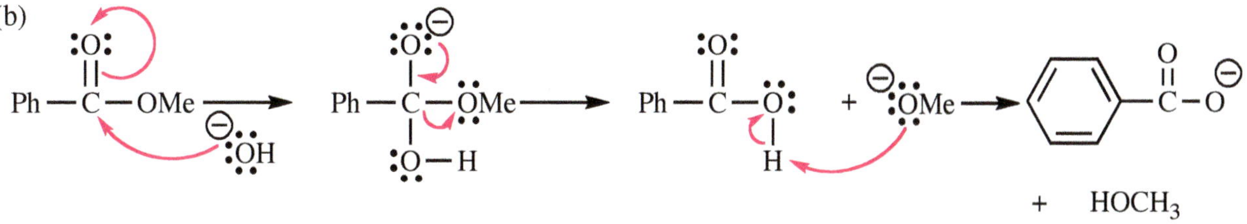

Ph—C(=O)—OMe + ⁻OH ⟶ Ph—C(O⁻)(OMe)—O—H ⟶ Ph—C(=O)—O—H + ⁻OMe ⟶ PhC(=O)—O⁻ + $HOCH_3$

(c)

Ph—C(=O)—OEt + H—A ⟶ {PhC(=O⁺—H)—OEt ⟷ PhC⁺(—O—H)—OEt ⟷ PhC(—O—H)=O⁺Et}

H_2O: ⟶ PhC(OH)(OEt)—O⁺H₂ ⟶ PhC(OH)(OEt)—O—H + H—A ⟶ PhC(OH)(O—H)—O⁺(H)—Et

– EtOH

{Ph—C⁺(O—H)(O—H) ⟷ Ph—C(O—H)=O⁺—H ⟷ Ph—C(=O⁺—H)(O—H)} + H_2O: ⟶ PhC(=O)—OH

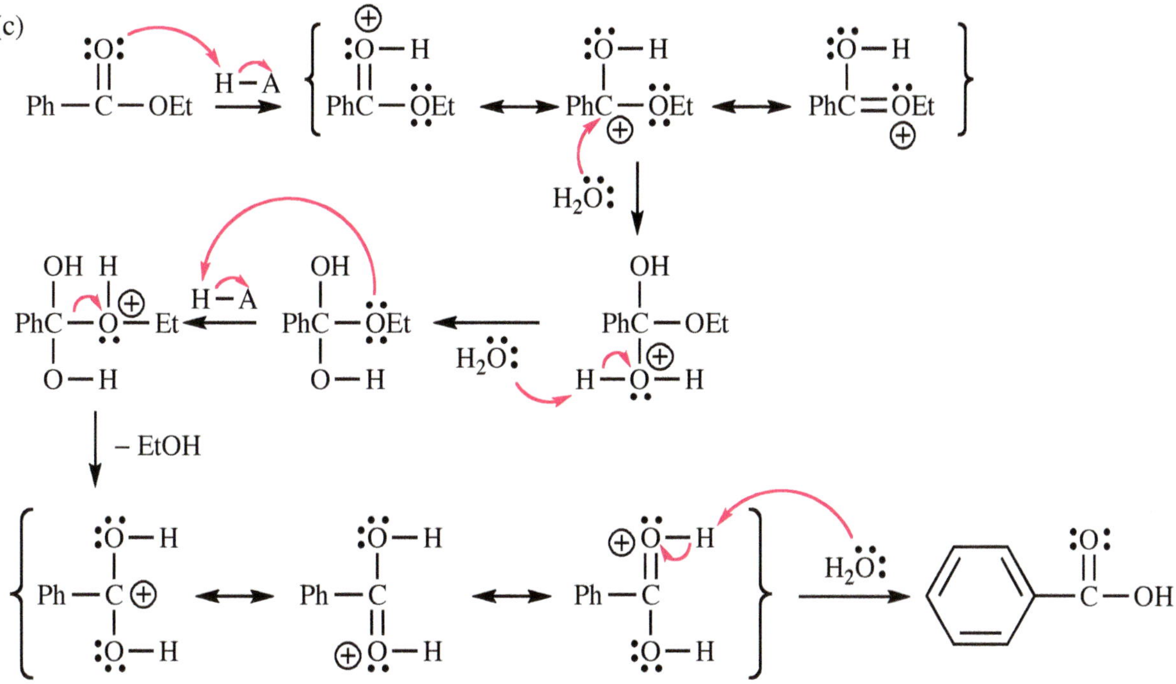

Note: species with positive charge on carbon adjacent to benzene also have resonance forms (not shown) with the positive charge distributed over the ring.

(d)

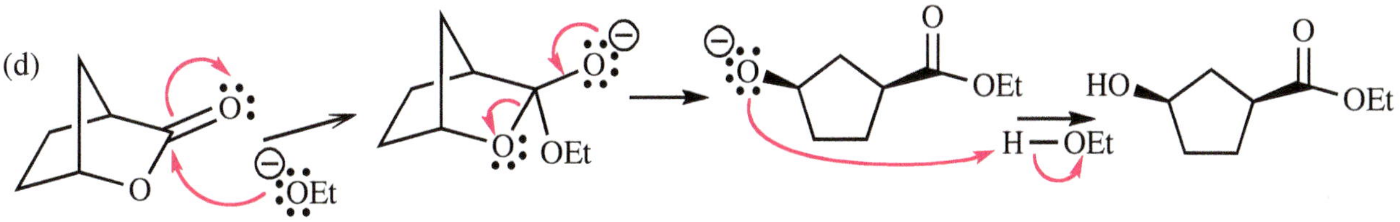

bicyclic lactone + ⁻OEt ⟶ tetrahedral intermediate (O⁻, OEt) ⟶ ⁻O—cyclopentyl—C(=O)OEt + H—OEt ⟶ HO—cyclopentyl—C(=O)OEt

21-49 continued

(e)

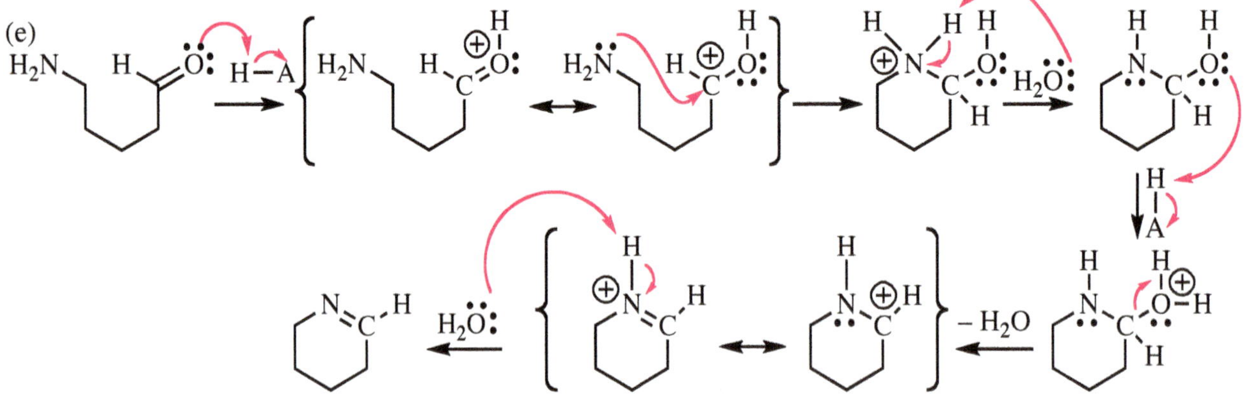

(f)

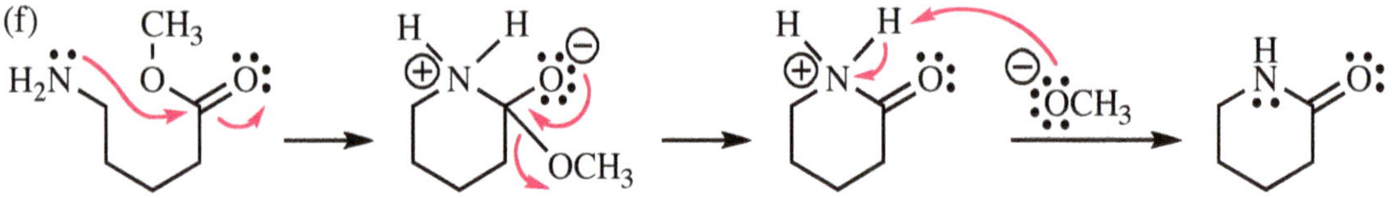

(g)

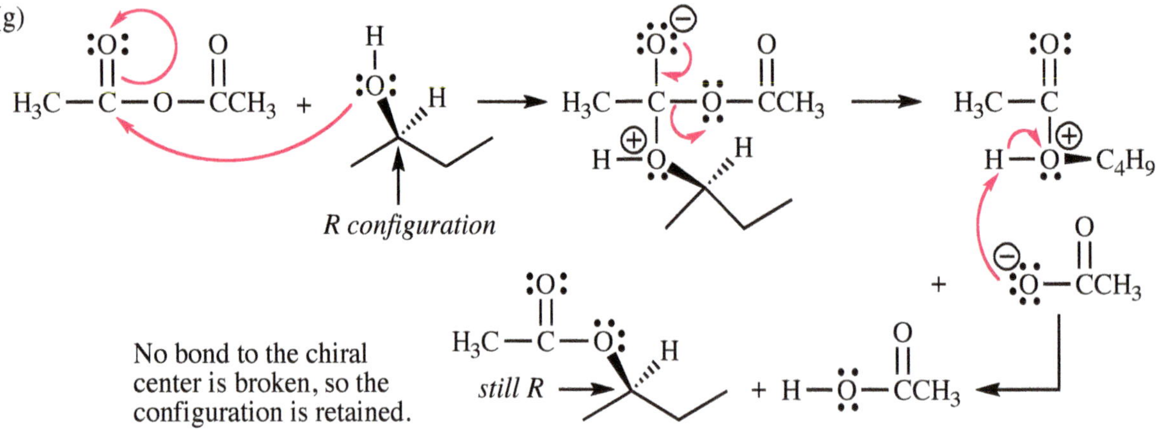

R configuration

No bond to the chiral center is broken, so the configuration is retained.

still R

21-50

(a) Ph–C(=O)–O–cyclohexyl

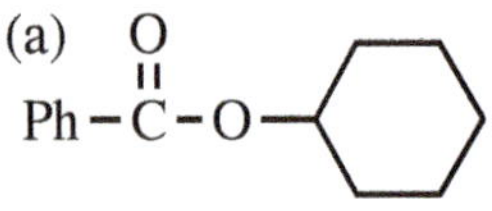

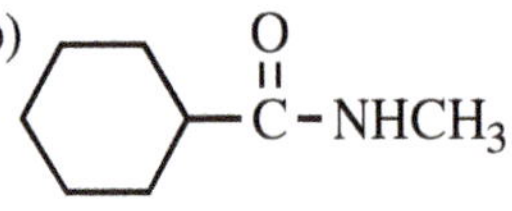

(b) cyclohexyl–C(=O)–NHCH$_3$

(c) Ph–C(=O)–N(pyrrolidine)

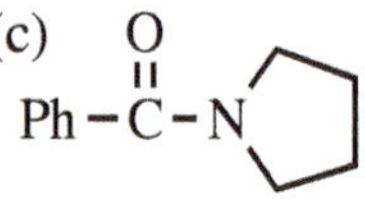

(d)

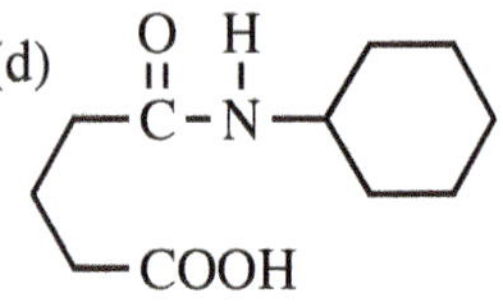

(e) PhCH$_2$OH

(f)

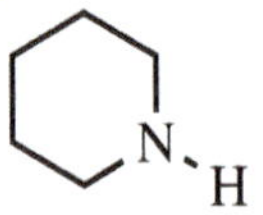

(g)

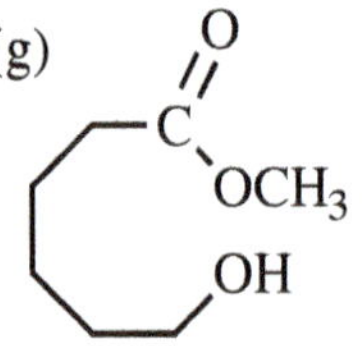

(h)

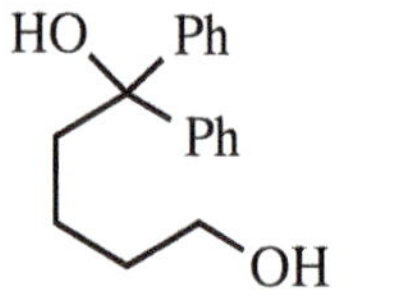

(i)

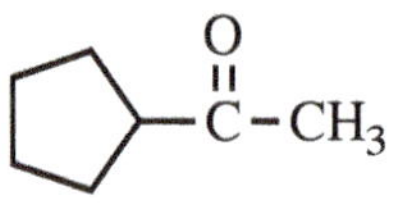

(j) Na$^+$

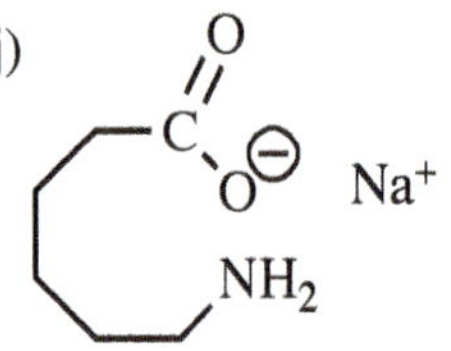

(k) PhCH$_2$CH$_2$CH$_2$C(CH$_3$)... CD$_2$NH$_2$

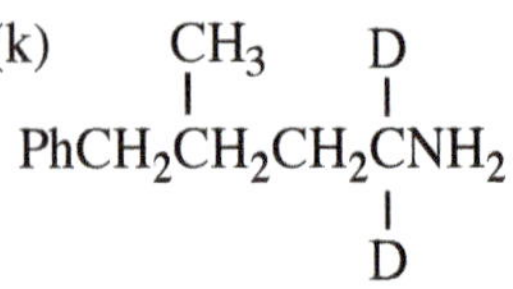

LiAlD$_4$ reduces the C=O; H_2O replaces the H on N.

(l) transesterification

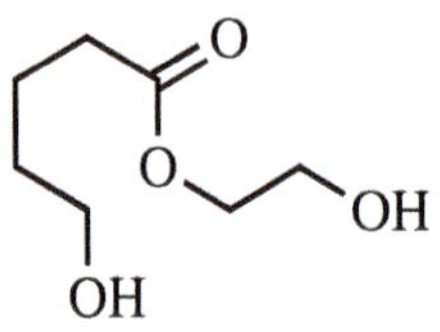

(m)

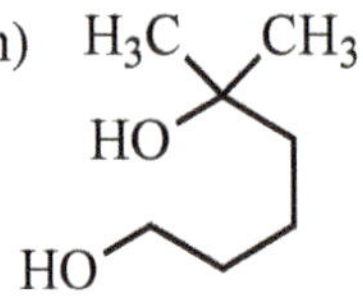

(n)

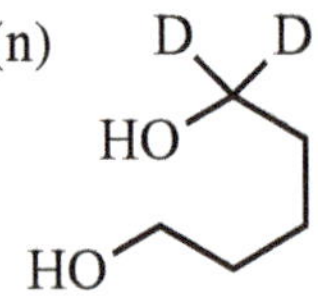

21-51

(a) 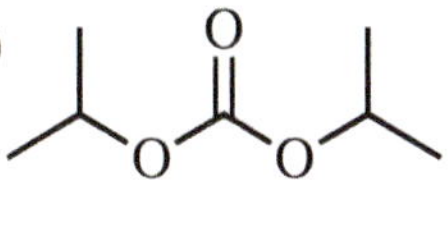 diisopropyl carbonate (b) 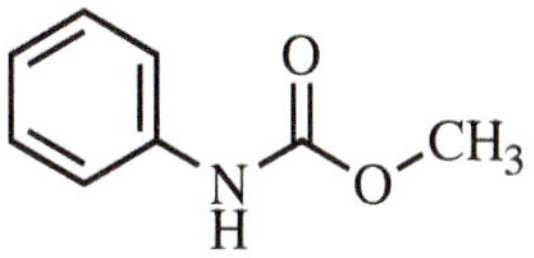 methyl N-phenylcarbamate (c) 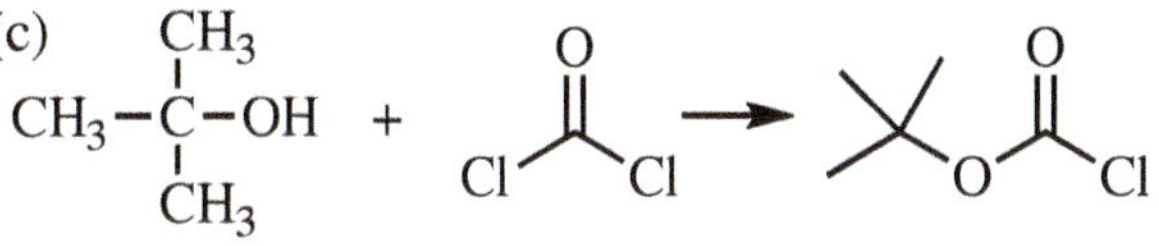 $CH_3-C(CH_3)_2-OH$ + $ClC(=O)Cl$ → tert-butyl chloroformate

(a) O, O, O

(b) O, N, H, O, CH_3

(c) CH_3, CH_3-C-OH, CH_3 + O, Cl, Cl → O, O, Cl

21-52

(a) 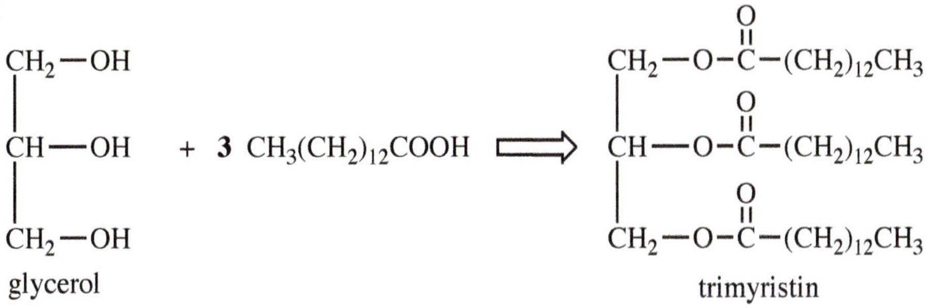 CH_2-OH / $CH-OH$ / CH_2-OH (glycerol) + **3** $CH_3(CH_2)_{12}COOH$ ⟹ $CH_2-O-C(=O)-(CH_2)_{12}CH_3$ / $CH-O-C(=O)-(CH_2)_{12}CH_3$ / $CH_2-O-C(=O)-(CH_2)_{12}CH_3$

glycerol trimyristin

(b) 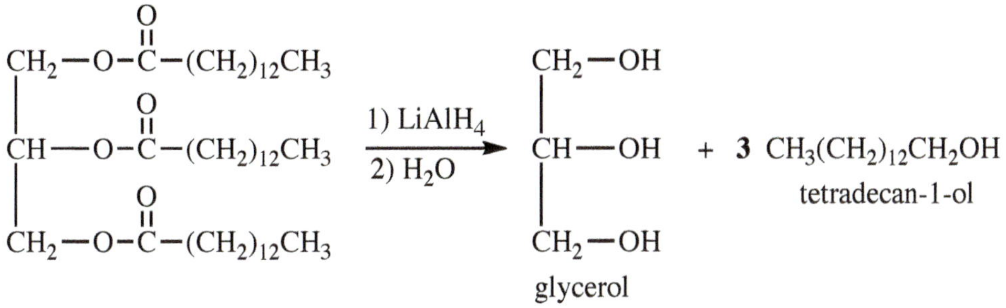 $CH_2-O-C(=O)-(CH_2)_{12}CH_3$ / $CH-O-C(=O)-(CH_2)_{12}CH_3$ / $CH_2-O-C(=O)-(CH_2)_{12}CH_3$ —1) $LiAlH_4$, 2) H_2O→ CH_2-OH / $CH-OH$ / CH_2-OH + **3** $CH_3(CH_2)_{12}CH_2OH$

tetradecan-1-ol

glycerol

21-53

(a) 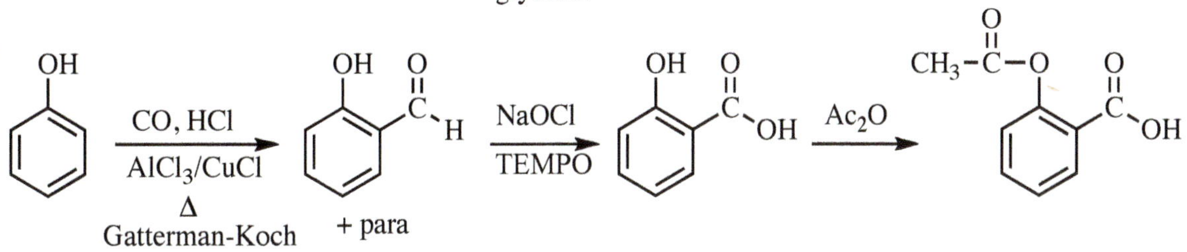 phenol (OH) —CO, HCl; $AlCl_3/CuCl$; Δ; Gatterman-Koch→ salicylaldehyde (OH, O, C, H) + para —NaOCl, TEMPO→ salicylic acid (OH, O, C, OH) —Ac_2O→ $CH_3-C(=O)-O$, O, C, OH (acetylsalicylic acid)

(b) 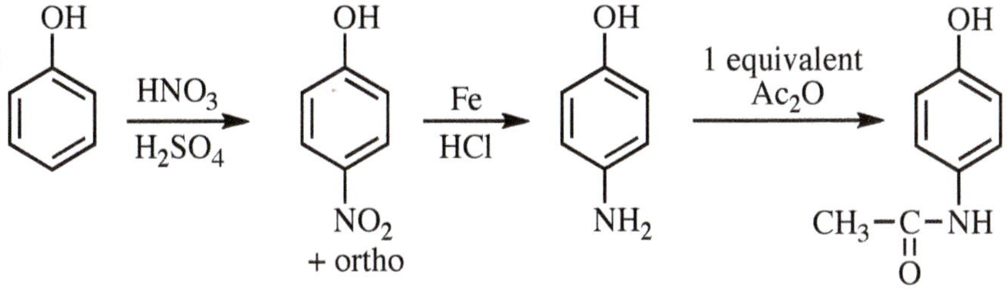 phenol (OH) —HNO_3, H_2SO_4→ OH, NO_2 + ortho —Fe, HCl→ OH, NH_2 —1 equivalent Ac_2O→ OH, $CH_3-C(=O)-NH$

NH_2 is more nucleophilic than OH, and an amide is more stable than an ester.

21-54

(a) 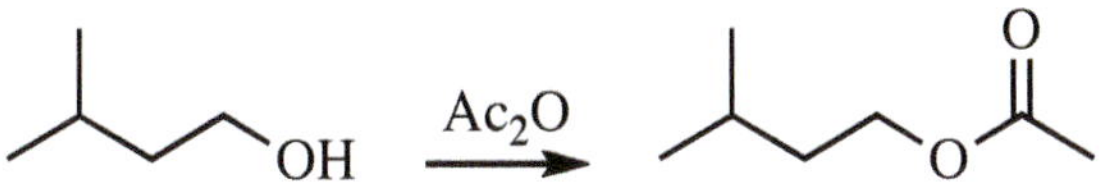 3-methylbutan-1-ol (OH) —Ac_2O→ O, O (3-methylbutyl acetate)

(b) 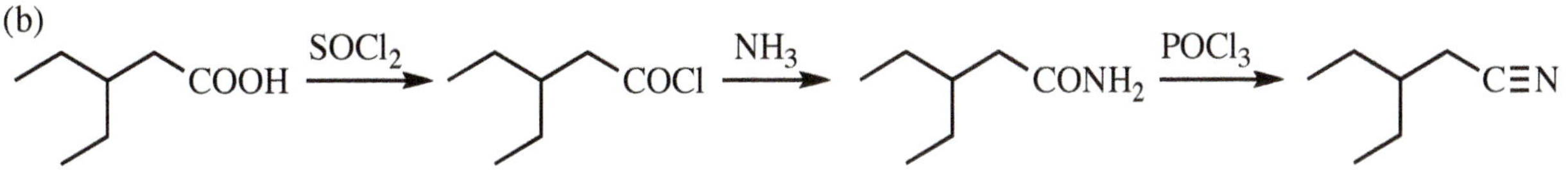 COOH —$SOCl_2$→ COCl —NH_3→ $CONH_2$ —$POCl_3$→ C≡N

21-54 continued

(c) (isobutylamine) NH_2 + $HC(=O)-O-C(=O)CH_3$ ⟶ (isobutyl)$-N(H)-C(=O)H$

(d) $CH_3C(=O)OCH_2CH_3$ + **2** CH_3CH_2MgBr ⟶ $\xrightarrow{H_3O^+}$ (3-methyl-3-pentanol, OH)

(e) (cyclohexyl)$-NH_2$ + Ac_2O ⟶ (cyclohexyl)$-N(H)-C(=O)-CH_3$

(f) (cyclohexyl)$-Br$ $\xrightarrow[\text{ether}]{\text{Mg}}$ (cyclohexyl)$-MgBr$ + $H-C(=O)-OEt$ ⟶ $\xrightarrow{H_3O^+}$ (cyclohexyl)$-CH(OH)-$(cyclohexyl)

two equivalents

(g) $CH_3O-C(=O)-C(=O)-OCH_3$ + $H_2N-CH_2CH_2-NH_2$ ⟶ (piperazine-2,3-dione, N–H)

(h) (cyclopentyl)$-CH_2OH$ $\xrightarrow[\text{TEMPO}]{\text{excess NaOCl}}$ (cyclopentyl)$-COOH$ $\xrightarrow{SOCl_2}$ (cyclopentyl)$-COCl$ $\xrightarrow{NH_3}$ (cyclopentyl)$-CONH_2$ $\xrightarrow{POCl_3}$ (cyclopentyl)$-C{\equiv}N$

21-55 Diethyl carbonate has *two* leaving groups on the carbonyl. It can undergo *two* nucleophilic acyl substitutions, followed by one nucleophilic addition.

(a) $EtO-C(=O)-OEt$ $\xrightarrow[\text{subst. 1}]{\text{PhMgBr}}$ $[Ph-C(=O)-OEt]$ $\xrightarrow[\text{subst. 2}]{\text{PhMgBr}}$ $[Ph-C(=O)-Ph]$ $\xrightarrow[\text{addition}]{\text{PhMgBr}}$ $\xrightarrow{H_3O^+}$ $Ph-C(OH)(Ph)-Ph$

(b) CH_3CH_2Br $\xrightarrow[\text{ether}]{\text{Mg}}$ CH_3CH_2MgBr

3 CH_3CH_2MgBr + $EtO-C(=O)-OEt$ ⟶ $\xrightarrow{H_3O^+}$ $CH_3CH_2-C(OH)(CH_2CH_3)-CH_2CH_3$

21-55 continued

(c) Lexan® is a polycarbonate that makes bisphenol A (BPA) into its carbonate ester. Phosgene will react faster than diethyl carbonate because acid chlorides react faster than esters, but the chemistry is the same. An acid catalyst would speed this transesterification.

BPA + EtO–C(=O)–OEt $\xrightarrow{HA}$ Lexan® polycarbonate

Sevin® insecticide can be made from diethyl carbonate by sequential addition of methylamine and 1-naphthol. The sequence is important: make the amide first because the N is a poor leaving group.

EtO–C(=O)–OEt + H_2NCH_3 $\xrightarrow{\Delta}$ H_3C–NH–C(=O)–OEt + 1-naphthol $\xrightarrow{H^+}$ Sevin®

21-56 Triethylamine is nucleophilic, but it has no H on nitrogen to lose, so it forms a salt instead of a stable amide.

$CH_3C(=O)$–Cl + :NEt_3 ⟶ $CH_3C(O^-)(Cl)$–$\overset{+}{N}Et_3$ $\xrightarrow{-\,Cl^-}$ $CH_3C(=O)$–$\overset{+}{N}$(Et)(Et)–Et

When ethanol is added, it attacks the carbonyl of the salt, with triethylamine as the leaving group.

$CH_3C(=O)$–$\overset{+}{N}Et_3$ + HÖ–CH_2CH_3 ⟶ $CH_3C(O^-)(\overset{+}{N}Et_3)$–$\overset{+}{O}H$–$CH_2CH_3$ ⟶ NEt_3 + $CH_3C(=O)$–$\overset{+}{O}H$–CH_2CH_3

↓

CH_3–C(=O)–O–CH_2CH_3 + $Et_3\overset{+}{N}H$ Cl^-

21-57

(a) H_2C=CH–$(CH_2)_4$–OH $\xrightarrow[\text{2) } Me_2S]{\text{1) } O_3,\ -78\ °C}$ O=CH–$(CH_2)_4$–OH

↓ Ag^+, NH_3 (aq)

^-O–C(=O)–$(CH_2)_4$–OH $\xrightarrow[\Delta]{HA}$ ε-caprolactone (seven-membered lactone)

21-57 continued

(b) CH_3O–C$_6H_4$ →(Br_2) CH_3O–C$_6H_4$–Br →(Mg, ether) →(CO_2) →(H_3O^+) CH_3O–C$_6H_4$–COOH →($SOCl_2$) CH_3O–C$_6H_4$–COCl →(NH_3) CH_3O–C$_6H_4$–$CONH_2$

(c) C$_6H_5$–CH_2Br →(KCN) C$_6H_5$–CH_2CN →(1) $LiAlH_4$; 2) H_2O) C$_6H_5$–$CH_2CH_2NH_2$

(d) 3,4,5-trihydroxybenzoic acid (COOH; HO, OH, OH) →(1) NaOH excess; 2) CH_3I excess) 3,4,5-trimethoxybenzoic acid (COOH; CH_3O, OCH_3, OCH_3) →(1) $LiAlH_4$; 2) H_2O) CH_2OH (CH_3O, OCH_3, OCH_3) →(TsCl, pyridine) CH_2OTs (CH_3O, OCH_3, OCH_3) →(KCN) CH_2CN (CH_3O, OCH_3, OCH_3) →(1) $LiAlH_4$; 2) H_2O) $CH_2CH_2NH_2$ (CH_3O, OCH_3, OCH_3)

Any methyl ester formed will be reduced with LAH.

21-58 The rate of a reaction depends on its activation energy, that is, the difference in energy between starting material and the transition state. The transition state in saponification is similar in structure, and therefore in energy, to the tetrahedral intermediate:

C$_6H_5$–C(O^-)(OH)–OCH_3

tetrahedral intermediate

The tetrahedral carbon has no resonance overlap with the benzene ring, so any resonance effect of a substituent on the ring will have very little influence on the energy of the transition state.

What will have a big influence on the activation energy is whether a substituent stabilizes or destabilizes the starting material. Anything that stabilizes the starting material will therefore increase the activation energy, slowing the reaction; anything that destabilizes the starting material will decrease the activation energy, speeding the reaction.

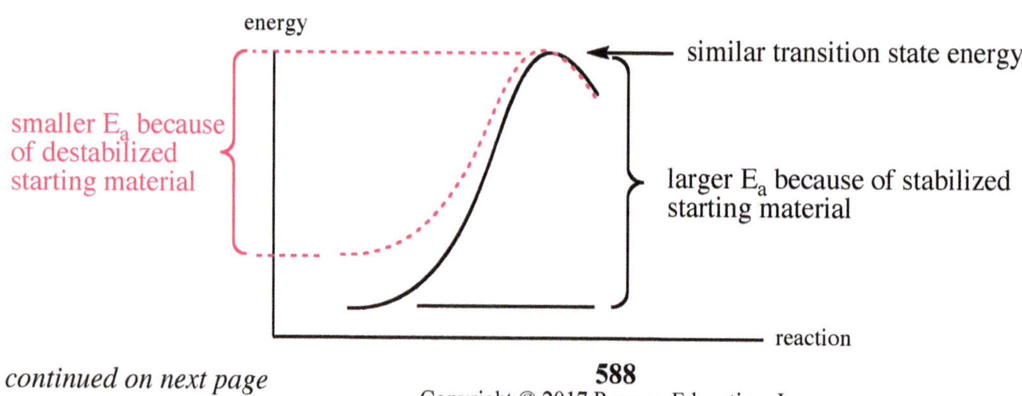

continued on next page

21-58 continued

(a) One of the resonance forms of methyl *p*-nitrobenzoate has a positive charge on the benzene carbon adjacent to the positive carbonyl carbon. This resonance form destabilizes the starting material, decreasing the activation energy, speeding the reaction.

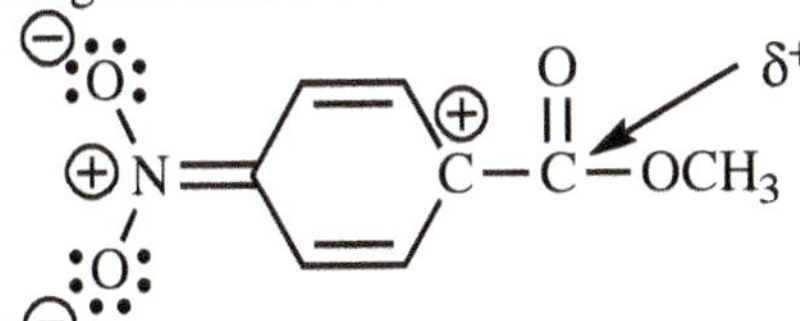

poor resonance contributor, destabilizing the starting material; no effect in the transition state

(b) One of the resonance forms of methyl *p*-methoxybenzoate has all atoms with full octets, and negative charge on the most electronegative atom. This resonance form stabilizes the starting material, increasing the activation energy, slowing the reaction.

good resonance contributor, stabilizing the starting material; no effect in the transition state

21-59 Text Figure 21-9 tells most of the story: order of reactivity: acid chlorides > anhydrides > esters > amides. Add steric and electronic factors to get the orders below.

(a) D < B < C < A

D *slowest—conjugated amide*; **B** *aliphatic amide*; **C** *ester*; **A** *fastest—anhydride*

(b) E < F < G < H

E *slowest—hindered amide*; **F** *aliphatic amide*; **G** *ester*; **H** *fastest—anhydride*

21-60

To compare two functional groups like ester and ketone, consider two options:
A) if the ester is more reactive than the ketone, then the one equivalent of Grignard reagent will react with all of the ester before it reacted with any of the ketone, and the product should be a full equivalent of ketone. (This would also be an effective method to prepare ketones--if it worked this way.)

OR B) if the ketone is more reactive than the ester, then after the first molecule of Grignard reacts with ester to produce a molecule of ketone, the second molecule of Grignard will react with the ketone before it reacts with another ester. By the time half of the ester is consumed, all of the Grignard will have been consumed in producing one-half equivalent of tertiary alcohol.

The experimental data show that Option B must be correct. Ketones are more reactive than esters.

21-61 COOH + NH_3 → O^- NH_4^+ —($-H_2O$, Δ)→ $C(=O)NH_2$ **B** —(Δ, $-H_2O$)→ C≡N **A**

21-62

(a) $CH_3{-}N{=}C{=}O + H_2O \longrightarrow H_3C{-}NH{-}C(=O){-}OH \longrightarrow CH_3NH_2\,(g) + CO_2\,(g)$

methyl isocyanate; a carbamic acid—unstable

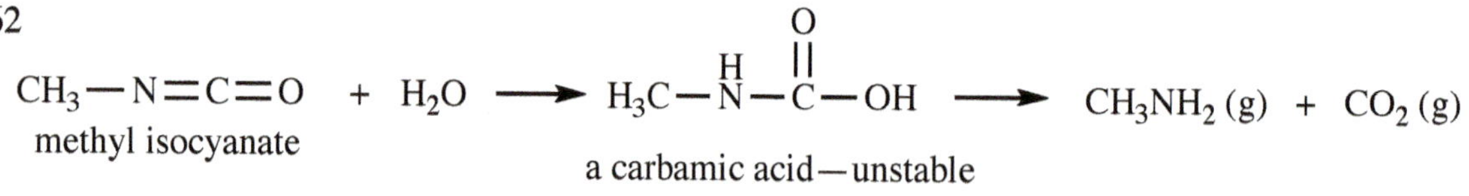

Both of these reactions are exothermic. In a closed vessel like an industrial reactor, the production of gaseous products causes a large pressure increase, risking an explosion.

(b)

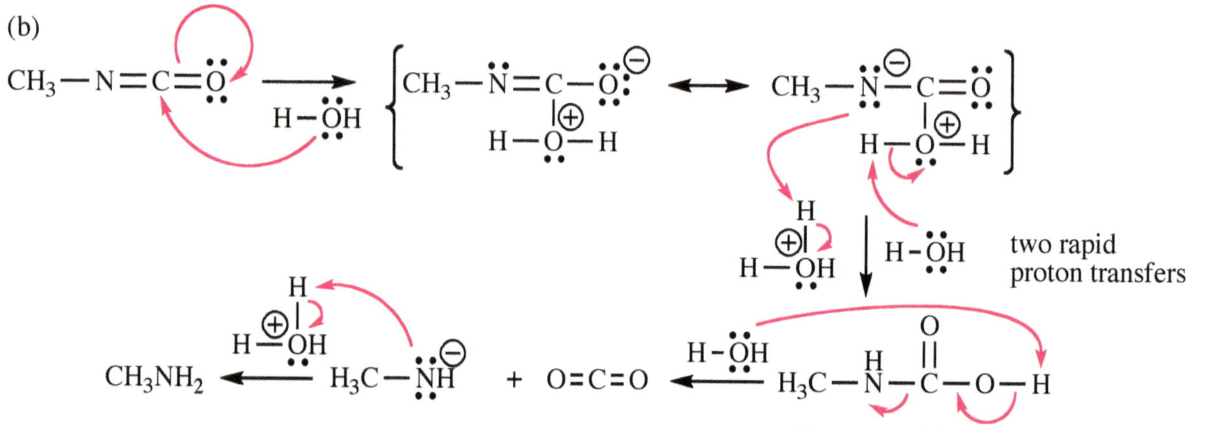

$CH_3{-}N{=}C{=}O$ + H−OH → { $CH_3{-}N{=}C(-O^{\ominus})(-O^{\oplus}H_2)$ ↔ $CH_3{-}N^{\ominus}{-}C(=O)(-O^{\oplus}H_2)$ }

two rapid proton transfers (H−OH, H_3O^+)

→ $H_3C{-}NH{-}C(=O){-}O{-}H$

Decomposition could be proposed as either acid- or base-catalyzed.

$H_3C{-}NH{-}C(=O){-}O{-}H$ + H−OH → O=C=O + $H_3C{-}NH^{\ominus}$ + $H_3O^{\oplus}$ → CH_3NH_2

(c) Alternatively, the sequence of reactions could be swapped.

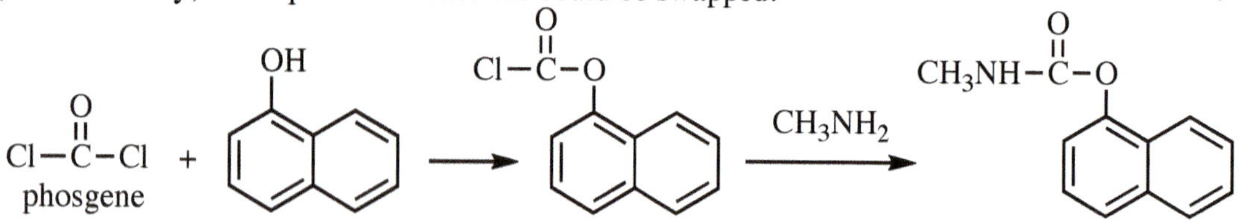

Cl−C(=O)−Cl (phosgene) + 1-naphthol (OH) → Cl−C(=O)−O−(1-naphthyl) $\xrightarrow{CH_3NH_2}$ $CH_3NH{-}C(=O){-}O{-}$(1-naphthyl)

21-63

(a) (i) The repeating functional group is an ester, so the polymer is a polyester (named Kodel®).

(ii) hydrolysis products:

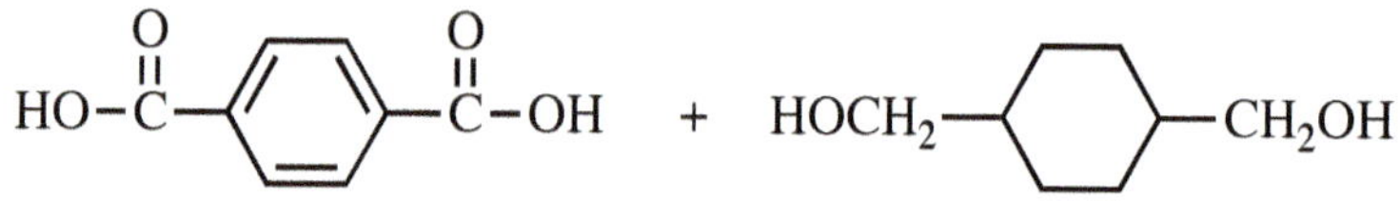

HO−C(=O)−C_6H_4−C(=O)−OH + $HOCH_2$−C_6H_{10}−CH_2OH

(iii) The monomers could be the same as the hydrolysis products, or else some reactive derivative of the dicarboxylic acid, like an acid chloride or an ester derivative.

(b)(i) The repeating functional group is an amide, so the polymer is a polyamide (named Nylon 6).

(ii) hydrolysis product:

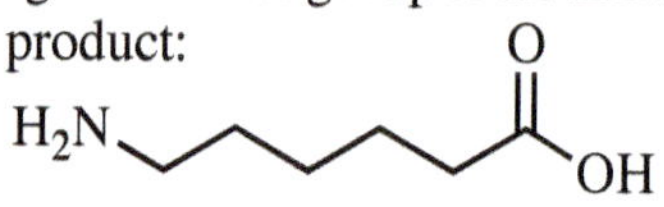

$H_2N{-}(CH_2)_5{-}C(=O)OH$

(iii) The monomer could be the same as the hydrolysis product, but in the polymer industry, the actual monomer used is the lactam shown at the right.

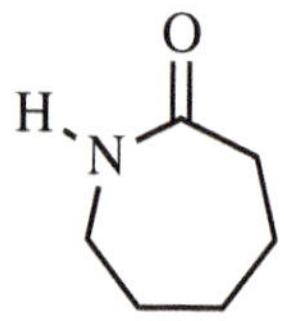

(H−N, C=O seven-membered lactam)

21-63 continued

(c) (i) The repeating functional group is a carbonate, so the polymer is a polycarbonate (named Lexan®).
(ii) hydrolysis products:

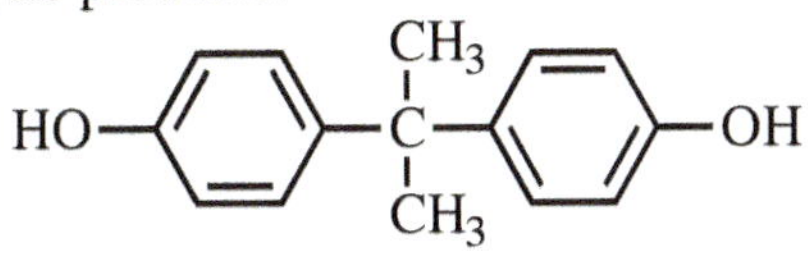

HO—C₆H₄—C(CH_3)₂—C₆H₄—OH + CO_2

(iii) The phenol monomer would be the same as the hydrolysis product; phosgene or a carbonate ester would be the other monomer.

(d) (i) The repeating functional group is an amide, so the polymer is a polyamide.
(ii) hydrolysis product:

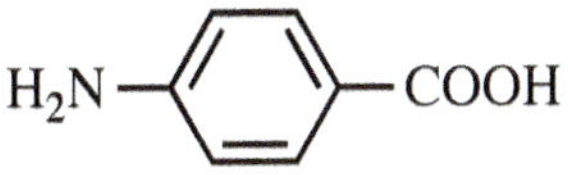

H_2N—C₆H₄—COOH *p*-aminobenzoic acid, PABA, used in sunscreens

(iii) The monomer could be the same as the hydrolysis product; a reactive derivative of the acid such as an ester could also be used.

21-64 A singlet at δ 2.15 is H on carbon next to carbonyl, the only type of proton in the compound. The IR spectrum shows no OH, and shows two carbonyl absorptions at high frequency, characteristic of an anhydride. The mass of the molecular ion at 102 proves that the anhydride must be acetic anhydride, a reagent commonly used in aspirin synthesis.

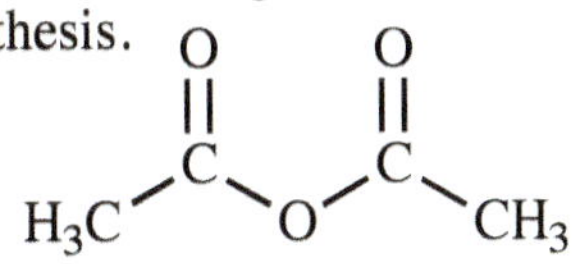

H_3C—C(=O)—O—C(=O)—CH_3

Acetic anhydride can be disposed of by hydrolyzing (carefully! exothermic!) and neutralizing in aqueous base.

21-65
The McLafferty rearrangement includes a six-membered cyclic transition state. When the chain is long enough on both sides of the C=O, the rearrangement can occur twice.

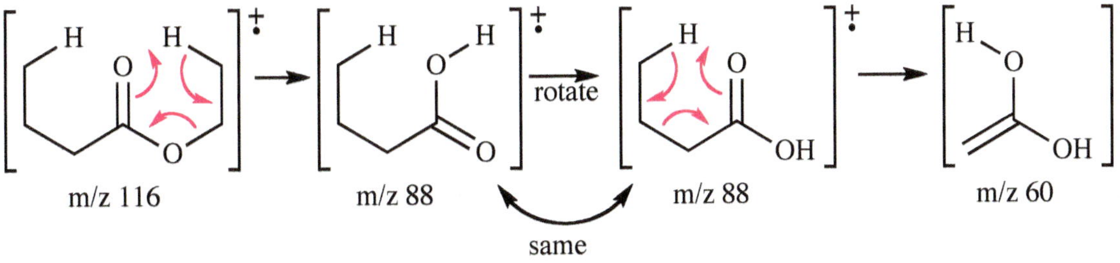

m/z 116 → m/z 88 —rotate→ m/z 88 → m/z 60

same

21-66
IR spectrum:
—sharp spike at 2250 cm^{-1} ⇒ C≡N
—1750 cm^{-1} ⇒ C=O } maybe an ester
—1200 cm^{-1} ⇒ C—O }

NMR spectrum:
—triplet and quartet ⇒ CH_3CH_2
—this quartet at δ 4.3 ⇒ CH_3CH_2O
—2H singlet at δ 3.5 ⇒ isolated CH_2

CH_3CH_2O + C(=O) + CH_2 + C≡N sum of the masses is 113, consistent with the MS

continued on next page

21-66 continued

The fragments can be combined in only two possible ways:

$$\underset{\textbf{A}}{CH_3CH_2O-\overset{\overset{O}{||}}{C}-CH_2C\equiv N} \qquad \underset{\textbf{B}}{CH_3CH_2OCH_2-\overset{\overset{O}{||}}{C}-C\equiv N}$$

The NMR proves the structure to be **A**. If the structure were **B**, the CH_2 between oxygen and the carbonyl would come farther downfield than the CH_2 of the ethyl (deshielded by oxygen and carbonyl instead of by oxygen alone). As this is not the case, the structure cannot be **B**.

The peak in the mass spectrum at m/z 68 is due to α-cleavage of the ester:

$$\left[CH_3CH_2O + \overset{\overset{O}{||}}{C}-CH_2C\equiv N\right]^{+\cdot} \longrightarrow \left\{ \overset{\overset{:O:}{||}}{\overset{\oplus}{C}}CH_2CN \longleftrightarrow \overset{\overset{:O^{\oplus}}{|||}}{C}CH_2CN \right\}$$

68 (cleavage); m/z 113 → m/z 68 + $CH_3CH_2O\bullet$ mass 45

21-67

IR spectrum: A strong carbonyl peak at 1720 cm^{-1}, in conjunction with the C—O peak at 1200 cm^{-1}, suggests the presence of an ester. An alkene peak appears at 1660 cm^{-1}.

$$C=C \qquad \overset{\overset{O}{||}}{C}-O-C$$

HNMR spectrum: The typical ethyl pattern stands out: 3H triplet at δ 1.25 and 2H quartet at δ 4.2. The chemical shift of the CH_2 suggests it is bonded to an oxygen. The other groups are: a 3H doublet at δ 1.8, likely to be a CH_3 next to one H; a 1H doublet at δ 5.8, a vinyl hydrogen with one neighboring H; and a 1H multiplet at δ 6.9, another vinyl H with many neighbors, the far downfield chemical shift suggests that it is beta to the carbonyl. The large coupling constant in the doublet at δ 5.8 shows that the two vinyl hydrogens are *trans*.

$$H-C=C-H\ (trans) \ + \ CH_3-\underset{H}{\underset{|}{C}} \ + \ OCH_2CH_3 \ + \ \overset{\overset{O}{||}}{C}-O-C$$

There is only one possible way to assemble these pieces:

$$\begin{array}{ccc} \delta\,1.8,\text{d}\ \ H_3C & & H\ \ \delta\,5.8,\text{d} \\ & C=C & \\ \delta\,6.9,\text{m}\ \ H & & \underset{\underset{O}{||}}{C}-OCH_2CH_3\ \ \delta\,1.25,\text{t} \\ & & \uparrow\ \delta\,4.2,\text{q} \end{array}$$

common name: ethyl crotonate

CNMR spectrum: The six unique carbons are unmistakable: the C=O of the ester at δ 166; the two vinyl carbons at δ 144 (beta to C=O) and at δ 123 (alpha to C=O); the CH_2—O of the ester at δ 60; and the two methyls at δ 18 and at δ 14.

Mass spectrum: This structure has a mass of 114, consistent with the molecular ion.
Major fragmentations shown on the next page.

continued on next page

21-67 continued

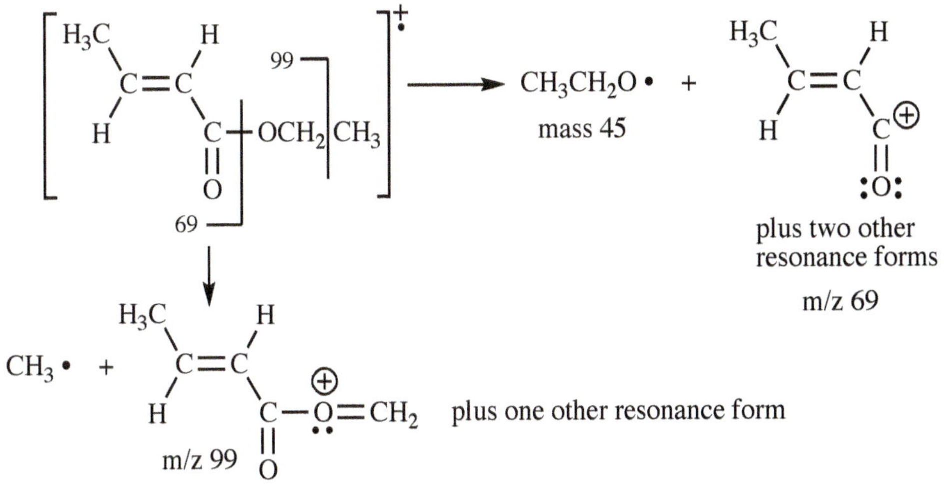

$$\left[\begin{array}{c} H_3C\text{—}CH\text{=}CH\text{—}C(\text{=}O)\text{—}OCH_2\text{—}CH_3 \end{array}\right]^{+\bullet} \longrightarrow CH_3CH_2O\bullet + H_3C\text{—}CH\text{=}CH\text{—}C^{\oplus}\text{=}\ddot{O}\text{:}$$

99 — 69 (cleavage positions)

$CH_3CH_2O\bullet$ mass 45

plus two other resonance forms

m/z 69

$CH_3\bullet$ + $H_3C\text{—}CH\text{=}CH\text{—}C(\text{=}O)\text{—}\overset{\oplus}{O}\text{=}CH_2$ plus one other resonance form

m/z 99

21-68 If you solved this problem, put a gold star on your forehead.

The formula $C_6H_8O_3$ indicates 3 elements of unsaturation.

IR spectrum: The absence of strong OH peaks shows that the compound is neither an alcohol nor a carboxylic acid. There are two carbonyl absorptions: the one about 1770 cm^{-1} is likely a strained cyclic ester (reinforced with the C—O peak around 1150 cm^{-1}), while the one at 1720 cm^{-1} is probably a ketone. (An anhydride also has two peaks, but they are of higher frequency than the ones in this spectrum.)

$$\overset{O}{\overset{\|}{C}}\text{—}O\text{—}C \qquad C\text{—}\overset{O}{\overset{\|}{C}}\text{—}C$$

HNMR spectrum: The NMR shows four types of protons. The 2H multiplet at δ 4.3 is a CH_2 group next to an oxygen on one side. The 1H multiplet at δ 3.7 is also strongly deshielded (probably by two carbonyls), a CH next to a CH_2. The 3H singlet at δ 2.45 is a CH_3 on one of the carbonyls. The remaining two hydrogens are highly coupled, a CH_2 where the two hydrogens are not equivalent. There are no vinyl hydrogens (and no alkene carbons in the carbon NMR), so the remaining element of unsaturation must be a ring.

Assemble the pieces: $\overset{O}{\overset{\|}{C}}\text{—}O\text{—}CH_2CH_2CH$ + $C\text{—}\overset{O}{\overset{\|}{C}}\text{—}CH_3$ + 1 ring

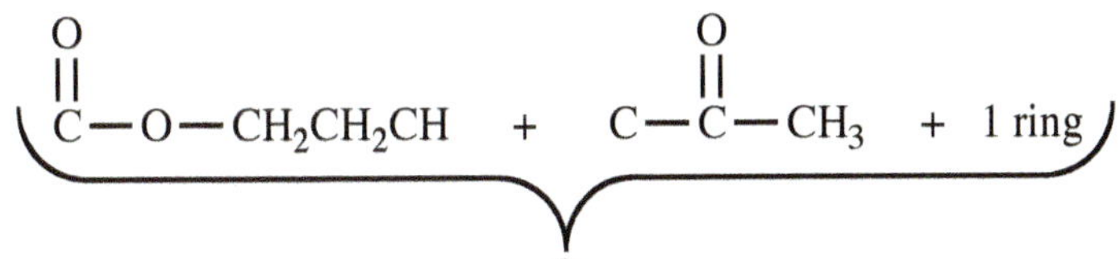

On each carbon, the "up" hydrogen is *cis* to the acetyl group, while the "down" hydrogen is *trans*. Thus, the two hydrogens on each of these carbons are not equivalent, leading to complex splitting.

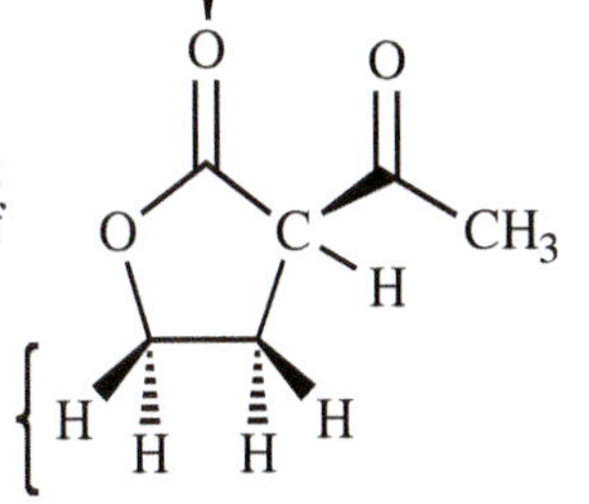

Carbon NMR:

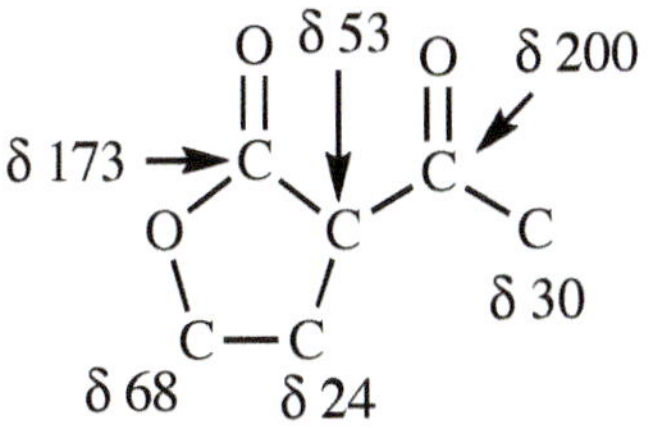

δ 173 (C=O ester), δ 53 (CH), δ 200 (C=O ketone), δ 30 (CH_3), δ 68 (CH_2O), δ 24 (CH_2)

21-69

The formula C_5H_9NO has 2 elements of unsaturation.

IR spectrum: The strongest peak at 1670 cm^{-1} comes low in the carbonyl region; in the absence of conjugation (no alkene peak observed), a carbonyl this low is almost certainly an amide. There is one broad peak in the NH/OH region, hinting at the likelihood of a secondary amide.

HNMR spectrum: The broad peak at δ 7.55 is exchangeable with D_2O; this is an amide proton. A broad, 2H peak at δ 3.3 is a CH_2 next to nitrogen. A broad, 2H peak at δ 2.4 is a CH_2 next to carbonyl. The 4H peak at δ 1.8 is probably two more CH_2 groups. There appears to be coupling among these protons but it is not resolved enough to be useful for interpretation. This is often the case when the compound is cyclic, with restricted rotation around carbon-carbon bonds, giving *non-equivalent* (axial and equatorial) hydrogens on the same carbon.

CNMR spectrum: The peak at δ 175 is the C=O of the amide. All of the peaks between δ 25 and δ 50 are aliphatic sp^3 carbons, no sp^2 carbons, so the remaining element of unsaturation cannot be a C=C; it must be a ring. The carbon peak farthest downfield is the carbon adjacent to N.

—C(=O)—N(H)—

CH_2—C(=O)—N(H)—CH_2

+ CH_2 + CH_2

The most consistent structure:

δ-valerolactam

21-70

There are multiple plausible mechanisms for production of **Z**. In all, an oxygen must eventually bridge between the two carbonyl carbons, and ethoxy must come from a molecule of ethanol.

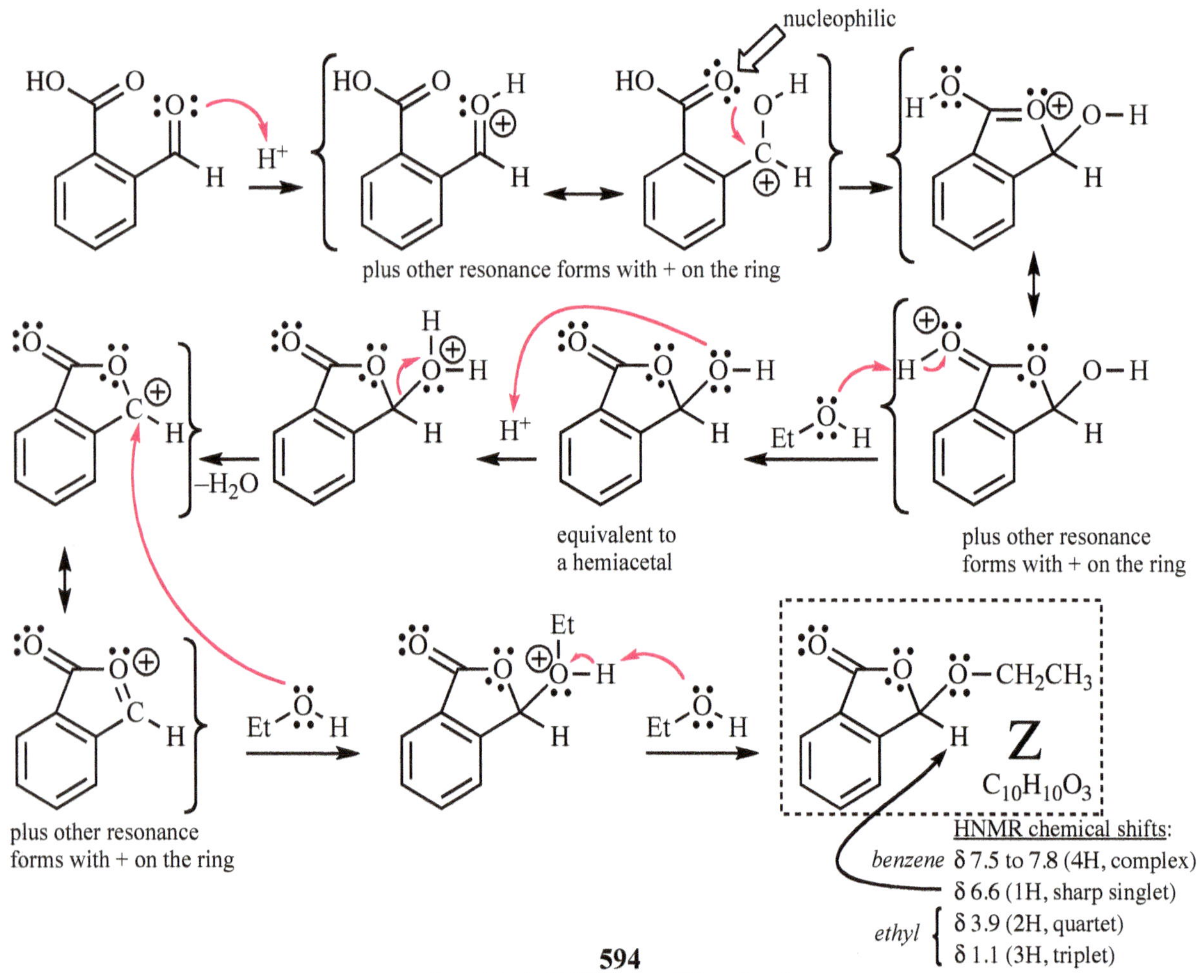

21-71
Where the wedge or dashed bonds are shown, these chiral centers are of known chirality, primarily by NMR experiments.

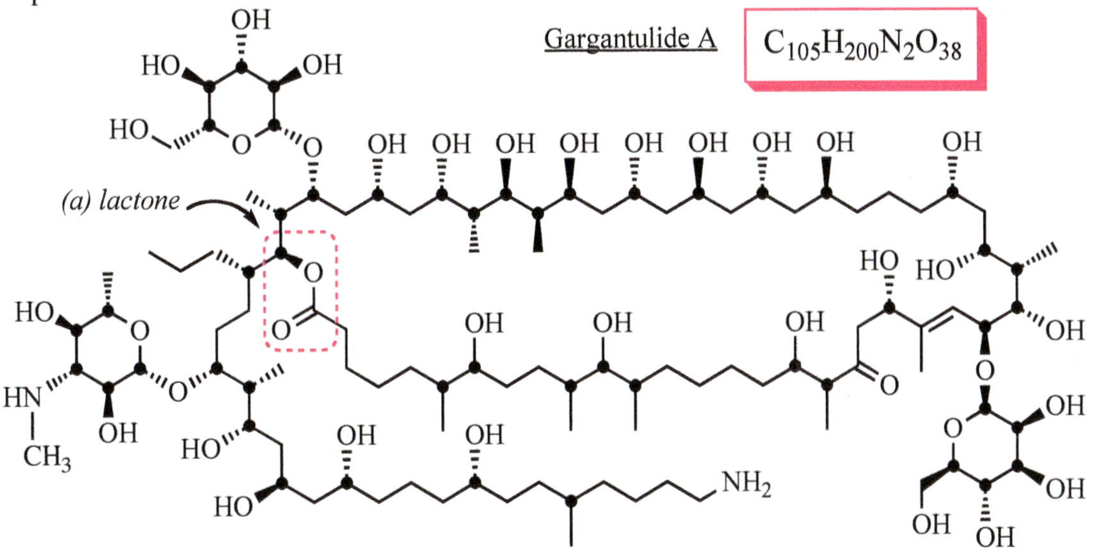

Gargantulide A $C_{105}H_{200}N_2O_{38}$

(a) lactone

(b) Macrolide ring contains 49 carbons and 1 oxygen. There are three external rings each containing 5 carbons and 1 oxygen.

(c) chiral centers are noted by a dot •:
15 in three external rings
8 on side chain
26 in macrolide ring
----
49 total chiral centers

CHAPTER 22—
CONDENSATIONS AND ALPHA SUBSTITUTIONS OF CARBONYL COMPOUNDS

22-1

(a)

$C_6H_5-CH=C(OH)-CH_3$ **1**

$C_6H_5-CH_2-C(OH)=CH_2$ **2**

(b) Enol **1** will predominate at equilibrium as its double bond is conjugated with the benzene ring, making it more stable than **2**.

(c)

basic conditions forming enol **1**

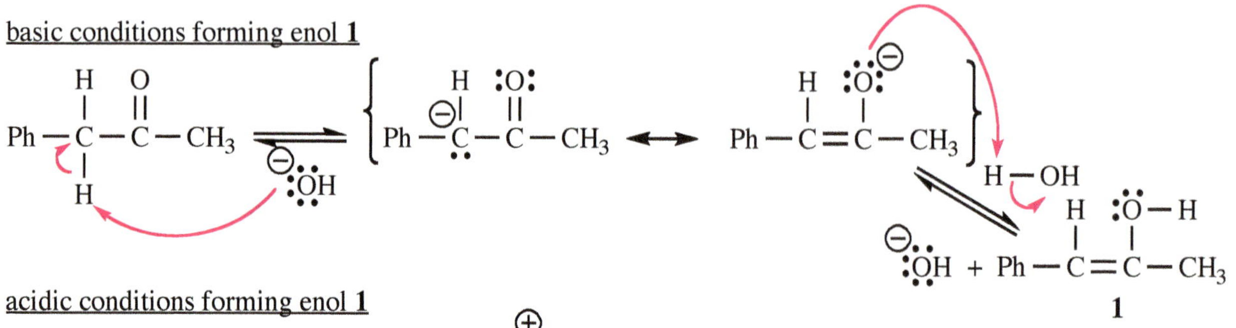

$Ph-CH_2-C(=O)-CH_3 + {}^{-}OH \rightleftharpoons \{Ph-\overset{-}{C}H-C(=O)-CH_3 \leftrightarrow Ph-CH=C(O^-)-CH_3\}$

$\xrightarrow{H-OH}$ $^{-}OH + Ph-CH=C(OH)-CH_3$ **1**

acidic conditions forming enol **1**

$Ph-CH_2-C(=O)-CH_3 + H-A \rightleftharpoons \{Ph-CH_2-C(=\overset{+}{O}H)-CH_3 \leftrightarrow Ph-CH_2-\overset{+}{C}(OH)-CH_3\}$

$\xrightarrow{{}^{-}A:}$ $Ph-CH=C(OH)-CH_3$ **1**

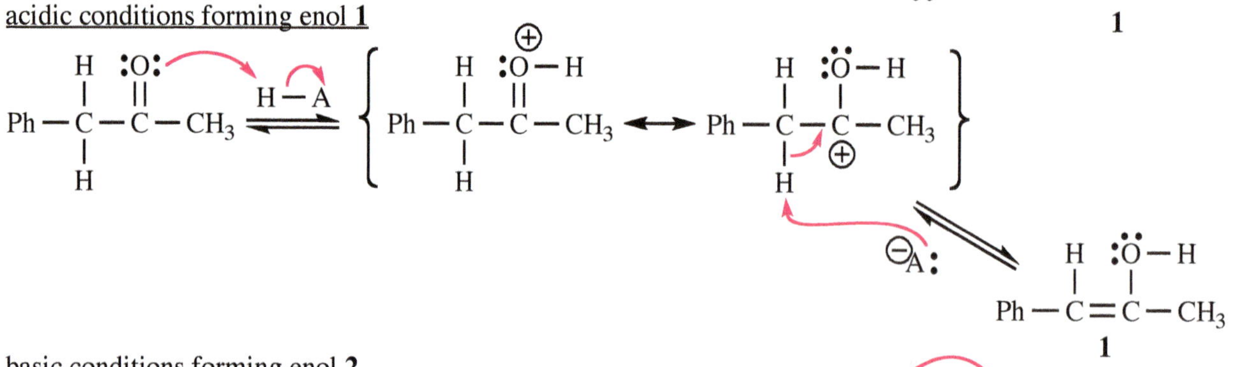

basic conditions forming enol **2**

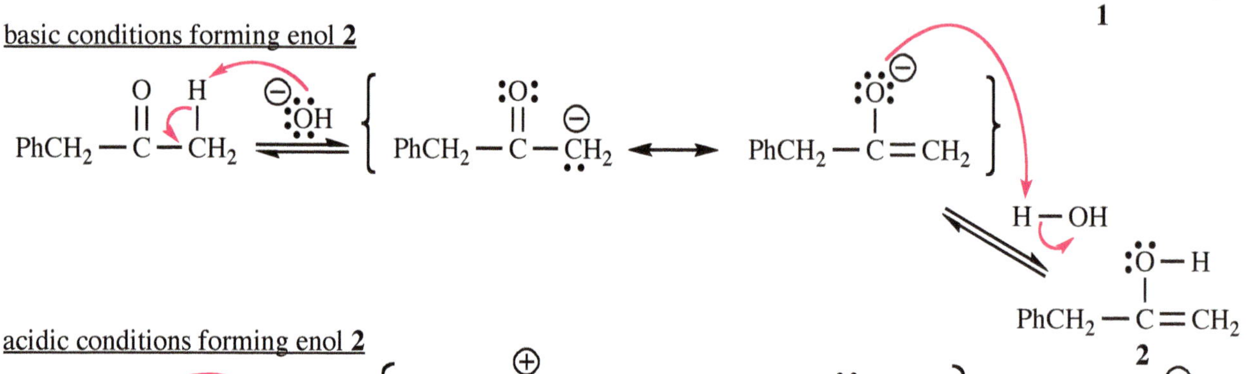

$PhCH_2-C(=O)-CH_3 + {}^{-}OH \rightleftharpoons \{PhCH_2-C(=O)-\overset{-}{C}H_2 \leftrightarrow PhCH_2-C(O^-)=CH_2\}$

$\xrightarrow{H-OH}$ $PhCH_2-C(OH)=CH_2$ **2** $+ {}^{-}OH$

acidic conditions forming enol **2**

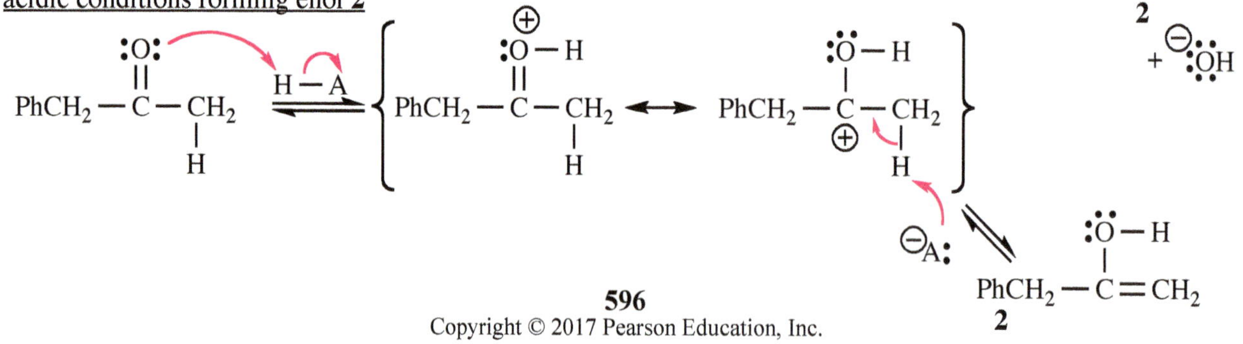

$PhCH_2-C(=O)-CH_3 + H-A \rightleftharpoons \{PhCH_2-C(=\overset{+}{O}H)-CH_3 \leftrightarrow PhCH_2-\overset{+}{C}(OH)-CH_3\}$

$\xrightarrow{{}^{-}A:}$ $PhCH_2-C(OH)=CH_2$ **2**

22-2

(a)

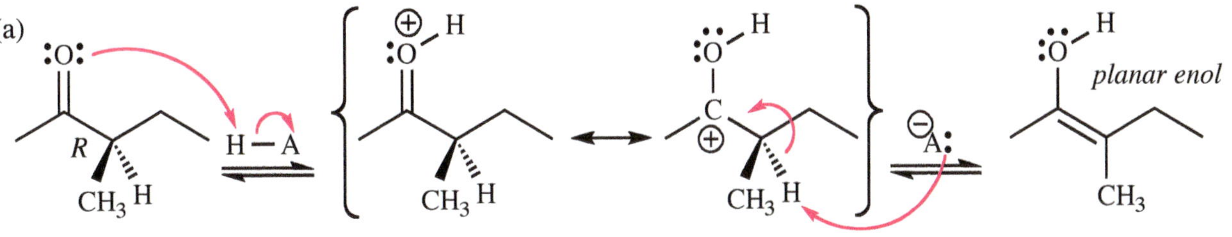

This planar enol intermediate has lost all chirality. Protonation can occur with equal probability at either face of the pi bond leading to racemic product.

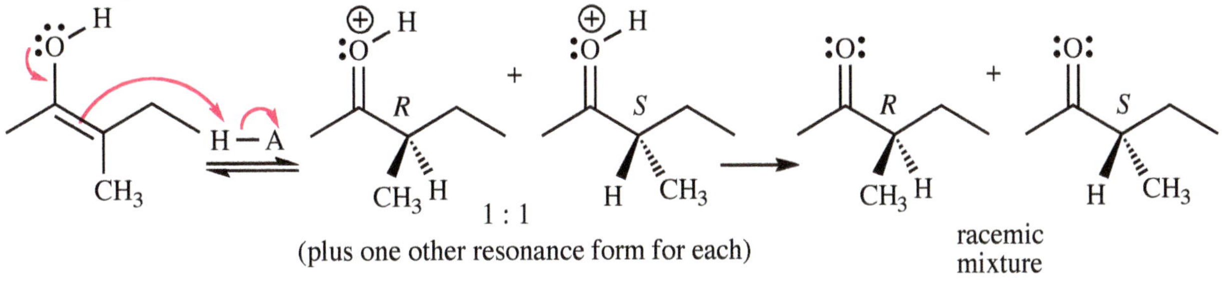

planar enol

1 : 1

(plus one other resonance form for each)

racemic mixture

(b)

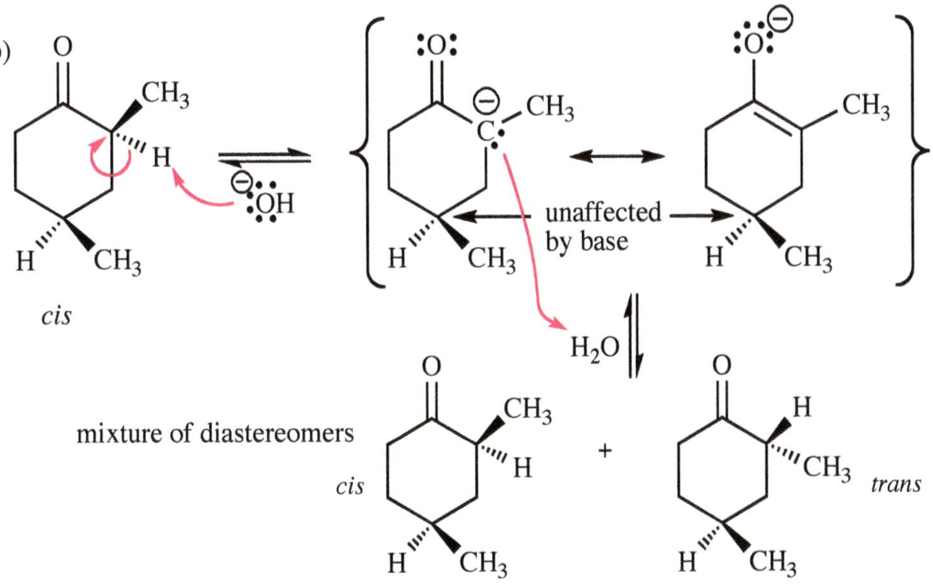

cis

unaffected by base

H_2O

mixture of diastereomers

cis

trans

22-3 If the ketone is not symmetric, more than one enolate may be formed.

(a)

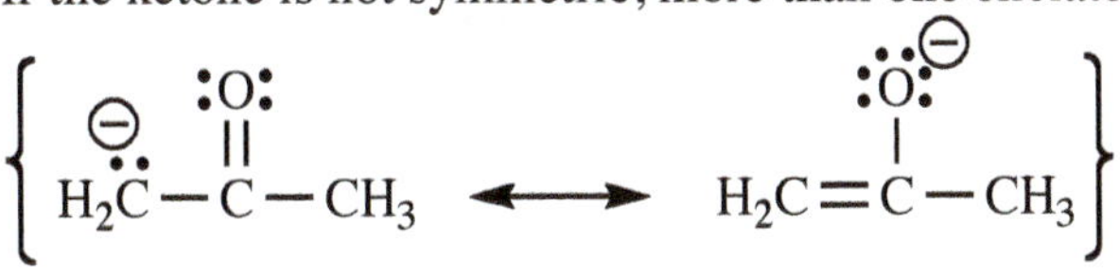

$$\left\{ H_2\overset{\ominus}{C}-\overset{O}{\overset{\|}{C}}-CH_3 \longleftrightarrow H_2C=\overset{O^{\ominus}}{\overset{|}{C}}-CH_3 \right\}$$

(b)

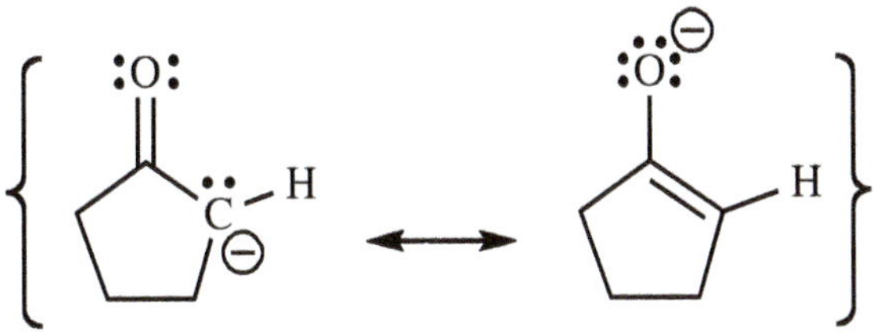

22-3 continued

(c)

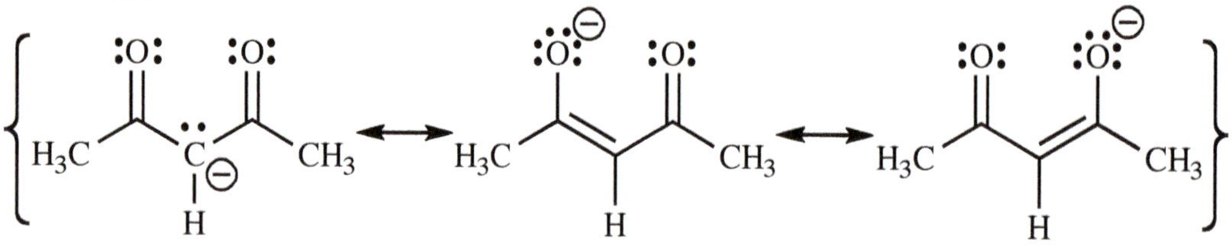

(d) Two enolates are possible.

and

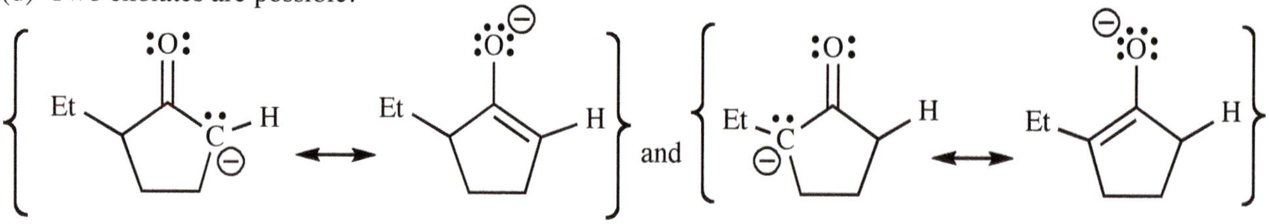

(e) Two enolates are possible.

and

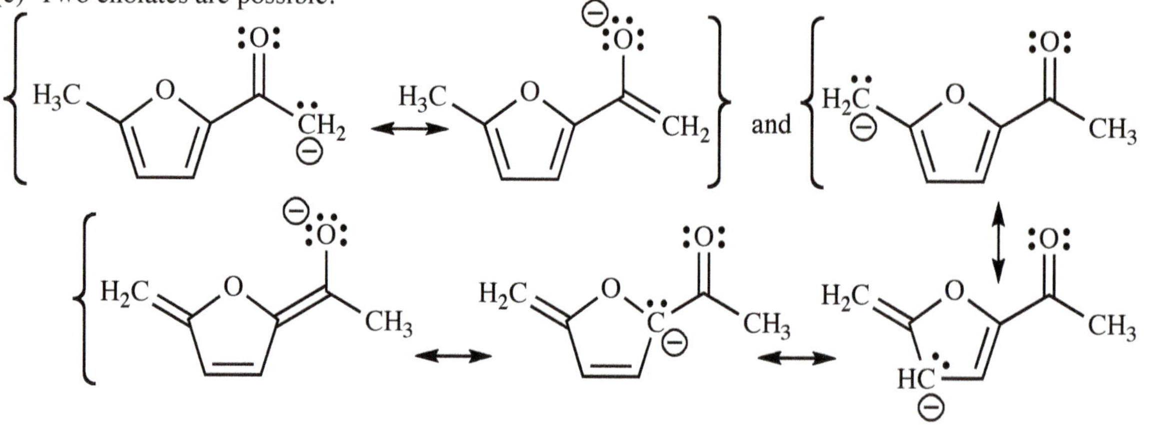

(f) Three enolates are possible.

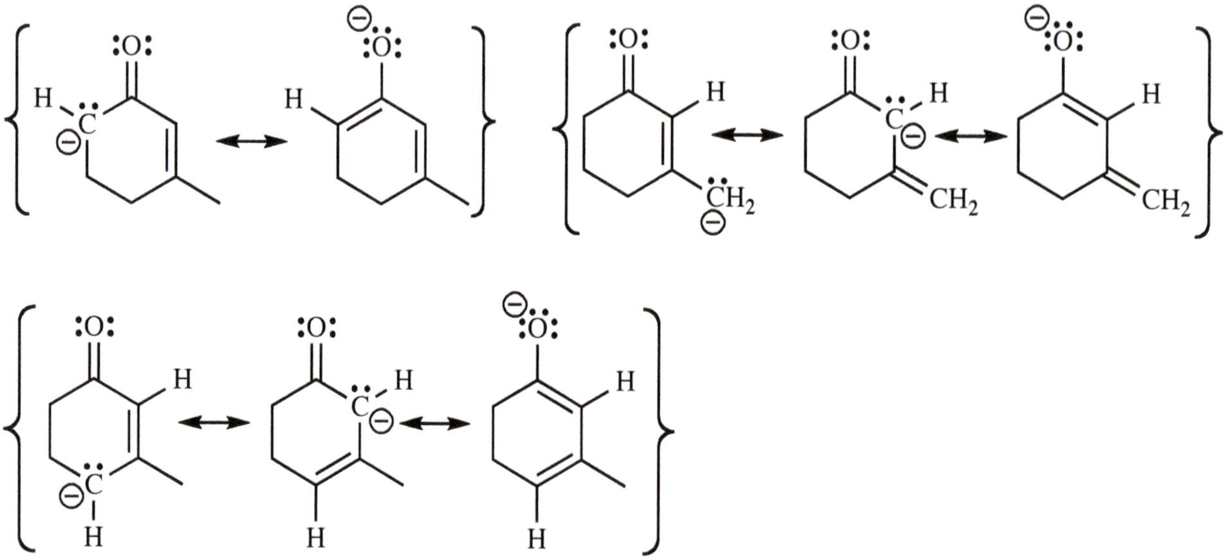

22-4

(a) Sodium ethoxide is not a strong enough base to deprotonate cyclohexanone more than a tiny amount:

cyclohexanone (pKa ≈ 20) + NaOEt ⇌ enolate ($\ominus$CH, Na^+) + HOEt (pKa ≈ 16)

A difference of 4 pK_a units means an equilibrium constant of ≈ 10^{-4}, meaning about 0.01% of the cyclohexanone is deprotonated.

Since ≈ 99.99% of the NaOEt is still present when benzyl bromide is added, these two react by Williamson ether synthesis. Unreacted cyclohexanone will remain in the reaction solution.

$$C_6H_5CH_2Br + NaOEt \longrightarrow C_6H_5CH_2OCH_2CH_3 + NaBr$$

(b) Two changes are needed to make the desired reaction feasible: (1) use a strong, non-nucleophilic base such as LDA that will completely deprotonate cyclohexanone, and (2) use an aprotic solvent such as THF or diethyl ether so that cyclohexanone cannot find a proton from solvent. These conditions will work fine.

22-5 New carbon-carbon bonds are shown in bold.

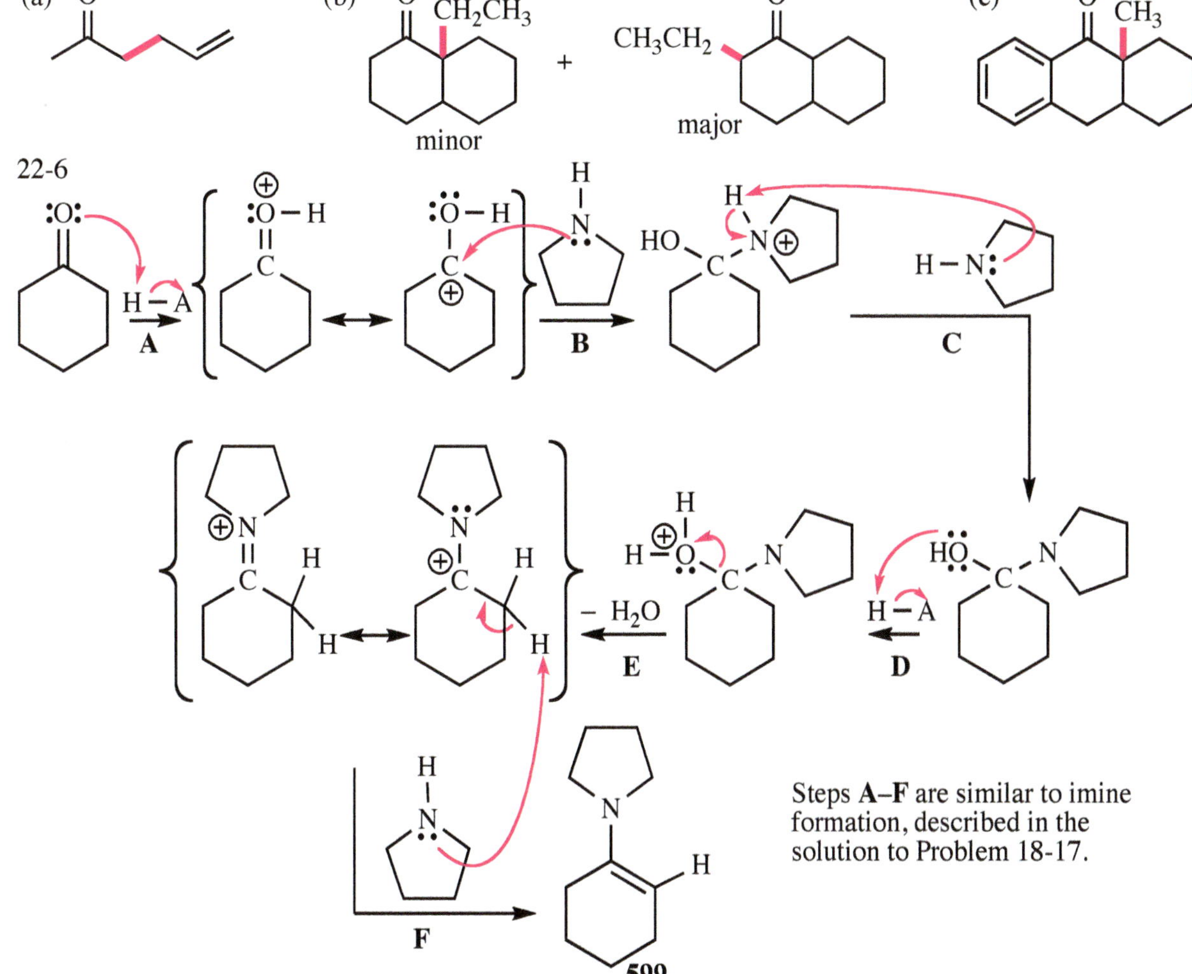

22-7

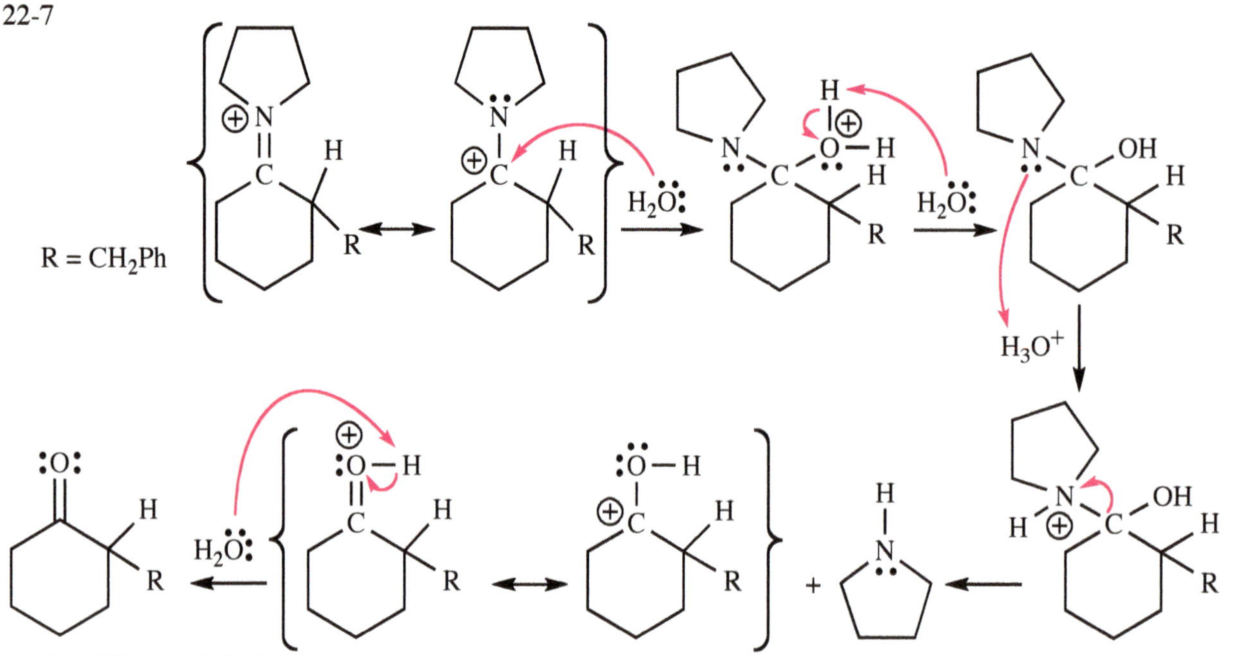

R = CH_2Ph

Pyrrolidine could also function as the base in the last step.

22-8

(a) 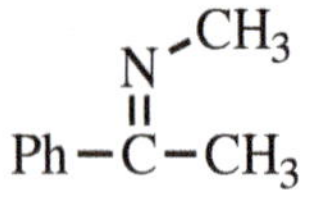 (b) 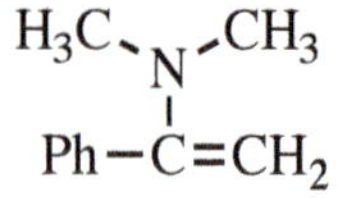 (c) 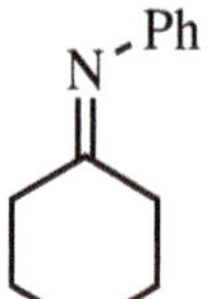 (d)

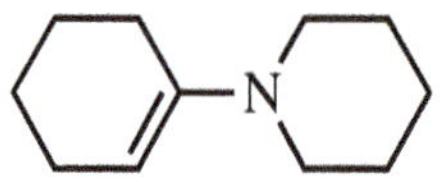

22-9 Any 2° aliphatic amines can be used to make enamines; pyrrolidine is a typical one.

(a)

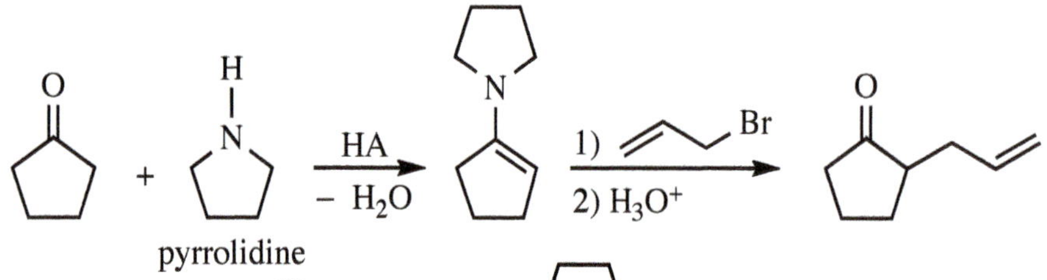

$\xrightarrow[-\ H_2O]{HA}$ $\xrightarrow[2)\ H_3O^+]{1)}$

pyrrolidine

(b)

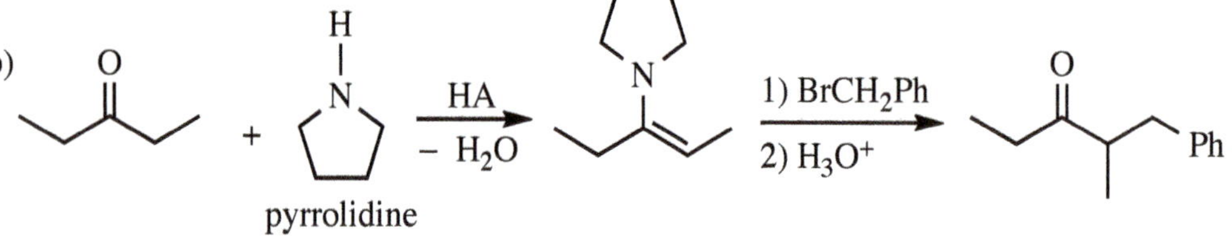

$\xrightarrow[-\ H_2O]{HA}$ $\xrightarrow[2)\ H_3O^+]{1)\ BrCH_2Ph}$

pyrrolidine

(c)

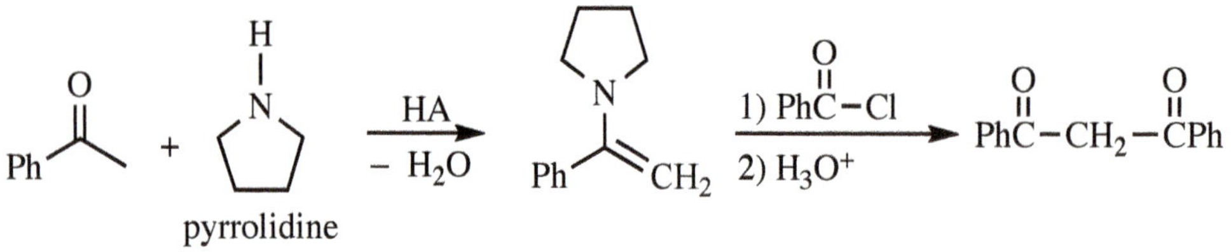

$\xrightarrow[-\ H_2O]{HA}$ $\xrightarrow[2)\ H_3O^+]{1)\ PhC(=O)-Cl}$ $PhC(=O)-CH_2-C(=O)Ph$

pyrrolidine

22-10

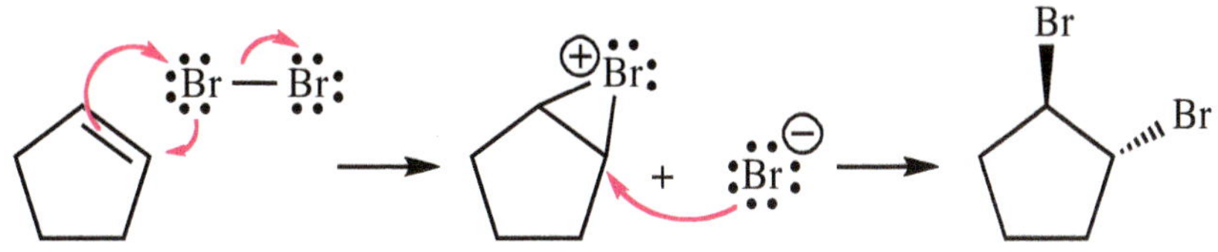

+ the enantiomer

Phenol is like an enol but is an even weaker nucleophile than an enol, yet it readily reacts with bromine. Only the ortho attack is shown here; the para would be abundant too.

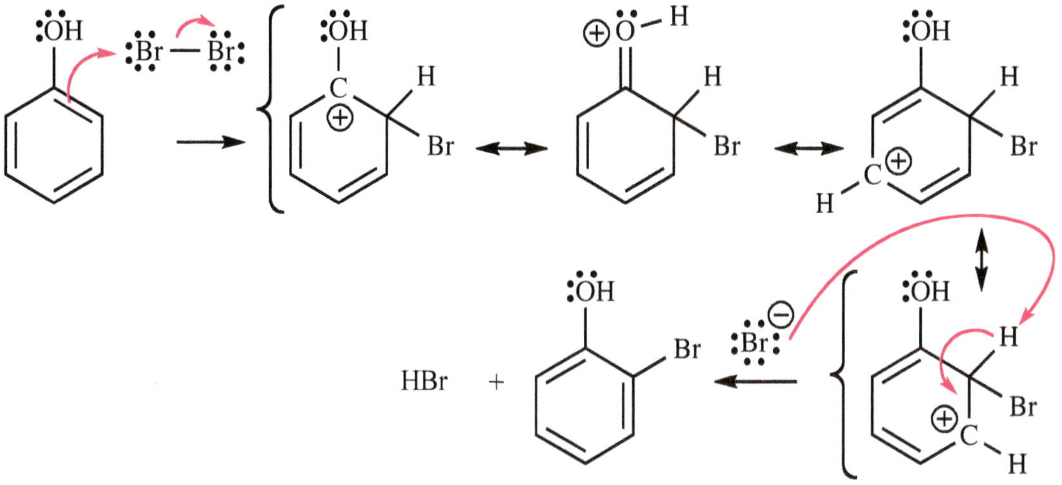

HBr +

22-11

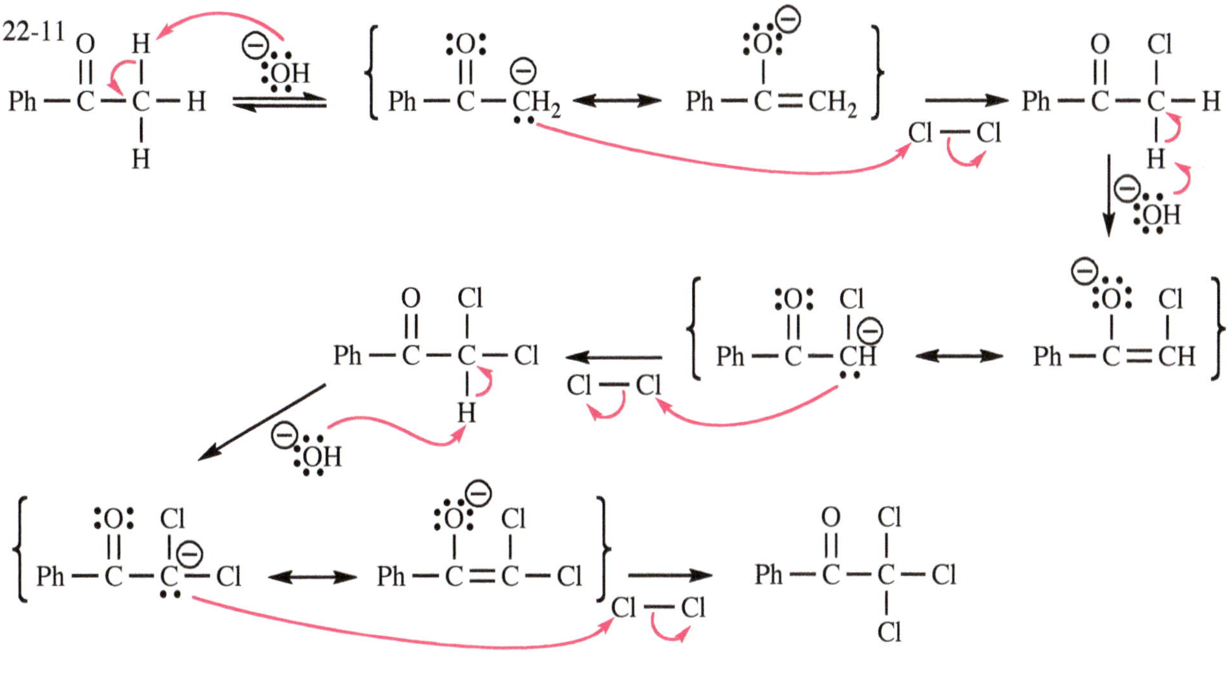

22-12 The cyclohexyl group is abbreviated "c-Hx".

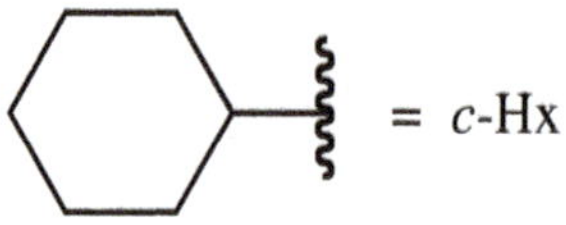

= c-Hx

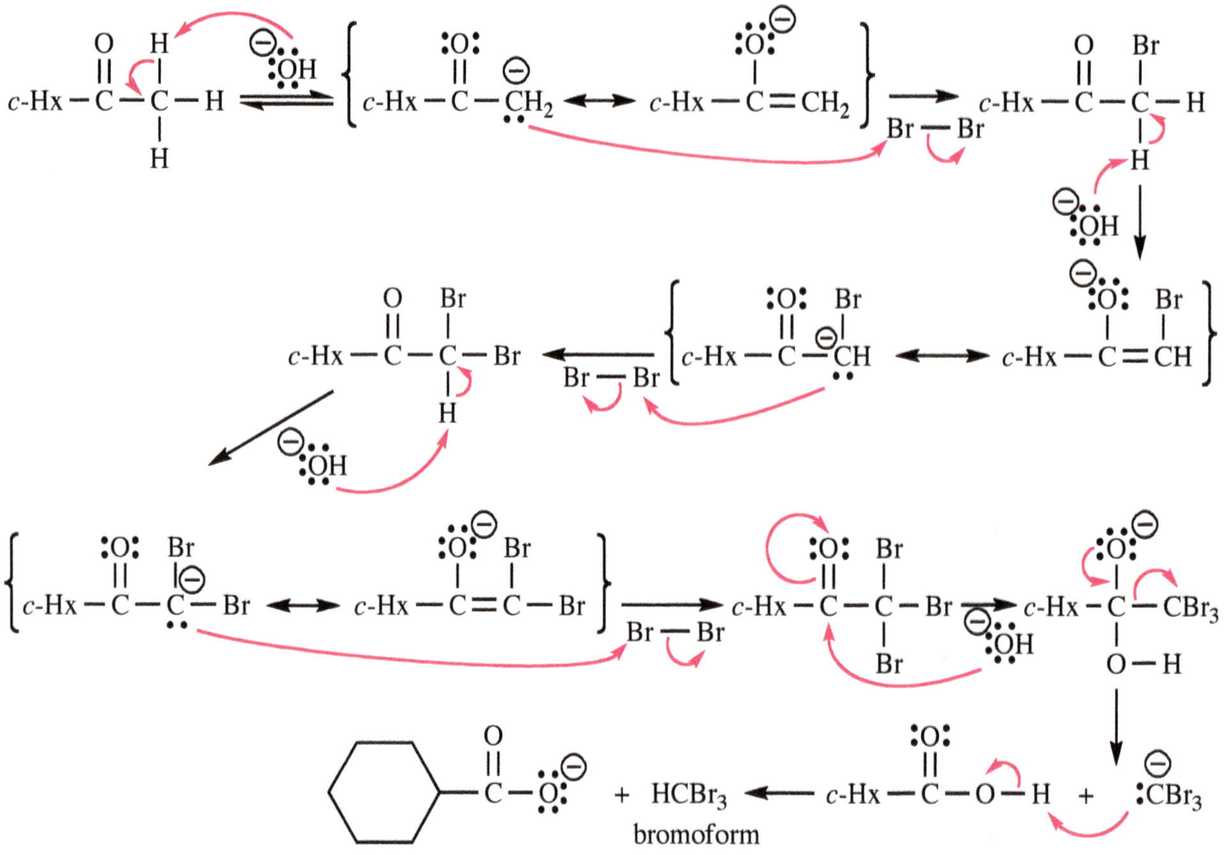

$c\text{-Hx}-C(=O)-CH_3 \rightleftharpoons$ (with $^{\ominus}OH$) $\{c\text{-Hx}-C(=O)-\overset{\ominus}{C}H_2 \longleftrightarrow c\text{-Hx}-C(-O^{\ominus})=CH_2\}$ $\xrightarrow{Br-Br}$ $c\text{-Hx}-C(=O)-CH_2Br$

$c\text{-Hx}-C(=O)-CH_2Br$ (with $^{\ominus}OH$) $\longrightarrow$ $\{c\text{-Hx}-C(=O)-\overset{\ominus}{C}HBr \longleftrightarrow c\text{-Hx}-C(-O^{\ominus})=CHBr\}$ $\xrightarrow{Br-Br}$ $c\text{-Hx}-C(=O)-CHBr_2$

$c\text{-Hx}-C(=O)-CHBr_2$ (with $^{\ominus}OH$) $\longrightarrow$ $\{c\text{-Hx}-C(=O)-\overset{\ominus}{C}Br_2 \longleftrightarrow c\text{-Hx}-C(-O^{\ominus})=CBr_2\}$ $\xrightarrow{Br-Br}$ $c\text{-Hx}-C(=O)-CBr_3$ $\xrightarrow{^{\ominus}OH}$ $c\text{-Hx}-C(O^{\ominus})(OH)-CBr_3$

$\longrightarrow c\text{-Hx}-C(=O)-O-H + {:}\overset{\ominus}{C}Br_3 \longrightarrow$ (cyclohexyl)$-C(=O)-O^{\ominus}$ + $HCBr_3$

bromoform

22-13

(a) cyclopentyl$-COO^{\ominus}$ Na^+ + $CHCl_3$

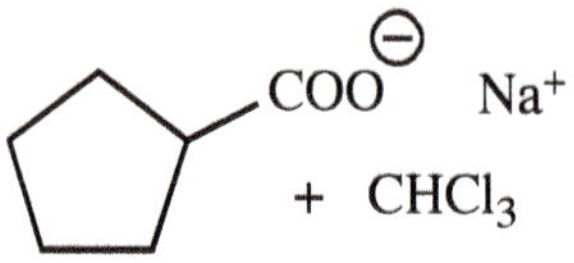

(b) cyclopentyl$-COO^{\ominus}$ Na^+ + CHI_3 (precipitate)

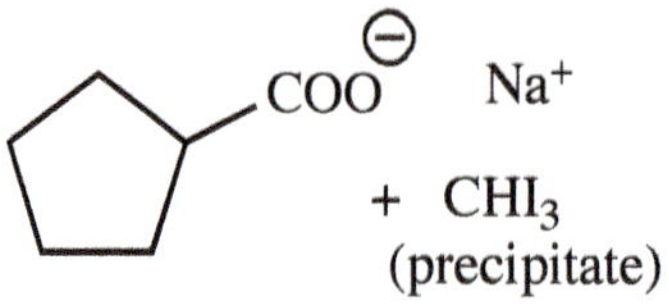

(c) $Ph-C(=O)-CBr_2-CH_3$

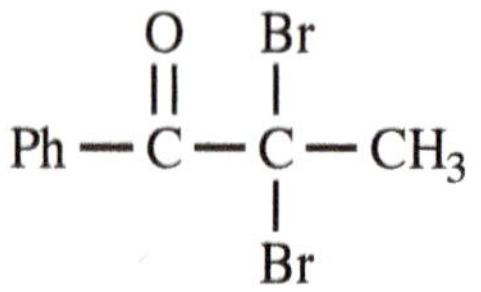

22-14 Methyl ketones, and those alcohols which can be oxidized to methyl ketones, will give a positive iodoform test. All of the compounds in this problem except pentan-3-one (part (d)) will give a positive iodoform test.

22-15

H—OAc

AcO⁻

Br — Br

HBr +

22-16

Cl_2 / CH_3COOH — KOH / Δ

from Solved Problem 22-2 — E2 elimination

This is a common and effective way to make carbon-3 reactive where it was not in the starting material.

22-17 Monobromination at the alpha carbon is typical of the HVZ reaction.

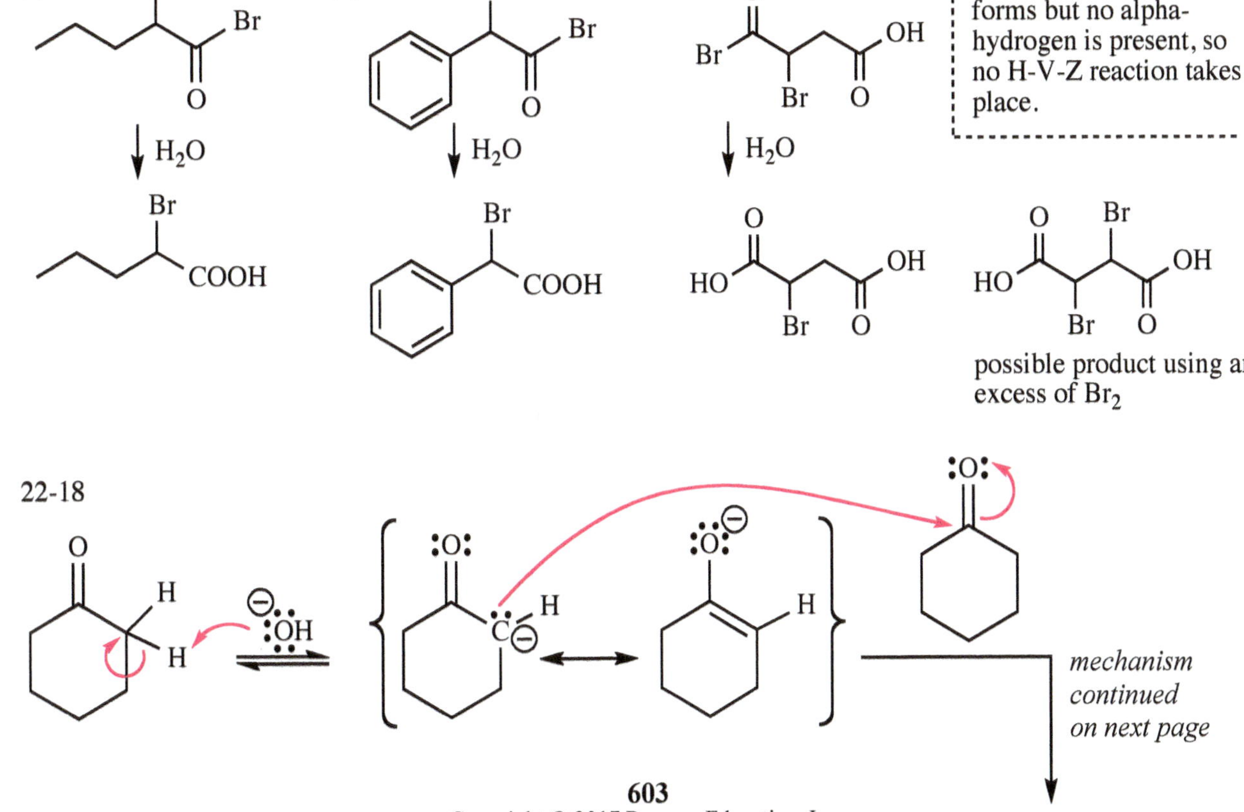

(a)
(b)
(c)
(d) An acyl bromide forms but no alpha-hydrogen is present, so no H-V-Z reaction takes place.
H_2O
COOH
possible product using an excess of Br_2
22-18
mechanism continued on next page

22-18 continued

In general, the equilibrium in aldol condensations of ketones favors reactants rather than products. There is significant steric hindrance at both carbons with new bonds, so it is reasonable to conclude that this reaction of cyclohexanone would also favor reactants at equilibrium.

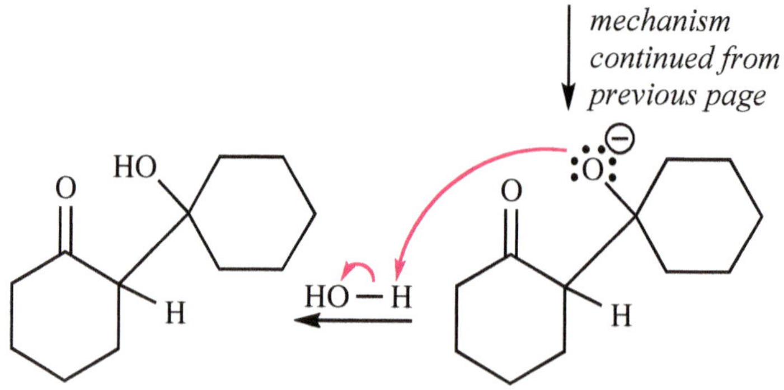

mechanism continued from previous page
HO
O
H
HO—H
O
H

22-19 New carbon-carbon bonds are shown in bold.

(a)

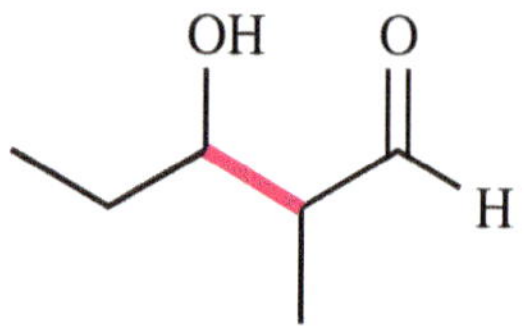

OH
O
H

(b)

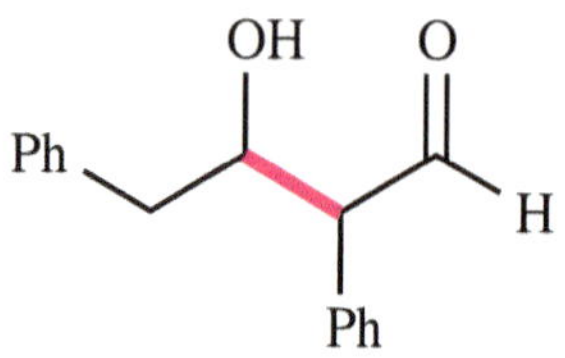

OH
O
Ph
H
Ph

(c)

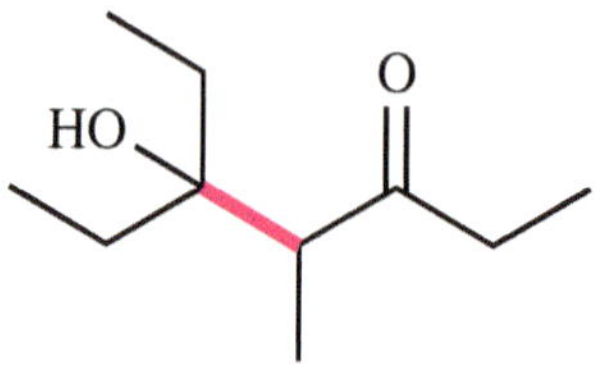

HO
O

22-20 All the steps in the aldol condensation are reversible. Adding base to diacetone alcohol promoted the reverse aldol reaction. The equilibrium greatly favors acetone.

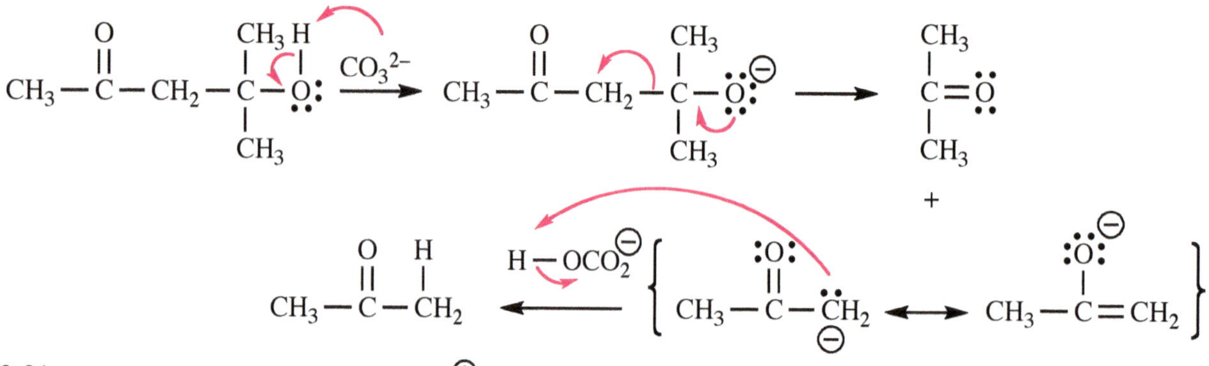

22-21

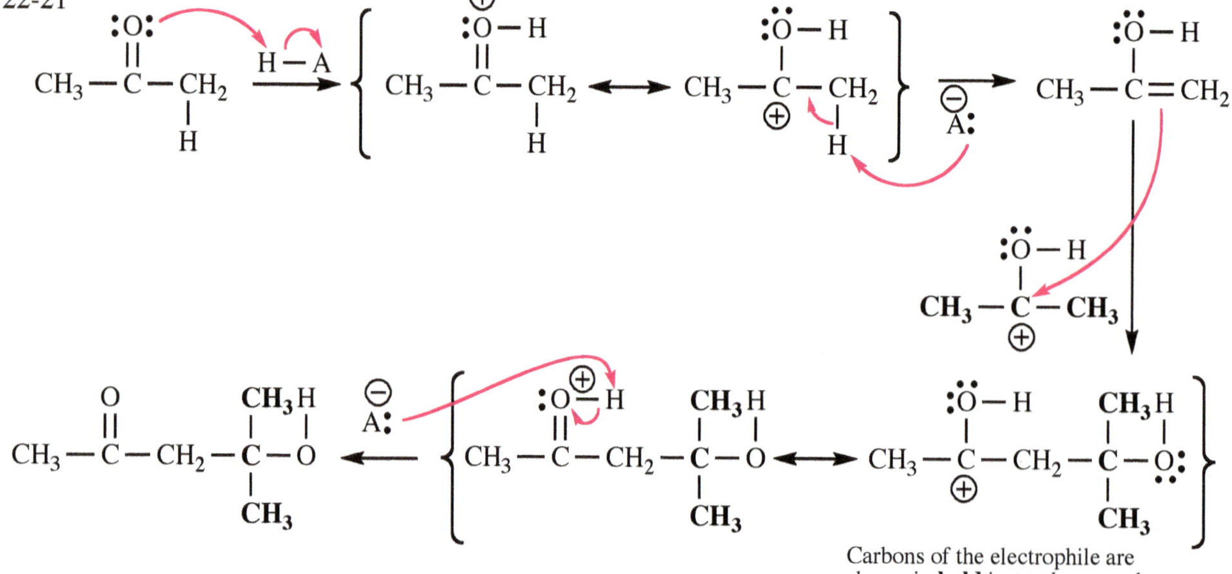

Carbons of the electrophile are shown in **bold** just to keep track of which carbons come from which molecule.

22-22

(a) acidic conditions

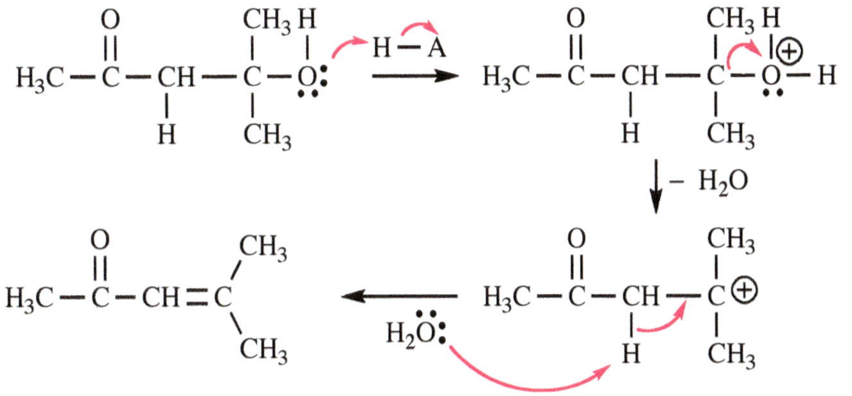

$$H_3C-C(=O)-CH(H)-C(CH_3)_2-\ddot{O}H \xrightarrow{H-A} H_3C-C(=O)-CH(H)-C(CH_3)_2-\overset{\oplus}{O}H_2$$

$$\xrightarrow{-H_2O} H_3C-C(=O)-CH(H)-\overset{\oplus}{C}(CH_3)_2 \xrightarrow{H_2\ddot{O}:} H_3C-C(=O)-CH=C(CH_3)_2$$

(b) basic conditions

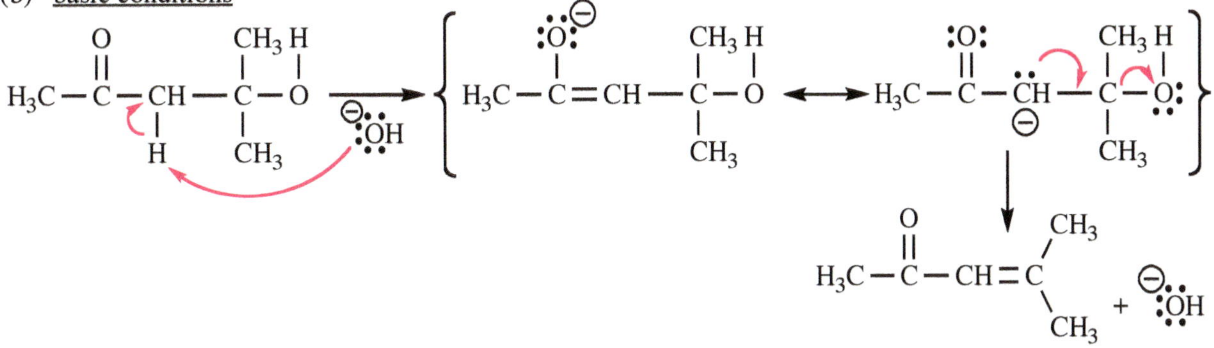

$$H_3C-C(=O)-CH(H)-C(CH_3)_2-OH \xrightarrow{^{\ominus}:\ddot{O}H} \left\{ H_3C-C(O^{\ominus})=CH-C(CH_3)_2-OH \longleftrightarrow H_3C-C(=O)-\overset{\ominus}{\ddot{C}}H-C(CH_3)_2-OH \right\}$$

$$\longrightarrow H_3C-C(=O)-CH=C(CH_3)_2 + {}^{\ominus}:\ddot{O}H$$

22-23

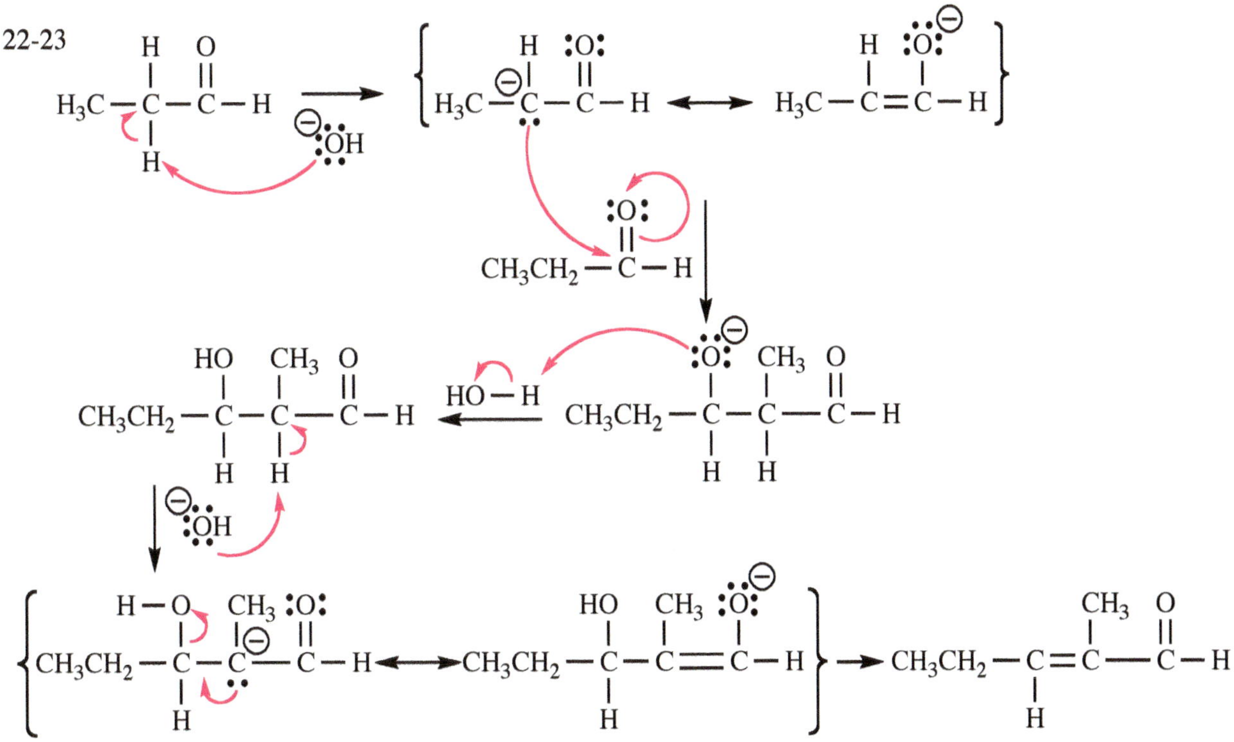

$$H_3C-CH(H)-C(=O)-H \xrightarrow{^{\ominus}:\ddot{O}H} \left\{ H_3C-\overset{\ominus}{\ddot{C}}H-C(=O)-H \longleftrightarrow H_3C-CH=C(O^{\ominus})-H \right\}$$

$$\xrightarrow{CH_3CH_2-C(=O)-H} CH_3CH_2-CH(O^{\ominus})-CH(CH_3)-C(=O)-H \xrightarrow{HO-H} CH_3CH_2-CH(OH)-CH(CH_3)-C(=O)-H$$

$$\xrightarrow{^{\ominus}:\ddot{O}H} \left\{ CH_3CH_2-CH(OH)-\overset{\ominus}{\ddot{C}}(CH_3)-C(=O)-H \longleftrightarrow CH_3CH_2-CH(OH)-C(CH_3)=C(O^{\ominus})-H \right\} \longrightarrow CH_3CH_2-CH=C(CH_3)-C(=O)-H$$

22-24 The carbon-carbon double bond is the new bond in each case.

(a) $E + Z$

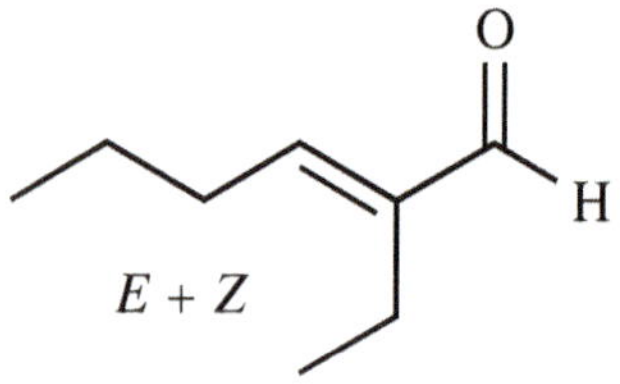

(b) $E + Z$

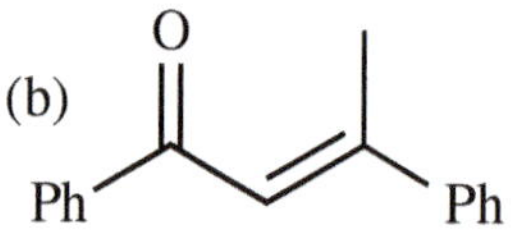

(c)

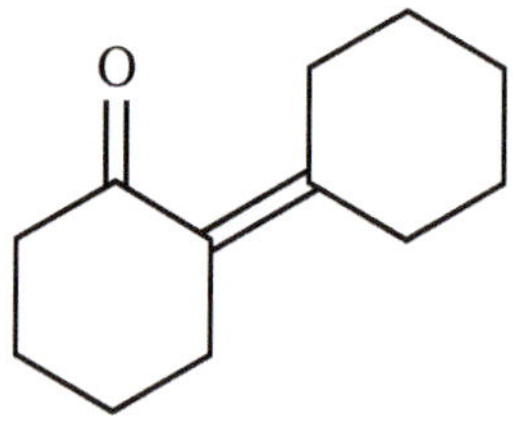

22-25

(a)

Step 1: carbon skeletons

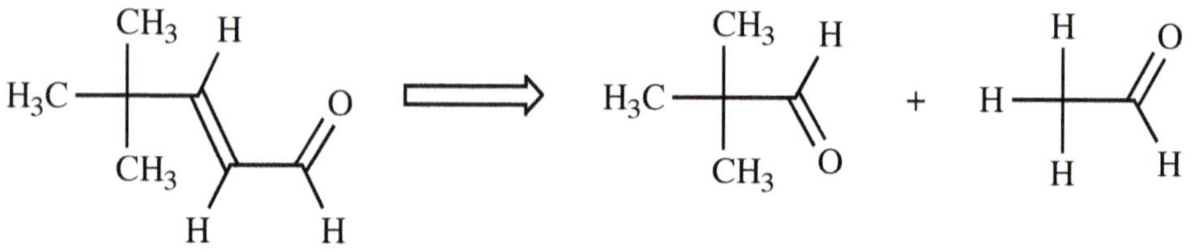

Step 2: nucleophile generation

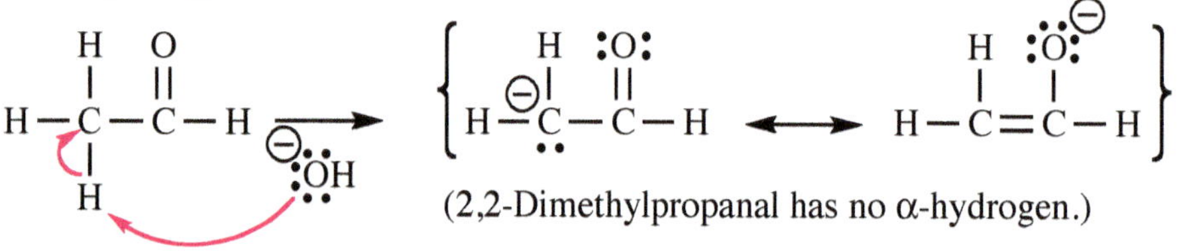

(2,2-Dimethylpropanal has no α-hydrogen.)

Step 3: nucleophilic attack

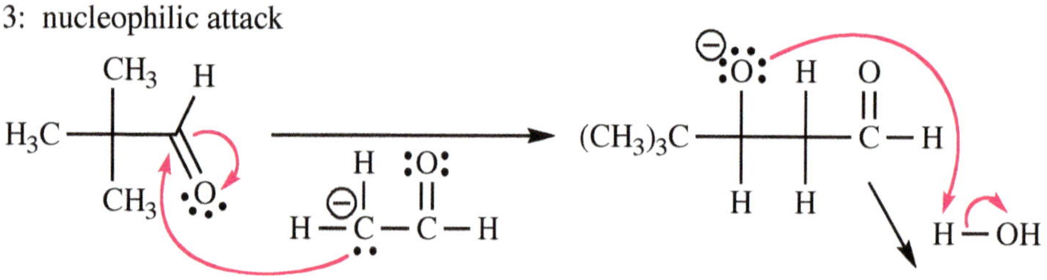

Step 4: dehydration to final product

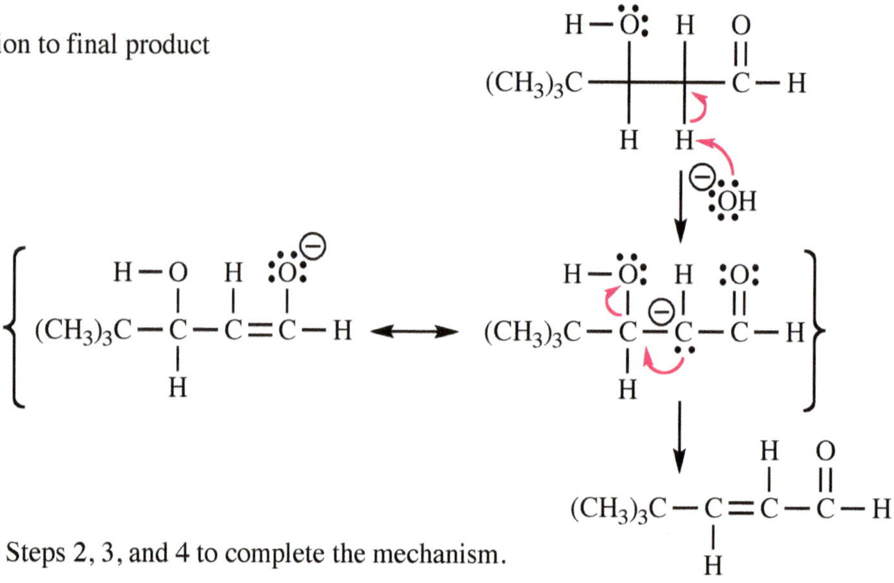

Step 5: Combine Steps 2, 3, and 4 to complete the mechanism.

22-25 continued

(b)

Step 1: carbon skeletons

Step 2: nucleophile generation

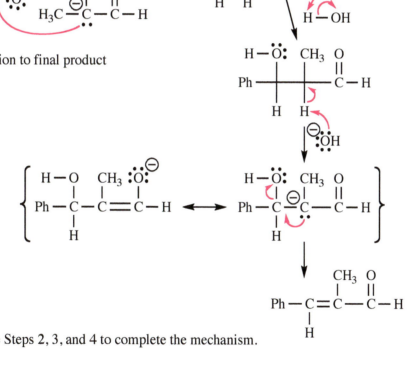

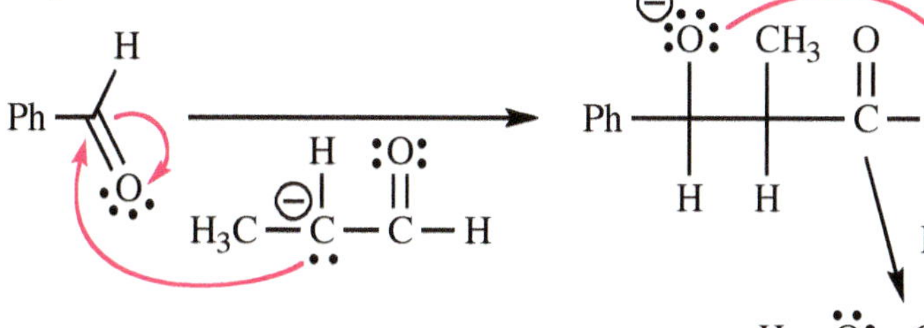

(Benzaldehyde has no α-hydrogen.)

Step 3: nucleophilic attack

Step 4: dehydration to final product

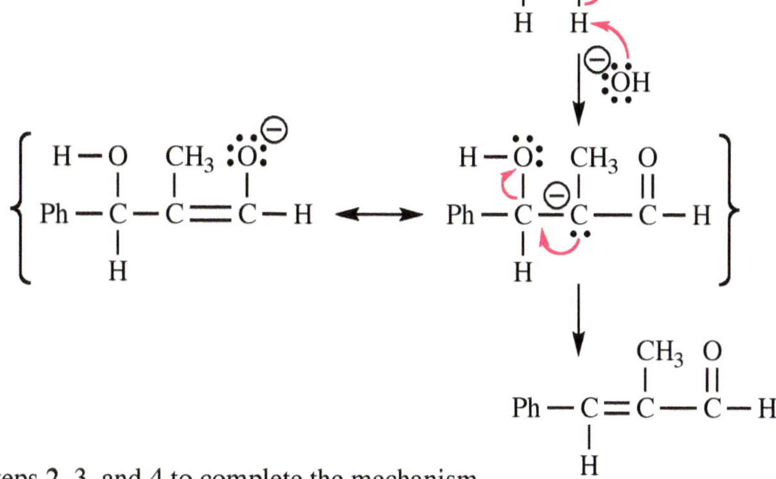

Step 5: Combine Steps 2, 3, and 4 to complete the mechanism.

22-26 This solution presents the sequence of reactions leading to the product, following the format of the Problem-Solving feature. **This is not a complete mechanism.**

Step 2: generation of the nucleophile

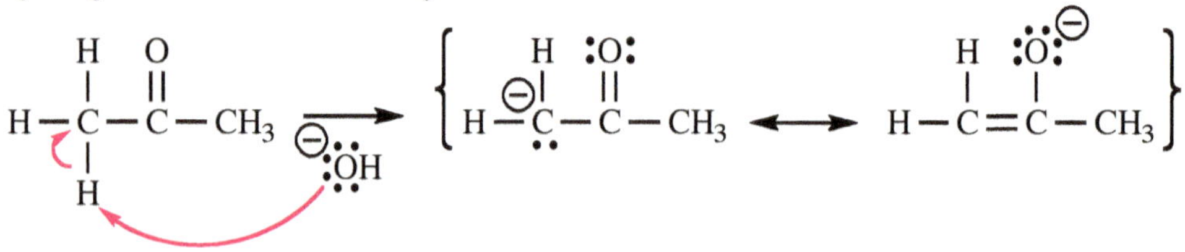

$$H-CH_2-C(=O)-CH_3 \xrightarrow{^-OH} \left\{ H-\overset{-}{C}H-C(=O)-CH_3 \longleftrightarrow H-CH=C(-O^-)-CH_3 \right\}$$

Step 3: nucleophilic attack

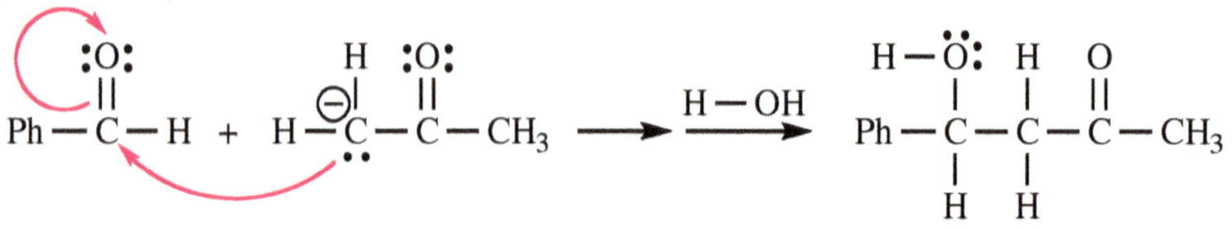

$$Ph-C(=O)-H + H-\overset{-}{C}H-C(=O)-CH_3 \longrightarrow \xrightarrow{H-OH} Ph-CH(OH)-CH_2-C(=O)-CH_3$$

Step 4: dehydration

$$Ph-CH(OH)-CH_2-C(=O)-CH_3 \xrightarrow{^-OH} Ph-CH=CH-C(=O)-CH_3$$

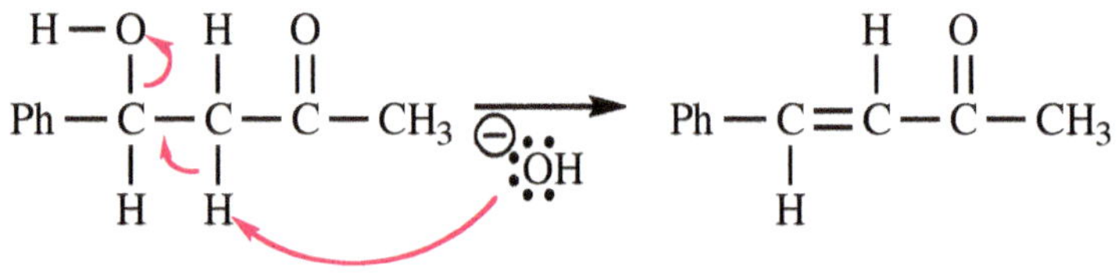

Remember that this is not an E2 reaction; it goes through an enolate intermediate.

The same sequence of steps subsequently occurs on the other side.

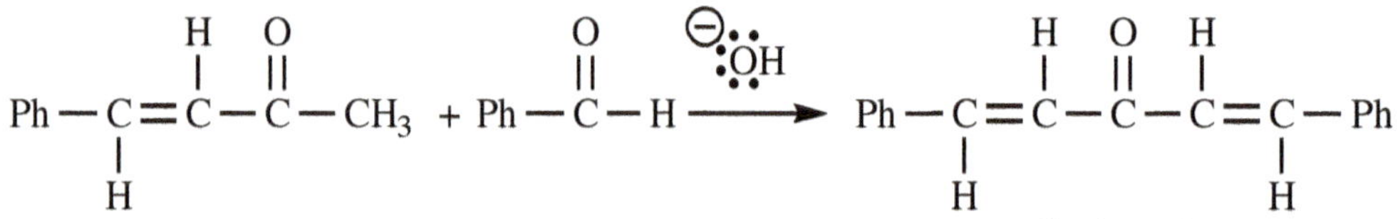

$$Ph-CH=CH-C(=O)-CH_3 + Ph-C(=O)-H \xrightarrow{^-OH} Ph-CH=CH-C(=O)-CH=CH-Ph$$

final product

22-27

A: 2-methylcyclohexanone + ^-OEt ⇌ {enolate: carbanion at C-2 (CH_3) ⟷ enolate O^- with C=C}

enolate + HC(=O)—Ph ⇌ B (alkoxide: 2-(CH(O⁻)-Ph)-2-methylcyclohexanone)

B + EtO—H ⇌ C (2-(CH(OH)-Ph)-2-methylcyclohexanone)

A

C

B

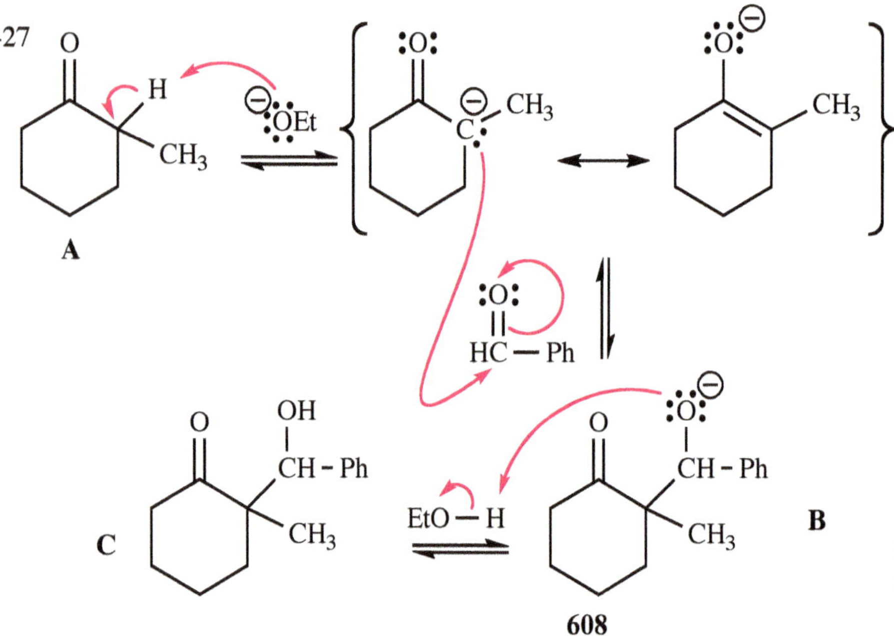

DISCUSSION ON NEXT PAGE

22-27 continued

There are three problems with the reaction as shown on the previous page:

1. Hydrogen on a 3° carbon (structure **A**) is less acidic than hydrogen on a 2° carbon. The 3° hydrogen will be removed at a slower rate than the 2° hydrogen.
2. Nucleophilic attack by the 3° carbon will be more hindered, and therefore slower, than attack by the 2° carbon. Structure **B** is quite hindered.
3. Once a normal aldol product is formed, dehydration gives a conjugated system that has great stability. The aldol product **C** cannot dehydrate because no α-hydrogen remains. Some **C** will form, but eventually the reverse-aldol process will return **C** to starting materials which, in turn, will react at the other α-carbon to produce the conjugated system. (This reason is the Kiss of Death for **C**.)

22-28

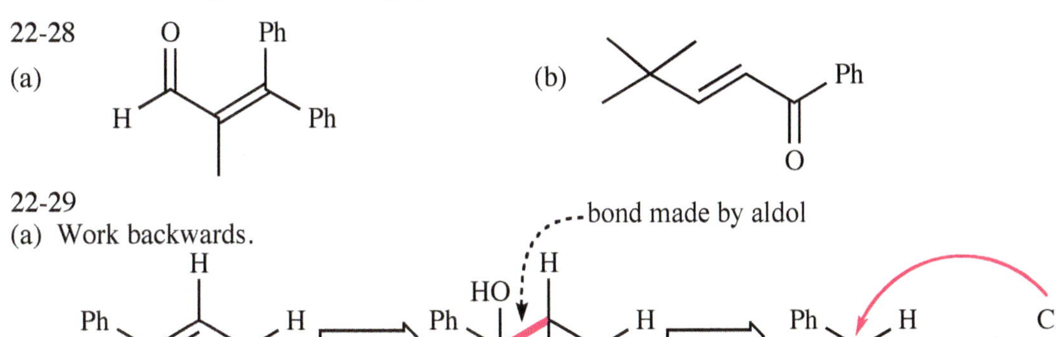

22-29

(a) Work backwards.

bond made by aldol

(b)

By adding the ketone to LDA, the enolate is generated rapidly, but it has nothing to react with.

By adding LDA to the ketone, the enolate is generated in a small amount in a sea of remaining ketone, allowing the aldol reaction to proceed rapidly.

aldol product ($E + Z$)

22-30

The formation of a 7-membered ring is unfavorable for entropy reasons: the farther apart the nucleophile and the electrophile, the harder time they will have finding each other. If the molecule has a possibility of forming a 5- or a 7-membered ring, it will almost always prefer to form the 5-membered ring.

22-31

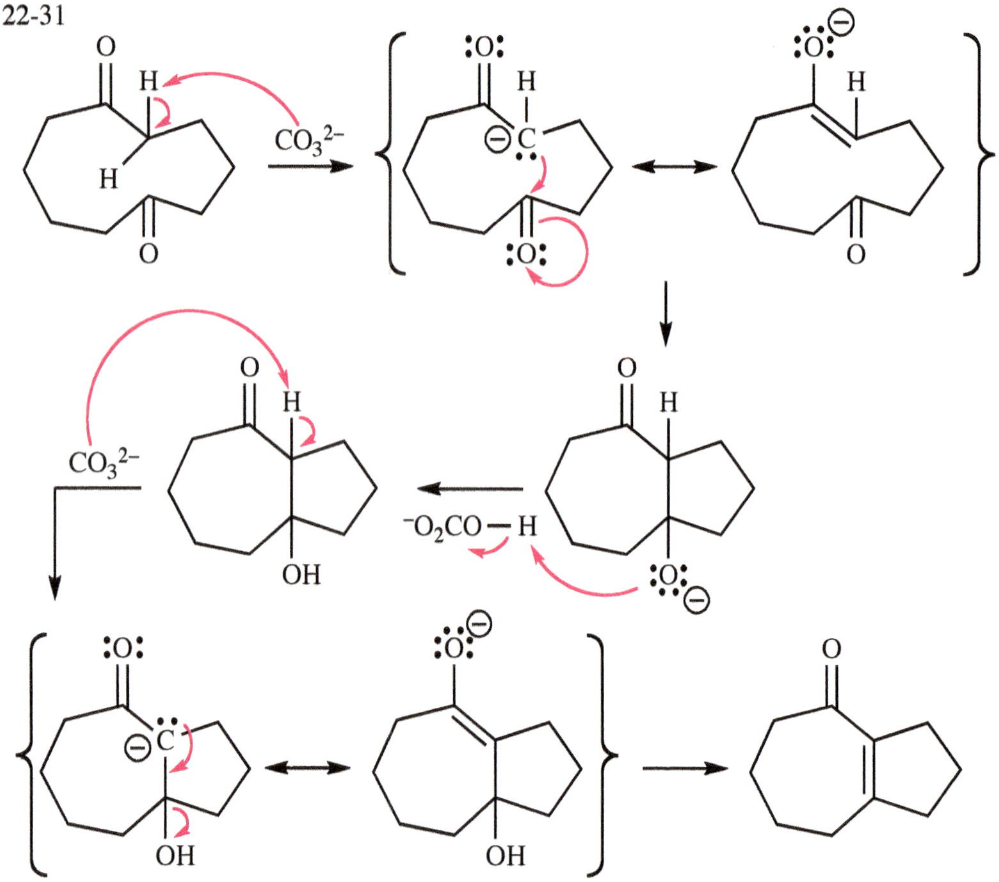

CO_3^{2-}

$^-O_2CO-H$

CO_3^{2-}

22-32

(a) $CH_3CH_2CH_2\overset{O}{\overset{||}{C}}-H$ + $CH_2-\overset{O}{\overset{||}{C}}H$ ($CH_2CH_2CH_3$) Not feasible: requires condensation of two different aldehydes, each with α-hydrogens.

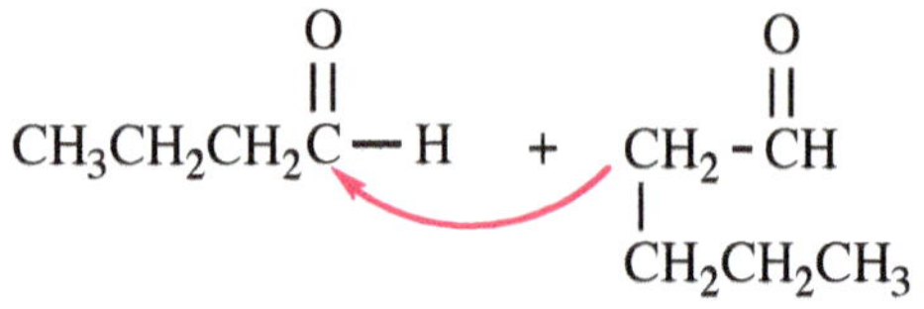

(b) $Ph\overset{O}{\overset{||}{C}}-CH_2CH_3$ + $CH_3CH_2\overset{O}{\overset{||}{C}}-Ph$ Feasible: this is a self-condensation.

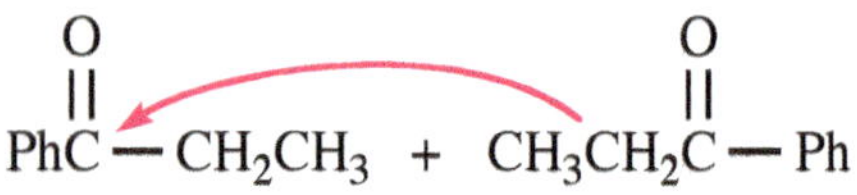

(c) $Ph\overset{O}{\overset{||}{C}}-H$ + $H_3C-\overset{O}{\overset{||}{C}}-CH_3$ Feasible: only one reactant has α-hydrogen; cannot use excess benzaldehyde as acetone has two reactive sites.

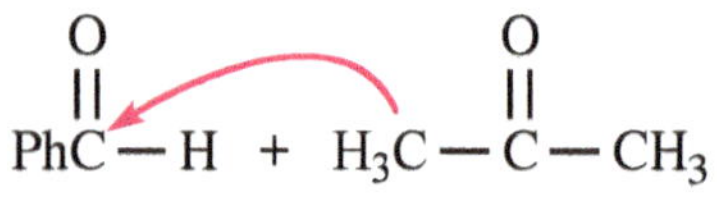

(d) Feasible; however, the cyclization from the carbon α to the aldehyde to the ketone carbonyl is also possible.

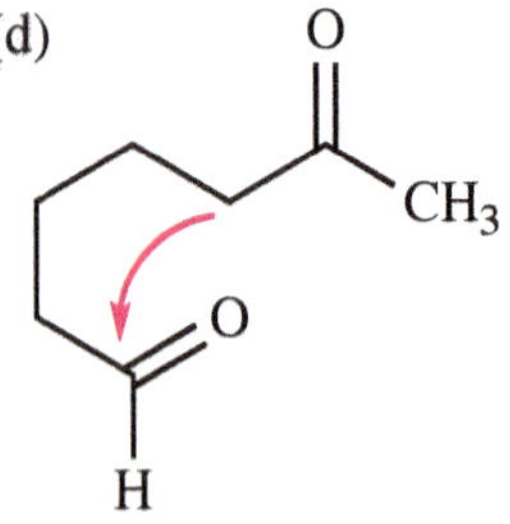

(e) Feasible: symmetric reagent will give the same product in either direction of cyclization.

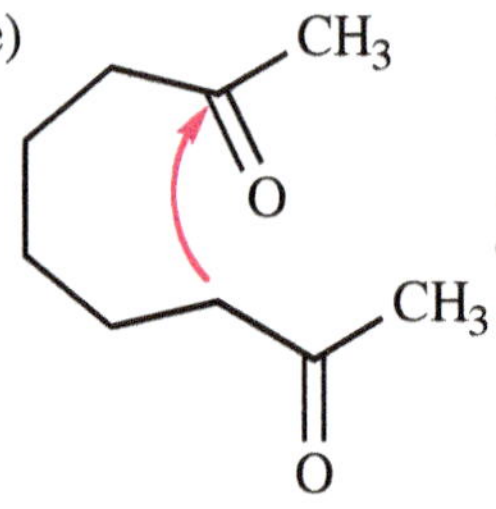

22-33 (a) and (b)

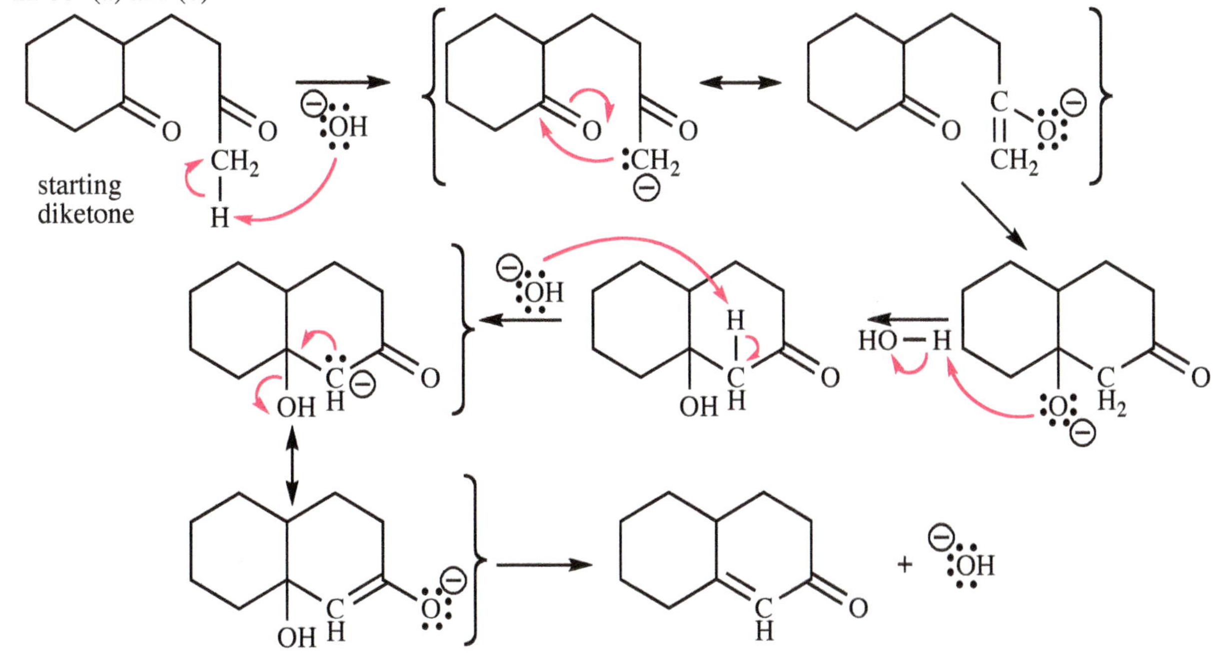

starting diketone

22-34

(a) The side reaction with sodium methoxide is transesterification. The starting material, and therefore the product, would be a mixture of methyl and ethyl esters.

$$H_3C-\overset{O}{\overset{\|}{C}}-OCH_2CH_3 \;+\; NaOCH_3 \rightleftharpoons H_3C-\overset{O}{\overset{\|}{C}}-OCH_3 \;+\; NaOCH_2CH_3$$

(b) Sodium hydroxide would irreversibly saponify the ester, completely stopping the Claisen condensation as the carbonyl no longer has a leaving group attached to it.

$$H_3C-\overset{O}{\overset{\|}{C}}-OCH_2CH_3 \;+\; NaOH \longrightarrow H_3C-\overset{O}{\overset{\|}{C}}-O^{\ominus}\ Na^+ \;+\; HOCH_2CH_3$$

22-35

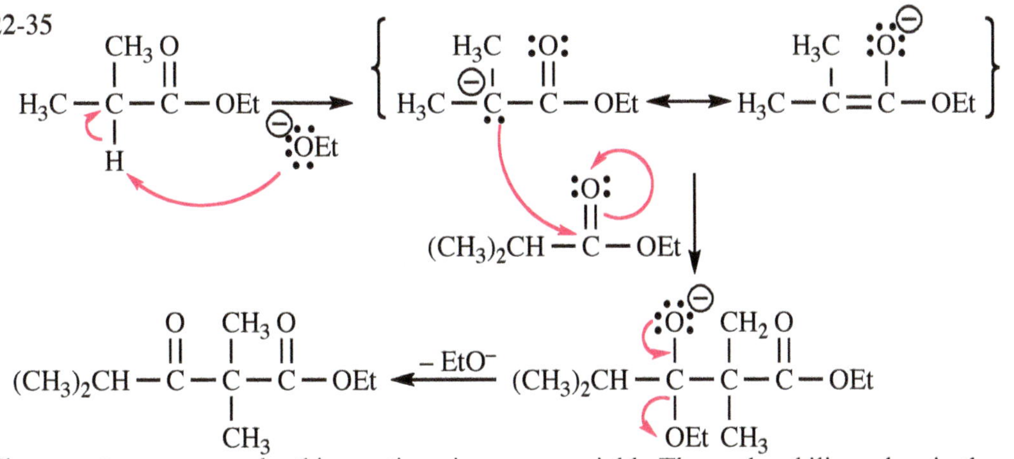

$(CH_3)_2CH-C(=O)-C(CH_3)_2-C(=O)-OEt$

There are two reasons why this reaction gives a poor yield. The nucleophilic carbon in the enolate is 3° and attack is hindered. More important, the final product has no hydrogen on the α-carbon, so the deprotonation by base that is the driving force in other Claisen condensations cannot occur here. What is produced is an *equilibrium mixture* of product and starting materials; the conversion to product is low.

22-36 Products after mild acid workup are shown. New carbon-carbon bonds are shown in bold.

(a) O, O, OCH_3

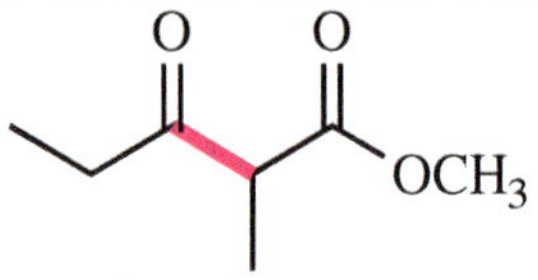

(b) Ph, O, O, OCH_2CH_3, Ph

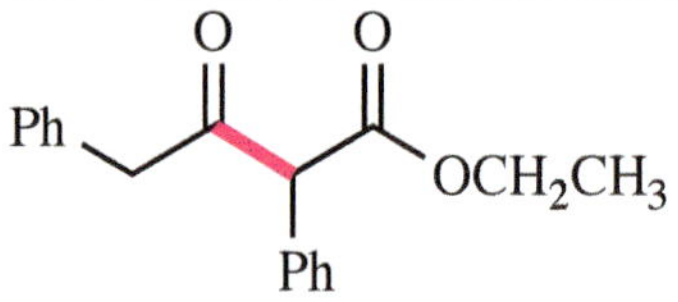

(c) O, O, OCH_3

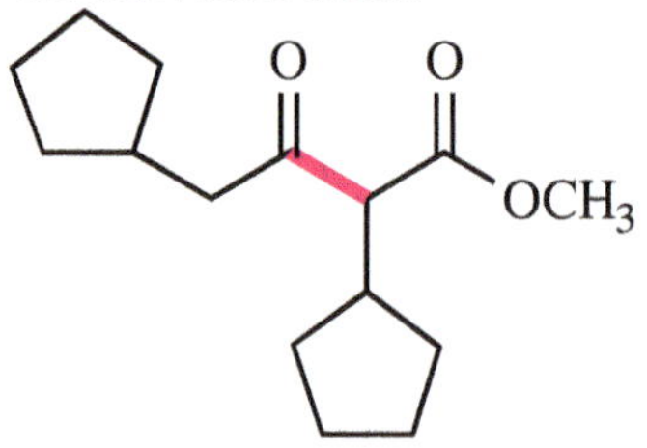

(d) O, O, O, OEt, O

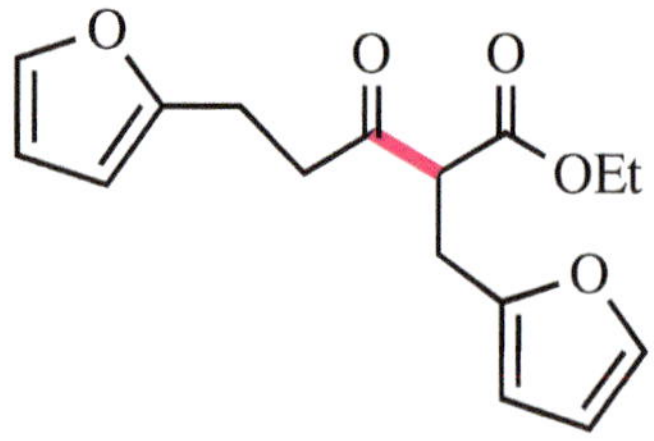

22-37

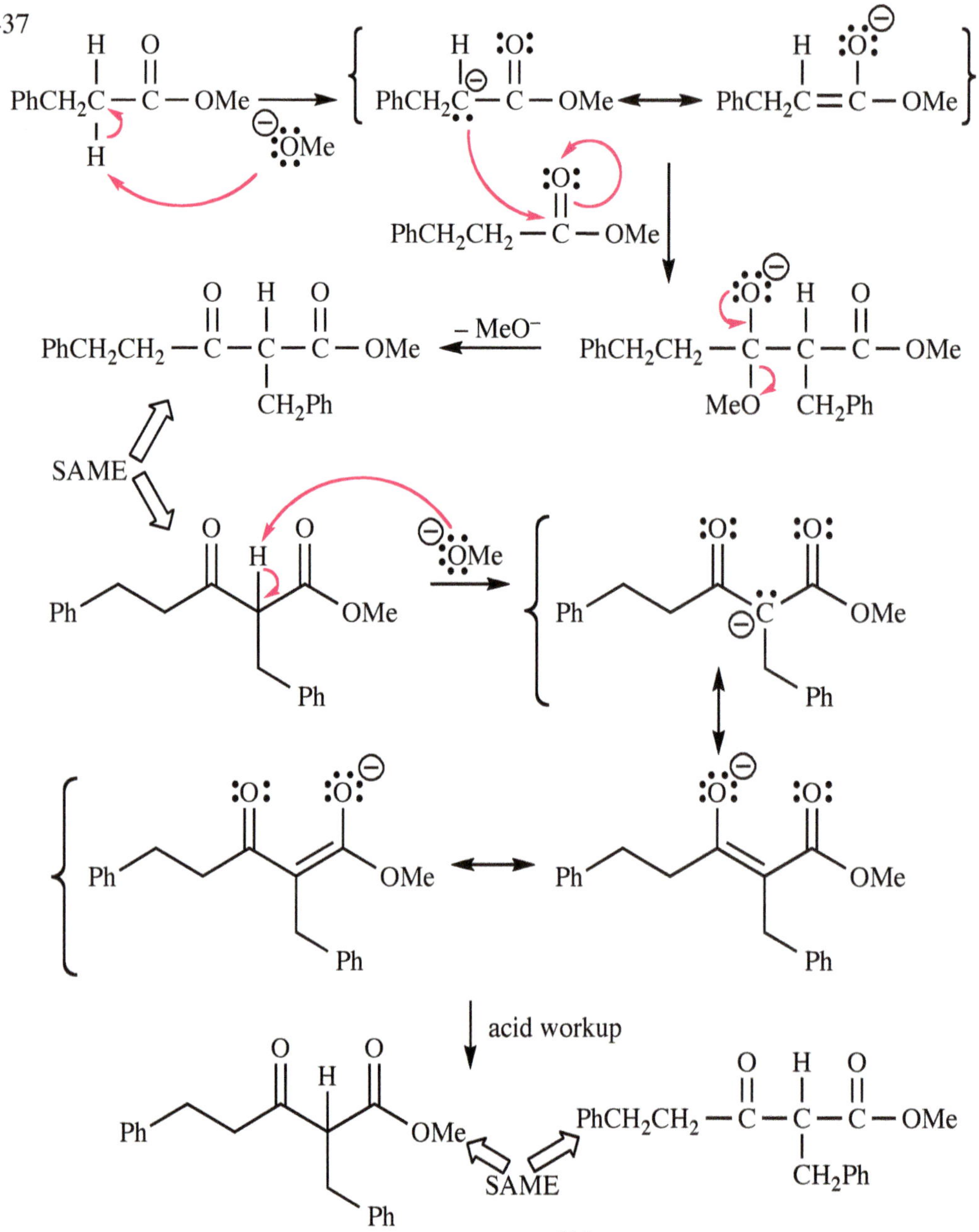

$PhCH_2C(H)(H)-C(=O)-OMe$ + $^{\ominus}$:OMe → { $PhCH_2\overset{\ominus}{C}H-C(=O)-OMe$ ⟷ $PhCH_2C(H)=C(O^{\ominus})-OMe$ }

+ $PhCH_2CH_2-C(=O)-OMe$ ↓

$PhCH_2CH_2-C(O^{\ominus})(MeO)-C(H)(CH_2Ph)-C(=O)-OMe$ —(– MeO⁻)→ $PhCH_2CH_2-C(=O)-C(H)(CH_2Ph)-C(=O)-OMe$

SAME

Ph, O, H, O, OMe, Ph + $^{\ominus}$:OMe → { Ph, :O:, :O:, C⊖, OMe, Ph ⟷ Ph, :O:⊖, :O:, OMe, Ph ⟷ Ph, :O:, :O:⊖, OMe, Ph }

↓ acid workup

Ph, O, H, O, OMe, Ph SAME $PhCH_2CH_2-C(=O)-C(H)(CH_2Ph)-C(=O)-OMe$

22-38 Condensed structural formulas and line formulas are shown.

(a) 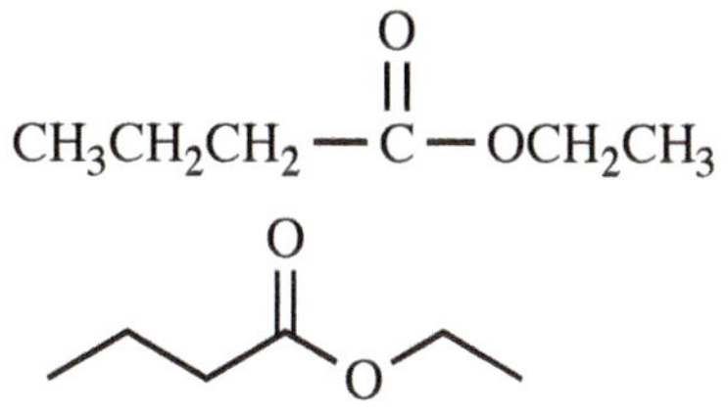 $CH_3CH_2CH_2-C(=O)-OCH_2CH_3$

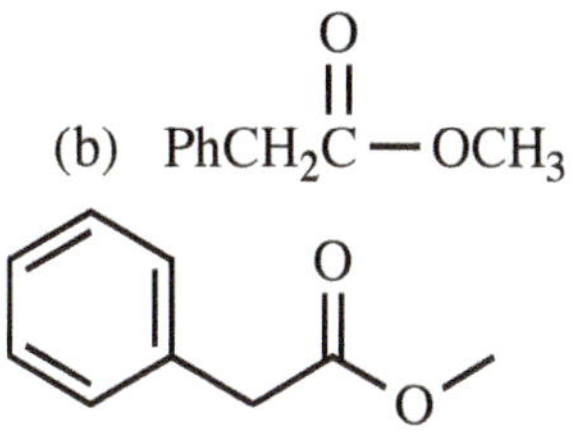

(b) $PhCH_2C(=O)-OCH_3$

(c) $CH_3CH(CH_3)CH_2-C(=O)-OCH_2CH_3$

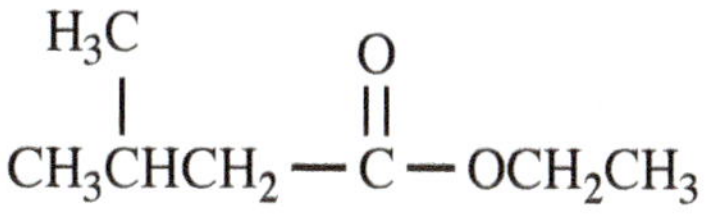

This one would be difficult because the alpha-carbon is hindered.

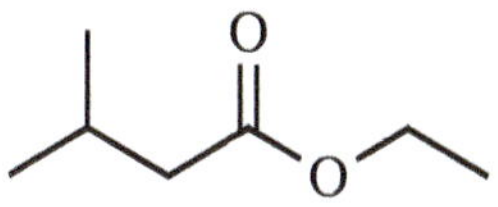

22-39

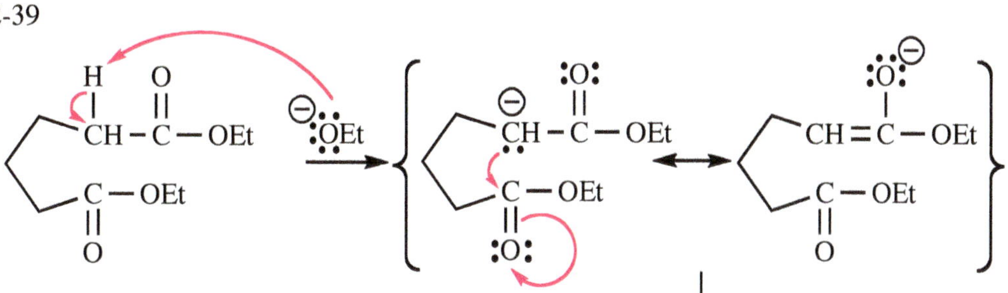

This is the final product, after removal of the α-hydrogen by ethoxide, followed by reprotonation during the workup.

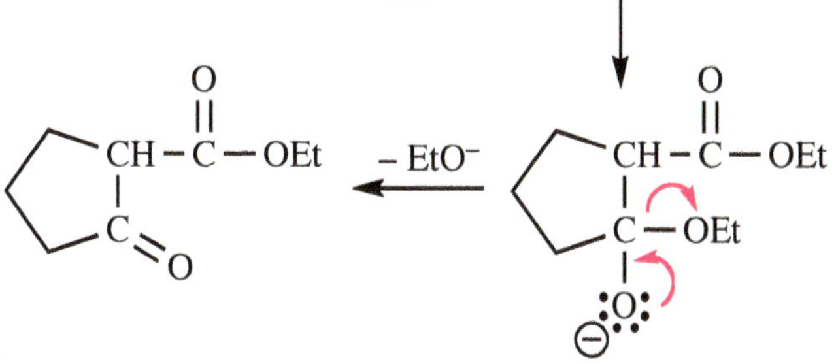

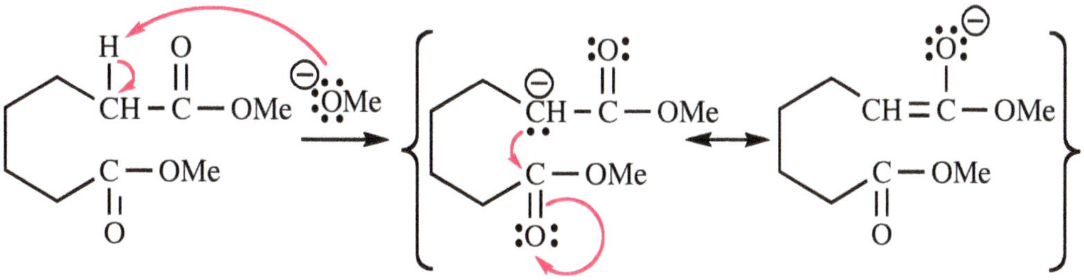

– EtO⁻

This is the final product, after removal of the α-hydrogen by methoxide, followed by reprotonation during the workup.

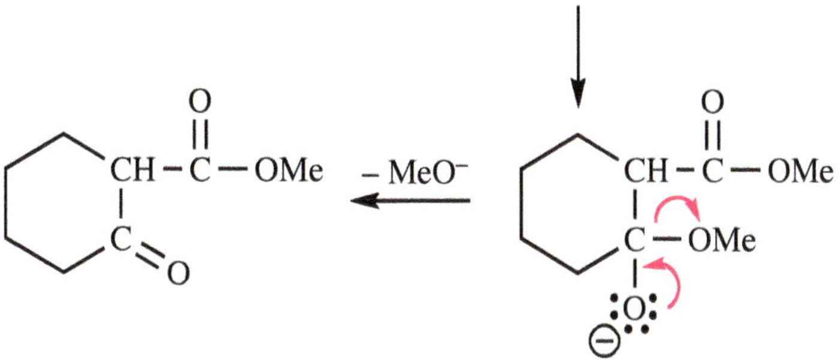

– MeO⁻

22-40

(a) This structure cannot be formed by a Dieckmann as it is not a β-keto ester

(b) 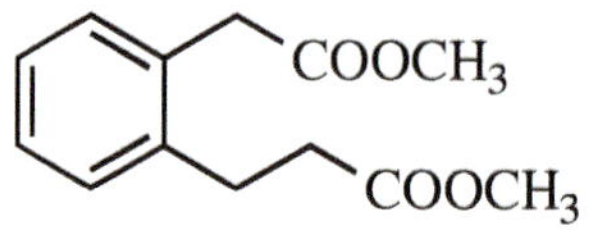 [benzene ring with CH_2COOCH_3 and $CH_2CH_2COOCH_3$ substituents] + $NaOCH_3$ (mixture of products results)

(c) 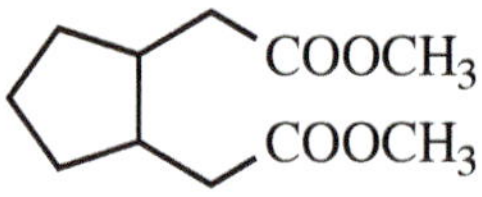 [cyclopentane with CH_2COOCH_3 and CH_2COOCH_3 substituents] + $NaOCH_3$

(d) 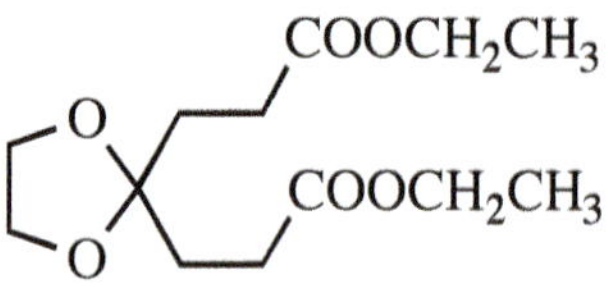 [1,3-dioxolane ketal bearing two $CH_2CH_2COOCH_2CH_3$ chains] + $NaOCH_2CH_3$

The protecting group is necessary to prevent aldol condensation. Aqueous acid workup removes the protecting group.

22-41

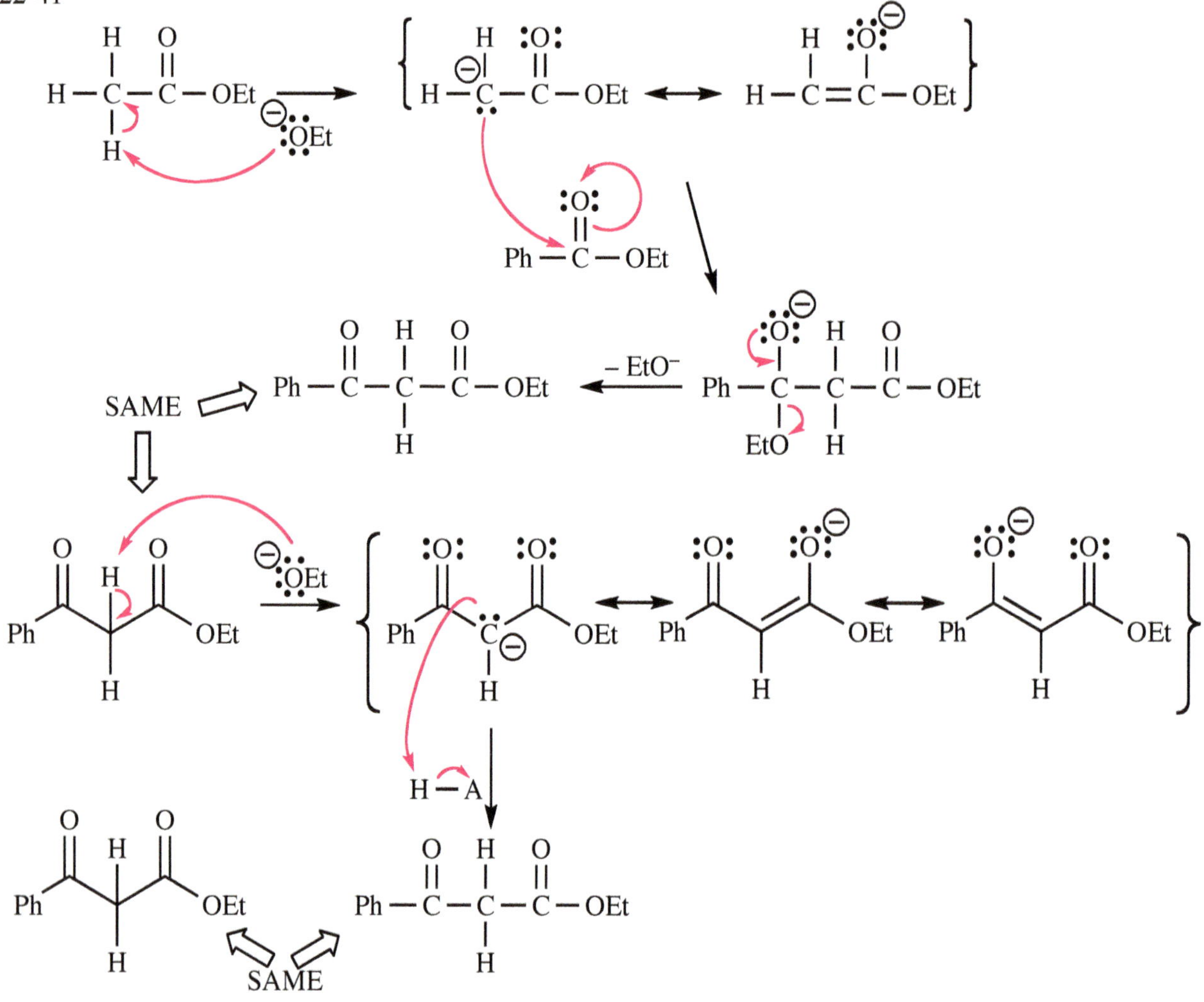

$H-CH_2-C(=O)-OEt$ + $^{\ominus}OEt$ → { $H-\overset{\ominus}{C}H-C(=O)-OEt$ ↔ $H-CH=C(-O^{\ominus})-OEt$ }

+ $Ph-C(=O)-OEt$ → $Ph-C(O^{\ominus})(EtO)-CH_2-C(=O)-OEt$ —(– EtO^-)→ $Ph-C(=O)-CH_2-C(=O)-OEt$

SAME

$PhC(=O)CH_2C(=O)OEt$ + $^{\ominus}OEt$ → { $PhC(=O)\overset{\ominus}{C}HC(=O)OEt$ ↔ $PhC(=O)CH=C(O^{\ominus})OEt$ ↔ $PhC(O^{\ominus})=CHC(=O)OEt$ }

H—A

$PhC(=O)CH_2C(=O)OEt$ $Ph-C(=O)-CH_2-C(=O)-OEt$

SAME

22-42 New carbon-carbon bonds are shown in bold.

(a) 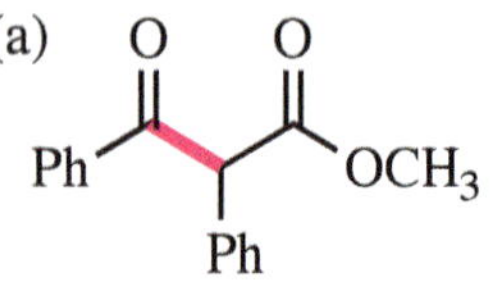 $PhC(=O)CH(Ph)C(=O)OCH_3$

(b) 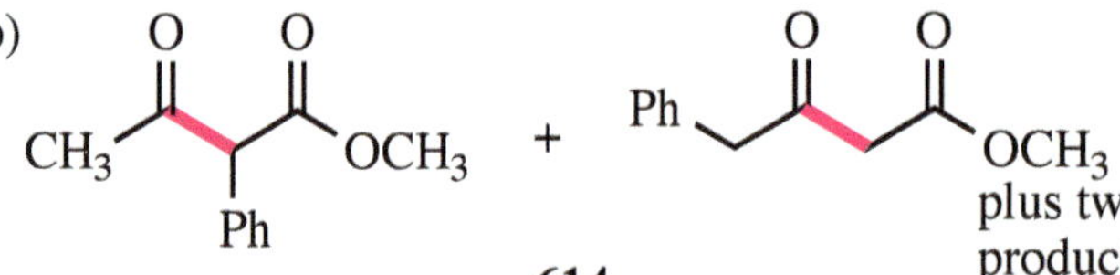 $CH_3C(=O)CH(Ph)C(=O)OCH_3$ + $PhCH_2C(=O)CH_2C(=O)OCH_3$

plus two self-condensation products—a poor choice because both esters have α-hydrogens

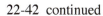

22-42 continued

(c) EtO–C(=O)–C(=O)–CH$_2$–C(=O)–OEt

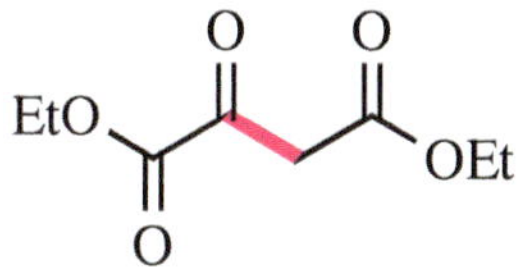

(d) EtO–C(=O)–CH(CH$_3$)–C(=O)–OEt

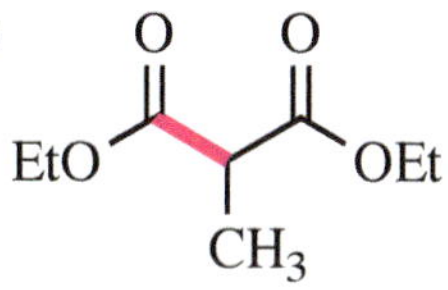

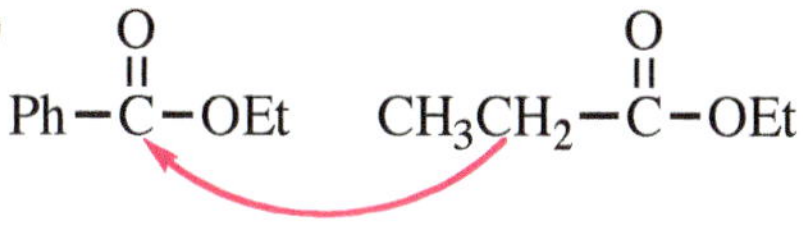

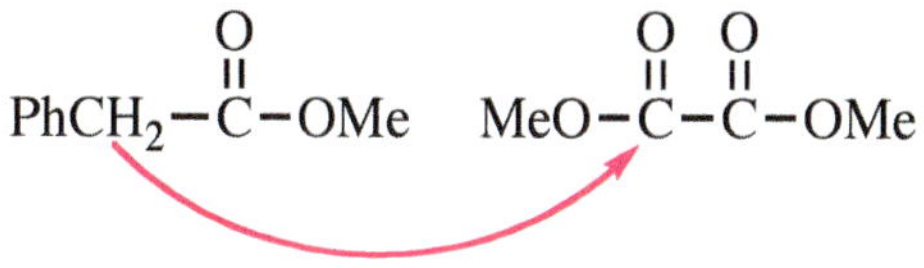

22-43

(a) $Ph-\overset{O}{\overset{||}{C}}-OEt$ $\quad CH_3CH_2-\overset{O}{\overset{||}{C}}-OEt$

(b) $PhCH_2-\overset{O}{\overset{||}{C}}-OMe$ $\quad MeO-\overset{O}{\overset{||}{C}}-\overset{O}{\overset{||}{C}}-OMe$

(c) $EtO-\overset{O}{\overset{||}{C}}-CH_2Ph$ $\quad EtO-\overset{O}{\overset{||}{C}}-OEt$

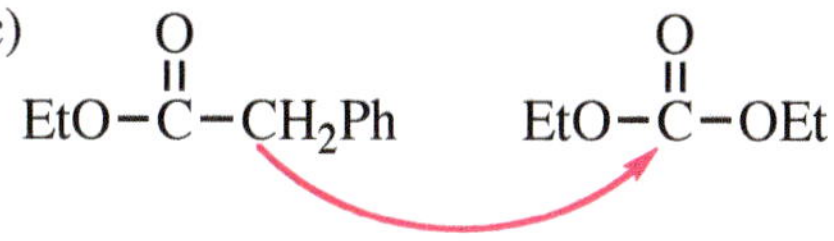

(d) $(CH_3)_3C-\overset{O}{\overset{||}{C}}-OMe$ $\quad CH_3CH_2CH_2CH_2-\overset{O}{\overset{||}{C}}-OMe$

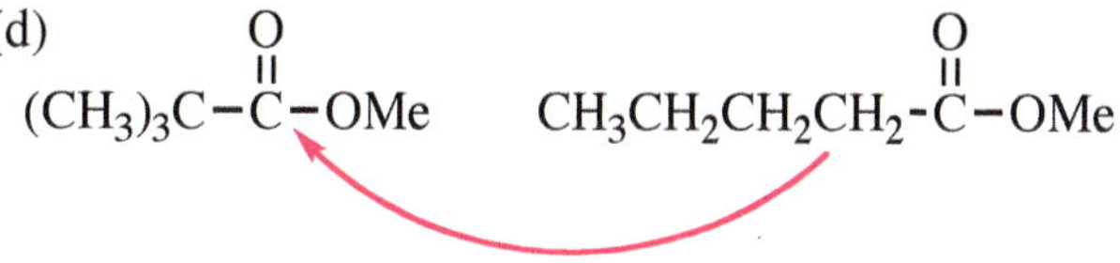

22-44 These Claisen products would likely be accompanied by aldol by-products.

(a) O O Ph

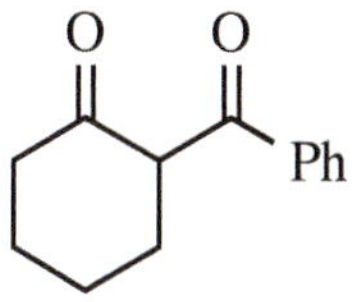

(b) O O + O O

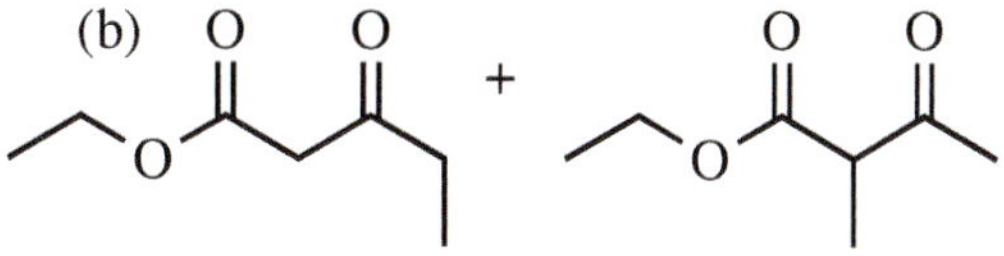

Unsymmetrical ketones can form two different enolates.

(c) O O

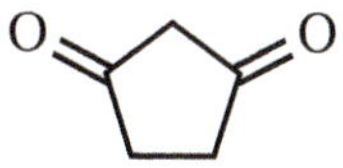

22-45

(a) two ways: O + $CH_3O-\overset{O}{\overset{||}{C}}-Ph$ OR O OCH$_3$ Ph O

(b) $CH_3CH_2-\overset{O}{\overset{||}{C}}-CH_2CH_3 + CH_3CH_2O-\overset{O}{\overset{||}{C}}-\overset{O}{\overset{||}{C}}-OCH_2CH_3$

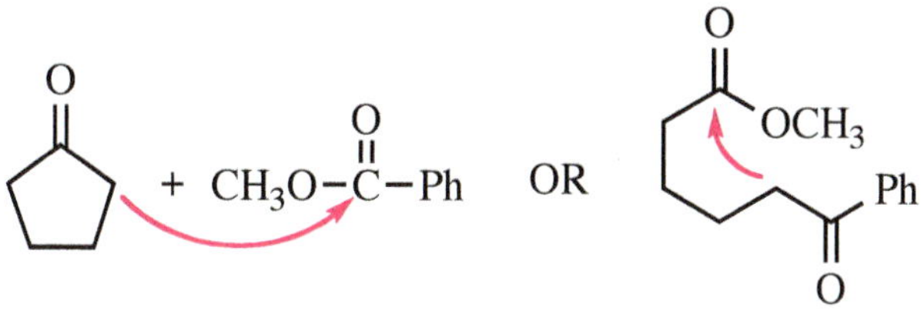

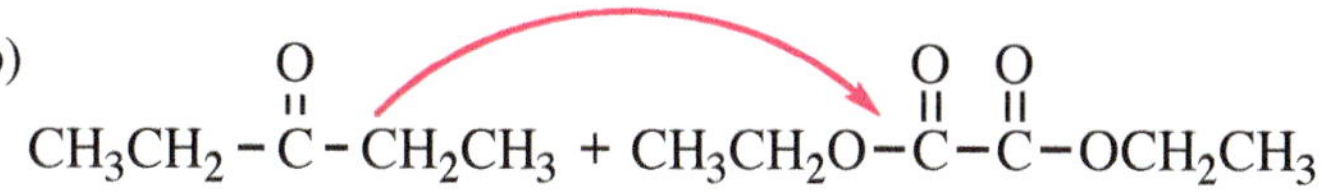

(c) O O OCH$_2$CH$_3$

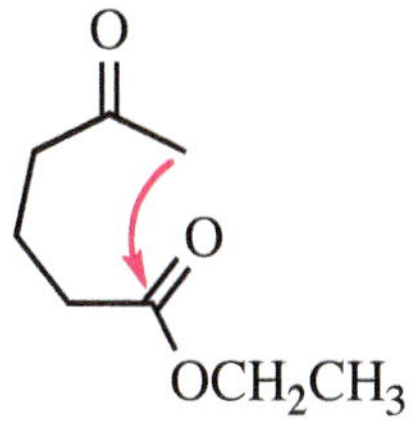

(d) two ways: O O + $CH_3CH_2O-\overset{O}{\overset{||}{C}}-OCH_2CH_3$ OR O O OCH$_2$CH$_3$ O OCH$_2$CH$_3$

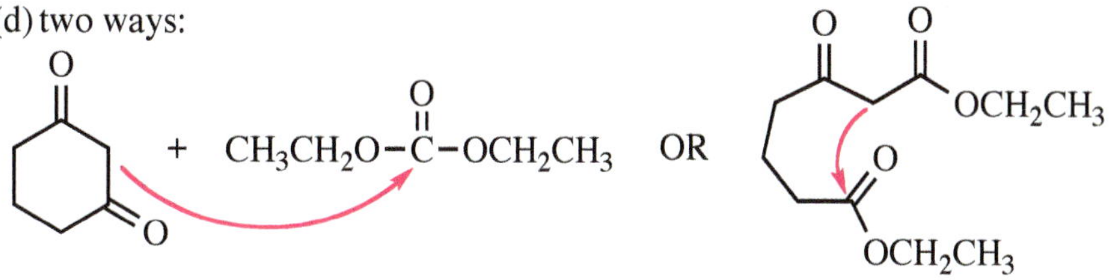

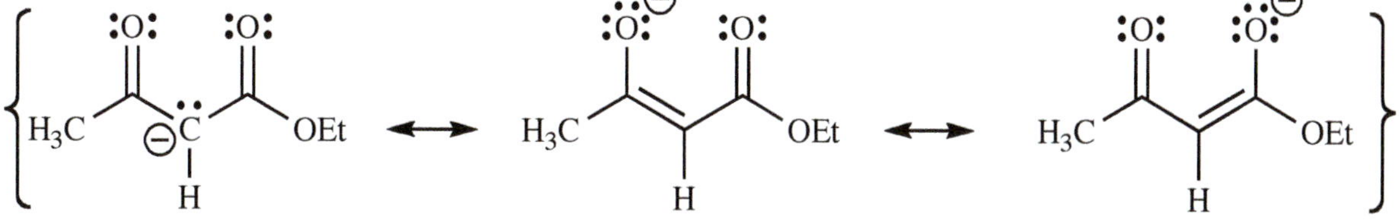

22-46

(a) { H$_3$C–C(=O)–C$^{\ominus}$H–C(=O)–OEt ⟷ H$_3$C–C(–O$^{\ominus}$)=CH–C(=O)–OEt ⟷ H$_3$C–C(=O)–CH=C(–O$^{\ominus}$)–OEt }

22-46 continued

(b)

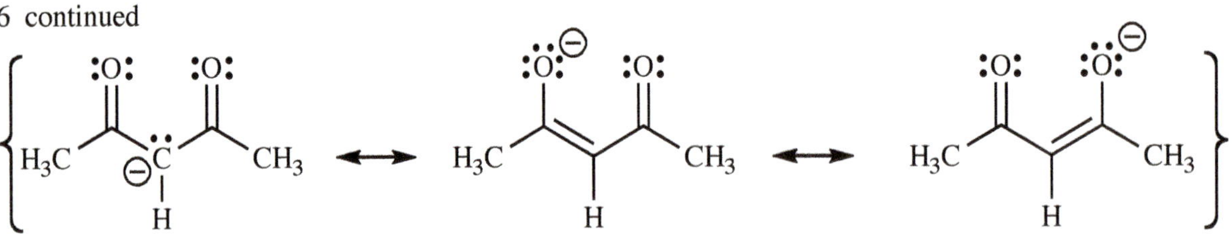

(c)

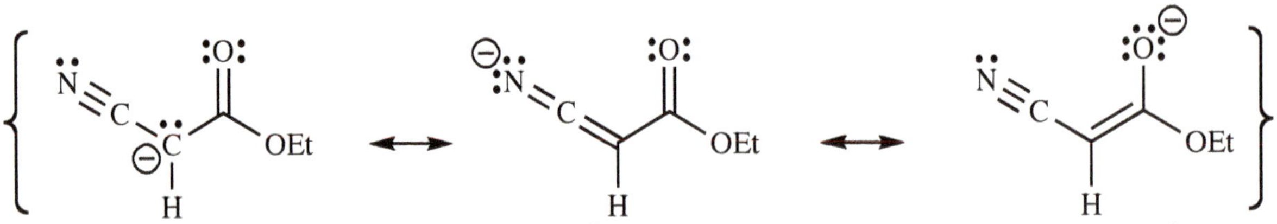

(d)

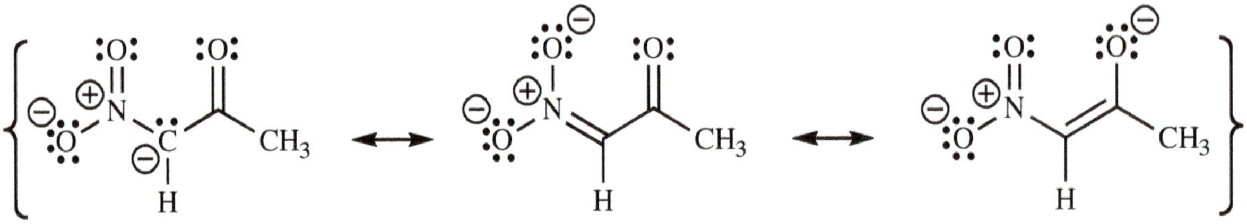

(Other resonance forms of the nitro group are not shown.)

22-47 In the products, the wavy lines cross the bonds that must be made by alkylation, before hydrolysis and decarboxylation produce the substituted acetic acid.

(a) EtO(C=O)CH$_2$(C=O)OEt —1) NaOEt, 2) PhCH$_2$Br→ EtO(C=O)CH(CH$_2$Ph)(C=O)OEt —H_3O^+, Δ→ PhCH$_2$CH$_2$COOH + CO_2 + 2 EtOH

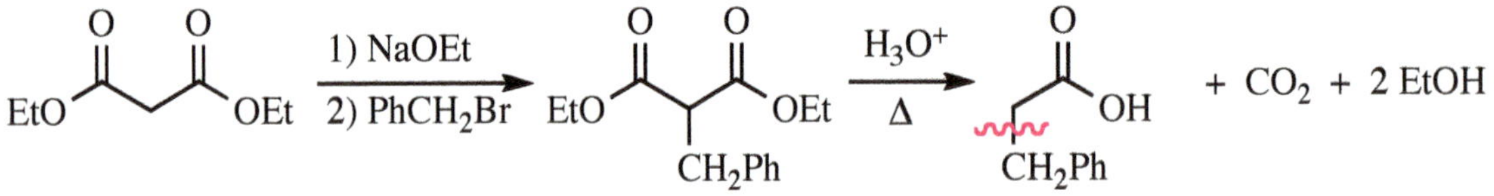

(b) EtO(C=O)CH$_2$(C=O)OEt —1) NaOEt, 2) CH_3I→ —1) NaOEt, 2) CH_3I→ EtO(C=O)C(CH$_3$)$_2$(C=O)OEt —H_3O^+, Δ→ (CH$_3$)$_2$CHCOOH + CO_2 + 2 EtOH

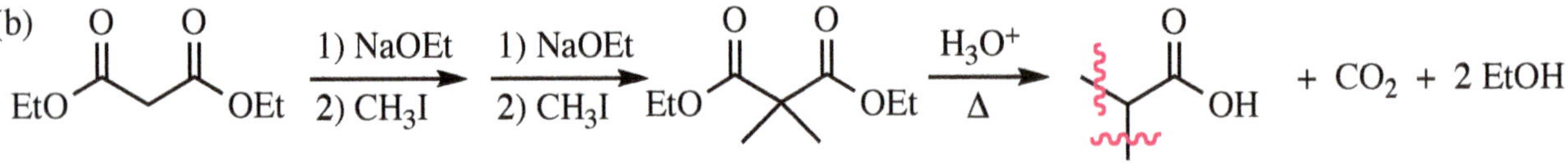

(c) EtO(C=O)CH$_2$(C=O)OEt —1) NaOEt, 2) Ph(CH$_2$)$_2$Br→ EtO(C=O)CH(CH$_2$CH$_2$Ph)(C=O)OEt —H_3O^+, Δ→ Ph(CH$_2$)$_3$COOH + CO_2 + 2 EtOH

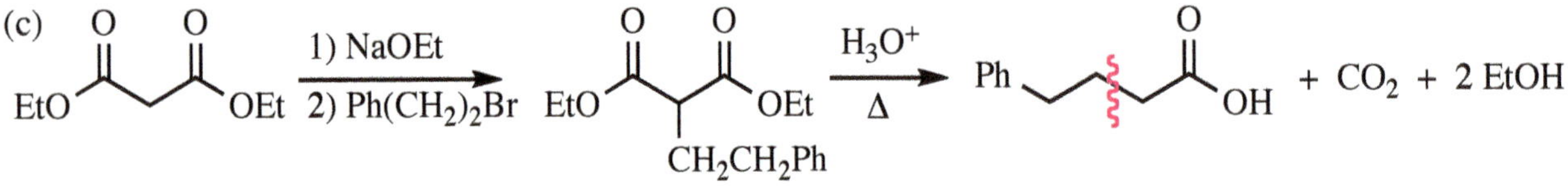

(d) EtO(C=O)CH$_2$(C=O)OEt —1) 2 NaOEt, 2) Br(CH$_2$)$_4$Br→ EtO(C=O)C(cyclopentane)(C=O)OEt —H_3O^+, Δ→ cyclopentane-COOH + CO_2 + 2 EtOH

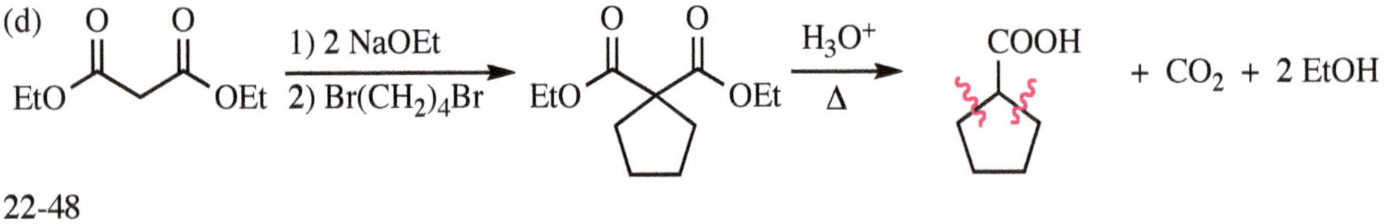

22-48

(a) Only two substituent groups plus a hydrogen atom can appear on the alpha carbon after decarboxylation at the end of the malonic ester synthesis. The product shown has three alkyl groups, so it cannot be made by malonic ester synthesis.

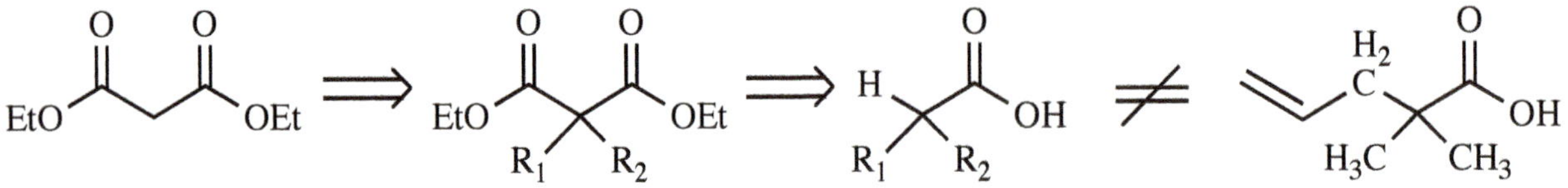

EtO(C=O)CH$_2$(C=O)OEt ⟹ EtO(C=O)C(R$_1$)(R$_2$)(C=O)OEt ⟹ H–C(R$_1$)(R$_2$)–COOH ≠ CH$_2$=CH–CH$_2$–C(CH$_3$)$_2$–COOH

desired product

22-48 continued

(b) H_3C, CH_3 ester (pK$_a$ 24) + LDA ⟶ HN (pK$_a$ ≈40) + {enolate ⟷ enolate resonance form, Li^+}

pK$_a$ 24

LDA

pK$_a$ ≈40

With such a large difference in pK$_a$ values, products are favored >> 99%.

(c) methyl 2-methylpropanoate $\xrightarrow{\text{LDA}}$ enolate $\xrightarrow[S_N2]{\text{allyl Br}}$ methyl 2,2-dimethylpent-4-enoate $\xrightarrow[\Delta]{H_3O^+}$ 2,2-dimethylpent-4-enoic acid

plus resonance form as shown in part (b)

Saponification in NaOH with an acid workup is another option here.

22-49

(a) $PhCH_2CH_2-\overset{O}{\overset{\|}{C}}-CH_3$ + CO_2 + EtOH

(b) cyclobutyl methyl ketone + CO_2 + EtOH

(c) cyclopentanone + CO_2 + EtOH

22-50 In the products, the wavy lines indicate the bonds that must be made by alkylation, before hydrolysis and decarboxylation produce the substituted acetone.

(a) ethyl acetoacetate $\xrightarrow[\text{2) } PhCH_2Br]{\text{1) NaOEt}}$ EtO-ester with CH_2Ph $\xrightarrow[\Delta]{H_3O^+}$ CH_2Ph ketone + CO_2 + EtOH

(b) ethyl acetoacetate $\xrightarrow[\text{2) } Br(CH_2)_4Br]{\text{1) 2 NaOEt}}$ EtO-ester with cyclopentane ring $\xrightarrow[\Delta]{H_3O^+}$ cyclopentyl methyl ketone + CO_2 + EtOH

(c) ethyl acetoacetate $\xrightarrow[\text{2) } PhCH_2Br]{\text{1) NaOEt}}$ EtO-ester with CH_2Ph $\xrightarrow[\text{2) Br-allyl}]{\text{1) NaOEt}}$ EtO-ester with CH_2Ph and allyl $\xrightarrow{\Delta \mid H_3O^+}$ CO_2 + EtOH + allyl/benzyl substituted ketone (Ph)

22-51

(a) There are two problems with attempting to make this compound by acetoacetic ester synthesis. The acetone "core" of the product is shown in the box. This product would require alkylation at BOTH carbons of the acetone "core" of acetoacetic ester; in reality, only one carbon undergoes alkylation in the acetoacetic ester synthesis. Second, it is not possible to do an S_N2 type reaction on an unsubstituted benzene ring, so neither benzene could be attached by acetoacetic ester synthesis.

(b)

LDA

Br

S_N2

plus resonance forms showing delocalization of e^- into ring and C=O

(c)

H^+

Br

H_3O^+

S_N2

22-52

β γ

α

came from

Michael acceptor

Michael donor

forward direction

NaOEt

COOEt

temporary ester group

resonance-stabilized

H_3O^+

CO_2 + EtOH +

H_3O^+

Δ

22-53 First, you might wonder why this sequence does not make the desired product:

MVK

poor yield

resonance-stabilized

The poor yield in this conjugate addition is due primarily to the numerous competing reactions: the ketone enolate can self-condense (aldol), can condense with the ketone of MVK (aldol), or can deprotonate the methyl of MVK to generate a new nucleophile. The complex mixture of products makes this route practically useless. (continued on next page)

22-53 continued

What permits enamines (or other stabilized enolates) to work are: a) the certainty of which atom is the nucleophile, and b) the lack of self-condensation. Enamines can also do conjugate addition:

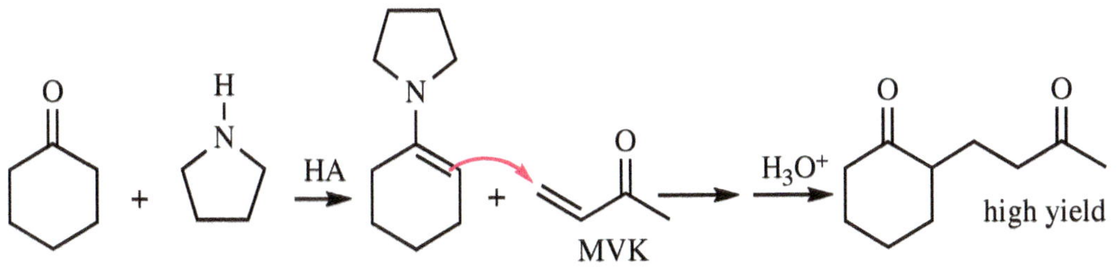

O + H–N (pyrrolidine) →(HA) N-enamine + MVK → →(H_3O^+) high yield

22-54 The enolate of acetoacetic ester can be used in a Michael addition to an α,β-unsaturated ketone like MVK.

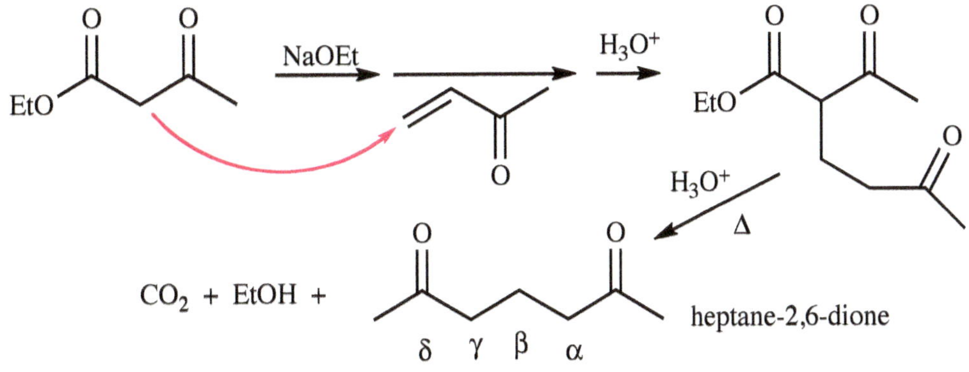

EtO (acetoacetic ester) →(NaOEt) → →(H_3O^+) EtO

→(H_3O^+, Δ) CO_2 + EtOH + heptane-2,6-dione

δ γ β α

22-55 { :N≡C–CH=CH$_2$ ⟷ N=C–CH=CH$_2$ ⟷ N=C=CH–CH$_2$ }

acrylonitrile

Nuc:

N≡C–CH$_2$–CH$_2$–Nuc ← Nuc—H { :N≡C–CH–CH$_2$–Nuc ⟷ N=C=CH–CH$_2$–Nuc }

{ O–N–CH=CH$_2$ ⟷ O–N=CH–CH$_2$ } → Nuc: { O–N=CH–CH$_2$–Nuc }

nitroethylene

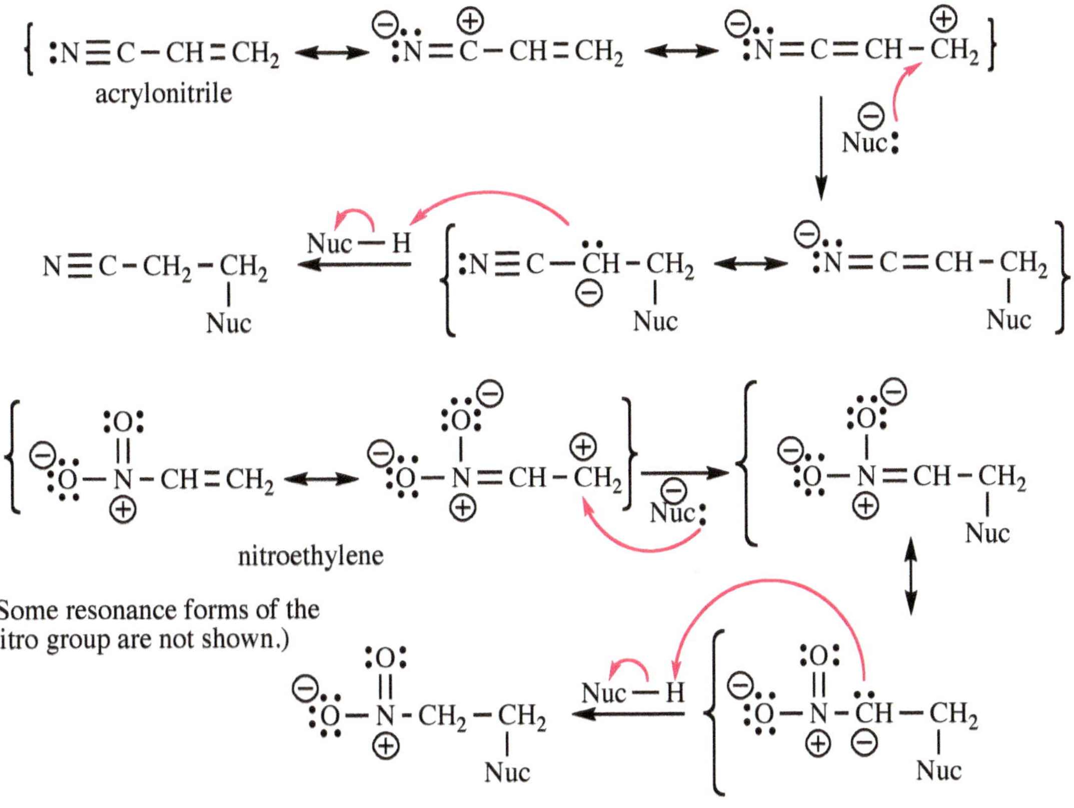

(Some resonance forms of the nitro group are not shown.)

O–N–CH$_2$–CH$_2$–Nuc ← Nuc—H { O–N–CH–CH$_2$–Nuc }

22-56

(a) $PhCH{=}CH{-}\overset{O}{\overset{\|}{C}}{-}OEt$

EtO, OEt (diethyl malonate)

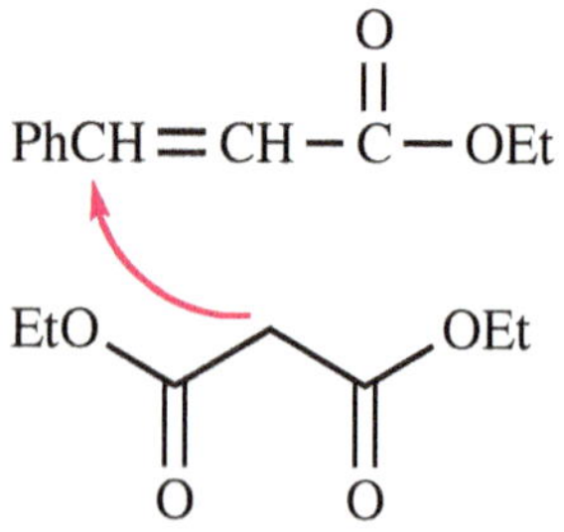

(b) $CH_2{=}CH{-}C{\equiv}N$

EtO (ethyl acetoacetate)

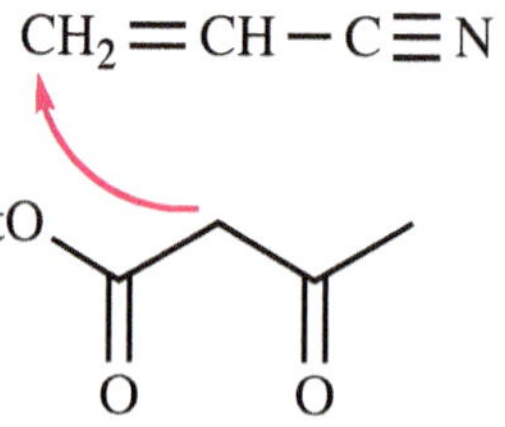

followed by hydrolysis and decarboxylation

(c) two ways

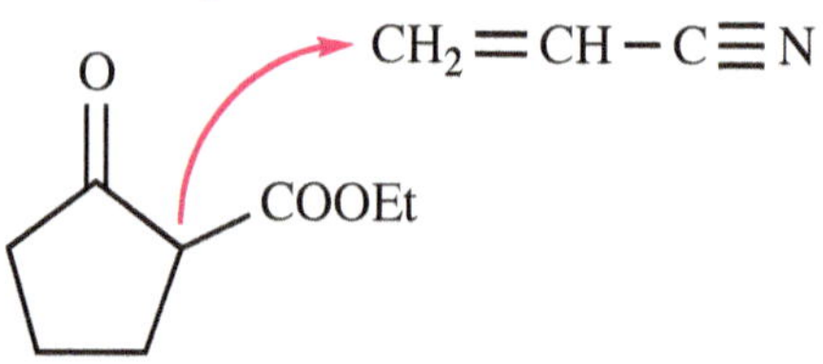

$CH_2{=}CH{-}C{\equiv}N$

O

COOEt

followed by hydrolysis and decarboxylation

OR

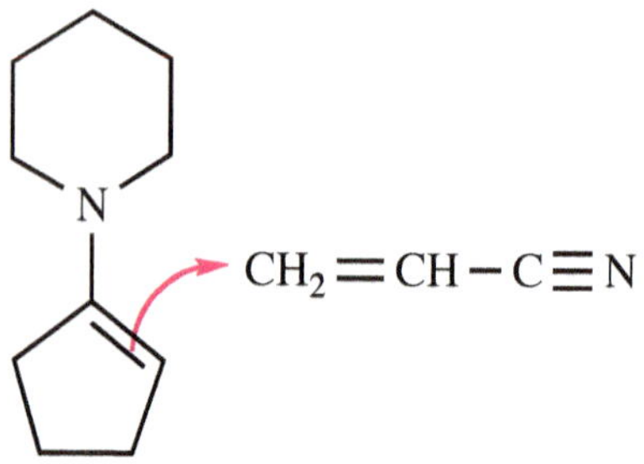

N

$CH_2{=}CH{-}C{\equiv}N$

followed by hydrolysis

(d) H_3C CH_3 N

$CH_2{=}CH{-}\overset{O}{\overset{\|}{C}}{-}Ph$

CH_3

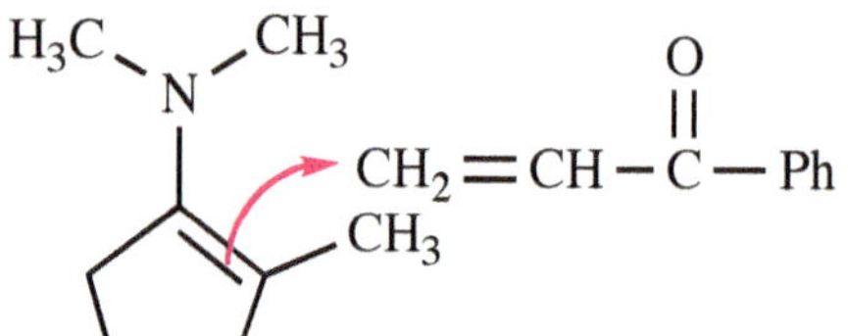

followed by hydrolysis

(Here is a dose of reality: The accuracy checker has correctly pointed out that the reaction in part (d) is more challenging than it first appears. The enamine shown is actually the minor product from formation of the enamine (about 40%, whereas the double bond in the less substituted position is about 60%). The two isomers can be separated, with the desired isomer being used for this reaction. As an alternative, formation of an enolate is equally problematic; deprotonation of the ketone with LDA gives the less substituted enolate in about 99%! Conclusion: the reaction shown can be used, but there is more to the real-life story.)

(e) O O OEt

1) NaOEt 2) CH_3I

O O OEt + O

NaOEt

Δ H_3O^+

O O + CO_2 + EtOH

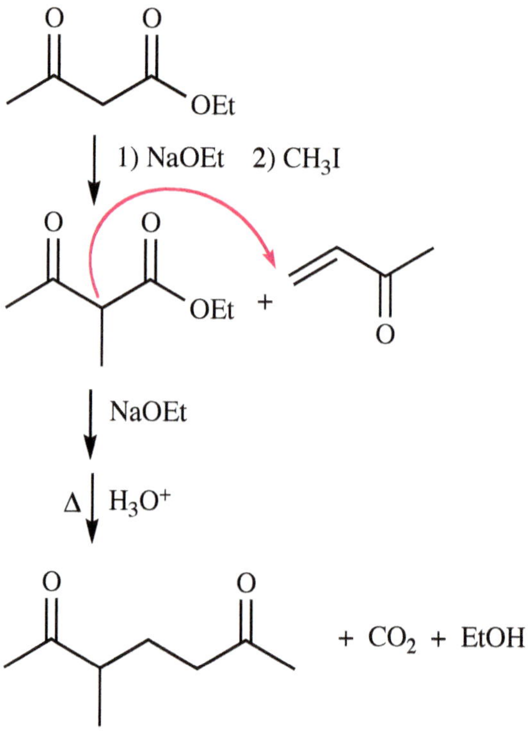

(Could also be synthesized by the Stork enamine reactions.)

(f) O

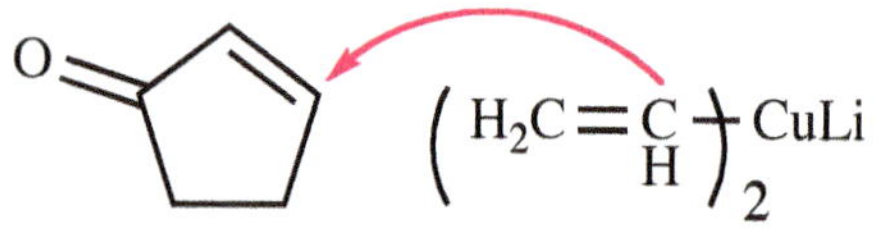

$\left(H_2C{=}\underset{H}{C}\right)_2 CuLi$

22-57

Step 1: carbon skeleton

comes from

Step 2: nucleophile generation

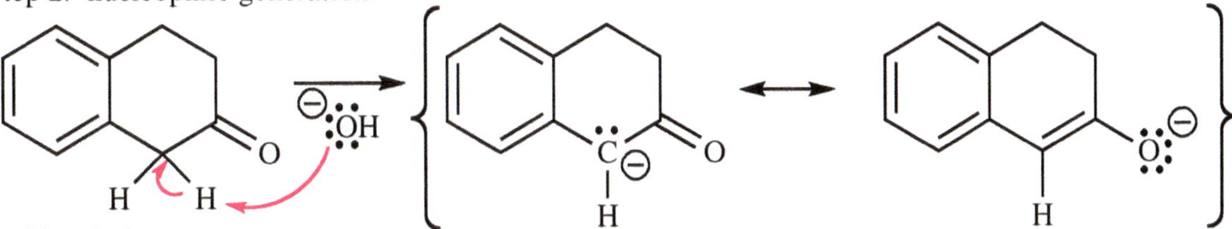

These hydrogen atoms are more acidic than the other alpha hydrogens because the anion can be stabilized by the benzene ring.

plus resonance forms with negative charge on the benzene ring

Step 3: nucleophilic attack (Michael addition)

continued on next page

22-57 continued

Step 4: conversion to final product

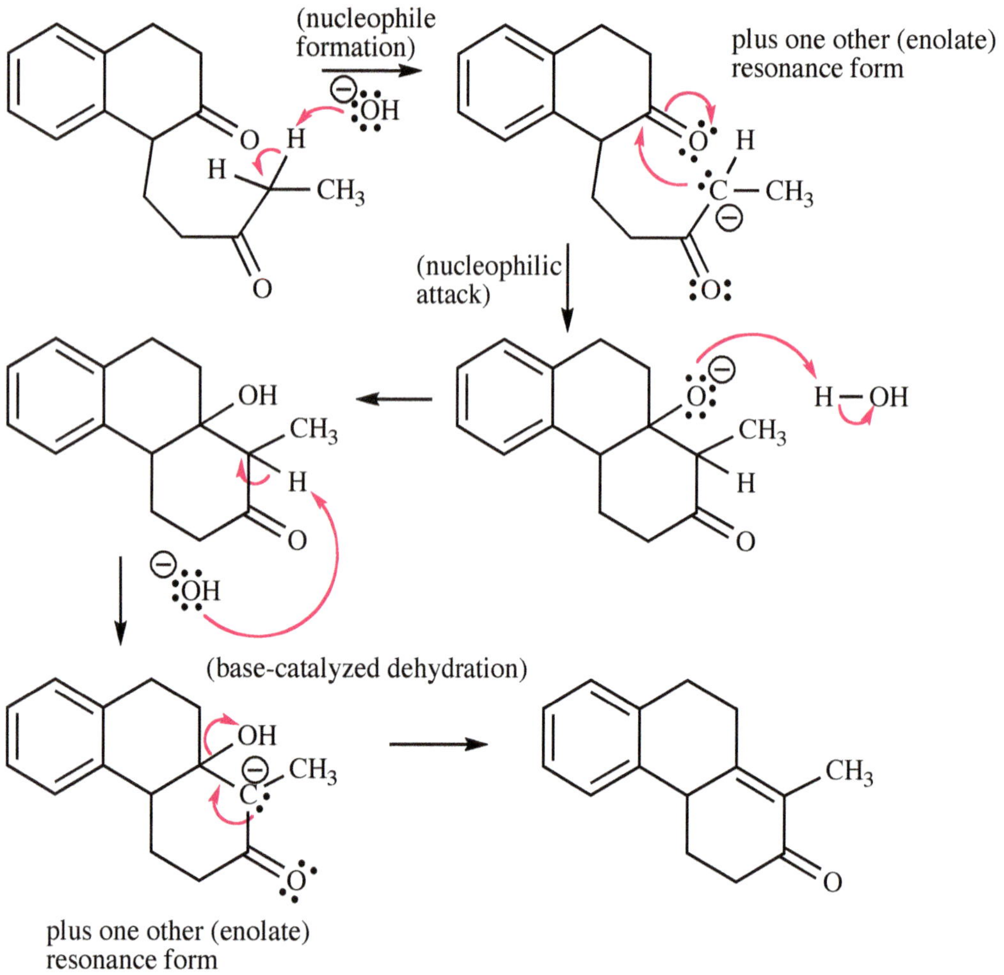

(nucleophile formation)

plus one other (enolate) resonance form

(nucleophilic attack)

(base-catalyzed dehydration)

plus one other (enolate) resonance form

Step 5: The complete mechanism is the combination of Steps 2, 3, and 4. Notice that this mechanism is simply described by:
1) Enolate formation, followed by Michael addition;
2) Aldol condensation, followed by dehydration.

22-58

Step 1: carbon skeleton

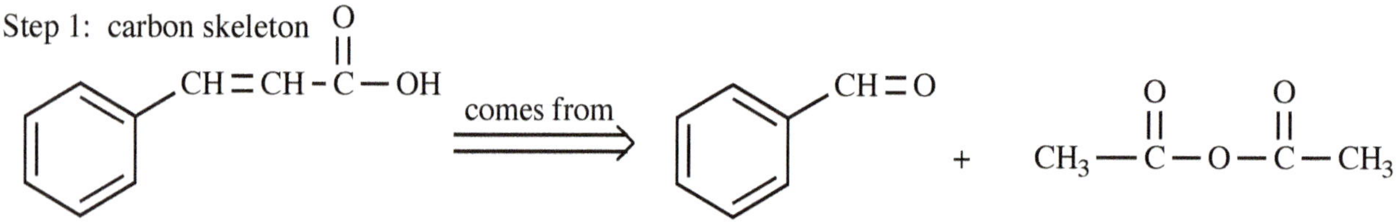

$C_6H_5-CH=CH-C(=O)-OH$ comes from $C_6H_5-CH=O$ + $CH_3-C(=O)-O-C(=O)-CH_3$

Step 2: nucleophile generation

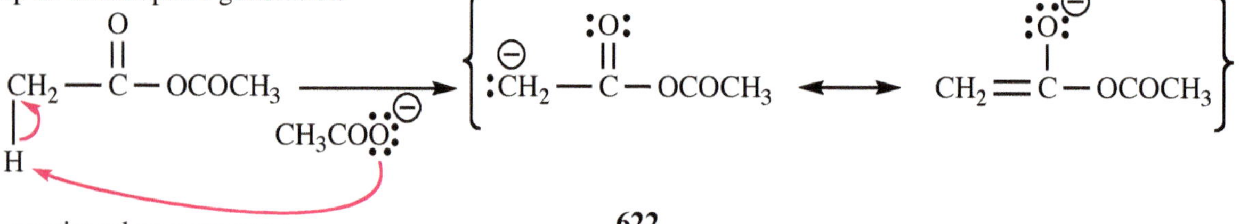

$CH_2(H)-C(=O)-OCOCH_3$ + CH_3COO^{-} $\longrightarrow$ $\{ {}^{-}:CH_2-C(=O)-OCOCH_3 \longleftrightarrow CH_2=C(O^{-})-OCOCH_3 \}$

continued on next page

22-58 continued

Step 3: nucleophilic attack

$$Ph-\overset{\overset{\displaystyle :\ddot{O}:}{\|}}{C}-H \;+\; \overset{\ominus}{:}CH_2-\overset{\overset{\displaystyle O}{\|}}{C}-OCOCH_3 \longrightarrow Ph-\overset{\overset{\displaystyle \ominus:\ddot{O}:}{|}}{CH}-CH_2-\overset{\overset{\displaystyle O}{\|}}{C}-OCOCH_3 \xrightarrow{H-OAc} Ph-\overset{\overset{\displaystyle OH}{|}}{CH}-CH_2-\overset{\overset{\displaystyle O}{\|}}{C}-OCOCH_3$$

Step 4: conversion to final product

$$Ph-\overset{\overset{\displaystyle OH}{|}}{CH}-\underset{\underset{\displaystyle H}{|}}{CH}-\overset{\overset{\displaystyle O}{\|}}{C}-OCOCH_3 \xrightarrow{CH_3CO\ddot{O}:^{\ominus}} Ph-\overset{\overset{\displaystyle OH}{|}}{CH}-\overset{\ominus}{\ddot{C}H}-\overset{\overset{\displaystyle :\ddot{O}:}{\|}}{C}-OCOCH_3$$

plus one other (enolate) resonance form

$$\downarrow$$

$$Ph-CH=CH-\overset{\overset{\displaystyle O}{\|}}{C}-OCOCH_3$$

<u>hydrolysis</u>

$$PhCH=CH-\overset{\overset{\displaystyle :\ddot{O}:}{\|}}{C}-O-\overset{\overset{\displaystyle O}{\|}}{C}-CH_3 \xrightarrow{H-A} \left\{ PhCH=CH-\overset{\overset{\displaystyle :\overset{\oplus}{O}-H}{\|}}{C}-O-\overset{\overset{\displaystyle O}{\|}}{C}-CH_3 \longleftrightarrow PhCH=CH-\overset{\overset{\displaystyle :\ddot{O}-H}{|}}{\underset{\oplus}{C}}-O-\overset{\overset{\displaystyle O}{\|}}{C}-CH_3 \right.$$

plus other resonance forms with:
a) positive on the other oxygen;
b) positive on the benzylic carbon;
c) positive on the benzene ring

$$\xrightarrow{H_2\ddot{O}:} PhCH=CH-\overset{\overset{\displaystyle OH}{|}}{\underset{\underset{\displaystyle H-\overset{\oplus}{\ddot{O}}-H}{|}}{C}}-O-\overset{\overset{\displaystyle :\ddot{O}:}{\|}}{C}-CH_3 \quad (H-A;\ H_2\ddot{O}:)$$

two fast proton transfers

$$PhCH=CH-\overset{\overset{\displaystyle OH}{|}}{\underset{\underset{\displaystyle OH}{|}}{C}}-O-\overset{\overset{\displaystyle :\overset{\oplus}{O}-H}{\|}}{C}-CH_3 \longrightarrow O=\overset{\overset{\displaystyle O-H}{|}}{\underset{\underset{\displaystyle CH_3}{|}}{C}} \;+\; \left\{ PhCH=CH-\overset{\overset{\displaystyle H-\ddot{O}:}{|}}{\underset{\underset{\displaystyle :\ddot{O}H}{|}}{C^{\oplus}}} \longleftrightarrow PhCH=CH-\overset{\overset{\displaystyle H-\overset{\oplus}{O}:}{\|}}{\underset{\underset{\displaystyle :\ddot{O}H}{|}}{C}} \right.$$

plus two other resonance forms

plus other resonance forms with:
a) positive on the other oxygen;
b) positive on the benzylic carbon;
c) positive on the benzene ring

$$\xrightarrow{H_2\ddot{O}:} PhCH=CH-\overset{\overset{\displaystyle O}{\|}}{C}-OH$$

Step 5: The complete mechanism is the combination of Steps 2, 3, and 4.

22-59 The Robinson annulation consists of a Michael addition followed by aldol cyclization with dehydration. In the retrosynthetic direction, disconnect the alkene formed in the aldol/dehydration, then disconnect the Michael addition to discover the reactants.

(a) Aldol and dehydration form the α,β double bond:

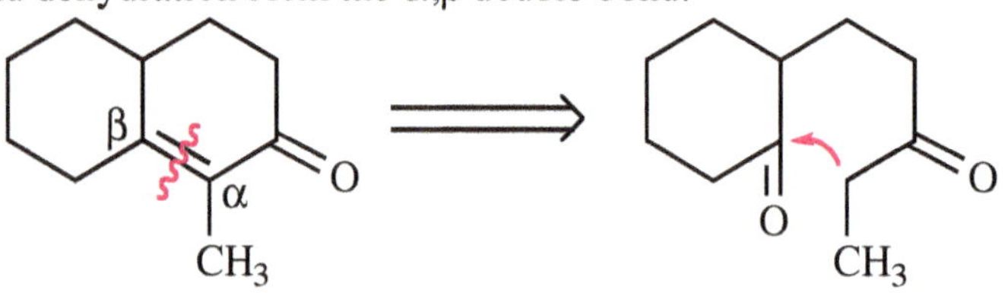

β α O CH_3 ⟹ O O CH_3

Michael addition forms a bond to the β' carbon:

β' α' O O CH_3 ⟹ O + O CH_3

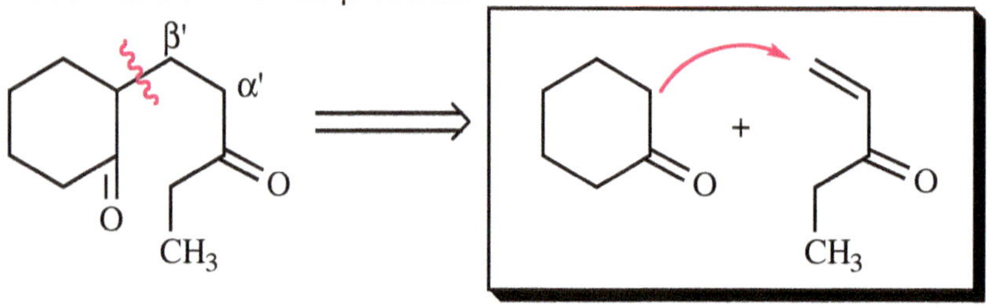

(b) Aldol and dehydration form the α,β double bond:

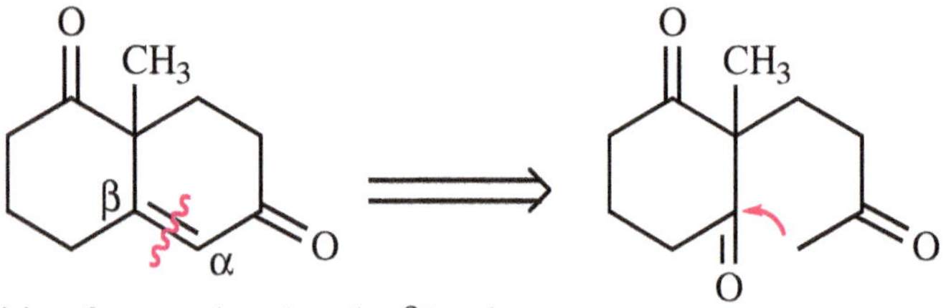

O CH_3 β α O ⟹ O CH_3 O O

Michael addition forms a bond to the β' carbon:

O CH_3 β' α' O O ⟹ O CH_3 O + O

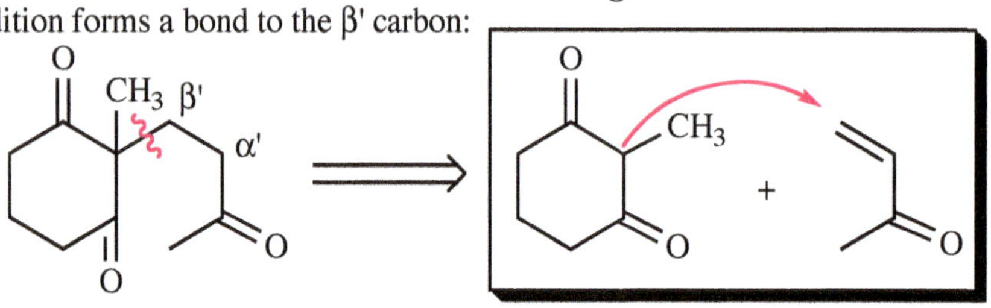

22-60 The most acidic hydrogens are shown as **H**.

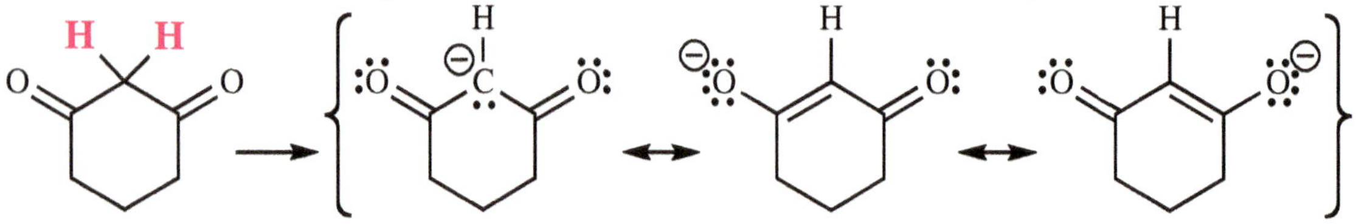

(a) O O H → { :O: :Ö:⊖ ↔ :Ö:⊖ :Ö: }

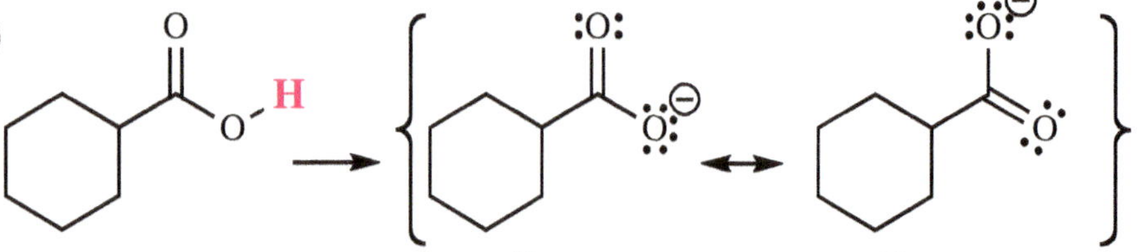

(b) H H O O → { :Ö: ⊖C H :Ö: ↔ ⊖:Ö: H :Ö: ↔ :Ö: H :Ö:⊖ }

22-60 continued

(c)

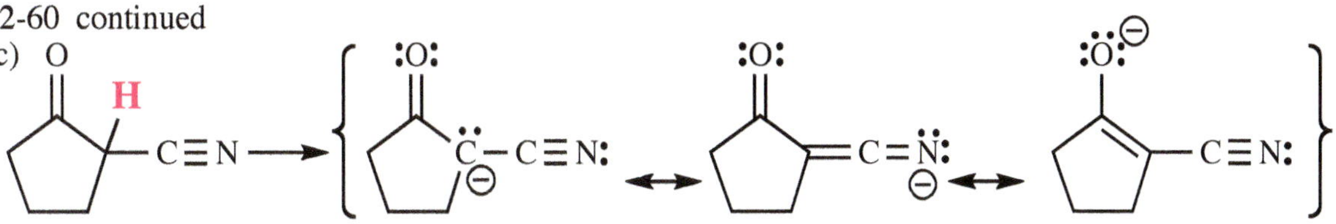

(d)

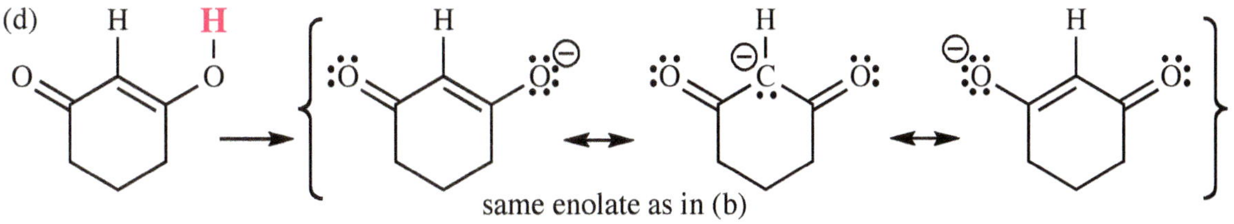

same enolate as in (b)

(e)

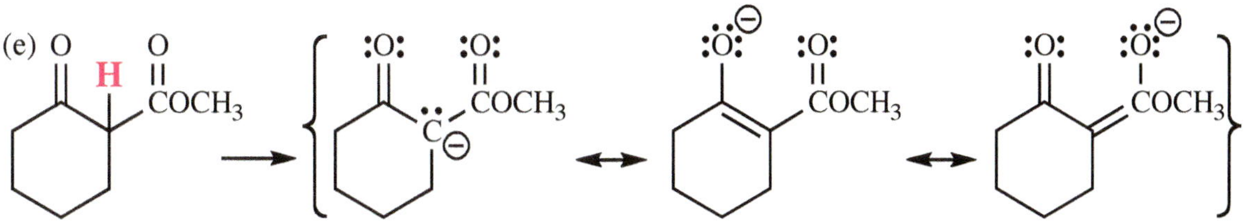

(f)

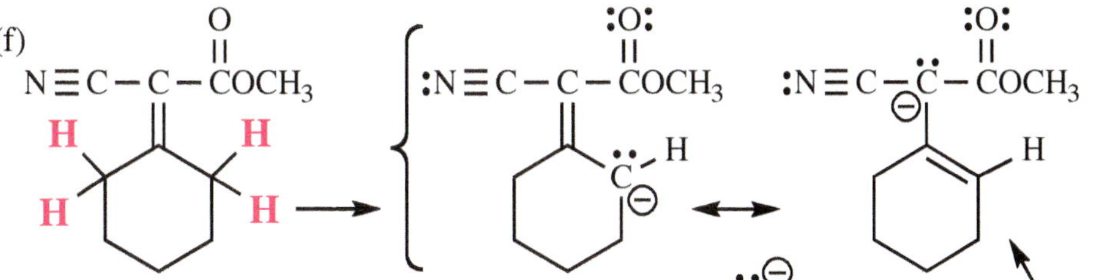

Because of the stereochemistry of the double bond, the H atoms on the left side of the ring are not identical to those on the right side. However, the two enolates will be equivalent except for the double bond geometry.

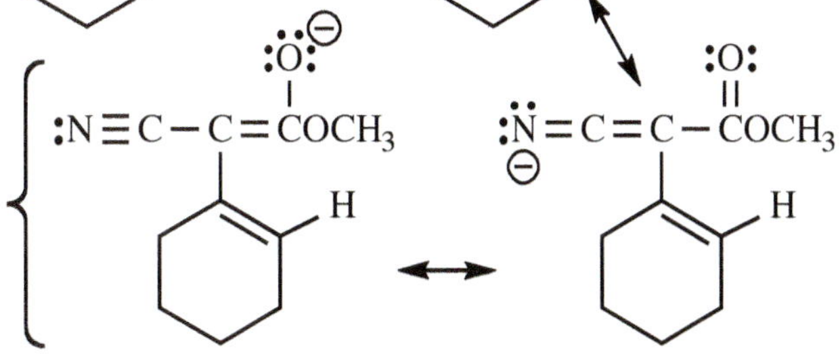

(g)

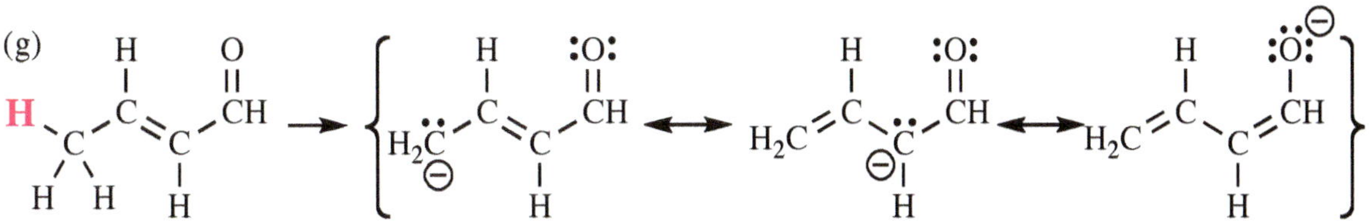

(h)

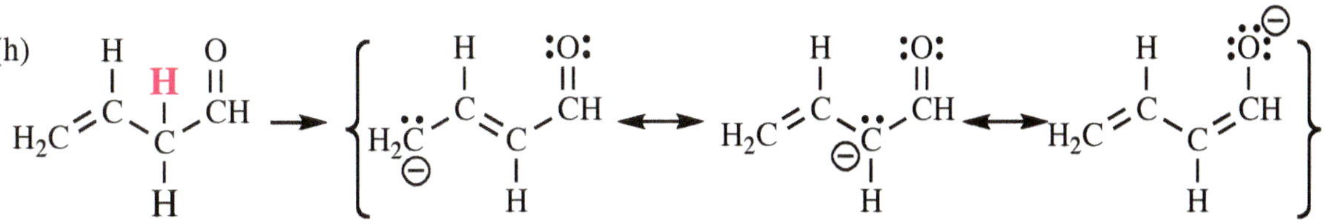

same enolate as in (g)

22-61 In order of increasing acidity. The most acidic protons are shown as H. (The approximate pK_a values are shown for comparison.)

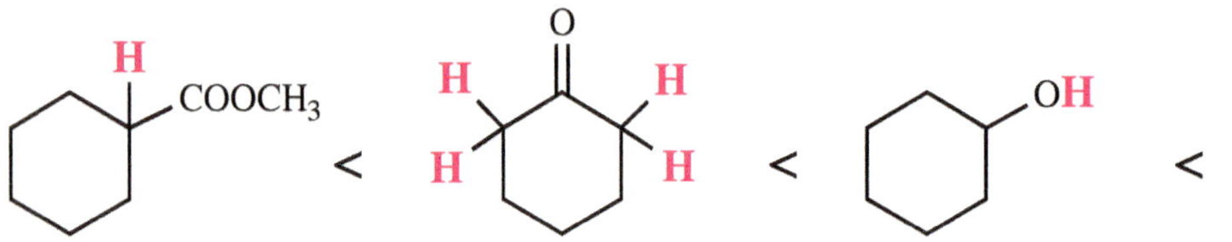

(g) pK_a 25 least acidic < (b) pK_a 20 < (f) pK_a 17-18 <

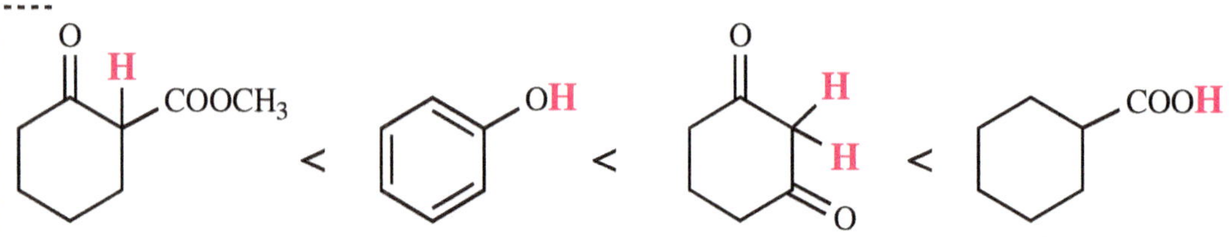

(a) pK_a 13 < (e) pK_a 10 < (c) pK_a 9 < (d) pK_a 5 most acidic

fully deprotonated by ethoxide ion

22-62 The wavy line lies across the bond formed in the aldol condensation.

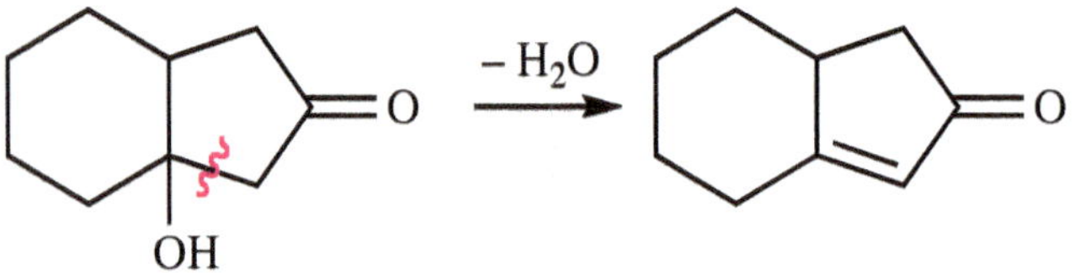

(a) $\xrightarrow{-H_2O}$

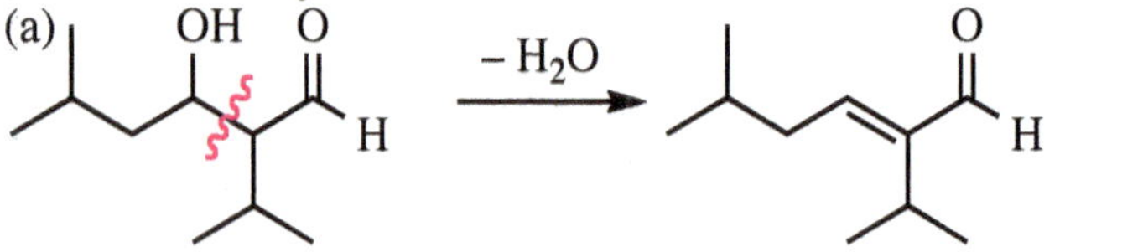

(b) $\xrightarrow{-H_2O}$

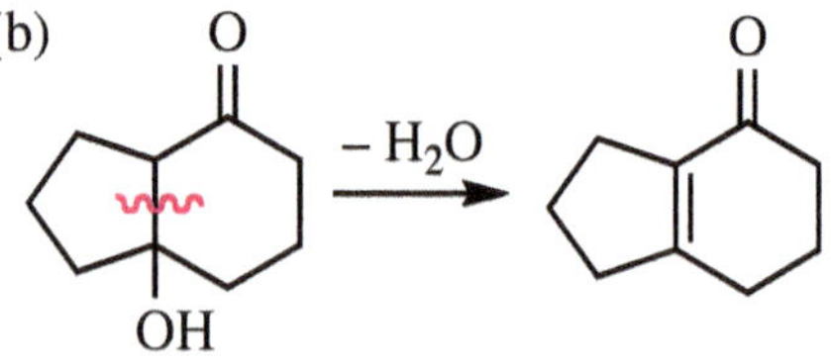

(c) $\xrightarrow{-2\,H_2O}$

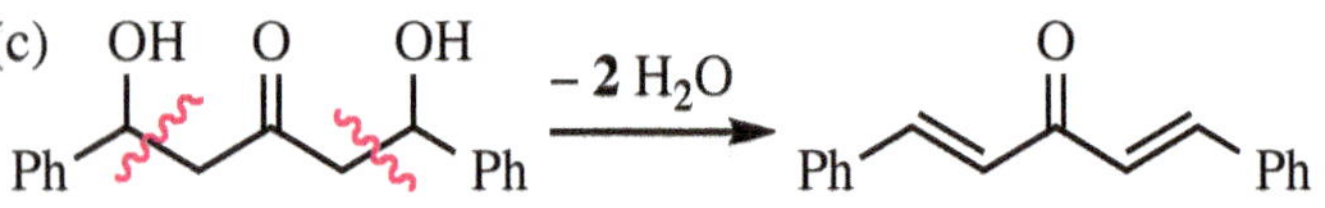

(d) $\xrightarrow{-H_2O}$

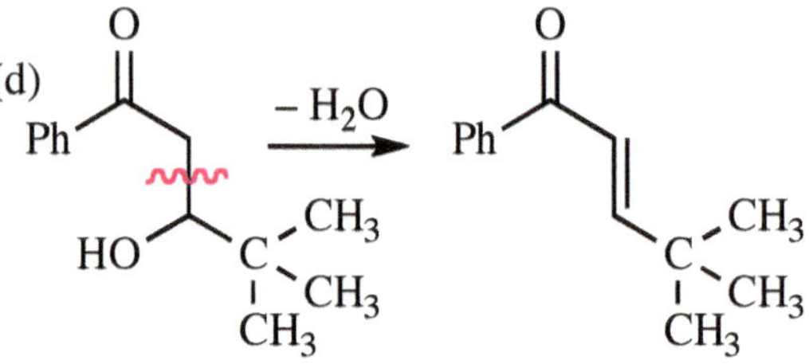

(e) $\xrightarrow{-H_2O}$

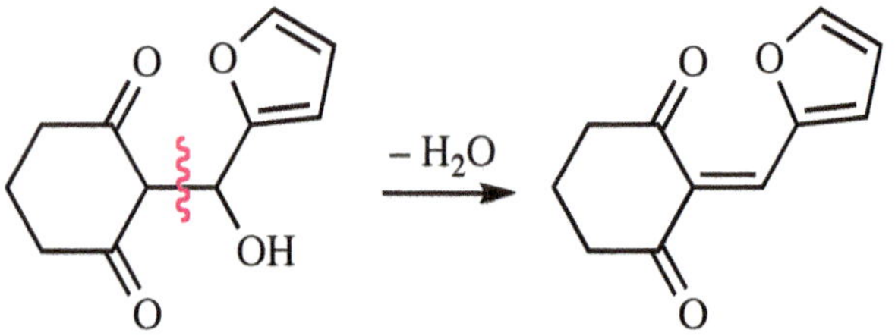

(f) $\xrightarrow{-H_2O}$

22-63 The bold bond is formed in the Claisen condensation.

(a) (b) 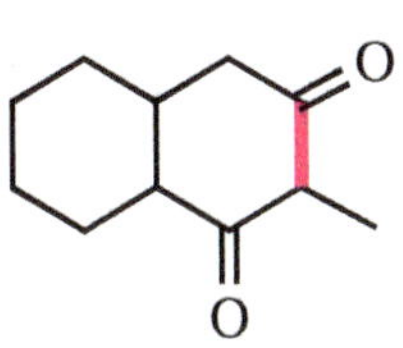 (c) 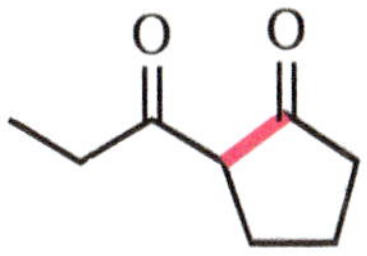 (d)

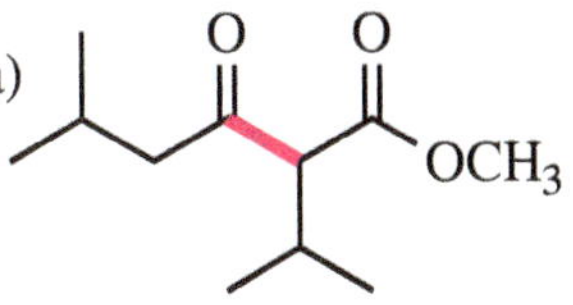

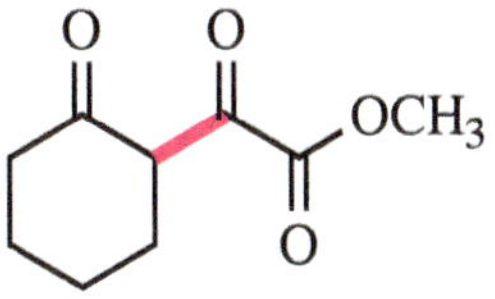

(e)

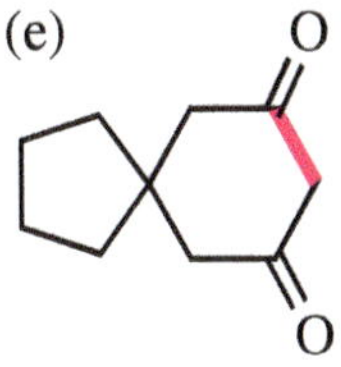

22-64

(a) mechanism of aldol condensation in problem 22-62(a)

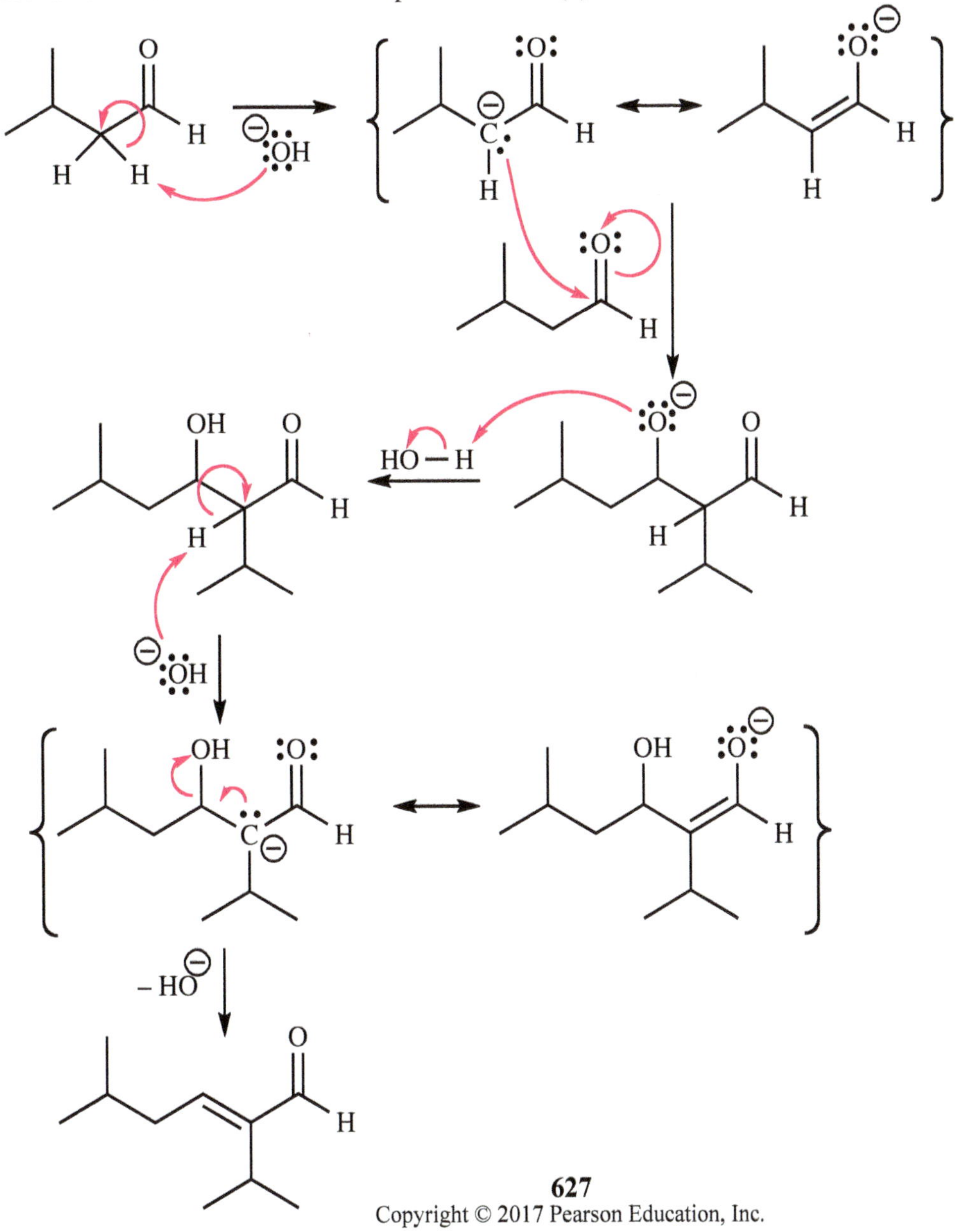

22-64 continued

(b) mechanism of aldol condensation in problem 22-62(b)

22-64 continued

(c) mechanism of Claisen condensation in problem 22-63(a)

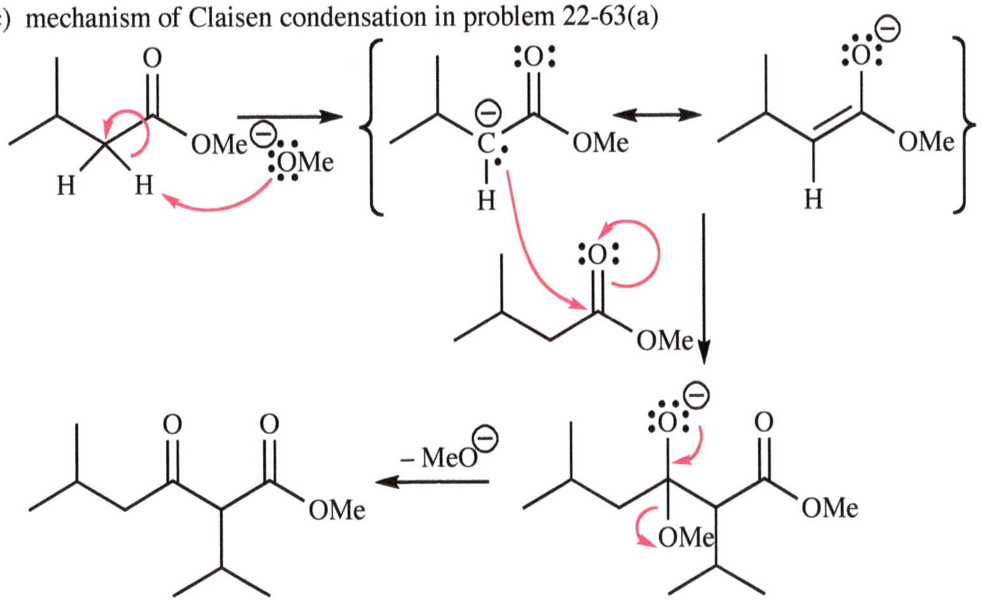

OMe ⊖:OMe

H H

:O:

C

H

OMe

:O:⊖

OMe

H

:O:

OMe

:O:⊖

O

OMe

OMe

– MeO⊖

O O

OMe

(This product will be deprotonated by methoxide but regenerated upon acidic workup.)

(d) mechanism of Claisen condensation in problem 22-63(b)

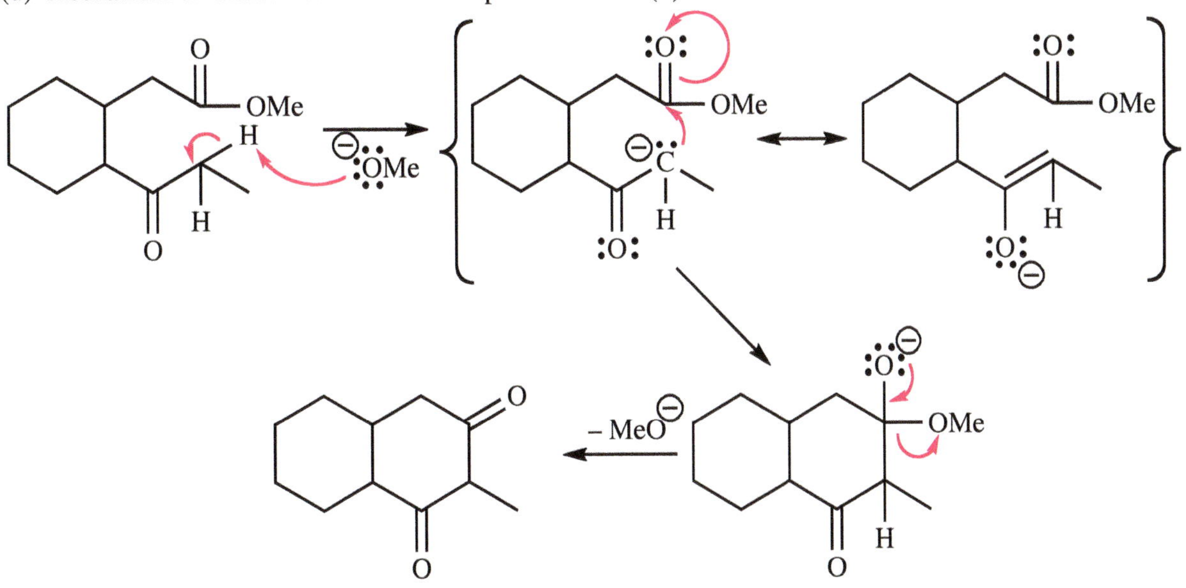

O

OMe

H

H

O

⊖:OMe

:O:

OMe

⊖C

H

:O:

:O:

OMe

H

:O:⊖

:O:⊖

OMe

H

O

– MeO⊖

O

O

(This product will be deprotonated by methoxide but regenerated upon acidic workup.)

22-65

keto ⇌ enol

The enol form is stable because of the conjugation and because of intramolecular hydrogen-bonding in a six-membered ring.

In dicarbonyl compounds in general, the weaker the electron-donating ability of the group G, the more it will exist in the enol form: aldehydes (G = H) are almost completely enolized, then ketones (G = R group), esters (G = OR), and finally amides (G = NR_2) which have virtually no enol content.

keto ⇌ enol

22-66

(a) Rank these compounds in order of increasing acid strength.

D < C < A < B

D *weakest* *diester—pK_a ≈13*

C *ketoester—pK_a ≈11*

A *diketone—pK_a ≈9*

B *strongest* *diketone with two electron-withdrawing groups—pK_a < 9*

(b) Rank these compounds in order of increasing enol content. In each case, draw the most stable enol.

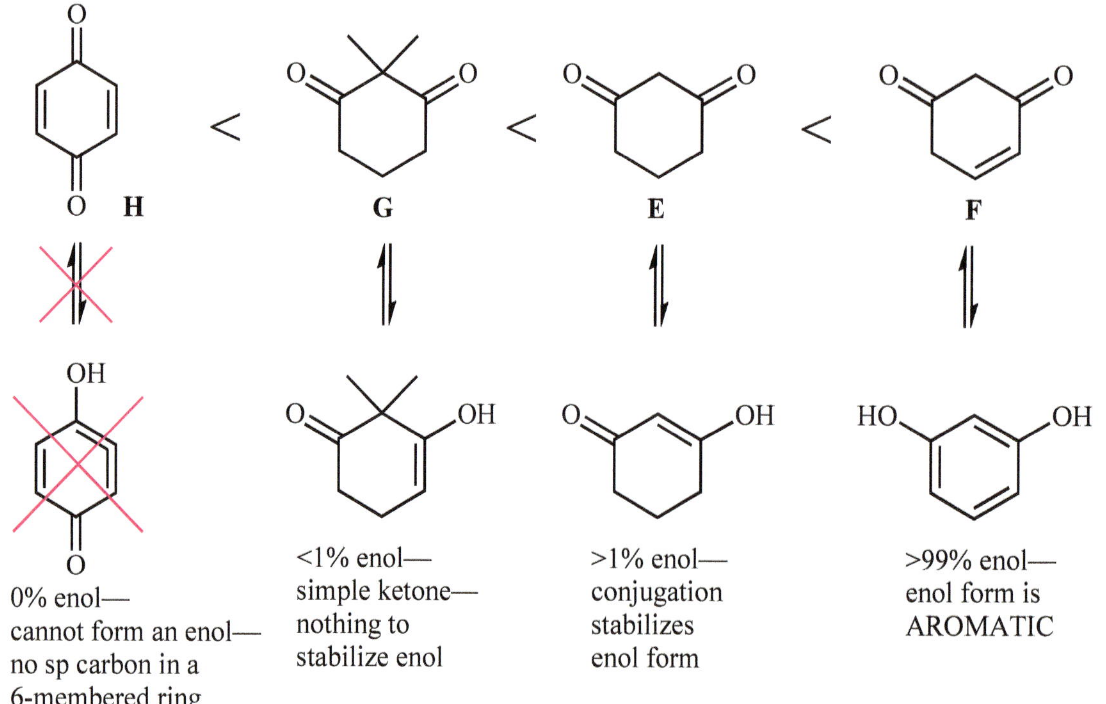

22-67 All of these Robinson annulations are catalyzed by NaOH.

(a) CH_3 ... Ph ... O + ... O

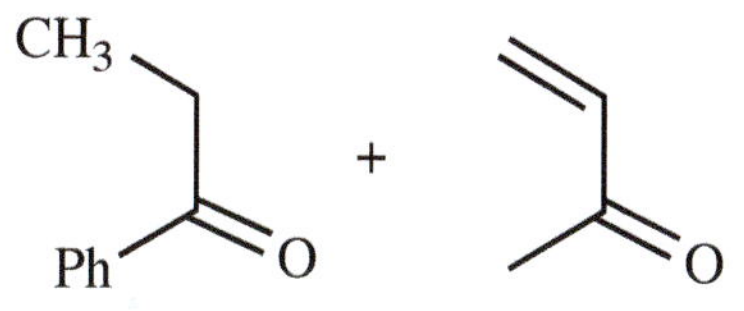

(b) ... O + ... O ... CH_3

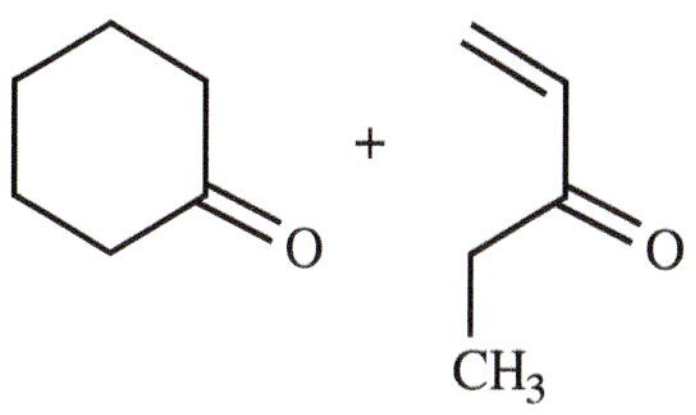

(c) ... O + ... O

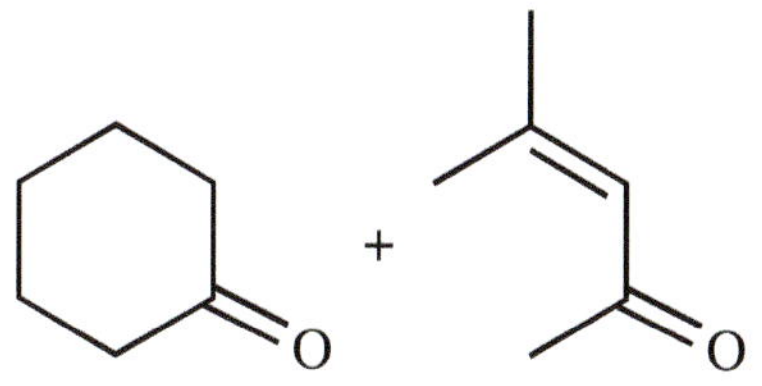

22-68 All products shown are after acidic workup.

(a) aldol self-condensation

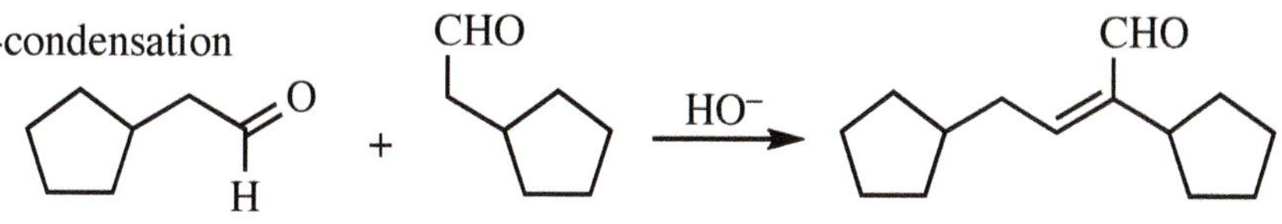

CHO ... O ... H + ... $\xrightarrow{HO^-}$ ... CHO

(b) Claisen self-condensation

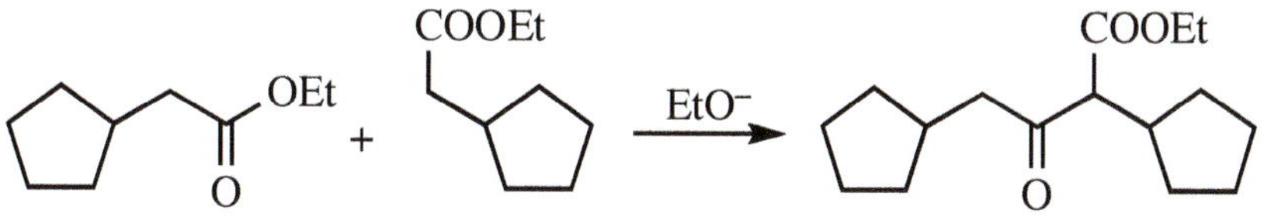

COOEt ... OEt ... O + ... $\xrightarrow{EtO^-}$ ... COOEt ... O

(c) aldol cyclization

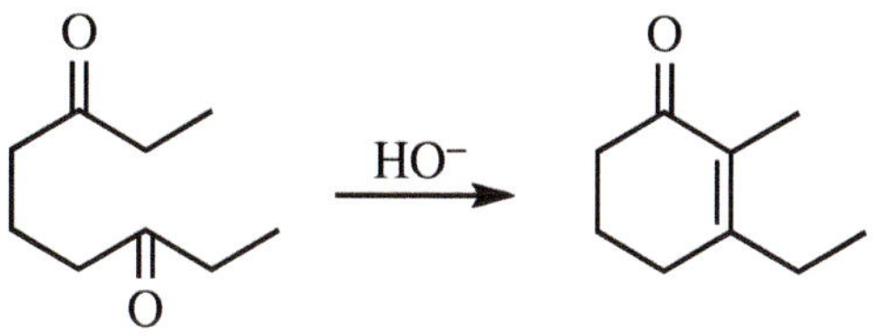

O ... O $\xrightarrow{HO^-}$ O

(d) mixed Claisen

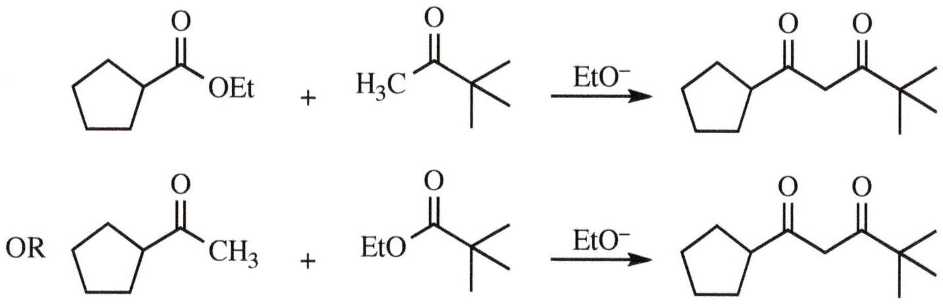

O ... OEt + H_3C ... O $\xrightarrow{EtO^-}$ O O

OR O ... CH_3 + EtO ... O $\xrightarrow{EtO^-}$ O O

(e) mixed aldol

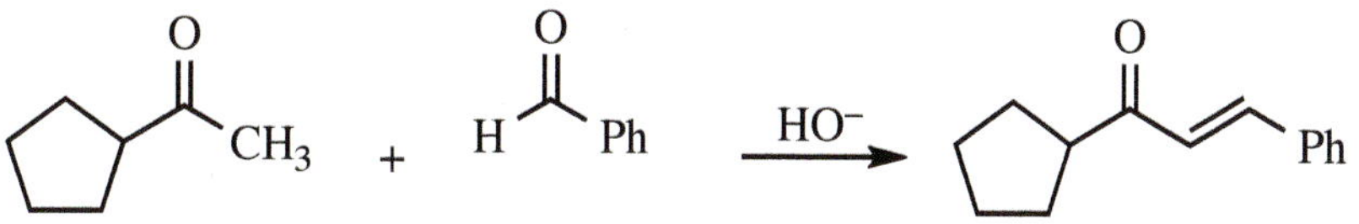

O ... CH_3 + H ... O ... Ph $\xrightarrow{HO^-}$ O ... Ph

22-68 continued

(f) enamine acylation or mixed Claisen

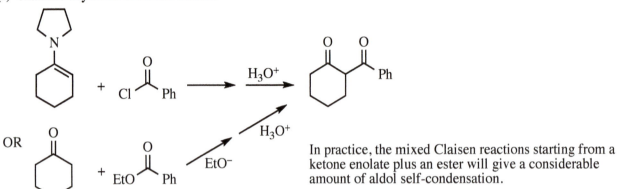

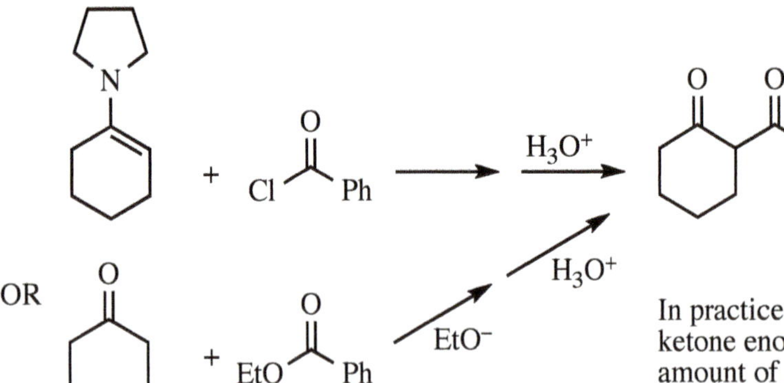

In practice, the mixed Claisen reactions starting from a ketone enolate plus an ester will give a considerable amount of aldol self-condensation.

22-69

(a) O Br

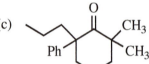

(b) COOH + CHI_3 after acid workup

(c) O CH_3 CH_3 Ph

(d) O H

(e) O Ph

(f) O $CH_2CH_2CH_2CH_3$ OCH_3 O

(g) O $CH_2CH_2CH_2CH_3$

(h) O O O → O + CO_2 + CH_3OH

initial product

(i) O O O OEt → O O + CO_2 + CH_3CH_2OH

initial product

22-70. Predict the products of these reactions.

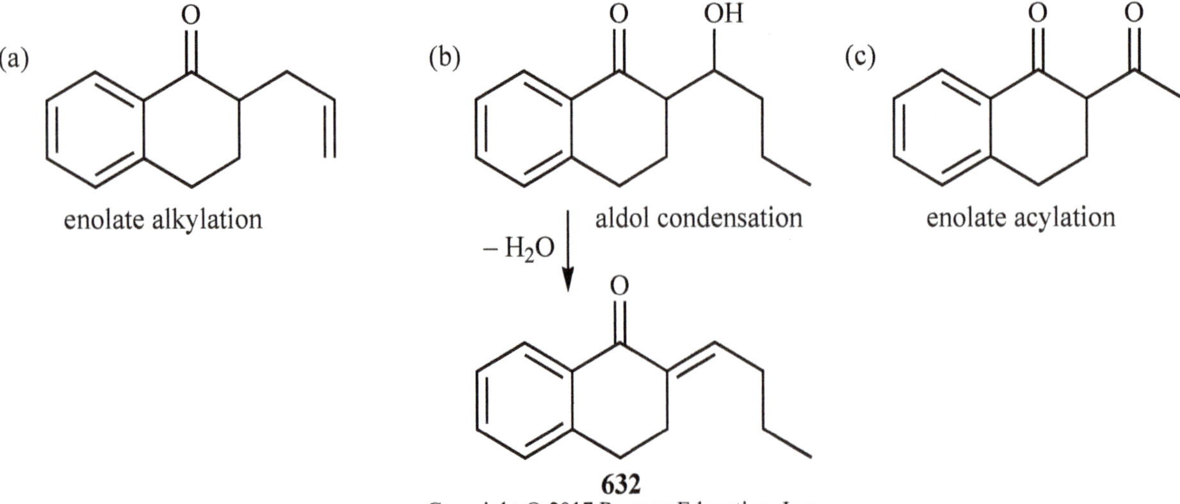

22-71

(a) reagents: Br_2, HOAc

(b) reagents: Br_2, PBr_3, followed by H_2O

(c) reagents: excess I_2 (or Br_2 or Cl_2), NaOH

(d) $Ph-C(=O)-H + Ph_3\overset{\oplus}{P}-\overset{\ominus}{C}HCH_3 \longrightarrow PhCH{=}CHCH_3 + Ph_3P{=}O$

(e) furan-2-carbaldehyde + propiophenone $\xrightarrow{NaOH}$ furyl–CH=C(CH_3)–C(=O)Ph

(f) cyclopentanone $\xrightarrow[HA]{Et_2NH}$ enamine (Et_2N–cyclopentene) + Cl–C(=O)–(N-CH_3 pyrrol-2-yl) $\longrightarrow \xrightarrow{H_2O}$ 2-acylcyclopentanone (N-CH_3 pyrrole)

An enamine avoids self-condensation. Any 2° amine could be chosen to form the enamine.

22-72 In the products, the wavy lines indicate the bonds that must be made by alkylation, before hydrolysis and decarboxylation produce the substituted acetic acid.

(a) $EtO_2CCH_2CO_2Et$ $\xrightarrow[2)\ PhCH_2Br]{1)\ NaOEt}$ $EtO_2CCH(CH_2Ph)CO_2Et$ $\xrightarrow[\Delta]{H_3O^+}$ $PhCH_2CH_2COOH$ + CO_2 + 2 EtOH

(b) $EtO_2CCH_2CO_2Et$ $\xrightarrow[2)\ CH_3CH_2Br]{1)\ NaOEt}$ $EtO_2CCH(CH_2CH_3)CO_2Et$ $\xrightarrow[2)\ \text{cyclopentylmethyl bromide}]{1)\ NaOEt}$ $EtO_2CC(CH_2CH_3)(CH_2\text{-cyclopentyl})CO_2Et$ $\xrightarrow[H_3O^+]{\Delta}$ CO_2 + 2 EtOH + cyclopentyl–CH_2–$CH(CH_2CH_3)$–COOH

(c) $EtO_2CCH_2CO_2Et$ $\xrightarrow[2)\ Br(CH_2)_5Br]{1)\ 2\ NaOEt}$ diethyl cyclohexane-1,1-dicarboxylate $\xrightarrow[\Delta]{H_3O^+}$ cyclohexane–COOH + CO_2 + 2 EtOH

22-73 In the products, the wavy lines indicate the bonds that must be made by alkylation, before hydrolysis and decarboxylation produce the substituted acetone.

(a) $EtO_2CCH_2COCH_3$ $\xrightarrow[2)\ EtBr]{1)\ NaOEt}$ $EtO_2CCH(CH_2CH_3)COCH_3$ $\xrightarrow[2)\ PhCH_2Br]{1)\ NaOEt}$ $EtO_2CC(CH_2CH_3)(CH_2Ph)COCH_3$ $\xrightarrow[\Delta]{H_3O^+}$ $PhCH_2CH(CH_2CH_3)COCH_3$ + CO_2 + EtOH

22-73 continued

(b) Ethyl acetoacetate $\xrightarrow[\text{2) } Br(CH_2)_4Br]{\text{1) 2 NaOEt}}$ ethyl 1-acetylcyclopentanecarboxylate $\xrightarrow[\Delta]{H_3O^+}$ cyclopentyl methyl ketone + CO_2 + EtOH

(c) The acetoacetic ester synthesis makes substituted acetone, so where is the acetone in this product?

substituted acetone

The single bond to this substituted acetone can be made by the acetoacetic ester synthesis. How can we make the α,β double bond? Aldol condensation!

EtOOC

make by conjugate addition

make by aldol cyclization/dehydration

Ethyl acetoacetate + methyl vinyl ketone $\xrightarrow{\text{NaOEt}}$ Michael adduct $\xrightarrow[\Delta]{H_3O^+}$ 2,6-heptanedione $\xrightarrow{\text{NaOH}}$ 3-methylcyclohex-2-enone

22-74 These compounds are made by aldol condensations followed by other reactions. The key is to find the skeleton made by the aldol.

(a) Where is the possible α,β-unsaturated carbonyl in this skeleton?

$$PhCH_2CH_2-CH(OH)Ph \Longrightarrow PhCH=CH-C(=O)-Ph \overset{\text{reverse aldol}}{\Longrightarrow} Ph-C(=O)-H + CH_3-C(=O)-Ph$$

<u>forward synthesis</u>

$$Ph-C(=O)-H + H_3C-C(=O)-Ph \xrightarrow{\text{NaOH}} PhCH=CH-C(=O)-Ph \xrightarrow{H_2,\ Pt} PhCH_2CH_2-CH(OH)Ph$$

(b) The aldol skeleton is not immediately apparent in this formidable product. What can we see from it? Most obvious is the β-dicarbonyl (β-ketoester), which we know to be a good nucleophile, capable of substitution or Michael addition. In this case, Michael addition is most likely, as the site of attack is β to another carbonyl.

Ph

OCH_3

β-ketoester

$\Longrightarrow$ 3-methylcyclohex-2-enone + PhCOCH$_2$COOCH$_3$

Ph

OCH_3

AHA! The aldol product reveals itself. (See the solution to 22-73 (c).)

continued on next page

22-74 continued

forward synthesis

NaOH; + ; 1) $NaOCH_3$ 2) HA

(c) The key in this product is the α-nitroketone, the equivalent of a β-dicarbonyl system, capable of doing Michael addition to the β-carbon of the other α,β-unsaturated system.

reverse Michael $\Longrightarrow$; reverse aldol $\Longrightarrow$

forward synthesis

+ ; NaOH ; Na_2CO_3 ; HA

$Z + E$—each will give the same final product

22-75 Starting materials are shown.

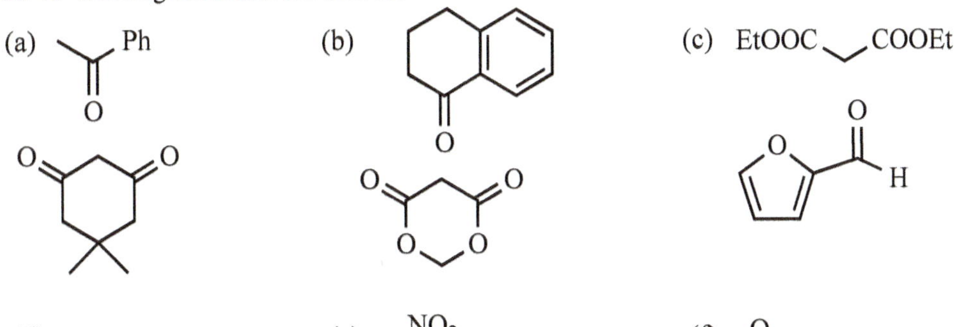

(d) EtO CN

(e) NO_2 O_2N

(f)

22-76 This common sequence demonstrates aldol-type condensation, followed by conjugate addition and alkylation of the enolate intermediate.

NC CN; base; $-H_2O$ → **X** ; $(CH_3CH_2)_2CuLi$ (no workup) → **Y** ; CH_3I → **Z**

22-77 A reminder about how to approach mechanisms from the Problem-Solving Strategy:
1. Analyze the carbon skeleton of the product to discern what pieces have to come together, and how.
2. Figure out which piece has to be the nucelophile, and how to generate a strong-enough nucleophile.
3. Consider the most electrophilic site on the other fragment, or within the same molecule in some cases.
4. Determine if the initial product needs to be converted to the desired final product. In this chapter, the final reaction is often a dehydration.
5. Put all the steps together in a continuous flow, one step at a time.

(a) O H H :OH⊖ → { :O: C⊖ H ↔ :O:⊖ H } PhC—H → O :O:⊖ HC—Ph H ↓ HO—H

O H C—Ph E + Z ← − HO⊖ { :O:⊖ :OH HC—Ph ↔ :O: :OH HC—Ph C⊖ } ← ⊖:OH O :OH HC—Ph H

(b) O H H ⊖:OMe → { :O: H C⊖ ↔ :O:⊖ H } :O: PhC—OCH₃ → O :O:⊖ PhC—OCH₃ H ← − CH₃O⊖ ← HA O O C—Ph H

22-77 continued

(c) Robinson annulations are explained most easily by remembering that the first step is a Michael addition, followed by aldol cyclization with dehydration.

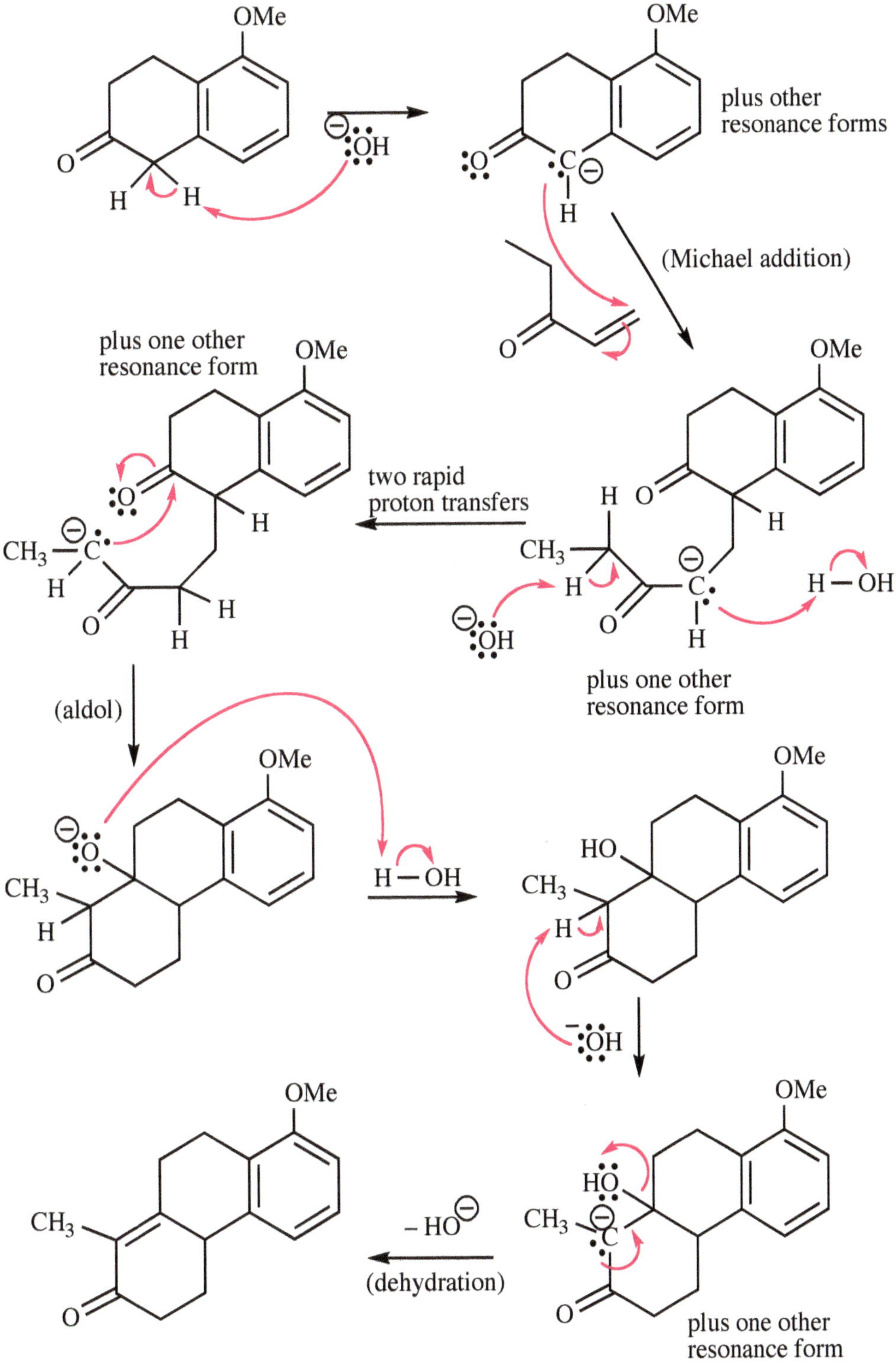

22-77 continued

(d)

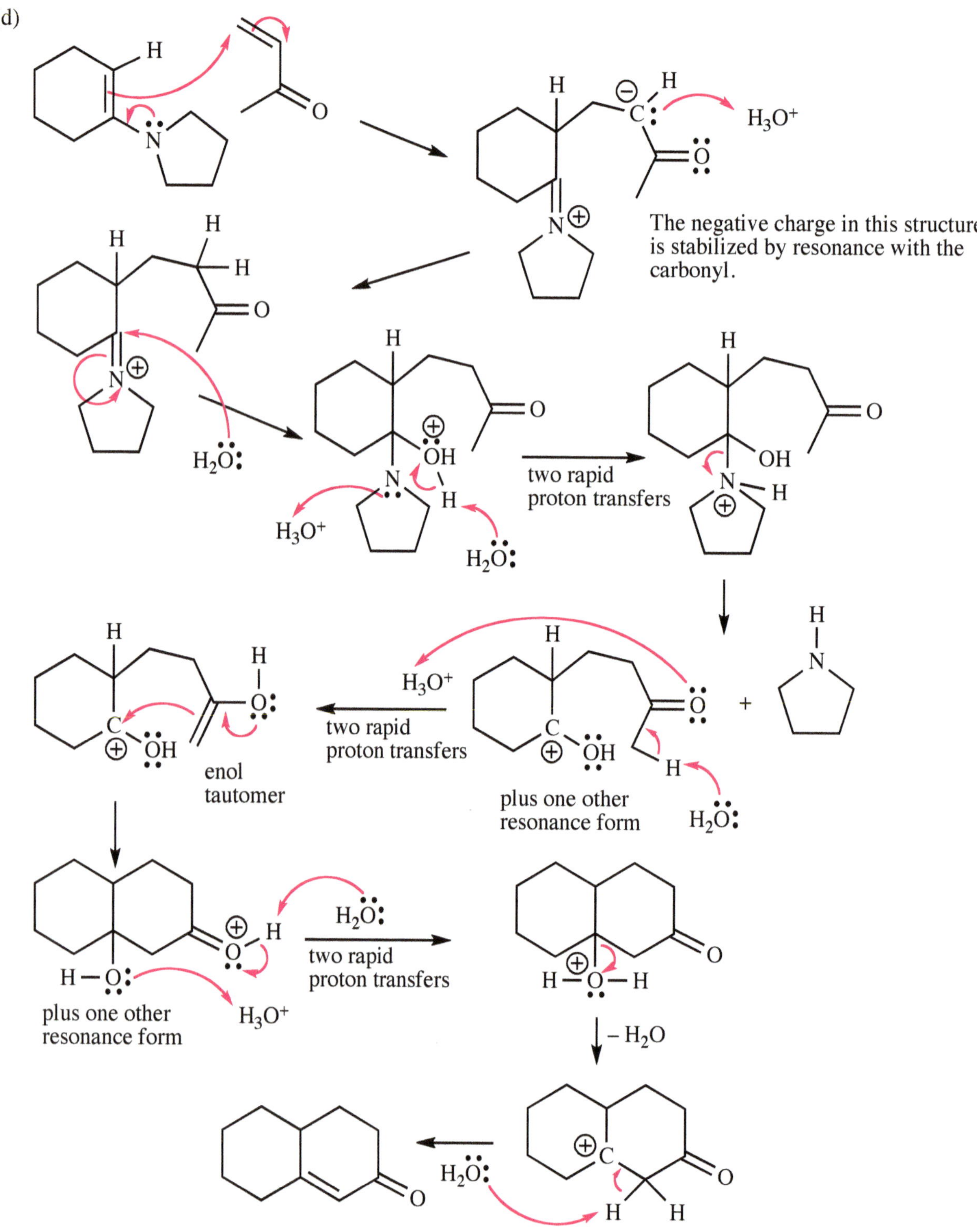

22-77 continued

(e) For simplicity, the benzene ring is abbreviated as Ph; consequently, for the resonance forms with the (+) on the carbon of the carbonyl, the delocalization of that (+) onto the benzene ring is not shown—but remember that the benzene ring still plays an important role in stabilizing the (+) charge.

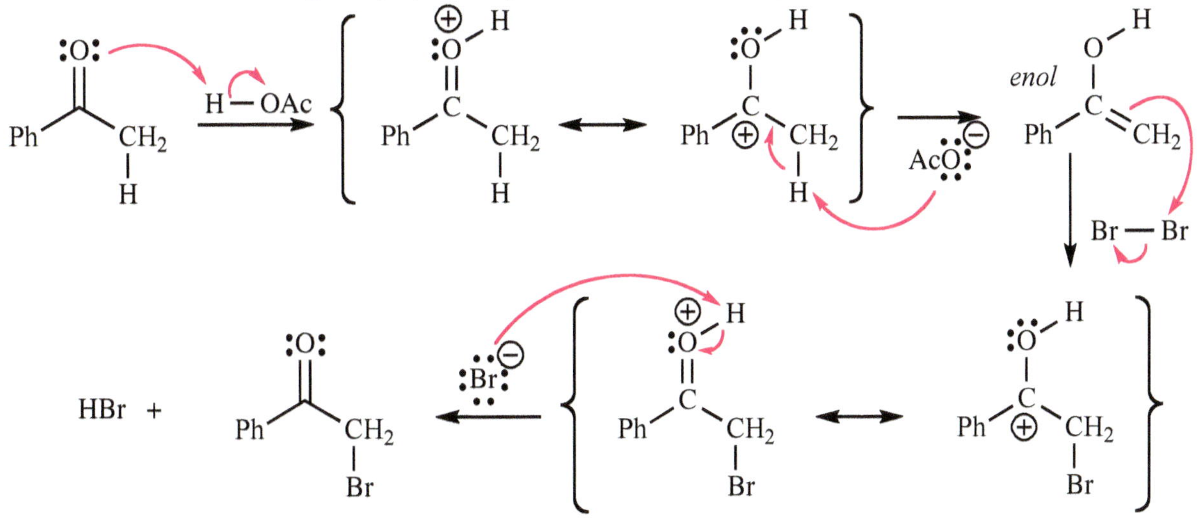

H—OAc; enol; AcO⁻; Br—Br; HBr +

22-78 In this problem, structures of intermediates with positive charge on double-bonded N will have a less-significant resonance contributor with the positive charge on C without an octet of electrons.

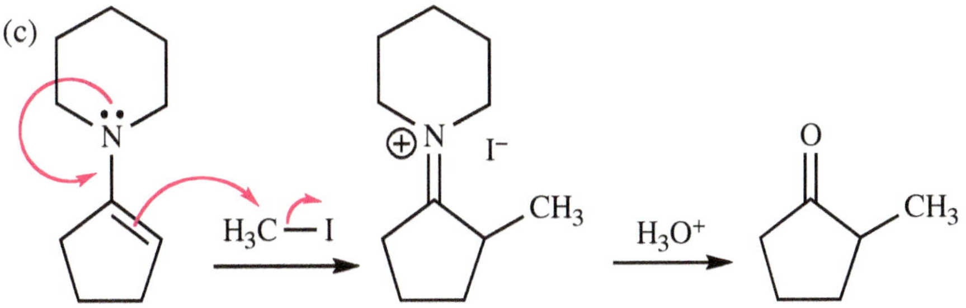

(a) Cl^- H_3O^+

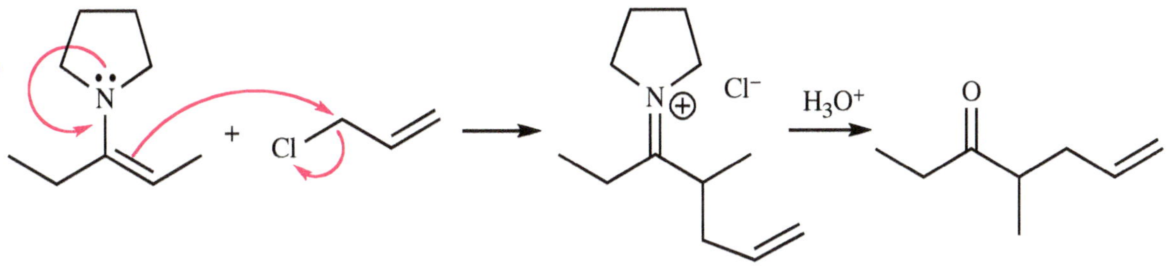

(b) Cl^- H_3O^+

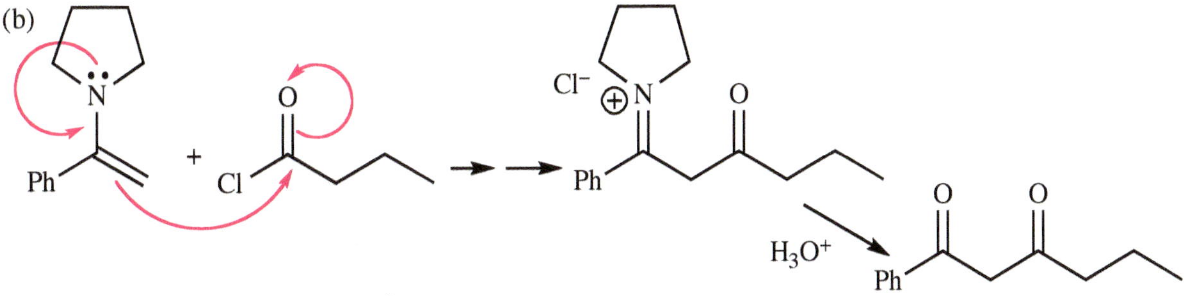

(c) $H_3C—I$ I^- CH_3 H_3O^+ CH_3

22-78 continued

(d)

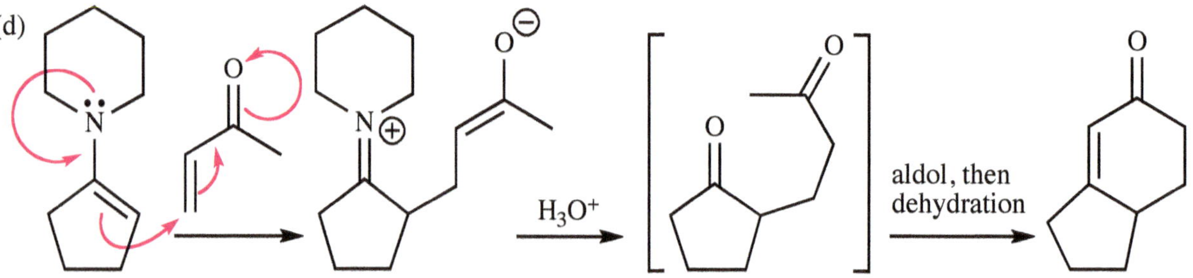

H_3O^+

aldol, then dehydration

22-79

(a)

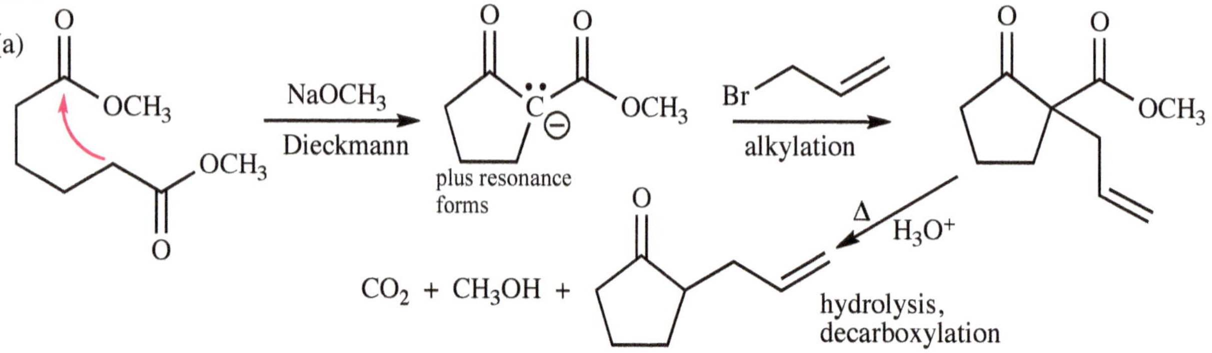

OCH_3

OCH_3

$NaOCH_3$

Dieckmann

OCH_3

plus resonance forms

Br

alkylation

OCH_3

Δ

H_3O^+

hydrolysis, decarboxylation

CO_2 + CH_3OH +

(b)

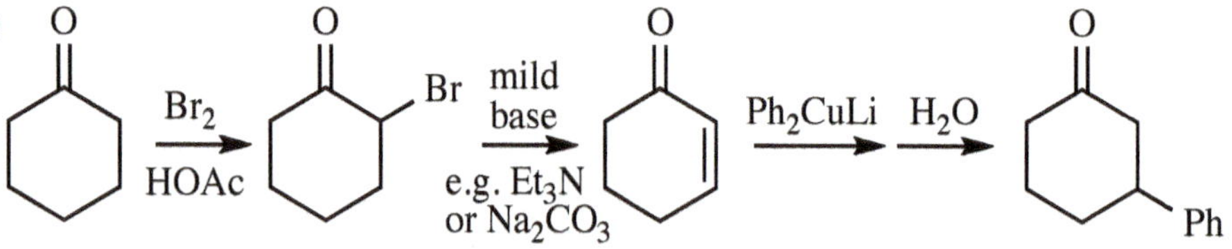

Br_2

HOAc

Br

mild base

e.g. Et_3N or Na_2CO_3

Ph_2CuLi H_2O

Ph

(c)

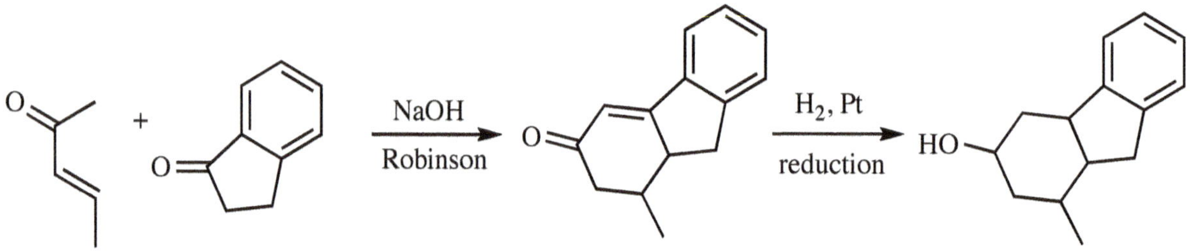

NaOH

Robinson

H_2, Pt

reduction

HO

(d)

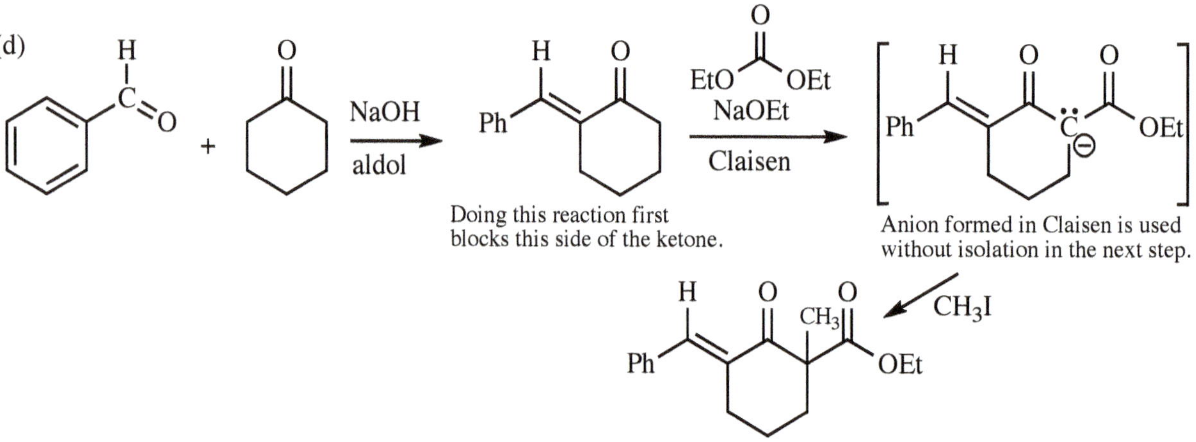

NaOH

aldol

Ph

Doing this reaction first blocks this side of the ketone.

EtO OEt

NaOEt

Claisen

Ph OEt

Anion formed in Claisen is used without isolation in the next step.

CH_3I

Ph CH_3 OEt

22-80

(a)

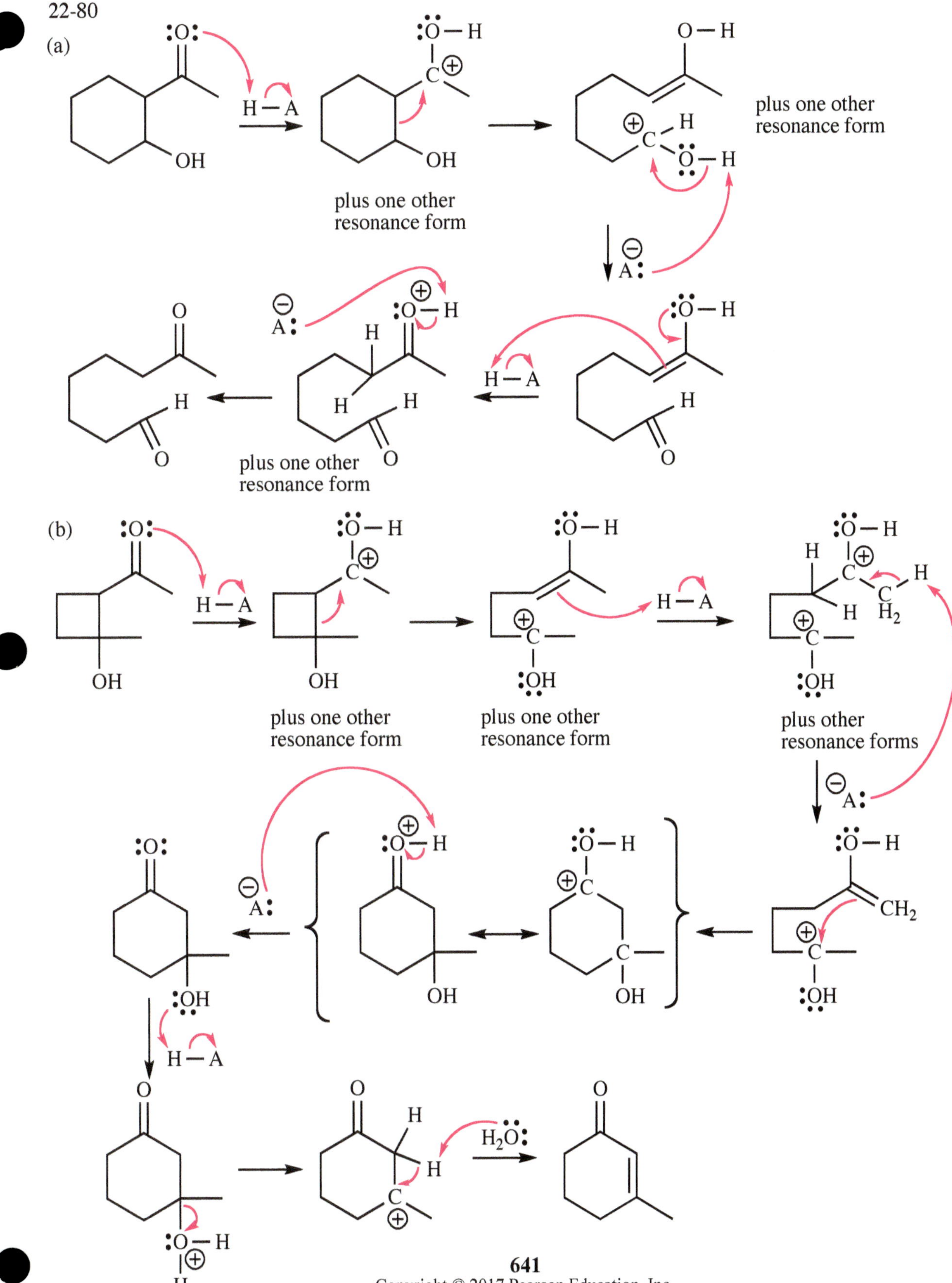

plus one other resonance form
plus one other resonance form
plus one other resonance form
(b)
plus one other resonance form
plus one other resonance form
plus other resonance forms

22-80 continued

(c)

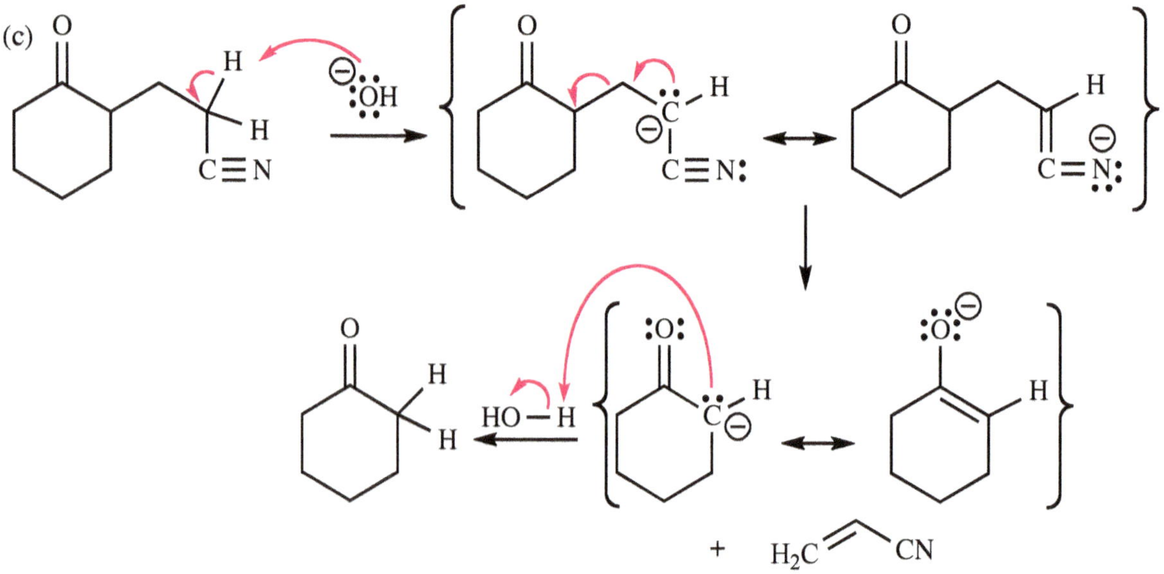

+ H_2C=CH–CN

(d)

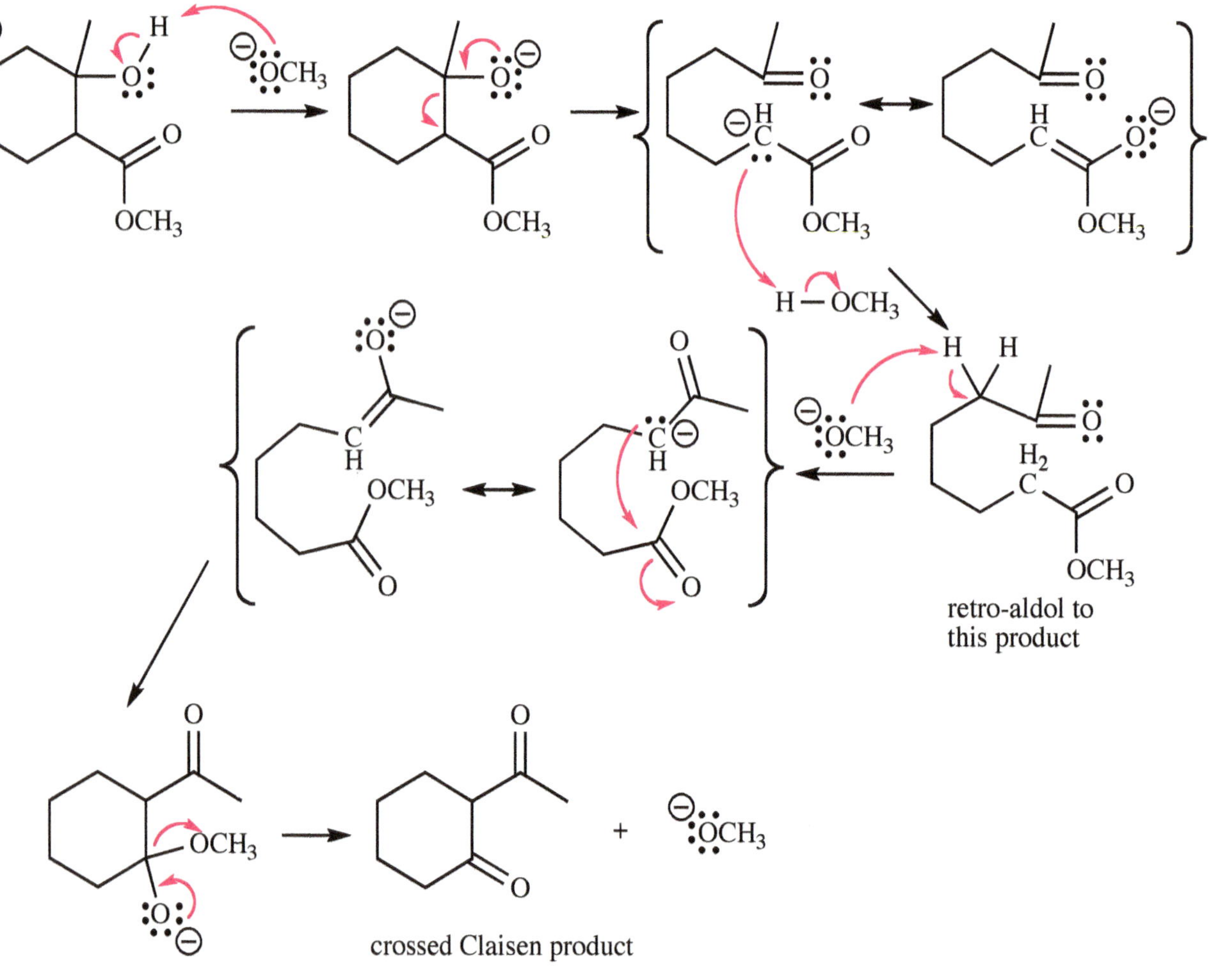

retro-aldol to this product

crossed Claisen product

22-81

Reaction of two molecules of Grignard reagent at an ester is expected. The unexpected product arises from a conjugate addition.

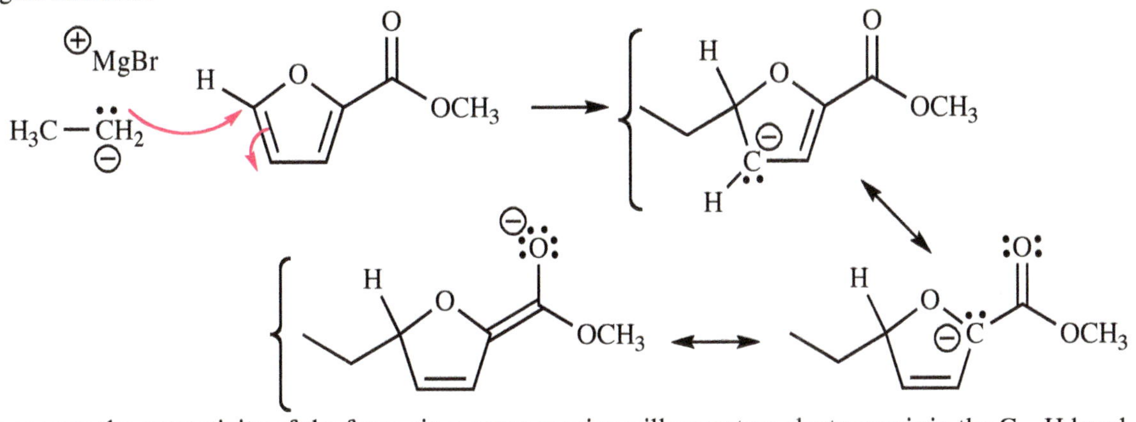

To regenerate the aromaticity of the furan ring, some species will accept an electron pair in the C—H bond. The most likely electron acceptor (the definition of an oxidizing agent) is the magnesium ion.

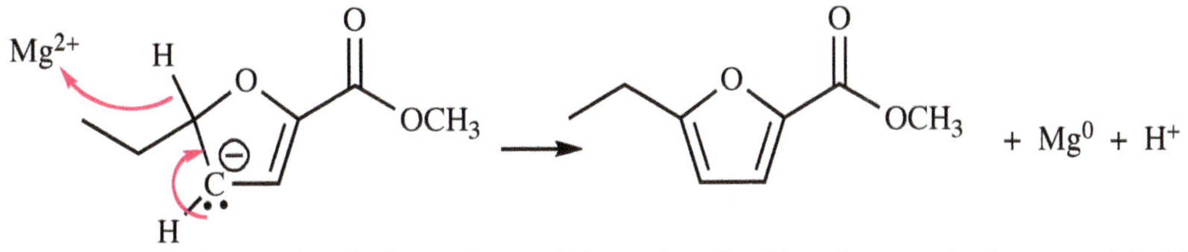

$+ Mg^0 + H^+$

Once this conjugate addition product is formed, two Grignards will add to the ester in the normal fashion. See the solution to Problem 10-18(a), or text section 10-9D, to review the mechanism of Grignard addition to an ester.

22-82

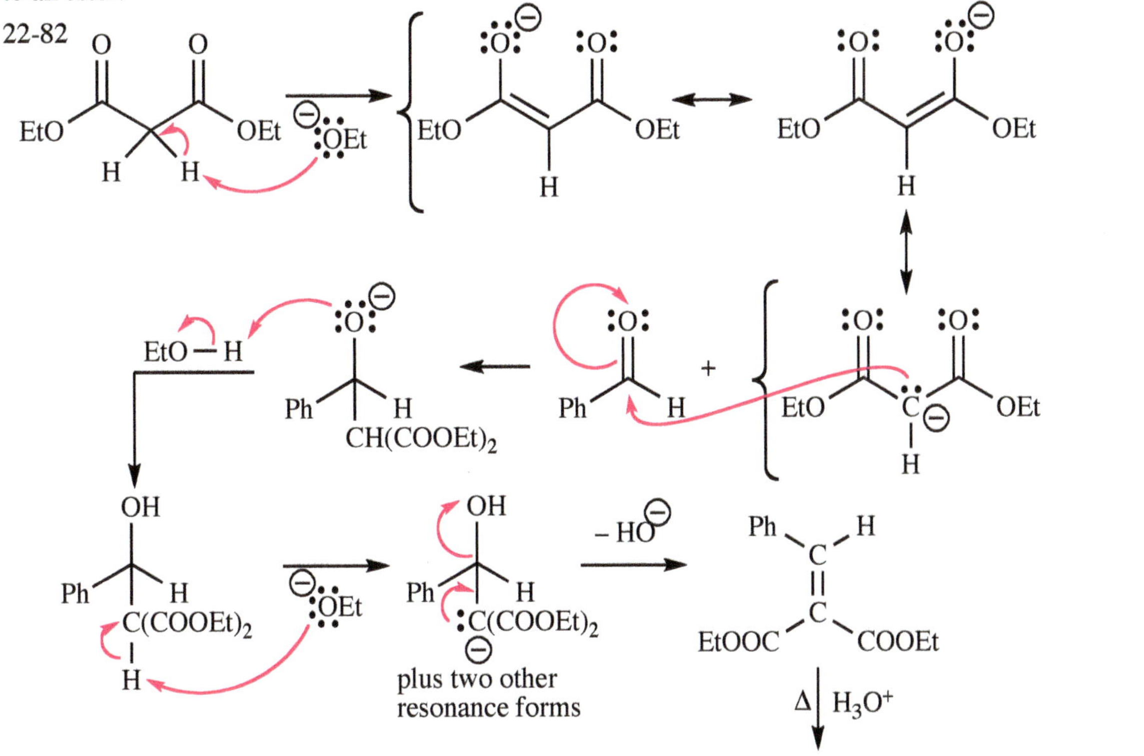

plus two other resonance forms

mechanism continued on next page

22-82 continued

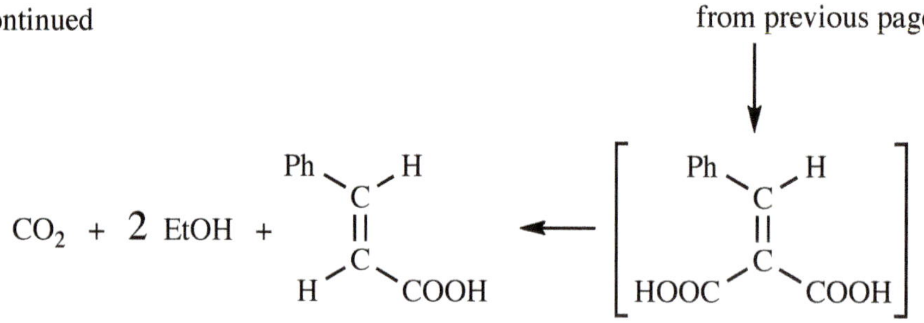

from previous page

CO_2 + 2 EtOH + (Ph)(H)C=C(H)(COOH) ← [(Ph)(H)C=C(HOOC)(COOH)]

22-83

$CH_2OPO_3^{2-}$–C=O–HO–CH–HC–OH–HC–OH–$CH_2OPO_3^{2-}$ → (with $^{-}$OH) $CH_2OPHO_3^{2-}$–C=O–HO–CH–HC–O^{-}–HC–OH–$CH_2OPO_3^{2-}$ → $CH_2OPO_3^{2-}$–C=O–HO–CH^{-} (plus one other resonance form) + HC=O–HC–OH–$CH_2OPO_3^{2-}$

HO–H

$CH_2OPO_3^{2-}$–C=O–HO–CH_2

dihydroxyacetone phosphate

glyceraldehyde-3-phosphate

To identify the retro-aldol, we must first locate the HO that is beta to the C=O.

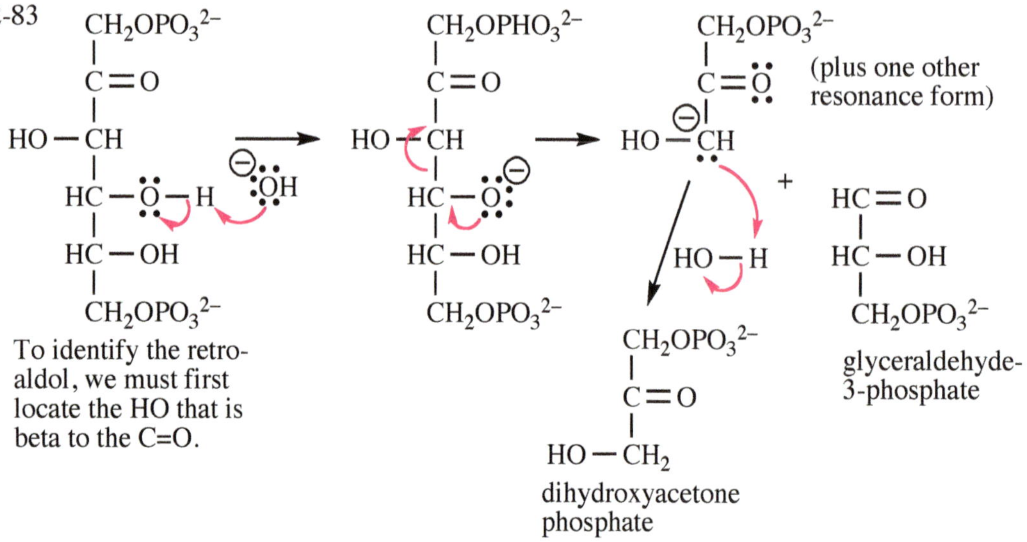

22-84 This is an aldol condensation. **P** stands for a protein chain in this problem.

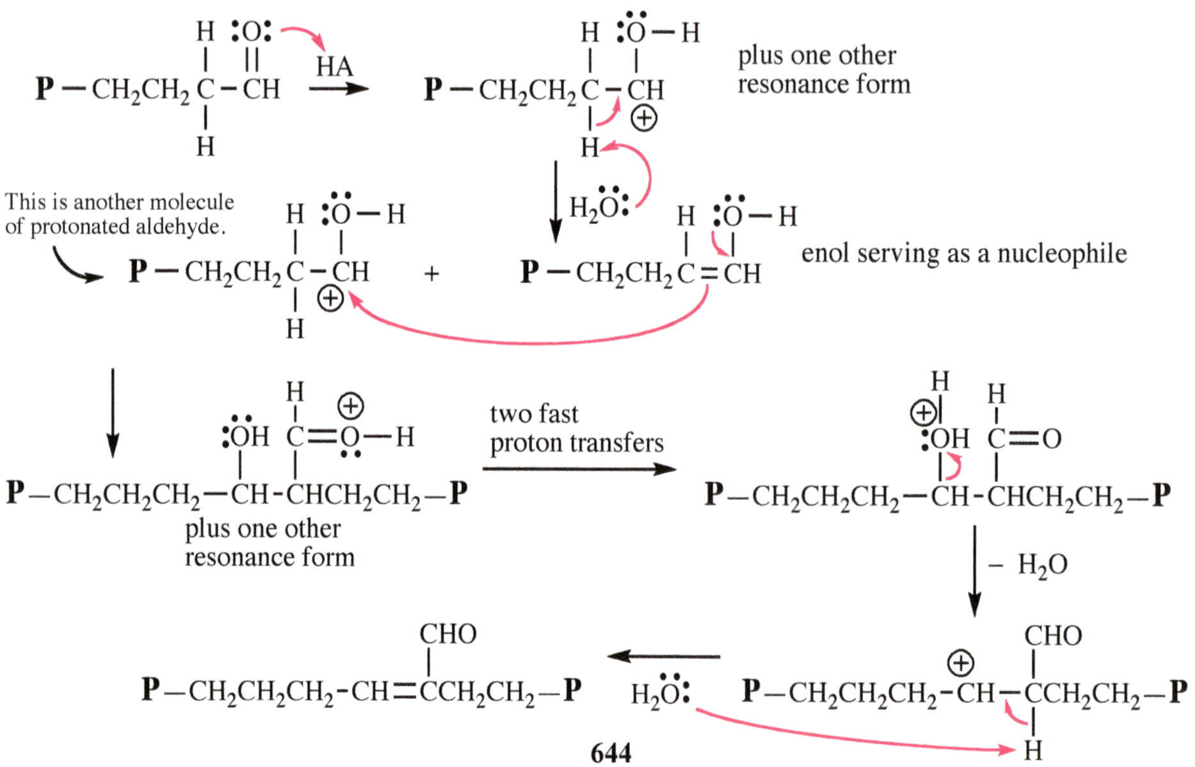

P–CH_2CH_2CH(H)–CH=O →(HA) **P**–CH_2CH_2CH(H)–CH^{+}–OH   plus one other resonance form

$H_2\ddot{O}$:

This is another molecule of protonated aldehyde.

P–CH_2CH_2CH(H)–CH^{+}–OH + **P**–CH_2CH_2C=CH–OH   enol serving as a nucleophile

P–$CH_2CH_2CH_2$–CH(OH)–CH(CH=O^{+}–H)CH_2CH_2–**P**   plus one other resonance form

two fast proton transfers

P–$CH_2CH_2CH_2$–CH(OH_2^{+})–CH(CHO)CH_2CH_2–**P**

– H_2O

P–$CH_2CH_2CH_2$–CH=C(CHO)CH_2CH_2–**P** ← $H_2\ddot{O}$: **P**–$CH_2CH_2CH_2$–CH^{+}–C(H)(CHO)CH_2CH_2–**P**

22-85

(a) aldol, Michael, aldol cyclization, decarboxylation

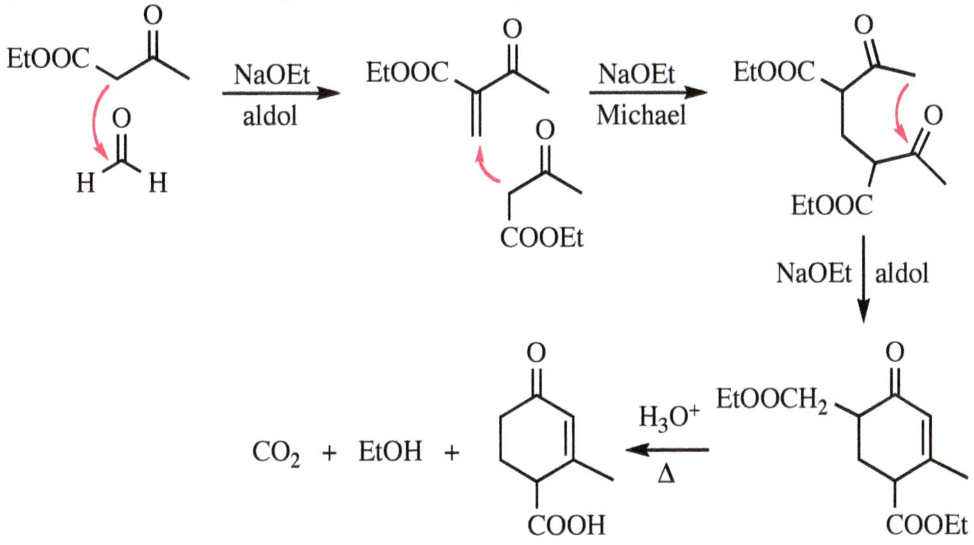

(b) Michael followed by Claisen condensation; hydrolysis and decarboxylation

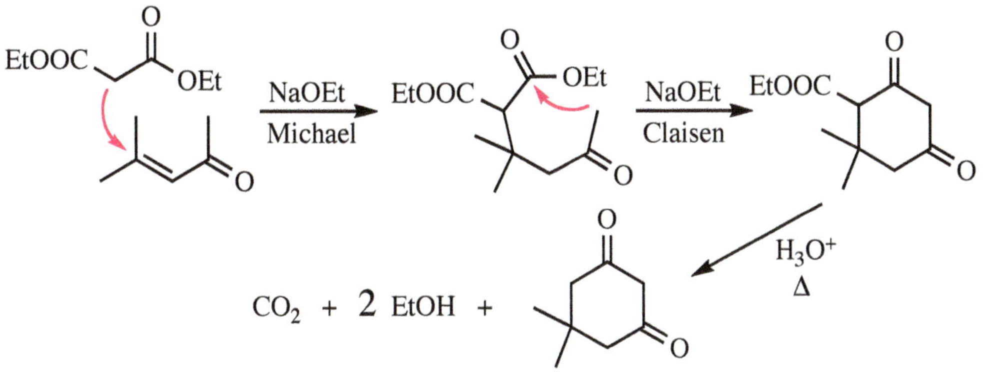

CHAPTER 23—CARBOHYDRATES AND NUCLEIC ACIDS

Remember about Fischer projections, first introduced in Chapter 5, section 5-10: vertical bonds are equivalent to dashed bonds, going behind the plane of the paper, and horizontal bonds are equivalent to wedge bonds, coming toward the viewer.

Fischer projection: CHO; HO—H; HO—H; CH_2OH ≡ O=C—H; HO▸C◂H; HO▸C◂H; CH_2OH

Fischer projections are always drawn with the highest priority functional group at the top.

view from left side ⇐ A; X—Y; X—Y; X—Y; X—Y; B ⇒ view from right side

23-1

CHO; H—OH; HO—H; H—OH; H—OH; CH_2OH — glucose

CHO; HO—H; H—OH; HO—H; HO—H; CH_2OH — mirror image

CH_2OH; C=O; HO—H; H—OH; H—OH; CH_2OH — fructose

CH_2OH; O=C; H—OH; HO—H; HO—H; CH_2OH — mirror image

All four of these compounds are chiral and optically active.

23-2

(a)

CHO; H—OH; H—OH; CH_2OH

CHO; HO—H; HO—H; CH_2OH

CHO; H—OH; HO—H; CH_2OH

CHO; HO—H; H—OH; CH_2OH

two asymmetric carbons ⇒ four stereoisomers (two pairs of enantiomers) if none are meso

(b)

CH_2OH; C=O; H—OH; CH_2OH

CH_2OH; O=C; HO—H; CH_2OH

one chiral center ⇒ two stereoisomers (enantiomers)

(c) An aldohexose has four chiral carbons and sixteen stereoisomers. A ketohexose has three chiral carbons and eight stereoisomers.

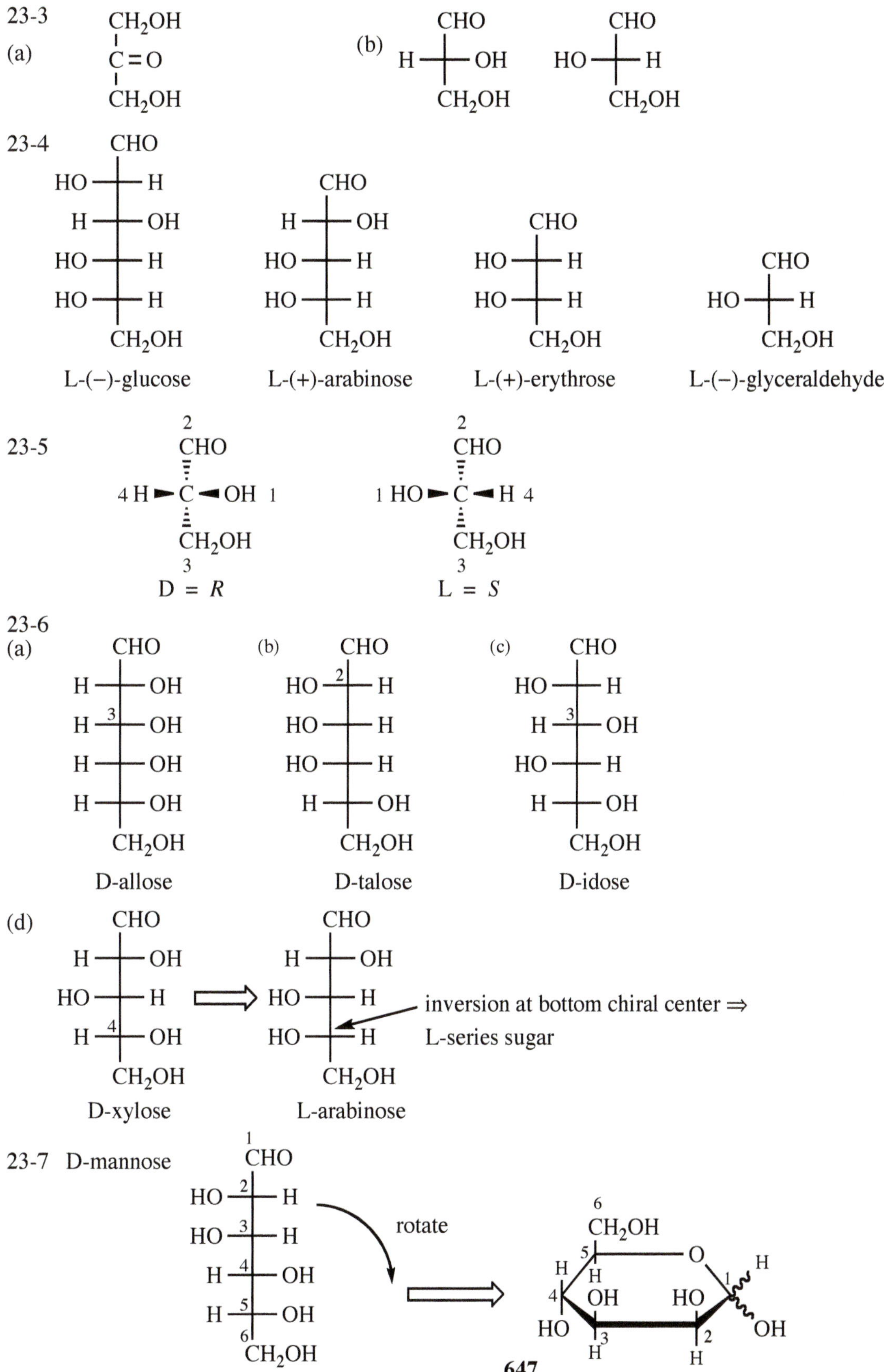
23-3
(a)
(b)
23-4
L-(−)-glucose
L-(+)-arabinose
L-(+)-erythrose
L-(−)-glyceraldehyde
23-5
D = R
L = S
23-6
(a)
(b)
(c)
D-allose
D-talose
D-idose
(d)
D-xylose
L-arabinose
inversion at bottom chiral center ⇒
L-series sugar
23-7 D-mannose
rotate

23-8 D-allose

OH at C-3 is axial.

23-9 D-talopyranose

OH groups at C-2 and C-4 are axial.

23-10

(a) D-arabinofuranose

(b) D-ribofuranose

23-11

23-12

(a) α-D-mannopyranose

axial = α

(b) β-D-galactopyranose

equatorial = β

(c) β-D-allopyranose

equatorial = β

(d) α-D-arabinofuranose

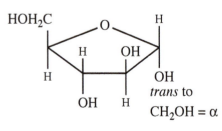

trans to CH_2OH = α

(e) β-D-ribofuranose

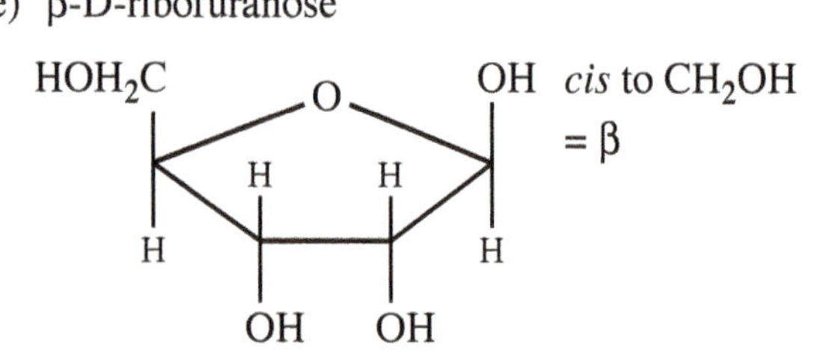

Students: This concept map intends to clarify the difference between an acetal and a ketal as mentioned in text section 23-4. The terms *hemiketal* and *ketal* have been resurrected by IUPAC for organic nomenclature, now considered subsets of *hemiacetal* and *acetal*. Biochemists still use these terms (as do many organic chemists!) to differentiate between monosaccharides containing the ketone or aldehyde functional group, *i.e.*, ketoses and aldoses.

To summarize: an aldehyde reacts with one molecule of alcohol to form a *hemiacetal* and with two molecules of alcohol to form an *acetal*; a ketone reacts with one molecule of alcohol to form a *hemiketal* and with two molecules of alcohol to form a *ketal*.

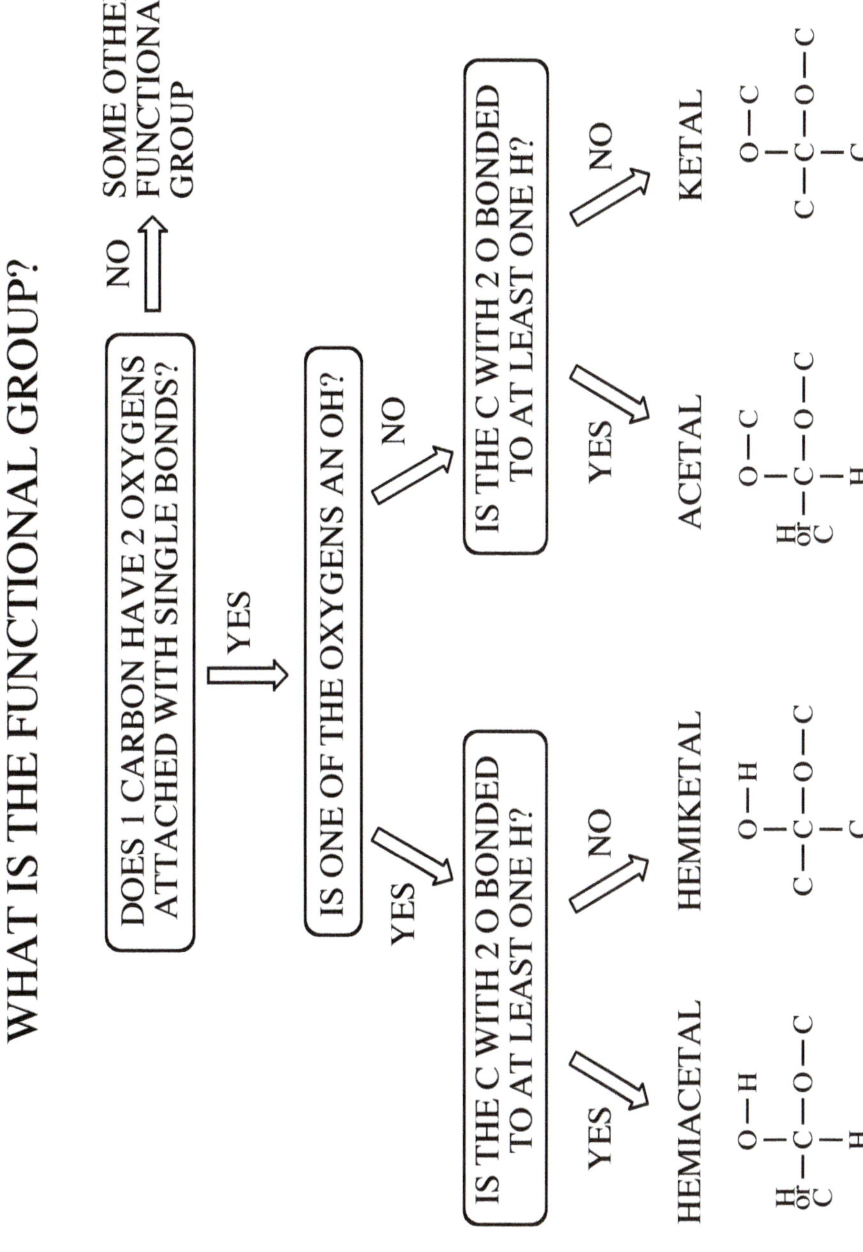

23-13 a = fraction of galactose as the α anomer; b = fraction of galactose as the β anomer

$$a(+150.7^\circ) + b(+52.8^\circ) = +80.2^\circ$$

$$a + b = 1; \quad b = 1 - a$$

$$a(+150.7^\circ) + (1-a)(+52.8^\circ) = +80.2^\circ \xrightarrow{\text{solve for "a"}} a = 0.28; \quad b = 0.72$$

The equilibrium mixture contains 28% of the α anomer and 72% of the β anomer.

23-14

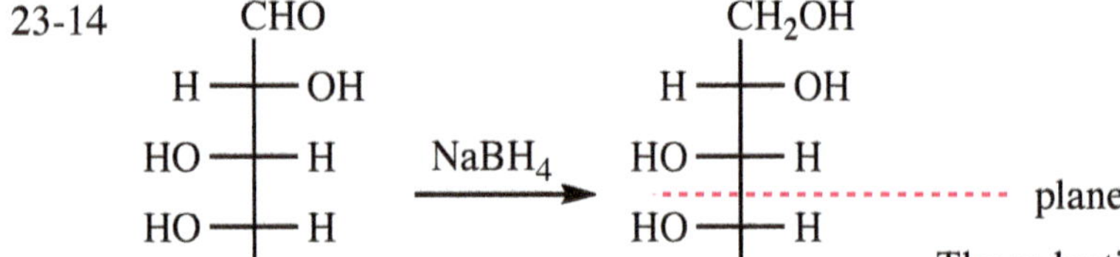

D-galactose

The reduction product of D-galactose still has four chiral centers but also has a plane of symmetry; it is a meso compound and is therefore optically inactive.

23-15

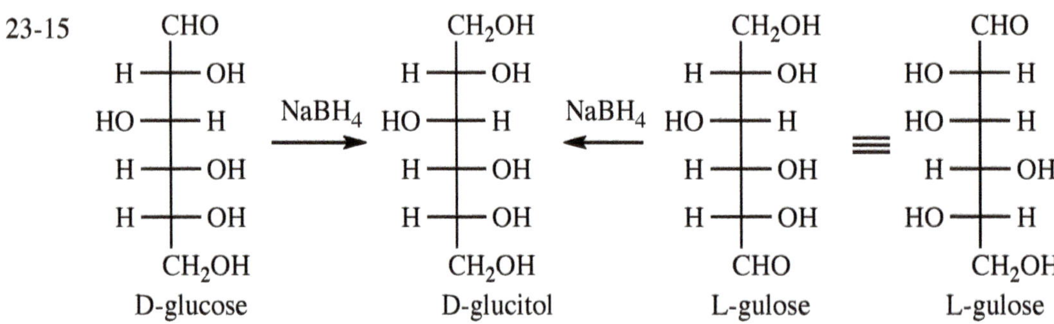

D-glucose D-glucitol L-gulose L-gulose

23-16

(a)

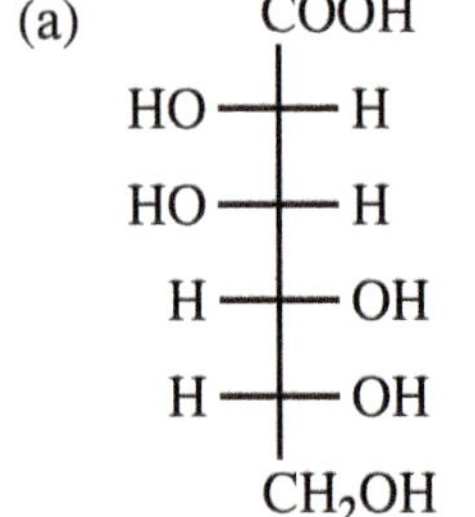

D-mannonic acid

(b)

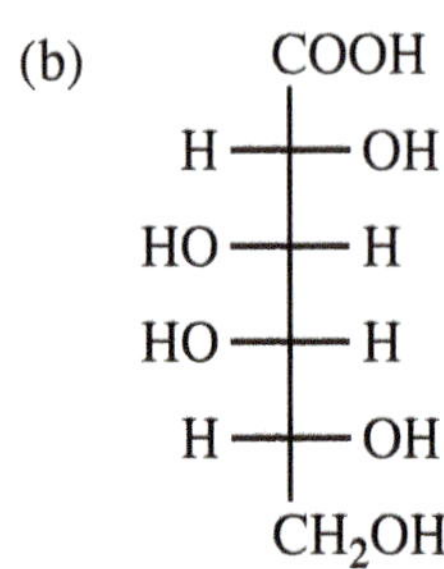

D-galactonic acid

(c) No reaction— D-fructose is a ketohexose; only aldoses react.

23-17

(a)

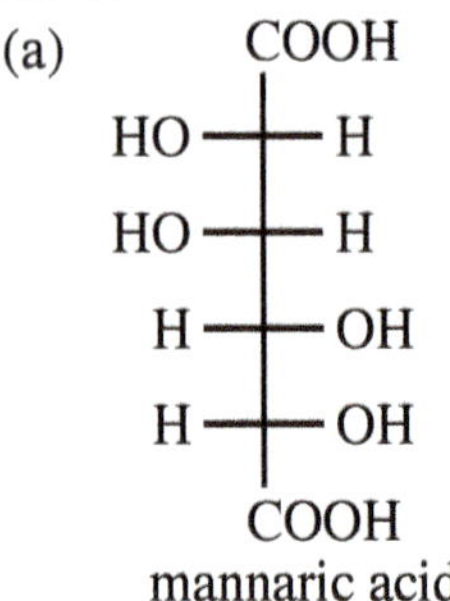

mannaric acid

(b)

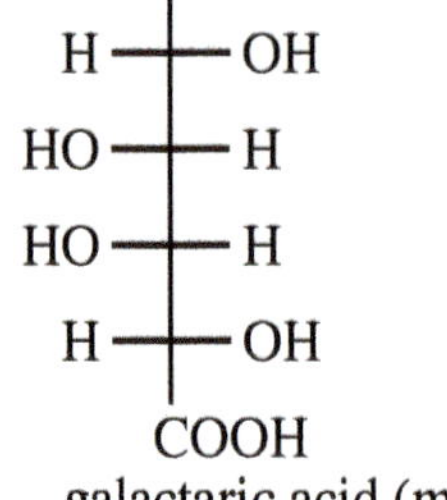

galactaric acid (meso)

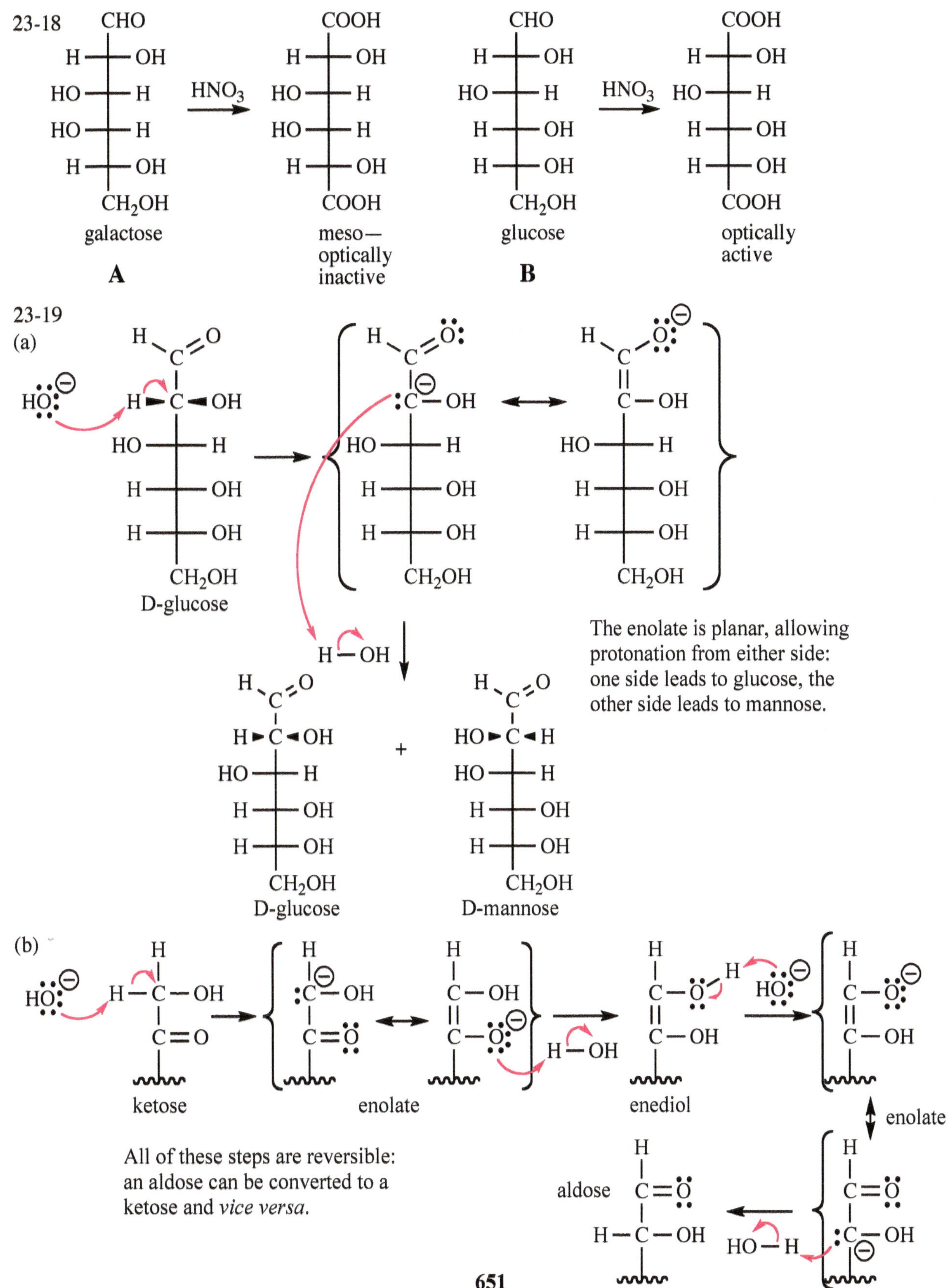

23-18
CHO
H OH
HO H
HO H
H OH
CH2OH
galactose
A
HNO3
COOH
H OH
HO H
HO H
H OH
COOH
meso—
optically
inactive
CHO
H OH
HO H
H OH
H OH
CH2OH
glucose
B
HNO3
COOH
H OH
HO H
H OH
H OH
COOH
optically
active
23-19
(a)
D-glucose
D-glucose
D-mannose
The enolate is planar, allowing protonation from either side: one side leads to glucose, the other side leads to mannose.
(b)
ketose
enolate
enediol
enolate
aldose
All of these steps are reversible: an aldose can be converted to a ketose and vice versa.

23-20

(a) not reducing: an acetal ending in "oside"
(b) reducing: a hemiacetal ending in "ose"
(c) reducing: a hemiacetal ending in "ose"
(d) not reducing: an acetal ending in "oside"
(e) reducing: one of the rings has a hemiacetal.
(f) not reducing: all anomeric carbons are in acetal form.

Like many common names, the name "sucrose" is misleading: the "ose" ending suggests a hemiacetal structure, but sucrose is a glycoside, so technically its name should end in "oside".

23-21 This problem asks for structures from 23-20 parts (a), (c), and (d).

(a) axial = α

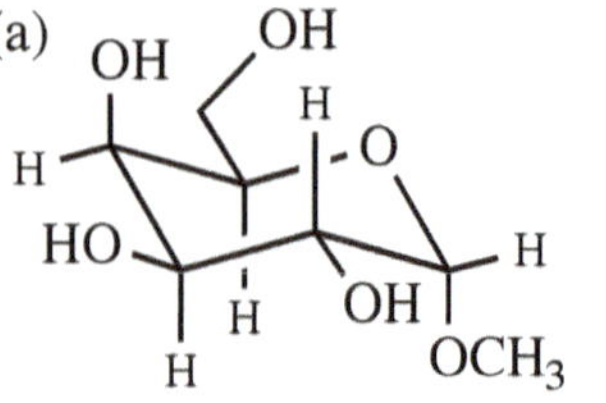

(c) axial = α

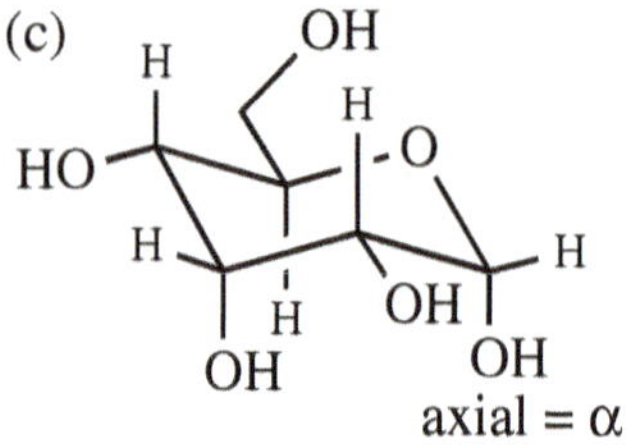

(d) HOH_2C ... OCH_2CH_3 *cis* to CH_2OH = β

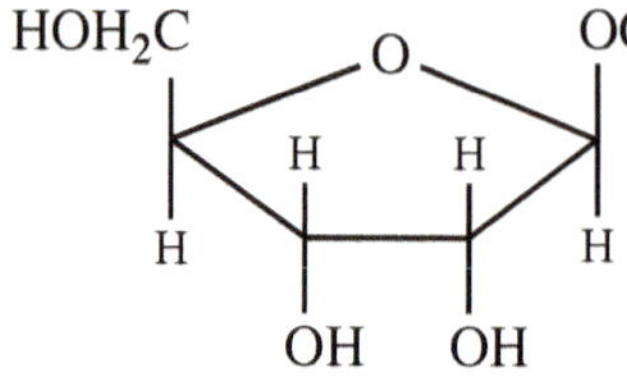

23-22

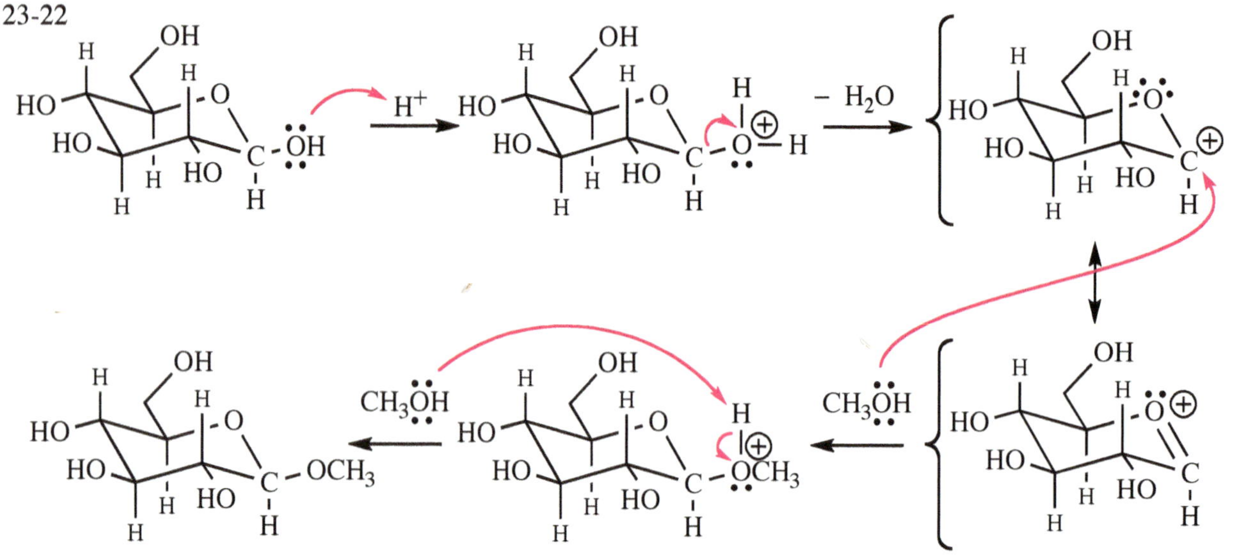

H^+

$- H_2O$

$CH_3\ddot{O}H$

$CH_3\ddot{O}H$

23-23

H_3O^+ → 2 glucose

\+ HO—CHPh (CN) benzaldehyde cyanohydrin

Ph—CH(=O) + HCN

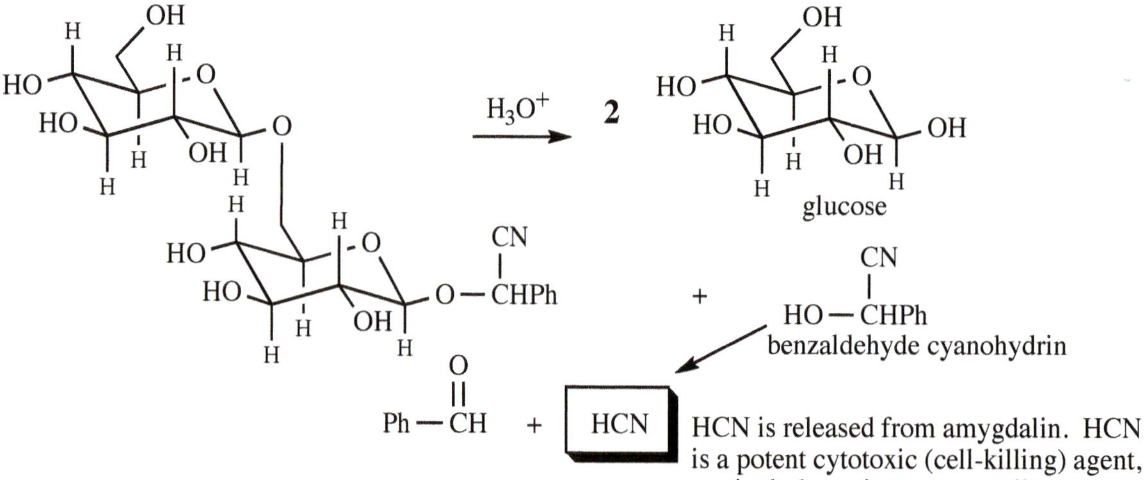

HCN is released from amygdalin. HCN is a potent cytotoxic (cell-killing) agent, particularly toxic to nerve cells.

23-24

α and β-D-fructofuranose $\xrightarrow[\text{HCl} \; - H_2O]{CH_3CH_2OH}$ ethyl β-D-fructofuranoside + ethyl α-D-fructofuranoside

The aglycone in each product is circled.

23-25

(a) D-fructose $\xrightarrow[\text{excess } CH_3I]{Ag_2O}$

(b) acidic hydrolysis → H_3O^+ → CH_3OH + 1,3,4,6-tetra-O-methyl open-chain fructose

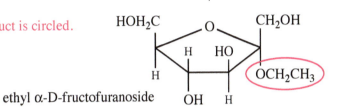

(c) Determining that the open chain form of fructose is a ketone at C-2 with a free OH at C-5 shows that fructose exists as a furanose hemiacetal.

23-26

$CH_3-OSO_3CH_3$ → + $^{-}OSO_3CH_3$

23-27 Excess CH_3I is needed to form methyl ethers at all OH groups.

(a)

(b)

23-28 Excess acetic anyhydride is needed to acetylate all OH groups.

(a) H, OAc, H, O, AcO, AcO, H, H AcO, H, OAc

(b) $AcOH_2C$, O, OAc, H, H, H, H, AcO, OAc

23-29 Reagents for the Ruff degradation are: 1. Br_2, H_2O; 2. $H_2O_2, Fe_2(SO_4)_3$.

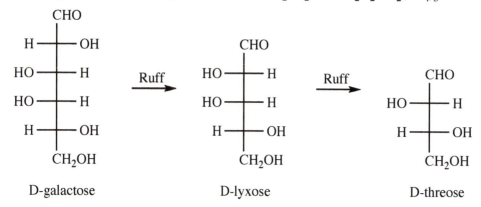

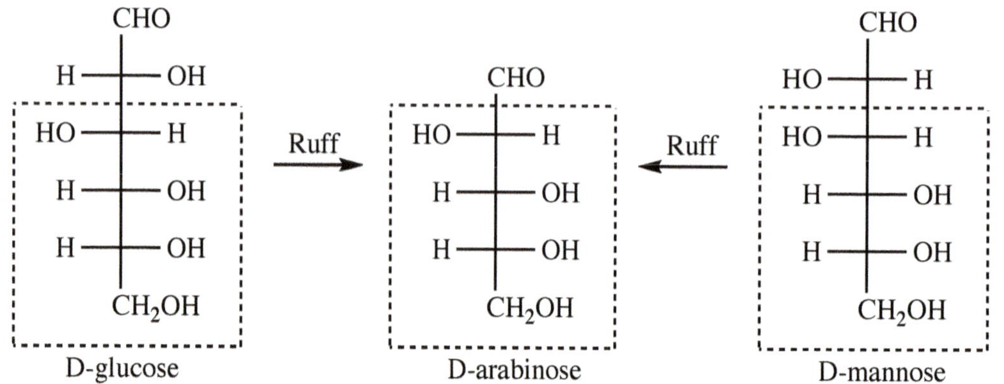

D-glucose D-arabinose D-mannose

23-30 Reagents for the Ruff degradation are: 1. Br_2, H_2O; 2. $H_2O_2, Fe_2(SO_4)_3$.

CHO
H — OH
HO — H
HO — H
H — OH
CH_2OH

Ruff →

CHO
HO — H
HO — H
H — OH
CH_2OH

Ruff →

CHO
HO — H
H — OH
CH_2OH

D-galactose D-lyxose D-threose

23-31 Reagents for the Ruff degradation are: 1. Br_2, H_2O; 2. $H_2O_2, Fe_2(SO_4)_3$.

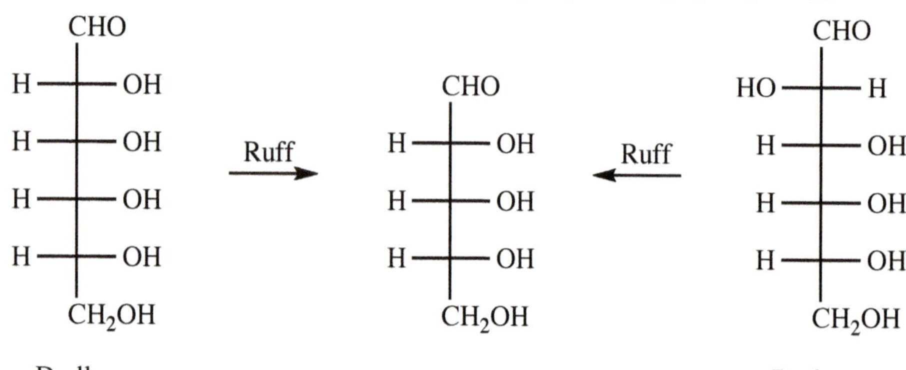

D-allose D-altrose

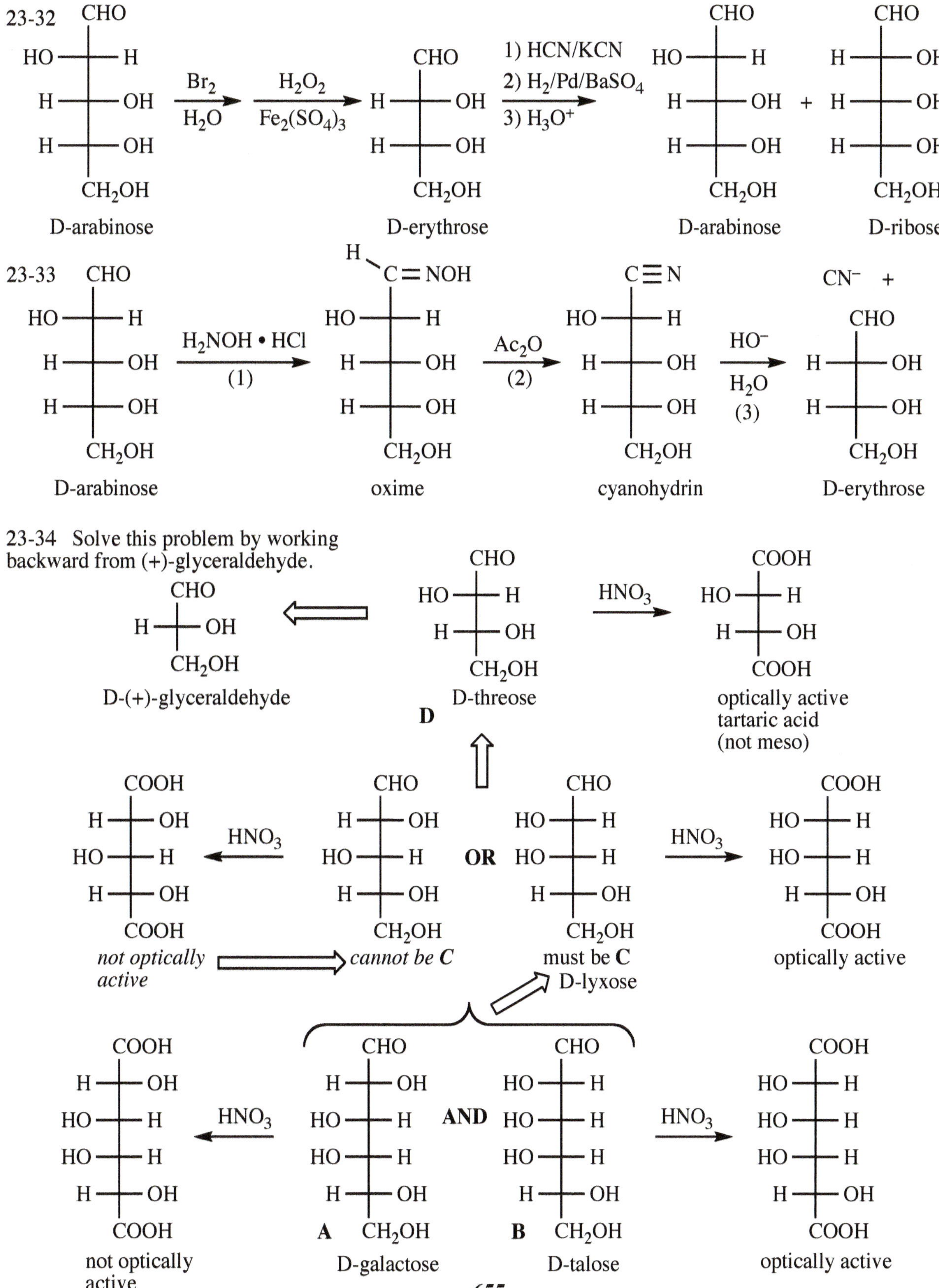
23-32
CHO
HO
H
OH
CH2OH
D-arabinose
Br2
H2O
H2O2
Fe2(SO4)3
D-erythrose
1) HCN/KCN
2) H2/Pd/BaSO4
3) H3O+
D-arabinose
+
D-ribose
23-33
D-arabinose
H2NOH • HCl
(1)
H
C=NOH
oxime
Ac2O
(2)
C≡N
cyanohydrin
HO−
H2O
(3)
CN− +
D-erythrose
23-34 Solve this problem by working backward from (+)-glyceraldehyde.
D-(+)-glyceraldehyde
D-threose
D
HNO3
COOH
optically active
tartaric acid
(not meso)
not optically active
cannot be C
OR
must be C
D-lyxose
optically active
AND
A
D-galactose
B
D-talose
not optically active
optically active

23-35 This problem requires logic, inference, and working backwards, just as it did for Emil Fischer.

(a) Glucose and mannose degrade by one carbon to give arabinose. Parts (a) and (b) determine the structure of arabinose.

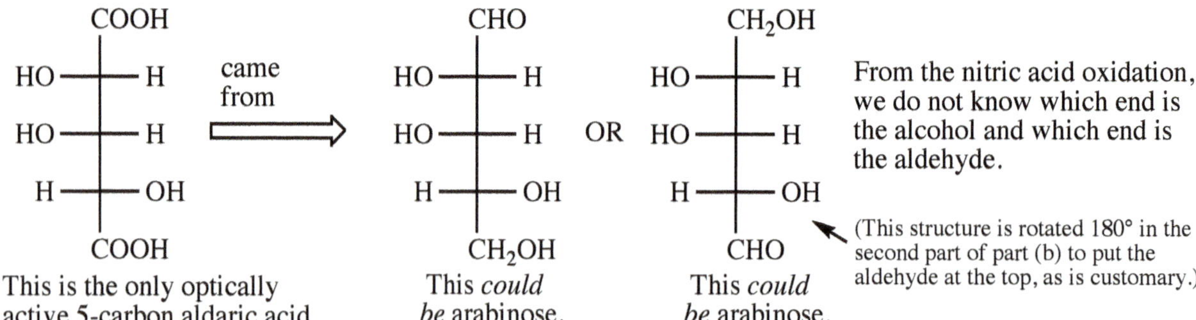

COOH
HO — H
HO — H
H — OH
COOH
This is the only optically active 5-carbon aldaric acid.
came from
CHO
HO — H
HO — H
H — OH
CH_2OH
This *could be* arabinose.
OR
CH_2OH
HO — H
HO — H
H — OH
CHO
This *could be* arabinose.
From the nitric acid oxidation, we do not know which end is the alcohol and which end is the aldehyde.
(This structure is rotated 180° in the second part of part (b) to put the aldehyde at the top, as is customary.)

(b) Once we get to the 4-carbon sugars, then we can tell which 5-carbon sugar must be arabinose.

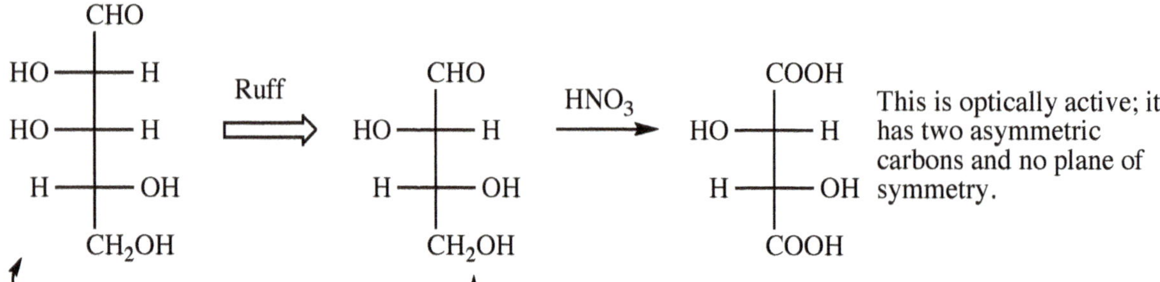

CHO
HO — H
HO — H
H — OH
CH_2OH
Ruff
CHO
HO — H
H — OH
CH_2OH
HNO_3
COOH
HO — H
H — OH
COOH
This is optically active; it has two asymmetric carbons and no plane of symmetry.

This possible arabinose degrades to this 4-carbon sugar, oxidation of which gives an **optically active** aldaric acid. Conclusion: this starting material is not arabinose.

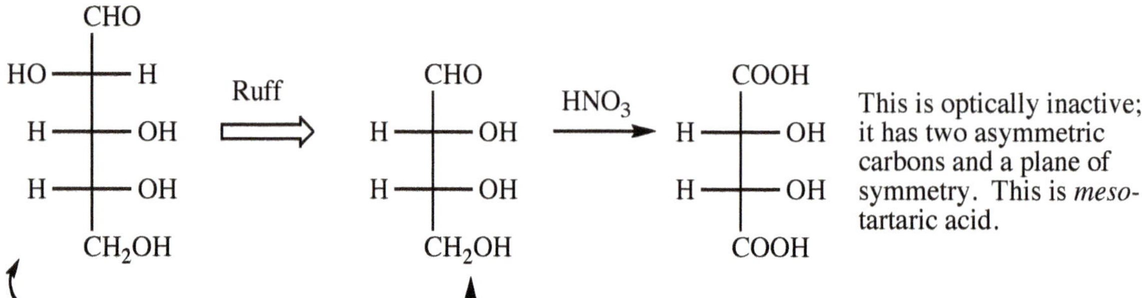

CHO
HO — H
H — OH
H — OH
CH_2OH
Ruff
CHO
H — OH
H — OH
CH_2OH
HNO_3
COOH
H — OH
H — OH
COOH
This is optically inactive; it has two asymmetric carbons and a plane of symmetry. This is *meso*-tartaric acid.

This possible arabinose degrades to this 4-carbon sugar, oxidation of which gives an **optically inactive** aldaric acid. Conclusion: this starting material is arabinose, and the 4-carbon sugar is erythrose.

CONCLUSIONS from (a) and (b):

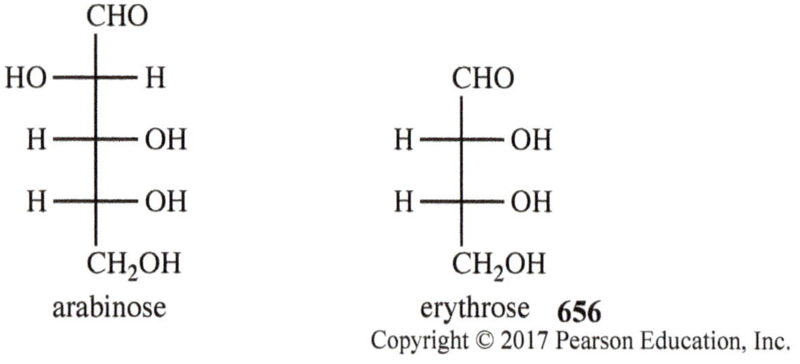

CHO
HO — H
H — OH
H — OH
CH_2OH
arabinose
CHO
H — OH
H — OH
CH_2OH
erythrose

23-35 continued

(c) Arabinose came from Ruff degradation of glucose or mannose.

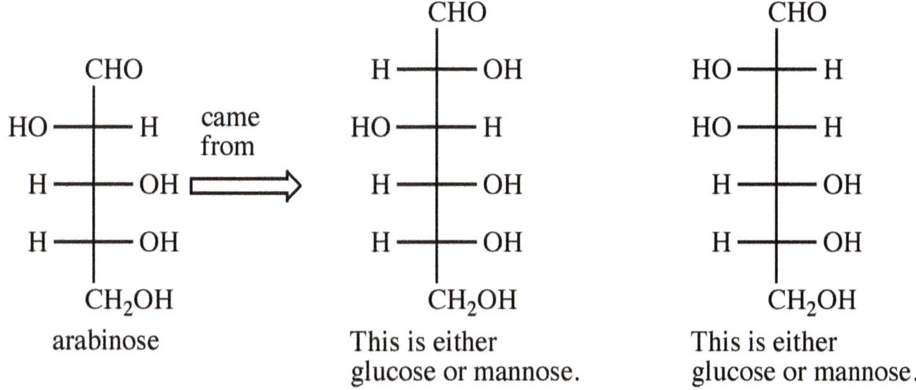

(d) Chemically interchange the aldehyde with the alcohol at C-6. Mannose produces mannose.

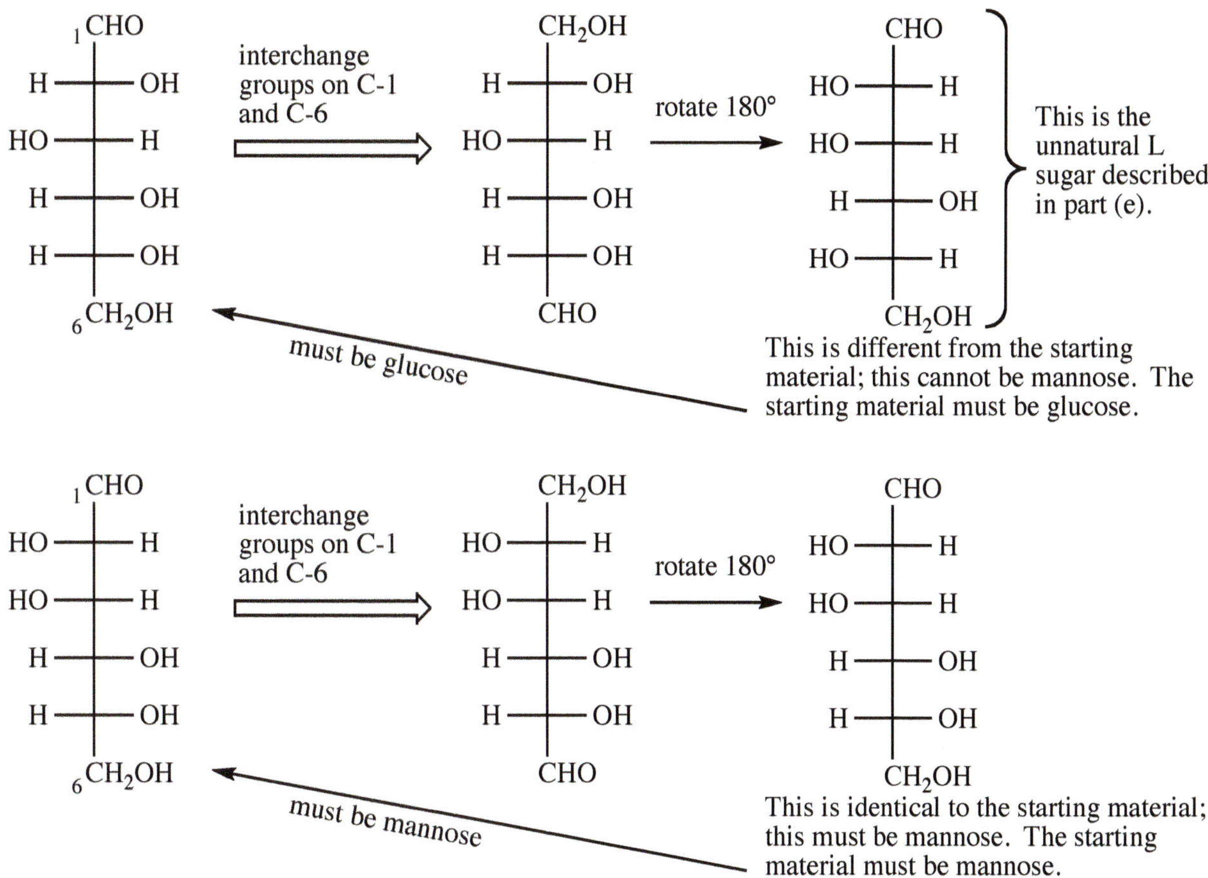

(e) The unnatural L sugar is shown at the end of the first sequence in part (d). This completes Fischer's proof of glucose.

It is hard for us in the 21st century to appreciate the challenges that Fischer and his students faced while doing this work: they had no spectroscopy, no chromatography—and since the incandescent light bulb had been invented only a decade earlier, they might not have had any electric lights! They might have had to interpret their NMRs by candlelight!

23-36

The glycoside linkage does not change—only the stereochemistry at the hemiacetal anomeric carbon.

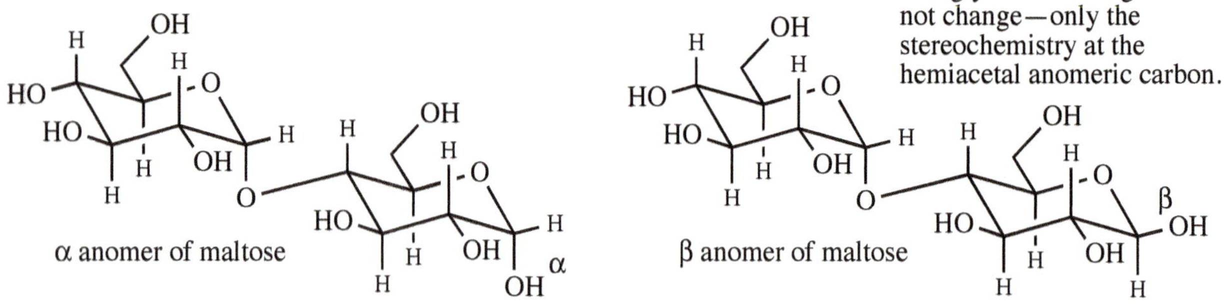

α anomer of maltose

β anomer of maltose

23-37

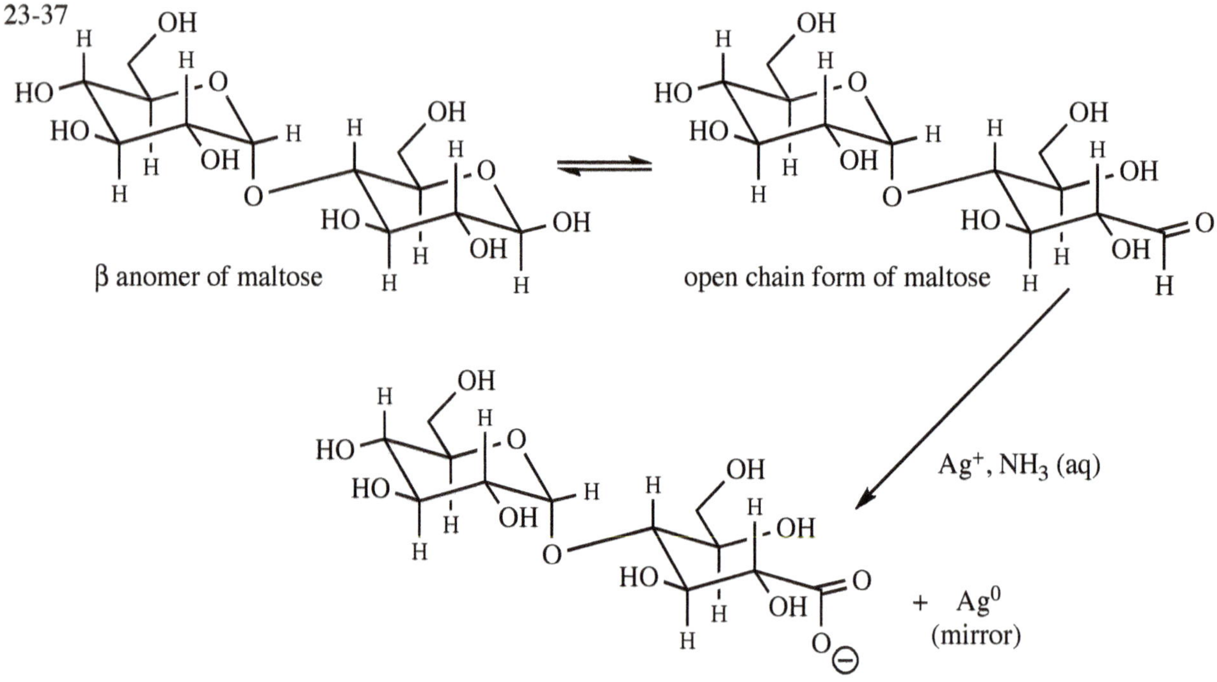

β anomer of maltose ⇌ open chain form of maltose

Ag^+, NH_3 (aq)

+ Ag^0 (mirror)

23-38 Lactose is a hemiacetal; in water, the hemiacetal is in equilibrium with the open-chain form and can react as an aldehyde. Therefore, lactose can mutarotate and is a reducing sugar.

α anomer of lactose

β anomer of lactose

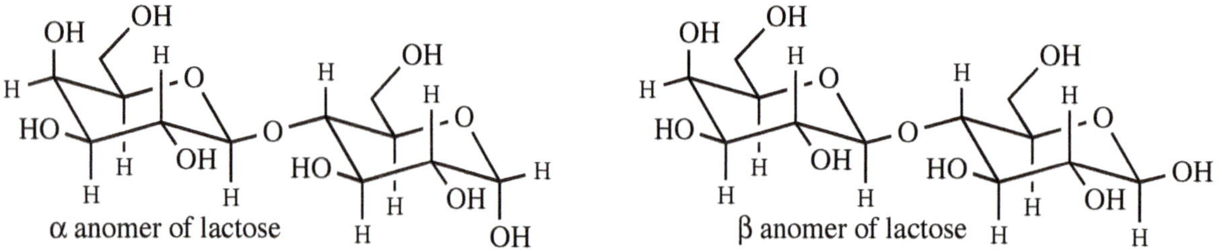

23-39 Gentiobiose is a hemiacetal; in water, the hemiacetal is in equilibrium with the open-chain form and can react as an aldehyde. Gentiobiose can mutarotate and is a reducing sugar.

23-40 Trehalose must be two glucose molecules connected by an α-1,1'-glycoside. Wedge and dash bonds are shown at the anomeric carbons for clarity.

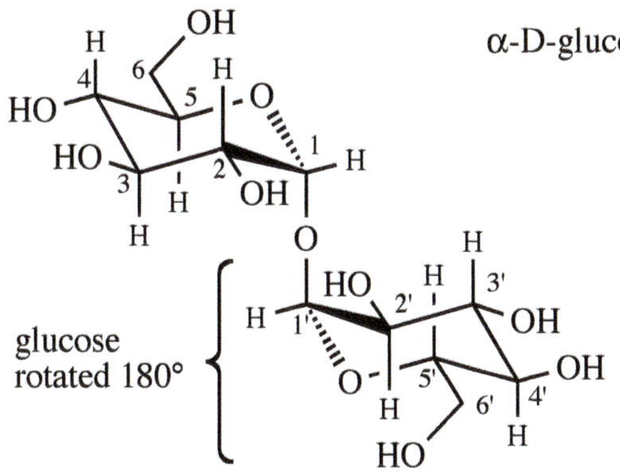

α-D-glucopyranosyl-α-D-glucopyranoside

glucose rotated 180°

23-41 raffinose

melibiose = 6-O-(α-D-galactopyranosyl)-D-glucopyranose

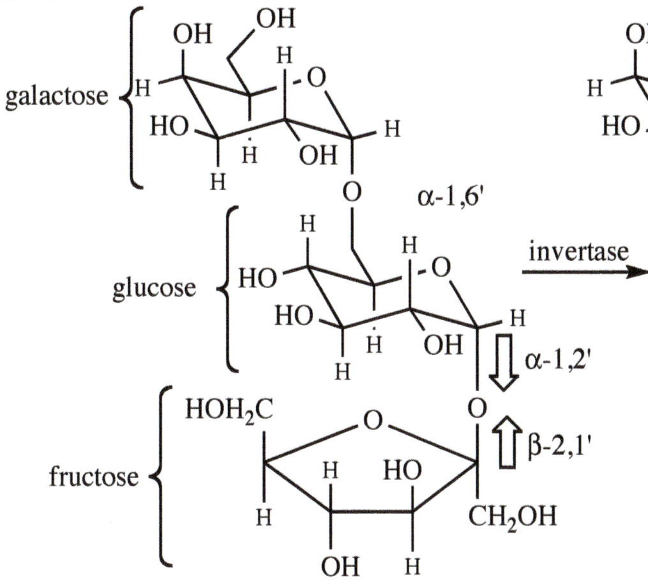

galactose

glucose

fructose

α-1,6'

invertase

α-1,2'

β-2,1'

The lower glycoside linkage is α-1,2' from the glucose point of view, but β-2,1' from the fructose point of view.

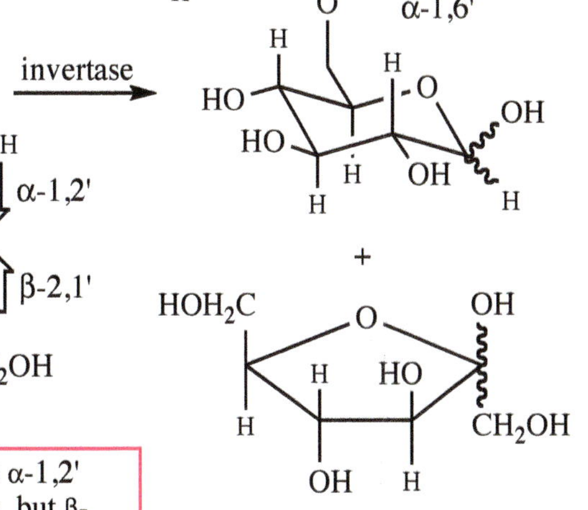

fructose
(D-fructofuranose)

23-42 cellulose acetate

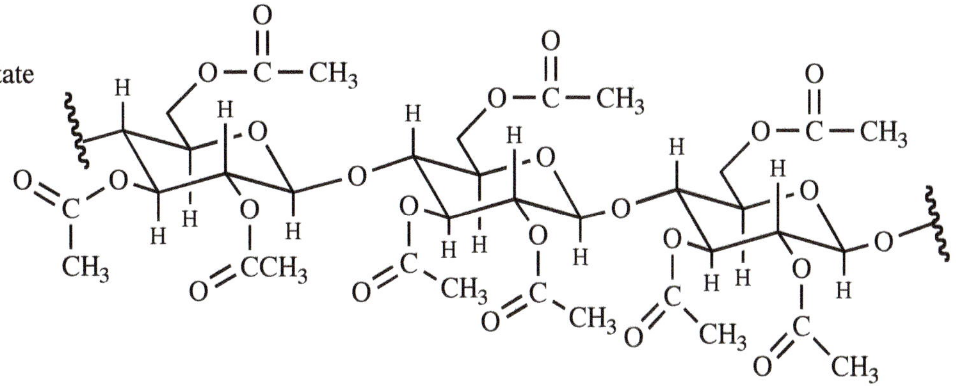

23-43 cytosine:

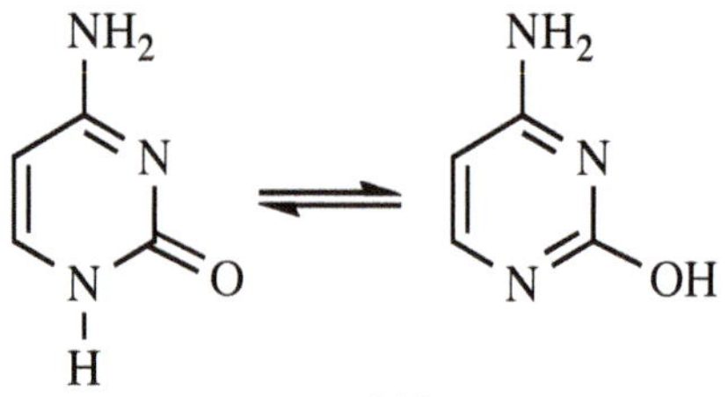

uracil:

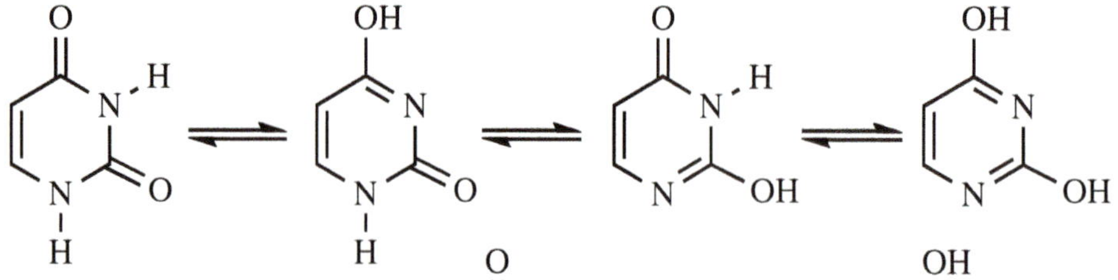

guanine:

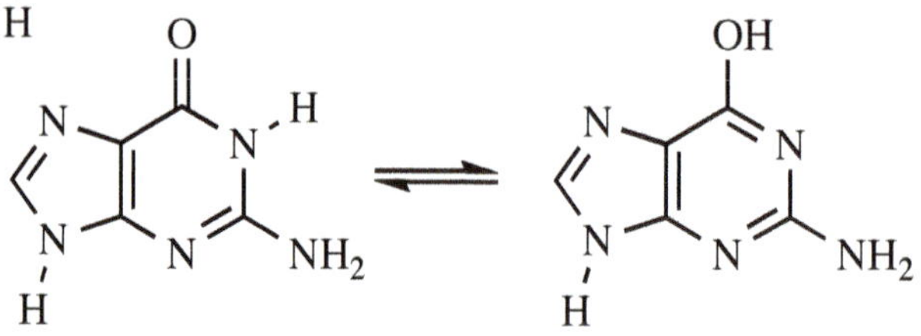

23-44 Aminoglycosides, including nucleosides, are similar to acetals: stable to base, cleaved by acid.

(a)

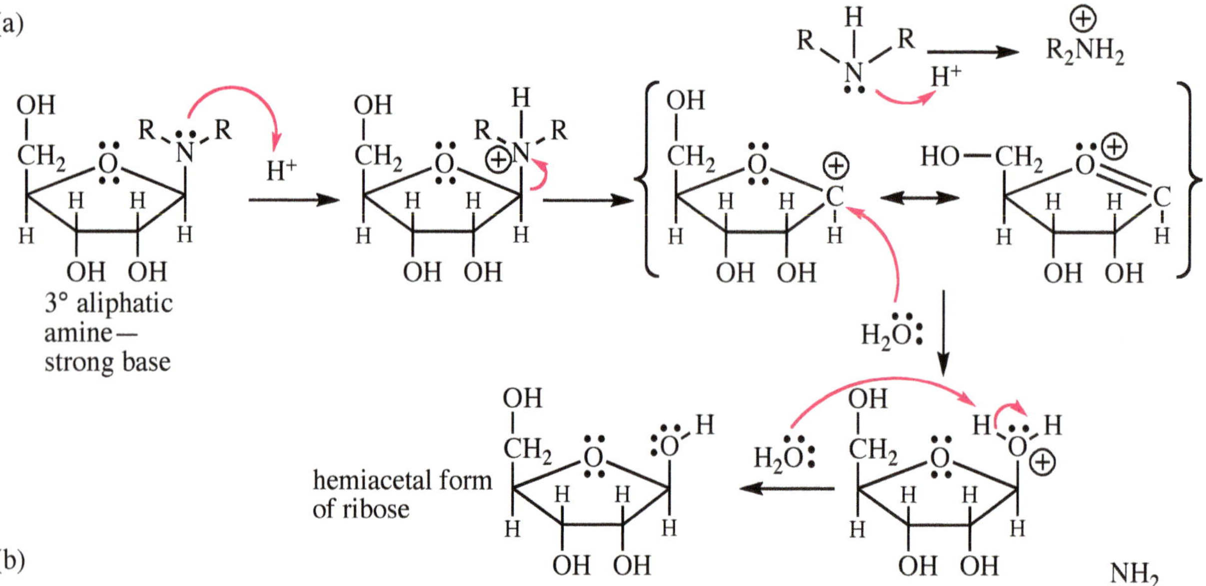

3° aliphatic amine—strong base

hemiacetal form of ribose

(b)

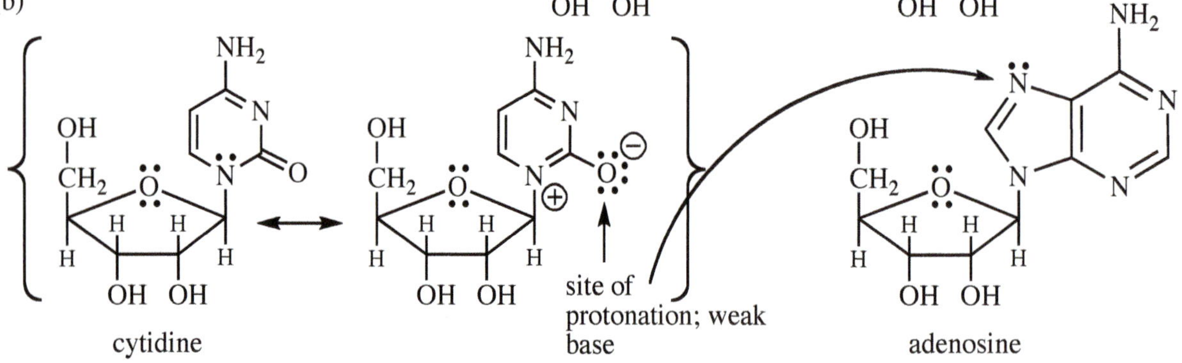

cytidine

site of protonation; weak base

adenosine

See discussion on next page.

23-44(b) continued

Nucleosides are less rapidly hydrolyzed in aqueous acid because the site of protonation (the N in adenosine, and in cytidine, the oxygen shown with the negative charge in the second resonance form) is much less basic than the aliphatic amine in an aminoglycoside as shown in part (a). Nucleosides require stronger acid, or longer time and higher temperature, to be hydrolyzed.

This is important in living systems, as it would cause genetic damage or even death of an organism if its DNA or RNA were too easily decomposed. Organisms go to great lengths and expend considerable energy to maintain the structural integrity of their DNA.

23-45

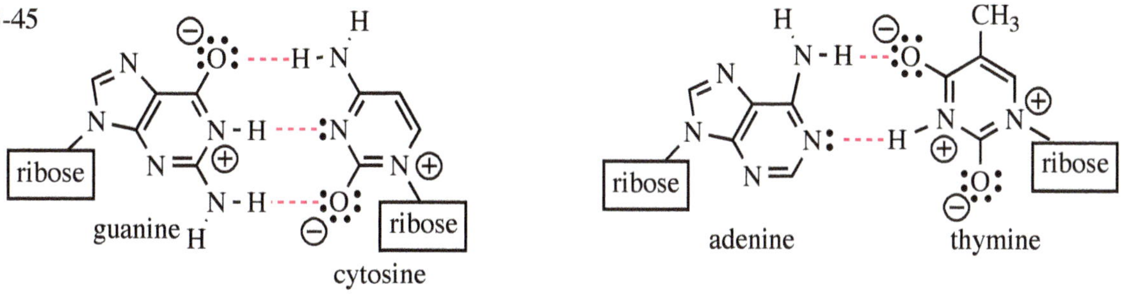

guanine cytosine adenine thymine

The polar resonance forms show how the hydrogen bonds are particularly strong. Each oxygen has significant negative charge, and in each pair, one H—N is polarized more strongly because the N has positive charge.

23-46

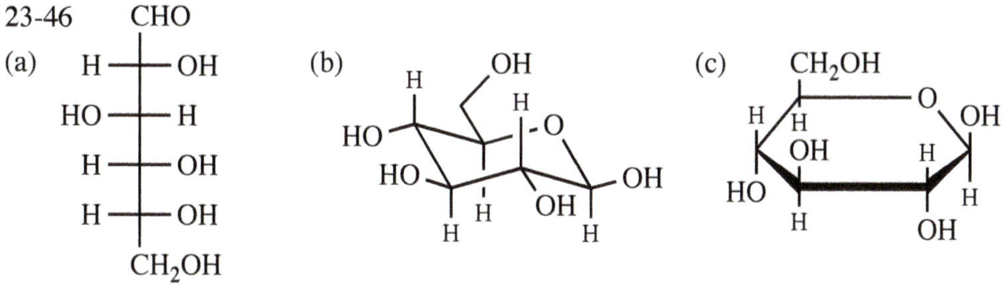

(a) (b) (c)

23-47

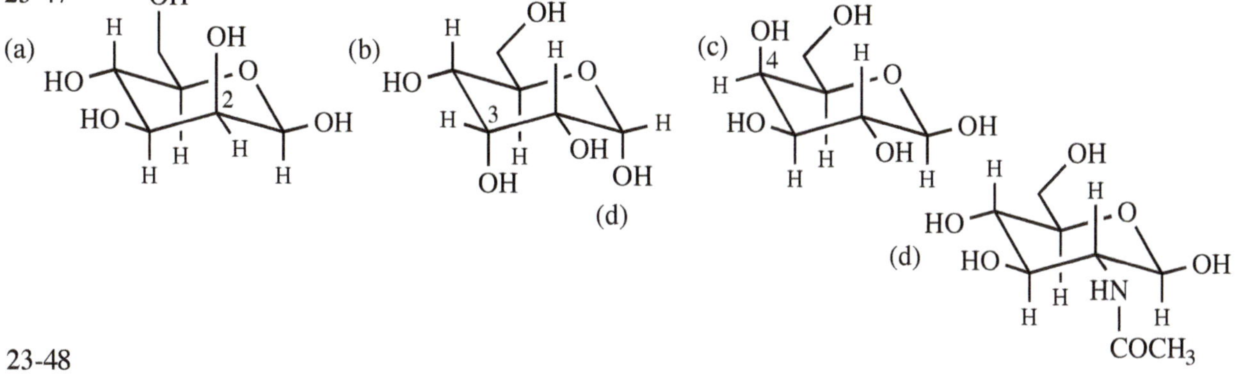

(a) (b) (c) (d)

23-48

(a) D-(−)-ribose (b) D-(+)-altrose (c) L-(+)-erythrose (d) L-(−)-galactose (e) L-(+)-idose

23-49

(a) D-aldohexose (D configuration, aldehyde, 6 carbons)
(b) D-aldopentose (D configuration, aldehyde, 5 carbons)
(c) L-ketohexose (L configuration, ketone, 6 carbons)
(d) L-aldohexose (L configuration, aldehyde, 6 carbons)
(e) D-ketopentose (D configuration, ketone, 5 carbons)
(f) L-aldotetrose (L configuration, aldehyde, 4 carbons)
(g) 2-acetamido-D-aldohexose (D configuration, aldehyde, 6 carbons in chain, with acetamido group at C-2)

23-50

(a)

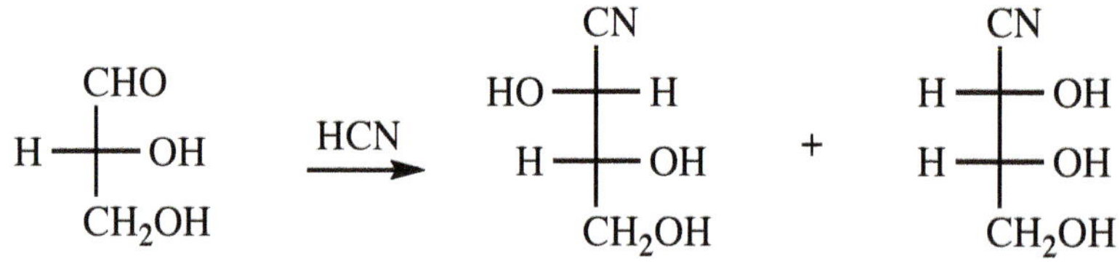

(b) The products are diastereomers with different physical properties. They could be separated by crystallization, distillation, or chromatography.

(c) Both products are optically active. Each has two chiral centers and no plane of symmetry.

23-51

(a)

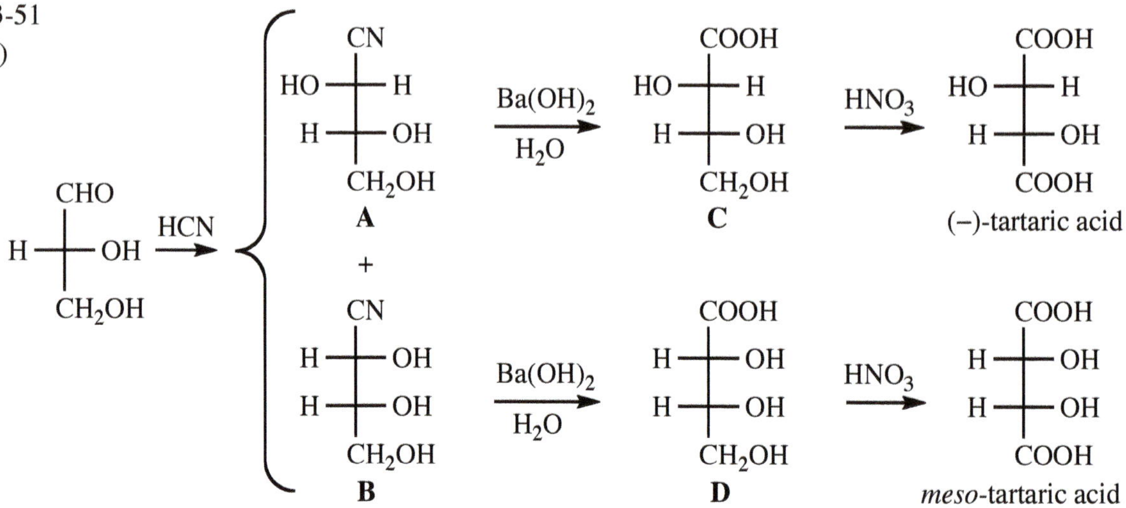

(b)

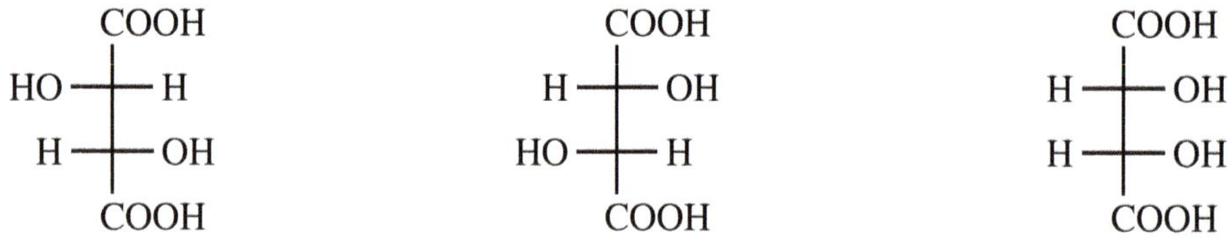

(*S*,*S*)-(−)-tartaric acid (*R*,*R*)-(+)-tartaric acid (*R*,*S*)-*meso*-tartaric acid

23-52

(a)

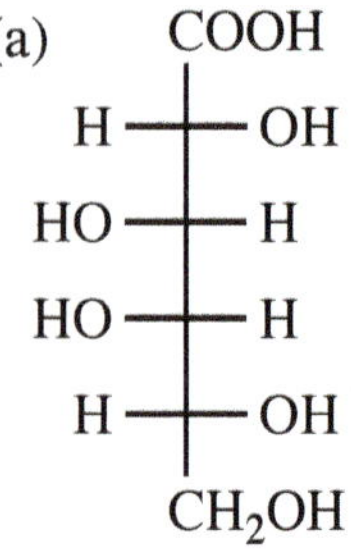

(b)

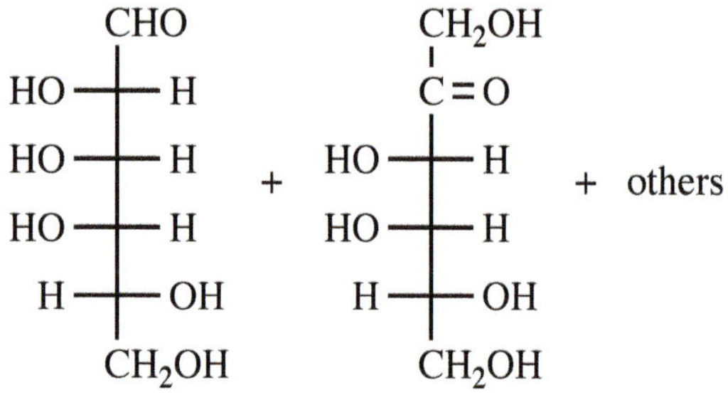

(c)

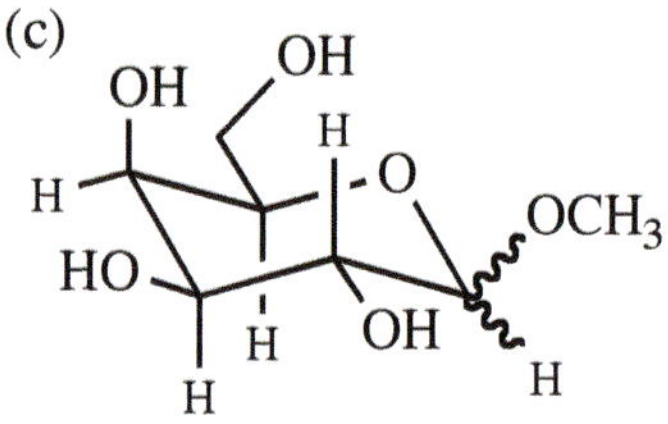

23-52 continued

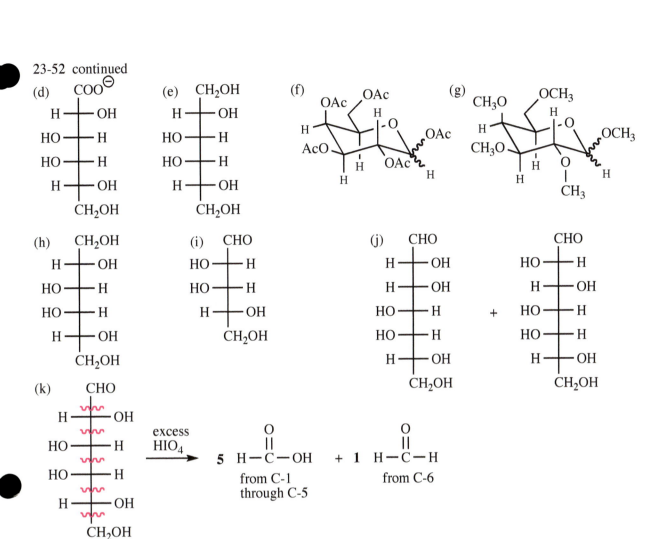

(d)

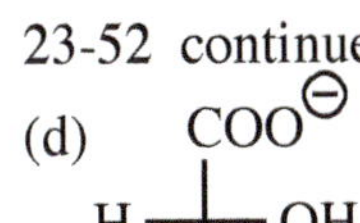

	$COO^{\ominus}$	
H		OH
HO		H
HO		H
H		OH
	CH_2OH	

(e)

	CH_2OH	
H		OH
HO		H
HO		H
H		OH
	CH_2OH	

(f) OAc, OAc, H, O, H, OAc, AcO, H, OAc, H, H

(g) CH_3O, OCH_3, H, O, H, OCH_3, CH_3O, H, O, H, CH_3, H

(h)

	CH_2OH	
H		OH
HO		H
HO		H
H		OH
	CH_2OH	

(i)

	CHO	
HO		H
HO		H
H		OH
	CH_2OH	

(j)

	CHO	
H		OH
H		OH
HO		H
HO		H
H		OH
	CH_2OH	

+

	CHO	
HO		H
H		OH
HO		H
HO		H
H		OH
	CH_2OH	

(k)

	CHO	
H		OH
HO		H
HO		H
H		OH
	CH_2OH	

$\xrightarrow{\text{excess } HIO_4}$ **5** H—C(=O)—OH (from C-1 through C-5) + **1** H—C(=O)—H (from C-6)

23-53

(a) H, OH, H, O, HO, HO, H, OH, OCH_3, H, H

(b) H, OCH_3, OCH_3, O, CH_3O, CH_3O, H, H, OH, H, H

(c) CH_3OH_2C, O, OH, H CH_3O, H, OH, H, CH_2OCH_3

(d) CH_3O, OCH_3, H, O, H, CH_3O, H, O, OCH_3, H, H, CH_3

23-54

(a) H, OH, H, O, HO, HO, H, OH, H, H, O, OH, H, O, H, OH, HO, H, OH, H, H

23-54 continued

(b) 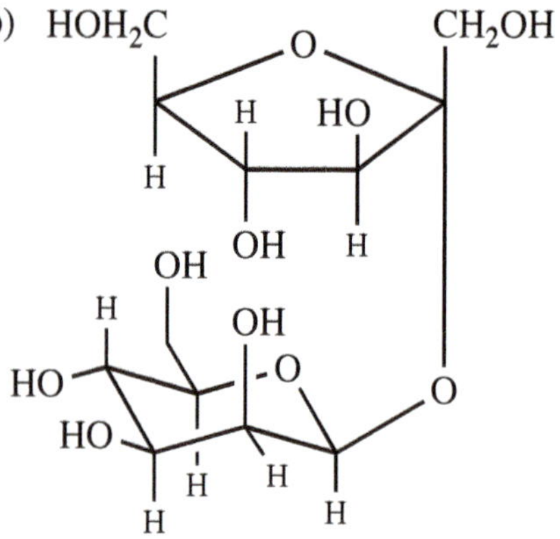 HOH₂C ... CH₂OH

(c)

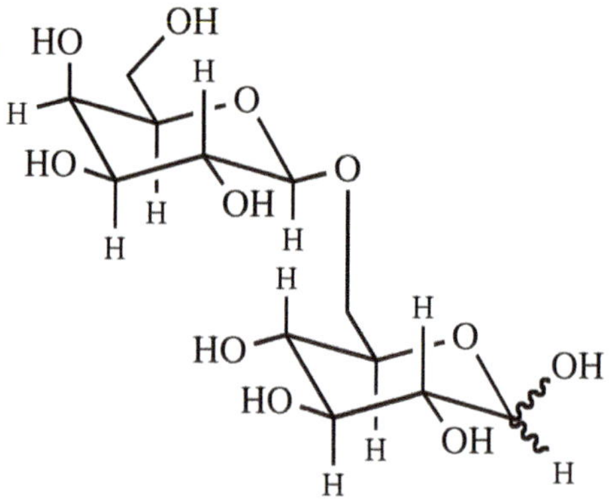

23-55

(a) No, there is no relation between the amount of G and A.
(b) Yes, this must be true mathematically.
(c) Chargaff's rule must apply only to double-stranded DNA. For each G in one strand, there is a complementary C in the opposing strand, but there is no correlation between G and C *in the same strand.*

23-56 These are reducing sugars and would undergo mutarotation. The required structural feature is the presence of a hemiacetal or a hemiketal that could equilibrate with the open-chain aldehyde or hydroxy ketone. —in problem 23-53: (b) and (c);
—in problem 23-54: (a) and (c).

23-57

(a) 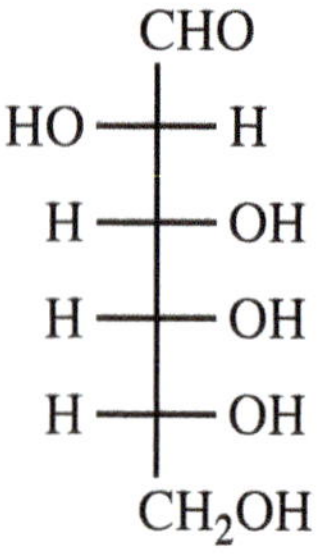 D-altrose $C_6H_{12}O_6$

(4) Not symmetric so aldaric acid is optically active.
(5) Removal of top carbon gives a pentose; the aldaric acid is symmetric so not optically active.

(b) 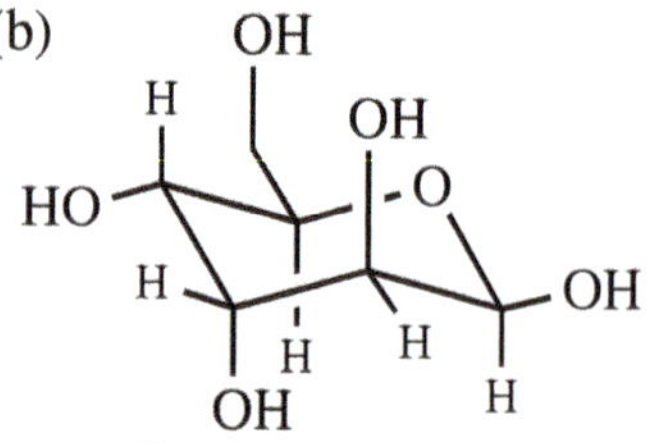 β-D-altropyranose

(7) Free OH at C5 shows the pyranose cyclic form.

23-58

galactose

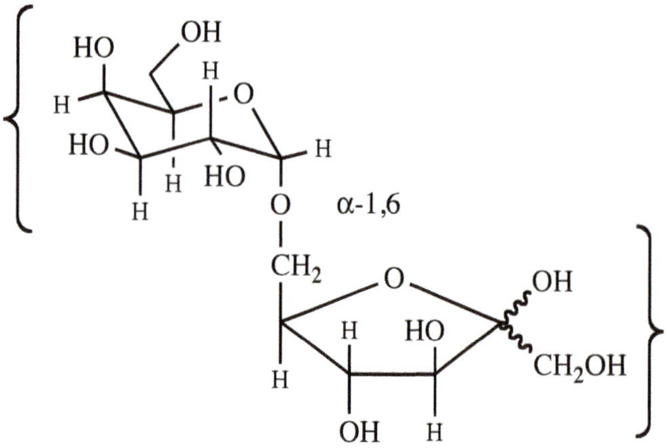

α-1,6

fructose

6-O-(α-D-galactopyranosyl)-D-fructofuranose

23-59

(a) These two D-aldopentoses will give optically active aldaric acids.

D-arabinose: CHO; HO–H; H–OH; H–OH; CH_2OH

D-lyxose: CHO; HO–H; HO–H; H–OH; CH_2OH

(b) Only D-threose (of the aldotetroses) will give an optically active aldaric acid.

D-threose: CHO; HO–H; H–OH; CH_2OH

(c) **X** is D-galactose.

X: CHO; H–OH; HO–H; HO–H; H–OH; CH_2OH —HNO_3→ COOH; H–OH; HO–H; HO–H; H–OH; COOH optically inactive

↓ Ruff

CHO; HO–H; HO–H; H–OH; CH_2OH —HNO_3→ COOH; HO–H; HO–H; H–OH; COOH optically active

The other aldohexose that gives an optically inactive aldaric acid is D-allose, with all OH groups on the right side of the Fischer projection. Ruff degradation followed by nitric acid gives an optically *inactive* aldaric acid, however, so **X** cannot be D-allose.

(d) The optically active, five-carbon aldaric acid comes from the optically active pentose, not from the optically inactive, six-carbon aldaric acid. The principle is not violated.

(e) CHO; HO–H; HO–H; H–OH; CH_2OH —Ruff→ CHO; HO–H; H–OH; CH_2OH (D-threose) —HNO_3→ COOH; HO–H; H–OH; COOH (*S*,*S*)-tartaric acid optically active

23-60 Tagatose is a monosaccharide, a ketohexose, that is found in the pyranose form.

(a) CH_2OH; C=O; HO–H; HO–H; H–OH; CH_2OH

(b) [pyranose chair structure: ring O; C1 CH_2OH; C2 OH; C3 OH, H; C4 HO, H; C5 HO, H; C6]

23-61

(a) acetal

(b) ⇒ benzaldehyde

β-D-glucose

The 4- and 6-OH groups of glucose reacted with benzaldehyde to form the acetal.

(c)

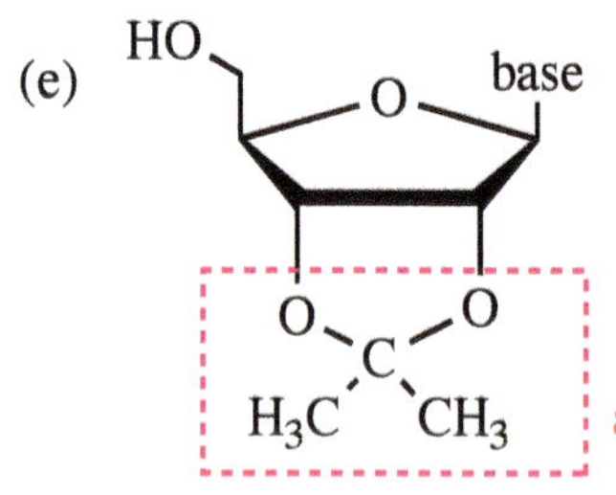

(c) and (d) The chiral center is marked with (*). This stereoisomer is a diastereomer since only one chiral center is inverted. This diastereomer puts the phenyl group in an axial position, definitely less stable than the structure shown in part (a). Only the product shown in (a) will be isolated from this reaction.

(e)

Only the 2'- and 3'-OH groups are close enough to form this cyclic acetal (ketal) from acetone, called an acetonide.

acetonide

23-62

(a) The less hindered 1° OH will react first.

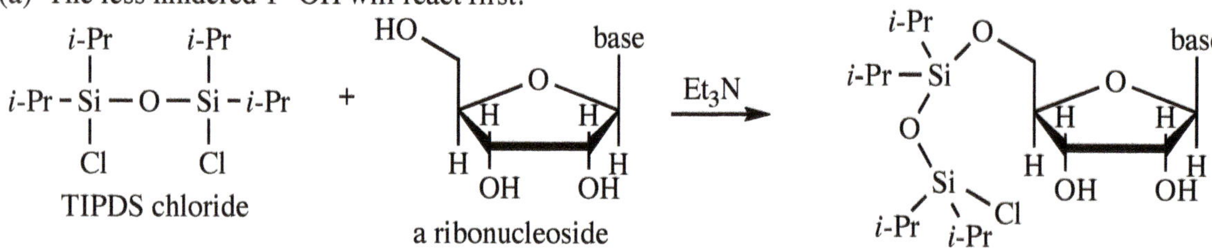

(b)

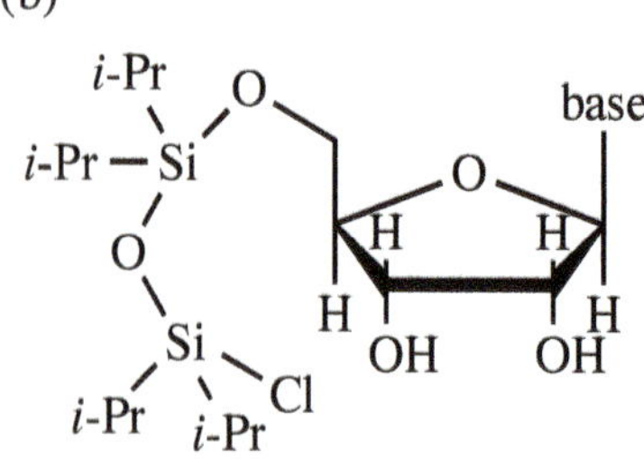

Et_3N

Because Si is larger than C, the Si—O bond length is around 1.6 Angstrom units, whereas a typical C—O bond is about 1.4 Angstrom units.

23-62 continued

(c)

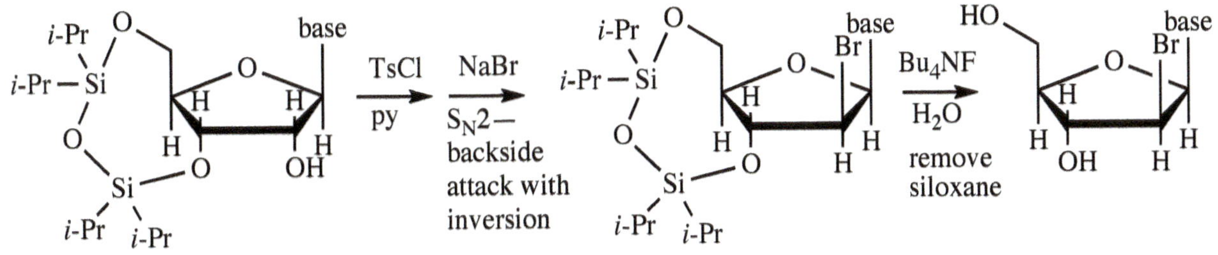

TsCl, py → NaBr, S_N2— backside attack with inversion → Bu_4NF, H_2O, remove siloxane

23-63

(a)

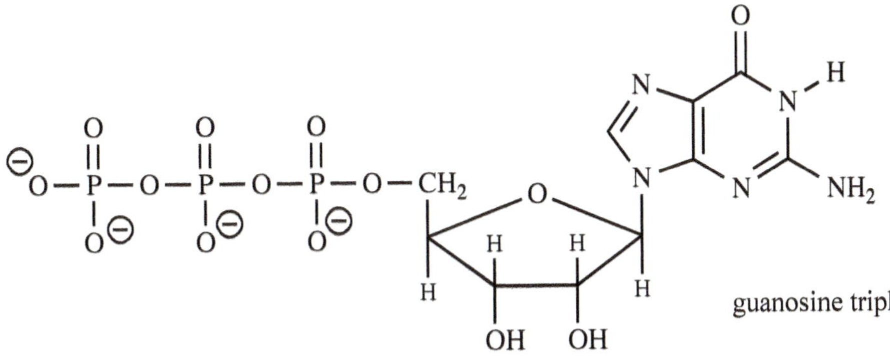

guanosine triphosphate (GTP)

(b)

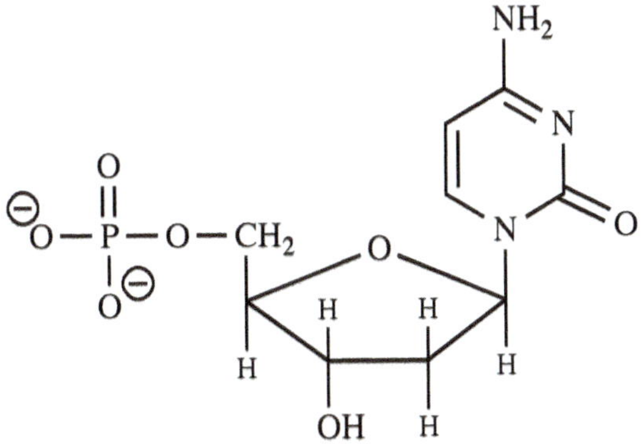

deoxycytidine monophosphate (dCMP)

(c)

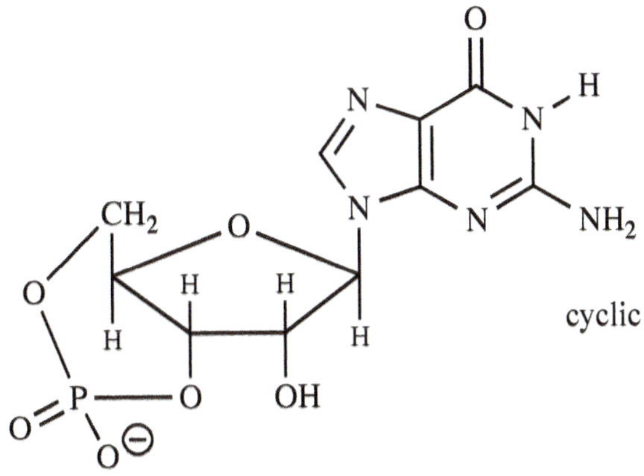

cyclic guanosine monophosphate (cGMP)

23-64 Bonds from C to H are omitted for simplicity.

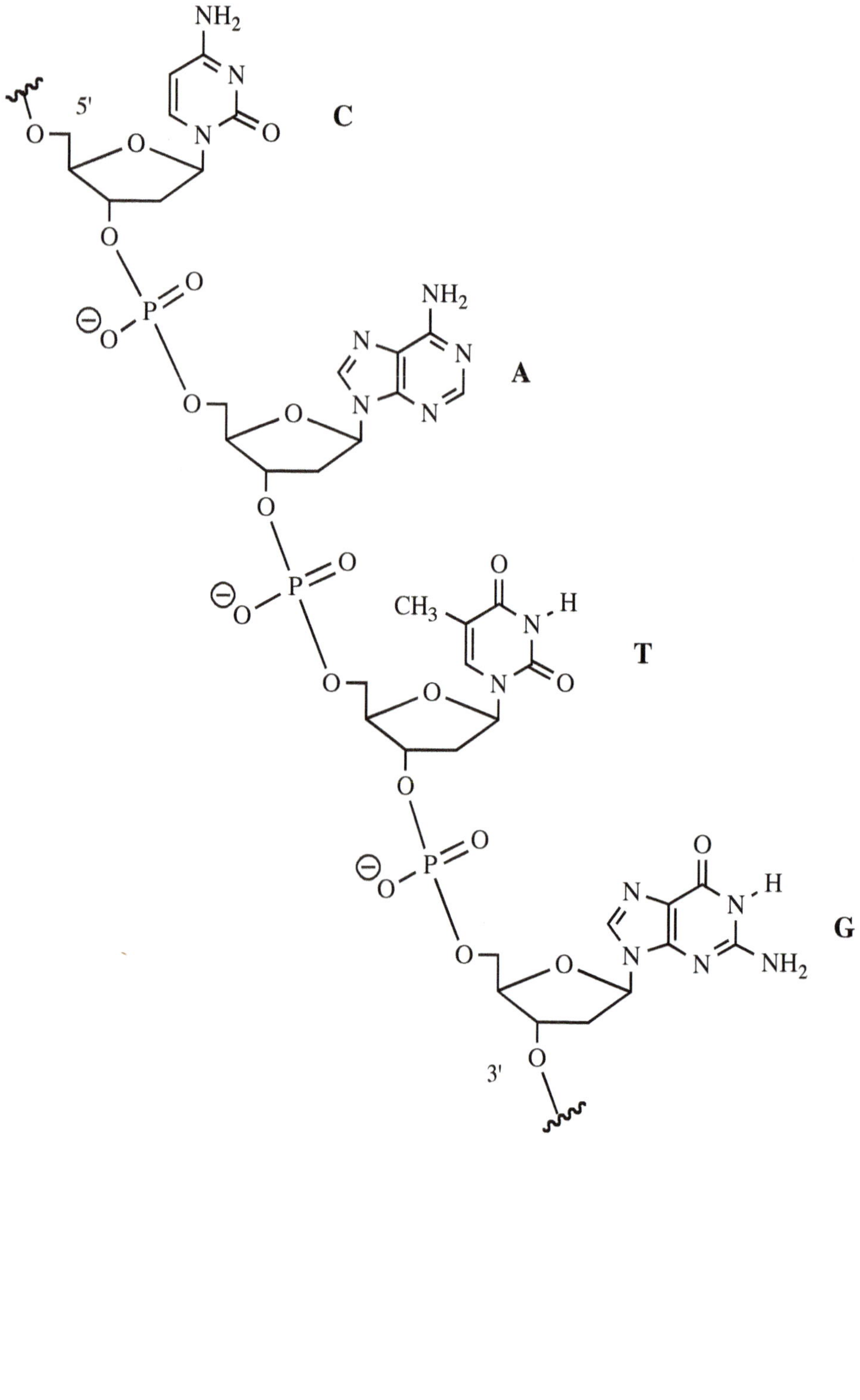

23-65 Bonds from C to H are omitted for simplicity.

(a)

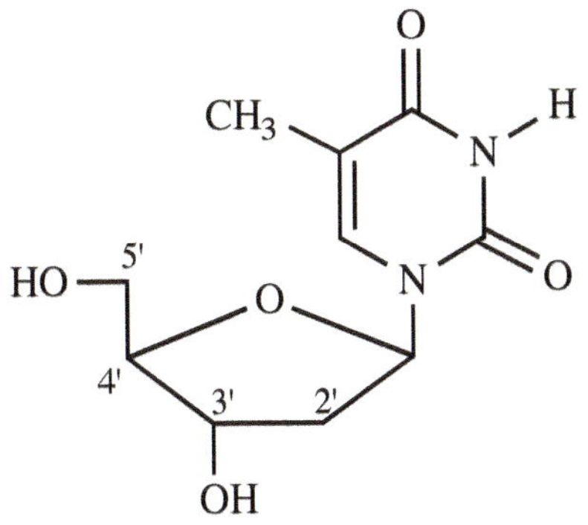

2'-deoxythymidine (abbreviated dT)

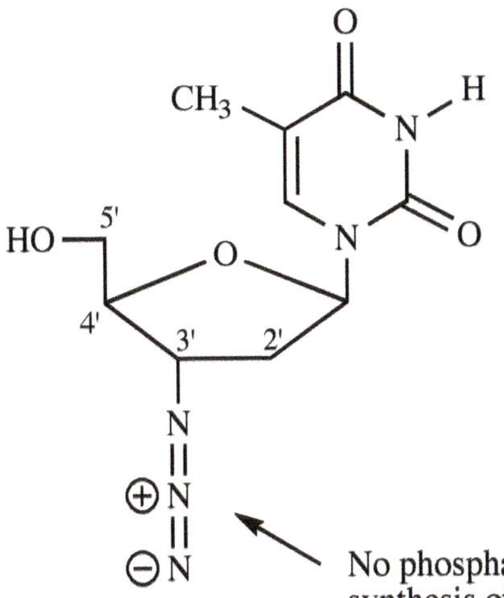

3'-azido-2',3'-dideoxythymidine (AZT)

No phosphate can attach to the azide group, so synthesis of the DNA chain is terminated.

(b)

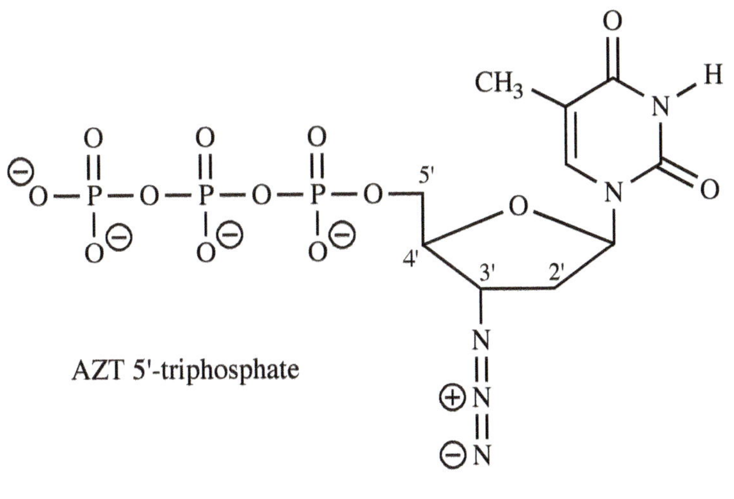

AZT 5'-triphosphate

23-66 Recall from text section 19-16 that nitrous acid is unstable, generating nitrosonium ion.

(a) $H{-}O{-}N{=}O + H^+ \longrightarrow H_2O + \overset{\oplus}{N}{=}O$

cytosine (C)

two rapid proton transfers

$H-A$

H_2O

$-H_2O$

$-N_2$ gas

No mechanism for substitution of diazonium ions was presented in Chapter 19. What is shown here is one possibility but not the only one.

If this species exists, its lifetime is very short as water will attack quickly.

enol tautomer of uracil

uracil (U)

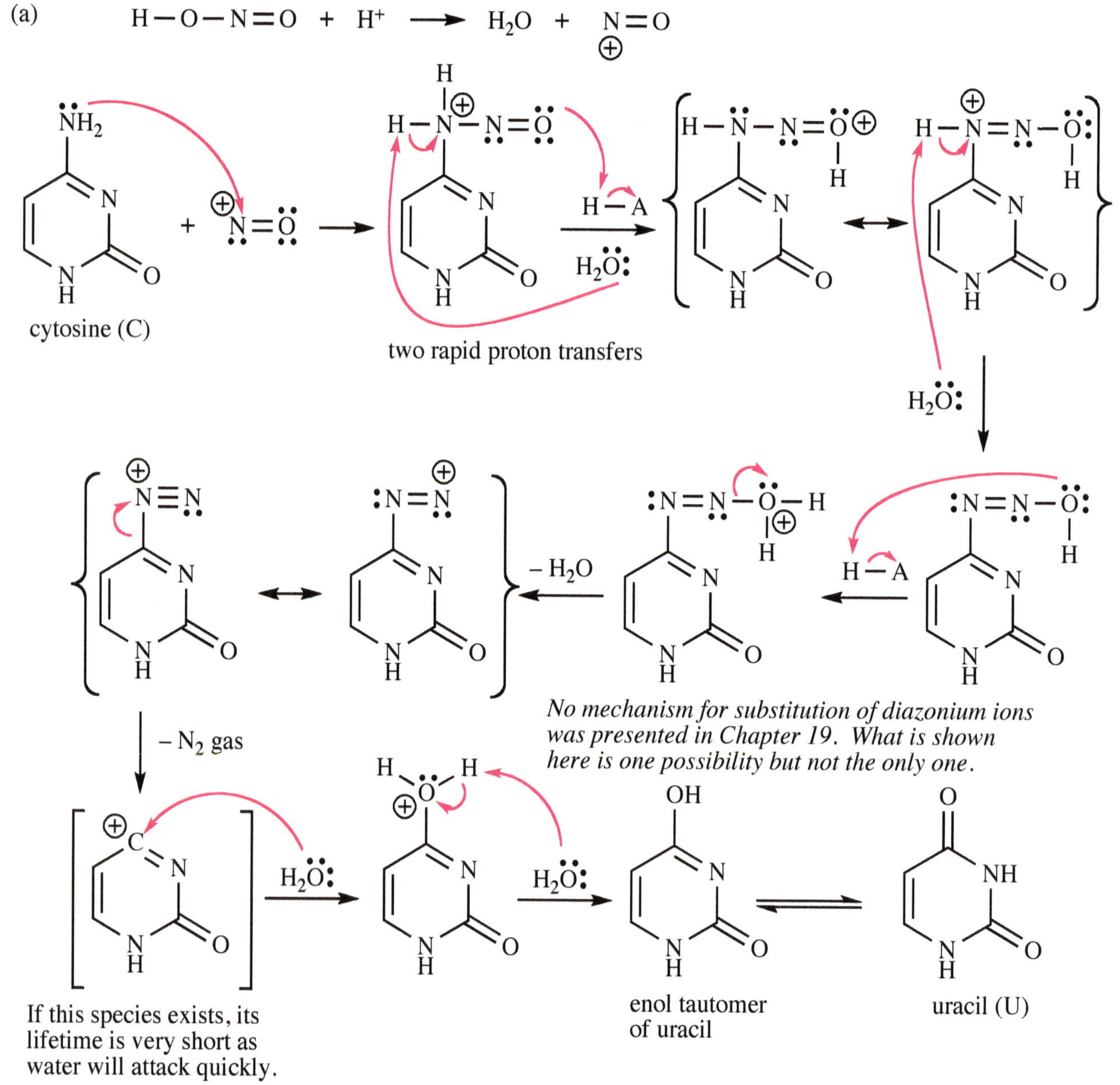

(b) In base pairing, cytosine pairs with guanine. If cytosine is converted to uracil, however, each replication will not carry the complement of cytosine (guanine) but instead will carry the complement of uracil (adenine). This is the definition of a mutation, where the wrong base is inserted in a nucleic acid chain.

(c) In RNA, the transformation of cytosine (C) to uracil (U) is not detected as a problem because U is a base normally found in RNA so it goes unrepaired. In DNA, however, thymine (with an extra methyl group) is used instead of uracil. If cytosine is diazotized to uracil, the DNA repair enzymes detect it as a mutation and correct it.

23-67

(a)

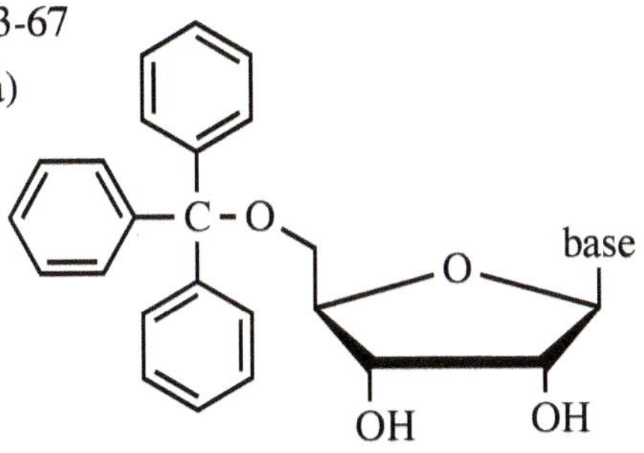

(b) Trityl groups are specific for 1° alcohols for steric reasons: the trityl group is so big that even a 2° alcohol is too crowded to react at the central carbon. It is possible for a trityl to go on a 2° alcohol, but the reaction is exceedingly slow and in the presence of a 1° alcohol, the reaction is done at the 1° alcohol long before the 2° alcohol gets started.

(c) Reactions happen faster when the product or intermediate is stabilized. Each OCH_3 group stabilizes the carbocation by resonance as shown here; two OCH_3 groups stabilize more than just one, increasing the rate of removal. The color comes from the extended conjugation through all three rings and out onto the OCH_3 groups. Compare the DMT structure with that of phenolphthalein, the most common acid base indicator, that turns pink in its ring open form shown here. Phenolphthalein is simply another trityl group with two OH substituents instead of the two OCH_3 groups as in DMT.

One of the major resonance contributors shows the delocalization of the positive charge on the oxygen.

One of the resonance contributors shows the delocalization of the positive charge onto the central carbon.

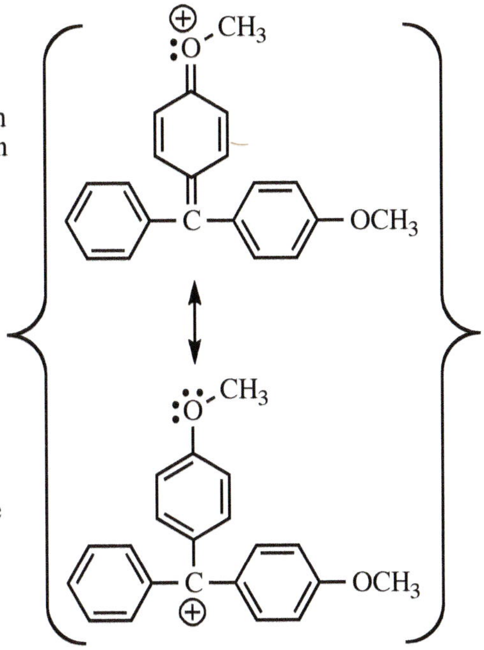

phenolphthalein

plus many more resonance forms

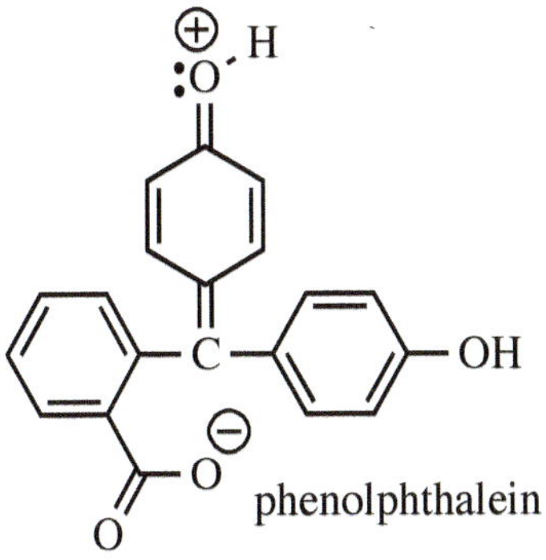

CHAPTER 24—AMINO ACIDS, PEPTIDES, AND PROTEINS

24-1

(a) COOH, H_2N, H

(b) COOH, H_2N, H, NH, N

(c) COOH, H_2N, H, CH_2OH

(d) COOH, H_2N, CH_2, H, NH

24-2

(a) The configurations around the asymmetric carbons of (*R*)-cysteine and (*S*)-alanine are the same. The *designation* of configuration changes because sulfur changes the priorities of the side chain and the COOH.

(*S*)-alanine with group priorities shown: ^{2} COOH; ^{1} H_2N; ^{3} CH_3; $_4$H

^{3} COOH; ^{1} H_2N; ^{2} CH_2SH; $_4$H — (*R*)-Cysteine with group priorities shown; note that the COOH and the CH_2SH priorities are reversed compared with (*S*)-alanine.

(b) Fischer projections show that both (*S*)-alanine and (*R*)-cysteine are L-amino acids.

COOH; H_2N—H; CH_3 — (*S*)-alanine, L-alanine

COOH; H_2N—H; CH_2SH — (*R*)-cysteine, L-cysteine

24-3 In their evolution, plants have needed to be more resourceful than animals in developing biochemical mechanisms for survival. Thus, plants make more of their own required compounds than animals do. The amino acid phenylalanine is produced by plants but required in the diet of mammals. To interfere with a plant's production of phenylalanine is fatal to the plant, but since humans do not produce phenylalanine, glyphosate is virtually nontoxic to us.

24-4 Here is a simple way of determining if a group will be protonated: *at solution pH below the group's pK_a value, the group will be protonated; at pH higher than the group's pK_a value, it will not be protonated.*

(a) $H_2N-CH(CH(CH_3)CH_2CH_3)-COO^{\ominus}$ pH 11 ⇨ Both groups are deprotonated.

(b) $H-N^{\oplus}H$ (pyrrolidine ring)—COOH pH 2 ⇨ Both groups are protonated.

(c) $H_3\overset{\oplus}{N}-CH-COO^{\ominus}$ with side chain $CH_2CH_2CH_2NH-C(=\overset{\oplus}{N}H_2)-NH_2$ pH 7 ⇨ Only the amines remain protonated.

(d) $H_3\overset{\oplus}{N}-CH-COO^{\ominus}$ with side chain $CH_2CH_2COO^{\ominus}$ pH 7 ⇨ Only the amine remains protonated.

24-4 continued

(e)

	alanine	lysine	aspartic acid
(i) pH 6	$H_3\overset{+}{N}-CH(CH_3)-COO^-$	$H_3\overset{+}{N}-CH((CH_2)_4\overset{+}{N}H_3)-COO^-$	$H_3\overset{+}{N}-CH(CH_2COO^-)-COO^-$
(ii) pH 11	$H_2N-CH(CH_3)-COO^-$	$H_2N-CH((CH_2)_4NH_2)-COO^-$	$H_2N-CH(CH_2COO^-)-COO^-$
(iii) pH 2	$H_3\overset{+}{N}-CH(CH_3)-COOH$	$H_3\overset{+}{N}-CH((CH_2)_4\overset{+}{N}H_3)-COOH$	$H_3\overset{+}{N}-CH(CH_2COOH)-COOH$

24-5

$$\left\{ (H_2N)C(=\overset{+}{N}H_2)(NH_2) \longleftrightarrow (H_2N)\overset{+}{C}(NH_2)(NH_2) \longleftrightarrow (H_2N)C(NH_2)(=\overset{+}{N}H_2) \longleftrightarrow H_2\overset{+}{N}=C(NH_2)(NH_2) \right\}$$

Protonation of the guanidino group gives a resonance-stabilized cation with all octets filled and the positive charge delocalized over three nitrogen atoms. Arginine's strongly basic isoelectric point reflects the unusual basicity of the guanidino group due to this resonance stabilization in the protonated form. (See Problems 2-40 and 19-54(a).)

24-6 tryptophan — histidine

tryptophan: $H_2N-CH(COOH)-CH_2$–(indol-3-yl); ring N–H: not basic ⟶ (like pyrrole)

histidine: $H_2N-CH(COOH)-CH_2$–(imidazol-4-yl); ring N: basic ⟶ (like pyridine); ring NH: not basic (like pyrrole)

The basicity of any nitrogen depends on its electron pair's availability for bonding with a proton. In tryptophan, the nitrogen's electron pair is part of the aromatic π system; without this electron pair in the π system, the molecule would not be aromatic. Using this electron pair for bonding to a proton would therefore destroy the aromaticity—not a favorable process.

In the imidazole ring of histidine, the electron pair of one nitrogen is also part of the aromatic π system and is unavailable for bonding; this nitrogen is not basic. The electron pair on the other nitrogen, however, is in an sp^2 orbital available for bonding, and is about as basic as pyridine.

24-7 At pH 9.7, alanine (isoelectric point (pI) 6.0) has a charge of –1 and will migrate to the anode. Lysine (pI 9.7) is at its isoelectric point and will not move. Aspartic acid (pI 2.8) has a charge of –2 and will also migrate to the anode, faster than alanine.

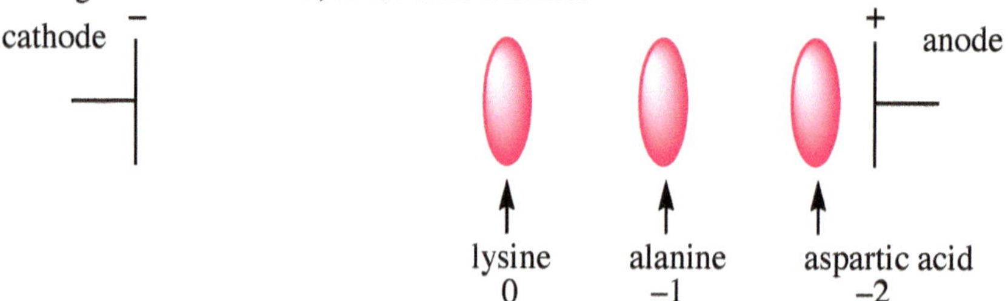

24-8 At pH 6.0, tryptophan (pI 5.9) has a charge of zero and will not migrate. Cysteine (pI 5.0) has a partial negative charge and will move toward the anode. Histidine (pI 7.6) has a partial positive charge and will move toward the cathode.

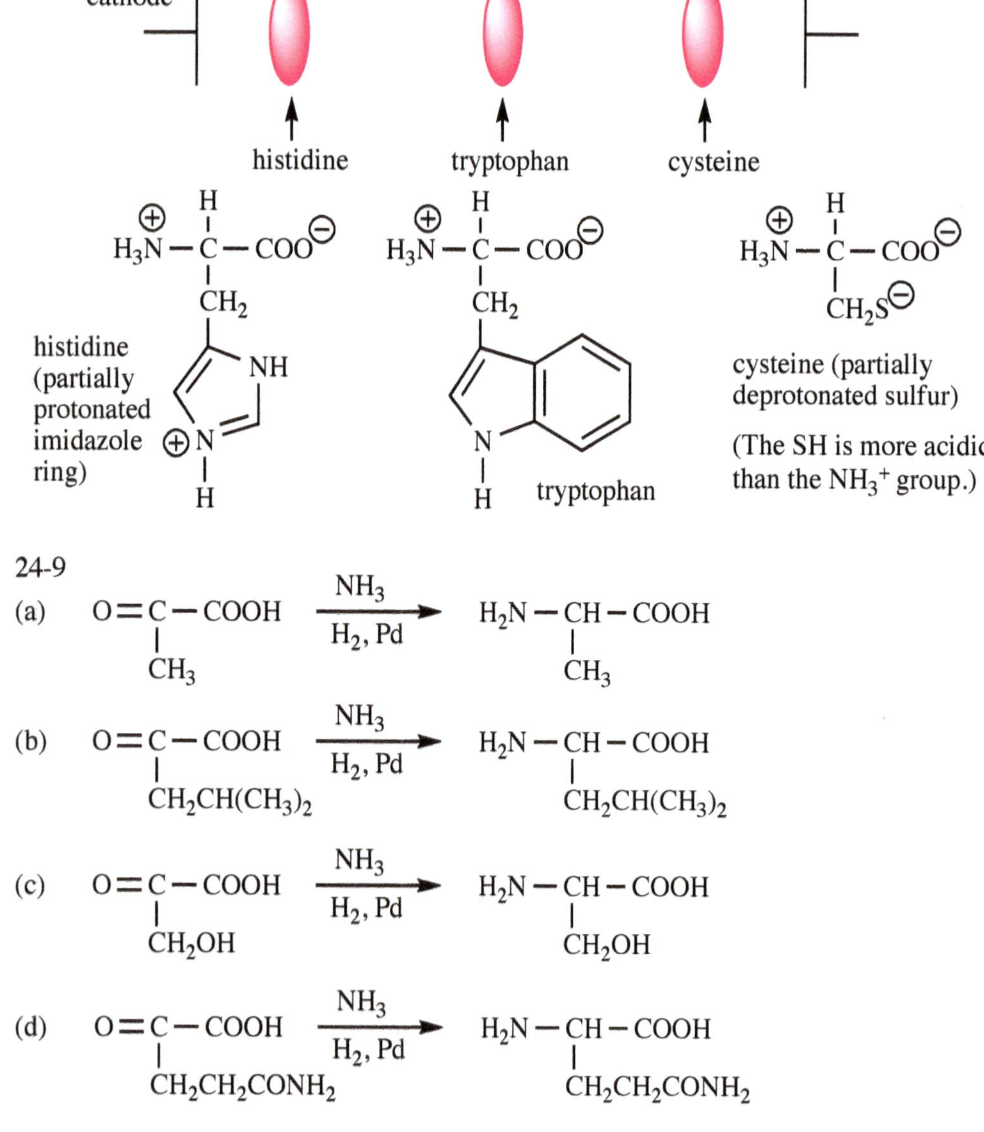

24-10 All of these reactions use: first arrow: (1) Br_2/PBr_3, followed by H_2O workup; second arrow: (2) excess NH_3, followed by neutralizing workup.

(a) $H_2C(H)-COOH \xrightarrow{(1)} Br-CH(H)-COOH \xrightarrow{(2)} H_2N-CH(H)-COOH$

(b) $H_2C(CH_2CH(CH_3)_2)-COOH \xrightarrow{(1)} Br-CH(CH_2CH(CH_3)_2)-COOH \xrightarrow{(2)} H_2N-CH(CH_2CH(CH_3)_2)-COOH$

(c) $H_2C(CH_2CH_2COOH)-COOH \xrightarrow{(1)} Br-CH(CH_2CH_2COOH)-COOH \xrightarrow{(2)} H_2N-CH(CH_2CH_2COOH)-COOH$

In part (c), care must be taken to avoid reaction α to the other COOH. In practice, this would be accomplished by using less than one-half mole of bromine per mole of the diacid.

24-11

(a) $PhCH_2-CH(=O) \xrightarrow[H_2O]{NH_3,\ HCN} PhCH_2-CH(NH_2)(C\equiv N) \xrightarrow[\Delta]{H_3O^+} H_3\overset{\oplus}{N}-CH(CH_2Ph)-COOH$

(b) While the solvent for the Strecker synthesis is water, the proton acceptor is ammonia and the proton donor is ammonium ion.

$PhCH_2-CH(=\ddot{O}:) + :NH_3 \longrightarrow PhCH_2-CH(-\ddot{O}:^{\ominus})(-\overset{\oplus}{N}H_2-H)$ (with $:NH_3$ and $H-\overset{\oplus}{N}H_3$)

two fast proton transfers

$\longrightarrow PhCH_2-CH(-\ddot{O}-H)(NH_2)$

$PhCH_2-CH(-\ddot{O}-H)(NH_2) + H_3\overset{\oplus}{N}-H \longrightarrow PhCH_2-CH(-\overset{\oplus}{O}H_2)(NH_2) \xrightarrow{-H_2O}$

$\left\{ PhCH_2-\overset{\oplus}{C}H(-\ddot{N}H_2) \longleftrightarrow PhCH_2-CH(=\overset{\oplus}{N}H_2) \right\}$

This protonated imine is rapidly attacked by cyanide nucleophile.

$\xrightarrow{:\overset{\ominus}{C}\equiv N} PhCH_2-CH(C\equiv N)(NH_2)$

(abbreviate as $R-C\equiv N$ on the next page)

continued on next page

24-11(b) continued

mechanism of acid hydrolysis of the nitrile

$$R-C\equiv N: \xrightarrow{H^+} \left\{ R-C\equiv \overset{\oplus}{N}-H \longleftrightarrow R-\overset{\oplus}{C}=\ddot{N}-H \right\} \xrightarrow{H_2\ddot{O}:} R-C(=\ddot{N}-H)-\overset{\oplus}{O}H_2$$

$$\xrightarrow{H_2\ddot{O}:} R-C(=\ddot{N}-H)-\ddot{O}-H \xrightarrow{H^+} \left\{ R-C(=\overset{\oplus}{N}H-H)-\ddot{O}-H \longleftrightarrow R-\overset{\oplus}{C}(-\ddot{N}H-H)-\ddot{O}-H \longleftrightarrow R-C(-\ddot{N}H-H)=\overset{\oplus}{O}-H \right\}$$

$$\xrightarrow{H_2\ddot{O}:} R-C(-\overset{\oplus}{O}H-H)(-HO)-\ddot{N}H-H \xrightarrow[\text{proton transfers}]{H_2\ddot{O}:\ \text{two rapid}} R-C(OH)(HO)-\overset{\oplus}{N}H_2-H + H^+$$

$$\xrightarrow{-NH_3} \left\{ R-\overset{\oplus}{C}(-\ddot{O}-H)-\ddot{O}H \longleftrightarrow R-C(-\ddot{O}-H)=\overset{\oplus}{O}H \longleftrightarrow R-C(=\overset{\oplus}{O}-H)-\ddot{O}H \right\} \xrightarrow{H_2\ddot{O}:} H_3\overset{\oplus}{N}-CH(CH_2Ph)-COOH$$

24-12

(a)
$$O=C(H)-CH_2CH(CH_3)_2 \xrightarrow[H_2O]{NH_3,\ HCN} H_2N-CH(CH_2CH(CH_3)_2)-CN \xrightarrow[\Delta]{H_3O^+} H_3\overset{\oplus}{N}-CH(CH_2CH(CH_3)_2)-COOH$$

(b)
$$O=C(H)-CH(CH_3)_2 \xrightarrow[H_2O]{NH_3,\ HCN} H_2N-CH(CH(CH_3)_2)-CN \xrightarrow[\Delta]{H_3O^+} H_3\overset{\oplus}{N}-CH(CH(CH_3)_2)-COOH$$

(c)
$$O=C(H)-CH_2COOH \xrightarrow[H_2O]{NH_3,\ HCN} H_2N-CH(CH_2COO^{\ominus})-CN \xrightarrow[\Delta]{H_3O^+} H_3\overset{\oplus}{N}-CH(CH_2COOH)-COOH$$

before acidification

24-13 In acid solution, the free amino acid will be protonated, with a positive charge, and probably soluble in water as are other organic ions. The acylated amino acid, however, is not basic since the nitrogen is present as an amide. In acid solution, the acylated amino acid is neutral and not soluble in water. Water extraction or ion-exchange chromatography (Figure 24-12) would be practical techniques to separate these compounds.

24-14 abbreviate $H_3\overset{\oplus}{N}-CH(CH_2Ph)-C(=O)OEt$ as $R-C(=O)OEt$

$$R-C(=O)-OEt \xrightarrow{H^+} RC^{\oplus}(OH)-OEt \xrightarrow{H_2O} RC(OH)(OEt)-\overset{\oplus}{O}H_2 \xrightarrow{H_2O} RC(OH)_2-OEt \xrightarrow{H^+} RC(OH)_2-\overset{\oplus}{O}(H)-Et \xrightarrow{-\,EtOH} RC(=\overset{\oplus}{O}H)-OH \xrightarrow{H_2O} H_3\overset{\oplus}{N}-CH(CH_2Ph)-C(=O)OH$$

plus two other resonance forms

plus two other resonance forms

(protonated form in acid solution)

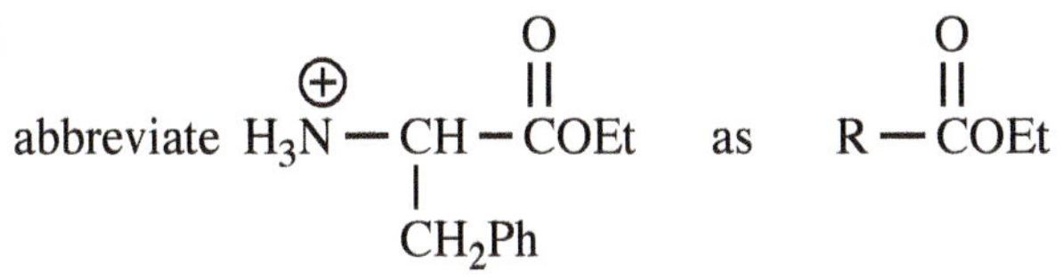

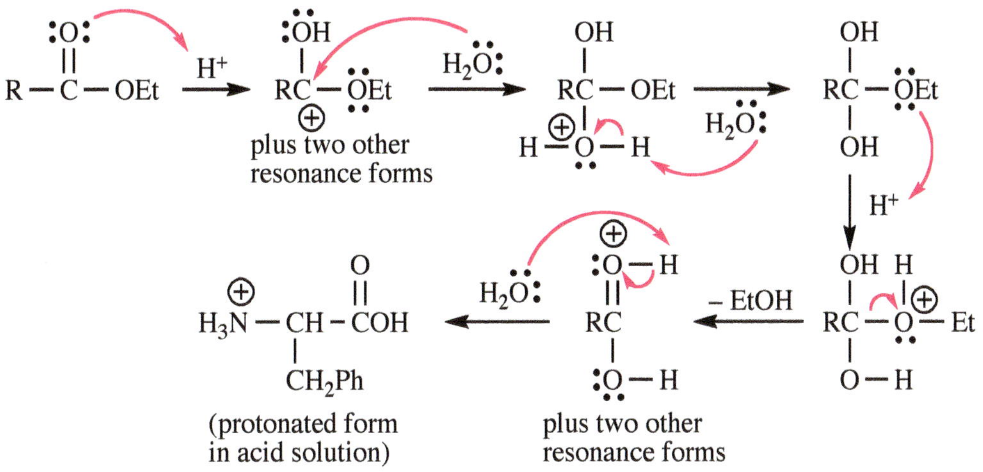

24-15

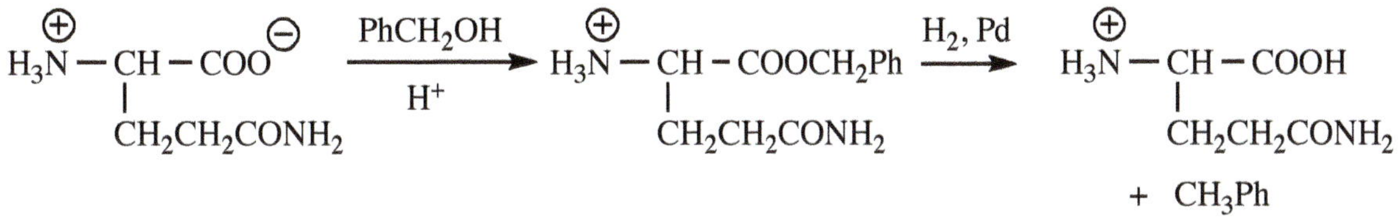

$$H_3\overset{\oplus}{N}-CH(CH_2CH_2CONH_2)-COO^{\ominus} \xrightarrow[H^+]{PhCH_2OH} H_3\overset{\oplus}{N}-CH(CH_2CH_2CONH_2)-COOCH_2Ph \xrightarrow{H_2,\ Pd} H_3\overset{\oplus}{N}-CH(CH_2CH_2CONH_2)-COOH + CH_3Ph$$

24-16

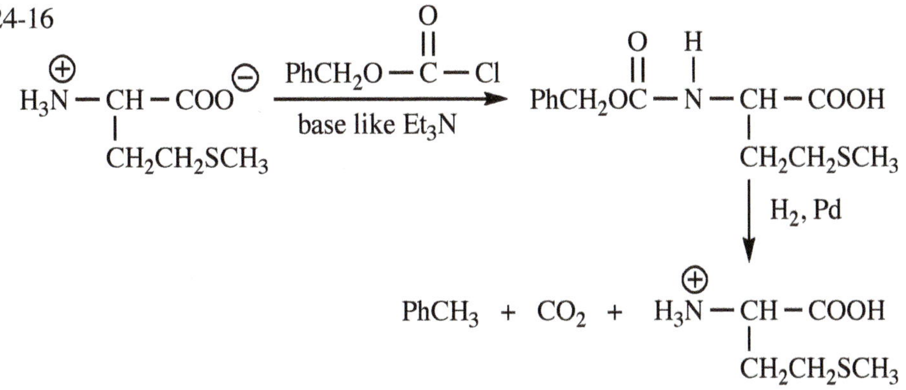

$$H_3\overset{\oplus}{N}-CH(CH_2CH_2SCH_3)-COO^{\ominus} \xrightarrow[\text{base like } Et_3N]{PhCH_2O-C(=O)-Cl} PhCH_2OC(=O)-NH-CH(CH_2CH_2SCH_3)-COOH$$

$$\xrightarrow{H_2,\ Pd} PhCH_3 + CO_2 + H_3\overset{\oplus}{N}-CH(CH_2CH_2SCH_3)-COOH$$

24-17

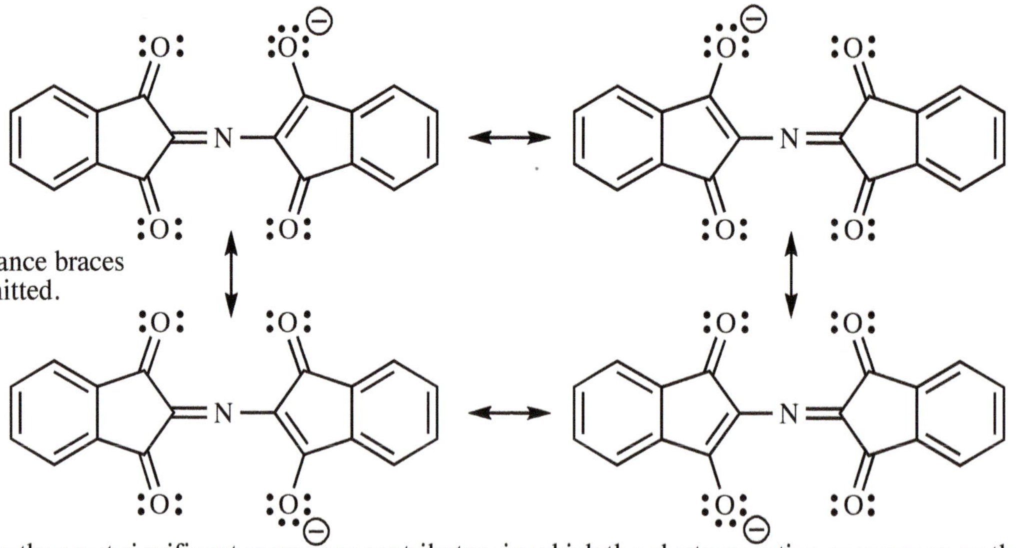

Resonance braces are omitted.

These are the most significant resonance contributors in which the electronegative oxygens carry the negative charge. There are also two other forms in which the negative charge is on the carbons bonded to the nitrogen, plus the usual resonance forms involving the alternate Kekulé structures of the benzene rings.

24-18

(a) $\overset{\oplus}{H_3N}$—CH(CH(OH)CH_3)—C(=O)—NH—CH(CH_2Ph)—C(=O)—NH—CH($CH_2CH_2SCH_3$)—C(=O)—$O^{\ominus}$

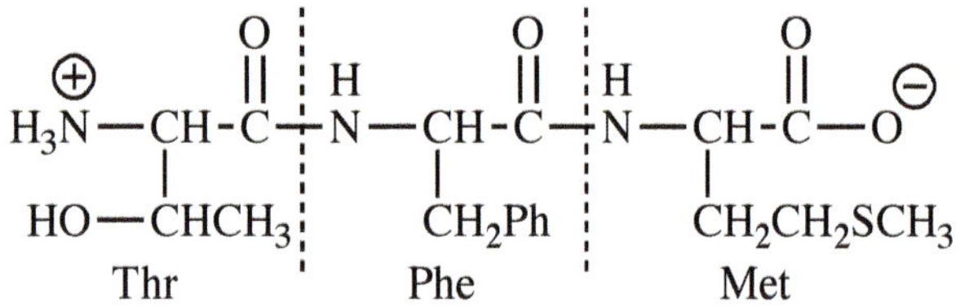

Thr Phe Met

(b) $\overset{\oplus}{H_3N}$—CH(CH_2OH)—C(=O)—NH—CH($(CH_2)_3$—HN—C(=NH)—NH_2)—C(=O)—NH—CH(H)—C(=O)—NH—CH(CH_2Ph)—C(=O)—$O^{\ominus}$

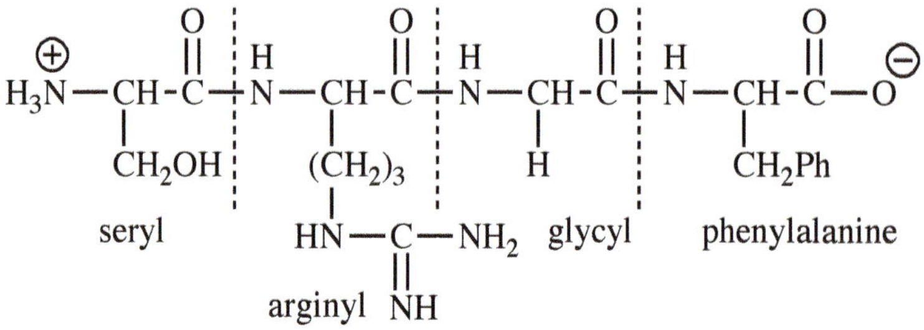

seryl arginyl glycyl phenylalanine

(c) $\overset{\oplus}{H_3N}$—CH($CH_3CH_2CH_2CH_3$)—C(=O)—NH—CH($(CH_2)_2SCH_3$)—C(=O)—NH—CH($(CH_2)_2CONH_2$)—C(=O)—NH—CH(CH_2COOH)—C(=O)—NH—CH($(CH_2)_4NH_2$)—C(=O)—$O^{\ominus}$

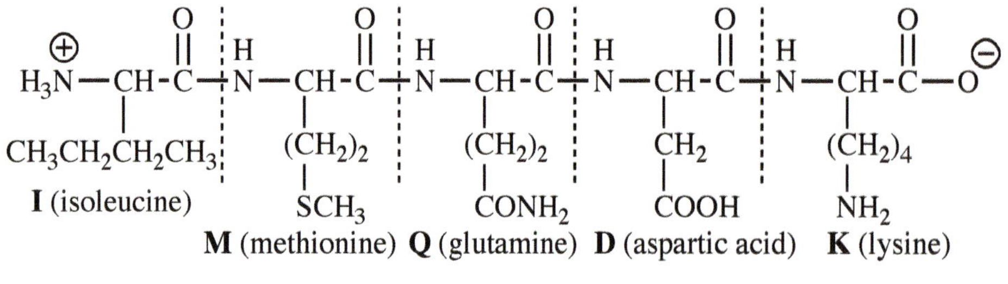

I (isoleucine) **M** (methionine) **Q** (glutamine) **D** (aspartic acid) **K** (lysine)

(d) $\overset{\oplus}{H_3N}$—CH($(CH_2)_2$COOH)—C(=O)—NH—CH(CH_2CHMe_2)—C(=O)—NH—CH($CHMe_2$)—C(=O)—NH—CH(CHCH_3—CH_2CH_3)—C(=O)—NH—CH(CH_2OH)—C(=O)—$O^{\ominus}$

E (glutamic acid) **L** (leucine) **V** (valine) **I** (isoleucine) **S** (serine)

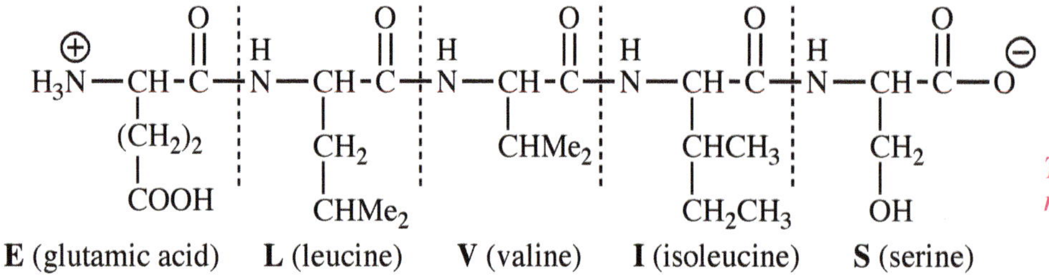

Try spelling your name in peptides!

24-19

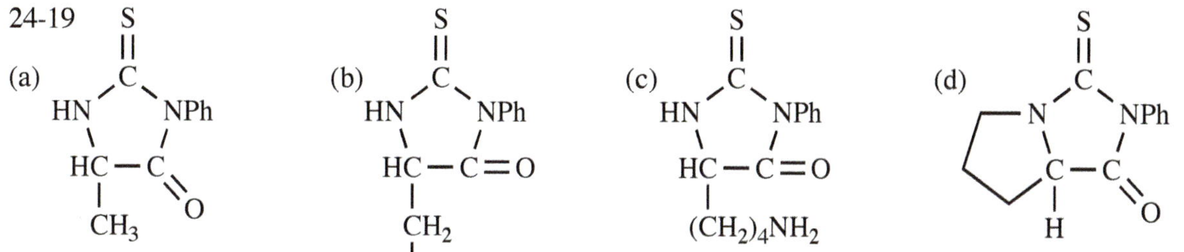

(a) [PTH structure: thiohydantoin ring with CH_3 side chain]

(b) [PTH structure: thiohydantoin ring with CH_2-indole side chain]

(c) [PTH structure: thiohydantoin ring with $(CH_2)_4NH_2$ side chain]

(d) [PTH structure: proline-fused thiohydantoin]

24-20

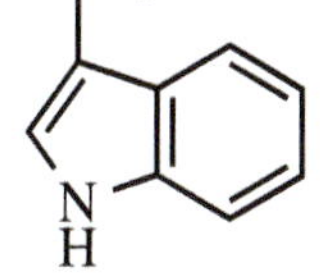

Step 3

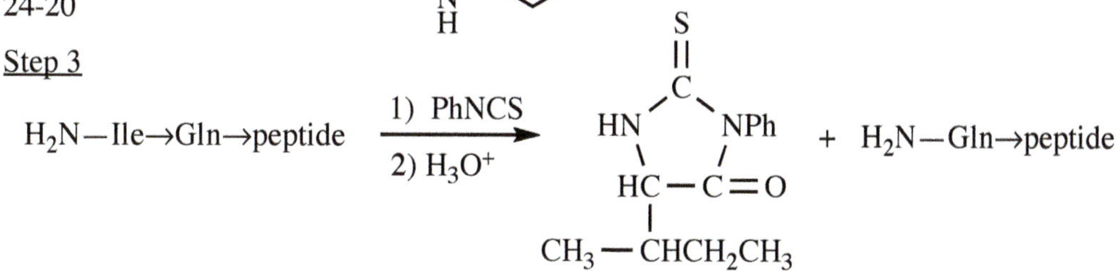

$H_2N-Ile \rightarrow Gln \rightarrow peptide \xrightarrow[2)\ H_3O^+]{1)\ PhNCS}$ PTH with side chain $CH_3-CHCH_2CH_3$ + $H_2N-Gln \rightarrow peptide$

Step 4

$H_2N-Gln \rightarrow peptide \xrightarrow[2)\ H_3O^+]{1)\ PhNCS}$ PTH with side chain $CH_2CH_2CONH_2$ + $H_2N-peptide$

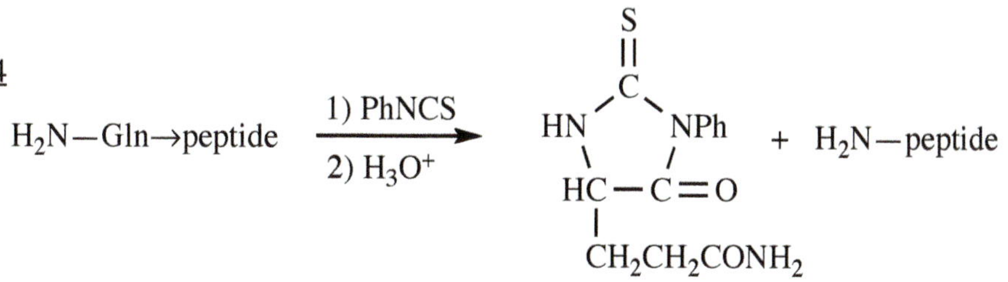

24-21 Abbreviate the N-terminus of the peptide chain as NH_2R .

(a) This is a nucleophilic aromatic substitution by the addition-elimination mechanism. The presence of two nitro groups makes this reaction feasible under mild conditions.

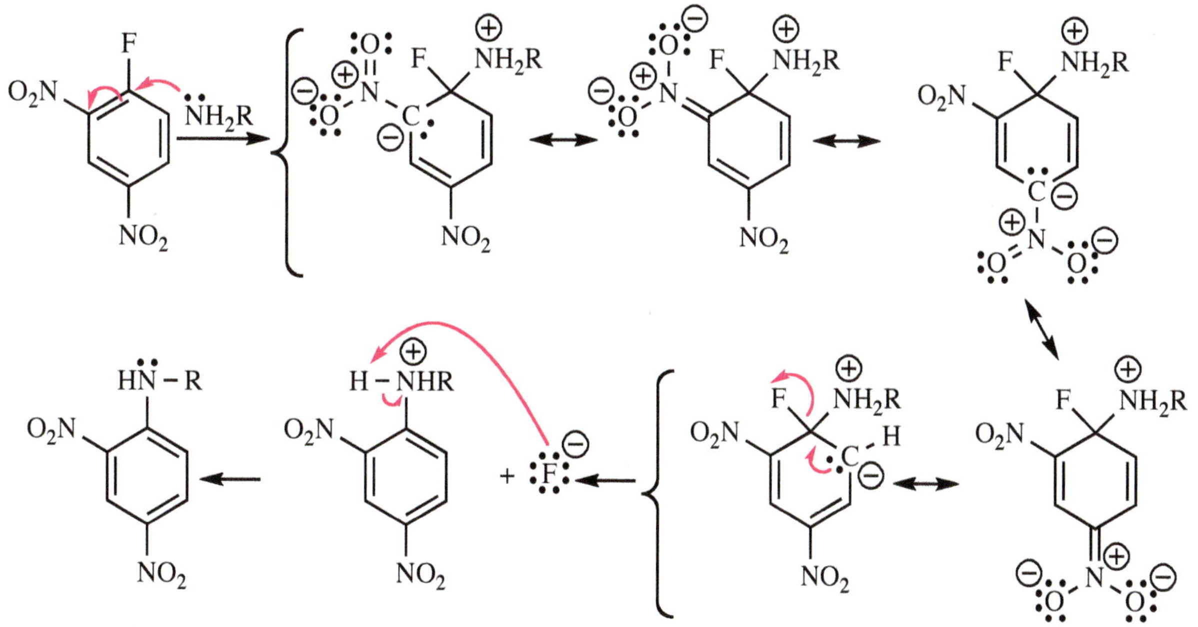

[Mechanism: 1-fluoro-2,4-dinitrobenzene + NH_2R → resonance-stabilized anionic intermediate → loss of F^- → $H-NHR^+$ → 2,4-dinitrophenyl-NHR]

(b) The main drawback of the Sanger method is that only one amino acid is analyzed per sample of protein. The Edman degradation can usually analyze more than 20 amino acids per sample of protein.

24-22

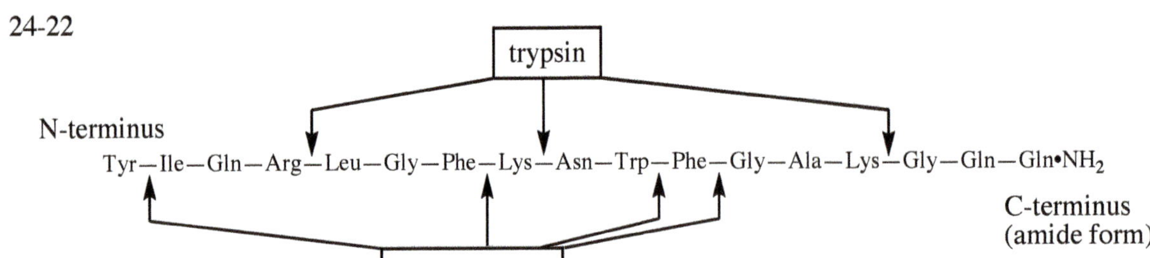

24-23

Phe—Gln—Asn

Pro—Arg—Gly•NH_2

Cys—Tyr—Phe

Asn—Cys—Pro—Arg

Tyr—Phe—Gln—Asn

Cys—Tyr—Phe—Gln—Asn—Cys—Pro—Arg—Gly•NH_2

24-24 We use "*c*-Hx" to stand for "cyclohexyl".

N=C=N ⟹ *c*-Hx—N=C=N—*c*-Hx

DiCyclohexylCarbodiimide

DCC

<u>mechanism</u>

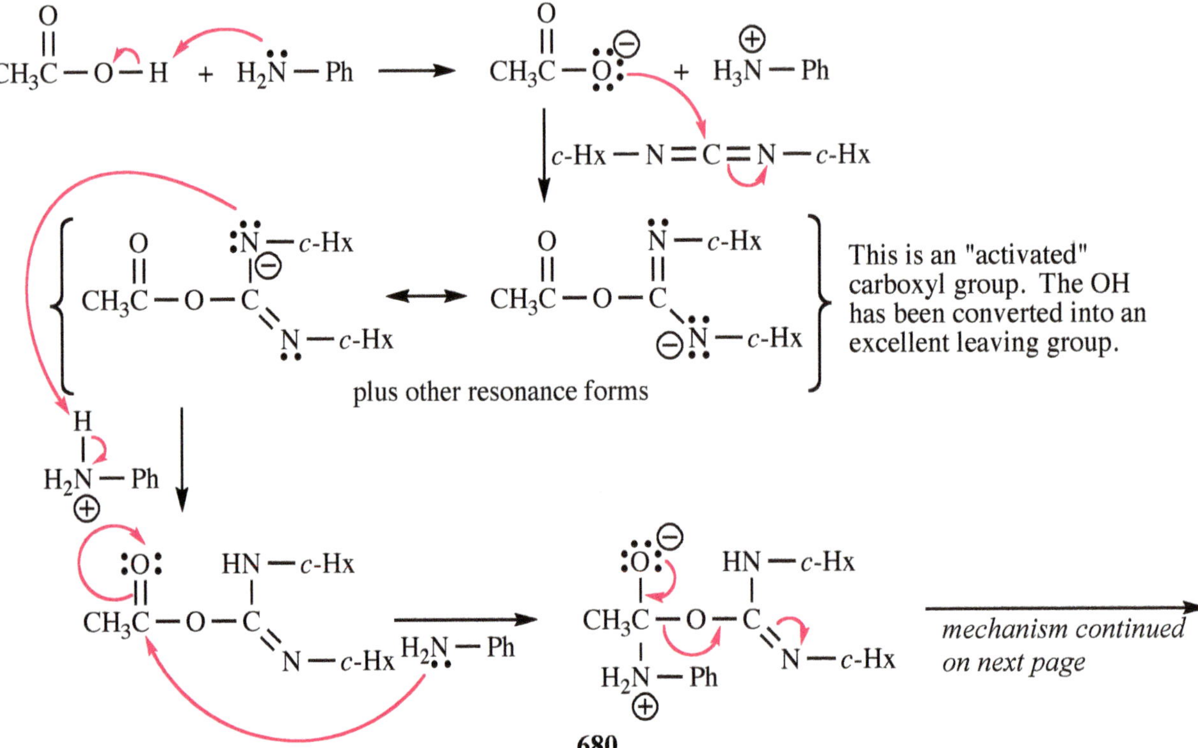

24-24 continued

$$\longrightarrow CH_3\overset{O}{\overset{\|}{C}}-\overset{\oplus}{N}HPh(H) + c\text{-}Hx-\overset{\ominus}{\ddot{N}}-\overset{O}{\overset{\|}{C}}-NH-c\text{-}Hx \longrightarrow CH_3\overset{O}{\overset{\|}{C}}-NHPh + c\text{-}Hx-NH-\overset{O}{\overset{\|}{C}}-NH-c\text{-}Hx$$

resonance-stabilized

DiCyclohexyl**U**rea, DCU

24-25

Fmoc: fluorenylmethyl–CH_2–O–C(=O)–

Fmoc | Ala | Val | Phe

$$\text{Fmoc}-NH-CH(CH_3)-C(=O)-NH-CH(CH(CH_3)_2)-C(=O)-NH-CH(CH_2Ph)-C(=O)-O-\text{(P)}$$

polymer

↓ piperidine/DMF

piperidine

$$\overset{\oplus}{H_3N}-CH(CH_3)-C(=O)-NH-CH(CH(CH_3)_2)-C(=O)-NH-CH(CH_2Ph)-C(=O)-O-\text{(P)}$$

↓ DCC, Fmoc — $NHCH_2COOH$ (Fmoc—glycine)

$$\text{Fmoc}-NH-CH(H)-C(=O)-NH-CH(CH_3)-C(=O)-NH-CH(CH(CH_3)_2)-C(=O)-NH-CH(CH_2Ph)-C(=O)-O-\text{(P)}$$

↓ piperidine/DMF

$$\overset{\oplus}{H_3N}-CH(H)-C(=O)-NH-CH(CH_3)-C(=O)-NH-CH(CH(CH_3)_2)-C(=O)-NH-CH(CH_2Ph)-C(=O)-O-\text{(P)}$$

↓ DCC, Fmoc — NH — CH(CH_2CHMe_2)·COOH (Fmoc—leucine)

$$\text{Fmoc}-NH-CH(CH_2CHMe_2)-C(=O)-NH-CH(H)-C(=O)-NH-CH(CH_3)-C(=O)-NH-CH(CH(CH_3)_2)-C(=O)-NH-CH(CH_2Ph)-C(=O)-O-\text{(P)}$$

↓ HF

$$\overset{\oplus}{H_3N}-CH(CH_2CHMe_2)-C(=O)-NH-CH(H)-C(=O)-NH-CH(CH_3)-C(=O)-NH-CH(CH(CH_3)_2)-C(=O)-NH-CH(CH_2Ph)-C(=O)-OH$$

24-26

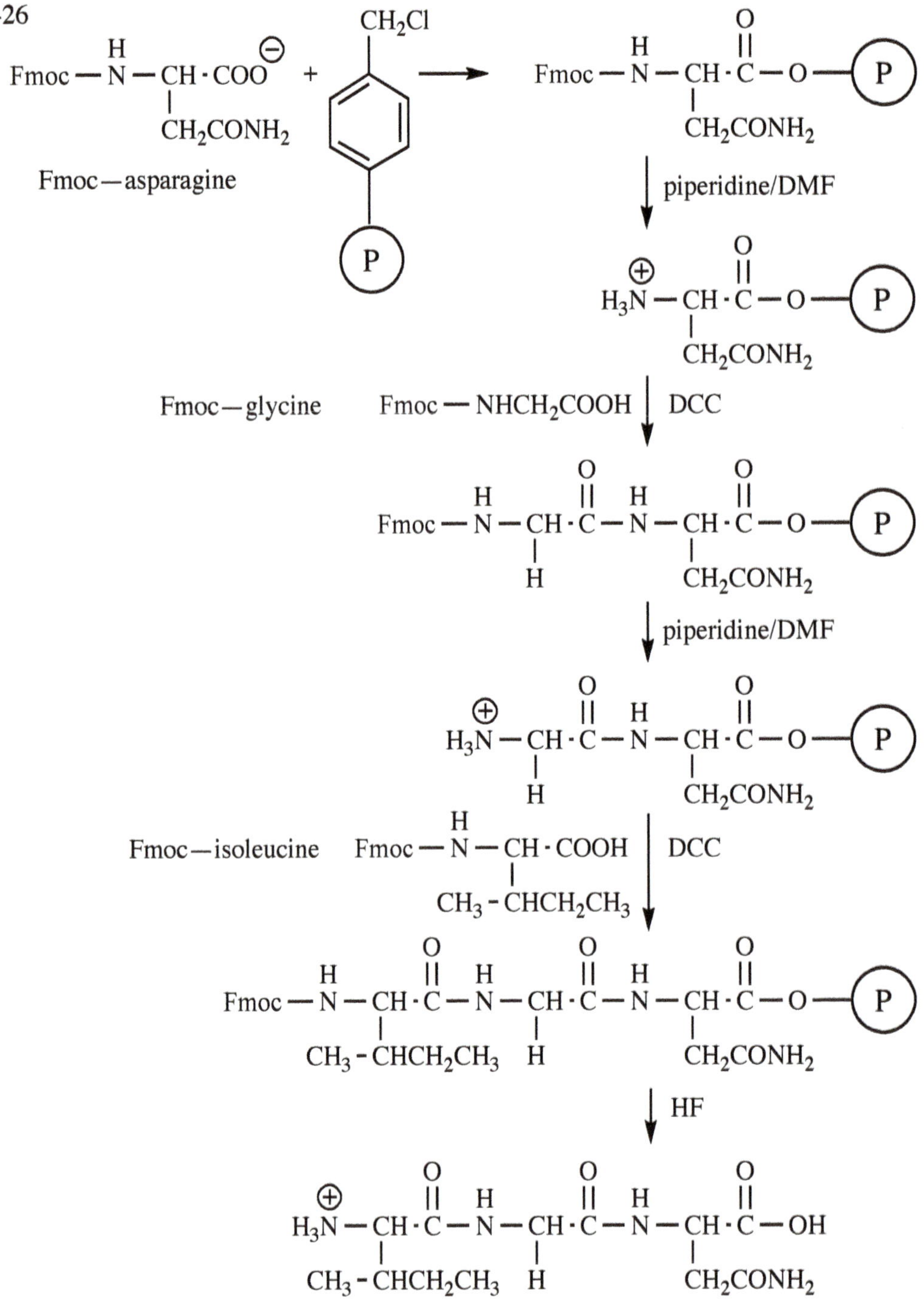

$$\text{Fmoc—NH—CH(CH}_2\text{CONH}_2\text{)·COO}^{\ominus} + \text{ClCH}_2\text{—C}_6\text{H}_4\text{—(P)} \longrightarrow \text{Fmoc—NH—CH(CH}_2\text{CONH}_2\text{)·C(=O)—O—(P)}$$

Fmoc—asparagine

piperidine/DMF

$$\overset{\oplus}{\text{H}_3\text{N}}\text{—CH(CH}_2\text{CONH}_2\text{)·C(=O)—O—(P)}$$

Fmoc—glycine Fmoc — $NHCH_2COOH$ DCC

$$\text{Fmoc—NH—CH(H)·C(=O)—NH—CH(CH}_2\text{CONH}_2\text{)·C(=O)—O—(P)}$$

piperidine/DMF

$$\overset{\oplus}{\text{H}_3\text{N}}\text{—CH(H)·C(=O)—NH—CH(CH}_2\text{CONH}_2\text{)·C(=O)—O—(P)}$$

Fmoc—isoleucine Fmoc—NH—CH(CH_3-$CHCH_2CH_3$)·COOH DCC

$$\text{Fmoc—NH—CH(CH}_3\text{-CHCH}_2\text{CH}_3\text{)·C(=O)—NH—CH(H)·C(=O)—NH—CH(CH}_2\text{CONH}_2\text{)·C(=O)—O—(P)}$$

HF

$$\overset{\oplus}{\text{H}_3\text{N}}\text{—CH(CH}_3\text{-CHCH}_2\text{CH}_3\text{)·C(=O)—NH—CH(H)·C(=O)—NH—CH(CH}_2\text{CONH}_2\text{)·C(=O)—OH}$$

24-27

(a) phenylalanine:

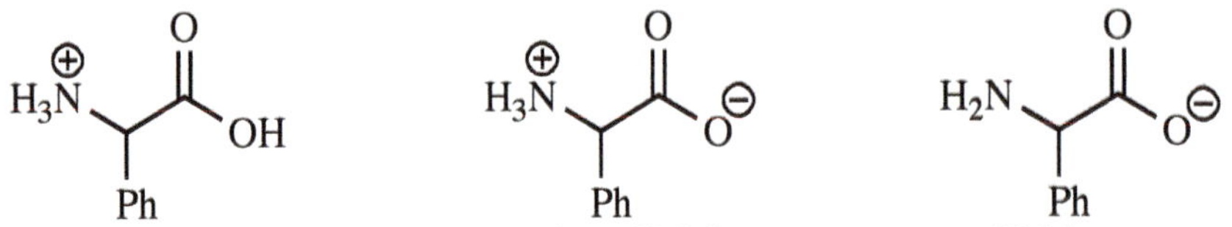

$\overset{\oplus}{H_3N}$–CH(Ph)–C(=O)OH
at pH 1
fully protonated

$\overset{\oplus}{H_3N}$–CH(Ph)–C(=O)$O^{\ominus}$
at pI = pH 5.5
charges balanced

H_2N–CH(Ph)–C(=O)$O^{\ominus}$
at pH 11
fully deprotonated

24-27 continued

(b) histidine:

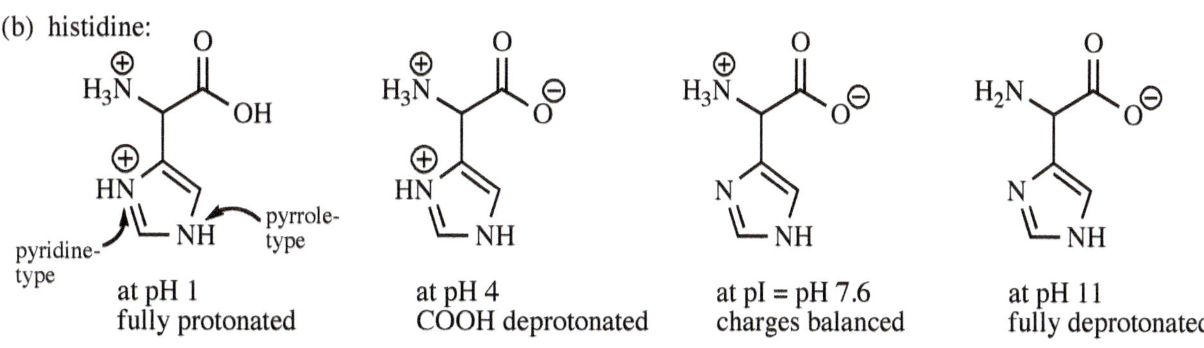

The imidazole ring is aromatic, with a pyrrole-type N that is not basic and a pyridine-type N that is weakly basic. Aromatic amines like pyridine are weaker bases than aliphatic amines by several pK units.

(c) glutamic acid:

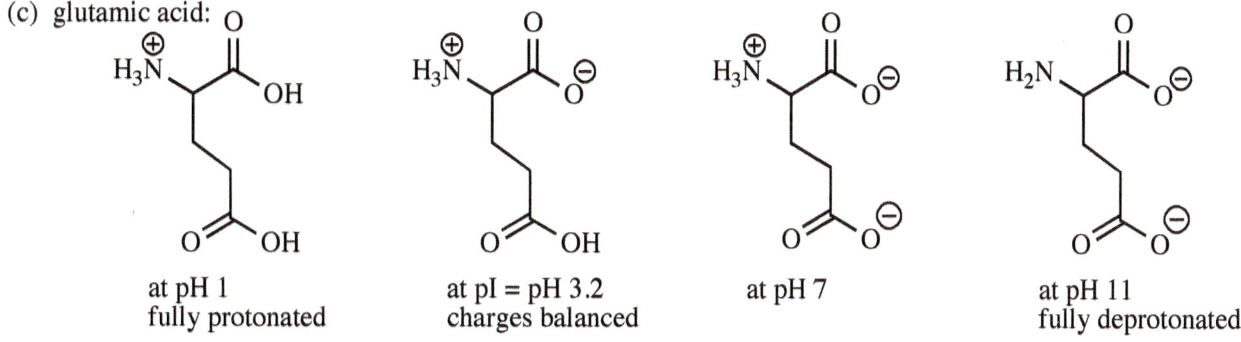

The NH_3^+ group is a strong electron-withdrawing group and strengthens the nearby COOH by 2–3 pK_a units as compared to the side chain COOH, which is not affected by induction in this molecule.

24-28

serine-glutamine-methionine

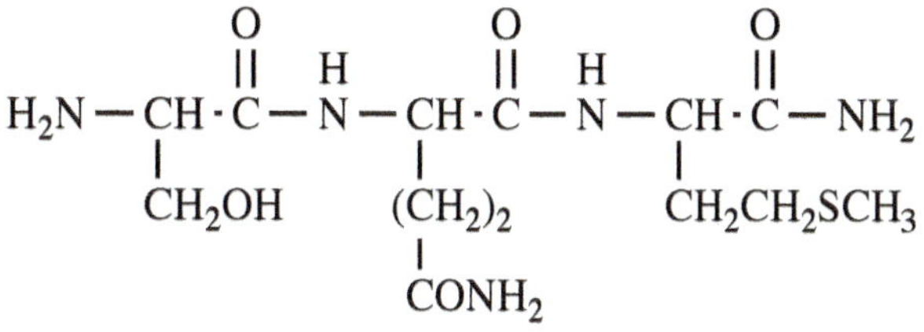

$H_2N—CH(CH_2OH)—C(=O)—NH—CH((CH_2)_2CONH_2)—C(=O)—NH—CH(CH_2CH_2SCH_3)—C(=O)—NH_2$

24-29

(a) + CO_2 + CHO

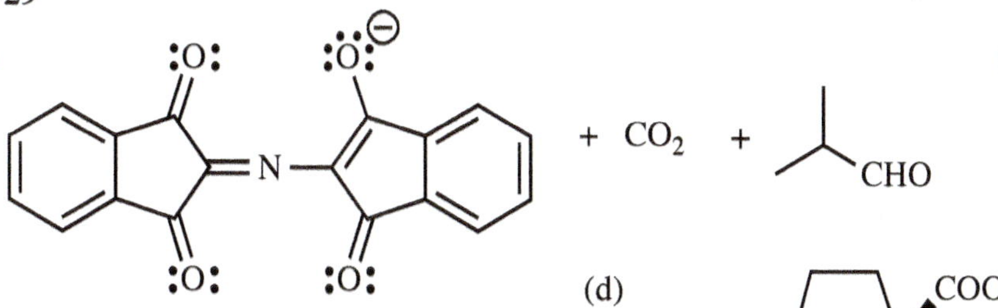

(b) $H_2N—CH(CH_3)COOH$

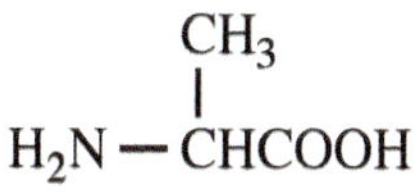

(c) $CH_3C(=O)—NH·CH·COOH$ with side chain $(CH_2)_4NHCOCH_3$

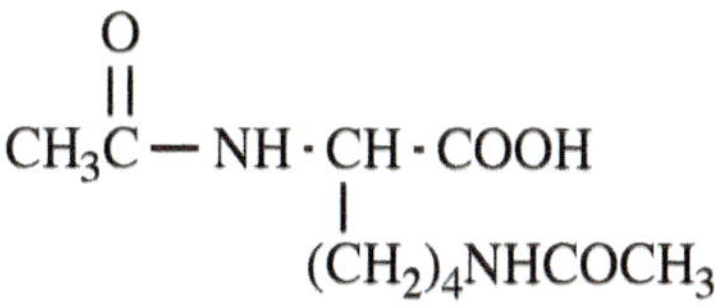

(d) The enzyme does not recognize D amino acids.

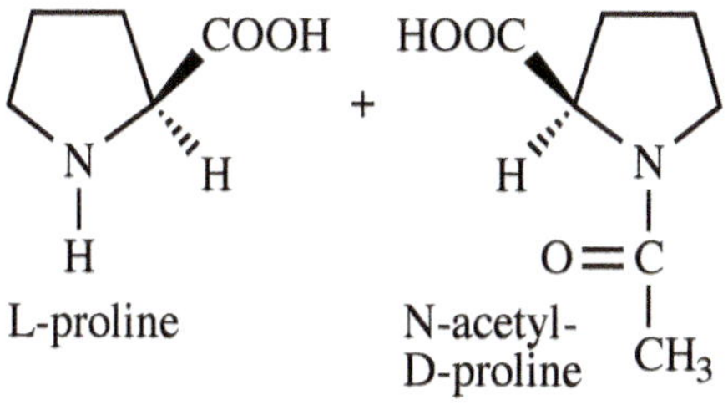

L-proline + N-acetyl-D-proline

24-29 continued

(e) $H_2N-C(H)(CN)-CH(CH_3)CH_2CH_3$

$$H_2N-\underset{\substack{|\\ H_3C-CHCH_2CH_3}}{\overset{\substack{H\\|}}{C}}-CN$$

(f) $$\overset{\oplus}{H_3N}-\underset{\substack{|\\ H_3C-CHCH_2CH_3}}{\overset{\substack{H\\|}}{C}}-COOH$$

isoleucine

(g) $$Br-\underset{\substack{|\\ CH_2CH(CH_3)_2}}{\overset{\substack{H\\|}}{C}}-\overset{\substack{O\\||}}{C}-Br$$

With a water workup, the acid bromide would become COOH.

(h) $$H_2N-\underset{\substack{|\\ CH_2CH(CH_3)_2}}{\overset{\substack{H\\|}}{C}}-COO^{\ominus}\ \overset{\oplus}{N}H_4 \quad \text{OR} \quad H_2N-\underset{\substack{|\\ CH_2CH(CH_3)_2}}{\overset{\substack{H\\|}}{C}}-\overset{\substack{O\\||}}{C}-NH_2$$

leucine

This is another possible answer, depending on whether the acid bromide is hydrolyzed before adding ammonia.

24-30

(a) $$\underset{\substack{|\\ CH(CH_3)_2}}{\overset{\substack{O\\||}}{C}}-COOH \xrightarrow[H_2,\ Pd]{NH_3} \overset{\oplus}{H_3N}-\underset{\substack{|\\ CH(CH_3)_2}}{CH}-COOH$$

valine after acid workup

(b) $$\underset{\substack{|\\ H_3C-CHCH_2CH_3}}{H_2C}-COOH \xrightarrow[PBr_3]{Br_2} \xrightarrow{H_2O} Br-\underset{\substack{|\\ H_3C-CHCH_2CH_3}}{CH}-COOH \xrightarrow{\text{excess } NH_3} \overset{\oplus}{H_3N}-\underset{\substack{|\\ H_3C-CHCH_2CH_3}}{CH}-COOH$$

isoleucine after acid workup

(c) $$\underset{\substack{|\\ CH_2CH(CH_3)_2}}{\overset{\substack{O\\||}}{C}}-H \xrightarrow[H_2O]{NH_3,\ HCN} H_2N-\underset{\substack{|\\ CH_2CH(CH_3)_2}}{CH}-CN \xrightarrow[\Delta]{H_3O^+} \overset{\oplus}{H_3N}-\underset{\substack{|\\ CH_2CH(CH_3)_2}}{CH}-COOH$$

leucine

24-31

(a) $$H_2N-\underset{\substack{|\\ CH_3}}{CH}\cdot COOH + HOCH(CH_3)_2 \xrightarrow{H^+} \overset{\oplus}{H_3N}-\underset{\substack{|\\ CH_3}}{CH}-COOCH(CH_3)_2$$

(b) $$H_2N-\underset{\substack{|\\ CH_3}}{CH}\cdot COOH + Ph\overset{\substack{O\\||}}{C}-Cl \xrightarrow{\text{(pyridine)}} Ph\overset{\substack{O\\||}}{C}-\overset{H}{N}-\underset{\substack{|\\ CH_3}}{CH}-COOH$$

(c) $$H_2N-\underset{\substack{|\\ CH_3}}{CH}\cdot COOH + PhCH_2O\overset{\substack{O\\||}}{C}-Cl \xrightarrow{Et_3N} PhCH_2O\overset{\substack{O\\||}}{C}-\overset{H}{N}-\underset{\substack{|\\ CH_3}}{CH}-COOH$$

(d) $$H_2N-\underset{\substack{|\\ CH_3}}{CH}\cdot COOH + Me_3CO\overset{\substack{O\\||}}{C}-O-\overset{\substack{O\\||}}{C}OCMe_3 \longrightarrow Me_3CO\overset{\substack{O\\||}}{C}-\overset{H}{N}-\underset{\substack{|\\ CH_3}}{CH}-COOH$$

24-32

$$\text{HO}\blacktriangleright\underset{\text{CH}_3}{\overset{\text{COOH}}{\text{C}}}\blacktriangleleft\text{H} \xrightarrow[\text{pyridine}]{\text{TsCl}} \text{TsO}\blacktriangleright\underset{\text{CH}_3}{\overset{\text{COOH}}{\text{C}}}\blacktriangleleft\text{H} \xrightarrow{\text{excess NH}_3} \text{H}\blacktriangleright\underset{\text{CH}_3}{\overset{\text{COO}^{\ominus}}{\text{C}}}\blacktriangleleft\text{NH}_2$$

D-alanine

24-33

$$\text{(indol-3-yl)CH}_2\text{—CHO} \xrightarrow[\text{H}_2\text{O}]{\text{NH}_3,\ \text{HCN}} \text{H}_2\text{N—CH(CH}_2\text{-indol-3-yl)—CN} \xrightarrow[\Delta]{\text{H}_3\text{O}^+} \text{H}_3\overset{\oplus}{\text{N}}\text{—CH(CH}_2\text{-indol-3-yl)—COOH}$$

racemic tryptophan

24-34

(a) $H_2N-CH(CH_2CH_2SCH_3)-C(=O)-NH-CH(CHCH_3OH)-C(=O)-OH$ neutral

(b) $H_2N-CH(CHCH_3OH)-C(=O)-NH-CH(CH_2CH_2SCH_3)-C(=O)-OH$ neutral

(c) $H_2N-CH((CH_2)_3-HN-C(=NH)NH_2)-C(=O)-NH-CH(CH_2COOH)-C(=O)-NH-CH((CH_2)_4NH_2)-C(=O)-OH$

basic: two basic side chains and one acidic side chain

(d) $H_2N-CH(CH_2CH_2COOH)-C(=O)-NH-CH(CH_2SH)-C(=O)-NH-CH(CH_2CH_2CONH_2)-C(=O)-OH$

acidic: carboxylic acid side chain, and weakly acidic SH

24-35

(a), (b), (c)

$$\underbrace{CH_3CH_2-CH(CH_3)-CH(CONH_2)-NH}_{\text{isoleucine}}\overset{*}{-}\underbrace{C(=O)-CH(NH\overset{*}{-}CO-CH_2NH_2)-CH_2CH_2-C(=O)-NH_2}_{\text{glutamine}}$$

C-terminus → $CONH_2$; $CO-CH_2NH_2$ (glycine) ← N-terminus

Peptide bonds are denoted with asterisks (*).

(d) glycylglutaminylisoleucinamide; Gly—Gln—Ile • NH_2

24-36

Aspartame: $H_2N-CH(CH_2COOH)-C(=O)-NH-CH(CH_2Ph)-C(=O)-OCH_3$ } methyl ester

aspartic acid (from Edman degradation) | phenylalanine

no free COOH ⇒ no reaction with carboxypeptidase

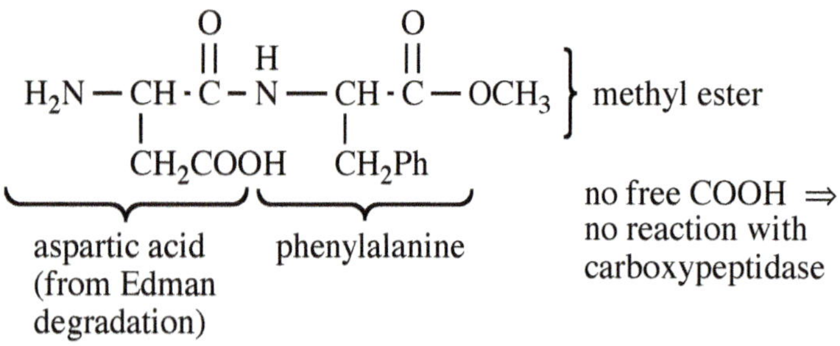

Aspartame is aspartylphenylalanine methyl ester.

24-37

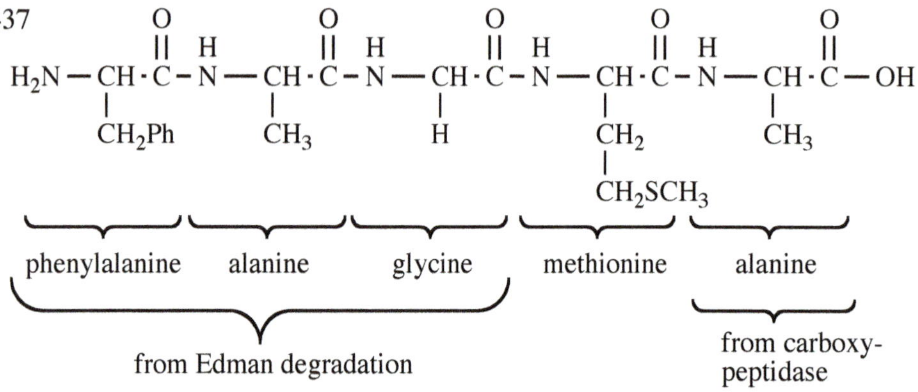

$H_2N-CH(CH_2Ph)-C(=O)-NH-CH(CH_3)-C(=O)-NH-CH(H)-C(=O)-NH-CH(CH_2CH_2SCH_3)-C(=O)-NH-CH(CH_3)-C(=O)-OH$

phenylalanine — alanine — glycine — methionine — alanine

phenylalanine, alanine, glycine: from Edman degradation

alanine: from carboxypeptidase

24-38

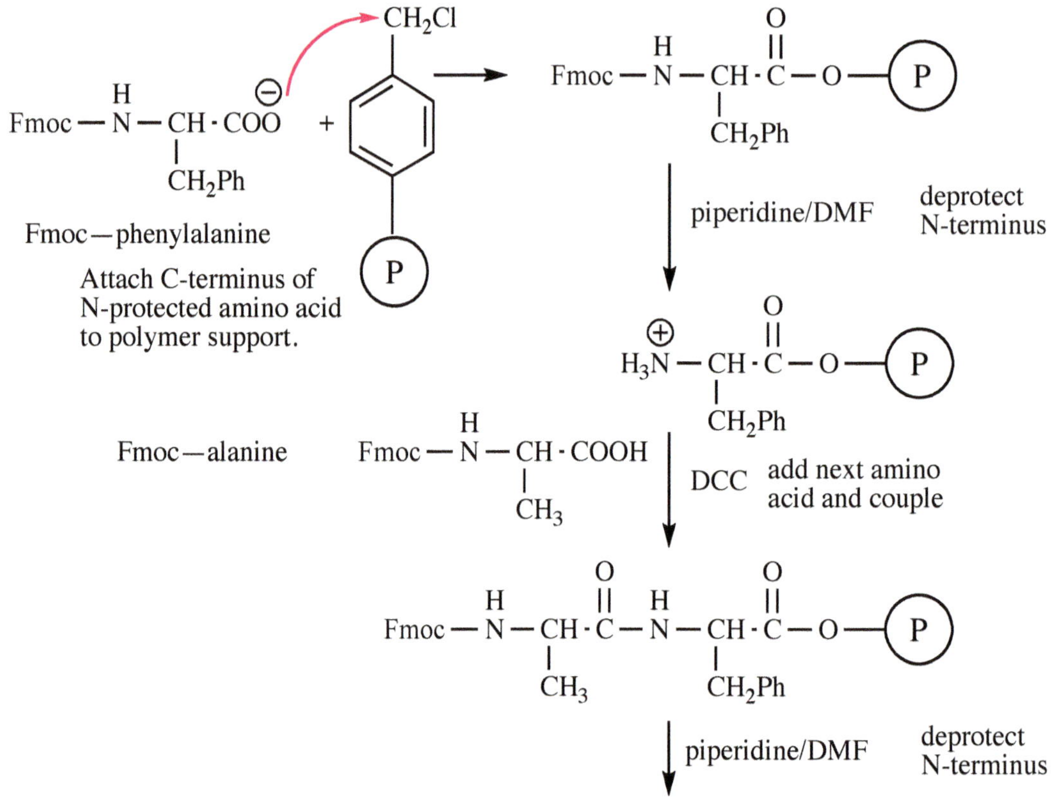

$Fmoc-NH-CH(CH_2Ph)-COO^{\ominus}$ + CH_2Cl–C₆H₄–(P) → $Fmoc-NH-CH(CH_2Ph)-C(=O)-O-$(P)

Fmoc—phenylalanine

Attach C-terminus of N-protected amino acid to polymer support.

↓ piperidine/DMF — deprotect N-terminus

$H_3\overset{\oplus}{N}-CH(CH_2Ph)-C(=O)-O-$(P)

Fmoc—alanine $Fmoc-NH-CH(CH_3)-COOH$

↓ DCC — add next amino acid and couple

$Fmoc-NH-CH(CH_3)-C(=O)-NH-CH(CH_2Ph)-C(=O)-O-$(P)

↓ piperidine/DMF — deprotect N-terminus

continued on next page

24-38 continued

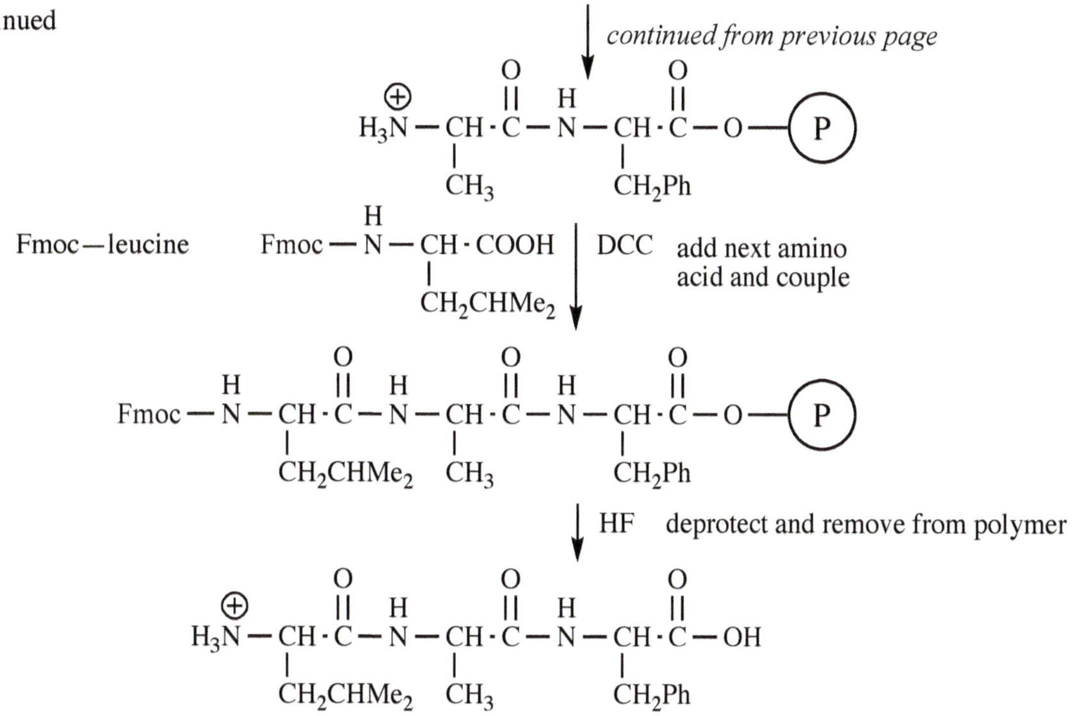

continued from previous page

$H_3\overset{\oplus}{N}-CH(CH_3)-C(=O)-NH-CH(CH_2Ph)-C(=O)-O-$ (P)

Fmoc—leucine $\quad$ $Fmoc-NH-CH(CH_2CHMe_2)-COOH$

DCC $\quad$ add next amino acid and couple

$Fmoc-NH-CH(CH_2CHMe_2)-C(=O)-NH-CH(CH_3)-C(=O)-NH-CH(CH_2Ph)-C(=O)-O-$ (P)

HF $\quad$ deprotect and remove from polymer

$H_3\overset{\oplus}{N}-CH(CH_2CHMe_2)-C(=O)-NH-CH(CH_3)-C(=O)-NH-CH(CH_2Ph)-C(=O)-OH$

24-39

(a) There are two possible sources of ammonia in the hydrolysate. The C-terminus could have been present as the amide instead of the carboxyl, or the glutamic acid could have been present as its amide, glutamine.

(b) The C-terminus is present as the amide. The N-terminus is present as the lactam (cyclic amide) combining the amino group with the carboxyl group of the glutamic acid side chain.

(c) The fact that hydrolysis does not release ammonia implies that the C-terminus is not an amide. Yet, carboxypeptidase treatment gives no reaction, showing that the C-terminus is not a free carboxyl group. Also, treatment with phenyl isothiocyanate gives no reaction, suggesting no free amine at the N-terminus. The most plausible explanation is that the N-terminus has reacted with the C-terminus to produce a cyclic amide, a lactam. (These large rings, called macrocycles, are often found in nature as hormones or antibiotics.)

24-40

(a) Lipoic acid is a mild oxidizing agent. In the process of oxidizing another reactant, lipoic acid is reduced.

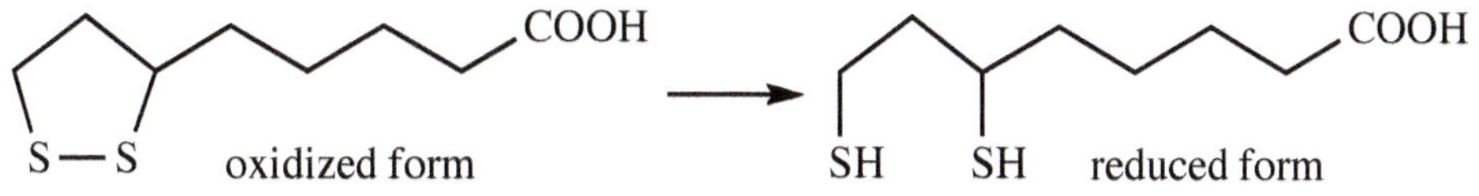

COOH (S—S ring) oxidized form $\longrightarrow$ COOH (SH SH) reduced form

(b) $O=C$ / $HC-CH_2SH$ / HN $\quad$ $HSCH_2-CH$ / $C=O$ / NH $\quad$ + S—S (ring) R $\longrightarrow$ $O=C$ / $HC-CH_2S-SCH_2-CH$ / HN $\quad$ $C=O$ / NH $\quad$ + SH SH R

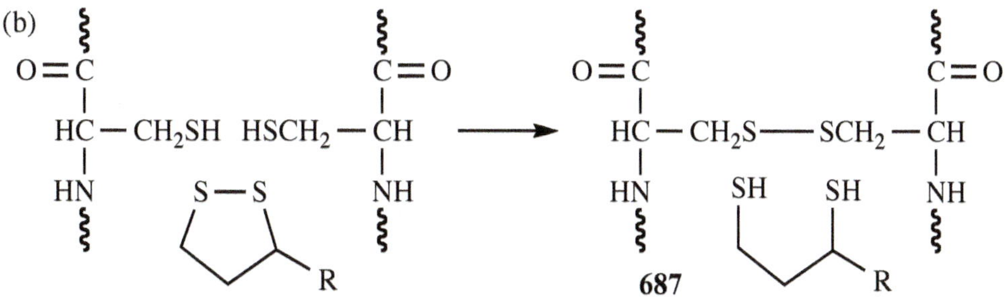

24-40 continued

(c)

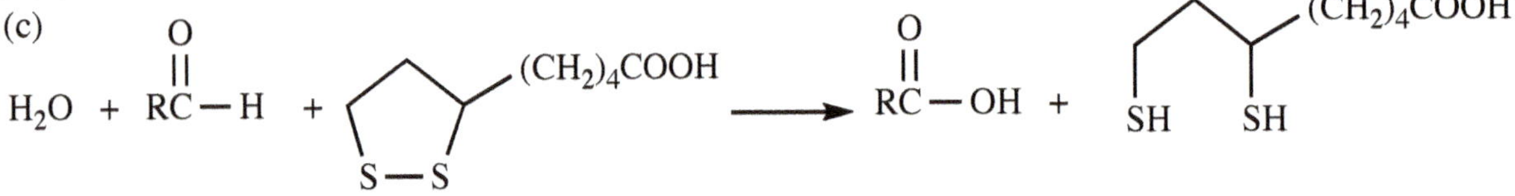

$$H_2O + RC(=O)-H + \text{(1,2-dithiolane)}-(CH_2)_4COOH \longrightarrow RC(=O)-OH + HS-CH_2CH_2-CH(SH)-(CH_2)_4COOH$$

24-41

(a) histidine: $H_2N-CH\cdot COOH$ with CH_2 side chain to the imidazole ring

basic → N (pyridine-type N)

not basic: pyrrole-type N (NH)

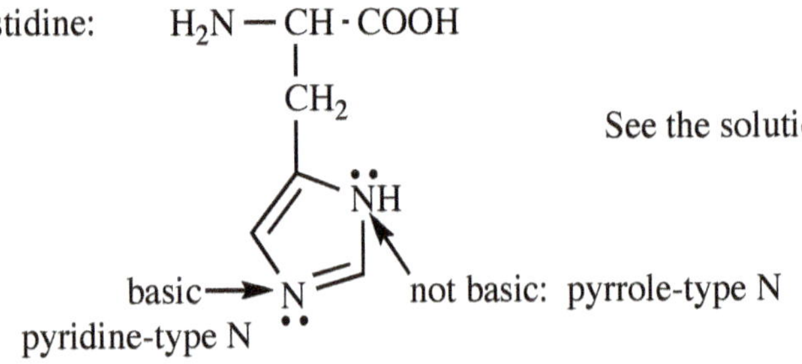

See the solution to Problem 24-6.

(b) Resonance forms of protonated imidazole: $\oplus$N–H / N–H … C$\oplus$ … N–H / $\oplus$N–H, each with H on both nitrogens.

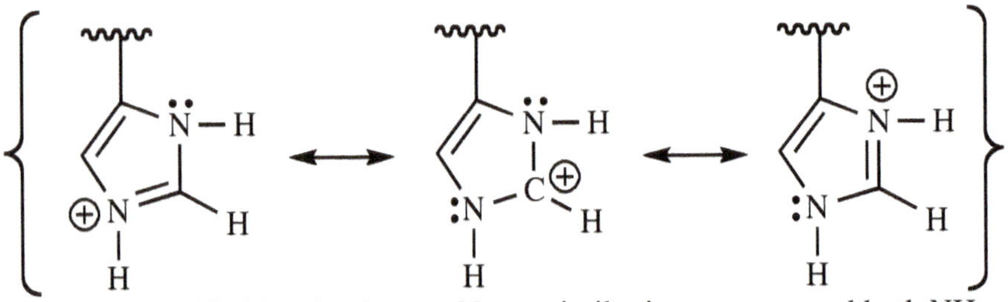

In the protonated imidazole, the two Ns are similar in structure, and both NH groups are acidic.

(c) Imidazole $\xrightarrow{H^+}$ protonated imidazole (resonance-stabilized) $\xrightarrow{-H^+}$ imidazole with H on the other N

resonance-stabilized

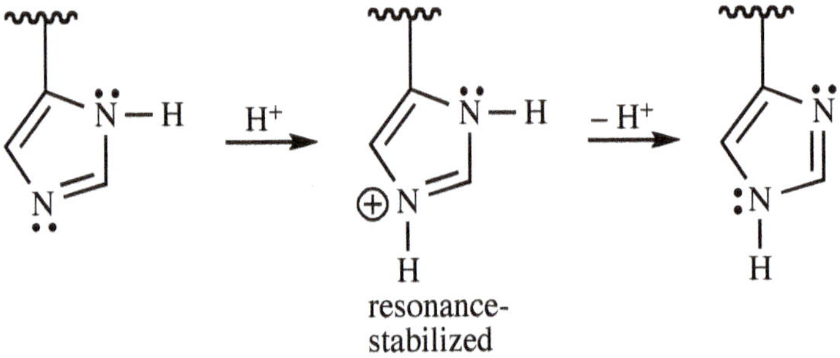

We usually think of protonation-deprotonation reactions occurring in solution where protons can move with solvent molecules. In an enzyme active site, there is no "solvent", so there must be another mechanism for movement of protons. Often, conformational changes in the protein will move atoms closer or farther. Histidine serves the function of moving a proton toward or away from a particular site by using its different nitrogens in concert as a proton acceptor and a proton donor.

24-42 The high isoelectric point suggests a strongly basic side chain as in lysine. The N—CH_2 bond in the side chain of arginine is likely to have remained intact during the metabolism. Can you propose a likely mechanism for this reaction? Think of this as a nucleophilic acyl substitution on C=NH instead of C=O.

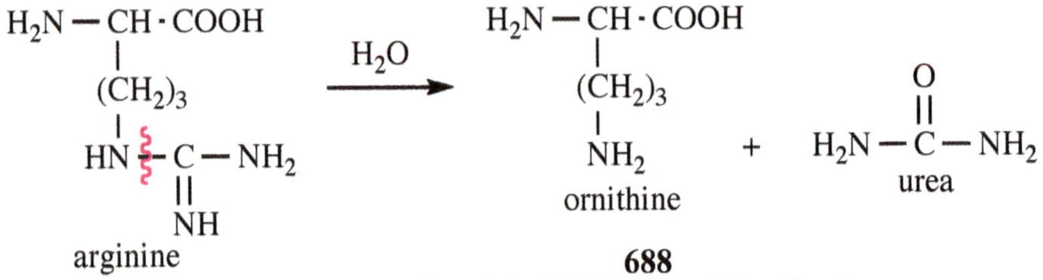

$$\underset{\text{arginine}}{H_2N-CH(COOH)-(CH_2)_3-HN-C(=NH)-NH_2} \xrightarrow{H_2O} \underset{\text{ornithine}}{H_2N-CH(COOH)-(CH_2)_3-NH_2} + \underset{\text{urea}}{H_2N-C(=O)-NH_2}$$

24-43

(a) glutathione:

Notice that the first peptide bond uses the COOH at carbon-5 of glutamic acid. Edman degradation will not cleave the peptide bond with this unusual bonding.

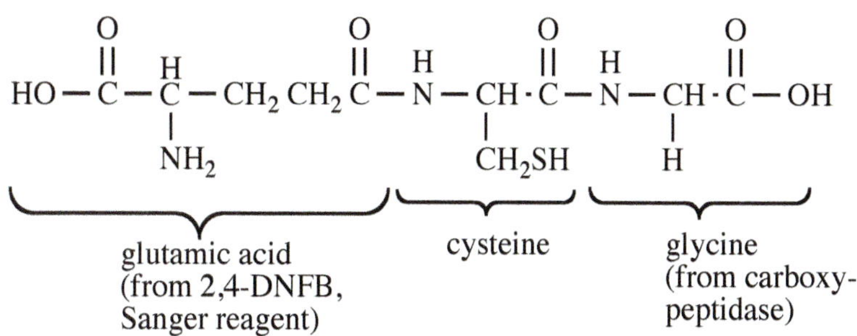

(b) reaction: **2** glutathione + H_2O_2 ⟶ glutathione disulfide + **2** H_2O

structure of glutathione disulfide:

```
      O          O        O         O
      ||  H      ||  H    ||  H     ||
HO — C — C — CH2 CH2 C — N — CH·C — N — CH·C — OH
          |                  |         |
         NH2               H2C — S     H
                                 |  <==  new S–S bond
         NH2               H2C — S     H
          |              H   |     H   |
HO — C — C — CH2 CH2 C — N — CH·C — N — CH·C — OH
      ||  H          ||         ||        ||
      O              O          O         O
```

24-44 Peptide A analysis gave NH_3 in addition to the amino acids, so the Glu in the analysis must have been Gln in the original peptide.

end groups: N-terminus Ala ———————— Ile C-terminus

chymotrypsin fragments

A Glu—Gly—Tyr (middle)

B Ala—Lys—Phe

N-terminus

C Arg—(Ser? Leu?)—Ile

C-terminus

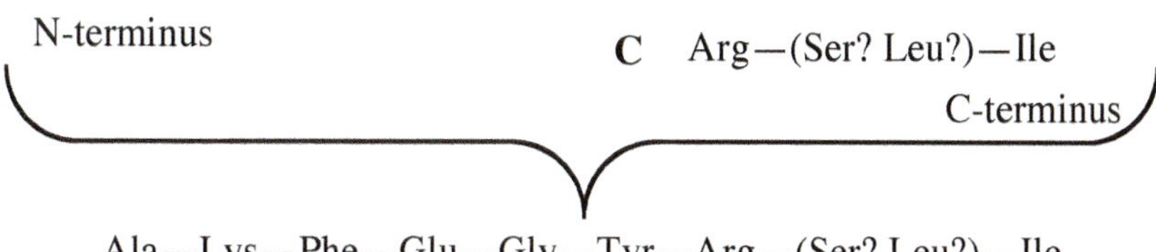

Ala—Lys—Phe—Glu—Gly—Tyr—Arg—(Ser? Leu?)—Ile

trypsin fragments

D Ala—Lys

E Phe—Glu—Gly—Tyr—Arg

F Ser—Leu—Ile

Ala—Lys—Phe—Gln—Gly—Tyr—Arg—Ser—Leu—Ile

24-45

(a) One important factor in ester reactivity is the ability of the alkoxy group to leave, and the main factor in determining leaving group ability is the stabilization of the anion. An NHS ester is more reactive than an alkyl ester because the anion R_2NO^- has an electron-withdrawing group on the O^-, thereby distributing the negative charge over two atoms instead of just one. A simple alkyl ester has the full negative charge on the oxygen with nowhere to go, RO^-.

This resonance form of the leaving group shows the (+) charge on N that stabilizes the (–) on oxygen.

(b) $RCOOH + Et_3N \longrightarrow Et_3NH^+ + \{RCOO^- \leftrightarrow RCOO^-\}$

abbreviate O–Succ

tetrahedral intermediate

reactive anhydride

NHS ester

tetrahedral intermediate

CF_3COO^- + NHS ester

NHS trifluoroacetate is an amazing reagent. Not only does it activate the carboxylic acid through a mixed anhydride to form an ester under mild conditions, but the NHS leaving group is also the nucleophile that forms an ester that is both stable enough to work with, yet easily reactive when it needs to be. Perfect!

(c) $RCO–O–Succ + H_2NR \longrightarrow$ tetrahedral intermediate $\longrightarrow RCONH_2R^+ + {}^-O–Succ \longrightarrow RCONHR + HO–Succ$

24-46

(a)

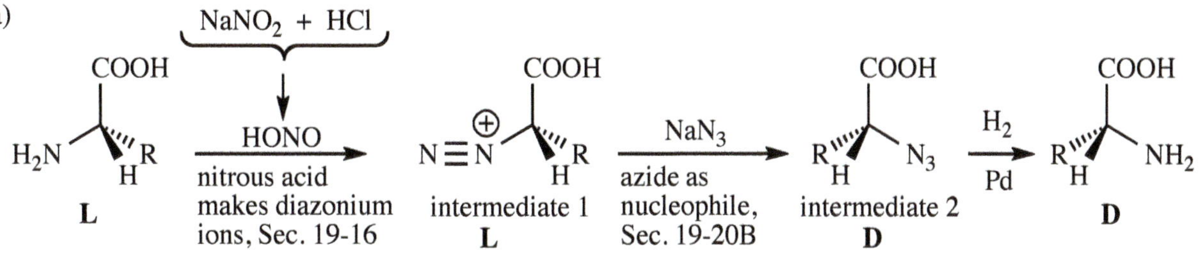

NaNO$_2$ + HCl → HONO

L (COOH, H_2N, R, H) —HONO: nitrous acid makes diazonium ions, Sec. 19-16→ intermediate 1 L ($N\equiv\overset{+}{N}$, COOH, R, H) —NaN_3: azide as nucleophile, Sec. 19-20B→ intermediate 2 D (COOH, R, H, N_3) —H_2/Pd→ D (COOH, R, H, NH_2)

(b) The product has the opposite configuration from the starting material because of the stereochemistry of the reactions. Diazotization does not break a bond to the chiral center so the L configuration is retained in Reaction 1. It is well documented that azide substitution is an S_N2 process that proceeds with inversion of configuration; this is why this process works. The third reaction does not break or form a bond to the chiral center, so the D configuration is retained.

24-47

(a) and (c) In D_2O, protons on COOH and amine NH_2 exchange rapidly. When H is replaced with D, it does not generate a signal in the HNMR. (In contrast, amide N-H protons exchange very slowly.)

(b)

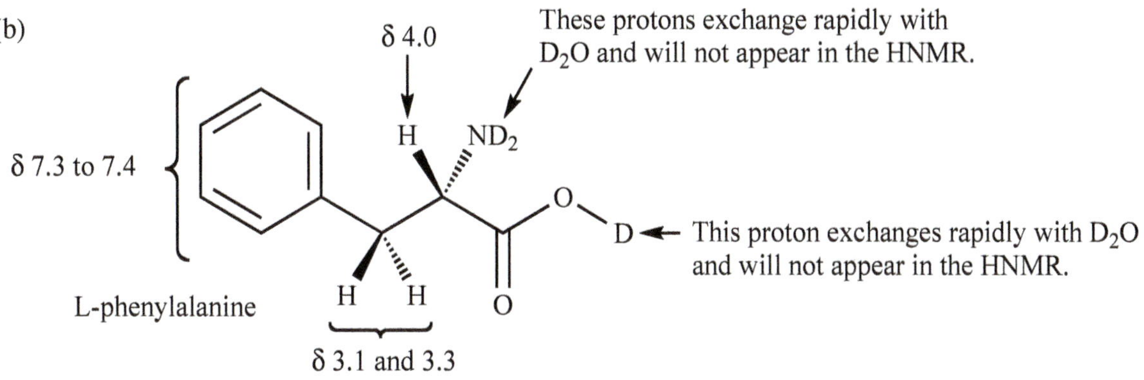

δ 4.0

These protons exchange rapidly with D_2O and will not appear in the HNMR.

δ 7.3 to 7.4

H ND$_2$ O D

This proton exchanges rapidly with D_2O and will not appear in the HNMR.

L-phenylalanine

H H O

δ 3.1 and 3.3

(d) In compounds with chiral centers as in amino acids, neighboring protons are not in identical environments. These protons are called *diastereotopic*, and they are distinguishable in the HNMR, often exhibiting different chemical shifts as above. See text section 13-10.

CHAPTER 25—LIPIDS

25-1

trimyristin

$$\begin{array}{l} CH_2-O-\overset{O}{\overset{\|}{C}}-(CH_2)_{12}CH_3 \\ | \\ CH-O-\overset{O}{\overset{\|}{C}}-(CH_2)_{12}CH_3 \\ | \\ CH_2-O-\overset{O}{\overset{\|}{C}}-(CH_2)_{12}CH_3 \end{array}$$

Trimyristin is easily isolated from nutmeg by stirring the nutmeg in some acetone, filtering, allowing the acetone to evaporate, leaving a waxy residue of trimyristin.

25-2

all *cis*

H_2C-O / $HC-O$ / H_2C-O (trilinolein) $\xrightarrow[\text{Ni}]{\text{excess } H_2}$

$$\begin{array}{l} CH_2-O-\overset{O}{\overset{\|}{C}}-(CH_2)_{16}CH_3 \\ | \\ CH-O-\overset{O}{\overset{\|}{C}}-(CH_2)_{16}CH_3 \\ | \\ CH_2-O-\overset{O}{\overset{\|}{C}}-(CH_2)_{16}CH_3 \end{array}$$

trilinolein, m.p. < –4 °C
(liquid at room temperature)

tristearin, m.p. 72 °C
(solid at room temperature)

You could predict that the melting point of trilinolein is lower than triolein (–4 °C) because more double bonds lower the melting point. Sources differ on the m.p. of trilinolein, ranging from –17 °C to –43 °C.

25-3

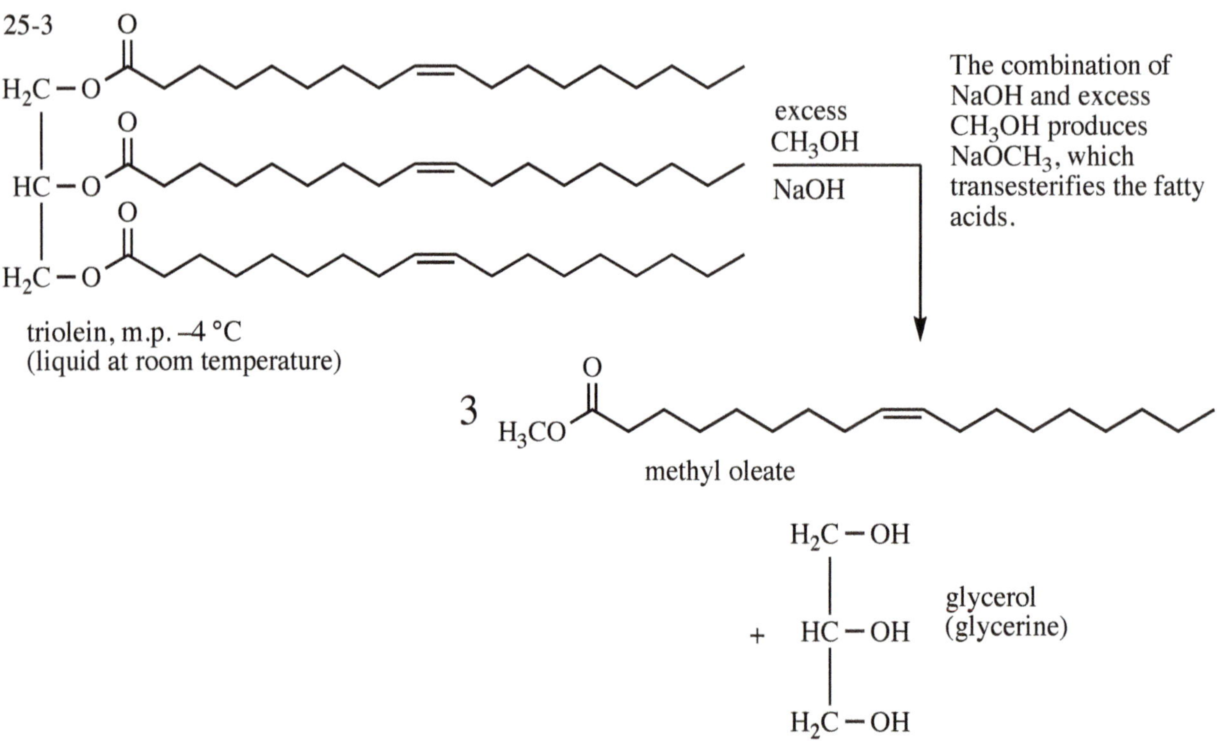

25-4

(a)

$$2\ CH_3(CH_2)_{16}-\overset{\overset{\large O}{\|}}{C}-O^-\ Na^+ + Ca^{2+} \longrightarrow \left[CH_3(CH_2)_{16}-\overset{\overset{\large O}{\|}}{C}-O\right]_2 Ca \ + \ 2\ Na^+$$

(b)

$$2\ CH_3(CH_2)_{16}-\overset{\overset{\large O}{\|}}{C}-O^-\ Na^+ + Mg^{2+} \longrightarrow \left[CH_3(CH_2)_{16}-\overset{\overset{\large O}{\|}}{C}-O\right]_2 Mg \ + \ 2\ Na^+$$

(c)

$$3\ CH_3(CH_2)_{16}-\overset{\overset{\large O}{\|}}{C}-O^-\ Na^+ + Fe^{3+} \longrightarrow \left[CH_3(CH_2)_{16}-\overset{\overset{\large O}{\|}}{C}-O\right]_3 Fe \ + \ 3\ Na^+$$

25-5

(a) Both sodium carbonate (its old name is "washing soda") and sodium phosphate will increase the pH above 6, so that the carboxyl group of the soap molecule will remain ionized, thus preventing precipitation.

(b) In the presence of calcium, magnesium, and ferric ions, the carboxylate group of soap will form precipitates called "hard-water scum", or as scientists label it, "bathtub ring". Both carbonate and phosphate ions will form complexes or precipitates with these cations, thereby preventing precipitation of the soap from solution.

25-6 In each structure, the hydrophilic portion is circled. The uncircled part is hydrophobic.

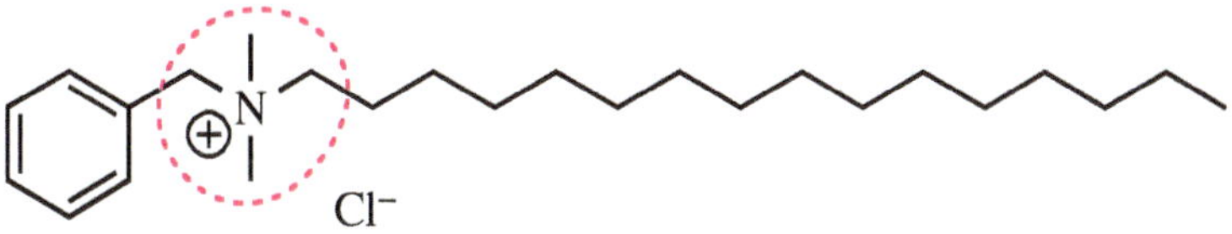

N
Cl^-
benzalkonium chloride

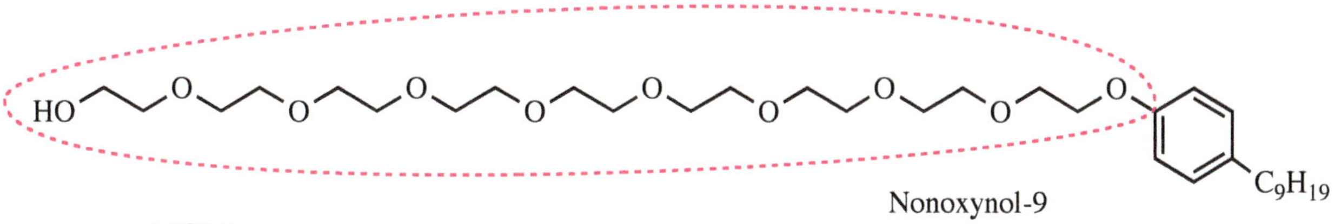

HO O O O O O O O O
C_9H_{19}
Nonoxynol-9

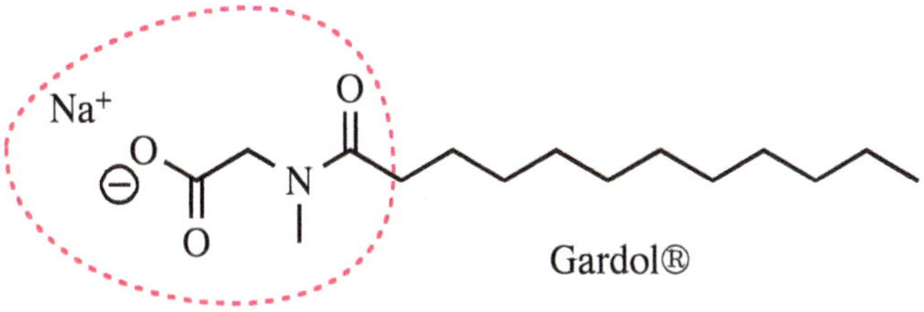

Na^+
O O
N
O
Gardol®

25-7

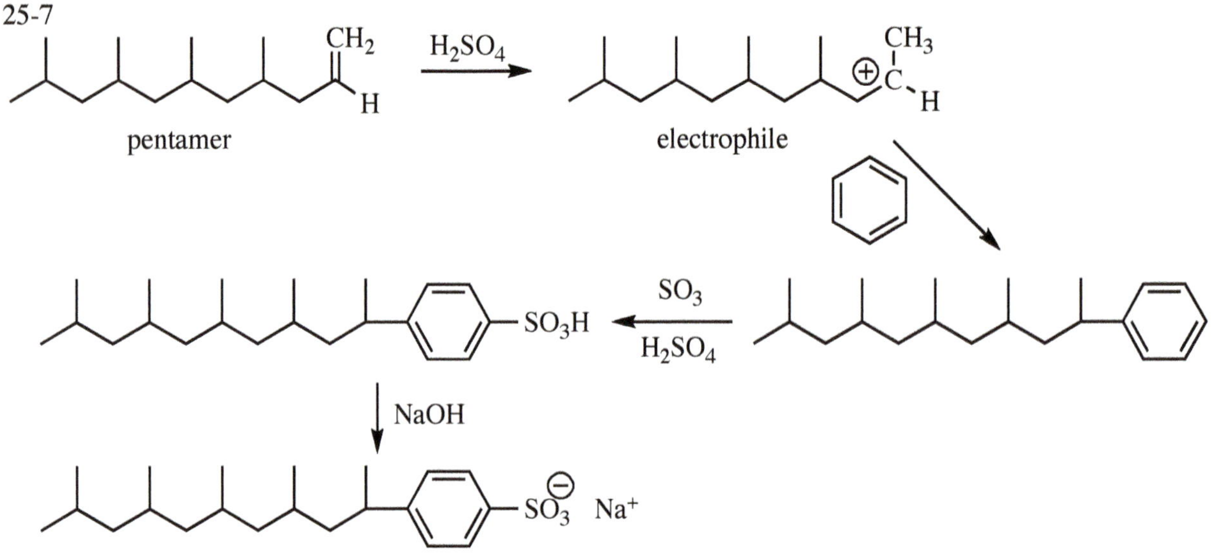

pentamer

electrophile

25-8

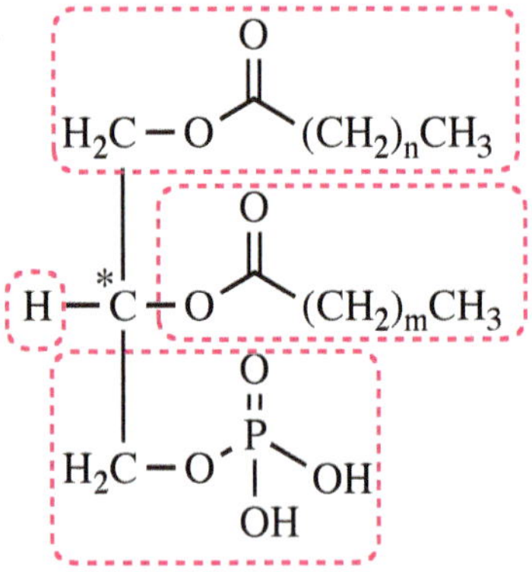

A phosphatidic acid has four different groups on the center carbon of the 3-carbon glycerol portion as shown in the dashed boxes, making that central carbon an asymmetric carbon atom.

25-9 Estradiol is a phenol and can be ionized with aqueous NaOH. Testosterone does not have any hydrogens acidic enough to react with NaOH. Treatment of a solution of estradiol and testosterone in organic solvent with aqueous base will extract the phenoxide form of estradiol into the aqueous layer, leaving testosterone in the organic layer. Acidification of the aqueous base will precipitate estradiol, which can be filtered. Evaporation of the organic solvent will leave testosterone.

25-10 Models may help. Abbreviations: "ax" = axial; "eq" = equatorial. Note that substituents at *cis*-fused ring junctures are axial to one ring and equatorial to another.

(a)

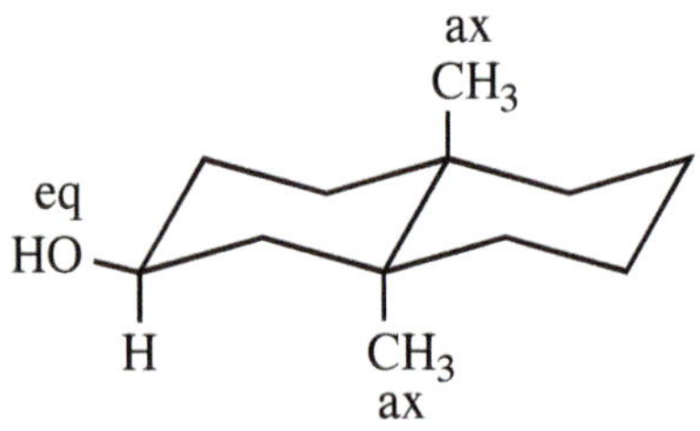

(b)

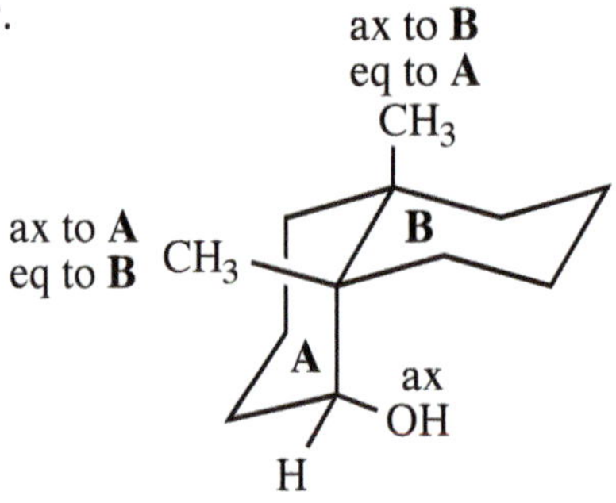

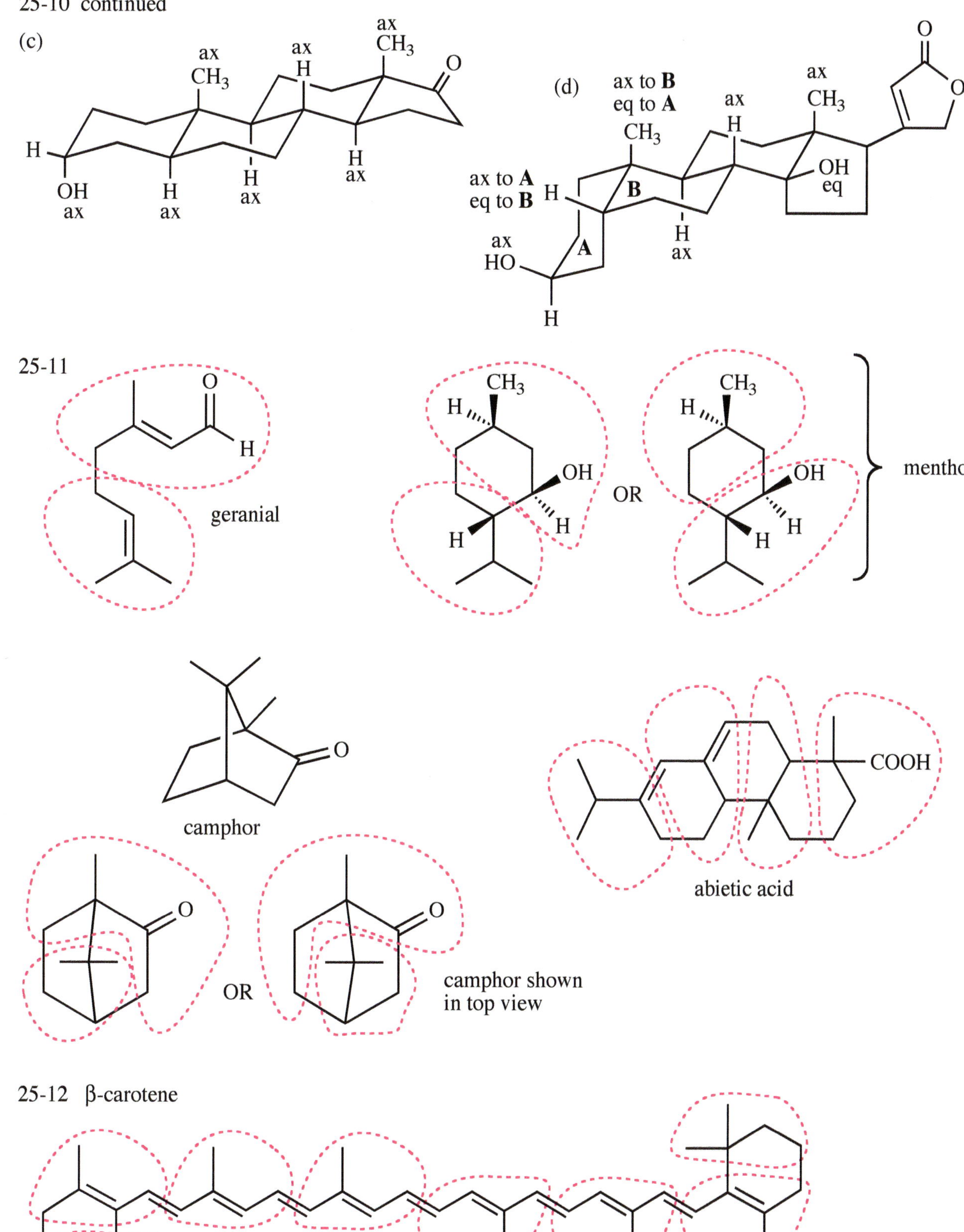
25-10 continued
(c)
ax
CH3
ax
H
ax
CH3
O
H
OH
ax
H
ax
H
ax
H
ax
(d)
ax to B
eq to A
CH3
ax
H
ax
CH3
O
O
OH
eq
ax to A
eq to B
H
B
A
ax
HO
H
H
ax
25-11
O
H
geranial
CH3
H
OH
H
H
OR
CH3
H
OH
H
H
menthol
O
camphor
COOH
abietic acid
O
OR
O
camphor shown
in top view
25-12 β-carotene

25-13

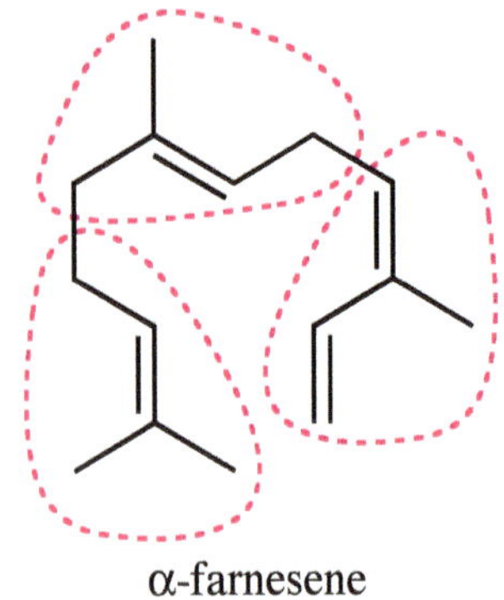

α-farnesene
sesquiterpene

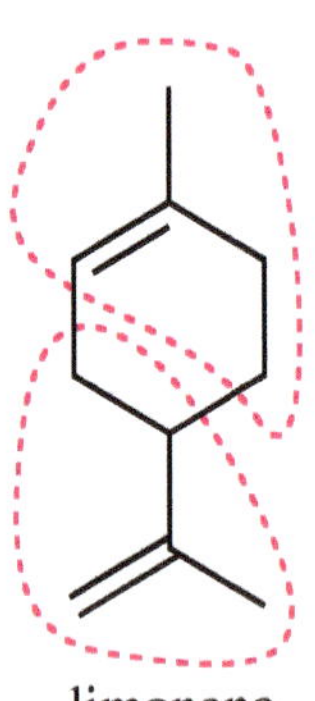

limonene
monoterpene

(Limonene can also be circled in the other direction around the ring—see menthol in the solution to problem 25-11.)

OR

α-pinene
monoterpene

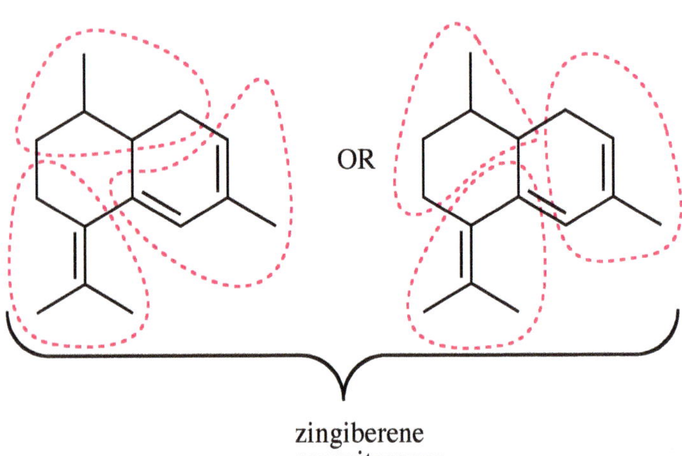

zingiberene
sesquiterpene

25-14

(a) A saturated fat is a solid triglyceride in which all three chains are saturated fatty acids with an even number of carbons, most commonly between 12 and 20 carbons.

$$\begin{array}{l} CH_2-O-C(=O)-(CH_2)_{12}CH_3 \\ | \\ CH-O-C(=O)-(CH_2)_{12}CH_3 \\ | \\ CH_2-O-C(=O)-(CH_2)_{12}CH_3 \end{array}$$

(b) A polyunsaturated oil is a liquid triglyceride in which all three fatty acid chains contain an even number of carbons, most commonly between 12 and 20 carbons, with two or more double bonds distributed among the three chains.

H_2C-O, $HC-O$, H_2C-O

25-14 continued

(c) A wax is an ester composed of a long-chain alcohol and a long-chain fatty acid, for example:

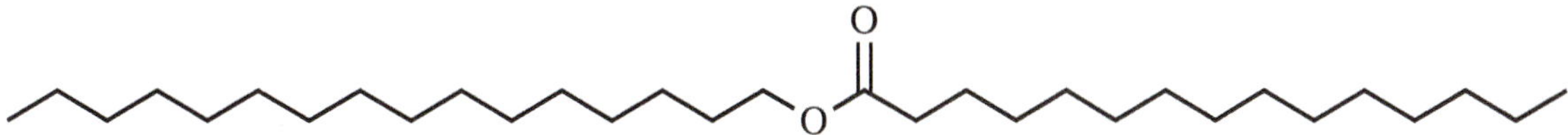

(d) A soap is the sodium or potassium salt of a fatty acid.

O

O ⊖ Na^+ or K^+

(e) A detergent can be anionic (the original category), cationic, or nonionic.

⊕ N

Cl^-

benzalkonium chloride—
a cationic detergent

HO O O O O O O O O O C_9H_{19}

Nonoxynol-9—
a nonionic detergent

O SO_3^- Na^+

SDS, sodium dodecyl sulfate, or
SLS, sodium lauryl sulfate,
an anionic detergent

(f) A phospholipid is a class of lipid, usually a triglyceride, that contains a phosphate group, for example:

$$\begin{array}{l} CH_2-O-C(=O)-(CH_2)_{16}CH_3 \\ | \\ CH-O-C(=O)-(CH_2)_{16}CH_3 \\ | \\ CH_2-O-P(=O)(O^{\ominus})-O^{\ominus} \end{array}$$

a phosphatidic acid
in ionized form

(g) A prostaglandin is a metabolite of arachidonic acid, a 20-carbon polyunsaturated fatty acid, containing a 5-membered ring. Many prostaglandins are potent mammalian hormones.

HO

COOH

HO

OH

$PGF_{2\alpha}$

$C_{20}H_{34}O_5$

(h) A steroid is a tetracyclic compound originally discovered in mammalian systems, but now found in almost all plants and animals. All contain this four-ring structural unit.

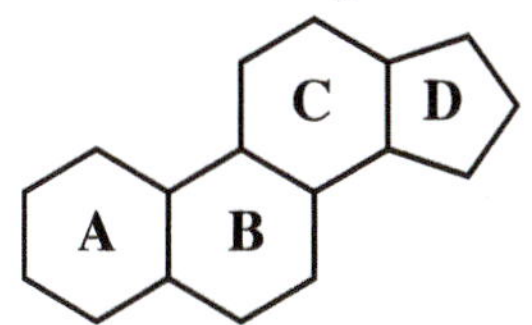

substructure of all steroids

(i) A sesquiterpene is a 15-carbon compound composed of three isoprene units.

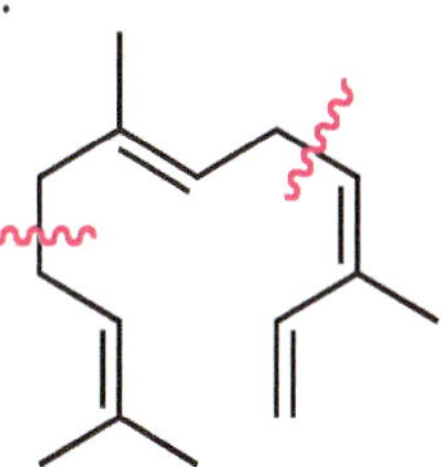

α-farnesene
a sesquiterpene

25-15

(a) **3** $CH_3(CH_2)_7CH{=}C(CH_2)_7COO^-\ Na^+$ + $HOCH(CH_2OH)_2$

soap glycerol

(b)

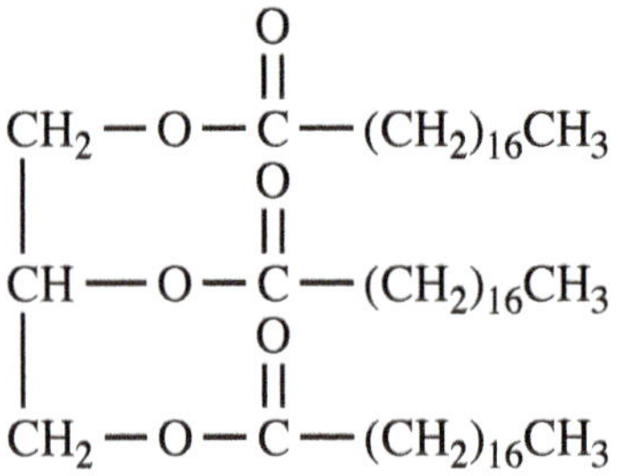

$$\begin{array}{l} CH_2-O-\overset{\overset{\large O}{\|}}{C}-(CH_2)_{16}CH_3 \\ | \\ CH-O-\overset{\overset{\large O}{\|}}{C}-(CH_2)_{16}CH_3 \\ | \\ CH_2-O-\overset{\overset{\large O}{\|}}{C}-(CH_2)_{16}CH_3 \end{array}$$

tristearin

(c)

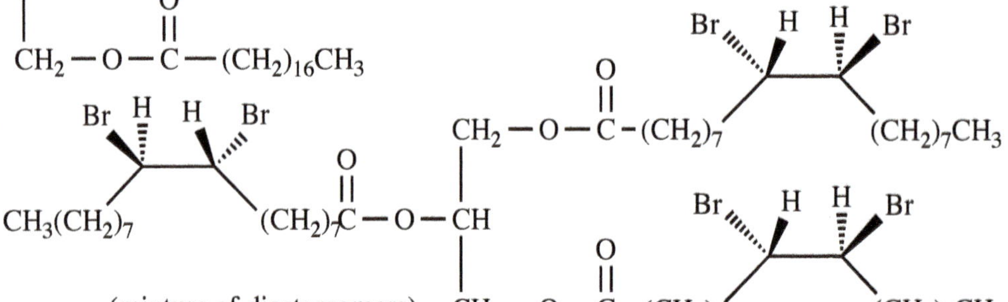

Br H H Br

$CH_3(CH_2)_7$ — CHBr–CHBr — $(CH_2)_7C(=O)-O-CH$

$CH_2-O-C(=O)-(CH_2)_7$ — CHBr–CHBr — $(CH_2)_7CH_3$

$CH_2-O-C(=O)-(CH_2)_7$ — CHBr–CHBr — $(CH_2)_7CH_3$

(mixture of diastereomers)

(d) **3** $CH_3(CH_2)_7CHO$ +

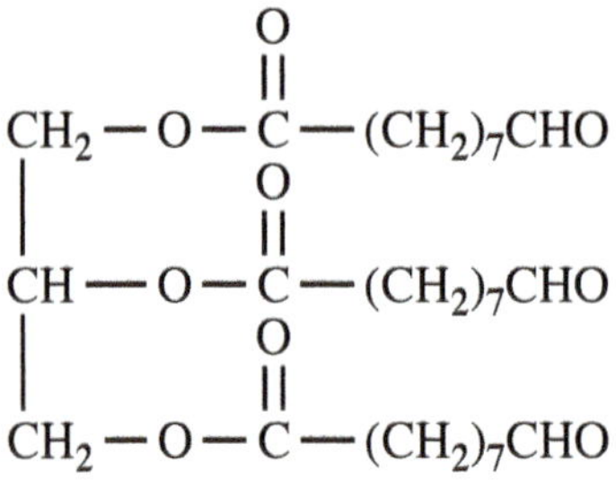

$$\begin{array}{l} CH_2-O-\overset{\overset{\large O}{\|}}{C}-(CH_2)_7CHO \\ | \\ CH-O-\overset{\overset{\large O}{\|}}{C}-(CH_2)_7CHO \\ | \\ CH_2-O-\overset{\overset{\large O}{\|}}{C}-(CH_2)_7CHO \end{array}$$

(e) **3** $CH_3(CH_2)_7COOH$ +

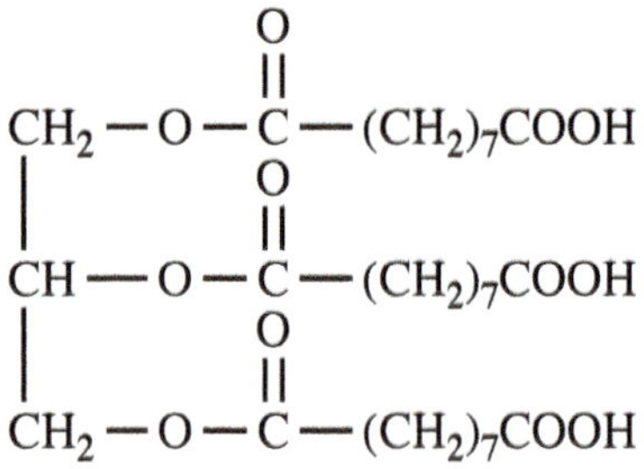

$$\begin{array}{l} CH_2-O-\overset{\overset{\large O}{\|}}{C}-(CH_2)_7COOH \\ | \\ CH-O-\overset{\overset{\large O}{\|}}{C}-(CH_2)_7COOH \\ | \\ CH_2-O-\overset{\overset{\large O}{\|}}{C}-(CH_2)_7COOH \end{array}$$

(f)

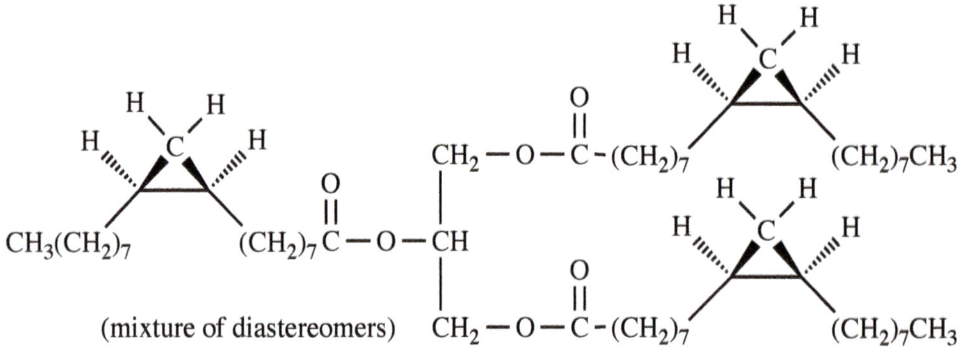

$CH_3(CH_2)_7$ — cyclopropane (CH_2; H, H) — $(CH_2)_7C(=O)-O-CH$

$CH_2-O-C(=O)-(CH_2)_7$ — cyclopropane (CH_2; H, H) — $(CH_2)_7CH_3$

$CH_2-O-C(=O)-(CH_2)_7$ — cyclopropane (CH_2; H, H) — $(CH_2)_7CH_3$

(mixture of diastereomers)

25-16

(a) $CH_3(CH_2)_7CH{=}CH(CH_2)_7COOH \xrightarrow[Ni]{H_2} \xrightarrow[2)\ H_3O^+]{1)\ LiAlH_4} CH_3(CH_2)_{16}CH_2OH$

(b) $CH_3(CH_2)_7CH{=}CH(CH_2)_7COOH \xrightarrow[Ni]{H_2} CH_3(CH_2)_{16}COOH$

(c) $CH_3(CH_2)_{16}COOH$ + $HOCH_2(CH_2)_{16}CH_3$ $\xrightarrow[\Delta]{H^+}$ $CH_3(CH_2)_{16}COOCH_2(CH_2)_{16}CH_3$

from (b) from (a)

25-16 continued

(d) $CH_3(CH_2)_7CH{=}CH(CH_2)_7COOH \xrightarrow[\text{2) } Me_2S]{\text{1) } O_3, -78\ °C} CH_3(CH_2)_7CH{=}O + O{=}CH(CH_2)_7COOH$

nonanal

(e) $CH_3(CH_2)_7CH{=}CH(CH_2)_7COOH \xrightarrow[H_2O,\ \Delta]{KMnO_4} CH_3(CH_2)_7COOH + HOOC(CH_2)_7COOH$

nonanedioic acid

(f) $CH_3(CH_2)_7CH{=}CH(CH_2)_7COOH \xrightarrow[PBr_3]{\text{excess } Br_2} \xrightarrow{H_2O}$ $CH_3(CH_2)_7$–CH(Br)(H)–CH(H)(Br)–$(CH_2)_6CH(Br)COOH$

Br H H Br, Br

25-17 (a) triglyceride (b) synthetic detergent (c) wax (d) sesquiterpene
(e) prostaglandin (f) steroid

25-18

(a)

$H_2C{-}O{-}C({=}O){-}(CH_2)_{16}CH_3$
$HC{-}O{-}C({=}O){-}(CH_2)_7CH{=}CH(CH_2)_7CH_3$
$H_2C{-}O{-}P({=}O)(O^{\ominus}){-}OCH_2CH_2\overset{\oplus}{N}H_3$

$\xrightarrow[\Delta]{NaOH}$

$H_2C{-}OH$, $HC{-}OH$, $H_2C{-}OH$ glycerol

$Na^+\ {}^{\ominus}O{-}C({=}O){-}(CH_2)_{16}CH_3$
$Na^+\ {}^{\ominus}O{-}C({=}O){-}(CH_2)_7CH{=}CH(CH_2)_7CH_3$
$Na^+\ {}^{\ominus}O{-}P({=}O)(O^{\ominus})\ Na^+{-}OCH_2CH_2NH_2$

NaOH removes proton from NH_3.

(b)

$H_2C{-}O{-}C({=}O){-}(CH_2)_{14}CH_3$
$HC{-}O{-}C({=}O){-}(CH_2)_{14}CH_3$
$H_2C{-}O{-}P({=}O)(O^{\ominus}){-}OCH_2CH_2\overset{\oplus}{N}(CH_3)_3$

$\xrightarrow[\Delta]{NaOH}$

$H_2C{-}OH$, $HC{-}OH$, $H_2C{-}OH$ glycerol

2 $Na^+\ {}^{\ominus}O{-}C({=}O){-}(CH_2)_{14}CH_3$

$Na^+\ {}^{\ominus}O{-}P({=}O)(O^{\ominus})\ Na^+{-}OCH_2CH_2\overset{\oplus}{N}(CH_3)_3$

25-19

$CH_2{-}O{-}C({=}O){-}(CH_2)_{16}CH_3$
$CH{-}O{-}C({=}O){-}(CH_2)_{16}CH_3$
$CH_2{-}O{-}C({=}O){-}(CH_2)_{16}CH_3$

$\xrightarrow[\text{2) } H_3O^+]{\text{1) } LiAlH_4}$ $CH_2OH{-}CHOH{-}CH_2OH$ + **3** $CH_3(CH_2)_{16}CH_2OH$

$CH_3(CH_2)_{16}CH_2OH \xrightarrow{SO_3,\ H_2SO_4\ (cold)} CH_3(CH_2)_{16}CH_2OSO_3H$

$CH_3(CH_2)_{16}CH_2OSO_3H \xrightarrow{NaOH} CH_3(CH_2)_{16}CH_2OSO_3^-\ Na^+$

25-20 Reagents in parts (a), (b), and (d) would react with alkenes. If both samples contained alkenes, these reagents could not distinguish the samples. Saponification (part (c)), however, is a reaction of an ester, so only the vegetable oil would react, not the hydrocarbon oil mixture. With saponification of a vegetable oil, as NaOH is consumed, the pH would gradually drop; with a petroleum oil, the pH would not change.

25-21 (a) Add an aqueous solution of calcium ion or magnesium ion. Sodium stearate will produce a precipitate, while the sulfonate will not precipitate.
(b) Beeswax, an ester, can be saponified with NaOH. Paraffin wax is a solid mixture of alkanes and will not react.
(c) Myristic acid will dissolve (or be emulsified) in dilute aqueous base. Trimyristin will remain unaffected.
(d) Triolein (an unsaturated oil) will decolorize bromine in CCl_4, but trimyristin (a saturated fat) will not.

25-22

(a)

$$\begin{array}{l} CH_2-O-\overset{O}{\overset{\|}{C}}-(CH_2)_{12}CH_3 \\ | \\ \overset{*}{C}H-O-\overset{O}{\overset{\|}{C}}-(CH_2)_7CH{=}CH(CH_2)_7CH_3 \\ | \\ CH_2-O-\overset{O}{\overset{\|}{C}}-(CH_2)_7CH{=}CH(CH_2)_7CH_3 \end{array}$$

asymmetric * carbon — optically active

(b)

$$\begin{array}{l} CH_2-O-\overset{O}{\overset{\|}{C}}-(CH_2)_7CH{=}CH(CH_2)_7CH_3 \\ | \\ CH-O-\overset{O}{\overset{\|}{C}}-(CH_2)_{12}CH_3 \\ | \\ CH_2-O-\overset{O}{\overset{\|}{C}}-(CH_2)_7CH{=}CH(CH_2)_7CH_3 \end{array}$$

not optically active; symmetric

25-23

$$\begin{array}{l} CH_2-O-\overset{O}{\overset{\|}{C}}-(CH_2)_{16}CH_3 \\ | \\ \overset{*}{C}H-O-\overset{O}{\overset{\|}{C}}-(CH_2)_7CH{=}CH(CH_2)_7CH_3 \\ | \\ CH_2-O-\overset{O}{\overset{\|}{C}}-(CH_2)_7CH{=}CH(CH_2)_7CH_3 \end{array}$$

asymmetric * carbon — optically active

(a)

$$\xrightarrow[Ni]{H_2}\quad \begin{array}{l} CH_2-O-\overset{O}{\overset{\|}{C}}-(CH_2)_{16}CH_3 \\ | \\ CH-O-\overset{O}{\overset{\|}{C}}-(CH_2)_{16}CH_3 \\ | \\ CH_2-O-\overset{O}{\overset{\|}{C}}-(CH_2)_{16}CH_3 \end{array}$$

not optically active

25-23 continued

(b)

$\xrightarrow[CCl_4]{Br_2}$ $CH_3(CH_2)_7$–CHBr–CHBr–$(CH_2)_7C(=O)$–O–$\overset{*}{C}H$(CH_2–O–$C(=O)(CH_2)_{16}CH_3$)(CH_2–O–$C(=O)$–$(CH_2)_7$–CHBr–CHBr–$(CH_2)_7CH_3$)

optically active
(mixture of diastereomers)

(c) Products are not optically active.

$HOCH(CH_2OH)_2$ + $Na^+\ ^-OOC(CH_2)_{16}CH_3$ + **2** $Na^+\ ^-OOC(CH_2)_7CH{=}CH(CH_2)_7CH_3$
glycerol

(d)

$\xrightarrow[2)\ Me_2S]{1)\ O_3,\ -78\ °C}$ CH_2–O–$C(=O)$–$(CH_2)_{16}CH_3$ / $\overset{*}{C}H$–O–$C(=O)$–$(CH_2)_7CHO$ / CH_2–O–$C(=O)$–$(CH_2)_7CHO$ + **2** $O{=}CH(CH_2)_7CH_3$

optically active

not optically active

25-24 The products in (b), (c) and (f) are mixtures of stereoisomers.

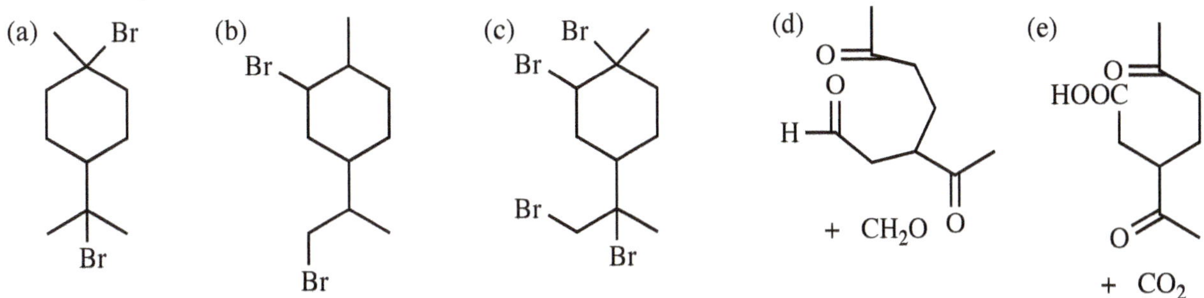

(f) H, CH_3, H, HO, HO

25-25 Is it any wonder that Olestra cannot be digested!

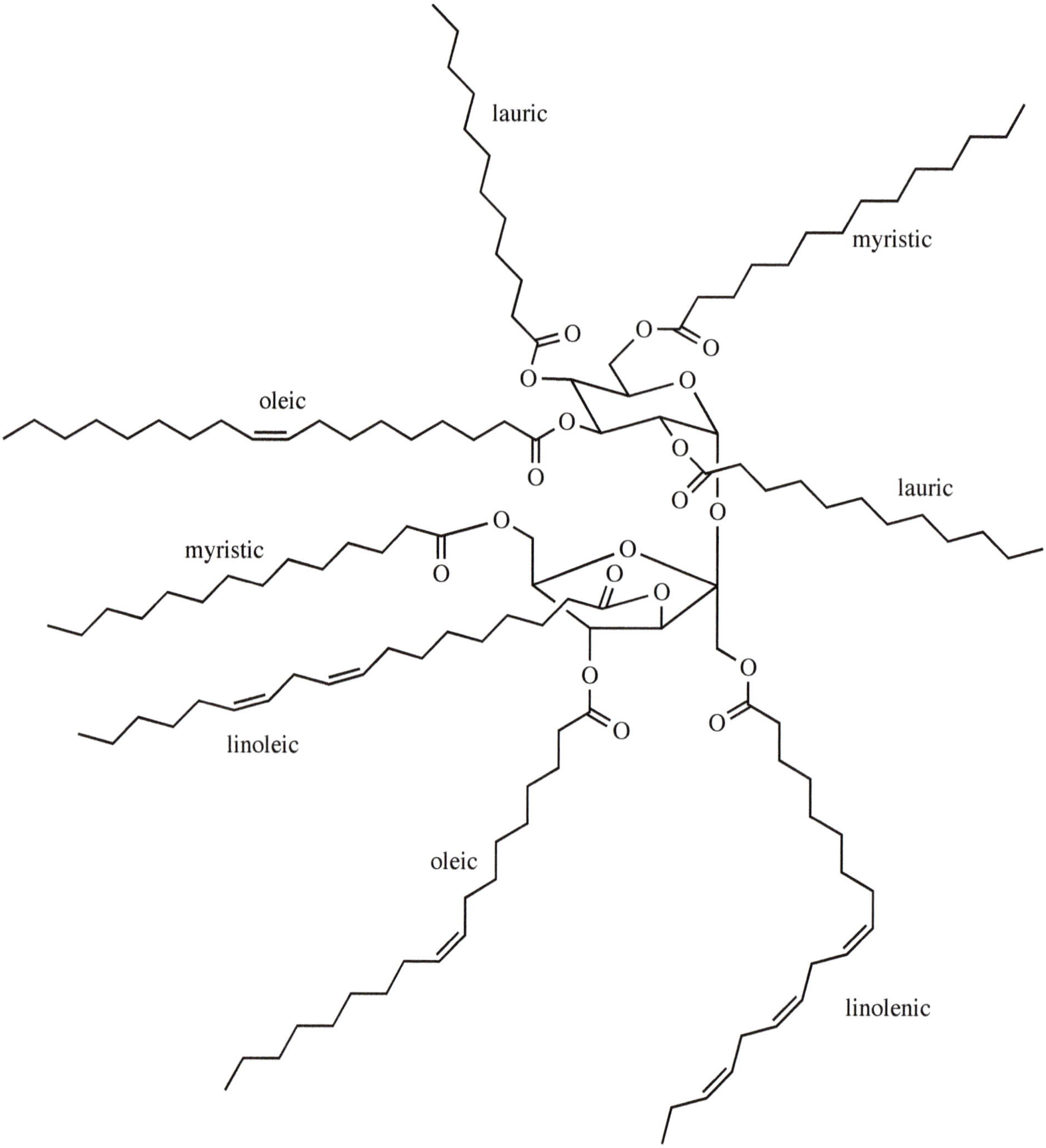

lauric

myristic

oleic

lauric

myristic

linoleic

oleic

linolenic

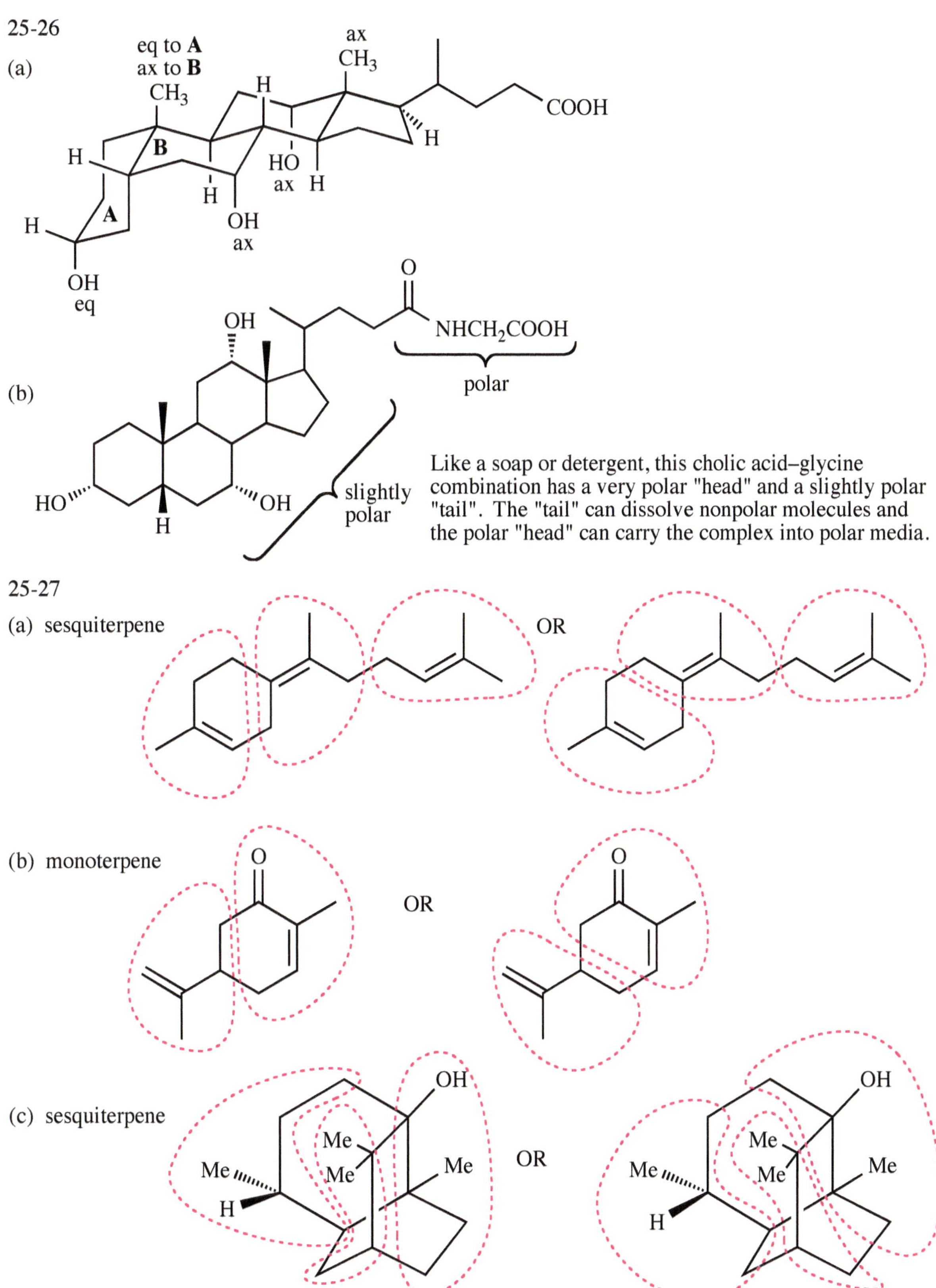
25-26
(a)
eq to A
ax to B
CH3
H
ax
CH3
COOH
H
B
H
HO
ax H
H
A
H
OH
ax
OH
eq
(b)
O
OH
NHCH2COOH
polar
HO
OH
H
slightly
polar
Like a soap or detergent, this cholic acid–glycine combination has a very polar "head" and a slightly polar "tail". The "tail" can dissolve nonpolar molecules and the polar "head" can carry the complex into polar media.
25-27
(a) sesquiterpene
OR
(b) monoterpene
O
OR
O
(c) sesquiterpene
OH
Me
Me
Me
Me
H
OR
OH
Me
Me
Me
Me
H

25-27 continued

(d) sesquiterpene

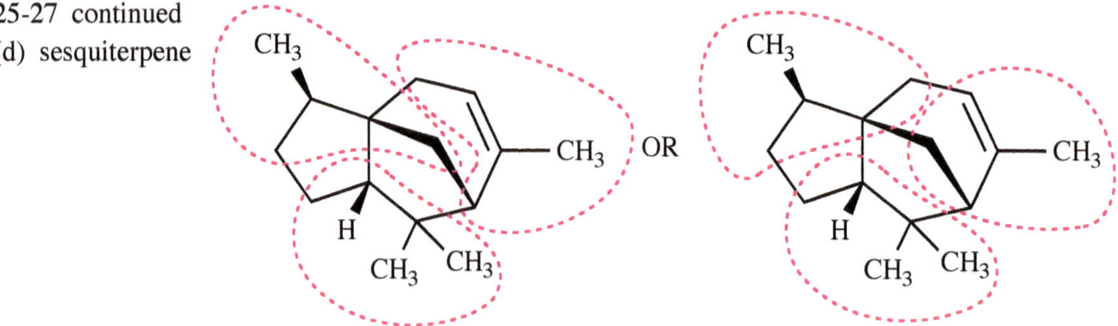

25-28 The formula $C_{18}H_{34}O_2$ has two elements of unsaturation; one is the carbonyl, so the other must be an alkene or a ring. Catalytic hydrogenation gives stearic acid, so the carbon cannot include a ring; it must contain an alkene. The products from $KMnO_4$ oxidation determine the location of the alkene:

$$HOOC(CH_2)_4COOH \;+\; HOOC(CH_2)_{10}CH_3 \;\Rightarrow\; HOOC(CH_2)_4CH{=}CH(CH_2)_{10}CH_3$$

If the alkene were *trans*, the coupling constant for the vinyl protons would be about 15 Hz; a 10 Hz coupling constant indicates a *cis* alkene. The 7 Hz coupling is from the vinyl Hs to the neighboring CH_2 groups.

COOH

petroselenic acid H H

J = 10 Hz

25-29

COOH linolenic acid

Ni ↓ 1 H_2

COOH linoleic acid—known

COOH NEW

COOH NEW

Ni ↓ 1 H_2

COOH NEW

COOH NEW

COOH oleic acid—known

In addition to the new isomers with different positions of the double bonds, there has been growing concern over the *trans* fatty acids created by isomerizing the naturally occurring *cis* double bonds. More manufacturers are now listing the percent of "*trans* fat" on their food labels.

25-30

(a) Of the two, only nepetalactone is a terpene. The other has only 8 carbons and terpenes must be in multiples of 5 carbons.

(b) Of the two, only the second is aromatic, as can be readily seen in one of the resonance forms.

like benzene like furan

(c) Each compound is cleaved with NaOH (aq) to give an enolate that tautomerizes to the more stable keto form.

NaOH H_2O enolate

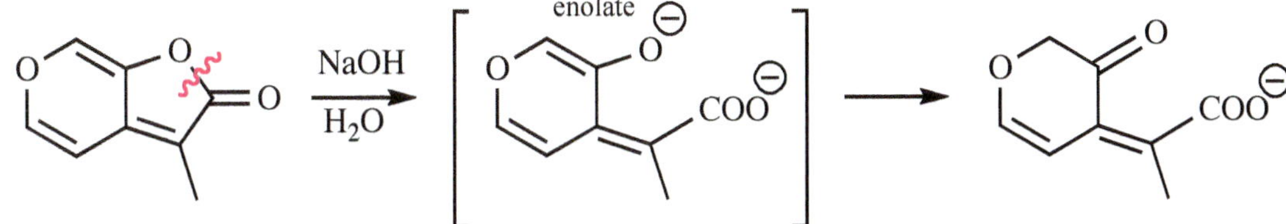

25-31

(a) Four of these components in Vicks Vapo-Rub® are terpenes. The only one that is not is decane; it is not composed of isoprene units despite having the correct number of carbons for a monoterpene.

Three of the four terpenes have had their isoprene units indicated in previous problems; in each of those three, there are two possible ways to assign the isoprene units, so those pictures will not be duplicated here.

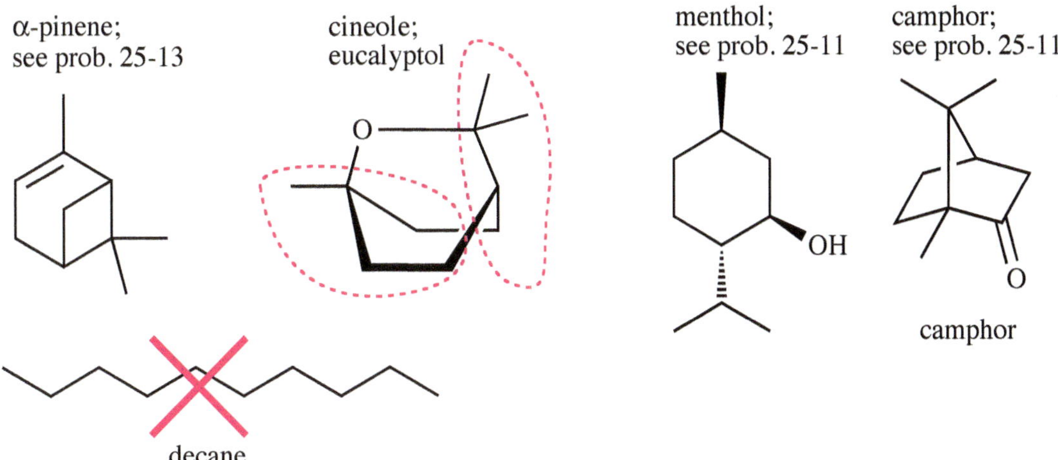

(b) Vicks Vapo-Rub® must be optically active, as it contains four optically active terpenes.

25-32

(a) High temperature and the diradical O_2 molecule strongly suggest a radical mechanism.

(b) Radical stability follows the order: benzylic > allylic > 3° > 2° > 1°. A radical at C-11 would be *doubly* allylic, making it a prime site for radical reaction.

(c)

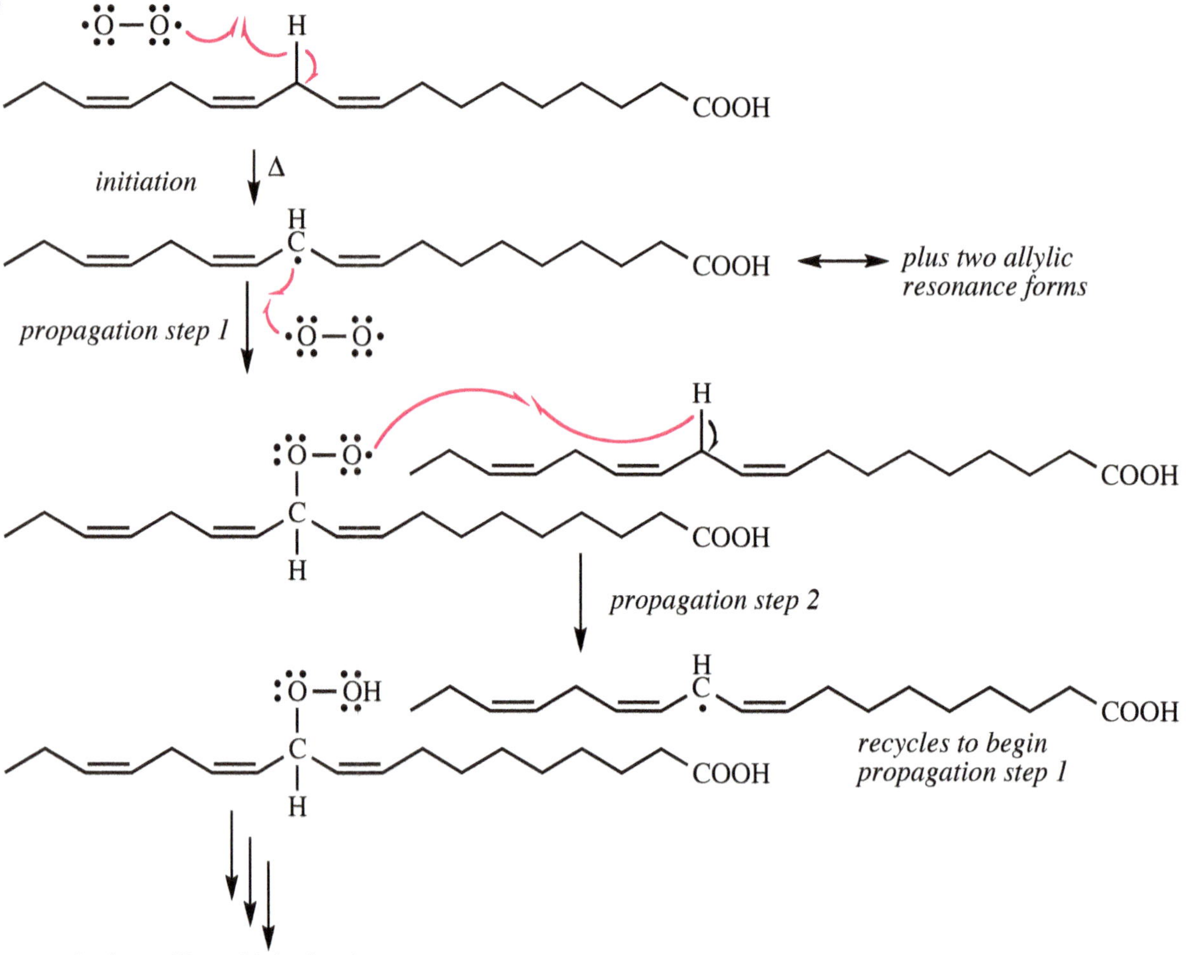

H

COOH

initiation Δ

H
C

COOH ⟷ *plus two allylic resonance forms*

propagation step 1

H

COOH

C

COOH

H

propagation step 2

OH

H
C

COOH

C

COOH

recycles to begin propagation step 1

H

foul-smelling aldehydes, ketones, and carboxylic acids = "rancidity"

(d) Antioxidants are molecules that stop the free radical chain mechanism. In each of the two cases here, abstraction of the phenolic H makes an oxygen radical that is highly stabilized by resonance, so stable that it does not continue the free radical chain process. It only takes a small amount of antioxidant to prevent the chain mechanism. Interestingly, in breakfast cereals, the antioxidant is usually put in the plastic bag that the cereal is packaged in, rather than in the food itself. BHA and BHT can also be used directly in food as there is no evidence that they are harmful to humans.

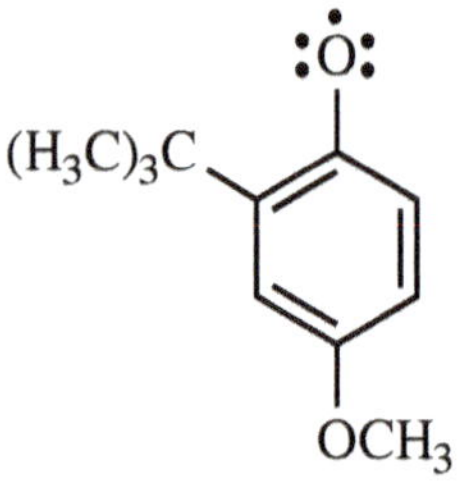

$(H_3C)_3C$ OCH_3

BHA radical

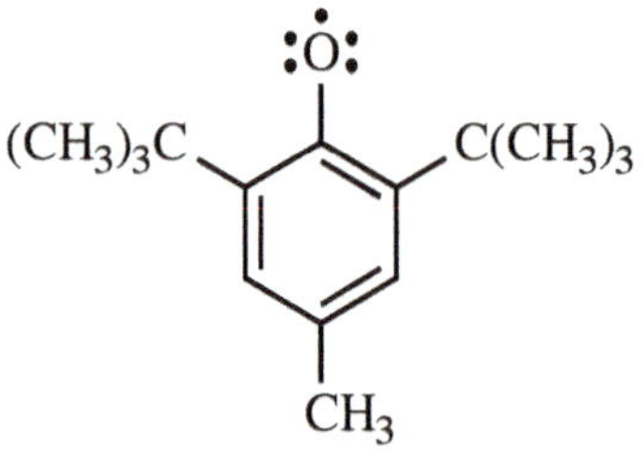

$(CH_3)_3C$ $C(CH_3)_3$ CH_3

BHT radical

CHAPTER 26—SYNTHETIC POLYMERS

Note: In this chapter, the "wavy bond" symbol means the continuation of a polymer chain.

26-1

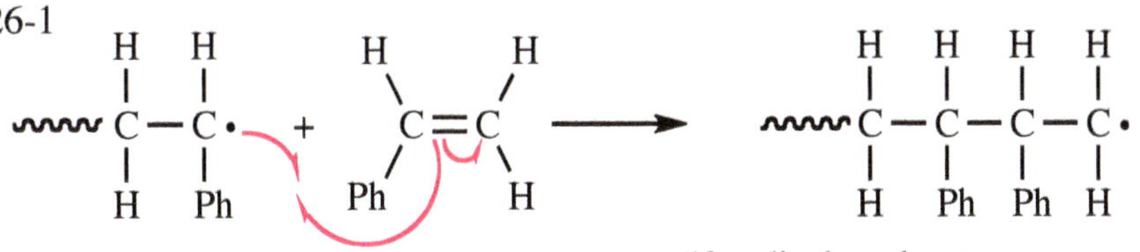

$$\text{~~~}CH_2\text{—}\dot{C}H(Ph) + CH(H)\text{=}C(H)(Ph) \longrightarrow \text{~~~}CH_2\text{—}CH(Ph)\text{—}CH(Ph)\text{—}\dot{C}H_2$$

1° radical, and *not* resonance-stabilized—
This orientation is not observed.

Orientation of addition typically generates the more stable intermediate; the energy difference between a 1° radical (shown above) and a benzylic radical is huge. Moreover, this energy difference accumulates with each repetition of the propagation step. The phenyl substituents must necessarily be on alternating carbons because the orientation of attack is always the same—not a random process.

26-2

$$PhCOO\text{—}OOCPh \xrightarrow{\Delta} 2\ Ph\bullet + 2\ CO_2$$

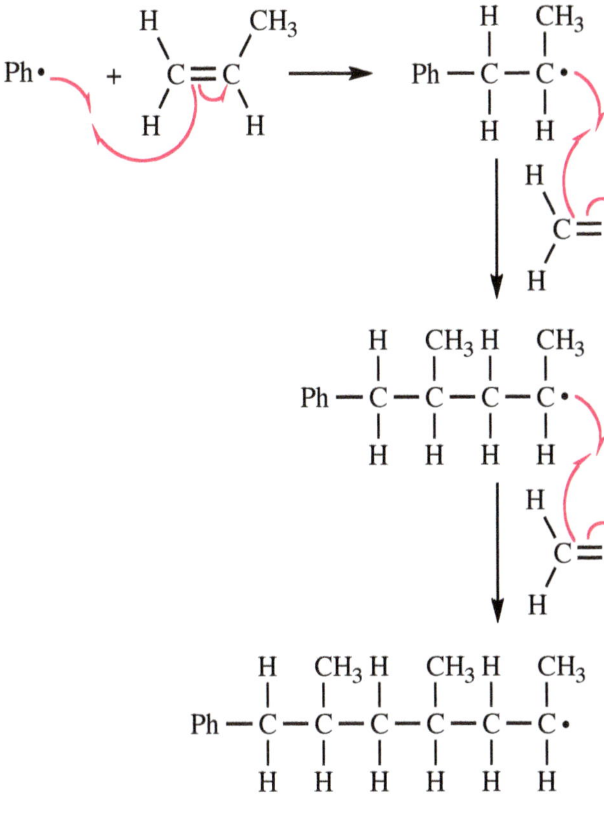

$$Ph\bullet + CH_2\text{=}CH(CH_3) \longrightarrow Ph\text{—}CH_2\text{—}\dot{C}H(CH_3)$$

$$\downarrow + CH_2\text{=}CH(CH_3)$$

$$Ph\text{—}CH_2\text{—}CH(CH_3)\text{—}CH_2\text{—}\dot{C}H(CH_3)$$

$$\downarrow + CH_2\text{=}CH(CH_3)$$

$$Ph\text{—}CH_2\text{—}CH(CH_3)\text{—}CH_2\text{—}CH(CH_3)\text{—}CH_2\text{—}\dot{C}H(CH_3) \quad \text{etc.}$$

26-3 The benzylic hydrogen will be abstracted in preference to a 2° hydrogen because the benzylic radical is both 3° and resonance-stabilized, and the 2° radical is neither.

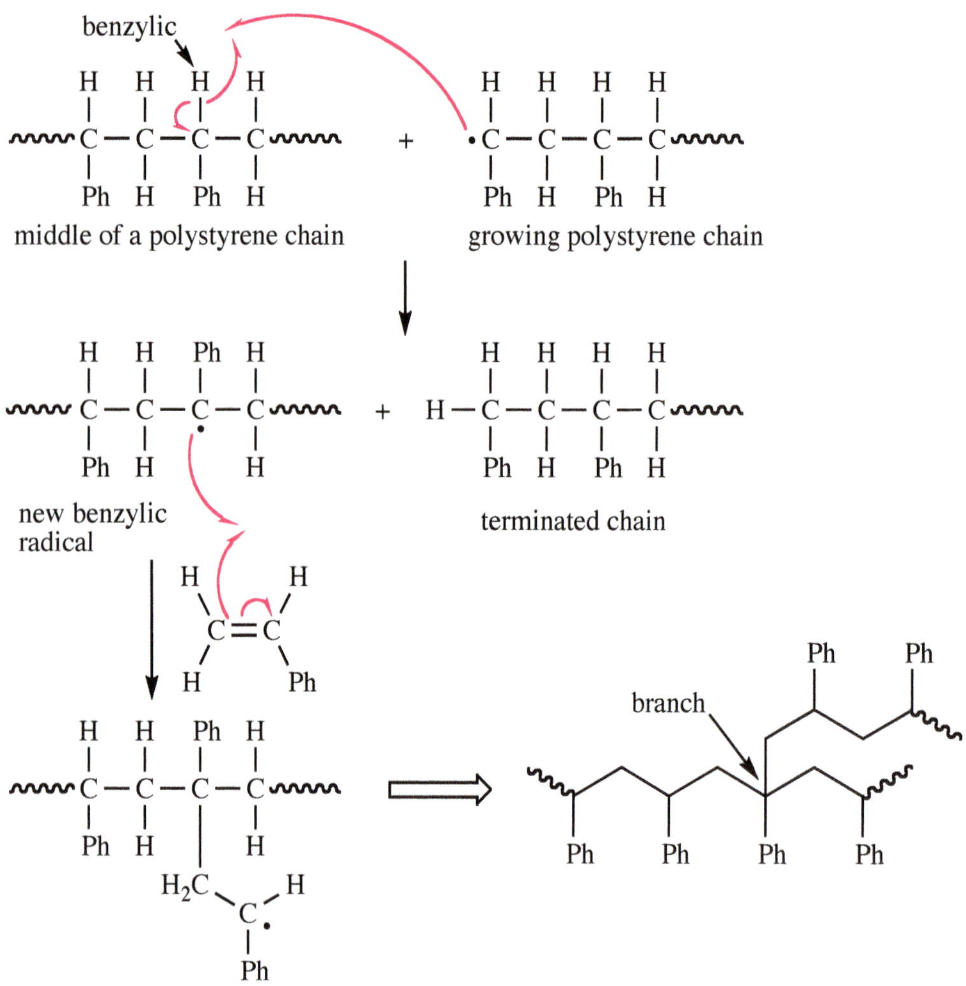

26-4 Addition occurs with the orientation giving the more stable intermediate. In the case of isobutylene, the growing chain will bond at the less substituted carbon to generate the more highly substituted carbocation.

3° carbocation—favored

OR

1° carbocation—disfavored (also more steric hindrance)

26-5

(a) Chlorine can stabilize a carbocation intermediate by resonance.

26-5 continued

(b) The —$OCOCH_3$ group can stabilize the carbocation intermediate by resonance.

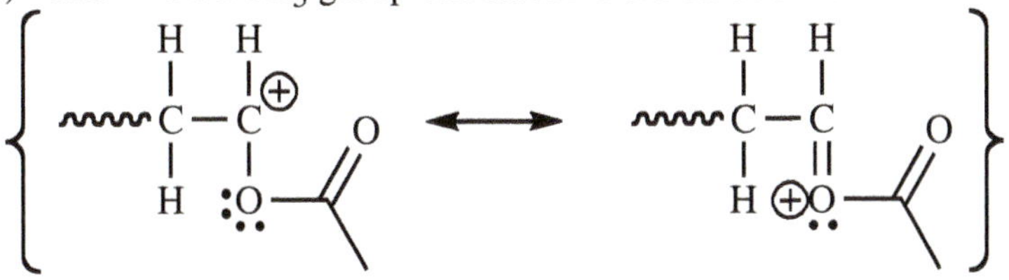

(c) Terrible for cationic polymerization: both substituents are electron-withdrawing and would *destabilize* the carbocation intermediate.

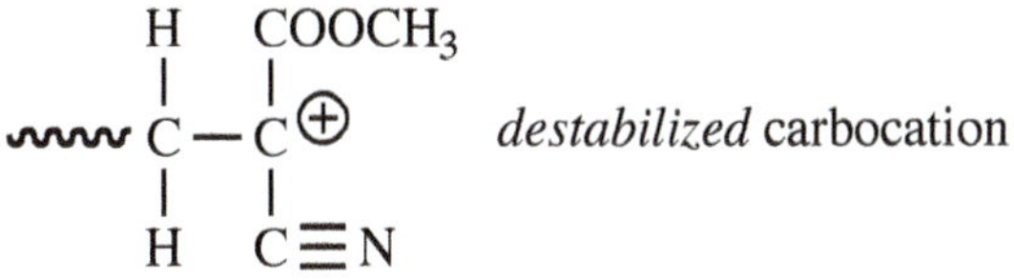

H COOCH$_3$

destabilized carbocation

26-6 benzylic

middle of a polystyrene chain

growing polystyrene chain

hydride transfer

terminated chain

new benzylic cation

branch

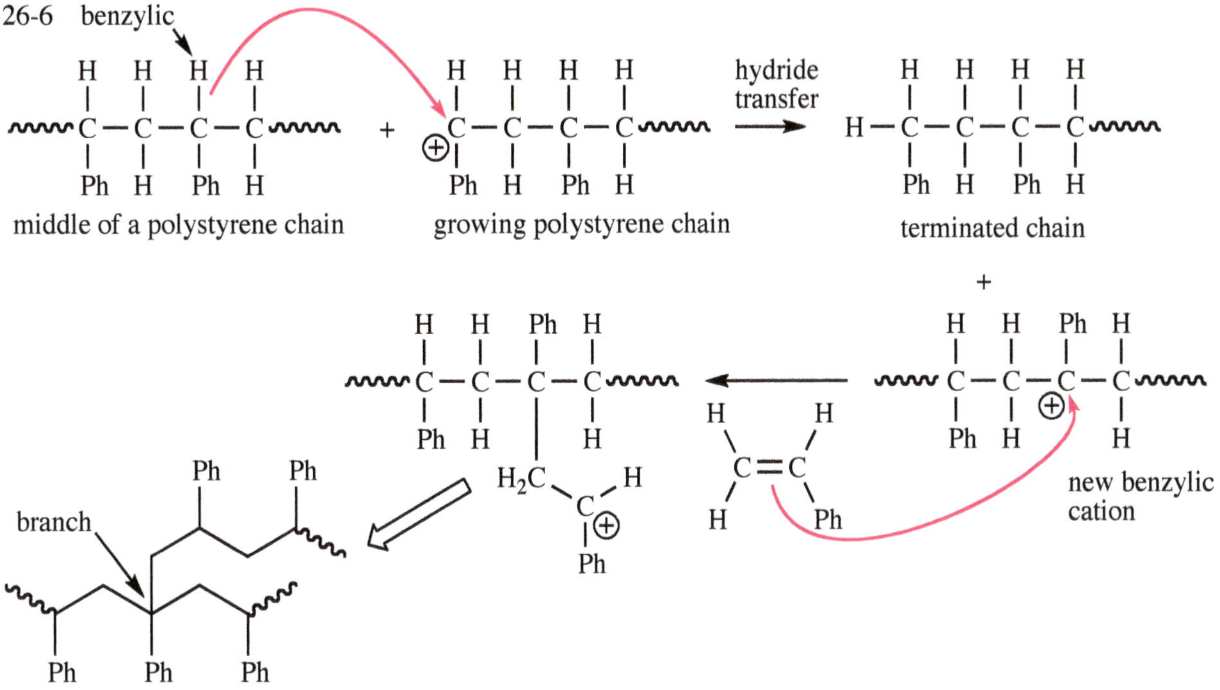

Polystyrene is particularly susceptible to branching because the 3° benzylic cation produced by a hydride transfer is so stable. In poly(isobutylene), there is no hydrogen on the carbon with the stabilizing substituents; any hydride transfer would generate a 2° carbocation at the expense of a 3° carbocation at the end of a growing chain—this is an increase in energy and therefore unfavorable.

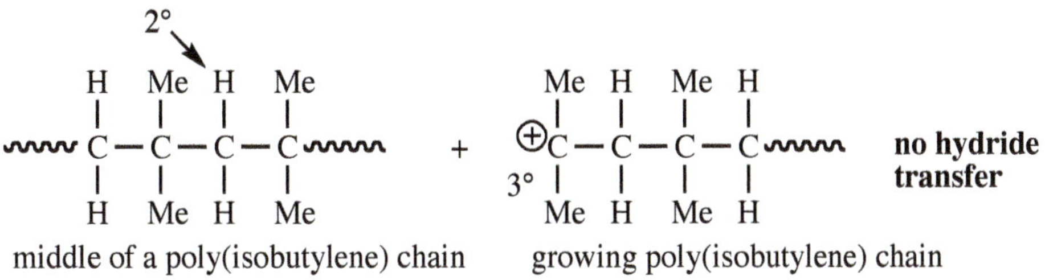

2°

3°

no hydride transfer

middle of a poly(isobutylene) chain

growing poly(isobutylene) chain

26-7

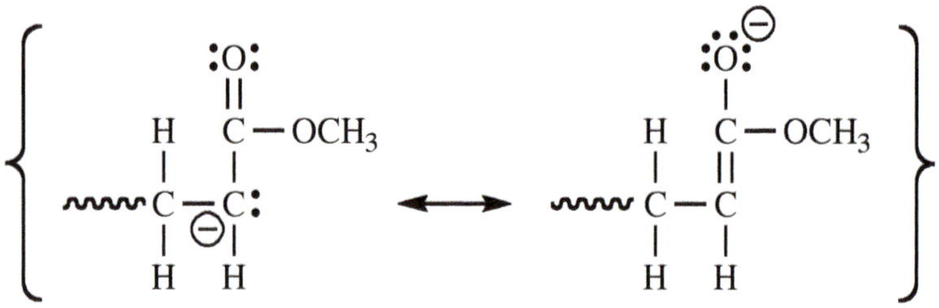

26-8

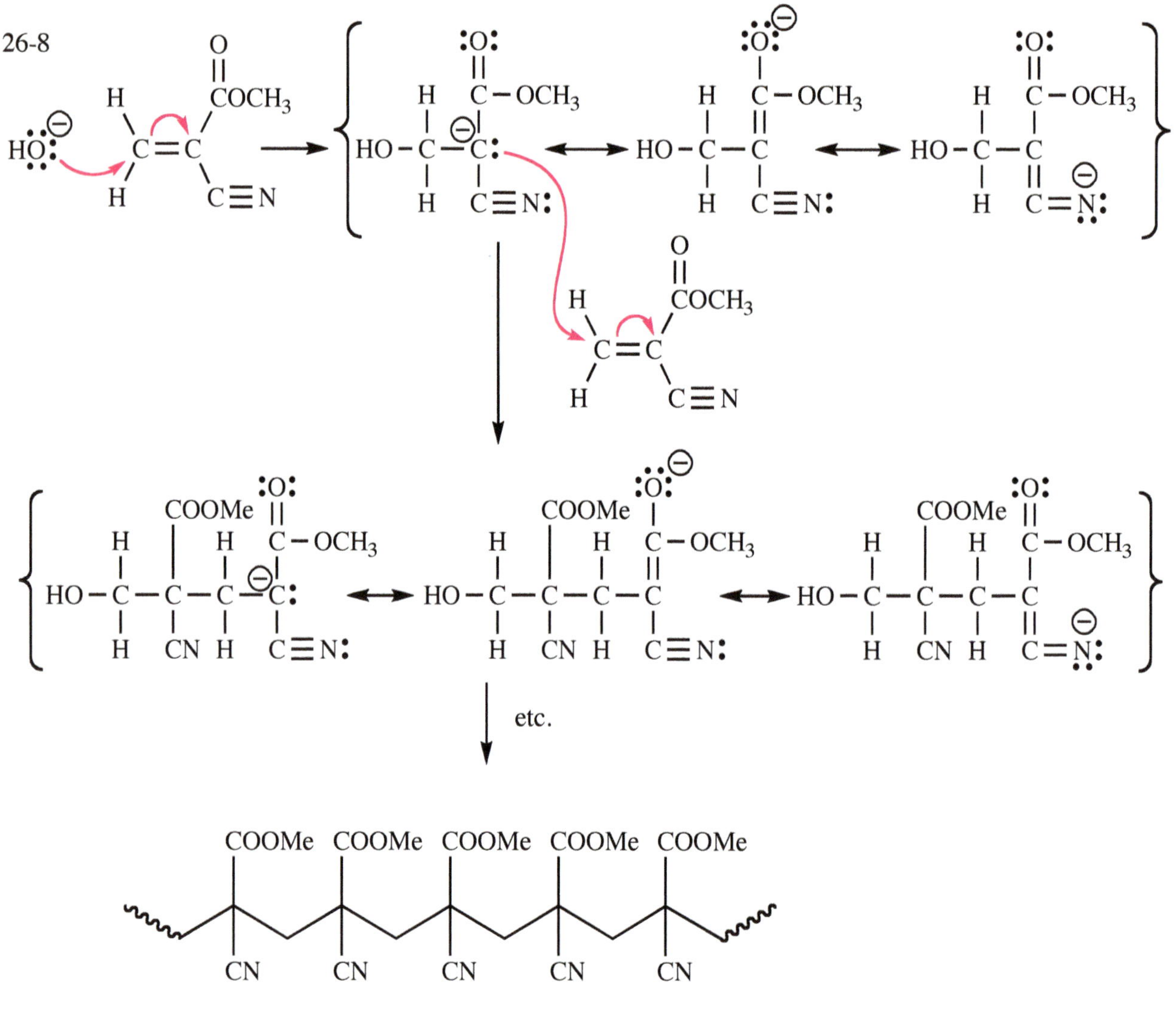

This polymerization goes so quickly because the anionic intermediate is highly resonance stabilized by the carbonyl and the cyano groups. A stable intermediate suggests a low activation energy, which translates to a fast reaction.

26-9

(a)

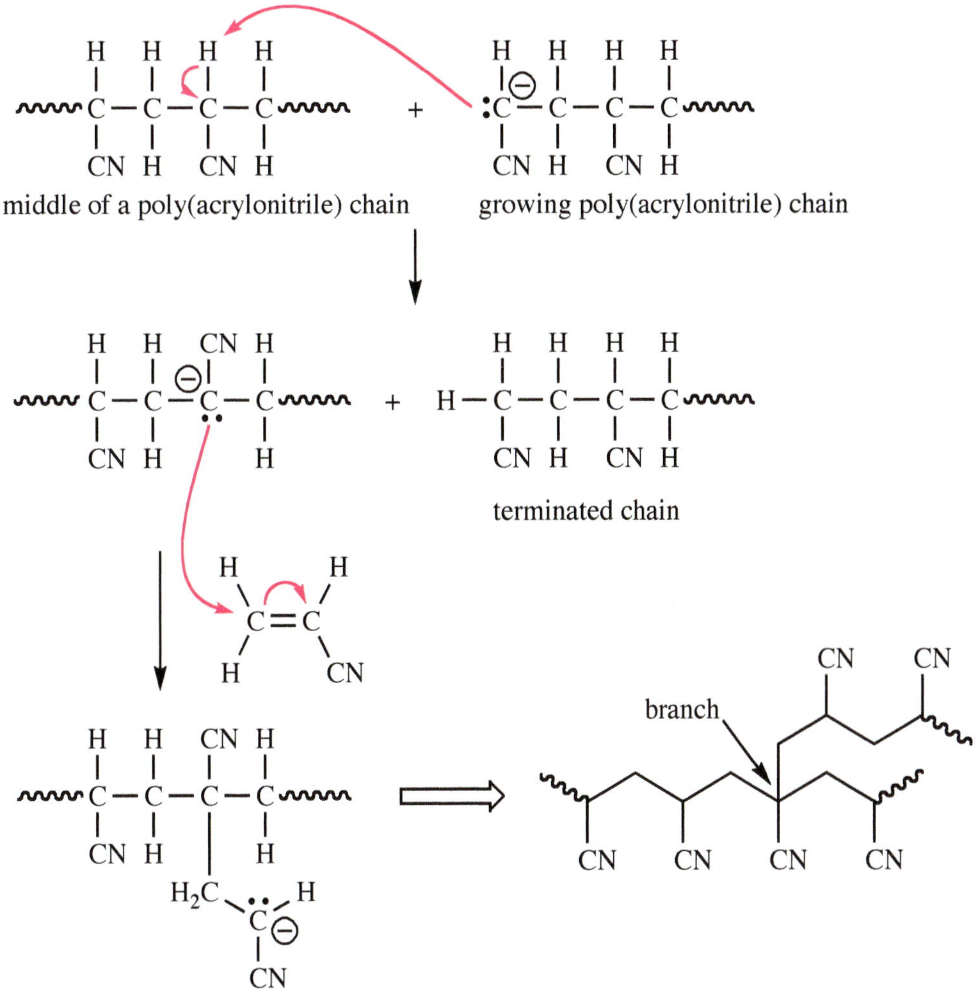

middle of a poly(acrylonitrile) chain

growing poly(acrylonitrile) chain

terminated chain

branch

(b) The chain-branching hydride transfer (from a cationic mechanism) or proton transfer (from an anionic mechanism) ends a less-highly-substituted end of a chain and generates an intermediate on a more-highly-substituted middle of a chain (a 3° carbon in these mechanisms). This stabilizes a carbocation, but greater substitution *destabilizes* a carbanion. Branching can and does happen in anionic mechanisms, but it is less likely than in cationic mechanisms.

26-10 isotactic poly(acrylonitrile)

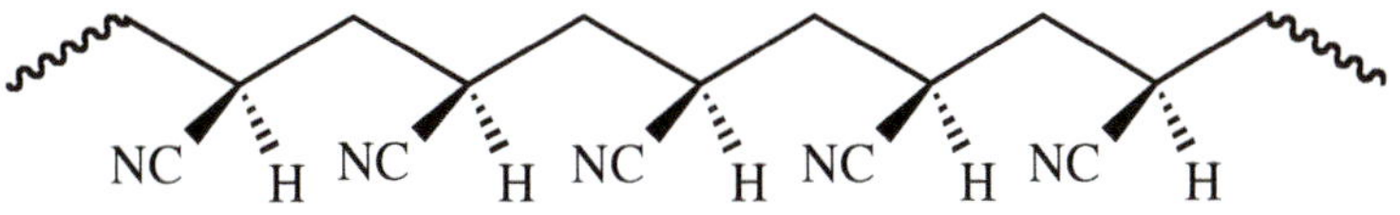

syndiotactic polystyrene

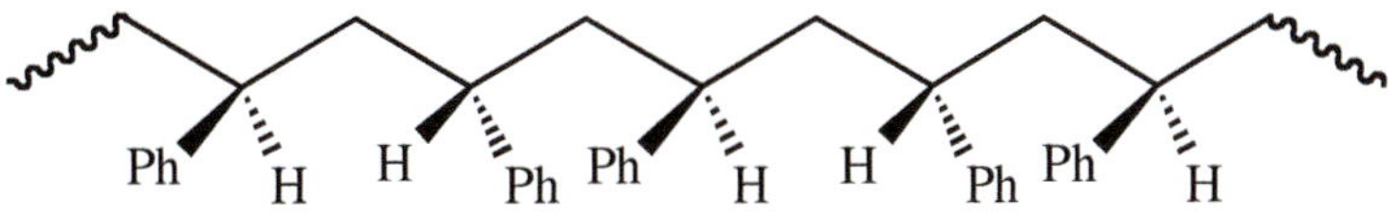

26-11

(a) all *trans*

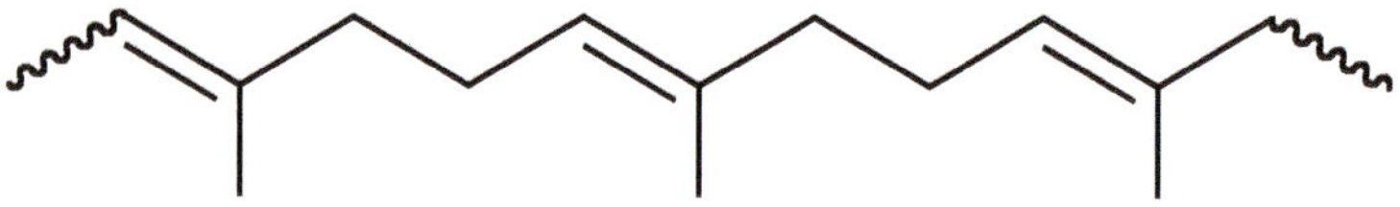

(b) The *trans* double bonds in gutta-percha allow for more ordered packing of the chains, that is, a higher degree of crystallinity. (Recall how *cis* double bonds in fats and oils lower the melting points because the *cis* orientation disrupts the ordering of the packing of the chains.) The more crystalline a polymer is, the less elastic it is.

26-12 Whether the alkene is *cis* or *trans* is not specified.

(a) isobutylene isoprene

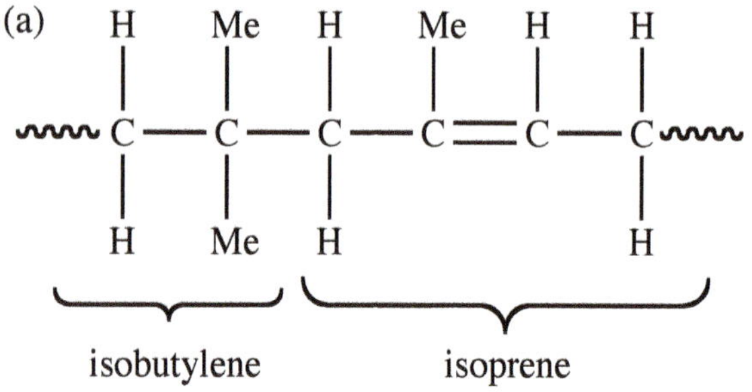

(b) styrene butadiene

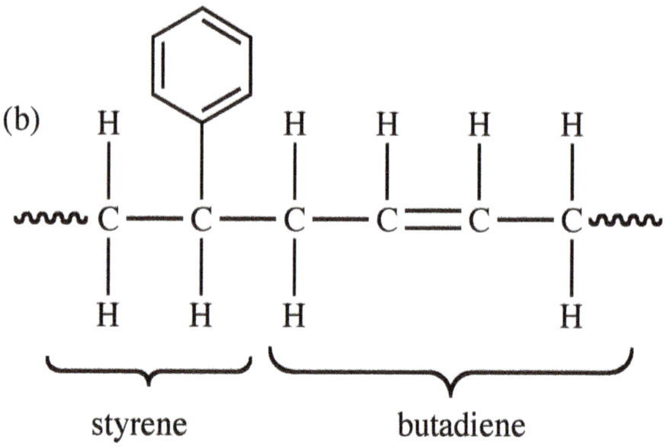

26-13 The repeating unit in each polymer is boxed.

(a) Nomex®

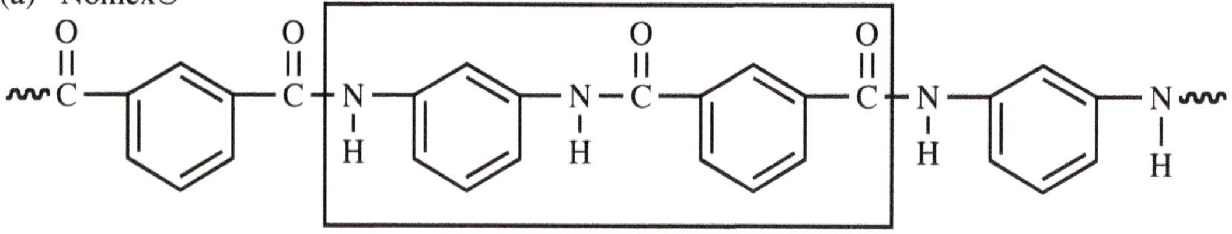

(b) Kevlar®

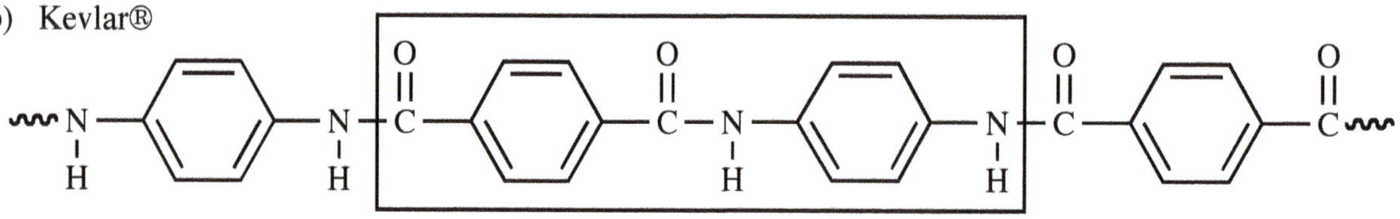

26-14 Kodel® polyester (only one repeating unit shown)

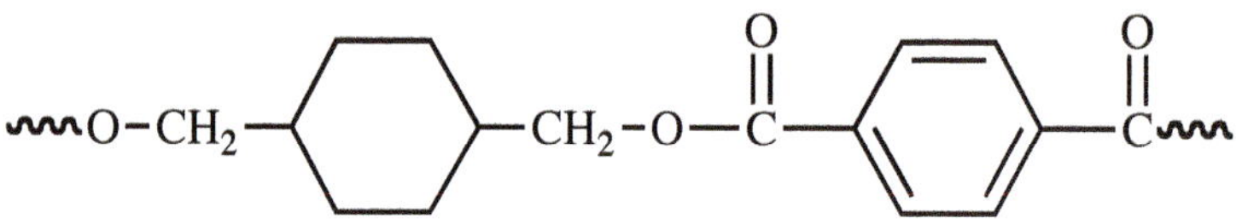

26-15

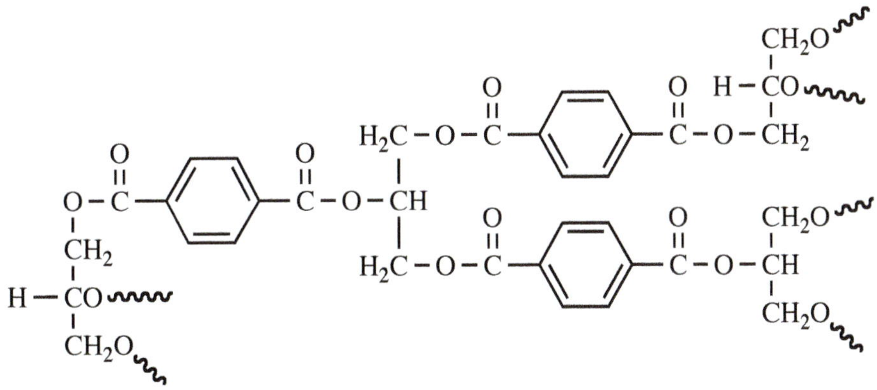

Glycerol is a trifunctional molecule, so not only does it grow in two directions to make a chain, it grows in three directions. All of its chains are cross-linked, forming a three-dimensional lattice with very little motion possible. The more cross-linked the polymer is, the more rigid it is.

26-16 For simplicity in this problem, bisphenol A will be abbreviated as a substituted phenol.

(a)

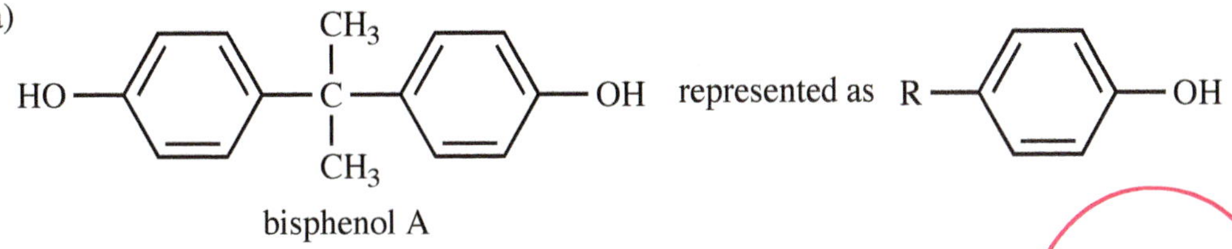

bisphenol A represented as R—C6H4—OH

mechanism

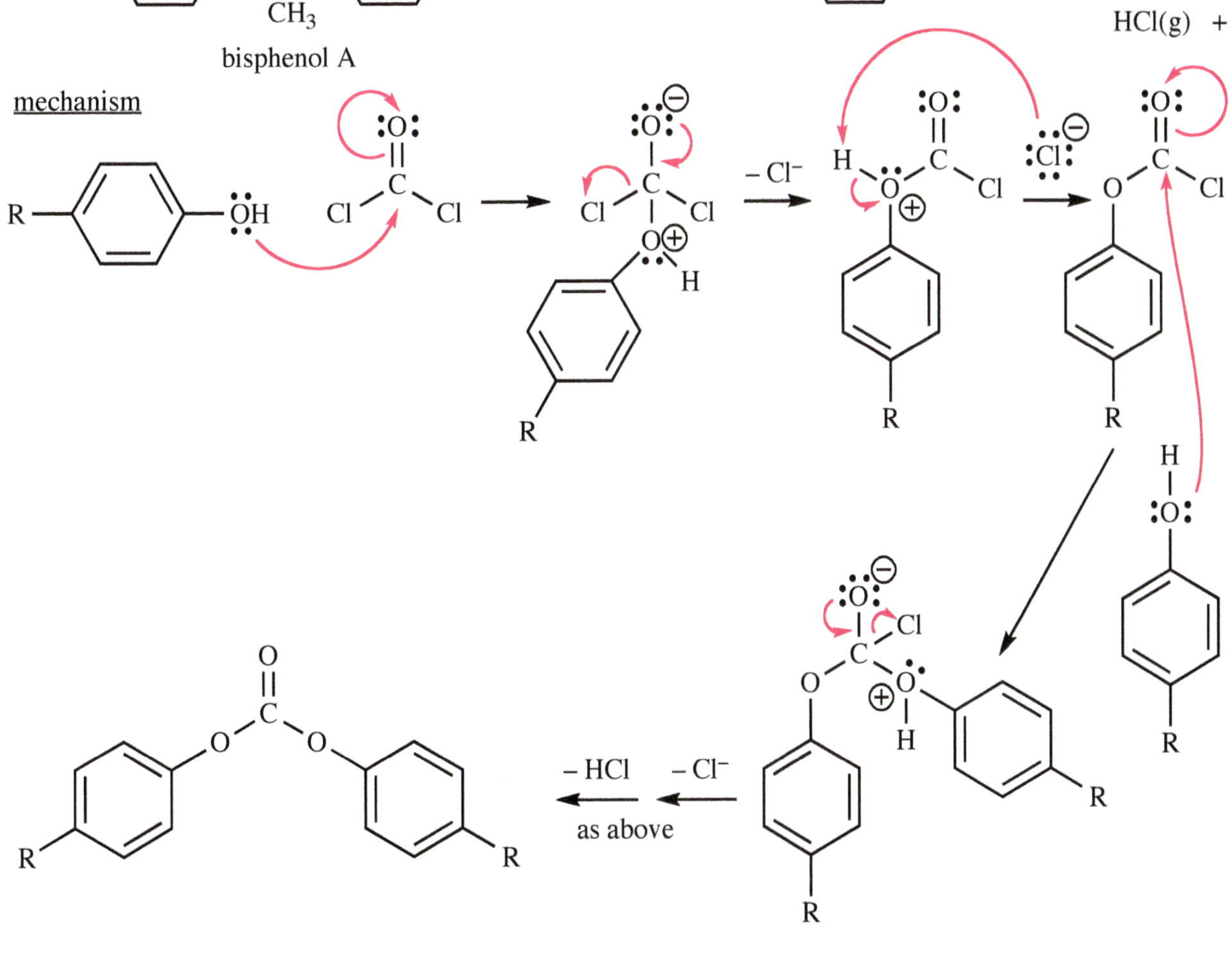

$HCl(g)$ +

– Cl⁻

– HCl – Cl⁻ as above

26-16 continued

(b)

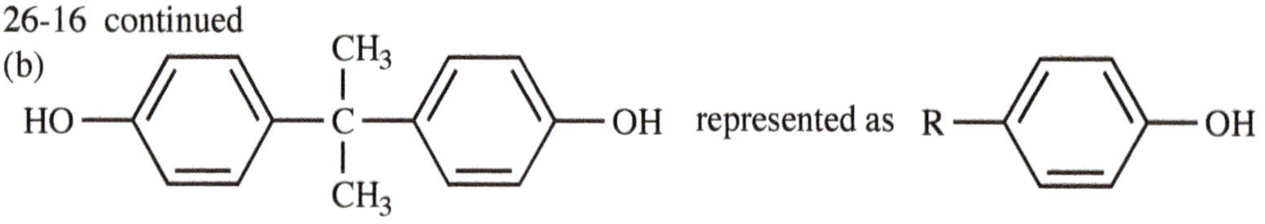

HO—C(CH3)2—OH represented as R—C6H4—OH

bisphenol A

R—C6H4—O—H + EtO⁻ ⇌ R—C6H4—O⁻ + EtOH

mechanism

R—C6H4—O⁻ + EtO—C(=O)—OEt → EtO—C(O⁻)(OEt)—O—C6H4—R → (− EtO⁻) → ArO—C(=O)—OEt

− EtO⁻

R

R

R

R

R

R

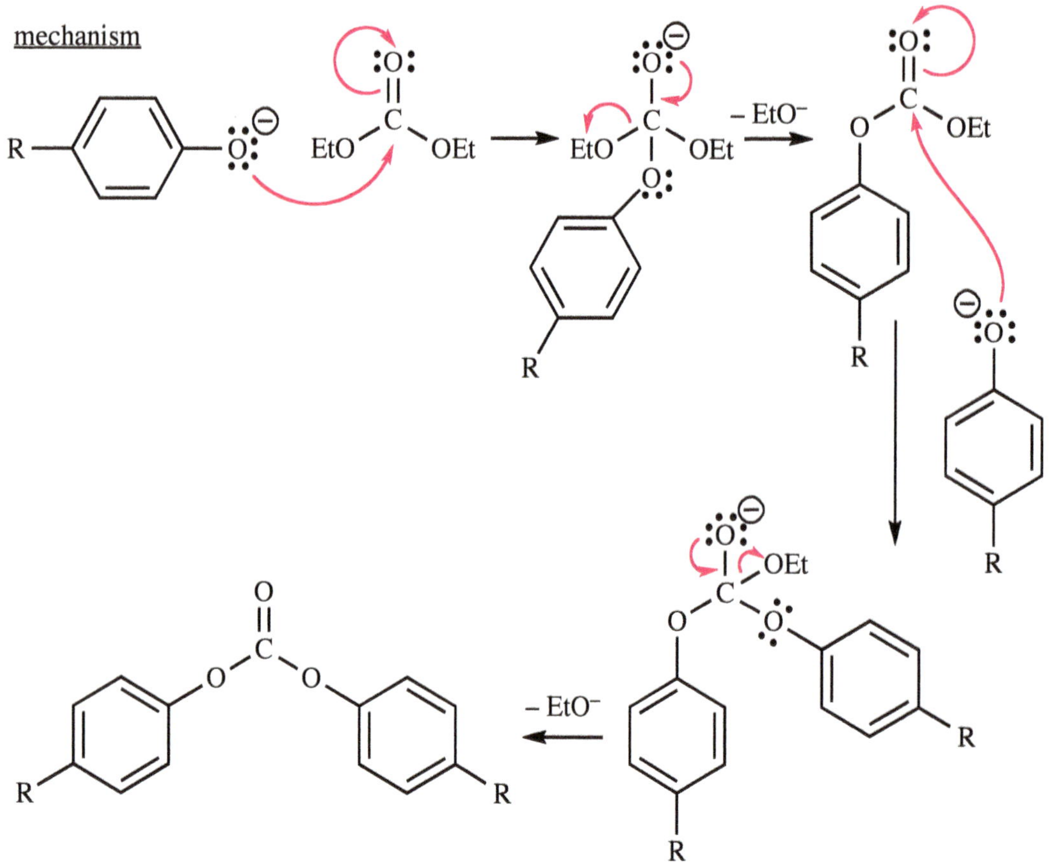

When ethoxide leaves from diethyl carbonate, it immediately deprotonates another phenol, generating ethanol as the small molecule produced in this condensation.

26-17 Bisphenol A is made by condensing two molecules of phenol with one molecule of acetone, with the loss of a molecule of water. This is an electrophilic aromatic substitution (more specifically, a Friedel-Crafts alkylation), and would require an acid catalyst to generate the carbocation. While a Lewis acid could be used, the mechanism below shows a protic acid.

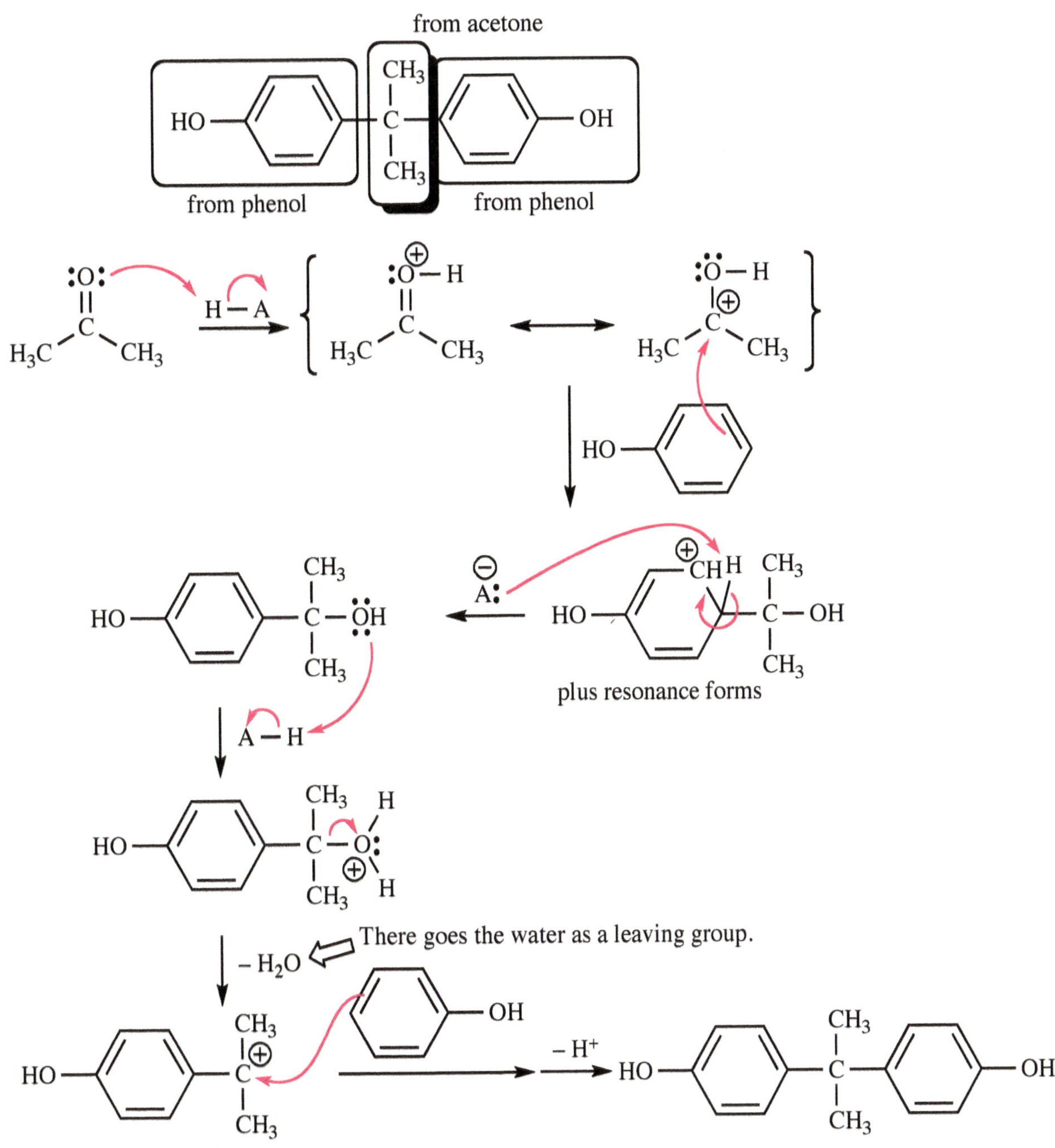

26-18

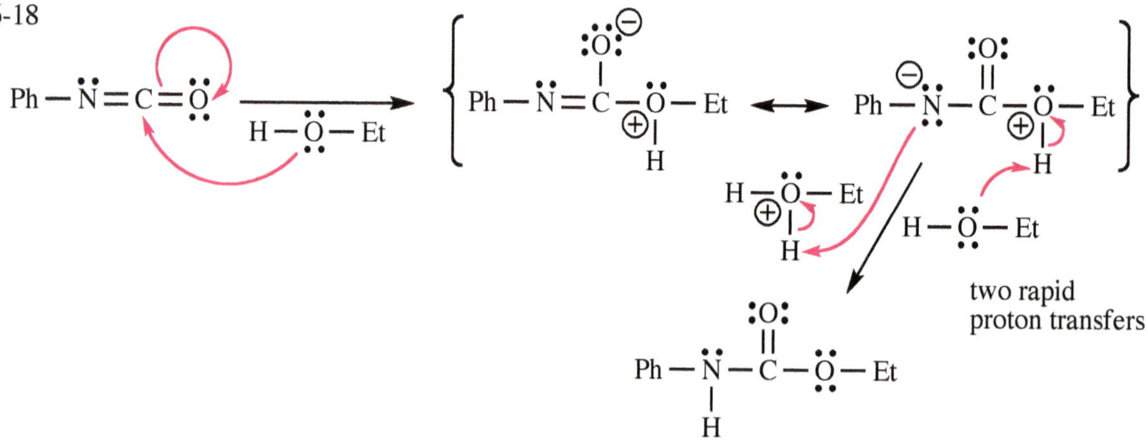

$$Ph-\ddot{N}=C=\ddot{O}: \xrightarrow{H-\ddot{O}-Et} \left\{ Ph-\ddot{N}=C(-O^-)-\overset{+}{O}(H)-Et \longleftrightarrow Ph-\ddot{N}^--C(=O)-\overset{+}{O}(H)-Et \right\}$$

two rapid proton transfers

$$Ph-\ddot{N}(H)-C(=O)-\ddot{O}-Et$$

26-19 Glycerol is a trifunctional alcohol. It uses two of its OH groups in a growing chain. The third OH group cross-links with another chain. The more cross-linked a polymer, the more rigid it is.

26-20

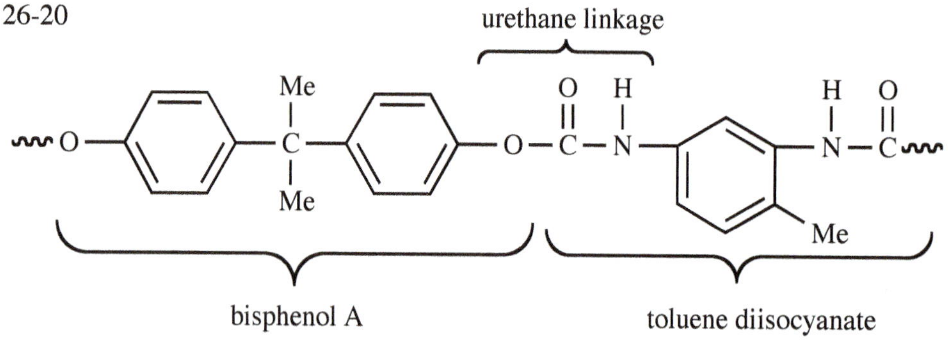

urethane linkage

$$\sim O-C_6H_4-C(Me)_2-C_6H_4-O-C(=O)-N(H)-C_6H_3(Me)-N(H)-C(=O)\sim$$

bisphenol A — toluene diisocyanate

26-21

(a)

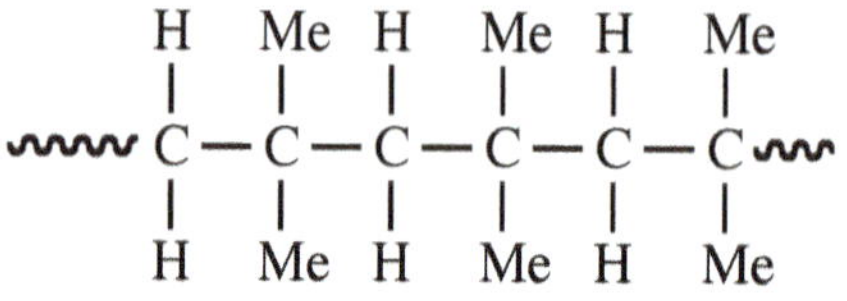

$$\sim C(H)(H)-C(Me)(Me)-C(H)(H)-C(Me)(Me)-C(H)(H)-C(Me)(Me)\sim$$

(b) Polyisobutylene is an addition polymer. No small molecule is lost, so this cannot be a condensation polymer.

(c) Either cationic polymerization or free-radical polymerization would be appropriate. The carbocation or free-radical intermediate would be 3° and therefore relatively stable. Anionic polymerization would be inappropriate as there is no electron-withdrawing group to stabilize the anion.

26-22

(a) Polychloroprene (Neoprene®) is an addition polymer.

(b) Polychloroprene comes from the diene, chloroprene, just as natural rubber comes from isoprene:

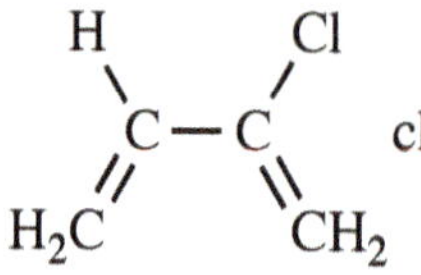

$$H_2C=C(H)-C(Cl)=CH_2$$

chloroprene

26-23

(a) It is a polyurethane.

(b) As with all polyurethanes, it is a step-growth polymer.

26-23 continued

(c) $\sim\sim CH_2CH_2CH_2-N(H)-C(=O)-O\sim\sim \xrightarrow{H_2O} HOCH_2CH_2CH_2NH_2 + CO_2$

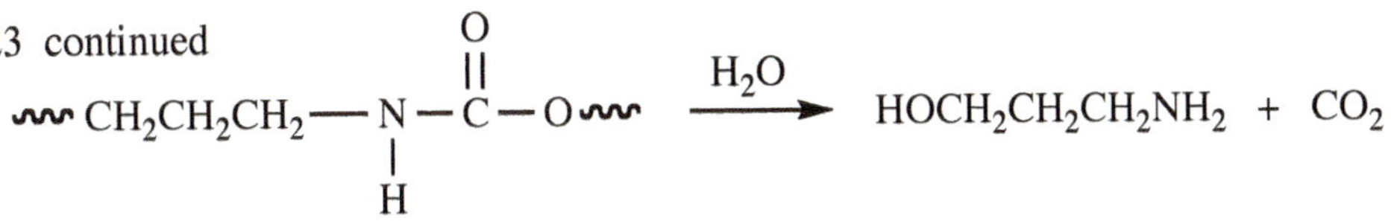

26-24

(a) It is a polyester.

(b) As with all polyesters, it is a step-growth (condensation) polymer.

(c)

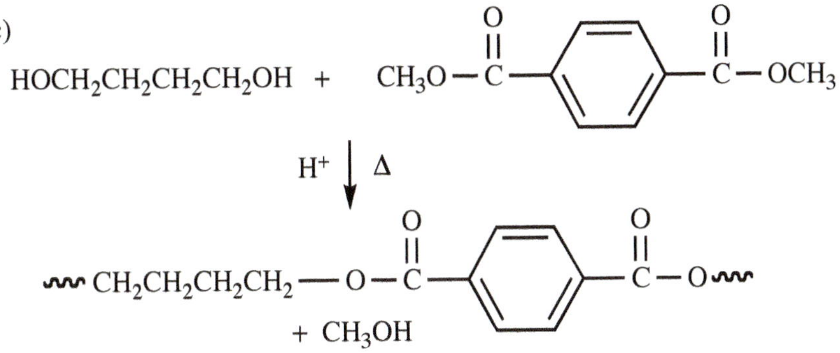

$HOCH_2CH_2CH_2CH_2OH + CH_3O-C(=O)-C_6H_4-C(=O)-OCH_3$

$\downarrow$ H^+, Δ

$\sim\sim CH_2CH_2CH_2CH_2-O-C(=O)-C_6H_4-C(=O)-O\sim\sim$

$+ CH_3OH$

Using the dicarboxylic acid instead of the ester would produce water as the small neutral molecule lost in this condensation.

26-25

(a) Urylon® is a polyurea.

(b) A polyurea is a step-growth polymer.

(c) $\sim\sim (CH_2)_9-N(H)-C(=O)-N(H)\sim\sim \xrightarrow{H_2O} H_2N(CH_2)_9NH_2 + CO_2$

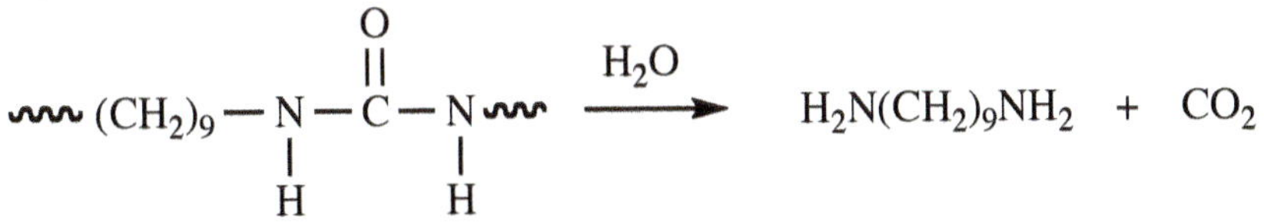

26-26

(a) Polyethylene glycol, abbreviated PEG, is a polyether.

(b) PEG is usually made from ethylene oxide (first reaction shown). In theory, PEG could also be made by intermolecular dehydration of ethylene glycol (second reaction shown), but the yields are low and the chains are short.

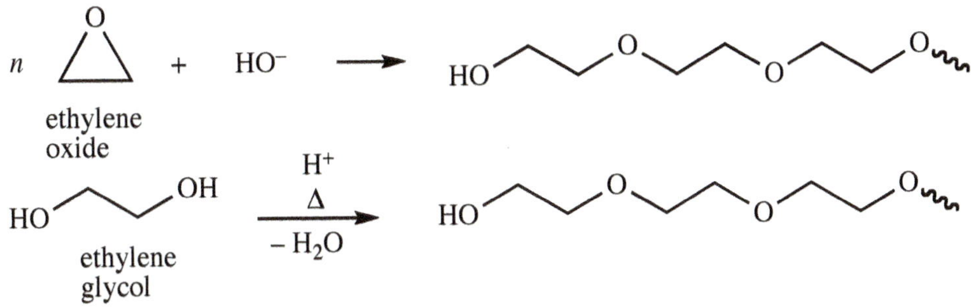

n (ethylene oxide) + HO^- $\longrightarrow$ $HO-CH_2CH_2-O-CH_2CH_2-O-CH_2CH_2-O\sim\sim$

ethylene oxide

$HOCH_2CH_2OH$ (ethylene glycol) $\xrightarrow[-\ H_2O]{H^+,\ \Delta}$ $HO-CH_2CH_2-O-CH_2CH_2-O-CH_2CH_2-O\sim\sim$

ethylene glycol

(c) Basic catalysts are most likely as they open the epoxide to generate a new nucleophile. Acid catalysts are possible but they risk dehydration and ether cleavage.

26-26 continued

(d) Mechanism of ethylene oxide polymerization (showing hydroxide as the base):

etc.

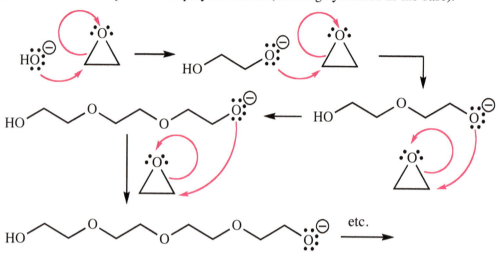

26-27

The key to determining the starting monomer for a ring-opening metathesis polymer (ROMP) is to "reconnect" the two carbons of the repeating unit. This process is similar to determining the starting material in an ozonolysis problem, where the two new C=O were reconnected as an alkene.

(a) 4 C in repeating unit ⇨

(a) cyclohexane plus 2 C in repeating unit ⇨

redraw

bicyclo[2.2.2]octene

26-28

(a) ∿ CH_2—O—CH_2—O—CH_2—O—CH_2—O ∿ Delrin® (polyformaldehyde)

(b) All of these intermediates are resonance-stabilized.

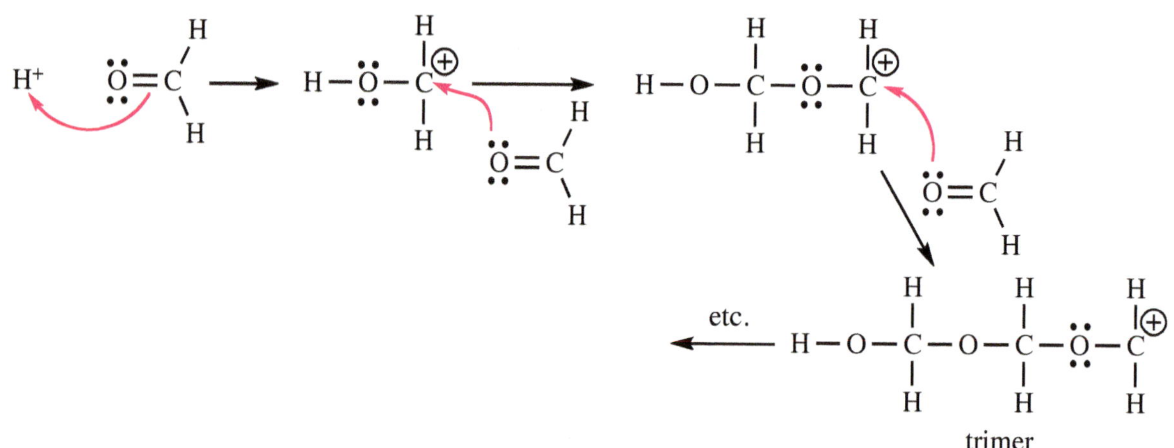

(c) Delrin® is an addition polymer; instead of adding across the double bond of an alkene, addition occurs across the double bond of a carbonyl group.

26-29

(a) *cis*

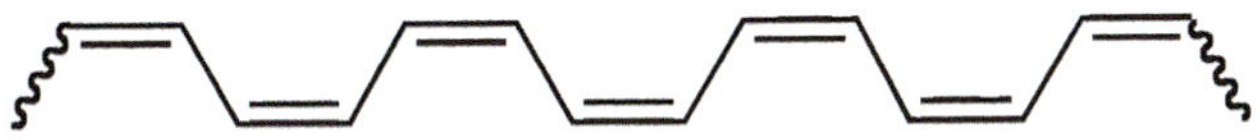

trans

(b) Each structure has a fully conjugated chain. It is reasonable to expect electrons to be able to be transferred through the π system, just as resonance effects can work over long distances through conjugated systems.

(c) It is not surprising that the conductivity is directional. Electrons must flow along the π system of the chain, so if the chains were aligned, conductivity would be greater in the direction parallel to the polymer chains. (It is possible, though less likely, that electrons could pass from the π system of one chain to the π system of another, that is, perpendicular to the direction of the chain; we would expect reduced conductivity in that direction.)

26-30

(a) A Nylon is a polyamide. Amides can be hydrolyzed in aqueous acid, cleaving the polymer chain in the process.

$$\sim N(H)-C(=O)\sim C(=O)-N(H)\sim \xrightarrow{H_3O^+} \sim \overset{\oplus}{N}H_3 + HO-C(=O)\sim C(=O)-OH + H_3\overset{\oplus}{N}\sim$$

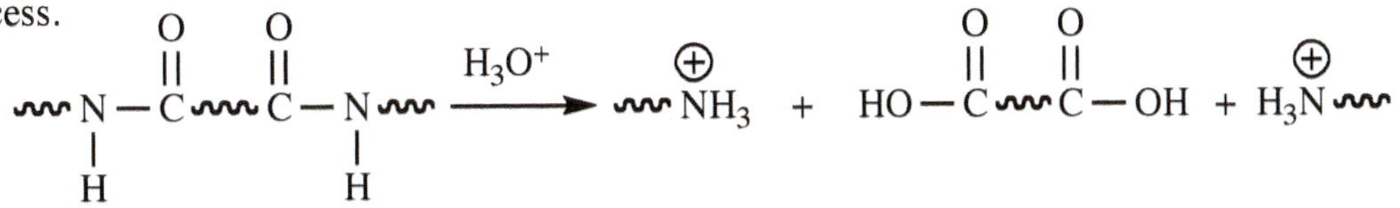

(b) A polyester can be saponified in aqueous base, cleaving the polymer chain in the process.

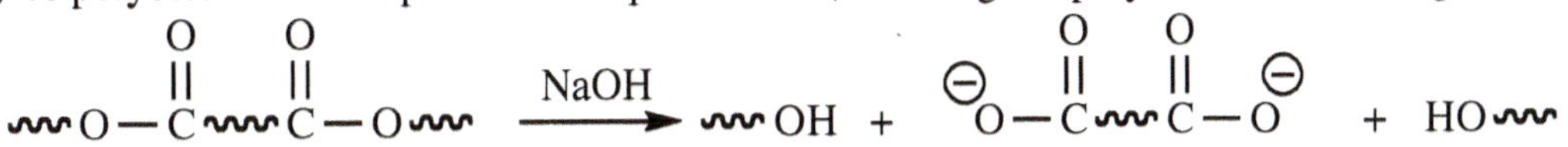

$$\sim O-C(=O)\sim C(=O)-O\sim \xrightarrow{NaOH} \sim OH + {}^{\ominus}O-C(=O)\sim C(=O)-O^{\ominus} + HO\sim$$

26-31

(a)

$$\sim[CH_2-CH(OC(=O)CH_3)]_3\sim \xrightarrow[H^+ \text{ or } HO^-]{H_2O} \sim[CH_2-CH(OH)]_3\sim$$

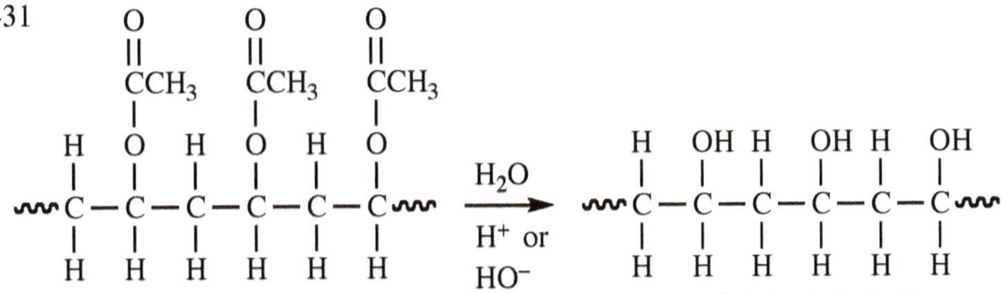

poly(vinyl acetate) poly(vinyl alcohol)

(b) A polyester is a condensation polymer in which monomer units are linked through ester groups as part of the polymer chain. Poly(vinyl acetate) is really a substituted polyethylene, an **addition** polymer, with only carbons in the chain; the ester groups are in the side chains, not in the polymer backbone.

(c) Hydrolysis of the esters in poly(vinyl acetate) does not affect the chain because the ester groups do not occur in the chain as they do in Dacron®.

(d) Vinyl alcohol cannot be polymerized because it is unstable, tautomerizing to acetaldehyde.

$$H_2C=CH(OH) \rightleftharpoons CH_3-CH(=O)$$

26-32

(a)

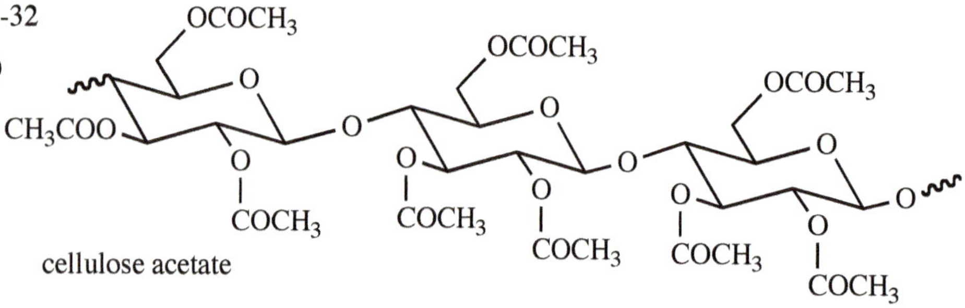

OCOCH3
OCOCH3
OCOCH3
CH3COO
O
COCH3
COCH3
COCH3
COCH3
COCH3

cellulose acetate

(b) Cellulose has three OH groups per glucose monomer, which form hydrogen bonds with other polar groups. Transforming these OH groups into acetates makes the polymer much less polar and therefore more soluble in organic solvents.
(c) The acetone dissolved the cellulose acetate in the fibers. As the acetone evaporated, the cellulose acetate remained but no longer had the fibrous, woven structure of cloth. It recrystallized as white fluff.
(d) Any article of clothing made from synthetic fibers is susceptible to the ravages of organic solvents. The structure of the shoe may disintegrate, and the solvent may penetrate more quickly.

26-33

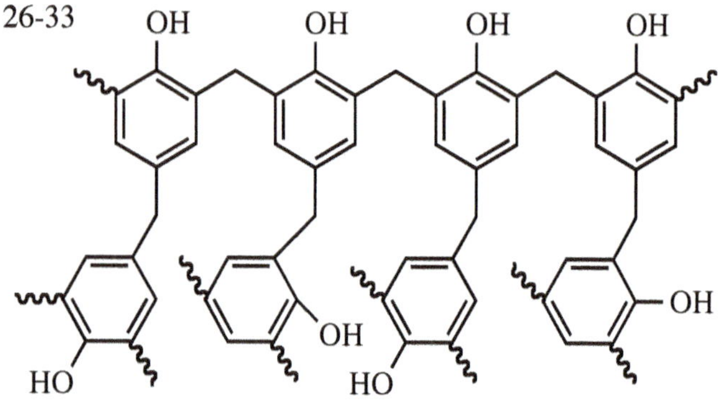

OH
HO

Bakelite® is highly cross-linked through the ortho and para positions of phenol; each phenol can form a chain at two ring positions, then form a branch at the third position.

mechanism

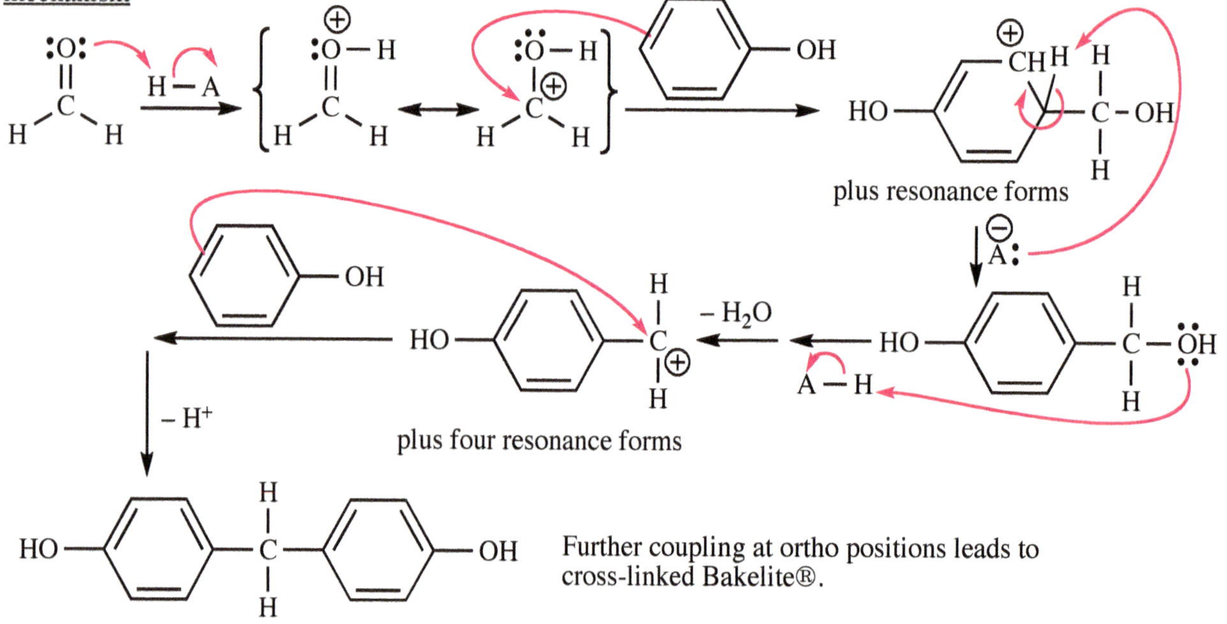

H—A
OH
HO
plus resonance forms
– H2O
A—H
plus four resonance forms
– H+

Further coupling at ortho positions leads to cross-linked Bakelite®.

26-34

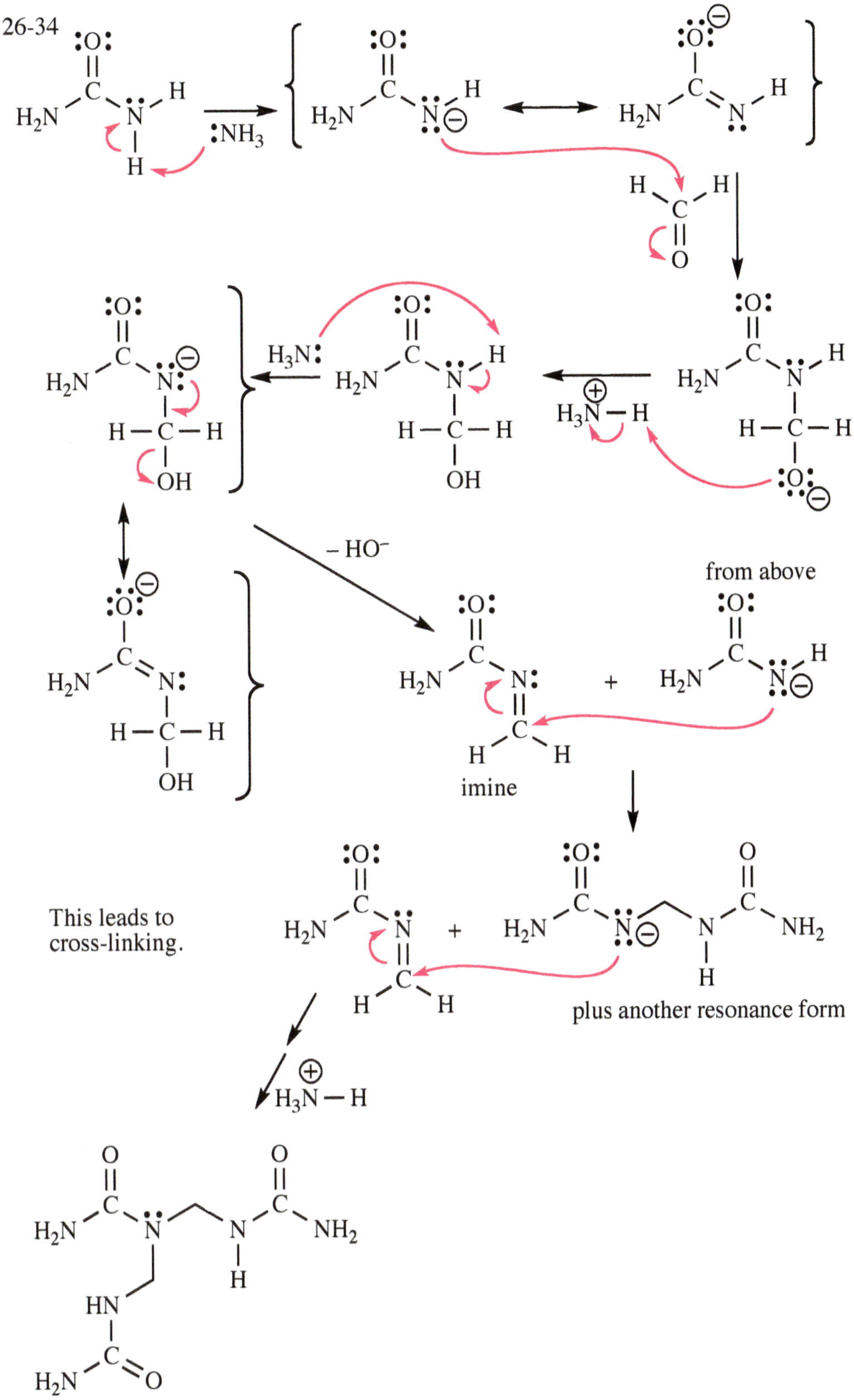

H₂N–C(=O)–NH₂ + :NH₃ → {H₂N–C(=O)–NH⁻ ↔ H₂N–C(–O⁻)=N–H}

H₂C=O

H₂N–C(=O)–NH–CH₂–O⁻ + H₃N⁺–H → H₂N–C(=O)–NH–CH₂–OH + H₃N:

{H₂N–C(=O)–N⁻–CH₂–OH ↔ H₂N–C(–O⁻)=N–CH₂–OH}

– HO⁻

from above

H₂N–C(=O)–N=CH₂ + H₂N–C(=O)–NH⁻

imine

This leads to cross-linking.

H₂N–C(=O)–N=CH₂ + H₂N–C(=O)–N⁻–CH₂–NH–C(=O)–NH₂

plus another resonance form

H₃N⁺–H

H₂N–C(=O)–N(–CH₂–NH–C(=O)–NH₂)–CH₂–NH–C(=O)–NH₂

26-35

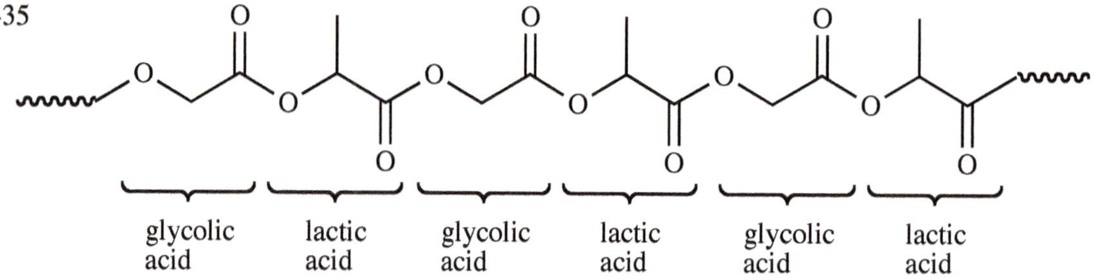

glycolic acid | lactic acid | glycolic acid | lactic acid | glycolic acid | lactic acid

26-36

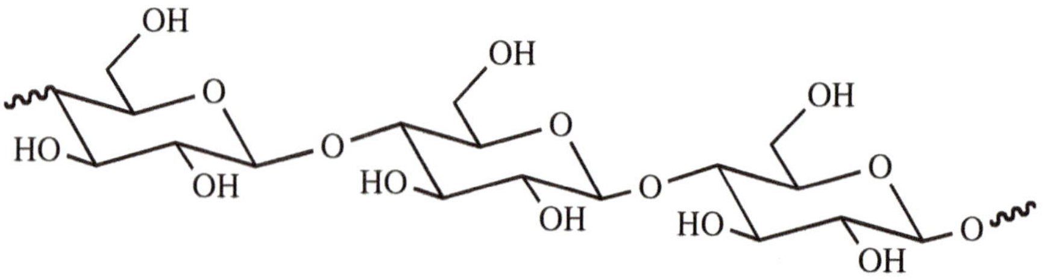

cellulose = cotton

polypropylene

As we have seen repeatedly through this presentation of organic chemistry, physical and chemical behavior depend on *structure*. The structure of cotton, i.e., cellulose, has multiple oxygen atoms that form hydrogen bonds with water. When cotton gets wet, it holds onto the water tightly, as you have seen if you have put cotton clothes in a clothes dryer—it takes a long time to dry. Polypropylene is a hydrocarbon with no hydrogen-bonding groups; the fiber feels dry because it cannot hold the water the way cotton can. Athletic garments are increasingly using polypropylene because they allow evaporation and cooling during periods of exertion, exactly the opposite of cotton.

Note to the student: BON VOYAGE!
I hope you have enjoyed your travels
through the wonders of organic chemistry.
Jan William Simek

Notes

Notes

Notes

Notes

Notes

Notes

Notes

Notes

Notes

Notes

Notes

Notes